DYNAMICS of MECHANICAL SYSTEMS

DYNAMICS of MECHANICAL SYSTEMS

Harold Josephs
Ronald L. Huston

CRC PRESS

Boca Raton London New York Washington, D.C.

Library of Congress Cataloging-in-Publication Data

Josephs, Harold.
Dynamics of mechanical systems / by Harold Josephs and Ronald L. Huston.
p. ; cm.
Includes bibliographical references and index.
ISBN 0-8493-0593-4 (alk. paper)
1. Mechanical engineering. I. Huston, Ronald L., 1937- II. Title.

TJ145 .J67 2002.
621—dc21 2002276809
CIP

Direct all inquiries to CRC Press LLC, 2000 N.W. Corporate Blvd., Boca Raton, Florida 33431.

Visit the CRC Press Web site at www.crcpress.com

International Standard Book Number 0-8493-0593-4
Library of Congress Card Number 2002276809
Printed in the United States of America 1 2 3 4 5 6 7 8 9 0
Printed on acid-free paper

Preface

This is a textbook intended for mid- to upper-level undergraduate students in engineering and physics. The objective of the book is to give readers a working knowledge of dynamics, enabling them to analyze mechanical systems ranging from elementary and fundamental systems such as planar mechanisms to more advanced systems such as robots, space mechanisms, and human body models. The emphasis of the book is upon the fundamental procedures underlying these dynamic analyses. Readers are expected to obtain skills ranging from the ability to perform insightful hand analyses to the ability to develop algorithms for numerical/computer analyses. In this latter regard, the book is also intended to serve as an independent study text and as a reference book for beginning graduate students and for practicing engineers.

Mechanical systems are becoming increasingly sophisticated, with applications requiring greater precision, improved reliability, and extended life. These enhanced requirements are spurred by a demand for advanced land, air, and space vehicles; by a corresponding demand for advanced mechanisms, manipulators, and robotics systems; and by a need to have a better understanding of the dynamics of biosystems. The book is intended to enable its readers to make engineering advances in each of these areas. The authors believe that the skills needed to make such advances are best obtained by illustratively studying fundamental mechanical components such as pendulums, gears, cams, and mechanisms while reviewing the principles of vibrations, stability, and balancing. The study of these subjects is facilitated by a knowledge of kinematics and skill in the use of Newton's laws, energy methods, Lagrange's equations, and Kane's equations. The book is intended to provide a means for mastering all of these concepts.

The book is written to be readily accessible to students and readers having a background in elementary physics, mathematics through calculus and differential equations, and elementary mechanics. The book itself is divided into 20 chapters, with the first two chapters providing introductory remarks and a review of vector algebra. The next three chapters are devoted to kinematics, with the last of these focusing upon planar kinematics. Chapter 6 discusses forces and force systems, and Chapter 7 provides a comprehensive review of inertia including inertia dyadics and procedures for obtaining the principal moments of inertia and the corresponding principal axes of inertia.

Fundamental principles of dynamics (Newton's laws and d'Alembert's principle) are presented in Chapter 8, and the use of impulse–momentum and work–energy principles is presented in the next two chapters with application to accident reconstruction. Chapters 11 and 12 introduce generalized dynamics and the use of Lagrange's equation and Kane's equations with application to multiple rod pendulum problems. The next five chapters are devoted to applications that involve the study of vibration, stability, balancing, cams, and gears, including procedures for studying nonlinear vibrations and engine balancing. The last three chapters present an introduction to multibody dynamics with application to robotics and biosystems.

Application and illustrative examples are discussed and presented in each chapter, and exercises and problems are provided at the end of each chapter. In addition, each chapter has its own list of references for additional study. Although the earlier chapters provide the basis for the latter chapters, each chapter is written to be as self-contained as possible, with excerpts from earlier chapters provided as needed.

Acknowledgments

The book is an outgrowth of notes the authors have compiled over the past three decades in teaching various courses using the subject material. These notes, in turn, are based upon information contained in various texts used in these courses and upon the authors' independent study and research.

The authors acknowledge the inspiration for a clearly defined procedural study of dynamics by Professor T. R. Kane at the University of Pennsylvania, now nearly 50 years ago. The authors particularly acknowledge the administrative support and assistance of Charlotte Better in typing and preparing the entire text through several revisions. The work of Xiaobo Liu and Doug Provine for preparation of many of the figures is also acknowledged.

The Authors

Harold Josephs, Ph.D., P.E., has been a professor in the Department of Mechanical Engineering at Lawrence Technological University in Southfield, MI, since 1984, subsequent to working in industry for General Electric and Ford Motor Company. Dr. Josephs is the author of numerous publications, holds nine patents, and has presented numerous seminars to industry in the field of safety, bolting, and joining. Dr. Josephs maintains an active consultant practice in safety, ergonomics, and accident reconstruction. His research interests are in fastening and joining, human factors, ergonomics, and safety. Dr. Josephs received his B.S. degree from the University of Pennsylvania, his M.S. degree from Villanova University, and his Ph.D. from the Union Institute. He is a licensed Professional Engineer, Certified Safety Professional, Certified Professional Ergonomist, Certified Quality Engineer, Fellow of the Michigan Society of Engineers, and a Fellow of the National Academy of Forensic Engineers.

Ronald L. Huston, Ph.D., P.E., is distinguished research professor and professor of mechanics in the Department of Mechanical, Industrial, and Nuclear Engineering at the University of Cincinnati. He is also a Herman Schneider chair professor. Dr. Huston has been at the University of Cincinnati since 1962. In 1978, he served as a visiting professor at Stanford University, and from 1979 to 1980 he was division director of civil and mechanical engineering at the National Science Foundation. From 1990 to 1996, Dr. Huston was a director of the Monarch Research Foundation. He is the author of over 140 journal articles, 142 conference papers, 4 books, and 65 book reviews and is a technical editor of *Applied Mechanics Reviews*, and book review editor of the *International Journal of Industrial Engineering*. Dr. Huston is an active consultant in safety, biomechanics, and accident reconstruction. His research interests are in multibody dynamics, human factors, biomechanics, and ergonomics and safety. Dr. Huston received his B.S. degree (1959), M.S. degree (1961), and Ph.D. (1962) from the University of Pennsylvania, Philadelphia. He is a Licensed Professional Engineer and a Fellow of the American Society of Mechanical Engineers.

Contents

Chapter 1 Introduction....1
1.1 Approach to the Subject....1
1.2 Subject Matter....1
1.3 Fundamental Concepts and Assumptions....2
1.4 Basic Terminology in Mechanical Systems....3
1.5 Vector Review....5
1.6 Reference Frames and Coordinate Systems....6
1.7 Systems of Units....9
1.8 Closure....11
References....11
Problems....12

Chapter 2 Review of Vector Algebra....15
2.1 Introduction....15
2.2 Equality of Vectors, Fixed and Free Vectors....15
2.3 Vector Addition....16
2.4 Vector Components....19
2.5 Angle Between Two Vectors....23
2.6 Vector Multiplication: Scalar Product....23
2.7 Vector Multiplication: Vector Product....28
2.8 Vector Multiplication: Triple Products....33
2.9 Use of the Index Summation Convention....37
2.10 Review of Matrix Procedures....38
2.11 Reference Frames and Unit Vector Sets....41
2.12 Closure....44
References....44
Problems....45

Chapter 3 Kinematics of a Particle....57
3.1 Introduction....57
3.2 Vector Differentiation....57
3.3 Position, Velocity, and Acceleration....59
3.4 Relative Velocity and Relative Acceleration....61
3.5 Differentiation of Rotating Unit Vectors....63
3.6 Geometric Interpretation of Acceleration....66
3.7 Motion on a Circle....66
3.8 Motion in a Plane....68
3.9 Closure....71
References....71
Problems....71

Chapter 4 Kinematics of a Rigid Body....77
4.1 Introduction....77
4.2 Orientation of Rigid Bodies....77

4.3 Configuration Graphs....79
4.4 Simple Angular Velocity and Simple Angular Acceleration....83
4.5 General Angular Velocity....85
4.6 Differentiation in Different Reference Frames....87
4.7 Addition Theorem for Angular Velocity....90
4.8 Angular Acceleration....93
4.9 Relative Velocity and Relative Acceleration of Two Points on a Rigid Body....97
4.10 Points Moving on a Rigid Body....103
4.11 Rolling Bodies....106
4.12 The Rolling Disk and Rolling Wheel....107
4.13 A Conical Thrust Bearing....110
4.14 Closure....113
References....113
Problems....114

Chapter 5 Planar Motion of Rigid Bodies — Methods of Analysis....125
5.1 Introduction....125
5.2 Coordinates, Constraints, Degrees of Freedom....125
5.3 Planar Motion of a Rigid Body....128
5.3.1 Translation....129
5.3.2 Rotation....130
5.3.3 General Plane Motion....130
5.4 Instant Center, Points of Zero Velocity....133
5.5 Illustrative Example: A Four-Bar Linkage....136
5.6 Chains of Bodies....142
5.7 Instant Center, Analytical Considerations....147
5.8 Instant Center of Zero Acceleration....150
Problems....156

Chapter 6 Forces and Force Systems....163
6.1 Introduction....163
6.2 Forces and Moments....163
6.3 Systems of Forces....165
6.4 Zero Force Systems....170
6.5 Couples....170
6.6 Wrenches....173
6.7 Physical Forces: Applied (Active) Forces....177
6.7.1 Gravitational Forces....177
6.7.2 Spring Forces....178
6.7.3 Contact Forces....180
6.7.4 Action–Reaction....181
6.8 First Moments....182
6.9 Physical Forces: Inertia (Passive) Forces....184
References....187
Problems....187

Chapter 7 Inertia, Second Moment Vectors, Moments and Products of Inertia, Inertia Dyadics....199
7.1 Introduction....199
7.2 Second-Moment Vectors....199

7.3 Moments and Products of Inertia 200
7.4 Inertia Dyadics 203
7.5 Transformation Rules 205
7.6 Parallel Axis Theorems 206
7.7 Principal Axes, Principal Moments of Inertia: Concepts 208
7.8 Principal Axes, Principal Moments of Inertia: Example 211
7.9 Principal Axes, Principal Moments of Inertia: Discussion 215
7.10 Maximum and Minimum Moments and Products of Inertia 223
7.11 Inertia Ellipsoid 228
7.12 Application: Inertia Torques 228
References 230
Problems 230

Chapter 8 Principles of Dynamics: Newton's Laws and d'Alembert's Principle 241
8.1 Introduction 241
8.2 Principles of Dynamics 242
8.3 d'Alembert's Principle 243
8.4 The Simple Pendulum 245
8.5 A Smooth Particle Moving Inside a Vertical Rotating Tube 246
8.6 Inertia Forces on a Rigid Body 249
8.7 Projectile Motion 251
8.8 A Rotating Circular Disk 253
8.9 The Rod Pendulum 255
8.10 Double-Rod Pendulum 258
8.11 The Triple-Rod and *N*-Rod Pendulums 260
8.12 A Rotating Pinned Rod 263
8.13 The Rolling Circular Disk 267
8.14 Closure 270
References 270
Problems 271

Chapter 9 Principles of Impulse and Momentum 279
9.1 Introduction 279
9.2 Impulse 279
9.3 Linear Momentum 280
9.4 Angular Momentum 282
9.5 Principle of Linear Impulse and Momentum 285
9.6 Principle of Angular Impulse and Momentum 288
9.7 Conservation of Momentum Principles 294
9.8 Examples 295
9.9 Additional Examples: Conservation of Momentum 301
9.10 Impact: Coefficient of Restitution 303
9.11 Oblique Impact 306
9.12 Seizure of a Spinning, Diagonally Supported, Square Plate 309
9.13 Closure 310
Problems 311

Chapter 10 Introduction to Energy Methods 321
10.1 Introduction 321
10.2 Work 321
10.3 Work Done by a Couple 326

10.4 Power 327
10.5 Kinetic Energy 327
10.6 Work–Energy Principles 329
10.7 Elementary Example: A Falling Object 332
10.8 Elementary Example: The Simple Pendulum 333
10.9 Elementary Example — A Mass–Spring System 336
10.10 Skidding Vehicle Speeds: Accident Reconstruction Analysis 338
10.11 A Wheel Rolling Over a Step 341
10.12 The Spinning Diagonally Supported Square Plate 342
10.13 Closure 344
References (Accident Reconstruction) 344
Problems 344

Chapter 11 Generalized Dynamics: Kinematics and Kinetics 353
11.1 Introduction 353
11.2 Coordinates, Constraints, and Degrees of Freedom 353
11.3 Holonomic and Nonholonomic Constraints 357
11.4 Vector Functions, Partial Velocity, and Partial Angular Velocity 359
11.5 Generalized Forces: Applied (Active) Forces 363
11.6 Generalized Forces: Gravity and Spring Forces 367
11.7 Example: Spring-Supported Particles in a Rotating Tube 369
11.8 Forces That Do Not Contribute to the Generalized Forces 375
11.9 Generalized Forces: Inertia (Passive) Forces 377
11.10 Examples 379
11.11 Potential Energy 389
11.12 Use of Kinetic Energy to Obtain Generalized Inertia Forces 394
11.13 Closure 401
References 401
Problems 402

Chapter 12 Generalized Dynamics: Kane's Equations and Lagrange's Equations 415
12.1 Introduction 415
12.2 Kane's Equations 415
12.3 Lagrange's Equations 423
12.4 The Triple-Rod Pendulum 429
12.5 The *N*-Rod Pendulum 433
12.6 Closure 435
References 436
Problems 436

Chapter 13 Introduction to Vibrations 439
13.1 Introduction 439
13.2 Solutions of Second-Order Differential Equations 439
13.3 The Undamped Linear Oscillator 444
13.4 Forced Vibration of an Undamped Oscillator 446
13.5 Damped Linear Oscillator 447
13.6 Forced Vibration of a Damped Linear Oscillator 449
13.7 Systems with Several Degrees of Freedom 450
13.8 Analysis and Discussion of Three-Particle Movement: Modes of Vibration 455

13.9 Nonlinear Vibrations 458
13.10 The Method of Krylov and Bogoliuboff....463
13.11 Closure 466
References....466
Problems 467

Chapter 14 Stability 479
14.1 Introduction....479
14.2 Infinitesimal Stability....479
14.3 A Particle Moving in a Vertical Rotating Tube 482
14.4 A Freely Rotating Body....485
14.5 The Rolling/Pivoting Circular Disk 488
14.6 Pivoting Disk with a Concentrated Mass on the Rim 493
14.6.1 Rim Mass in the Uppermost Position 498
14.6.2 Rim Mass in the Lowermost Position 502
14.7 Discussion: Routh–Hurwitz Criteria....505
14.8 Closure 509
References....509
Problems 510

Chapter 15 Balancing....513
15.1 Introduction....513
15.2 Static Balancing....513
15.3 Dynamic Balancing: A Rotating Shaft 514
15.4 Dynamic Balancing: The General Case 516
15.5 Application: Balancing of Reciprocating Machines....520
15.6 Lanchester Balancing Mechanism 525
15.7 Balancing of Multicylinder Engines....526
15.8 Four-Stroke Cycle Engines 528
15.9 Balancing of Four-Cylinder Engines....529
15.10 Eight-Cylinder Engines: The Straight-Eight and the V-8 532
15.11 Closure 534
References....534
Problems 534

Chapter 16 Mechanical Components: Cams 539
16.1 Introduction....539
16.2 A Survey of Cam Pair Types....540
16.3 Nomenclature and Terminology for Typical Rotating Radial Cams with Translating Followers....541
16.4 Graphical Constructions: The Follower Rise Function....543
16.5 Graphical Constructions: Cam Profiles 544
16.6 Graphical Construction: Effects of Cam–Follower Design 545
16.7 Comments on Graphical Construction of Cam Profiles....549
16.8 Analytical Construction of Cam Profiles 550
16.9 Dwell and Linear Rise of the Follower 551
16.10 Use of Singularity Functions....553
16.11 Parabolic Rise Function....557
16.12 Sinusoidal Rise Function....560
16.13 Cycloidal Rise Function 563
16.14 Summary: Listing of Follower Rise Functions....566

16.15 Closure 568
References 568
Problems 569

Chapter 17 Mechanical Components: Gears 573
17.1 Introduction 573
17.2 Preliminary and Fundamental Concepts: Rolling Wheels 573
17.3 Preliminary and Fundamental Concepts: Conjugate Action 575
17.4 Preliminary and Fundamental Concepts: Involute Curve Geometry 578
17.5 Spur Gear Nomenclature 581
17.6 Kinematics of Meshing Involute Spur Gear Teeth 584
17.7 Kinetics of Meshing Involute Spur Gear Teeth 588
17.8 Sliding and Rubbing between Contacting Involute Spur Gear Teeth 589
17.9 Involute Rack 591
17.10 Gear Drives and Gear Trains 592
17.11 Helical, Bevel, Spiral Bevel, and Worm Gears 595
17.12 Helical Gears 595
17.13 Bevel Gears 596
17.14 Hypoid and Worm Gears 597
17.15 Closure 599
17.16 Glossary of Gearing Terms 599
References 601
Problems 602

Chapter 18 Introduction to Multibody Dynamics 605
18.1 Introduction 605
18.2 Connection Configuration: Lower Body Arrays 605
18.3 A Pair of Typical Adjoining Bodies: Transformation Matrices 609
18.4 Transformation Matrix Derivatives 612
18.5 Euler Parameters 613
18.6 Rotation Dyadics 617
18.7 Transformation Matrices, Angular Velocity Components, and Euler Parameters 623
18.8 Degrees of Freedom, Coordinates, and Generalized Speeds 628
18.9 Transformations between Absolute and Relative Coordinates 632
18.10 Angular Velocity 635
18.11 Angular Acceleration 640
18.12 Joint and Mass Center Positions 643
18.13 Mass Center Velocities 645
18.14 Mass Center Accelerations 647
18.15 Kinetics: Applied (Active) Forces 647
18.16 Kinetics: Inertia (Passive) Forces 648
18.17 Multibody Dynamics 650
18.18 Closure 651
References 651
Problems 652

Chapter 19 Introduction to Robot Dynamics 661
19.1 Introduction 661
19.2 Geometry, Configuration, and Degrees of Freedom 661
19.3 Transformation Matrices and Configuration Graphs 663

19.4 Angular Velocity of Robot Links 665
19.5 Partial Angular Velocities 667
19.6 Transformation Matrix Derivatives 668
19.7 Angular Acceleration of the Robot Links 668
19.8 Joint and Mass Center Position 669
19.9 Mass Center Velocities 671
19.10 Mass Center Partial Velocities 673
19.11 Mass Center Accelerations 673
19.12 End Effector Kinematics 674
19.13 Kinetics: Applied (Active) Forces 677
19.14 Kinetics: Passive (Inertia) Forces 680
19.15 Dynamics: Equations of Motion 681
19.16 Redundant Robots 682
19.17 Constraint Equations and Constraint Forces 684
19.18 Governing Equation Reduction and Solution: Use of Orthogonal Complement Arrays 687
19.19 Discussion, Concluding Remarks, and Closure 689
References 691
Problems 691

Chapter 20 Application with Biosystems, Human Body Dynamics 701
20.1 Introduction 701
20.2 Human Body Modeling 702
20.3 A Whole-Body Model: Preliminary Considerations 703
20.4 Kinematics: Coordinates 706
20.5 Kinematics: Velocities and Acceleration 709
20.6 Kinetics: Active Forces 715
20.7 Kinetics: Muscle and Joint Forces 716
20.8 Kinetics: Inertia Forces 719
20.9 Dynamics: Equations of Motion 721
20.10 Constrained Motion 722
20.11 Solutions of the Governing Equations 724
20.12 Discussion: Application and Future Development 727
References 730
Problems 731

Appendix I Centroid and Mass Center Location for Commonly Shaped Bodies with Uniform Mass Distribution 735

Appendix II Inertia Properties (Moments and Products of Inertia) for Commonly Shaped Bodies with Uniform Mass Distribution 743

Index 753

1

Introduction

1.1 Approach to the Subject

This book presents an introduction to the dynamics of mechanical systems; it is based upon the principles of elementary mechanics. Although the book is intended to be self-contained, with minimal prerequisites, readers are assumed to have a working knowledge of fundamental mechanics' principles and a familiarity with vector and matrix methods. The readers are also assumed to have knowledge of elementary physics and calculus. In this introductory chapter, we will review some basic assumptions and axioms and other preliminary considerations. We will also begin a review of vector methods, which we will continue and expand in Chapter 2.

Our procedure throughout the book will be to develop a general methodology which we will then simplify and specialize to topics of interest. We will attempt to illustrate the concepts through examples and exercise problems. The reader is encouraged to solve as many problems as possible. Indeed, it is our belief that a basic understanding of the concepts and an intuitive grasp of the subject are best obtained through solving the exercise problems.

1.2 Subject Matter

Dynamics is a subject in the general field of mechanics, which in turn is a discipline of classical physics. Mechanics can be divided into two divisions: solid mechanics and fluid mechanics. Solid mechanics may be further divided into flexible mechanics and rigid mechanics. Flexible mechanics includes such subjects as strength of materials, elasticity, viscoelasticity, plasticity, and continuum mechanics. Alternatively, aside from statics, dynamics is the essence of rigid mechanics. Figure 1.2.1 contains a chart showing these subjects and their relations to one another.

Statics is a study of the behavior of rigid body systems when there is no motion. Statics is concerned primarily with the analysis of forces and force systems and the determination of equilibrium configurations. In contrast, *dynamics* is a study of the behavior of moving rigid body systems. As seen in Figure 1.2.1, dynamics may be subdivided into three sub-subjects: kinematics, inertia, and kinetics.

Kinematics is a study of motion without regard to the cause of the motion. Kinematics includes an analysis of the positions, displacements, trajectories, velocities, and accelerations of the members of the system. *Inertia* is a study of the mass properties of the bodies of a system and of the system as a whole in various configurations. *Kinetics* is a study of forces. Forces are generally divided into two classes: *applied* (or "active") forces and *inertia* (or "passive") forces. Applied forces arise from contact between bodies and from gravity; inertia forces occur due to the motion of the system.

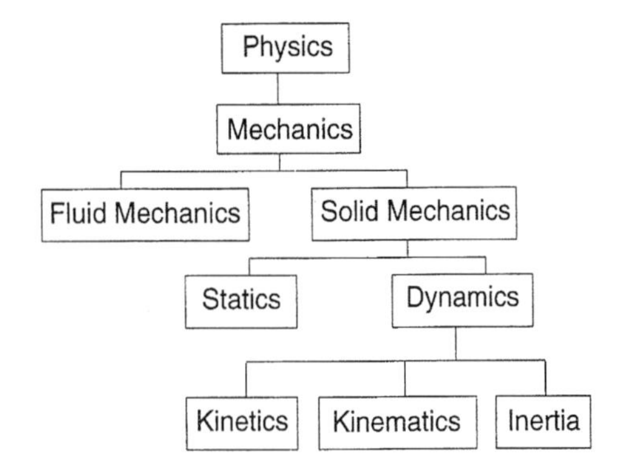

FIGURE 1.2.1
Subdivisions of mechanics.

1.3 Fundamental Concepts and Assumptions

The study of dynamics is based upon several fundamental concepts and basic assumptions that are intuitive and based upon common experience: time, space, force, and mass. *Time* is a measure of change or a measure of a process of events; in dynamics, time is assumed to be a continually increasing, non-negative quantity. *Space* is a geometric region where events occur; in the study of dynamics, space is usually defined by reference frames or coordinate systems. *Force* is intuitively described as a push or a pull. The effect of a force depends upon the magnitude, direction, and point of application of the push or pull; a force is thus ideally suited for representation by a vector. *Mass* is a measure of inertia representing a resistance to change in motion; mass is the source of gravitational attraction and thus also the source of weight forces.

In our study we will assume the existence of an *inertial reference frame*, which is simply a reference frame where Newton's laws are valid. More specifically, we will assume the Earth to be an inertial reference frame for the range of systems and problems considered in this book.

Newton's laws may be briefly stated as follows:

1. In the absence of applied forces, a particle at rest remains at rest and a particle in motion remains in motion, moving at a constant speed along a straight line.
2. A particle subjected to an applied force will accelerate in the direction of the force, and the acceleration will be proportional to the magnitude of the force and inversely proportional to the mass of the particle. Analytically, this may be expressed as

$$\mathbf{F} = m\mathbf{a} \tag{1.3.1}$$

   where $\mathbf{F}$ is the force (a vector), m is the particle mass, and $\mathbf{a}$ is the resulting acceleration (also a vector).
3. Within a mechanical system, interactive forces occur in pairs with equal magnitudes but opposite directions (the law of action and reaction).

1.4 Basic Terminology in Mechanical Systems

Particular terminology is associated with dynamics, and specifically with mechanical system dynamics, which we will use in the text. We will attempt to define the terms as we need them, but it might also be helpful to mention some of them here:

- A *space* is a region or geometric entity occupied by particles where, for our purposes, dynamic events will occur.
- A *reference frame* may be regarded as a coordinate axis system containing and locating the points of a space. Typical reference frames employ Cartesian axes systems.
- A *particle* is a small body whose dimensions are either negligible or irrelevant in the description of its motion and of its response to forces applied to it. "Small" is, of course, a relative term. A body considered as a particle may be small in some contexts but not in others (for example, an Earth satellite or an automobile). Particles are generally identified with points in space, and they generally have finite masses.
- A *rigid body* is a set of particles whose distances from one another remain fixed, or constant, such as a sandstone. The number of particles in a body is usually quite large. A reference frame may be regarded, for kinematic purposes, as a rigid body whose particles have zero masses.
- A *degree of freedom* is defined as a way in which a particle, body, or system can move. The number of degrees of freedom possessed by a particle, body, or system is defined as the number of geometric parameters (for example, coordinates, distances, or angles) needed to uniquely describe the location, orientation, and/or configuration of the particle, body, or system.
- A *constraint* is a restriction on the motion of a particle, body, or system. Constraints can be either geometric (holonomic) or kinematic (nonholonomic).
- A *machine* is an arrangement of a system of bodies designed for applying, transmitting, and/or changing forces and motion.
- A *mechanism* is a machine intended primarily for the transmission of motion. The three general categories of machines are:
  1. *Gear systems*, which are toothed bodies in contact whose objectives are to transmit motion between rotating shafts.
  2. *Cam systems*, which are bodies with curved profiles in contact whose objectives are to transmit motion between a rotating member and a nonrotating member. The term "cam" is sometimes also used to describe a gear tooth.
  3. *Linkages*, which are multibody systems intended to provide either a desired motion of a rigid body or the motion of a point of a body along a curve.
- A *link* is a connective member of a machine or a mechanism. A link maintains a constant distance between two points of a mechanism, although links may be one way, such as cables.
- A *driver* is an "input" link that stimulates a motion.
- A *follower* is an "output" link that responds to the input stimulus of the driver.

- A *joint* is a connective member of a mechanism, usually bringing together two elements of a mechanism. Two elements brought together by a joint are sometimes called *kinematic pairs*. Figure 1.4.1 shows a number of commonly used joints (or kinematic pairs).
- A *kinematic chain* is a series of links that are either joined together or are in contact with one another. A kinematic chain may contain one or more loops. A loop is a chain whose ends are connected. An *open chain* (or "open tree") contains no loops, a *simple chain* contains one loop, and a *complex chain* involves more than one loop or one loop with open branches.

Joint Name	Schematic Representation	Degrees of Freedom
Revolute (pin)	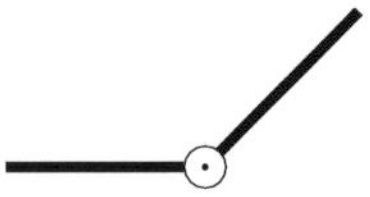	1
Prismatic (slider)		1
Helix (screw)	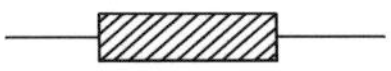	1
Cylinder (sliding pin)	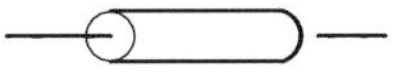	2
Spherical (ball and socket)	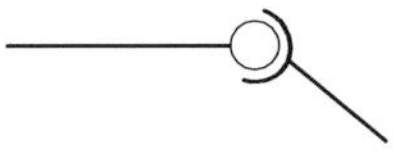	3
Planar	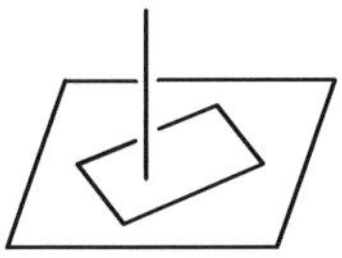	2
Universal (hook)	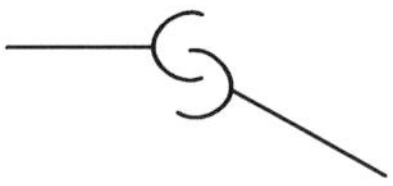	2
Spur gear (rollers)	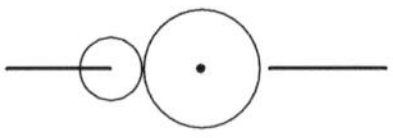	1
Cam	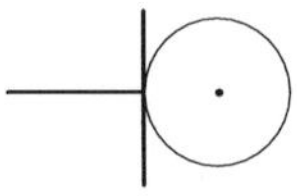	1

FIGURE 1.4.1
Commonly used joints and kinematics pairs.

1.5 Vector Review

Because vectors are used extensively in the text, it is helpful to review a few of their fundamental concepts. We will expand this review in Chapter 2. Mathematically, a vector may be defined as an element of a vector space (see, for example, References 1.1 to 1.3). For our purposes, we may think of a vector simply as a directed line segment.

Figure 1.5.1 shows some examples of vectors as directed line segments. In this context, vectors are seen to have several characteristics: magnitude, orientation, and sense. The *magnitude* of a vector is simply its length; hence, in a graphical representation as in Figure 1.5.1, the magnitude is simply the geometrical length. (Observe, for example, that vector **2B** has a length and magnitude twice that of vector **B**.) The *orientation* of a vector refers to its inclination in space; this inclination is usually measured relative to some fixed coordinate system. The *sense* of a vector is designated by the position of the arrowhead in the geometrical representation. Observe, for example, in Figure 1.5.1 that vectors **A** and –**A** have opposite sense. The combined characteristics of orientation and sense are sometimes called the *direction* of a vector.

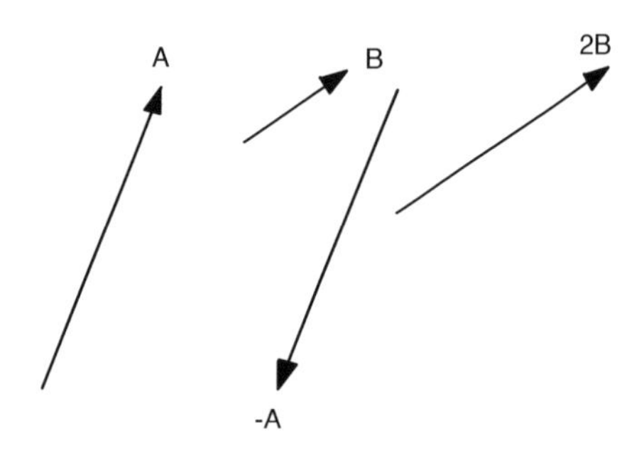

FIGURE 1.5.1
Vectors depicted as directed line segments.

In this book, we will use vectors to represent forces, velocities, and accelerations. We will also use them to locate points and to indicate directions. The units of a vector are those of its magnitude. In turn, the units of the magnitude depend upon the quantity the vector represents. For example, if a vector represents a force, its magnitude is measured in force units such as Newtons (N) or pounds (lb). Alternatively, if a vector represents velocity, its units might be meters per second (m/sec) or feet per second (ft/sec). Hence, vectors representing different quantities will have graphical representations with different length scales. (A review of specific systems of units is presented in Section 1.7.)

Because vectors have the characteristics of magnitude and direction they are distinct from scalars, which are simply elements of a real or complex number system. For example, the magnitude of a vector is a scalar; the direction of a vector is not a scalar. To distinguish vectors from scalars, vectors are printed in bold-face type, such as **V**. Also, because the magnitude of a vector is never negative (length is never negative), absolute-value signs are used to designate the magnitude, such as $|\mathbf{V}|$.

In the next chapter, we will review algebraic operations of vectors, such as the addition and multiplication of vectors. In preparation for this, it is helpful to review the concept of multiplication of vectors by scalars. Specifically, if a vector **V** is multiplied by a scalar s, the product, written as $s\mathbf{V}$, is a vector whose magnitude is $|s|\,|\mathbf{V}|$, **where** $|s|$ **is the absolute value of the scalar** s. The direction of $s\mathbf{V}$ is the same as that of **V** if s is positive and opposite that of **V** if s is negative. Figure 1.5.2 shows some examples of products of scalars and vectors.

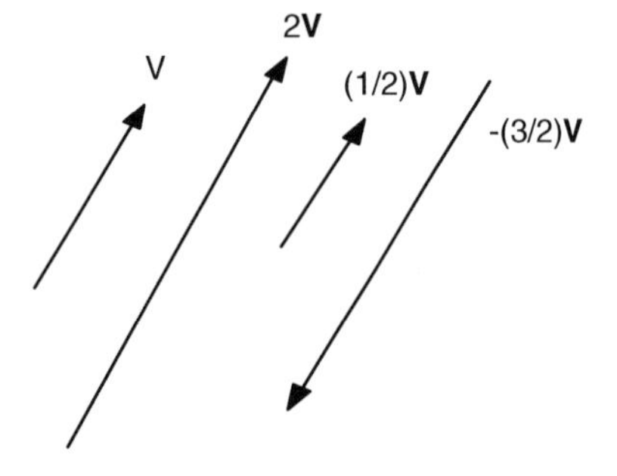

FIGURE 1.5.2
Examples of products of scalars and a vector **V**.

Two kinds of vectors occur so frequently that they deserve special attention: zero vectors and unit vectors. A *zero vector* is simply a vector with magnitude zero. A *unit vector* is a vector with magnitude one; unit vectors have no dimensions or units.

Zero vectors are useful in equations involving vectors. Unit vectors are useful for separating the characteristics of vectors. That is, every vector **V** may be expressed as the product of a scalar and a unit vector. In such a product, the scalar represents the magnitude of the vector and the unit vector represents the direction. Specifically, if **V** is written as:

$$\mathbf{V} = s\mathbf{n} \tag{1.5.1}$$

where s is a scalar and **n** is a unit vector, then s and **n** are:

$$s = |\mathbf{V}| \quad \text{and} \quad \mathbf{n} = \mathbf{V}/|\mathbf{V}| \tag{1.5.2}$$

This means that given any non-zero vector **V** we can always find a unit vector **n** with the same direction as V; thus, **n** represents the direction of **V**.

1.6 Reference Frames and Coordinate Systems

We can represent a reference frame by identifying it with a coordinate–axes system such as a Cartesian coordinate system. Specifically, we have three mutually perpendicular lines, called *axes*, which intersect at a point O called the *origin*, as in Figure 1.6.1. The space is then filled with "points" that are located relative to O by distances from O to P measured along lines parallel to the axes. These distances form sets of three numbers, called the *coordinates* of the points. Each point is then associated with its coordinates.

The points in space may also be located relative to O by introducing additional lines conveniently associated with the points together with the angles these lines make with the mutually perpendicular axes. The coordinates of the points may then involve these angles.

To illustrate these concepts, consider first the Cartesian coordinate system shown in Figure 1.6.2, where the axes are called X, Y, and Z. Let P be a typical point in space. Then the coordinates of P are the distances x, y, and z from P to the planes Y–Z, Z–X, and X–Y, respectively.

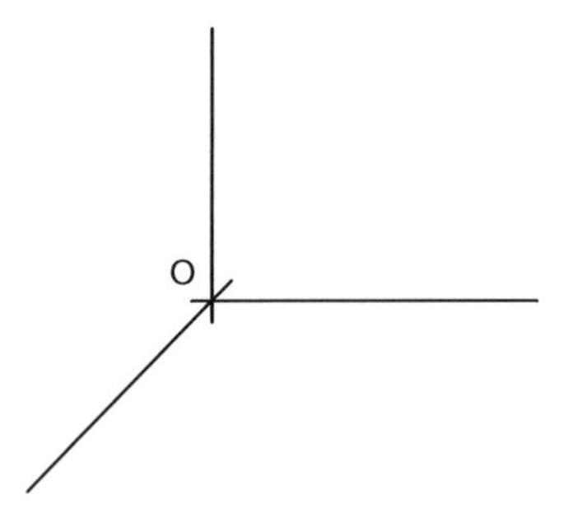

FIGURE 1.6.1
A reference frame with origin O.

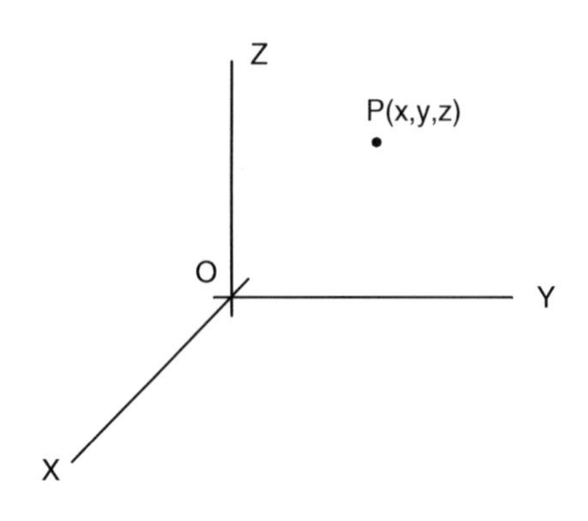

FIGURE 1.6.2
Cartesian coordinate system.

Consider the task of locating P relative to the origin O by moving from O to P along lines parallel to X, Y, and Z, as shown in Figure 1.6.3. The coordinates may then be interpreted as the distances along these lines. The distance d from O to P is then given by the Pythagorean relation:

$$d = \left(x^2 + y^2 + z^2\right)^{1/2} \tag{1.6.1}$$

Finally, the point P is identified by either the name P or the set of three numbers (x, y, z).

To illustrate the use of additional lines and angles, consider the cylindrical coordinate system shown in Figure 1.6.4. In this system, a typical point P is located relative to the origin O by the coordinates (r, θ, z) measuring: (1) the distance r along the newly introduced radial line, (2) the inclination angle θ between the radial line and the X-axis, and (3) the distance z along the line parallel to the Z-axis, as shown in Figure 1.6.4.

By comparing Figures 1.6.3 and 1.6.4 we can readily obtain expressions relating Cartesian and cylindrical coordinates. Specifically, we obtain the relations:

$$\begin{aligned} x &= r \cos\theta \\ y &= r \sin\theta \\ z &= z \end{aligned} \tag{1.6.2}$$

and

$$\begin{aligned} r &= \left(x^2 + y^2\right)^{1/2} \\ \theta &= \tan^{-1}(y / x) \\ z &= z \end{aligned} \tag{1.6.3}$$

As a third illustration, consider the coordinate system shown in Figure 1.6.5. In this case, a typical point P is located relative to the origin O by the coordinates (ρ, ϕ, θ) measuring the distance and angles as shown in Figure 1.6.5. Such a system is called a *spherical* coordinate system.

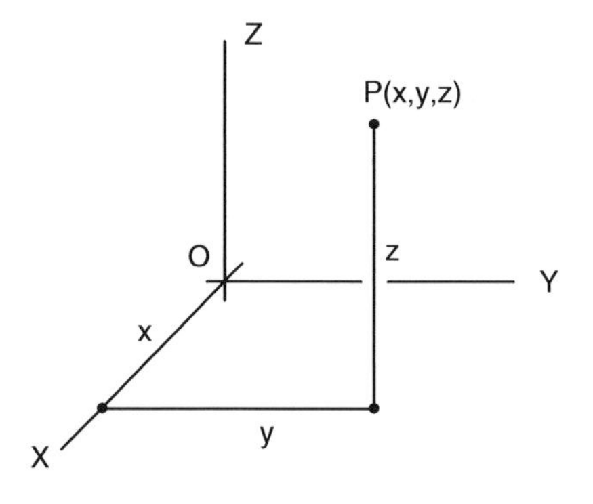

FIGURE 1.6.3
Location of P relative to O.

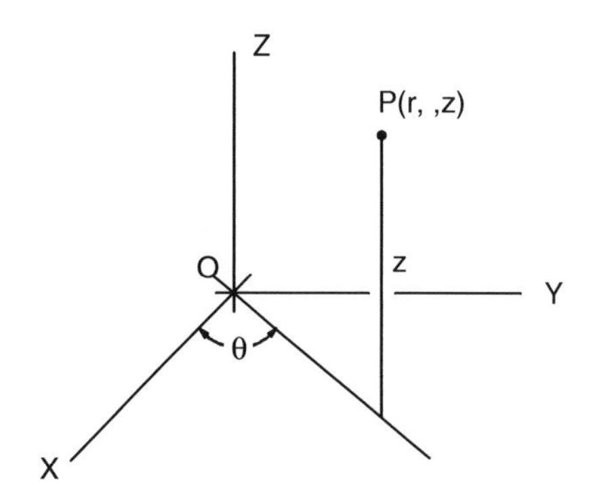

FIGURE 1.6.4
Cylindrical coordinate system.

By comparing Figures 1.6.3 and 1.6.5, we obtain the following relations between the Cartesian and spherical coordinates:

$$\begin{aligned} x &= \rho \sin\phi \cos\theta \\ y &= \rho \sin\phi \sin\theta \\ z &= \rho \cos\phi \end{aligned} \tag{1.6.4}$$

and

$$\begin{aligned} \rho &= \left(x^2 + y^2 + z^2\right)^{1/2} \\ \phi &= \cos^{-1}\left[z / \left(x^2 + y^2 + z^2\right)^{1/2}\right] \\ \theta &= \tan^{-1}(y / x) \end{aligned} \tag{1.6.5}$$

The uses of vectors and coordinate systems are closely related. To illustrate this, consider again the Cartesian coordinate system shown in Figure 1.6.6. Let the unit vectors $\mathbf{n}_x$, $\mathbf{n}_y$, and $\mathbf{n}_z$ be parallel to the X-, Y-, and Z-axes, as shown. Let $\mathbf{p}$ be a position vector locating P relative to O (that is, $\mathbf{p}$ is $\boldsymbol{OP}$). Then it is readily seen that $\mathbf{p}$ may be expressed as the vector sum (see details in the next chapter):

$$\mathbf{p} = x\mathbf{n}_x + y\mathbf{n}_y + z\mathbf{n}_z \tag{1.6.6}$$

Also, the magnitude of $\mathbf{p}$ is:

$$|\mathbf{p}| = \left(x^2 + y^2 + z^2\right)^{1/2} \tag{1.6.7}$$

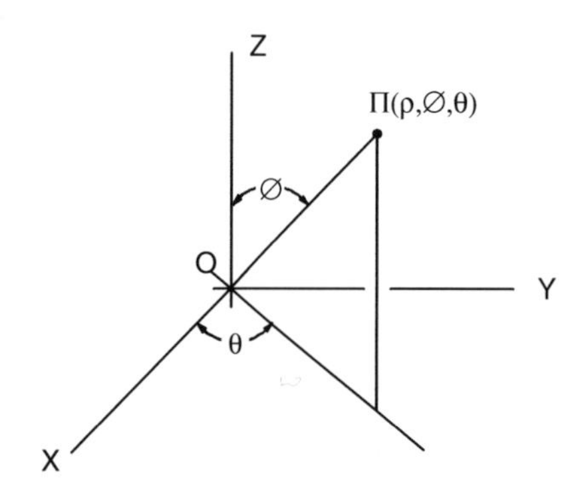

FIGURE 1.6.5
Spherical coordinate system.

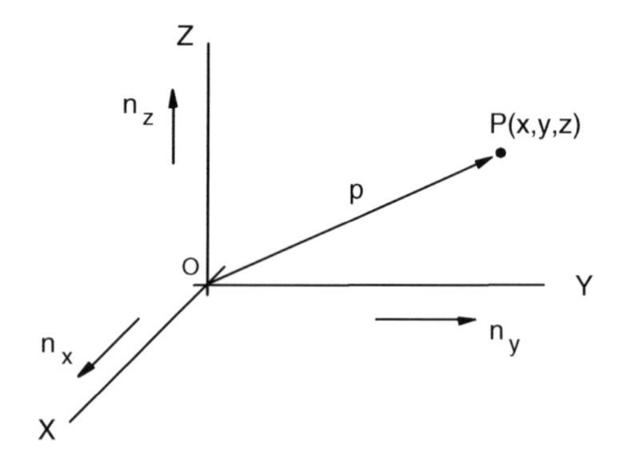

FIGURE 1.6.6
Position vector in a Cartesian coordinate system.

1.7 Systems of Units

In this book, we will use both the English and the International unit systems. On occasion, we will want to make conversions between them. Table 1.7.1 presents a listing of unit conversion factors for commonly occurring quantities in mechanical system dynamics.

TABLE 1.7.1

Conversion Factors between English and International Unit Systems

	To Convert		Multiply by
	from	to	
Acceleration (L/T^2)	ft/sec^2	m/sec^2	3.048 000* × 10^{-1}
	ft/sec^2	cm/sec^2	3.048 000* × 10^{1}
	ft/sec^2	in./sec^2	1.200 000* × 10^{1}
	ft/sec^2	g	3.105 590 × 10^{-2}
	m/sec^2	ft/sec^2	3.280 840 × 10^{0}
	m/sec^2	in./sec^2	3.937 008 × 10^{1}
	m/sec^2	cm/sec^2	1.000 000* × 10^{2}
	m/sec^2	g	1.018 894 × 10^{-1}
	in./sec^2	ft/sec^2	8.333 333 × 10^{-2}
	in./sec^2	m/sec^2	2.540 000* × 10^{-2}
	in./sec^2	cm/sec^2	2.540 000* × 10^{0}
	in./sec^2	g	2.587 992 × 10^{-3}
	cm/sec^2	ft/sec^2	3.280 840 × 10^{-2}
	cm/sec^2	in./sec^2	3.937 008 × 10^{-1}
	cm/sec^2	m/sec^2	1.000 000* × 10^{-2}
	cm/sec^2	g	1.018 894 × 10^{-3}
	g	ft/sec^2	3.220 000* × 10^{1}
	g	in./sec^2	3.864 000* × 10^{2}
	g	m/sec^2	9.814 564 × 10^{0}
	g	cm/sec^2	9.814 564 × 10^{-2}
Angular velocity ($1/T$)	rpm	rad/sec	1.047 198 × 10^{-1}
	rad/sec	rpm	9.549 297 × 10^{0}
Angle	deg	rad	1.745 329 × 10^{-2}
	rad	deg	5.729 578 × 10^{1}
Area	ft^2	m^2	9.290 304 × 10^{-2}
	ft^2	cm^2	9.290 304 × 10^{2}
	in.^{2}	m^2	6.451 600* × 10^{-4}
	in.^{2}	cm^2	6.451 600* × 10^{0}
	ft^2	in.^{2}	1.440 000* × 10^{2}
	in.^{2}	ft^2	6.944 444 × 10^{-3}
	m^2	ft^2	1.076 391 × 10^{1}
	m^2	in.^{2}	1.550 003 × 10^{3}
	m^2	cm^2	1.000 000* × 10^{4}
	cm^2	ft^2	1.076 391 × 10^{-3}
	cm^2	in.^{2}	1.550 003 × 10^{-1}
	cm^2	m^2	1.000 000* × 10^{-4}
Energy/work	ft lb	Nm (or J)	1.355 818 × 10^{0}
	in. lb	Nm (or J)	1.129 848 × 10^{-1}
	ft lb	in. lb	1.200 000* × 10^{1}
	in. lb	ft lb	8.333 333 × 10^{-2}
	Nm (or J)	ft lb	7.375 621 × 10^{-1}
	Nm (or J)	in. lb	8.850 745 × 10^{0}
Force	lb	N	4.448 222 × 10^{0}
	N	lb	2.248 089 × 10^{-1}

TABLE 1.7.1 (CONTINUED)

Conversion Factors between English and International Unit Systems

	To Convert		Multiply by
	from	to	
Length	ft	m	$3.048\ 000 \times 10^{-1}$
	ft	cm	$3.048\ 000 \times 10^{1}$
	ft	in.	$1.200\ 000^{*} \times 10^{1}$
	in.	m	$2.540\ 000^{*} \times 10^{-2}$
	in.	cm	$2.540\ 000^{*} \times 10^{0}$
	in.	ft	$8.333\ 333 \times 10^{-2}$
	m	ft	$3.280\ 839 \times 10^{0}$
	m	in.	$3.937\ 008 \times 10^{1}$
	m	cm	$1.000\ 000^{*} \times 10^{2}$
	cm	ft	$3.280\ 840 \times 10^{-2}$
	cm	in.	$3.937\ 008 \times 10^{-1}$
	cm	m	$1.000\ 000 \times 10^{-2}$
	mi	ft	$5.280\ 000^{*} \times 10^{3}$
	mi	km	$1.609\ 344 \times 10^{0}$
	km	mi	$6.213\ 712 \times 10^{-1}$
Mass	slug	kg	$1.459\ 390 \times 10^{1}$
	kg	slug	$6.852\ 178 \times 10^{-2}$
Mass density	slug/ft^3	kg/m^3	$5.153\ 788 \times 10^{2}$
	slug/ft^3	g/cm^3	$5.153\ 788 \times 10^{-1}$
	kg/m^3	slug/ft^3	$1.940\ 320 \times 10^{-3}$
	kg/m^3	g/cm^3	$1.000\ 000^{*} \times 10^{-3}$
	g/cm^3	slug/ft^3	$1.940\ 320 \times 10^{0}$
	g/cm^3	kg/m^3	$1.000\ 000^{*} \times 10^{3}$
Moment or torque	ft lb	Nm	$1.355\ 818 \times 10^{0}$
	ft lb	in. lb	$1.200\ 000^{*} \times 10^{1}$
	in. lb	Nm	$1.129\ 848 \times 10^{-1}$
	in. lb	ft lb	$8.333\ 333 \times 10^{-2}$
	Nm	ft lb	$7.375\ 621 \times 10^{-1}$
	Nm	in. lb	$8.850\ 745 \times 10^{0}$
Moments and products of inertia	slug ft^2	kg m^2	$1.355\ 818 \times 10^{0}$
	kg m^2	slug ft^2	$7.375\ 623 \times 10^{-1}$
Power or energy/work rate	ft lb/sec	HP	$1.818\ 182 \times 10^{-3}$
	ft lb/sec	Nm/sec (or W)	$1.355\ 818 \times 10^{0}$
	ft lb/min	HP	$3.030\ 303 \times 10^{-5}$
	ft lb/min	Nm/sec (or W)	$2.259\ 697 \times 10^{-2}$
	HP	Nm/sec (or W)	$7.456\ 999 \times 10^{2}$
	HP	ft lb/sec	$7.375\ 621 \times 10^{-1}$
	HP	ft lb/min	$4.425\ 373 \times 10^{1}$
	Nm/sec (or W)	ft lb/sec	$1.341\ 022 \times 10^{-3}$
	Nm/sec (or W)	ft lb/min	$5.500\ 000^{*} \times 10^{2}$
	Nm/sec (or W)	HP	$3.300\ 000^{*} \times 10^{4}$
Pressure or stress	lb/in.^{2} (or psi)	N/m^2 (or Pa)	$6.894\ 757 \times 10^{3}$
	lb/in.^{2} (or psi)	lb/ft^2	$1.440\ 000^{*} \times 10^{2}$
	lb/ft^2	N/m^2 (or Pa)	$4.788\ 026 \times 10^{1}$
	lb/ft^2	lb/in.^{2} (or psi)	$6.944\ 444 \times 10^{-3}$
	N/m^2 (or Pa)	lb/in.^{2} (or psi)	$1.450\ 377 \times 10^{-4}$
	N/m^2 (or Pa)	lb/ft^2	$2.088\ 543 \times 10^{-2}$
Velocity	ft/sec	m/sec	$3.048\ 000^{*} \times 10^{-1}$
	ft/sec	cm/sec	$3.048\ 000^{*} \times 10^{1}$
	ft/sec	km/hr	$1.097\ 280^{*} \times 10^{0}$
	ft/sec	in./sec	$1.200\ 000^{*} \times 10^{1}$
	ft/sec	mi/hr (or mph)	$6.818\ 182 \times 10^{-1}$
	in./sec	m/sec	$2.540\ 000^{*} \times 10^{-2}$
	in./sec	cm/sec	$2.540\ 000^{*} \times 10^{0}$
	in./sec	km/hr	$9.144\ 000^{*} \times 10^{-2}$

TABLE 1.7.1 (CONTINUED)

Conversion Factors between English and International Unit Systems

	To Convert		Multiply by
	from	to	
Velocity	in./sec	ft/sec	$8.333\ 333 \times 10^{-2}$
	in./sec	mi/hr (or mph)	$5.681\ 818 \times 10^{-2}$
	mi/hr (or mph)	m/sec	$4.470\ 400^{*} \times 10^{-1}$
	mi/hr (or mph)	km/hr	$1.609\ 344 \times 10^{0}$
	mi/hr (or mph)	ft/sec	$1.466\ 667 \times 10^{0}$
	mi/hr (or mph)	in./sec	$1.760\ 000^{*} \times 10^{1}$
	m/sec	ft/sec	$3.280\ 840 \times 10^{0}$
	m/sec	in./sec	$3.937\ 008 \times 10^{1}$
	m/sec	mi/hr (or mph)	$2.236\ 936 \times 10^{0}$
	m/sec	cm/sec	$1.000\ 000 \times 10^{2}$
	m/sec	km/hr	$3.600\ 000^{*} \times 10^{0}$
	km/hr	ft/sec	$9.113\ 444 \times 10^{-1}$
	km/hr	in./sec	$1.093\ 613 \times 10^{1}$
	km/hr	mi/hr (or mph)	$6.213\ 712 \times 10^{-1}$
	km/hr	m/sec	$2.777\ 777 \times 10^{-1}$
Weight density	lb/ft^3	N/m^3	$1.570\ 670 \times 10^{2}$
	lb/ft^3	$lb/in.^3$	$5.787\ 037 \times 10^{-4}$
	$lb/in.^3$	N/m^3	$9.089\ 525 \times 10^{-2}$
	$lb/in.^3$	lb/ft^3	$1.728\ 000^{*} \times 10^{3}$
	N/m^3	lb/ft^3	$6.366\ 671 \times 10^{-3}$
	N/m^3	$lb/in.^3$	$1.100\ 167 \times 10^{1}$

Note: cm, centimeters; deg, degrees; ft, feet; g, grams; *g*, gravity acceleration (taken as 32.2 ft/sec^2); HP, horsepower; in., inches; J, Joules; kg, kilograms; lb, pounds; m, meters; mi, miles; mph, miles per hour; N, Newtons; Pa, Pascals; psi, pounds per square inch; rad, radius; rpm, revolutions per minute; sec, seconds; W, watts.

* Exact by definition.

1.8 Closure

In this introduction to our study of mechanical system dynamics, we have focused upon terminology and procedures that are believed to be useful in the sequel. As we proceed, we will continue to introduce terminology and procedures as needed. In this regard, we will expand and elaborate upon our review of vector methods in Chapter 2. Throughout the text we will attempt to illustrate the subject matter under discussion by examples and by providing exercises (or problems) for the reader. These problems, appearing at the ends of the chapters, are not intended to be burdensome but instead to serve as a learning aid for the reader. In addition, references will be provided for parallel study and for more in-depth study.

References

1.1. Noble, B., *Applied Linear Algebra*, Prentice Hall, Englewood Cliffs, NJ, 1969, pp. 104, 461.
1.2. Usmani, R. A., *Applied Linear Algebra*, Dekker, New York, 1987, chap. 1.
1.3. Shields, P. C., *Elementary Linear Algebra*, Worth Publishers, New York, 1968, pp. 31–32.
1.4. Baumeister, T., Avallone, E. A., and Baumeister, T., III, Eds., *Marks' Standard Handbook for Mechanical Engineers*, 8th ed., McGraw-Hill, 1978, pp. 1-33–1-39.

1.5. Hsu, H. P., *Vector Analysis*, Simon & Schuster Technical Outlines, New York, 1969.
1.6. Brand, L., *Vector and Tensor Analysis*, Wiley, New York, 1964.
1.7. Kane, T. R., *Analytical Elements of Mechanics*, Vol. 2, Academic Press, New York, 1961.
1.8. Kane, T. R., and Levinson, D. A., *Dynamics: Theory and Applications*, McGraw-Hill, New York, 1985, pp. 361–371.
1.9. Likins, P. W., *Elements of Engineering Mechanics*, McGraw-Hill, New York, 1973.
1.10. Beer, F. P., and Johnston, E. R., Jr., *Vector Mechanics for Engineers*, 6th ed., McGraw-Hill, New York, 1996.
1.11. Yeh, H., and Abrams, J. I., *Principles of Mechanics of Solids and Fluids*, Vol. 1, *Particle and Rigid-Body Mechanics*, McGraw-Hill, New York, 1960.
1.12. Haug, E. J., *Intermediate Dynamics*, Prentice Hall, Englewood Cliffs, NJ, 1992.
1.13. Meriam, J. L., and Kraige, L. G., *Engineering Mechanics*, Vol. 2, *Dynamics*, 3rd ed., John Wiley & Sons, New York, 1992.
1.14. Hibbler, R. C., *Engineering Mechanics: Statics and Dynamics*, Macmillan, New York, 1974.
1.15. Shelley, J. P., *Vector Mechanics for Engineers*, Vol. II, *Dynamics*, Schaum's Solved Problems Series, McGraw-Hill, New York, 1991.
1.16. Jong, I. C., and Rogers, B. G., *Engineering Mechanics, Statics and Dynamics*, Saunders College Publishing, Holt, Rinehart & Winston, Philadelphia, PA, 1991.
1.17. Huston, R. L., *Multibody Dynamics*, Butterworth–Heinemann, Stoneham, MA, 1990.
1.18. Higdon, A., and Stiles, W. B., *Engineering Mechanics*, Vol. II, *Dynamics*, 3rd ed., Prentice Hall, Englewood Cliffs, NJ, 1968.
1.19. Meirovitch, L., *Methods of Analytical Dynamics*, McGraw-Hill, New York, 1970.
1.20. Paul, B., *Kinematics and Dynamics of Planar Machinery*, Prentice Hall, Englewood Cliffs, NJ, 1979.
1.21. Sneck, H. J., *Machine Dynamics*, Prentice Hall, Englewood Cliffs, NJ, 1991.
1.22. Mabie, H. H., and Reinholtz, C. F., *Mechanisms and Dynamics of Machinery*, 4th ed., Wiley, New York, 1987.
1.23. Ginsberg, J. H., *Advanced Engineering Dynamics*, Harper & Row, New York, 1988.
1.24. Shames, I. H., *Engineering Mechanics*, Prentice Hall, Englewood Cliffs, NJ, 1980.
1.25. Liu, C. Q., and Huston, R. L., *Formulas for Dynamic Analyses*, Marcel Dekker, New York, 1999.

Problems

Section 1.5 Vector Review

P1.5.1: Suppose a velocity vector **V** is expressed in the form:

$$\mathbf{V} = 3\mathbf{i} + 4\mathbf{j} + 12\mathbf{k} \text{ ft/sec}$$

where **i**, **j**, and **k** are mutually perpendicular unit vectors.

a. Determine the magnitude of **V**.
b. Find a unit vector **n** parallel to **V** and having the same sense as **V**.

Section 1.6 Reference Frames and Coordinate Systems

P1.6.1: Suppose the Cartesian coordinates (x, y, z) of a point P are (3, 1, 4).

a. Find the cylindrical coordinates (r, θ, z) of P.
b. Find the spherical coordinates (ρ, ϕ, θ) of P.

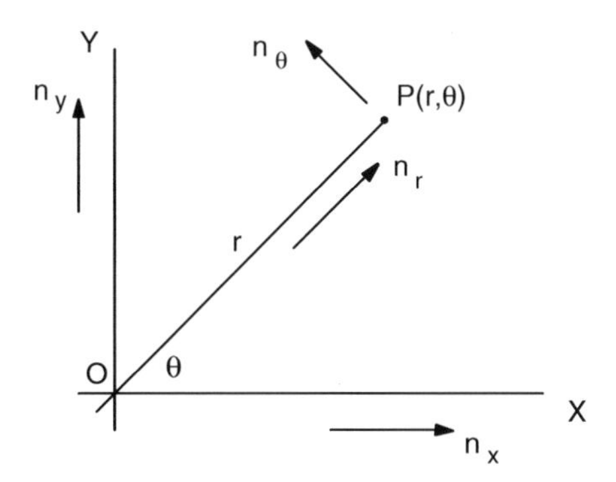

FIGURE P1.6.1
Polar coordinate system.

P1.6.2: Suppose the Cartesian coordinates (x, y, z) of a point P are $(-1, 2, -5)$.

a. Find the cylindrical coordinates (r, θ, z) of P.
b. Find the spherical coordinates (ρ, ϕ, θ) of P.

P1.6.3: Suppose the cylindrical coordinates (r, θ, z) of a point P are $(4, \pi/6, 2)$.

a. Find the Cartesian coordinates (x, y, z) of P.
b. Find the Spherical coordinates (ρ, ϕ, θ) of P.

P1.6.4: Suppose the spherical coordinates (ρ, ϕ, θ) of a point P are $(7, \pi/4, \pi/3)$.

a. Find the Cartesian coordinates (x, y, z) of P.
b. Find the cylindrical coordinates (r, θ, z) of P.

P1.6.5: Consider the cylindrical coordinate system (r, θ, z) with z identically zero. This system then reduces to the "polar coordinate" system as shown in Figure P1.6.1.

a. Express the coordinates (x, y) in terms of (r, θ).
b. Express the coordinates (r, θ) in terms of (x, y).

P1.6.6: See Problem 1.6.5. Let n_x and n_y be unit vectors parallel to the X- and Y-axes, as shown. Let n_r and n_θ be unit vectors parallel and perpendicular to the radial line as shown.

a. Express n_r and n_θ in terms of n_x and n_y.
b. Express n_x and n_y in terms of n_r and n_θ.

Section 1.7 Systems of Units

P1.7.1: An automobile A is traveling at 60 mph.

a. Express the speed of A in ft/sec.
b. Express the speed of A in km/sec.

P1.7.2: A person weighs 150 lb.

a. What is the person's weight in N?
b. What is the person's mass in slug?
c. What is the person's mass in kg?

P1.7.3: An automobile A accelerates from a stop at 3 mph/sec.

a. Express the acceleration in ft/sec^2.
b. Express the acceleration in m/sec^2.
c. Express the acceleration in g.

2

Review of Vector Algebra

2.1 Introduction

In Chapter 1, we reviewed the basic concepts of vectors. We considered vectors, scalars, and the multiplication of vectors and scalars. We also examined zero vectors and unit vectors. In this chapter, we will build upon these ideas as we develop a review of vector algebra. Specifically, we will review the concepts of vector equality, vector addition, and vector multiplication. We will also review the concepts of reference frames and unit vector sets. Finally, we will review the elementary procedures of matrix algebra.

2.2 Equality of Vectors, Fixed and Free Vectors

Recall that the characteristics of a vector are its *magnitude*, its *orientation*, and its *sense*. Indeed, we could say that a vector is defined by its characteristics. The concept of vector equality follows from this definition: Specifically, two vectors are *equal* if (and only if) they have the same characteristics. Vector equality is fundamental to the development of vector algebra. For example, if vectors are equal, they may be interchanged in vector equations, which enables us to simplify expressions. It should be noted, however, that vector equality does not necessarily denote physical equality, particularly when the vectors model physical quantities. This occurs, for example, with forces. We will explore this concept later.

Two fundamental ideas useful in relating mathematical and physical quantities are the concepts of *fixed* and *free* vectors. A *fixed* vector has its location restricted to a line fixed in space. To illustrate this, consider the fixed line L as shown in Figure 2.2.1. Let $\mathbf{v}$ be a vector whose location is restricted to L, and let the location of $\mathbf{v}$ along L be arbitrary. Then $\mathbf{v}$ is a *fixed* vector. Because the location of $\mathbf{v}$ along L is arbitrary, $\mathbf{v}$ might even be called a *sliding vector.*

Alternatively, a *free* vector is a vector that may be placed anywhere in space if its characteristics are maintained. Unit vectors such as $\mathbf{n}_x$, $\mathbf{n}_y$, and $\mathbf{n}_z$ shown in Figure 2.2.1 are examples of free vectors. Most vectors in our analyses will be free vectors. Indeed, we will assume that vectors are free vectors unless otherwise stated.

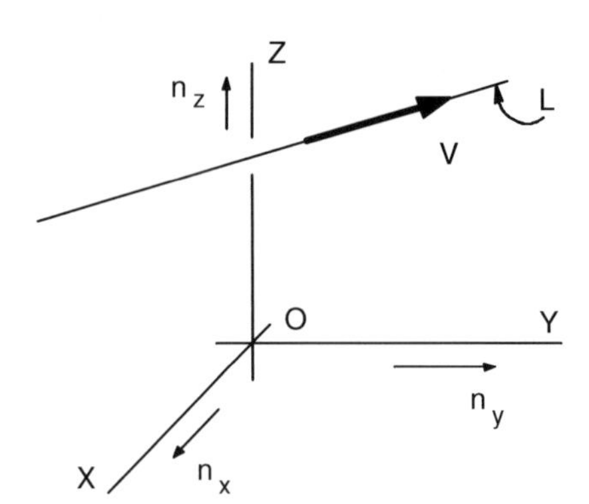

FIGURE 2.2.1
A fixed line L and a fixed vector **V**.

2.3 Vector Addition

Vectors obey the parallelogram law of addition. This is a simple geometric algorithm for understanding and exhibiting the powerful analytical utility of vectors. To see this, consider two vectors **A** and **B** as in Figure 2.3.1. Let **A** and **B** be free vectors. To add **A** and **B**, let them be connected "head-to-tail" (without changing their characteristics) as in Figure 2.3.2. That is, relocate **B** so that its tail is at the head of **A**. Then, the sum of **A** and **B**, called the *resultant*, is the vector **R** connecting the tail of **A** to the head of **B**, as in Figure 2.3.3. That is,

$$\mathbf{R} = \mathbf{A} + \mathbf{B} \tag{2.3.1}$$

The vectors **A** and **B** are called the *components* of **R**.

The reason for the name "parallelogram law" is that the same result is obtained if the head of **B** is connected to the tail of **A**, as in Figure 2.3.4. The two ways of adding **A** and **B** produce a parallelogram, as shown. The order of the addition — that is, which vector is taken first and which is taken second — is therefore unimportant; hence, vector addition is *commutative*. That is,

$$\mathbf{A} + \mathbf{B} = \mathbf{B} + \mathbf{A} \tag{2.3.2}$$

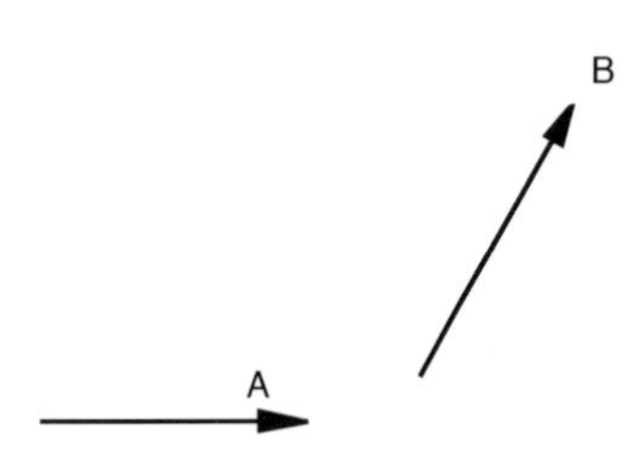

FIGURE 2.3.1
Two vectors **A** and **B** to be added.

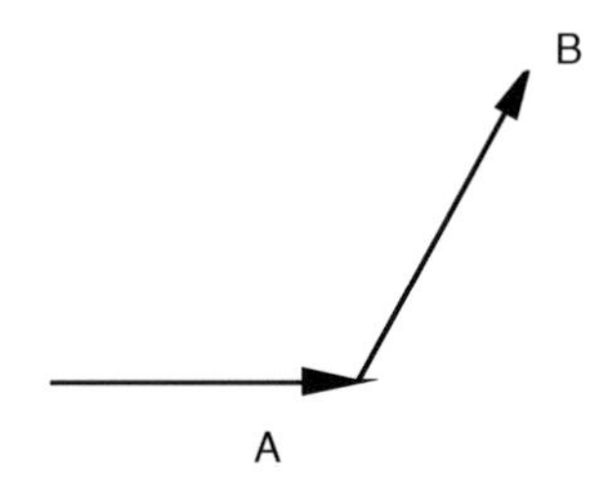

FIGURE 2.3.2
Vectors **A** and **B** connected head to tail.

Vector ***subtraction*** may be defined from vector addition. Specifically, the difference of two vectors **A** and **B**, written as **A** – **B**, is simply the sum of **A** with the negative of **B**. That is,

$$\mathbf{A}-\mathbf{B}=\mathbf{A}+(-\mathbf{B}) \tag{2.3.3}$$

An item of interest in vector addition is the magnitude of the resultant, which may be determined using the geometry of the parallelogram and the law of cosines. For example, in Figure 2.3.5, let θ be the angle between **A** and **B**, as shown. Then, the magnitude of the resultant **R** is given by:

$$|\mathbf{R}|=|\mathbf{A}|^2+|\mathbf{B}|^2-2|\mathbf{A}|+|\mathbf{B}|\cos(\pi-\theta) \tag{2.3.4}$$

Example 2.3.1: Resultant Magnitude

To illustrate the use of Eq. (2.3.4), suppose the magnitude of **A** is 15 N, the magnitude of **B** is 12 N, and the angle θ between **A** and **B** is 60°. Then, the magnitude of the resultant **R** is:

$$|\mathbf{R}|=(15)^2+(12)^2-(2)(15)(12)\cos(2\pi/3)^{1/2}=23.43N \tag{2.3.5}$$

Observe from Eq. (2.3.4) that if we double the magnitude of both **A** and **B**, the magnitude of the resultant **R** is also doubled. Indeed, if we multiply **A** and **B** by any scalar s, **R** will also be multiplied by s. That is,

$$s\mathbf{A}+s\mathbf{B}=s\mathbf{R}=s(\mathbf{A}+\mathbf{B}) \tag{2.3.6}$$

This means that vector addition is *distributive* with respect to scalar multiplication.

Next, suppose we have *three* vectors **A**, **B**, and **C**, and suppose we wish to find their resultant. Suppose further that the vectors are *not* parallel to the same plane, as, for example, in Figure 2.3.6. The resultant **R** is obtained in the same manner as before. That is, the vectors are connected head to tail, as depicted in Figure 2.3.7. Then, the resultant **R** is obtained by connecting the tail of the first vector **A** to the head of the third vector **C** as in Figure 2.3.7. That is,

$$\mathbf{R}=\mathbf{A}+\mathbf{B}+\mathbf{C} \tag{2.3.7}$$

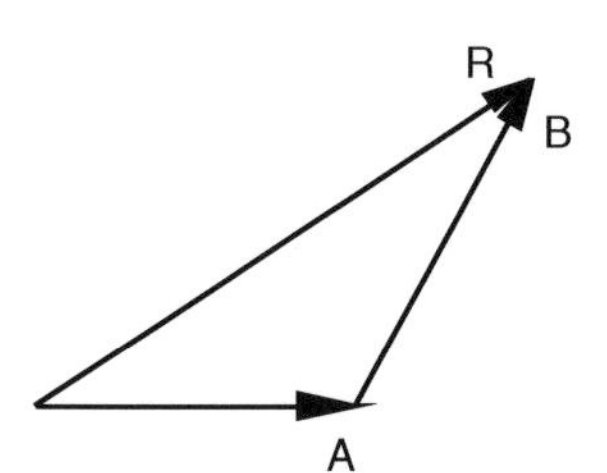

FIGURE 2.3.3
Resultant (sum) **R** of vectors **A** and **B**.

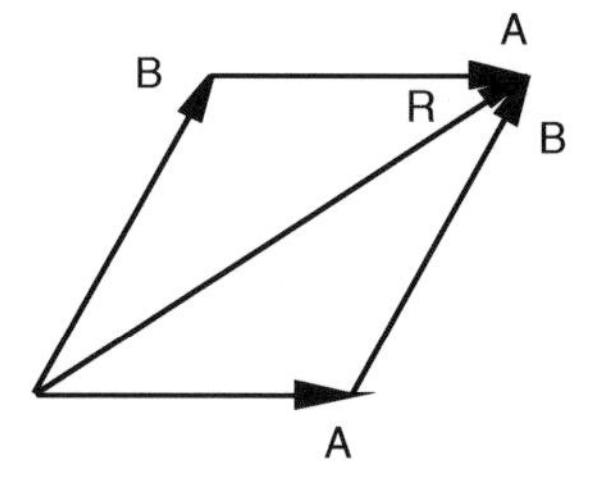

FIGURE 2.3.4
Two ways of adding vectors **A** and **B**.

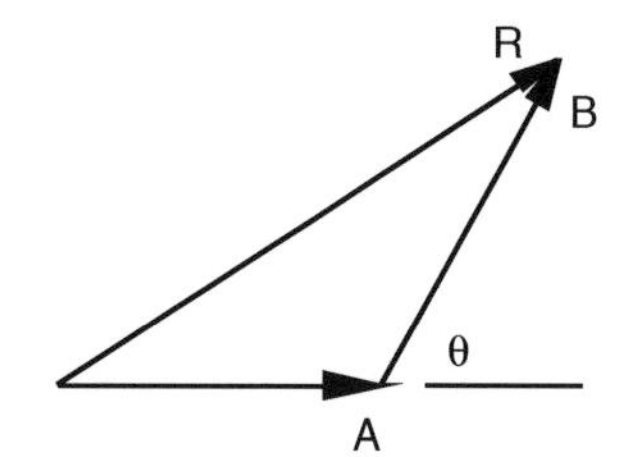

FIGURE 2.3.5
Vector triangle geometry.

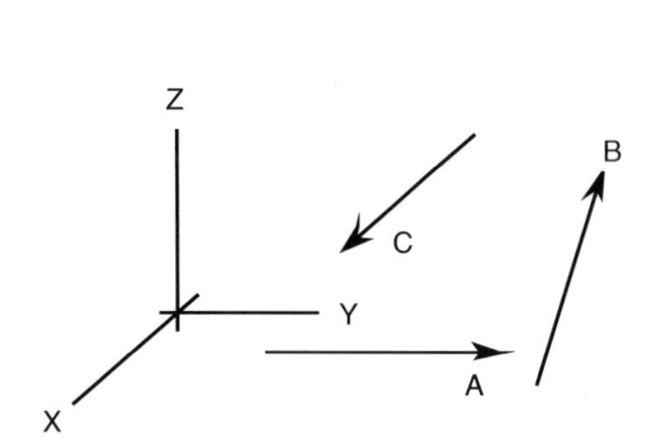

FIGURE 2.3.6
Vectors **A**, **B**, and **C** to be added.

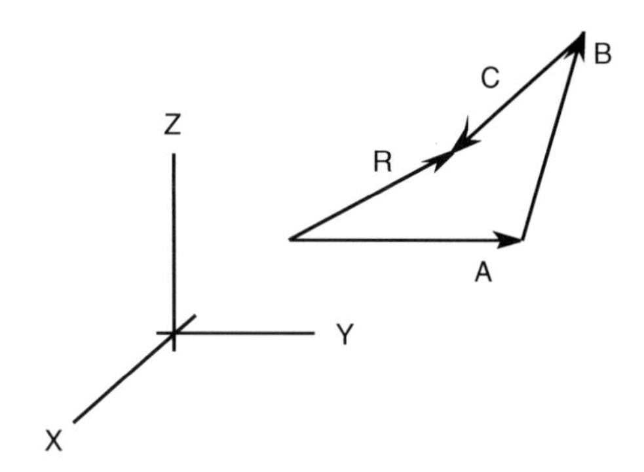

FIGURE 2.3.7
Resultant of vectors **A**, **B**, and **C** of Figure 2.3.6.

Example 2.3.2: Resultant Magnitude in Three Dimensions

As an illustration, suppose **C** is perpendicular to the plane of vectors **A** and **B** of Example 2.3.1 and suppose the magnitude of **C** is 10 N. Then, from the results of Eq. (2.3.5), the magnitude of **R** is:

$$|\mathbf{R}| = \left[(23.43)^2 + (10)^2\right]^{1/2} = 25.47N \tag{2.3.8}$$

This procedure has several remarkable features. First, as before, the order of the components in Eq. (2.3.7) is unimportant. That is,

$$\begin{aligned} &\mathbf{A}+\mathbf{B}+\mathbf{C}=\mathbf{C}+\mathbf{A}+\mathbf{B}=\mathbf{B}+\mathbf{C}+\mathbf{A} \\ &=\mathbf{A}+\mathbf{C}+\mathbf{B}=\mathbf{B}+\mathbf{A}+\mathbf{C}=\mathbf{C}+\mathbf{B}+\mathbf{A}=\mathbf{R} \end{aligned} \tag{2.3.9}$$

Second, to obtain the resultant **R** we may first add any two of the vectors, say **A** and **B**, and then add the resultant of this sum to **C**. This means the summation in Eq. (2.3.7) is *associative*. That is,

$$\mathbf{R}=\mathbf{A}+\mathbf{B}+\mathbf{C}=(\mathbf{A}+\mathbf{B})+\mathbf{C}=\mathbf{A}+(\mathbf{B}+\mathbf{C}) \tag{2.3.10}$$

This feature may be used to obtain the magnitude of the resultant by repeated use of the law of cosines as before.

Third, observe that configurations exist where the magnitude of the resultant is less than the magnitude of the individual components. Indeed, the magnitude of the resultant could be zero.

Finally, with three or more components, the procedure of finding the magnitude of the resultant by repeated use of the law of cosines is cumbersome and tedious.

An attractive feature of vector addition is that the resultant magnitude may be obtained by strictly analytical means — that is, without regard to triangle geometry. This is the original reason for using vectors in analysis. We will explore this further in the next section.

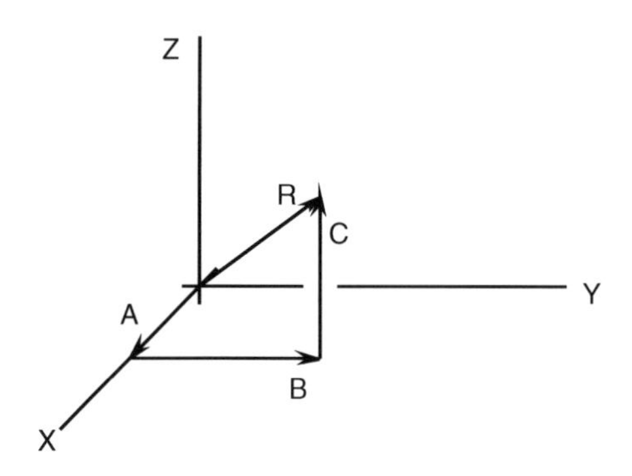

FIGURE 2.4.1
Vector **R** with mutually perpendicular components.

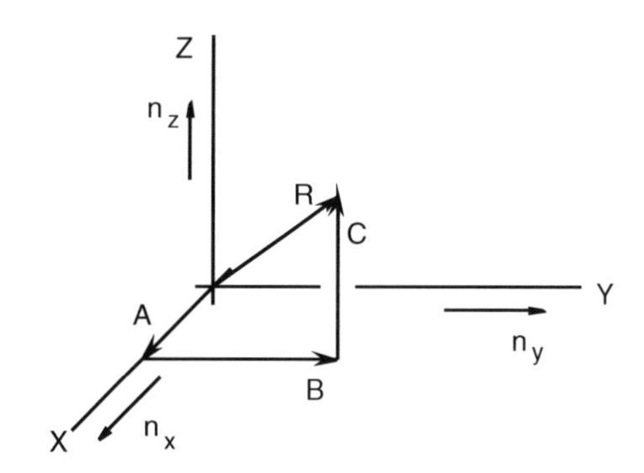

FIGURE 2.4.2
Vectors **A**, **B**, and **C** and unit vectors $\mathbf{n}_x$, $\mathbf{n}_y$, and $\mathbf{n}_z$.

2.4 Vector Components

Consider again Eq. (2.3.7) where we have the vector sum:

$$\mathbf{R} = \mathbf{A} + \mathbf{B} + \mathbf{C} \tag{2.4.1}$$

Instead of thinking of this expression as a sum of components, consider it as a representation of the vector **R**. Suppose further that the components **A**, **B**, and **C** happen to be mutually perpendicular and parallel to coordinate axes, as shown in Figure 2.4.1. Then, by the Pythagoras theorem, the magnitude of **R** is simply:

$$|\mathbf{R}| = \left(|\mathbf{A}|^2 + |\mathbf{B}|^2 + |\mathbf{C}|^2\right)^{1/2} \tag{2.4.2}$$

To develop these ideas still further, suppose that $\mathbf{n}_x$, $\mathbf{n}_y$, and $\mathbf{n}_z$ are unit vectors parallel to *X*, *Y*, and *Z*, as in Figure 2.4.2. Then, from our discussion in Chapter 1, we see that **A**, **B**, and **C** can be expressed in the forms:

$$\begin{aligned} \mathbf{A} &= |\mathbf{A}|\mathbf{n}_x = a\mathbf{n}_x \\ \mathbf{B} &= |\mathbf{B}|\mathbf{n}_y = b\mathbf{n}_y \\ \mathbf{C} &= |\mathbf{C}|\mathbf{n}_z = c\mathbf{n}_z \end{aligned} \tag{2.4.3}$$

where *a*, *b*, and *c* are scalars representing the magnitudes of **A**, **B**, and **C**. Hence, **R** may be expressed as:

$$\mathbf{R} = a\mathbf{n}_x + b\mathbf{n}_y + c\mathbf{n}_z \tag{2.4.4}$$

Then, the magnitude of **R** is:

$$|\mathbf{R}| = \left(a^2 + b^2 + c^2\right)^{1/2} \tag{2.4.5}$$

A question that arises is what if **A**, **B**, and **C** are not mutually perpendicular? How then can we find the magnitude of the resultant? A powerful feature of the vector method is that the same general procedure can be used regardless of the directions of the components. All that is required is to express the components as sums of vectors parallel to the *X*-, *Y*-, and *Z*-axes (or other convenient mutually perpendicular directions). For example, suppose a vector **A** is inclined relative to the *X*-, *Y*-, and *Z*-axes, as in Figure 2.4.3. Let θ_x, θ_y, and θ_z be the angles that **A** makes with the axes. Next, let us express **A** in the desired form:

$$\mathbf{A} = \mathbf{A}_x + \mathbf{A}_y + \mathbf{A}_z \tag{2.4.6}$$

where $\mathbf{A}_x$, $\mathbf{A}_y$, and $\mathbf{A}_z$ are vector components of **A** parallel to *X*, *Y*, and *Z*. $\mathbf{A}_x$, $\mathbf{A}_y$, and $\mathbf{A}_z$ may be considered as "projections" of **A** along the *X*-, *Y*-, and *Z*-axes. Their magnitudes are proportional to the magnitude of **A** and the cosines of the angles θ_x, θ_y, and θ_z. That is,

$$\begin{aligned} |\mathbf{A}_x| &= |\mathbf{A}|\cos\theta_x = a_x \\ |\mathbf{A}_y| &= |\mathbf{A}|\cos\theta_y = a_y \\ |\mathbf{A}_z| &= |\mathbf{A}|\cos\theta_z = a_z \end{aligned} \tag{2.4.7}$$

where a_x, a_y, and a_z are defined as given in the equations. As before, let $\mathbf{n}_x$, $\mathbf{n}_y$, and $\mathbf{n}_z$ be unit vectors parallel to *X*, *Y*, and *Z*. Then, $\mathbf{A}_x$, $\mathbf{A}_y$, and $\mathbf{A}_z$ can be expressed as:

$$\begin{aligned} \mathbf{A}_x &= a_x\mathbf{n}_x = |\mathbf{A}|\cos\theta_x\mathbf{n}_x \\ \mathbf{A}_y &= a_y\mathbf{n}_y = |\mathbf{A}|\cos\theta_y\mathbf{n}_y \\ \mathbf{A}_z &= a_z\mathbf{n}_z = |\mathbf{A}|\cos\theta_z\mathbf{n}_z \end{aligned} \tag{2.4.8}$$

Finally, by substituting into Eq. (2.4.6), **A** may be expressed in the form:

$$\mathbf{A} = a_x\mathbf{n}_x + a_y\mathbf{n}_y + a_z\mathbf{n}_z = |\mathbf{A}|\cos\theta_x\mathbf{n}_x + |\mathbf{A}|\cos\theta_y\mathbf{n}_y + |\mathbf{A}|\cos\theta_z\mathbf{n}_z \tag{2.4.9}$$

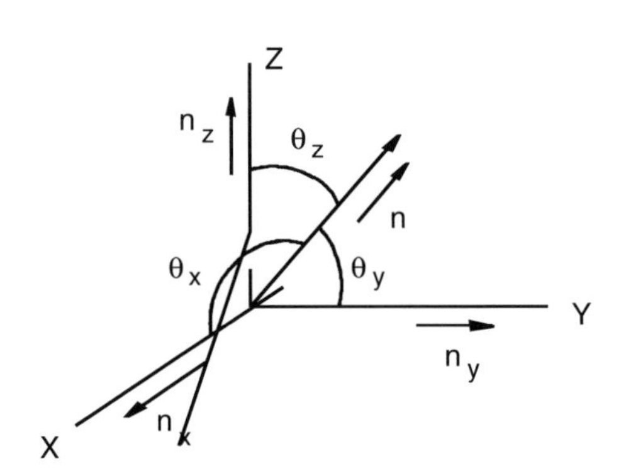

FIGURE 2.4.3
A vector **A** inclined relative to the coordinate axes.

Then, the magnitude of **A** is simply:

$$|\mathbf{A}| = \left(a_x^2 + a_y^2 + a_z^2\right)^{1/2} \quad (2.4.10)$$

Next, suppose that **n** is a unit vector parallel to **A**, as in Figure 2.4.3. Then, by following the same procedure as in Eq. (2.4.9), **n** may be expressed as:

$$\mathbf{n} = \cos\theta_x \mathbf{n}_x + \cos\theta_y \mathbf{n}_y + \cos\theta_z \mathbf{n}_z \quad (2.4.11)$$

Because the magnitude of **n** is 1, we then have:

$$1 = \cos^2\theta_x + \cos^2\theta_y + \cos^2\theta_z \quad (2.4.12)$$

Example 2.4.1: Vector Addition Using Components

To illustrate the use of these ideas, consider again the vectors of Examples 2.3.1 and 2.3.2 (see Figures 2.3.6 and 2.3.7). Specifically, let **A** be parallel to the *Y*-axis with a magnitude of 15 N. Let **B** be parallel to the *Y–Z* plane and inclined at 60° relative to the *Y*-axis. Let the magnitude of **B** be 12 N. Let **C** be parallel to the *X*-axis with a magnitude of 10 N (Figure 2.4.4). As before let $\mathbf{n}_x$, $\mathbf{n}_y$, and $\mathbf{n}_z$ be unit vectors parallel to the *X*-, *Y*-, and *Z*-axes. Then, **A**, **B**, and **C** may be expressed as:

$$\begin{aligned} &\mathbf{A} = 15\mathbf{n}_x N \\ &\mathbf{B} = 12\cos 60°\mathbf{n}_y + 12\sin 60°\mathbf{n}_z = 6\mathbf{n}_y + 10.392\mathbf{n}_z N \\ &\mathbf{C} = 10\mathbf{n}_x N \end{aligned} \quad (2.4.13)$$

Hence, **R** becomes:

$$\begin{aligned} \mathbf{R} &= \mathbf{A} + \mathbf{B} + \mathbf{C} = 15\mathbf{n}_y + \left(6\mathbf{n}_y + 10.392\mathbf{n}_z\right) + 10\mathbf{n}_x N \\ &= 10\mathbf{n}_x + 21\mathbf{n}_y + 10.392\mathbf{n}_z N \end{aligned} \quad (2.4.14)$$

Then, the magnitude of **R** is:

$$|\mathbf{R}| = \left[(10)^2 + (21)^2 + (10.392)^2\right]^{1/2} N = 25.47N \quad (2.4.15)$$

This is the same result as in Example 2.3.2 (see Eq. (2.3.8)). Here, however, we obtained the result without using the law of cosines as in Eq. (2.3.4).

Example 2.4.2: Direction Cosines

A particle *P* is observed to move on a curve *C* in a Cartesian reference frame *R*, as shown in Figure 2.4.5. The coordinates of *P* are functions of time *t*. Suppose that at an instant of interest *x*, *y*, and *z* have the values 8 m, 12 m, and 7 m, respectively. Determine the orientation angles of the line of sight of an observer of *P* if the observer is at the origin *O*. Specifically, determine the angles θ_x, θ_y, and θ_z of *OP* with the *X*-, *Y*-, and *Z*-axes.

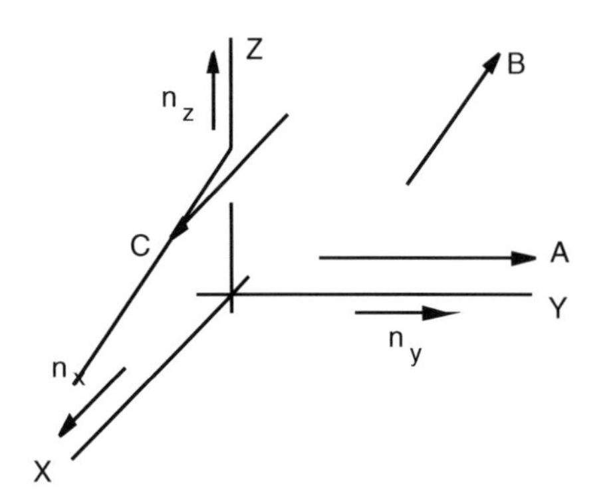

FIGURE 2.4.4
The system of Figure 2.3.6.

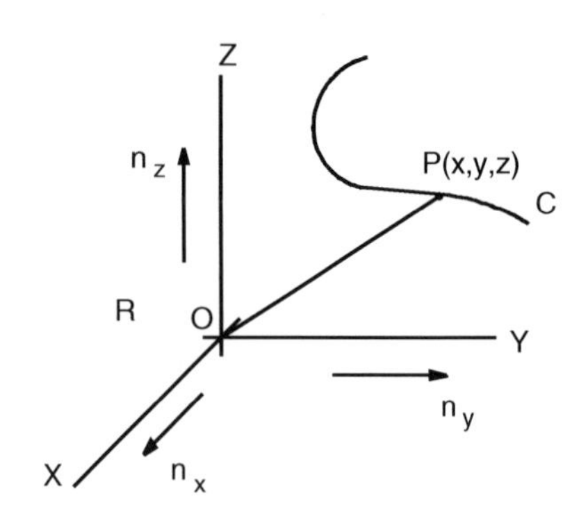

FIGURE 2.4.5
A particle P moving in a reference frame R.

Solution: The line of sight OP may be represented by the position vector **p** of Figure 2.4.5. In terms of unit vectors $\mathbf{n}_x$, $\mathbf{n}_y$, and $\mathbf{n}_z$ parallel to X, Y, and Z, **p** may be expressed as:

$$\mathbf{p} = x\mathbf{n}_x + y\mathbf{n}_y + z\mathbf{n}_z = 8\mathbf{n}_x + 12\mathbf{n}_y + 7\mathbf{n}_z \ \ m \tag{2.4.16}$$

Then, the magnitude of **p** is:

$$|\mathbf{p}| = \left[(8)^2 + (12)^2 + (7)^2\right]^{1/2} = 16.03\,\text{m} \tag{2.4.17}$$

Therefore, a unit vector **n** parallel to **p** is:

$$\mathbf{n} = \mathbf{p}/|\mathbf{p}| = 0.499\mathbf{n}_x + 0.749\mathbf{n}_y + 0.437\mathbf{n}_z \tag{2.4.18}$$

Then, from Eq. (2.4.11), the direction cosines are:

$$\cos\theta_x = 0.499, \ \cos\theta_y = 0.749, \ \cos\theta_z = 0.437 \tag{2.4.19}$$

Hence, θ_x, θ_y, and θ_z are:

$$\theta_x = 60.6\,\text{deg} \quad \theta_y = 41.54\,\text{deg} \quad \theta_z = 64.11\,\text{deg} \tag{2.4.20}$$

Observe that the functional representation of the coordinates x, y, and z of P as:

$$x = x(t), \quad y = y(t), \quad z = z(t) \tag{2.4.21}$$

forms a set of parametric equations defining C.

Finally, if a vector **V** is expressed in the form:

$$\mathbf{V} = v_x\mathbf{n}_x + v_y\mathbf{n}_y + v_z\mathbf{n}_z \tag{2.4.22}$$

where $\mathbf{n}_x$, $\mathbf{n}_y$, and $\mathbf{n}_z$ are mutually perpendicular unit vectors, then v_x, v_y, and v_z are called the scalar components of **V** relative to $\mathbf{n}_x$, $\mathbf{n}_y$, and $\mathbf{n}_z$. Then, from Eq. (2.4.10), the magnitude of **V** is:

$$|\mathbf{V}| = \left(v_x^2 + v_y^2 + v_z^2\right)^{1/2} \tag{2.4.23}$$

Observe that if **V** is zero, then $|\mathbf{V}|$ is zero, and each of the scalar components is also zero. This is the basis for force equilibrium procedures of elementary mechanics.

2.5 Angle Between Two Vectors

The concept of the angle between two vectors is useful in developing the procedures of vector multiplication. We already used this idea in Section 2.3 with the law of cosines (see Figure 2.3.5). The angle between two vectors is defined as follows: Let **A** and **B** be any nonzero vectors as in Figure 2.5.1. Let the vectors be connected tail to tail, as in Figure 2.5.2. Then the angle as shown is defined as the angle between the vectors. Observe that θ always has values between 0 and 180°.

2.6 Vector Multiplication: Scalar Product

Multiplying vectors can be accomplished in several ways, which we will review in this and the following sections. Consider first the *scalar* product, so called because the result is a scalar: Given any two vectors **A** and **B**, the scalar product, written as $\mathbf{A} \cdot \mathbf{B}$, is defined as:

$$\mathbf{A} \cdot \mathbf{B} = |\mathbf{A}||\mathbf{B}| \cos\theta \tag{2.6.1}$$

where θ is the angle between **A** and **B** (see Section 2.5.1). Because a dot is placed between the vectors, the operation is often called the *dot product*.

Observe in the definition of Eq. (2.6.1) that, if we interchange the positions of **A** and **B**, the result remains the same. That is,

$$\mathbf{A} \cdot \mathbf{B} = |\mathbf{A}||\mathbf{B}| \cos\theta = |\mathbf{B}||\mathbf{A}| \cos\theta = \mathbf{B} \cdot \mathbf{A} \tag{2.6.2}$$

Hence, the scalar product is *commutative.*

Consider some special cases. First, observe that the scalar product of two perpendicular vectors is zero, as the $\cos\theta$ is zero. Next, consider the scalar product of a vector **A** with itself:

$$\mathbf{A} \cdot \mathbf{A} = |\mathbf{A}||\mathbf{A}| \cos\theta \tag{2.6.3}$$

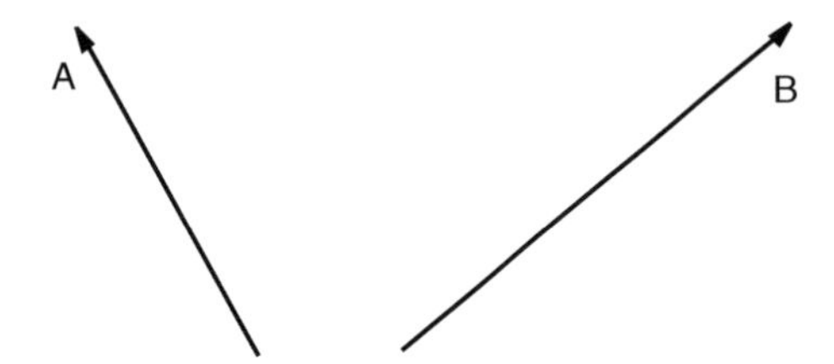

FIGURE 2.5.1
Two nonzero vectors.

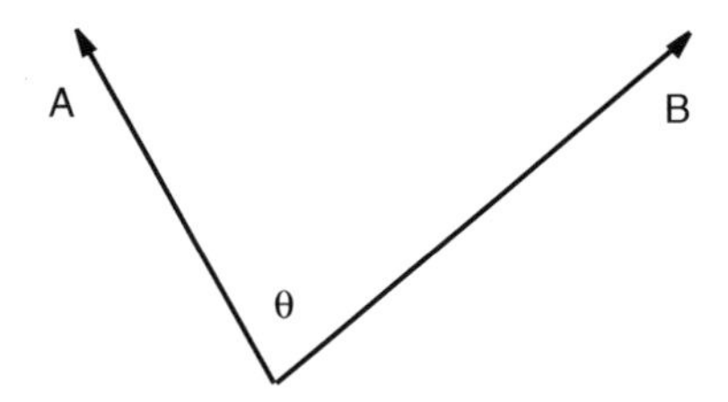

FIGURE 2.5.2
Vectors **A** and **B** connected tail to tail.

However, the angle that a vector makes with itself is zero (see Problem P2.5.1); hence, we have:

$$\mathbf{A}\cdot\mathbf{A} = |\mathbf{A}||\mathbf{A}| = |\mathbf{A}|^2 = \mathbf{A}^2 \tag{2.6.4}$$

where the last equality is a definition of $\mathbf{A}^2$.

Suppose in this last case that **A** is a unit vector, say **n**. Then,

$$\mathbf{n}\cdot\mathbf{n} = \mathbf{n}^2 = |\mathbf{n}|^2 = 1 \tag{2.6.5}$$

Next, suppose that $\mathbf{n}_1$, $\mathbf{n}_2$, and $\mathbf{n}_3$ are mutually perpendicular unit vectors parallel to the *X*-, *Y*-, and *Z*-axes as shown in Figure 2.6.1. Then, the various scalar products of these unit vectors may be expressed as:

$$\mathbf{n}_1\cdot\mathbf{n}_1 = 1, \quad \mathbf{n}_2\cdot\mathbf{n}_2 = 1, \quad \mathbf{n}_3\cdot\mathbf{n}_3 = 1 \tag{2.6.6}$$

$$\mathbf{n}_1\cdot\mathbf{n}_2 = \mathbf{n}_2\cdot\mathbf{n}_1 = \mathbf{n}_1\cdot\mathbf{n}_3 = \mathbf{n}_3\cdot\mathbf{n}_1 = \mathbf{n}_2\cdot\mathbf{n}_3 = \mathbf{n}_3\cdot\mathbf{n}_2 = 0 \tag{2.6.7}$$

These results may be expressed in the compact form:

$$\mathbf{n}_i\cdot\mathbf{n}_j = \delta_{ij} = \begin{cases} 1 & i = j \\ 0 & i \neq j \end{cases} \tag{2.6.8}$$

where δ_{ij} is often called Kronecker's delta function.

Next, suppose that **n** is a unit vector parallel to a line *L* and that **A** is a nonzero vector, as in Figure 2.6.2. Then, $\mathbf{A} \bullet \mathbf{n}$ is:

$$\mathbf{A}\cdot\mathbf{n} = |\mathbf{A}||\mathbf{n}|\cos\theta = |\mathbf{A}|\cos\theta \tag{2.6.9}$$

We can interpret this result as the "projection" of **A** onto *L*. Indeed, suppose we express **A** in terms of two components: one parallel to *L*, called $\mathbf{A}_{||}$, and the other perpendicular to *L*, called $\mathbf{A}_{\perp}$. Then,

$$\mathbf{A} = \mathbf{A}_{||} + \mathbf{A}_{\perp} \tag{2.6.10}$$

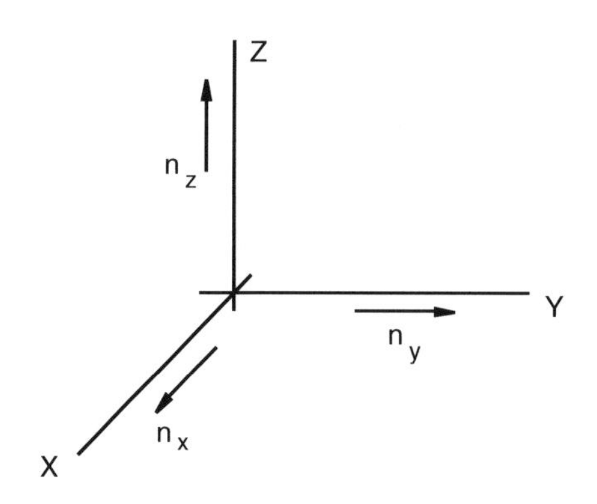

FIGURE 2.6.1
Mutually perpendicular unit vectors.

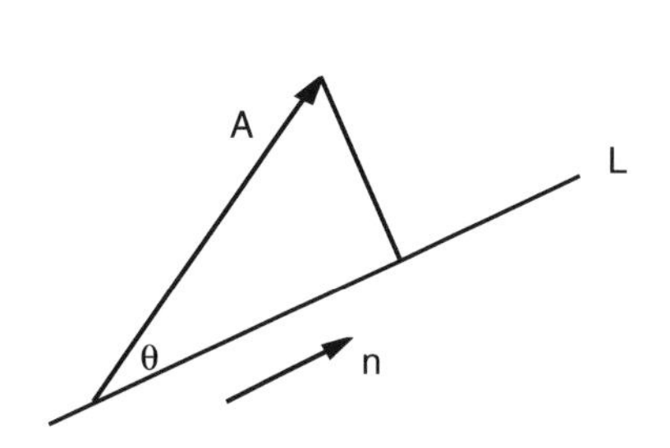

FIGURE 2.6.2
Vector **A**, line *L*, and unit vector **n**.

where $A_{||}$ is:

$$\mathbf{A}_{||} = (\mathbf{A} \cdot \mathbf{n})\mathbf{n} = (|\mathbf{A}|\cos\theta)\mathbf{n} \tag{2.6.11}$$

To further develop these components, let a vector **C** be the resultant (sum) of vectors **A** and **B**. That is,

$$\mathbf{C} = \mathbf{A} + \mathbf{B} \tag{2.6.12}$$

Let *L* be a line passing through the tail *O* of **C**, as in Figure 2.6.3, and let **n** be a unit vector parallel to *L*. Let *A* and *B* be the projection points of the heads of **A** and **B** onto *L*, as shown. Then, from Eq. (2.6.9), the lengths of the line segments *OA*, *AB*, and *OB* may be expressed as:

$$OA = \mathbf{n} \cdot \mathbf{A} \quad AB = \mathbf{n} \cdot \mathrm{B} \quad OB = \mathbf{n} \cdot \mathbf{C} \tag{2.6.13}$$

However, from Figure 2.6.3, we see that:

$$OB = OA + AB \tag{2.6.14}$$

Hence,

$$\mathbf{n} \cdot \mathbf{C} = \mathbf{n} \cdot \mathbf{A} + \mathbf{n} \cdot \mathbf{B} \tag{2.6.15}$$

Therefore, from Eq. (2.6.12), we have the *distributive law*:

$$\mathbf{n} \cdot (\mathbf{A} + \mathbf{B}) = \mathbf{n} \cdot \mathbf{A} + \mathbf{n} \cdot \mathbf{B} \tag{2.6.16}$$

Continuing in this manner, suppose *s* is a scalar. Then, from the definition of Eq. (2.6.1), we have:

$$s\mathbf{A} \cdot \mathbf{B} = (s\mathbf{A}) \cdot \mathbf{B} + \mathbf{A} \cdot (s\mathbf{B}) = \mathbf{A} \cdot s\mathbf{B} = \mathbf{A} \cdot \mathbf{B}s \tag{2.6.17}$$

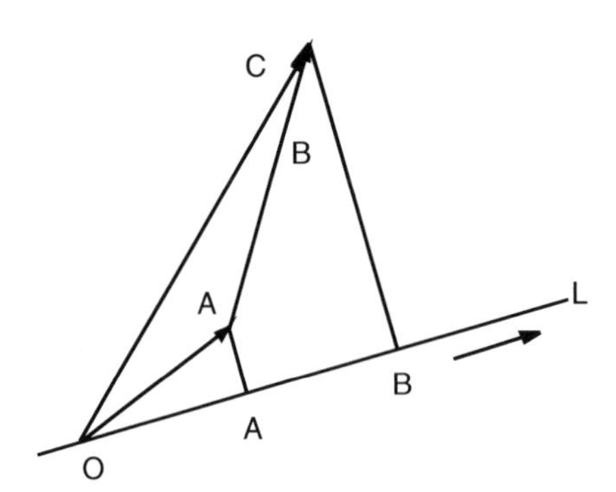

FIGURE 2.6.3
A vector sum projected onto a line L.

That is, the scalar may be placed anywhere in the product. The parentheses are unnecessary. The dot, however, must remain between the vectors.

Next, suppose $\mathbf{n}_x$, $\mathbf{n}_y$, and $\mathbf{n}_z$ are mutually perpendicular unit vectors parallel to X-, Y-, and Z-axes, and suppose $\mathbf{V}$ is a nonzero vector, as in Figure 2.6.4. Let θ_x, θ_y, and θ_z be the angles between $\mathbf{V}$ and the X-, Y-, and Z-axes. Then, the projections of $\mathbf{V}$ onto X, Y, and Z are:

$$\mathbf{V}\cdot\mathbf{n}_x = |\mathbf{V}|\cos\theta_x \quad \mathbf{V}\cdot\mathbf{n}_y = |\mathbf{V}|\cos\theta_y \quad \mathbf{V}\cdot\mathbf{n}_z = |\mathbf{V}|\cos\theta_z \tag{2.6.18}$$

Then, from Eq. (2.6.11), if $\mathbf{V}_x$, $\mathbf{V}_y$, and $\mathbf{V}_z$ are mutually perpendicular components of $\mathbf{V}$ parallel to X, Y, and Z, we can express $\mathbf{V}_x$, $\mathbf{V}_y$, and $\mathbf{V}_z$ as:

$$\mathbf{V}_x = (\mathbf{V}\cdot\mathbf{n}_x)\mathbf{n}_x \quad \mathbf{V}_y = (\mathbf{V}\cdot\mathbf{n}_y)\mathbf{n}_y \quad \mathbf{V}_z = (\mathbf{V}\cdot\mathbf{n}_z)\mathbf{n}_z \tag{2.6.19}$$

Then, we also may express $\mathbf{V}$ as:

$$\begin{aligned}\mathbf{V} = \mathbf{V}_x + \mathbf{V}_y + \mathbf{V}_z &= (\mathbf{V}\cdot\mathbf{n}_x)\mathbf{n}_x + (\mathbf{V}\cdot\mathbf{n}_y)\mathbf{n}_y + (\mathbf{V}\cdot\mathbf{n}_z)\mathbf{n}_z \\ &= V_x\mathbf{n}_x + V_y\mathbf{n}_y + V_z\mathbf{n}_z\end{aligned} \tag{2.6.20}$$

where V_x, V_y, and V_z are the scalar components of $\mathbf{V}$ along X, Y, and Z.

As a final example, suppose $\mathbf{n}_1$, $\mathbf{n}_2$, and $\mathbf{n}_3$ are mutually perpendicular unit vectors, and suppose $\mathbf{A}$ and $\mathbf{B}$ can be expressed as:

$$\mathbf{A} = \mathbf{A}_1\mathbf{n}_1 + \mathbf{A}_2\mathbf{n}_2 + \mathbf{A}_3\mathbf{n}_3 \tag{2.6.21}$$

$$\mathbf{B} = \mathbf{B}_1\mathbf{n}_1 + \mathbf{B}_2\mathbf{n}_2 + \mathbf{B}_3\mathbf{n}_3 \tag{2.6.22}$$

Then, by using Eqs. (2.6.8), (2.6.16), and (2.6.17), we can express $\mathbf{A}\cdot\mathbf{B}$ as:

$$\begin{aligned}A\cdot B &= (A_1\mathbf{n}_1 + A_2\mathbf{n}_2 + A_3\mathbf{n}_3)\cdot(B_1\mathbf{n}_1 + B_2\mathbf{n}_2 + B_3\mathbf{n}_3) \\ &= A_1B_1\mathbf{n}_1\cdot\mathbf{n}_1 + A_1B_2\mathbf{n}_1\cdot\mathbf{n}_2 + A_1B_3\mathbf{n}_1\cdot\mathbf{n}_3 \\ &\quad + A_2B_1\mathbf{n}_2\cdot\mathbf{n}_1 + A_3B_2\mathbf{n}_3\cdot\mathbf{n}_2 + A_2B_3\mathbf{n}_2\cdot\mathbf{n}_3 \\ &\quad + A_3B_1\mathbf{n}_3\cdot\mathbf{n}_1 + A_3B_2\mathbf{n}_3\cdot\mathbf{n}_2 + A_3B_3\mathbf{n}_3\cdot\mathbf{n}_3 \\ &= A_1B_1 + A_2B_2 + A_3B_3\end{aligned} \tag{2.6.23}$$

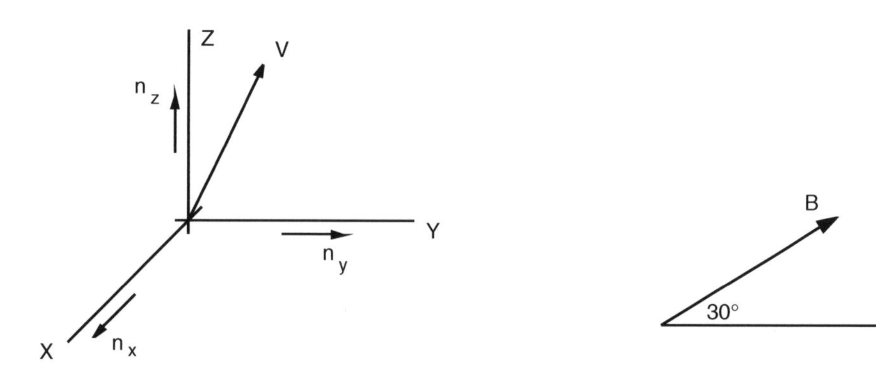

FIGURE 2.6.4
A vector **V** and coordinate axes *X*, *Y*, and *Z*.

FIGURE 2.6.5
Vectors **A** and **B**.

This result will prove to be useful in many ways. It demonstrates the analytical advantage of vector algebra. For example, if we want to obtain the magnitude of a vector, it is simply:

$$|\mathbf{A}| = \sqrt{|\mathbf{A}|^2} = \sqrt{\mathbf{A} \cdot \mathbf{A}} = \sqrt{A_1^2 + A_2^2 + A_3^2} \tag{2.6.24}$$

Also, if we want to find the angle θ between two vectors, we have:

$$\cos\theta = \frac{\mathbf{A} \cdot \mathbf{B}}{|\mathbf{A}||\mathbf{B}|} = \frac{A_1B_1 + A_2B_2 + A_3B_3}{\left(A_1^2 + A_2^2 + A_3^2\right)^{\frac{1}{2}}\left(B_1^2 + B_2^2 + B_3^2\right)^{\frac{1}{2}}} \tag{2.6.25}$$

The application and utility of the scalar product may be further illustrated by a few examples, as in the following paragraphs.

Example 2.6.1: Scalar Product Definition

Consider the vectors **A** and **B,** shown in Figure 2.6.5. Let the magnitude of **A** and **B** be 4 and 3, respectively. Evaluate the scalar product.

Solution: From Figure 2.6.5, the angle between **A** and **B** is 30°; therefore, from Eq. (2.6.1) the scalar product is:

$$\mathbf{A} \cdot \mathbf{B} = |\mathbf{A}||\mathbf{B}|\cos\theta = (4)(3)\left(\sqrt{3}/2\right) = 6\sqrt{3} \tag{2.6.26}$$

Example 2.6.2: Magnitude of a Vector

Let the vector **A** be expressed in terms of mutually perpendicular unit vectors $\mathbf{n}_1$, $\mathbf{n}_2$, and $\mathbf{n}_3$ as:

$$\mathbf{A} = 5\mathbf{n}_1 - 2\mathbf{n}_2 + 4\mathbf{n}_3 \tag{2.6.27}$$

Determine the magnitude of **A**.

Solution: From Eqs. (2.6.4) and (2.6.23) we have:

$$|\mathbf{A}|^2 = \mathbf{A}^2 = \mathbf{A} \cdot \mathbf{A} = (5)^2 + (-2)^2 + (4)^2 = 45 \tag{2.6.28}$$

Hence,

$$|\mathbf{A}| = \sqrt{45} \tag{2.6.29}$$

Example 2.6.3: Angle Between Two Vectors

Let vectors **A** and **B** be expressed in terms of unit vectors $\mathbf{n}_1$, $\mathbf{n}_2$, and $\mathbf{n}_3$ as:

$$\mathbf{A} = 3\mathbf{n} - \mathbf{n}_2 + 4\mathbf{n}_3 \quad \text{and} \quad \mathbf{B} = -5\mathbf{n}_1 + 2\mathbf{n}_2 + 7\mathbf{n}_3 \tag{2.6.30}$$

Determine the angle between **A** and **B**.

Solution: From Eq. (2.6.25), cosθ is:

$$\cos\theta = \frac{\mathbf{A}\cdot\mathbf{B}}{|\mathbf{A}||\mathbf{B}|} = \frac{(3)(-5)+(-1)(2)+(4)(7)}{\sqrt{26}\sqrt{78}} = 0.244 \tag{2.6.31}$$

Then, θ is:

$$\theta = \cos^{-1}(0.244) = 75.86\,\text{deg} \tag{2.6.32}$$

Example 2.6.4: Projection of a Vector Along a Line

Consider a velocity vector **V** and points *P* and *Q* of a Cartesian coordinate system *R* as shown in Figure 2.6.6. Let **V** be expressed in terms of mutually perpendicular unit vectors **i**, **j**, and **k** as:

$$\mathbf{V} = 3\mathbf{i} + 4\mathbf{j} + 4\mathbf{k}\ \text{ft/sec} \tag{2.6.33}$$

Let the coordinates of *P* and *Q* be as shown and let *L* be a line passing through *P* and *Q*. Determine the projection of **V** along *L*.

Solution: From Eq. (2.6.9), the projection of **V** along *L* is simply where **n** is a unit vector along *L*. Because *L* passes through *P* and *Q*, **n** is:

$$\mathbf{n} = \mathbf{PQ}/|\mathbf{PQ}| = (-\mathbf{i} + 5\mathbf{j} - 2\mathbf{k})/\sqrt{29} \tag{2.6.34}$$

Therefore, the projection of **V** along (or onto) *L* is:

$$\begin{aligned} \mathbf{v}\cdot\mathbf{n} &= (3\mathbf{i} + 4\mathbf{j} + 4\mathbf{k})\cdot(-\mathbf{i} + 5\mathbf{j} - 2\mathbf{k})/\sqrt{29} \\ &= [(3)(-1) + (4)(5) + (4)(-2)]/\sqrt{29} = 1.67\ \text{ft/sec} \end{aligned} \tag{2.6.35}$$

2.7 Vector Multiplication: Vector Product

Next, consider the "vector" product, so called because the result is a vector. Given any vectors **A** and **B**, the vector product, written **A** × **B**, is defined as:

$$\mathbf{A}\times\mathbf{B} = |\mathbf{A}||\mathbf{B}|\sin\theta\ \mathbf{n} \tag{2.7.1}$$

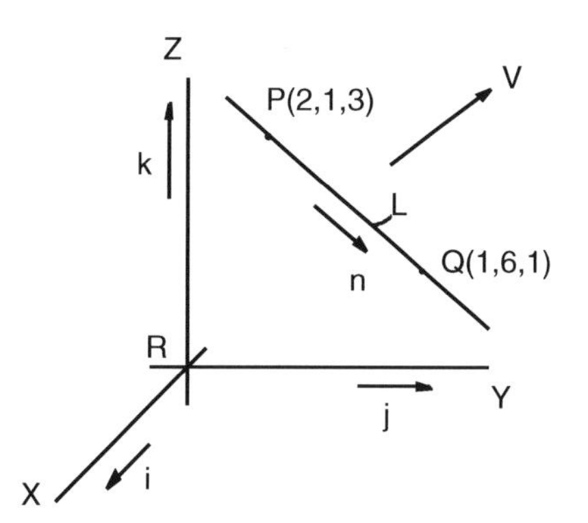

FIGURE 2.6.6
A vector **V** and line L and a Cartesian coordinate system R.

where, as before, θ is the angle between **A** and **B** and **n** is a unit vector perpendicular to both **A** and **B**. The sense of **n** is determined by the "right-hand rule": if **A** is rotated toward **B** so that θ is diminished, the sense of **n** is in the direction of advance of a right-hand screw, with axis perpendicular to **A** and **B**, and rotated in the same way (**A** to **B**). Because a cross is placed between the vectors, the operation is often called the *cross product*.

Observe in the definition of **n** that the sense of **n** depends upon which vector occurs first in the product; hence, the vector product is anticommutative. That is,

$$\mathbf{A}\times\mathbf{B}=-\mathbf{B}\times\mathbf{A} \tag{2.7.2}$$

Observe further from the definition that, if s is a scalar, then:

$$s\mathbf{A}\times\mathbf{B}=(s\mathbf{A})\times\mathbf{B}=\mathbf{A}\times(s\mathbf{B})=\mathbf{A}\times\mathbf{B}s \tag{2.7.3}$$

That is, the scalar can be placed at any position in the operation; the parentheses are unnecessary. The cross, however, must remain between the vectors.

Consider some examples: First, the vector product of a vector with itself is zero because the angle a vector makes with itself is zero. That is,

$$\mathbf{A}\times\mathbf{A}=0 \tag{2.7.4}$$

Next, consider the vector product of mutually perpendicular unit vectors: Let $\mathbf{n}_1$, $\mathbf{n}_2$, and $\mathbf{n}_3$ be parallel to the X-, Y-, and Z-axes, as shown in Figure 2.7.1. Then, from the definition Eq. (2.7.1), we have the following results:

$$\begin{array}{lll} \mathbf{n}_1\times\mathbf{n}_1=0 & \mathbf{n}_1\times\mathbf{n}_2=\mathbf{n}_3 & \mathbf{n}_1\times\mathbf{n}_3=-\mathbf{n}_2 \\ \mathbf{n}_2\times\mathbf{n}_1=-\mathbf{n}_3 & \mathbf{n}_2\times\mathbf{n}_2=0 & \mathbf{n}_2\times\mathbf{n}_3=\mathbf{n}_1 \\ \mathbf{n}_3\times\mathbf{n}_1=\mathbf{n}_2 & \mathbf{n}_3\times\mathbf{n}_2=-\mathbf{n}_1 & \mathbf{n}_3\times\mathbf{n}_3=0 \end{array} \tag{2.7.5}$$

Observe that these results may be summarized by the expression:

$$\mathbf{n}_i\times\mathbf{n}_j=\sum_{i=1}^{3}e_{ijk}\mathbf{n}_k \tag{2.7.6}$$

where e_{ijk} is the permutation symbol defined as:

$$e_{ijk} = \begin{cases} 1 & \textit{if } i, \ j, \text{ and } k \text{ are distinct and cyclic (for example: 1, 2, 3)} \\ -1 & \textit{if } i, \ j, \text{ and } k \text{ are distinct and anticyclic (for example: 1, 3, 2)} \\ 0 & \textit{if } i, \ j, \text{ and } k \text{ are not distinct and cyclic (for example: 1, 1, 3)} \end{cases} \tag{2.7.7}$$

This definition is readily seen to be equivalent to the expression:

$$e_{ijk} = 1/2(i-j)(j-k)(k-i) \tag{2.7.8}$$

When the unit vectors are arranged as in Figure 2.7.1, so that minus signs do *not* appear in the equations with cyclic indices, the system is said to be "right-handed" or *dextral*. Alternatively, when the indices are *anticyclic* and minus signs do not occur, the system is said to be "left-handed" or *sinistral*. Figure 2.7.2 shows an example of a sinistral system. In this book we will always use right-handed systems.

Next, consider the vector product of a vector **A** with a sum of vectors **B** + **C**. Let **D** be the resultant of **B** and **C**, and let $\mathbf{n}_A$ be a unit vector parallel to and with the same sense as **A**. Then, from Eq. (2.6.15), we have:

$$\mathbf{n}_A \cdot \mathbf{D} = \mathbf{n}_A \cdot (\mathbf{B} + \mathbf{C}) = \mathbf{n}_A \cdot \mathbf{B} + \mathbf{n}_A \cdot \mathbf{C} \tag{2.7.9}$$

Let $\mathbf{D}_{||}$, $\mathbf{B}_{||}$, and $\mathbf{C}_{||}$ be defined as:

$$\mathbf{D}_{||} = (\mathbf{n}_A \cdot \mathbf{D})\mathbf{n}_A, \quad \mathbf{B}_{||} = (\mathbf{n}_A \cdot \mathbf{B})\mathbf{n}_A, \text{ and } \mathbf{C}_{||} = (\mathbf{n}_A \cdot \mathbf{C})\mathbf{n}_A \tag{2.7.10}$$

Then, from Eq. (2.7.9), we have:

$$\mathbf{D}_{||} = \mathbf{B}_{||} + \mathbf{C}_{||} \tag{2.7.11}$$

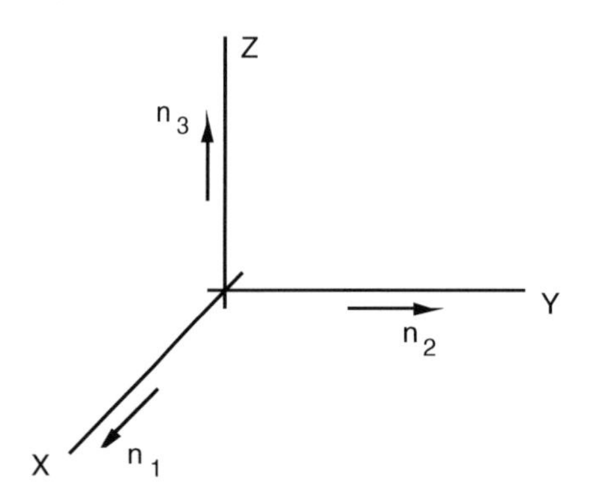

FIGURE 2.7.1
Mutually perpendicular unit vectors and coordinate axes.

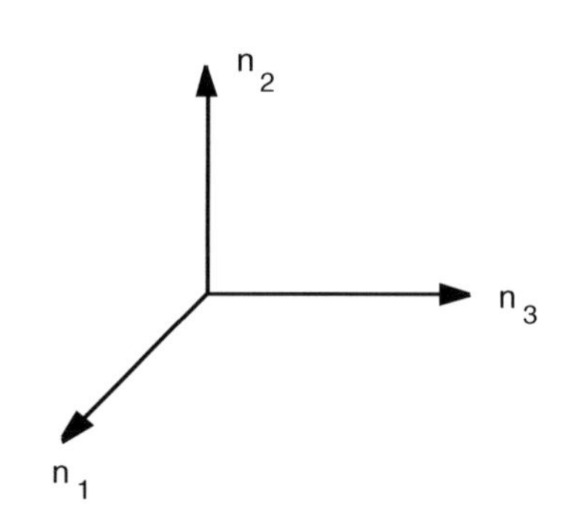

FIGURE 2.7.2
A sinistral unit vector system.

$\mathbf{D}_{||}$, $\mathbf{B}_{||}$, and $\mathbf{C}_{||}$ are the projections of **D**, **B**, and **C** along **A**; hence, let $\mathbf{D}_\perp$, $\mathbf{B}_\perp$, and $\mathbf{C}_\perp$ be defined as:

$$\mathbf{D}_\perp = \mathbf{D} - \mathbf{D}_{||},\ \mathbf{B}_\perp = \mathbf{B} - \mathbf{B}_{||},\ \text{and } \mathbf{C}_\perp = \mathbf{C} - \mathbf{C}_{||} \tag{2.7.12}$$

$\mathbf{D}_\perp$, $\mathbf{B}_\perp$, and $\mathbf{C}_\perp$ are the components of **D**, **B**, and **C** perpendicular to **A**. Because **D** is the resultant of **B** and **C** (**D** = **B** + **C**), Eq. (2.7.11) shows that:

$$\mathbf{D}_\perp = \mathbf{B}_\perp + \mathbf{C}_\perp \tag{2.7.13}$$

Hence, let $\mathbf{D}_\perp$, $\mathbf{B}_\perp$, and $\mathbf{C}_\perp$ be depicted as in Figure 2.7.3.

Consider the products $\mathbf{n}_A \times \mathbf{D}_\perp$, $\mathbf{n}_A \times \mathbf{B}_\perp$, and $\mathbf{n}_A \times \mathbf{C}_\perp$: because $|\mathbf{n}_A|$ is 1.0, and because $\mathbf{n}_A$ is perpendicular to $\mathbf{D}_\perp$, $\mathbf{B}_\perp$, and $\mathbf{C}_\perp$, the definition of Eq. (2.7.1) shows that $\mathbf{n}_A \times \mathbf{D}_\perp$, $\mathbf{n}_A \times \mathbf{B}_\perp$, and $\mathbf{n}_A \times \mathbf{C}_\perp$ are vectors in the plane of $\mathbf{D}_\perp$, $\mathbf{B}_\perp$, and $\mathbf{C}_\perp$ with the same magnitudes as $\mathbf{D}_\perp$, $\mathbf{B}_\perp$, and $\mathbf{C}_\perp$ and with directions perpendicular to $\mathbf{D}_\perp$, $\mathbf{B}_\perp$, and $\mathbf{C}_\perp$. Let $\hat{\mathbf{D}}_\perp$, $\hat{\mathbf{B}}_\perp$, and $\hat{\mathbf{C}}_\perp$ represent $\mathbf{n}_A \times \mathbf{D}_\perp$, $\mathbf{n}_A \times \mathbf{B}_\perp$, and $\mathbf{n}_A \times \mathbf{C}_\perp$. Then, in view of Figure 2.7.3, $\hat{\mathbf{D}}_\perp$, $\hat{\mathbf{B}}_\perp$, and $\hat{\mathbf{C}}_\perp$ may be represented as in Figure 2.7.4. Hence, we have:

$$\hat{\mathbf{D}}_\perp = \hat{\mathbf{B}}_\perp + \hat{\mathbf{C}}_\perp \tag{2.7.14}$$

or,

$$\mathbf{n}_A \times \mathbf{D}_\perp = \mathbf{n}_A \times \mathbf{B}_\perp + \mathbf{n}_A \times \mathbf{C}_\perp \tag{2.7.15}$$

Then, from Eq. (2.7.13) we have:

$$\mathbf{n}_A \times (\mathbf{B}_\perp + \mathbf{C}_\perp) = \mathbf{n}_A \times \mathbf{B}_\perp + \mathbf{n}_A \times \mathbf{C}_\perp \tag{2.7.16}$$

By multiplying by $|\mathbf{A}|$ (a scalar) we then obtain (see Eq. (2.3.6)):

$$\mathbf{A} \times (\mathbf{B}_\perp + \mathbf{C}_\perp) = \mathbf{A} \times \mathbf{B}_\perp + \mathbf{A} \times \mathbf{C}_\perp \tag{2.7.17}$$

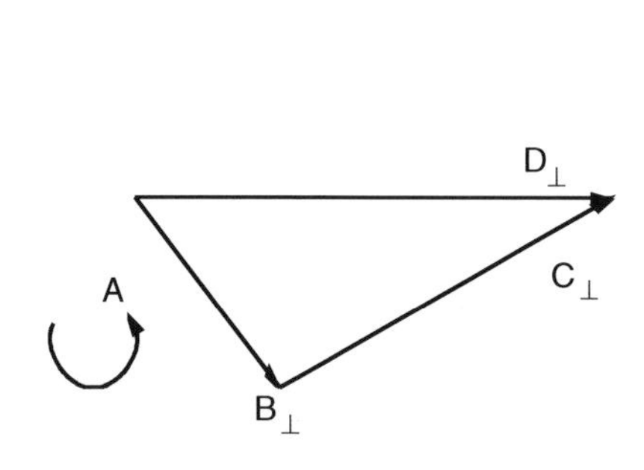

FIGURE 2.7.3
A representation of vectors perpendicular to a vector **A**.

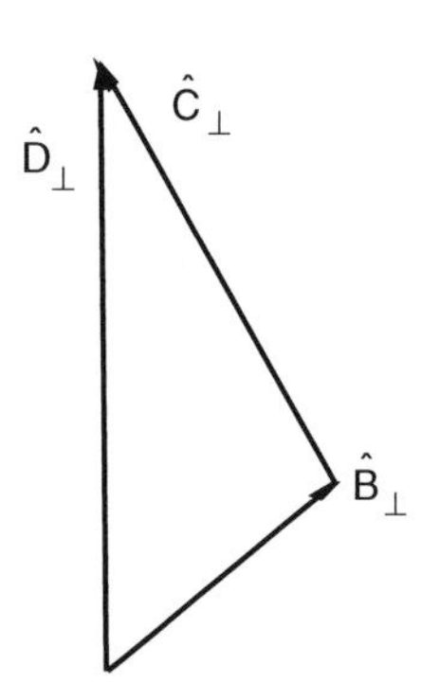

FIGURE 2.7.4
A representation of $\hat{\mathbf{D}}_\perp$ ($= \mathbf{n}_A \times \mathbf{D}_\perp$), $\hat{\mathbf{B}}_\perp$ ($= \mathbf{n}_A \times \mathbf{B}_\perp$), and $\hat{\mathbf{C}}_\perp$ ($= \mathbf{n}_A \times \mathbf{C}_\perp$).

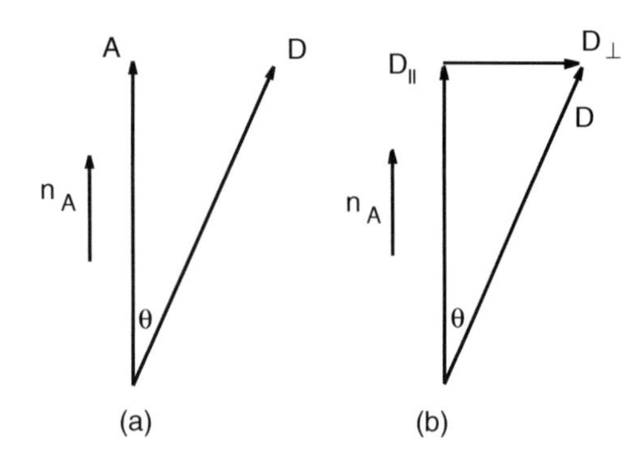

FIGURE 2.7.5
Vectors $\mathbf{A}$, $\mathbf{D}$, $\mathbf{n}_A$, $\mathbf{D}_{||}$, and $\mathbf{D}_\perp$.

Returning to the product $\mathbf{A} \times \mathbf{D}$, let $\mathbf{A}$ and $\mathbf{D}$ be depicted as in Figure 2.7.5a, where θ is the angle between the vectors. As before, let $\mathbf{n}_A$ be a unit vector parallel to $\mathbf{A}$. Then, $\mathbf{n}_A \times \mathbf{D}$ is a vector perpendicular to $\mathbf{n}_A$ and with magnitude $|\mathbf{D}|\,|\sin\theta|$. From Figure 2.7.5b, we see that:

$$|\mathbf{D}||\sin\theta| = |\mathbf{D}_\perp| \tag{2.7.18}$$

By similar reasoning we have:

$$\mathbf{A} \times \mathbf{B} = \mathbf{A} \times \mathbf{B}_\perp \quad \text{and} \quad \mathbf{A} \times \mathbf{C} = \mathbf{A} \times \mathbf{C}_\perp \tag{2.7.19}$$

Therefore, by comparing Eqs. (2.7.17) and (2.7.19), we have:

$$\mathbf{A} \times \mathbf{D} = \mathbf{A} \times (\mathbf{B} + \mathbf{C}) = \mathbf{A} \times \mathbf{B} + \mathbf{A} \times \mathbf{C} \tag{2.7.20}$$

This establishes the *distributive* law.

Finally, suppose that $\mathbf{n}_1$, $\mathbf{n}_2$, and $\mathbf{n}_3$ are mutually perpendicular unit vectors, and suppose that vectors $\mathbf{A}$ and $\mathbf{B}$ are expressed in the forms:

$$\mathbf{A} = A_1\mathbf{n}_1 + A_2\mathbf{n}_2 + A_3\mathbf{n}_3 \quad \text{and} \quad B = B_1\mathbf{n}_1 + B_2\mathbf{n}_2 + B_3\mathbf{n}_3 \tag{2.7.21}$$

Then, by repeated use of Eqs. (2.3.6), (2.7.5), and (2.7.20) we see that $\mathbf{A} \times \mathbf{B}$ may be expressed as:

$$\begin{aligned} \mathbf{A} \times \mathbf{B} &= (A_2B_3 - A_3B_2)\mathbf{n}_1 + (A_3B_1 - A_3B_1)\mathbf{n}_2 + (A_1B_2 - A_2B_1)\mathbf{n}_3 \\ &= \sum_{i=1}^{3}\sum_{j=1}^{3}\sum_{k=1}^{3} e_{ijk} A_i B_i \mathbf{n}_k \end{aligned} \tag{2.7.22}$$

By recalling the elementary rules for expanding determinants, we see that Eq. (2.7.22) may be written as:

$$\mathbf{A} \times \mathbf{B} = \begin{vmatrix} n_1 & n_2 & n_3 \\ A_1 & A_2 & A_3 \\ B_1 & B_2 & B_3 \end{vmatrix} \tag{2.7.23}$$

This is a useful algorithm for computing the vector product.

Example 2.7.1: Vector Product Computation and Geometric Properties of the Vector Product

Let vectors **A** and **B** be expressed in terms of mutually perpendicular dextral unit vectors $\mathbf{n}_1$, $\mathbf{n}_2$, and $\mathbf{n}_3$ as:

$$\mathbf{A} = 7\mathbf{n}_1 - 2\mathbf{n}_2 + 4\mathbf{n}_3 \quad \text{and} \quad \mathbf{B} = \mathbf{n}_1 + 3\mathbf{n}_2 - 8\mathbf{n}_3 \tag{2.7.24}$$

Let **C** be the vector product $\mathbf{A} \times \mathbf{B}$.

a. Find **C**.
b. Show that **C** is perpendicular to both **A** and **B**.
c. Show that $\mathbf{B} \times \mathbf{A} = -\mathbf{C}$.

Solution:

a. From Eq. (2.7.24), **C** is:

$$\mathbf{C} = \begin{vmatrix} \mathbf{n}_1 & \mathbf{n}_2 & \mathbf{n}_3 \\ 7 & -2 & 4 \\ 1 & 3 & -8 \end{vmatrix} = 4\mathbf{n}_1 + 60\mathbf{n}_2 + 23\mathbf{n}_3 \tag{2.7.25}$$

b. If **C** is perpendicular to **A**, with the angle θ between **C** and **A** being 90°, $\mathbf{C} \bullet \mathbf{A}$ is zero because cosθ is zero. Conversely, if $\mathbf{C} \bullet \mathbf{A}$ is zero, and neither **C** nor **A** is zero, then cosθ is zero, making **C** perpendicular to **A**. From Eq. (2.6.21), $\mathbf{C} \bullet \mathbf{A}$ is:

$$\mathbf{C} \cdot \mathbf{A} = (4)(7) + (60)(-2) + (23)(4) = 0 \tag{2.7.26}$$

Similarly, $\mathbf{C} \cdot \mathbf{B}$ is

$$\mathbf{C} \cdot \mathbf{B} = (4)(1) + (60)(3) + (23)(-8) = 0 \tag{2.7.27}$$

c. From Eq. (2.7.23), $\mathbf{B} \times \mathbf{A}$ is:

$$\mathbf{B} \times \mathbf{A} = \begin{vmatrix} \mathbf{n}_1 & \mathbf{n}_2 & \mathbf{n}_3 \\ 1 & 3 & -8 \\ 7 & -2 & 4 \end{vmatrix} = -4\mathbf{n}_1 - 60\mathbf{n}_2 - 23\mathbf{n}_3 \tag{2.7.28}$$

which is seen to be from Eq. (2.7.25).

2.8 Vector Multiplication: Triple Products

On many occasions it is necessary to consider the product of *three* vectors. Such products are called "triple" products. Two triple products that will be helpful to use are the *scalar triple product* and the *vector triple product*.

Given three vectors **A**, **B**, and **C**, the scalar triple product has the form $\mathbf{A} \cdot (\mathbf{B} \times \mathbf{C})$. The result is a scalar. The scalar triple product is seen to be a combination of a vector product and a scalar product.

Recall from Eqs. (2.6.21) and (2.7.23) that if $\mathbf{n}_1$, $\mathbf{n}_2$, and $\mathbf{n}_3$ are mutually perpendicular unit vectors, the algorithms for evaluating the scalar and vector products are:

$$\mathbf{A} \cdot \mathbf{B} = A_1 B_1 + A_2 B_2 + A_3 B_3 \tag{2.8.1}$$

and

$$A \times B = \begin{vmatrix} \mathbf{n}_1 & \mathbf{n}_2 & \mathbf{n}_3 \\ A_1 & A_2 & A_3 \\ B_1 & B_2 & B_3 \end{vmatrix} \tag{2.8.2}$$

where the A_i and B_i ($i = 1, 2, 3$) are the $\mathbf{n}_i$ components of **A** and **B**.

By comparing Eqs. (2.8.1) and (2.8.2), we see that the scalar triple product $\mathbf{A} \cdot (\mathbf{B} \times \mathbf{C})$ may be obtained by replacing $\mathbf{n}_1$, $\mathbf{n}_2$, and $\mathbf{n}_3$ in:

$$\mathbf{A} \cdot (\mathbf{B} \times \mathbf{C}) = \begin{vmatrix} A_1 & A_2 & A_3 \\ B_1 & B_2 & B_3 \\ C_1 & C_2 & C_3 \end{vmatrix} \tag{2.8.3}$$

Recall from the elementary rules of evaluating determinants that the rows and columns may be interchanged without changing the value of the determinant. Also, the rows and columns may be cyclically permuted without changing the determinant value. If the rows or columns are anticyclically permuted, the value of the determinant changes sign. Hence, we can rewrite Eq. (2.8.3) in the forms:

$$\begin{aligned} \mathbf{A} \cdot (\mathbf{B} \times \mathbf{C}) &= \begin{vmatrix} A_1 & A_2 & A_3 \\ B_1 & B_2 & B_3 \\ C_1 & C_2 & C_3 \end{vmatrix} = \begin{vmatrix} B_1 & B_2 & B_3 \\ C_1 & C_2 & C_3 \\ A_1 & A_2 & A_3 \end{vmatrix} = \begin{vmatrix} C_1 & C_2 & C_3 \\ A_1 & A_2 & A_3 mp \\ B_1 & B_2 & B_3 \end{vmatrix} \\ &= -\begin{vmatrix} B_1 & B_2 & B_3 \\ A_1 & A_2 & A_3 \\ C_1 & C_2 & C_3 \end{vmatrix} = -\begin{vmatrix} A_1 & A_2 & A_3 \\ C_1 & C_2 & C_3 \\ B_1 & B_2 & B_3 \end{vmatrix} = -\begin{vmatrix} C_1 & C_2 & C_3 \\ B_1 & B_2 & B_3 \\ A_1 & A_2 & A_3 \end{vmatrix} \end{aligned} \tag{2.8.4}$$

By examining Eq. (2.8.4), we see that the dot and the cross may be interchanged in the product. Also, the parentheses are unnecessary. Finally, the vectors may be cyclically permuted without changing the value of the product. An anticyclic permutation changes the sign of the result. Specifically,

$$\begin{aligned} \mathbf{A} \cdot (\mathbf{B} \times \mathbf{C}) &= \mathbf{A} \cdot \mathbf{B} \times \mathbf{C} = \mathbf{A} \times \mathbf{B} \cdot \mathbf{C} = \mathbf{C} \cdot \mathbf{A} \times \mathbf{B} \\ &= \mathbf{B} \cdot \mathbf{C} \times \mathbf{A} = -\mathbf{B} \cdot \mathbf{A} \times \mathbf{C} = -\mathbf{C} \cdot \mathbf{B} \times \mathbf{A} = -\mathbf{A} \cdot \mathbf{C} \times \mathbf{B} \end{aligned} \tag{2.8.5}$$

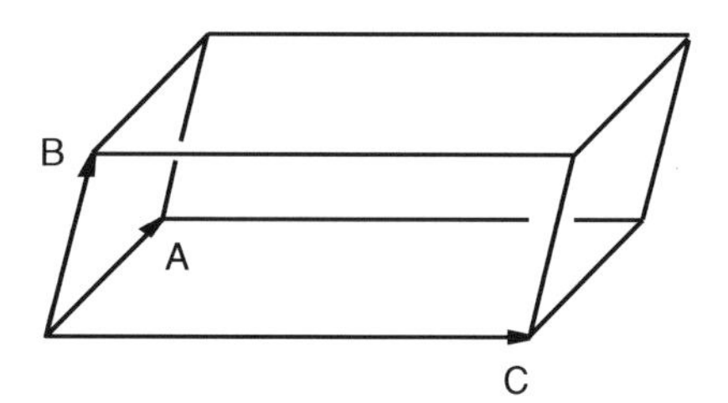

FIGURE 2.8.1
A parallelepiped with vectors **A**, **B**, and **C** along the edges.

If the vectors **A**, **B**, and **C** coincide with the edges of a parallelepiped as in Figure 2.8.1, the scalar triple product of **A**, **B**, and **C** is seen to be equal to the *volume* of the parallelepiped. That is, the volume V is:

$$V = \mathbf{A} \times \mathbf{B} \cdot \mathbf{C} \tag{2.8.6}$$

Example 2.8.1: Verification of Interchangeability of Terms of Triple Scalar Products

Let vectors **A**, **B**, and **C** be expressed in terms of mutually perpendicular unit vectors $\mathbf{n}_1$, $\mathbf{n}_2$, and $\mathbf{n}_3$ as:

$$\mathbf{A} = -3\mathbf{n}_1 - \mathbf{n}_2 + \mathbf{n}_3, \quad \mathbf{B} = 2\mathbf{n}_1 + 4\mathbf{n}_2 - 7\mathbf{n}_3, \quad \mathbf{C} = -\mathbf{n}_1 + 3\mathbf{n}_2 - 5\mathbf{n}_3 \tag{2.8.7}$$

Verify the equalities of Eq. (2.8.5).

Solution: From Eq. (2.8.2), the vector products $\mathbf{A} \times \mathbf{B}$ and $\mathbf{B} \times \mathbf{C}$ are:

$$\mathbf{A} \times \mathbf{B} = \begin{vmatrix} \mathbf{n}_1 & \mathbf{n}_2 & \mathbf{n}_3 \\ 3 & -1 & 1 \\ 2 & 4 & -7 \end{vmatrix} = 3\mathbf{n}_1 + 23\mathbf{n}_2 + 14\mathbf{n}_3 \tag{2.8.8}$$

$$\mathbf{B} \times \mathbf{C} = \begin{vmatrix} \mathbf{n}_1 & \mathbf{n}_2 & \mathbf{n}_3 \\ 2 & 4 & -7 \\ -1 & 3 & -5 \end{vmatrix} = \mathbf{n}_1 + 17\mathbf{n}_2 + 10\mathbf{n}_3 \tag{2.8.9}$$

Then, from Eq. (2.6.22), the triple scalar products $\mathbf{A} \times \mathbf{B} \cdot \mathbf{C}$ and $\mathbf{B} \times \mathbf{C} \cdot \mathbf{A}$ are:

$$\mathbf{A} \times \mathbf{B} \cdot \mathbf{C} = (3)(-1) + (23)(3) + (14)(-5) = -4 \tag{2.8.10}$$

and

$$\mathbf{B} \times \mathbf{C} \cdot \mathbf{A} = (1)(3) + (17)(-1) + (10)(1) = -4 \tag{2.8.11}$$

The other equalities are verified similarly (see Problem P2.8.1).

Next, the vector triple product has one of the two forms: $\mathbf{A} \times (\mathbf{B} \times \mathbf{C})$ or $(\mathbf{A} \times \mathbf{B}) \times \mathbf{C}$. The result is a vector. The position of the parentheses is important, as the two forms generally produce different results.

To explore this, let the vectors **A**, **B**, and **C** be expressed in terms of mutually perpendicular unit vectors $\mathbf{n}_i$ with scalar components $\mathbf{A}_i$, $\mathbf{B}_i$, and $\mathbf{C}_i$ ($i = 1, 2, 3$). Then, by using the algorithms of Eqs. (2.8.1) and (2.8.2), we see that the vector triple products may be expressed as:

$$\mathbf{A} \times (\mathbf{B} \cdot \mathbf{C}) = (\mathbf{A} \cdot \mathbf{C})\mathbf{B} - (\mathbf{A} \cdot \mathbf{B})\mathbf{C} \tag{2.8.12}$$

and

$$(\mathbf{A} \times \mathbf{B}) \times \mathbf{C} = (\mathbf{A} \cdot \mathbf{C})\mathbf{B} - (\mathbf{B} \cdot \mathbf{C})\mathbf{A} \tag{2.8.13}$$

Observe that the last terms in these expressions are different.

Example 2.8.2: Validity of Eqs. (2.8.6) and (2.8.7) and the Necessity of Parentheses

Verify Eqs. (2.8.6) and (2.8.7) using the vectors of Eq. (2.8.7).

Solution: From Eqs. (2.8.2) and (2.8.9), $\mathbf{A} \times (\mathbf{B} \times \mathbf{C})$ is:

$$\mathbf{A} \times (\mathbf{B} \times \mathbf{C}) = \begin{vmatrix} \mathbf{n}_1 & \mathbf{n}_2 & \mathbf{n}_3 \\ 3 & -1 & 1 \\ 1 & 17 & 10 \end{vmatrix} = -27\mathbf{n}_1 - 29\mathbf{n}_2 + 52\mathbf{n}_3 \tag{2.8.14}$$

From Eq. (2.6.22), $(\mathbf{A} \cdot \mathbf{C})\mathbf{B} - (\mathbf{A} \cdot \mathbf{B})\mathbf{C}$ is:

$$\begin{aligned} (\mathbf{A} \cdot \mathbf{C})\mathbf{B} - (\mathbf{A} \cdot \mathbf{C})\mathbf{C} &= [(3)(-1) + (-1)(3) + (1)(-5)](2\mathbf{n}_1 + 4\mathbf{n}_2 - 7\mathbf{n}_3) \\ &\quad - [(3)(2) + (-1)(4) + (1)(-7)](-\mathbf{n}_1 + 3\mathbf{n}_2 - 5\mathbf{n}_3) \\ &= (-11)(2\mathbf{n}_1 + 4\mathbf{n}_2 - 7\mathbf{n}_3) \\ &\quad - (-5)(-\mathbf{n}_1 + 3\mathbf{n}_2 - 5\mathbf{n}_3) \\ &= -27\mathbf{n}_1 - 29\mathbf{n}_2 + 52\mathbf{n}_3 \end{aligned} \tag{2.8.15}$$

Similarly, $(\mathbf{A} \times \mathbf{B}) \times \mathbf{C}$ and $(\mathbf{A} \cdot \mathbf{C})\mathbf{B} - (\mathbf{B} \cdot \mathbf{C})\mathbf{A}$ are:

$$(\mathbf{A} \times \mathbf{B}) \times \mathbf{C} = \begin{vmatrix} \mathbf{n}_1 & \mathbf{n}_2 & \mathbf{n}_3 \\ 3 & 23 & 14 \\ -1 & 3 & -5 \end{vmatrix} = -157\mathbf{n}_1 + \mathbf{n}_2 + 32\mathbf{n}_3 \tag{2.8.16}$$

and

$$\begin{aligned} (\mathbf{A} \cdot \mathbf{C})\mathbf{B} - (\mathbf{B} \cdot \mathbf{C})\mathbf{A} &= [(3)(-1) + (-1)(3) + (1)(-5)](2\mathbf{n}_1 + 4\mathbf{n}_2 - 7\mathbf{n}_3) \\ &\quad - [(2)(-1) + (4)(3) + (-7)(-5)](3\mathbf{n}_1 - \mathbf{n}_2 + \mathbf{n}_3) \\ &= (-11)(2\mathbf{n}_1 + 4\mathbf{n}_2 - 7\mathbf{n}_3) \\ &\quad - (45)(3\mathbf{n}_1 - \mathbf{n}_2 + \mathbf{n}_3) \\ &= -157\mathbf{n}_1 + \mathbf{n}_2 + 32\mathbf{n}_3 \end{aligned} \tag{2.8.17}$$

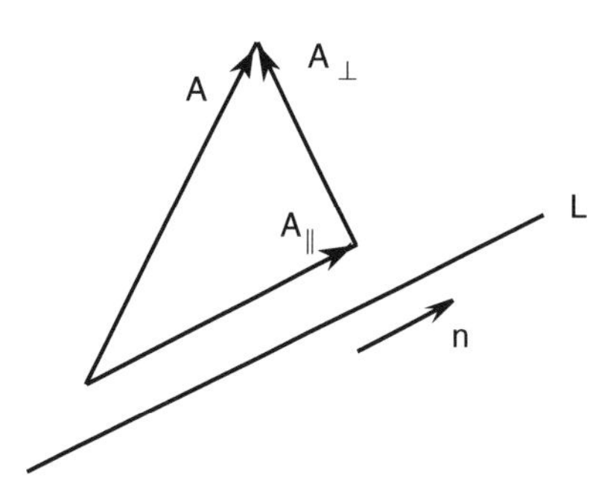

FIGURE 2.8.2
Projection of a vector A parallel and perpendicular to a line.

Observe that the results in Eqs. (2.8.14) and (2.8.15) are identical and thus consistent with Eq. (2.8.12). Similarly, the results of Eqs. (2.8.16) and (2.8.17) are the same, thus verifying Eq. (2.8.13). Finally, observe that the results of Eqs. (2.8.14) and (2.8.16) are different, thus demonstrating the necessity for parentheses on the left sides of Eqs. (2.8.12) and (2.8.13).

Recall from Eq. (2.6.10) that the projection $\mathbf{A}_{||}$ of a vector **A** along a line *L* is:

$$\mathbf{A}_{||} = (\mathbf{A} \cdot \mathbf{n})\mathbf{n} \tag{2.8.18}$$

where **n** is a unit vector parallel to *L*, as in Figure 2.8.2. The vector triple product may be used to obtain $\mathbf{A}_{\perp}$, the component of perpendicular to *L*. That is,

$$\mathbf{A}_{\perp} = \mathbf{n} \times (\mathbf{A} \times \mathbf{n}) \tag{2.8.19}$$

To see this, use Eqs. (2.8.6) and (2.8.8) to expand the product. That is,

$$\mathbf{n} \times (\mathbf{A} \times \mathbf{n}) = \mathbf{A} - (\mathbf{n} \cdot \mathrm{A})\mathbf{n} = \mathbf{A} - \mathbf{A}_{||} = \mathbf{A}_{\perp} \tag{2.8.20}$$

2.9 Use of the Index Summation Convention

Observe in the previous sections that expressing a vector in terms of mutually perpendicular unit vectors results in a sum of products of the scalar components and the unit vectors. Specifically, if **v** is any vector and if $\mathbf{n}_1$, $\mathbf{n}_2$, and $\mathbf{n}_3$ are mutually perpendicular unit vectors, we can express **v** in the form:

$$\mathbf{v} = -v_1\mathbf{n}_1 + v_2\mathbf{n}_1 + v_3\mathbf{n}_3 = \sum_{i=1}^{3} v_i\mathbf{n}_i \tag{2.9.1}$$

Because these sums occur so frequently, and because the pattern of the indices is similar for sums of products, it is convenient to introduce the "summation convention": if an index is repeated, there is a sum over the range of the index (usually 1 to 3). This means, for example, in Eq. (2.9.1), that the summation sign may be deleted. That is, **v** may be expressed as:

$$\mathbf{V} = v_i\mathbf{n}_i \tag{2.9.2}$$

In using this convention, several rules are useful. First, in an equation or expression, a repeated index may not be repeated more than one time. Second, any letter may be used for a repeated index. For example, in Eq. (2.9.2) we may write:

$$\mathbf{v} = v_i \mathbf{n}_i = v_j \mathbf{n}_j = v_k \mathbf{n}_k \tag{2.9.3}$$

Finally, if an index is not repeated it is a "free" index. In an equation, there must be a correspondence of free indices on both sides of the equation and in each term of the equation.

In using the summation convention, we can express the scalar and vector products as follows: If $\mathbf{n}_1$, $\mathbf{n}_2$, and $\mathbf{n}_3$ are mutually perpendicular unit vectors and if vectors **a** and **b** are expressed as:

$$\mathbf{a} = a_i \mathbf{n}_i \quad \text{and} \quad \mathbf{b} = b_i \mathbf{n}_i \tag{2.9.4}$$

then the products $\mathbf{a} \cdot \mathbf{b}$ and $\mathbf{a} \times \mathbf{b}$ are:

$$\mathbf{a} \cdot \mathbf{b} = a_i b_i \quad \textit{and} \quad \mathbf{a} \times \mathbf{b} = e_{ijk} a_i b_j \mathbf{n}_k \tag{2.9.5}$$

where, as before, e_{ijk} is the permutation symbol (see Eq. (2.7.7)).

With a little practice, the summation convention becomes natural in analysis procedures. We will employ it when it is convenient.

Example 2.9.1: Kronecker Delta Function Interpreted as a Substitution Symbol

Consider the expression $\delta_{ij} V_j$ where the δ_{ij} $(i, j = 1, 2, 3)$ are components of the Kronecker delta function defined by Eq. (2.6.7) and where V_j are the components of a vector **V** relative to a mutually perpendicular unit vector set $\mathbf{n}_i$ $(i = 1, 2, 3)$. From the summation convention, we have:

$$\delta_{ij} V_j = \delta_{i1} V_1 + \delta_{i2} V_2 + \delta_{i3} V_3 \quad (i = 1,\ 2,\ 3) \tag{2.9.6}$$

In this equation, i has one of the values 1, 2, or 3. If i is 1, the right side of the equation reduces to V_1; if i is 2, the right side becomes V_2; and, if i is 3, the right side is V_3. Therefore, the right side is simply V_i. That is,

$$\delta_{ij} V_j = V_i \tag{2.9.7}$$

The Kronecker delta function may then be interpreted as an index operator, substituting an i for the j, thus the name *substitution symbol.*

2.10 Review of Matrix Procedures

In continuing our review of vector algebra, it is helpful to recall the elementary procedures in matrix algebra. For illustration purposes, we will focus our attention primarily on square matrices and on row and column arrays. Recall that a matrix A is simply an array of numbers a_{ij} ($i = 1,\ldots, m_i$; $j = 1,\ldots, n$) arranged in m rows and n columns as:

$$A = \begin{bmatrix} a_{11} & a_{12} & \cdots & a_1 n \\ a_{21} & a_{22} & \cdots & a_2 n \\ \cdots & & & \\ a_{m1} & a_{2n} & \cdots & a_{mn} \end{bmatrix} \tag{2.10.1}$$

The entries a_{ij} are usually called the *elements* of the matrix. The first subscript indicates the row position, and the second subscript designates the column position.

Two matrices A and B are said to be equal if they have equal elements. That is,

$$A = B \quad \text{if and only if} \quad a_{ij} = b_{ij} \quad i = 1,\ldots,n; j = 1,\ldots,m \tag{2.10.2}$$

If all the elements of matrix are zero, it is called a *zero matrix*. If a matrix has only one row, it is called a *row matrix* or *row array*. If a matrix has only one column, it is called a *column matrix* or *column array*. A matrix with an equal number of rows and columns is a *square* matrix. If all the elements of a square matrix are zero except for the diagonal elements, the matrix is called a *diagonal matrix*. If all the diagonal elements of a diagonal matrix have the value 1, the matrix is called an *identity matrix*. If a square matrix has a zero determinant, it is said to be a *singular matrix*. The *transpose* of a matrix A (written A^T) is the matrix obtained by interchanging the rows and columns of A. If a matrix and its transpose are equal, the matrix is said to be *symmetric*. If a matrix is equal to the negative of its transpose, it is said to be *antisymmetric*.

Recall that the algebra of matrices is based upon a few simple rules: First, the multiplication of a matrix A by a scalar s produces a matrix B whose elements are equal to the elements of A multiplied by s. That is,

$$B = sA \quad \text{if and only if} \quad b_{ij} = sa_{ij} \tag{2.10.3}$$

Next, the sum of two matrices A and B is a matrix C whose elements are equal to the sum of the respective elements of A and B. That is,

$$C = A + B \quad \text{if and only if} \quad c_{ij} = a_{ij} + b_{ij} \tag{2.10.4}$$

Matrix subtraction is defined similarly.

The product of matrices is defined through the "row–column" product algorithm. The product of two matrices A and B (written AB) is possible only if the number of columns in the first matrix A is equal to the number of rows of the second matrix B. When this occurs, the matrices are said to be *conformable*. If C is the product AB, then the element c_{ij} is the sum of products of the elements of the ith row of A with the corresponding elements of the jth column of B. Specifically,

$$c_{ij} = \sum_{k=1}^{n} a_{ik} b_{kj} \tag{2.10.5}$$

where n is the number of columns of A and the number of rows of B.

Matrix products are distributive over addition and subtraction. That is,

$$A(B+C)=AB+AC \tag{2.10.6}$$

and

$$(B+C)A=BA+CA \tag{2.10.7}$$

Matrix products are also associative. That is, for conformable matrices A, B, and C, we have:

$$(AB)C=A(BC)=ABC \tag{2.10.8}$$

Hence, the parentheses are unnecessary.

Next, it is readily seen that the transpose of a product is the product of the transposes in reverse order. That is:

$$(AB)^T=B^TA^T \tag{2.10.9}$$

Finally, if A is a nonsingular square matrix, the *inverse* of A, written as A^{-1}, is the matrix such that:

$$AA^{-1}=A^{-1}A=I \tag{2.10.10}$$

where I is the identity matrix. A^{-1} may be determined as follows: Let M_{ij} be the *minor* of the element a_{ij} defined as the determinant of the matrix occurring when the ith row of A and the jth column of a are deleted. Let $\hat{A}_{ij}$ be the *adjoint* of a_{ij} defined as:

$$\hat{A}_{ij}=(-1)^{i+j}M_{ij} \tag{2.10.11}$$

Then A^{-1} is the matrix with elements:

$$A^{-1}=\left[\hat{A}_{ij}\right]^T\Big/\det A \tag{2.10.12}$$

where detA designates the determinant of A and $[\hat{A}_{ij}]^T$ is the transpose of the matrix of adjoints.

If it happens that:

$$A^{-1}=A^T \tag{2.10.13}$$

then A is said to be *orthogonal*. In this case, the rows and columns of A may be considered as components of mutually perpendicular unit vectors.

2.11 Reference Frames and Unit Vector Sets

In the analysis of dynamical systems, it is frequently useful to express kinematical and dynamical quantities in several different reference frames. Orthogonal transformation matrices (as discussed in this section) are useful in obtaining relationships between the representations of the quantities in the different reference frames.

To explore these ideas consider two unit vector sets $\mathbf{n}_i$ and $\hat{\mathbf{n}}_i$ and an arbitrary $\mathbf{V}$ as in Figure 2.11.1. Let the sets be inclined relative to each other as shown. Recall from Eq. (2.6.19) that $\mathbf{V}$ may be expressed in terms of the $\mathbf{n}_i$ as:

$$\mathbf{V} = (\mathbf{V}\cdot\mathbf{n}_1)\mathbf{n}_1 + (\mathbf{V}\cdot\mathbf{n}_2)\mathbf{n}_2 + (\mathbf{V}\cdot\mathbf{n}_3)\mathbf{n}_3 = (\mathbf{V}\cdot\mathbf{n}_i)\mathbf{n}_i = \mathbf{V}_i\mathbf{n}_i \tag{2.11.1}$$

where the $\mathbf{V}_i$ are the scalar components of $\mathbf{V}$. Similarly, $\mathbf{V}$ may be expressed in terms of the $\hat{\mathbf{n}}_i$ as:

$$\mathbf{V} = (\mathbf{V}\cdot\hat{\mathbf{n}}_i)\hat{\mathbf{n}}_i = \hat{\mathbf{V}}_i\hat{\mathbf{n}}_i \tag{2.11.2}$$

Given the relative inclination of the unit vector sets, our objective is to obtain relations between the $\mathbf{V}_i$ and the $\hat{\mathbf{V}}_i$. To that end, let S be a matrix with elements S_{ij} defined as:

$$S_{ij} = \mathbf{n}_i \cdot \hat{\mathbf{n}}_j \tag{2.11.3}$$

Consider the $\mathbf{n}_i$: from Eq. (2.11.2), we can express $\mathbf{n}_i$ as:

$$\mathbf{n}_1 = (\mathbf{n}_1\cdot\hat{\mathbf{n}}_i)\hat{\mathbf{n}}_i = S_{1i}\hat{\mathbf{n}}_i \tag{2.11.4}$$

Similarly, $\mathbf{n}_2$ and $\mathbf{n}_3$ may be expressed as:

$$\mathbf{n}_2 = S_{2i}\hat{\mathbf{n}}_i \quad \text{and} \quad \mathbf{n}_3 = S_{3i}\hat{\mathbf{n}}_i \tag{2.11.5}$$

Thus, in general, we have:

$$\mathbf{n}_i = S_{ij}\hat{\mathbf{n}}_j \tag{2.11.6}$$

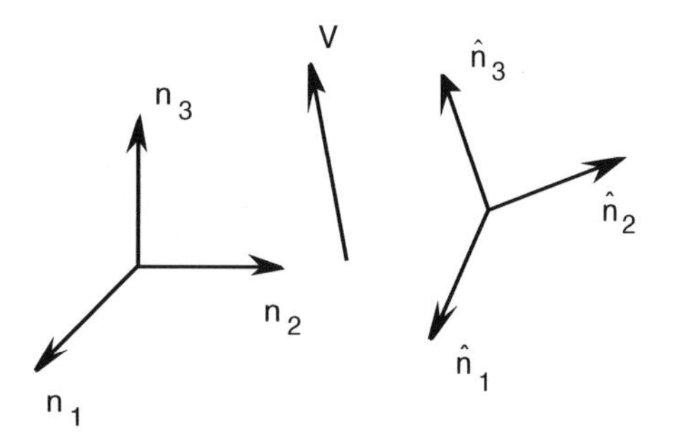

FIGURE 2.11.1
Two unit vector sets.

Similarly, if we express $\hat{\mathbf{n}}_1$, $\hat{\mathbf{n}}_2$, and $\hat{\mathbf{n}}_3$ in terms of the $\hat{\mathbf{n}}_i$, we have:

$$\hat{\mathbf{n}}_i = S_{ij}\mathbf{n}_j \tag{2.11.7}$$

Observe the difference between Eqs. (2.11.6) and (2.11.7): in Eq. (2.11.6), the sum is taken over the *second* index of S_{ij}, whereas in Eq. (2.11.7) it is taken over the *first* index. This is consistent with Eq. (2.11.3), where we see that the first index is associated with the $\mathbf{n}_i$ and the second with the $\hat{\mathbf{n}}_i$. Observe the same pattern in Eqs. (2.11.6) and (2.11.7).

By substituting from Eqs. (2.11.6) and (2.11.7) into Eqs. (2.11.1) and (2.11.2), we obtain:

$$\mathbf{V} = V_i\hat{\mathbf{n}}_i = V_iS_{ij}\hat{\mathbf{n}}_j = \hat{V}_j\hat{\mathbf{n}}_j \tag{2.11.8}$$

and

$$\mathbf{V} = \hat{V}_i\hat{\mathbf{n}}_i = \hat{V}_iS_{ij}\mathbf{n}_j = V_j\mathbf{n}_j \tag{2.11.9}$$

Hence, we have:

$$V_i = S_{ij}\hat{V}_j \quad \text{and} \quad \hat{V}_i = S_{ji}V_j \tag{2.11.10}$$

Observe that Eq. (2.11.10) has the same form as Eqs. (2.11.6) and (2.11.7).

By substituting from Eqs. (2.11.6) and (2.11.7), we obtain the expression:

$$S_{ij}S_{kj} = \delta_{ik} \quad \text{and} \quad S_{ji}S_{jk} = \delta_{ik} \tag{2.11.11}$$

where, as in Eq. (2.6.7), δ_{ik} is Kronecker's delta function. In matrix form, this may be written as:

$$SS^T = S^TS = I \quad \text{or} \quad S^T = S^{-1} \tag{2.11.12}$$

where, as before, I is the identity matrix. Hence, S is an orthogonal transformation matrix (see Eq. (2.10.12)).

To illustrate these ideas, imagine the unit vector sets $\mathbf{n}_i$ and $\hat{\mathbf{n}}_i$ to be aligned with each other such that $\mathbf{n}_i$ and $\hat{\mathbf{n}}_i$ are parallel ($i = 1, 2, 3$). Next, let the $\hat{\mathbf{n}}_i$ be rotated relative to the $\mathbf{n}_i$ and $\hat{\mathbf{n}}_i$ so that the angle between $\hat{\mathbf{n}}_2$ and $\mathbf{n}_2$ (and also between $\hat{\mathbf{n}}_3$ and $\mathbf{n}_3$) is α, as shown in Figure 2.11.2. Then, by inspection of the figure, $\mathbf{n}_i$ and $\hat{\mathbf{n}}_i$ are related by the expressions:

$$\begin{aligned} \mathbf{n}_1 &= \hat{\mathbf{n}}_1 & \hat{\mathbf{n}}_1 &= \mathbf{n}_1 \\ \mathbf{n}_2 &= c_\alpha\hat{\mathbf{n}}_2 + s_\alpha\mathbf{n}_3 \quad \text{and} & \hat{\mathbf{n}}_2 &= c_\alpha\mathbf{n}_2 + s_\alpha\mathbf{n}_3 \\ \mathbf{n}_3 &= s_\alpha\hat{\mathbf{n}}_2 + c_\alpha\hat{\mathbf{n}}_3 & \hat{\mathbf{n}}_3 &= -s_\alpha\mathbf{n}_2 + c_\alpha\mathbf{n}_3 \end{aligned} \tag{2.11.13}$$

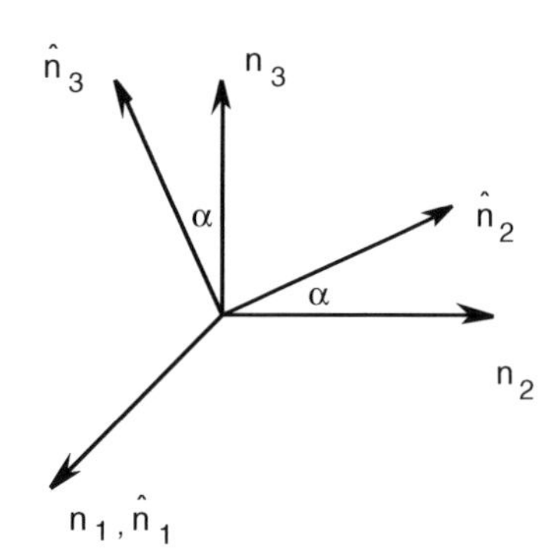

FIGURE 2.11.2
Unit vector sets $\mathbf{n}_i$ and $\hat{\mathbf{n}}_i$.

where s_α and c_α represent sinα and cosα. Hence, from Eq. (2.11.3), the matrix S has the elements:

$$S = \left[S_{ij}\right] = \begin{bmatrix} 1 & 0 & 0 \\ 0 & c_\alpha & -s_\alpha \\ 0 & s_\alpha & c_\alpha \end{bmatrix} \tag{2.11.14}$$

Observe that SS^T and S^TS are:

$$SS^T = \begin{bmatrix} 1 & 0 & 0 \\ 0 & c_\alpha & -s_\alpha \\ 0 & s_\alpha & c_\alpha \end{bmatrix} \begin{bmatrix} 1 & 0 & 0 \\ 0 & c_\alpha & s_\alpha \\ 0 & -s_\alpha & c_\alpha \end{bmatrix} = \begin{bmatrix} 1 & 0 & 0 \\ 0 & 1 & 0 \\ 0 & 0 & 1 \end{bmatrix} \tag{2.11.15}$$

and

$$S^TS = \begin{bmatrix} 1 & 0 & 0 \\ 0 & c_\alpha & s_\alpha \\ 0 & -s_\alpha & c_\alpha \end{bmatrix} \begin{bmatrix} 1 & 0 & 0 \\ 0 & c_\alpha & -s_\alpha \\ 0 & s_\alpha & c_\alpha \end{bmatrix} = \begin{bmatrix} 1 & 0 & 0 \\ 0 & 1 & 0 \\ 0 & 0 & 1 \end{bmatrix} \tag{2.11.16}$$

Observe in Figure 2.11.2 the rotation of the $\hat{\mathbf{n}}_i$ and $\hat{\mathbf{n}}_1$ through the angle α is a dextral rotation. Imagine analogous dextral rotations of the $\hat{\mathbf{n}}_i$ about $\hat{\mathbf{n}}_2$ and $\hat{\mathbf{n}}_3$ through the angles β and γ. Then, it is readily seen that the transformation matrices analogous to S of Eq. (2.11.14) are:

$$B = \begin{bmatrix} c_\beta & 0 & s_\beta \\ 0 & 1 & 0 \\ -s_\beta & 0 & c_\beta \end{bmatrix} \quad \text{and} \quad C = \begin{bmatrix} c_\gamma & -s_\gamma & 0 \\ s_\gamma & c_\gamma & 0 \\ 0 & 0 & 1 \end{bmatrix} \tag{2.11.17}$$

(In this context, the transformation matrix of Eq. (2.11.15) might be called A.)

Finally, let the $\hat{\mathbf{n}}_i$ have a general inclination relative to the $\mathbf{n}_i$ as shown in Figure 2.11.1. The $\hat{\mathbf{n}}_i$ may be brought into this configuration by initially aligning them with the $\mathbf{n}_i$ and then by performing *three* successive dextral rotations of the $\hat{\mathbf{n}}_i$ about $\hat{\mathbf{n}}_1$, $\hat{\mathbf{n}}_2$, and $\hat{\mathbf{n}}_3$ through the angles α, β, and γ. The transformation matrix S between the $\mathbf{n}_i$ and the $\hat{\mathbf{n}}_i$ is, then:

$$S = ABC = \begin{bmatrix} 1 & 0 & 0 \\ 0 & c_\alpha & -s_\alpha \\ 0 & s_\alpha & c_\alpha \end{bmatrix} \begin{bmatrix} c_\beta & 0 & s_\beta \\ 0 & 1 & 0 \\ -s_\beta & 0 & c_\beta \end{bmatrix} \begin{bmatrix} c_\gamma & -s_\gamma & 0 \\ s_\gamma & c_\gamma & 0 \\ 0 & 0 & 1 \end{bmatrix} \tag{2.11.18}$$

or,

$$S = \begin{bmatrix} c_\beta c_\gamma & -c_\beta s_\gamma & s_\beta \\ (c_\alpha s_\gamma + s_\alpha s_\beta c_\gamma) & (c_\alpha c_\gamma - s_\alpha s_\beta s_\gamma) & -s_\alpha c_\beta \\ (s_\alpha s_\gamma - c_\alpha s_\beta c_\gamma) & (s_\alpha c_\gamma + c_\alpha s_\beta s_\gamma) & c_\alpha c_\beta \end{bmatrix} \tag{2.11.19}$$

Also, it is readily seen that:

$$S^T = C^T B^T A^T \tag{2.11.20}$$

and that:

$$SS^T = S^T S = I \tag{2.11.21}$$

2.12 Closure

The foregoing discussion is intended primarily as a review of basic concepts of vector and matrix algebra. Readers who are either unfamiliar with these concepts or who want to explore the concepts in greater depth may want to consult a mathematics or vector mechanics text as provided in the References. We will freely employ these concepts throughout this text.

References

2.1. Hinchey, F. A., *Vectors and Tensors for Engineers and Scientists*, Wiley, New York, 1976.
2.2. Hsu, H. P., *Vector Analysis*, Simon & Schuster Technical Outlines, New York, 1969.
2.3. Haskell, R. E., *Introduction to Vectors and Cartesian Tensors*, Prentice Hall, Englewood Cliffs, NJ, 1972.
2.4. Shields, P. C., *Elementary Linear Algebra*, Worth Publishers, New York, 1968.
2.5. Ayers, F., *Theory and Problems of Matrices*, Schawn Outline Series, McGraw-Hill, New York, 1962.
2.6. Pettofrezzo, A. J., *Elements of Linear Algebra*, Prentice Hall, Englewood Cliffs, NJ, 1970.
2.7. Usamani, R. A., *Applied Linear Algebra*, Marcel Dekker, New York, 1987.
2.8. Borisenko, A. I., and Tarapov, I. E., *Vector and Tensor Analysis with Applications* (translated by R. A. Silverman), Prentice Hall, Englewood Cliffs, NJ, 1968.

Problems

Section 2.3 Vector Addition

P2.3.1: Given the vectors **A**, **B**, **C**, and **D** as shown in Figure P2.3.1, let the magnitudes of these vectors be:

$$|\mathbf{A}| = 8 \text{ lb}, \ |\mathbf{B}| = 6 \text{ lb}, \ |\mathbf{C}| = 12 \text{ lb}, \ |\mathbf{D}| = 5 \text{ lb}$$

Find the magnitudes of the following sums:

a. **A** + **B**
b. **B** + **A**
c. 2**A** + 2**B**
d. **A** – **B**
e. **A** + **C**
f. **B** + **C**
g. **A** + **B** + **C**
h. **C** + **D**
i. **B** + **D**
j. **A** + **B** + **C** + **D**
k. 3**A** – 2**B** + **C** – 4**D**
l. **D** – 4**A** + **B**

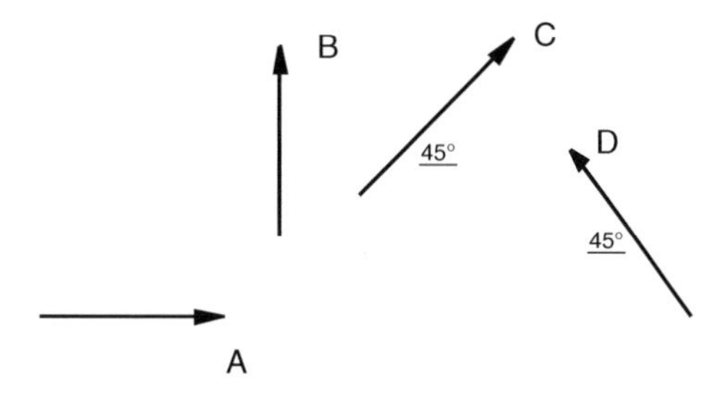

FIGURE P2.3.1
A set of four vectors.

Section 2.4 Vector Components

P2.4.1: Given the unit vectors **i** and **j** and the vectors **A** and **B** as shown in Figure P2.4.1, let the magnitudes of **A** and **B** be 10 N and 15 N, respectively. Express **A** and **B** in terms of **i** and **j**.

P2.4.2: See Problem P2.4.1. Let **C** be the resultant (or sum) of **A** and **B**.

a. Express **C** in terms of the unit vectors **i** and **j**.
b. Find the magnitude of **C**.

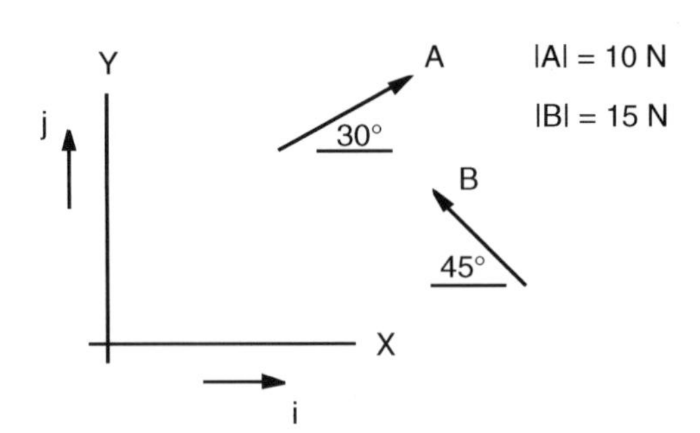

FIGURE P2.4.1
Vectors **A** and **B** and a unit vector.

P2.4.3: Given the unit vectors **i** and **j** and the vectors **A** and **B** as shown in Figure P2.4.3, let the magnitudes of **A** and **B** be 15 lb and 26 lb, respectively. Express **A** and **B** in terms of **i** and **j**.

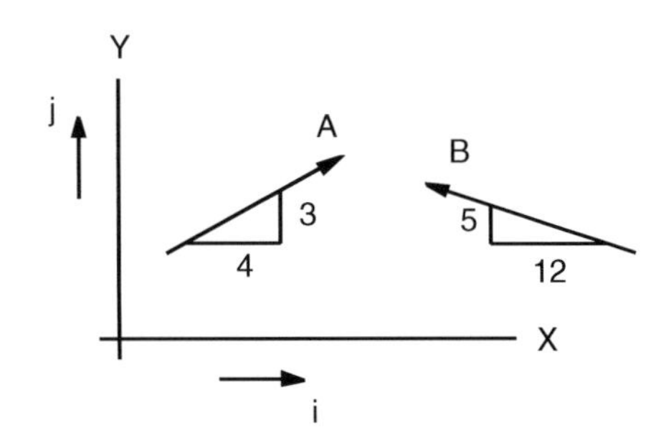

FIGURE P2.4.3
Vectors **A** and **B** and unit vectors.

P2.4.4: See Problem 2.4.3. Let **C** be the resultant of **A** and **B**.

a. Express **C** in terms of **i** and **j**.
b. Find the magnitude of **C**.

P2.4.5: Given the Cartesian coordinate system as shown in Figure P2.4.5, let A, B, and C be points in space as shown. Let $\mathbf{n}_x$, $\mathbf{n}_y$, and $\mathbf{n}_z$ be unit vectors parallel to X, Y, and Z.

a. Express the position vectors **AB**, **BC**, and **CA** in terms of $\mathbf{n}_x$, $\mathbf{n}_y$, and $\mathbf{n}_z$.
b. Find unit vectors parallel to **AB**, **BC**, and **CA**.

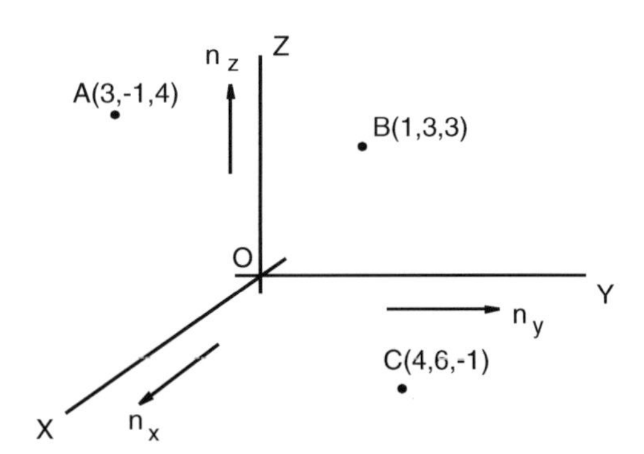

FIGURE P2.4.5
A Cartesian coordinate system with points *A*, *B*, and *C*.

P2.4.6: See Problem P2.4.5. Let **R** be the resultant of **AB**, **BC**, and **CA**. Express **R** in terms of $\mathbf{n}_x$, $\mathbf{n}_y$, and $\mathbf{n}_z$.

P2.4.7: See Figure P2.4.7. Forces $\mathbf{F}_1,\ldots,\mathbf{F}_6$ act along the edges and diagonals of a parallelepiped (or box) as shown. Let **i**, **j**, and **k** be mutually perpendicular unit vectors parallel to the edges of the box. Let the box be 12 inches long, 3 inches high, and 4 inches deep. Let the magnitudes of $\mathbf{F}_1,\ldots,\mathbf{F}_6$ be:

$$|\mathbf{F}_1| = 15\,\text{lb} \quad |\mathbf{F}_2| = 18\,\text{lb} \quad |\mathbf{F}_3| = 18\,\text{lb}$$

$$|\mathbf{F}_4| = 20\,\text{lb} \quad |\mathbf{F}_5| = 17\,\text{lb} \quad |\mathbf{F}_6| = 26\,\text{lb}$$

a. Find unit vectors $\mathbf{n}_1,\ldots,\mathbf{n}_6$ parallel to the forces $\mathbf{F}_1,\ldots,\mathbf{F}_6$. Express these unit vectors in terms of **i**, **j**, and **k**.
b. Express $\mathbf{F}_1,\ldots,\mathbf{F}_6$ in terms of unit vectors $\mathbf{n}_1,\ldots,\mathbf{n}_6$.
c. Express $\mathbf{F}_1,\ldots,\mathbf{F}_6$ in terms of **i**, **j**, and **k**.
d. Find the resultant **R** of $\mathbf{F}_1$. Express **R** in terms of **i**, **j**, and **k**.
e. Find the magnitude of **R**.

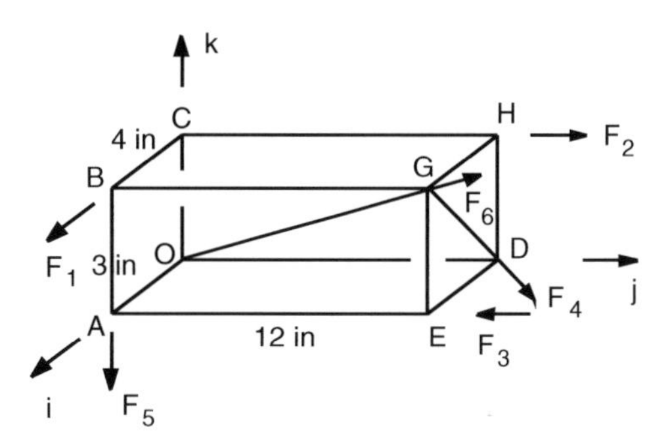

FIGURE P2.4.7
Forces acting along the edges and diagonals of a parallelepiped.

P2.4.8: Let forces be given as:

$$\mathbf{F}_1 = 3\mathbf{n}_1 - 2\mathbf{n}_2 + \mathbf{n}_3 \quad \text{N}$$

$$\mathbf{F}_2 = 4\mathbf{n}_1 + 6\mathbf{n}_2 - 2\mathbf{n}_3 \quad \text{N}$$

$$\mathbf{F}_3 = -\mathbf{n}_1 + 3\mathbf{n}_2 + 4\mathbf{n}_3 \quad \text{N}$$

$$\mathbf{F}_4 = 7\mathbf{n}_1 - 8\mathbf{n}_2 + 3\mathbf{n}_3 \quad \text{N}$$

$$\mathbf{F}_5 = -5\mathbf{n}_1 + 4\mathbf{n}_2 - 9\mathbf{n}_3 \quad \text{N}$$

where $\mathbf{n}_1$, $\mathbf{n}_2$, and $\mathbf{n}_3$ are mutually perpendicular unit vectors.

a. Find a force **F** such that the resultant of **F** and the forces $\mathbf{F}_1,\ldots,\mathbf{F}_6$ is zero. Express **F** in terms of $\mathbf{n}_1$, $\mathbf{n}_2$, and $\mathbf{n}_3$.
b. Determine the magnitude of **F**.

P2.4.9: See Figure P2.4.9. Let P and Q be points of a Cartesian coordinate system with coordinates as shown. Let $\mathbf{n}_x$, $\mathbf{n}_y$, and $\mathbf{n}_z$ be unit vectors parallel to the coordinate axes as shown.

a. Construct the position vectors **OP**, **OQ**, and **PQ** and express the results in terms of $\mathbf{n}_x$, $\mathbf{n}_y$, and $\mathbf{n}_z$.
b. Determine the distance between P and Q if the coordinates are expressed in meters.
c. Find the angles between **OP** and the X-, Y-, and Z-axes.
d. Find the angles between **OQ** and the X-, Y-, and Z-axes.
e. Find the angles between **PQ** and the X-, Y-, and Z-axes.

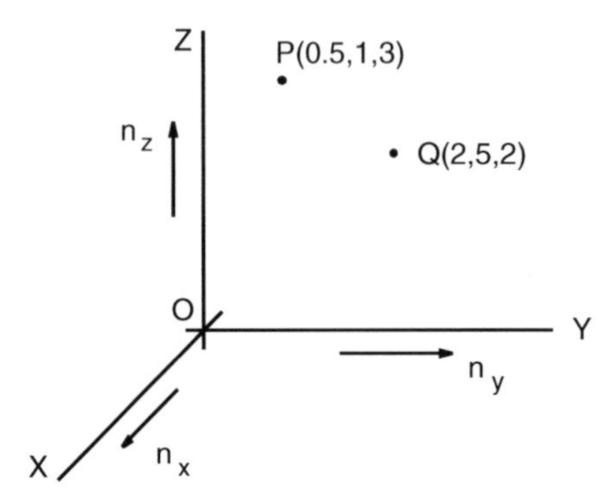

FIGURE P2.4.9
A Cartesian coordinate system with points P and Q.

Section 2.5 Angle Between Two Vectors

P2.5.1: From the definition in Section 2.5, what is the angle between two parallel vectors with the same sense? What is the angle if the vectors have opposite sense? What is the angle between a vector and itself?

Section 2.6 Vector Multiplication: Scalar Product

P.2.6.1: Consider the vectors **A** and **B** shown in Figure P2.6.1. Let the magnitudes of **A** and **B** be 8 and 5, respectively. Evaluate the scalar product $\mathbf{A} \cdot \mathbf{B}$.

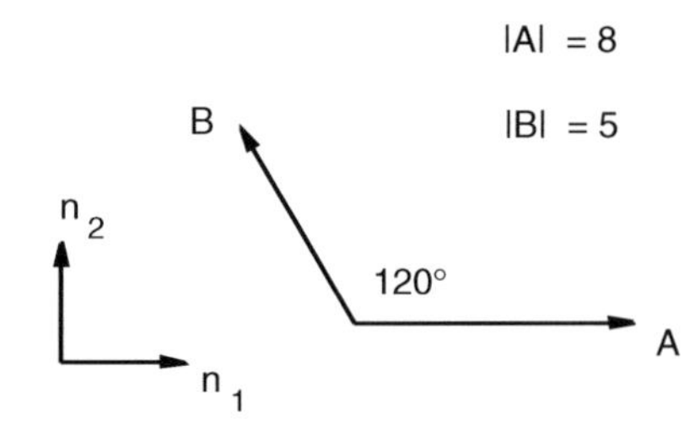

FIGURE P2.6.1
Vectors **A** and **B**.

P2.6.2: See Problem P2.6.1 and Figure P2.6.1. Express **A** and **B** in terms of the unit vectors $\mathbf{n}_1$ and $\mathbf{n}_2$ as shown in the figure. Use Eq. (2.6.2) to evaluate $\mathbf{A} \cdot \mathbf{B}$.

P2.6.3: See Problems P2.6.1 and P2.6.2. Let **C** be the resultant of **A** and **B**. Find the magnitude of **C**.

P2.6.4: Let $\mathbf{n}_1$, $\mathbf{n}_2$, and $\mathbf{n}_3$ be mutually perpendicular unit vectors. Let vectors **A** and **B** be expressed in terms of $\mathbf{n}_1$, $\mathbf{n}_2$, and $\mathbf{n}_3$ as:

$$\mathbf{A} = -3\mathbf{n}_1 + 5\mathbf{n}_2 + 6\mathbf{n}_3 \quad \text{and} \quad \mathbf{B} = 4\mathbf{n}_1 - 2\mathbf{n}_2 + 7\mathbf{n}_3$$

a. Determine the magnitudes of **A** and **B**.
b. Find the angle between **A** and **B**.

P2.6.5: Let vectors **A** and **B** be expressed in terms of mutually perpendicular unit vectors as:

$$\mathbf{A} = 4\mathbf{n}_1 + \mathbf{n}_2 - 5\mathbf{n}_3 \quad \text{and} \quad \mathbf{B} = 2\mathbf{n}_1 + \mathbf{B}_y\mathbf{n}_2 + \mathbf{n}_3$$

Find B_y so that the angle between **A** and **B** is 90°.

P2.6.6: See Figure P2.6.6. Let the vector **V** be expressed in terms of mutually perpendicular unit vectors $\mathbf{n}_x$, $\mathbf{n}_y$, and $\mathbf{n}_z$, as:

$$\mathbf{V} = 3\mathbf{n}_x + 4\mathbf{n}_y - 2\mathbf{n}_z$$

Let the line *L* pass through points *A* and *B* where the coordinates of *A* and *B* are as shown. Find the projection of **V** along *L*.

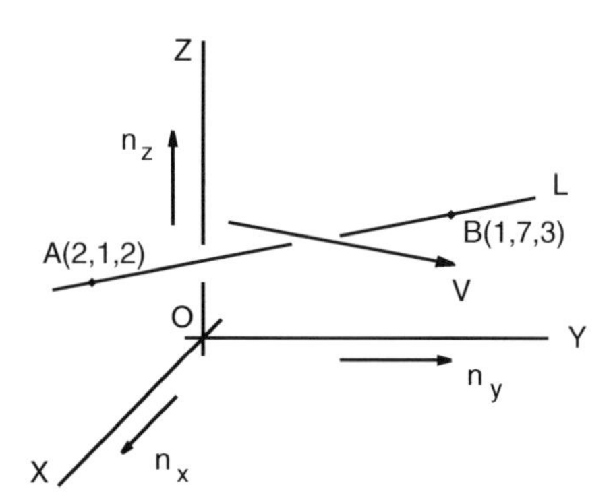

FIGURE P2.6.6
A vector **V** and a line *L*.

P2.6.7: A force **F** is expressed in terms of mutually perpendicular unit vectors $\mathbf{n}_x$, $\mathbf{n}_y$, and $\mathbf{n}_z$ as:

$$\mathbf{F} = -4\mathbf{n}_x + 2\mathbf{n}_y - 7\mathbf{n}_z \text{ lb}$$

If **F** moves a particle P from point *A* (1, –2, 4) to point *B* (–3, 4, –5), find the projection of **F** along the line *AB*, where the coordinates (in feet) of *A* and *B* are referred to an *X*, *Y*, *Z* Cartesian system and where $\mathbf{n}_x$, $\mathbf{n}_y$, and $\mathbf{n}_z$ are parallel to *X*, *Y*, and *Z*.

P2.6.8: See Problem P2.6.7. Let the work *W* done by **F** be defined as the product of the projection of **F** along *AB* and distance between *A* and *B*. Find *W*.

Section 2.7 Vector Multiplication: Vector Product

P2.7.1: Consider the vectors **A** and **B** shown in Figure P2.7.1. Using Eq. (2.7.1), determine the magnitude of the vector product **A** × **B**.

a. If **A** and **B** are in the *X*–*Y* plane, what is the direction of **A** × **B**?
b. What is the direction of **B** × **A**?

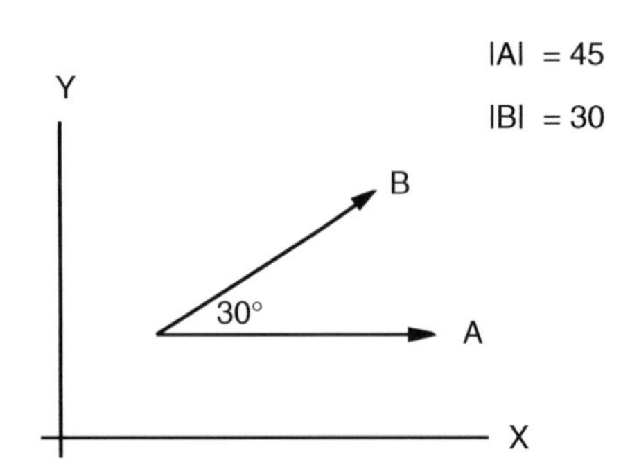

FIGURE P2.7.1
Vectors **A** and **B** in the *X*–*Y* plane.

P2.7.2: See Problem P2.7.1. What is the magnitude of 6**A** × 2**B**?

P2.7.3: Consider the vectors **A**, **B**, and **C** shown in Figure P2.7.3. Let **i**, **j**, and **k** be mutually perpendicular unit vectors as shown, with **k** being along the *Z*-axis.

a. Evaluate the sum **B** + **C**.
b. Determine the vector products **A** × **B** and **A** × **C**.
c. Use the results of (b) to determine the sum of **A** × **B** and **A** × **C**.
d. Use the results of (a) to determine the product **A** × (**B** + **C**). Compare the result with that of (c).

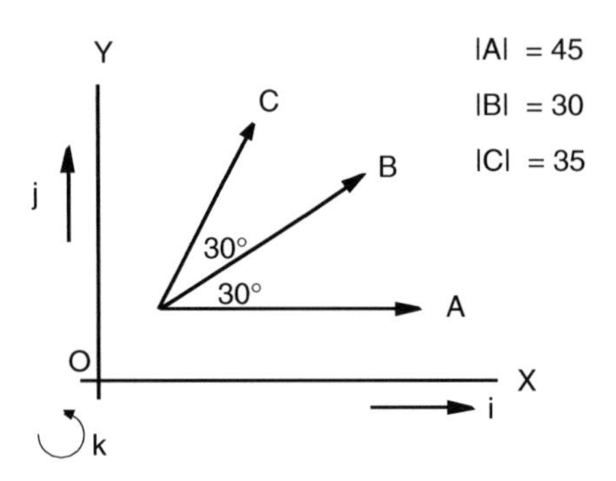

FIGURE P2.7.3
Vectors **A**, **B**, and **C** in the *X*–*Y* plane.

P.2.7.4: Let $\mathbf{n}_1$, $\mathbf{n}_2$, and $\mathbf{n}_3$ be mutually perpendicular (dextral) unit vectors, and let vectors **A**, **B**, and **C** be expressed in terms of $\mathbf{n}_1$, $\mathbf{n}_2$, and $\mathbf{n}_3$ as:

$$\mathbf{A} = 6\mathbf{n}_1 - \mathbf{n}_2 + \mathbf{n}_3$$
$$\mathbf{B} = -5\mathbf{n}_1 + 4\mathbf{n}_2 - 2\mathbf{n}_2$$
$$\mathbf{C} = 3\mathbf{n}_1 + 5\mathbf{n}_2 - 7\mathbf{n}_3$$

a. Evaluate the sum **B** + **C**.
b. Use Eq. (2.7.24) to determine the vector products **A** × **B** and **A** × **C**.
c. Use the results of (b) to determine the sum of **A** × **B** and **A** × **C.**
d. Use the results of (a) to determine the product **A** × (**B** + **C**). Compare the result with that of (c).

P2.7.5: Let **A** and **B** be expressed in terms of mutually perpendicular unit vectors $\mathbf{n}_1$, $\mathbf{n}_2$, and $\mathbf{n}_3$ as:

$$\mathbf{A} = 3\mathbf{n}_1 - 6\mathbf{n}_2 + 8\mathbf{n}_3$$

$$\mathbf{B} = 4\mathbf{n}_1 - \mathbf{n}_2 - 5\mathbf{n}_3$$

Find a unit vector perpendicular to **A** and **B**.

P2.7.6: See Figure P2.7.6. Let X, Y, and Z be mutually perpendicular coordinate axes, and let A, B, and C be points with coordinates relative to X, Y, and Z as shown.

a. Form the position vectors **AB**, **BC**, and **CA**, and express the results in terms of the unit vectors $\mathbf{n}_x$, $\mathbf{n}_y$, and $\mathbf{n}_z$.
b. Evaluate the vector products $\mathbf{AB} \times \mathbf{BC}$, $\mathbf{BC} \times \mathbf{CA}$, and $\mathbf{CA} \times \mathbf{AB}$.
c. Find a unit vector **n** perpendicular to the triangle ABC.

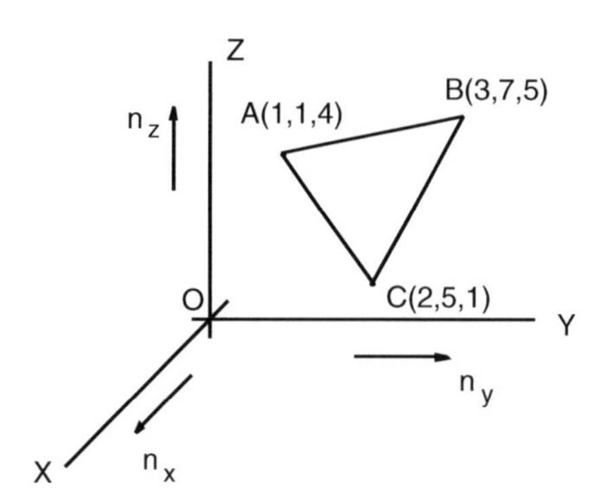

FIGURE P2.7.6
A triangle ABC in an X, Y, Z reference frame.

P2.7.7: See Figure P2.7.7. Let LP and LQ be lines passing through points P_1, P_2 and Q_1, Q_2 as shown. Let the coordinates of P_1, P_2, Q_1, and Q_2 relative to the X, Y, Z system be as shown. Find a unit vector **n** perpendicular to LP and LQ, and express the results in terms of the unit vectors $\mathbf{n}_x$, $\mathbf{n}_y$, and $\mathbf{n}_z$ shown in Figure P2.7.7.

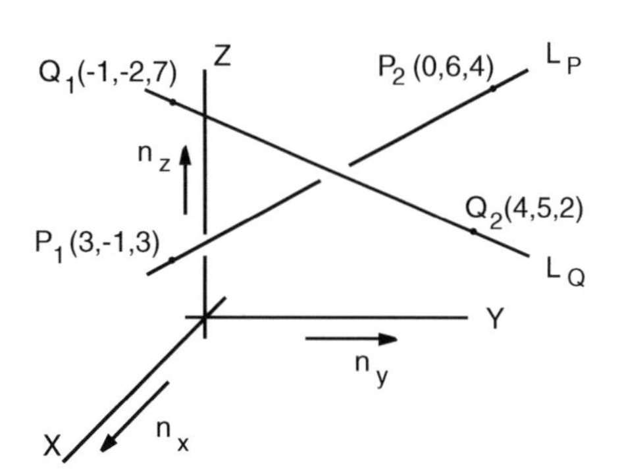

FIGURE P2.7.7
Lines LP and LQ in an X, Y, Z reference frame.

P2.7.8: See Figure P2.7.8. Let points P and Q have the coordinates (in feet) as shown. Let L be a line passing through P and Q, and let **F** be a force acting along L. Let the magnitude of **F** be 7 lb.

a. Find a unit vector **n** parallel to L, in the direction of P to Q. Express the results in terms of the unit vectors $\mathbf{n}_x$, $\mathbf{n}_y$, and $\mathbf{n}_z$ shown in Figure P2.7.7.
b. Express **F** in terms of $\mathbf{n}_x$, $\mathbf{n}_y$, and $\mathbf{n}_z$.
c. Form the vectors **OP** and **OQ**, calculate $\mathbf{OP} \times \mathbf{F}$ and $\mathbf{OQ} \times \mathbf{F}$, and express the results in terms of $\mathbf{n}_x$, $\mathbf{n}_y$, and $\mathbf{n}_z$. Compare the results.

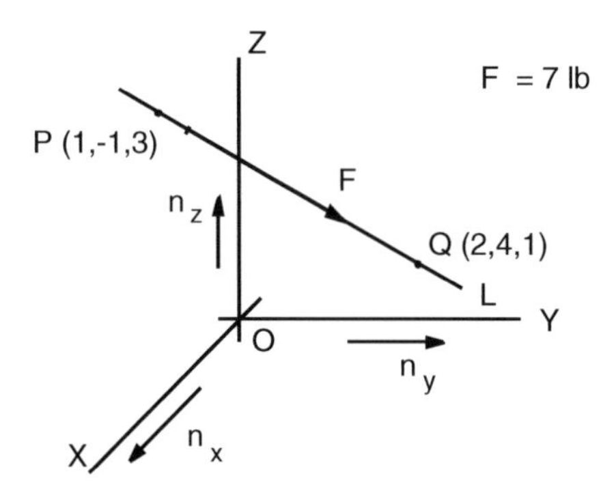

FIGURE P2.7.8
Coordinate system X, Y, Z with points P and Q, line L, and force **F**.

P2.7.9: Let e_{ijk} and δ_{jk} be the permutation symbol and the Kronecker delta symbol as in Eqs. (2.7.7) and (2.6.7). Evaluate the sums:

$$\sum_{j=1}^{3}\sum_{k=1}^{3} e_{ijk}\delta_{jk} \quad i = 1,2,3$$

Section 2.8 Vector Multiplication: Triple Products

P2.8.1: See Example 2.8.1. Verify the remaining equalities of Eq. (2.8.5) for the vectors **A**, **B**, and **C** of Eq. (2.8.7).

P2.8.2: Use Eq. (2.8.6) to find the volume of the parallelepiped shown in Figure P2.8.2, where the coordinates are measured in meters.

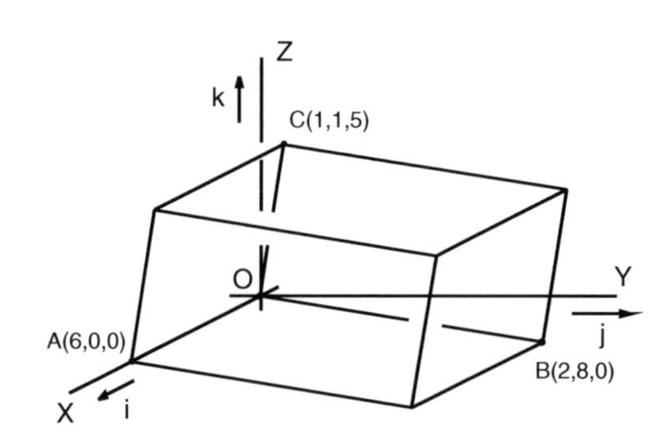

FIGURE P2.8.2
Parallelepiped.

P2.8.3: The triple scalar product is useful in determining the distance d between two non-parallel, non-intersecting lines. Specifically, d is the projection of a vector $\mathbf{P}_1\mathbf{P}_2$, which connects any point P_1 on one of the lines with any point P_2 on the other line, onto the common perpendicular between the lines. Thus, if **n** is a unit vector parallel to the common perpendicular, d is given by:

$$d = \mathbf{P}_1\mathbf{P}_2 \cdot \mathbf{n}$$

Using these concepts, find the distance d between the lines L_1 and L_2 shown in Figure P2.8.3, where the coordinates are expressed in feet. (Observe that **n** may be obtained from the vector product $(\mathbf{P}_1\mathbf{Q}_1 \times \mathbf{P}_2\mathbf{Q}_2)/|\mathbf{P}_1\mathbf{Q}_1 \times \mathbf{P}_2\mathbf{Q}_2|$.)

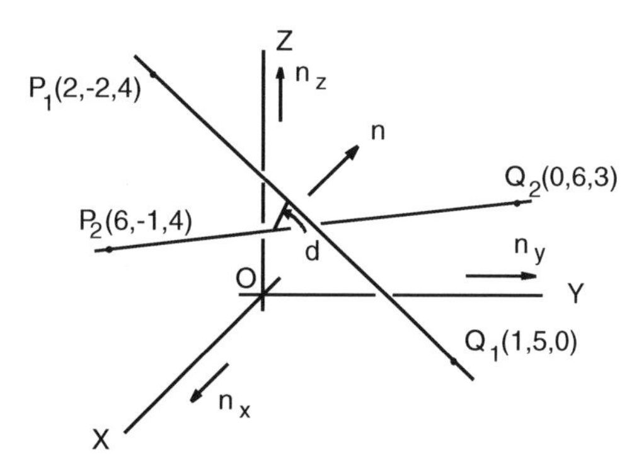

FIGURE P2.8.3
Distance d between non-parallel, non-intersecting lines.

P2.8.4: The triple vector product is useful in determining second-moment vectors. Let P be a particle with mass m located at a point P with coordinates (x, y, z) relative to a Cartesian coordinate frame X, Y, Z, as in Figure P2.8.4. (If a particle is small, it may be identified by a point.) Let position vector **p** locate P relative to the origin O. Let $\mathbf{n}_a$ be an arbitrarily directed unit vector and let $\mathbf{n}_x$, $\mathbf{n}_y$, and $\mathbf{n}_z$ be unit vectors parallel to axes X, Y, and Z as shown. The second moment vector $\mathbf{I}_a$ of P for O for the direction $\mathbf{n}_a$ is then defined as:

$$\mathbf{I}_a = m\mathbf{p} \times (\mathbf{n}_a \times \mathbf{p})$$

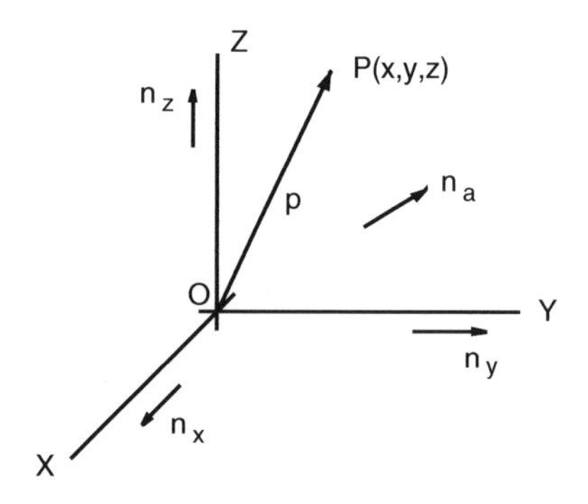

FIGURE P2.8.4
A particle P in a Cartesian reference frame.

a. Show that in this expression the parentheses are unnecessary; that is, unlike the general triple vector products in Eqs. (2.8.12) and (2.8.13), we have here:

$$m\mathbf{p} \times (\mathbf{n}_a \times \mathbf{p}) = m(\mathbf{p} \times n_a) \times \mathbf{p}$$

b. Observing that **p** may be expressed as:

$$\mathbf{p} = x\mathbf{n}_x + y\mathbf{n}_y + z\mathbf{n}_z$$

find $\mathbf{I}_x$, $\mathbf{I}_y$, and $\mathbf{I}_z$, the second moment vectors for the directions $\mathbf{n}_x$, $\mathbf{n}_y$, and $\mathbf{n}_z$.

c. Let unit vector $\mathbf{n}_a$ be expressed in terms of $\mathbf{n}_x$, $\mathbf{n}_y$, and $\mathbf{n}_z$ as:

$$\mathbf{n}_a = a_x\mathbf{n}_x + a_y\mathbf{n}_y + a_z\mathbf{n}_z$$

show that $\mathbf{I}_a$ may then be expressed as:

$$\mathbf{I}_a = a_x\mathbf{I}_x + a_y\mathbf{I}_y + a_z\mathbf{I}_z$$

Section 2.9 Use of the Index Summation Convention

P2.9.1: Evaluate and/or expand the following terms:

a. δ_{kk}
b. $e_{ijk}\delta_{jk}$
c. $a_{ij}b_{jk}$

Section 2.10 Review of Matrix Procedures

P2.10.1: Given the matrices A, B, and C:

$$A = \begin{bmatrix} 1 & -2 & 4 \\ 5 & 8 & 3 \\ 7 & -6 & 9 \end{bmatrix} \quad B = \begin{bmatrix} 4 & 8 & -3 \\ 5 & -6 & 9 \\ 0 & 5 & 2 \end{bmatrix} \quad C = \begin{bmatrix} 7 & -1 & -7 \\ 3 & 6 & 8 \\ 4 & -2 & 9 \end{bmatrix}$$

a. Compute the products AB and BC.
b. Compute the products (AB)C and A(BC) and compare the results (see Eq. (2.8.10)).
c. Find B^T, A^T, and the product B^TA^T.
d. Find $(AB)^T$ and compare with B^TA^T (see Eq. (2.10.9)).
e. Find $\hat{A}$, the matrix of adjoints for A (see Eq. (2.10.11)).
f. Compute A^{-1}.
g. Compute the products $A^{-1}A$ and AA^{-1}.

P2.10.2: Given the matrix S (an orthogonal matrix):

$$S = \begin{bmatrix} 1/2 & \sqrt{6}/4 & \sqrt{6}/4 \\ -\sqrt{3}/2 & \sqrt{2}/4 & \sqrt{2}/4 \\ 0 & -\sqrt{2}/2 & \sqrt{2}/2 \end{bmatrix}$$

a. Find detS.
b. Find S^{-1}.
c. Find S^T.
d. Compare S^{-1} and S^T.

Section 2.11 Reference Frames and Unit Vector Sets

P2.11.1: Determine the numerical values of the transformation matrix elements of Eq. (2.11.19) if α, β, and γ have the values:

$$\alpha = 30°, \quad \beta = 45°, \quad \gamma = 60°$$

P2.11.2: See Problem P2.11.1. Show that $S^T = S^{-1}$ and that $\det S = 1$.

3

Kinematics of a Particle

3.1 Introduction

Kinematics is a study of motion without regard to the cause of the motion. Often this motion occurs in three dimensions. For such cases, and even when the motion is restricted to two dimensions, it is convenient to describe the motion using vector quantities. Indeed, the principal kinematic quantities of position, velocity, and acceleration are vector quantities.

In this chapter, we will study the kinematics of particles. We will think of a "particle" as an object that is sufficiently small that it can be identified by and represented by a point. Hence, we can study the kinematics of particles by studying the movement of points. In the next chapter, we will extend our study to rigid bodies and will think of a rigid body as being simply a collection of particles. We begin our study with a discussion of vector differentiation.

3.2 Vector Differentiation

Consider a vector **V** whose characteristics (magnitude and direction) are dependent upon a parameter t (time). Let the functional dependence of **V** on t be expressed as:

$$\mathbf{V} = \mathbf{V}(t) \tag{3.2.1}$$

Then, as with scalar functions, the derivative of **V** with respect to t is defined as:

$$d\mathbf{V}/dt \overset{D}{=} \lim_{\Delta t \to 0} \frac{\mathbf{V}(t + \Delta t) - \mathbf{V}(t)}{\Delta t} \tag{3.2.2}$$

The manner in which **V** depends upon t depends in turn upon the reference frame in which **V** is observed. For example, if **V** is fixed in a reference frame $\hat{R}$, then in $\hat{R}$, **V** is independent of t. If, however, $\hat{R}$ moves relative to a second reference frame, R, then in R **V** depends upon t (time). Hence, even though the rate of change of **V** relative to an observer in $\hat{R}$ is zero, the rate of change of **V** relative to an observer in R is not necessarily zero. Therefore, in general, the derivative of a vector function will depend upon the reference frame in which that derivative is calculated. Hence, to avoid ambiguity, a superscript is usually added to the derivative symbol to designate the reference frame in which the derivative is calculated. Thus, instead of simply writing $d\mathbf{V}/dt$ we will write ${}^{R}d\mathbf{V}/dt$, unless the reference frame is clearly understood.

To explore the properties of vector differentiation, let $\mathbf{n}_1$, $\mathbf{n}_2$, and $\mathbf{n}_3$ be mutually perpendicular unit vectors fixed in a reference frame R. Let $\mathbf{V}$ be a vector function of t and let $\mathbf{V}$ be expressed in terms of $\mathbf{n}_1$, $\mathbf{n}_2$, and $\mathbf{n}_3$ as:

$$\mathbf{V} = v_i\mathbf{n}_i = v_1(t)\mathbf{n}_1 + v_2(t)\mathbf{n}_2 + \mathbf{n}_2 + v_3(t)\mathbf{n} \tag{3.2.3}$$

By substituting into Eq. (3.2.2), the derivative of $\mathbf{V}$ with respect to t in R is then:

$$\begin{aligned} {}^{R}d\mathbf{V}/dt &= \underset{\Delta t \to 0}{\mathbf{Lim}}\Big\{\big[v_1(t+\Delta t)\mathbf{n}_1 + v_2(t+\Delta t)\mathbf{n}_2 + v_3(t+\Delta t)\mathbf{n}_3\big] \\ &\quad -\big[v_1(t)\mathbf{n}_1 + v_2(t)\mathbf{n}_2 + v_3(t)\mathbf{n}_3\big]\Big\} / \Delta t \end{aligned} \tag{3.2.4}$$

Because the $\mathbf{n}_i$ (i = 1, 2, 3) are fixed in R, they are independent of t in R and thus are constant in R. From elementary calculus, we recall that if the limits exist, the limit of a sum is equal to the sum of the limits. Also, if the limits exist, the limit of a constant multiplied by a function is equal to the constant multiplied by the limit of the function. Hence, we can rewrite Eq. (3.2.4) as:

$$\begin{aligned} {}^{R}d\mathbf{V}/dt &= \mathbf{n}_1 \underset{\Delta t \to 0}{\mathbf{Lim}}\big[v_1(t+\Delta t) - v_1(t)\big] / \Delta t + \mathbf{n}_2\big[v_2(t+\Delta t) - v_2(t)\big] / \Delta t \\ &\quad + \mathbf{n}_3 \underset{\Delta t \to 0}{\mathbf{Lim}}\big[v_3(t+\Delta t) - v_3(t)\big] / \Delta t \\ &= \mathbf{n}_1 dv_1/dt + \mathbf{n}_2\, dv_2/dt + \mathbf{n}_3\, dv_3/dt \end{aligned} \tag{3.2.5}$$

This expression shows that vector derivatives are simply linear combinations of scalar derivatives. Specifically, to calculate the derivative of a vector function in a reference frame R we simply need to express $\mathbf{V}$ in terms of unit vectors fixed in R and then differentiate the scalar components. Although this procedure always works, it may not be the most convenient in all circumstances. We will explore and develop other procedures in Section 3.5.

Example 3.2.1: Differentiation of a Position Vector

To illustrate these ideas, consider the following example: Let the position vector $\mathbf{p}$ locating a moving point P relative to the origin of a Cartesian reference frame R be expressed as:

$$\mathbf{p} = x\mathbf{n}_1 + y\mathbf{n}_2 + z\mathbf{n}_3 \tag{3.2.6}$$

where $\mathbf{n}_1$, $\mathbf{n}_2$, and $\mathbf{n}_3$ are mutually perpendicular unit vectors fixed in R. Let x, y, and z be functions of time t given by:

$$x = (2t - 2), \quad y = t^3 - 3t, \quad z = -t^4 + 4t \tag{3.2.7}$$

where x, y, and z are measured in meters, and t is measured in seconds. Then, the derivative of $\mathbf{p}$ with respect to t in R is:

$$d\mathbf{p}/dt = (2t-2)\mathbf{n}_1 + \left(3t^2 - 3\right)\mathbf{n}_2 + \left(-4t^3 + 4\right)\mathbf{n}_3 \text{ m/sec} \tag{3.2.8}$$

Observe that when t is zero, $\mathbf{p}$ and $d\mathbf{p}/dt$ are:

$$\mathbf{p} = 0 \quad \text{and} \quad d\mathbf{p}/dt = -2\mathbf{n}_1 - 3\mathbf{n}_2 + 4\mathbf{n}_3 \text{ m/sec} \quad (t = 0) \tag{3.2.9}$$

Alternatively, when t is one, $\mathbf{p}$ and $d\mathbf{p}/dt$ are:

$$\mathbf{p} = -\mathbf{n}_1 - 2\mathbf{n}_2 + 3\mathbf{n}_3 \text{ m} \quad \text{and} \quad d\mathbf{p}/dt = 0 \quad (t = 1 \text{ sec}) \tag{3.2.10}$$

This shows that a vector may be zero while its derivative is not zero, and conversely the derivative may be zero while the vector is not zero.

As noted in Eq. (3.2.5), vector derivatives may be expressed as linear combinations of scalar derivatives. This means that the algebra associated with scalar derivatives is applicable with vector derivatives as well. For example, the rules for differentiating sums and products are the same for scalars and vectors. Also, the chain rule is the same. Specifically, let $\mathbf{V}$ and $\mathbf{W}$ be vector functions of t, and let s be a scalar function of t. Then from Eq. (3.2.5) it is readily seen that:

$$d(\mathbf{V} + \mathbf{W}) / dt = d\mathbf{V}/dt + d\mathbf{W}/dt \tag{3.2.11}$$

$$d(s\mathbf{V}/dt) = (ds/dt)\mathbf{V} + s(d\mathbf{V}/dt) \tag{3.2.12}$$

$$d(\mathbf{V} \cdot \mathbf{W}) / dt = (d\mathbf{V}/dt) \cdot \mathbf{W} + \mathbf{V} \cdot (d\mathbf{W}/dt) \tag{3.2.13}$$

and

$$d(\mathbf{V} \times \mathbf{W}/dt = d\mathbf{V}/dt) \times \mathbf{W} + \mathbf{V} \times (d\mathbf{W}/dt) \tag{3.2.14}$$

Finally, if $\mathbf{V}$ is a function of s and if s is a function of t, we have:

$$d\mathbf{V}/dt = (d\mathbf{V}/ds)(ds/dt) \tag{3.2.15}$$

3.3 Position, Velocity, and Acceleration

Consider the movement of a particle P along a curve C as in Figure 3.3.1. Let P be a point representing particle P. Let $\mathbf{p}$ locate P relative to a fixed point O of a reference frame R as shown. The *position* of P is designated by the vector $\mathbf{p}$. The *velocity* of **P**, $\mathbf{V}^P$, is defined as the time derivative of $\mathbf{p}$ in R. That is,

$$\mathbf{V}^P = {}^R d\mathbf{p}/dt \tag{3.3.1}$$

Similarly, the *acceleration* of P is defined as the time derivative of $\mathbf{V}^P$ in R. That is,

$$\mathbf{a}^P = {}^R d\mathbf{V}^P/dt = {}^R d^2\mathbf{p}/dt^2 \tag{3.3.2}$$

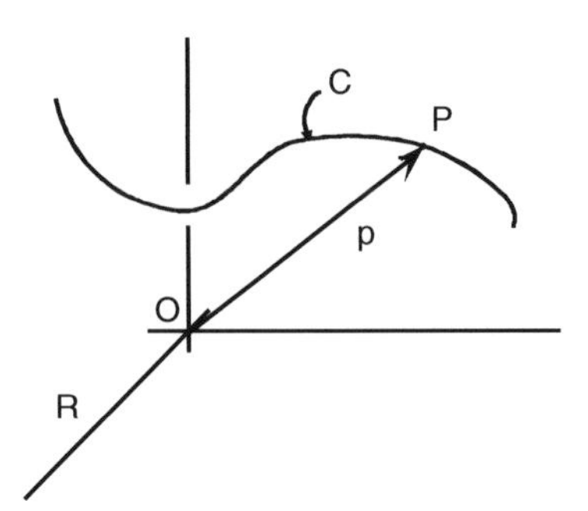

FIGURE 3.3.1
Point P moving along a curve C in a reference frame R.

Consider the characteristics of $\mathbf{V}^P$. From Eq. (3.3.1) and the definition of the derivative, we have:

$$\mathbf{V}^P = \underset{\Delta t \to 0}{\mathbf{Lim}} \left[\mathbf{p}(t+\Delta t) - \mathbf{p}(t)\right] / \Delta t = \underset{\Delta t \to 0}{\mathbf{Lim}} \frac{\Delta \mathbf{p}}{\Delta t} \tag{3.3.3}$$

where $\Delta\mathbf{p}$ is the difference between $\mathbf{p}(t + \Delta t)$ and $\mathbf{p}(t)$. That is,

$$\mathbf{p}(t+\Delta t) = \mathbf{p}(t) + \Delta\mathbf{p} \tag{3.3.4}$$

If P is at P_1 at time t and at P_2 at time $t + \Delta t$, then $\Delta\mathbf{p}$ locates P_2 relative to P_1 as in Figure 3.3.2. Observe that as t approaches zero, P_2 approaches P_1 and $\mathbf{p}$ becomes tangent to C. In this latter regard, $\Delta\mathbf{p}$ is a *chord vector* on C. Recall from elementary calculus that the slope of a tangent to a curve is the limiting slope of a chord as its end points (in this case, P_1 and P_2) approach each other. From Eq. (3.3.3), in the limiting process the velocity has the same direction as $\Delta\mathbf{p}$. This means that $\mathbf{V}^P$ is tangent to C at P. Observe that the acceleration of P is, in general, *not* tangent to C (see Problems 3.3.2 and 3.3.3).

Suppose now that X, Y, and Z are Cartesian coordinate axes fixed in R. Let $\mathbf{N}_X$, $\mathbf{N}_Y$, and $\mathbf{N}_Z$ be unit vectors parallel to X, Y, and Z, as shown in Figure 3.3.3. Let x, y, and z be the

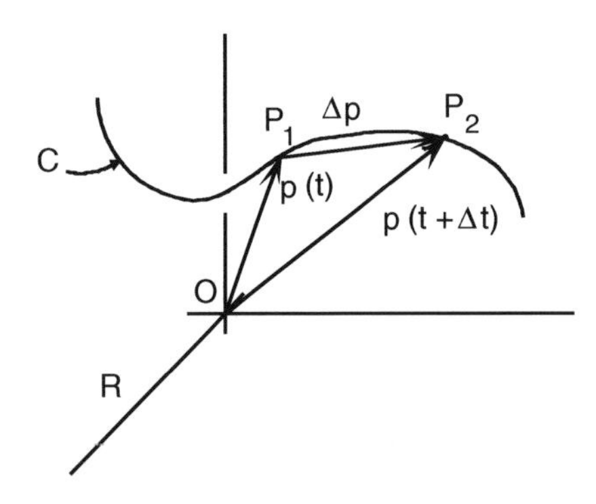

FIGURE 3.3.2
Position of P at times t and $t + \Delta t$.

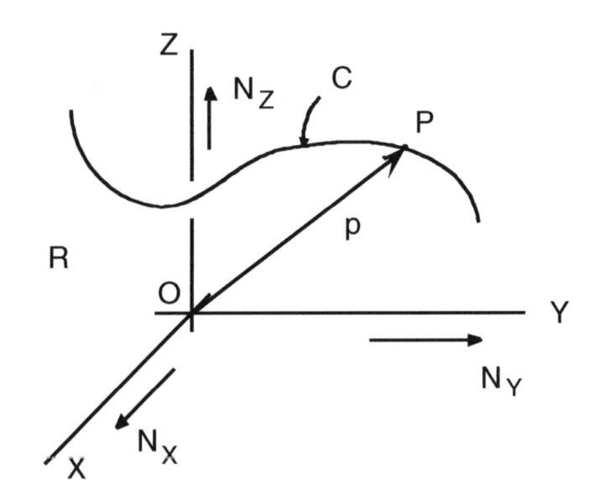

FIGURE 3.3.3
Coordinate axes and unit vectors.

coordinates of P relative to X, Y, and Z. Then, the position velocity and acceleration of P may be expressed in terms of $\mathbf{N}_X$, $\mathbf{N}_Y$, and $\mathbf{N}_Z$ as:

$$\mathbf{p} = x\mathbf{N}_X + y\mathbf{N}_Y + z\mathbf{N}_Z$$

$$\mathbf{V}^P = \dot{x}\mathbf{N}_X + \dot{y}\mathbf{N}_y + \dot{z}\mathbf{N}_Z$$

and

$$\mathbf{a}^P = \ddot{x}\mathbf{N}_X + \ddot{y}\mathbf{N}_y + \ddot{z}\mathbf{N}_Z \tag{3.3.5}$$

3.4 Relative Velocity and Relative Acceleration

Consider again the movement of a particle P represented by a point P in a reference frame R. As before, let the position vector $\mathbf{p}$ locate P relative to a fixed point O in R as in Figure 3.4.1. From Eq. (3.3.1), the derivative of $\mathbf{p}$ in R is the velocity of P in R. That is,

$$\mathbf{V}^P = d\mathbf{p}/dt \tag{3.4.1}$$

where we have deleted the superscript R in $d\mathbf{p}/dt$ because the reference frame is clearly understood. Because O is fixed in R, $\mathbf{V}^P$ is often called the *absolute velocity* of P in R.

Consider now a second particle Q represented by point Q also moving in R. Let $\mathbf{q}$ locate Q relative to O as in Figure 3.4.2. Then, as with P, the velocity of Q in R is:

$$\mathbf{V}^Q = d\mathbf{q}/dt \tag{3.4.2}$$

Again, because O is fixed in R, $\mathbf{V}^Q$ is often called the absolute velocity of Q in R.

The *difference* in the absolute velocities of P and Q is called the *relative velocity* of P with respect to Q in R and is written as ${}^P\mathbf{V}^{P/Q}$ or as $\mathbf{V}^{P/Q}$. That is,

$$\mathbf{V}^{P/Q} = \mathbf{V}^P - \mathbf{V}^Q \tag{3.4.3}$$

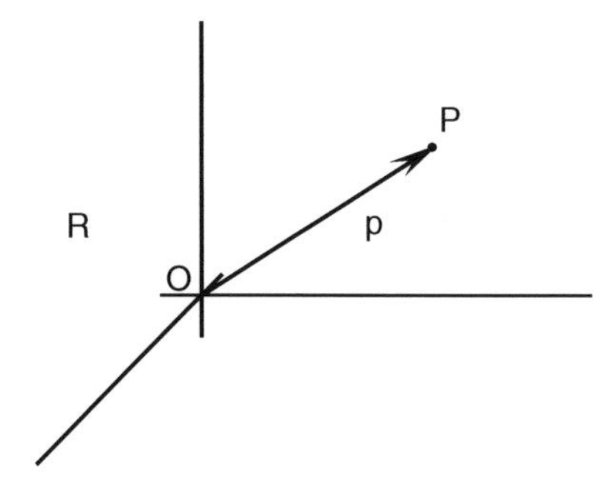

FIGURE 3.4.1
A point P moving in a reference frame R.

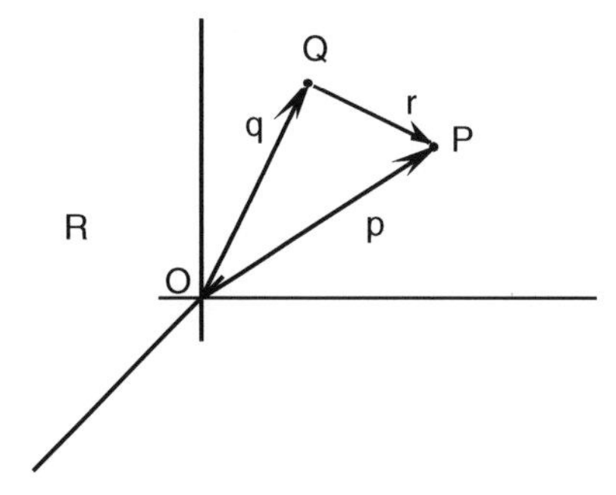

FIGURE 3.4.2
Points P and Q moving in reference frame R.

Consider again Figure 3.4.2. Let **r** be the position vector locating P relative to Q. Then, from the rules of vector addition, we have:

$$\mathbf{p} = \mathbf{q} + \mathbf{r} \quad \text{op} \quad \mathbf{r} = \mathbf{p} - \mathbf{q} \tag{3.4.4}$$

By differentiating, we have:

$$d\mathbf{r}/dt = d\mathbf{p}/dt - d\mathbf{q}/dt = \mathbf{V}^P - \mathbf{V}^Q - \mathbf{V}^{P/Q} \tag{3.4.5}$$

Therefore, we see that the relative velocity $\mathbf{V}^{P/Q}$ is simply the derivative of the position vector locating P with respect to Q. Hence, $\mathbf{V}^{P/Q}$ may be interpreted as the movement of P as seen by an observer moving with Q.

From an analytical perspective, Eq. (3.4.3) is generally written in the form:

$$\mathbf{V}^P = \mathbf{V}^Q + \mathbf{V}^{P/Q} \tag{3.4.6}$$

Thus, if the velocity of Q is known and if the velocity of P relative to Q is known, then the velocity of P may be calculated.

The same concepts hold for accelerations. Specifically, by differentiating in Eqs. (3.4.5) and (3.4.6), we have:

$$\mathbf{a}^{P/Q} = \mathbf{a}^P - \mathbf{a}^Q \quad \text{or} \quad \mathbf{a}^P = \mathbf{a}^Q + \mathbf{a}^{P/Q} \tag{3.4.7}$$

where $\mathbf{a}^{P/Q}$ is called the acceleration of P *relative* to Q in R, and $\mathbf{a}^P$ and $\mathbf{a}^Q$ are called *absolute* accelerations of P and Q in R.

Example 3.4.1: Relative Velocity of Two Particles of a Rigid Body

Consider a rigid body B moving in a reference frame R, and let P and Q be particles of B represented by points P and Q. Let P be located relative to Q by the vector **r** given by:

$$\mathbf{r} = 2\mathbf{n}_1 - 3\mathbf{n}_2 + 7\mathbf{n}_3 \text{ ft} \tag{3.4.8}$$

where $\mathbf{n}_1$, $\mathbf{n}_2$, and $\mathbf{n}_3$ are unit vectors fixed in B as in Figure 3.4.3. Let the angular velocity of B in R be given as:

$$\boldsymbol{\omega} = -2\mathbf{n}_1 + 3\mathbf{n}_2 - 4\mathbf{n}_3 \text{ rad/sec} \tag{3.4.9}$$

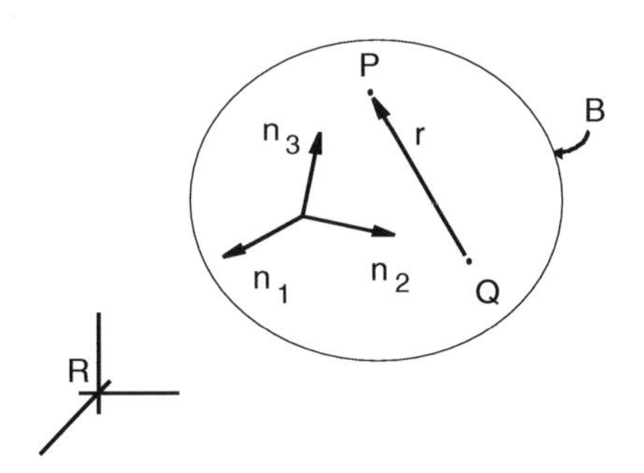

FIGURE 3.4.3
A body B with particles P and Q.

It can be shown that the velocity of P relative to Q in R may be expressed as:

$$^{R}\mathbf{V}^{P/Q} = \boldsymbol{\omega} \times \mathbf{r} \tag{3.4.10}$$

Assuming this to be correct, and if the velocity of Q in R is also known as $5\mathbf{n}_1 - 8\mathbf{n}_2 + 3\mathbf{n}_3$ ft/sec, find the velocity of P relative to Q in R and the absolute velocity of P in R.

Solution: From Eqs. (3.4.8) to (3.4.10), the relative velocity is:

$$\begin{aligned} ^{R}\mathbf{V}^{P/Q} &= (-2\mathbf{n}_1 + 3\mathbf{n}_2 - 4\mathbf{n}_3) \times (2\mathbf{n}_1 - 3\mathbf{n}_2 + 7\mathbf{n}_3) \\ &= 33\mathbf{n}_1 + 6\mathbf{n}_2 \text{ ft/sec} \end{aligned} \tag{3.4.11}$$

Then, from Eq. (3.4.6), the velocity of P is:

$$\begin{aligned} \mathbf{V}^{P} &= (5\mathbf{n}_1 - 8\mathbf{n}_2 + 3\mathbf{n}_3) + (33\mathbf{n}_1 + 6\mathbf{n}_2) \\ &= 38\mathbf{n}_1 + 2\mathbf{n}_2 + 3\mathbf{n}_3 \end{aligned} \tag{3.4.12}$$

Observe that even though the distance between P and Q is constant, the relative velocity of P and Q in R is not zero. Observe further, however, that relative to an observer in B the relative velocity is zero. That is,

$$^{B}\mathbf{V}^{P/Q} = 0 \tag{3.4.13}$$

3.5 Differentiation of Rotating Unit Vectors

In Section 3.2, we observed that we can calculate the derivative of a vector in a reference frame R by first expressing the vector in terms of unit vectors fixed in R and then by differentiating the components. In this procedure, the vector derivative is expressed in terms of scalar derivatives. Although in principle this procedure always works, it may not always be the most convenient way to obtain the derivative. Consider, for example, the velocity of a point P moving along a fixed curve C, as discussed in Section 3.3. The velocity vector $\mathbf{V}^P$ is tangent to C as in Figure 3.5.1. Then, if $\mathbf{n}$ is a unit vector parallel to $\mathbf{V}^P$, we may express $\mathbf{V}^P$ as:

$$\mathbf{V}^{P} = |\mathbf{V}^{P}|\mathbf{n} = v\mathbf{n} \tag{3.5.1}$$

where v is defined as $|\mathbf{V}^P|$ **to simplify the notation;** v is often called the *speed* of P.

Suppose now that we want to calculate the acceleration of P. By differentiating in Eq. (3.5.1), we have:

$$\mathbf{a}^{P} = (dv/dt)\mathbf{n} + v^{R}\, d\mathbf{n}/dt \tag{3.5.2}$$

The first term involves a scalar derivative and presents no difficulty. The second term involves the derivative of a unit vector $\mathbf{n}$ that is ***not*** fixed in R. As with any other vector,

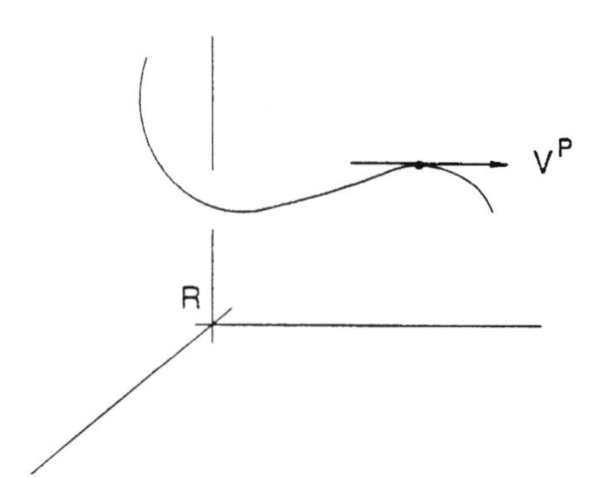

FIGURE 3.5.1
A point P moving on a curve C in a reference frame R.

we can calculate the derivative of **n** by expressing it in terms of unit vectors fixed in R. As we will see presently, however, there are more convenient procedures.

We focus our attention on unit vectors because they define the direction characteristics of vectors, which distinguish vectors from scalars. It is quickly seen that the derivative of a unit vector is a vector perpendicular to the unit vector. To see this, recall that the magnitude of a unit vector is 1. Therefore,

$$\mathbf{n} \cdot \mathbf{n} = 1 \tag{3.5.3}$$

Then, by differentiating, we have:

$$(d\mathbf{n}/dt) \cdot \mathbf{n} + \mathbf{n} \cdot d\mathbf{n}/dt = 0 \tag{3.5.4}$$

The dot product, however, is commutative (see Eq. (2.6.2)). That is,

$$(d\mathbf{n}/dt) \cdot \mathbf{n} = \mathbf{n} \cdot (d\mathbf{n}/dt) \tag{3.5.5}$$

Hence, Eq. (3.5.4) becomes:

$$2\mathbf{n} \cdot (d\mathbf{n}/dt) = 0 \quad \text{or} \quad \mathbf{n} \cdot (d\mathbf{n}/dt) = 0 \tag{3.5.6}$$

This shows that if $d\mathbf{n}/dt$ is not zero, it is perpendicular to **n**.

Next, consider the case of a unit vector that moves in such a way that it always remains perpendicular to a fixed line. Specifically, let Z be a line fixed in a reference frame R, and let L be a line that intersects Z and is perpendicular to Z. Let the unit vector **n** be parallel to L. Finally, let X and Y be lines fixed in R such that X, Y, and Z form a mutually perpendicular set. Let L rotate in the X–Y plane (see Figure 3.5.2). Let θ be the angle between L and X as shown. As L rotates, θ changes. Thus, θ is a function of time t and is a measure of the rotation of L and of **n**, as well. Because **n** rotates with L, it is called a *rotating unit vector*.

Let $\mathbf{n}_x$, $\mathbf{n}_y$, and $\mathbf{n}_z$ be unit vectors parallel to X, Y, and Z, as shown in Figure 3.5.2. Then, the derivative of **n** in R is simply:

$$d\mathbf{n}/dt = \mathbf{n}_z \times \mathbf{n}\, d\theta/dt \tag{3.5.7}$$

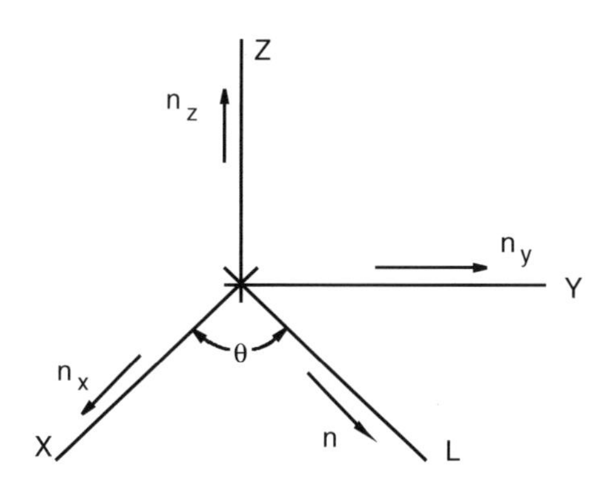

FIGURE 3.5.2
Fixed axes *X*, *Y*, and *Z*; a rotating line *L*; and unit vector parallel to *L*.

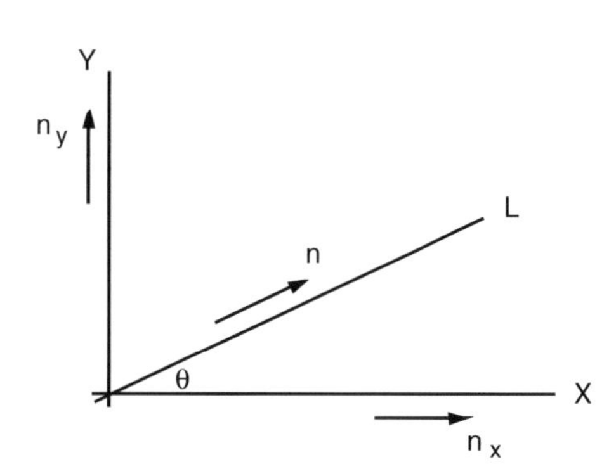

FIGURE 3.5.3
A view of the *X*–*Y* plane of Figure 3.5.2.

To verify this, we can evaluate $d\mathbf{n}/dt$ by expressing $\mathbf{n}$ in terms of $\mathbf{n}_x$, $\mathbf{n}_y$, and $\mathbf{n}_z$ and then differentiate the scalar components as noted earlier. To this end, consider the *X*–*Y* plane of Figure 3.5.2 as redrawn and shown in Figure 3.5.3. We see that $\mathbf{n}$ may be expressed as:

$$\mathbf{n} = \cos\theta\,\mathbf{n}_x + \sin\theta\,\mathbf{n}_y \tag{3.5.8}$$

Then, $d\mathbf{n}/dt$ is:

$$d\mathbf{n}/dt = (d\mathbf{n}/d\theta)(d\theta/dt) = (-\sin\theta\,\mathbf{n}_x + \cos\theta\mathbf{n}_y)(d\theta/dt) \tag{3.5.9}$$

Observe, however, from Eq. (3.5.8) that $\mathbf{n}_z \times \mathbf{n}$ is:

$$\mathbf{n}_z \times \mathbf{n} = -\sin\theta\,\mathbf{n}_y + \cos\theta\,\mathbf{n}_y \tag{3.5.10}$$

By comparing Eqs. (3.5.9) and (3.5.10), Eq. (3.5.7) is established (see also References 3.1 and 3.2).

Equation (3.5.7) is an algorithm for computing the derivative of a rotating unit vector. It is a convenient computational procedure in that it avoids the task of expressing $\mathbf{n}$ in terms of fixed unit vectors. Moreover, it shows that we can calculate a vector derivative in terms of a multiplication. This latter concept has profound implications for digital computation. Indeed, a digital computer is ideally suited for performing the arithmetic operations of addition, subtraction, multiplication, and division. Equation (3.5.7) thus extends the capability to include differentiating without resorting to difference equations as in numerical scalar differentiation. Equation (3.5.7) also forms a basis for the concept of angular velocity as developed in Chapter 4. We will explore the application of Eq. (3.5.7) in the remaining sections of this chapter.

3.6 Geometric Interpretation of Acceleration

Consider again a point P moving along curve C as in Figure 3.6.1. As before, let C be fixed in reference frame R. Then, from Eqs. (3.5.1) and (3.5.2) the velocity and acceleration of P in R are:

$$\mathbf{V}^P = v\mathbf{n} \quad \text{and} \quad \mathbf{a}^P = \frac{dv}{dt}\mathbf{n} + v\frac{d\mathbf{n}}{dt} \tag{3.6.1}$$

where $\mathbf{n}$ is a unit vector tangent to C at P and where v is the magnitude of $\mathbf{V}^P$. From Eq. (3.5.6), we see that $d\mathbf{n}/dt$ is perpendicular to $\mathbf{n}$. Therefore, the two terms of $\mathbf{a}^P$ in Eq. (3.6.1) are perpendicular to each other. The first term arises from a change in the speed of P. The second arises from a change in the direction of P.

Recall that a vector may be characterized as having magnitude and direction; hence, changes in a vector can arise from changes in either the magnitude of the vector or the direction of the vector, or both. Therefore, because $\mathbf{a}^P$ is the time rate of change of v^P, the terms of $\mathbf{a}^P$ arise due to changes in the magnitude of $\mathbf{V}^P$ (that is, $dv/dt\ \mathbf{n}$) and due to changes in the direction of $\mathbf{V}^P$ (that is, $v\ d\mathbf{n}/dt$).

3.7 Motion on a Circle

To illustrate these concepts, consider the special case of a point moving on a circle. Specifically, let C be a circle with radius r and center O. Let P be a point moving on C as in Figure 3.7.1. Let θ measure the inclination of the radial line OP. Let $\mathbf{n}_r$ and $\mathbf{n}_\theta$ be unit vectors parallel to OP and parallel to the tangent of C at P, as shown. Let $\mathbf{n}_z$ be a unit vector normal to the plane of C such that:

$$\mathbf{n}_z = \mathbf{n}_r \times \mathbf{n}_\theta \tag{3.7.1}$$

The position vector $\mathbf{p}$ locating P relative to O is:

$$\mathbf{p} = r\mathbf{n}_r \tag{3.7.2}$$

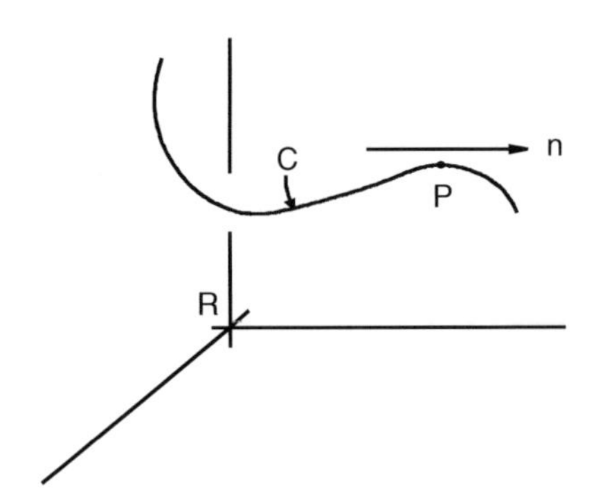

FIGURE 3.6.1
A point moving on a curve.

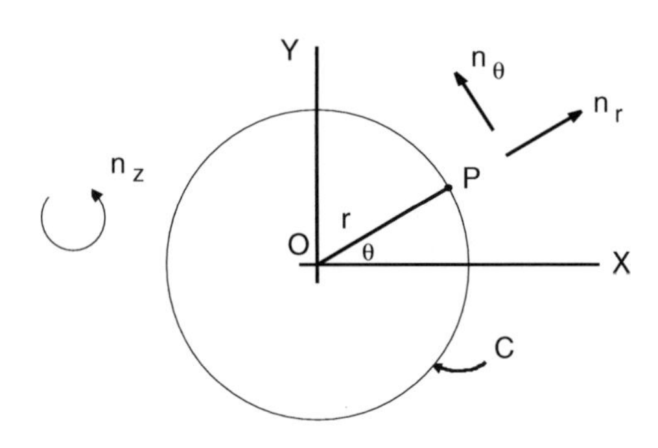

FIGURE 3.7.1
A point P moving on a circle C.

Hence, by differentiating, the velocity of P becomes:

$$\begin{aligned} \mathbf{V}^P &= d\mathbf{p}/dt = d(r\mathbf{n}_r)/dt \\ &= r\,d\mathbf{n}_r/dt = r\mathbf{n}_z \times \mathbf{n}_r(d\theta/dt) \\ &= r(d\theta/dt)\mathbf{n}_z \end{aligned} \tag{3.7.3}$$

where the fourth equality is obtained from Eq. (3.5.7). By differentiating again, we find the acceleration of P to be:

$$\begin{aligned} \mathbf{a}^P &= d\mathbf{V}^P/dt = d\big[r(d\theta/dt)\mathbf{n}_\theta\big]/dt \\ &= r\big[d^2 r/d\theta^2 + (d\theta/dt)\,d\mathbf{n}_\theta/dt\big] \\ &= r\big[d^2 r/d\theta^2 + (d\theta/dt)^2\mathbf{n}_z \times \mathbf{n}_\theta\big] \\ &= r\big[(d^2\theta/dt^2)\mathbf{n}_\theta - (d\theta/dt)^2\mathbf{n}_r\big] \end{aligned} \tag{3.7.4}$$

where again we have used Eq. (3.5.7) to obtain $d\mathbf{n}_\theta/dt$.

Observe in Eq. (3.7.3) that $\mathbf{V}^P$ is tangent to C as expected. Alternatively, observe in Eq. (3.7.4) that $\mathbf{a}^P$ has two components: one tangent to C and the other normal to C. The tangential component of $\mathbf{a}^P$ arises from a change in the speed of P along C. The normal component of $\mathbf{a}^P$ arises from a change in the direction of P as it moves along C.

To see these results more clearly let ω, v, and α be defined as:

$$\omega = d\theta/dt, \quad v = r d\theta/dt = r\omega, \quad \text{and} \quad \alpha = d\omega/dt \tag{3.7.5}$$

In terms of v, ω, and α, $\mathbf{V}^P$ and $\mathbf{a}^P$ are then:

$$\mathbf{V}^P = v\mathbf{n}_\theta = r\omega\mathbf{n}_\theta \tag{3.7.6}$$

and

$$\mathbf{a}^P = r\alpha\mathbf{n}_\theta - r\omega^2\mathbf{n}_r = (dv/dt)\mathbf{n}_\theta - (v^2/r)\mathbf{n}_r \tag{3.7.7}$$

These expressions will be useful to us throughout the text. We consider a generalization of these concepts in the next section.

Example 3.7.1: Movement of a Particle on the Rim of a Rotor

Consider a rotor R with radius r rotating about its axis with an angular speed and an angular acceleration as shown in Figure 3.7.2. Let P be a particle on the rim of R which is located by the angular coordinate θ as shown. If $\mathbf{n}_r$ and $\mathbf{n}_\theta$ are radial and tangential unit vectors, as shown, express the velocity and acceleration of P in terms of $\mathbf{n}_r$ and $\mathbf{n}_\theta$.

Solution: The angular speed is simply the time derivative of θ. Hence, we have:

$$\omega = \dot{\theta} \quad \text{and} \quad \alpha = \dot{\omega} = \dot{\theta} \tag{3.7.8}$$

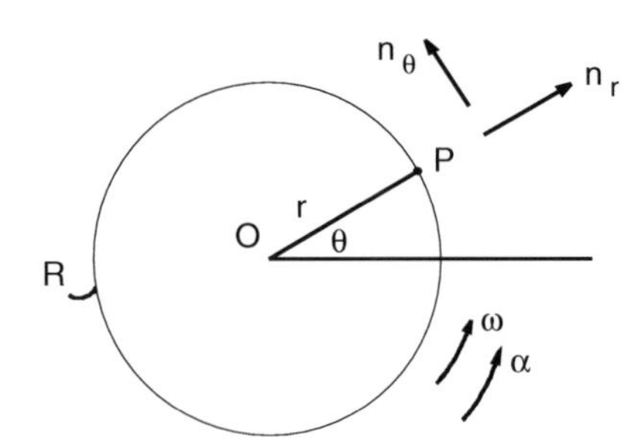

FIGURE 3.7.2
A particle P on the rim of a rotor.

Then, from Eq. (3.7.6), the velocity of P may be expressed as:

$$\mathbf{V}^P = r\omega\mathbf{n}_\theta = r\ddot{\theta}\mathbf{n}_\theta = v\mathbf{n}_\theta \tag{3.7.9}$$

Similarly, from Eq. (3.7.7), the acceleration of P may be expressed as:

$$\begin{aligned} \mathbf{a}^P &= r\alpha\mathbf{n}_\theta - r\omega^2\mathbf{n}_r = (dv/dt)\mathbf{n}_\theta - (v^2/r)\mathbf{n}_r \\ &= r\ddot{\theta}\mathbf{n}_\theta - r\ddot{\theta}^2\mathbf{n}_r \end{aligned} \tag{3.7.10}$$

3.8 Motion in a Plane

Let a point P move in a plane. Let the location of P be defined by the polar coordinates r and θ as in Figure 3.8.1. As before, let $\mathbf{n}_r$ and $\mathbf{n}_\theta$ be unit vectors parallel to the radial line and parallel to the transverse direction. Let $\mathbf{n}_z$ be normal to the plane as shown such that:

$$\mathbf{n}_z = \mathbf{n}_r \times \mathbf{n}_\theta \tag{3.8.1}$$

The position vector locating P relative to the fixed point O is then:

$$\mathbf{p} = r\mathbf{n}_r \tag{3.8.2}$$

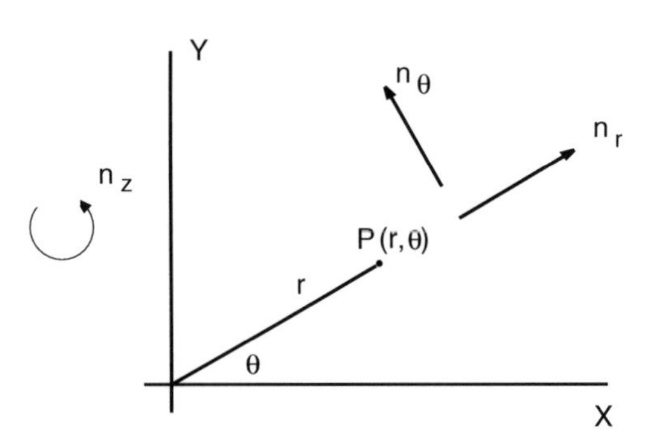

FIGURE 3.8.1
Motion of a point in a plane.

By differentiating, the velocity of p in a reference frame R containing O is:

$$\begin{aligned}\mathbf{V}^P &= d\mathbf{p}/dt = (dr/dt)\mathbf{n}_r + r(d\mathbf{n}_r/dt) \\ &= (dr/dt)\mathbf{n}_r + r\mathbf{n}_z \times \mathbf{n}_r(d\theta/dt) \\ &= (dr/dt)\mathbf{n}_r + r(d\theta/dt)\mathbf{n}_\theta\end{aligned} \tag{3.8.3}$$

where, as before, $d\mathbf{n}_r/dt$ is evaluated using Eq. (3.5.7).

Similarly, by differentiating again we obtain the acceleration of P as:

$$\begin{aligned}\mathbf{a}^P &= d\mathbf{V}^P/dt = d/dt[(dr/dt)\mathbf{n}_r + r(d\theta/dt)\mathbf{n}_\theta] \\ &= (d^2r/dt^2)\mathbf{n}_r + (dr/dt)(d\mathbf{n}_r/dt) + (dr/dt)(d\theta/dt)\mathbf{n}_\theta \\ &\quad + rd^2\theta/dt^2\mathbf{n}_\theta + r(d\theta/dt)(d\mathbf{n}_\theta/dt) \\ &= d^2r/dt^2\mathbf{n}_r + (dr/dt)(d\theta/dt)\mathbf{n}_z \times \mathbf{n}_r + (dr/d\theta)(d\theta/dt)\mathbf{n}_\theta \\ &\quad + r(d^2\theta/dt^2)\mathbf{n}_\theta + r(d\theta/dt)^2\mathbf{n}_z \times \mathbf{n}_\theta \\ &= (d^2r/dt^2)\mathbf{n}_r + (dr/dt)(d\theta/dt)\mathbf{n}_\theta + (dr/dt)(d\theta/dt)\mathbf{n}_\theta \\ &\quad + r(d^2\theta/dt^2)\mathbf{n}_\theta - r(d\theta/dt)^2\mathbf{n}_r\end{aligned}$$

or

$$\mathbf{a}_p = \left[d^2r/dt^2 - r(d\theta/dt)^2\right]\mathbf{n}_r + \left[rd^2\theta/dt^2 + 2(dr/dt)(d\theta/dt)\mathbf{n}_\theta\right] \tag{3.8.4}$$

Suppose, as before, that we define ω and α to be:

$$\omega = d\theta/dt \quad \text{and} \quad \alpha = d\omega/dt = d^2\theta/dt^2 \tag{3.8.5}$$

Thus, in terms of α and ω $\mathbf{V}^P$ and $\mathbf{a}^P$ are:

$$\mathbf{V}^P = r\omega n_\theta \tag{3.8.6}$$

and

$$\mathbf{a}^P = \left[d^2r/dt^2 - r\omega^2\right]\mathbf{n}_r + \left[r\alpha + 2\omega(dr/dt)\right]\mathbf{n}_\theta \tag{3.8.7}$$

or, alternatively,

$$\mathbf{a}^P = \left[\ddot{r} - r\omega^2\right]\mathbf{n}_r + \left[r\alpha + 2\omega\dot{r}\right]\mathbf{n}_\theta \tag{3.8.8}$$

where the overdot represents differentiation with respect to time t.

Observe that, if r is a constant, P moves on a circle and Eq. (3.8.6) has the same form as Eq. (3.7.6). Similarly, when r is a constant Eq.s (3.8.7) and (3.7.7) have identical forms.

Finally, it might be instructive to note that when r is not a constant and when $d\theta/dt$ is not zero, the last term in Eq. (3.8.7) is not zero. This term, called the *Coriolis acceleration*, often produces anti-intuitive results in the analysis of problems with moving bodies. We will discuss the kinematics of rigid bodies in the next chapter.

Example 3.8.1: Motion of a Cam Follower

Let a cam have a profile given by the equation:

$$r = a + b \sin\theta \quad a > b \tag{3.8.9}$$

Let P be a "follower" of the cam profile as depicted in Figure 3.8.2. If the cam is fixed and if the follower moves along the profile so that the angular rate $\dot{\theta}$ is constant, determine the radial $\mathbf{n}_r$ and transverse $\mathbf{n}_\theta$ components of the velocity and acceleration of P.

Solution: From Eq. (3.8.3), the velocity of P is:

$$\mathbf{V}^P = \dot{r}\mathbf{n}_r + r\dot{\theta}\mathbf{n}_\theta \tag{3.8.10}$$

Then by substituting from Eq. (3.8.9) we obtain:

$$\mathbf{V}^P = (b\cos\theta)\dot{\theta}\mathbf{n}_r + (a + b\sin\theta)\dot{\theta}\mathbf{n}_\theta \tag{3.8.11}$$

Similarly, from Eq. (3.8.4) the acceleration of P is:

$$\mathbf{a}^P = \left(\ddot{r} - r\dot{\theta}^2\right)\mathbf{n}_r + \left(r\ddot{\theta} + 2\dot{r}\dot{\theta}\right)\mathbf{n}_\theta \tag{3.8.12}$$

Then, by substituting from Eq. (3.8.9) we obtain:

$$\mathbf{a}^P = (-a - 2b\sin\theta)\dot{\theta}^2\,\mathbf{n}_r + (2b\cos\theta)\dot{\theta}^2\,\mathbf{n}_\theta \tag{3.8.13}$$

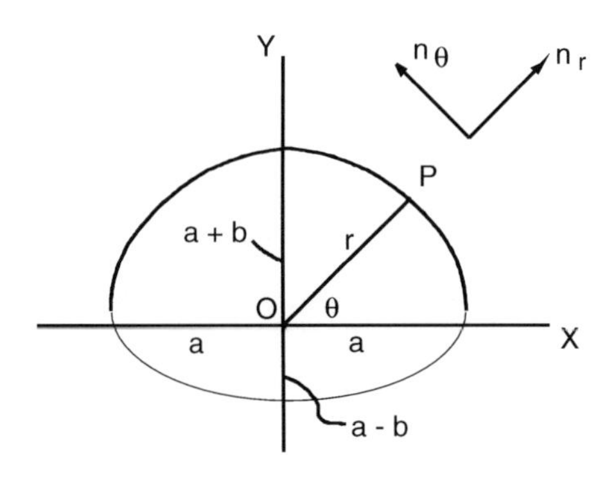

FIGURE 3.8.2
A fixed cam and follower P.

3.9 Closure

In the foregoing sections, we have briefly discussed and reviewed the kinematics of points (or particles). The emphasis has been the movement in a plane. In the following two chapters, we will extend these concepts to study the motion of rigid bodies and of points moving relative to rigid bodies.

References

3.1. Kane, T. R., *Analytical Elements of Mechanics*, Vol. 2, *Dynamics*, Academic Press, 1961, pp. 20–21.
3.2. Huston, R. L., *Multibody Dynamics*, Butterworth-Heineman, Stoneham, MA, 1990, pp. 14–16.

Problems

Section 3.2 Vector Differentiation

P3.2.1: Let the position vector **p** locating a point P relative to the origin O of a Cartesian reference frame R be expressed as:

$$\mathbf{p} = (2t - t^2)\mathbf{n}_x + (2 - 2t)^2\mathbf{n}_y + (t^3 - 3t)\mathbf{n}_z \text{ m}$$

where $\mathbf{n}_x$, $\mathbf{n}_y$, and $\mathbf{n}_z$ are unit vectors parallel to the X-, Y-, and Z-axes of R and where t is time in seconds.

a. Let $\mathbf{V} = d\mathbf{p}/dt$ (velocity of P in R). Find **V** and $|\mathbf{V}|$ **at times 0, 2, and 4 seconds.**
b. Let $a = d\mathbf{V}/dt = d^2\mathbf{p}/dt^2$ (acceleration of P in R). Find **a** and $|\mathbf{a}|$ **at times 0, 2, and 4 seconds.**
c. Find the time t^* such that $|\mathbf{V}|$ **= 0.**
d. See (c). Find **p** and **a** at t^*. (Observe that even if **V** is zero, **a** is not necessarily zero.)

P3.2.2: Let scalar s and vectors **V** and **W** be given as:

$$s = \sin 2\pi t$$

$$\mathbf{V} = t\mathbf{i} - t^2\mathbf{j} + 3(t-4)\mathbf{k}$$

$$\mathbf{W} = -t^2\mathbf{i} + 4\mathbf{j} - t\mathbf{k}$$

where **i**, **j**, and **k** are mutually perpendicular constant unit vectors. Verify Eqs. (3.2.11) through (3.2.14) for these expressions.

Section 3.3 Position, Velocity, and Acceleration

P3.3.1: The position vector **p** locating a particle P relative to a point O of a reference frame R is:

$$\mathbf{p} = t^2\mathbf{n}_x + (2t-3)\mathbf{n}_y + t^3\mathbf{n}_z \text{ ft}$$

where $\mathbf{n}_x$, $\mathbf{n}_y$, and $\mathbf{n}_z$ are unit vectors parallel to the X-, Y-, and Z-axes of a Cartesian coordinate system fixed in R.

a. Find expressions **v** and **a** for the velocity and acceleration of P in R.
b. Evaluate the magnitudes of the velocity and acceleration of P where $t = 2$ sec.

P3.3.2: See Problem P3.3.1, and find a unit vector tangent to the velocity of P when $t = 1$ sec.

P3.3.3: See Problems P3.3.1 and 3.3.2. Suppose the acceleration of P is expressed in the form:

$$\mathbf{a} = dV/dt\ \boldsymbol{\tau} + \left(V^2/\rho\right)\mathbf{n}$$

where V is the magnitude ("speed") of P, ρ is the radius of curvature of the curve C on which P moves, and $\boldsymbol{\tau}$ and **n** are unit vectors tangent and normal to C at the position of P. Find ρ, dV/dt, and **n** at $t = 1$. (Express **n** in terms of $\mathbf{n}_x$, $\mathbf{n}_y$, and $\mathbf{n}_z$.)

P3.3.4: A particle P moves in a plane. Let the position vector **p** locating P relative to the origin O of X–Y Cartesian axes be:

$$\mathbf{p} = 7\cos\pi t\,\mathbf{i} + 7\sin\pi t\,\mathbf{j} \text{ m}$$

where **i** and **j** are unit vectors parallel to X and Y.

a. Find expressions for the velocity and acceleration of P in terms of t, **i**, and **j**.
b. Find the velocity and acceleration of P at times $t = 0$, 0.5, 1, and 2 sec.

P3.3.5: See Problem 3.3.4, and let the position vector **p** have the general form:

$$\mathbf{p} = r\cos\omega t\,\mathbf{i} + r\sin\omega t\,\mathbf{j}$$

a. Find expressions for the velocity and acceleration of P.
b. Show that P moves on a circle with radius r.
c. Show that the velocity of P is tangent to the circular path of P.
d. Show that the acceleration of P is radial, directed toward the origin.
e. Show that the velocity and acceleration of P are perpendicular.
f. Find the magnitudes of the velocity and acceleration of P.

P3.3.6: A particle P moves in a straight line with a constant acceleration a_0. Find expressions for the speed v and displacements s of P as functions of time t. Let v and s when $t = 0$ be v_0 and s_0.

P3.3.7: See Problem 3.3.6. Suppose an automobile traveling at 45 mph on a straight, level roadway is suddenly braked so that its acceleration (or deceleration) is $-\mu g$, where μ is a friction coefficient valued at 0.75 and g is the gravity acceleration (32.2 ft/sec²). Determine the distance d traveled by the automobile before it stops and the time t that it takes to stop.

Section 3.4 Relative Velocity and Relative Acceleration

P3.4.1: See Example 3.4.1. Consider again rigid body B moving in reference frame R with P and Q being two particles of B represented by points P and Q, as shown in Figure P3.4.1.

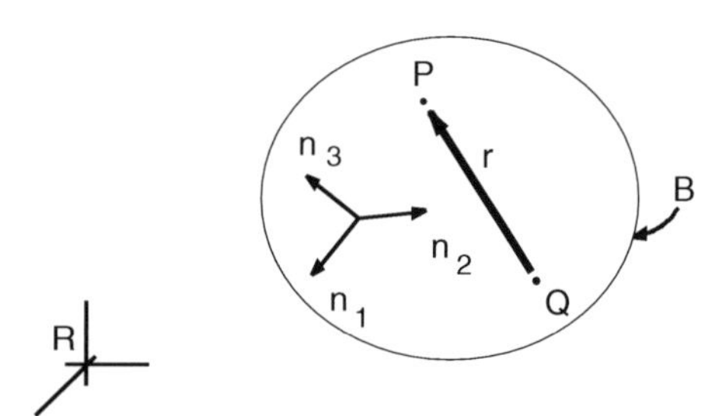

FIGURE P3.4.1
A body B with particles P and Q.

As before, let P be located relative to Q by the vector **r** given by:

$$\mathbf{r} = 2\mathbf{n}_1 - 3\mathbf{n}_2 + 7\mathbf{n}_3 \text{ ft}$$

where $\mathbf{n}_1$, $\mathbf{n}_2$, and $\mathbf{n}_3$ are unit vectors fixed in B. Let the angular velocity ω and the angular acceleration α of B in R be given as:

$$\boldsymbol{\omega} = -2\mathbf{n}_1 + 3\mathbf{n}_2 - 4\mathbf{n}_3 \text{ rad/sec}$$

and

$$\boldsymbol{\alpha} = 4\mathbf{n}_1 + 6\mathbf{n}_2 - 8\mathbf{n}_3 \text{ rad/sec}^2$$

It can be shown that the acceleration of P relative to Q in R may be expressed as:

$${}^R\mathbf{a}^{P/Q} = \boldsymbol{\alpha} \times \mathbf{r} + \omega \times (\boldsymbol{\omega} \times \mathbf{r})$$

Assuming this to be the case and if the acceleration of Q in R is also known as $-10\mathbf{n}_1 + 9\mathbf{n}_2 + 12\mathbf{n}_3$ ft/sec², find the acceleration of P relative to Q in R and the absolute acceleration of P in R.

P3.4.2: A southbound motorist traveling at 45 mph approaches an intersection with an east–west highway. When the motorist is 150 feet from the intersection, a traffic light controlling southbound traffic turns red, and the motorist brakes, causing the car to decelerate at 16.1 ft/sec² (.5 g). At the same time, an eastbound motorist operating a pickup truck on the east–west highway is traveling at 35 mph and is 200 feet west of the intersection with the north–south highway. At that point, the traffic light turns green for the pickup operator, causing the operator to accelerate the truck at the rate of 3 mph per second.

a. Determine the velocity of the pickup truck relative to the automobile. Express the answer in both feet per second and miles per hour and in terms of unit vectors **i** and **j** where **i** is directed east and **j** is directed north.
b. Determine the acceleration of the pickup truck relative to the automobile. Express the answer in feet per second squared (ft/sec^2) and in terms of the unit vectors **i** and **j** defined in (a).

P3.4.3: Car A is traveling west at a constant speed of 30 mph. As car A crosses an intersection (see Figure P3.4.3), car *B* starts from rest 120 feet south of the intersection and moves north with a constant acceleration of 4.4 ft/sec^2. Determine the position, velocity, and acceleration of *B* relative to A, 6 seconds after A crosses the intersection.

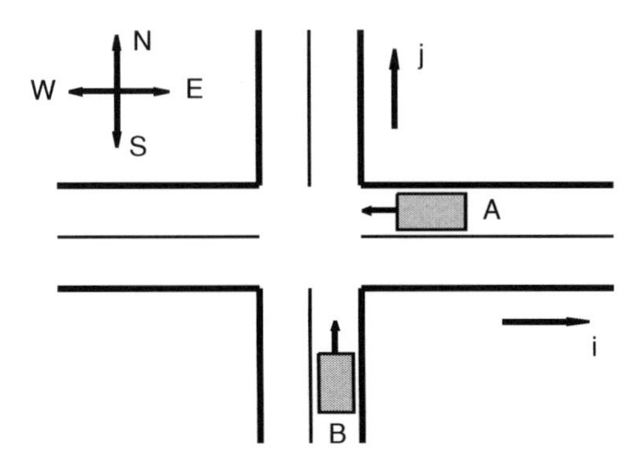

FIGURE P3.4.3
Cars *A* and *B* passing through an intersection.

Section 3.5 Differentiation of Rotating Unit Vectors

P3.5.1: Consider a point *P* moving on a circle at constant speed *V* as represented in Figure P3.5.1. Let the circle have radius *r* and let **p** locate *P* relative to the center *O*, as shown, with θ being the angle between **p** and the *X*-axis.

a. Express **p** in terms of the unit vectors $\mathbf{n}_x$ and $\mathbf{n}_y$ shown in Figure P3.5.1.
b. Compute the velocity **V** of *P*. Show that **V** is perpendicular to **p**.
c. Express the speed *V* in terms of *r* and θ.
d. Compute the velocity **V** of *P* using Eq. (3.5.9).
e. Compute the acceleration **a** of *P* by differentiating the velocity **V** obtained in (b). Show that **a** is directed opposite to **p**.
f. Compute **a** by differentiating **V** obtained in (d) by using Eq. (3.5.9).

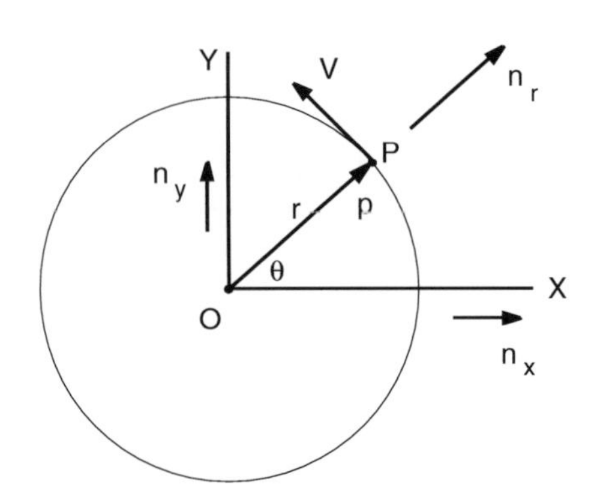

FIGURE P3.5.1
A point *P* moving on a circle at constant speed.

P3.5.2: Consider the Earth to be a sphere with a radius of 3960 miles. Let *P* be a point on the surface of the Earth at a latitude of 40°, as represented in Figure P3.5.2. With the Earth rotating on its axis one revolution per day, determine the magnitudes of the velocity and acceleration of *P*.

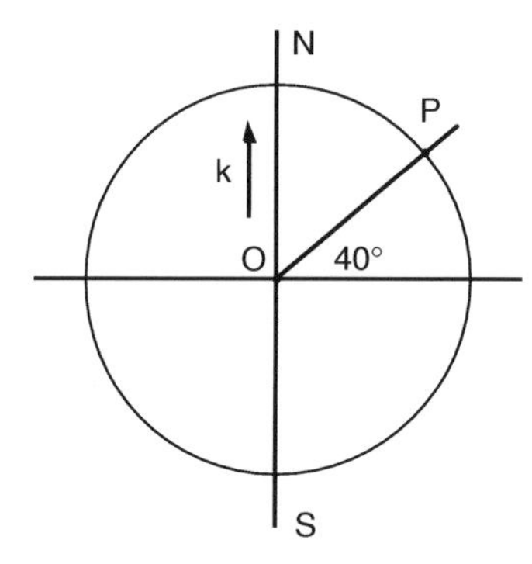

FIGURE P3.5.2
A point *P* on the surface of the Earth.

Section 3.7 Motion on a Circle

P3.7.1: A particle *P* is located on the rim of a flywheel as shown in Figure P3.7.1. If the radius of the flywheel is 300 mm, and if the angular displacement θ, the angular speed ω, and the angular acceleration α are as shown in the figure, determine the $\mathbf{n}_x$ and $\mathbf{n}_y$ components of the velocity and acceleration of *P*.

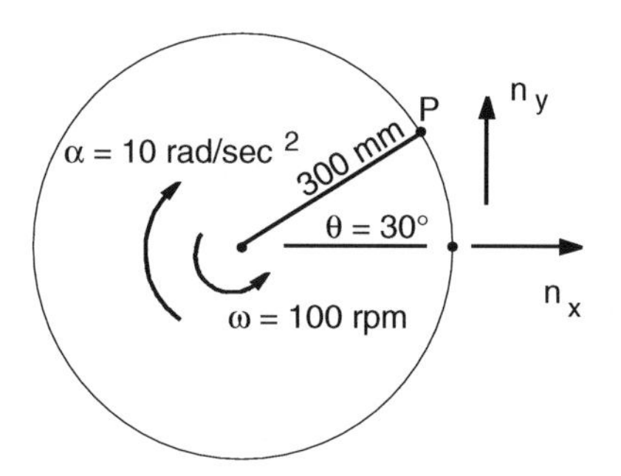

FIGURE P3.7.1
A particle *P* on the rim of a flywheel.

P3.7.2: An automobile A is traveling around a horizontal curve with a radius of 1000 feet. If the maximum radial acceleration is 0.6 *g* (where $g = 32.2$ ft/sec^2), determine the maximum speed *V* of A.

P3.7.3: A pickup truck T travels around a 600-ft-radius curve at 35 mph. If the driver brakes the truck at 5 ft/sec^2, determine the magnitude of the acceleration **a** of T. What is the angle between **a** and the direction of travel of T?

Section 3.8 Motion in a Plane

P3.8.1: See Example 3.8.1, and suppose that for the fixed cam and follower of Figure 3.8.1 the following geometric and kinematic data are given:

$$a = 150 \text{ mm} \quad b = 100 \text{ mm} \quad \text{and} \quad \dot{\theta} = 40 \text{ rpm}$$

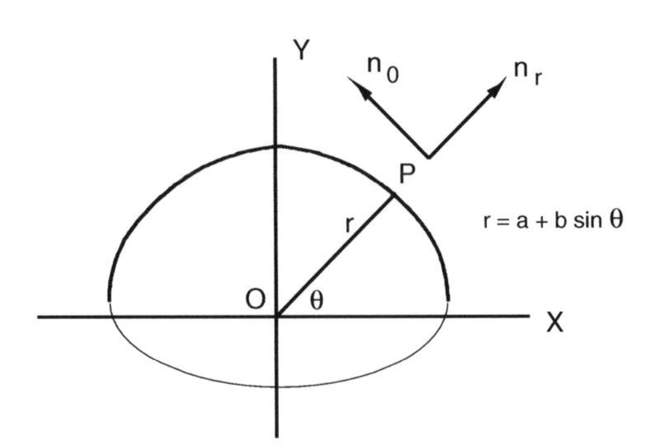

FIGURE P3.8.1
A fixed cam and follower *P*.

Determine the velocity and acceleration of *P* when θ is (a) 0°; (b) 30°; (c) 90°; and (d) 180°.

P3.8.2: Solve Problem P3.8.1 if $\dot{\theta}$, instead of being constant, is:

$$\dot{\theta} = 40 + 10 \sin 2\pi t \text{ rpm}$$

P3.8.3: Suppose in Problem P3.8.1 at $t = 0$ that the cam is rotating clockwise at a constant rate of 40 rpm, and that the radial line *OP* has a fixed inclination of 30°. Determine the velocity and acceleration of *P* relative to the cam when the cam has turned through (a) 0°; (b) 30°; (c) 90°; and (d) 180°.

P3.8.4: Repeat Problem P3.8.3 for a counterclockwise rotation of the cam at 40 rpm.

P3.8.5: In Problem P3.8.3 and P3.8.4, find the velocity and acceleration of *P* relative to the *X*–Y frame.

P3.8.6: An insect is crawling along a radial line of an old record turntable. If the turntable rotates clockwise at 33 1/3 rpm and if the insect crawls at 4 in./sec, determine the magnitude of the acceleration of the insect as it crosses the center of the turntable.

4

Kinematics of a Rigid Body

4.1 Introduction

The majority of machine elements can be modeled as rigid bodies. The same is true for mechanical systems in general. Therefore, in this chapter we will review the kinematics of rigid bodies. Although the motion of many machines and mechanical systems is restricted to a plane, we will initially develop a general (three-dimensional) analysis. Application with planar motion will then be simple, direct, and well founded. In the course of our discussions we will also develop several useful expressions regarding the relative motion of particles and rigid bodies.

4.2 Orientation of Rigid Bodies

Consider a rigid body B moving in a reference frame R as in Figure 4.2.1. Let $\mathbf{n}_1$, $\mathbf{n}_2$, $\mathbf{n}_3$ and $\mathbf{N}_1$, $\mathbf{N}_2$, $\mathbf{N}_3$ be mutually perpendicular dextral unit vector sets in B and R as shown. Then, the *orientation* of B in R can be defined in terms of the relative inclinations of the unit vector sets. To explore and develop this, recall from Section 2.11 that a transformation matrix S between the unit vector sets has the elements:

$$S_{ij} = \mathbf{N}_i \cdot \mathbf{n}_j \tag{4.2.1}$$

Recall that S has the convenient property of being *orthogonal*. That is, the inverse and transpose are equal:

$$S^{-1} = S^T \tag{4.2.2}$$

Recall further that the inclinations of the $\mathbf{n}_j$ relative to the $\mathbf{N}_i$ can be described in terms of three angles α, β, and γ, defined as follows: Let the $\mathbf{n}_j$ be mutually aligned with the $\mathbf{N}_i$. Then, three successive dextral rotations of B about $\mathbf{n}_1$, $\mathbf{n}_2$, and $\mathbf{n}_3$ through the angles α, β, and γ bring the $\mathbf{n}_j$ (and, hence, B itself) into a general inclination relative to the $\mathbf{N}_i$ (and, hence, relative to R). In this procedure, the transformation matrix S may be expressed as (see Eqs. (2.11.18) and (2.11.19)):

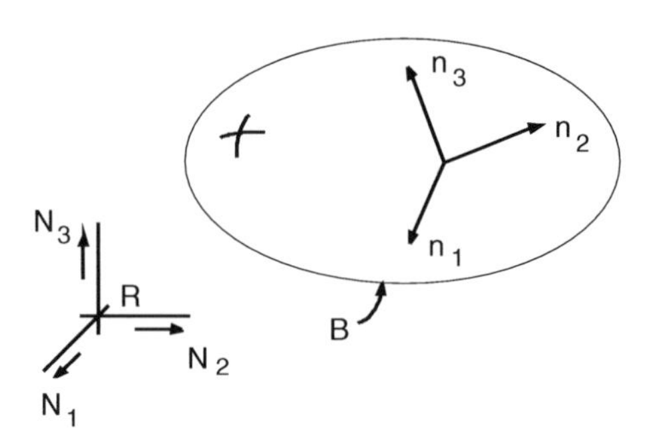

FIGURE 4.2.1
A rigid body moving in a reference frame.

$$
\begin{aligned}
S = ABC &= \begin{bmatrix} 1 & 0 & 0 \\ 0 & c_\alpha & -s_\alpha \\ 0 & s_\alpha & c_\alpha \end{bmatrix} \begin{bmatrix} c_\beta & 0 & s_\beta \\ 0 & 1 & 0 \\ -s_\beta & 0 & c_\beta \end{bmatrix} \begin{bmatrix} c_\gamma & -s_\gamma & 0 \\ s_\gamma & c_\gamma & 0 \\ 0 & 0 & 1 \end{bmatrix} \\
&= \begin{bmatrix} c_\beta c_\gamma & -c_\beta s_\gamma & s_\beta \\ (c_\alpha s_\gamma + s_\alpha s_\beta c_\gamma) & (c_\alpha c_\gamma - s_\alpha s_\beta s_\gamma) & -s_\alpha c_\beta \\ (s_\alpha s_\gamma - c_\alpha s_\beta c_\gamma) & (s_\alpha c_\gamma + c_\alpha s_\beta s_\gamma) & c_\alpha c_\beta \end{bmatrix}
\end{aligned}
\tag{4.2.3}
$$

Example 4.2.1: Use of Transformation Matrices

Suppose in Eq. (4.2.3) that α, β, and γ have the values 30, 60, and 45 degrees, respectively. Determine expressions relating the unit vector sets $\mathbf{N}_i$ and $\mathbf{n}_i$.

Solution: By substituting for α, β, and γ in Eq. (4.2.3), we obtain the transformation matrix S as:

$$
S = \begin{bmatrix} 0.354 & -0.354 & 0.866 \\ 0.918 & 0.306 & -0.250 \\ -0.176 & 0.884 & 0.433 \end{bmatrix} \tag{4.2.4}
$$

From Eq. (4.2.1) we have:

$$
\mathbf{N}_i = S_{ij}\mathbf{n}_j \quad \text{and} \quad \mathbf{n}_i = S_{ji}\mathbf{N}_j \tag{4.2.5}
$$

These expressions in turn may be written in the matrix forms:

$$
\mathbf{N} = S\mathbf{n} \quad \text{and} \quad \mathbf{n} = S^T\mathbf{N} \tag{4.2.6}
$$

where **N** and **n** are column arrays of the unit vectors $\mathbf{N}_i$ and $\mathbf{n}_i$. Hence, by comparing Eqs. (4.2.4) to (4.2.6), we obtain the desired relations:

$$
\begin{aligned}
\mathbf{N}_1 &= 0.354\mathbf{n}_1 - 0.354\mathbf{n}_2 + 0.866\mathbf{n}_3 \\
\mathbf{N}_2 &= 0.918\mathbf{n}_1 + 0.306\mathbf{n}_2 - 0.250\mathbf{n}_3 \\
\mathbf{N}_3 &= -0.176\mathbf{n}_1 + 0.884\mathbf{n}_2 + 0.433\mathbf{n}_3
\end{aligned}
\tag{4.2.7}
$$

and

$$\mathbf{n}_1 = 0.354\mathbf{N}_1 + 0.918\mathbf{N}_2 - 0.176\mathbf{N}_3$$
$$\mathbf{n}_2 = -0.354\mathbf{N}_1 + 0.306\mathbf{N}_2 + 0.884\mathbf{N}_3 \quad (4.2.8)$$
$$\mathbf{n}_3 = 0.866\mathbf{N}_1 - 0.250\mathbf{N}_2 + 0.433\mathbf{N}_3$$

Just as S defines the relative inclinations of the $\mathbf{n}_j$ and the $\mathbf{N}_{i,}$ it also defines the relative inclination or *orientation* of B and R. In this context the angles α, β, and γ are called *orientation angles*. We will find these concepts to be useful in our discussion about angular velocity. Before we consider that, however, it is useful to consider a procedure for determining the transformation matrices for various orientation angle sets. We do this in the following section.

4.3 Configuration Graphs

Consider again the problem of relating unit vector sets to each other — that is, expressing the vectors of one set in terms of the vectors in the other set (see Section 2.11). As before, let $\mathbf{N}_i$ ($i = 1, 2, 3$) and $\hat{\mathbf{N}}_j$ ($j = 1, 2, 3$) be mutually perpendicular dextral unit vector sets, and let $\mathbf{N}_i$ and $\hat{\mathbf{N}}_1$ be aligned with each other. Let the remaining vectors be inclined relative to one another as in Figure 4.3.1.

Observe that when the orientation angle is zero the unit vector sets are coincident and mutually aligned. Recall further that the equations relating the unit vectors are (see Eq. (2.11.14)):

$$\begin{aligned} \mathbf{N}_1 &= \hat{\mathbf{N}}_1 & & & \hat{\mathbf{N}}_1 &= \mathbf{N}_1 \\ \mathbf{N}_2 &= c_\alpha \hat{\mathbf{N}}_2 - s_\alpha \hat{\mathbf{N}}_3 & &\text{and} & \hat{\mathbf{N}}_2 &= c_\alpha \mathbf{N}_2 + s_\alpha \mathbf{N}_3 \\ \mathbf{N}_3 &= s_\alpha \hat{\mathbf{N}}_2 + c_\alpha \hat{\mathbf{N}}_3 & &\text{and} & \hat{\mathbf{N}}_3 &= -s_\alpha \mathbf{n}_2 + c_\alpha \mathbf{N}_3 \end{aligned} \quad (4.3.1)$$

Hence, we have the matrix expressions:

$$\begin{bmatrix} \mathbf{N}_1 \\ \mathbf{N}_2 \\ \mathbf{N}_3 \end{bmatrix} = \begin{bmatrix} 1 & 0 & 0 \\ 0 & c_\alpha & -s_\alpha \\ 0 & s_\alpha & c_\alpha \end{bmatrix} \begin{bmatrix} \hat{\mathbf{N}}_1 \\ \hat{\mathbf{N}}_2 \\ \hat{\mathbf{N}}_3 \end{bmatrix} \quad \text{and} \quad \begin{bmatrix} \hat{\mathbf{N}}_1 \\ \hat{\mathbf{N}}_2 \\ \hat{\mathbf{N}}_3 \end{bmatrix} = \begin{bmatrix} 1 & 0 & 0 \\ 0 & c_\alpha & s_\alpha \\ 0 & -s_\alpha & c_\alpha \end{bmatrix} \begin{bmatrix} \mathbf{N}_1 \\ \mathbf{N}_2 \\ \mathbf{N}_3 \end{bmatrix} \quad (4.3.2)$$

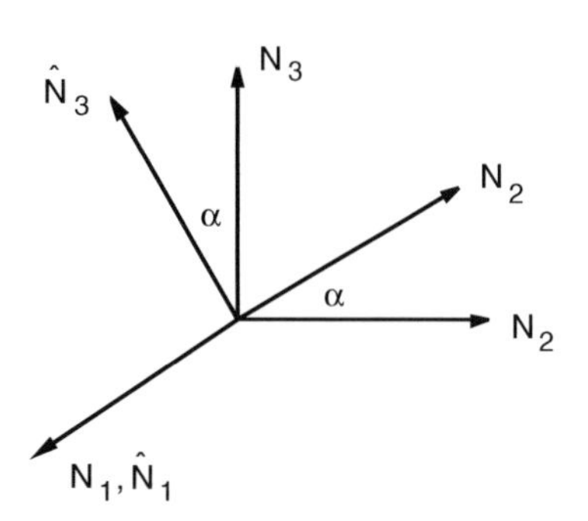

FIGURE 4.3.1
Unit vector sets $\mathbf{N}_i$ and $\hat{\mathbf{N}}_i$.

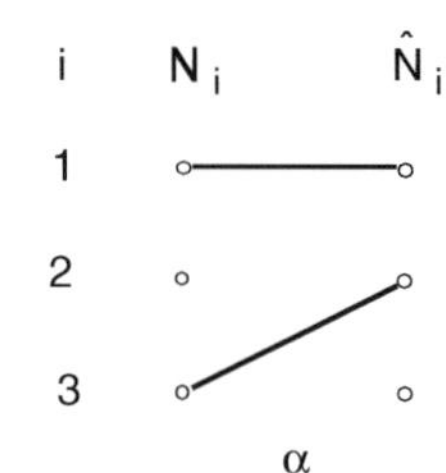

FIGURE 4.3.2
Configuration graph for the $\mathbf{N}_i$ and $\hat{\mathbf{N}}_i$.

The information of Eqs. (4.3.1) and (4.3.2) may be condensed into a diagram called a *configuration graph*, as shown in Figure 4.3.2. In this diagram, each dot, or node, represents one of the unit vectors as indicated. The horizontal line connecting the nodes of $\mathbf{N}_1$ and $\hat{\mathbf{N}}_1$ indicates that $\mathbf{N}_1$ and $\hat{\mathbf{N}}_1$ are equal. The inclined line connecting the nodes of $\mathbf{N}_3$ and $\hat{\mathbf{N}}_2$ means that $\mathbf{N}_3$ and $\hat{\mathbf{N}}_2$ are "inside" vectors in the alignment of Figure 4.3.1. Specifically, consider the plane of vectors that are perpendicular to $\mathbf{N}_1$ and $\hat{\mathbf{N}}_1$, as in Figure 4.3.3. Observe that $\mathbf{N}_3$ and $\hat{\mathbf{N}}_2$ lie between $\hat{\mathbf{N}}_3$ and $\mathbf{N}_2$; hence, $\mathbf{N}_3$ and $\hat{\mathbf{N}}_2$ are called "outside" vectors. In the configuration graph of Figure 4.3.2, the outside vectors are not connected by either a horizontal or inclined line.

Next, consider the relationship of $\mathbf{N}_2$, $\hat{\mathbf{N}}_2$, $\mathbf{N}_3$, and $\hat{\mathbf{N}}_3$ of Eq. (4.3.1):

$$\begin{aligned} \mathbf{N}_2 &= c_\alpha \hat{\mathbf{N}}_2 - s_\alpha \hat{\mathbf{N}}_3 \\ \mathbf{N}_3 &= s_\alpha \hat{\mathbf{N}}_2 + c_\alpha \hat{\mathbf{N}}_3 \end{aligned} \quad \text{and} \quad \begin{aligned} \hat{\mathbf{N}}_2 &= c_\alpha \mathbf{N}_2 + s_\alpha \mathbf{N}_3 \\ \mathbf{N}_3 &= -s_\alpha \mathbf{N}_2 + c_\alpha \mathbf{N}_3 \end{aligned} \tag{4.3.3}$$

Observe in these expressions that vectors with the same index are related by a positive cosine term, and vectors with different indices are related by either a positive or negative sine term. Specifically, those vectors related with a positive sine term are the *inside* vectors and those related with the negative sine term are the *outside* vectors. Therefore, by examining the configuration graph of Figure 4.3.2 we can immediately construct Eqs. (4.3.1) and (4.3.2).

To illustrate this concept further, consider the configuration graph of Figure 4.3.4. In this graph, the horizontal line indicates that $\hat{\mathbf{N}}_2$ and $\hat{\mathbf{n}}_2$ are equal. The inclined line indicates that $\hat{\mathbf{N}}_1$ and $\hat{\mathbf{n}}_3$ are inside vectors, and the absence of a line connecting $\hat{\mathbf{N}}_3$ and $\hat{\mathbf{n}}_1$ means that $\hat{\mathbf{N}}_3$ and $\hat{\mathbf{n}}_1$ are outside vectors. The orientation angle is β (at the bottom of the graph). From the graph, the vectors might be depicted as in Figure 4.3.5. Therefore the relations between the $\hat{\mathbf{N}}_1$ and $\mathbf{n}_1$ are:

$$\begin{aligned} \hat{\mathbf{N}}_1 &= c_\beta \hat{\mathbf{n}}_1 + s_\beta \hat{\mathbf{n}}_3 \\ \hat{\mathbf{N}}_2 &= \hat{\mathbf{n}}_2 \\ \hat{\mathbf{N}}_3 &= -s_\beta \hat{\mathbf{n}}_1 + c_\beta \hat{\mathbf{n}}_1 \end{aligned} \quad \text{and} \quad \begin{aligned} \hat{\mathbf{n}}_1 &= -s_\beta \hat{\mathbf{N}}_1 + c_\beta \hat{\mathbf{N}}_3 \\ \hat{\mathbf{n}}_2 &= \hat{\mathbf{N}}_2 \\ \hat{\mathbf{n}}_3 &= c_\beta \hat{\mathbf{N}}_3 + s_\beta \hat{\mathbf{N}}_1 \end{aligned} \tag{4.3.4}$$

Finally, consider the configuration graph of Figure 4.3.6 relating unit vector sets $\hat{\mathbf{n}}_i$ and $\mathbf{n}_i$. Using the same procedures as above, the vectors are related by the expressions:

$$\begin{aligned} \hat{\mathbf{n}}_1 &= c_\gamma \mathbf{n}_1 - s_\gamma \mathbf{n}_2 \\ \hat{\mathbf{n}}_2 &= s_\gamma \mathbf{n}_1 + c_\gamma \mathbf{n}_2 \\ \hat{\mathbf{n}}_3 &= \mathbf{n}_3 \end{aligned} \quad \text{and} \quad \begin{aligned} \mathbf{n}_1 &= c_\gamma \hat{\mathbf{n}}_1 + s_\gamma \hat{\mathbf{n}}_2 \\ \mathbf{n}_2 &= -s_\gamma \hat{\mathbf{n}}_1 + c_\gamma \hat{\mathbf{n}}_2 \\ \mathbf{n}_3 &= \hat{\mathbf{n}}_3 \end{aligned} \tag{4.3.5}$$

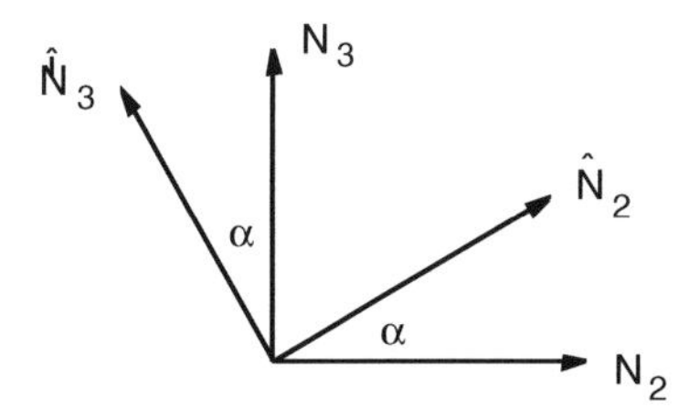

FIGURE 4.3.3
Plane of unit vectors normal to $\mathbf{N}_1$ and $\hat{\mathbf{N}}_i$.

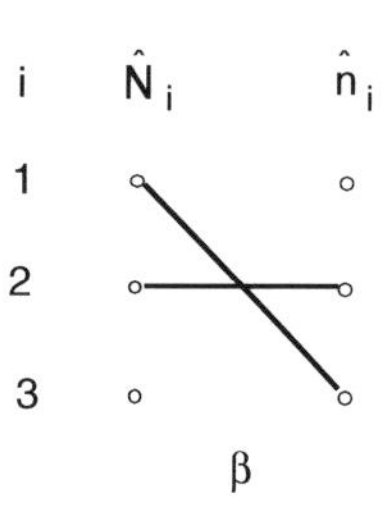

FIGURE 4.3.4
Configuration graph relating unit vector sets $\hat{\mathbf{N}}_1$ and $\hat{\mathbf{n}}_i$.

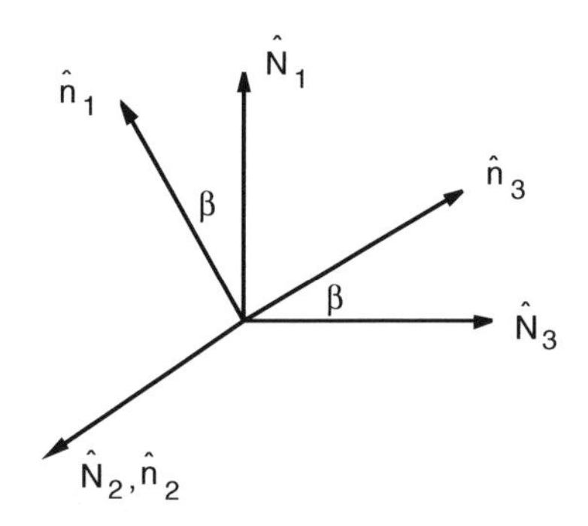

FIGURE 4.3.5
Unit vectors of the graph of Figure 4.3.4.

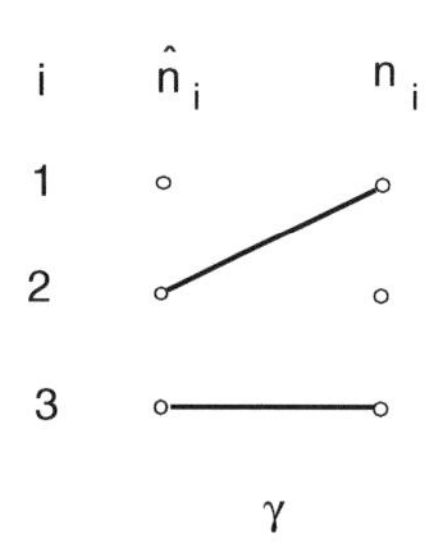

FIGURE 4.3.6
Configurations graph between unit vector sets $\mathbf{n}_i$ and $\hat{\mathbf{n}}_i$.

Configuration graphs may be combined to produce equations describing general orientations of the unit vector sets. To illustrate this suppose that $\mathbf{n}_1$, $\mathbf{n}_2$, and $\mathbf{n}_3$ are fixed in a body B. Let B have a general orientation in a reference frame R with unit vectors $\mathbf{N}_1$, $\mathbf{N}_2$, and $\mathbf{N}_3$, as in Figure 4.3.7 (and Figure 4.2.2). B may be brought into this general orientation by initially mutually aligning the unit vectors $\mathbf{N}_i$ and $\mathbf{n}_i$ (i = 1, 2, 3). Then, successive rotation of B about $\mathbf{n}_1$, $\mathbf{n}_2$, and $\mathbf{n}_3$ through the angles α, β, and γ brings B into its general orientation. An expanded configuration graph describing this orientation may be constructed by adjoining the configuration graphs of Figures 4.3.2, 4.3.4, and 4.3.6. Figure 4.3.8 shows the expanded graph. By using the rules stated above we can relate the $\mathbf{n}_i$ and $\mathbf{N}_i$ by the expressions:

$$\begin{aligned}
\mathbf{N}_1 &= c_\beta c_\gamma \mathbf{n}_1 - c_\beta s_\gamma \mathbf{n}_2 + s_\beta \mathbf{n}_3 \\
\mathbf{N}_2 &= \left(c_\alpha s_\gamma + s_\alpha s_\beta c_\gamma\right)\mathbf{n}_1 + \left(c_\alpha c_\gamma - s_\alpha s_\beta s_\gamma\right)\mathbf{n}_2 - s_\alpha c_\beta \mathbf{n}_3 \\
\mathbf{N}_3 &= \left(s_\alpha s_\gamma - s_\beta c_\alpha c_\gamma\right)\mathbf{n}_1 + \left(s_\alpha c_\gamma + c_\alpha s_\beta s_\gamma\right)\mathbf{n}_2 + c_\alpha c_\beta \mathbf{n}_3
\end{aligned} \tag{4.3.6}$$

and

$$\begin{aligned}
\mathbf{n}_1 &= c_\beta c_\gamma \mathbf{N}_1 + \left(c_\alpha s_\gamma + s_\alpha s_\beta c_\gamma\right)\mathbf{N}_2 + \left(s_\alpha s_\gamma - s_\beta c_\alpha c_\gamma\right)\mathbf{N}_3 \\
\mathbf{n}_2 &= -c_\beta s_\gamma \mathbf{N}_1 + \left(c_\alpha c_\gamma - s_\alpha s_\beta s_\gamma\right)\mathbf{N}_2 + \left(s_\alpha c_\gamma + c_\alpha s_\beta s_\gamma\right)\mathbf{N}_3 \\
\mathbf{n}_3 &= s_\beta \mathbf{N}_1 - s_\alpha c_\beta \mathbf{N}_2 + c_\alpha c_\beta \mathbf{N}_3
\end{aligned} \tag{4.3.7}$$

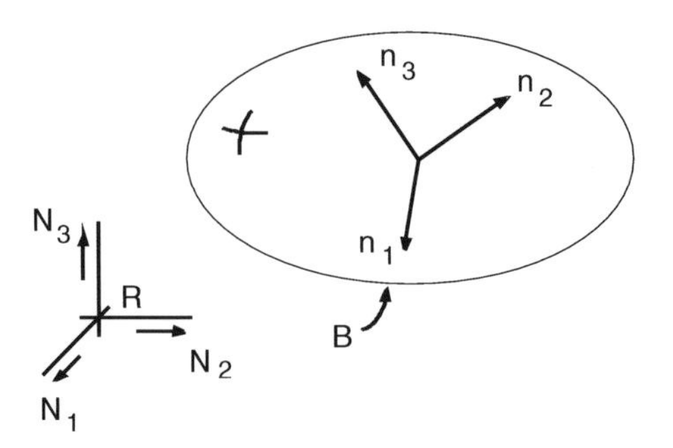

FIGURE 4.3.7
A rigid body B with a general orientation in a reference frame R.

In matrix form these equations may be expressed as:

$$\mathbf{N} = S\mathbf{n} \quad \text{and} \quad \mathbf{n} = S^T\mathbf{N} \tag{4.3.8}$$

where **N** and **n** represent the columns of the $\mathbf{N}_i$ and $\mathbf{n}_i$, and S is the matrix triple product:

$$S = ABC \tag{4.3.9}$$

where, as before, A, B, and C are:

$$A = \begin{bmatrix} 1 & 0 & 0 \\ 0 & c_\alpha & -s_\alpha \\ 0 & s_\alpha & c_\alpha \end{bmatrix}, \quad B = \begin{bmatrix} c_\beta & 0 & s_\beta \\ 0 & 1 & 0 \\ -s_\beta & 0 & c_\beta \end{bmatrix}, \quad C = \begin{bmatrix} c_\gamma & -s_\gamma & 0 \\ s_\gamma & c_\gamma & 0 \\ 0 & 0 & 1 \end{bmatrix} \tag{4.3.10}$$

Observe that by carrying out the product of Eq. (4.3.9) with A, B, and C given by Eq. (4.3.10) leads to Eq. (4.2.3) (see Problem 4.3.1).

Finally, observe that a body B may be brought into a general orientation in a reference frame R by successively rotating B an arbitrary sequence of vectors as illustrated in the following example.

Example 4.3.1: A 1–3–1 (Euler Angle) Rotation Sequence

Consider rotating B about $\mathbf{n}_1$, then $\mathbf{n}_3$, and then $\mathbf{n}_1$ again through angles θ_1, θ_2, and θ_3. In this case, the configuration graph takes the form as shown in Figure 4.3.9. With the rotation angles being θ_1, θ_2, and θ_3, the transformation matrices are:

$$A = \begin{bmatrix} 1 & 0 & 0 \\ 0 & c_1 & -s_1 \\ 0 & s_1 & c_1 \end{bmatrix}, B = \begin{bmatrix} c_2 & s_2 & 0 \\ -s_2 & c_2 & 0 \\ 0 & 0 & 1 \end{bmatrix}, C = \begin{bmatrix} 1 & 0 & 0 \\ 0 & c_3 & -s_3 \\ 0 & s_3 & c_3 \end{bmatrix} \tag{4.3.11}$$

and the general transformation matrix becomes:

$$S = ABC = \begin{bmatrix} c_2 & s_2c_3 & -s_2s_3 \\ -c_1s_2 & (c_1c_2c_3 - s_1s_3) & (-c_1c_2c_3 - s_1c_3) \\ -s_1s_2 & (s_1c_2c_3 + c_1s_3) & (-s_1c_2s_3 + c_1c_3) \end{bmatrix} \tag{4.3.12}$$

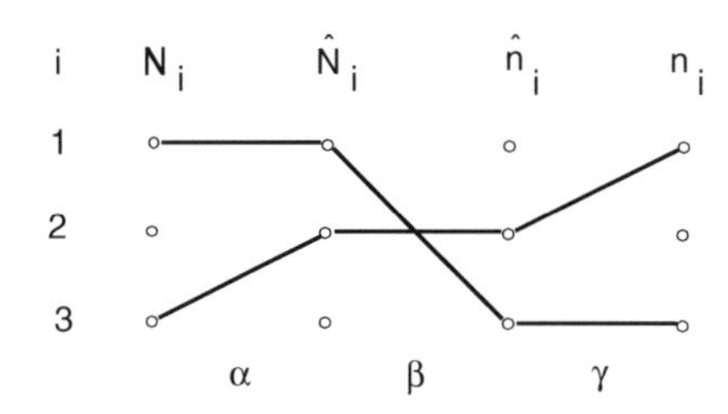

FIGURE 4.3.8
Configuration graph defining the orientation of B in R (see Figure 4.3.7).

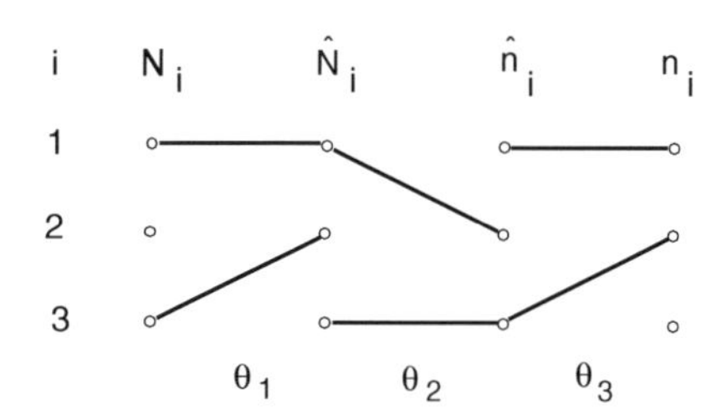

FIGURE 4.3.9
Configuration graph for a 1–3–1 rotation.

Hence, the unit vectors are related by the expressions:

$$\begin{aligned}
\mathbf{N}_1 &= c_2\mathbf{n}_1 + s_2c_3\mathbf{n}_2 - s_2s_3\mathbf{n}_3 \\
\mathbf{N}_2 &= -c_1s_2\mathbf{n}_1 + (c_1c_2c_3 - s_1s_3)\mathbf{n}_2 + (-c_1c_2c_3 - s_1c_3)\mathbf{n}_3 \\
\mathbf{N}_3 &= -s_1s_2\mathbf{n}_1 + (s_1c_2c_3 + c_1s_3)\mathbf{n}_2 + (-s_1c_2s_3 + c_1c_3)\mathbf{n}_3
\end{aligned} \tag{4.3.13}$$

and

$$\begin{aligned}
\mathbf{n}_1 &= c_2\mathbf{N}_1 + c_1s_2\mathbf{N}_2 - s_1s_2\mathbf{N}_3 \\
\mathbf{n}_2 &= s_2c_3\mathbf{N}_1 + (c_1c_2c_3 - s_1s_3)\mathbf{N}_2 + (s_1c_2c_3 + c_1s_3)\mathbf{N}_3 \\
\mathbf{n}_3 &= -s_2s_3\mathbf{N}_1 + (-c_1c_2s_3 - s_1c_3)\mathbf{N}_2 + (-s_1c_2s_3 + c_1c_3)\mathbf{N}_3
\end{aligned} \tag{4.3.14}$$

The angles θ_1, θ_2, and θ_3 are *Euler orientation angles* and the rotation sequence is referred to as a 1–3–1 sequence. (The angles α, β, and γ are called *dextral orientation angles* or *Bryan orientation angles* and the rotation sequence is a 1–2–3 sequence.)

4.4 Simple Angular Velocity and Simple Angular Acceleration

Of all kinematic quantities, angular velocity is the most fundamental and the most useful in studying the motion of rigid bodies. In this and the following three sections we will study angular velocity and its applications.

We begin with a study of simple angular velocity, where a body rotates about a fixed axis. Specifically, let B be a rigid body rotating about an axis Z–Z fixed in both B and a reference frame R as in Figure 4.4.1. Let **n** be a unit vector parallel to Z–Z as shown. Simple angular velocity is then defined to be a vector parallel to **n** measuring the rotation rate of B in R.

To quantify this further, consider an end view of B and of axis Z–Z as in Figure 4.4.2. Let X and Y be axes fixed in R and let L be a line fixed in B and parallel to the X–Y plane. Let θ be an angle measuring the inclination of L relative to the X-axis as shown. Then, the angular velocity ω (simple angular velocity) of B in R is defined to be:

$$\boldsymbol{\omega} \overset{D}{=} \dot{\theta}\text{n} \tag{4.4.1}$$

where $\dot{\theta}$ is sometimes called the *angular speed* of B in R.

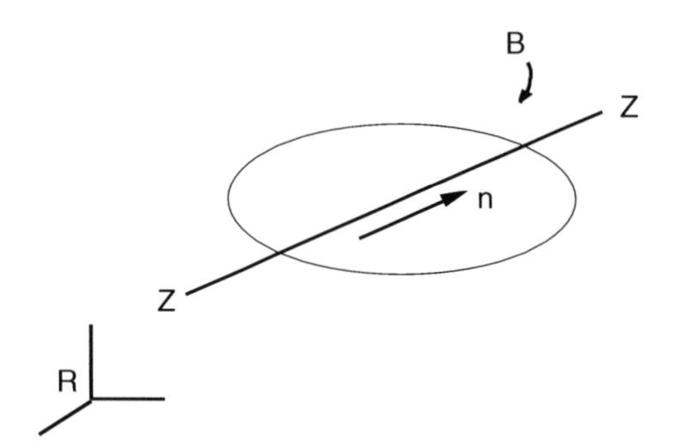

FIGURE 4.4.1
A body B rotating in a reference frame R.

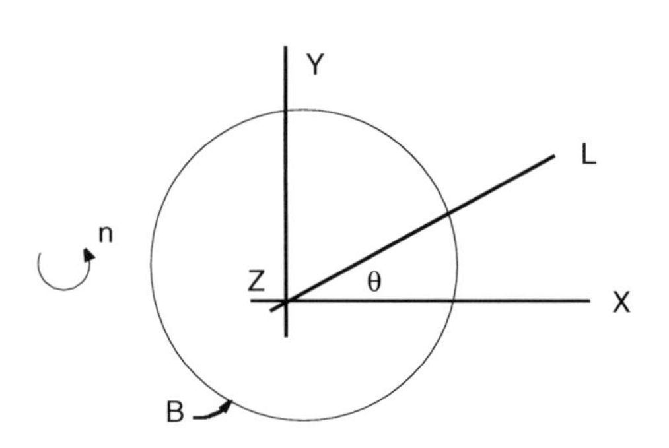

FIGURE 4.4.2
End view of body B.

Observe that θ measures the rotation of B in R. It is also a measure of the orientation of B in R. In this context, the angular velocity of B in R is a measure of *the rate of change of orientation* of B in R.

The simple angular *acceleration* α of B in R is then defined as the time rate of change of the angular velocity. That is,

$$\boldsymbol{\alpha} = d\boldsymbol{\omega}/dt = \ddot{\theta}\mathbf{n} \tag{4.4.2}$$

If we express α and β in the forms:

$$\boldsymbol{\alpha} = \alpha\mathbf{n} \quad \text{and} \quad \omega = \boldsymbol{\omega}\mathbf{n} \tag{4.4.3}$$

then α, ω, and θ are related by the expressions:

$$\omega = \dot{\theta} \quad \text{and} \quad \alpha = \dot{\omega} = \ddot{\theta} \tag{4.4.4}$$

By integrating we have the relations:

$$\theta = \int \omega dt + \theta_o \quad \text{and} \quad \omega = \int \alpha dt + \omega_o \tag{4.4.5}$$

where θ_o and ω_o are values of θ and ω at some initial time t_o.

From the chain rule for differentiation we have:

$$\alpha = \dot{\omega} = d\omega/dt = (d\omega/d\theta)(d\theta/dt) = \omega d\omega/d\theta = d(\omega^2/2)/dt \tag{4.4.6}$$

Then by integrating we obtain:

$$\omega^2 = 2\int \alpha d\theta + \omega_o^2 \tag{4.4.7}$$

where, as before, ω_o is an initial value of ω when θ is θ_o. If α is constant, we have the familiar relation:

$$\omega^2 = \omega_o^2 + 2\alpha\theta \tag{4.4.8}$$

Example 4.4.1: Revolutions Turned Through During Braking

Suppose a rotor is rotating at an angular speed of 100 rpm. Suppose further that the rotor is braked to rest with a constant angular deceleration of 5 rad/sec². Find the number N of revolutions turned through during braking.

Solution: From Eq. (4.4.8), when the rotor is braked to rest, its angular speed ω is zero. The angle θ turned through during braking is, then,

$$\theta = -\omega_o^2/2\alpha = -[(2\pi)(100)/60]^2/[(2)(-5)]$$
$$= 10.966\,\text{radians} = 628.3\,\text{degrees}$$

Hence, the number of revolutions turned through is:

$$\mathrm{N} = 1.754$$

4.5 General Angular Velocity

Angular velocity may be defined intuitively as the time rate of change of orientation. Generally, however, no single quantity defines the orientation for a rigid body. Hence, unlike velocity, angular velocity cannot be considered as the derivative of a single quantity. Nevertheless, it is possible to define the angular velocity in terms of derivatives of a set of unit vectors fixed in the body. These unit vector derivatives thus provide a measure of the rate of change of orientation of the body.

Specifically, let B be a body whose orientation is changing in a reference frame R, as depicted in Figure 4.5.1. Let $\mathbf{n}_1$, $\mathbf{n}_2$, and $\mathbf{n}_3$ be mutually perpendicular unit vectors as shown. Then, the angular velocity of B in R, written as ${}^R\omega^B$, is defined as:

$$ {}^R\boldsymbol{\omega}^B \overset{D}{=} \left(\frac{d\mathbf{n}_2}{dt}\cdot\mathbf{n}_3\right)\mathbf{n}_1 + \left(\frac{d\mathbf{n}_3}{dt}\cdot\mathbf{n}_1\right)\mathbf{n}_2 + \left(\frac{d\mathbf{n}_1}{dt}\cdot\mathbf{n}_2\right)\mathbf{n}_3 \tag{4.5.1}$$

The angular velocity vector has several properties that are useful in dynamical analyses. Perhaps the most important is the property of producing derivatives by vector multiplication: specifically, let $\mathbf{c}$ be any vector fixed in B. Then, the derivative of $\mathbf{c}$ in R is given by the single expression:

$$d\mathbf{c}/dt = {}^R\boldsymbol{\omega}^B \times \mathbf{c} \tag{4.5.2}$$

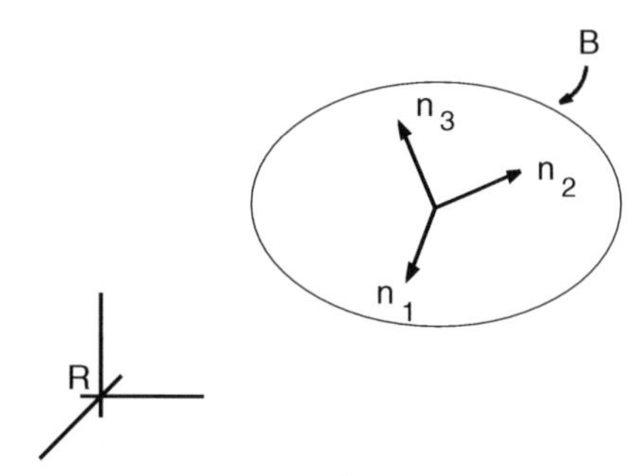

FIGURE 4.5.1
A rigid body changing orientation in a reference frame.

To see this, let **c** be expressed in terms of the unit vectors $\mathbf{n}_1$, $\mathbf{n}_2$, and $\mathbf{n}_3$. That is, let

$$\mathbf{c} = c_1\mathbf{n}_1 + c_2\mathbf{n}_2 + c_3\mathbf{n}_3 \tag{4.5.3}$$

Because **c** is fixed in B, the scalar components c_i (i = 1, 2, 3) are constants; hence, the derivative of **c** in R is:

$$d\mathbf{c}/dt = c_1 d\mathbf{n}_1/dt + c_2\mathbf{n}_2/dt + c_3\mathbf{n}_3/dt \tag{4.5.4}$$

Next, consider the product ${}^R\omega^B \times \mathbf{n}_1$:

$$\begin{aligned} {}^R\boldsymbol{\omega}^B \times \mathbf{n}_1 &= \left[\left(\frac{d\mathbf{n}_2}{dt}\cdot\mathbf{n}_3\right)\mathbf{n}_1 + \left(\frac{d\mathbf{n}_3}{dt}\cdot\mathbf{n}_1\right)\mathbf{n}_2 + \left(\frac{d\mathbf{n}_1}{dt}\cdot\mathbf{n}_2\right)\mathbf{n}_3\right]\times\mathbf{n}_1 \\ &= -\left(\frac{d\mathbf{n}_3}{dt}\cdot\mathbf{n}_1\right)\mathbf{n}_3 + \left(\frac{d\mathbf{n}_1}{dt}\cdot\mathbf{n}_2\right)\mathbf{n}_2 \end{aligned} \tag{4.5.5}$$

Recall that:

$$\mathbf{n}_1\cdot\mathbf{n}_1 = 1 \quad \text{and} \quad \mathbf{n}_1\cdot\mathbf{n}_3 = 0 \tag{4.5.6}$$

By differentiating these expressions we have:

$$\frac{d\mathbf{n}_1}{dt}\cdot\mathbf{n}_1 = 0 \quad \text{and} \quad \frac{d\mathbf{n}_1}{dt}\cdot\mathbf{n}_3 = -\mathbf{n}_1\cdot\frac{d\mathbf{n}_3}{dt} \tag{4.5.7}$$

Hence, Eq. (4.5.5) may be written as:

$${}^R\boldsymbol{\omega}^B \times \mathbf{n}_1 = \left(\frac{d\mathbf{n}_1}{dt}\cdot\mathbf{n}_1\right)\mathbf{n}_1 + \left(\frac{d\mathbf{n}_1}{dt}\cdot\mathbf{n}_2\right)\mathbf{n}_2 + \left(\frac{d\mathbf{n}_1}{dt}\cdot\mathbf{n}_3\right)\mathbf{n}_3 \tag{4.5.8}$$

Because *any* vector **V** may be expressed as $(\mathbf{V}\cdot\mathbf{n}_1)\mathbf{n}_1 + (\mathbf{V}\cdot\mathbf{n}_2)\mathbf{n}_2 + (\mathbf{V}\cdot\mathbf{n}_3)\mathbf{n}_3$, we see that the right side of Eq. (4.5.8) is an identity for $d\mathbf{n}_1/dt$. Thus, we have the result:

$${}^R\boldsymbol{\omega}^B \times \mathbf{n}_1 = d\mathbf{n}_1/dt \tag{4.5.9}$$

Similarly, we have the companion results:

$${}^R\boldsymbol{\omega}^B \times \mathbf{n}_2 = d\mathbf{n}_2/dt \quad \text{and} \quad {}^R\boldsymbol{\omega}^B \times \mathbf{n}_3 = d\mathbf{n}_3/dt \tag{4.5.10}$$

Finally, by using these results in Eq. (4.5.4) we have:

$$\begin{aligned} d\mathbf{c}/dt &= c_1\,{}^R\boldsymbol{\omega}^B \times \mathbf{n}_1 + c_2\,{}^R\boldsymbol{\omega}^B \times \mathbf{n}_2 + c_3\,{}^R\boldsymbol{\omega}^B \times \mathbf{n}_3 \\ &= {}^R\boldsymbol{\omega}^B \times (c_1\mathbf{n}_1 + c_2\mathbf{n}_2 + c_3\mathbf{n}_3) \\ &= {}^R\boldsymbol{\omega}^B \times \mathbf{c} \end{aligned} \tag{4.5.11}$$

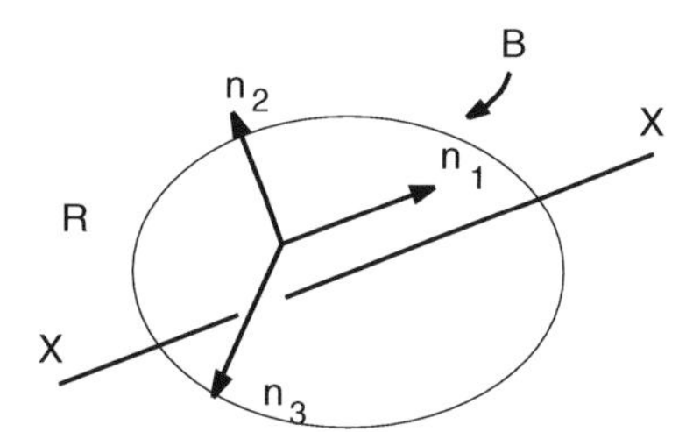

FIGURE 4.5.2
Rotation of a body about a fixed axis X–X.

Observe the pattern of the terms in Eq. (4.5.1). They all have the same form, and they may be developed from one another by simply permuting the indices.

Example 4.5.1: Simple Angular Velocity

We may also observe that Eq. (4.5.1) is consistent with our earlier results on simple angular velocity. To see this, let B rotate in R about an axis parallel to, say, $\mathbf{n}_1$, as shown in Figure 4.5.2. Let X–X be fixed in both B and R. Then, from Eq. (4.4.1), the angular velocity of B in R is:

$$^R\boldsymbol{\omega}^B = \dot{\theta}\mathbf{n}_1 \tag{4.5.12}$$

where, as before, θ measures the rotation angle. From Eq. (4.5.1), we see that $^R\omega^B$ may be expressed as:

$$^R\boldsymbol{\omega}^B = \left(\frac{d\mathbf{n}_2}{dt}\cdot\mathbf{n}_3\right)\mathbf{n}_1 + \left(\frac{d\mathbf{n}_3}{dt}\cdot\mathbf{n}_1\right)\mathbf{n}_2 + \left(\frac{d\mathbf{n}_1}{dt}\cdot\mathbf{n}_2\right)\mathbf{n}_3 \tag{4.5.13}$$

Because $\mathbf{n}_1$ is fixed, parallel to axis X–X, its derivative is zero; hence, the third term in Eq. (4.5.13) is zero. The first two terms may be evaluated using Eq. (3.5.7). Specifically,

$$d\mathbf{n}_2/dt = \dot{\theta}\mathbf{n}_1 \times \mathbf{n}_2 = \dot{\theta}\mathbf{n}_3 \quad \text{and} \quad d\mathbf{n}_3/dt = \dot{\theta}\mathbf{n}_1 \times \mathbf{n}_3 = -\dot{\theta}\mathbf{n}_2 \tag{4.5.14}$$

By substituting into Eq. (4.5.13), we have:

$$^R\boldsymbol{\omega}^B = \dot{\theta}\mathbf{n}_1 \tag{4.5.15}$$

which is identical to Eq. (4.5.11).

4.6 Differentiation in Different Reference Frames

Consider next the differentiation of a vector with respect to different reference frames. Specifically, let $\mathbf{V}$ be the vector and let R and $\hat{R}$ be two distinct reference frames. Let $\hat{\mathbf{n}}_i$ be mutually perpendicular unit vectors fixed in $\hat{R}$, as represented in Figure 4.6.1. Let $\mathbf{V}$ be expressed in terms of the $\hat{\mathbf{n}}_i$ as:

$$\mathbf{V} = \hat{V}_1\hat{\mathbf{n}}_1 + \hat{V}_2\hat{\mathbf{n}}_2 + \hat{V}_3\hat{\mathbf{n}}_3 \tag{4.6.1}$$

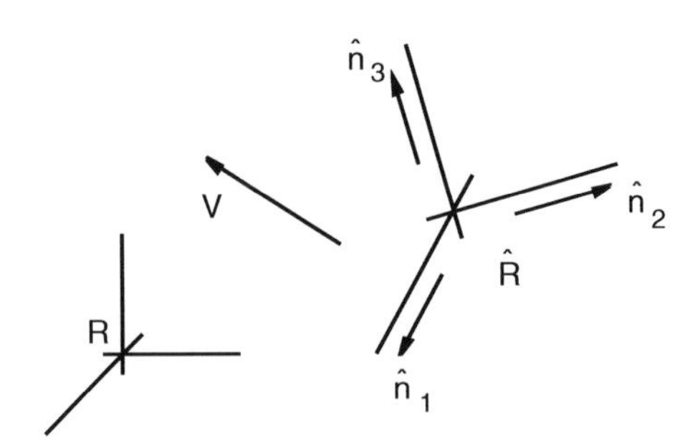

FIGURE 4.6.1
Vector V and reference frames R and $\hat{R}$.

Because the $\hat{\mathbf{n}}_i$ are fixed in $\hat{R}$, the derivative of **V** in $\hat{R}$ is obtained by simply differentiating the scalar components in Eq. (4.6.1). That is,

$$ {}^{\hat{R}}d\mathbf{V}/dt = \left(d\hat{V}_1/dt\right)\hat{\mathbf{n}}_1 + \left(d\hat{V}_2/dt\right)\hat{\mathbf{n}}_2 + \left(d\hat{V}_3/dt\right)\hat{\mathbf{n}}_3 \tag{4.6.2} $$

Next, relative to reference frame R, the derivative of **V** is:

$$ \begin{aligned} {}^{R}d\mathbf{V}/dt &= \left(d\hat{V}_1/dt\right)\hat{\mathbf{n}}_1 + \hat{V}_1 {}^{R}d\hat{\mathbf{n}}_1/dt + \left(d\hat{V}_2/dt\right)\hat{\mathbf{n}}_2 \\ &\quad + \hat{V}_2 {}^{R}d\hat{\mathbf{n}}_3/dt + \left(d\hat{V}_3/dt\right)\hat{\mathbf{n}}_3 + \hat{V}_3 {}^{R}d\hat{\mathbf{n}}_3/dt \\ &= {}^{\hat{R}}d\mathbf{V}/dt + \hat{V}_1 {}^{R}d\hat{\mathbf{n}}_1/dt + \hat{V}_2 {}^{R}d\mathbf{n}_2/dt + \hat{V}_3 {}^{R}d\hat{\mathbf{n}}_3/dt \end{aligned} \tag{4.6.3} $$

where the second equality is determined from Eq. (4.6.2). Because the $\hat{\mathbf{n}}_i$ are fixed in $\hat{R}$, the derivatives in R are:

$$ {}^{R}d\hat{\mathbf{n}}_1/dt = {}^{R}\boldsymbol{\omega}^{\hat{R}} \times \hat{\mathbf{n}}_i \quad (i = 1,2,3) \tag{4.6.4} $$

Hence, ${}^{R}d\mathbf{V}/dt$ becomes:

$$ \begin{aligned} {}^{R}d\mathbf{V}/dt &= {}^{\hat{R}}d\mathbf{V}/dt + V_1 {}^{R}\boldsymbol{\omega}^{\hat{R}} \times \hat{\mathbf{n}}_1 + V_2 {}^{R}\boldsymbol{\omega}^{\hat{R}} \times \hat{\mathbf{n}}_2 \\ &\quad + V_3 {}^{R}\boldsymbol{\omega}^{\hat{R}} \times \hat{\mathbf{n}}_3 \\ &= {}^{\hat{R}}d\mathbf{V}/dt + {}^{R}\boldsymbol{\omega}^{\hat{R}} \times \left(\hat{V}_1\hat{\mathbf{n}}_1 + \hat{V}_2\hat{\mathbf{n}}_2 + \hat{V}_3\hat{\mathbf{n}}_3\right) \end{aligned} $$

or

$$ {}^{R}d\mathbf{V}/dt = {}^{\hat{R}}d\mathbf{V}/dt + {}^{R}\boldsymbol{\omega}^{\hat{R}} \times \mathbf{V} \tag{4.6.5} $$

Because there were no restrictions on **V**, Eq. (4.6.5) may be written as:

$$ {}^{R}d(\ )/dt = {}^{\hat{R}}d(\ )/dt + {}^{R}\boldsymbol{\omega}^{\hat{R}} \times (\ ) \tag{4.6.6} $$

where any vector quantity may be inserted in the parentheses.

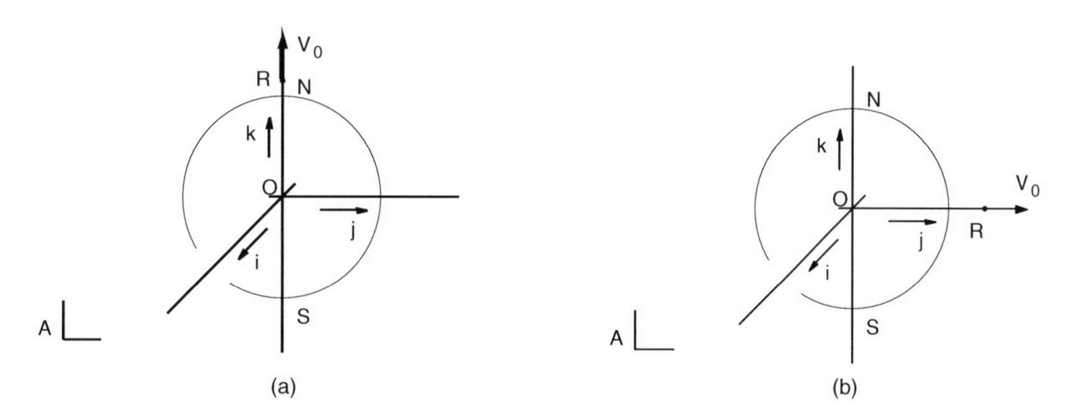

FIGURE 4.6.2
Vertical launch of a rocket from the surface of the Earth. (a) Launch at the North Pole; (b) launch at the Equator.

Example 4.6.1: Effect of Earth Rotation

Equation (4.6.6) is useful for developing expressions for velocity and acceleration of particles moving relative to rotating bodies. To illustrate this, consider the case of a rocket R launched vertically from the Earth's surface. Let the vertical speed of R relative to the Earth (designated as E) be V_0. Suppose we want to determine the velocity of R in an astronomical reference frame A in which E rotates for the cases when R is launched from (a) the North Pole, and (b) the equator.

Solution: Consider a representation of R, E, and A as in Figure 4.6.2a and b, where **i**, **j**, and **k** are mutually perpendicular unit vectors fixed in E, with **k** being along the north/south axis, the axis of rotation of E. In both cases the velocity of R in A may be expressed as:

$$ {}^{A}\mathbf{V}^{R} = {}^{A}d\mathbf{OR}/dt \tag{4.6.7} $$

where O is the center of the Earth which is also fixed in A.

a. For launch at the North Pole, the position vector **OR** is simply $(r + h)\mathbf{k}$, where r is the radius of E (approximately 3960 miles) and h is the height of R above the surface of E. Thus, from Eq. (4.6.6), ${}^{A}\mathbf{V}^{R}$ is:

$$ {}^{A}\mathbf{V}^{R} = {}^{E}d\big[(r+h)\mathbf{k}\big]/dt + {}^{A}\boldsymbol{\omega}^{E} \times (r+h)\mathbf{k} \tag{4.6.8} $$

Observe that angular velocity of E in A is along **k** with a speed of one revolution per day. That is,

$$ {}^{A}\boldsymbol{\omega}^{E} = \omega\mathbf{k} = \frac{2\pi}{(24)(3600)}\mathbf{k} = 7.27 \times 10^{-5}\mathbf{k}\,\text{rad/sec} \tag{4.6.9} $$

Hence, by substituting into Eq. (4.6.8), we have:

$$ {}^{A}\mathbf{V}^{R} = \dot{h}\mathrm{k} = \mathbf{V}_{o}\mathrm{k} \tag{4.6.10} $$

b. For the launch at the Equator, the position vector **OR** is $(r + h)\mathbf{i}$. In this case, Eq. (4.6.7) becomes:

$$
\begin{aligned}
{}^{A}\mathbf{V}^{R} &= {}^{E}d(r+h)\mathbf{i}/dt + {}^{A}\omega^{E} \times (r+h)\mathbf{i} \\
&= \dot{h}\mathbf{i} + (r+h)\omega\mathbf{j} \\
&= V_o\mathbf{i} + [(3960)(5280) + h](7.27\times 10^{-5})\mathbf{j}
\end{aligned}
$$

or

$$
{}^{A}\mathbf{V}^{R} = V_o\mathbf{i} + (1520 + h\omega)\mathbf{j}\,\text{ft/sec} \tag{4.6.11}
$$

Observe that h is small (at least, initially) compared with r. Thus, a reasonable approximation to ${}^{A}\mathbf{V}^{R}$ is:

$$
{}^{A}\mathbf{V}^{R} = V_O\mathbf{i} + 1520\mathbf{j}\,\text{ft/sec} \tag{4.6.12}
$$

Observe also the differences in the results of Eqs. (4.6.10) and (4.6.11).

4.7 Addition Theorem for Angular Velocity

Equation (4.6.6) is useful for establishing the addition theorem for angular velocity — one of the most important equations of rigid body kinematics. Consider a body B moving in a reference frame $\hat{R}$, which in turn is moving in a reference frame R as depicted in Figure 4.7.1. Let **V** be an arbitrary vector fixed in B. Using Eq. (4.6.6), the derivative of **V** in R is:

$$
{}^{R}d\mathbf{V}/dt = {}^{\hat{R}}d\mathbf{V}/dt + {}^{R}\boldsymbol{\omega}^{\hat{R}} \times \mathbf{V} \tag{4.7.1}
$$

Because **V** is fixed in B its derivatives in R and $\hat{R}$ may be expressed in the forms (see Eq. (4.5.2)):

$$
{}^{R}d\mathbf{V}/dt = {}^{R}\boldsymbol{\omega}^{R} \times \mathbf{V} \quad \text{and} \quad {}^{\hat{R}}d\mathbf{V}/dt = {}^{\hat{R}}\boldsymbol{\omega}^{B} \times \mathbf{V} \tag{4.7.2}
$$

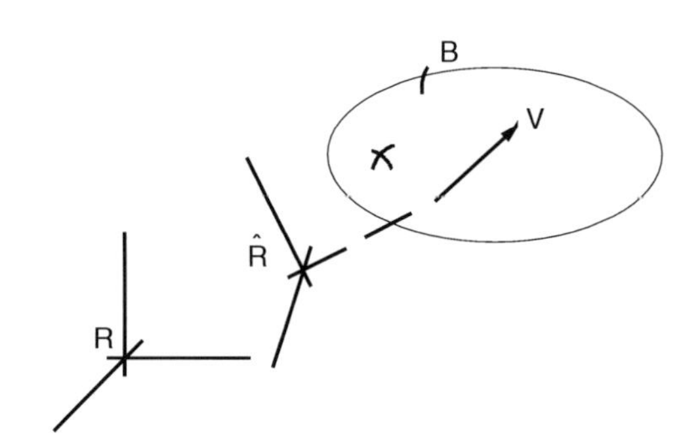

FIGURE 4.7.1
Body B moving in reference frame R and $\hat{R}$.

Hence, by substituting into Eq. (4.7.1) we have:

$$^{R}\boldsymbol{\omega}^{B} \times \mathbf{V} = {}^{\hat{R}}\boldsymbol{\omega}^{B} \times \mathbf{V} + {}^{R}\boldsymbol{\omega}^{\hat{R}} \times \mathbf{V}$$

or

$$\left({}^{R}\boldsymbol{\omega}^{B} - {}^{\hat{R}}\boldsymbol{\omega}^{B} - {}^{R}\boldsymbol{\omega}^{\hat{R}}\right) \times \mathbf{V} = 0 \tag{4.7.3}$$

Because **V** is arbitrary, we thus have:

$$^{R}\boldsymbol{\omega}^{B} - {}^{\hat{R}}\boldsymbol{\omega}^{B} - {}^{R}\boldsymbol{\omega}^{\hat{R}} = 0$$

or

$$^{R}\boldsymbol{\omega}^{B} = {}^{\hat{R}}\boldsymbol{\omega}^{B} + {}^{R}\boldsymbol{\omega}^{\hat{R}} \tag{4.7.4}$$

Equation (4.7.4) is an expression of the addition theorem for angular velocity. Because body B may itself be considered as a reference frame, Eq. (4.7.4) may be rewritten in the form:

$$^{R_0}\boldsymbol{\omega}^{R_2} = {}^{R_0}\boldsymbol{\omega}^{R_1} + {}^{R_1}\boldsymbol{\omega}^{R_2} \tag{4.7.5}$$

Equation (4.7.5) may be generalized to include reference frames. That is, suppose a reference frame R_n is moving in a reference frame R_0 and suppose that there are (n –1) intermediate reference frames, as depicted in Figure 4.7.2. Then, by repeated use of Eq. (4.7.5), we have:

$$^{R_0}\boldsymbol{\omega}^{R_n} = {}^{R_0}\boldsymbol{\omega}^{R_1} + {}^{R_1}\boldsymbol{\omega}^{R_2} + \cdots + {}^{R_{n-1}}\boldsymbol{\omega}^{R_n} \tag{4.7.6}$$

The addition theorem together with the configuration graphs of Section 4.3 are useful for obtaining more insight into the nature of angular velocity. Consider again a body B moving in a reference frame R as in Figure 4.7.3. Let the orientation of B in R be described by dextral orientation angles α, β, and γ. Let $\mathbf{n}_i$ and $\mathbf{N}_i$ (i = 1, 2, 3) be unit vector sets fixed in B and R, respectively. Then, from Figure 4.3.8, the configuration graph relating $\mathbf{n}_i$ and

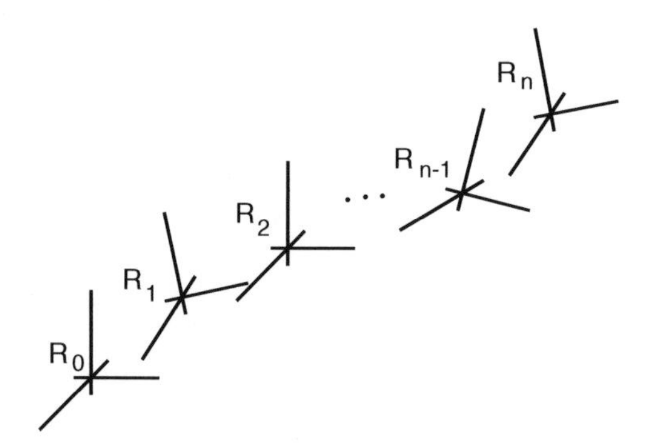

FIGURE 4.7.2
A set of $n + 1$ reference frames.

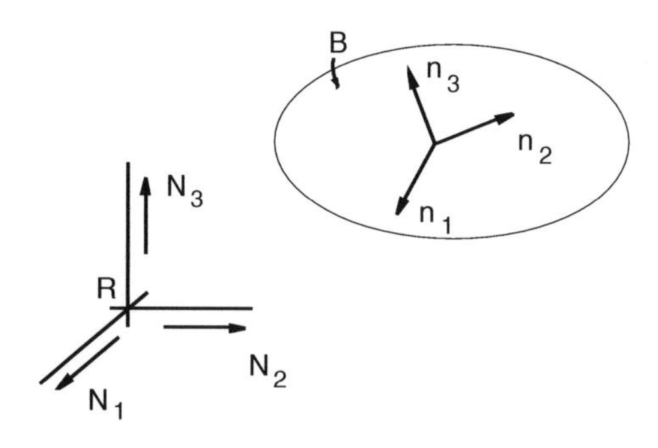

FIGURE 4.7.3
A body B moving in a reference frame R.

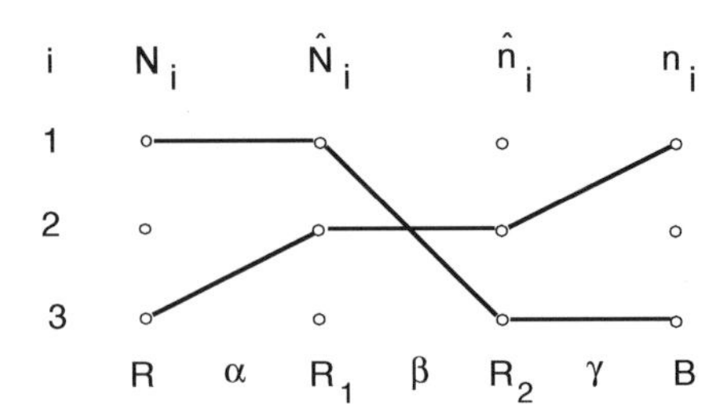

FIGURE 4.7.4
Configuration graph relating $\mathbf{n}_i$ and $\mathbf{N}_i$.

$\mathbf{N}_i$ is as reproduced in Figure 4.7.4. Recall that the unit vector sets $\hat{\mathbf{N}}_i$ and $\hat{\mathbf{n}}_i$ are fixed in intermediate reference frames R_1 and R_2 used in the development of the configuration graph. The horizontal lines in the graph signify axes of simple angular velocity (see Section 4.4) of adjoining reference frames. The respective rotation angles are written at the base of the graph. Therefore, from Eq. (4.7.6) we have:

$$ {}^R\boldsymbol{\omega}^B = {}^R\boldsymbol{\omega}^{R_1} + {}^{R_1}\boldsymbol{\omega}^{R_2} + {}^{R_2}\boldsymbol{\omega}^B \tag{4.7.7} $$

From the configuration graph of Figure 4.7.4 we have (see Eq. (4.4.1)):

$$ {}^R\boldsymbol{\omega}^{R_1} = \dot{\alpha}\mathbf{N}_1 = \dot{\alpha}\hat{\mathbf{N}}_1,\ {}^{R_1}\boldsymbol{\omega}^{R_2} = \dot{\beta}\hat{\mathbf{N}}_2 = \dot{\beta}\hat{\mathbf{n}}_2,\ {}^R\boldsymbol{\omega}^B = \dot{\gamma}\hat{\mathbf{n}}_3 = \dot{\gamma}\mathbf{n}_3 \tag{4.7.8} $$

Hence, ${}^R\omega^B$ is:

$$ {}^R\boldsymbol{\omega}^B = \dot{\alpha}\mathbf{N}_1 + \dot{\beta}\hat{\mathbf{N}}_2 + \dot{\gamma}\hat{\mathbf{n}}_3 = \dot{\alpha}\hat{\mathbf{N}}_1 + \dot{\beta}\hat{\mathbf{n}}_2 + \dot{\gamma}\mathbf{n}_3 \tag{4.7.9} $$

Finally, using the analysis associated with the configuration graph (see Eqs. (4.3.6) and (4.3.7)), we can express ${}^R\omega^B$ as:

$$ {}^R\boldsymbol{\omega}^B = \left(\dot{\alpha} + \dot{\gamma}s_\beta\right)\mathbf{N}_1 + \left(\dot{\beta}c_\alpha - \dot{\gamma}c_\beta s_\alpha\right)\mathbf{N}_2 + \left(\dot{\beta}s_\alpha - \dot{\gamma}c_\beta c_\alpha\right)\mathbf{N}_3 \tag{4.7.10} $$

or alternatively as:

$$ {}^R\boldsymbol{\omega}^B = \left(\dot{\alpha}c_\beta c_\gamma + \dot{\beta}s_\gamma\right)\mathbf{n}_1 + \left(\dot{\beta}c_\gamma - \dot{\alpha}c_\beta s_\gamma\right)\mathbf{n}_2 + \left(\dot{\alpha}s_\beta + \dot{\gamma}\right)\mathbf{n}_3 \tag{4.7.11} $$

Example 4.7.1: A Simple Gyro

(See Reference 4.1.) A circular disk (or gyro), *D*, is mounted in a yoke (or gimbal), *Y*, as shown in Figure 4.7.5. *Y* is mounted on a shaft *S* and is free to rotate about an axis that is perpendicular to *S*. *S* in turn can rotate about its own axis in bearings fixed in a reference frame *R*. Let *D* have an angular speed Ω relative to *Y*; let the rotation of *Y* in *S* be measured by the angle ϕ; and let the rotation of *S* in *R* be measured by the angle ψ. Let $\mathbf{N}_{Yi}$ and $\mathbf{N}_{Ri}$ be configured so that when *Y* is vertical and when the plane of *D* is parallel to *S*, the unit vector sets are mutually aligned. Also in this configuration, let the orientation angles ϕ and ψ be zero. Determine the angular velocity of *D* in *R* by constructing a configuration graph as in Figure 4.7.4 and by using the addition theorem of Eq. (4.7.6).

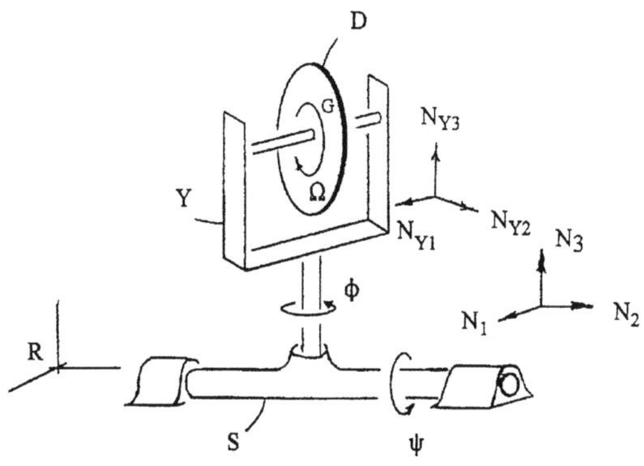

FIGURE 4.7.5
A spinning disk D and a supporting yoke Y and shaft S.

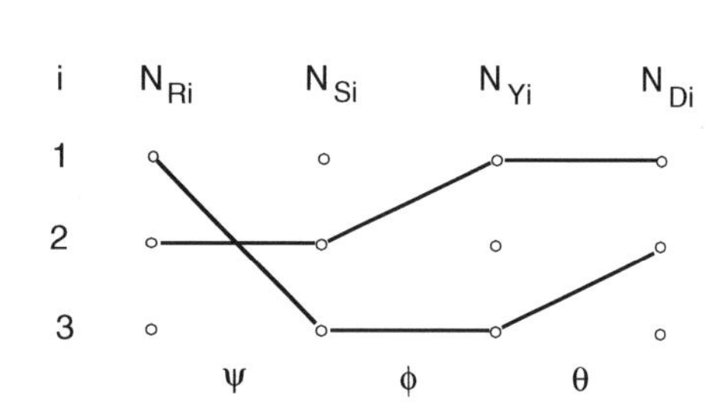

FIGURE 4.7.6
Configuration graph for the system of Figure 4.7.5.

Solution: Let Ω be the derivative of a rotation angle θ of D in Y; let $\mathbf{N}_{Di}$ ($i = 1, 2, 3$) be mutually perpendicular unit vectors fixed in D; and let the $\mathbf{N}_{Di}$ be mutually aligned with the $\mathbf{N}_{Yi}$ when θ is zero. Similarly, let $\mathbf{N}_{Si}$ ($i = 1, 2, 3$) be mutually perpendicular unit vectors fixed in S which are aligned with the $\mathbf{N}_{Ri}$ when ψ is zero. Then, a configuration graph representing the various unit vector sets and the orientation angles can be constructed, as shown in Figure 4.7.6. From this graph and from Eq. (4.7.6), the angular velocity of D in R may be expressed as:

$$^{R}\boldsymbol{\omega}^{D} = \dot{\psi}\mathbf{N}_{R2} + \dot{\phi}\mathbf{N}_{S3} + \dot{\theta}\mathbf{N}_{Y1} \tag{4.7.12}$$

The configuration graph is useful for expressing $^{R}\omega^{D}$ solely in terms of one of the unit vector sets. For example, in terms of the yoke unit vectors $\mathbf{N}_{Yi}$, $^{R}\omega^{D}$ is:

$$^{R}\boldsymbol{\omega}^{D} = \left(\Omega + \dot{\psi}s_{\phi}\right)\mathbf{N}_{Y1} + \dot{\psi}c_{\phi}\mathbf{N}_{Y2} + \dot{\phi}\mathbf{N}_{Y3} \tag{4.7.13}$$

where Ω is $\dot{\theta}$ (see also Problem P4.7.1).

4.8 Angular Acceleration

The angular acceleration of a body B in a reference frame R is defined as the derivative in R of the angular velocity of B in R. Specifically,

$$^{R}\boldsymbol{\alpha}^{B} = {}^{R}d\,{}^{R}\boldsymbol{\omega}^{B}/dt \quad \text{or} \quad \boldsymbol{\alpha} = d\boldsymbol{\omega}/dt \tag{4.8.1}$$

Unfortunately, there is not an addition theorem for angular acceleration analogous to that for angular velocity. That is, in general,

$$^{R_0}\boldsymbol{\alpha}^{R_n} \neq {}^{R_0}\boldsymbol{\alpha}^{R_1} + {}^{R_1}\boldsymbol{\alpha}^{R_2} + \ldots + {}^{R_{n-1}}\boldsymbol{\alpha}^{R_n} \tag{4.8.2}$$

To see this, consider three reference frames R_0, R_1, and R_2 as in Eq. (4.7.5):

$$^{R_0}\boldsymbol{\omega}^{R_2} = {}^{R_0}\boldsymbol{\omega}^{R_1} + {}^{R_1}\boldsymbol{\omega}^{R_2} \tag{4.8.3}$$

By differentiating in R_0 we obtain:

$$^{R_0}d\,{}^{R_0}\boldsymbol{\omega}^{R_2}/dt = {}^{R_0}d\,{}^{R_0}\boldsymbol{\omega}^{R_1}/dt + {}^{R_0}d\,{}^{R_1}\boldsymbol{\omega}^{R_2}/dt$$

or

$$^{R_0}\boldsymbol{\alpha}^{R_2} = {}^{R_0}\boldsymbol{\alpha}^{R_1} + {}^{R_0}d\,{}^{R_1}\boldsymbol{\omega}^{R_2}/dt \tag{4.8.4}$$

Consider the last term: From Eq. (4.6.6) we have:

$$\begin{aligned} ^{R_0}d\,{}^{R_1}\boldsymbol{\omega}^{R_2}/dt &= {}^{R_1}d\,{}^{R_1}\boldsymbol{\omega}^{R_2}/dt + {}^{R_0}\boldsymbol{\omega}^{R_1} \times {}^{R_1}\boldsymbol{\omega}^{R_2} \\ &= {}^{R_1}\boldsymbol{\alpha}^{R_2} + {}^{R_0}\boldsymbol{\omega}^{R_1} \times {}^{R_1}\boldsymbol{\omega}^{R_2} \end{aligned} \tag{4.8.5}$$

Then, by substituting into Eq. (4.8.4) we have:

$$^{R_0}\boldsymbol{\alpha}^{R_2} = {}^{R_0}\boldsymbol{\alpha}^{R_1} + {}^{R_1}\boldsymbol{\alpha}^{R_2} + {}^{R_0}\boldsymbol{\omega}^{R_1} \times {}^{R_1}\boldsymbol{\omega}^{R_2} \tag{4.8.6}$$

Hence, unless ${}^{R_0}\boldsymbol{\omega}^{R_1} \times {}^{R_1}\boldsymbol{\omega}^{R_2}$ is zero, the addition theorem for angular acceleration does not hold as it does for angular velocity. However, if ${}^{R_0}\boldsymbol{\omega}^{R_1} \times {}^{R_1}\boldsymbol{\omega}^{R_2}$ is zero (for example, if ${}^{R_0}\boldsymbol{\omega}^{R_1}$ and ${}^{R_1}\boldsymbol{\omega}^{R_2}$ are parallel), then the addition theorem for angular acceleration holds as well. This occurs if all rotations of the reference frames are parallel, or, equivalently, if all the bodies (treated as reference frames) move parallel to the same plane.

In some occasions the angular acceleration of a body B in a reference frame R (${}^R\alpha^B$) may be computed by simply differentiating the scalar components of the angular velocity of B in R (${}^R\omega^B$). This occurs if ${}^R\omega^B$ is expressed either in terms of unit vectors fixed in R or in terms of unit vectors fixed in B. To see this, consider first the case where the limit vectors of ${}^R\omega^B$ are fixed in R. That is, let $\mathbf{N}_i$ (i = 1, 2, 3) be mutually perpendicular unit vectors fixed in R and let ${}^R\omega^B$ be expressed as:

$$^R\boldsymbol{\omega}^B = \Omega_1\mathbf{N}_1 + \Omega_2\mathbf{N}_2 + \Omega_3\mathbf{N}_3 \tag{4.8.7}$$

Then, because the $\mathbf{N}_i$ are fixed in R, their derivatives in R are zero. Hence, in this case, ${}^R\alpha^B$ becomes:

$$^R\boldsymbol{\alpha}^B = \dot{\Omega}_1\mathbf{N}_1 + \dot{\Omega}_2\mathbf{N}_2 + \dot{\Omega}_3\mathbf{N}_3 \tag{4.8.8}$$

Next, consider the case where the unit vectors of ${}^R\omega^B$ are fixed in B. That is, let $\mathbf{n}_i$ (i = 1, 2, 3) be mutually perpendicular unit vectors fixed in B and let ${}^R\omega^B$ be expressed as:

$$^R\boldsymbol{\omega}^B = \omega_1\mathbf{n}_1 + \omega_2\mathbf{n}_2 + \omega_3\mathbf{n}_3 \tag{4.8.9}$$

Then ${}^R\alpha^B$ is:

$$\begin{aligned}{}^R\boldsymbol{\alpha}^B &= {}^R d\,{}^R\boldsymbol{\omega}^B/dt = \dot{\omega}_1\mathbf{n}_1 + w_1\,{}^R d\mathbf{n}_1/dt + \dot{\omega}_2\mathbf{n}_2 \\ &\quad + \omega_2\,{}^R d\mathbf{n}_2/dt + \dot{\omega}_3\mathbf{n}_3 + \omega_3\,{}^R d\mathbf{n}_3/dt \\ &= \dot{\omega}_1\mathbf{n}_1 + \dot{\omega}_2\mathbf{n}_2 + \dot{\omega}_3\mathbf{n}_3 \\ &\quad + \omega_1\,{}^R d\mathbf{n}_1/dt + \omega_2\,{}^R d\mathbf{n}_2/dt + \omega_3\,{}^R d\mathbf{n}_3/dt\end{aligned} \tag{4.8.10}$$

Observe that the last three terms each involve derivatives of unit vectors fixed in B. From Eq. (4.5.2) these derivatives may be expressed as:

$${}^R d\mathbf{n}_i/dt = {}^R\boldsymbol{\omega}^B \times \mathbf{n}_i \quad (i = 1,2,3) \tag{4.8.11}$$

Hence, the last three terms of Eq. (4.8.10) may be expressed as:

$$\begin{aligned}&\omega_1\,{}^R d\mathbf{n}_1/dt + \omega_2\,{}^R d\mathbf{n}_2/dt + \omega_3\,{}^R d\mathbf{n}_3/dt \\ &\quad = \omega_1\,{}^R\boldsymbol{\omega}^B \times \mathbf{n}_1 + \omega_2\,{}^R\boldsymbol{\omega}^B \times \mathbf{n}_2 + \omega_3\,{}^R\boldsymbol{\omega}^B \times \mathbf{n}_3 \\ &\quad = {}^R\boldsymbol{\omega}^B \times \omega_1\mathbf{n}_1 + {}^R\boldsymbol{\omega}^B \times \omega_2\mathbf{n}_2 + {}^R\boldsymbol{\omega}^B \times \omega_3\mathbf{n}_3 \\ &\quad = {}^R\boldsymbol{\omega}^B \times (\omega_1\mathbf{n}_1 + \omega_2\mathbf{n}_2 + \omega_3\mathbf{n}_3) \\ &\quad = {}^R\boldsymbol{\omega}^B \times {}^R\boldsymbol{\omega}^B = 0\end{aligned} \tag{4.8.12}$$

Therefore, ${}^R\boldsymbol{\alpha}^B$ becomes:

$${}^R\boldsymbol{\alpha}^B = \dot{\omega}_1\mathbf{n}_1 + \dot{\omega}_2\mathbf{n}_2 + \dot{\omega}_3\mathbf{n}_3 \tag{4.8.13}$$

Example 4.8.1: A Simple Gyro

See Example 4.7.1. Consider again the simple gyro of Figure 4.7.5 and as shown in Figure 4.8.1. Recall from Eq. (4.7.13) that the angular velocity of the gyro D in the fixed frame R may be expressed as:

$${}^R\boldsymbol{\omega}^D = (\Omega + \dot{\psi}s_\phi)\mathbf{N}_{Y1} + \dot{\psi}c_\phi\mathbf{N}_{Y2} + \dot{\phi}\mathbf{N}_{Y3} \tag{4.8.14}$$

where as before Ω (the gyro spin) is $\dot{\theta}$; ϕ and ψ are orientation angles as shown in Figure 4.8.1; and $\mathbf{N}_{Y1}$, $\mathbf{N}_{Y2}$, and $\mathbf{N}_{Y3}$ are unit vectors fixed in the yoke Y. Determine the angular acceleration of D in R.

Solution: By differentiating in Eq. (4.4.14) we obtain:

$$\begin{aligned}{}^R\boldsymbol{\alpha}^D &= (\dot{\Omega} + \ddot{\psi}s_\phi + \dot{\psi}\dot{\phi}c_\phi)\mathbf{N}_{Y1} + (\Omega + \dot{\psi}s_\phi)\,{}^R d\mathbf{N}_{Y1}/dt \\ &\quad + (\ddot{\psi}c_\phi - \dot{\psi}\dot{\phi}s_\phi)\mathbf{N}_{Y2} + \dot{\psi}c_\phi\,{}^R d\mathbf{N}_{Y2}/dt \\ &\quad + \ddot{\phi}\mathbf{N}_{Y3} + \dot{\phi}\,{}^R d\mathbf{N}_{Y3}/dt\end{aligned} \tag{4.8.15}$$

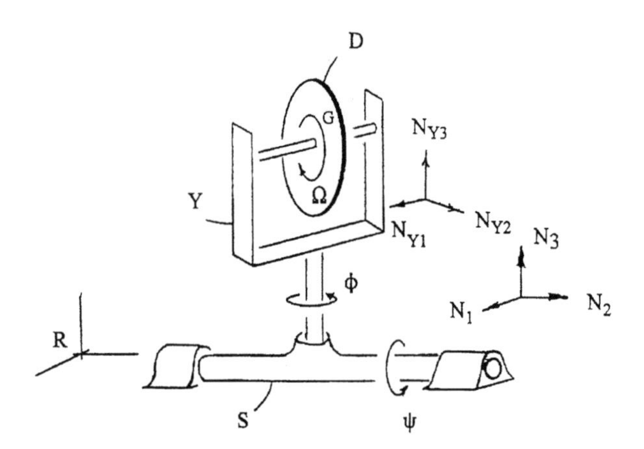

FIGURE 4.8.1
The gyro of Figure 4.7.5.

Because the $\mathbf{N}_{Yi}$ (i = 1, 2, 3) are fixed in the yoke Y, their derivatives may be evaluated from the expression:

$$^{R}d\mathbf{N}_{Yi}/dt = {}^{R}\boldsymbol{\omega}^{Y} \times \mathbf{N}_{Yi} \quad (i = 1,2,3) \tag{4.8.16}$$

where from Eqs. (4.7.12) and (4.7.13) $^{R}\boldsymbol{\omega}^{Y}$ is:

$$^{R}\boldsymbol{\omega}^{Y} = \dot{\psi}\mathbf{N}_{R2} + \dot{\phi}\mathbf{N}_{S3} = \dot{\psi}s_{\phi}\mathbf{N}_{Y1} + \dot{\psi}c_{\phi}\mathbf{N}_{Y2} + \dot{\phi}\mathbf{N}_{Y3} \tag{4.8 17}$$

Hence, the $^{R}d\mathbf{N}_{Yi}/dt$ are:

$$\begin{aligned} ^{R}d\mathbf{N}_{Y1}/dt &= \dot{\phi}\mathbf{N}_{Y2} - \dot{\psi}c_{\phi}\mathbf{N}_{Y3} \\ ^{R}d\mathbf{N}_{Y2}/dt &= -\dot{\phi}\mathbf{N}_{Y1} + \dot{\psi}s_{\phi}\mathbf{N}_{Y3} \\ ^{R}d\mathbf{N}_{Y3}/dt &= \dot{\psi}c_{\phi}\mathbf{N}_{Y1} - \dot{\psi}s_{\phi}\mathbf{N}_{Y2} \end{aligned} \tag{4.18}$$

and, by substitution into Eq. (4.8.15), $^{R}\alpha^{D}$ becomes:

$$\begin{aligned} ^{R}\boldsymbol{\alpha}^{D} &= \left(\dot{\Omega} + \ddot{\psi}s_{\phi} + \dot{\phi}\dot{\psi}c_{\phi}\right)\mathbf{N}_{Y1} + \left(\ddot{\psi}c_{\phi} + \Omega\dot{\phi} - \dot{\psi}\dot{\phi}s_{\phi}\right)\mathbf{N}_{Y2} \\ &+ \left(\ddot{\phi} - \Omega\dot{\psi}c_{\phi}\right)\mathbf{N}_{Y3} \end{aligned} \tag{4.19}$$

Comment: Observe that by comparing Eqs. (4.8.14) and (4.8.19) we see that $^{R}\boldsymbol{\alpha}^{D}$ is considerably longer and more detailed than $^{R}\boldsymbol{\omega}^{D}$. This raises the question as to whether a different unit vector set might produce a simpler expression for $^{R}\boldsymbol{\alpha}^{D}$. Unfortunately, this is not the case, as seen in Problem P4.8.2. It happens that the $\mathbf{N}_{Yi}$ are the preferred unit vectors for simplicity.

4.9 Relative Velocity and Relative Acceleration of Two Points on a Rigid Body

Consider a body *B* moving in a reference frame *R* as in Figure 4.9.1. Let *P* and *Q* be points fixed in *B*. Consider the velocity and acceleration of *P* and *Q* and their *relative* velocity and acceleration in reference frame *R*. Let *O* be a fixed point (the origin) of *R*. Let vectors **p** and **q** locate *P* and *Q* relative to *O* and let vector **r** locate *P* relative to *Q*. Then, from Figure 4.9.1, these vectors are related by the expression:

$$\mathbf{p} = \mathbf{q} + \mathbf{r} \tag{4.9.1}$$

By differentiating we have:

$${}^R\mathbf{V}^P = {}^R\mathbf{V}^Q + {}^R d\mathbf{r}/dt \quad \text{or} \quad {}^R\mathbf{V}^P - {}^R\mathbf{V}^Q = {}^R d\mathbf{r}/dt = {}^R\mathbf{V}^{P/Q} \tag{4.9.2}$$

Because *P* and *Q* are fixed in *B*, **r** is fixed in *B*. Hence, from Eq. (4.5.2), we have:

$${}^R d\mathbf{r}/dt = {}^R\boldsymbol{\omega}^B \times \mathbf{r} \tag{4.9.3}$$

Then, by substituting into Eq. (4.9.2), we have:

$${}^R\mathbf{V}^{P/Q} = {}^R\boldsymbol{\omega}^B \times \mathbf{r} \quad \text{or} \quad {}^R\mathbf{V}^P = {}^R\mathbf{V}^Q + {}^R\boldsymbol{\omega}^Q \times \mathbf{r} \tag{4.9.4}$$

By differentiating in Eq. (4.9.4), we obtain the following relations for accelerations:

$${}^R\mathbf{a}^{P/Q} = {}^R\boldsymbol{\alpha}^B \times \mathbf{r} + {}^R\boldsymbol{\omega}^B \times \left({}^R\boldsymbol{\omega}^B \times \mathbf{r}\right) \tag{4.9.5}$$

and

$${}^R\mathbf{a}^P = {}^R\mathbf{a}^Q + {}^R\boldsymbol{\alpha}^B \times \mathbf{r} + {}^R\boldsymbol{\omega}^B \times \left({}^R\boldsymbol{\omega}^B \times \mathbf{r}\right) \tag{4.9.6}$$

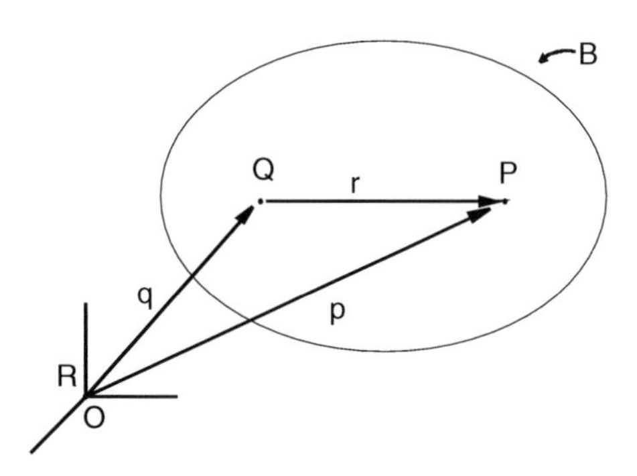

FIGURE 4.9.1
A body *B* moving in a reference frame *R* with points *P* and *Q* fixed in *B*.

Equations (4.9.4) and (4.9.6) are sometimes used to provide an interpretation of rigid body motion as being a superposition of translation and rotation. Specifically, at any instant, from Eq. (4.9.4) we may envision a body as translating with a velocity $\mathbf{V}^Q$ and rotating about Q with an angular velocity ${}^R\boldsymbol{\omega}^B$.

Example 4.9.1: Relative Movement of a Sports Car Operator's Hands During a Simple Turn

To illustrate the application of Eqs. (4.9.4) and (4.9.5), suppose a sports car operator traveling at a constant speed of 15 miles per hour, with hands on the steering wheel at 10 o'clock and 2 o'clock, is making a turn to the right with a turning radius of 25 feet. Suppose the steering wheel has a diameter of 12 inches and is in the vertical plane. Suppose further that the operator is turning the steering wheel at one revolution per second. Find the velocity and acceleration of the motorist's left hand relative to the right hand.

Solution: Let the movement of the car be represented as in Figure 4.9.2, and let the steering wheel be represented as in Figure 4.9.3, where $\mathbf{n}_x$, $\mathbf{n}_y$, and $\mathbf{n}_z$ are mutually perpendicular unit vectors fixed in the car. Let $\mathbf{n}_x$ be forward; $\mathbf{n}_z$ be vertical, directed up; and $\mathbf{n}_y$ be to the left. Let W represent the steering wheel, C represent the car, and S the road surface (fixed frame). Let O represent the center of W, and let L and R represent the motorist's left and right hands. The desired relative velocity and acceleration (${}^S\mathbf{V}^{L/R}$ and ${}^S\mathbf{a}^{L/R}$) may be obtained directly from Eqs. (4.9.4) and (4.9.5) once the angular velocity and the angular acceleration of W are known. From the addition theorem for angular velocity (Eq. (4.7.6)), the angular velocity of the steering wheel W in S is:

$$ {}^S\boldsymbol{\omega}^W = {}^C\boldsymbol{\omega}^W + {}^S\boldsymbol{\omega}^C \tag{4.9.7} $$

From the data presented, the angular velocity of the car C in S is:

$$ {}^S\boldsymbol{\omega}^C = -(V/\rho)\mathbf{n}_z = -(22/25)\mathbf{n}_z \text{ rad/sec} \tag{4.9.8} $$

where V is the speed of C (15 mph or 22 ft/sec) and ρ is the turn radius. Similarly, the angular velocity of the steering wheel W in C is:

$$ {}^C\boldsymbol{\omega}^W = 2\pi\mathbf{n}_x \text{ rad/sec} \tag{4.9.9} $$

Hence, ${}^S\boldsymbol{\omega}^W$ is:

$$ {}^S\boldsymbol{\omega}^W = 2\pi\mathbf{n}_x - (22/25)\mathbf{n}_z \text{ rad/sec} \tag{4.9.10} $$

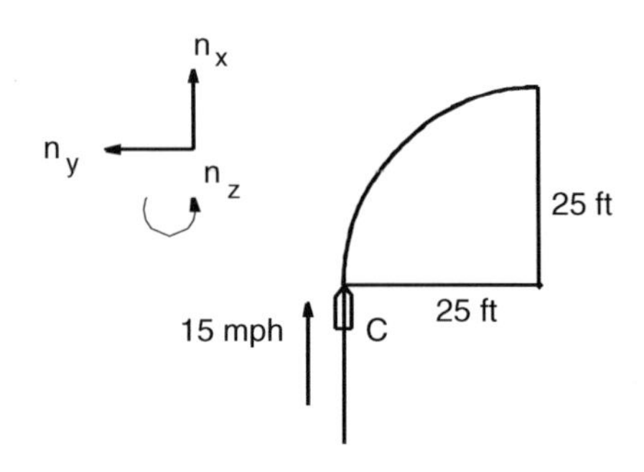

FIGURE 4.9.2
Sports car C entering a turn to the right.

By differentiating in Eq. (4.9.10), the angular acceleration of W in S is:

$$ {}^{S}\boldsymbol{\alpha}^{W} = 2\pi\, {}^{S}d\mathbf{n}_x/dt + 0 = 2\pi\, {}^{S}\boldsymbol{\omega}^{C} \times \mathbf{n}_x = -(2\pi)(22/25)\mathbf{n}_y \tag{4.9.11} $$

Finally, from Eqs. (4.9.4) and (4.9.5), the desired relative velocity and acceleration expressions are:

$$ \begin{aligned} {}^{S}\mathbf{V}^{L/R} &= {}^{S}\boldsymbol{\omega}^{W} \times \mathbf{RL} = \left[2\pi\mathbf{n}_x - (22/25)\mathbf{n}_z\right] \times \left(\sqrt{3}/2\right)\mathbf{n}_y \\ &= 0.762\mathbf{n}_x + 5.44\mathbf{n}_z \text{ ft/sec} \end{aligned} \tag{4.9.12} $$

and

$$ \begin{aligned} {}^{S}\mathrm{a}^{L/R} &= {}^{S}\boldsymbol{\alpha}^{W} \times \mathrm{RL} + {}^{S}\boldsymbol{\omega}^{W} \times \left({}^{S}\boldsymbol{\omega}^{W} \times \mathrm{RL}\right) \\ &= -2\pi(22/25)\mathbf{n}_y \times \left(\sqrt{3}/2\right)\mathbf{n}_y + \left[2\pi\mathbf{n}_x - (22/25)\mathbf{n}_z\right] \times \left(0.762\mathbf{n}_x + 5.44\mathbf{n}_z\right) \\ &= -33.5\mathbf{n}_y \text{ ft/sec}^2 \end{aligned} \tag{4.9.13} $$

Example 4.9.2: Absolute Movement of Sports Car Operator's Hands

To further illustrate the application of Eqs. (4.9.4) to (4.9.6), suppose we are interested in determining the velocity and acceleration of the sports car operator's left hand in the previous example. Specifically, find the velocity and acceleration of L in S.

Solution: Observe that the steering wheel hub O is fixed in both the steering wheel W and the car C. Observe further that O moves on the 25-ft-radius curve (or circle) at 15 mph (or 22 ft/sec). Therefore, the velocity and acceleration of O are:

$$ {}^{S}\mathbf{V}^{O} = 22\mathbf{n}_x \text{ ft/sec} \quad \text{and} \quad {}^{S}\mathbf{a}^{O} = -(22)^2/25\mathbf{n}_y = -19.36\mathbf{n}_y \tag{4.9.14} $$

By knowing ${}^{S}\mathbf{V}^{O}$ and ${}^{S}\mathbf{a}^{O}$, we can use Eqs. (4.9.4) and (4.9.6) to determine ${}^{S}\mathbf{V}^{L}$ and ${}^{S}\mathbf{a}^{L}$. That is,

$$ {}^{S}\mathbf{V}^{L} = {}^{S}\mathbf{V}^{O} + {}^{S}\boldsymbol{\omega}^{W} \times OL $$

and

$$ {}^{S}\mathbf{a}^{L} = {}^{S}\mathbf{a}^{O} + {}^{S}\boldsymbol{\alpha}^{W} \times \mathbf{OL} + {}^{S}\boldsymbol{\omega}^{W} \times \left({}^{S}\boldsymbol{\omega}^{W} \times \mathbf{OL}\right) \tag{4.9.15} $$

where from Figure 4.9.3 the position vector **OL** is:

$$ \mathbf{OL} = (0.5)\left(0.866\mathbf{n}_y + 0.5\mathbf{n}_z\right) = 0.433\mathbf{n}_y + 0.25\mathbf{n}_z \tag{4.9.16} $$

Hence, from Eqs. (4.9.10), (4.9.11), and (4.9.14), ${}^{S}\mathbf{V}^{L}$ and ${}^{S}\mathbf{a}^{L}$ are seen to be:

$$ {}^{S}\mathbf{V}^{L} = 22\mathbf{n}_x + \left(2\pi\mathbf{n}_x - 22/25\mathbf{n}_z\right) \times \left(0.433\mathbf{n}_y + 0.25\mathbf{n}_z\right) $$

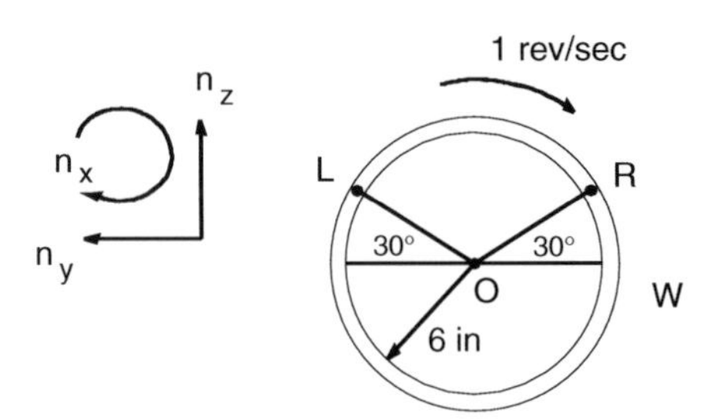

FIGURE 4.9.3
Sports car steering wheel W and left and right hands L and R.

or

$$^{S}\mathbf{V}^{L} = 22.381\mathbf{n}_x - 1.571\mathbf{n}_y + 2.721\mathbf{n}_z \text{ ft/sec} \tag{4.9.17}$$

and

$$\begin{aligned} ^{S}\mathbf{a}^{L} &= -19.36\mathbf{n}_y + (-5.529)\mathbf{n}_y \times (0.433\mathbf{n}_y + 0.25\mathbf{n}_z) \\ &\quad + (6.283\mathbf{n}_x - 0.88\mathbf{n}_z) \times [(6.283\mathbf{n}_x - 0.88\mathbf{n}_z) \times (0.433\mathbf{n}_y + 0.25\mathbf{n}_z)] \\ &= -2.765\mathbf{n}_x - 36.79\mathbf{n}_y - 9.87\mathbf{n}_z \text{ ft/sec}^2 \end{aligned} \tag{4.9.18}$$

Example 4.9.3: A Rod with Constrained End Movements

As a third example illustrating the use of Eqs. (4.9.4) and (4.9.6) consider a rod whose ends A and B are restricted to movement on horizontal and vertical lines as in Figure 4.9.4. Let the rod have length 13 m. Let the velocity and acceleration of end A be:

$$\mathbf{V}^A = 6\mathbf{n}_x \text{ m/s} \quad \text{and} \quad \mathbf{a}^A = -3\mathbf{n}_x \text{ m/s}^2 \tag{4.9.19}$$

where $\mathbf{n}_x$, $\mathbf{n}_y$, and $\mathbf{n}_z$ are unit vectors parallel to the coordinate axes as shown. Let the angular velocity and angular acceleration of the rod along the rod itself be zero. That is,

$$\boldsymbol{\omega}^{AB} \cdot \mathbf{n} = 0 \quad \text{and} \quad \boldsymbol{\alpha}^{AB} \cdot \mathbf{n} = 0 \tag{4.9.20}$$

where $\mathbf{n}$ is a unit vector along the rod. Find the velocity and acceleration of end B for the rod position shown in Figure 4.9.4.

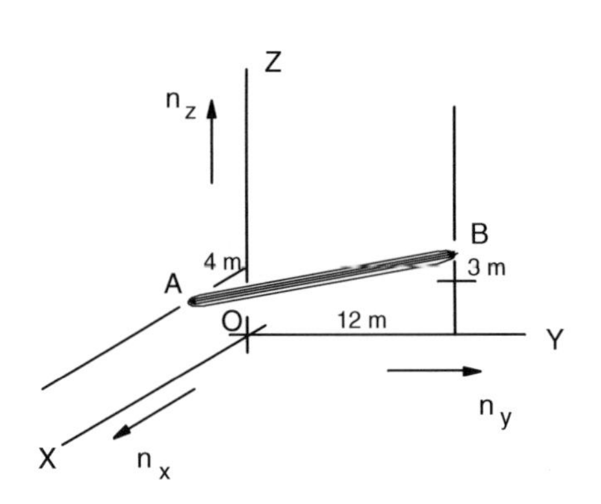

FIGURE 4.9.4
A rod AB with constrained motion of its ends.

Solution: From Eq. (4.9.4), the velocity of end B may be expressed as:

$$\mathbf{V}^B = \mathbf{V}^A + \boldsymbol{\omega}^{AB} \times \mathbf{AB} \tag{4.9.21}$$

For the rod position of Figure 4.9.4, **AB** is:

$$\mathbf{AB} = -4\mathbf{n}_x + 12\mathbf{n}_y + 3\mathbf{n}_z \text{ m} \tag{4.9.22}$$

Let $\boldsymbol{\omega}^{AB}$ be expressed in the form:

$$\boldsymbol{\omega}^{AB} = \omega_x \mathbf{n}_x + \omega_y \mathbf{n}_y + \omega_z \mathbf{n}_z \tag{4.9.23}$$

Let $\mathbf{V}^B$ be expressed as:

$$\mathbf{V}^B = v^B \mathbf{n}_z \tag{4.9.24}$$

Then, by substituting from Eqs. (4.9.19), (4.9.22), (4.9.23), and (4.9.24) we have:

$$\begin{aligned} v^B \mathbf{n}_z = {} & 6\mathbf{n}_x + (3\omega_y - 12\omega_z)\mathbf{n}_x + (-3\omega_x - 4\omega_z)\mathbf{n}_y \\ & + (12\omega_x + 4\omega_y)\mathbf{n}_z \end{aligned} \tag{4.9.25}$$

Also, from Eq. (4.9.20) we have:

$$-4\omega_x + 12\omega_y + 3\omega_z = 0 \tag{4.9.26}$$

Equations (4.9.25) and (4.9.26) are equivalent to the following four scalar equations for ω_x, ω_y, ω_z, and v^B:

$$\begin{aligned} 0\omega_x + 3\omega_y - 12\omega_z + 0v^B &= -6 \\ -3\omega_x + 0\omega_y - 4\omega_z + 0v^B &= 0 \\ 12\omega_x + 4\omega_y + 0\omega_z - v^B &= 0 \\ -4\omega_x + 12\omega_y + 3\omega_z + 0v^B &= 0 \end{aligned} \tag{4.9.27}$$

Solving for ω_x, ω_y, ω_z, and v^B, we obtain:

$$\omega_x = -0.568 \text{ rad/sec}, \; \omega_y = -0.296 \text{ rad/sec}, \; \omega_3 = -0.426 \text{ rad/sec} \tag{4.9.28}$$

and

$$v^B = -8 \text{ m/s} \tag{4.9.29}$$

Similarly, from Eq. (4.9.6), the acceleration of end A may be expressed as:

$$\mathbf{a}^A = \mathbf{a}^B + \boldsymbol{\alpha}^{AB} \times \mathbf{AB} + \boldsymbol{\omega}^{AB} \times \left(\boldsymbol{\omega}^{AB} \times \mathbf{AB}\right) \tag{4.9.30}$$

Let α^{AB} be expressed in the form:

$$\boldsymbol{\alpha}^{AB} = \alpha_x \mathbf{n}_x + \alpha_y \mathbf{n}_y + \alpha_z \mathbf{n}_z \tag{4.9.31}$$

Let $\mathbf{a}^B$ be expressed as:

$$\mathbf{a}^B = a^B \mathbf{n}_z \tag{4.9.32}$$

Then, by substituting from Eqs. (4.9.19), (4.9.22), (4.9.23), (4.9.31), and (4.9.32) into (4.9.30), we have:

$$\begin{aligned} a^B \mathbf{n}_z = &-3\mathbf{n}_x + \left(3\alpha_y - 12\alpha_z\right)\mathbf{n}_x + \left(-3\alpha_x - 4\alpha_z\right)\mathbf{n}_y \\ &+ \left(12\alpha_x + 4\alpha_y\right)\mathbf{n}_z - 8\omega_y \mathbf{n}_x + \left(8\omega_x - 6\omega_z\right)\mathbf{n}_y + 6\omega_y \mathbf{n}_z \end{aligned} \tag{4.9.33}$$

Also, from Eq. (4.9.20) we have:

$$-4\alpha_x + 12\alpha_y + 3\alpha_z = 0 \tag{4.9.34}$$

Using Eq. (4.9.29) for values of ω_x, ω_y, and ω_z, Eqs. (4.9.33) and (4.9.34) are equivalent to the following four scalar equations for α_x, α_y, α_z, and a^B:

$$\begin{aligned} 0\alpha_x + 3\alpha_y - 12\alpha_z + 0a^B &= 0.632 \\ -3\alpha_x + 0\alpha_y - 4\alpha_z + 0a^B &= 7.1 \\ 12\alpha_x + 4\alpha_y + 0\alpha_z - a^B &= 1.776 \\ -4\alpha_x + 12\alpha_y + 3\alpha_z + 0a^B &= 0 \end{aligned} \tag{4.9.35}$$

Solving for α_x, α_y, α_z, and a^B, we obtain:

$$\alpha_x = -2.083\,\text{rad/sec}^2, \quad \alpha_y = -0.641\,\text{rad/sec}^2, \quad \alpha_z = -0.213\,\text{rad/sec}^2 \tag{4.9.36}$$

and

$$a^B = -29.33\,\text{m/s} \tag{4.9.37}$$

4.10 Points Moving on a Rigid Body

Consider next a point P moving on a body B which in turn is moving in a reference frame R as depicted in Figure 4.10.1. Let Q be a point fixed in B. Let $\mathbf{p}$ and $\mathbf{q}$ be position vectors locating P and Q relative to a fixed point O in R. Let vector $\mathbf{r}$ locate P relative to Q. Then, from Figure 4.10.1, we have:

$$\mathbf{p} = \mathbf{q} + \mathbf{r} \tag{4.10.1}$$

By differentiating, we obtain:

$${}^R\mathbf{V}^P = {}^R\mathbf{V}^Q + {}^R d\mathbf{r}/dt \tag{4.10.2}$$

where, as before, ${}^R\mathbf{V}^P$ and ${}^R\mathbf{V}^Q$ represent the velocities of P and Q in R. From Eq. (4.6.6), ${}^R d\mathbf{r}/dt$ is:

$${}^R d\mathbf{r}/dt = {}^B d\mathbf{r}/dt + {}^R\boldsymbol{\omega}^R \times \mathbf{r} \tag{4.10.3}$$

Because Q is fixed in B, and because $\mathbf{r}$ locates P relative to Q, ${}^R d\mathbf{r}/dt$ is the velocity of P in B. Hence, ${}^R\mathbf{V}^P$ becomes:

$${}^R\mathbf{V}^P = {}^R\mathbf{V}^Q + {}^B\mathbf{V}^P + {}^R\boldsymbol{\omega}^B \times \mathbf{r} \tag{4.10.4}$$

Suppose that an instant of interest P happens to coincide with Q. Then, at that instant, $\mathbf{r}$ is zero and ${}^R\mathbf{V}^P$ is:

$${}^R\mathbf{V}^P = {}^B\mathbf{V}^P + {}^R\mathbf{V}^Q \tag{4.10.5}$$

Observe that at any instant there *always* exists a point P^*, fixed in B, which coincides with P. Therefore, Eq. (4.10.5) may be rewritten as:

$${}^R\mathbf{V}^P = {}^B\mathbf{V}^P + {}^R\mathbf{V}^{P^*} \tag{4.10.6}$$

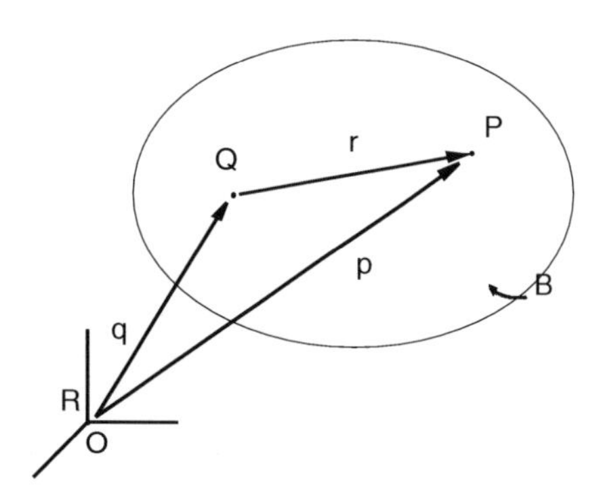

FIGURE 4.10.1
A point P moving on a body B, moving in reference frame R.

By differentiating in Eq. (4.10.4), we can obtain a relation determining the acceleration of P. That is,

$$^{R}\mathbf{a}^{P} = {}^{R}\mathbf{a}^{Q} + {}^{R}d^{B}\mathbf{V}^{P}/dt + {}^{R}\boldsymbol{\alpha}^{B} \times \mathbf{r} + {}^{R}\boldsymbol{\omega}^{B} \times {}^{R}d\mathbf{r}/dt \tag{4.10.7}$$

where, as before, $^{R}\mathbf{a}^{P}$ and $^{R}\mathbf{a}^{Q}$ are the acceleration of P and Q in R, and $^{R}\alpha^{B}$ is the acceleration of B in R. By using Eq. (4.6.6), we can express $^{R}d^{B}\mathbf{V}^{P}/dt$ as:

$$^{R}d^{B}\mathbf{V}^{P}/dt = {}^{B}d^{B}\mathbf{V}^{P}/dt + {}^{R}\boldsymbol{\omega}^{B} \times {}^{B}\mathbf{V}^{P} = {}^{B}\mathbf{a}^{P} + {}^{R}\boldsymbol{\omega}^{B} \times {}^{B}\mathbf{V}^{P} \tag{4.10.8}$$

Then, by using Eq. (4.10.3), $^{R}\mathbf{a}^{P}$ may be written as:

$$^{R}\mathbf{a}^{P} = {}^{R}\mathbf{a}^{Q} + {}^{B}\mathbf{a}^{P} + {}^{R}\boldsymbol{\omega}^{B} + {}^{B}\mathbf{V}^{P} + {}^{R}\boldsymbol{\alpha}^{B} \times \mathbf{r} + {}^{R}\boldsymbol{\omega}^{B} \times \left[{}^{B}\mathbf{V}^{P} \times \left({}^{R}\boldsymbol{\omega}^{B} \times \mathbf{r}\right)\right]$$

or

$$^{R}\mathbf{a}^{P} = {}^{B}\mathbf{a}^{P} + {}^{R}\mathbf{a}^{Q} + 2\,{}^{R}\boldsymbol{\omega}^{B} \times {}^{B}\mathbf{V}^{P} + {}^{R}\boldsymbol{\alpha}^{B} \times \mathbf{r} + {}^{R}\boldsymbol{\omega}^{B} \times \left({}^{R}\boldsymbol{\omega}^{B} \times \mathbf{r}\right) \tag{4.10.9}$$

Suppose again that at an instant of interest P happens to coincide with Q. Then, $\mathbf{r}$ is zero and $^{R}\mathbf{a}^{P}$ is:

$$^{R}\mathbf{a}^{P} = {}^{B}\mathbf{a}^{P} + {}^{R}\mathbf{a}^{Q} + 2\,{}^{R}\boldsymbol{\omega}^{B} \times {}^{B}\mathbf{V}^{P} \tag{4.10.10}$$

Hence, in general, if P^* is the point of B coinciding with P we have:

$$^{R}\mathrm{a}^{P} = {}^{B}\mathrm{a}^{P} + {}^{R}\mathrm{a}^{P^*} + 2\,{}^{R}\boldsymbol{\omega}^{B} \times {}^{B}\mathrm{V}^{P} \tag{4.10.11}$$

The term $2\,{}^{R}\boldsymbol{\omega}^{B} \times {}^{B}\mathbf{V}^{P}$ is called the *Coriolis acceleration*, after the French mechanician de Coriolis (1792–1843) who is credited with being the first to discover it. We have already seen this term in our analysis of the movement of a point in a plane in Chapter 3 (see Eq. (3.8.7)). This term is not generally intuitive, and it often gives rise to surprising and unexpected effects.

Example 4.10.1: Movement of Sports Car Operator's Hands

Equations (4.10.6) and (4.10.11) may also be used to determine the velocity and acceleration of the sports car operator's left hand of Example 4.9.2. Recall that the sports car is making a turn to the right at 15 mph with a turn radius of 25 feet, and that the operator's left hand is at 10 o'clock on a 12-in.-diameter vertical steering wheel, as in Figures 4.10.2 and 4.10.3. Recall further that the operator is turning the wheel clockwise at one revolution per second, as in Figure 4.10.3.

Solution: From Eq. (4.10.6), the velocity of the left hand L may be expressed as:

$$^{S}\mathbf{V}^{L} = {}^{C}\mathbf{V}^{L} + {}^{S}\mathbf{V}^{L^*} \tag{4.10.12}$$

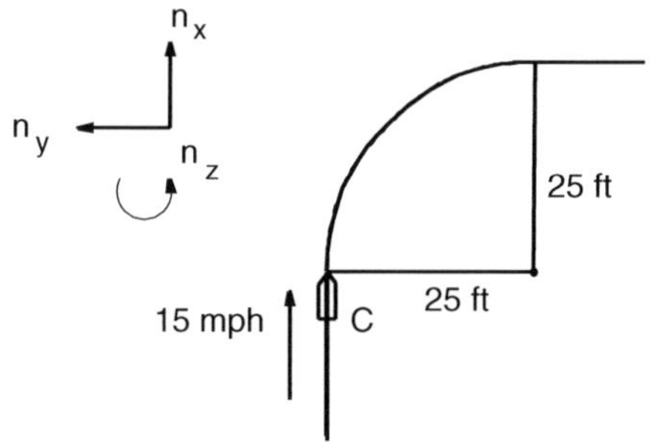
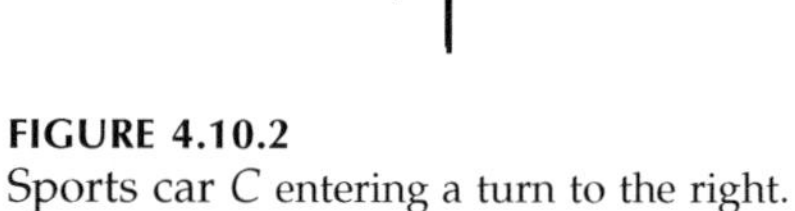

FIGURE 4.10.2
Sports car C entering a turn to the right.

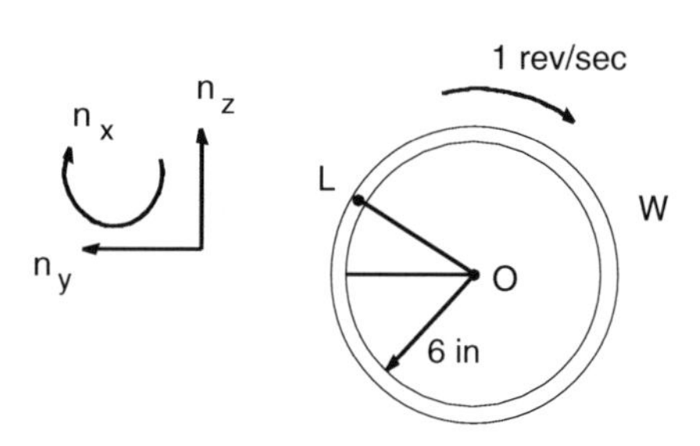

FIGURE 4.10.3
Sports car steering wheel W and left hand L.

where L^* is that point of the car C where L is located at the instant of interest and where, as before, S is the fixed road surface. From Figure 4.10.3, ${}^C\mathbf{V}^L$ is:

$$\begin{aligned} {}^C\mathbf{V}^L &= {}^C\boldsymbol{\omega}^W \times \mathbf{OL} = 2\pi\mathbf{n}_x \times (0.433\mathbf{n}_y + 0.25\mathbf{n}_z) \\ &= -1.571\mathbf{n}_y + 2.71\mathbf{n}_z \text{ ft/sec} \end{aligned} \tag{4.10.13}$$

From Figure 4.10.2, ${}^S\mathbf{V}^{L^*}$ is:

$$\begin{aligned} {}^S\mathbf{V}^{L^*} &= {}^S\mathbf{V}^O + {}^S\boldsymbol{\omega}^C \times \mathbf{OL} \\ &= 22\mathbf{n}_x + 0.88\mathbf{n}_z \times (0.433\mathbf{n}_y + 0.25\mathbf{n}_z) \\ &= 22.381\mathbf{n}_x \text{ ft/sec} \end{aligned} \tag{4.10.14}$$

Therefore, by substituting into Eq. (4.10.12), the velocity of L is:

$${}^S\mathbf{V}^L = 22.381\mathbf{n}_x - 1.571\mathbf{n}_y + 2.71\mathbf{n}_z \text{ ft/sec} \tag{4.10.15}$$

In like manner, from Eq. (4.10.11), the acceleration of L is:

$${}^S\mathbf{a}^L = {}^C\mathbf{a}^L + {}^S\mathbf{a}^{L^*} + 2\,{}^S\boldsymbol{\omega}^C \times {}^C\mathbf{V}^L \tag{4.10.16}$$

From Figure 4.10.3, ${}^C\mathbf{a}^L$ is:

$$\begin{aligned} {}^C\mathbf{a}^L &= {}^C\boldsymbol{\omega}^W \times ({}^C\boldsymbol{\omega}^W \times \mathbf{OL}) = 6.238\mathbf{n}_x \times [6.283\mathbf{n}_x \times (0.433\mathbf{n}_y + 0.25\mathbf{n}_z)] \\ &= -17.095\mathbf{n}_y - 9.87\mathbf{n}_z \text{ ft/sec}^2 \end{aligned} \tag{4.10.17}$$

From Figure 4.10.2, ${}^S\mathbf{a}^{L^*}$ is:

$$\begin{aligned} {}^S\mathbf{a}^{L^*} &= {}^S\mathbf{a}^O + {}^S\boldsymbol{\omega}^C \times \left({}^S\boldsymbol{\omega}^C \times \mathbf{OL}\right) \\ &= -19.36\mathbf{n}_y + \left(-0.88\mathbf{n}_z\right) \times \left[\left(-0.88\mathbf{n}_z\right) \times \left(0.433\mathbf{n}_y + 0.25\mathbf{n}_z\right)\right] \\ &= -19.7\mathbf{n}_y \text{ ft/sec}^2 \end{aligned} \tag{4.10.18}$$

Finally, from both figures $2{}^S\omega^C \times {}^C\mathbf{V}^L$ is:

$$\begin{aligned} 2{}^S\boldsymbol{\omega}^C \times {}^C\mathbf{V}^L &= 2\left(-0.88\mathbf{n}_z\right) \times \left(-1.571\mathbf{n}_y + 2.71\mathbf{n}_z\right) \\ &= -2.765\mathbf{n}_x \text{ ft/sec}^2 \end{aligned} \tag{4.10.19}$$

Therefore, by substituting into Eq. (4.10.16), the acceleration of L is:

$${}^S\mathbf{a}^L = -2.765\mathbf{n}_x - 36.79\mathbf{n}_y - 9.87\mathbf{n}_z \text{ ft/sec}^2 \tag{4.10.20}$$

4.11 Rolling Bodies

Rolling motion is an important special case in the kinematics of rigid bodies. It is particularly important in machine kinematics. Rolling can occur between two bodies or between a body and a surface. Rolling between two bodies occurs when the bodies are moving relative to each other but still are in contact with each other, with the contacting points having zero relative velocity. Similarly, a body rolls on a surface when it is moving relative to the surface but is still in contact with the surface with the contacting point (or points) having zero velocity relative to the surface.

Rolling may be defined analytically as follows: Let S be a surface and let B be a body that rolls on S as depicted in Figure 4.11.1. (S could be a portion of a body upon which B rolls.) Let B and S be counterformal so that they are in contact at a single point. Let C be the point of B that is in contact with S. *Rolling* then occurs when:

$${}^S\mathbf{V}^C = 0 \tag{4.11.1}$$

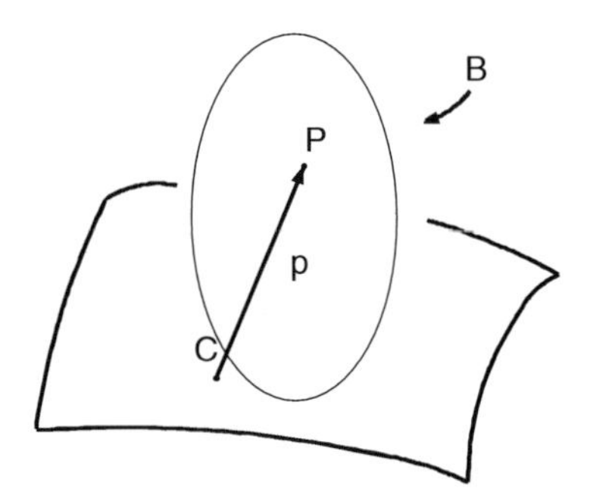

FIGURE 4.11.1
A body B rolling on a surface S.

Rolling motion is governed by the magnitude and direction of the angular velocity of B in S. Let $\mathbf{n}$ be a unit vector normal to S at C. Then B has *pure rolling* in S if the angular velocity of B in S is perpendicular to $\mathbf{n}$. That is,

$$\text{Pure Rolling: } {}^S\boldsymbol{\omega}^B \cdot \mathbf{n} = 0 \tag{4.11.2}$$

B is *pivoting* in S if ${}^S\omega^B$ is parallel to $\mathbf{n}$. That is,

$$\text{Pivoting: } {}^S\boldsymbol{\omega}^B = \left|{}^S\boldsymbol{\omega}^B\right|\mathbf{n} = \omega\mathbf{n} \tag{4.11.3}$$

B is at *rest* relative to S if ${}^S\omega^B$ is zero:

$$\text{Rest: } {}^S\boldsymbol{\omega}^B = 0 \tag{4.11.4}$$

Finally, B has *general rolling* in S if B is neither at rest nor pivoting or has pure rolling in S. (Pure rolling is desired in machine elements to reduce the wear of the rolling surfaces.)

Consider again Figure 4.11.1. Let P be an arbitrary point of B. Because C is fixed in B and because ${}^S\mathbf{V}^C$ is zero, Eqs. (3.4.6), (4.5.2), and (4.9.4) show that the velocity of P in S is simply:

$${}^S\mathbf{V}^P = {}^S\boldsymbol{\omega}^B \times \mathbf{p} \tag{4.11.5}$$

where $\mathbf{p}$ is the position vector locating P relative to C.

The acceleration of P in S may be obtained by differentiating in Eq. (4.11.5).

4.12 The Rolling Disk and Rolling Wheel

As an illustration of these ideas, consider a uniform circular disk D rolling on a horizontal flat surface S as depicted in Figure 4.12.1 (see References 4.1 to 4.4). Let C be the contact point between D and S, and let G be the center of D. Let axes X, Y, and Z form a Cartesian reference frame fixed relative to S with the Z-axis being normal to S. Let $\mathbf{N}_1$, $\mathbf{N}_2$, and $\mathbf{N}_3$ be unit vectors parallel to X, Y, and Z. Let T be a line in the X–Y plane (the plane of S) which is at all times tangent to D at C. Then, the angle ϕ between T and the X-axis defines the *turning* of D. Let L_R be a radial line fixed in D. Let L_D be a diametral line passing through G and C. Then, the angle ψ between L_R and L_D measures the *roll* of D. Finally, the angle θ between L_D and a vertical line measures the *lean* of D.

Because D rolls on S, the kinematics of D in S can be expressed in terms of θ, ϕ, and ψ. To see this, consider the angular velocity of D in S. Let $\mathbf{n}_1$, $\mathbf{n}_2$, and $\mathbf{n}_3$ be unit vectors parallel to T, to the axis of D, and to L_D as shown. Let $\hat{\mathbf{n}}_1$, $\hat{\mathbf{n}}_2$, and $\mathbf{n}_3$ be mutually perpendicular unit vectors with $\hat{\mathbf{n}}_1$ parallel to $\mathbf{n}_1$, $\hat{\mathbf{n}}_2$ in the plane of S and perpendicular to $\mathbf{n}_1$ (as shown), and $\hat{\mathbf{n}}_3$ parallel to $\mathbf{N}_3$. Finally, let $\mathbf{d}_1$, $\mathbf{d}_2$, and $\mathbf{d}_3$ be mutually perpendicular unit vectors fixed in D with $\mathbf{d}_2$ parallel to $\mathbf{n}_2$ and $\mathbf{d}_3$ parallel to L_R.

Let S ($\mathbf{N}_1$, $\mathbf{N}_2$, $\mathbf{N}_3$), R_1 ($\hat{\mathbf{n}}_1$, $\hat{\mathbf{n}}_2$, $\hat{\mathbf{n}}_3$), R_2 ($\mathbf{n}_1$, $\mathbf{n}_2$, $\mathbf{n}_3$), and D ($\mathbf{d}_1$, $\mathbf{d}_2$, $\mathbf{d}_3$) be reference frames containing the unit vector sets as indicated. Then, the configuration graph defining the orientation of D in S is shown in Figure 4.12.2.

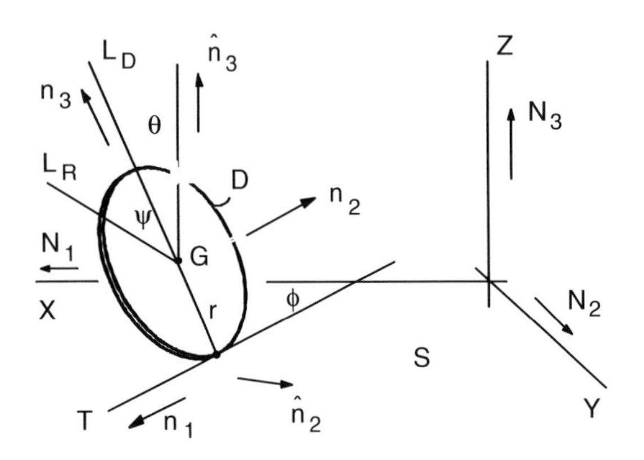

FIGURE 4.12.1
A circular disk rolling on a surface *S*. (Reprinted from Huston, R. L., and Liu, C. Q., *Formulas for Dynamic Analysis*, p. 258, by courtesy of Marcel Dekker, Inc., 2001.)

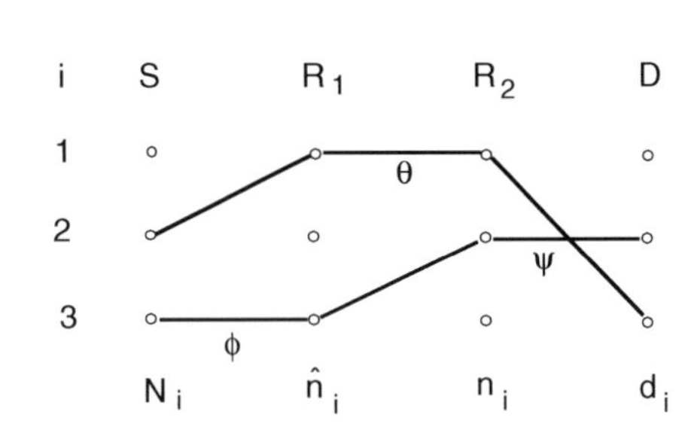

FIGURE 4.12.2
Configuration graph orienting *D* in *S*.

From the addition theorem for angular velocity (Eq. (4.7.6)), the angular velocity of *D* in *S* may be expressed as:

$$^S\boldsymbol{\omega}^D = \dot{\phi}\mathbf{N}_3 + \dot{\theta}\hat{\mathbf{n}}_1 + \dot{\psi}\mathbf{n}_2 \tag{4.12.1}$$

From the configuration graph of Figure 4.12.2 we may express $^S\omega^D$ entirely in terms of one of the unit vector sets. In terms of the $\mathbf{n}_i$ (fixed in R_2) $^S\omega^D$ is:

$$^S\boldsymbol{\omega}^D = \dot{\theta}\mathbf{n}_1 + \left(\dot{\psi} + \dot{\phi}s_\theta\right)\mathbf{n}_2 + \dot{\phi}c_\theta\mathbf{n}_3 \tag{4.12.2}$$

where s_θ and c_θ represent sinθ and cosθ.

From Eqs. (4.11.5) and (4.12.2), the velocity of *G* is:

$$\mathbf{V}^G = {}^S\boldsymbol{\omega}^D \times r\mathbf{n}_3 = r\left(\dot{\psi} + \dot{\phi}s_\theta\right)\mathbf{n}_1 - r\dot{\theta}\mathbf{n}_2 \tag{4.12.3}$$

where *r* is the radius of D.

The angular acceleration of *D* in *S* and the acceleration of *G* in *S* may be obtained by differentiation of Eqs. (4.12.2) and (4.12.3). To perform this differentiation, it is necessary

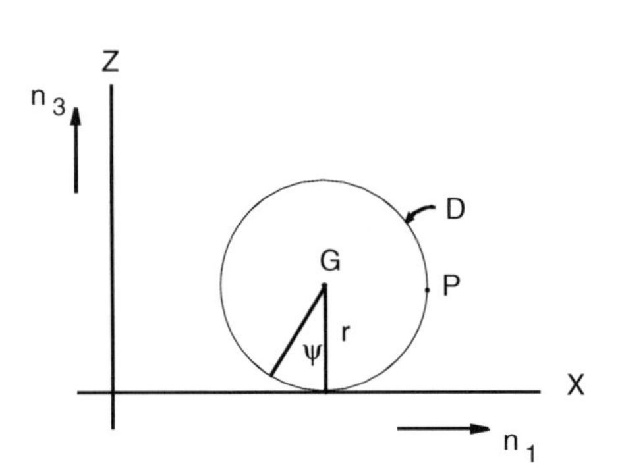

FIGURE 4.12.3
A circular disk rolling in a straight line.

to know the derivative of the $\mathbf{n}_i$ in S. These derivatives may be obtained by noting that the $\mathbf{n}_i$ are fixed in R_2 and by using Eq. (4.5.2). That is,

$$^S d\mathbf{n}_i/dt = {}^S\boldsymbol{\omega}^{R_2} \times \mathbf{n}_i \quad (i = 1,2,3) \tag{4.12.4}$$

From the configuration graph of Figure 4.12.2, we see that $^S\boldsymbol{\omega}^{R_2}$ is:

$$^S\boldsymbol{\omega}^{R_2} = \dot{\theta}\mathbf{n}_1 + \dot{\phi}s_\theta\mathbf{n}_2 + \dot{\phi}c_\theta\mathbf{n}_3 \tag{4.12.5}$$

Hence, the derivatives in S of the $\mathbf{n}_i$ are:

$$\begin{aligned} ^S d\mathbf{n}_1/dt &= \dot{\phi}c_\theta\mathbf{n}_2 - \dot{\phi}s_\theta\mathbf{n}_3, \quad ^S d\mathbf{n}_2/dt = -\dot{\phi}c_\theta\mathbf{n}_1 + \dot{\theta}\mathbf{n}_3 \\ ^S d\mathbf{n}_3/dt &= \dot{\phi}s_\theta\mathbf{n}_1 - \dot{\theta}\mathbf{n}_2 \end{aligned} \tag{4.12.6}$$

Using these results the angular acceleration of D in S and the acceleration of G in S are found to be:

$$\begin{aligned} ^S\boldsymbol{\alpha}^D &= \left(\ddot{\theta} - \dot{\psi}\dot{\phi}c_\theta\right)\mathrm{n}_1 + \left(\ddot{\psi} + \ddot{\phi}s_\theta + \dot{\theta}\dot{\phi}c_\theta\right)\mathrm{n}_2 \\ &+ \left(\ddot{\phi}c_\theta - \dot{\phi}\dot{\theta}s_\theta + \dot{\psi}\dot{\theta}\right)\mathrm{n}_3 \end{aligned} \tag{4.12.7}$$

and

$$\begin{aligned} ^S\mathbf{a}^G &= r\left(\ddot{\psi} + \ddot{\phi}s_\theta + 2\dot{\phi}\dot{\theta}c_\theta\right)\mathbf{n}_1 + r\left(-\ddot{\theta} + \dot{\psi}\dot{\phi}c_\theta + \dot{\phi}^2 s_\theta c_\theta\right)\mathbf{n}_2 \\ &+ r\left(-\dot{\psi}\dot{\phi}s_\theta - \dot{\phi}^2 s_\theta^2 - \dot{\theta}^2\right)\mathbf{n}_3 \end{aligned} \tag{4.12.8}$$

Finally, consider the special case of straight-line rolling of a disk or wheel as depicted in Figure 4.11.4. As before, let the disk have radius r, center G, and contact point C. Let P be a point on the rim of D such that GP is perpendicular to GC, as in Figure 4.11.4. We can obtain expressions for the kinematics of D and G directly from the expressions of Eqs. (4.12.2), (4.12.3), (4.2.7), and (4.12.8) by setting θ and ϕ equal to zero. That is:

$$\begin{aligned} ^S\boldsymbol{\omega}^D &= \dot{\psi}\mathbf{n}_2, \quad ^S\mathbf{V}^G = r\dot{\psi}\mathbf{n}_1 \\ ^S\boldsymbol{\alpha}^D &= \ddot{\psi}\mathbf{n}_2, \quad ^S\mathbf{a}^G = r\ddot{\psi}\mathbf{n}_1 \end{aligned} \tag{4.12.9}$$

Observe the relative simplicity of these expressions. Indeed, these expressions may be recognized as those learned in elementary mechanics.

Finally, using Eqs. (4.11.1), (4.11.5), (4.9.4), and (4.9.6), we can evaluate the velocity and acceleration of P and C. The results are:

$$\mathbf{V}^P = r\dot{\psi}\mathbf{n}_1 - r\dot{\psi}\mathbf{n}_3, \quad a^p = \left(r\ddot{\psi} - r\dot{\psi}^2\right)\mathbf{n}_1 - r\ddot{\psi}\mathbf{n}_3$$

and

$$\mathbf{V}^C = 0, \ \mathbf{a}^C = r\dot{\psi}^2\mathbf{n}_3 \tag{4.12.10}$$

Observe that even though the velocity of the contact point is zero, its acceleration is not zero.

4.13 A Conical Thrust Bearing

As another example illustrating rolling kinematics, consider a thrust bearing* consisting of a cylindrical shaft with a conical end rolling on four spheres (or balls) as depicted in Figure 4.13.1. Let the spheres roll in a cylindrical race R as shown. Let S refer to the shaft, and let B refer to a typical ball. Let C_1, C_2, and C_3 be contact points between S and B and between B and R, as shown. Let G be the center of B. Let θ be the half-angle of the conical end of S, and let a be the distance between G and the axis of S. Let r be the radius of B.

As noted above, B rolls on both R and S. Suppose we desire B to have *pure rolling* on S (see Eq. (4.11.2)). One might ask if there is a relationship between a, r, and θ that will produce pure rolling between B and S while maintaining rolling between B and R. To answer this question, consider an analysis of the kinematics of B and S: Figure 4.13.2 shows an enlarged view of the contact region between B and S. Let $\mathbf{n}_1$, $\mathbf{n}_2$, and $\mathbf{n}_3$ be a set of mutually perpendicular unit vectors such that $\mathbf{n}_2$ is parallel to the axis of S and $\mathbf{n}_1$ is parallel to a line passing through G and C_3 and intersecting the axis of S. Let $\mathbf{n}_\perp$ be a unit vector parallel to GC and thus normal to the contacting surfaces of B and S at C_1. Let $\mathbf{n}_{||}$ be perpendicular to $\mathbf{n}_\perp$ and $\mathbf{n}_3$. Thus, $n_{||}$ is parallel to a cone element passing through C_1.

To study the kinematics and especially the rolling, it is helpful to first obtain expressions for the angular velocities of S and B: Let Ω be the angular speed of the shaft S as indicated in Figure 4.13.2. Then, the angular velocity of S in the reference frame R of the race is:

$$^R\boldsymbol{\omega}^S = \Omega\mathbf{n}_2 \tag{4.13.1}$$

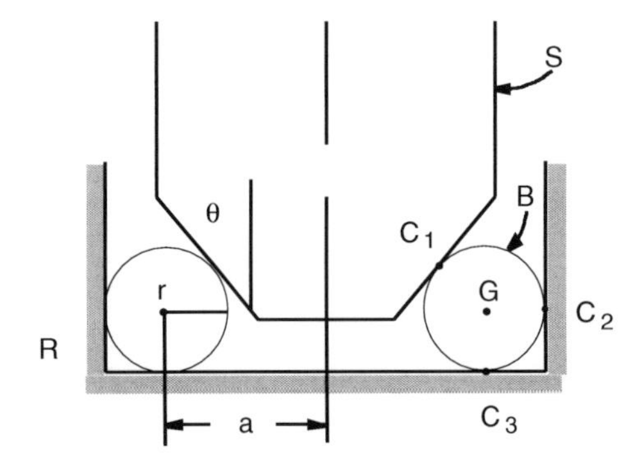

FIGURE 4.13.1
A conical thrust ball bearing.

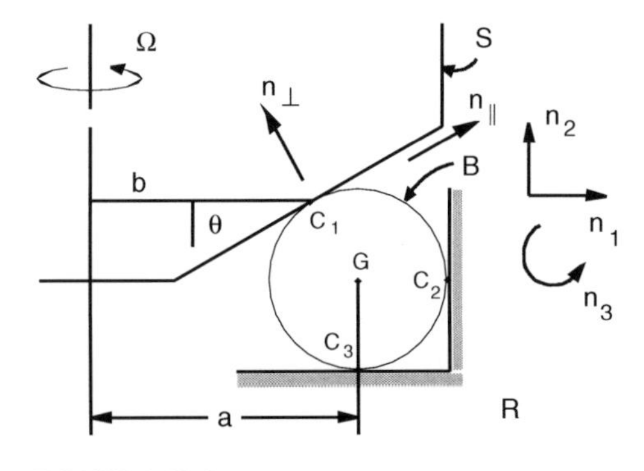

FIGURE 4.13.2
Ball and shaft geometry.

* This problem is discussed by Kane (4.1, 4.2) and Cabannes (4.5); see also Ramsey (4.6).

Let the angular velocity of ball B in R be expressed in terms of $\mathbf{n}_1$, $\mathbf{n}_2$, and $\mathbf{n}_3$ as:

$$^R\boldsymbol{\omega}^B = \omega_1\mathbf{n}_1 + \omega_2\mathbf{n}_2 + \omega_3\mathbf{n}_3 \tag{4.13.2}$$

where the scalar components ω_1, ω_2, and ω_3 are to be determined. To this end, observe that because B rolls on R at both C_2 and C_3, the velocity of its center G may be expressed both as:

$$^R\mathbf{V}^G = {}^R\boldsymbol{\omega}^B \times r\mathbf{n}_2 \quad \text{and as} \quad {}^R\mathbf{V}^G = {}^R\boldsymbol{\omega}^B \times (-r\mathbf{n}_1) \tag{4.13.3}$$

By substituting from Eq. (4.13.2) into (4.13.3) and by comparing the results, we obtain the relation:

$$r\omega_1\mathbf{n}_3 - r\omega_3\mathbf{n}_1 = r\omega_2\mathbf{n}_3 - r\omega_3\mathbf{n}_2 \tag{4.13.4}$$

Hence, we have:

$$\omega_3 = 0 \quad \text{and} \quad \omega_1 = \omega_2 \stackrel{D}{=} \omega \tag{4.13.5}$$

Therefore, $^R\omega^B$ becomes:

$$^R\boldsymbol{\omega}^B = \omega\mathbf{n}_1 + \omega\mathbf{n}_2 \tag{4.13.6}$$

Next, consider the velocity of C_1: If C_1 is considered to be a point of the shaft S, it is seen to move on a circle about the shaft axis. Its velocity is thus:

$$\mathbf{V}^{C_1} = {}^R\boldsymbol{\omega}^S \times b\mathbf{n}_1 = -b\Omega\mathbf{n}_3 \tag{4.13.7}$$

where the last equality is obtained by using Eq. (4.13.1) for $^R\omega^S$. Alternatively, if C_1 is considered to be a point of the ball B, its velocity can be obtained from the "rolling" Eq. (4.11.5). That is,

$$\mathbf{V}^{C_1} = {}^R\boldsymbol{\omega}^B \times \mathbf{C_2C_1} \tag{4.13.8}$$

where the position vector $\mathbf{C_2C_1}$ may be expressed as:

$$\mathbf{C_2C_1} = r\mathbf{n}_2 + r\mathbf{n}_\perp = -r\cos\theta\mathbf{n}_1 + r(1+\sin\theta)\mathbf{n}_2 \tag{4.13.9}$$

By using Eq. (4.13.6) for $^R\boldsymbol{\omega}^B$, by substituting into Eq. (4.13.8), by evaluating the vector product, and by comparing the result with that of Eq. (4.13.7), we obtain:

$$-b\Omega = r\omega(1+\sin\theta) + r\omega\cos\theta \tag{4.13.10}$$

From Figure 4.13.2, however, it is seen that b may be expressed in terms of a, r, and θ as:

$$b = a - r\cos\theta \tag{4.13.11}$$

By substituting from Eq. (4.13.11) into (4.13.10), we can solve for the angular speed ratio Ω/ω as:

$$\Omega/\omega = -[r(1+\sin\theta+\cos\theta)]/(a - r\cos\theta) \tag{4.13.12}$$

Recall that our objective is to determine the relationship between a, r, and θ so that there will be pure rolling between B and S at C_1. We have not yet used the pure rolling criterion. To invoke the criterion, we recall from Eq. (4.11.2) that to have *pure* rolling at C_1, we must have:

$${}^B\boldsymbol{\omega}^S \cdot \mathbf{n}_\perp = 0 \tag{4.13.13}$$

where ${}^B\omega^S$ is the angular velocity of the shaft S relative to the ball B. From Eqs. (4.7.5), (4.13.1), and (4.13.6), ${}^B\omega^S$ is:

$${}^B\boldsymbol{\omega}^S = {}^R\boldsymbol{\omega}^S - {}^R\boldsymbol{\omega}^B = \Omega\mathbf{n}_2 - \omega\mathbf{n}_1 - \omega\mathbf{n}_3 = -\omega\mathbf{n}_1 + (\Omega - \omega)\mathbf{n}_2 \tag{4.13.14}$$

From Figure 4.13.2, we see that $\mathbf{n}_\perp$ may be expressed in terms of $\mathbf{n}_1$ and $\mathbf{n}_2$ as:

$$\mathbf{n}_\perp = -\cos\theta\mathbf{n}_1 + \sin\theta\mathbf{n}_2 \tag{4.13.15}$$

Hence, the pure rolling criterion of Eq. (4.13.13) becomes:

$$\omega\cos\theta + (\Omega - \omega)\sin\theta = 0 \tag{4.13.16}$$

By solving Eq. (4.13.16) for the angular speed ratio Ω/ω, we obtain:

$$\Omega/\omega = (\sin\theta - \cos\theta)/\sin\theta \tag{4.13.17}$$

Finally, by equating expressions for Ω/ω of Eqs. (4.13.12) and (4.13.17), we obtain an equation involving only a, r, and θ. By solving this equation for a/r we obtain:

$$a/r = (1+\sin\theta)/(\cos\theta - \sin\theta) \tag{4.13.18}$$

This is the desired geometrical relation that will ensure pure rolling at C_1.

This result can also be obtained by using a rather insightful geometrical argument: Specifically, to have pure rolling at C_1, ball B may be considered as being part of an imaginary cone with axis along C_2C_3 and rolling on the conical end of S. This is depicted in Figure 4.13.3, where O is at the apex of the rolling cones. Let h measure the elevation of C_1 above O, as shown. Then, from the figure, we see that b and h are:

$$b = a - r\cos\theta = h\tan\theta \tag{4.13.19}$$

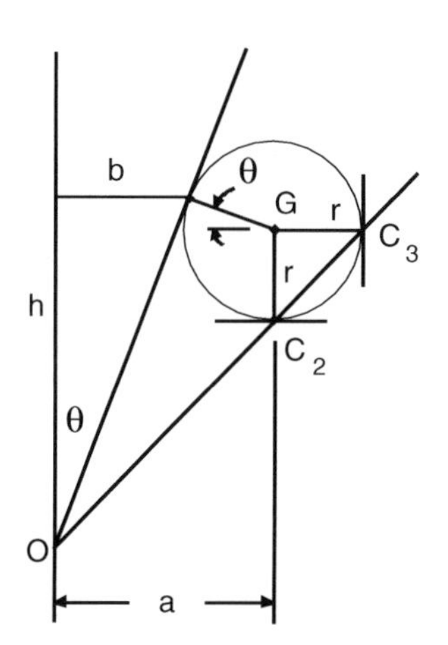

FIGURE 4.13.3
Geometry of rolling cones.

and

$$h = a + r + r\sin\theta \tag{4.13.20}$$

(Note that line OC_2 is inclined at 45°.) By eliminating h between these two expressions and then solving for a/r we obtain:

$$a/r = (1+\sin\theta)/(\cos\theta - \sin\theta) \tag{4.13.21}$$

This is seen to be identical to the result of Eq. (4.13.18).

4.14 Closure

This completes our study and review of the fundamentals of kinematics. In the following chapters, we will specialize and apply these kinematic principles to mechanisms, machines, and multibody systems.

References

4.1. Kane, T. R., *Analytical Elements of Mechanics*, Vol. 2, Academic Press, New York, 1961, pp. 66, 97, 205, 318.
4.2. Kane, T. R., *Dynamics*, Holt, New York, 1968, pp. 63, 66.
4.3. Kane, T. R., and Levinson, D. A., *Dynamics: Theory and Applications*, McGraw-Hill, New York, 1985.
4.4. Huston, R. L., and Liu, C. Q., *Formulas for Dynamic Analysis*, Marcel Dekker, New York, 2001.
4.5. Cabannes, H., *General Mechanics*, Blaisdell, Boston, 1968, pp. 384–387.
4.6. Ramsey, A. S., *Dynamics*, Part II, Cambridge University Press, London, 1937, p. 160.

Problems

Section 4.2 Orientation of Rigid Bodies

P4.2.1: In Eq. (4.2.3), suppose that the orientation angles are $\alpha = 30°$, $\beta = -45°$, and $\gamma = 60°$. Determine the transformation matrix S and verify that S is orthogonal; that is, the inverse S_{-1} is equal to the transpose.

Section 4.3 Configuration Graphs

P4.3.1: Consider the configuration graph of Figure P4.3.1. Find the relations between the $\mathbf{N}_i$ and the $\mathbf{n}_i$. Express the results in the matrix forms:

$$\mathbf{N} = S\mathbf{n} \quad \text{and} \quad \mathbf{n} = T\mathbf{N}$$

Show that:

$$S = T^T = S_{-1}$$

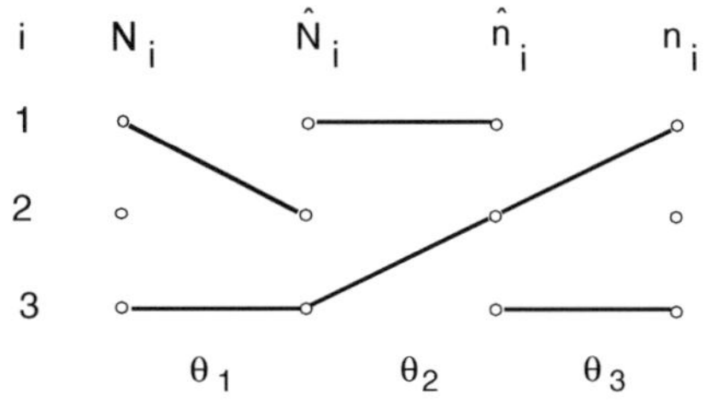

FIGURE P4.3.1
Configuration graph.

P4.3.2: Repeat Problem P4.3.1 for the configuration graph of Figure P4.3.2.

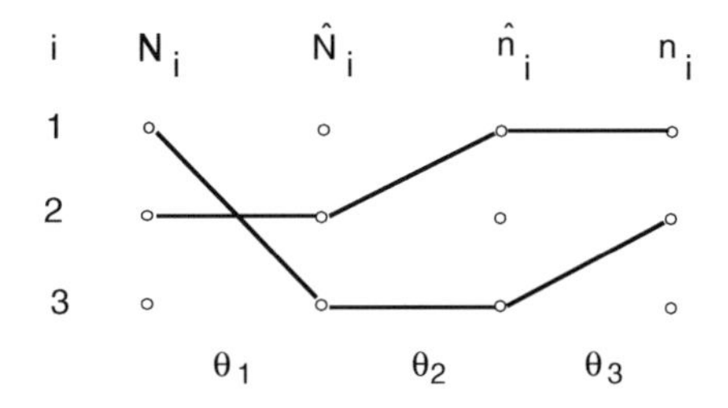

FIGURE P4.3.2
Configuration graph.

Section 4.4 Simple Angular Velocity and Simple Angular Acceleration

P4.4.1: A flywheel rotating at 400 rpm is brought to rest in 3 minutes by a uniform braking moment. Determine the angular deceleration and the number of revolutions turned through during the braking.

P4.4.2: Consider the rotating disk D of Figure P4.4.2 where θ is the rotation angle of D. Let AB be a line scribed on D, and let ϕ be the angle AB makes with a fixed horizontal

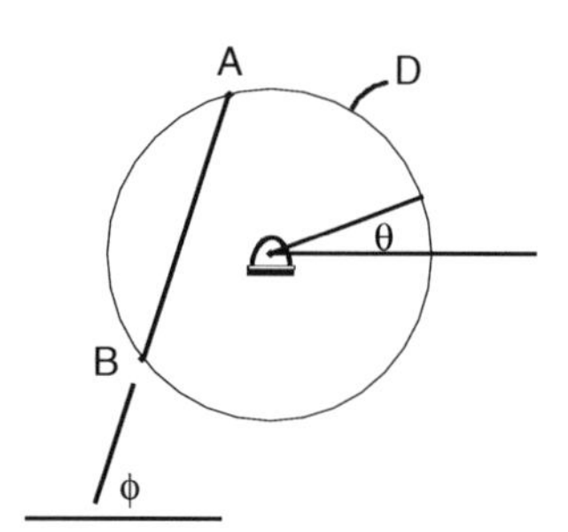

FIGURE P4.4.2
Rotating disk D, rotation angle θ, and scribed line AB.

line. Show that $d\phi/dt = d\theta/dt$ and that, therefore, the angular speed and angular acceleration of D may be expressed as:

$$\omega = d\phi/dt \quad \text{and} \quad \alpha = d^2\phi/dt^2$$

P4.4.3: Refer to Problem P4.4.2. Suppose ϕ is a function of time t given by:

$$\phi = 5 + 4t + 8t^2 \text{ deg}$$

where t is in seconds. Find the angular speed ω and angular acceleration α of D for t = (a) 1 sec, (b) 2 sec, and (c) 5 sec.
P4.4.4: See Problem P4.4.3. Find the angle turned through by D from $t = 2$ to $t = 5$ sec.

Section 4.5 General Angular Velocity

P4.5.1: Let a body B be moving in a reference frame R such that the angular velocity of B in R is given by:

$$^R\boldsymbol{\omega}^B = 8\mathbf{n}_x + 4\mathbf{n}_y - 5\mathbf{n}_z \text{ rad/sec}$$

where $\mathbf{n}_x$, $\mathbf{n}_y$, and $\mathbf{n}_z$ are fixed in B. Let $\mathbf{n}_x$, $\mathbf{n}_y$, and $\mathbf{n}_z$ be parallel to the X, Y, Z coordinate axes fixed in B as in Figure P4.5.1. Let P and Q be particles of B with coordinates $P(1, 2, -3)$ and $Q(-2, 4, 7)$, measured in meters. Using Eq. (4.5.2), find:

a. The velocity of P relative to O in R.
b. The velocity of Q relative to O in R.
c. The velocity of P relative to Q in R.

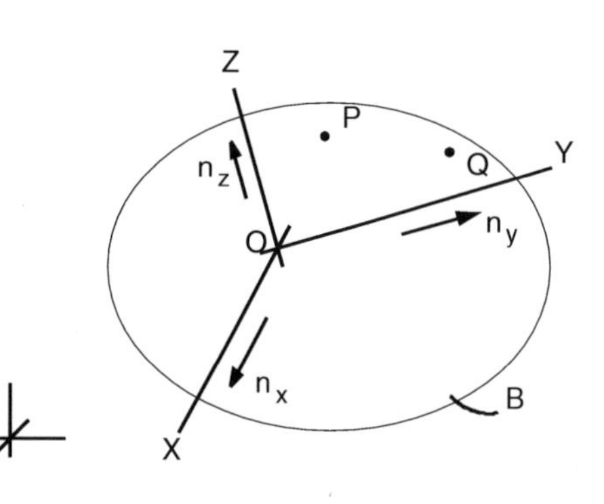

FIGURE P4.5.1
A body B moving in a reference frame R.

Section 4.6 Differentiation in Different Reference Frames

P4.6.1: Refer again to Problem P4.5.1. Let the origin O of the reference frame fixed in body B have a velocity in reference frame R given by:

$$^R\mathbf{V}^O = 8\mathbf{n}_x - 6\mathbf{n}_y + 5\mathbf{n}_z \text{ m/s}$$

Next, let $\hat{P}$ be a point that moves relative to B such that the location of $\hat{P}$ relative to O is given by the vector $\mathbf{O\hat{P}}$ as:

$$\mathbf{O\hat{P}} = 4t\mathbf{n}_x - 3t^2\mathbf{n}_y + 9t^3\mathbf{n}_z \text{ m}$$

Find the velocity of $\hat{P}$ in R at times $t = 0, 1,$ and 2 sec.

P4.6.2: See Example 4.6.1. Suppose the rocket is launched from the Earth at a latitude of 40° North. Find the velocity of R in the astronomical frame A.

Section 4.7 Addition Theorem for Angular Velocity

P4.7.1: See Example 4.7.1. Using the configuration graph of Figure 4.7.6, express $^R\boldsymbol{\omega}^D$ of Eq. (4.7.12) in terms of (a) $\mathbf{N}_{Ri}$, (b) $\mathbf{N}_{Si}$, and (c) $\mathbf{N}_{Di}$.

P4.7.2: Using Eq. (4.7.6), show that:

$$^R\boldsymbol{\omega}^B = -{}^B\boldsymbol{\omega}^R$$

for a body B moving in a reference frame R.

P4.7.3: An end-effector, or gripper, E is mounted on the end of a shaft S as shown in Figure P4.7.3. The rotation axis of E is perpendicular to the axis of S with the half closing angle being β as shown. S in turn is mounted in a cylinder C having a common axis with C and a rotation angle about its axis through an angle ψ. C is hinged to a turret platform T as shown, with ϕ being the angle between the axis of C and the horizontal. Finally, T rotates about a vertical axis through an angle θ, in a fixed reference frame R. Let $\mathbf{n}_1$, $\mathbf{n}_2$, and $\mathbf{n}_3$ be unit vectors fixed in T, with $\mathbf{n}_3$ being vertical and $\mathbf{n}_1$ parallel to the hinge axis with C. Find the angular velocity of E in R, and express the result in terms of $\mathbf{n}_1$, $\mathbf{n}_2$, and $\mathbf{n}_3$.

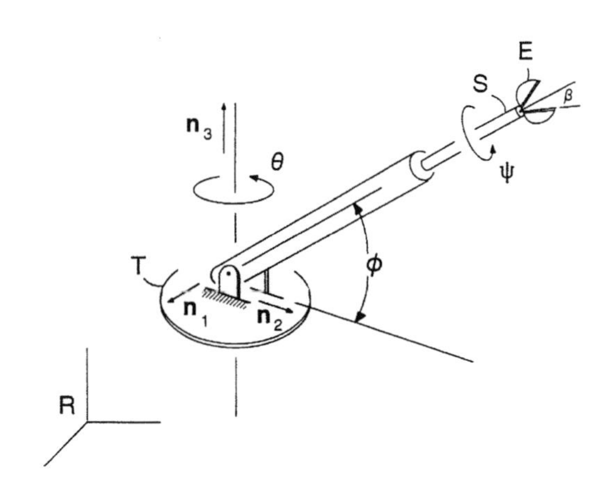

FIGURE P4.7.3
An end-effector E at the extremity of a manipulator shaft.

P4.7.4: A sport utility vehicle traveling at 45 mph on a straight level roadway suddenly goes out of control, spinning counterclockwise (looking from above) and becoming perpendicular to the direction of travel in 0.25 sec. If the wheel diameter is 26 in., determine the angular velocity of the right rear wheel relative to the road surface. Express the result in terms of unit vectors $\mathbf{n}_x$, $\mathbf{n}_y$, and $\mathbf{n}_z$ fixed in the vehicle, with $\mathbf{n}_x$ being forward and $\mathbf{n}_z$ being up. (*Hint:* The angular velocity of a wheel relative to the vehicle is equal to the speed divided by the wheel radius.)

Section 4.8 Angular Acceleration

P4.8.1: Use Eq. (4.6.6) to validate Eq. (4.4.13).

P4.8.2: See Examples 4.7.1 and 4.8.1 and also Problem 4.7.1. Determine ${}^R\boldsymbol{\alpha}^D$ in terms of unit vectors $\mathbf{N}_{Ri}$, $\mathbf{N}_{Si}$, and $\mathbf{N}_{Di}$ (i = 1, 2, 3) by using the result of Eq. (4.8.19) and the configuration graph of Figure 4.7.6. Check the results by differentiating the expression for ${}^R\boldsymbol{\omega}^D$ as determined from Problem 4.7.1.

P4.8.3: See Problem P4.8.2. Find ${}^Y\boldsymbol{\alpha}^D$, ${}^S\boldsymbol{\alpha}^D$, and ${}^R\boldsymbol{\alpha}^S$. Show that ${}^R\boldsymbol{\alpha}^D \neq {}^R\boldsymbol{\alpha}^S + {}^S\boldsymbol{\alpha}^Y + {}^Y\boldsymbol{\alpha}^D$.

Section 4.9 Relative Velocity and Relative Acceleration of Two Points on a Rigid Body

P4.9.1: A body B moves in a reference frame R with angular velocity in R given by:

$$ {}^R\boldsymbol{\omega}^B = 6\mathbf{n}_x - 4\mathbf{n}_y + 7\mathbf{n}_z \text{ rad/sec} $$

where $\mathbf{n}_x$, $\mathbf{n}_y$, and $\mathbf{n}_z$ are unit vectors parallel to an X-, Y-, Z-axis system fixed in B as in Figure P4.9.1. Let O be the origin of X, Y, Z and let the velocity of O in R be:

$$ {}^R\mathbf{V}^O = -10\mathbf{n}_x + 2\mathbf{n}_y - 8\mathbf{n}_z \text{ m/s} $$

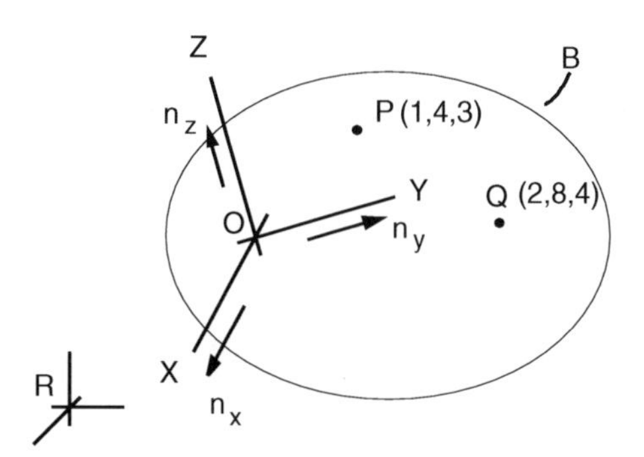

FIGURE P4.9.1
A body B moving in a reference frame R.

Finally, let P and Q be particles of B with coordinates (1, 4, 3) and (2, 8, 4), respectively, relative to X, Y, Z as shown. Find: (a) ${}^R\mathbf{V}^P$; (b) ${}^R\mathbf{V}^Q$; (c) ${}^R\mathbf{V}^{P/Q}$. Check the results of (c) by constructing a vector $\mathbf{r}$ from Q to P and then evaluating ${}^R d\mathbf{r}/dt$.

P4.9.2: See Problem P4.9.1. Let the angular acceleration of B in R be given by:

$$ {}^R\boldsymbol{\alpha}^B = -2\mathbf{n}_x + 4\mathbf{n}_y - \mathbf{n}_z \text{ rad/sec}^2 $$

and let the acceleration of O in R be:

$$^{R}\mathbf{a}^{O} = -6\mathbf{n}_x - 8\mathbf{n}_y + 2\mathbf{n}_z \text{ m/sec}^2$$

Find: (a) $^{R}\mathbf{a}^{P}$, (b) $^{R}\mathbf{a}^{Q}$, and (c) $^{R}\mathbf{a}^{P/Q}$. Check the results of (c) by evaluating where, as before, $\mathbf{r}$ locates P relative to Q.

P4.9.3: A rod pendulum has a movable support as represented in Figure P4.9.3. Let the pendulum length ℓ be 3 ft and let the horizontal displacement x of the support O and the angular displacement θ of the rod be:

$$x = 2\sin(\pi/4)t \text{ ft}$$

and

$$\theta = \pi/3\cos(\pi/6)t$$

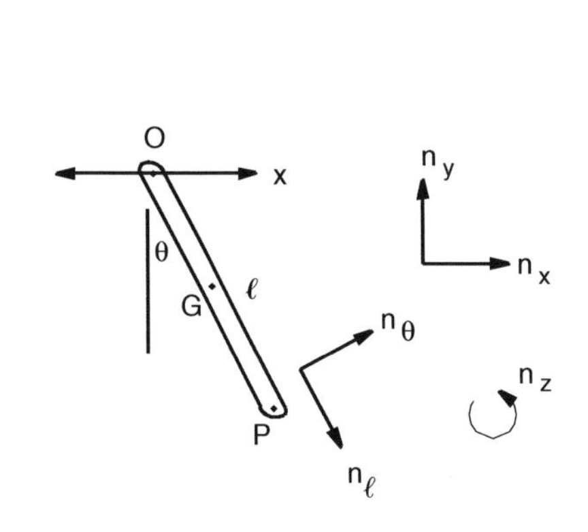

FIGURE P4.9.3
A rod pendulum with a horizontally moving support.

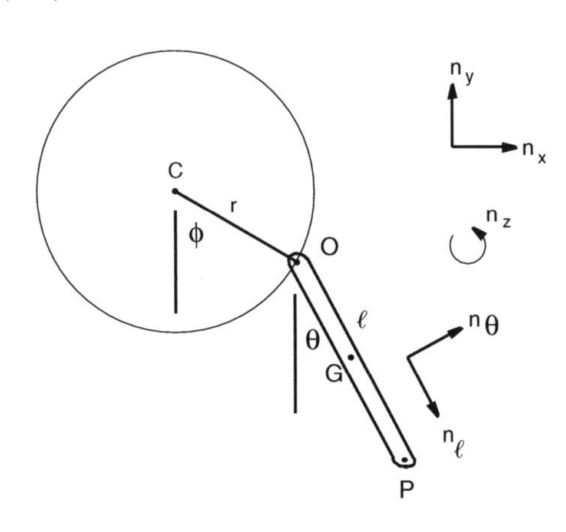

FIGURE P4.9.4
A rod pendulum with a support moving vertically on a circle.

Find the velocity and acceleration of the rod center G and the rod extremity P for time t equal to (a) 0, (b) 1, and (c) 2 sec. Express the results in terms of the unit vectors sets $\mathbf{n}_x$, $\mathbf{n}_y$, $\mathbf{n}_z$ and $\mathbf{n}_\ell$, $\mathbf{n}_\theta$, $\mathbf{n}_z$, as shown in Figure 4.9.4.

P4.9.4: See Problem 4.9.3. Instead of moving on a horizontal line, let the pendulum support O move on a vertical circle as in Figure P4.9.4. As before, let the pendulum length ℓ be 3 ft. Let the circle on which O moves have a radius of 1.25 ft. Let the inclination angles θ and ϕ of Figure P4.9.4 be:

$$\theta = (\pi/3)\cos(\pi/6)t$$

and

$$\phi = \pi\sin(\pi/3)t$$

Find the velocity and acceleration of the rod center G and the rod extremity P for time t equal to (a) 0, (b) 1, and (c) 2 sec, and express the results in terms of the unit vector sets $\mathbf{n}_x$, $\mathbf{n}_y$, $\mathbf{n}_z$ and $\mathbf{n}_\ell$, $\mathbf{n}_\theta$, $\mathbf{n}_z$, shown in Figure 4.9.4.

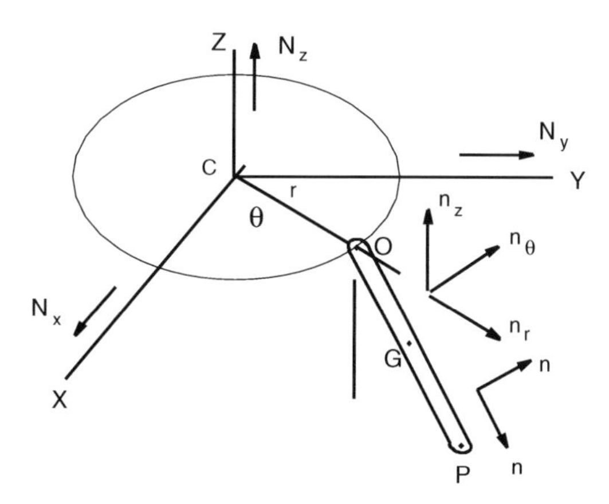

FIGURE P4.9.5
A rod pendulum with support moving horizontally on a circle, and rotating about a radial line.

P4.9.5: See Problems P4.9.3 and P4.9.4. Instead of the pendulum moving on a vertical circle, let it support O moving on a horizontal circle as in Figure P4.9.5. As before, let the pendulum length ℓ be 3 ft. Let the circle on which O moves have a radius of 1.25 ft. Let the inclination angles θ and ϕ of Figure P4.9.5 be:

$$\theta = (\pi/3)\cos(\pi/6)t$$

and

$$\phi = \pi \sin(\pi/3)t$$

where θ now defines the rotation of the rod about a radial line. Find the velocity and acceleration of the rod center G and the rod extremity P for time t equal to (a) 0, (b) 1, and (c) 2 sec. Express the results in terms of the unit vector sets $\mathbf{n}\ell$, $\mathbf{n}_\theta$, $\mathbf{n}_r$; $\mathbf{n}_r$, $\mathbf{n}_\phi$, $\mathbf{n}_z$; and $\mathbf{N}_x$, $\mathbf{N}_y$, $\mathbf{N}_z$.

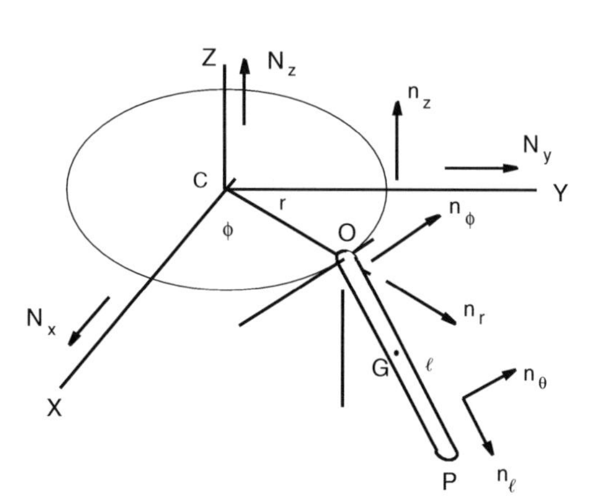

FIGURE P4.9.6
A rod pendulum with support moving horizontally on a circle, and rotating about a tangential line.

P4.9.6: See Problems P4.9.3, P4.9.4, and P4.9.5. As in Problem P4.9.4, let the pendulum support O move on a horizontal circle, but let the rod rotate about a tangential line as shown in Figure P4.9.6. As before, let the pendulum length ℓ be 3 ft. Let the circle on which O moves have a radius of 1.25 ft. Let the inclination angles θ and ϕ of Figure P4.9.6 be:

$$\theta = (\pi/3)\cos(\pi/6)t$$

and

$$\phi = \pi \sin(\pi/3)t$$

where θ now defines the rotation of the rod about a tangential line. Find the velocity and acceleration of the rod center *G* and the rod extremity *P* for time *t* equal to (a) 0, (b) 1, and (c) 2 sec. Express the results in terms of the unit vector sets $\mathbf{n}_\ell$, $\mathbf{n}_\phi$, $\mathbf{n}_\theta$; $\mathbf{n}_r$, $\mathbf{n}_\phi$, $\mathbf{n}_z$; and $\mathbf{N}_x$, $\mathbf{N}_y$, $\mathbf{N}_z$.

P4.9.7: See Example 4.9.3. Suppose end *B* of rod *AB* is constrained to move on an inclined line in the *Y–Z* plane as in Figure P4.9.7. As in Example 4.9.3, let *AB* have length 13 m, with end *A* having velocity of $6\mathbf{n}_x$ and an acceleration of $-3\mathbf{n}_x$ m/s^2. Let the projections of the angular velocity and angular acceleration of *AB* along *AB* itself be zero, as in Eqs. (4.9.21). Then, for the position shown in Figure P4.9.7, find the velocity and acceleration of end *B*.

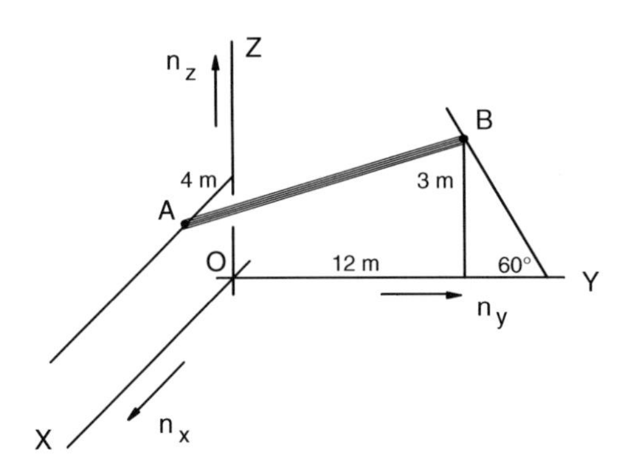

FIGURE P4.9.7
A rod *AB* with constrained motions of its ends.

P4.9.8: The box shown in Figure P4.9.8 has an angular velocity relative to a reference frame *R* given by:

$${}^R\omega^{Box} = t^2\mathbf{n}_1 + (2+t)\mathbf{n}_2 + (4+t^2)\mathbf{n}_3 \text{ rad/sec}$$

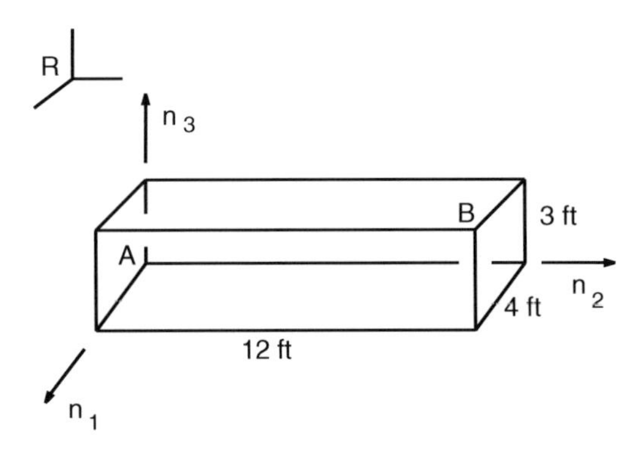

FIGURE P4.9.8
A box moving in a reference frame *R*.

where t is time and $\mathbf{n}_1$, $\mathbf{n}_2$, and $\mathbf{n}_3$ are mutually perpendicular unit vectors fixed relative to the box as shown. The velocity and acceleration of corner A of the box in R are:

$$^R\mathbf{V}^A = 2\mathbf{n}_1 + 5\mathbf{n}_2 - 6\mathbf{n}_3 \text{ ft/sec}$$

and

$$^R\mathbf{a}^A = 2\mathbf{n}_1 - 8\mathbf{n}_2 - 4\mathbf{n}_3 \text{ ft/sec}^2$$

Find at times $t = 0$ and $t = 1$:

a. The angular acceleration of the box in R.
b. The velocity of corner B in R.
c. The acceleration of corner A in R.

Section 4.10 Points Moving on a Rigid Body

P4.10.1: A particle P moves along a diametral slot of a rotating disk D as in Figure P4.10.1. Let the speed v of P be constant at 2 ft/sec and directed from A to B. Let the angular speed ω and angular acceleration α of D be 3 rad/sec and 5 rad/sec², respectively, with directions as indicated in Figure P4.10.1. Finally let the radius of D be 2 ft. Find the velocity and acceleration of P, relative to a fixed reference frame R in which D rotates, when P is at (a) A, (b) O, and (c) B. Express the results in terms of the radial and tangential unit vectors $\mathbf{n}_r$ and $\mathbf{n}_\theta$.

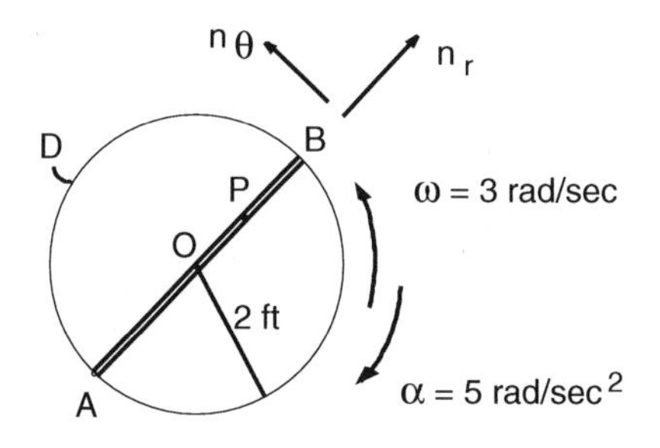

FIGURE P4.10.1
A particle moving along a diametral slot of a rotating disk.

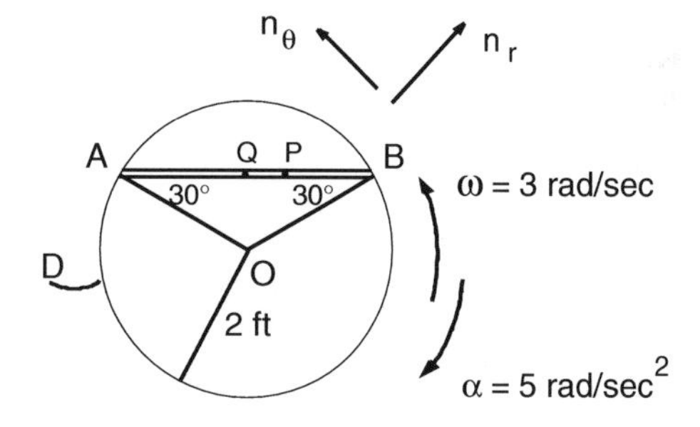

FIGURE P4.10.2
A particle P moving along a chord slot of a rotating disk.

P4.10.2: Repeat Problem P4.10.1, where the slot AB is now along a chord of the disk D as in Figure P4.10.2. As before, let the speed of P be uniform at 2 ft/sec and directed from A to B. Find the velocity and acceleration of P, relative to a fixed reference frame R in which D rotates, when P is at (a) A, (b) B, and (c) Q, the midpoint of AB. Express the results in terms of the radial and tangential unit vectors $\mathbf{n}_r$ and n_θ.

P4.10.3: See Problem P4.9.8. The box shown in Figure P4.10.3 has an angular velocity relative to a reference frame R given by:

$$^R\boldsymbol{\omega}^{Box} = t^2\mathbf{n}_1 + (2+t)\mathbf{n}_2 + \left(4+t^2\right)\mathbf{n}_3$$

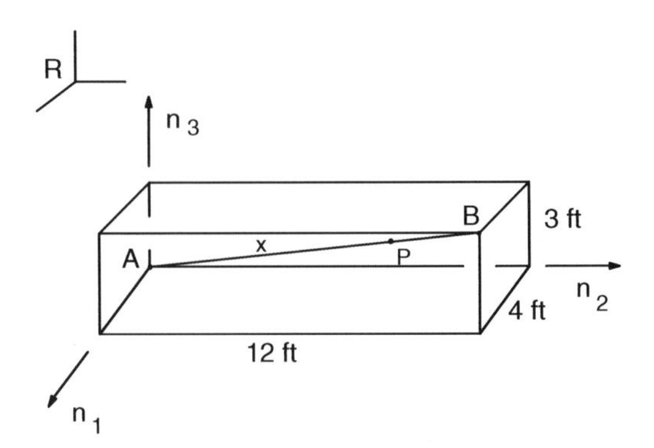

FIGURE P4.10.3
A box moving in a reference frame *R*.

where, as before, *t* is time, and $\mathbf{n}_1$, $\mathbf{n}_2$, and $\mathbf{n}_3$ are mutually perpendicular unit vectors fixed relative to the box as shown. Also, as before, the velocity and acceleration of corner *A* of the box in *R* are:

$$^R\mathbf{V}^A = 2\mathbf{n}_1 + 5\mathbf{n}_2 - 6\mathbf{n}_3 \text{ ft/sec}$$

and

$$^R\mathbf{a}^A = 2\mathbf{n}_1 - 8\mathbf{n}_2 - 4\mathbf{n}_3 \text{ ft/sec}^2$$

A particle *P* moves along the diagonal *AB* of the box with the distance *x* from corner *A* given by:

$$x = 13 - 13\sin(\pi t/2)\text{ ft}$$

Find at times $t = 0$ and $t = 1$:

a. The velocity of *P* relative to the box.
b. The velocity of *P* relative to *R*.
c. The acceleration of *P* relative to the box.
d. The acceleration of *P* relative to R.

Section 4.11 Rolling Bodies

P4.11.1: A circular disk *D* rolls at a constant speed on a straight line as indicated in Figure P4.11.1. Let *D* roll to the right and remain in a vertical plane. Let the radius of *D* be 0.3 m, and let the velocity of the center *O* at *D* be 10 m/sec (a constant). For the instant shown in the figure, find: (a) the angular speed ω of *D*, and (b) the velocity and acceleration of points *P*, *Q*, and *C* of *D*. Express the velocities and accelerations in terms of unit vectors $\mathbf{n}_x$ and $\mathbf{n}_y$.

P4.11.2: See Problem P4.11.1. Let the disk *D* be slowing down so that the acceleration (or deceleration) of *O* is to the left at 5 m/sec². For the configuration shown in Figure P4.11.2, with V_O being 5 m/sec, find: a) the acceleration α of *D*, and (b) the acceleration of *P*, *Q*, and *C*.

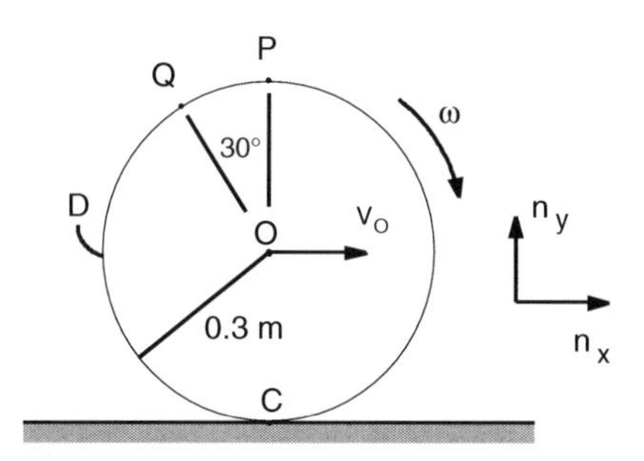

FIGURE P4.11.1
A circular disk rolling on a straight line.

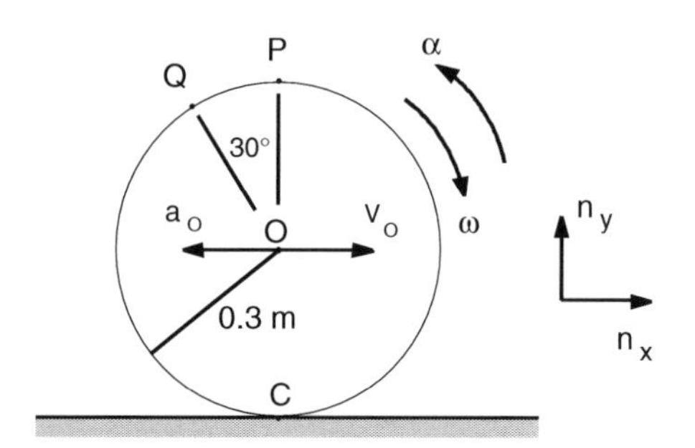

FIGURE P4.11.2
A rolling, decelerating circular disk.

P4.11.3: An automobile A traveling at 30 mph goes around a curve as represented in Figure P4.11.3. Consider the left rear wheel W_{LR}. Let the radius of the curve, approximated as a circle, upon which W_{LR} travels be 100 ft. If the diameter of W_{LR} is 26 in., find the angular velocity of W_{LR}. Express the result in terms of unit vectors $\mathbf{n}_x$, $\mathbf{n}_y$, and $\mathbf{n}_z$ fixed relative to A, with $\mathbf{n}_x$ being forward, $\mathbf{n}_y$ left, and $\mathbf{n}_z$ vertical up.

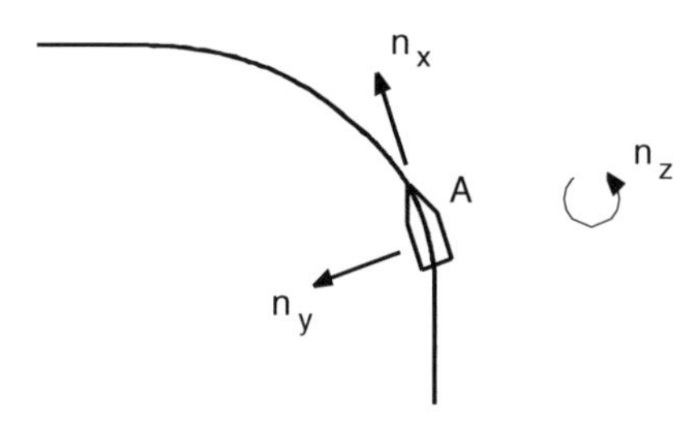

FIGURE 4.11.3
An automobile going around a turn to the left.

P4.11.4: See Problem P4.11.3. Let the rear axle of A be 54 in. long. Repeat Problem 4.11.3 for the right rear wheel W_{RR}.

P4.11.5: See Problems P4.11.3 and P4.11.4. Suppose A is going around the curve at a constant speed. Find the angular accelerations of W_{LR} and W_{RR}. As before, express the results in terms of $\mathbf{n}_x$, $\mathbf{n}_y$, and $\mathbf{n}_z$.

P4.11.6: See Problem P4.11.5. Suppose that instead of going at a constant speed around the curve A is slowing at the rate of 8 ft/sec². Determine the angular acceleration of W_{LR} and W_{RR}.

P4.11.7: A circular disk D with radius r rolls on a circle of radius R with a *lean angle* θ toward the inside of the circle as shown in Figure P4.11.7. Let $\mathbf{n}_r$, $\mathbf{n}_t$, and $\mathbf{n}_z$ be mutually perpendicular unit vectors, with $\mathbf{n}_t$ being tangent to the circle at the contact point C between D and the circle; $\mathbf{n}_r$ is a radial unit vector directed toward the center of the circle; and $\mathbf{n}_z$ is a vertical unit vector perpendicular to the circle. Let the center O of D have a velocity and tangential acceleration relative to a reference frame in which the circle is fixed given by:

$$\mathbf{V}^O = V\mathbf{n}_t \quad \text{and} \quad \mathbf{a}^O = a\mathbf{n}_t$$

If θ is a constant, determine expressions for the velocity and acceleration of point P at the top of D as in Figure P4.11.7.

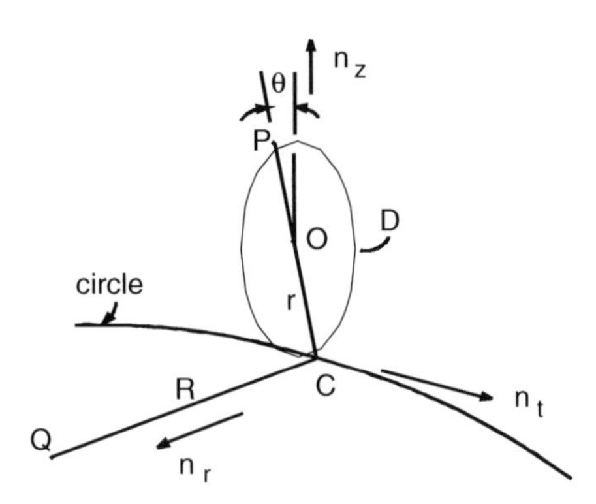

FIGURE P4.11.7
A leaning disk rolling on a circle.

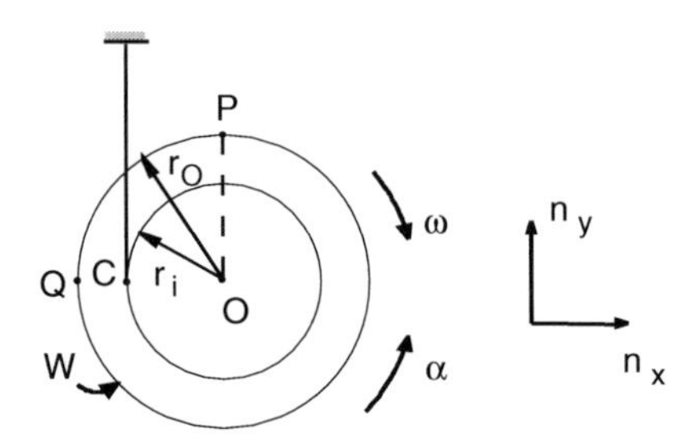

FIGURE P4.11.8
An unwinding differential pulley.

P4.11.8 A differential pulley *W* is being unwound as in Figure P4.11.8. Let the inner and outer radii of *W* be 12 cm and 18 cm, and let the angular speed and angular acceleration of *W* be 15 rad/sec and 3 rad/sec^2, respectively, in the directions shown. Determine the velocities and accelerations of points *O*, *C*, *P*, and *Q* of *W* when they are in the positions shown. Express the results in terms of unit vectors $\mathbf{n}_x$ and $\mathbf{n}_y$.

5

Planar Motion of Rigid Bodies — Methods of Analysis

5.1 Introduction

In this chapter, we consider planar motion — an important special case of the kinematics of rigid bodies. Planar motion characterizes the movement of the vast majority of machine elements and mechanisms. When a body has planar motion, the description of that motion is greatly simplified. Special methods of analysis can be used that provide insight not usually obtained in three-dimensional analyses. We begin our study with a general discussion of coordinates, constraints, and degrees of freedom. We then consider the planar motion of a body and the special methods of analysis that are applicable.

5.2 Coordinates, Constraints, Degrees of Freedom

In our discussion, we will use the term *coordinate* to refer to a parameter locating a particle or to a parameter defining the orientation of a body. In this sense, a coordinate is similar to the measurement used in elementary mathematics to locate a point or to orientate a line. We can bridge the difference between mathematical and physical objects by simply identifying particles with points and bodies with line segments.

Coordinates are not unique. For example, a point P in a plane is commonly located either by Cartesian coordinates or polar coordinates as shown in Figure 5.2.1.

In Cartesian coordinates, P is located by distances (x, y) to coordinate axes. In polar coordinates, P is located by the distance r to the origin (or pole) and by the inclination θ of the line connecting P with the pole. In both coordinate systems, *two* independent parameters are needed to locate P.

If P is free to move in the plane, the values of the coordinates will change as P moves. Because these changes can occur independently for each coordinate, P is said to have two *degrees of freedom*. If, however, P is restricted in its movement so that it must remain on, say, a curve C, then P is said to be *constrained*. The coordinates of P are then no longer independent.

Suppose, for example, that P is required to move on a circle C, centered at O with radius a, as in Figure 5.2.2. The coordinates of P are then restricted by a *constraint equation*. Using Cartesian coordinates, the constraint equation is:

$$x^2 + y^2 = a^2 \tag{5.2.1}$$

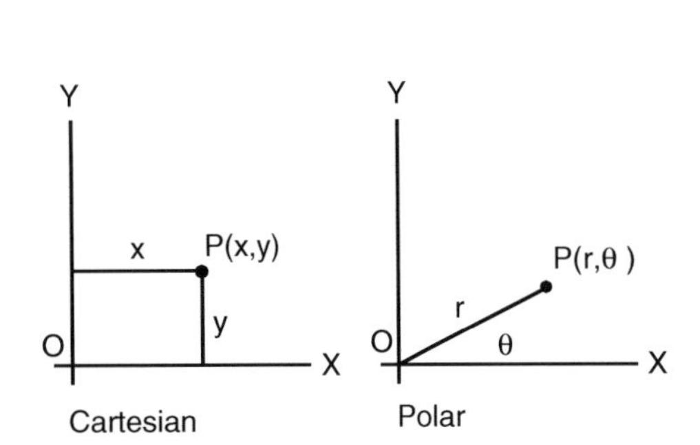

FIGURE 5.2.1
Cartesian and polar coordinates of a point.

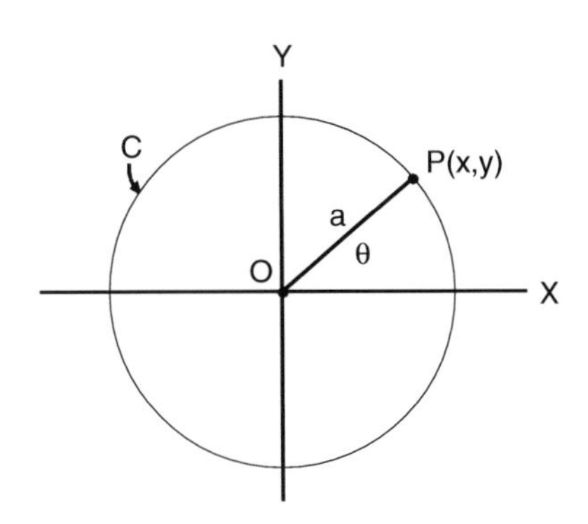

FIGURE 5.2.2
A point restricted to move on a circle.

Using polar coordinates the constraint equation is:

$$\mathbf{r} = a \tag{5.2.2}$$

The number of degrees of freedom of a *mechanical system* is sometimes defined as the difference between the number of *coordinates* needed to define the location of the particles of the system and the number of *constraint equations* needed to define the restrictions on the movement of the particles.

In three dimensions, it is generally known that an unrestrained particle has three degrees of freedom and that an unrestrained body has six degrees of freedom (three in *translation* and three in *rotation*). To discuss and examine this further, consider a particle P, free to move in space as depicted in Figure 5.2.3. Then, in a Cartesian coordinate system, P may be located by the coordinates (x, y, z) defining the distances of P from the coordinate planes. Alternatively, P may be located by the position vector $\mathbf{p}$ expressed as:

$$\mathbf{p} = x\mathbf{n}_x + y\mathbf{n}_y + z\mathbf{n}_z \tag{5.2.3}$$

where $\mathbf{n}_x$, $\mathbf{n}_y$, and $\mathbf{n}_z$ are unit vectors parallel to the coordinate axes as in Figure 5.2.3.

Constraints on the movement of P are often expressed in terms of the position vector $\mathbf{p}$. For example, if P is restricted to planar motion, say in the X–Y plane, then this restriction (or constraint) may be expressed as:

$$\mathbf{p} \cdot \mathbf{n}_z = 0 \quad \text{or as} \quad z = 0 \tag{5.2.4}$$

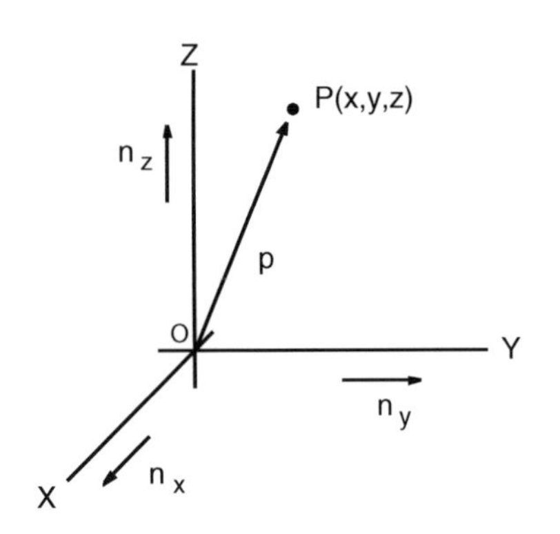

FIGURE 5.2.3
A particle P moving in space.

If *two* particles are moving freely in three dimensions, then each particle will have three degrees of freedom so that the *system* of two particles will have *six* degrees of freedom. Thus, for a system of N freely moving particles (in three dimensions), there will be $3N$ degrees of freedom.

Suppose a system of two particles, moving in three dimensions, is restricted by the requirement that the particles remain a fixed distance d from each other. Let the particles be called P_1 and P_2 and let them have coordinates (x_1, y_1, z_1) and (x_2, y_2, z_2) and position vectors $\mathbf{p}_1$ and $\mathbf{p}_2$ as in Figure 5.2.4. Then, the fixed-distance constraint may be expressed as:

$$|\mathbf{p}_1 - \mathbf{p}_2| = d \quad \text{or} \quad (\mathbf{p}_1 - \mathbf{p}_2)^2 = d^2$$

or as:

$$(x_1 - x_2)^2 + (y_1 - y_2)^2 + (z_1 - z_2)^2 = d^2 \tag{5.2.5}$$

Because this constraint may be expressed by a single equation (any one of Eqs. (5.2.5)), the system of particles has (6 – 1), or five, degrees of freedom.

A rigid body may be considered to be a system of particles whose distances are fixed relative to each other — such as a sandstone (see Figure 5.2.5). Suppose a body is considered to have N particles p_i ($i = 1,\ldots, N$). Then, the fixed distances of the first three particles relative to each other may be expressed by the equations:

$$(\mathbf{p}_1 - \mathbf{p}_2)^2 = a^2, \; (\mathbf{p}_2 - \mathbf{p}_3)^2 = b^2, \; (\mathbf{p}_3 - \mathbf{p}_1)^2 = c^2 \tag{5.2.6}$$

where a, b, and c are constants and $\mathbf{p}_i$ locates P_i relative to a fixed point (or origin) O. If P_1, P_2, and P_3 are not colinear, the remaining particles P_i ($i = 4,\ldots, N$) will remain a fixed distance from P_1, P_2, and P_3 and from each other if the following expressions are satisfied:

$$(\mathbf{p}_i - \mathbf{p}_1)^2 = a_i^2, \; (\mathbf{p}_i - \mathbf{p}_2)^2 = b_i^2, \; (\mathbf{p}_i - \mathbf{p}_3)^2 = c_i^2 \; (i = 4,\ldots,N) \tag{5.2.7}$$

where a_i, b_i, and c_i are constants.

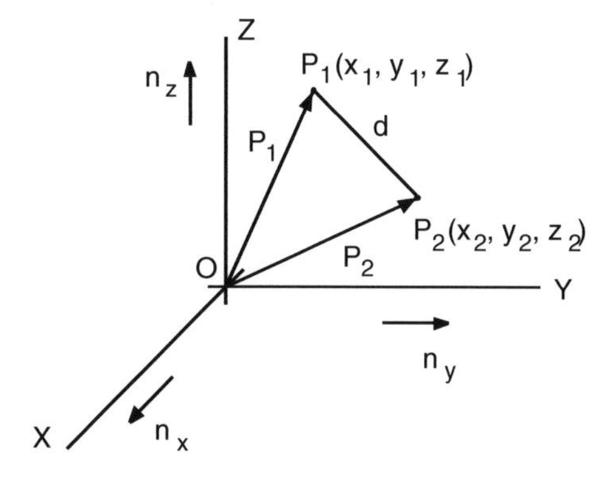

FIGURE 5.2.4
Two particles separated by a fixed distance.

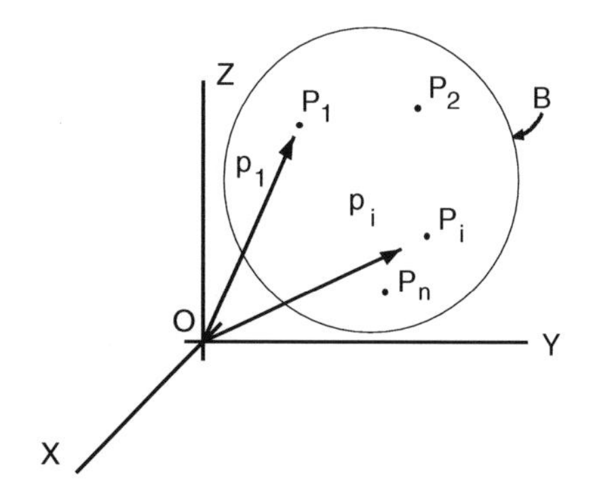

FIGURE 5.2.5
A rigid body considered a set of particles.

Equations (5.2.6) and (5.2.7) form a system of (3 + 3(N – 3)) or (3N – 6) constraint equations. Because N unrestrained particles (in three dimensions) require 6N coordinates to define their locations, the (3N – 6) constraint equations reduce the number of degrees of freedom for the rigid body to 6N – (3N – 6), or six.

In like manner it is seen that, if a rigid body is restricted to *planar movement*, it has three degrees of freedom.

5.3 Planar Motion of a Rigid Body

When a body has planar motion, each particle of the body moves in a plane; however, all of the particles do not move in the same plane. Instead, they move in planes that are parallel to each other, as depicted in Figure 5.3.1. In the figure, P_1 and P_2 are typical particles of a body B. They move in parallel planes π_1 and π_2. If the particles happen to be on the same normal line N of the planes, they have identical motions. This means that if we consider the movements of the particles of B in one of the planes (parallel to the planes of motion), we are in effect considering the motion of *all* the particles of B. That is, any particle of B not in our considered plane of motion can be identified with a particle in that plane. Hence, the motion of B can be studied entirely in a plane.

Because a body with planar motion has at most three degrees of freedom, the kinematic analysis is greatly simplified from that of general three-dimensional motion. Many of the kinematical quantities are then more conveniently described with scalars than with vectors. For example, with planar motion the angular velocity of a body is always directed normal to the plane of motion. Hence, a vector is not needed to define its direction.

To demonstrate this, consider the definition of angular velocity (Eq. (4.5.1)):

$$\boldsymbol{\omega} = \left[(d\mathbf{n}_2/dt)\cdot\mathbf{n}_3\right]\mathbf{n}_1 + \left[(d\mathbf{n}_3/dt)\cdot\mathbf{n}_1\right]\mathbf{n}_2 + \left[(d\mathbf{n}_1/dt)\cdot\mathbf{n}_2\right]\mathbf{n}_3 \tag{5.3.1}$$

where $\mathbf{n}_1$, $\mathbf{n}_2$, and $\mathbf{n}_3$ are mutually perpendicular unit vectors fixed in the body. Let $\mathbf{n}_3$ be parallel to line N and thus perpendicular to the plane of motion of the body B (see Figure 5.3.1). Let particles P_1 and P_2 of B lie on N, a distance d apart. Because the orientation of N is fixed, $\mathbf{n}_3$ has fixed orientation and is thus constant. Hence, $d\mathbf{n}_3/dt$ is zero. Also, because $\mathbf{n}_1$ and $\mathbf{n}_2$ are perpendicular to $\mathbf{n}_3$ they remain parallel to the planes of motion of B. Their

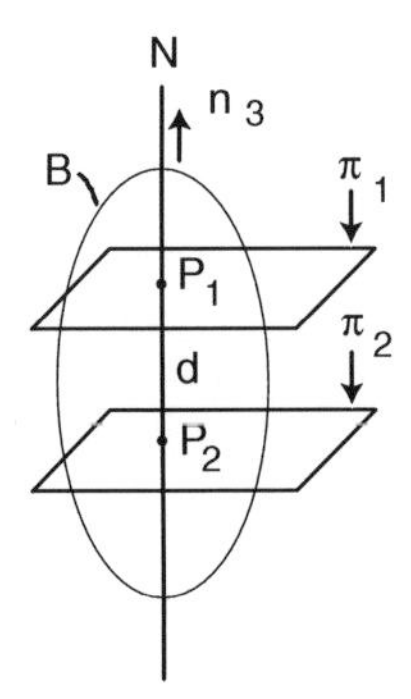

FIGURE 5.3.1
A body with planar motion.

derivatives are thus parallel to the motion planes and therefore perpendicular to $\mathbf{n}_3$ (see Section 4.5). Hence, $(d\mathbf{n}_2/dt) \cdot \mathbf{n}_3$ is zero. Equation (5.3.1) thus becomes:

$$\boldsymbol{\omega} = 0\mathbf{n}_1 + 0\mathbf{n}_2 + \left[(d\mathbf{n}_1/dt)\cdot \mathbf{n}_2\right]\mathbf{n}_3 \overset{D}{=} \omega \mathbf{n}_3 \tag{5.3.2}$$

where by inspection the angular speed ω is defined as $(d\mathbf{n}_1/dt) \bullet \mathbf{n}_2$.

Equation (5.3.2) shows that the angular velocity ω of B is always parallel to $\mathbf{n}_3$, a fixed unit vector. Therefore, $\boldsymbol{\omega}$ is characterized by the angular speed ω. Hence, in planar motion, ω is sometimes called the *angular velocity* of B.

By differentiating in Eq. (5.3.2), the angular acceleration α of B takes the simple form:

$$\boldsymbol{\alpha} = d\boldsymbol{\omega}/dt = (d\omega/dt)\mathbf{n}_3 \overset{D}{=} \alpha \mathbf{n}_3 \tag{5.3.3}$$

where the scalar α is called the *angular acceleration* of B and is often written as $\dot{\omega}$.

Because the angular velocity and angular acceleration of a body in planar motion are defined by the two scalars ω and α, respectively, the objectives of kinematic analyses are often reduced to finding expressions for velocities and accelerations of points on the body. In such analyses, it is often convenient to categorize the motion of the body as being translation, rotation, or general motion.

Translation occurs when *all* particles of the body have equal velocities. If, in addition to this, all the particles have straight-line motion, the body is said to have *rectilinear motion.*

Rotation occurs when the particles of the body move in concentric circles (this is sometimes called *pure rotation*).

General plane motion occurs when a body has planar motion that is neither translation nor rotation. General plane motion may be considered as a superposition of translation and rotation. (Translation and rotation are thus special cases of general plane motion.)

5.3.1 Translation

The kinematic analysis of a body in translation is relatively simple. The angular velocity ω and the angular acceleration α are zero:

$$\omega = \alpha = 0 \tag{5.3.4}$$

If P is a typical particle of the body B with velocity $\mathbf{v}^P$, and if Q is any other particle, then:

$$\mathbf{v}^Q = \mathbf{v}^P \tag{5.3.5}$$

Then, by differentiation, we have:

$$\mathbf{a}^Q = \mathbf{a}^P \tag{5.3.6}$$

If the translating body also has rectilinear motion, the direction of the velocity and acceleration of the particles is constant. The particle velocity and acceleration can then be defined in terms of scalars v and a as:

$$\mathbf{v}^P = \mathbf{v}^Q = v\mathbf{n} \quad \text{and} \quad \mathbf{a}^P = \mathbf{a}^Q = a\mathbf{n} \tag{5.3.7}$$

where **n** is a unit vector parallel to the line of motion.

5.3.2 Rotation

If a body is in pure rotation, with its particles moving in circles, the kinematics can be developed using the procedures of Section 3.7 describing the motion of points on circles. To illustrate this, consider the body B depicted in Figure 5.3.2. Let P be a typical particle of B. Let P move on a circle with radius r_P and center O as shown. Then, O has zero velocity (otherwise, P would not move in a circle). From Eqs. (3.7.6) and (3.7.7), the velocity and acceleration of P may then be expressed as:

$$\mathbf{v}^P = r_p\omega\mathbf{n}_\theta \quad \text{and} \quad \mathbf{a}^P = r_p\alpha\mathbf{n}_\theta - r_p\omega^2\mathbf{n}_r \tag{5.3.8}$$

where ω and α represent the angular velocity and angular acceleration, respectively, of B and where $\mathbf{n}_r$ and $\mathbf{n}_\theta$ are radial and tangential unit vectors, respectively, as shown in Figure 5.3.2.

Let Q be any other point of B. If Q is located a distance r_Q from O, then the velocity and acceleration of Q can be expressed as:

$$\mathbf{v}_Q = r_Q\omega\mathbf{n}_\theta \quad \text{and} \quad \mathbf{a}_Q = r_Q\alpha\mathbf{n}_\theta - r_Q\omega^2\mathbf{n}_r \tag{5.3.9}$$

By comparing Eqs. (5.3.8) and (5.3.9), we see that the magnitudes of the velocities and accelerations are directly proportional to the radii of the circles on which the particles move. Also, we see that if we know the velocity and acceleration of one point, say P, we can find the angular velocity and angular acceleration of the body and, hence, the velocity and acceleration of any and all other points.

5.3.3 General Plane Motion

General plane motion may be considered as a superposition of translation and rotation. The velocity and acceleration of a typical particle P of a body in general plane motion may be obtained from Eqs. (4.9.4) and (4.9.6). Specifically, if Q is a particle of B in the plane of motion of P, the velocity and acceleration of P may be expressed as:

$$\mathbf{v}^P = \mathbf{v}^Q + \boldsymbol{\omega} + \mathbf{r} \tag{5.3.10}$$

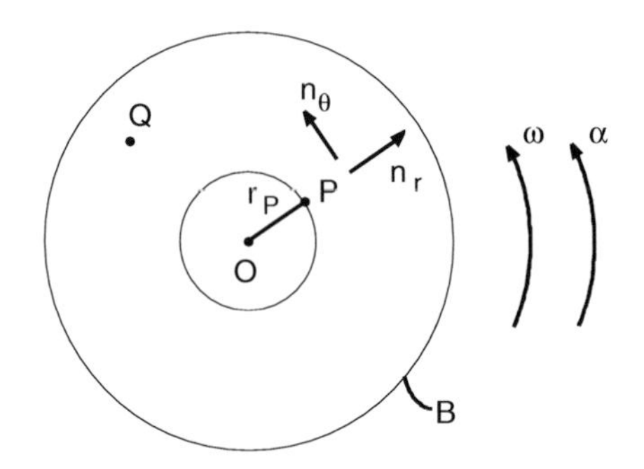

FIGURE 5.3.2
A body B in pure rotation.

and

$$\mathbf{a}^P = \mathbf{a}^Q + \boldsymbol{\alpha} \times \mathbf{r} + \boldsymbol{\omega} \times (\boldsymbol{\omega} \times \mathbf{r}) \tag{5.3.11}$$

Example 5.3.1: Motion of a Piston, Connecting Rod, Crank Arm

To illustrate these concepts, consider the piston, connecting rod, and crank arm shown in Figure 5.3.3. The crank arm OQ is pinned at O and thus has pure rotation about O. The motion of the piston P is translation. The connecting rod QP has general plane motion.

Let the length of the crank arm and the connecting rod be r and ℓ and let the angles that they make with the horizontal (the piston/cylinder axis) be θ and ϕ as shown in Figure 5.3.3. Let the crank turn at a uniform angular speed Ω. Determine the velocity and acceleration of the piston P.

Solution: Let unit vectors $\mathbf{n}_r$, $\mathbf{n}_\theta$, $\mathbf{n}_x$, and $\mathbf{n}_y$ be introduced parallel and perpendicular to the crank arm and horizontal and vertical as in Figure 5.3.4. From Eq. (5.3.9), the velocity and acceleration of Q are:

$$\mathbf{v}_Q = r\Omega\mathbf{n}_\theta \quad \text{and} \quad \mathbf{a}_Q = -r\Omega^2\mathbf{n}_r \tag{5.3.12}$$

where Ω is $\dot{\theta}$. In terms of $\mathbf{n}_x$ and $\mathbf{n}_y$, $\mathbf{v}_Q$ and $\mathbf{a}_Q$ are:

$$\mathbf{v}_Q = r\Omega\sin\theta\,\mathbf{n}_x + r\Omega\cos\theta\,\mathbf{n}_y \tag{5.3.13}$$

and

$$\mathbf{a}_Q = -r\Omega^2\cos\theta\,\mathbf{n}_x - r\Omega^2\sin\theta\,\mathbf{n}_y \tag{5.3.14}$$

From Eqs. (5.3.10) and (5.3.11), the velocity and acceleration of P are:

$$\mathbf{v}_P = \mathbf{v}_Q + \boldsymbol{\omega}_{QP} \times \mathbf{QP} \tag{5.3.15}$$

and

$$\mathbf{a}_P = \mathbf{a}_Q + \boldsymbol{\alpha}_{QP} \times \mathbf{QP} + \boldsymbol{\omega}_{QP} \times \left(\boldsymbol{\omega}_{QP} \times \mathbf{QP}\right) \tag{5.3.16}$$

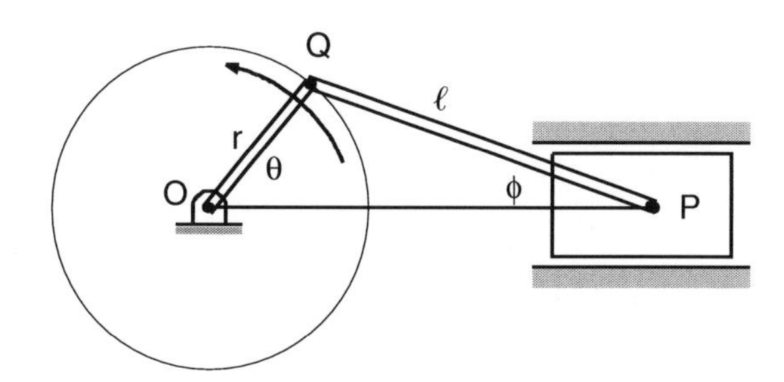

FIGURE 5.3.3
A piston, connecting rod, and crank arm.

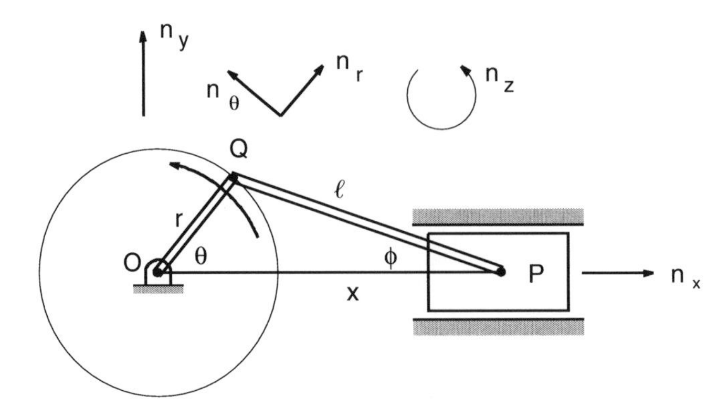

FIGURE 5.3.4
Geometrical parameters and unit vectors of the piston, connecting rod, and crank arm system.

where $\boldsymbol{\omega}_{QP}$ and $\boldsymbol{\alpha}_{QP}$ are the angular velocity and angular acceleration, respectively, of the connecting rod QP. Because QP has planar motion, we see from Figure 5.3.4 that $\boldsymbol{\omega}_{QP}$ and $\boldsymbol{\alpha}_{QP}$ may be expressed as:

$$\boldsymbol{\omega}_{QP} = -\dot{\phi}\mathbf{n}_z \quad \text{and} \quad \boldsymbol{\alpha}_{QP} = -\ddot{\phi}\mathbf{n}_z \tag{5.3.17}$$

where $\mathbf{n}_z$ ($= \mathbf{n}_x \times \mathbf{n}_y$) is perpendicular to the plane of motion. Also, the position vector $\mathbf{QP}$ may be expressed as:

$$\mathbf{QP} = \ell \cos\phi\, \mathbf{n}_x - \ell \sin\phi\, \mathbf{n}_y \tag{5.3.18}$$

By carrying out the indicated operations of Eqs. (5.3.15) and (5.3.16) and by using Eqs. (5.3.13) and (5.3.14), $\mathbf{v}_P$ and $\mathbf{a}_P$ become:

$$\mathbf{v}_P = \left(-r\Omega \sin\theta - \ell\dot{\phi}\sin\phi\right)\mathbf{n}_x + \left(r\Omega\cos\theta - \ell\dot{\phi}\cos\right)\mathbf{n}_y \tag{5.3.19}$$

and

$$\begin{aligned} \mathbf{a}_P = &\left(-r\Omega^2 \cos\theta - \ell\ddot{\phi}\sin\phi - \ell\dot{\phi}^2\cos\phi\right)\mathbf{n}_x \\ &+\left(-r\Omega^2 \sin\theta - \ell\ddot{\phi}\cos\phi + \ell\dot{\phi}^2\sin\phi\right)\mathbf{n}_y \end{aligned} \tag{5.3.20}$$

Observe from Figure 5.3.4 that P moves in translation in the $\mathbf{n}_x$ direction. Therefore, the velocity and acceleration of P may be expressed simply as:

$$\mathbf{v}_P = \dot{x}\mathbf{n}_x \quad \text{and} \quad \mathbf{a}_P = \ddot{x}\mathbf{n}_x \tag{5.3.21}$$

where x is the distance OP. By comparing Eqs. (5.3.19) and (5.3.20) with (5.3.21), we have:

$$\dot{x} = -r\Omega\sin\theta - \ell\dot{\phi}\sin\phi \tag{5.3.22}$$

$$0 = r\Omega\cos\theta - \ell\dot{\phi}\cos\phi \tag{5.3.23}$$

$$\ddot{x} = -r\Omega^2\cos\theta - \ell\ddot{\phi}\sin\phi - \ell\dot{\phi}^2\cos\phi \tag{5.3.24}$$

and

$$0 = -r\Omega^2\sin\theta - \ell\ddot{\phi}\cos\phi + \ell\dot{\phi}^2\sin\phi \tag{5.3.25}$$

Observe further from Figure 5.3.4 that x may be expressed as:

$$x = r\cos\theta + \ell\cos\phi \tag{5.3.26}$$

and that from the law of sines we have:

$$\frac{\ell}{\sin\theta} = \frac{r}{\sin\phi} \tag{5.3.27}$$

By differentiating in Eqs. (5.3.26) and (5.3.27), we immediately obtain Eqs. (5.3.22) and (5.3.23). Finally, observe that by differentiating in Eqs. (5.3.22) and (5.3.23) we obtain Eqs. (5.3.24) and (5.3.25).

5.4 Instant Center, Points of Zero Velocity

If a point O of a body B with planar motion has zero velocity, then O is called a *center of zero velocity.* If O has zero velocity throughout the motion of B, it is called a *permanent center of zero velocity.* If O has zero velocity during only a part of the motion of B, or even for only an instant during the motion of B, then O is called an *instant center of zero velocity.*

For example, if a body B undergoes pure rotation, then points on the axis of rotation are permanent centers of zero velocity. If, however, B is in translation, then there are no points of B with zero velocity. If B has general plane motion, there may or may not be points of B with zero velocity. However, as we will see, if B has no centers of zero velocity within itself, it is always possible to mathematically extend B to include such points. In this latter context, bodies in translation are seen to have centers of zero velocity at infinity (that is, infinitely far away).

The advantage, or utility, of knowing the location of a center of zero velocity can be seen from Eq. (5.3.10):

$$\mathbf{v}^P = \mathbf{v}^Q + \boldsymbol{\omega}\times\mathbf{r} \tag{5.4.1}$$

where P and Q are points of a body B and where $\mathbf{r}$ locates P relative to Q. If Q is a center of zero velocity, then $\mathbf{v}^Q$ is zero and $\mathbf{v}^P$ is simply:

$$\mathbf{v}^P = \boldsymbol{\omega}\times\mathbf{r} \tag{5.4.2}$$

This means that during the time that Q is a center of zero velocity, P moves in a circle about Q. Indeed, during the time that Q is a center of zero velocity, B may be considered to be in pure rotation about Q. Finally, observe in Eq. (5.4.2) that ω is normal to the plane of motion. Hence, we have:

$$|\mathbf{v}^P| = |\boldsymbol{\omega}|\,|\mathbf{r}| \quad \text{or} \quad v^P = r^P\omega \tag{5.4.3}$$

where the notation is defined by inspection.

If a body B is at rest, then every point of B is a center of zero velocity. If B is not at rest but has planar motion, then at most one point of B, in a given plane of motion, is a center of zero velocity.

To prove this last assertion, suppose B has two particles, say O and Q, with zero velocity. Then, from Eq. (5.3.10), we have:

$$\mathbf{v}^O = \mathbf{v}^Q + \boldsymbol{\omega} \times \mathbf{r} \tag{5.4.4}$$

where $\mathbf{r}$ locates O relative to Q. If both O and Q have zero velocity then,

$$\boldsymbol{\omega} \times \mathbf{r} = 0 \tag{5.4.5}$$

If O and Q are distinct particles, then $\mathbf{r}$ is not zero. Because $\boldsymbol{\omega}$ is perpendicular to $\mathbf{r}$, Eq. (5.4.5) is then satisfied only if $\boldsymbol{\omega}$ is zero. The body is then in translation and all points have the same velocity. Therefore, because both O and Q are to have zero velocity, all points have zero velocity, and the body is at rest — a contradiction to the assumption of a moving body. That is, the only way that a body can have more than one center of zero velocity, in a plane of motion, is when the body is at rest.

We can demonstrate these concepts by graphical construction. That is, we can show that if a body has planar motion, then there exists a particle of the body (or of the mathematical extension of the body) with zero velocity. We can also develop a graphical procedure for finding the point.

To this end, consider the body B in general plane motion as depicted in Figure 5.4.1. Let P and Q be two typical distinct particles of B. Then, if B has a center O of zero velocity, P and Q may be considered to be moving in a circle about O. Suppose the velocities of P and Q are represented by vectors, or line segments, as in Figure 5.4.2. Then, if P and Q rotate about a center O of zero velocity, the velocity vectors of P and Q will be perpendicular to lines through O and P and Q, as in Figure 5.4.3.

Observe that unless the velocities of P and Q are parallel, the lines through P and Q perpendicular to these velocities will always intersect. Hence, with non-parallel velocities, the center O of zero velocity always exists.

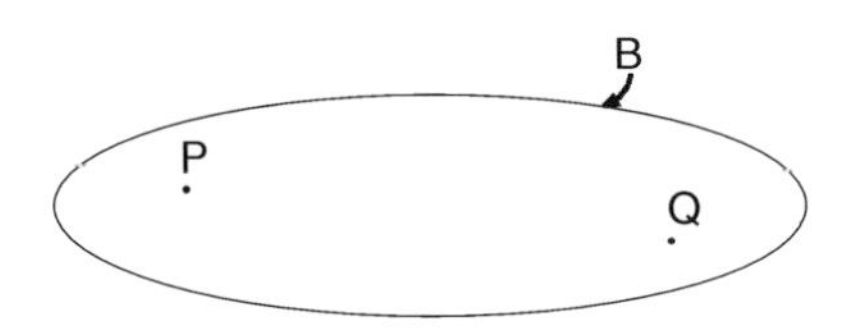

FIGURE 5.4.1
A body B in general plane motion.

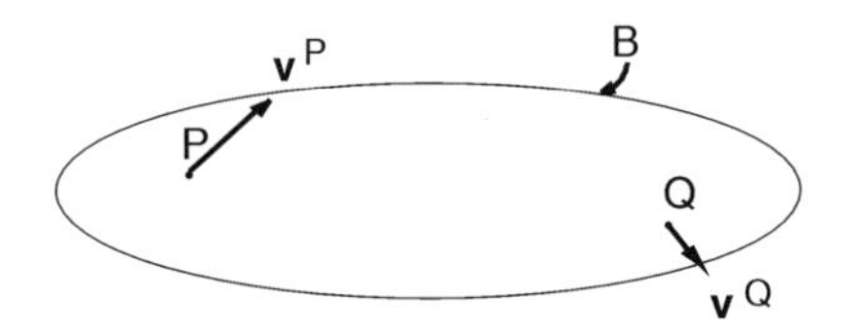

FIGURE 5.4.2
Vector representations of the velocities of particles P and Q.

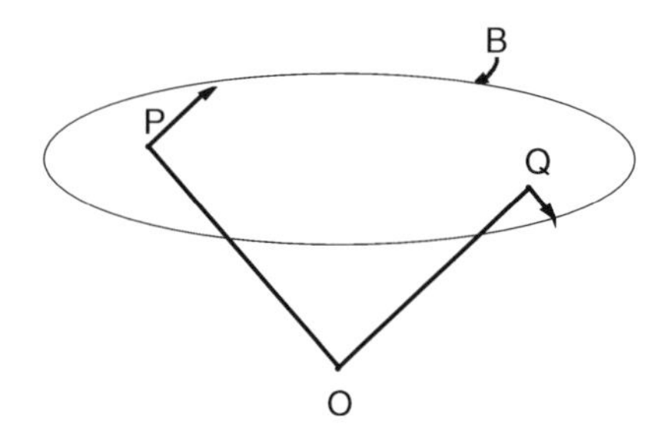

FIGURE 5.4.3
Location of center O of zero velocity by the intersection of lines perpendicular to velocity vectors.

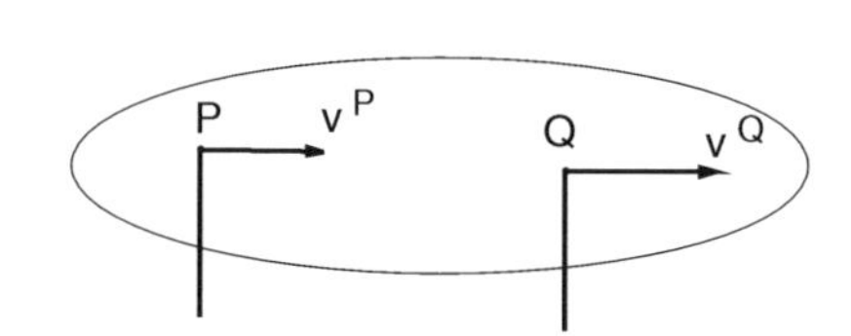

FIGURE 5.4.4
A body in translation with equal velocity particles and center of zero velocity infinitely far away.

If the velocities of P and Q are parallel with equal magnitudes, and the same sense, then they are equal. That is,

$$\mathbf{v}^P = \mathbf{v}^Q \quad \text{and} \quad \boldsymbol{\omega} \times \mathbf{r} = 0 \tag{5.4.6}$$

where the last equality follows from Eq. (5.3.10) with $\mathbf{r}$ locating P relative to Q. If P and Q are distinct, $\mathbf{r}$ is not zero; hence, $\boldsymbol{\omega}$ is zero. B is then in translation. Lines through P and Q perpendicular to $\mathbf{v}^P$ and $\mathbf{v}^Q$ will then be parallel to each other and thus not intersect (except at infinity). That is, the center of zero velocity is infinitely far away (see Figure 5.4.4).

If the velocities of P and Q are parallel with non-equal magnitudes, then the center of zero velocity will occur on the line connecting P and Q. To see this, first observe that the relative velocities of P and Q will have zero projection along the line connecting P and Q: That is, from Eq. (5.3.10), we have:

$$\mathbf{v}^P = \mathbf{v}^Q + \boldsymbol{\omega} \times \mathbf{r} \quad \text{or} \quad \mathbf{v}^{P/Q} = \boldsymbol{\omega} \times \mathbf{r} \tag{5.4.7}$$

Hence, $\mathbf{v}^{P/Q}$ must be perpendicular to $\mathbf{r}$. (This simply means that P and Q cannot approach or depart from each other; otherwise, the rigidity of B would be violated.) Next, observe that if $\mathbf{v}^P$ and $\mathbf{v}^Q$ are parallel, their directions may be defined by a common unit vector $\mathbf{n}$. That is,

$$\mathbf{v}^P = \mathbf{v}^P \mathbf{n}, \quad \mathbf{v}^Q = v^Q \mathbf{n}, \quad \text{and} \quad \mathbf{v}^{P/Q} = v^{P/Q} \mathbf{n} \tag{5.4.8}$$

where $\mathbf{v}^P$, $\mathbf{v}^Q$, and $\mathbf{v}^{P/Q}$ are appropriate scalars. By comparing Eqs. (5.4.7) and (5.4.8) we see that $\mathbf{n}$ must be perpendicular to $\mathbf{r}$. Hence, when $\mathbf{v}^P$ and $\mathbf{v}^Q$ are parallel but with non-equal magnitudes, their directions must be perpendicular to the line connecting P and Q. Therefore, lines through P and Q and perpendicular to $\mathbf{v}^P$ and $\mathbf{v}^Q$ will coincide with each other and with the line connecting P and Q (see Figure 5.4.5).

Next, observe from Eqs. (5.3.10) and (5.4.3) that, if O is the center of zero velocity, then the magnitude of $\mathbf{v}^P$ is proportional to the distance between O and P. Similarly, the magnitude of $\mathbf{v}^Q$ is proportional to the distance between O and Q. These observations enable us to locate O on the line connecting P and Q. Specifically, from Eq. (5.4.3), the distance from P to O is simply $|\mathbf{v}^P|/\omega$.

From a graphical perspective, O can be located as in Figure 5.4.6. Similar triangles are formed by O, P, Q and the "arrow ends" of $\mathbf{v}^P$ and $\mathbf{v}^Q$.

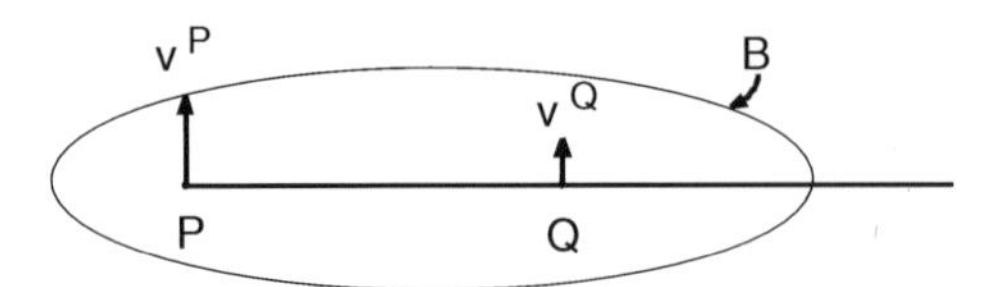

FIGURE 5.4.5
A body B with particles P and Q having parallel velocities with non-equal magnitudes.

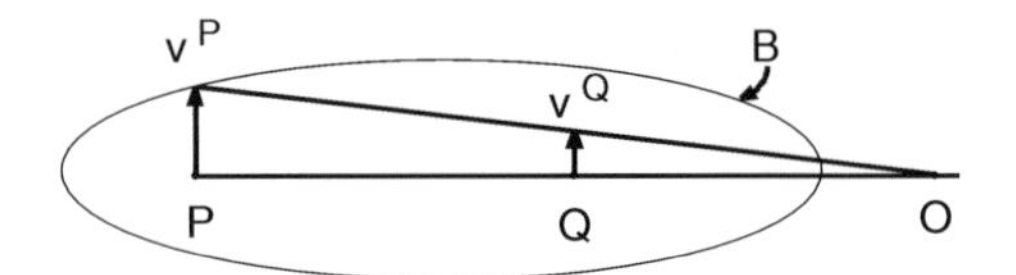

FIGURE 5.4.6
Location of the center of zero velocity for a body having distinct particles with parallel but unequal velocities.

To summarize, we see that if a body has planar motion, there exists a unique point O of the body (or the body extended) that has zero velocity. O may be located at the intersection of lines through two points that are perpendicular to the velocity vectors of the points. Alternatively, O may be located on a line perpendicular to the velocity vector of a single point P of the body at a distance $|\mathbf{v}^P|/\omega$ from P (see Figure 5.4.7). Finally, when the zero velocity center O is located, the body may be considered to be rotating about O. Then, the velocity of any point P of the body is proportional to the distance from O to P and is directed parallel to the plane of motion of B and perpendicular to the line segment OP.

5.5 Illustrative Example: A Four-Bar Linkage

Consider the planar linkage shown in Figure 5.5.1. It consists of three links, or bars (B_1, B_2, and B_3), and four joints (O, P, Q, and R). Joints O and R are fixed while joints P and Q may move in the plane of the linkage. The ends of each bar are connected to a joint; thus, the bars may be identified (or labeled) by their joint ends. That is, B_1 is OP, B_2 is PQ, B_3 is QR. In this context, we may also imagine a fourth bar B_4 connecting the fixed joints O and R. The system then has four bars and is thus referred to as a four-bar linkage.

The four-bar linkage may be used to model many physical systems employed in mechanisms and machines, particularly cranks and connecting rods. The four-bar linkage is thus an excellent practical example for illustrating the concepts of the foregoing sections. In this context, observe that bars $OP(B_1)$ and $RQ(B_3)$ undergo pure rotation, while bar $PQ(B_2)$ undergoes general plane motion, and bar $OR(B_4)$ is fixed, or at rest.

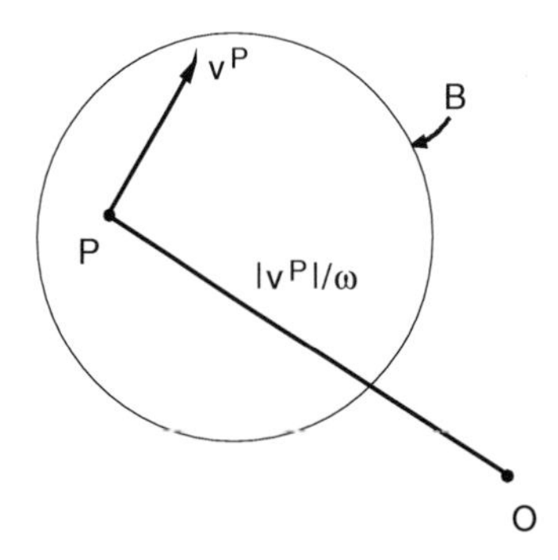

FIGURE 5.4.7
Location of the center for zero velocity knowing the velocity of one particle and the angular speed.

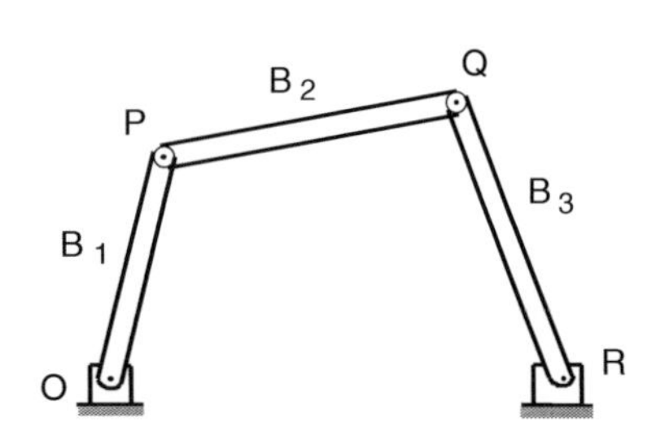

FIGURE 5.5.1
A four-bar linkage.

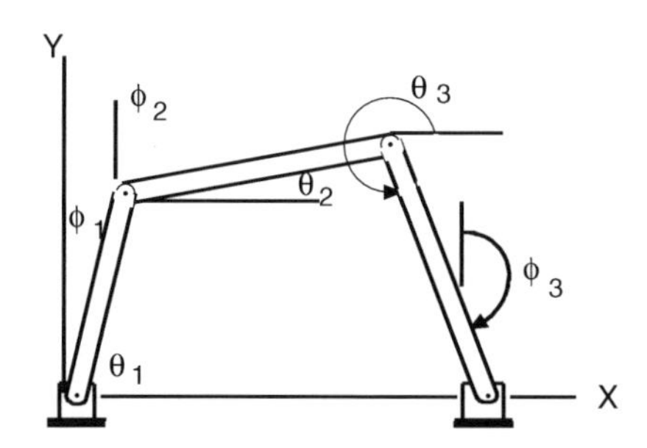

FIGURE 5.5.2
Absolute orientation angles.

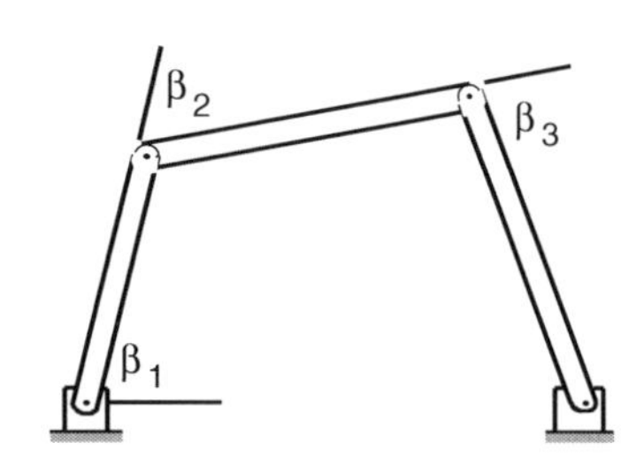

FIGURE 5.5.3
Relative orientation angles.

The system of Figure 5.5.1 has one degree of freedom: The rotations of bars $OP(B_1)$ and $RQ(B_3)$ each require two coordinates, and the general motion of bar $PQ(B_2)$ requires an additional three coordinates for a total of *five* coordinates. Nevertheless, requiring joint P to be connected to both B_1 and B_2 and joint Q to be connected to both B_2 and B_3 produces *four* position (or coordinate) constraints. Thus, there is but *one* degree of freedom.

A task encountered in the kinematic analyses of linkages is that of describing the orientation of the individual bars. One method is to define the orientations of the bars in terms of angles that the bars make with the horizontal (or X-axis) such as θ_1, θ_2, and θ_3 as in Figure 5.5.2. Another method is to define the orientation in terms of angles that the bars make with the vertical (or Y-axis) such as ϕ_1, ϕ_2, and ϕ_3 as in Figure 5.5.2. A third method is to define the orientations in terms of angles that the bars make with each other, as in Figure 5.5.3. The latter angles are generally called *relative orientation angles* whereas the former are called *absolute orientation angles.*

Relative orientation angles are usually more meaningful in describing the configuration of a physical system. Absolute orientation angles are usually easier to work with in the analysis of the problem. In our example, we will use the first set of absolute angles θ_1, θ_2, and θ_3.

Because the system has only one degree of freedom, the orientation angles are not independent. They may be related to each other by constraint equations obtained by considering the linkage of four bars as a closed loop: Specifically, consider the position vector equation:

$$\mathbf{OP} + \mathbf{PQ} + \mathbf{QR} + \mathbf{RO} = 0 \tag{5.5.1}$$

This equation locates O relative to itself through position vectors taken around the loop of the mechanism. It is called the *loop closure equation.*

Let ℓ_1, ℓ_2, ℓ_3, and ℓ_4 be the lengths of bars B_1, B_2, B_3, and B_4. Then, Eq. (5.5.1) may be written as:

$$\ell_1\boldsymbol{\lambda}_1 + \ell_2\boldsymbol{\lambda}_2 + \ell_3\boldsymbol{\lambda}_3 + \ell_4\boldsymbol{\lambda}_4 = 0 \tag{5.5.2}$$

where $\boldsymbol{\lambda}_1$, $\boldsymbol{\lambda}_2$, $\boldsymbol{\lambda}_3$, and $\boldsymbol{\lambda}_4$ are unit vectors parallel to the rods as shown in Figure 5.5.4. These vectors may be expressed in terms of horizontal and vertical unit vectors $\mathbf{n}_x$ and $\mathbf{n}_y$ as:

$$\begin{aligned} \boldsymbol{\lambda}_1 &= \cos\theta_1\mathbf{n}_x + \sin\theta_1\mathbf{n}_y, \quad & \boldsymbol{\lambda}_2 &= \cos\theta_2\mathbf{n}_x + \sin\theta_2\mathbf{n}_y \\ \boldsymbol{\lambda}_3 &= \cos\theta_3\mathbf{n}_x - \sin\theta_3\mathbf{n}_4, \quad & \boldsymbol{\lambda}_4 &= -\mathbf{n}_x \end{aligned} \tag{5.5.3}$$

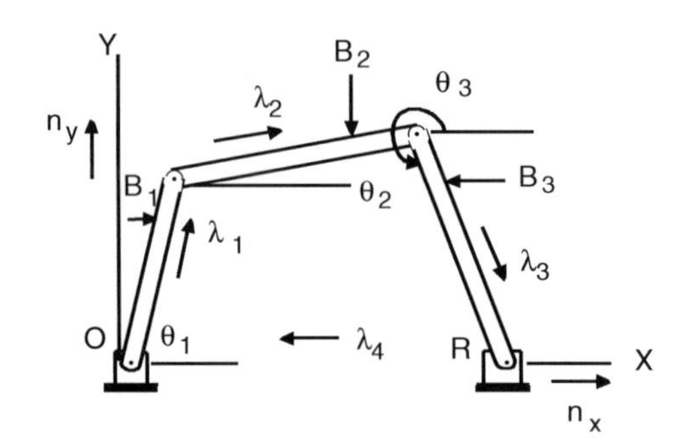

FIGURE 5.5.4
Linkage geometry and unit vectors.

Hence, by substituting into Eq. (5.5.2) we have:

$$\begin{aligned}&\left(\ell_1\cos\theta_1+\ell_2\cos\theta_2+\ell_3\cos\theta_3-\ell_4\right)\mathbf{n}_x\\&+\left(\ell_1\sin\theta_1+\ell_2\sin\theta_2+\ell_3\sin\theta_3\right)\mathbf{n}_y=0\end{aligned} \tag{5.5.4}$$

This immediately leads to two scalar constraint equations relating θ_1, θ_2, and θ_3:

$$\ell_1\cos\theta_1+\ell_2\cos\theta_2+\ell_3\cos\theta_3=\ell_4 \tag{5.5.5}$$

and

$$\ell_1\sin\theta_1+\ell_2\sin\theta_2+\ell_3\sin\theta_3=0 \tag{5.5.6}$$

The objective in a kinematic analysis of a four-bar linkage is to determine the velocity and acceleration of the various points of the linkage and to determine the angular velocities and angular accelerations of the bars of the linkage. In such an analysis, the motion of one of the three moving bars, say B_1, is generally given. The objective is then to determine the motion of bars B_1 and B_2. In this case, B_1 is the *driver*, and B_2 and B_3 are *followers*.

The procedures of Section 5.4 may be used to meet these objectives. To illustrate the details, consider the specific linkage shown in Figure 5.5.5. The bar lengths and orientations are given in the figure. Also given in Figure 5.5.5 are the angular velocity and angular acceleration of $OP(B_1)$. B_1 is thus a driver bar and $PQ(B_2)$ and $QR(B_3)$ are follower bars. Our objective, then, is to find the angular velocities and angular accelerations of B_2 and B_3 and the velocity and acceleration of P and Q.

To begin the analysis, first observe that, by comparing Figures 5.5.4 and 5.5.5, the angles and lengths of Figure 5.5.5 satisfy Eqs. (5.5.5) and (5.5.6). To see this, observe that ℓ_1, ℓ_2, ℓ_3, ℓ_4, θ_1, θ_2, θ_3, and θ_4 have the values:

$$\begin{aligned}&\ell_1=2.0\text{ m},\ \ \ell_2=3.0\text{ m},\ \ \ell_3=4.95\text{ m},\ \ \ell_4=6.098\text{ m}\\&\theta_1=90°,\ \ \theta_2=30°,\ \ \theta_3=315°\ (\text{or } -45°)\end{aligned} \tag{5.5.7}$$

Then Eqs. (5.5.5) and (5.5.6) become:

$$(2.0)\cos 90+(3.0)\cos 30+4.95\cos 315=6.098 \tag{5.5.8}$$

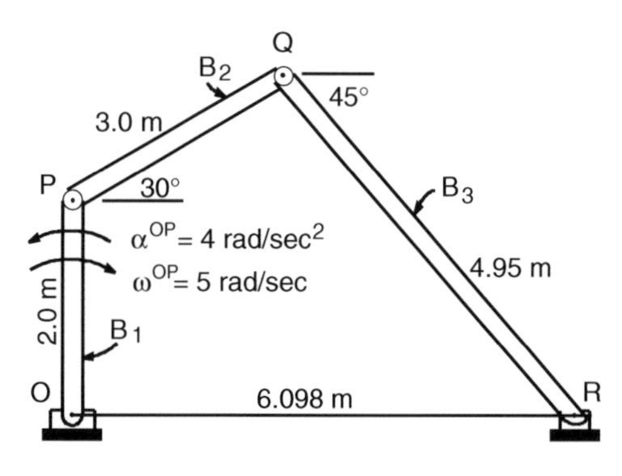

FIGURE 5.5.5
Example four-bar linkage.

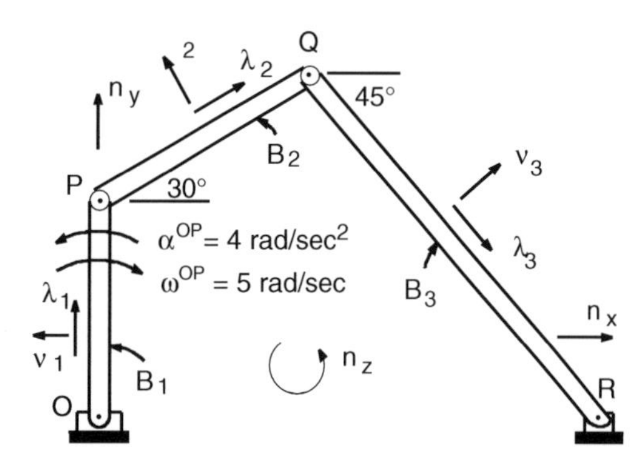

FIGURE 5.5.6
Unit vectors for the analysis of the linkage of Figure 5.5.5.

and

$$(2.0)\sin 90 + (3.0)\sin 30 + 4.95\sin 315 = 0 \tag{5.5.9}$$

Next, recall that B_1 and B_3 have pure rotation about points O and R, respectively, and that B_2 has general plane motion.

Third, let us introduce unit vectors $\boldsymbol{\lambda}_i$ and $\mathbf{v}_i$ (i = 1, 2, 3) parallel and perpendicular to the bars as in Figure 5.5.6. Then, in the configuration shown, the $\boldsymbol{\lambda}_i$ and $\mathbf{v}_i$ may be expressed in terms of horizontal and vertical unit vectors $\mathbf{n}_x$ and $\mathbf{n}_y$ as:

$$\boldsymbol{\lambda}_1 = \mathbf{n}_y \quad \text{and} \quad \mathbf{v}_1 = -\mathbf{n}_x \tag{5.5.10}$$

$$\boldsymbol{\lambda}_2 = \left(\sqrt{3}/2\right)\mathbf{n}_x + (1/2)\mathbf{n}_y \quad \text{and} \quad \mathbf{v}_2 = -(1/2)\mathbf{n}_x + \left(\sqrt{3}/2\right)\mathbf{n}_y \tag{5.5.11}$$

$$\boldsymbol{\lambda}_3 = \left(\sqrt{2}/2\right)\mathbf{n}_x - \left(\sqrt{2}/2\right)\mathbf{n}_y \quad \text{and} \quad \mathbf{v}_3 = \left(\sqrt{2}/2\right)\mathbf{n}_x + \left(\sqrt{2}/2\right)\mathbf{n}_y \tag{5.5.12}$$

Consider the velocity analysis: because B_1 has pure rotation, its angular velocity is:

$$\boldsymbol{\omega}^{OP} \overset{\text{D}}{=} \boldsymbol{\omega}_1 = -5\mathbf{n}_z \ \text{rad/sec} \tag{5.5.13}$$

The velocity of joint P is then:

$$\begin{aligned} \mathbf{v}^P &= \boldsymbol{\omega}_1 \times \mathbf{OP} = -5\mathbf{n}_z \times 2.0\boldsymbol{\lambda}_1 \\ &= -10\mathbf{v}_1 = 10\mathbf{n}_x \ \text{m/sec} \end{aligned} \tag{5.5.14}$$

(Recall that O is a center of zero velocity of B_1 and that P moves in a circle about O.)

Because B_2 has general plane motion, the velocity of Q may be expressed as:

$$\begin{aligned}\mathbf{v}^Q &= \mathbf{v}^P + \boldsymbol{\omega}_2 \times \mathbf{PQ} = 10\mathbf{n}_y + \omega_2 \mathbf{n}_z \times (3.0)\boldsymbol{\lambda}_2 = 10\mathbf{n}_x + 3\omega_2 \mathbf{v}_2 \\ &= 10\mathbf{n}_x + 3\omega_2 \left[\left(-1/2\, \mathbf{n}_x + \left(\sqrt{3}/2 \right) \mathbf{n}_y \right] \\ &= \left[10 - (3/2)\omega_2 \right] \mathbf{n}_x + \left(\sqrt{3}/2 \right) \omega_2 \mathbf{n}_y \end{aligned} \tag{5.5.15}$$

where ω_2 is the angular speed of B_2. Note that Q is fixed in both B_2 and B_3.

Because B_3 has pure rotation with center R, Q moves in a circle about R. Hence, $\mathbf{v}^Q$ may be expressed as:

$$\begin{aligned}\mathbf{v}^Q &= \boldsymbol{\omega}_3 \times \mathbf{RQ} = \omega_3 \mathbf{n}_z \times \left(-4.95\,\omega_3 \right) \mathbf{v}_3 \\ &= -4.95\,\omega_3 \left[\left(\sqrt{2}/2 \right) \mathbf{n}_x + \left(\sqrt{2}/2 \right) \mathbf{n}_y \right] = -3.5\,\omega_2 \mathbf{n}_x - 3.5\,\omega_3 \mathbf{n}_y \end{aligned} \tag{5.5.16}$$

where ω_3 is the angular speed of B_3.

Comparing Eqs. (5.5.15) and (5.5.16) we have the scalar equations:

$$10 - 1.5\,\omega_2 = -3.5\,\omega_3 \tag{5.5.17}$$

and

$$\left(3\sqrt{3}/2 \right) \omega_2 = -3.5\,\omega_3 \tag{5.5.18}$$

Solving for ω_2 and ω_3 we obtain:

$$\omega_2 = 2.44 \text{ rad/sec} \quad \text{and} \quad \omega_3 = -1.81 \text{ rad/sec} \tag{5.5.19}$$

Hence, $\mathbf{v}^Q$ becomes:

$$\mathbf{v}^Q = 6.34\,\mathbf{n}_x + 6.34\,\mathbf{n}_y \text{ m/sec} \tag{5.5.20}$$

Observe that in calculating the angular speeds of B_2 and B_3 we could also use an analysis of the instant centers as discussed in Section 5.4. Because the velocities of P and Q are perpendicular to, respectively, $B_1(OP)$ and $B_3(QR)$, we can construct the diagram shown in Figure 5.5.7 to obtain $\boldsymbol{\omega}_2$, $\boldsymbol{\omega}_3$, and $\mathbf{v}^Q$. By extending OP and RQ until they intersect, we obtain the instant center of zero velocity of B_2. Then, IP and IQ are perpendicular to, respectively, $\mathbf{v}^P$ and $\mathbf{v}^Q$. Triangle IOR forms a 45° right triangle. Hence, the distance between I and P is (6.098 – 2.0) m, or 4.098 m. Because $|\mathbf{v}^P|$ is 10 m/sec, ω_2 is:

$$\omega_2 = |\mathbf{v}^P| / |IP| = 10/4.098 = 2.44 \text{ rad/sec} \tag{5.5.21}$$

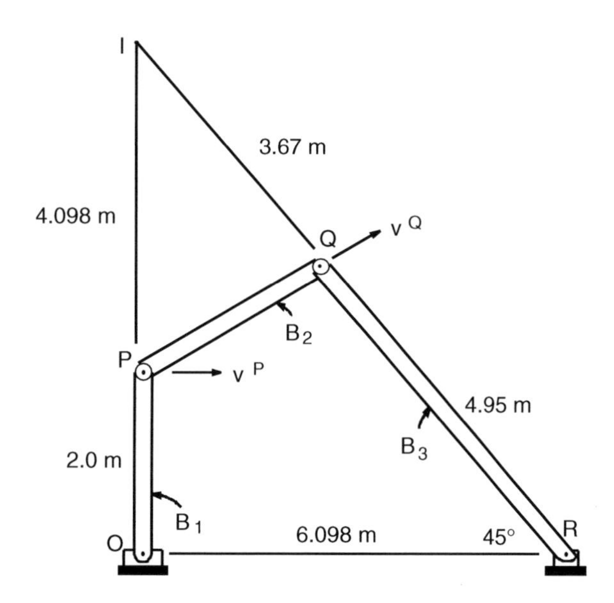

FIGURE 5.5.7
Instant center of zero velocity of B_2.

Similarly, the distance IQ is 3.67 m, and $\mathbf{v}^Q$ is, then,

$$\mathbf{v}^Q = |IQ|\omega_2\mathbf{v}_3 = (3.67)(2.44)\,\mathbf{v}_3 = 8.95\,\mathbf{v}_3 = 6.34\,\mathbf{n}_x + 6.34\,\mathbf{n}_y \tag{5.5.22}$$

Then ω_3 becomes:

$$\omega_3 = -|\mathbf{v}^Q| / |\mathbf{QR}| = -8.95/4.95 = -1.81 \text{ rad/sec} \tag{5.5.23}$$

Next, consider an acceleration analysis. Because B_1 has pure rotation, P moves in a circle about O and its acceleration is:

$$\begin{aligned}\mathbf{a}^P &= \boldsymbol{\alpha}_1 \times \mathbf{OP} + \boldsymbol{\omega}_1 \times (\boldsymbol{\omega}_1 \times \mathbf{OP}) = 4\mathbf{n}_3 \times 2\boldsymbol{\lambda}_1 + (-5\mathbf{n}_z) \times [(-5\mathbf{n}_z) \times 2\boldsymbol{\lambda}_1] \\ &= 8\mathbf{v}_1 - 50\boldsymbol{\lambda}_1 = -8\mathbf{n}_x - 50\mathbf{n}_y \text{ m/sec}^2\end{aligned} \tag{5.5.24}$$

Because P and Q are both fixed on B_2, the acceleration of Q may be expressed as:

$$\begin{aligned}\mathbf{a}^Q &= \mathbf{a}^P + \boldsymbol{\alpha}_2 \times \mathbf{PQ} + \boldsymbol{\omega}_2 \times (\boldsymbol{\omega}_2 \times \mathbf{PQ}) \\ &= -8\mathbf{n}_x - 50\mathbf{n}_y + \alpha_2\mathbf{n}_z \times (3.0\boldsymbol{\lambda}_2) + (2.44\mathbf{n}_2) \times [(2.44\mathbf{n}_z) \times (3.0\boldsymbol{\lambda}_2)] \\ &\quad - 8\mathbf{n}_x - 50\mathbf{n}_y + 3.0\alpha_2\mathbf{v}_2 - 17.86\boldsymbol{\lambda}_2 \\ &= -8\mathbf{n}_x - 50\mathbf{n}_y + 3.0\alpha_2\left[(-1/2)\mathbf{n}_x + (\sqrt{3}/2)\mathbf{n}_y\right] - 17.86\left[(\sqrt{3}/2\mathbf{n}_x + (1/2)\mathbf{n}_y)\right] \\ &= (-23.467 - 1.5\alpha_2)\mathbf{n}_x + (-58.93 + 2.6\alpha_2)\mathbf{n}_y\end{aligned} \tag{5.5.25}$$

Because Q also moves in a circle about R, we have:

$$\begin{aligned}
\mathbf{a}^Q &= \boldsymbol{\alpha}_3 \times \mathbf{RQ} + \boldsymbol{\omega}_3 \times (\boldsymbol{\omega}_3 \times \mathbf{RQ}) \\
&= \alpha_3 \mathbf{n}_z \times (-4.95\boldsymbol{\lambda}_3) + (-1.81\mathbf{n}_z) \times [(-1.81\mathbf{n}_z) \times (-4.95\boldsymbol{\lambda}_3)] \\
&= -4.95\alpha_3 \mathbf{v}_3 + 16.21\boldsymbol{\lambda}_3 \\
&= -4.95\boldsymbol{\alpha}_3 \left[(\sqrt{2}/2)\mathbf{n}_x + (\sqrt{2}/2)\mathbf{n}_y\right] + 16.21\left[(\sqrt{2}/2)\mathbf{n}_x - (\sqrt{2}/2)\mathbf{n}_y\right] \\
&= -3.5\boldsymbol{\alpha}_3 \mathbf{n}_x - 3.5\boldsymbol{\alpha}_3 \mathbf{n}_y + 11.46\mathbf{n}_x - 11.46\mathbf{n}_y \\
&= (11.46 - 3.5\boldsymbol{\alpha}_3)\mathbf{n}_x + (-11.46 - 3.5\boldsymbol{\alpha}_3)\mathbf{n}_y
\end{aligned} \tag{5.5.26}$$

Comparing Eqs. (5.5.25) and (5.5.26), we have:

$$-23.467 - 1.5\alpha_2 = 11.46 - 3.5\alpha_3 \tag{5.5.27}$$

and

$$-58.93 + 2.6\alpha_2 = -11.46 - 3.5\alpha_3 \tag{5.5.28}$$

Solving for α_2 and α_3 we obtain:

$$\alpha_2 = 3.06 \text{ rad/sec}^2 \quad \text{and} \quad \alpha_3 = 11.28 \text{ rad/sec}^2 \tag{5.5.29}$$

Hence, the acceleration of Q is:

$$\mathbf{a}^Q = -28.02\mathbf{n}_x - 50.94\mathbf{n}_y \text{ ft/sec}^2 \tag{5.5.30}$$

Observe how much more effort is required to obtain accelerations than velocities.

5.6 Chains of Bodies

Consider next a chain of identical pin-connected bars moving in a vertical plane and supported at one end as represented in Figure 5.6.1. Let the chain have N bars, and let their orientations be measured by N angles θ_i ($i = 1,\ldots, N$) that the bars make with the vertical Z-axis as in Figure 5.6.1. Because N angles are required to define the configuration and positioning of the system, the system has N degrees of freedom. A chain may be considered to be a *finite-segment* model of a cable; hence, an analysis of the system of Figure 5.6.1 can provide insight into the behavior of cable and tether systems.

A kinematical analysis of a chain generally involves determining the velocities and accelerations of the connecting joints and the centers of the bars and also the angular velocities and angular accelerations of the bars. To determine these quantities, it is easier to use the *absolute* orientation angles of Figure 5.6.1 than the *relative* orientation angles

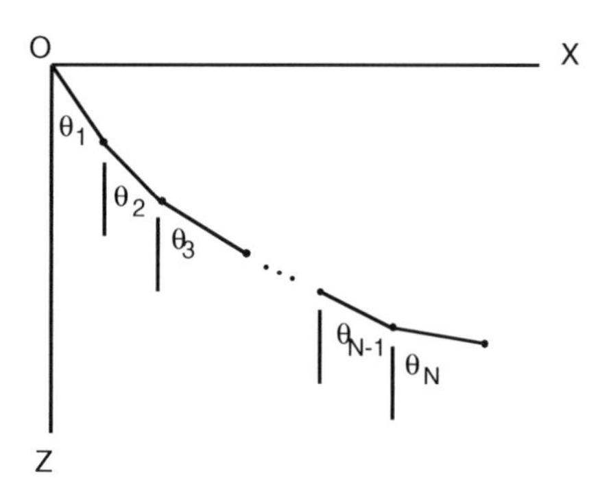

FIGURE 5.6.1
A chain of N bars.

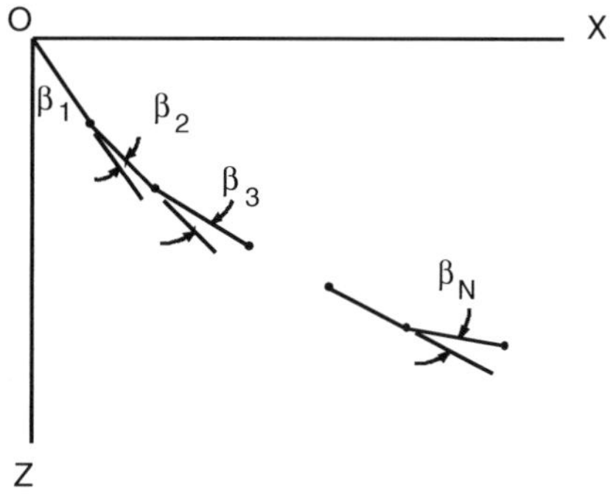

FIGURE 5.6.2
Relative orientation angles.

shown in Figure 5.6.2. The relative angles have the advantage of being more intuitive in their description of the inclination of the bars.

In our discussion we will use absolute angles because of their simplicity in analysis. To begin, consider a typical pair of adjoining bars such as B_j and B_k as in Figure 5.6.3. Let the connecting joints of the bars be O_j, O_k, and O_ℓ as shown, and let G_j and G_k be the centers of the bars. Let $\mathbf{n}_{j3}$, $\mathbf{n}_{k3}$ and $\mathbf{n}_{j1}$, $\mathbf{n}_{k1}$ be unit vectors parallel and perpendicular, respectively, to the bars in the plane of motion.

By using this notation, the system may be numbered and labeled serially from the support pin O as in Figure 5.6.4. Let $\mathbf{N}_x$, $\mathbf{N}_y$, and $\mathbf{N}_z$ be unit vectors parallel to the X-, Y-, and Z-axes, as shown.

Because the X–Z plane is the plane of motion, the angular velocity and angular acceleration vectors will be perpendicular to the X–Z plane and, thus, parallel to the Y-axis. Specifically, the angular velocities and angular accelerations may be expressed as:

$$\omega_k = \dot{\theta}_k \mathbf{N}_Y \quad \text{and} \quad \boldsymbol{\alpha}_k = \ddot{\theta}_k \mathbf{N}_Y \quad (k = 1, \ldots, N) \tag{5.6.1}$$

Next, the velocity and acceleration of G_1, the center of B_1, may be readily obtained by noting that G_1 moves on a circle. Thus, we have:

$$\mathbf{v}^{G1} = (\ell/2)\dot{\theta}_1 \mathbf{n}_{11} \quad \text{and} \quad \mathbf{a}^{G1} = (\ell/2)\ddot{\theta}_1 \mathbf{n}_{11} - (\ell/2)\dot{\theta}_1^2 \mathbf{n}_{13} \tag{5.6.2}$$

In terms of $\mathbf{N}_X$ and $\mathbf{N}_Z$, these expressions become:

$$\mathbf{v}^{G1} = (\ell/2)\dot{\theta}_1 (\cos\theta_1 \mathbf{N}_X - \sin\theta_1 \mathbf{N}_Z) \tag{5.6.3}$$

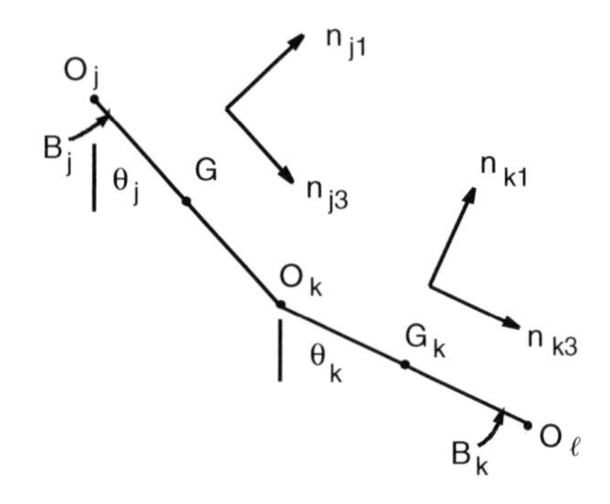

FIGURE 5.6.3
Two typical adjoining bars.

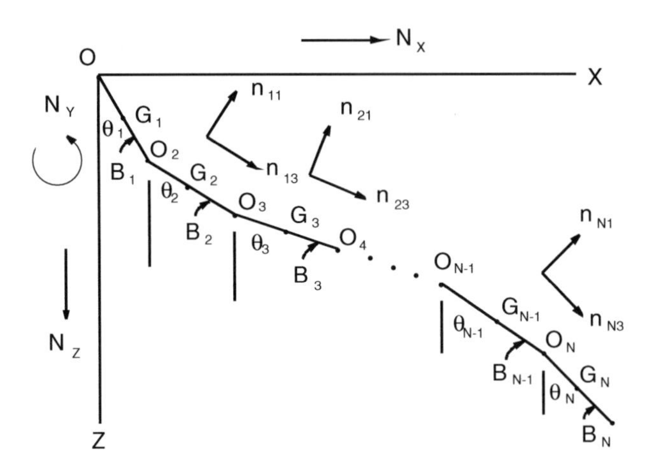

FIGURE 5.6.4
Numbering and labeling of the systems.

and

$$\mathbf{a}^{G1} = (\ell/2)\left[\left(\ddot{\theta}_1\cos\theta_1 - \dot{\theta}_1^2\sin\theta_1\right)\mathbf{N}_X + \left(-\ddot{\theta}_1\sin\theta_1 - \dot{\theta}_1^2\cos\theta_1\right)\mathbf{N}_Z\right] \tag{5.6.4}$$

Similarly, the velocity and acceleration of O_2 are:

$$\mathbf{v}^{O_2} = \ell\dot{\theta}_1\mathbf{n}_{11} \quad \text{and} \quad \mathbf{a}^{O_2} = \ell\ddot{\theta}_1\mathbf{n}_{11} - \ell\dot{\theta}_1^2\mathbf{n}_{13} \tag{5.6.5}$$

and in terms of $\mathbf{N}_X$ and $\mathbf{N}_Z$, they are:

$$\mathbf{v}^{O_2} = \ell\dot{\theta}_1\left(\cos\theta_1\,\mathbf{N}_X - \sin\theta_1\mathbf{N}_Z\right) \tag{5.6.6}$$

and

$$\mathbf{a}^{O_2} = \ell\left[\left(\ddot{\theta}_1\cos\theta_1 - \dot{\theta}_1^2\sin\theta_1\right)\mathbf{N}_X + \left(-\ddot{\theta}_1\sin\theta_1 - \dot{\theta}_1^2\cos\theta_1\right)\mathbf{N}_Z\right] \tag{5.6.7}$$

Observe how much simpler the expressions are when the *local* (as opposed to *global*) unit vectors are used.

Consider next the velocity and acceleration of the center G_2 and the distal joint O_3 of B_2. From the relative velocity and acceleration formulas, we have (see Eqs. (3.4.6) and (3.4.7)):

$$\mathbf{v}^{G_2} = \mathbf{v}^{O_2} + \mathbf{v}^{G_2/O_2} \quad \text{and} \quad \mathbf{a}^{G_2} = \mathbf{a}^{O_2} + \mathbf{a}^{G_2/O_2} \tag{5.6.8}$$

Because O_2 and G_2 are both fixed on B_2, we have:

$$\mathbf{v}^{G_2/O_2} = \boldsymbol{\omega}_2 \times (\ell/2)\mathbf{n}_{23} = (\ell/2)\dot{\theta}_2\mathbf{n}_{21} \tag{5.6.9}$$

and

$$\begin{aligned} \mathbf{a}^{G_2/O_2} &= \boldsymbol{\alpha}_2 \times (\ell/2)\mathbf{n}_{23} + \boldsymbol{\omega}_2 \times \left[\boldsymbol{\omega}_2 \times (\ell/2)\mathbf{n}_{23}\right] \\ &= (\ell/2)\ddot{\theta}_2\mathbf{n}_{21} - (\ell/2)\dot{\theta}_2^2\mathbf{n}_{23} \end{aligned} \tag{5.6.10}$$

(G_2 may be viewed as moving on a circle about O_2.) Hence, by substituting into Eq. (5.6.8), we have:

$$\mathbf{v}^{G_2} = \ell\dot{\theta}_1\mathbf{n}_{11} + (\ell/2)\dot{\theta}_2\mathbf{n}_{21} \tag{5.6.11}$$

and

$$\mathbf{a}^{G_2} = \ell\ddot{\theta}_1\mathbf{n}_{11} - \ell\dot{\theta}_1^2\mathbf{n}_{13} + (\ell/2)\ddot{\theta}_2\mathbf{n}_{21} - (\ell/1)\dot{\theta}_2^2\mathbf{n}_{23} \tag{5.6.12}$$

In terms of $\mathbf{N}_X$ and $\mathbf{N}_Z$, these expressions become:

$$\begin{aligned} \mathbf{v}^{G_2} &= \left[\ell\dot{\theta}_1 \cos\theta_1 + (\ell/2)\dot{\theta}_2 \cos\theta_2\right]\mathbf{N}_X \\ &\quad + \left[-\ell\dot{\theta}_1 \sin\theta_1 - (\ell/2)\dot{\theta}_2 \sin\theta_2\right]\mathbf{N}_Z \end{aligned} \tag{5.6.13}$$

and

$$\begin{aligned} \mathbf{a}^{G_2} &= \left[\ell\ddot{\theta}_1 \cos\theta_1 - \ell\dot{\theta}_1^2 \sin\theta_1 + (\ell/2)\ddot{\theta}_2 \cos\theta_2 - (\ell/2)\dot{\theta}_2^2 \sin\theta_2\right]\mathbf{N}_X \\ &\quad + \left[-\ell\theta_1 \sin\ddot{\theta}_1 - \ell\dot{\theta}_1^2 \cos\theta_1 - (\ell/2)\ddot{\theta}_2 \sin\theta_2 - (\ell/2)\dot{\theta}_2^2 \cos\theta_2\right]\mathbf{N}_Z \end{aligned} \tag{5.6.14}$$

Similarly, the velocity of acceleration of O_3 is:

$$\mathbf{v}^{O_3} = \ell\dot{\theta}_1\mathbf{n}_{11} + \ell\dot{\theta}_2\mathbf{n}_{21} \tag{5.6.15}$$

and

$$\mathbf{a}^{O_3} = \ell\ddot{\theta}_1\mathbf{n}_{11} - \ell\dot{\theta}_1^2\mathbf{n}_{13} + \ell\ddot{\theta}_2\mathbf{n}_{21} - \ell\dot{\theta}_2^2\mathbf{n}_{23} \tag{5.6.16}$$

In terms of $\mathbf{N}_X$ and $\mathbf{N}_Z$, these expressions become:

$$\begin{aligned} \mathbf{v}^{O_3} &= \left[\ell\dot{\theta}_1 \cos\theta_1 + \ell\dot{\theta}_2 \cos\theta_2\right]\mathbf{N}_X \\ &\quad + \left[-\ell\dot{\theta}_1 \sin\theta_1 - \ell\dot{\theta}_2 \sin\theta_2\right]\mathbf{N}_Z \end{aligned} \tag{5.6.17}$$

and

$$\mathbf{a}^{O_3} = \left[\ell\ddot{\theta}\cos\theta_1 - \ell\dot{\theta}_1^2\sin\theta_1 + \ell\ddot{\theta}_2\cos\theta_2 - \ell\dot{\theta}_2^2\sin\theta_2\right]\mathbf{N}_X$$
$$+\left[-\ell\ddot{\theta}_1\sin\theta_1 - \ell\dot{\theta}_1^2\cos\theta_1 - \ell\ddot{\theta}_2\sin\theta_2 - \ell\dot{\theta}_2^2\cos\theta_2\right]\mathbf{N}_Z \tag{5.6.18}$$

Observe that using the local unit vectors again leads to simpler expressions (compare Eqs. (5.6.11) and (5.6.12) with Eqs. (5.6.13) and (5.6.14)). Nevertheless, with the use of the local unit vectors we have mixed sets in the individual equations. For example, in Eq. (5.6.11), the unit vectors are neither parallel nor perpendicular; hence, the components are not readily added. Therefore, for computational purposes, the use of the global unit vectors is preferred.

The velocities and accelerations of the remaining points of the system may be obtained similarly. Indeed, we can inductively determine the velocity and acceleration of the center of a typical bar B_k as:

$$\mathbf{v}^{G_k} = \ell\dot{\theta}_1\mathbf{n}_{11} + \ell\dot{\theta}_2\mathbf{n}_{21} + \ldots + \ell\dot{\theta}_j\mathbf{n}_{j1} + (\ell/2)\dot{\theta}_k\mathbf{n}_{k1} \tag{5.6.19}$$

and

$$\mathbf{a}^{G_k} = \ell\ddot{\theta}_1\mathbf{n}_{11} + \ell\ddot{\theta}_2\mathbf{n}_{21} + \ldots + \ell\ddot{\theta}_j\mathbf{n}_{j1} + (\ell/2)\ddot{\theta}_k\mathbf{n}_{k1}$$
$$- \ell\dot{\theta}_1^2\mathbf{n}_{13} - \ell\dot{\theta}_2^2\mathbf{n}_{23} - \ldots - \ell\dot{\theta}_j^2\mathbf{n}_{j3} - (\ell/2)\dot{\theta}_k^2\mathbf{n}_{k3} \tag{5.6.20}$$

where $\mathbf{n}_{j1}$, $\mathbf{n}_{j3}$, and θ_j are associated with the bar B_j, immediately preceding B_k. In terms of $\mathbf{N}_X$ and $\mathbf{N}_Z$, these expressions become:

$$\mathbf{v}^{G_K} = \left[\ell\dot{\theta}_1\cos\theta_1 + \ell\dot{\theta}_2\cos\theta_2 + \ldots + \ell\dot{\theta}_j\cos\theta_j + (\ell/2)\dot{\theta}_k\cos\theta_k\right]\mathbf{N}_X$$
$$+\left[-\ell\dot{\theta}_1\sin\theta_1 - \ell\dot{\theta}_2\sin\theta_2 - \ldots - \ell\dot{\theta}_j\sin\theta_j - (\ell/2)\dot{\theta}_k\sin\theta_k\right]\mathbf{N}_Z \tag{5.6.21}$$

and

$$\mathbf{a}^{G_k} = \left[\ell\ddot{\theta}_1\cos\theta_1 + \ell\ddot{\theta}_2\cos\theta_2 + \ldots + \ell\ddot{\theta}_j\cos\theta_j + (\ell/2)\ddot{\theta}_k\cos\theta_k\right.$$
$$\left.- \ell\dot{\theta}_1^2\sin\theta_1 - \ell\dot{\theta}_2^2\sin\theta_2 - \ldots - \ell\dot{\theta}_j^2\sin\theta_j - (\ell/2)\dot{\theta}_k^2\sin\theta_k\right]\mathbf{N}_X$$
$$+\left[-\ell\ddot{\theta}_1\sin\theta_1 - \ell\ddot{\theta}_2\sin\theta_2 - \ldots - \ell\ddot{\theta}_j\sin\theta_j - (\ell/2)\ddot{\theta}_k\sin\theta_k\right.$$
$$\left.- \ell\dot{\theta}_1^2\cos\theta_1 - \ell\dot{\theta}_2^2\cos\theta_2 - \ldots - \ell\dot{\theta}_j^2\cos\theta_j - (\ell/2)\dot{\theta}_k^2\cos\theta_k\right]\mathbf{N}_Z \tag{5.6.22}$$

The velocity and acceleration of O_3 may be obtained from these latter expressions by simply replacing the fraction $(\ell/2)$ by ℓ.

5.7 Instant Center, Analytical Considerations

In Section 5.4, we developed an intuitive and geometrical description of centers of zero velocity. Here, we examine these concepts again, but this time from a more analytical perspective. Consider again a body B moving in planar motion as represented in Figure 5.7.1. Let the X–Y plane be a plane of motion of B. Let P be a typical point of B, and let C be a center of zero velocity of B. (That is, C is that point of B [or B extended] that has zero velocity.) Finally, let (x_P, y_P) and (x_C, y_C) be the X–Y coordinates of P and C, and let P and C be located relative to the origin O, and relative to each other, by the vectors $\mathbf{p}_P$, $\mathbf{p}_C$, and $\mathbf{r}$, as shown.

If $\mathbf{n}_x$ and $\mathbf{n}_y$ are unit vectors parallel to the X- and Y-axes, respectively, $\mathbf{p}_P$ and $\mathbf{p}_C$ may be expressed as:

$$\mathbf{p}_P = x_P\mathbf{n}_x + y_P\mathbf{n}_y \quad \text{and} \quad \mathbf{p}_C = x_C\mathbf{n}_x + y_C\mathbf{n}_y \tag{5.7.1}$$

Then $\mathbf{r}$, which locates C relative to P, may be expressed as:

$$\mathbf{r} = \mathbf{p}_C - \mathbf{p}_P = (x_C - x_P)\mathbf{n}_x + (y_C - y_P)\mathbf{n}_y = r\cos\theta\mathbf{n}_x + r\sin\theta\mathbf{n}_y \tag{5.7.2}$$

where r is the magnitude of $\mathbf{r}$ and θ is the inclination of $\mathbf{r}$ relative to the X-axis.

From Eq. (4.9.4), the velocities of P and C are related by the expression:

$$\mathbf{v}^C = \mathbf{v}^P + \boldsymbol{\omega} \times \mathbf{r} \tag{5.7.3}$$

where $\boldsymbol{\omega}$ is the angular velocity of B. Because C is a center of zero velocity, we have:

$$\mathbf{v}^C \quad \text{and thus} \quad \mathbf{v}^P = -\boldsymbol{\omega} \times \mathbf{r} \tag{5.7.4}$$

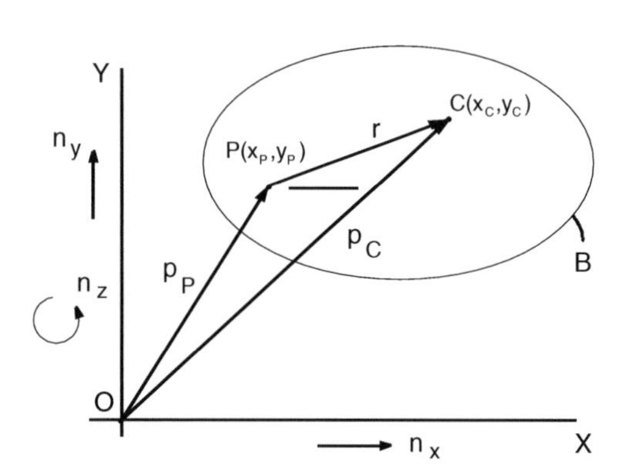

FIGURE 5.7.1
A body in plane motion with center for zero velocity C.

Let $\mathbf{n}_z$ be a unit vector normal to the X–Y plane generated by $\mathbf{n}_x \times \mathbf{n}_y$. Then, ω may be expressed as:

$$\boldsymbol{\omega} = \omega \mathbf{n}_z \tag{5.7.5}$$

where ω is positive when B rotates counterclockwise, as viewed in Figure 5.7.1.

Using Eqs. (5.7.1) to (5.7.5), $\mathbf{v}^P$ may be expressed as:

$$\begin{aligned} \mathbf{v}^P &= \dot{x}_P \mathbf{n}_x + \dot{y}_P \mathbf{n}_y = -\boldsymbol{\omega} \times \mathbf{r} \\ &= -\begin{vmatrix} \mathbf{n}_x & \mathbf{n}_y & \mathbf{n}_z \\ 0 & 0 & \omega \\ r\cos\theta & r\sin\theta & 0 \end{vmatrix} \\ &= r\omega \sin\theta \, \mathbf{n}_x - r\omega \cos\theta \, \mathbf{n}_y \end{aligned} \tag{5.7.6}$$

By comparing components, we obtain:

$$r\sin\theta = \dot{x}_P/\omega \quad \text{and} \quad r\cos\theta = -\dot{y}_P/\omega \tag{5.7.7}$$

We can readily locate C using these results: from Eqs. (5.7.1), (5.7.2), and (5.7.7), we have:

$$\mathbf{p}_C = \mathbf{p}_P + \mathbf{r} = x_C \mathbf{n}_x + y_C \mathbf{n}_y = (x_P + r\cos\theta)\mathbf{n}_x + (y_P + r\sin\theta)\mathbf{n}_y \tag{5.7.8}$$

Therefore, by comparing components, we have:

$$x_C = x_P - \dot{y}_P/\omega \quad \text{and} \quad y_C = y_P + \dot{x}_P/\omega \tag{5.7.9}$$

Equation (5.7.9) shows that if we know the location of a typical point P of B, the velocity of P, and the angular speed of B, we can locate the center C of zero velocity of B.

Let Q be a second typical point of B (distinct from P). Then, from Eq. (5.7.9) we have:

$$x_C = x_Q - \dot{y}_Q/\omega \quad \text{and} \quad y_C = y_Q + \dot{x}_Q/\omega \tag{5.7.10}$$

By comparing the terms of Eqs. (5.7.9) and (5.7.10), we have:

$$x_P - \dot{y}_P/\omega = x_Q - \dot{y}_Q/\omega \quad \text{and} \quad y_P + \dot{x}_P/\omega = y_Q + \dot{x}_Q/\omega \tag{5.7.11}$$

Solving for ω we obtain:

$$\omega = \frac{\dot{x}_Q - \dot{x}_P}{y_P - y_Q} \quad \text{and} \quad \omega = \frac{\dot{y}_P - \dot{y}_Q}{x_P - x_Q} \tag{5.7.12}$$

Equation (5.7.12) shows that if we know the velocities of two points of B we can determine the angular velocity of B. Then, from Eq. (5.7.9), the coordinates (x_C, y_C) of the center of zero velocity can be determined. That is,

$$x_C = x_p - \dot{y}_P\left(\frac{x_P - x_Q}{\dot{y}_P - \dot{y}_Q}\right) \quad \text{and} \quad y_C = y_P + \dot{x}_P\left(\frac{x_P - x_Q}{\dot{y}_P - \dot{y}_Q}\right) \tag{5.7.13}$$

We can verify these expressions using the geometric procedures of Section 5.4. Consider, for example, a body B moving in the X–Y plane with center of zero velocity C as in Figure 5.7.2. Let P and Q be two points of B whose positions and velocities are known. Then, the magnitude of their velocities designated by v_P and v_Q are related to the angular speed ω of B as:

$$v_P = \left|\mathbf{v}^P\right| = a\omega \quad \text{and} \quad v_Q = \left|\mathbf{v}^Q\right| = (a+b)\omega \tag{5.7.14}$$

where a and b are the distances shown in Figure 5.7.2. By comparing and combining these expressions, we have:

$$v_Q = v_P + b\omega \quad \text{or} \quad \omega = \left(v_Q - v_P\right)/b \tag{5.7.15}$$

From the geometry of Figure 5.7.2 we see that:

$$v_P = \dot{y}_P/\cos\theta, \quad v_Q = \dot{y}_Q/\cos\theta, \quad b = \left(x_Q - x_P\right)/\cos\theta \tag{5.7.16}$$

By substituting into Eq. (5.7.15) we obtain:

$$\omega = \frac{\dot{y}_Q - \dot{y}_P}{x_Q - x_P} \tag{5.7.17}$$

This verifies the second equation of Eq. (5.7.12); the first expression of Eq. (5.7.12) can be verified similarly.

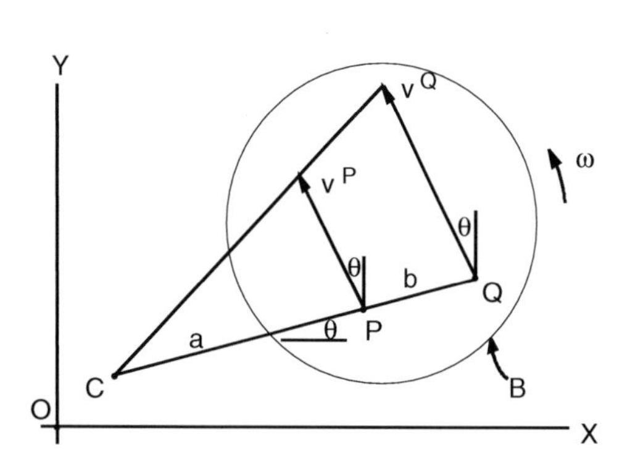

FIGURE 5.7.2
A body B with zero velocity center C and typical points P and Q.

5.8 Instant Center of Zero Acceleration

We can extend and generalize these procedures to obtain a center of zero acceleration — that is, a point of a body (or the body extended) that has zero acceleration. To this end, consider again a body *B* moving in planar motion as depicted in Figure 5.8.1. As before, let *P* and *Q* be typical points of *B*, and let *C* be the sought-after center of zero acceleration. Let (x_P, y_P), (x_Q, y_Q), and (x_C, y_C) be the *X–Y* coordinates of *P*, *Q*, and *C*. Let **r** locate *C* relative to P. Let **r** have magnitude *r* and inclination θ relative to the *X*-axis as shown in the figure. Finally, let ω and α represent the angular speed and angular acceleration of *B*.

Because *P* and *C* are fixed in *B*, their accelerations are related by the expression (see Eq. (4.9.6)):

$$\mathbf{a}^C = \mathbf{a}^P + \boldsymbol{\alpha}\times\mathbf{r} + \boldsymbol{\omega}\times(\boldsymbol{\omega}\times\mathbf{r}) \tag{5.8.1}$$

Therefore, if the acceleration of *C* is zero, then the acceleration of *P* is:

$$\mathbf{a}^P = -\boldsymbol{\alpha}\times\mathbf{r} - \boldsymbol{\omega}\times(\boldsymbol{\omega}\times\mathbf{r}) \tag{5.8.2}$$

If $\mathbf{n}_z$ is a unit vector normal to the *X–Y* plane, then the angular velocity and angular acceleration vectors may be expressed as (see Eq. (5.7.5)):

$$\boldsymbol{\omega} = \omega\mathbf{n}_z \quad \text{and} \quad \boldsymbol{\alpha} = \alpha\mathbf{n}_z \tag{5.8.3}$$

Also, from Figure 5.8.1, the position vector **r** may be written as:

$$\mathbf{r} = r\cos\theta\,\mathbf{n}_x + r\sin\theta\,\mathbf{n}_y \tag{5.8.4}$$

Then terms $\alpha \times \mathbf{r}$ and $\omega \times (\omega \times \mathbf{r})$ in Eq. (5.8.2) are:

$$\boldsymbol{\alpha}\times\mathbf{r} = -r\alpha\sin\theta\,\mathbf{n}_x + r\alpha\cos\theta\,\mathbf{n}_y \tag{5.8.5}$$

and

$$\boldsymbol{\omega}\times(\boldsymbol{\omega}\times\mathbf{r}) = -r\omega^2\cos\theta\,\mathbf{n}_x - r\omega^2\sin\theta\,\mathbf{n}_y \tag{5.8.6}$$

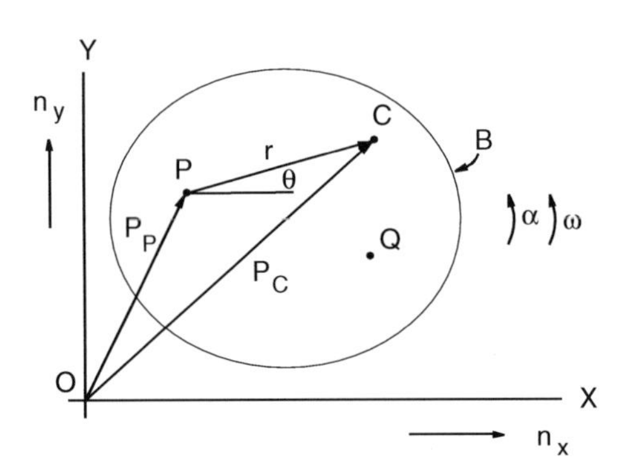

FIGURE 5.8.1
A body *B* in planar motion with center *C* of zero acceleration.

Hence, Eq. (5.8.2) becomes:

$$\mathbf{a}^P = \ddot{x}_P\mathbf{n}_x + \ddot{y}_P\mathbf{n}_y = \left(r\alpha\sin\theta + r\omega^2\cos\theta\right)\mathbf{n}_x + \left(-r\alpha\cos\theta + r\omega^2\sin\theta\right)\mathbf{n}_y \tag{5.8.7}$$

Then, $\ddot{x}_P$ and $\ddot{y}_P$ are:

$$\ddot{x}_P = r\alpha\sin\theta + r\omega^2\cos\theta$$

and

$$\ddot{y}_P = -r\alpha\cos\theta + r\omega^2\sin\theta \tag{5.8.8}$$

Solving for r sinθ and r cosθ we obtain:

$$r\sin\theta = \frac{\alpha\ddot{x}_P + \omega^2\ddot{y}_P}{\alpha^2 + \omega^4}$$

and

$$r\cos\theta = \frac{\omega^2\ddot{x}_P - \alpha\ddot{y}_P}{\alpha^2 + \omega^4} \tag{5.8.9}$$

From Figure 5.8.1, we see that C may be located relative to O by the equation:

$$\mathbf{p}_C = x_C\mathbf{n}_x + y_C\mathbf{n}_y = \mathbf{p}_P + \mathbf{r} = x_P\mathbf{n}_x + y_P\mathbf{n}_y + r\cos\theta\,\mathbf{n}_x + r\sin\theta\,\mathbf{n}_y \tag{5.8.10}$$

This leads to the component and coordinate expressions:

$$x_C = x_P + r\cos\theta = x_P + \frac{\omega^2\ddot{x}_P - \alpha\ddot{y}_P}{\alpha^2 + \omega^4} \tag{5.8.11}$$

and

$$y_C = y_P + r\sin\theta = y_P + \frac{\alpha\ddot{x}_P + \omega^2\ddot{y}_P}{\alpha^2 + \omega^4} \tag{5.8.12}$$

Equations (5.8.11) and (5.8.12) can be used to locate C if we know the position and acceleration of a typical point P of B and the angular speed and angular acceleration of B. Then, once C is located, the acceleration of any other point Q may be obtained from the expression:

$$\mathbf{a}^Q = \boldsymbol{\alpha}\times\mathbf{q} + \boldsymbol{\omega}\times(\boldsymbol{\omega}\times\mathbf{q}) \tag{5.8.13}$$

where $\mathbf{q}$ is a vector locating Q relative to C.

Alternatively, Eqs. (5.8.11) and (5.8.12) may be used to obtain the angular speed ω and the angular acceleration α of B if the acceleration of typical points P and Q are known. To see this, observe first that for point Q expressions analogous to Eqs. (5.8.11) and (5.8.12) are:

$$x_C = x_Q + \frac{\omega^2 \ddot{x}_Q - \alpha \ddot{y}_Q}{\alpha^2 + \omega^4} \tag{5.8.14}$$

and

$$y_C = y_Q + \frac{\alpha \ddot{x}_Q + \omega^2 \ddot{y}_Q}{\alpha^2 + \omega^4} \tag{5.8.15}$$

Next, by subtracting these expressions from Eqs. (5.8.11) and (5.8.12) we have:

$$x_P - x_Q = \frac{\alpha\left(\ddot{y}_P - \ddot{y}_Q\right) + \omega^2\left(\ddot{x}_Q - \ddot{x}_P\right)}{\alpha^2 + \omega^4} \tag{5.8.16}$$

$$y_P - y_Q = \frac{\alpha\left(\ddot{x}_Q - \ddot{x}_P\right) + \omega^2\left(\ddot{y}_Q - \ddot{y}_P\right)}{\alpha^2 + \omega^4} \tag{5.8.17}$$

The expressions may be solved for ω and α as follows: Let ξ and η be defined as:

$$\xi \overset{D}{=} \frac{\alpha}{\alpha^2 + \omega^4} \quad \text{and} \quad \eta \overset{D}{=} \frac{\omega^2}{\alpha^2 + \omega^4} \tag{5.8.18}$$

Then, Eqs. (5.8.16) and (5.8.17) become:

$$\left(\ddot{y}_P - \ddot{y}_Q\right)\xi + \left(\ddot{x}_Q - \ddot{x}_P\right)\eta = x_P - x_Q \tag{5.8.19}$$

and

$$\left(\ddot{x}_Q - \ddot{x}_P\right)\xi + \left(\ddot{y}_Q - \ddot{y}_P\right)\eta = y_P - y_Q \tag{5.8.20}$$

Solving for ξ and η we obtain:

$$\xi = \frac{1}{\Delta}\left[\left(x_P - x_Q\right)\left(\ddot{y}_Q - \ddot{y}_P\right) - \left(y_P - y_Q\right)\left(\ddot{x}_Q - \ddot{x}_P\right)\right] \tag{5.8.21}$$

$$\eta = \frac{1}{\Delta}\left[\left(y_P - y_Q\right)\left(\ddot{y}_P - \ddot{y}_Q\right) - \left(x_P - x_Q\right)\left(\ddot{x}_Q - \ddot{x}_P\right)\right] \tag{5.8.22}$$

where Δ, the determinant of the coefficients, is:

$$\Delta = -\left[\left(\ddot{y}_P - \ddot{y}_Q\right)^2 + \left(\ddot{x}_Q - \ddot{x}_P\right)^2\right] \tag{5.8.23}$$

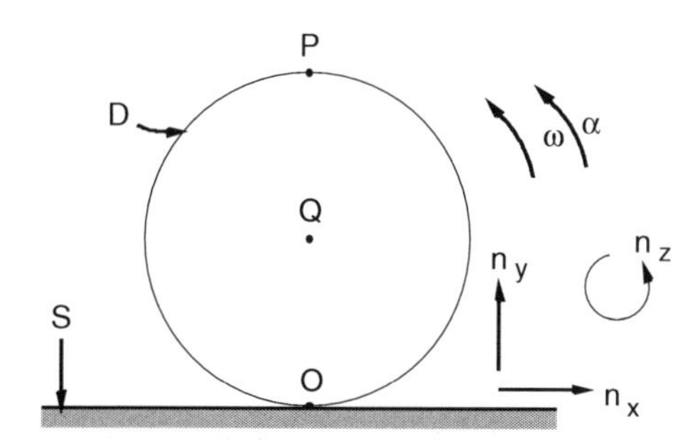

FIGURE 5.8.2
A rolling disk.

Finally, from Eq. (5.8.18) we have:

$$\xi^2 = \frac{\alpha^2}{\left(\alpha^2 + \omega^4\right)^2}, \quad \eta^2 = \frac{\omega^4}{\left(\alpha^2 + \omega^4\right)^2}, \quad \text{and} \quad \xi^2 + \eta^2 = \frac{1}{\alpha^2 + \omega^4} \tag{5.8.24}$$

Hence, α and ω^2 are:

$$\alpha = \xi / \left(\xi^2 + \eta^2\right) \quad \text{and} \quad \omega^2 = \eta / \left(\xi^2 + \eta^2\right) \tag{5.8.25}$$

To illustrate the application of these ideas, consider a circular disk D rolling to the left in a straight line on a surface S as in Figure 5.8.2. Let Q be the center of D, let O be the contact point (instant center of zero velocity) of D with S, and let P be a point on the periphery or rim of D. Finally, let D have radius r, angular speed ω, and angular acceleration α, as indicated.

Because Q moves on a straight line, its velocity may be expressed as:

$$\mathbf{v}^Q = -r\omega\mathbf{n}_x \tag{5.8.26}$$

Then, by differentiating, the acceleration of Q is:

$$\mathbf{a}^Q = d\mathbf{v}^Q/dt = -r\alpha\mathbf{n}_x \tag{5.8.27}$$

Because O and P are also fixed on D, their velocities and acceleration may be obtained from the expressions:

$$\mathbf{v}^O = \mathbf{v}^Q + \boldsymbol{\omega} \times \left(-r\mathbf{n}_y\right), \quad \mathbf{v}^P = \mathbf{v}^Q + \boldsymbol{\omega} \times \left(r\mathbf{n}_y\right) \tag{5.8.28}$$

and

$$\mathbf{a}^O = \mathbf{a}^Q + \boldsymbol{\alpha} \times \left(-r\mathbf{n}_y\right) + \boldsymbol{\omega} \times \left[\boldsymbol{\omega} \times \left(-r\mathbf{n}_y\right)\right], \quad \mathbf{a}^P = \mathbf{a}^Q + \boldsymbol{\alpha} \times \left(r\mathbf{n}_y\right) + \boldsymbol{\omega} \times \left[\boldsymbol{\omega} \times \left(r\mathbf{n}_y\right)\right] \tag{5.8.29}$$

By substituting from Eqs. (5.8.26) and (5.8.27), by recognizing that $\boldsymbol{\omega}$ and $\boldsymbol{\alpha}$ are $\omega\mathbf{n}_z$ and $\alpha\mathbf{n}_z$, and by carrying out the indicated operations, we obtain:

$$\mathbf{v}^O = -r\omega\mathbf{n}_x + \omega\mathbf{n}_z \times \left(-r\mathbf{n}_y\right) = 0 \tag{5.8.30}$$

$$\mathbf{v}^P = -r\omega\mathbf{n}_x + \omega\mathbf{n}_z \times \left(r\mathbf{n}_y\right) = -2r\omega\mathbf{n}_x \tag{5.8.31}$$

$$\mathbf{a}^O = -r\alpha\mathbf{n}_x + \alpha\mathbf{n}_z \times \left(-r\mathbf{n}_y\right) + \omega\mathbf{n}_z \times \left[\omega\mathbf{n}_z \times \left(-r\mathbf{n}_y\right)\right] = r\omega^2\mathbf{n}_y \tag{5.8.32}$$

$$\begin{aligned} \mathbf{a}^P &= -r\alpha\mathbf{n}_x + \alpha\mathbf{n}_z \times \left(r\mathbf{n}_y\right) + \omega\mathbf{n}_z \times \left[\omega\mathbf{n}_z \times \left(r\mathbf{n}_y\right)\right] \\ &= -2r\alpha\mathbf{n}_x - r\omega^2\mathbf{n}_y \end{aligned} \tag{5.8.33}$$

Observe that the velocity of O is zero, as expected, but the acceleration of O is *not* zero. To find the point C with zero acceleration, we can use Eqs. (5.8.11) and (5.8.12): specifically, for a Cartesian (X–Y) axes system with origin at O, we find the coordinates of C to be:

$$x_C = x_Q + \frac{\omega^2\ddot{x}_Q - \alpha\ddot{y}_Q}{\alpha^2 + \omega^4} = 0 + \frac{\omega^2(-r\alpha) - \alpha(0)}{\alpha^2 + \omega^4} = -\frac{r\alpha\omega^2}{\alpha^2 + \omega^4} \tag{5.8.34}$$

and

$$y_C = y_Q + \frac{\alpha\ddot{x}_Q + \omega^2\ddot{y}_Q}{\alpha^2 + \omega^4} = r + \frac{\alpha(-r\alpha) + \omega^2(0)}{\alpha^2 + \omega^4} = \frac{r\omega^4}{\alpha^2 + \omega^4} \tag{5.8.35}$$

For positive values of ω and α, the position of C is depicted in Figure 5.8.3.

To verify the results, consider calculating the acceleration of C using the expression:

$$\mathbf{a}^C = \mathbf{a}^Q + \boldsymbol{\alpha} \times \mathbf{QC} + (\boldsymbol{\omega} \times \mathbf{QC}) \tag{5.8.36}$$

where $\mathbf{QC}$ is:

$$\mathbf{QC} = -\frac{r\alpha\omega^2}{\alpha^2 + \omega^4}\mathbf{n}_x - \left[r - \left(\frac{r\omega^4}{\alpha^2 + \omega^4}\right)\right]\mathbf{n}_y \tag{5.8.37}$$

Using Eq. (5.8.27), $\mathbf{a}^C$ becomes:

$$\begin{aligned} \mathbf{a}^C &= -r\alpha\mathbf{n}_x - \alpha\frac{r\alpha\omega^2}{\alpha^2 + \omega^4}\mathbf{n}_y + \alpha\left[r - \left(\frac{r\omega^4}{\alpha^2 + \omega^4}\right)\right]\mathbf{n}_x \\ &+ \omega^2\frac{r\alpha\omega^2}{\alpha^2 + \omega^4}\mathbf{n}_x + \omega^2\left[r - \left(\frac{r\omega^4}{\alpha^2 + \omega^4}\right)\right]\mathbf{n}_y = 0 \end{aligned} \tag{5.8.38}$$

Finally, we can check the consistency of Eqs. (5.8.18), (5.8.21), and (5.8.22): from Eqs. (5.8.21) and (5.8.22), ξ and η are:

$$\xi = \frac{1}{\Delta}\left[\left(x_P - x_Q\right)\left(\ddot{y}_Q - \ddot{y}_P\right) - \left(y_P - y_Q\right)\left(\ddot{x}_Q - \ddot{x}_P\right)\right] \tag{5.8.39}$$

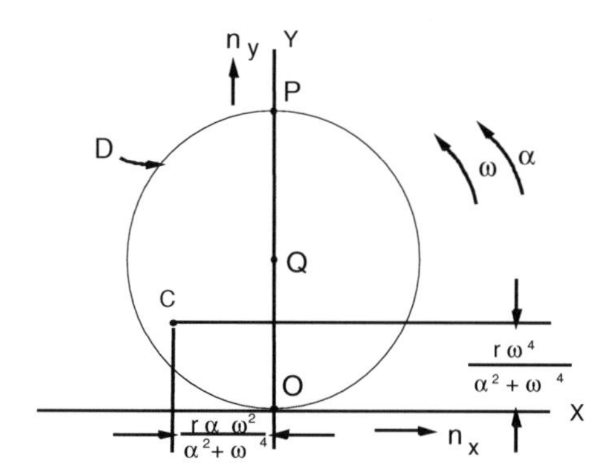

FIGURE 5.8.3
Location of center C of zero acceleration for rolling disk of Figure 5.8.2.

and

$$\eta = \frac{1}{\Delta}\left[\left(y_P - y_Q\right)\left(\ddot{y}_P - \ddot{y}_Q\right) - \left(x_P - x_Q\right)\left(\ddot{x}_Q - \ddot{x}_P\right)\right] \tag{5.8.40}$$

where, from Eq. (5.8.23), Δ is:

$$\Delta = -\left[\left(\ddot{y}_P - \ddot{y}_Q\right)^2 + \left(\ddot{x}_Q - \ddot{x}_P\right)^2\right] \tag{5.8.41}$$

From Eqs. (5.8.27) and (5.8.33) and from Figure 5.8.3 we have:

$$\begin{aligned} x_P &= 0, & y_P &= 2r \\ x_Q &= 0, & y_Q &= r \\ \ddot{x}_P &= -2r\alpha, & \ddot{y}_P &= -r\omega^2 \\ \ddot{x}_Q &= -r\alpha, & \ddot{y}_Q &= 0 \end{aligned} \tag{5.8.42}$$

Then, Δ becomes:

$$\Delta = -\left[\left(-r\omega^2 - 0\right)^2 + (-r\alpha + 2r\alpha)^2\right] = -r^2\left(\omega^4 + \alpha^2\right) \tag{5.8.43}$$

Hence, ξ and η are:

$$\xi = \frac{-1}{r^2\left(\omega^4 + \alpha^2\right)}\left[0 - (2r - r)(-r\alpha + 2r\alpha)\right] = \frac{\alpha}{\alpha^2 + \omega^4} \tag{5.8.44}$$

$$\eta = \frac{-1}{r^2\left(\omega^4 + \alpha^2\right)}\left[(2r - r)\left(-r\omega^2 - 0\right) - 0\right] = \frac{\omega^2}{\alpha^2 + \omega^4} \tag{5.8.45}$$

These expressions are identical with those of Eq. (5.8.18).

Problems

Section 5.2 Coordinates, Constraints, Degrees of Freedom

P5.2.1: Consider a pair of eyeglasses to be composed of a frame containing the lenses and two rods hinged to the frame for fitting over the ears. How many degrees of freedom do the eyeglasses have?

P5.2.2: Let a simple model of the human arm consist of three bodies representing the upper arm, the lower arm, and the hands. Let the upper arm have a spherical (ball-and-socket) connection with the chest, let the elbow be represented as a pin (or hinge), and let the hand movement be governed by a twist of the lower arm and vertical and horizontal rotations. How many degrees of freedom does the model have?

P5.2.3: How many degrees of freedom does a vice, as commonly found in a workshop, have? (Include the axial rotation of the adjustment handle about its long axis and the potential rotation of the vice itself about a vertical axis.)

P5.2.4: See Figure P5.2.4. A wheel *W*, having planar motion, rolls without slipping in a straight line. Let *C* be the contact point between *W* and the rolling surface *S*. How many degrees of freedom does *W* have? What are the constraint equations?

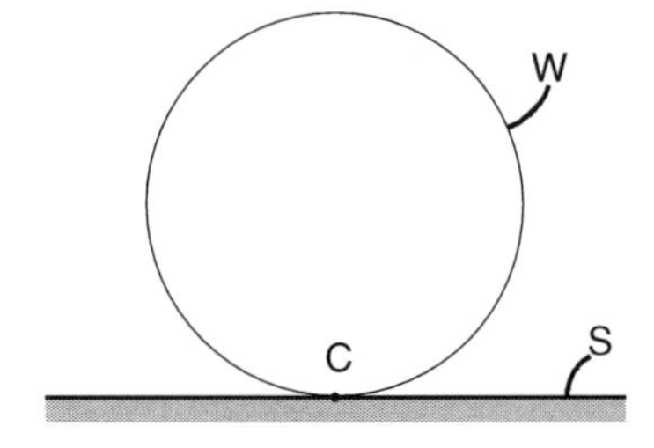

FIGURE P5.2.4
A wheel rolling in a straight line.

P5.2.5: See Problem P5.2.4. Suppose *W* is allowed to slip along *S*. How many degrees of freedom does *W* then have?

P5.2.6: How many degrees of freedom are there in a child's tricycle whose wheels roll without slipping on a flat horizontal surface? (Neglect the rotation of the pedals about their individual axes.)

Section 5.3 Planar Motion of a Rigid Body

P5.3.1: Classify the movement of the following bodies as being (1) translation, (2) rotation, and/or (3) general plane motion.

a. Eraser on a chalk board
b. Table-saw blade
c. Radial-arm-saw blade
d. Bicycle wheel of a bicycle moving in a straight line
e. Seat of a bicycle moving in a straight line
f. Foot pedal of a bicyclist moving in a straight line

P5.3.2: Consider again the piston, connecting rod, and crank arm of Figure 5.3.3 and as shown again in Figure P5.3.2. Let the length of the crank arm OQ be 6 in., and let the connecting rod be 15 in. long. Let the crank arm be rotating counterclockwise at 750 rpm. Determine the velocity and acceleration of Q and P when θ has the values 0, 30, 45, 60, and 90 degrees. Express the results in terms of unit vectors $\mathbf{n}_x$ and $\mathbf{n}_y$ as shown in Figure P5.3.2.

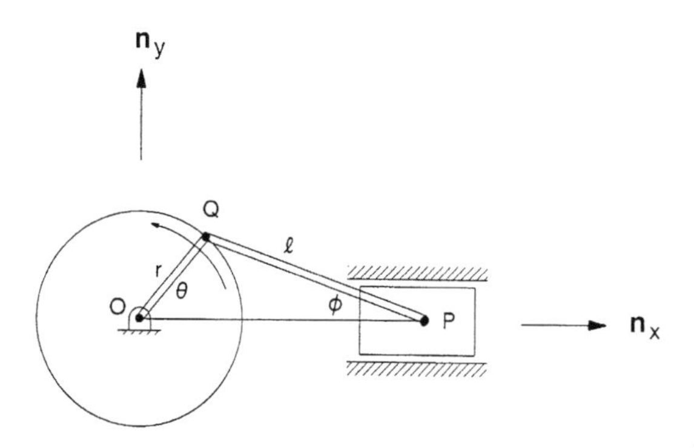

FIGURE P5.3.2
A piston, connecting rod, and crank arm.

P5.3.3: Repeat Problem P5.3.2 if the crank arm is rotating clockwise at 1000 rpm.

P5.3.4: Review the analysis of Example 5.3.1. Develop equations analogous to Eqs. (5.3.24) to (5.3.27) (i.e., the crank arm OQ has an angular acceleration α).

P5.3.5: See Problems P5.3.2 and P5.3.4. Repeat Problem P5.3.2 if the crank arm OQ is increasing its rotation rate at 50 rpm per second.

P5.3.6: A wheel W having planar motion rolls on an inner rim as represented in Figure P5.3.6. Suppose the inner rim radius r_i is 8 cm and the outer radius r_O is 14 cm. Suppose further that the velocity and acceleration of the center O of W are $24\mathbf{n}_x$ cm/sec and $-28\mathbf{n}_x$ cm/sec^2, respectively, where $\mathbf{n}_x$ and $\mathbf{n}_y$ are horizontal and vertical unit vectors, respectively, as shown. Determine the velocity and acceleration of points P, Q, C, and R of W when they are in the positions shown in Figure P5.3.6.

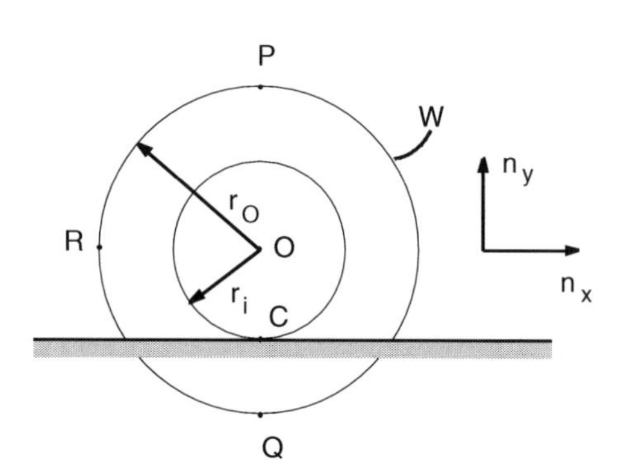

FIGURE P5.3.6
A wheel rolling on an inner rim.

P5.3.7: A wheel with a cable wrapped around its hub is rolling to the left (without slipping) as shown in Figure P5.3.7. Let the velocity and acceleration of the center O of W be $-5\mathbf{n}_1$ ft/sec and $-3\mathbf{n}_1$ ft/sec^2, respectively. Let the wheel radius r_O be 1.4 ft, and let the hub radius r_i be 0.8 ft. Let the cable pass over a pulley and support a block B as shown.

a. Find the velocity and acceleration of B.
b. Find the velocity and acceleration of points P and Q on the rim of W in the positions shown in Figure P5.3.7.

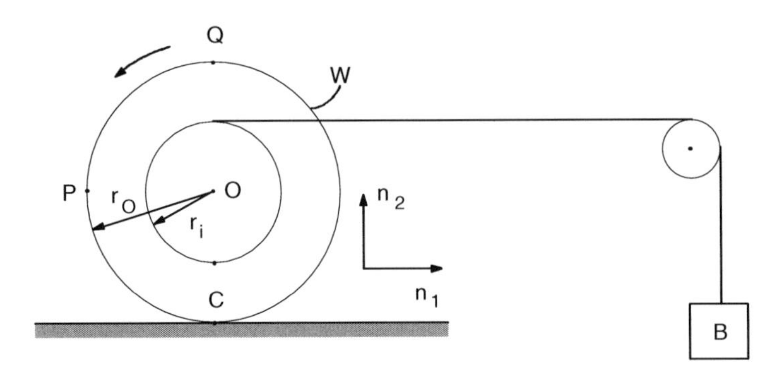

FIGURE P5.3.7
A wheel with a hub/cable lifting a block *B*.

P5.3.8: For the linkage shown in Figure P5.3.8 and for the position shown, determine the velocity and acceleration of points *A* and *B*.

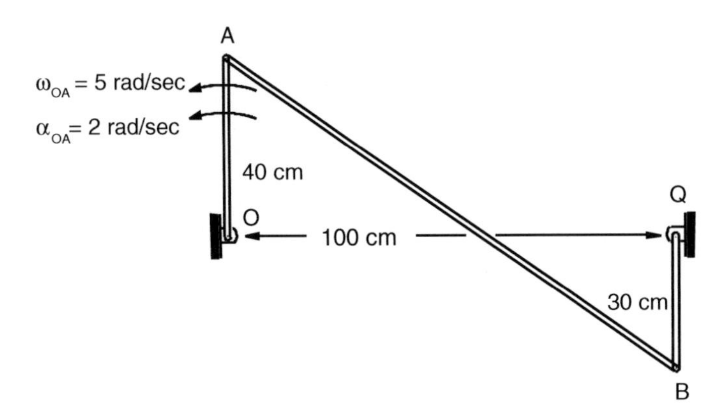

FIGURE P5.3.8
A four-bar linkage.

P5.3.9: Repeat Problem P5.3.8 for the linkage shown in Figure P5.3.9.

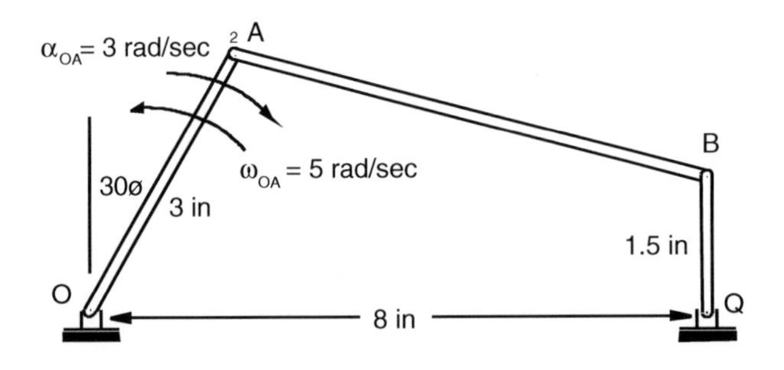

FIGURE P5.3.9
A four-bar linkage.

Section 5.4 Instant Center, Points of Zero Velocity

P5.4.1: A 2-ft-radius wheel rolls to the right so that its center *O* has a velocity of 3 ft/sec, as presented in Figure P5.4.1.

a. Locate (or identify) the instant center of zero velocity.
b. Determine the angular velocity of the wheel.
c. Determine the velocities of points *A*, *Q*, *P*, and *B* on the rim of the wheel (see Figure P5.4.1). Express the results in terms of horizontal and vertical unit vectors $\mathbf{n}_x$ and $\mathbf{n}_y$, respectively.

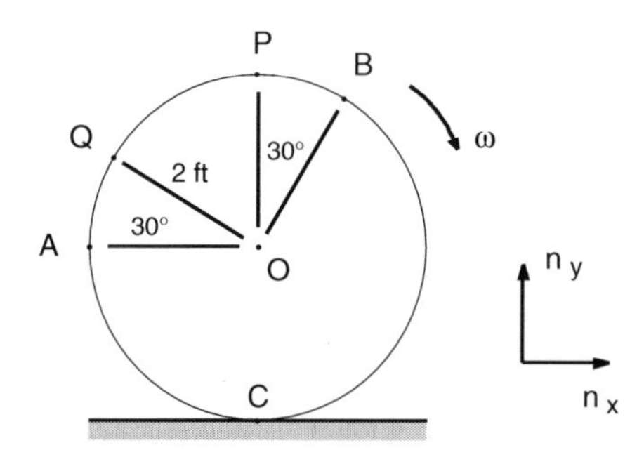

FIGURE P5.4.1
A wheel rolling to the right.

P5.4.2: Repeat Problem P5.4.1 if the wheel is rolling to the left with *O* having a velocity of 3 ft/sec to the left.

P5.4.3: Consider the four-bar linkage shown in Figure P5.4.3. Let link *AB* have a counterclockwise angular speed of 5 rad/sec, as shown.

a. Locate or identify the instant centers of zero velocity for links *AB*, *BC*, and *CD* for the position shown in Figure P5.4.3.
b. Use the results of (a) to determine the velocities of joints *B* and *C* and the angular speeds of links *BC* and *CD*.
c. Express the velocities of *B* and *C* in terms of unit vectors and of Figure P5.4.3.

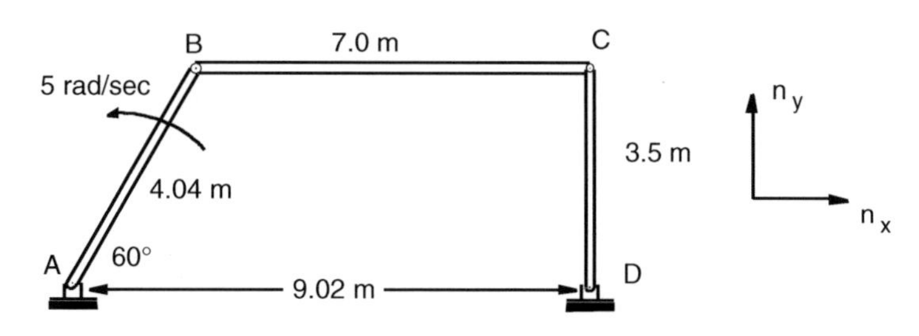

FIGURE P5.4.3
A four-bar linkage.

P5.4.4: Repeat Problem P5.4.3 if the angular speed of link *AB* is 5 rad/sec clockwise.

P5.4.5: Consider the crank, connecting rod, and piston system shown in Figure P5.4.5. Let the crank *OA* have a clockwise angular speed of 300 rpm as shown.

a. Locate or identify the instant center of zero velocity of the crank OA, the connecting rod AP, and the piston P.
b. Use the results of (a) to determine the velocities of A and P and the angular velocity of connecting rod AP.
c. Express the velocities of A and P in terms of the unit vectors $\mathbf{n}_x$ and $\mathbf{n}_y$ of Figure P5.4.5.

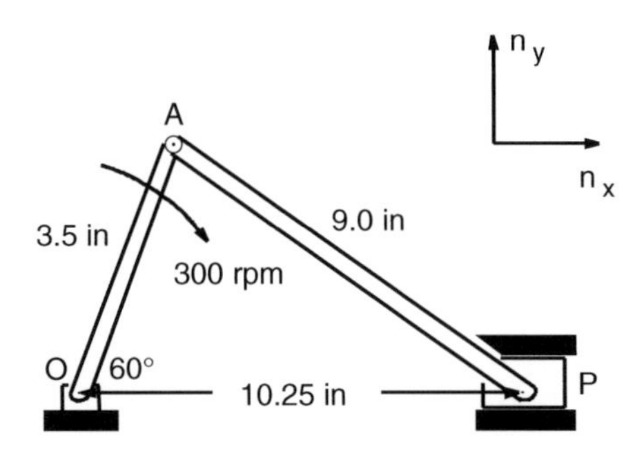

FIGURE P5.4.5
A crank, connecting rod, and piston system.

Section 5.7 Instant Center, Analytical Considerations

P5.7.1: A body B having planar motion has a counterclockwise angular velocity of 7 rad/sec, as represented in Figure P5.7.1. Suppose a point P with coordinates (3, 4) (in meters) relative to X–Y coordinates in a reference frame R has the velocity:

$$\mathbf{V}^P = -6\sqrt{2}\mathbf{n}_x - 6\sqrt{2}\mathbf{n}_y \text{ m/sec}$$

where $\mathbf{n}_x$ and $\mathbf{n}_y$ are the horizontal and vertical unit vectors as in Figure P5.7.1. Determine the coordinates of the instant center C of B and the velocity of point Q of B whose coordinates are (2, 4). Express the velocity of Q in terms of $\mathbf{n}_x$ and $\mathbf{n}_y$.

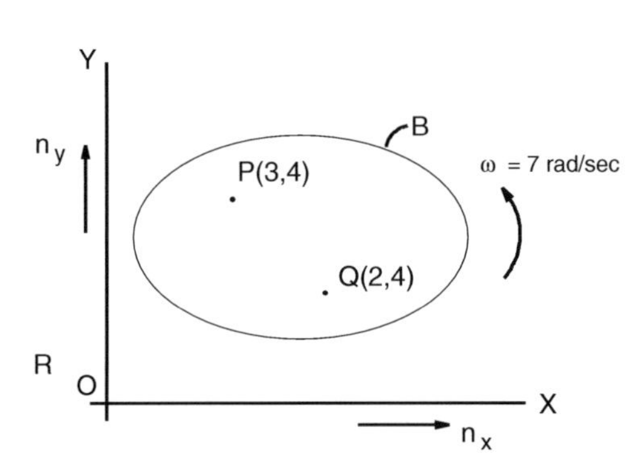

FIGURE P5.7.1
A body B in planar motion with known angular velocity and velocity of one point.

P5.7.2: A body B having planar motion has particle P and Q whose coordinates relative to the X- and Y-axes are (13, 8) and (25, 13) (in centimeters), as in Figure P5.7.2. Let the velocities of P and Q be:

$$\mathbf{V}^P = -25\mathbf{n}_x + 60\mathbf{n}_y \text{ cen/sec}$$

and

$$\mathbf{V}^Q = -50\mathbf{n}_x + 120\mathbf{n}_y \text{ cen/sec}$$

where $\mathbf{n}_x$ and $\mathbf{n}_y$ are horizontal and vertical unit vectors as in Figure P5.7.2. Determine (a) the coordinates of the instant center *C* of *B*, (b) the angular velocity ω of *B*, and (c) the velocity of point *A* of *B* (in terms of $\mathbf{n}_x$ and $\mathbf{n}_y$), whose coordinates are (30, 10).

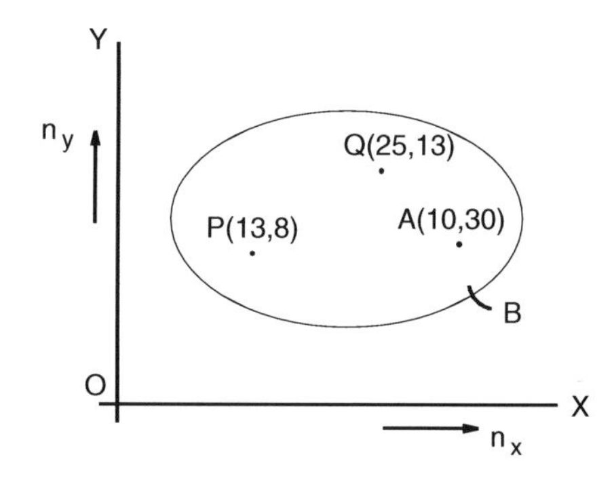

FIGURE P5.7.2
A body *B* in planar motion with known velocities of two points.

Section 5.8 Instant Center of Zero Acceleration

P5.8.1: A 2-ft-radius circular disc rolls to the right, with its center *O* having a speed of 3.5 ft/sec, as in Figure P5.8.1. The disc, however, is decelerating so that *O* has an acceleration to the left of 1.5 ft/sec². Determine the *X*–*Y* coordinates of the center at zero acceleration.

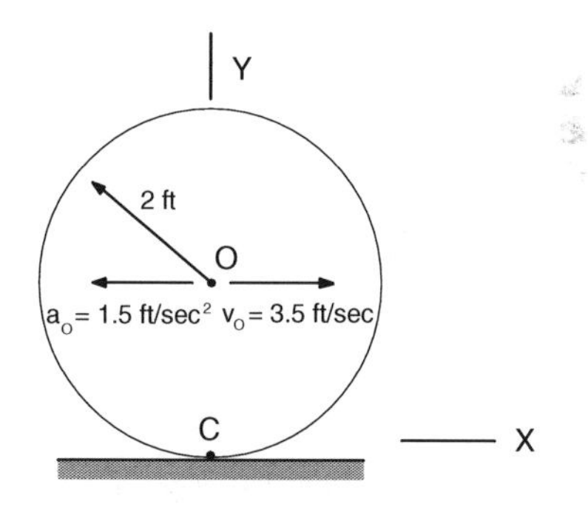

FIGURE P5.8.1
A rolling but decelerating circular disc.

6

Forces and Force Systems

6.1 Introduction

Kinematics is a study of motion without regard to the causes of the motion. Alternatively, *kinetics* is a study of forces causing a motion. While knowledge of forces is based upon experience and intuition and upon studies in elementary physics and statics, it is helpful to review the fundamental concepts again, as they are essential in studying the dynamics of mechanical systems. Unlike unit vectors, position vectors, or velocity vectors, force vectors are generally associated with a point of application. Force vectors are thus *bound* vectors, whereas the other vectors are *free* vectors.

To illustrate the importance of the point of application, consider the rod shown in Figure 6.1.1. Let a force **F** be applied perpendicular to the rod, first at end *A* and then at end *B*. The effect upon the rod of the different points of application is that in the first case (end *A*) the rod will rotate clockwise, whereas in the second case (end *B*) the rod will rotate counterclockwise.

In this chapter, we review and discuss the analysis of forces and sets of forces (force systems) as they are applied with mechanical systems. We will consider various representations of force systems, various kinds of force systems, and various methods of analysis.

6.2 Forces and Moments

If we intuitively define a force as a push or pull applied in some direction at a point, the force may be represented as a *bound vector*. The force then has (1) a magnitude, (2) a line of action (defining its orientation), (3) a sense (*push* or *pull*), and (4) a point of application. Figure 6.2.1 depicts a force **F** acting at a point *P* with line of action *L*.

A *moment* is defined as the rotational effect of a force about a point. This point is generally distinct from the points on the line of action of the force. Specifically, the moment $\mathbf{M}_O$ of a force **F** about a point *O* is defined as:

$$\mathbf{M}_O \stackrel{D}{=} \mathbf{p} \times \mathbf{F} \tag{6.2.1}$$

where **p** is a position vector locating any point *Q*, on the line of action of **F**, relative to *O*. Figure 6.2.2 depicts point *O*, a typical point *Q*, vectors **p** and **F**, and the line of action *L* of **F**.

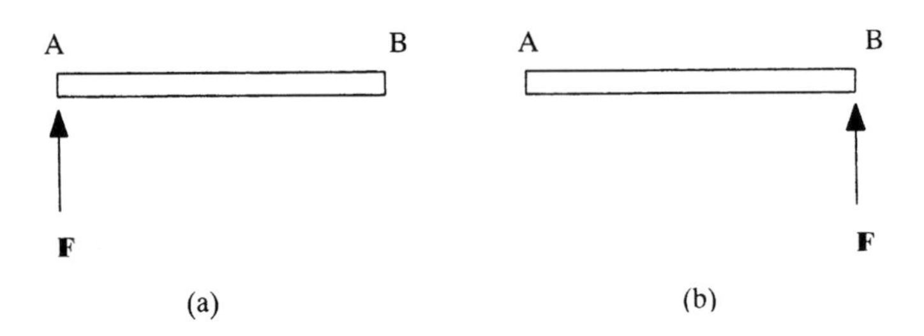

FIGURE 6.1.1
A force applied at different ends of a rod.

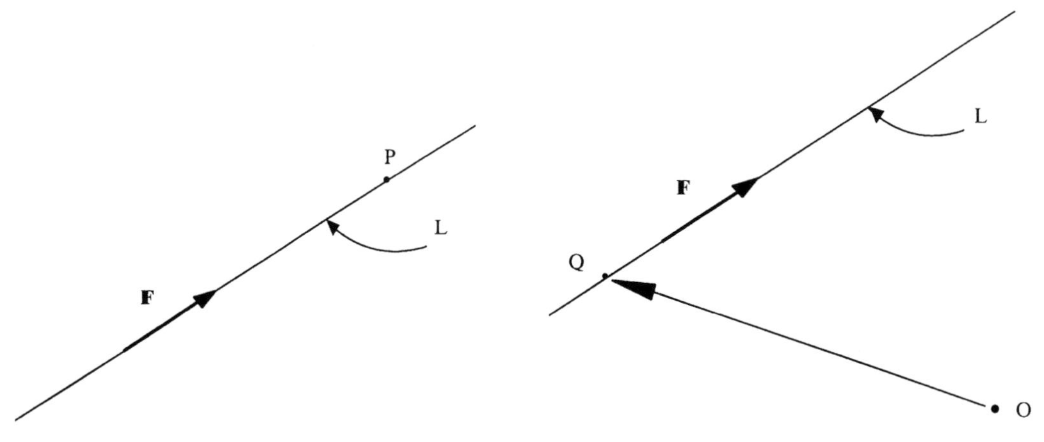

FIGURE 6.2.1
A force **F**, its line of action *L*, and point of application *P*.

FIGURE 6.2.2
A force **F** with line of action *L*, a point *O*, and an arbitrary point *Q* on *L*.

Observe that because *Q* is *any* point on the line of action *L* of **F**, the position vector **p** of Eq. (6.2.1) is not unique. Hence, if $\hat{Q}$ is some point of *L*, distinct from *Q*, the position vector $\hat{\mathbf{p}}$ locating $\hat{Q}$ relative to *O*, could be used in place of **p** in Eq. (6.2.1). That is,

$$\mathbf{M}_O = \mathbf{p} \times \mathbf{F} \quad \text{or alternatively} \quad \mathbf{M}_O = \hat{\mathbf{p}} \times \mathbf{F} \tag{6.2.2}$$

It happens that whether *Q* or $\hat{Q}$ is selected on *L* or whether **p** or $\hat{\mathbf{p}}$ is used in Eq. (6.2.2), the resulting moment $\mathbf{M}_O$ is the same. To see this, consider Figure 6.2.3 depicting points *Q* and $\hat{Q}$ and vectors **p** and $\hat{\mathbf{p}}$. Observe that p, $\hat{\mathbf{p}}$, and the vector $\mathbf{Q\hat{Q}}$ are related by the expression:

$$\hat{\mathbf{p}} = \mathbf{p} + \mathbf{Q\hat{Q}} \tag{6.2.3}$$

Observe further that $\mathbf{Q\hat{Q}}$ is parallel to the force F. Therefore, $\mathbf{M}_O$ becomes:

$$\mathbf{M}_O = \hat{\mathbf{p}} \times \mathbf{F} = \left(\mathbf{p} + \mathbf{Q\hat{Q}}\right) \times \mathbf{F} = \mathbf{p} \times \mathbf{F} = 0 \tag{6.2.4}$$

This expression shows that M_O is unique even though the choice of *Q* on *L* is arbitrary. Because the choice is arbitrary, we can select the position of *Q* so as to simplify the

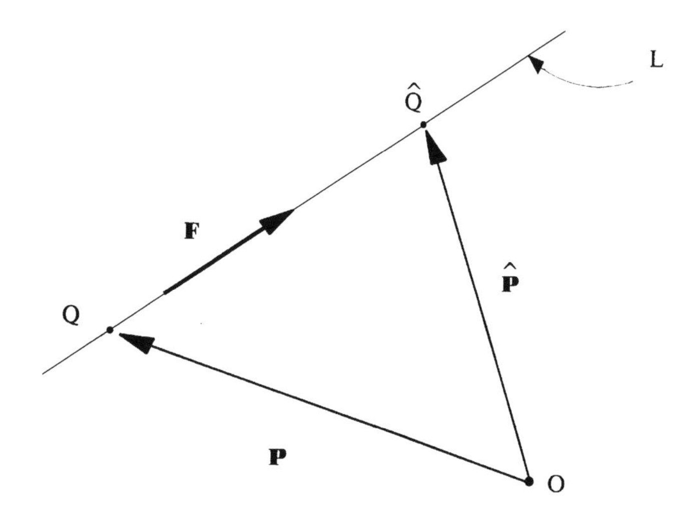

FIGURE 6.2.3
Points Q and Q on L and position vectors $\mathbf{p}$ and $\hat{\mathbf{p}}$.

computation of $\mathbf{M}_O$. For example, Q may be chosen to be the point at the intersection of the line through O and perpendicular to L.

Finally, observe that if O is a point on L, then $\mathbf{M}_O$ is zero.

6.3 Systems of Forces

Consider a mechanical system V consisting of particles and rigid bodies as depicted in Figure 6.3.1. Let S be a set of forces acting on V. That is, let S consist of forces $\mathbf{F}_i$ ($i = 1,\ldots, N$) with lines of action passing through points P_i of the particles and bodies of V as in Figure 6.3.2. Then, S is called a *system of forces*.

Systems of forces — particularly systems with many forces — are conveniently characterized by (and thus represented by) two vectors: the *resultant* and the *moment* about some point O.

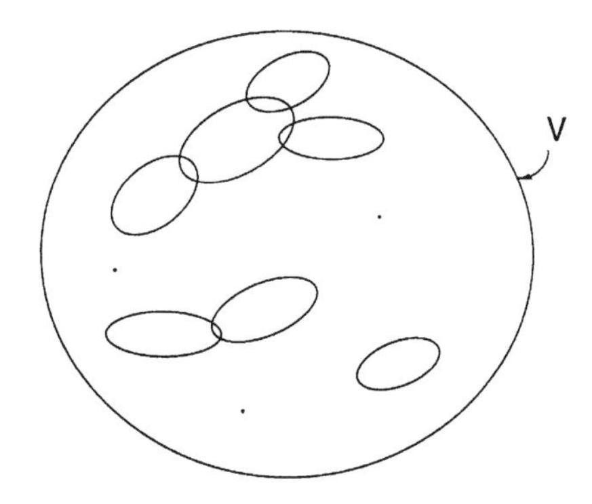

FIGURE 6.3.1
A mechanical system.

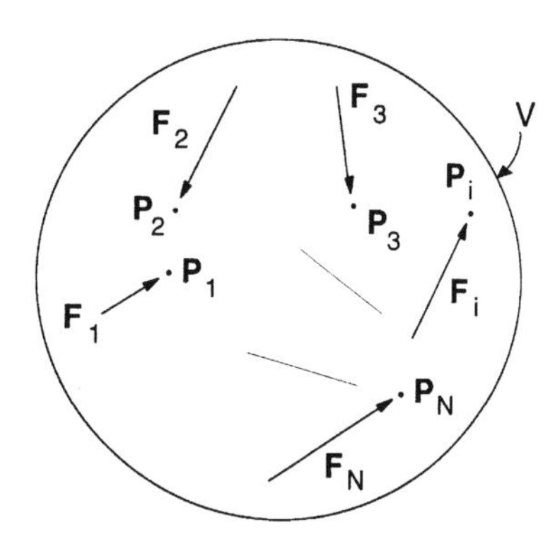

FIGURE 6.3.2
A set of forces acting on a mechanical system.

The resultant **R** of a system of forces S is simply the sum of the forces. That is,

$$\mathbf{R} = \sum_{i=1}^{N} \mathbf{F}_i \tag{6.3.1}$$

Unlike the individual forces of S, the resultant of S is a free vector and is not associated with any particular point of V.

The moment of a system S of forces about a point O, designated by $\mathbf{M}^{S/O}$ or $\mathbf{M}_O$, is simply the sum of the moments of the individual forces of S about O. That is,

$$\mathbf{M}^{S/O} = \mathbf{M}_O = \sum_{i=1}^{N} \mathbf{p}_i \times \mathbf{F}_i \tag{6.3.2}$$

where $\mathbf{p}_i$ is the position vector locating an arbitrary point, typically P_i, on the line of action of $\mathbf{F}_i$, relative to O (see Figure 6.3.3). Like **R**, $\mathbf{M}_O$ is also a free vector.

The resultant of a system of forces is unique. The moment of the system about an arbitrary point O, however, depends upon the location of O. If some other point, say $\hat{O}$, is chosen instead of O, the moment $\mathbf{M}^{S/\hat{O}}$ or $\mathbf{M}_{\hat{O}}$ is generally different than $\mathbf{M}_O$. But, this raises the question: What is the relationship between $\mathbf{M}_{\hat{O}}$ and $\mathbf{M}_O$? To answer this question consider Figure 6.3.4 depicting a force system S, points O and $\hat{O}$, and position vectors $\mathbf{p}_i$ and $\hat{\mathbf{p}}_i$, locating a point on the line of action of typical force $\mathbf{F}_i$, relative to O and $\hat{O}$. Then, from the definition of Eq. (6.3.2) and $\mathbf{M}_O$ and $\mathbf{M}_{\hat{O}}$ are:

$$\mathbf{M}_O = \sum_{i=1}^{N} \mathbf{p}_i \times \mathbf{F}_i \quad \text{and} \quad \mathbf{M}_{\hat{O}} = \sum_{i=1}^{N} \hat{\mathbf{p}}_i \times \mathbf{F}_i \tag{6.3.3}$$

From Figure 6.3.4, we see that $\mathbf{p}_i$ and $\hat{\mathbf{p}}_i$ are related by:

$$\mathbf{p}_i = \hat{\mathbf{p}}_i + \mathbf{O}\hat{\mathbf{O}} \tag{6.3.4}$$

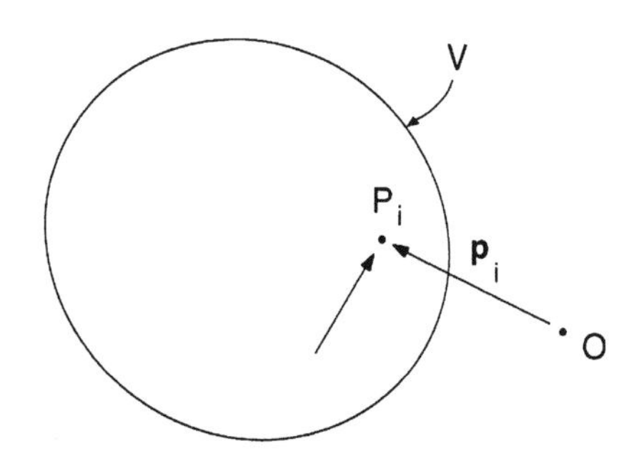

FIGURE 6.3.3
Position vector $\mathbf{p}_i$ for moment calculation.

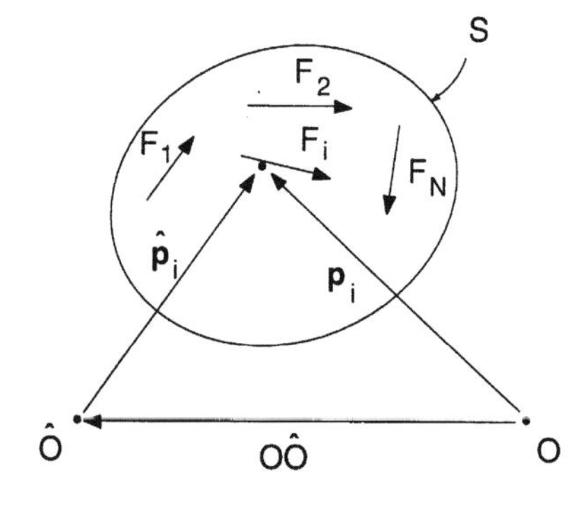

FIGURE 6.3.4
A force system S and points O and $\hat{O}$.

Hence, by substitution into Eq. (6.3.3) we have:

$$\begin{aligned}\mathbf{M}_O &= \sum_{i=1}^{N} \mathbf{p}_i \times \mathbf{F}_i = \sum_{i=1}^{N} \left(\hat{\mathbf{p}}_i + \mathbf{O\hat{O}}\right) \times \mathbf{F}_i = \sum_{i=1}^{N} \hat{\mathbf{p}}_i \times \mathbf{F}_i + \sum_{i=1}^{N} \mathbf{O\hat{O}} \times \mathbf{F}_i \\ &= \mathbf{M}_{\hat{O}} + \mathbf{O\hat{O}} \times \sum_{i=1}^{N} \mathbf{F}_i = \mathbf{M}_{\hat{O}} + \mathbf{O\hat{O}} \times \mathbf{R}\end{aligned} \tag{6.3.5}$$

That is,

$$\mathbf{M} = \mathbf{M}_{\hat{O}} + \mathbf{O\hat{O}} \times \mathbf{R} \quad \text{and then} \quad \mathbf{M}_{\hat{O}} = \mathbf{M}_O + \mathbf{\hat{O}O} \times \mathbf{R} \tag{6.3.6}$$

Equation (6.3.6) shows that if we know the resultant **R** and the moment of S about some point O we can readily find the moment of S about any other point $\hat{O}$.

To illustrate these ideas, consider the box or parallelepiped shown in Figure 6.3.5. Let the vertices of the box be lettered as shown and let the dimensions be $12 \times 4 \times 3$ feet. Let forces be applied to the box as in Figure 6.3.6. Let the forces be directed as shown and let them have magnitudes as:

$$\begin{aligned}&F_1 = 10\,\text{lb},\ F_2 = 10\,\text{lb},\ F_3 = 15\,\text{lb},\ F_4 = 8\,\text{lb} \\ &F_5 = 8\,\text{lb},\ \ F_6 = 26\,\text{lb},\ F_7 = 20\,\text{lb}\end{aligned} \tag{6.3.7}$$

The objective is to find: (1) the resultant of the system of forces; (2) the moment of the system about O; (3) the moment about Q; and (4) the relation between the moments about O and Q.

To perform the analysis, let the forces be expressed in terms of the unit vectors $\mathbf{n}_1$, $\mathbf{n}_2$, and $\mathbf{n}_3$. Because the forces $\mathbf{F}_1, \ldots, \mathbf{F}_5$ are along the box edges, they may readily be expressed in terms of the unit vectors as:

$$\begin{aligned}&\mathbf{F}_1 = 10\mathbf{n}_3\,\text{lb},\ \mathbf{F}_2 = -10\mathbf{n}_3\,\text{lb},\ \mathbf{F}_3 = -15\mathbf{n}_1\,\text{lb} \\ &\mathbf{F}_4 = -8\mathbf{n}_2\,\text{lb},\ \mathbf{F}_5 = 8\mathbf{n}_2\,\text{lb}\end{aligned} \tag{6.3.8}$$

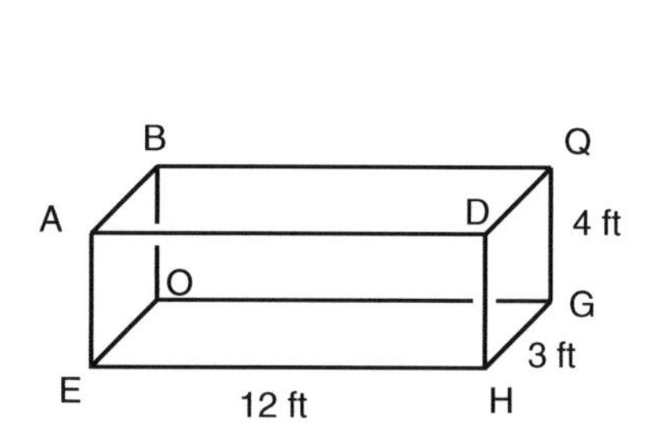

FIGURE 6.3.5
A box loaded as in Figure 6.3.6.

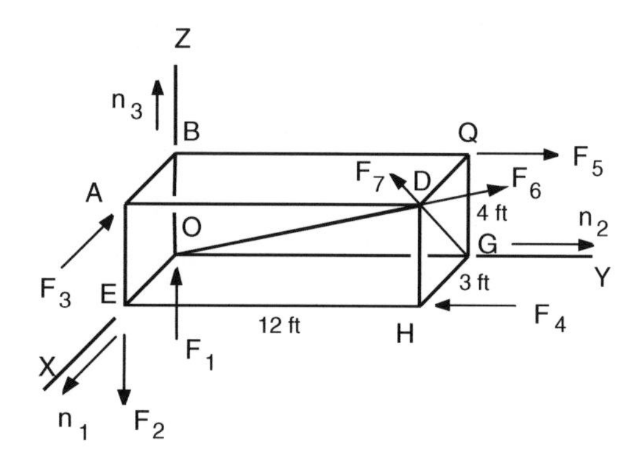

FIGURE 6.3.6
Forces applied to the box.

The forces $\mathbf{F}_6$ and $\mathbf{F}_7$ may be expressed in terms of $\mathbf{n}_1$, $\mathbf{n}_2$, and $\mathbf{n}_3$ by first expressing them in terms of unit vectors $\boldsymbol{\mu}$ and $\mathbf{v}$ along the box diagonals as:

$$\mathbf{F}_6 = 26\boldsymbol{\mu} \quad \text{and} \quad \mathbf{F}_7 = 20\mathbf{v} \tag{6.3.9}$$

where $\boldsymbol{\mu}$ and $\mathbf{v}$ are directed along OQ and GQ, respectively. Then, $\boldsymbol{\mu}$ and $\mathbf{v}$ are:

$$\boldsymbol{\mu} = OQ/|OQ| = (3\mathbf{n}_1 + 12\mathbf{n}_2 + 4\mathbf{n}_3)/13 \tag{6.3.10}$$

and

$$\mathbf{v} = GQ/|GQ| = (3\mathbf{n}_1 + 4\mathbf{n}_3)/5 \tag{6.3.11}$$

Hence, $\mathbf{F}_6$ and $\mathbf{F}_7$ are:

$$\mathbf{F}_6 = 6\mathbf{n}_1 + 24\mathbf{n}_2 + 8\mathbf{n}_3 \text{ lb} \tag{6.3.12}$$

and

$$\mathbf{F}_7 = 12\mathbf{n}_1 + 16\mathbf{n}_3 \text{ lb} \tag{6.3.13}$$

These results may be tabulated as in Table 6.3.1.

The components of the resultant $\mathbf{R}$ are then obtained by adding the columns in Table 6.3.1. That is,

$$\mathbf{R} = 3\mathbf{n}_1 + 24\mathbf{n}_2 + 24\mathbf{n}_3 \text{ lb} \tag{6.3.14}$$

The moment of the system about O may be calculated using Eq. (6.3.2): using convenient position vectors $\mathbf{M}_O$ becomes:

$$\begin{aligned}
\mathbf{M}_O &= OO \times \mathbf{F}_1 + OC \times \mathbf{F}_2 + OB \times \mathbf{F}_3 + \mathbf{F}_3 + OC \times \mathbf{F}_4 \\
&\quad + OB \times \mathbf{F}_5 + OO \times \mathbf{F}_6 + OG \times \mathbf{F}_7 \\
&= 0 + 3\mathbf{n}_1 \times (-10\mathbf{n}_3) + 4\mathbf{n}_3 \times (-15\mathbf{n}_1) + 3\mathbf{n}_1 \times (-8\mathbf{n}_2) \\
&\quad + 4\mathbf{n}_3 \times (8\mathbf{n}_2) + 0 + 12\mathbf{n}_2 \times (12\mathbf{n}_1 + 16\mathbf{n}_3) \\
&= 30\mathbf{n}_2 = 60\mathbf{n}_2 - 24\mathbf{n}_3 - 32\mathbf{n}_1 - 144\mathbf{n}_3 + 192\mathbf{n}_1
\end{aligned}$$

or:

$$\mathbf{M}_O = 160\mathbf{n}_1 - 30\mathbf{n}_2 - 168\mathbf{n}_3 \text{ ft lb} \tag{6.3.15}$$

TABLE 6.3.1

A Component Listing of the Forces of Figure 6.3.6

$\mathbf{F}_i$	$\lvert\mathbf{F}_i\rvert$ (lb)	$\mathbf{n}_1$	$\mathbf{n}_2$	$\mathbf{n}_3$
$\mathbf{F}_1$	10	0	0	10
$\mathbf{F}_2$	10	0	0	–10
$\mathbf{F}_3$	15	–15	0	0
$\mathbf{F}_4$	8	0	–8	0
$\mathbf{F}_5$	8	0	8	0
$\mathbf{F}_6$	26	6	24	8
$\mathbf{F}_7$	20	12	0	16

Similarly, this moment about Q is:

$$\begin{aligned}\mathbf{M}_Q &= QO\times\mathbf{F}_1 + QA\times\mathbf{F}_2 + QB\times\mathbf{F}_3 + QH\times\mathbf{F}_4\\ &\quad + QQ\times\mathbf{F}_5 + QD\times\mathbf{F}_6 + QD\times\mathbf{F}_7\\ &= (-4\mathbf{n}_3 - 12\mathbf{n}_2)\times(10\mathbf{n}_3) + (-12\mathbf{n}_2 + 3\mathbf{n}_1)\times(-10\mathbf{n}_3)\\ &\quad + (-12\mathbf{n}_2)\times(-15\mathbf{n}_1) + (3\mathbf{n}_1 - 4\mathbf{n}_3)\times(-8\mathbf{n}_2)\\ &\quad + O + 3\mathbf{n}_1\times(6\mathbf{n}_1 + 24\mathbf{n}_2 + 8\mathbf{n}_3) + 3\mathbf{n}_1\times(12\mathbf{n}_3 + 16\mathbf{n}_3)\\ &= -120\mathbf{n}_1 + 120\mathbf{n}_1 + 30\mathbf{n}_2 - 180\mathbf{n}_3 - 24\mathbf{n}_3 - 32\mathbf{n}_1\\ &\quad + 72\mathbf{n}_3 - 24\mathbf{n}_2 - 48\mathbf{n}_2\end{aligned}$$

or

$$\mathbf{M}_Q = -32\mathbf{n}_1 - 42\mathbf{n}_2 - 132\mathbf{n}_3 \text{ ft lb} \tag{6.3.16}$$

According to Eq. (6.3.6), $\mathbf{M}_O$ and $\mathbf{M}_Q$ are related by the expression:

$$\mathbf{M}_O = \mathbf{M}_Q + \mathbf{OQ}\times\mathbf{R} \tag{6.3.17}$$

By substituting from Eqs. (6.3.14), (6.3.15), and (6.3.16), we can check the validity of Eq. (6.3.17):

$$\begin{aligned}160\mathbf{n}_1 - 30\mathbf{n}_2 - 168\mathbf{n}_3 &\stackrel{?}{=} -32\mathbf{n}_1 - 42\mathbf{n}_2 - 132\mathbf{n}_3\\ &\quad + (12\mathbf{n}_2 + 4\mathbf{n}_3)\times(3\mathbf{n}_1 + 24\mathbf{n}_2 + 24\mathbf{n}_3)\\ &= -32\mathbf{n}_1 - 42\mathbf{n}_2 - 132\mathbf{n}_3 - 36\mathbf{n}_3 + 288\mathbf{n}_1 + 12\mathbf{n}_2 - 96\mathbf{n}_1\\ &= 160\mathbf{n}_1 - 30\mathbf{n}_2 - 168\mathbf{n}_3\end{aligned}$$

Equation (6.3.17) is thus verified.

6.4 Zero Force Systems

If a system of forces has a zero resultant and a zero moment about some point O, it is called a *zero force system*. Recall from Eq. (6.3.6) the moment of the system about a point $\hat{O}$, distinct from O, is:

$$\mathbf{M}_{\hat{O}} = \mathbf{M}_O + \hat{\mathbf{O}}\mathbf{O} \times \mathbf{R} \tag{6.4.1}$$

Thus, if $\mathbf{R}$ is zero, $\mathbf{M}_O$ and $\mathbf{M}_{\hat{O}}$ are equal. This means that if the resultant is zero, the moment of the force system about all points is the same. Then, if $\mathbf{M}_O$ is also zero, the moment of the force system about all points is zero.

The concept of a zero force system is the basic principle of static analysis: If a mechanical system is in static equilibrium, the set of all the forces exerted on the system is a zero force system.

6.5 Couples

Closely related to zero force systems is the concept of a *couple*. Specifically, a force system that has a zero resultant but a *non-zero* moment about some point O is called a couple. Observe again from Eq. (6.4.1) that if $\mathbf{R}$ is zero, the force system has the same moment about all points. Thus, a couple has the same moment about all points. This moment is called the *torque* of the couple.

A couple may have a large number of forces or it may have as few as two forces (see Figure 6.4.1). If a couple has only two forces it is called a *simple couple*. If a force system S has many forces, it is often tedious and time consuming to conduct an analysis of the effect of S on a physical system. This difficulty can be overcome by the use of *equivalent force systems*. Two force systems are said to be *equivalent* if they have equal resultants and equal moments about some point.

Observe from Eq. (6.3.6) that if two force systems (say S and $\hat{S}$) have equal resultants ($\mathbf{R} = \hat{\mathbf{R}}$) and equal moments about some point O, then they have equal moments about all points. To see this, let Q be any point distinct from O. Then, from Eq. (6.3.6) we have:

$$\mathbf{M}^{S/Q} = \mathbf{M}^{S/O} + \mathbf{QO} \times \mathbf{R} \quad \text{and} \quad \mathbf{M}^{\hat{S}/Q} = \mathbf{M}^{\hat{S}/O} + \mathbf{QO} \times \hat{\mathbf{R}} \tag{6.5.1}$$

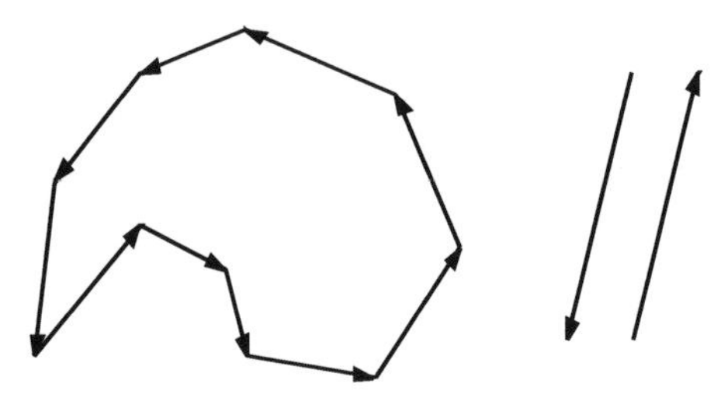

FIGURE 6.4.1
Couples.

With the right sides of these equations being equal, we have:

$$\mathbf{M}^{S/Q} = \mathbf{M}^{\hat{S}/Q} \tag{6.5.2}$$

If two force systems are equivalent and if one of the force systems contains fewer forces than the other, it would follow that the force system with the fewer forces would be easier to work with than the one with the greater number of forces.

To illustrate these concepts consider again the box of Section 6.3 loaded with seven forces as in Figure 6.5.1. Recall that the resultant of these seven forces and the moment of the forces about point O were found to be (see Eqs. (6.3.14) and (6.3.15)):

$$\mathbf{R} = 3\mathbf{n}_1 + 24\mathbf{n}_2 + 24\mathbf{n}_3 \text{ lb} \tag{6.5.3}$$

and

$$\mathbf{M}_O = 160\mathbf{n}_1 - 30\mathbf{n}_2 - 168\mathbf{n}_3 \text{ ft lb} \tag{6.5.4}$$

Consider now the same box but with the force systems as depicted in Figure 6.5.2. **F** is a force equal to **R** with line of action through O and where C represents a couple whose torque T is equal to $\mathbf{M}_O$. This force system is readily seen to be equivalent to the force system of Figure 6.5.1. That is, the resultant of each force system is **R** and the moment of each force system about O is $\mathbf{M}_O$.

The major advantage of equivalent force systems is that they may be interchanged in analyses of physical systems containing rigid bodies. If two force systems S and $\hat{S}$ are equivalent and if S has fewer forces than $\hat{S}$, then S is said to be a *reduction* of $\hat{S}$. As noted earlier, a reduced force system is generally more convenient in analyses simply because there are fewer vectors to consider.

Equivalent force systems cannot be interchanged when physical systems have flexible or deformable bodies. Indeed, for nonrigid bodies, equivalent force systems may produce vastly different effects. Consider for example the extensible rod shown in Figure 6.5.3. Let this rod be subjected to two different but equivalent force systems as in Figure 6.5.4. The two force systems are seen to be equivalent because they are both zero systems; however, the one produces tension and extension of the bar while the other produces compression and contraction of the bar.

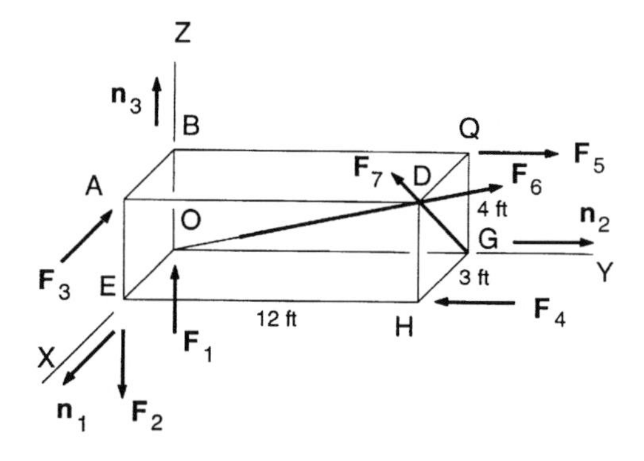

FIGURE 6.5.1
Forces applied to the box.

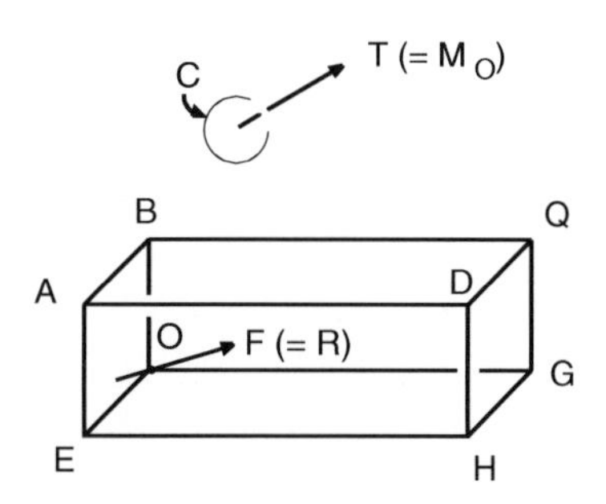

FIGURE 6.5.2
Equivalent force system.

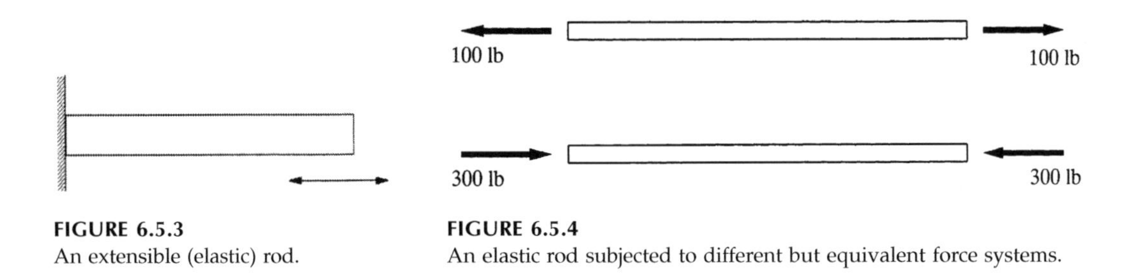

FIGURE 6.5.3
An extensible (elastic) rod.

FIGURE 6.5.4
An elastic rod subjected to different but equivalent force systems.

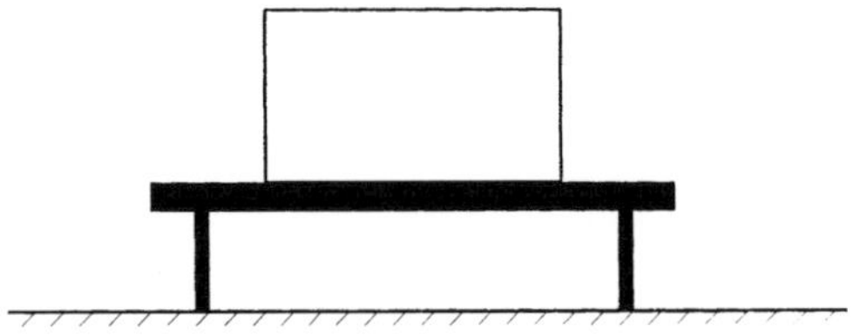

FIGURE 6.5.5
A homogeneous block resting on a table.

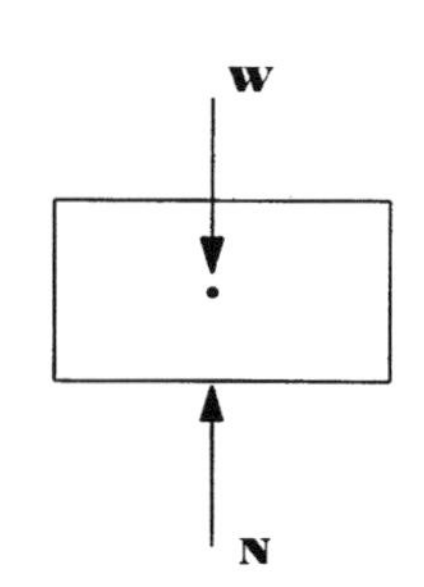

FIGURE 6.5.6
A free-body diagram of a homogeneous block.

As a second example consider the force analysis of a homogeneous block resting on a table as in Figure 6.5.5. An initial step in a force analysis of the block is to construct a *free-body diagram* of the block — that is, a sketch of the block with all forces acting on the block shown in the sketch. A typical free-body diagram of the block of Figure 6.5.5 might be that as shown in Figure 6.5.6, where W represents the weight or gravitational force on the block and N represents the supporting contact force from the table on the block.

A little reflection reveals that W and N are not the *actual* physical forces on the block but instead are simply a representation of the physical forces. Indeed, for a homogeneous block, gravitational forces are acting on each particle of the block. These forces are distributed uniformly throughout the volume occupied by the block. Similarly, for a smooth table surface with ideal (flat) surface geometry, the contact forces are distributed uniformly across the surface. Thus, the actual physical gravitational and contact forces are very large in number. The forces W and N are forces *equivalent* to the large number of gravitational and contact forces. If the block is rigid (and thus infinitely strong), the effect on the block of the physical forces and of W and N are the same. That is, W and N provide an accurate representation of the physical forces. Also, by being much fewer in number, W and N lead to a simple analysis. That is,

$$W = N \tag{6.5.5}$$

If, however, the block is not rigid, then, with W and N being concentrated forces, infinite stresses occur at their points of application, leading to deformation of the block.

This raises questions as to when equivalent forces can have meaningful application with actual physical systems, because no physical systems are perfectly rigid or infinitely strong. To answer this question, consider that even though physical bodies, such as structural members, are not rigid, they often behave like rigid bodies in that their geometry does not appreciably change when they are subject to customary loads. This observation forms

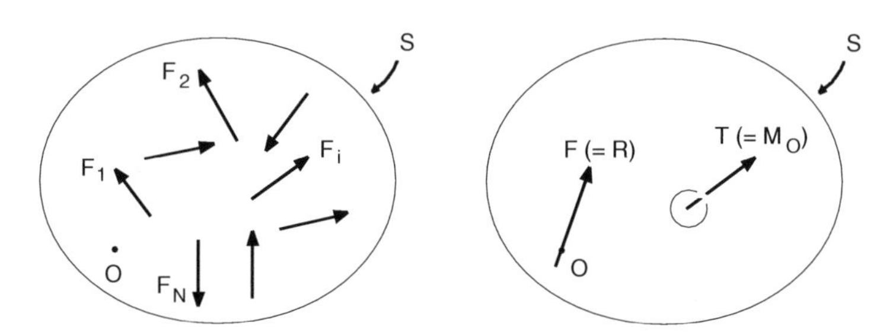

FIGURE 6.5.7
Equivalent force systems.

the basis of Saint Venant's principle, which states that if two equivalent force systems are applied to a physical body, the stress distributions resulting from the force systems are different, but the difference is only significant in the regions of application of the force systems. At points distant from the regions of force applications, the stress distributions are essentially the same. That is, equivalent but distinct force systems produce distinct effects locally, but the same effects globally. This means that even for nonrigid, nonideal bodies equivalent force systems may often be interchanged for the purpose of global analysis.

Finally, observe that for every force system there is an equivalent force system consisting of a single force passing through an arbitrary point together with a couple. That is, for rigid (or nearly rigid) bodies every force system may be replaced by an equivalent force system consisting of a single force and a couple.

To see this let S be a given force system. Let $\mathbf{R}$ be the resultant of S and let $\mathbf{M}_O$ be the moment of S about some point O. Let $\hat{S}$ be a force system consisting of a single force $\mathbf{F}$, equal to $\mathbf{R}$, with the line of action passing through O together with a couple with torque $\mathbf{T}$ equal to $\mathbf{M}_O$ (see Figure 6.5.7). Then, S and $\hat{S}$ are equivalent because they have equal resultants and equal moments about point O.

6.6 Wrenches

Consider again the example of the equivalent force system of the forces on the box of the foregoing section (see Figures 6.5.1 and 6.5.2). Recall, in that example, that we replaced the original force system by a force system consisting of a single force $\mathbf{F}$ passing through O together with a couple with torque $\mathbf{T}$ (see Figure 6.5.2). Suppose that instead of having the line of action of $\mathbf{F}$ pass through O we pass it through Q as in Figure 6.6.1. Then, for the force systems to be equivalent, we must adjust the torque of the couple, so that now the torque is $\mathbf{M}_Q$, which is the moment of the original system of forces about Q.

Observe that aside from the placement of the force $\mathbf{F}$ the only difference in the two equivalent force systems is in the torques $\mathbf{T}$ and $\hat{\mathbf{T}}$ of the accompanying couples. For the system of Figure 6.5.2, the torque $\mathbf{T}$ is $\mathbf{M}_O$, whereas here (Figure 6.6.1) the torque $\hat{\mathbf{T}}$ is $\mathbf{M}_Q$. From Eqs. (6.3.15) and (6.3.16), $\mathbf{M}_O$ and $\mathbf{M}_Q$ are:

$$\mathbf{M}_O = 160\mathbf{n}_1 - 30\mathbf{n}_2 - 168\mathbf{n}_3 \text{ ft lb} \quad \text{and} \quad \mathbf{M}_Q = -32\mathbf{n}_1 - 42\mathbf{n}_2 - 132\mathbf{n}_3 \text{ ft lb} \tag{6.6.1}$$

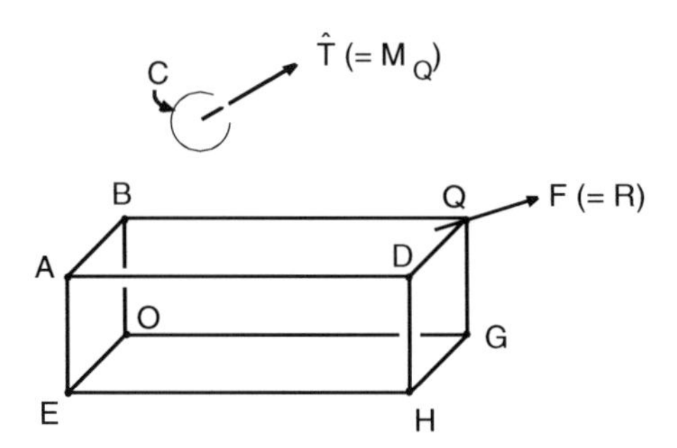

FIGURE 6.6.1
Alternative equivalent force system on box (see Figure 6.5.2).

Their magnitudes are:

$$|\mathbf{M}_O| = |\mathbf{T}| = 243\ \text{ft lb} \quad \text{and} \quad |\mathbf{M}_Q| = |\hat{\mathbf{T}}| = 142\ \text{ft lb} \tag{6.6.2}$$

Because the magnitudes of $\mathbf{M}_O$ and $\mathbf{M}_Q$ are different, the question arises as to whether there is a point Q^* such that, if the line of action of $\mathbf{F}$ is placed through Q^*, then the magnitude of the torque is a minimum.

$$\mathbf{M}_O = \mathbf{M}_{\hat{O}} + \mathbf{O\hat{O}} \times \mathbf{R} \tag{6.6.3}$$

To answer this question, consider again Eq. (6.3.6), which provides a relationship between force system moments about distinct points: let $\mathbf{n}_R$ be a unit vector parallel to $\mathbf{R}$ and consider the projection of $\mathbf{M}_O$ and $\mathbf{M}_{\hat{O}}$ along $\mathbf{n}_R$:

$$\mathbf{n}_R \cdot \mathbf{M}_O = \mathbf{n}_R \cdot \mathbf{M}_{\hat{O}} + \mathbf{n}_R \cdot \mathbf{O\hat{O}} \times \mathbf{R} \tag{6.6.4}$$

where the last term is zero because $\mathbf{N}_R$ is parallel to $\mathbf{R}$.

Eq. (6.6.4) shows that the projections of $\mathbf{M}_O$ and $M_{\hat{O}}$ parallel to $\mathbf{n}_R$ are equal. This means that, for a given force system, the projections in the $\mathbf{R}$ direction of the moments about all points are the same. Therefore, if Q is any point, $\mathbf{M}_Q$ may be expressed as:

$$\mathbf{M}_Q = \mathbf{M}^* \mathbf{n}_R + \mathbf{M}_\perp \mathbf{n}_\perp \tag{6.6.5}$$

where $n_\perp$ is a unit vector perpendicular to $\mathbf{n}_R$ (and $\mathbf{R}$), and M^* and $M_\perp$ are scalar components of $\mathbf{M}_Q$ in the $\mathbf{n}_R$ and $\mathbf{n}_\perp$ directions. The component M^* is the same for all points Q. That is, if $\hat{Q}$ is any other point, then $\mathbf{M}_{\hat{Q}}$ has the form:

$$\mathbf{M}_{\hat{Q}} = \mathbf{M}^* \mathbf{n}_R + \hat{\mathbf{M}}_\perp \mathbf{n}_\perp \tag{6.6.6}$$

From Eqs. (6.6.5) and (6.6.6) we see that the squares of the magnitudes of $\mathbf{M}_Q$ and $\mathbf{M}_{\hat{Q}}$ are:

$$|\mathbf{M}_Q|^2 = \mathbf{M}^{*2} + \mathbf{M}_\perp^2 \quad \text{and} \quad |\mathbf{M}_{\hat{Q}}|^2 = \mathbf{M}^{*2} + \hat{\mathbf{M}}_\perp^2 \tag{6.6.7}$$

These expressions show that if there is some point Q^* such that the component of $\mathbf{M}_{Q^*}$ along $n_\perp$ is zero, then $|\mathbf{M}_{Q^*}|$ will be smaller or, at most, equal to the magnitude of the moment about any other point. That is, if Q is any point, then:

$$|\mathbf{M}_Q| \geq |\mathbf{M}_{Q^*}| = \mathbf{M}^* \tag{6.6.8}$$

The question that remains, however, is do such points Q^* exist? To answer this question, observe that if Q^* exists, then for any point O we have:

$$\mathbf{M}_{Q^*} = \mathbf{M}_O + \mathbf{Q}^*\mathbf{O} \times \mathbf{R} = \mathbf{M}^*\mathbf{n}_R \tag{6.6.9}$$

If we consider the vector product of $\mathbf{n}_R$ with the terms of this equation, we have:

$$\mathbf{n}_R \times \mathbf{M}_O + \mathbf{n}_R \times (\mathbf{Q}^*\mathbf{O} \times \mathbf{R}) = 0 \tag{6.6.10}$$

By expanding the triple product and by dividing by $|\mathbf{R}|$, we obtain:

$$\frac{(\mathbf{n}_R \times \mathbf{M}_O)}{|\mathbf{R}|} + \mathbf{Q}^*\mathbf{O} - (\mathbf{n}_R \cdot \mathbf{Q}^*\mathbf{O})\mathbf{n}_R \tag{6.6.11}$$

Solving for $\mathbf{OQ}^*$, we obtain:

$$\mathbf{OQ}^* = -\mathbf{Q}^*\mathbf{O} = \frac{\mathbf{R} \times \mathbf{M}_O}{\mathbf{R}^2} + (\mathbf{OQ}^* \cdot \mathbf{n}_R)\mathbf{n}_R \tag{6.6.12}$$

Observe that the terms $(\mathbf{R} \times \mathbf{M}_O)/\mathbf{R}^2$ and $(\mathbf{OQ}^* \bullet \mathbf{n}_R)\mathbf{n}_R$ are perpendicular ($\mathbf{R} \times \mathbf{M}_O$ is perpendicular to $\mathbf{R}$, and $\mathbf{n}_R$ is parallel to $\mathbf{R}$). Hence, let Q^* be selected such that:

$$\mathbf{OQ}^* = (\mathbf{R} \times \mathbf{M}_O)/\mathbf{R}^2 \tag{6.6.13}$$

Then, from Eq. (6.6.9), we have:

$$\begin{aligned}
\mathbf{M}_Q^* &= \mathbf{M}_O + \mathbf{Q}^*\mathbf{O} \times \mathbf{R} = \mathbf{M}_O - \mathbf{OQ}^* \times \mathbf{R} \\
&= \mathbf{M}_O - \frac{(\mathbf{R} \times \mathbf{M}_O)}{\mathbf{R}^2} \times \mathbf{R} \\
&= \mathbf{M}_O - \mathbf{M}_O + \frac{\mathbf{M}_O \cdot \mathbf{R}}{\mathbf{R}^2} \\
&= (\mathbf{M}_O \cdot \mathbf{n}_R)\mathbf{n}_R = M^*\mathbf{n}_R
\end{aligned} \tag{6.6.14}$$

Therefore, if Q^* is located according to Eq. (6.6.13), then $|\mathbf{M}_Q^*|$ is smaller than (or at most equal to) the magnitude of the moment about any other point. That is, $\mathbf{M}_Q^*$ is a *minimum moment*. Moreover, upon examination of Eqs. (6.6.11) to (6.6.14) we see that if L is a line through Q^* and parallel to $\mathbf{n}_R$, then all points on L produce minimum moments.

To summarize: if S is a given force system, and if O is an arbitrarily chosen point, then there exists an equivalent force system S^* consisting of a single force $\mathbf{F}^*$ passing through a point Q^* together with a couple with torque $\mathbf{M}^*$ where $\mathbf{F}^*$ and $\mathbf{M}^*$ are:

$$\mathbf{F}^* = \mathbf{R}(\text{resultant of S}) \tag{6.6.15}$$

and

$$\mathbf{M}^* = (\mathbf{M}_O \cdot \mathbf{n}_R)\mathbf{n}_R \quad (\mathbf{n}_R = \mathbf{R}/|\mathbf{R}|) \tag{6.6.16}$$

and where Q^* is located relative to O by the expression:

$$\mathbf{OQ} = (\mathbf{R} \times \mathbf{M}_O)/\mathbf{R}^2 \tag{6.6.17}$$

The magnitude of $\mathbf{M}^*$ will be smaller (or, at most, equal to) the magnitude of the torque of any other equivalent force–couple system.

Observe that $\mathbf{F}^*$ is parallel to $\mathbf{M}^*$. This equivalent force system thus has the characteristics of a push or a pull with a twist about the line of action of the push or pull. Such a system is called a *wrench* (although it might be better be described as a *screwdriver*).

As an example, consider again the force system on the box as in Figures 6.3.6 and 6.5.1 and reproduced in Figure 6.6.2. Recall from Eqs. (6.3.14) and (6.3.15) that the resultant and the moment of the system about point O are:

$$\mathbf{R} = 3\mathbf{n}_1 + 24\mathbf{n}_2 + 24\mathbf{n}_3 \text{ lb} \tag{6.6.18}$$

and

$$\mathbf{M}_O = 160\mathbf{n}_1 - 30\mathbf{n}_2 - 168\mathbf{n}_3 \text{ ft lb} \tag{6.6.19}$$

Then, from Eqs. (6.6.15) and (6.6.16), the wrench equivalent to the force system consists of the single force $\mathbf{F}^*$ and couple with the torque $\mathbf{M}^*$ where $\mathbf{F}^*$ and $\mathbf{M}^*$ are:

$$\mathbf{F}^* = \mathbf{R} = 3\mathbf{n}_1 + 24\mathbf{n}_2 + 24\mathbf{n}_3 \text{ lb} \tag{6.6.20}$$

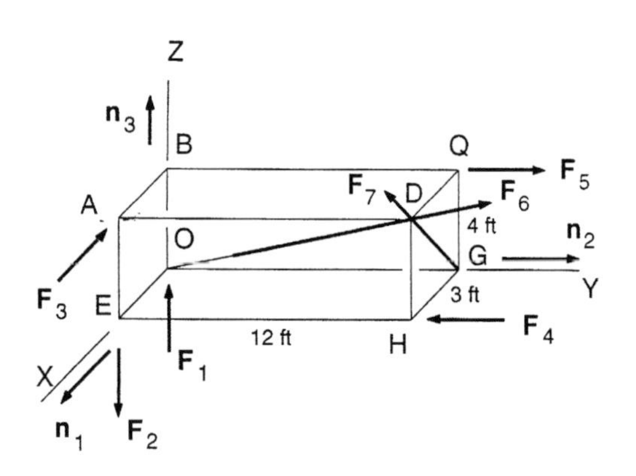

FIGURE 6.6.2
Force system applied to a box.

and

$$\mathbf{M}^* = (\mathbf{M}_O \cdot \mathbf{n}_R)\mathbf{n}_R = -11.03m_1 - 88.22m_2 - 88.22m_3 \text{ ft lb} \tag{6.6.21}$$

$\mathbf{F}^*$ passes through point Q located relative to O by the expression:

$$\mathbf{OQ}^* = (\mathbf{R} \times \mathbf{M}_O)/\mathbf{R}^2 = -2.853\mathbf{n}_1 + 3.742\mathbf{n}_2 - 3.385\mathbf{n}_3 \tag{6.6.22}$$

Observe that Q^* is not a geometrically significant point of the box. This means that wrenches, although physically simple, may not be especially convenient in practical problems.

6.7 Physical Forces: Applied (Active) Forces

The forces acting upon a mechanical system may be divided into two categories: *applied forces* and *inertia forces*, or alternatively into *active forces* and *passive forces*, respectively. Applied (or active) forces arise externally to the system. They are generally composed of gravity, spring, and/or contact forces. Occasionally, they will arise from electrical, magnetic, or radiation fields. Inertia (or passive) forces arise due to motion (that is, acceleration) of the system.

For ideal mechanical systems with rigid bodies, we may replace the forces by equivalent force systems. For nonideal systems, but for systems with nearly rigid bodies whose geometry does not change significantly during the application of the forces, we may still replace the forces by an equivalent force system if we neglect local effects at the points of application (see Section 6.5).

6.7.1 Gravitational Forces

For gravitational (or *weight*) forces, we may replace the forces on bodies by a single force passing through the *center of gravity* of the body. For bodies near the surface of the Earth, having dimensions small compared with the radius of the Earth, the center of gravity may be identified with the *mass center* of the body (see Section 6.8). For homogeneous bodies, the mass center is at the centroid of the geometrical figure occupied by the body. The magnitude of the single force is simply the weight of the body (the product of the mass and the gravity acceleration). The line of action of the force passes through the mass center and the Earth center, and the force is directed toward the Earth center.

For nonhomogeneous bodies, the mass center must be located using definitions and procedures presented in Section 6.8. For systems remote from the Earth's surface, if the gravitational forces are replaced by a single force passing through the mass center, there must be an accompanying couple. The torque of this couple (called the *gravitational moment*) is generally non-zero (although in many cases its magnitude may be insignificant). For systems with a significant gravitational moment, it is also possible to find a point through which the gravitational forces may be replaced by an equivalent single force. This point is then the center of gravity. In the sequel, we will not consider such systems. We will therefore make no distinction between the mass center and the center of gravity.

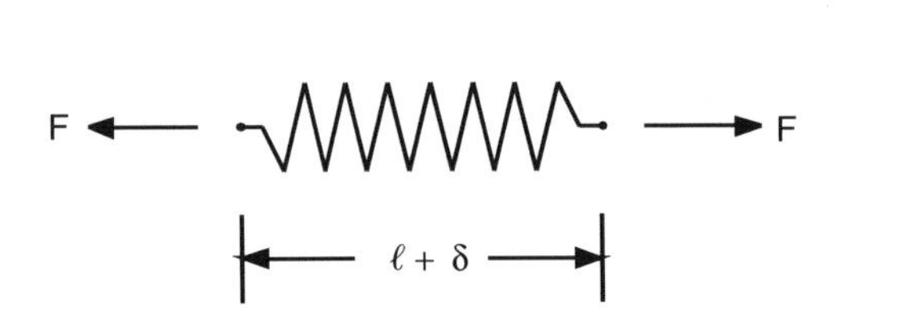

FIGURE 6.7.1
A coil spring subjected to tension forces.

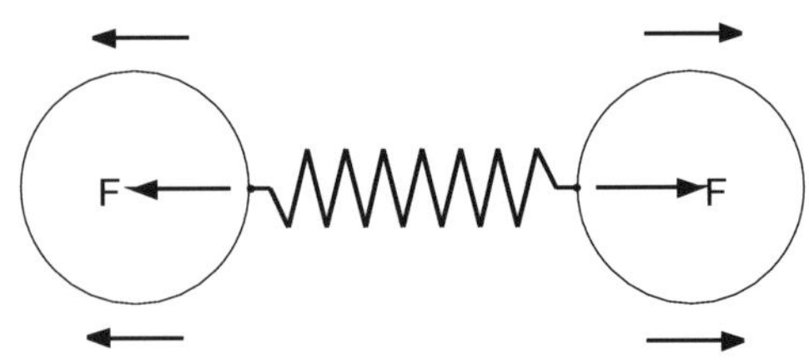

FIGURE 6.7.2
Separating bodies, elongating a spring.

6.7.2 Spring Forces

Spring forces and moments are used to model the effects of spring components within or external to a mechanical system. These spring components may be coil springs, leaf springs, torsion springs, elastic bands, bumper stops, or even flexible/elastic bodies. Spring forces and moments are usually approximated as proportional to the deformation of the spring component being considered. For example, a coil spring might be represented as in Figure 6.7.1 where F is a tension force applied along the axis of the spring, ℓ is the natural (unstretched) length of the spring, and δ is the elongation caused by the tension force. Then, F and δ are approximately related by the expression:

$$F = k\delta \tag{6.7.1}$$

where k is a positive constant called the *spring modulus*.

If the coil spring of Figure 6.7.1 is subjected to axial compressive forces F, the spring will shorten, and, if δ measures the shortening, Eq. (6.7.1) also provides the approximate relation between F and δ. Tension forces and elongation are usually considered to be positive with compression forces and shortening regarded as negative. Observe, however, that if two bodies of a mechanical system exert tension forces on a spring so as to elongate the spring, as in Figure 6.7.2, then the spring will react by exerting forces on the bodies, tending to bring the bodies closer together.

Similarly, torsion springs might be represented as in Figure 6.7.3 where M is a moment (for example, a couple torque) applied along the axis of the spring, and θ is the resulting angular deformation of the spring. Then, M and θ are approximately related by the expression:

$$M = \kappa\theta \tag{6.7.2}$$

where κ is a positive constant.

Multiple springs, or combinations of springs, are often employed in mechanical systems. Specifically, springs occur in series and in parallel with one another. Consider first a series of two springs as in Figure 6.7.4. Let the springs have moduli k_1 and k_2 and unstretched lengths ℓ_1 and ℓ_2. If this spring series is then subjected to a tension force F, the spring combination will elongate or stretch. Let δ be the elongation. If δ_1 and δ_2 are the elongation of the individual springs, then δ is simply:

$$\delta = \delta_1 + \delta_2 \tag{6.7.3}$$

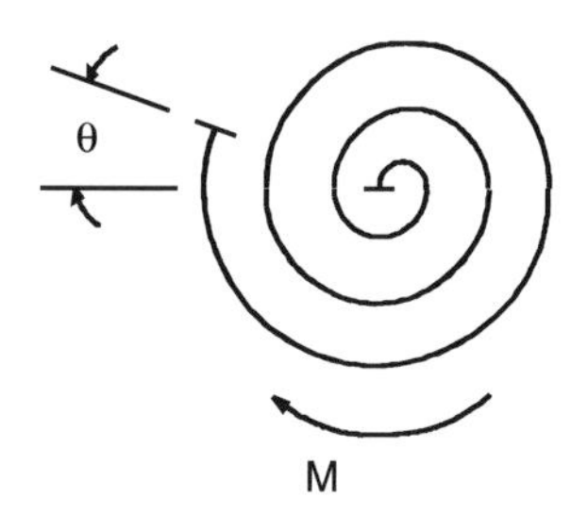

FIGURE 6.7.3
A representation of a torsional spring subjected to a turning moment.

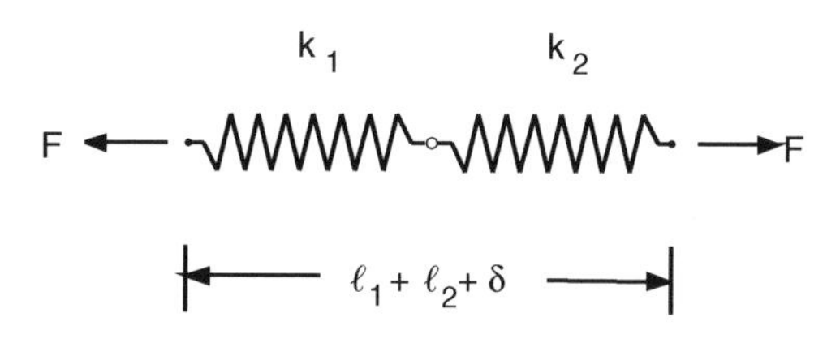

FIGURE 6.7.4
Two springs in series.

Because the springs are in series, the force is the same in each spring. Therefore, from Eq. (6.7.1) δ_1 and δ_2 are:

$$\delta_1 = F/k_1 \quad \text{and} \quad \delta_2 = F/k_2 \tag{6.7.4}$$

Hence, from Eq. (6.7.3) δ is:

$$\delta = F/k_1 + F/k_2 = F\left(\frac{1}{k_1} + \frac{1}{k_2}\right) \tag{6.7.5}$$

Solving for F we obtain:

$$F = \left(\frac{k_1 k_2}{k_1 + k_2}\right)\delta \tag{6.7.6}$$

where the ratio $k_1k_2/(k_1 + k_2)$ is the equivalent modulus k for the spring series.

Observe that Eq. (6.7.6) also holds if the spring series is compressed. Also, by repeated use of Eq. (6.7.6) we can obtain equivalent moduli for any number of springs in a series.

Consider next a pair of springs parallel as in Figure 6.7.5. As before, let k_1 and k_2 be the moduli of the springs. Let F be a tension force on the spring combination, let ℓ be the unstretched length of the springs, and let δ be the elongation of the springs. In this case, the springs elongate the same amount, whereas the forces in the springs are different. That is,

$$\delta = \delta_1 = \delta_2 \tag{6.7.7}$$

thus,

$$F_1/k_1 = F_2/k_2 = \delta \tag{6.7.8}$$

where F_1 and F_2 are the individual spring forces. From equilibrium considerations of the spring ends we have:

$$F = F_1 + F_2 \tag{6.7.9}$$

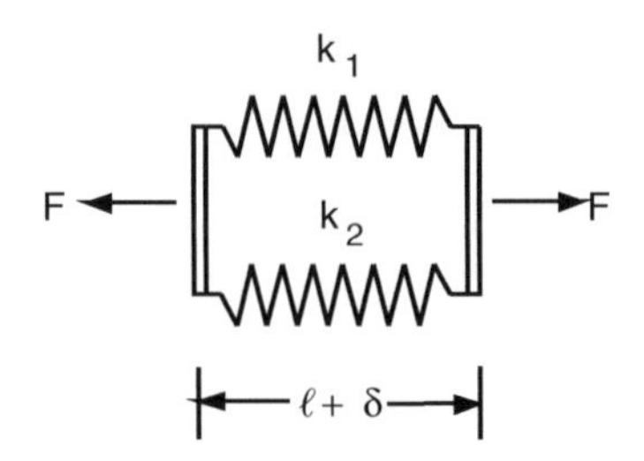

FIGURE 6.7.5
Two springs in parallel.

Therefore, by combining Eqs. (6.7.8) and (6.7.9), we obtain:

$$F = k_1\delta + k_2\delta = (k_1 + k_2)\delta \tag{6.7.10}$$

where the sum $k_1 + k_2$ is now the equivalent spring modulus.

Observe that Eq. (6.7.10) may be expanded to include any number of springs in parallel, and that the expression is also valid for springs in compression. Finally, observe that Eqs. (6.7.6) and (6.7.10) are also applicable with torsion springs.

6.7.3 Contact Forces

Contact forces are generally more difficult to model than gravity and spring forces. That is, it is usually more difficult to find equivalent force systems for contact forces. For many physical systems of practical importance, however, it is possible to represent the contact forces by a single force passing through a *point of contact*, defined as that point where the contacting bodies initially meet. To illustrate this, consider two bodies A and B coming into contact with one another, as in Figure 6.7.6. Let B be a body of a physical system, and let A be a body outside the system. Let A exert forces on B through the contact A has with the surface of B. Let C be the initial point of contact as represented in Figure 6.7.6b. Then, for nonrigid bodies, as A is pressed into B, the surfaces of the bodies will deform into a *contact patch*, as in Figure 6.7.6c.

For nearly rigid bodies, the dimensions of the contact patch are usually small compared with the dimensions of the bodies themselves. This means that if the forces exerted across the contact patch are replaced by a single force passing through a point of the contact, patched together with a couple, the torque of the couple is small and can generally be neglected.

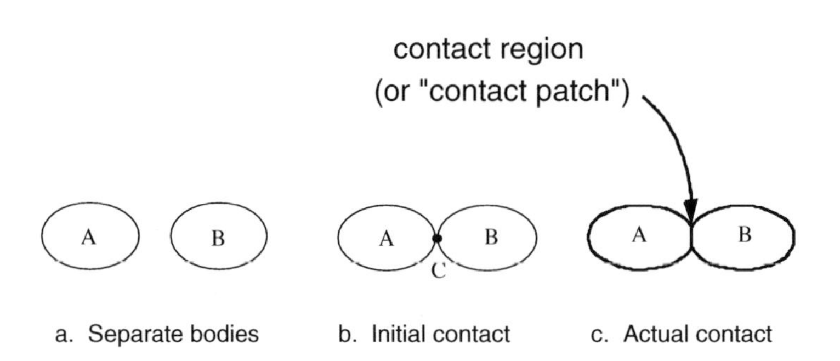

FIGURE 6.7.6
Contacting bodies.

When the contact forces are replaced by a nearly equivalent single force, it is often convenient to divide this single force into components: one normal to the contact surface and the other tangent to the contacting surface. The component tangent to the contacting surface is called the *friction force*. If there is sliding between the bodies, this friction force may often be regarded as proportional to the normal force, but with lesser magnitude than the normal force and directed opposite to the direction of sliding. That is, when there is sliding, the friction force is like a drag force and is represented by:

$$F = \mu N \quad \text{with} \quad 0 \le \mu \le 1 \tag{6.7.11}$$

where μ is called the *coefficient of friction* or *drag factor*. When μ is zero, the friction force is zero, and the contacting surfaces (or at least one of the contacting surfaces) are said to be *smooth*.

6.7.4 Action–Reaction

Forces exerted by one body on another (say, A or B) are responded to by forces exerted from the second body onto the first (B on A). These *reaction* forces are equal in magnitude, but are directed opposite to the original forces. To see this, consider two bodies A and B exerting forces on each other as in Figure 6.7.7. Also, consider free-body diagrams of A and B as in Figure 6.7.8. In the diagram for A, let S_A and $\hat{S}_A$ represent force systems exerted on A from bodies external to A (S_A) and from B ($\hat{S}_A$). Similarly, let S_B and $\hat{S}_B$ represent forces on B from bodies external to B (S_B) and from A ($\hat{S}_B$). For equilibrium (whether static or dynamic), the force systems on the individual and combined bodies must be zero systems (see Chapter 7). Therefore, we have:

$$\begin{aligned} S_A + S_B &= 0 \\ S_A + \hat{S}_A &= 0 \\ S_B + \hat{S}_B &= 0 \end{aligned} \tag{6.7.12}$$

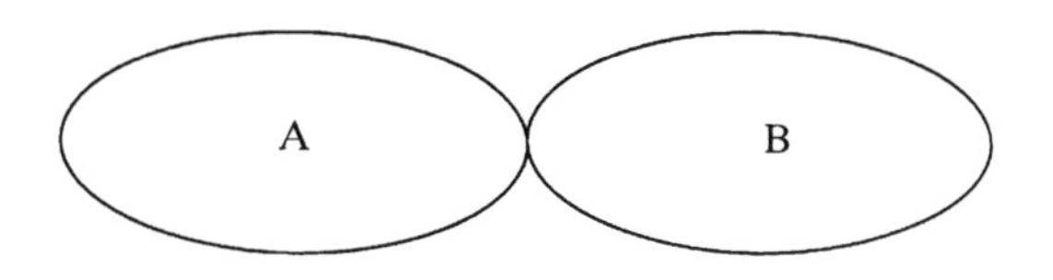

FIGURE 6.7.7
Bodies A and B exerting forces on each other.

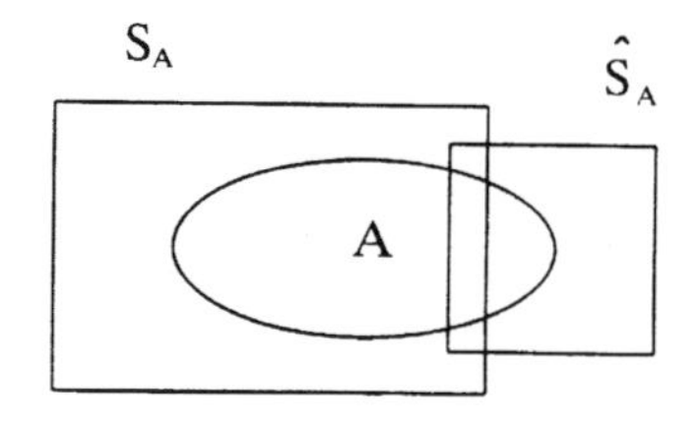

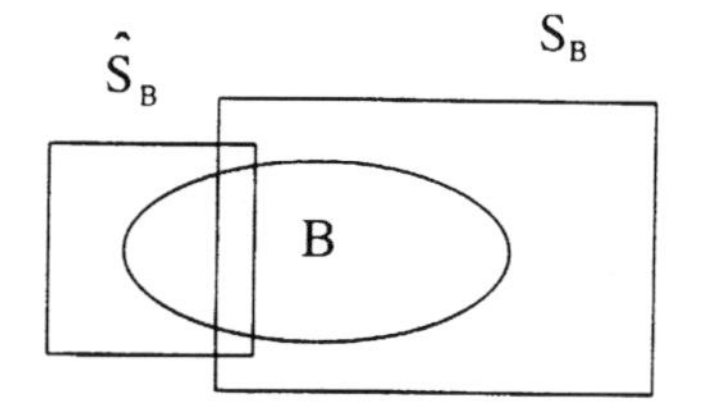

FIGURE 6.7.8
Free body diagrams of bodies A and B.

By eliminating S_A and S_B between these three equations, we have:

$$\hat{S}_A + \hat{S}_B = 0 \tag{6.7.13}$$

Now, suppose that $\hat{S}_A$ is represented by a single force, say $\mathbf{F}_A$, passing through some common point C of A and B (or A and B extended) together with a couple with torque $\mathbf{T}_A$. Similarly, let $\hat{S}_B$ be represented by a single force $\mathbf{F}_B$ passing through C together with a couple with torque $\mathbf{T}_B$. Then, because $\hat{S}_A$ and $\hat{S}_B$ taken together form a zero system (Eq. (6.7.3)), the resultant of $\hat{S}_A$ and $\hat{S}_B$ and the moment of $\hat{S}_A$ and $\hat{S}_B$ about C must be zero. That is,

$$\mathbf{F}_A + \mathbf{F}_B = 0 \quad \text{or} \quad \mathbf{F}_A = -\mathbf{F}_B \tag{6.7.14}$$

and

$$\mathbf{T}_A + \mathbf{T}_B = 0 \quad \text{or} \quad \mathbf{T}_A = -\mathbf{T}_B \tag{6.7.15}$$

Equations (6.7.13), (6.7.14), and (6.7.15), or the equivalent wording, represent the *law of action and reaction.*

6.8 First Moments

Consider a particle P with mass m (or, alternatively, a point P with associated mass m) as depicted in Figure 6.8.1. Let O be an arbitrary reference point, and let $\mathbf{p}$ be a position vector locating P relative to O. The *first moment* of P relative to O, $\phi^{P/O}$, is defined as:

$$\phi^{P/O} \stackrel{D}{=} m\mathbf{p} \tag{6.8.1}$$

Consider next a set S of N particles P_i ($i = 1,\ldots, N$) having masses as in Figure 6.8.2, where O is an arbitrary reference point. The first moment of S for O, $\phi^{S/O}$, is defined as the sum of the first moments of the individual particles of S for O. That is,

$$\phi^{S/O} \stackrel{D}{=} \mathbf{m}_1\mathbf{p}_1 + \mathbf{m}_2\mathbf{p}_2 + \ldots + \mathbf{m}_N\mathbf{p}_N = \sum_{i=1}^{N} \mathbf{m}_i\mathbf{p}_i \tag{6.8.2}$$

Observe that, in general, $\phi^{S/O}$ is not zero. However, if a point G can be found such that the first moment of S relative to G, $\phi^{S/G}$, is zero, then G is defined as the *mass center* of S.

Using this definition, the existence and location of G can be determined from Eq. (6.8.2). Specifically, if G is the mass center and if $\mathbf{r}_i$ locates $\mathbf{p}_i$ relative to G, as in Figure 6.8.3, then the first moment of S relative to G may be expressed as:

$$\phi^{S/G} = \sum_{i=1}^{N} m_i\mathbf{r}_i = 0 \tag{6.8.3}$$

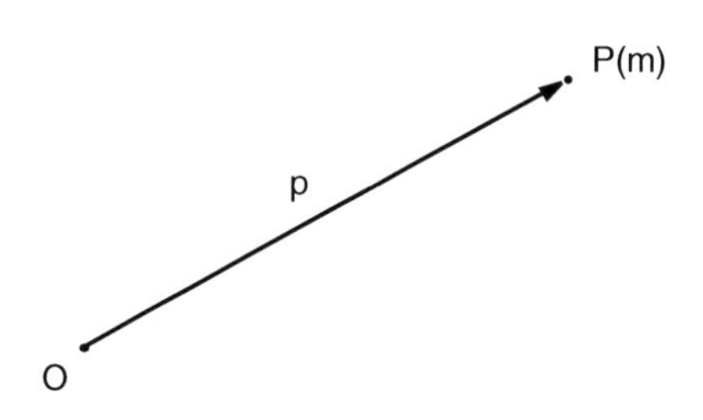

FIGURE 6.8.1
A particle P and reference point O.

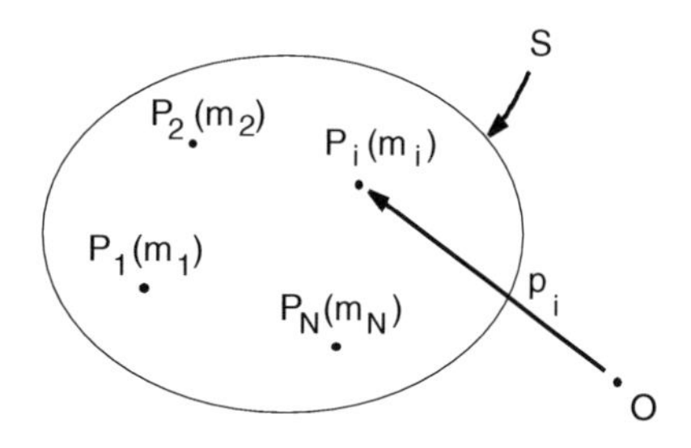

FIGURE 6.8.2
A set S of particles and a reference point O.

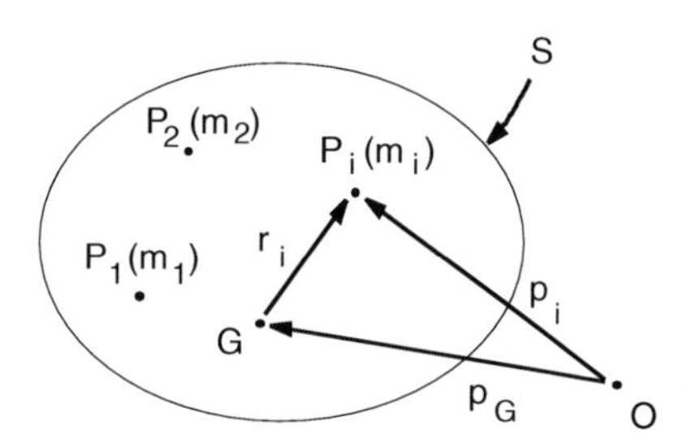

FIGURE 6.8.3
A set S of particles with mass center G.

From Figure 6.8.3, we have:

$$\mathbf{p}_i = \mathbf{p}_G + \mathbf{r}_i \quad \text{or} \quad \mathbf{r}_i = \mathbf{p}_i - \mathbf{p}_G \tag{6.8.4}$$

Hence, by substituting into Eq. (6.8.3), we obtain:

$$\begin{aligned}\sum_{i=1}^{N} m_i \mathbf{r}_i &= \sum_{i=1}^{N} m_i(\mathbf{p}_i - \mathbf{p}_G) \\ &= \sum_{i=1}^{N} m_i \mathbf{p}_i - \sum_{i=1}^{N} m_i \mathbf{p}_G \\ &= \sum_{i=1}^{N} m_i \mathbf{p}_i - \left(\sum_{i=1}^{N} m_i\right)\mathbf{p}_G \\ &= 0\end{aligned} \tag{6.8.5}$$

By solving the last equation for $\mathbf{p}_G$, we obtain:

$$\mathbf{p}_G = (1/M)\sum_{i=1}^{N} m_i \mathbf{p}_i \tag{6.8.6}$$

where M is the total mass of S. That is,

$$M = \sum_{i=1}^{N} m_i \tag{6.8.7}$$

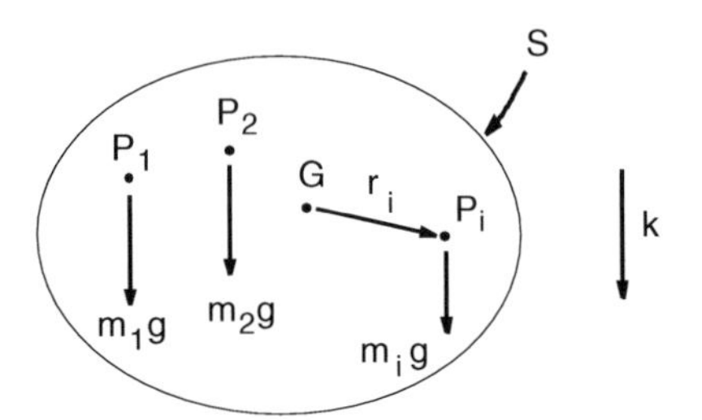

FIGURE 6.8.4
Gravitational forces on the particles of a body.

Equation (6.8.6) demonstrates the existence of G by providing an algorithm for its location.

We may think of a body as though it were composed of particles, just as a sandstone is composed of particles of sand. Then, the sums in Eqs. (6.8.2) through (6.8.7) become very large, and in the limit they may be replaced by integrals.

For homogeneous bodies, the mass is uniformly distributed throughout the region, or geometric figure, occupied by the body. The mass center location is then solely determined by the shape of the figure of the body. The mass center position is then said to be at the *centroid* of the geometric figure of the body. The centroid location for common and simple geometric figures may be determined by routine integration. The results of such integrations are listed in figurative form in Appendix I. As the name implies, a centroid is at the intuitive center or middle of a figure.

As an illustration of these concepts, consider the gravitational forces acting on a body B with an arbitrary shape. Let B be composed of N particles P_i having masses m_i ($i = 1,\ldots, N$), and let G be the mass center of B, as depicted in Figure 6.8.4.

Let the set of all the gravitational forces acting on B be replaced by a single force **W** passing through G together with a couple with torque **T**. Then, from the definition of equivalent force systems, **W** and **T** are:

$$\mathbf{W} = \sum_{i=1}^{N} m_i g\mathbf{k} = \left(\sum_{i=1}^{N} m_i\right) g\mathbf{k} = Mg\mathbf{k} \tag{6.8.8}$$

and

$$\mathbf{T} = \sum_{i=1}^{N} \mathbf{r}_i \times m_i g\mathbf{k} = \left(\sum_{i=1}^{N} m_i \mathbf{r}_i\right) \times g\mathbf{k} = 0 \tag{6.8.9}$$

where **k** is a downward-directed unit vector as in Figure 6.8.4 and M is the total mass of particles of B. The last equality of Eq. (6.8.9) follows from the definition of the mass center in Eq. (6.8.3).

Equations (6.8.7) and (6.8.8) show how dramatic the reduction in forces can be through the use of equivalent force systems.

6.9 Physical Forces: Inertia (Passive) Forces

Inertia forces arise due to acceleration of particles and their masses. Specifically, the inertia force on a particle is proportional to both the mass of the particle and the acceleration of the particle. However, the inertia force is directed opposite to the acceleration. Thus, if P is a particle with mass m and with acceleration $\mathbf{a}$ in an inertial reference frame R, then the inertia force $\mathbf{F}^*$ exerted on P is:

$$\mathbf{F}^* = -m\mathbf{a} \tag{6.9.1}$$

An *inertial* reference frame is defined as a reference frame in which Newton's laws of motion are valid. From elementary physics, we recall that from Newton's laws we have the expression:

$$\mathbf{F} = m\mathbf{a} \tag{6.9.2}$$

where $\mathbf{F}$ represents the resultant of all applied forces on a particle P having mass m and acceleration $\mathbf{a}$. By comparing Eqs. (6.9.1) and (6.9.2), we have:

$$\mathbf{F} + \mathbf{F}^* = 0 \tag{6.9.3}$$

Equation (6.9.3) is often referred to as *d'Alembert's principle*. That is, the sum of the applied and inertia forces on a particle is zero. Equation (6.9.3) thus also presents an algorithm or procedure for the analysis of dynamic systems as though they were static systems.

For rigid bodies, considered as sets of particles, the inertia force system is somewhat more involved than for that of a single particle due to the large number of particles making up a rigid body; however, we can accommodate the resulting large number of inertia forces by using equivalent force systems, as discussed in Section 6.5. To do this, consider the representation of a rigid body as a set of particles as depicted in Figure 6.9.1. As B moves in an inertial frame R, the particles of B will be accelerated and thus experience inertia forces; hence, the inertia force system exerted on B will be made up of the inertia forces on the particles of B. The inertia force exerted on a typical particle P_i of B is:

$$\mathbf{F}_i^* = -m_i\mathbf{a}_i \tag{6.9.4}$$

where m_i is the mass of P_i and A_i is the acceleration of P_i in R.

Using the procedures of Section 6.5, we can replace this system of many forces by a single force $\mathbf{F}^*$ passing through an arbitrary point, together with a couple having a torque

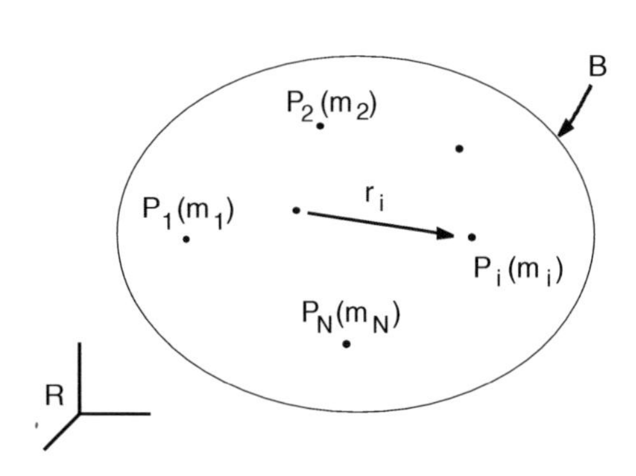

FIGURE 6.9.1
Representation of a rigid body as a set of particles.

$\mathbf{T}^*$. It is generally convenient to let $\mathbf{F}^*$ pass through the mass center G of the body. Then, $\mathbf{F}^*$ and $\mathbf{T}^*$ are:

$$\mathbf{F}^* = \sum_{i=1}^{N} \left(-m_i \mathbf{a}_i\right) \tag{6.9.5}$$

and:

$$\mathbf{T}^* = \sum_{i=1}^{N} \mathbf{r}_i \times \left(-m_i \mathbf{a}_i\right) \tag{6.9.6}$$

where $\mathbf{r}_i$ locates P_i relative to G.

Using Eq. (4.9.6), we see that because both P_i and G are fixed on B, $\mathbf{a}_i$ may be expressed as:

$$\mathbf{a}_i = \mathbf{a}_G + \boldsymbol{\alpha} \times \mathbf{r}_i + \boldsymbol{\omega} \times \left(\boldsymbol{\omega} \times \mathbf{r}_i\right) \tag{6.9.7}$$

where $\boldsymbol{\alpha}$ and $\boldsymbol{\omega}$ are the angular acceleration and angular velocity of B in R. Hence, by substituting into Eq. (6.9.5), $\mathbf{F}^*$ becomes:

$$\begin{aligned} \mathbf{F}^* &= \sum_{i=1}^{N} -m_i \left[\mathbf{a}_G + \boldsymbol{\alpha} \times \mathbf{r}_i + \boldsymbol{\omega} \times \left(\boldsymbol{\omega} \times \mathbf{r}_i\right)\right] \\ &= -\left(\sum_{i=1}^{N} m_i\right) \mathbf{a}_G - \boldsymbol{\alpha} \times \sum_{i=1}^{N} \left(m_i \mathbf{r}_i\right) - \boldsymbol{\omega} \times \left[\boldsymbol{\omega} \times \sum_{i=1}^{N} \left(m_i \mathbf{r}_i\right)\right] \\ &= -M\mathbf{a}_G - \boldsymbol{\alpha} \times 0 - \boldsymbol{\omega} \times \left(\boldsymbol{\omega} \times 0\right) \end{aligned} \tag{6.9.8}$$

or

$$\mathbf{F} = -M\mathbf{a}_G \tag{6.9.9}$$

where M is the total mass of B and where the last two terms of Eq. (6.9.8) are zero because G is the mass center of B (see Eq. (6.8.3)).

Similarly, by substituting for $\mathbf{a}_i$ in Eq. (6.9.6), $\mathbf{T}^*$ becomes:

$$\begin{aligned} \mathbf{T}^* &= \sum_{i=1}^{N} \mathbf{r}_i \times \left(-m_i\right) \left[\mathbf{a}_G + \boldsymbol{\alpha} \times \mathbf{r}_i + \boldsymbol{\omega} \times \left(\boldsymbol{\omega} \times \mathbf{r}_i\right)\right] \\ &= -\left(\sum_{i=1}^{N} m_i \mathbf{r}_i\right) \times \mathbf{a}_G - \sum_{i=1}^{N} m_i \mathbf{r}_i \times \left(\boldsymbol{\alpha} \times \mathbf{r}_i\right) - \sum_{i=1}^{N} m_i \mathbf{r}_i \times \left[\boldsymbol{\omega} \times \left(\boldsymbol{\omega} \times \mathbf{r}_i\right)\right] \\ &= -0 \times \mathbf{a}_G - \sum_{i=1}^{N} m_i \mathbf{r}_i \times \left(\boldsymbol{\alpha} \times \mathbf{r}_i\right) - \boldsymbol{\omega} \times \sum_{i=1}^{N} m_i \mathbf{r}_i \times \left(\boldsymbol{\omega} \times \mathbf{r}_i\right) \end{aligned} \tag{6.9.10}$$

or

$$\mathbf{T}^* = \left[\sum_{i=1}^{N} m_i \mathbf{r}_i \times (\boldsymbol{\alpha} \times \mathbf{r}_i)\right] - \boldsymbol{\omega} \times \left[\sum_{i=1}^{N} m_i \mathbf{r}_i \times (\boldsymbol{\omega} \times \mathbf{r}_i)\right] \tag{6.9.11}$$

where the first term of Eq. (6.9.10) is zero because G is the mass center, and the last term is obtained from its counterpart in the previous line by using the identity:

$$\mathbf{a} \times (\mathbf{b} \times \mathbf{c}) \equiv (\mathbf{a} \cdot \mathbf{c})\mathbf{b} - (\mathbf{a} \cdot \mathbf{b})\mathbf{c} \tag{6.9.12}$$

To see this, simply expand the triple products of Eq. (6.9.10) using the identity. Specifically,

$$\mathbf{r}_i \times [\boldsymbol{\omega} \times (\boldsymbol{\omega} \times \mathbf{r}_i)] = \mathbf{r}_i \cdot (\boldsymbol{\omega} \times \mathbf{r}_i)\boldsymbol{\omega} - (\mathbf{r}_i \cdot \boldsymbol{\omega})\boldsymbol{\omega} \times \mathbf{r}_i = -(\mathbf{r}_i \cdot \boldsymbol{\omega})\boldsymbol{\omega} \times \mathbf{r}_i \tag{6.9.13}$$

and

$$\boldsymbol{\omega} \times [\mathbf{r}_i \times (\boldsymbol{\omega} \times \mathbf{r}_i)] = \boldsymbol{\omega} \cdot (\boldsymbol{\omega} \times \mathbf{r}_i)\mathbf{r}_i - (\boldsymbol{\omega} \cdot \mathbf{r}_i)\boldsymbol{\omega} \times \mathbf{r}_i = -(\boldsymbol{\omega} \cdot \mathbf{r}_i)\boldsymbol{\omega} \times \mathbf{r}_i \tag{6.9.14}$$

where the first terms are zero because $\mathbf{r}_i$ and $\boldsymbol{\omega}$ are perpendicular to $\boldsymbol{\omega} \times \mathbf{r}_i$. Comparing Eqs. (6.9.13) and (6.9.14), we see the results are the same. That is,

$$\mathbf{r}_i \times [\boldsymbol{\omega} \times (\boldsymbol{\omega} \times \mathbf{r}_i)] = \boldsymbol{\omega} \times [\mathbf{r}_i \times (\boldsymbol{\omega} \times \mathbf{r}_i)] \tag{6.9.15}$$

Neither of the terms of Eq. (6.9.11) is in a form convenient for computation or analysis; however, the terms have similar forms. Moreover, we can express these forms in terms of the *inertia dyadic* of the body as discussed in the next chapter. This, in turn, will enable us to express $\mathbf{T}^*$ in terms of the moments and products of inertia of B for its mass center G.

References

6.1. Kane, T. R., *Analytical Elements of Mechanics*, Vol. 1, Academic Press, New York, 1961, p. 150.
6.2. Kane, T. R., *Dynamics*, Holt, Rinehart & Winston, New York, 1968, pp. 92–115.
6.3. Huston, R. L., *Multibody Dynamics*, Butterworth-Heinemann, Stoneham, MA, 1990, pp. 153–212.

Problems

Section 6.2 Forces and Moments

P6.2.1: A force $\mathbf{F}$ with magnitude 12 lb acts along a line L passing through points P and Q as in Figure P6.2.1. Let the coordinates of P and Q relative to a Cartesian reference frame be as shown, measured in feet. Express $\mathbf{F}$ in terms of unit vectors $\mathbf{n}_x$, $\mathbf{n}_y$, and $\mathbf{n}_z$, which are parallel to the X-, Y-, and Z-axes.

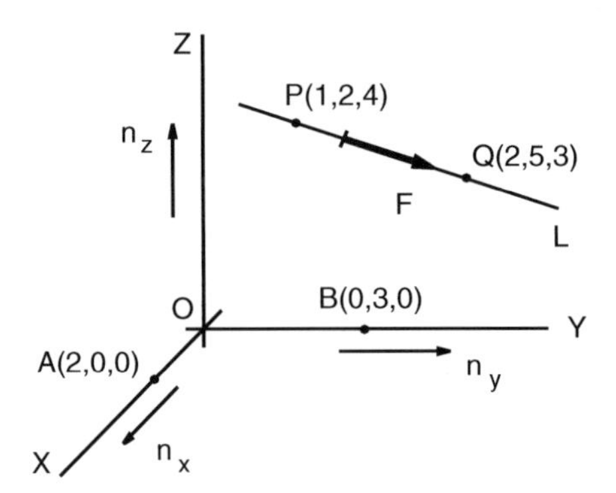

FIGURE P6.2.1
A force **F** acting along a line *L*.

P6.2.2: See Problem P6.2.1. Find the moment of **F** about points *O*, *A*, and *B* of Figure P6.2.1. Express the results in terms of $\mathbf{n}_x$, $\mathbf{n}_y$, and $\mathbf{n}_z$.

P6.2.3: A force **F** with magnitude 52 N acts along diagonal *OA* of a box with dimensions as represented in Figure P6.2.3. Express **F** in terms of the unit vectors $\mathbf{n}_1$, $\mathbf{n}_2$, and $\mathbf{n}_3$, which are parallel to the edges of the box as shown.

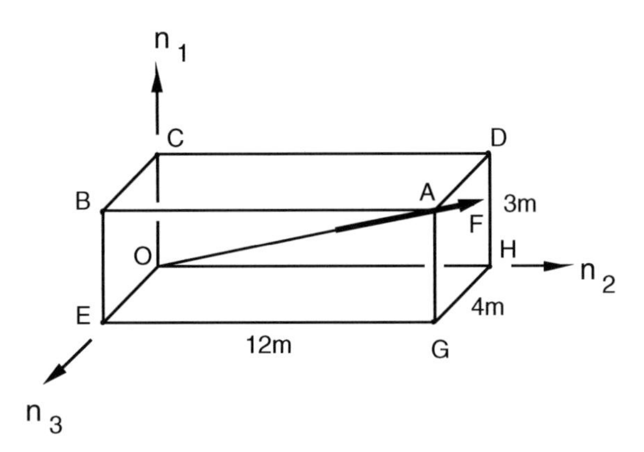

FIGURE P6.2.3
A force **F** acting along a box diagonal.

P6.2.4: See Problem P6.2.3. Find the moment of **F** about the corners A, B, C, D, E, G, and H of the box. Express the results in terms of $\mathbf{n}_1$, $\mathbf{n}_2$, and $\mathbf{n}_3$.

Section 6.3 Systems of Forces

P6.3.1: Consider the force system exerted on the box as represented in Figure P6.3.1. The force system consists of ten forces with lines of action and magnitudes as listed in Table P6.3.1. Let the dimensions of the box be 12 m, 4 m, and 3 m, as shown.

a. Express the forces $\mathbf{F}_1,\ldots, \mathbf{F}_{10}$ in terms of the unit vectors $\mathbf{n}_1$, $\mathbf{n}_2$, and $\mathbf{n}_3$ shown in Figure P6.3.1.
b. Find the resultant of the force system expressed in terms of $\mathbf{n}_1$, $\mathbf{n}_2$, and $\mathbf{n}_3$.
c. Find the moment of the force system about *O* (expressed in terms of $\mathbf{n}_1$, $\mathbf{n}_2$, and $\mathbf{n}_3$).

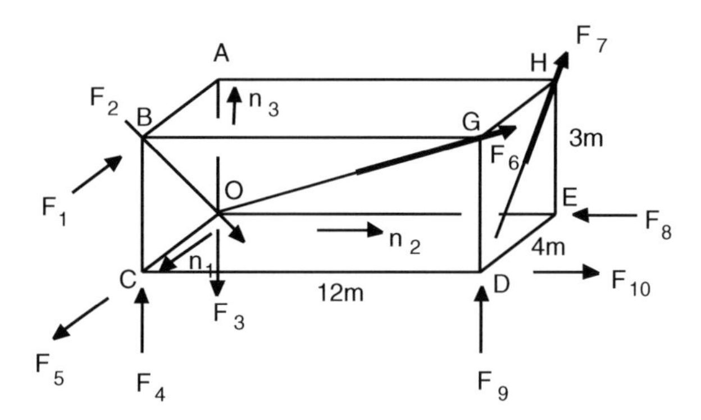

FIGURE P6.3.1
A force system exerted on a box.

TABLE P6.3.1
Forces, Their Magnitudes, and Lines of Action for the Force System of Figure P6.3.1

Force	Magnitude (N)	Line of Action
$\mathbf{F}_1$	18	*BA*
$\mathbf{F}_2$	25	*BO*
$\mathbf{F}_3$	30	*AO*
$\mathbf{F}_4$	10	*CB*
$\mathbf{F}_5$	12	*OC*
$\mathbf{F}_6$	26	*OG*
$\mathbf{F}_7$	20	*DH*
$\mathbf{F}_8$	25	*EO*
$\mathbf{F}_9$	16	*DG*
$\mathbf{F}_{10}$	24	*CD*

P6.3.2: See Problem 6.3.1, and (a) find the moment of the force system about points *C* and *H*, and (b) verify Eq. (6.3.6) for these results. That is, show that:

$$\mathbf{M}_C = \mathbf{M}_H + \mathbf{CH} \times \mathbf{R}$$

where **R** is the resultant of the force system.

P6.3.3: A cube with 2-ft sides has forces exerted upon it as shown in Figure P6.3.3. The magnitudes and lines of action of these forces are listed in Table P6.3.3.

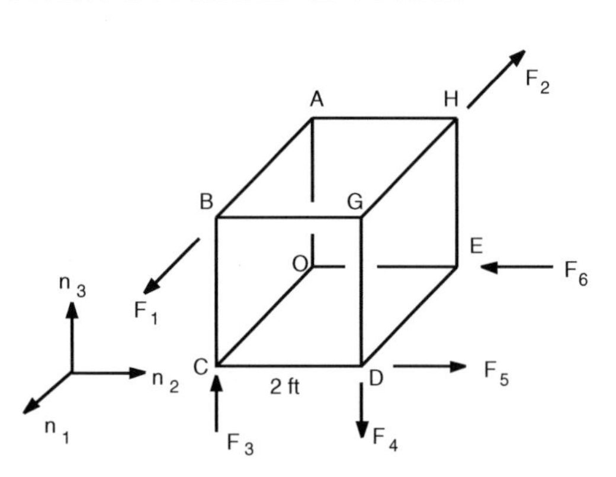

FIGURE P6.3.3
Forces exerted on a cube.

a. Determine the resultant **R** of this system of forces.
b. Find the moments of the force system about points C, H, E, and G.

Express the results in terms of the unit vectors $\mathbf{n}_1$, $\mathbf{n}_2$, and $\mathbf{n}_3$ shown in Figure P6.3.3.

TABLE P6.3.3

Forces, Their Magnitudes, and Lines of Action for the Force System of Figure P6.3.3

Force	Magnitude (lb)	Line of Action
$\mathbf{F}_1$	10	*AB*
$\mathbf{F}_2$	10	*GH*
$\mathbf{F}_3$	8	*CB*
$\mathbf{F}_4$	8	*GD*
$\mathbf{F}_5$	12	*CD*
$\mathbf{F}_6$	12	*EO*

6.4 Special Force Systems: Zero Force Systems and Couples

P6.4.1: In Figure P6.4.1, the box is subjected to forces as shown. The magnitudes and directions of the forces are listed in Table P6.4.1. Determine the magnitudes of forces $\mathbf{F}_1$, $\mathbf{F}_3$, $\mathbf{F}_4$, $\mathbf{F}_5$, $\mathbf{F}_6$, and $\mathbf{F}_{10}$ so that the system of forces on the box is a zero system.

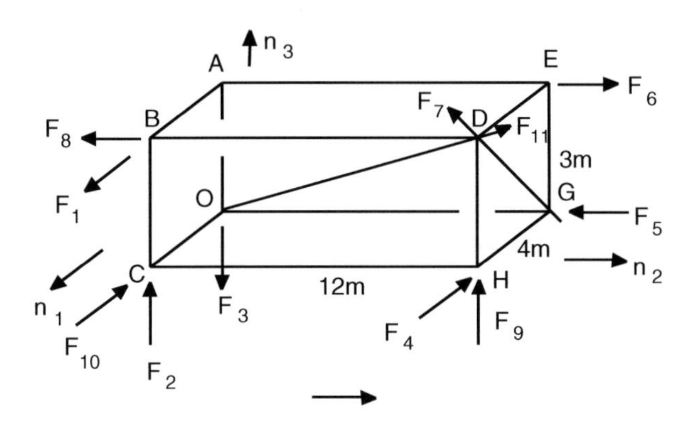

FIGURE P6.4.1
Force system exerted on a box.

TABLE P6.4.1

Forces, Their Magnitudes, and Lines of Action for the Forces of Figure P6.4.1

Force	Magnitude (N)	Line of Action
$\mathbf{F}_1$	10	*AB*
$\mathbf{F}_2$	8	*CB*
$\mathbf{F}_3$	8	*AO*
$\mathbf{F}_4$	6	*HG*
$\mathbf{F}_5$	7	*GO*
$\mathbf{F}_6$	5	*AE*
$\mathbf{F}_7$	10	*GD*
$\mathbf{F}_8$	16	*DB*
$\mathbf{F}_9$	12	*HD*
$\mathbf{F}_{10}$	9	*CO*
$\mathbf{F}_{11}$	26	*OD*

P6.4.2: Three cables support a 1500-lb load as depicted in Figure P6.4.2. By considering the connecting joint O of the cables to the load to be in static equilibrium and by recalling that forces in cables are directed along the cable, determine the forces in each of the cables. The location of the fixed ends of the cables (A, B, and C) are determined by their x, y, z coordinates (measured in feet) as shown in Figure P6.4.2.

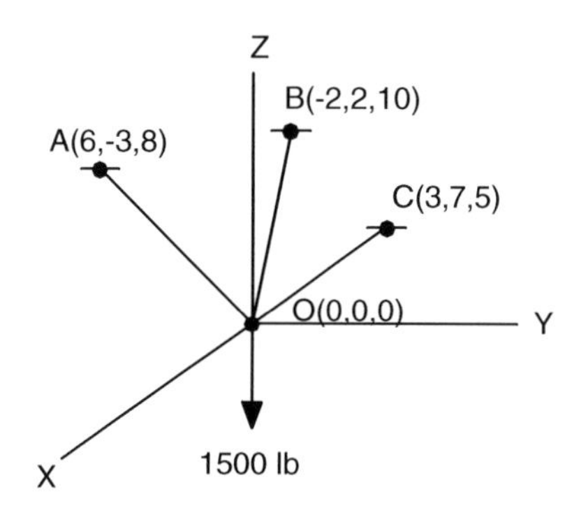

FIGURE P6.4.2
Cables OA, OB, and OC supporting a vertical load.

P6.4.3: Repeat Problem P6.4.2 if the load is 2000 lb.

P6.4.4: Repeat Problem P6.4.3 if the coordinates of the cable supports (A, B, and C) are given in meters instead of feet.

P6.4.5: Show that the force system exerted on the box shown in Figure P6.4.5 is a couple. The magnitudes and directions of the forces are listed in Table P6.4.5.

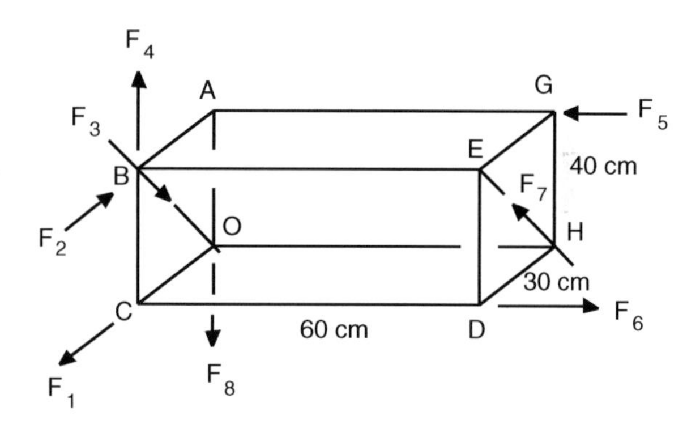

FIGURE P6.4.5
Forces system exerted on a box.

TABLE P6.4.5

Forces, Their Magnitudes, and Lines of Action for the Forces of Figure P6.4.5

Force	Magnitude (N)	Line of Action
$\mathbf{F}_1$	8	OC
$\mathbf{F}_2$	8	BA
$\mathbf{F}_3$	15	BO
$\mathbf{F}_4$	12	CB
$\mathbf{F}_5$	10	GA
$\mathbf{F}_6$	10	CD
$\mathbf{F}_7$	15	HE
$\mathbf{F}_8$	12	AO

P6.4.6: See Problem P6.4.5. Find the torque of the couple.

P6.4.7: See Problems P6.4.5 and P6.4.6. Find the moment of the force system about points O, A, D, and G.

P6.4.8: Show that the magnitude of the torque of a simple couple (two equal-magnitude but oppositely directed forces) is simply the product of the magnitude of one of the forces multiplied by the distance between the parallel lines of action of the forces. Show further that the orientation of the torque is perpendicular to the plane of the forces, with sense determined by the *right-hand rule.*

P6.4.9: Show that a set of simple couples is also a couple. Show further that the torque of this composite couple is then simply the resultant (sum) of the torques of the simple couples.

6.5 Equivalent Force Systems

P6.5.1: See Problem P6.3.1. Consider again the force system exerted on the box of Problem P6.3.1 as shown in Figure P6.5.1, where the magnitudes and lines of action of the forces are listed in Table P6.5.1. Suppose this force system is to be replaced by an equivalent force system consisting of a single force **F** passing through O together with a couple having torque **T**. Find **F** and **T**. (Express the results in terms of the unit vectors $\mathbf{n}_1$, $\mathbf{n}_2$, and $\mathbf{n}_3$ of Figure P6.5.1.)

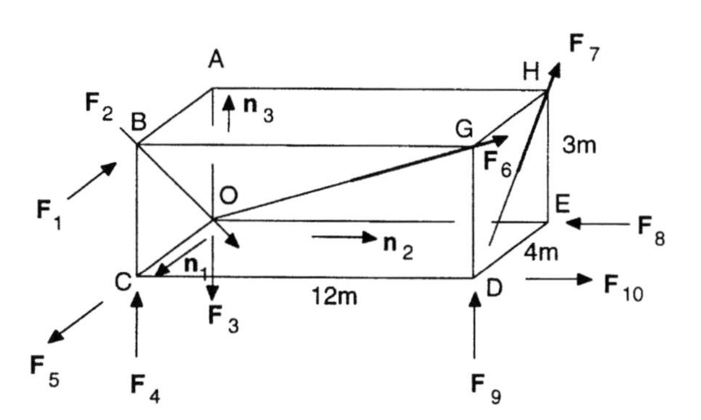

FIGURE P6.5.1
A force system exerted on a box.

TABLE P6.5.1

Forces, Their Magnitudes, and Lines of Action for the Force System of Figure P6.5.1

Force	Magnitude (N)	Line of Action
$\mathbf{F}_1$	18	*BA*
$\mathbf{F}_2$	25	*BO*
$\mathbf{F}_3$	30	*AO*
$\mathbf{F}_4$	10	*CB*
$\mathbf{F}_5$	12	*OC*
$\mathbf{F}_6$	26	*OG*
$\mathbf{F}_7$	20	*DH*
$\mathbf{F}_8$	25	*EO*
$\mathbf{F}_9$	16	*DG*
$\mathbf{F}_{10}$	24	*CD*

P6.5.2: Repeat Problem P6.5.1 with the single force **F** passing through G. Compare the magnitudes of the respective couple torques.

P6.5.3: Consider a homogeneous rectangular block with dimensions a, b, and c as represented in Figure P6.5.3. Let the gravitational forces acting on the block be replaced by an equivalent force system consisting of a single force **W**. If ρ is the uniform mass density of the block, find the magnitude and line of action of **W**.

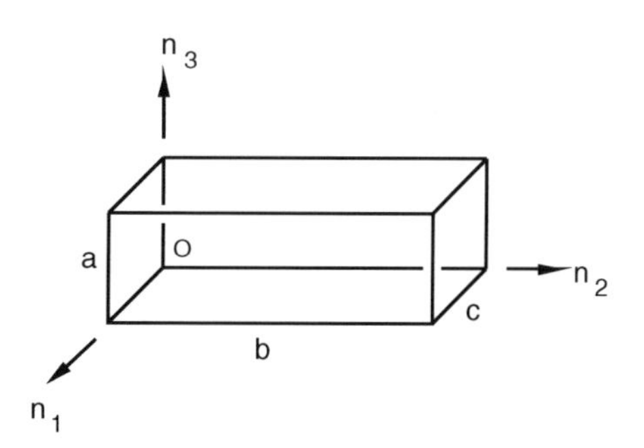

FIGURE P6.5.3
A homogeneous rectangular block.

P6.5.4: See Problem P6.5.3. Suppose the gravitational force system on the block is replaced by an equivalent force system consisting of a single force **W** passing through O together with a couple with torque **T**. Find **F** and **T**. Express the results in terms of the unit vectors $\mathbf{n}_1$, $\mathbf{n}_2$, and $\mathbf{n}_3$ shown in Figure P6.5.3.

6.6 Wrenches

P6.6.1: See Problems P6.5.1 and P6.3.1. For the force system of Problems P6.5.1 and P6.3.1, find the point Q^* for which there is a minimum moment of the force system. Specifically, find the position vector $\mathbf{OQ}^*$ locating Q^* relative to O. Express the result in terms of the unit vectors $\mathbf{n}_1$, $\mathbf{n}_2$, and $\mathbf{n}_3$ shown in Figure P6.5.1.

P6.6.2: See Problem P6.6.1. Find a wrench that is equivalent to the force system of Problems P6.3.1, P6.5.1, and P6.6.1. Express the results in terms of the unit vectors of Figure P6.5.1.

6.7 Physical Forces: Applied (Active) Forces

P6.7.1: Suppose the human arm is modeled by three bodies B_1, B_2, and B_3 representing the upper arm, lower arm, and hand, as shown in Figure P6.7.1A. Suppose further that the

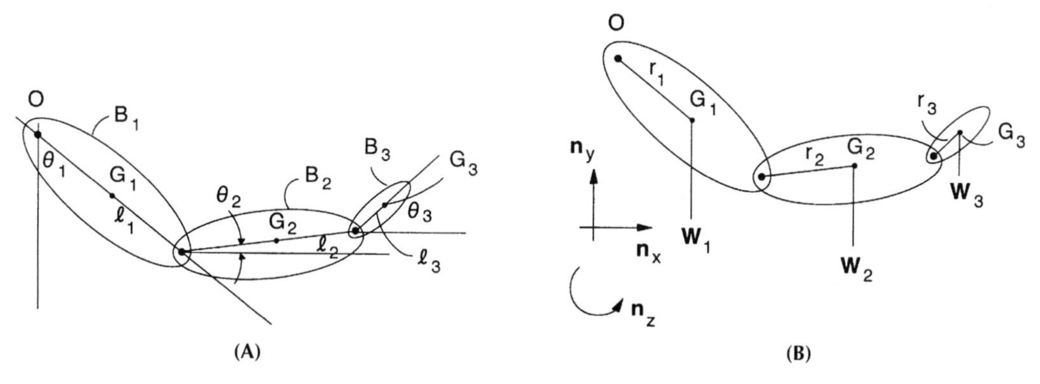

FIGURE P6.7.1
(A) A model of the human arm. (B) Equivalent gravity (weight) forces on the human arm model.

gravitational forces acting on these bodies are represented by equivalent gravity (weight) force systems consisting of single vertical forces W_1, W_2, and W_3 as in Figure P6.7.1B. Finally, suppose we want to find an equivalent gravity force system for the entire arm consisting of a single force **W** passing through the shoulder joint O together with a couple with torque **M**. Find **W** and **M**. Express the results in terms of the angles θ_1, θ_2, and θ_3; the distances r_1, r_2, r_3, ℓ_1, ℓ_2, and ℓ_3; the force magnitudes W_1, W_2, and W_3; and the unit vectors $\mathbf{n}_x$, $\mathbf{n}_y$, and $\mathbf{n}_z$ shown in Figures P6.7.1A and B.

P6.7.2: See Problem P6.7.1. Table P6.7.2 provides numerical values for the geometric quantities and weights of the arm model of Figures P6.7.1A and B. Using these values, determine the magnitudes of **W** and **M**.

TABLE P6.7.2

Numerical Values for the Geometric Parameters and Weights of Figures P6.7.1A and B

i	θ_i (°)	r_i (in.)	ℓ_i (in.)	W_i (lb)
1	45	4.45	11.7	5.0
2	15	6.5	14.5	3.0
3	30	2.5	6.0	1.15

P6.7.3: See Problems P6.7.1 and P6.7.2. Suppose the equivalent force system is to be a wrench, where the couple torque **M** is a minimum. Locate a point G on the line of action of the equivalent force. Find the magnitudes of the equivalent wrench force and minimum moment.

P6.7.4: Three springs are connected in series as in Figure P6.7.4. Find the elongation for the applied forces. The spring moduli and force magnitudes are listed in Table P6.7.4.

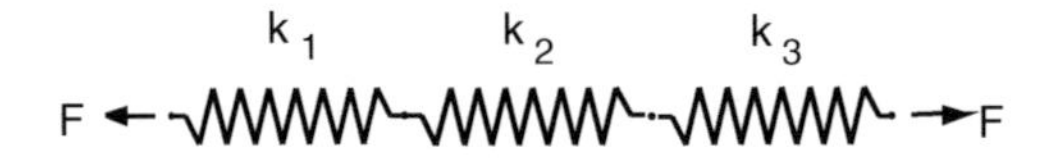

FIGURE P6.7.4
Three springs in series.

TABLE P6.7.4

Physical Data for the System of Figure P6.7.4

Force F Spring Moduli	12 lb (lb/in.)	50 N (N/mm)
k_1	6	10
k_2	5	12
k_3	8	7

P6.7.5: See Problem P6.7.4. Let the springs of Problem P6.7.4 be arranged in parallel as in Figure P6.7.5. Find the elongation δ for the applied forces. Assume that the springs are sufficiently close (or even coaxial) so that the rotation of the attachment plates can be ignored.

P6.7.6: A block is resting on an incline plane as shown in Figure P6.7.6. Let μ be the coefficient of friction between the block and the plane. Find the inclination angle θ of the incline where the block is on the verge of sliding down the plane.

P6.7.7: See Problem P6.7.6. Let the inclination angle θ be small. Let the *drag factor* (f) be defined as an effective coefficient of friction that accounts for the small slope. Find f in terms of μ and θ. What would be the value of f if the block is sliding up the plane?

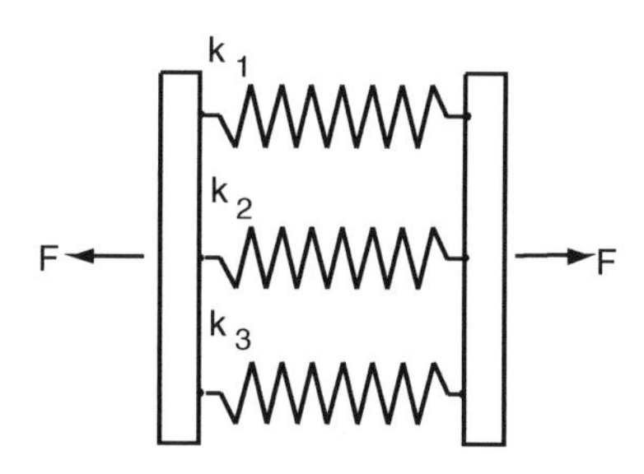

FIGURE P6.7.5
Three springs in parallel.

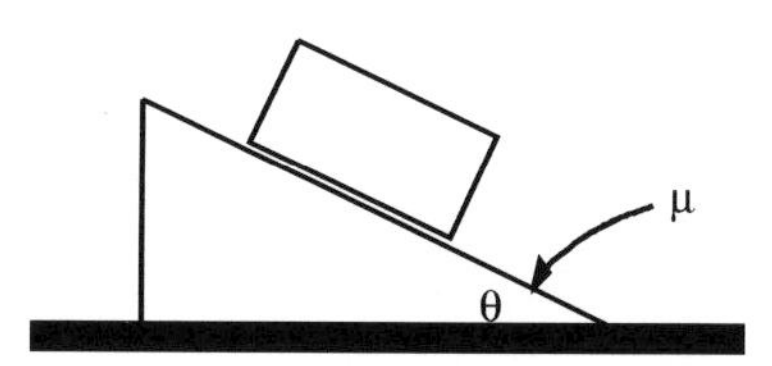

FIGURE P6.7.6
Block on an inclined plane.

P6.7.8: A homogeneous block is pushed to the right by a force P as in Figure P6.7.8. If the friction coefficient between the block and plane is μ, find the maximum elevation h above the surface where the force can be applied so that the block will slide and not tip.

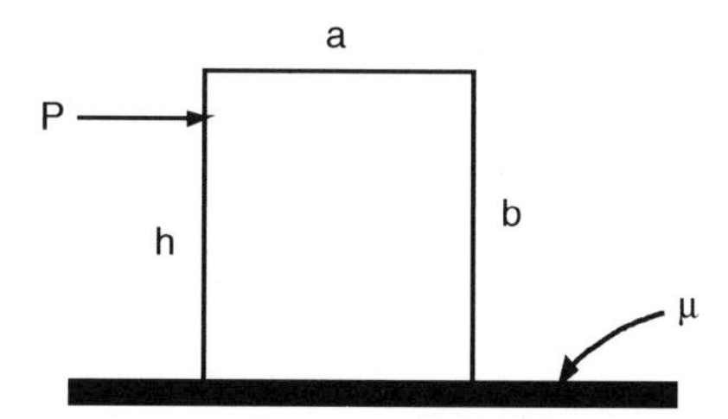

FIGURE P6.7.8
A block pushed along a surface.

P6.7.9: Figure P6.7.9 shows a schematic representation of a *short-shoe* drum brake. When a force F is applied to the brake lever, the friction between the brake shoe and the drum creates a braking force and braking moment on the drum. Let μ be the coefficient of friction between the shoe and the drum. If the force F is known, determine the braking moment M on the drum if the drum is rotating: (a) clockwise, and (b) counterclockwise. Express M in terms of μ, F, and the dimensions a, b, c, d, and r shown in Figure P6.7.9.

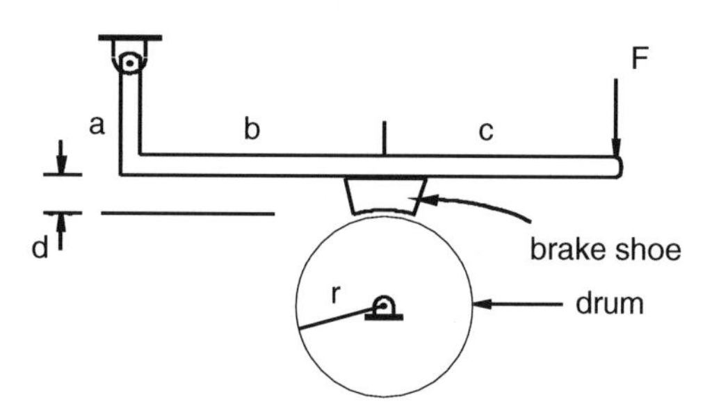

FIGURE P6.7.9
A schematic representation of a brake system.

P6.7.10: See Problem P6.7.9. Find a relation between μ and the dimensions of Figure P6.7.9 so that the brake is self-locking when the drum is rotating counterclockwise. (*Self-locking* means that a negligible force F is required to brake the drum.)

Section 6.8 Mass Center

P6.8.1: Let a system of 10 particles P_i ($i = 1,\ldots, 10$) have masses m_i and coordinates (x_i, y_i, z_i) relative to a Cartesian coordinate system as in Table P6.8.1. Find the coordinates $\bar{x}$, $\bar{y}$, $\bar{z}$ of the mass center of this set of particles if the m_i are expressed in kilograms and the x_i, y_i, z_i are in meters. How does the result change if the m_i are expressed in slug and the x_i, y_i, z_i in feet?

TABLE P6.8.1

Masses and Coordinates of a Set of Particles

P_i	m_i	x_i	y_i	z_i
P_1	6	–1	0	2
P_2	4	3	–7	4
P_3	3	4	–5	–5
P_4	8	8	0	7
P_5	1	–2	–3	–9
P_6	9	1	–2	0
P_7	4	7	–8	3
P_8	5	–3	4	–2
P_9	5	1	–8	9
P_{10}	2	6	–5	1

P6.8.2: See Problem P6.8.1. From the definition of mass center as expressed in Eq. (6.8.3) show that the coordinates of the mass center may be obtained by the simple expressions:

$$\bar{x} = (1/\text{m})\sum_{i=1}^{N} \text{m}_i x_i \quad \bar{y} = (1/\text{m})\sum_{i=1}^{N} \text{m}_i y_i \quad \bar{z} = (1/\text{m})\sum_{i=1}^{N} \text{m}_i z_i$$

where $\text{m} = \sum_{i=1}^{N} \text{m}_i$.

P6.8.3: Use the definition of mass center as expressed in Eq. (6.8.3) to show that the mass center of a system of bodies may be obtained by: (a) letting each body B_i be represented by a particle G_i located at the mass center of the body and having the mass m_i of the body; and (b) by then locating the mass center of this set of particles.

P6.8.4: See Problem P6.8.3. Consider a thin, uniform-density, sheet-metal panel with a circular hole as in Figure P6.8.4. Let the center O of the hole be on the diagonal BC 13 in. from corner B. Locate the mass center relative to corner A (that is, distance from A in the AB direction and distance from A in the AC direction).

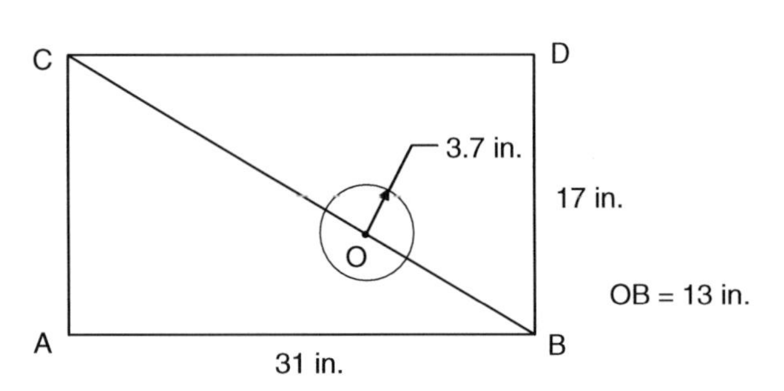

FIGURE P6.8.4
A thin, uniform-density panel with a circular hole.

P6.8.5: See Problem P6.8.4. Let the panel of Problem P6.8.4 be augmented by the addition of: (a) a triangular plate as in Figure P6.8.5A, and (b) by both a triangular plate and a semicircular plate as in Figure P6.8.5B. Locate the mass center in each case.

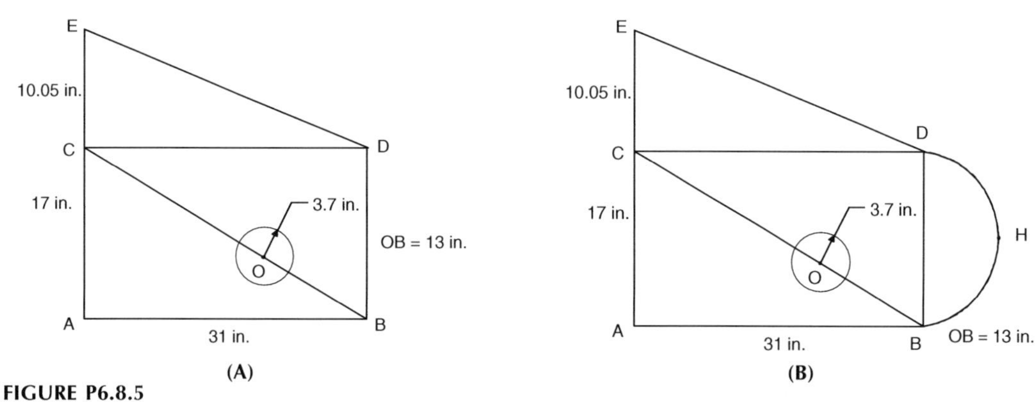

FIGURE P6.8.5
(A) A thin, uniform-density panel consisting of a triangular plate and a rectangular plate with a circular hole (see Figure P6.8.4). (B) A thin, uniform-density panel consisting of a triangular plate, a semicircular plate, and a rectangular plate with a circular hole (see Figures P6.8.4 and P6.8.5A).

P6.8.6: See Problem P6.8.5. Let the semicircular plate of Figure P6.8.5b be bent upward as depicted in a side view of the panel as in Figure P6.8.6. Locate the mass center parallel to the main panel relative to corner *A* as before and in terms of its distance, or elevation, above the main panel.

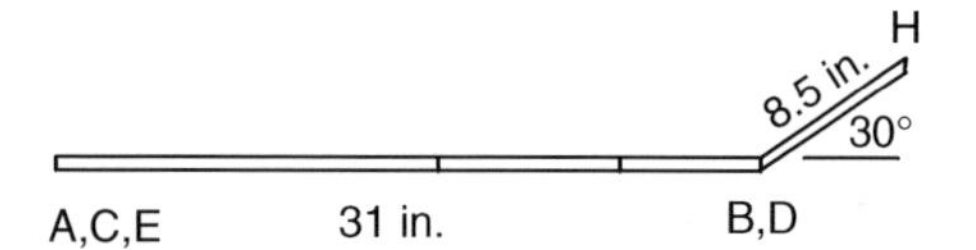

FIGURE P6.8.6
Side view of the panel of Figure P6.8.5B with bent upward semicircular plate (see Figure P6.8.5B).

P6.8.7: See Problems P6.7.1, P6.7.2, and P6.7.3. Consider again the model of the human arm as depicted in Figures P6.7.1A and B and in Figures P6.8.7A and B. Let the physical and geometrical data for the arm model be as in Table P6.7.2 and as listed again in Table P6.8.7. Determine the location of the mass center of the arm model relative to the shoulder joint *O* for the configuration of Table P6.8.7. Compare the results with those of Problem P6.7.3.

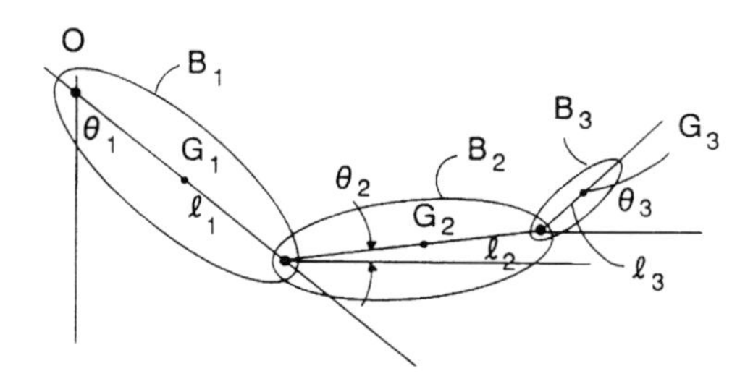

FIGURE P6.8.7
(A) A model of the human arm. (B) Equivalent gravity (weight) forces on the human arm model.

TABLE P6.8.7

Numerical Values for the Geometric Parameters and Weights of Figures P6.8.7A and B

i	θ_i (%)	r_i (in.)	ℓ_i (in.)	W_i (lb)
1	45	4.45	11.7	5.0
2	15	6.5	14.5	3.0
3	30	2.5	6.0	1.15

7

Inertia, Second Moment Vectors, Moments and Products of Inertia, Inertia Dyadics

7.1 Introduction

In this chapter we review various topics and concepts about inertia. Many readers will be familiar with a majority of these topics; however, some topics, particularly those concerned with three-dimensional aspects of inertia, may not be as well understood, yet these topics will be of most use to us in our continuing discussion of mechanical system dynamics. In a sense, we have already begun our review with our discussion of mass centers in the previous chapter. At the end of the chapter, however, we discovered that we need additional information to adequately describe the inertia torque of Eq. (6.9.11), shown again here:

$$\mathbf{T}^* = -\left[\sum_{i=1}^{N} m_i \mathbf{r}_i \times (\boldsymbol{\alpha} \times \mathbf{r}_i)\right] - \boldsymbol{\omega} \times \left[\sum_{i=1}^{N} m_i \mathbf{r}_i \times (\boldsymbol{\omega} \times \mathbf{r}_i)\right] \tag{7.1.1}$$

Indeed, the principal motivation for our review of inertia is to obtain simplified expressions for this torque. Our review will parallel the development in Reference 7.4 with a basis found in References 7.1 to 7.3. We begin with a discussion about second-moment vectors — a topic that will probably be unfamiliar to most readers. As we shall see, though, second-moment vectors provide a basis for the development of the more familiar topics, particularly moments and products of inertia.

7.2 Second-Moment Vectors

Consider a particle P with mass m and an arbitrary reference point O. Consider also an arbitrarily directed unit vector $\mathbf{n}_a$ as in Figure 7.2.1. Let $\mathbf{p}$ be a position vector locating P relative to O. The *second moment* of P relative to O for the direction $\mathbf{n}_a$ is defined as:

$$\mathbf{I}_a^{P/O} \stackrel{D}{=} m\mathbf{p} \times (\mathbf{n}_a \times \mathbf{p}) \tag{7.2.1}$$

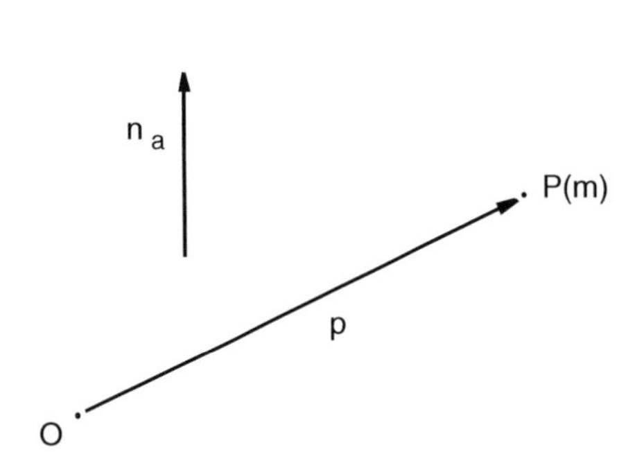

FIGURE 7.2.1
A particle P, reference point O, and unit vector $\mathbf{n}_a$.

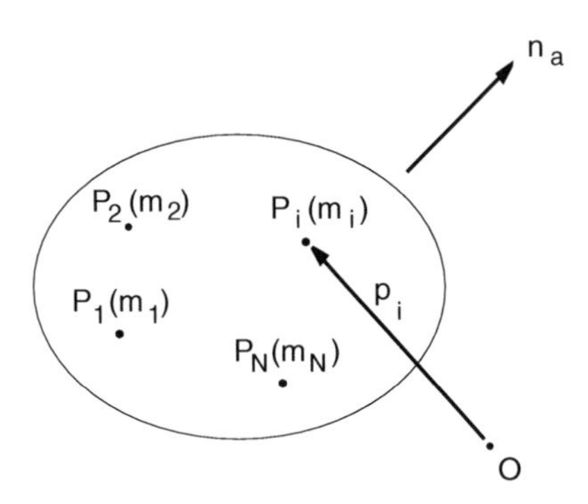

FIGURE 7.2.2
A set S of particles, reference point O and unit vector $\mathbf{n}_a$.

Observe that the second moment is somewhat more detailed than the first moment ($m\mathbf{p}$) defined in Eq. (6.8.1). The second moment depends upon the square of the distance of P from O and it also depends upon the direction of the unit vector $\mathbf{n}_a$.

The form of the definition of Eq. (7.2.1) is motivated by the form of the terms of the inertia torque of Eq. (7.1.1). Indeed, for a set S of particles, representing a rigid body B (Figure 7.2.2), the second moment is defined as the sum of the second moments of the individual particles. That is,

$$\mathbf{I}_a^{S/O} \overset{D}{=} \sum_{i=1}^{N} \mathbf{I}_a^{P_i/O} = \sum_{i=1}^{N} m_i \mathbf{p}_i \times (\mathbf{n}_a \times \mathbf{p}_i) \tag{7.2.2}$$

Then, except for the presence of $\mathbf{n}_a$ instead of $\boldsymbol{\alpha}$ or $\boldsymbol{\omega}$, the form of Eq. (7.2.2) is identical to the forms of Eq. (7.7.1). Hence, by examining the properties of the second moment vector, we can obtain insight into the properties of the inertia torque. We explore these properties in the following subsections.

7.3 Moments and Products of Inertia

Consider again a particle P, with mass m, a reference point O, unit vector $\mathbf{n}_a$, and a *second* unit vector $\mathbf{n}_b$ as in Figure 7.3.1. The moment and product of inertia of P relative to O for the directions $\mathbf{n}_a$ and $\mathbf{n}_b$ are defined as the scalar projections of the second moment vector (Eq. (7.2.1)) along $\mathbf{n}_a$ and $\mathbf{n}_b$. Specifically, the moment of inertia of P relative to O for the direction $\mathbf{n}_a$ is defined as:

$$I_{aa}^{P/O} \overset{D}{=} \mathbf{I}_a^{P/O} \cdot \mathbf{n}_a \tag{7.3.1}$$

Similarly, the product of inertia of P relative to O for the directions $\mathbf{n}_a$ and $\mathbf{n}_b$ is defined as:

$$I_{ab}^{P/O} \overset{D}{=} \mathbf{I}_a^{P/O} \cdot \mathbf{n}_b \tag{7.3.2}$$

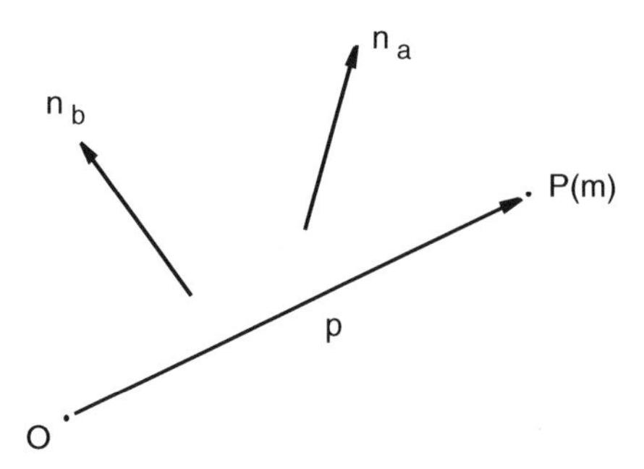

FIGURE 7.3.1
A particle P, reference point O, and unit vectors $\mathbf{n}_a$ and $\mathbf{n}_b$.

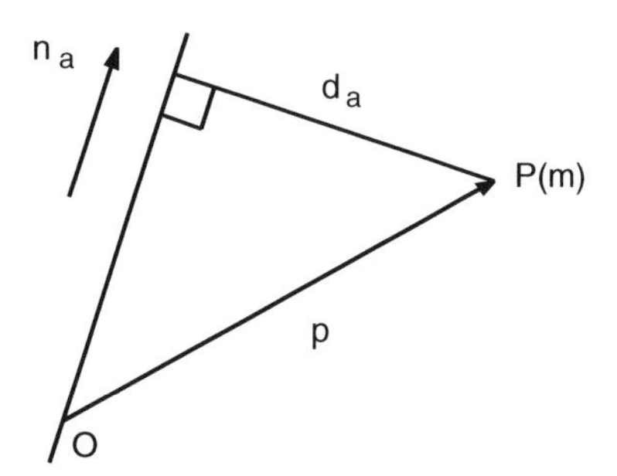

FIGURE 7.3.2
Distance from particle P to line through O and parallel to $\mathbf{n}_a$.

Observe that by substituting for $\mathbf{I}_a^{P/O}$ from Eq. (7.2.1) that $\mathbf{I}_{aa}^{P/O}$ and $\mathbf{I}_{ab}^{P/O}$ may be expressed in the form:

$$\mathbf{I}_{aa}^{P/O} = m\mathbf{p}\times(\mathbf{n}_a\times\mathbf{p})\cdot\mathbf{n}_a = m(\mathbf{p}\times\mathbf{n}_a)\cdot(\mathbf{p}\times\mathbf{n}_a) = m(\mathbf{p}\times\mathbf{n}_a)^2 \tag{7.3.3}$$

and

$$\mathbf{I}_{ab}^{P/O} = m\mathbf{p}\times(\mathbf{n}_a\times\mathbf{p})\cdot\mathbf{n}_b = m(\mathbf{p}\times\mathbf{n}_a)\cdot(\mathbf{p}\times\mathbf{n}_b) \tag{7.3.4}$$

Observe further that $(\mathbf{p}\times\mathbf{n}_a)^2$ may be identified with the square of the distance $\mathbf{d}_a$ from P to a line passing through O and parallel to $\mathbf{n}_a$ (see Figure 7.3.2). This distance is often referred to as the *radius of gyration* of P relative to O for the direction $\mathbf{n}_a$.

Observe also for the product of inertia of Eq. (7.3.3) that the unit vectors $\mathbf{n}_a$ and $\mathbf{n}_b$ may be interchanged. That is,

$$\mathbf{I}_{ab}^{P/O} = m(\mathbf{p}\times\mathbf{n}_a)\cdot(\mathbf{p}\times\mathbf{n}_b) = m(\mathbf{p}\times\mathbf{n}_a) = \mathbf{I}_{ba}^{P/O} \tag{7.3.5}$$

Note that no restrictions are placed upon the unit vectors $\mathbf{n}_a$ and $\mathbf{n}_b$. If, however, $\mathbf{n}_a$ and $\mathbf{n}_b$ are perpendicular, or, more generally, if we have three mutually perpendicular unit vectors, we can obtain additional geometric interpretations of moments and products of inertia. Specifically, consider a particle P with mass m located in a Cartesian reference frame R as in Figure 7.3.3. Let (x, y, z) be the coordinates of P in R. Then, from Eq. (7.2.1) the second moment vectors of P relative to origin O for the directions $\mathbf{n}_x$, $\mathbf{n}_y$, and $\mathbf{n}_z$ are:

$$\begin{aligned}\mathbf{I}_x^{P/O} &= m\mathbf{p}\times(\mathbf{n}_x\times\mathbf{p})\\ &= m(x\mathbf{n}_x + y\mathbf{n}_y + z\mathbf{n}_z)\times\left[\mathbf{n}_x\times(x\mathbf{n}_x + y\mathbf{n}_y + z\mathbf{n}_z)\right]\end{aligned}$$

or

$$\mathbf{I}_x^{P/O} = m\left[(y^2+z^2)\mathbf{n}_x - xy\mathbf{n}_y - xz\mathbf{n}_z\right] \tag{7.3.6}$$

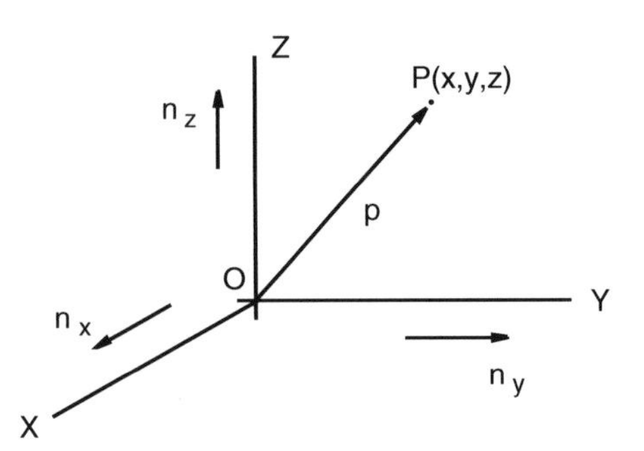

FIGURE 7.3.3
A particle P in a Cartesian reference frame.

and, similarly,

$$\mathbf{I}_y^{P/O} = m\left[-yx\mathbf{n}_x + \left(x^2 + z^2\right)\mathbf{n}_y - yz\mathbf{n}_z\right] \tag{7.3.7}$$

and

$$\mathbf{I}_z^{P/O} = m\left[-zx\mathbf{n}_x - zy\mathbf{n}_y + \left(x^2 + y^2\right)\mathbf{n}_z\right] \tag{7.3.8}$$

Using the definitions of Eqs. (7.3.1) and (7.3.2), we see that the various moments and products of inertia are then:

$$\begin{aligned} &\mathbf{I}_{xx}^{P/O} = m\left(x^2 + y^2\right), && \mathbf{I}_{xy}^{P/O} = -mxy, && \mathbf{I}_{xz}^{P/O} = -mxz \\ &\mathbf{I}_{yx}^{P/O} = -myx, && \mathbf{I}_{xy}^{P/O} = m\left(x^2 + y^2\right), && \mathbf{I}_{yz}^{P/O} = -myz \\ &\mathbf{I}_{zx}^{P/O} = -mzx, && \mathbf{I}_{zy}^{P/O} = -mzy, && \mathbf{I}_{zz}^{P/O} = m\left(y^2 + z^2\right) \end{aligned} \tag{7.3.9}$$

Observe that the moments of inertia are always nonnegative or zero, whereas the products of inertia may be positive, negative, or zero depending upon the position of P.

It is often convenient to normalize the moments and products of inertia by dividing by the mass m. Then, the normalized moment of inertia may be interpreted as a length squared, called the *radius of gyration* and defined as:

$$k_a \stackrel{D}{=} \left[\mathbf{I}_{aa}^{P/O}/m\right]^{1/2} \tag{7.3.10}$$

(See also Eq. (7.3.3).)

Finally, observe that the moments and products of inertia of Eqs. (7.3.9) may be conveniently listed in the matrix form:

$$\mathbf{I}_{ab}^{P/O} = m\begin{bmatrix} \left(y^2 + z^2\right) & -xy & -xz \\ -yx & \left(z^2 + x^2\right) & -yz \\ -zx & -zy & \left(x^2 + y^2\right) \end{bmatrix} \tag{7.3.11}$$

where a and b can be x, y, or z.

7.4 Inertia Dyadics

Comparing Eqs. (7.3.6), (7.3.7), and (7.3.8) with (7.3.9) we see that the second moment vectors may be expressed as:

$$\begin{aligned}\mathbf{I}_x^{P/O} &= I_{xx}\mathbf{n}_x + I_{xy}\mathbf{n}_y + I_{xz}\mathbf{n}_z \\ \mathbf{I}_y^{P/O} &= I_{yx}\mathbf{n}_x + I_{yy}\mathbf{n}_y + I_{yz}\mathbf{n}_z \\ \mathbf{I}_z^{P/O} &= I_{zx}\mathbf{n}_x + I_{zy}\mathbf{n}_y + I_{zz}\mathbf{n}_z\end{aligned} \tag{7.4.1}$$

where the superscripts on the moments and products of inertia have been deleted. We can simplify these expressions further by using the index notation introduced and developed in Chapter 2. Specifically, if the subscripts x, y, and z are replaced by the integers 1, 2, and 3, we have:

$$\begin{aligned}\mathbf{I}_1^{P/O} &= I_{11}\mathbf{n}_1 + I_{12}\mathbf{n}_2 + I_{13}\mathbf{n}_3 = I_{1j}\mathbf{n}_j \\ \mathbf{I}_2^{P/O} &= I_{21}\mathbf{n}_1 + I_{22}\mathbf{n}_2 + I_{23}\mathbf{n}_3 = I_{2j}\mathbf{n}_j \\ \mathbf{I}_3^{P/O} &= I_{31}\mathbf{n}_1 + I_{32}\mathbf{n}_2 + I_{33}\mathbf{n}_3 = I_{3j}\mathbf{n}_j\end{aligned} \tag{7.4.2}$$

or

$$\mathbf{I}_i^{P/O} = I_{ij}\mathbf{n}_i \quad (i = 1,2,3) \tag{7.4.3}$$

where the repeated index denotes a sum over the range of the index.

These expressions can be simplified even further by using the concept of a *dyadic*. A dyadic is the result of a product of vectors employing the usual rules of elementary algebra, except that the pre- and post-positions of the vectors are maintained. That is, if **a** and **b** are vectors expressed as:

$$\mathbf{a} = a_1\mathbf{n}_1 + a_2\mathbf{n}_2 + a_3\mathbf{n}_3 \quad \text{and} \quad \mathbf{b} = b_1\mathbf{n}_1 + b_2\mathbf{n}_2 + b_3\mathbf{n}_3 \tag{7.4.4}$$

then the dyadic product of **a** and **b** is defined through the operations:

$$\begin{aligned}\mathbf{ab} &\stackrel{D}{=} (a_1\mathbf{n}_1 + a_2\mathbf{n}_2 + a_3\mathbf{n}_3)(b_1\mathbf{n}_1 + b_2\mathbf{n}_2 + b_3\mathbf{n}_3) \\ &= a_1b_1\mathbf{n}_1\mathbf{n}_1 + a_1b_2\mathbf{n}_1\mathbf{n}_2 + a_1b_3\mathbf{n}_1\mathbf{n}_3 \\ &\quad + a_2b_1\mathbf{n}_2\mathbf{n}_1 + a_2b_2\mathbf{n}_2\mathbf{n}_2 + a_2b_3\mathbf{n}_2\mathbf{n}_3 \\ &\quad + a_3b_1\mathbf{n}_3\mathbf{n}_1 + a_3b_2\mathbf{n}_2\mathbf{n}_2 + a_3b_3\mathbf{n}_3\mathbf{n}_3\end{aligned} \tag{7.4.5}$$

where the unit vector products are called *dyads*. The nine dyads form the basis for a general dyadic (say, **D**), expressed as:

$$\begin{aligned}\mathbf{D} = & d_{11}\mathbf{n}_1\mathbf{n}_1 + d_{12}\mathbf{n}_1\mathbf{n}_2 + d_{13}\mathbf{n}_1\mathbf{n}_3 \\ & + d_{21}\mathbf{n}_2\mathbf{n}_1 + d_{22}\mathbf{n}_2\mathbf{n}_2 + d_{23}\mathbf{n}_2\mathbf{n}_3 \\ & + d_{31}\mathbf{n}_3\mathbf{n}_1 + d_{22}\mathbf{n}_3\mathbf{n}_2 + d_{33}\mathbf{n}_3\mathbf{n}_3 = d_{ij}\mathbf{n}_i\mathbf{n}_j\end{aligned} \tag{7.4.6}$$

Observe that a dyadic may be thought of as being a vector whose components are vectors; hence, dyadics are sometimes called *vector–vectors*. Observe further that the scalar components of $\mathbf{D}$ (the d_{ij} of Eq. (7.4.6)) can be considered as the elements of a 3×3 matrix and as the components of a second-order *tensor* (see References 7.5, 7.6, and 7.7).

In Eq. (7.3.12), we see that the moments and products of inertia may be assembled as elements of a matrix. In Eq. (7.4.3) these elements are seen to be I_{ij} $(i, j = 1, 2, 3)$. If these matrix elements are identified with dyadic components, we obtain the inertia dyadic defined as:

$$\begin{aligned}\mathbf{I}^{P/O} \overset{D}{=} & I_{11}\mathbf{n}_1\mathbf{n}_1 + I_{12}\mathbf{n}_1\mathbf{n}_2 + I_{13}\mathbf{n}_1\mathbf{n}_3 \\ & + I_{21}\mathbf{n}_2\mathbf{n}_1 + I_{22}\mathbf{n}_2\mathbf{n}_2 + I_{23}\mathbf{n}_2\mathbf{n}_3 \\ & + I_{31}\mathbf{n}_3\mathbf{n}_1 + I_{32}\mathbf{n}_3\mathbf{n}_2 + I_{33}\mathbf{n}_3\mathbf{n}_3\end{aligned} \tag{7.4.7}$$

or

$$\mathbf{I}^{P/O} = I_{ij}\mathbf{n}_i\mathbf{n}_j \tag{7.4.8}$$

By comparing Eqs. (7.4.2) and (7.4.7), we see that the inertia dyadic may also be expressed in the form:

$$\mathbf{I}^{P/O} = \mathbf{n}_1\mathbf{I}_1^{P/O} + \mathbf{n}_2\mathbf{I}_2^{P/O} + \mathbf{n}_3\mathbf{I}_3^{P/O} \tag{7.4.9}$$

Equation (7.3.5) shows that the matrix of moments and products of inertia is symmetric (that is, $I_{ij} = I_{ji}$). Then, by rearranging the terms of Eq. (7.4.7), we see that $\mathbf{I}^{P/O}$ may also be expressed as:

$$\mathbf{I}^{P/O} = \mathbf{I}_1^{P/O}\mathbf{n}_1 + \mathbf{I}_2^{P/O}\mathbf{n}_2 + \mathbf{I}_3^{P/O}\mathbf{n}_3 \tag{7.4.10}$$

The inertia dyadic may thus be interpreted as a vector whose components are second-moment vectors.

A principal advantage of using the inertia dyadic is that it can be used to generate second-moment vectors, moments of inertia, and products of inertia. Specifically, once the inertia dyadic is known, these other quantities may be obtained simply by dot product multiplication with unit vectors. That is,

$$\mathbf{I}_i^{P/O} = \mathbf{I}^{P/O} \cdot \mathbf{n}_i = \mathbf{n}_i \cdot \mathbf{I}^{P/O} \quad (i = 1,2,3) \tag{7.4.11}$$

and

$$\mathbf{I}_{ij}^{P/O} = \mathbf{n}_i \cdot \mathbf{I}^{P/O} \cdot \mathbf{n}_j \quad (i = 1,2,3) \tag{7.4.12}$$

Finally, for systems of particles or for rigid bodies, the inertia dyadic is developed from the contributions of the individual particles. That is, for a system S of N particles we have:

$$\mathbf{I}^{S/O} = \sum_{i=1}^{N} \mathbf{I}^{P_i/O} \tag{7.4.13}$$

7.5 Transformation Rules

Consider again the definition of the second moment vector of Eq. (7.2.1):

$$\mathbf{I}_a^{P/O} = m\mathbf{p}\times(\mathbf{n}_a \times \mathbf{p}) \tag{7.5.1}$$

Observe again the direct dependency of $\mathbf{I}_a^{P/O}$ upon $\mathbf{n}_a$. Suppose the unit vector $\mathbf{n}_a$ is expressed in terms of mutually perpendicular unit vectors $\mathbf{n}_i$ ($i = 1, 2, 3$) as:

$$\mathbf{n}_a = a_1\mathbf{n}_1 + a_2\mathbf{n}_2 + a_3\mathbf{n}_3 = a_i\mathbf{n}_i \tag{7.5.2}$$

Then, by substitution from Eq. (7.5.1) into (7.5.2), we have:

$$\mathbf{I}_a^{P/O} = m\mathbf{p}\times(a_i\mathbf{n}_i \times \mathbf{p}) = a_i m\mathbf{p}\times(\mathbf{n}_i \times \mathbf{p})$$

or

$$\mathbf{I}_a^{P/O} = a_i\mathbf{I}_i^{P/O} = a_1\mathbf{I}_1^{P/O} + a_2\mathbf{I}_2^{P/O} + a_3\mathbf{I}_3^{P/O} \tag{7.5.3}$$

Equation (7.5.3) shows that if we know the second moment vectors for each of three mutually perpendicular directions, we can obtain the second moment vector for any direction $\mathbf{n}_a$.

Similarly, suppose $\mathbf{n}_b$ is a second unit vector expressed as:

$$\mathbf{n}_b = b_1\mathbf{n}_1 + b_2\mathbf{n}_2 + b_3\mathbf{n}_3 = b_j\mathbf{n}_j \tag{7.5.4}$$

Then, by forming the projection of $\mathbf{I}_a^{P/O}$ onto $\mathbf{n}_b$ we obtain the product of inertia $\mathbf{I}_{ab}^{P/O}$, which in view of Eq. (7.5.3) can be expressed as:

$$\mathbf{I}_{ab}^{P/O} = \mathbf{n}_b \cdot \mathbf{I}_a^{P/O} = a_i b_j \mathbf{n}_j \cdot \mathbf{I}_i^{P/O}$$

or

$$\mathbf{I}_{ab}^{P/O} = a_i b_j \mathbf{I}_{ij}^{P/O} \tag{7.5.5}$$

Observe that the scalar components a_i and b_i of $\mathbf{n}_a$ and $\mathbf{n}_b$ of Eqs. (7.5.2) and (7.5.3) may be identified with transformation matrix components as in Eq. (2.11.3). Specifically, let $\hat{\mathbf{n}}_j$ (j = 1, 2, 3) be a set of mutually perpendicular unit vectors, distinct and noncollinear with the $\mathbf{n}_i$ (i = 1, 2, 3). Let the transformation matrix components be defined as:

$$S_{ij} = \mathbf{n}_i \cdot \hat{\mathbf{n}}_j \tag{7.5.6}$$

Then, in terms of the S_{ij}, Eq. (7.5.5) takes the form:

$$\hat{\mathbf{I}}_{k1}^{P/O} = S_{ik} S_{j1} I_{ij}^{P/O} \tag{7.5.7}$$

7.6 Parallel Axis Theorems

Consider once more the definition of the second moment vector of Eq. (7.2.1):

$$\mathbf{I}_a^{P/O} = m\mathbf{p} \times (\mathbf{n}_a \times \mathbf{p}) \tag{7.6.1}$$

Observe that just as $\mathbf{I}_a^{P/O}$ is directly dependent upon the direction of $\mathbf{n}_a$, it is also dependent upon the choice of the reference point O. The transformation rules discussed above enable us to evaluate the second-moment vector and other inertia functions as $\mathbf{n}_a$ changes. The parallel axis theorems discussed in this section will enable us to evaluate the second-moment vector and other inertia functions as the reference point O changes.

To see this, consider a set S of particles P_i (i = 1,…, N) with masses m_i as in Figure 7.6.1. Let G be the mass center of S and let O be a reference point. Let $\mathbf{p}$G locate G relative to O, let $\mathbf{p}_i$ locate P_i relative to O, and let $\mathbf{r}_i$ locate P_i relative to G. Finally, let $\mathbf{n}_a$ be an arbitrary unit vector. Then, from Eq. (7.2.2), the second moment of S for O for the direction of $\mathbf{n}_a$ is:

$$\mathbf{I}_a^{S/O} = \sum_{i=1}^{N} m_i \mathbf{p}_i \times (\mathbf{n}_a \times \mathbf{p}_i) \tag{7.6.2}$$

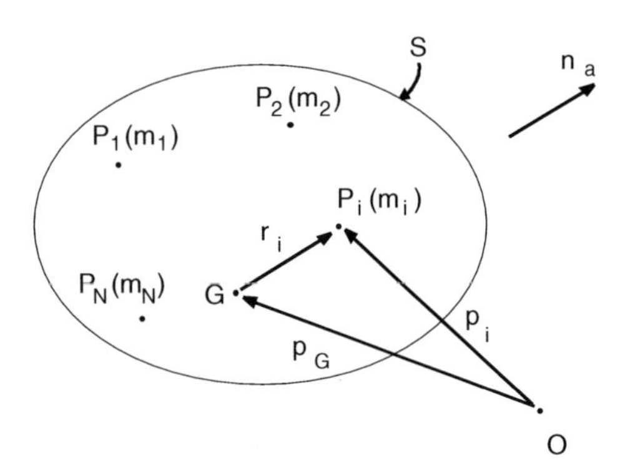

FIGURE 7.6.1
A set of particles with mass center G.

From Figure 7.6.1, we see that $\mathbf{p}_i$, $\mathbf{r}_i$, and $\mathbf{p}_G$ are related by:

$$\mathbf{p}_i = \mathbf{p}_G + \mathbf{r}_i \tag{7.6.3}$$

Then, by substituting into Eq. (7.6.2) we have:

$$\begin{aligned}
\mathbf{I}_a^{P/O} &= \sum_{i=1}^{N} m_i \{(\mathbf{p}_G + \mathbf{r}_i) \times [\mathbf{n}_a \times (\mathbf{p}_G + \mathbf{r}_i)]\} \\
&= \sum_{i=1}^{N} m_i \mathbf{p}_G \times (\mathbf{n}_a \times \mathbf{p}_G) + \sum_{i=1}^{N} m_i \mathbf{p}_G \times (\mathbf{n}_a \times \mathbf{r}_i) \\
&\quad + \sum_{i=1}^{N} m_i \mathbf{r}_i \times (\mathbf{n}_a \times \mathbf{p}_G) + \sum_{i=1}^{N} m_i \mathbf{r}_i \times (\mathbf{n}_a \times \mathbf{r}_i) \\
&= \left(\sum_{i=1}^{N} m_i\right) \mathbf{p}_G \times (\mathbf{n}_a \times \mathbf{p}_G) + \mathbf{p}_G \times \left(\mathbf{n}_a \times \cancelto{0}{\sum_{i=1}^{N} m_i \mathbf{r}_i}\right) \\
&\quad + \left(\cancelto{0}{\sum_{i=1}^{N} m_i \mathbf{r}_i}\right) \times (\mathbf{n}_a \times \mathbf{p}_G) + \sum_{i=1}^{N} m_i \mathbf{r}_i \times (\mathbf{n}_a \times \mathbf{r}_i)
\end{aligned}$$

or

$$\mathbf{I}_a^{S/O} = \mathbf{I}_a^{G/O} + \mathbf{I}_a^{S/G} \tag{7.6.4}$$

where $\mathbf{I}_a^{G/O}$ is defined as:

$$\mathbf{I}_a^{G/O} \stackrel{D}{=} M \mathbf{p}_G \times (\mathbf{n}_a \times \mathbf{p}_G) \tag{7.6.5}$$

where M is the total mass,

$$\left(\sum_{i=1}^{N} m_i\right),$$

of the particles of S; and $\mathbf{I}_a^{G/O}$ is the second moment of a particle located at G with a mass equal to the total mass M of S.

Eq. (7.6.4) is often called a *parallel axis theorem*. The reason for this name can be seen from the analogous equation for the moments of inertia: that is, by examining the projection of the terms of Eq. (7.6.4) along $\mathbf{n}_a$, we have:

$$\mathbf{I}_a^{S/O} \cdot \mathbf{n}_a = \mathbf{I}_a^{G/O} \cdot \mathbf{n}_a + \mathbf{I}_a^{S/G} \cdot \mathbf{n}_a$$

or

$$I_{aa}^{S/O} = I_{aa}^{G/O} + I_{aa}^{S/G} \tag{7.6.6}$$

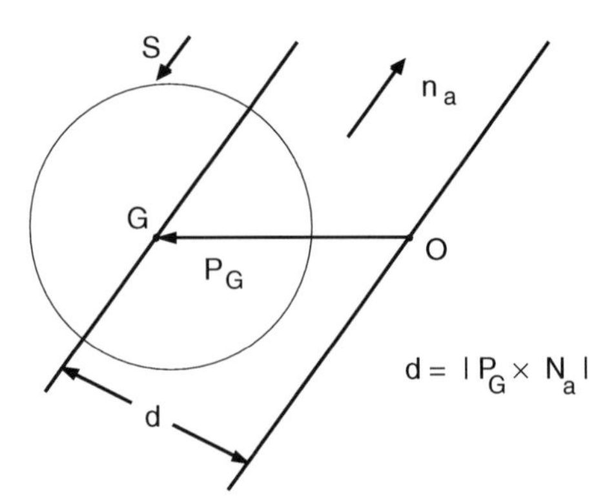

FIGURE 7.6.2
Parallel axes through O and G.

The first term on the right may be developed as:

$$I_{aa}^{G/O} = \mathbf{I}_a^{G/O} \cdot \mathbf{n}_a = M[\mathbf{p}_G \times (\mathbf{n}_a \times \mathbf{p}_G)] \cdot \mathbf{n}_a = M(\mathbf{p}_G \times \mathbf{n}_a)^2 = Md^2 \tag{7.6.7}$$

where d is $|\mathbf{p}G \times \mathbf{n}_a|$. Then, d is seen to be the distance between parallel lines passing through O and G and parallel to $\mathbf{n}_a$ (see Figure 7.6.2). Equation (7.6.6) may then be written:

$$I_{aa}^{S/O} = I_{aa}^{S/G} + Md^2 \tag{7.6.8}$$

By taking the projections of the terms of Eq. (7.6.4) along $\mathbf{n}_b$, a unit vector with a direction different than $\mathbf{n}_a$, we have for the products of inertia:

$$I_{ab}^{S/O} = I_{ab}^{G/O} + I_{ab}^{S/G} \tag{7.6.9}$$

Finally, by using the transformation rules of Eqs. (7.5.5) and (7.5.7) and by successively combining terms of Eqs. (7.6.6) and (7.6.8), we obtain the analogous equation for inertia dyadics:

$$\mathbf{I}^{S/O} = \mathbf{I}^{G/O} + \mathbf{I}^{S/G} \tag{7.6.10}$$

Equations (7.6.4), (7.6.6), (7.6.9), and (7.6.10) are versions of the parallel axis theorem for the second-moment vector, for the moments of inertia, for the products of inertia, and for the inertia dyadics. They show that if an inertia quantity is known relative to the mass center, then that quantity can readily be found relative to any other point.

Finally, inertia quantities computed relative to the mass center are called *central* inertia properties. Observe, then, in Eq. (7.6.8) that the central moment of inertia is the minimum moment of inertia for a given direction.

7.7 Principal Axes, Principal Moments of Inertia: Concepts

The foregoing paragraphs show that the second-moment vector may be used to generate the moments of inertia, the products of inertia, and the inertia dyadic. Recall that the second-moment vector is dependent upon the direction of the unit vector $\mathbf{n}_a$, used in the definition. That is, for a system of particles we have:

$$\mathbf{I}_a^{S/O} \stackrel{D}{=} \sum_{i=1}^{N} m_i \mathbf{p}_i \times (\mathbf{n}_a \times \mathbf{p}_i) \tag{7.7.1}$$

Observe that for an arbitrary unit vector, it is unlikely that $\mathbf{I}_a^{S/O}$ will be parallel to $\mathbf{n}_a$. If it happens, however, that $\mathbf{I}_a^{S/O}$ *is* parallel to $\mathbf{n}_a$, then $\mathbf{n}_a$ is said to be a *principal unit vector* or an *eigen unit vector.*

Observe that if $\mathbf{n}_a$ is a principal unit vector and if $\mathbf{n}_b$ is perpendicular to $\mathbf{n}_a$, then the product of inertia $\mathbf{I}_{ab}^{S/O}$ is zero. That is,

$$\mathbf{I}_{ab}^{S/O} = \mathbf{I}_a^{S/O} \cdot \mathbf{n}_b = 0 \tag{7.7.2}$$

Observe further that if $\mathbf{n}_a$ is a principal unit vector, the moment of inertia $\mathbf{I}_{aa}^{S/O}$ is the magnitude of $\mathbf{I}_a^{S/O}$. Then, we have:

$$\mathbf{I}_{aa}^{S/O} = \mathbf{I}_a^{S/O} \cdot \mathbf{n}_a = \left|\mathbf{I}_a^{S/O}\right| \quad \text{and} \quad \mathbf{I}_a^{S/O} = \mathbf{I}_{aa}^{S/O} \mathbf{n}_a \tag{7.7.3}$$

The direction of a principal unit vector is called a *principal direction*. The corresponding moment of inertia is called a *principal moment of inertia* or *eigenvalue of inertia.*

Consider again the inertia dyadic defined in Section 7.4. From Eq. (7.4.11) we have:

$$\mathbf{I}_a^{S/O} = \mathbf{I}^{S/O} \cdot \mathbf{n}_a \tag{7.7.4}$$

Hence, if $\mathbf{n}_a$ is a principal unit vector, we have:

$$\mathbf{I}^{S/O} \cdot \mathbf{n}_a = \mathbf{I}_{aa}^{S/O} \mathbf{n}_a \tag{7.7.5}$$

In view of these concepts and equations, the obvious questions that arise are do principal unit vectors exist and, if they exist, how can they be found? Equation (7.7.5) is often used as point of departure for an analysis to answer these questions and to discuss principal unit vectors in general.

To begin the analysis, let Eq. (7.7.5) be rewritten without the superscripts:

$$\mathbf{I} \cdot \mathbf{n}_a = \mathbf{I}_{aa} \mathbf{n}_a \tag{7.7.6}$$

Next, let $\mathbf{I}$ and $\mathbf{n}_a$ be expressed in terms of mutually perpendicular unit vectors $\mathbf{n}_i$ (i = 1, 2, 3) as:

$$\mathbf{I} = I_{ij} \mathbf{n}_i \mathbf{n}_j \quad \text{and} \quad \mathbf{n}_a = a_k \mathbf{n}_k \tag{7.7.7}$$

Then, Eq. (7.7.6) becomes:

$$\begin{aligned} \mathbf{I} \cdot \mathbf{n}_a &= I_{ij} \mathbf{n}_i \mathbf{n}_j \cdot a_k \mathbf{n}_k = I_{ij} \cdot a_k \mathbf{n}_i \mathbf{n}_j \cdot \mathbf{n}_k \\ &= I_{ij} a_k \mathbf{n}_i \delta_{jk} = I_{ij} a_j \mathbf{n}_i \\ &= I_{aa} \mathbf{n}_a = I_{aa} a_k \mathbf{n}_k \end{aligned}$$

or

$$I_{ij}a_j\mathbf{n}_i = I_{aa}a_i\mathbf{n}_i \tag{7.7.8}$$

Equation (7.7.8) is equivalent to the three scalar equations:

$$I_{ij}a_j = I_{aa}a_i \quad \text{or} \quad \left(I_{ij} - \lambda\delta_{ij}\right)a_i = 0 \;\left(i = 1,2,3\right) \tag{7.7.9}$$

where for simplicity in notation I_{aa} is replaced by λ. When the index sums of these equations are written explicitly, they become:

$$\begin{aligned} \left(I_{11} - \lambda\right)a_1 + I_{12}a_2 + I_{13}a_3 &= 0 \\ I_{21}a_1 + \left(I_{22} - \lambda\right)a_2 + I_{23}a_3 &= 0 \\ I_{31}a_1 + I_{32}a_2 + \left(I_{33} - \lambda\right)a_3 &= 0 \end{aligned} \tag{7.7.10}$$

Equations (7.7.10) form a set of three linear equations for the three scalar components a_i of $\mathbf{n}_a$. If we can determine the a_i we have in effect determined and found $\mathbf{n}_a$. However, Eqs. (7.7.10) are homogeneous — that is, each term on the left sides contains one and only one of the a_i, and the right sides are zero. This means that there is no solution (except the trivial solution $a_i = 0$), unless the determinant of the coefficients of the a_i is zero. That is, a nontrivial solution for the a_i exists only if:

$$\begin{vmatrix} \left(I_{11} - \lambda\right) & I_{12} & I_{13} \\ I_{21} & \left(I_{22} - \lambda\right) & I_{23} \\ I_{31} & I_{32} & \left(I_{33} - \lambda\right) \end{vmatrix} = 0 \tag{7.7.11}$$

By expanding the determinant, we obtain:

$$\lambda^3 - I_I\lambda^2 + I_{II}\lambda - I_{III} = 0 \tag{7.7.12}$$

where the coefficients I_I, I_{II}, and I_{III} are:

$$I_I = I_{11} + I_{22} + I_{33} \tag{7.7.13}$$

$$I_{II} = I_{22}I_{33} - I_{32}I_{23} + I_{11}I_{33} - I_{31}I_{13} + I_{11}I_{22} - I_{12}I_{21} \tag{7.7.14}$$

$$I_{III} = I_{11}I_{22}I_{33} - I_{11}I_{32}I_{23} + I_{12}I_{31}I_{23} - I_{12}I_{21}I_{33} + I_{21}I_{32}I_{13} - I_{31}I_{13}I_{22} \tag{7.7.15}$$

These coefficients are seen to be directly related to the elements I_{ij} of the inertia matrix. Indeed, they may be identified as:

I_I = sum of diagonal elements of I_{ij}
I_{II} = sum of diagonal elements of the cofactor matrix of I_{ij}
I_{III} = determinant of I_{ij}

Interestingly, these coefficients may be shown (see Reference 7.4) to be independent of the choice of unit vectors $\mathbf{n}_i$ in which $\mathbf{I}$ is expressed. As a consequence, I_I, I_{II}, and I_{III} are sometimes called *invariants* of **I**.

Equation (7.7.12) is a cubic equation for λ which in general has three distinct roots: λ_1, λ_2, and λ_3. This implies that there are at least three solutions for each at the a_1, a_2, and a_3 of Eq. (7.7.10). To obtain these solutions, we can select one of the roots (say, λ_1) and then substitute λ_1 for λ in Eq. (7.7.10). This produces three equations that are *dependent* because λ_1 is a solution of Eq. (7.7.11). Hence, only two of the three equations are independent. However, we can obtain a third independent equation by recalling that $\mathbf{n}_a$ is a unit vector and thus a_1, a_2, and a_3 also satisfy the equation:

$$a_1^2 + a_2^2 + a_3^2 = 1 \tag{7.7.16}$$

Then, by selecting any two equations from Eq. (7.7.10) together with Eq. (7.7.16), we can solve the resulting system of three equations and determine the values of a_1, a_2, and a_3.

Next, by letting λ be λ_2, we can repeat the process and find a second set of components (a_1, a_2, and a_3) of a principal unit vector $\mathbf{n}_a$. Similarly, for $\lambda = \lambda_3$, we obtain a third principal unit vector.

By following this procedure we obtain the components of three principal unit vectors defining three principal directions of inertia of the system. The values λ_1, λ_2, and λ_3 are then the corresponding principal moments of inertia. This procedure and these concepts are illustrated in the example in the following section.

7.8 Principal Axes, Principal Moments of Inertia: Example

Consider a system whose central inertia matrix relative to mutually perpendicular unit vectors $\mathbf{n}_1$, $\mathbf{n}_2$, and $\mathbf{n}_3$ is:

$$\left[I_{ij}\right] = \begin{bmatrix} I_{11} & I_{12} & I_{13} \\ I_{21} & I_{22} & I_{23} \\ I_{31} & I_{32} & I_{33} \end{bmatrix} = \begin{bmatrix} 9/2 & -3/4 & 3\sqrt{3}/4 \\ -3/4 & 39/8 & \sqrt{3}/8 \\ 3\sqrt{3}/4 & \sqrt{3}/8 & 37/8 \end{bmatrix} \tag{7.8.1}$$

To determine the principal values, the principal unit vectors, and the principal directions, we need to form Eq. (7.7.10). That is,

$$\begin{gathered}(9/2-\lambda)a_1 + (-3/4)a_2 + 3\sqrt{3}/4\,a_3 = 0 \\ (-3/4)a_1 + (39/8-\lambda)a_2 + \left(\sqrt{3}/8\right)a_3 = 0 \\ \left(3\sqrt{3}/4\right)a_1 + \left(\sqrt{3}/8\right)a_2 + (37/8-\lambda)a_3 = 0\end{gathered} \tag{7.8.2}$$

Because these equations are linear and homogeneous (right sides are zero), there is a nontrivial solution for the a_1, a_2, and a_3 only if the determinant of the coefficients is zero (Eq. (7.7.11)). Hence,

$$\begin{bmatrix} (9/2-\lambda) & (-3/4) & (3\sqrt{3}/4) \\ (-3/4) & (39/8-\lambda) & (\sqrt{3}/8) \\ (3\sqrt{3}/4) & (\sqrt{3}/8) & (37/8-\lambda) \end{bmatrix} = 0 \tag{7.8.3}$$

Expanding the determinant, we obtain:

$$\lambda^3 - 14\lambda^2 + 63\lambda - 90 = 0 \tag{7.8.4}$$

The coefficients of Eq. (7.8.4) are seen to be the parameters I_I, I_{II}, and I_{III} of Eqs. (7.7.13), (7.7.14), and (7.7.15).

By solving Eq. (7.8.4) we obtain:

$$\lambda = \lambda_1 = 3,\ \lambda = \lambda_2 = 5,\ \lambda = \lambda_3 = 6 \tag{7.8.5}$$

These are the principal moments of inertia. By substituting the first of these into Eq. (7.8.2), we obtain:

$$\begin{aligned} &(9/2-3)a_1 + (-3/4)a_2 + (3\sqrt{3}/4)a_3 = 0 \\ &(-3/4)a_1 + (39/8-3)a_2 + (\sqrt{3}/8)a_3 = 0 \\ &(3\sqrt{3}/4)a_1 + (\sqrt{3}/8)a_2 + (37/8-3)a_3 = 0 \end{aligned} \tag{7.8.6}$$

By eliminating fractions we have:

$$\begin{aligned} &2a_1 - a_2 + \sqrt{3}a_3 = 0 \\ &-6a_1 + 15a_2 + \sqrt{3}a_3 = 0 \\ &18a_1 + 3a_2 + 13\sqrt{3}a_3 = 0 \end{aligned} \tag{7.8.7}$$

These equations are seen to be dependent (multiplying the first by 12 and adding the result to the second produces the third); hence, only two of the equations are independent. A third independent equation may be obtained from Eq. (7.7.16) by requiring that $\mathbf{n}_a$ be a unit vector:

$$a_1^2 + a_2^2 + a_3^2 = 1 \tag{7.8.8}$$

Thus, by selecting any two of Eq. (7.8.7) and appending them to Eq. (7.8.8) we can determine a_1, a_2, and a_3. The result is:

$$a_1 = \sqrt{2}/2,\ a_2 = \sqrt{2}/4,\ a_3 = -\sqrt{6}/4 \tag{7.8.9}$$

Hence, the principal unit vector $\mathbf{n}_a$ corresponding to the principal value $\lambda = 3$ is:

$$\mathbf{n}_a = \sqrt{2}/2\mathbf{n}_1 + \sqrt{2}/4\mathbf{n}_2 - \sqrt{6}/4\mathbf{n}_3 \tag{7.8.10}$$

Similarly, by taking the second principal value $\lambda = 5$, substituting into Eq. (7.8.2), and using Eq. (7.8.8), we obtain the principal unit vector:

$$\mathbf{n}_b = \sqrt{3}/2\mathbf{n}_2 + 1/2\mathbf{n}_3 \tag{7.8.11}$$

Finally, for the third principal value $\lambda = 6$, the principal unit vector is:

$$\mathbf{n}_c = \sqrt{2}/2\mathbf{n}_1 - \sqrt{2}/4\mathbf{n}_2 + \sqrt{6}/4\mathbf{n}_3 \tag{7.8.12}$$

Observe that the principal unit vectors $\mathbf{n}_a$, $\mathbf{n}_b$, and $\mathbf{n}_c$ are mutually perpendicular. This is important because then the products of inertia relative to the directions of $\mathbf{n}_a$, $\mathbf{n}_b$, and $\mathbf{n}_c$ are zero. To explore this further, recall from Eq. (7.7.6) that because $\mathbf{n}_a$, $\mathbf{n}_b$, and $\mathbf{n}_c$ are principal unit vectors, we have:

$$\mathbf{I}\cdot\mathbf{n}_a = \lambda_1\mathbf{n}_a, \ \mathbf{I}\cdot\mathbf{n}_b = \lambda_2\mathbf{n}_b, \text{ and } \mathbf{I}\cdot\mathbf{n}_c = \lambda_3\mathbf{n}_c \tag{7.8.13}$$

Hence, we have:

$$\mathbf{n}_b\cdot\mathbf{I}\cdot\mathbf{n}_a = \mathbf{n}_c\cdot\mathbf{I}\cdot\mathbf{n}_b = \mathbf{n}_a\cdot\mathbf{I}\cdot\mathbf{n}_c = I_{ba} = I_{cb} = I_{ac} = 0 \tag{7.8.14}$$

Equation (7.8.14) shows that the components of the dyadic $\mathbf{I}$ in the "mixed" directions are zero; therefore, if we express $\mathbf{I}$ in terms of $\mathbf{n}_a$, $\mathbf{n}_b$, and $\mathbf{n}_c$ we have:

$$\begin{aligned}\mathbf{I} &= \lambda_1\mathbf{n}_a\mathbf{n}_a + \lambda_2\mathbf{n}_b\mathbf{n}_b + \lambda_3\mathbf{n}_c\mathbf{n}_c \\ &= 3\mathbf{n}_a\mathbf{n}_a + 5\mathbf{n}_b\mathbf{n}_b + 6\mathbf{n}_c\mathbf{n}_c\end{aligned} \tag{7.8.15}$$

In view of Eqs. (7.8.1) and (7.8.15), we see that the matrices of $\mathbf{I}$ referred to $\mathbf{n}_1$, $\mathbf{n}_2$, and $\mathbf{n}_3$ and to $\mathbf{n}_a$, $\mathbf{n}_b$, and $\mathbf{n}_c$ are vastly different. That is,

$$I_{ij} = \begin{bmatrix} I_{11} & I_{12} & I_{13} \\ I_{21} & I_{22} & I_{23} \\ I_{31} & I_{32} & I_{33} \end{bmatrix} = \begin{bmatrix} 9/2 & -3/4 & 3\sqrt{3}/4 \\ -3/4 & 39/8 & \sqrt{3}/8 \\ 3\sqrt{3}/4 & \sqrt{3}/8 & 37/8 \end{bmatrix} \tag{7.8.16}$$

and

$$I_{\alpha\beta} = \begin{bmatrix} I_{aa} & I_{ab} & I_{ac} \\ I_{ba} & I_{bb} & I_{bc} \\ I_{ca} & I_{cb} & I_{cc} \end{bmatrix} = \begin{bmatrix} 3 & 0 & 0 \\ 0 & 5 & 0 \\ 0 & 0 & 6 \end{bmatrix} \tag{7.8.17}$$

where the Greek subscripts α and β refer to the indices *a*, *b*, and *c*. We can obtain a direct relationship between these matrices by using the transformation matrices introduced in Chapter 2. Specifically, from Eq. (2.11.3) the elements of the transformation matrix between the n_i (i = 1, 2, 3) and the n_α (α = 1, b, c) are:

$$S_{i\alpha} = \mathbf{n}_i \cdot \mathbf{n}_\alpha \quad \text{and} \quad \mathbf{n}_\alpha = S_{i\alpha}\mathbf{n}_i \tag{7.8.18}$$

Let **I** be expressed in the forms:

$$\mathbf{I} = I_{ij}\mathbf{n}_i\mathbf{n}_j = I_{\alpha\beta}\mathbf{n}_\alpha\mathbf{n}_\beta \tag{7.8.19}$$

Then, the I_{ij} and the $I_{\alpha\beta}$ are related by the expressions:

$$I_{ij} = S_{i\alpha}S_{j\beta}I_{\alpha\beta} \quad \text{and} \quad I_{\alpha\beta} = S_{i\alpha}S_{j\beta}I_{ij} \tag{7.8.20}$$

From Eqs. (7.8.10), (7.8.11), (7.8.12), and (7.8.18), we see that for our example the $S_{i\alpha}$ are:

$$S_{i\alpha} = \begin{bmatrix} \sqrt{2}/2 & 0 & \sqrt{2}/2 \\ \sqrt{2}/4 & \sqrt{3}/2 & -\sqrt{2}/4 \\ -\sqrt{6}/4 & 1/2 & \sqrt{6}/4 \end{bmatrix} \tag{7.8.21}$$

Then, in matrix form, Eq. (7.8.20) becomes:

$$\begin{bmatrix} 9/2 & -3/4 & 3\sqrt{3}/4 \\ -3/4 & 39/8 & \sqrt{3}/8 \\ 3\sqrt{3}/4 & \sqrt{3}/8 & 37/8 \end{bmatrix} = \begin{bmatrix} \sqrt{2}/2 & 0 & \sqrt{2}/2 \\ \sqrt{2}/4 & \sqrt{3}/2 & -\sqrt{2}/4 \\ -\sqrt{6}/4 & 1/2 & \sqrt{6}/4 \end{bmatrix} \begin{bmatrix} 3 & 0 & 0 \\ 0 & 5 & 0 \\ 0 & 0 & 6 \end{bmatrix} \begin{bmatrix} \sqrt{2}/2 & \sqrt{2}/4 & -\sqrt{2}/4 \\ 0 & \sqrt{3}/2 & 1/2 \\ \sqrt{2}/2 & -\sqrt{2}/4 & \sqrt{6}/4 \end{bmatrix} \tag{7.8.22}$$

and

$$\begin{bmatrix} 3 & 0 & 0 \\ 0 & 5 & 0 \\ 0 & 0 & 6 \end{bmatrix} = \begin{bmatrix} \sqrt{2}/2 & \sqrt{2}/4 & -\sqrt{6}/4 \\ 0 & \sqrt{3}/2 & 1/2 \\ \sqrt{2}/2 & -\sqrt{2}/4 & \sqrt{6}/4 \end{bmatrix} \begin{bmatrix} 9/2 & -3/4 & 3\sqrt{3}/4 \\ -3/4 & 39/8 & \sqrt{3}/8 \\ 3\sqrt{3}/4 & \sqrt{3}/8 & 37/8 \end{bmatrix} \begin{bmatrix} \sqrt{2}/2 & 0 & \sqrt{2}/2 \\ \sqrt{2}/4 & \sqrt{3}/2 & -\sqrt{2}/4 \\ -\sqrt{6}/4 & 1/2 & \sqrt{6}/4 \end{bmatrix} \tag{7.8.23}$$

Finally, observe that the columns of the transformation matrix $[S_{i\alpha}]$ are the components of the principal unit vectors $\mathbf{n}_a$, $\mathbf{n}_b$, and $\mathbf{n}_c$.

7.9 Principal Axes, Principal Moments of Inertia: Discussion

In view of these results, several questions arise:

1. Are principal unit vectors always mutually perpendicular?
2. What if the roots of the cubic equation Eq. (7.7.12) are not all distinct?
3. What if the roots are not real?
4. Do we always need to solve a cubic equation to obtain the principal moments of inertia?
5. Will the procedures of the example always work?

To answer these questions consider again the definition of a principal unit vector, having the property expressed in Eq. (7.7.5). That is, if $\mathbf{n}_a$ is a principal unit vector, then:

$$\mathbf{I} \cdot \mathbf{n}_a = I_{aa}\mathbf{n}_a \quad (\text{no sum on } a) \tag{7.9.1}$$

Similarly if $\mathbf{n}_b$ is also a principal unit vector:

$$\mathbf{I} \cdot \mathbf{n}_b = I_{bb}\mathbf{n}_b \tag{7.9.2}$$

If we take the scalar product of the terms of Eqs. (7.9.1) and (7.9.2) with $\mathbf{n}_b$ and $\mathbf{n}_a$ and subtract the results, we obtain:

$$\begin{aligned} \mathbf{n}_b \cdot \mathbf{I} \cdot \mathbf{n}_a - \mathbf{n}_a \cdot \mathbf{I} \cdot \mathbf{n}_b &= I_{aa}\mathbf{b}_b \cdot \mathbf{n}_a - I_{aa}\mathbf{n}_a \cdot \mathbf{n}_b \\ &= \left(I_{aa} - I_{bb}\right)\mathbf{n}_a \cdot \mathbf{n}_b \end{aligned} \tag{7.9.3}$$

Because the inertia dyadic **I** is symmetric, $\mathbf{n}_a \cdot \mathbf{I} \cdot \mathbf{n}_b$ and $\mathbf{n}_b \cdot \mathbf{I} \cdot \mathbf{n}_a$ are equal, thus the left side of Eq. (7.9.3) is zero. Then, if the principal moments of inertia I_{aa} and I_{bb} are not equal, we have:

$$\mathbf{n}_a \cdot \mathbf{n}_b = 0 \tag{7.9.4}$$

This shows that principal unit vectors of distinct principal moments of inertia are mutually perpendicular.

Next, suppose that the principal moments of inertia are not all distinct. Suppose, for example, that I_{aa} and I_{bb} are equal. Are the associated principal unit vectors still perpendicular? To address this question, consider again Eq. (7.7.12):

$$\lambda^3 - I_I\lambda^2 + I_{II}\lambda - I_{III} = 0 \tag{7.9.5}$$

Recall from elementary algebra that if I_{aa}, I_{bb}, and I_{cc} are the roots of this equation then the equation may be written in the equivalent form:

$$\left(\lambda - I_{aa}\right)\left(\lambda - I_{bb}\right)\left(\lambda - I_{cc}\right) = 0 \tag{7.9.6}$$

By expanding Eq. (7.9.6) and by comparing the coefficients with those of Eq. (7.9.5), we discover that the roots are related to the coefficients (the invariants) by the expressions:

$$
\begin{aligned}
I_I &= I_{aa} + I_{bb} + I_{cc} \\
I_{II} &= I_{aa}I_{bb} + I_{bb}I_{cc} + I_{cc}I_{aa} \\
I_{III} &= I_{aa}I_{bb}I_{cc}
\end{aligned}
\tag{7.9.7}
$$

Now, if I_{cc} is distinct from I_{aa} (= I_{bb}), then by the reasoning of the foregoing paragraph the two corresponding principal unit vectors $\mathbf{n}_a$ and $\mathbf{n}_c$ will be distinct and perpendicular. That is,

$$
\mathbf{I}\cdot\mathbf{n}_a = I_{aa}\mathbf{n}_a, \quad \mathbf{I}\cdot\mathbf{n}_c = I_{cc}\mathbf{n}_c, \quad \text{and } \mathbf{n}_a\cdot\mathbf{n}_c = 0 \tag{7.9.8}
$$

Let $\mathbf{n}_b$ be a unit vector perpendicular to $\mathbf{n}_a$ and $\mathbf{n}_c$. Then, $\mathbf{I} \bullet \mathbf{n}_b$ will be some vector, say $\boldsymbol{\beta}$. However, because $\mathbf{n}_a$, $\mathbf{n}_b$, and $\mathbf{n}_c$ are mutually perpendicular, $\mathbf{I}$ can be expressed in the form:

$$
\mathbf{I} = (\mathbf{I}\cdot\mathbf{n}_a)\mathbf{n}_a + (\mathbf{I}\cdot\mathbf{n}_b)\mathbf{n}_b + (\mathbf{I}\cdot\mathbf{n}_c)\mathbf{n}_c \tag{7.9.9}
$$

Then, in view of Eq. (7.9.8) and if $\mathbf{I}\cdot\mathbf{n}_b$ is $\boldsymbol{\beta}$, we have:

$$
\mathbf{I} = I_{aa}\mathbf{n}_a\mathbf{n}_a + \boldsymbol{\beta}\mathbf{n}_b + I_{cc}\mathbf{n}_c\mathbf{n}_c \tag{7.9.10}
$$

Because $\mathbf{I}$ is also symmetric, we have:

$$
\boldsymbol{\beta}\mathbf{n}_b = \mathbf{n}_b\boldsymbol{\beta} \tag{7.9.11}
$$

Therefore, $\boldsymbol{\beta}$ must be parallel to $\mathbf{n}_b$. That is,

$$
\boldsymbol{\beta} = \beta\mathbf{n}_b \tag{7.9.12}
$$

where β is the magnitude of $\boldsymbol{\beta}$. Hence, $\mathbf{I}$ takes the form:

$$
\mathbf{I} = I_{aa}\mathbf{n}_a\mathbf{n}_a + \beta\mathbf{n}_b\mathbf{n}_b + I_{cc}\mathbf{n}_c\mathbf{n}_c \tag{7.9.13}
$$

Then, because I_I is the sum of the diagonal elements of the matrix of $\mathbf{I}$ (independently of the unit vectors in which $\mathbf{I}$ is expressed) and in view of the first of Eq. (7.9.7), we have:

$$
I_I = I_{aa} + \beta + I_{cc} = I_{aa} + I_{bb} + I_{cc} \tag{7.9.14}
$$

or

$$
\beta = I_{bb} \tag{7.9.15}
$$

Hence, by assuming that I_{bb} and I_{aa} are equal, Eq. (7.9.13) produces an expression for **I** in the form:

$$\begin{aligned} \mathbf{I} &= I_{aa}\mathbf{n}_a\mathbf{n}_a + I_{aa}\mathbf{n}_b\mathbf{n}_b + I_{cc}\mathbf{n}_c\mathbf{n}_c \\ &= I_{aa}\left(\mathbf{n}_a\mathbf{n}_a + \mathbf{n}_b\mathbf{n}_b\right) + I_{cc}\mathbf{n}_c\mathbf{n}_c \end{aligned} \tag{7.9.16}$$

Then, if $\boldsymbol{\eta}$ is any unit vector parallel to the plane of $\mathbf{n}_a$ and $\mathbf{n}_b$ (that is, perpendicular to $\mathbf{n}_c$), we have:

$$\begin{aligned} \mathbf{I}\cdot\boldsymbol{\eta} &= I_{aa}\left(\mathbf{n}_a\mathbf{n}_a + \mathbf{n}_b\mathbf{n}_b\right)\cdot\boldsymbol{\eta} + I_{cc}\mathbf{n}_c\mathbf{n}_c\cdot\boldsymbol{\eta} \\ &= I_{aa}\boldsymbol{\eta} + 0 \end{aligned} \tag{7.9.17}$$

Therefore, $\boldsymbol{\eta}$ is a principal unit vector (see Eq. (7.7.5)). This means that all unit vectors parallel to the plane of $\mathbf{n}_a$ and $\mathbf{n}_b$ are principal unit vectors. Thus, if two of the principal moments of inertia are equal, an infinite number of principal unit vectors are parallel to the plane that is normal to the principal unit vector associated with the distinct root of Eq. (7.7.12).

We can consider the case when all three of the roots of Eq. (7.7.12) are equal by similar reasoning. In this case, only one principal unit vector is found by the procedure of the previous sections. Let this vector be $\mathbf{n}_c$ and let $\mathbf{n}_a$ and $\mathbf{n}_b$ be unit vectors perpendicular to $\mathbf{n}_c$ such that $\mathbf{n}_a$, $\mathbf{n}_b$, and $\mathbf{n}_c$ form a mutually perpendicular set. Then, **I** has the form:

$$\mathbf{I} = \boldsymbol{\alpha}\mathbf{n}_a + \boldsymbol{\beta}\mathbf{n}_b + I\mathbf{n}_c\mathbf{n}_c \tag{7.9.18}$$

where $\boldsymbol{\alpha}$ and $\boldsymbol{\beta}$ are vectors to be determined and I is the triple root of Eq. (7.7.12). Because **I** is symmetric, we have:

$$\boldsymbol{\alpha}\mathbf{n}_a + \boldsymbol{\beta}\mathbf{n}_b = \mathbf{n}_a\boldsymbol{\alpha} + \mathbf{n}_b\boldsymbol{\beta} \tag{7.9.19}$$

or, equivalently, $\boldsymbol{\alpha}$ and $\boldsymbol{\beta}$ must be parallel to a plane normal to $\mathbf{n}_c$. Thus, $\boldsymbol{\alpha}$ and $\boldsymbol{\beta}$ have the forms:

$$\boldsymbol{\alpha} = \alpha_a\mathbf{n}_a + \alpha_b\mathbf{n}_b \quad \text{and} \quad \boldsymbol{\beta} = \beta_a\mathbf{n}_a + \beta_b\mathbf{n}_b \tag{7.9.20}$$

Hence, **I** becomes:

$$\mathbf{I} = \alpha_a\mathbf{n}_a\mathbf{n}_a + \beta_a\mathbf{n}_a\mathbf{n}_b + \alpha_b\mathbf{n}_b\mathbf{n}_a + \beta_b\mathbf{n}_b\mathbf{n}_b + I\mathbf{n}_c\mathbf{n}_c \tag{7.9.21}$$

The matrix of **I** relative to $\mathbf{n}_a$, $\mathbf{n}_b$, and $\mathbf{n}_c$ is:

$$\begin{bmatrix} \alpha_a & \beta_a & 0 \\ \alpha_b & \beta_b & 0 \\ 0 & 0 & 1 \end{bmatrix} \tag{7.9.22}$$

Because **I** is symmetric we have:

$$\alpha_b = \beta_a \tag{7.9.23}$$

From Eq. (7.9.7), the invariants I_I, I_{II}, and I_{III} are:

$$\begin{aligned} I_I &= 3I = \alpha_a + \beta_b + I \\ I_{II} &= 3I^2 = \beta_b I + \alpha_a I + \alpha_a\beta_b - \alpha_b^2 \\ I_{III} &= I^3 = I(\alpha_a\beta_b - \alpha_b^2) \end{aligned} \tag{7.9.24}$$

where the right sides of these expressions are obtained from the description of I_I, I_{II}, and I_{III} in Eqs. (7.7.13), (7.7.14), and (7.7.15) and from the matrix of **I** in Eq. (7.9.22). By solving for α_a, α_b, and β_b, we obtain:

$$\alpha_a = I, \ \ \alpha_b = \beta_a = 0, \ \ \beta_b = I \tag{7.9.25}$$

Hence, $\boldsymbol{\alpha}$ and $\boldsymbol{\beta}$ are:

$$\boldsymbol{\alpha}_a = I\mathbf{n}_a \quad \text{and} \quad \boldsymbol{\beta} = I\mathbf{n}_b \tag{7.9.26}$$

Finally, **I** becomes (see Eq. (7.9.21)):

$$\mathbf{I} = I\mathbf{n}_a\mathbf{n}_a + I\mathbf{n}_b\mathbf{n}_b + I\mathbf{n}_c\mathbf{n}_c \tag{7.9.27}$$

Then, if $\boldsymbol{\eta}$ is any vector:

$$\mathbf{I}\cdot\boldsymbol{\eta} = I\boldsymbol{\eta} \tag{7.9.28}$$

Hence, when there is a triple root of Eq. (7.7.12) *all* unit vectors are principal unit vectors.

Regarding the question of real roots of Eq. (7.7.12), suppose a root is not real. That is, let a root λ have the complex form:

$$\lambda = \alpha + i\beta \tag{7.9.29}$$

where α and β are real and i is the imaginary $\sqrt{-1}$. Let **n** be a principal unit vector associated with the complex root λ. Then, from Eq. (7.7.5), we have:

$$\mathbf{I}\cdot\mathbf{n} = \lambda\mathbf{n} \tag{7.9.30}$$

By following the procedures of the previous sections we can obtain **n** by knowing λ. Because λ is complex, the components of **n** will be complex. Then, **n** could be expressed in the form:

$$\mathbf{n} = \mathbf{n}_\alpha + i\mathbf{n}_\beta \tag{7.9.31}$$

where $\mathbf{n}_\alpha$ and $\mathbf{n}_\beta$ are real vectors. Hence, by substituting from Eqs. (7.9.29) and (7.9.31) into Eq. (7.9.30) (and recalling that **I** is real), we have:

$$\mathbf{I}\cdot\left(\mathbf{n}_\alpha + i\mathbf{n}_\beta\right) = (\alpha + i\beta)\left(\mathbf{n}_\alpha + i\mathbf{n}_\beta\right) \tag{7.9.32}$$

By expanding this equation and by matching real and imaginary terms, we obtain:

$$\mathbf{I}\cdot\mathbf{n}_\alpha = \alpha\mathbf{n}_\alpha - \beta\mathbf{n}_\beta \quad \text{and} \quad \mathbf{I}\cdot\mathbf{n}_\beta = \beta\mathbf{n}_\beta + \alpha\mathbf{n}_\beta \tag{7.9.33}$$

By multiplying the first of these by $\mathbf{n}_\beta$ and the second by $\mathbf{n}_\alpha$ and subtracting, we have:

$$\mathbf{n}_\beta\cdot I\cdot\mathbf{n}_\alpha - \mathbf{n}_\alpha\cdot I\cdot\mathbf{n}_\beta = \alpha\mathbf{n}_\beta\cdot\mathbf{n}_\alpha - \beta - \beta - \alpha\mathbf{n}_\alpha\cdot\mathbf{n}_\beta = -2\beta \tag{7.9.34}$$

Because **I** is symmetric, the left side is zero. Hence, we have the conclusion:

$$\beta = 0 \tag{7.9.35}$$

Therefore, because **I** is real and symmetric, the principal moments of inertia (the roots of Eq. (7.7.12)) are real.

Regarding the question as to whether we always need to solve a cubic equation (Eq. (7.7.12)), consider that if we know one of the roots (say, λ) of an equation, we can depress the equation and obtain a quadratic equation by dividing by $\lambda - \lambda_1$. We can obtain a root λ_1 if we know a principal unit vector (say, η_1), as then from Eq. (7.7.5) λ_1 is simply $|\mathbf{I}\cdot\mathbf{n}_1|$. The question then becomes: can a principal unit vector be found without solving Eq. (7.7.5) and the associated linear system of Eqs. (7.7.10)? The answer is yes, on occasion. Specifically, if Π is a plane of symmetry of a set S of particles, then the unit vector normal to Π is a principal unit vector for all reference points in Π.

To see this, recall that a plane of symmetry is such that for every particle on one side of the plane there is a corresponding particle on the other side having the same mass and at the same distance from the plane. Consider, for example, the plane Π depicted in Figure 7.9.1 with particles P_1 and P_2 of S equidistant from Π and on opposite sides of Π. Let the particles each have mass m and let their distances from Π be h. Let O be any reference point in Π, and let Q be the point of intersection of Π with the line connecting P_1 and P_2. Finally, let **n** be a unit vector normal to Π.

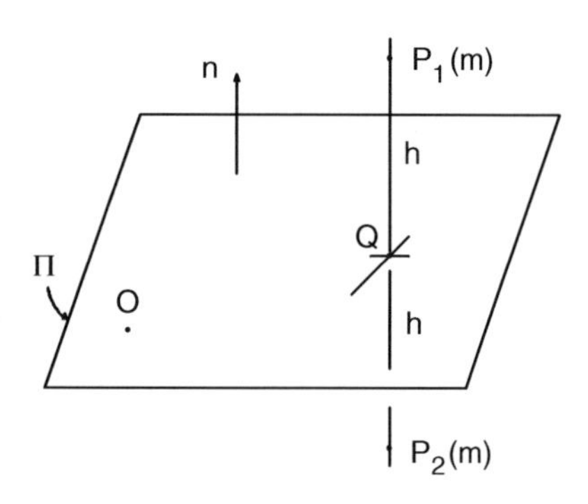

FIGURE 7.9.1
Particles on opposite sides of a plane of symmetry.

Let P_1 and P_2 be a typical pair of particles taken from a set S of particles for which Π is a plane of symmetry. Then, from Eq. (7.7.2) we see that the contributions of P_1 and P_2 to the second moment of S relative to O for the direction of **n** are represented by:

$$
\begin{aligned}
\mathbf{I}_n^{S/O} &= m\mathbf{p}_1 \times (\mathbf{n} \times \mathbf{p}_1) + m\mathbf{p}_2 \times (\mathbf{n} \times \mathbf{p}_2) + \ldots \\
&= m(\mathbf{OP} + h\mathbf{n}) \times [\mathbf{n} \times (\mathbf{OQ} + h\mathbf{n})] \\
&\quad + m(\mathbf{OQ} - h\mathbf{n}) \times [\mathbf{n} \times (\mathbf{OQ} - h\mathbf{n})] + \ldots \\
&= 2m|\mathbf{OQ}|^2 \mathbf{n} + \ldots
\end{aligned}
\tag{7.9.36}
$$

where the last equality follows from expanding the triple products of the previous equality and where the terms not written in Eq. (7.9.36) represent the contributions to $\mathbf{I}_a^{S/O}$ from the other pairs of particles of S. However, by analyses identical to those shown in Eq. (7.9.36), these contributions to $\mathbf{I}_a^{S/O}$ are also parallel to **n**. Therefore, $\mathbf{I}_n^{S/O}$ is parallel to **n**, and thus **n** is a principal unit vector of S for O.

A specialization of these ideas occurs if the system of particles all lie in the same plane, as with a planar body. A unit vector normal to the plane is then a principal unit vector.

There are other ways of determining principal unit vectors without solving Eqs. (7.7.10) and (7.7.12). For homogeneous bodies occupying common geometric shapes, we can simply refer to a table of results as in Appendix II. We can also make use of a number of theorems for principal unit vectors stated here, but without proof (see References 7.1 and 7.2 for additional information). If a line is parallel to a principal unit vector for a given reference point, then that line is called a *principal axis* for that reference point. A principal axis for the mass center as a reference point is called a *central principal axis*. Then, it can be shown that:

1. If a principal axis for a reference point other than the mass center also passes through the mass center, then it is also a central principal axis.
2. Mutually perpendicular principal axes that include a central principal axis for any point on the central principal axis are parallel to mutually perpendicular central principal axes.
3. A central principal axis is a principal axis for each of its points.

Finally, regarding the question as to whether the procedures of the example of the previous section will always produce principal unit vectors and principal moments of inertia, the answer is yes, but as seen above there may exist simpler procedures for any given problem. That is, a principal unit vector may often be identified by inspection — for example, as being normal to a plane of symmetry. In this case, the task of solving a cubic equation may be reduced to that of solving a quadratic equation.

To illustrate the procedure, suppose that a principal unit vector found by inspection is called $\mathbf{n}_3$. Then, let $\mathbf{n}_1$ and $\mathbf{n}_2$ be unit vectors perpendicular to $\mathbf{n}_3$ and also perpendicular to each other so that $\mathbf{n}_1$, $\mathbf{n}_2$, and $\mathbf{n}_3$ form a mutually perpendicular set. Then, because $\mathbf{n}_3$ is a principal unit vector, the inertia dyadic **I** expressed in terms of $\mathbf{n}_1$, $\mathbf{n}_2$, and $\mathbf{n}_3$ has a matrix of components in the form:

$$
I_{ij} = \begin{bmatrix} I_{11} & I_{12} & 0 \\ I_{21} & I_{22} & 0 \\ 0 & 0 & I_{33} \end{bmatrix}
\tag{7.9.37}
$$

From Eq. (7.7.5), if $\mathbf{n}_a$ is a principal unit vector, we have:

$$\mathbf{I}\cdot\mathbf{n}_a = I_{aa}\mathbf{n}_a \quad \text{no sum on } a \tag{7.9.38}$$

The associated scalar equations, Eq. (7.7.10), then become:

$$\begin{aligned}
&(I_{11}-I_{aa})a_1 + I_{12}a_2 = 0\\
&I_{21}a_1 + (I_{22}-I_{aa})a_2 = 0\\
&(I_{33}-I_{aa})a_3 = 0
\end{aligned} \tag{7.9.39}$$

where, as before, a_1, a_2, and a_3 are the components of $\mathbf{n}_a$ relative to $\mathbf{n}_1$, $\mathbf{n}_2$, and $\mathbf{n}_3$. By inspection, a solution to these equations is:

$$\mathbf{I}_{aa} = \mathbf{I}_{33}, \quad a_3 = 1, \quad a_1 = a_2 = 0 \tag{7.9.40}$$

If, however, I_{aa} is not equal to I_{33}, then a_3 is zero and the equations for a_1, a_2, and I_{aa} reduce to:

$$\begin{aligned}
&(I_{11}-I_{aa})a_1 + I_{12}a_2 = 0\\
&I_{21}a_1 + (I_{22}-I_{aa})a_2 = 0\\
&a_1^2 + a_2^2 = 1
\end{aligned} \tag{7.9.41}$$

Because the first two of these equations are linear and homogeneous, the third equation will be violated unless the determinant of the coefficients of the first two equations is zero. That is,

$$\begin{vmatrix} (I_{11}-I_{aa}) & I_{12} \\ I_{21} & (I_{22}-I_{aa}) \end{vmatrix} = 0 \tag{7.9.42}$$

Expanding the determinant we obtain:

$$I_{aa}^2 - I_{aa}(I_{11}+I_{22}) + I_{11}I_{22} - I_{12}^2 = 0 \tag{7.9.43}$$

Solving for I_{aa} we find:

$$I_{aa} = \frac{I_{11}+I_{22}}{2} \pm \left[\left(\frac{I_{11}-I_{22}}{2}\right)^2 + I_{12}^2\right]^{1/2} \tag{7.9.44}$$

When I_{aa} has the values as in Eq. (7.9.44), the first two equations of Eq. (7.9.41) become dependent. Hence, by taking the first and third of Eq. (7.9.41) we have a_1 and a_2 to be:

$$a_1 = I_{12}\Big/\left[I_{12}^2 + (I_{11}-I_{aa})^2\right]^{1/2} \tag{7.9.45}$$

and

$$a_2 = (I_{11} - I_{aa}) \Big/ \left[I_{12}^2 + (I_{11} - I_{aa})^2 \right]^{1/2} \tag{7.9.46}$$

When a plane of symmetry is identified or when a system or body is planar, the analysis is two dimensional, as above. In this case, the transformation between different unit vector sets is also simplified. Recall from Eqs. (7.5.5) and (7.5.7) that the moments and products of inertia referred to unit vectors $\hat{\mathbf{n}}_k$ and $\mathbf{n}_i$ are related by the expression:

$$\hat{I}_{k\ell} = S_{ik} S_{j\ell} I_{ij} \tag{7.9.47}$$

where the transformation matrix components are defined by Eq. (7.5.6) as:

$$S_{ik} = \mathbf{n}_i \cdot \hat{\mathbf{n}}_k \tag{7.9.48}$$

Suppose the unit vectors $\mathbf{n}_a$, $\mathbf{n}_b$, and $\mathbf{n}_c$ are principal unit vectors, with $\mathbf{n}_c$ being normal to a plane of symmetry. Let $\mathbf{n}_1$, $\mathbf{n}_2$, and $\mathbf{n}_3$ be mutually perpendicular unit vectors, with $\mathbf{n}_3$ being parallel to $\mathbf{n}_c$. Let θ be the angle between $\mathbf{n}_1$ and $\mathbf{n}_a$ as in Figure 7.9.2. Then, the transformation matrix components between the $\mathbf{n}_i$ ($i = 1, 2, 3$) and the $\mathbf{n}_\alpha$ ($\alpha = a, b, c$) are:

$$\mathbf{n}_i \cdot \mathbf{n}_\alpha = \begin{pmatrix} \cos\theta & \sin\theta & 0 \\ -\sin\theta & \cos\theta & 0 \\ 0 & 0 & 1 \end{pmatrix} \tag{7.9.49}$$

Using Eq. (7.9.47), the moments and products of inertia I_{ij}, relative to $\mathbf{n}_i$ and $\mathbf{n}_j$, then become:

$$I_{11} = \cos^2\theta I_{aa} + \sin^2\theta I_{bb} \tag{7.9.50}$$

$$I_{12} = I_{21} = -\sin\theta\cos\theta(I_{aa} - I_{bb}) \tag{7.9.51}$$

$$I_{22} = \sin^2\theta I_{aa} + \cos^2\theta I_{bb} \tag{7.9.52}$$

$$I_{13} = I_{31} = I_{23} = I_{32} = 0 \tag{7.9.53}$$

$$I_{33} = I_{cc} \tag{7.9.54}$$

Equations (7.9.50), (7.9.51), and (7.9.52) may be expressed in an even simpler form by using the trigonometric identities:

$$\begin{aligned} \sin^2\theta &= \frac{1}{2} - \frac{1}{2}\cos 2\theta \\ \cos^2 &\equiv \frac{1}{2} + \frac{1}{2}\cos 2\theta \\ \sin\theta\cos\theta &\equiv \frac{1}{2}\sin 2\theta \end{aligned} \tag{7.9.55}$$

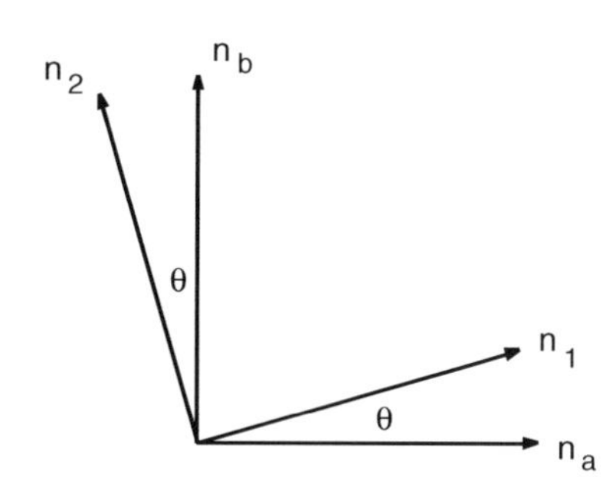

FIGURE 7.9.2
Unit vectors parallel to a plane of symmetry.

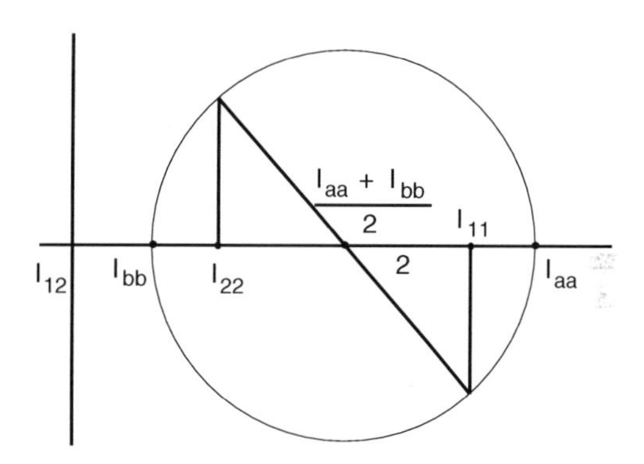

FIGURE 7.9.3
A Mohr circle diagram for moments and products of inertia.

Then, I_{11}, I_{12}, and I_{22} become:

$$I_{11} = \left(\frac{I_{aa} + I_{bb}}{2}\right) + \left(\frac{I_{aa} - I_{bb}}{2}\right)\cos 2\theta \qquad (7.9.56)$$

$$I_{12} = -\left(\frac{I_{aa} - I_{bb}}{2}\right)\sin 2\theta \qquad (7.9.57)$$

$$I_{22} = \left(\frac{I_{aa} + I_{bb}}{2}\right) - \left(\frac{I_{aa} - I_{bb}}{2}\right)\cos 2\theta \qquad (7.9.58)$$

Equations (7.9.56), (7.9.57), and (7.9.58) may be interpreted geometrically using a Mohr circle diagram as in Figure 7.9.3. In this diagram, the values of the moment of inertia I_{11} and I_{22} are found along the abscissa, while the value of the product of inertia I_{12} is found along the ordinate as shown.

7.10 Maximum and Minimum Moments and Products of Inertia

We will see presently and in the following chapters that the inertia dyadic and, specifically, the moments and products of inertia are the primary parameters of inertia torques. Therefore, the questions that arise are for which reference points and for which directions do the moments and products of inertia have maximum and minimum values? To address these questions, consider first the parallel axis theorem as expressed in Eq. (7.6.10):

$$\mathbf{I}^{S/O} = \mathbf{I}^{S/G} + \mathbf{I}^{G/O} \tag{7.10.1}$$

From this expression we see that if O is sufficiently far away from the mass center G, the components of $I^{G/O}$ become arbitrarily large; thus, there are no absolute maximum moments and products of inertia. However, if the reference point O is at the mass center G, the moments and products of inertia may have minimum values. Thus, we have the following question: if the mass center G is the reference point, for which directions do the moments and products of inertia have maximum and minimum values? To answer this question, consider the two-dimensional case and the Mohr circle analysis discussed at the end of the previous section. From Figure 7.9.3, we see that the maximum and minimum moments of inertia are the principal moments of inertia I_{aa} and I_{bb}. Also, the maximum absolute value of the product of inertia I_{12} occurs at 45° to the principal moment of inertia directions with value $(I_{aa} - I_{bb})/2$. The minimum absolute value of the product of inertia is zero. This occurs in directions corresponding to the directions of the principal moments of inertia. These results can also be seen by inspection of Eqs. (7.9.56), (7.9.57), and (7.9.58).

Do similar results hold, in general, in the three-dimensional case? The answer is yes, but the vertification is not as simple. Nevertheless, the results are readily obtained by using the *Lagrange multiplier method*. For readers not familiar with this method, it is sufficient to know that the method is used to find maximum and minimum values of functions of several variables which in turn are related to each other by *constraint* equations. For example, if the maximum or minimum value of a function $f(x, y, z)$ is to be found subject to a constraint:

$$g(x,y,z) = 0 \tag{7.10.2}$$

then a function $h(x, y, z)$ is defined as:

$$h(x,y,z,\lambda) \stackrel{D}{=} f(x,y,z) + \lambda g(x,y,z) \tag{7.10.3}$$

where λ is a parameter, called a *Lagrange multiplier,* to be determined along with x, y, and z such that h has maximum or minimum values. Specifically, if h is to be a maximum or minimum, the following equations need to be satisfied simultaneously:

$$\partial h/\partial x = 0, \quad \partial h/\partial y = 0, \quad \partial h/\partial z = 0, \quad \partial h/\partial \lambda = 0 \tag{7.10.4}$$

The last of these equations is identical with Eq. (7.10.2). Thus, Eq. (7.10.4) provides the necessary conditions for $f(x, y, z)$ to have maximum or minimum values. The solution of these equations then determines the values of x, y, z, and λ, producing the maximum or minimum values of f.

The advantage of the Lagrange multiplier method is that the constraint equation need not be solved independently in the analysis. The disadvantage is that an additional parameter λ is to be determined; however, the parameter λ may often be identified with an important physical or geometrical parameter in a given problem.

To apply this method in obtaining maximum and minimum central (mass center reference point) moments of inertia, let $\mathbf{I}$ be a central inertia dyadic with components I_{ij} relative to an arbitrary set of mutually perpendicular unit vectors $\mathbf{n}_i$. Then, if $\mathbf{n}_a$ is an arbitrarily directed unit vector, the moment of inertia relative to the $\mathbf{n}_a$ direction may be expressed as (see Eq. (7.5.5)):

$$I_{aa} = a_i a_j I_{ij} \tag{7.10.5}$$

where the a_i are the $\mathbf{n}_i$ components of $\mathbf{n}_a$ and where we are again employing the summation convention. Because $\mathbf{n}_a$ is a unit vector, the a_i must satisfy the relation:

$$a_i a_i = 1 \quad \text{or} \quad 1 - a_i a_i = 0 \tag{7.10.6}$$

This latter expression is a constraint equation analogous to Eq. (7.10.2). Thus, from Eq. (7.10.3), if we wish to find the direction $\mathbf{n}_a$ such that I_{aa} is a maximum or minimum, then we can form the function $h(a_i,\lambda)$ as:

$$h(a_i,\lambda) \overset{D}{=} a_i a_j I_{ij} + \lambda(1 - a_i a_i) \tag{7.10.7}$$

By setting the derivation of h with respect to the a_k equal to zero, as in Eq. (7.10.4), we have:

$$\begin{aligned} \partial h/\partial a_k &= (\partial a_i/\partial a_k) a_j I_{ij} + a_i (\partial a_j/\partial a_k) I_{ij} \\ &\quad - 2\lambda a_i \, \partial a_i/\partial a_k \\ &= \delta_{ik} a_i I_{ij} + a_i \delta_{jk} I_{ij} - 2\lambda a_i \delta_{ik} \\ &= 2 I_{kj} a_j - 2\lambda a_k = 0 \end{aligned}$$

or

$$I_{kj} a_j = \lambda a_k \quad (k = 1,2,3) \tag{7.10.8}$$

where we have used the symmetry of the inertia dyadic components. By setting the derivative of h with respect to λ equal to zero, we obtain Eq. (7.10.6):

$$a_k a_k = I \tag{7.10.9}$$

Equations (7.10.8) and (7.10.9) are identical to Eqs. (7.7.9) and (7.7.16), which determine the principal moments of inertia and their directions. Therefore, the directions producing the maximum and minimum central moments of inertia are the principal directions, and the maximum and minimum moments of inertia are to be found among the principal moments of inertia. Finally, the Lagrange multiplier λ is seen to be a principal moment of inertia.

We can use the same procedure to obtain the direction for the maximum products of inertia. From Eq. (7.5.5), the product of inertia relative to the directions of unit vectors $\mathbf{n}_a$ and $\mathbf{n}_b$ is:

$$I_{ab} = a_i b_j I_{ij} \tag{7.10.10}$$

where, as before, the a_i and b_i are the $\mathbf{n}_i$ components of $\mathbf{n}_a$ and $\mathbf{n}_b$. Then, because $\mathbf{n}_a$ and $\mathbf{n}_b$ are unit vectors, the a_i and b_i must satisfy the equations:

$$1-a_ia_i=0 \quad \text{and} \quad 1-b_ib_i=0 \tag{7.10.11}$$

If $\mathbf{n}_a$ and $\mathbf{n}_b$ are also perpendicular to each other, we have:

$$a_ib_i=0 \tag{7.10.12}$$

Equations (7.10.11) and (7.10.12) form *three* constraint equations to be satisfied while maximum values of I_{ab} of Eq. (7.10.10) are sought. Then, by generalizing the foregoing procedure, we form a function $h(a_i, b_i, \lambda, \mu, \nu)$ defined as:

$$h(a_i,b_i,\lambda,\mu,\nu)=a_ib_iI_{ij}+\lambda(1-a_ia_i)+\mu(1-b_ib_i)+\nu a_ib_i \tag{7.10.13}$$

where λ, μ, and ν are Lagrange multipliers. By setting the derivative of h with respect to a_k and b_k equal to zero, we obtain:

$$\partial h/\partial a_k = I_{kj}b_j - 2\lambda a_k + \nu b_k = 0 \tag{7.10.14}$$

and

$$\partial h/\partial b_k = I_{ik}a_i - 2\mu b_k + \nu a_k = 0 \tag{7.10.15}$$

If we multiply these equations by a_k and b_k, respectively (and sum over k), we obtain:

$$a_kI_{kj}b_j - 2\lambda = 0 \tag{7.10.16}$$

$$b_kI_{kj}b_j + \nu = 0 \tag{7.10.17}$$

$$a_kI_{ik}a_i + \nu = 0 \tag{7.10.18}$$

$$b_kI_{ih}a_i - 2\mu = 0 \tag{7.10.19}$$

where we have used Eqs. (7.10.11) and (7.10.12). By comparing Eqs. (7.10.16) and (7.10.19) and by recalling that I_{ik} is symmetric, we have:

$$\lambda = \mu \tag{7.10.20}$$

Then, Eqs. (7.10.14) and (7.10.15) may be rewritten in the forms:

$$I_{kj}b_j = 2\lambda a_k - \nu b_k \tag{7.10.21}$$

and

$$I_{kj}a_j = 2\lambda b_k - \nu a_k \tag{7.10.22}$$

By adding and subtracting these equations we have:

$$I_{kj}\left(a_j + b_j\right) = (2\lambda - \nu)\left(a_k + b_k\right) \tag{7.10.23}$$

and

$$I_{kj}\left(a_j - b_j\right) = -(2\lambda + \nu)\left(a_k - b_k\right) \tag{7.10.24}$$

These equations in turn may be expressed as:

$$\mathbf{I} \cdot \left(\mathbf{n}_a + \mathbf{n}_b\right) = (2\lambda - \nu)\left(\mathbf{n}_a + \mathbf{n}_b\right) \tag{7.10.25}$$

and

$$\mathbf{I} \cdot \left(\mathbf{n}_a - \mathbf{n}_b\right) = -(2\lambda + \nu)\left(\mathbf{n}_a - \mathbf{n}_b\right) \tag{7.10.26}$$

Thus, $(\mathbf{n}_a + \mathbf{n}_b)$ and $(\mathbf{n}_a - \mathbf{n}_b)$ are principal vectors of $\mathbf{I}$ corresponding to the principal values $(2\lambda - \nu)$ and $-(2\lambda + \nu)$. From our previous discussions, however, we know that for the general three-dimensional case with three distinct principal values, the principal values contain the maximum, the minimum, and intermediate values of the moments of inertia. For our discussion here, let these principal moments of inertia be designated as $I_{\alpha\alpha}$, $I_{\beta\beta}$, and $I_{\gamma\gamma}$, with principal unit vectors $\mathbf{n}_\alpha$, $\mathbf{n}_\beta$, and $\mathbf{n}_\gamma$. That is,

$$\mathbf{I} \cdot \mathbf{n}_\alpha = \mathbf{I}_{\alpha\alpha}\mathbf{n}_\alpha, \quad \mathbf{I} \cdot \mathbf{n}_\beta = \mathbf{I}_{\beta\beta}\mathbf{n}_\beta, \quad \mathbf{I} \cdot \mathbf{n}_\gamma = \mathbf{I}_{\gamma\gamma}\mathbf{n}_\gamma \quad \left(\text{no sum}\right) \tag{7.10.27}$$

Then, comparing Eqs. (7.10.25) and (7.10.26) with Eq. (7.10.27) we see that $(2\lambda - \nu)$ and $-(2\lambda + \nu)$ are to be identified with $I_{\alpha\alpha}$, $I_{\beta\beta}$, and $I_{\gamma\gamma}$ and $(\mathbf{n}_a + \mathbf{n}_b)$ and $(\mathbf{n}_a - \mathbf{n}_b)$ are in the direction of $\mathbf{n}_\alpha$, $\mathbf{n}_\beta$, and $\mathbf{n}_\gamma$. Specifically, if $(2\lambda - \nu)$ and $-(2\lambda + \nu)$ are identified with $I_{\alpha\alpha}$ and $I_{\beta\beta}$, we have:

$$2\lambda - \nu = I_{\beta\beta} \quad \text{and} \quad 2\lambda + \nu = -I_{\beta\beta} \tag{7.10.28}$$

Then, by adding these equations we have:

$$2\lambda = \left(I_{\alpha\alpha} - I_{\beta\beta}\right)/2 \tag{7.10.29}$$

However, from Eq. (7.10.16) we see that:

$$2\lambda = a_k I_{kj} b_j = \mathbf{n}_a \cdot \mathbf{I} \cdot \mathbf{n}_b = I_{ab} \tag{7.10.30}$$

Therefore, the maximum absolute values of the products of inertia are the half differences of the principal moments of inertia, and the directions where they occur are at 45° to the principal directions of the moments of inertia.

7.11 Inertia Ellipsoid

Geometrical interpretations are sometimes used to obtain insight into the nature of inertia quantities. Of these, one of the most extensively used is the inertia ellipsoid, which may be developed from Eq. (7.5.5). Specifically, for a given reference point O the moment of inertia of a system for the direction of a unit vector $\mathbf{n}_a$ is:

$$I_{aa} = a_i a_j I_{ij} \tag{7.11.1}$$

where the a_i and the I_{ij} are components of $\mathbf{n}_a$ and the inertia dyadic $\mathbf{I}$ relative to mutually perpendicular unit vectors $\mathbf{n}_i$ ($i = 1, 2, 3$). If the $\mathbf{n}_i$ are principal unit vectors, the products of inertia, I_{ij} ($i \neq j$), are zero. Eq. (7.11.1) then takes the simplified form:

$$I_{aa} = a_1^2 I_{11} + a_2^2 I_{22} + a_3^2 I_{33} \tag{7.11.2}$$

Then, by dividing by I_{aa}, we have:

$$1 = \frac{a_1^2}{\left(I_{aa}/I_{11}\right)} + \frac{a_2^2}{\left(I_{aa}/I_{22}\right)} + \frac{a_3^2}{\left(I_{aa}/I_{33}\right)} \tag{7.11.3}$$

Because the denominators of this expression are all positive, we may write it in the form:

$$1 = \frac{a_1^2}{a^2} + \frac{a_2^2}{b^2} + \frac{a_3^2}{c^2} \tag{7.11.4}$$

where a, b, and c are defined by inspection in comparison with Eq. (7.11.3). Eq. (7.11.4) is immediately seen to be the equation of an ellipsoid in the Cartesian coordinate space of (a_1, a_2, a_3). The distance from the ellipse center (at the origin O) to a point P on the surface is proportional to the moment of inertia of a system for the direction of a unit vector $\mathbf{n}_a$ parallel to **OP**. For example, if $\mathbf{n}_a$ is directed parallel to $\mathbf{n}_1$ such that (a_1, a_2, a_3) is (1, 0, 0), then I_{aa} is I_{11}, a principal moment of inertia. This property has led the ellipsoid of Eq. (7.11.4) to be called the *inertia ellipsoid*.

7.12 Application: Inertia Torques

Consider a rigid body B moving in an inertial reference frame R (see Section 6.9) as depicted in Figure 7.12.1. Let B be considered to be composed of a set of N particles P_i having masses m_i ($i = 1,\ldots, N$). Let G be the mass center of B. Then, from Eqs. (6.9.9) and (6.9.11), we recall that the system of inertia forces acting on B (through particles P_i) is equivalent to a single force $\mathbf{F}^*$ passing through G together with a couple having torque $\mathbf{T}^*$ where $\mathbf{F}^*$ and $\mathbf{T}^*$ are:

$$\mathbf{F}^* = -M\mathbf{a}_G \tag{7.12.1}$$

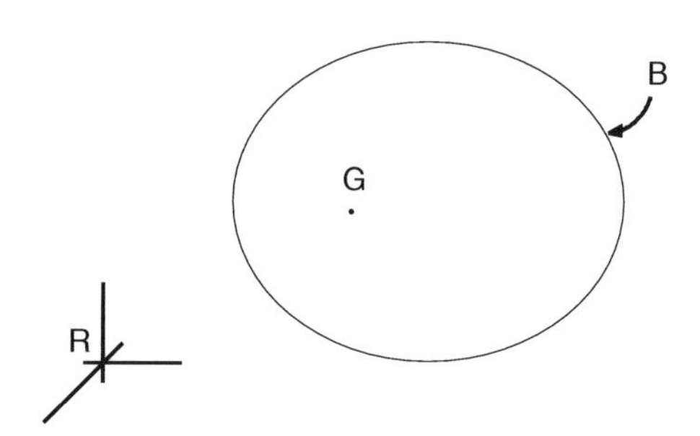

FIGURE 7.12.1
A body B moving in an inertial frame R.

and

$$\mathbf{T}^* = -\left[\sum_{i=1}^{N} m_i \mathbf{r}_i \times (\boldsymbol{\alpha} \times \mathbf{r}_i)\right] - \omega \times \left[\sum_{i=1}^{N} m_i \mathbf{r}_i \times (\boldsymbol{\omega} \times \mathbf{r}_i)\right] \tag{7.12.2}$$

where $\mathbf{a}_G$ is the acceleration of G in R; $\boldsymbol{\omega}$ and $\boldsymbol{\alpha}$ are the angular velocity and angular acceleration, respectively, of B in R; and $\mathbf{r}_i$ is a position vector locating P_i relative to G.

We are now in a position to further develop the terms of Eq. (7.12.2). Specifically, we can conveniently express them in terms of the inertia dyadic of Section 7.4. Recall first from Eq. (7.2.2) that the second moment of B for G, for a direction of a unit vector $\mathbf{n}_a$, is:

$$I_a^{B/G} = \sum_{i=1}^{N} m_i \mathbf{r}_i \times (\mathbf{n}_a \times r_i) \tag{7.12.3}$$

(Observe the similarity of the sum of Eq. (7.12.3) with those of Eq. (7.12.2).) Then, recall from Eq. (7.4.11) that the second moment may be expressed in terms of the inertia dyadic of B for G as:

$$I_a^{B/G} = \mathbf{n}_a \cdot \mathbf{I}^{B/G} \tag{7.12.4}$$

Consider now the terms of Eq. (7.12.2). Let $\boldsymbol{\alpha}$ and $\boldsymbol{\omega}$ be written in the forms:

$$\boldsymbol{\alpha} = \alpha \mathbf{n}_\alpha \quad \text{and} \quad \boldsymbol{\omega} = \omega \mathbf{n}_\omega \tag{7.12.5}$$

where $\mathbf{n}_\alpha$ and $\mathbf{n}_\omega$ are unit vectors in the directions of $\boldsymbol{\alpha}$ and $\boldsymbol{\omega}$ at any instant, with α and ω then being three magnitudes of $\boldsymbol{\alpha}$ and $\boldsymbol{\omega}$. Then, the first term of Eq. (7.12.2) may be expressed as:

$$\begin{aligned}
\sum_{i=1}^{N} m_i \mathbf{r}_i \times (\boldsymbol{\alpha} \times \mathbf{r}_i) &= \sum_{i=1}^{N} m_i \mathbf{r}_i \times (\alpha \mathbf{n}_\alpha \times \mathbf{r}_i) \\
&= \boldsymbol{\alpha} \sum_{i=1}^{N} m_i \mathbf{r}_i \times (\mathbf{n}_\alpha \times \mathbf{r}_i) \\
&= \alpha \mathbf{I}_\alpha^{B/G} \qquad (7.12.6) \\
&= \alpha \mathbf{n}_\alpha \cdot \mathbf{I}^{B/G} \\
&= \boldsymbol{\alpha} \cdot \mathbf{I}^{B/G}
\end{aligned}$$

where we have made use of Eqs. (7.12.4) and (7.12.5). Similarly, the second term of Eq. (7.12.2) may be expressed as:

$$\boldsymbol{\omega}\times\left[\sum_{i=1}^{N} m_i \mathbf{r}_i \times(\boldsymbol{\omega}\times\mathbf{r}_i)\right] = \boldsymbol{\omega}\times\left(\mathbf{I}^{B/G}\cdot\boldsymbol{\omega}\right) \tag{7.12.7}$$

Hence, $\mathbf{T}^*$ becomes:

$$\mathbf{T}^* = -\mathbf{I}^{B/G}\cdot\boldsymbol{\alpha} - \boldsymbol{\omega}\times\left(\mathbf{I}^{B/G}\cdot\boldsymbol{\omega}\right) \tag{7.12.8}$$

The scalar components of $\mathbf{T}^*$ are sometimes referred to as *Euler torques*. We will explore the significance of these terms in the next several chapters.

References

7.1. Kane, T. R., *Analytical Elements of Mechanics*, Vol. 2, Academic Press, New York, 1962.
7.2. Kane, T. R., *Dynamics*, Holt, Rinehart & Winston, New York, 1968.
7.3. Kane, T. R., and Levinson, D. A., *Dynamics: Theory and Applications*, McGraw-Hill, New York, 1985.
7.4. Huston, R. L., *Multibody Dynamics*, Butterworth-Heinemann, Stoneham, MA, 1990.
7.5. Hinchley, F. A., *Vectors and Tensors for Engineers and Scientists*, Wiley, New York, 1976.
7.6. Hsu, H. P., *Vector Analysis*, Simon & Schuster Technical Outlines, New York, 1969.
7.7. Haskell, R. E., *Introduction to Vectors and Cartesian Tensors*, Prentice Hall, Englewood Cliffs, NJ, 1972.

Problems

Section 7.2 Second Moment Vectors

P7.2.1: A particle P with mass 3 slug has coordinates (2, –1, 3), measured in feet, in a Cartesian coordinate system as represented in Figure P7.2.1. Determine the second moment of P relative to the origin O for the directions represented by the unit vectors $\mathbf{n}_x$, $\mathbf{n}_y$, $\mathbf{n}_z$, $\mathbf{n}_a$, and $\mathbf{n}_b$, where $\mathbf{n}_a$ and $\mathbf{n}_b$ are parallel to the X–Y plane, as shown in Figure P7.2.1. That is, find $\mathbf{I}_x^{P/O}$, $\mathbf{I}_y^{P/O}$, $\mathbf{I}_z^{P/O}$, $\mathbf{I}_a^{P/O}$, and $\mathbf{I}_b^{P/O}$.

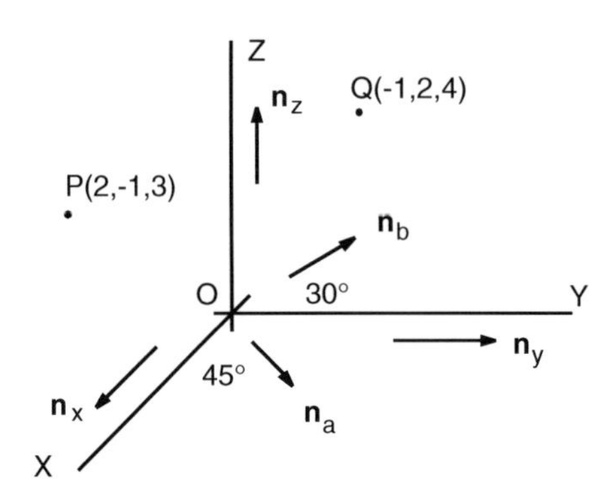

FIGURE P7.2.1
A particle P in a Cartesian reference frame.

P7.2.2: See Problem P7.2.1. Let point Q have coordinates (–1, 2, 4). Repeat Problem P7.2.1 with Q, instead of O, being the reference point. That is, find $\mathbf{I}_x^{P/Q}$, $\mathbf{I}_y^{P/Q}$, $\mathbf{I}_z^{P/Q}$, $\mathbf{I}_a^{P/Q}$, and $\mathbf{I}_b^{P/Q}$.

P7.2.3: Consider the system S of two particles P_1 and P_2 located on the X-axis of a Cartesian coordinate system as in Figure P7.2.3. Let the masses of P_1 and P_2 be m_1 and m_2, and let the distances of P_1 and P_2 from the origin O be x_1 and x_2. Find the second moment of P relative to O for the direction of $\mathbf{n}_y$. (Observe that $\mathbf{I}_y^{S/O}$ is parallel to $\mathbf{n}_y$.)

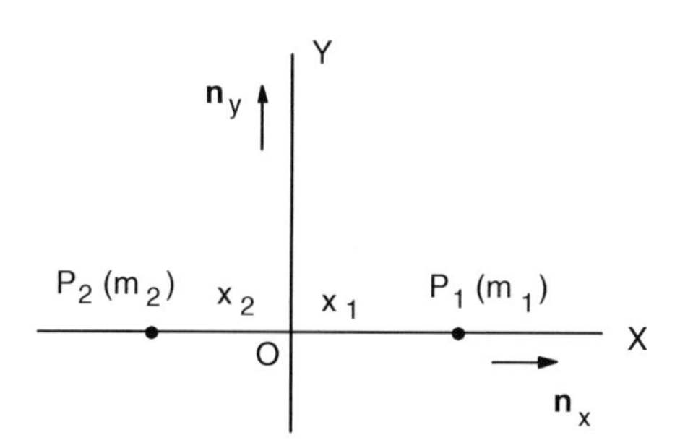

FIGURE P7.2.3
Particles P_1 and P_2 on the X-axis of a Cartesian reference frame.

P7.2.4: See Problems P7.2.1 and P7.2.3. Find a unit vector $\mathbf{n}$ such that $\mathbf{I}_n^{P/O}$ is parallel to $\mathbf{n}$.

P7.2.5: Let a set S of three particles P_1, P_2, and P_3 be located at the vertices of a triangle as shown in Figure P7.2.5. Let the particles have masses 2, 3, and 4 kg, respectively. Find $\mathbf{I}_x^{S/O}$, $\mathbf{I}_x^{S/O}$, and $\mathbf{I}_z^{S/O}$ where as before O is the origin of the x, y, z coordinate system of Figure P7.2.5, and $\mathbf{n}_x$, $\mathbf{n}_y$, and $\mathbf{n}_z$ are the unit vectors shown.

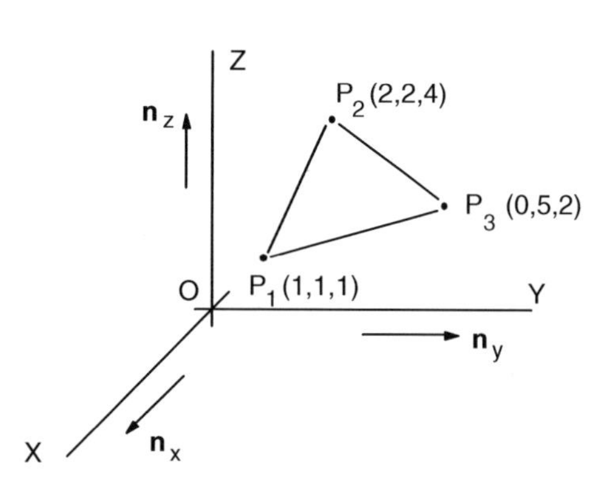

FIGURE P7.2.5
Particles at the vertices of a triangle.

P7.2.6: See Problem P7.2.5. Let $\mathbf{n}_a$ and $\mathbf{n}_b$ be unit vectors with coordinates relative to $\mathbf{n}_x$, $\mathbf{n}_y$, and $\mathbf{n}_z$ as:

$$\mathbf{n}_a = 0.75\mathbf{n}_x - 0.5\mathbf{n}_y + 0.433\mathbf{n}_z$$

and

$$\mathbf{n}_b = 0.655\mathbf{n}_y + 0.756\mathbf{n}_z$$

Determine $\mathbf{I}_a^{S/O}$ and $\mathbf{I}_b^{S/O}$.

P7.2.7: See Problems P7.2.5 and P7.2.6. Show that:

$$\mathbf{I}_a^{S/O} = 0.75\mathbf{I}_x^{S/O} - 0.50\mathbf{I}_y^{S/O} + 0.433\mathbf{I}_z^{S/O}$$

and

$$\mathbf{I}_b^{S/O} = 0.655\mathbf{I}_y^{S/O} + 0.756\mathbf{I}_z^{S/O}$$

P7.2.8: See Problem P7.2.5. Find the x, y, z coordinates of the mass center G of S. Find $\mathbf{I}_x^{S/G}$, $\mathbf{I}_y^{S/G}$, $\mathbf{I}_z^{S/G}$, $\mathbf{I}_a^{S/G}$, and $\mathbf{I}_b^{S/G}$.

P7.2.9: See Problem P7.2.8. Let G have an associated mass of 9 kg (equal to the sum of the masses of P_1, P_2, and P_3). Find $\mathbf{I}_x^{G/O}$, $\mathbf{I}_y^{G/O}$, $\mathbf{I}_z^{G/O}$, $\mathbf{I}_a^{G/O}$, and $\mathbf{I}_b^{G/O}$.

P7.2.10: See Problems P7.2.5, P7.2.6, and P7.2.9. Show that:

$$\mathbf{I}_x^{S/O} = \mathbf{I}_x^{S/G} + \mathbf{I}_x^{G/O}$$

$$\mathbf{I}_y^{S/O} = \mathbf{I}_y^{S/G} + \mathbf{I}_y^{G/O}$$

$$\mathbf{I}_z^{S/O} = \mathbf{I}_z^{S/G} + \mathbf{I}_z^{G/O}$$

$$\mathbf{I}_a^{S/O} = \mathbf{I}_a^{S/G} + \mathbf{I}_a^{G/O}$$

and

$$\mathbf{I}_b^{S/O} = \mathbf{I}_b^{S/G} + \mathbf{I}_b^{G/O}$$

P7.2.11: See Problem P7.2.5. Find a unit vector $\mathbf{n}$ perpendicular to the plane of P_1, P_2, and P_3. Find also $\mathbf{I}_\mathbf{n}^{S/O}$. Show that $\mathbf{I}_\mathbf{n}^{S/O}$ is parallel to $\mathbf{n}$.

Section 7.3 Moments and Products of Inertia

P7.3.1: See Problem P7.2.1. A particle P with mass of 3 slug has coordinates (2, –1, 3), measured in feet, in a Cartesian coordinate system as represented in Figure P7.3.1. Determine the following moments and products of inertia: $\mathbf{I}_{xx}^{P/O}$, $\mathbf{I}_{xy}^{P/O}$, $\mathbf{I}_{xz}^{P/O}$, $\mathbf{I}_{yy}^{P/O}$, $\mathbf{I}_{yz}^{P/O}$, $\mathbf{I}_{zz}^{P/O}$, $\mathbf{I}_{aa}^{P/O}$, $\mathbf{I}_{bb}^{P/O}$, $\mathbf{I}_{ab}^{P/O}$, and $\mathbf{I}_{ba}^{P/O}$.

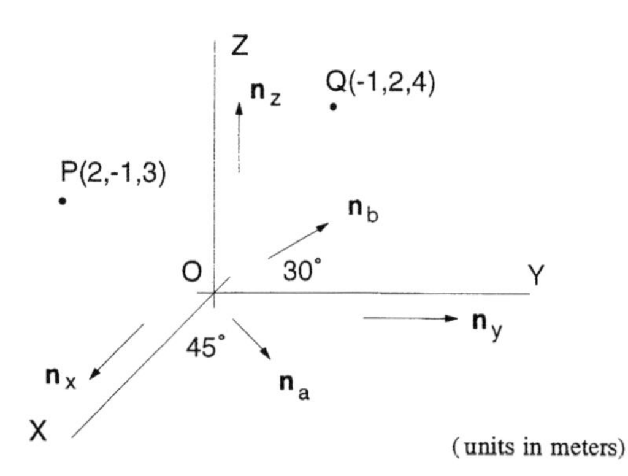

FIGURE P7.3.1
A particle P and a point Q.

P7.3.2: See Problems P7.2.2 and P7.3.1. Let Q have coordinates (–1, 2, 4). Repeat Problem P7.3.1 with Q, instead of O, being the reference point. That is, determine $\mathbf{I}_{xx}^{P/Q}$, $\mathbf{I}_{xy}^{P/Q}$, $\mathbf{I}_{xz}^{P/Q}$, $\mathbf{I}_{yy}^{P/Q}$, $\mathbf{I}_{yz}^{P/Q}$, $\mathbf{I}_{zz}^{P/Q}$, $\mathbf{I}_{aa}^{P/Q}$, $\mathbf{I}_{bb}^{P/Q}$, $\mathbf{I}_{ab}^{P/Q}$, and $\mathbf{I}_{ba}^{P/Q}$.

P7.3.3: See Problem P7.2.5. Let a set S of three particles P_1, P_2, and P_3 be located at the vertices of a triangle as shown in Figure P7.3.3. Let the particles have masses 2, 3, and

4 kg, respectively. Find the following moments and products of inertia of S relative to the origin O of the X-, Y-, Z-axis system of Figure P7.3.3: $\mathbf{I}_{xx}^{S/O}$, $\mathbf{I}_{xy}^{S/O}$, $\mathbf{I}_{xz}^{S/O}$, $\mathbf{I}_{yy}^{S/O}$, $\mathbf{I}_{yz}^{S/O}$, and $\mathbf{I}_{zz}^{S/O}$.

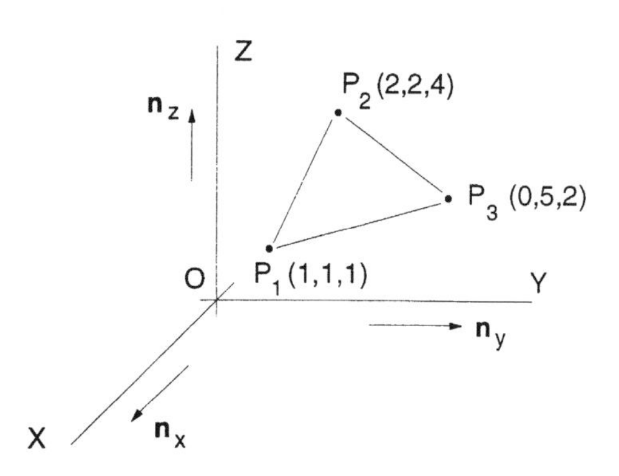

FIGURE P7.3.3
Particles at the vertices of a triangle.

P7.3.4: See Problems P7.2.5, P7.2.8, P7.2.9, and P7.3.3. For the system S shown in Figure P7.3.3, find the following moments and products of inertia: $\mathbf{I}_{xx}^{S/G}$, $\mathbf{I}_{xy}^{S/G}$, $\mathbf{I}_{xz}^{S/G}$, $\mathbf{I}_{yy}^{S/G}$, $\mathbf{I}_{yz}^{S/G}$, and $\mathbf{I}_{zz}^{S/G}$ where G is the mass center of S, as determined in Problem P7.2.8. (Compare the magnitudes of these results with those of Problem P7.3.3.)

P7.3.5: See Problems P7.2.9 and P7.3.3. For the system S shown in Figure P7.3.3, find the following moments and products of inertia: $\mathbf{I}_{xx}^{G/O}$, $\mathbf{I}_{xy}^{G/O}$, $\mathbf{I}_{xz}^{G/O}$, $\mathbf{I}_{yy}^{G/O}$, $\mathbf{I}_{yz}^{G/O}$, and $\mathbf{I}_{zz}^{G/O}$.

P7.3.6: See Problems P7.2.10, P7.3.3, P7.3.4, and P7.3.5. Show that:

$$\mathbf{I}_{ij}^{S/O} = \mathbf{I}_{ij}^{S/G} + \mathbf{I}_{ij}^{G/O} \quad (i, j = x, y, z)$$

P7.3.7: See Problems P7.2.5, P7.2.6, P7.2.7, and P7.3.3. As in Problem P7.2.6 let $\mathbf{n}_a$ and $\mathbf{n}_b$ be the unit vectors:

$$\mathbf{n}_a = 0.75\mathbf{n}_x - 0.5\mathbf{n}_y + 0.433\mathbf{n}_z$$

and

$$\mathbf{n}_b = 0.655\mathbf{n}_y + 0.756\mathbf{n}_z$$

Find $\mathbf{I}_{aa}^{S/O}$, $\mathbf{I}_{ab}^{S/O}$, and $\mathbf{I}_{bb}^{S/O}$.

Section 7.4 Inertia Dyadics

P7.4.1: Let vectors **a**, **b**, and **c** be expressed as:

$$\mathbf{a} = 6\mathbf{n}_1 - 3\mathbf{n}_2 + 4\mathbf{n}_3$$

$$\mathbf{b} = -5\mathbf{n}_1 + 4\mathbf{n}_2 - 7\mathbf{n}_3$$

$$\mathbf{c} = 3\mathbf{n}_1 - \mathbf{n}_2 + 9\mathbf{n}_3$$

where $\mathbf{n}_1$, $\mathbf{n}_2$, and $\mathbf{n}_3$ are mutually perpendicular unit vectors. Compute the following dyadic products: (a) **ab**, (b) **ba**, (c) **ca** + **cb**, (d) **c**(**a** + **b**), (e) (**a** + **b**)**c**, and (f) **ac** + **bc**.

P7.4.2: See Problem 7.2.1. A particle P with mass 3 slug has coordinates (2, –1, 3), measured in feet, in a Cartesian coordinate system as represented in Figure P7.4.2. Determine the inertia dyadic of P relative to the origin O, $\mathbf{I}^{P/O}$. Express the results in terms of the unit vectors $\mathbf{n}_x$, $\mathbf{n}_y$, and $\mathbf{n}_z$.

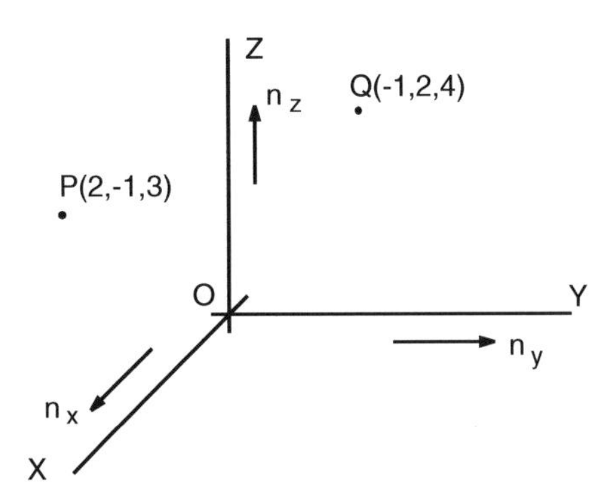

FIGURE P7.4.2
A particle P in a Cartesian reference frame.

P7.4.3: See Problem P7.2.2. Let Q have coordinates (–1, 2, 4). Repeat Problem P7.4.2 with Q instead of O being the reference point. That is, find $\mathbf{I}^{P/Q}$.

P7.4.4: See Problems P7.2.5 and P7.3.3. Let S be the set of three particles P_1, P_2, and P_3 located at the vertices of a triangle as shown in Figure P7.4.4. Let the particles have masses: 2, 3, and 4 kg, respectively. Find the inertia dyadic of S relative to O, $\mathbf{I}^{S/O}$. Express the results in terms of the unit vectors $\mathbf{n}_x$, $\mathbf{n}_y$, and $\mathbf{n}_z$.

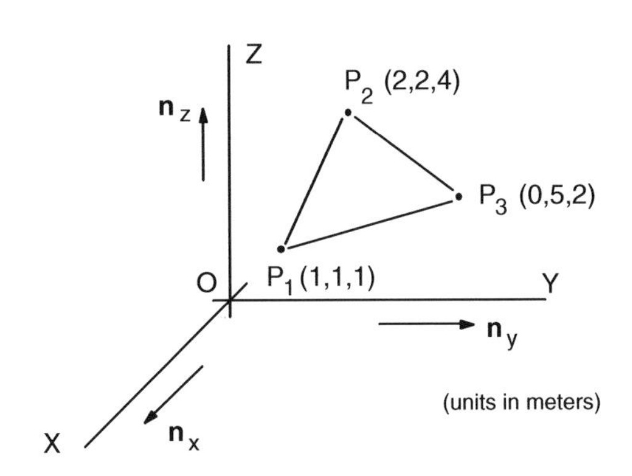

FIGURE P7.4.4
Particles at the vertices of a triangle.

P7.4.5: See Problems P7.2.8, P7.2.9, P7.3.4, and P7.4.4. Let G be the mass center of S. Find the inertia dyadic of S relative to G, $\mathbf{I}^{S/G}$. Express the results in terms of the unit vectors $\mathbf{n}_x$, $\mathbf{n}_y$, and $\mathbf{n}_z$.

P7.4.6: See Problems P7.4.4 and P7.4.5. Let G have an associated mass of 9 kg. Find the inertia dyadic of G relative to the origin O, $\mathbf{I}^{G/O}$. Express the result in terms of the unit vectors $\mathbf{n}_x$, $\mathbf{n}_y$, and $\mathbf{n}_z$.

P7.4.7: See Problems P7.4.5 and P7.4.6. Show that:

$$\mathbf{I}^{S/O} = \mathbf{I}^{S/G} + \mathbf{I}^{G/O}$$

P7.4.7: See Problems P7.2.5 and P7.4.4. Find the second moments of S relative to O for the directions of $\mathbf{n}_x$, $\mathbf{n}_y$, and $\mathbf{n}_z$.

P7.4.8: See Problems P7.3.3 and P7.4.4. Find the following moments and products of inertia of S for O: $\mathbf{I}_{xx}^{S/O}$, $\mathbf{I}_{xz}^{S/O}$, $\mathbf{I}_{yy}^{S/O}$, $\mathbf{I}_{yz}^{S/O}$, and $\mathbf{I}_{zz}^{S/O}$.

P7.4.9: See Problems P7.2.6 and P7.4.4. Let $\mathbf{n}_a$ and $\mathbf{n}_b$ be unit vectors with coordinates relative to $\mathbf{n}_x$, $\mathbf{n}_y$, and $\mathbf{n}_z$ as:

$$\mathbf{n}_a = 0.75\mathbf{n}_x - 0.5\mathbf{n}_y + 0.433\mathbf{n}_z$$

$$\mathbf{n}_b = 0.655\mathbf{n}_y + 0.756\mathbf{n}_z$$

Find the second moment vectors $\mathbf{I}_a^{S/O}$ and $\mathbf{I}_b^{S/O}$.

P7.4.10: See Problems P7.2.5, P7.2.6, P7.3.7, P7.4.4, and P7.4.9. Let $\mathbf{n}_a$ and $\mathbf{n}_b$ be the unit vectors of Problem P7.4.9. Find the following moments and products of inertia of S relative to O: $\mathbf{I}_{aa}^{S/O}$, $\mathbf{I}_{ab}^{S/O}$, and $\mathbf{I}_{bb}^{S/O}$.

Section 7.5 Transformation Rules

P7.5.1: Let S be a set of eight particles P_i ($i = 1,\ldots, 8$) located at the vertices of a cube as in Figure P7.5.1. Let the masses m_i of the P_i be as listed in the figure. Determine the second-moment vectors $\mathbf{I}_1^{S/O}$, $\mathbf{I}_2^{S/O}$, and $\mathbf{I}_3^{S/O}$ for the directions of the unit vectors $\mathbf{n}_1$, $\mathbf{n}_2$, and $\mathbf{n}_3$ shown in Figure P7.5.1.

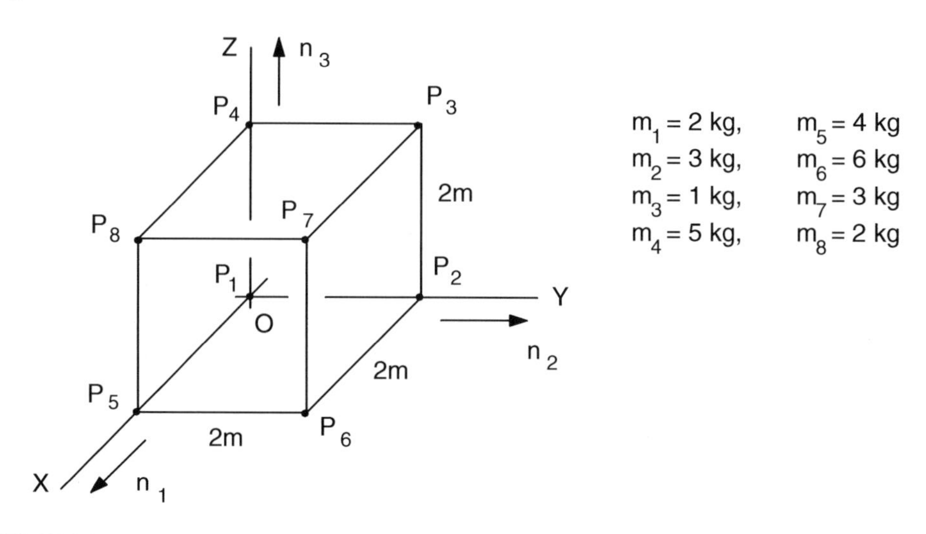

FIGURE P7.5.1
Particles at the vertices of a cube.

P7.5.2: See Problem P7.5.1. Let $\mathbf{n}_a$, $\mathbf{n}_b$, and $\mathbf{n}_c$ be unit vectors with components relative to $\mathbf{n}_1$, $\mathbf{n}_2$, and $\mathbf{n}_3$ as:

$$\mathbf{n}_a = 0.5\mathbf{n}_1 + 0.866\mathbf{n}_2$$

$$\mathbf{n}_b = -0.433\mathbf{n}_1 + 0.25\mathbf{n}_2 + 0.866\mathbf{n}_3$$

$$\mathbf{n}_c = 0.75\mathbf{n}_1 - 0.433\mathbf{n}_2 + 0.5\mathbf{n}_3$$

Determine the second moment vectors $\mathbf{I}_a^{S/O}$, $\mathbf{I}_b^{S/O}$, and $\mathbf{I}_c^{S/O}$.

P7.5.3: See Problem P7.5.1. Determine the moments and products of inertia $I_{ij}^{S/O}$ $(i, j = 1, 2, 3)$.

P7.5.4: See Problem P7.5.2. Let the transformation matrix between $\mathbf{n}_a$, $\mathbf{n}_b$, $\mathbf{n}_c$ and $\mathbf{n}_1$, $\mathbf{n}_2$, $\mathbf{n}_3$ have elements $S_{j\alpha}$ ($j = 1, 2, 3$; $\alpha = a, b, c$) defined as:

$$S_{j\alpha} = \mathbf{n}_j \cdot \mathbf{n}_\alpha$$

Find the $S_{j\alpha}$.

P7.5.5: See Problems P7.5.1 to P7.5.4. Find the moments and products of inertia $I_{\alpha\beta}^{S/O}$ $(\alpha, \beta = a, b, c)$. Also verify that:

$$I_{\alpha\beta}^{S/O} = S_{i\alpha} S_{j\beta} I_{ij}^{S/O}$$

and

$$I_{ij}^{S/O} = S_{i\alpha} S_{j\beta} I_{\alpha\beta}^{S/O}$$

P7.5.6: See Problem P7.5.3. Find the inertia dyadic $\mathbf{I}^{S/O}$. Express the results in terms of the unit vectors $\mathbf{n}_1$, $\mathbf{n}_2$, and $\mathbf{n}_3$ of Figure P7.5.1.

P7.5.7: See Problems P7.5.5 and P7.5.6. Verify that $I_{\alpha\beta}^{S/C}$ $(\alpha, \beta = a, b, c)$ is given by:

$$I_{\alpha\beta}^{S/O} = \mathbf{n}_\alpha \cdot \mathbf{I}^{S/O} \cdot \mathbf{n}_\beta$$

P7.5.8: See Problems P7.5.3 and P7.5.5. Verify that:

$$I_{11}^{S/O} + I_{22}^{S/O} + I_{33}^{S/O} = I_{aa}^{S/O} + I_{bb}^{S/O} + I_{cc}^{S/O}$$

P7.5.9: A 3-ft bar B weighs 18 pounds. Let the bar be homogeneous and uniform so that its mass center G is at the geometric center. Let the bar be placed on an X–Y plane so that it is inclined at 30° to the X-axis as shown in Figure P7.5.9. It is known that the moment of inertia of a homogeneous, uniform bar relative to its center is zero for directions parallel to the bar and $m\ell^2/12$ for directions perpendicular to the bar where m is the bar mass and ℓ is its length (see Appendix II). It is also known that the products of inertia for a bar for directions parallel and perpendicular to the bar are zero. Determine the moments and products of inertia: $I_{xx}^{B/G}$, $I_{yy}^{B/G}$, $I_{zz}^{B/G}$, $I_{xy}^{B/G}$, $I_{xz}^{B/G}$, and $I_{yz}^{B/G}$.

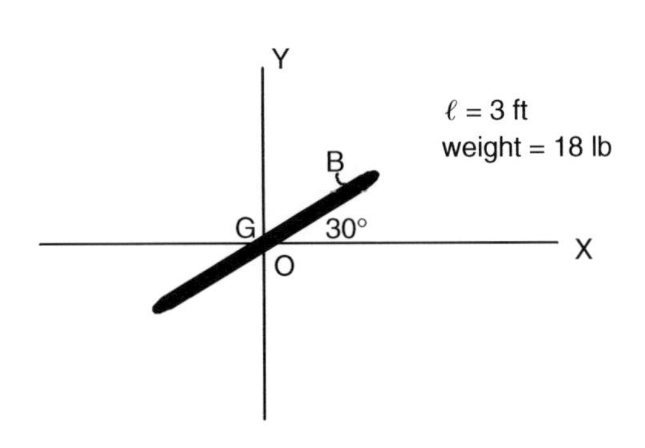

FIGURE P7.5.9
A homogeneous bar in the X–Y plane with center at the origin.

P7.5.10: A thin uniform circular disk D with mass m and radius r is mounted on a shaft S with a small misalignment, measured by the angle θ as represented in Figure P7.5.10. Knowing that the moments of inertia of D for its center for directions parallel to and perpendicular to its axis are $mr^2/2$ and $mr^2/4$, respectively, and that the corresponding products of inertia of D for its axis and diameter directions are zero (see Appendix II), find the moment of inertia of D for its center G for the shaft axis direction x: $\mathbf{I}_{xx}^{D/G}$.

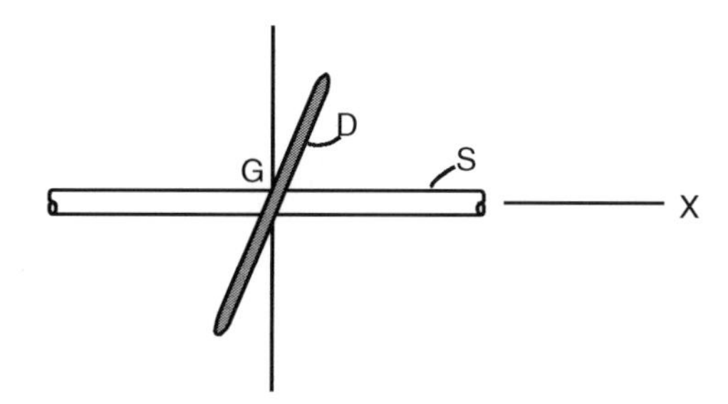

FIGURE P7.5.10
A misaligned circular disk on a shaft S.

Section 7.6 Parallel Axis Theorems

P7.6.1: Consider the homogeneous rectangular parallepiped (block) B shown in Figure P7.6.1. From Appendix II, we see that the moments of inertia of B for the mass center G for the X, Y, and Z directions are:

$$I_{xx}^{B/G} = \frac{1}{12}m(b^2 + c^2), \; I_{yy}^{B/G} = \frac{1}{12}m(a^2 + c^2), \; I_{zz}^{B/G} = \frac{1}{12}m(a^2 + b^2)$$

where m is the mass of B and a, b and c are the dimensions as shown in Figure P7.6.1. Let B have the following properties:

$$m = 12\,\text{kg}, \; a = 2\,\text{m}, \; b = 4\,\text{m}, \; c = 3\,\text{m}$$

Determine the moments of inertia of B relative to G for the directions of X, Y, and Z.

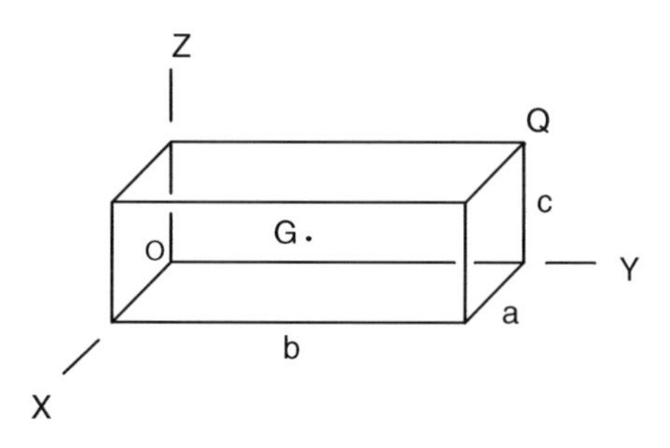

FIGURE P7.6.1
A homogeneous rectangular block.

P7.6.2: Repeat Problem P7.6.1 with B having the following properties:

$$m = 15\,\text{lb}, \; a = 2.5\,\text{ft}, \; b = 5\,\text{ft}, \; c = 3\,\text{ft}$$

P7.6.3: See Problems P7.6.1 and P7.6.2. For the properties of Problems P7.6.1 and P7.6.2, find the moments of inertia of B for Q for the direction X, Y, and Z where Q is a vertex of B with coordinates (a, b, c) as shown in Figure P7.6.1.

P7.6.4: A body B has mass center G with coordinates (1, 3, 2), in meters, in a Cartesian reference frame as represented in Figure P7.6.4. Let the mass of B be 0.5 kg. Let the inertia dyadic of B for the origin O have the matrix $\mathbf{I}_{ij}^{B/O}$ given by:

$$\mathbf{I}^{B/O} = I_{ij}^{B/O}\mathbf{n}_i\mathbf{n}_j, I_{ij}^{B/O} = \begin{bmatrix} 9 & -2.634 & -1 \\ -2.634 & 4 & -3 \\ -1 & -3 & -7 \end{bmatrix} \text{kg m}^2$$

where $\mathbf{n}_1$, $\mathbf{n}_2$, and $\mathbf{n}_3$ are parallel to the X-, Y-, and Z-axes. Determine the components of the inertia dyadic $\mathbf{I}_{ij}^{B/Q}$ of B for point Q, where the coordinates of Q are (2, 6, 3), in meters.

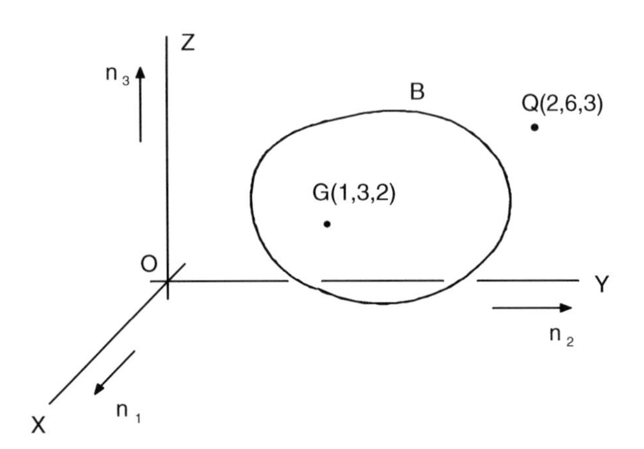

FIGURE P7.6.4
A body B in a Cartesian reference frame.

P7.6.5: A thin, rectangular plate P weighs 15 lb. The dimensions of the plate are 20 in. by 10 in. See Figure P7.6.5, and determine the moments of inertia of P relative to corner A for the X, Y, and Z directions (see Appendix II).

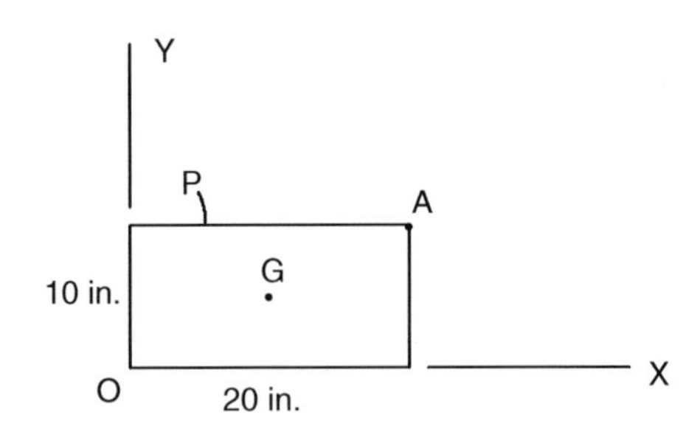

FIGURE P7.6.5
A rectangular plate in a Cartesian reference frame.

P7.6.6: Repeat Problem P7.6.5 for a plate with a 5-in.-diameter circular hole centered in the left half of the plate as represented in Figure P7.6.6.

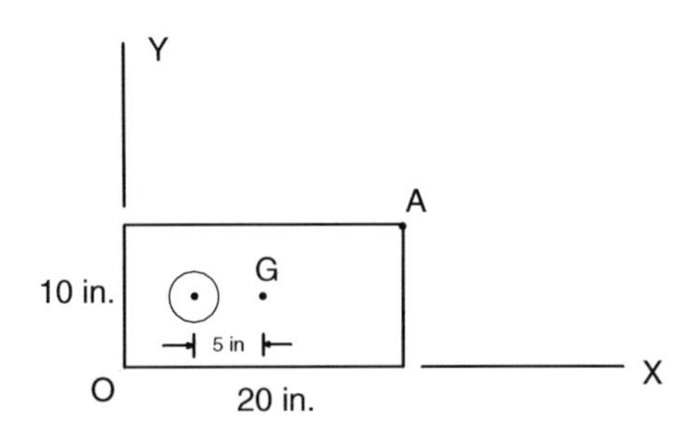

FIGURE 7.6.6
A rectangular plate with an offset circular hole.

Sections 7.7, 7.8, 7.9 Principal Moments of Inertia

P7.7.1: Review again the example of Section 7.8. Repeat the example for an inertia matrix given by:

$$I_{ij} = \begin{bmatrix} 32 & 2 & 6 \\ 2 & 32 & 9 \\ 6 & 9 & 16 \end{bmatrix} \text{slug ft}^2$$

P7.7.2: A 2 × 4-ft rectangular plate *OABC* is bonded to a 2-ft-square plate *CDEF*, forming a composite body *S* as in Figure P7.7.2. Let the rectangular plate weigh 40 lb and the square plate 20 lb.

a. Determine the *x*, *y*, *z* components of the mass center *G* of *S*.
b. Find the inertia dyadic of *S* for *G*. Express the results in terms of the unit vectors $\mathbf{n}_x$, $\mathbf{n}_y$, and $\mathbf{n}_z$ shown in Figure P7.7.2.
c. Find the principal moments of inertia of *S* for *G*.
d. Find the principal unit vectors of *S* for *G*. Express the results in terms of $\mathbf{n}_x$, $\mathbf{n}_y$, and $\mathbf{n}_z$.

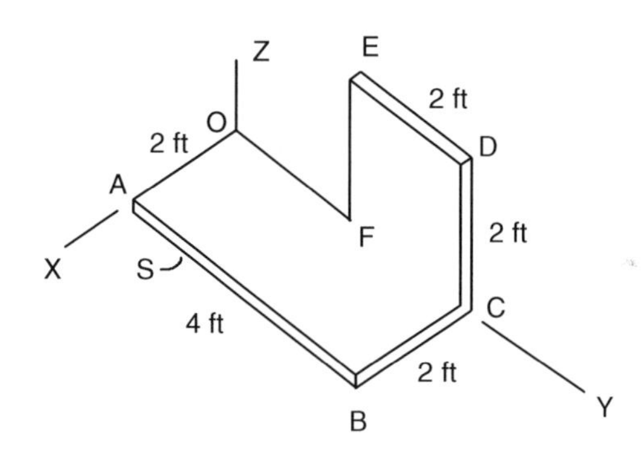

FIGURE P7.7.2
A composite plate body.

P7.7.3: Repeat Problem P7.7.2 if the square plate *CDEF* weighs 30 lb.

P7.7.4: Repeat Problem P7.7.3 if the square plate *CDEF* weighs 10 lb.

8

Principles of Dynamics: Newton's Laws and d'Alembert's Principle

8.1 Introduction

Dynamics is a combined study of motion (kinematics), forces (kinetics), and inertia (mass distributing). By using the principles of dynamics we can obtain mathematical models of the behavior of mechanical systems. In this chapter, and in subsequent chapters, we will explore the principles of dynamics and their applications.

The development of dynamics principles dates back to at least the 14th century, long before the development of calculus and other widely used analytical procedures. One of the earliest statements of a dynamics principle in the Western world is attributed to John Buridan in (1358) [8.1]:

> From this theory also appears the cause of why the natural motion of a heavy body downward is continually accelerated. For from the beginning only the gravity was moving it. Therefore, it moved more slowly, but in moving it impressed in the heavy body an impetus. This impetus now together with its gravity moves it. Therefore, the motion becomes faster, and by the amount it is faster so the impetus becomes more intense. Therefore, the movement evidently becomes continually faster.

While this statement seems to be intuitively reasonable, it is not strictly correct, as we now understand the physics of falling bodies. Moreover, the statement does not readily lead to a quantitative analysis.

The earliest principles that adequately describe the physics and lead to quantitative analysis are generally attributed to Isaac Newton. His principles, first published in 1687, are generally stated in three laws [8.2]:

First law (law of inertia): In the absence of forces applied to a particle, the particle will remain at rest or it will move along a straight line at constant velocity.

Second law (law of kinetics): If a force is applied to a particle, the particle accelerates in the direction of the force. The magnitude of the acceleration is proportional to the magnitude of the force and inversely proportional to the mass of the particle.

Third law (law of action–reaction): If two particles exert forces on each other, the respective forces are equal in magnitude and oppositely directed along the line joining the particles.

Recently, researchers have established that Newton's first law was known and stated in China in the third or fourth century BC. Under the leadership of Mo Tzu it was stated [8.3]:

> The cessation of motion is due to the opposing force. ...If there is no opposing force ... the motion will never stop. This is as true as that an ox is not a horse.

Newton's laws form the foundation for the principles of dynamics employed in modern analyses. We will briefly review some of these principles in the following section. We will then focus upon d'Alembert's principle in the remaining sections of the chapter and will illustrate use of the principle with several examples. We will consider other principles in subsequent chapters.

8.2 Principles of Dynamics

Newton's laws are almost universally accepted as the fundamental principles of mechanics. Newton's laws directly provide a means for studying dynamical systems. They also provide a means for developing other principles of dynamics. These other principles are often in forms that are more convenient than Newton's laws for the analysis of some classes of systems. Some of these other principles have been formulated independently of Newton's laws, but all of the principles are fundamentally equivalent.

The references for this chapter provide a brief survey of some of the principles of dynamics. They include (in addition to Newton's laws) Hamilton's principle, Lagrange's equations, d'Alembert's principle, Gibbs equations, Boltzmann–Hamel equations, Kane's equations, impulse–momentum, work–energy, and virtual work.

Hamilton's principle, which is widely used in structural analyses and in approximate analyses, states that the time integral of the difference of kinetic and potential energies of a mechanical system is a minimum. Hamilton's principle is thus an *energy* principle, which may be expressed analytically as:

$$\delta\left(\int_{t_1}^{t_2} L\,dt\right)=0 \tag{8.2.1}$$

where L, called the *Lagrangian*, is the difference in the kinetic and potential energies; δ represents a *variation operation*, as in the calculus of variations; and t_1 and t_2 are any two times during the motion of the system with $t_2 > t_1$.

From Hamilton's principle many dynamicists have developed Lagrange's equations, a very popular procedure for obtaining equations of motion for relatively simple systems. Lagrange's equations may be stated in the form:

$$\frac{d}{dt}\left(\frac{\partial K}{\partial \dot{q}_r}\right)-\frac{\partial K}{\partial q_r}=F_r \quad r=1,\ldots,n \tag{8.2.2}$$

where K is the kinetic energy; q_r $(r = 1,\ldots, n)$ are geometric variables, called *generalized coordinates*, which define the configuration of the system; n is the number of degrees of freedom of the system; and the F_r ($r = 1,\ldots, n$) are *generalized forces* exerted on the system.

Another procedure, similar to Lagrange's equations, is Gibbs equations, which state that:

$$\frac{\partial G}{\partial \ddot{q}_r} = F_r \quad (r = 1,\ldots,n) \tag{8.2.3}$$

where G, called the *Gibbs function*, is a *kinetic energy of acceleration* defined as:

$$G = \frac{1}{2}\sum_{i=1}^{N} m_i \left({}^{R}\mathbf{a}^{P_i}\right)^2 \tag{8.2.4}$$

where P_i is a typical particle of the mechanical system, m_i is the mass of P_i, ${}^{R}\mathbf{a}^{P_i}$ is the acceleration of P_i in an inertial reference frame R, and N is the number of particles of the system. An inertial reference frame is defined as a reference frame in which Newton's laws are valid.

A principle which we will examine and use in the remaining sections of this chapter, called *d'Alembert's principle*, is closely associated with Newton's laws. d'Alembert's principle introduces the concept of an *inertial force*, defined for a particle as:

$$\mathbf{F}_i^* \overset{D}{=} -m_i {}^{R}\mathbf{a}^{P_i} \tag{8.2.5}$$

Then, d'Alembert's principle states that the set of all applied and inertia forces on a mechanical system is a zero force system (see Section 6.4).

A relatively recent (1961) principle of dynamics, known as *Kane's equations*, states that the sum of the generalized applied and inertia forces on a mechanical system is zero for each generalized coordinate. That is,

$$F_r + F_r^* = 0 \quad r = 1,\ldots,n \tag{8.2.6}$$

Kane's equations combine the computational advantages of d'Alembert's principle and Lagrange's equations for a wide variety of mechanical systems. For this reason, Kane's equations were initially called *Lagrange's form of d'Alembert's principle.*

Finally, still other principles of dynamics include impulse–momentum, work–energy, virtual work, Boltzmann–Hamel equations, and Jourdain's principle. We will consider the impulse–momentum and work–energy principles in the next two chapters. The principles of virtual work and Jourdain's principle are similar to Kane's equations, and the Boltzmann–Hamel equations are similar to Lagrange's equations and Gibbs equations.

8.3 d'Alembert's Principle

Newton's second law, which is probably the best known of all dynamics principles, may be stated in analytical form as follows: Given a particle P with mass m and a force $\mathbf{F}$

applied to P, the acceleration of P in an inertial reference frame is related to $\mathbf{F}$ and m through the expression:

$$\mathbf{F} = m\mathbf{a} \tag{8.3.1}$$

As noted in the preceding section, an inertial reference frame (or a Newtonian reference frame) is defined as a reference frame in which Newton's laws are valid. This is a kind of circular definition that has led dynamics theoreticians and philosophers to contemplate and debate the existence of inertial or Newtonian reference frames. Intuitively, an inertial reference frame is a reference frame that is at rest relative to the universe (or relative to the "fixed stars"). Alternatively, an inertial reference frame is an axes system fixed in a rigid body having infinite mass. For the study of most mechanical systems of practical importance, the Earth may be considered to be an approximate inertial reference frame.

The analytical procedures of d'Alembert's principle may be developed from Eq. (8.3.1) by introducing the concept of an *inertia force* (see Section 6.9). Specifically, if a particle P with mass m has an acceleration $\mathbf{a}$ in an inertial reference frame R then the inertia force $\mathbf{F}^*$ on P in R is defined as:

$$\mathbf{F}^* \overset{D}{=} -m\mathbf{a} \tag{8.3.2}$$

Observe that the negative sign in this definition means that the inertia force will always be directed opposite to the acceleration. A familiar illustration of an inertia force is the radial thrust of a small object attached to a string and spun in a circle. Another illustration is the rearward thrust felt by an occupant of an automobile accelerating from rest.

By comparing Eqs. (8.3.1) and (8.3.2), the applied and inertia forces exerted on P are seen to be related by the simple expression:

$$\mathbf{F} + \mathbf{F}^* = 0 \tag{8.3.3}$$

Equation (8.3.3) is an analytical expression of d'Alembert's principle. Simply stated, the sum of the applied and inertia forces on a particle is zero.

When d'Alembert's principle is extended to a set of particles, or to rigid bodies, or to a system of particles and rigid bodies, the principle may be stated simply: the combined system of applied and inertia forces acting on a mechanical system is a zero system (see Section 6.4). When sets of particles, rigid bodies, or systems are considered, interactive forces, exerted between particles of the system on one another, cancel or "balance out" due to the law of action and reaction (see Reference 8.31).

Applied forces, which are generally gravity, contact, or electromagnetic forces, are sometimes called *active forces*. In that context, inertia forces are at times called *passive forces*.

d'Alembert's principle has analytical and computational advantages not enjoyed by Newton's laws. Specifically, with d'Alembert's principle, dynamical systems may be studied as though they are static systems. This means, for example, that free-body diagrams may be used to aid in the analysis. In such diagrams, inertia forces are simply included along with the applied forces. We will illustrate the use of d'Alembert's principle, with the accompanying free-body diagrams, in the next several sections.

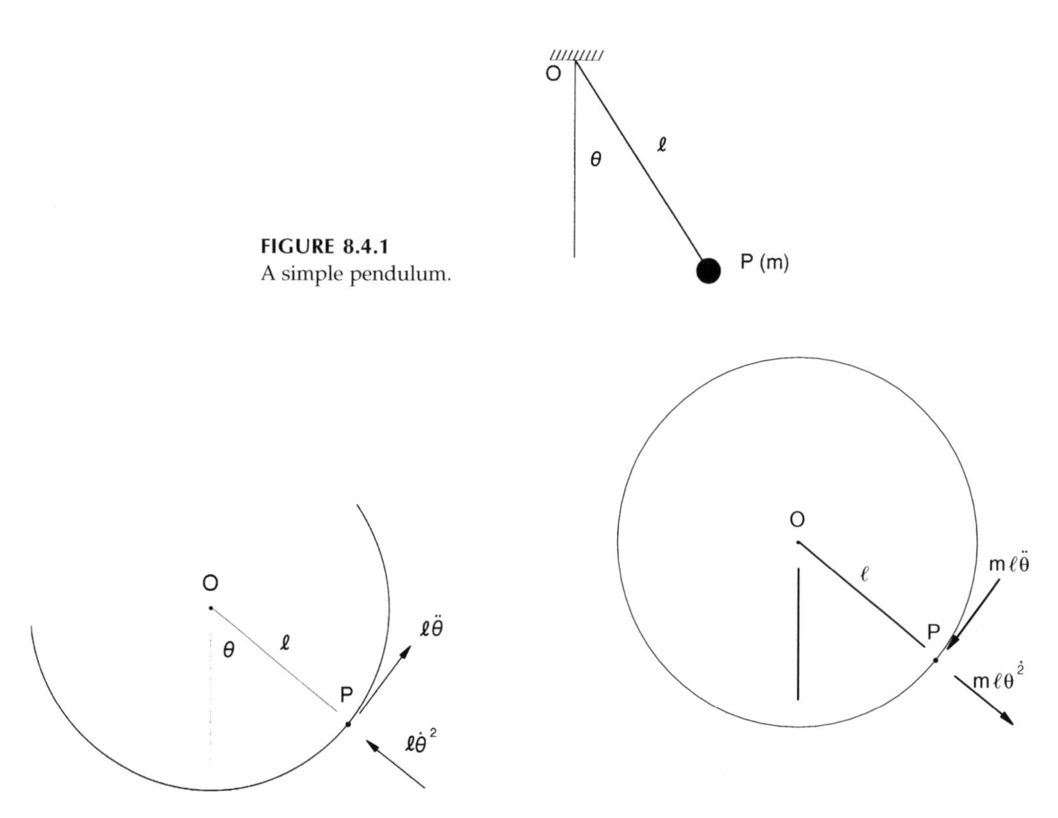

FIGURE 8.4.1
A simple pendulum.

FIGURE 8.4.2
Acceleration components of pendulum mass.

FIGURE 8.4.3
Inertia force components on the pendulum mass.

8.4 The Simple Pendulum

Consider the simple pendulum shown in Figure 8.4.1. It consists of a particle P of mass m attached to the end of a light (or massless) rod of length ℓ, which in turn is supported at its other end by a frictionless pin, at O. Let O be fixed in an inertial frame R. Under these conditions, P moves in a circle with radius ℓ and center O.

The acceleration of P may then be expressed in terms of radial and tangential components as in Figure 8.4.2 (see Section 3.7). Then, from Eq. (8.3.2), the inertia force exerted on P may be represented by components proportional to the acceleration components but oppositely directed, as in Figure 8.4.3.

In view of Figure 8.4.3, a free-body diagram of P may be constructed as in Figure 8.4.4 where T represents the tension in the connecting rod, and, as before, g is the gravity acceleration. Because the system of forces in a free-body diagram is a zero system (see Section 6.4), the forces must produce a zero resultant in all directions. Hence, by adding force components in the radial and tangential directions, we obtain:

$$T - \mathrm{mg}\cos\theta - m\ell\dot{\theta}^2 = 0 \tag{8.4.1}$$

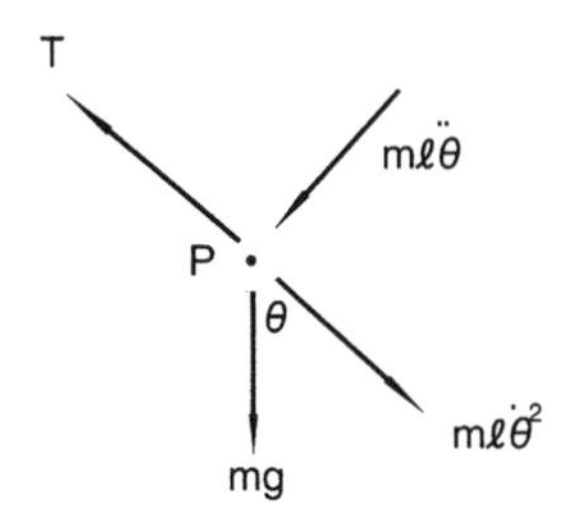

FIGURE 8.4.4
A free-body diagram of the pendulum mass.

and

$$m\ell\ddot{\theta} + \text{mg}\sin\theta = 0 \tag{8.4.2}$$

or, alternatively,

$$T = m\ell\dot{\theta}^2 + \text{mg}\cos\theta \tag{8.4.3}$$

and

$$\ddot{\theta} + (g/\ell)\sin\theta = 0 \tag{8.4.4}$$

Equation (8.4.4) is the classic pendulum equation. It is the governing equation for the orientation angle θ. Observe that it does not involve the pendulum mass m, but simply the length ℓ. This means that the pendulum motion is independent of its mass.

We will explore the solution of Eq. (8.4.4) in Chapter 13, where we will see that it is a nonlinear ordinary differential equation requiring approximate and numerical methods to obtain the solution. The nonlinearity occurs in the $\sin\theta$ term. If it happens that θ is "small" so that $\sin\theta$ may be closely approximated by θ, the equation takes the linear form:

$$\ddot{\theta} + (g/\ell)\theta = 0 \tag{8.4.5}$$

Equation (8.4.5) is called the *linear oscillator equation*. It usually forms the starting point for a study of vibrations (see Chapter 13). Once Eq. (8.4.4) is solved for θ, the result may be substituted into Eq. (8.4.3) to obtain the rod tension T.

Finally, it should be noted that dynamics principles such as d'Alembert's principle or Newton's laws simply lead to the governing equation. They do not lead to solutions of the equations, although some principles may produce equations that are in a form more suitable for easy solution than others.

8.5 A Smooth Particle Moving Inside a Vertical Rotating Tube

For a second example illustrating the use of d'Alembert's principle consider a circular tube T with radius r rotating with angular speed Ω about a vertical axis as depicted in Figure 8.5.1. Let T contain a smooth particle P with mass m which is free to slide within T. Let the position of P within T be defined by the angle θ as shown. Let $\mathbf{n}_r$ and $\mathbf{n}_\theta$ be radial and

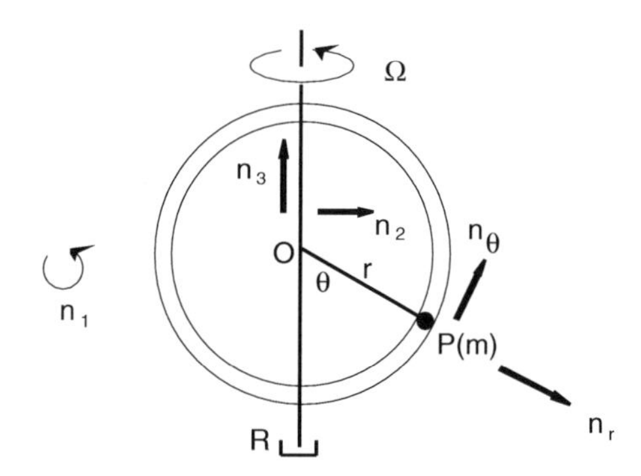

FIGURE 8.5.1
A rotating tube with an interior particle *P*.

tangential unit vectors, respectively, and let $\mathbf{n}_2$ and $\mathbf{n}_3$ be horizontal and vertical unit vectors, respectively, fixed in T.

Using the principles of kinematics of Chapter 4 we see that the acceleration of *P* in an inertial frame *R*, in which *T* is spinning, is (see Eq. (4.10.11)):

$$ {}^R\mathbf{a}^P = {}^T\mathbf{a}^P + {}^R\mathbf{a}^{P^*} + 2\,{}^R\boldsymbol{\omega}^T \times {}^T\mathbf{v}^P \tag{8.5.1} $$

where P^* is that point of *T* that coincides with *P*. P^* moves on a horizontal circle with radius r sinθ. The acceleration of P^* in *R* is, then,

$$ {}^R\mathbf{a}^{P^*} = -\dot{\Omega} r \sin\theta \mathbf{n}_1 - \Omega^2 r \sin\theta \mathbf{n}_2 \tag{8.5.2} $$

where $\mathbf{n}_1$ is a unit vector normal to the plane of *T*.

The velocity and acceleration of *P* in *T* are:

$$ {}^T\mathbf{v}^P = r\dot{\theta}\mathbf{n}_\theta \quad \text{and} \quad {}^T\mathbf{a}^P = -r\dot{\theta}^2\mathbf{n}_r + r\ddot{\theta}\mathbf{n}_\theta \tag{8.5.3} $$

By substituting from Eqs. (8.5.2) and (8.5.3) into Eq. (8.5.1) the acceleration of *P* in *R* becomes:

$$ \begin{aligned} {}^R\mathbf{a}^P &= -r\dot{\theta}^2\mathbf{n}_r + r\ddot{\theta}\mathbf{n}_\theta - r\dot{\Omega}\sin\theta\mathbf{n}_1 \\ &\quad - r\Omega^2\sin\theta\mathbf{n}_2 + 2\Omega\mathbf{n}_3 \times 4\dot{\theta} n_\theta \end{aligned} \tag{8.5.4} $$

where ${}^R\boldsymbol{\omega}^T$ is identified as being $\Omega\mathbf{n}_3$. The unit vectors $\mathbf{n}_2$ and $\mathbf{n}_3$ may be expressed in terms of $\mathbf{n}_r$ and $\mathbf{n}_\theta$ as:

$$ \begin{aligned} \mathbf{n}_2 &= \sin\theta\mathbf{n}_r + \cos\theta\mathbf{n}_\theta \\ \mathbf{n}_3 &= -\cos\theta\mathbf{n}_r + \sin\theta\mathbf{n}_\theta \end{aligned} \tag{8.5.5} $$

By substituting into Eq. (8.5.4) and by carrying out the indicated addition and multiplication, ${}^R\mathbf{a}^P$ becomes:

$$ \begin{aligned} {}^R\mathbf{a}^P &= \left(-r\dot{\Omega}\sin\theta - 2r\Omega\dot{\theta}\cos\theta\right)\mathbf{n}_1 + \left(-r\dot{\theta}^2 - r\Omega^2\sin^2\theta\right)\mathbf{n}_r \\ &\quad + \left(r\ddot{\theta} - r\Omega^2\sin\theta\cos\theta\right)\mathbf{n}_\theta \end{aligned} \tag{8.5.6} $$

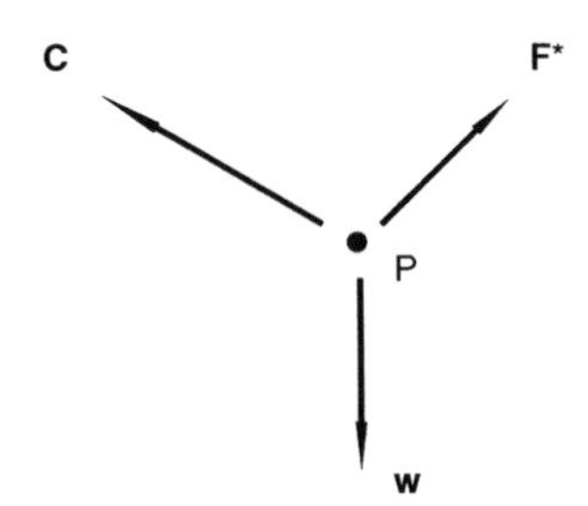

FIGURE 8.5.2
Free-body diagram of *P*.

Consider next the forces on *P*. The inertia force $\mathbf{F}^*$ on *P* is:

$$\mathbf{F}^* = -m\,{}^R\mathbf{a}^P \tag{8.5.7}$$

The applied forces on *P* consist of a vertical weight (or gravity) force $\mathbf{w}$ given by:

$$\mathbf{w} = -mg\mathbf{n}_3 \tag{8.5.8}$$

and a contact force $\mathbf{C}$ given by:

$$\mathbf{C} = N_1\mathbf{n}_1 + N_r\mathbf{n}_r \tag{8.5.9}$$

(Recall that *P* is smooth, thus there is no friction or contact force in the $\mathbf{n}_\theta$ direction.)

These forces on *P* are exhibited in the free-body diagram of Figure 8.5.2. Then, from d'Alembert's principle, we have:

$$\mathbf{C} + \mathbf{w} + \mathbf{F}^* = 0 \tag{8.5.10}$$

or

$$N_1\mathbf{n}_1 + N_r\mathbf{n}_r - mg\mathbf{n}_3 - m\,{}^R\mathbf{a}^P = 0 \tag{8.5.11}$$

By substituting from Eq. (8.5.6), and by using Eq. (8.5.5) to express $\mathbf{n}_3$ in terms of $\mathbf{n}_r$ and $\mathbf{n}_\theta$, the governing equation becomes:

$$\begin{aligned} &N_1\mathbf{n}_1 + N_r\mathbf{n}_r - mg\cos\theta\mathbf{n}_r - mg\sin\theta\mathbf{n}_q + \left(mr\dot{\Omega}\sin\theta + 2mr\Omega\dot{\theta}\cos\theta\right)\mathbf{n}_1 \\ &\quad + \left(mr\dot{\theta}^2 + mr\Omega^2\sin^2\theta\right)\mathbf{n}_r + \left(-mr\ddot{\theta} + mr\Omega^2\sin\theta\cos\theta\right)\mathbf{n}_\theta = 0 \end{aligned}$$

or

$$\begin{aligned} &\left(N_1 + mr\dot{\Omega}\sin\theta + 2mr\Omega\dot{\theta}\cos\theta\right)\mathbf{n}_1 \\ &+\left(N_r + mg\cos\theta + mr\dot{\theta}^2 + mr\Omega^2\sin^2\theta\right)\mathbf{n}_r \\ &+\left(-mr\ddot{\theta} + mr\Omega^2\sin\theta\cos\theta - mg\sin\theta\right)\mathbf{n}_\theta = 0 \end{aligned} \tag{8.5.12}$$

Therefore, the scalar governing equations are:

$$N_1 = -mr\dot{\Omega}\sin\theta - 2mr\Omega\dot{\theta}\cos\theta \tag{8.5.13}$$

$$N_r = -mg\cos\theta - mr\dot{\theta}^2 - mr\Omega^2\sin^2\theta \tag{8.5.14}$$

$$\ddot{\theta} - \Omega^2\sin\theta\cos\theta + (g/r)\sin\theta = 0 \tag{8.5.15}$$

Equations (8.5.13), (8.5.14), and (8.5.15) are three equations for the unknowns N_1, N_r, and θ. Observe that Eq. (8.5.15) involves only θ. Hence, by solving Eq. (8.5.15) for θ we can then substitute the result into Eqs. (8.5.13) and (8.5.14) to obtain N_1 and N_r. Observe further that if Ω is zero, Eq. (8.5.15) takes the same form as Eq. (8.4.4), the pendulum equation. We will consider the solution of Eq. (8.5.15) in Chapters 12 and 13.

8.6 Inertia Forces on a Rigid Body

For a more general example of an inertia force system, consider a rigid body B moving in an inertial frame R as depicted in Figure 8.6.1. Let G be the mass center of B and let P_i be a typical point of B. Then, from Eq. (4.9.6), the acceleration of P_i in R may be expressed as:

$${}^R\mathbf{a}^{P_i} = {}^R\mathbf{a}^G + \boldsymbol{\alpha}\times\mathbf{r}_i + \boldsymbol{\omega}\times(\boldsymbol{\omega}\times\mathbf{r}_i) \tag{8.6.1}$$

where G is the mass center of B, $\mathbf{r}_i$ is the position vector of P_i relative to G, $\boldsymbol{\alpha}$ is the angular acceleration of B in R, and $\boldsymbol{\omega}$ is the angular velocity of B in R.

Let B be considered to be composed of particles such as the crystals of a sandstone. Let P_i be a point of a typical particle having mass m_i. Then, from Eq. (8.2.5), the inertia force on the particle is:

$$F_i^* = -m_i\,{}^R\mathbf{a}^{P_i} \quad (\text{no sum on } i) \tag{8.6.2}$$

The inertia forces on B consist of the system of forces made up of the inertia forces on the particles of B. This system of forces (usually a very large number of forces) may be represented by an equivalent force system (see Section 6.5) consisting of a single force $\mathbf{F}^*$

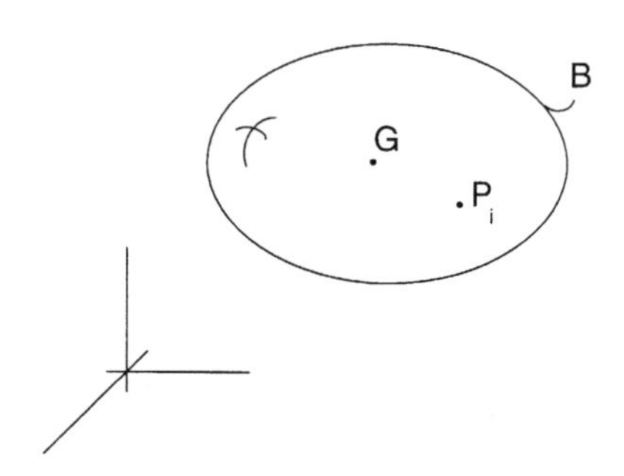

FIGURE 8.6.1
A rigid body moving in an inertial reference frame.

passing through an arbitrary point (say G) together with a couple with torque $\mathbf{T}^*$. Then, $\mathbf{F}^*$ and $\mathbf{T}^*$ are:

$$\mathbf{F}^* = \sum_{i=1}^{N} \mathbf{F}_{i^*} = \sum_{i=1}^{N} m_i {}^R\mathbf{a}^{P_i} \tag{8.6.3}$$

and

$$\mathbf{T}^* = \sum_{i=1}^{N} \mathbf{r}_i \times \mathbf{F}_{i^*} = -\sum_{i=1}^{N} m_i \mathbf{r}_i \times {}^R\mathbf{a}^{P_i} \tag{8.6.4}$$

where N is the number of particles of B. Recall that we already examined the summation in Eqs. (8.6.3) and (8.6.4) in Section 7.12. Specifically, by using the definitions of mass center and inertia dyadic we found that $\mathbf{F}^*$ and $\mathbf{T}^*$ could be expressed as (see Eqs. (6.9.9), (7.12.1), and (7.12.8)):

$$\mathbf{F}^* = -M {}^R\mathbf{a}^G \tag{8.6.5}$$

and

$$\mathbf{T}^* = -\mathbf{I}^{B/G} \cdot \boldsymbol{\alpha} - \boldsymbol{\omega} \times \left(\mathbf{I}^{B/G} \cdot \boldsymbol{\omega} \right) \tag{8.6.6}$$

where M is the total mass of B.

Consider the form of the inertia torque: Suppose $\mathbf{n}_1$, $\mathbf{n}_2$, and $\mathbf{n}_3$ are mutually perpendicular unit vectors parallel to central principal inertia axes of B. Then, the inertia dyadic $\mathbf{I}^{B/G}$ may be expressed as:

$$\mathbf{I}^{B/G} = I_{11}\mathbf{n}_1\mathbf{n}_1 + I_{22}\mathbf{n}_2\mathbf{n}_2 + I_{33}\mathbf{n}_3\mathbf{n}_3 \tag{8.6.7}$$

Let the angular acceleration and angular velocity of B be expressed as:

$$\begin{aligned} \boldsymbol{\alpha} &= \alpha_1\mathbf{n}_1 + \alpha_2\mathbf{n}_2 + \alpha_3\mathbf{n}_3 = \alpha_i\mathbf{n}_i \\ \text{and} \quad \boldsymbol{\omega} &= \omega_1\mathbf{n}_1 + \omega_2\mathbf{n}_2 + \omega_3\mathbf{n}_3 = \omega_i\mathbf{n}_i \end{aligned} \tag{8.6.8}$$

Then, in terms of the α_i, ω_i, I_{ii}, and the $\mathbf{n}_i$ ($i = 1, 2, 3$), the inertia torque $\mathbf{T}^*$ may be expressed as:

$$\mathbf{T}^* = T_1\mathbf{n}_1 + T_2\mathbf{n}_2 + T_3\mathbf{n}_3 = T_i\mathbf{n}_i \tag{8.6.9}$$

where the components T_i ($i = 1, 2, 3$) are:

$$T_1 = -\alpha_1 I_{11} + \omega_2\omega_3\left(I_{22} - I_{33}\right) \tag{8.6.10}$$

$$T_2 = -\alpha_2 I_{22} + \omega_3\omega_1\left(I_{33} - I_{11}\right) \tag{8.6.11}$$

$$T_3 = -\alpha_3 I_{33} + \omega_1\omega_2\left(I_{11} - I_{22}\right) \tag{8.6.12}$$

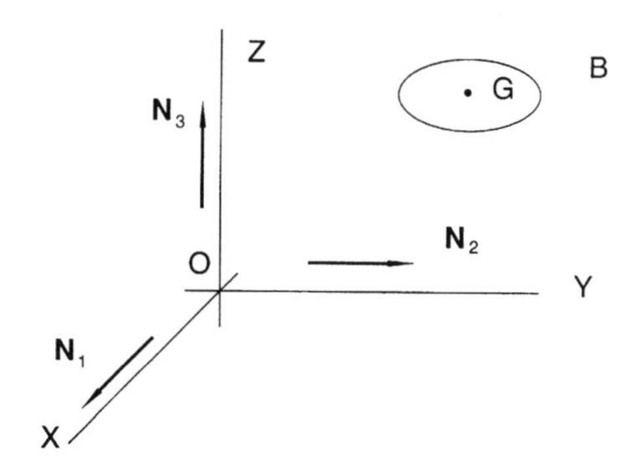

FIGURE 8.7.1
A body B moving as a projectile.

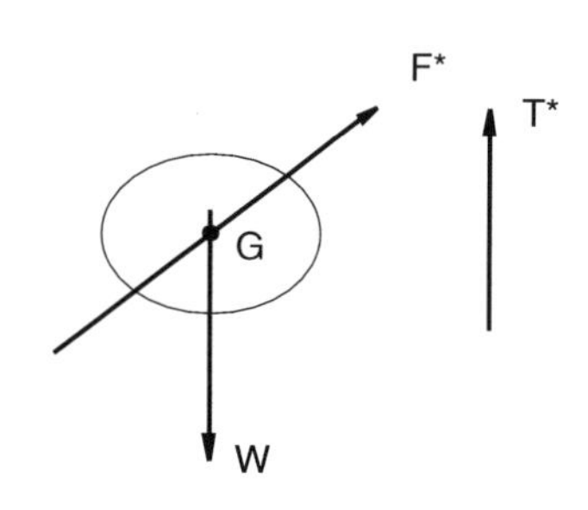

FIGURE 8.7.2
A free-body diagram of projectile B.

8.7 Projectile Motion

To illustrate the use of Eqs. (8.6.5) and (8.6.6), consider a body thrown into the air as a projectile as in Figure 8.7.1. Then, the only applied forces on B are due to gravity, which can be represented by the single weight force **W** passing through G as:

$$\mathbf{W} = -Mg\mathbf{N}_3 \tag{8.7.1}$$

where $\mathbf{N}_3$ is the vertical unit vector shown in Figure 8.7.1. Figure 8.7.2 shows a free-body diagram of B. Using d'Alembert's principle, the governing equations of motion of B are, then,

$$\mathbf{W} + \mathbf{F}^* = 0 \tag{8.7.2}$$

and

$$\mathbf{T}^* = 0 \tag{8.7.3}$$

or

$${}^R\mathbf{a}^G = -g\mathbf{N}_3 \tag{8.7.4}$$

and

$$T_1 = T_2 = T_3 = 0 \tag{8.7.5}$$

Suppose the acceleration of G is expressed in the form:

$${}^R\mathbf{a}^G = \ddot{x}\mathbf{N}_1 + \ddot{y}\mathbf{N}_2 + \ddot{z}\mathbf{N}_3 \tag{8.7.6}$$

where (x, y, z) are the coordinates of G relative to the X-, Y-, Z-axes system of Figure 8.7.1. Then, by substituting into Eq. (8.7.4), we obtain the scalar equations:

$$\ddot{x} = 0, \quad \ddot{y} = 0, \quad \ddot{z} = -g \tag{8.7.7}$$

These are differential equations governing the motion of a projectile. They are easy to solve given suitable initial conditions. For example, suppose that initially (at $t = 0$) we have G at the origin O and projected with speed V_O in the X–Z plane at an angle θ relative to the X-axis as shown in Figure 8.7.3. Specifically, at $t = 0$, let $x, y, z, \dot{x}, \dot{y}$, and $\dot{z}$ be:

$$x = y = z = 0, \quad \dot{x} = V_O \cos\theta, \quad \dot{y} = 0, \quad \dot{z} = V_O \sin\theta \tag{8.7.8}$$

Then, by integrating, we obtain the solutions of Eq. (8.6.19) in the forms:

$$x = (V_O \cos\theta)t \tag{8.7.9}$$

$$y = 0 \tag{8.7.10}$$

$$z = -gt^2/2 + (V_O \sin\theta)t \tag{8.7.11}$$

By eliminating t between Eqs. (8.7.9) and (8.7.11), we obtain:

$$(V_O^2 \cos^2\theta)z = (-g/2)x^2 + (V_O^2 \sin\theta\cos\theta)x \tag{8.7.12}$$

Equations (8.7.10) and (8.7.12) show that G moves in a plane, on a parabola. That is, a projectile always has planar motion and its mass center traces out a parabola.

From Eq. (8.7.11), we see that G is on the X-axis (that is, $z = 0$) when:

$$t = 0 \quad \text{and} \quad t = (2V_O/g)\sin\theta \tag{8.7.13}$$

The corresponding positions on the X-axis are:

$$x = 0 \quad \text{and} \quad x = d = (2V_O^2/g)\sin\theta\cos\theta \tag{8.7.14}$$

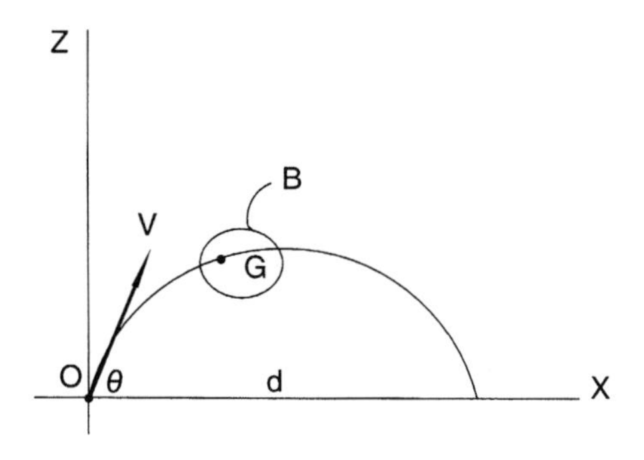

FIGURE 8.7.3
Projectile movement.

where d is the distance from the origin to where G returns again to the horizontal plane, or to the X-axis (see Figure 8.7.3). For a given V_O, Eq. (8.7.14) shows that d has a maximum value when θ is 45°.

Next, consider Eq. (8.7.5). Suppose that the unit vectors $\mathbf{n}_i$ ($i = 1, 2, 3$) are not only parallel to principal inertia axes but are also fixed in B. Then, from Eq. (4.6.6), we have:

$$\alpha_1 = \dot{\omega}_1, \;\; \alpha_2 = \dot{\omega}_2, \;\; \alpha_3 = \dot{\omega}_3 \tag{8.7.15}$$

Equation (8.7.5) then takes the form (see Eqs. (8.6.10) to (8.6.12)):

$$\dot{\omega}_1 = \omega_2\omega_3(I_{22} - I_{33})/I_{11} \tag{8.7.16}$$

$$\dot{\omega}_2 = \omega_3\omega_1(I_{33} - I_{11})/I_{22} \tag{8.7.17}$$

$$\dot{\omega}_3 = \omega_1\omega_2(I_{11} - I_{22})/I_{33} \tag{8.7.18}$$

These equations form a system of nonlinear differential equations. A simple solution of the equations is seen to be:

$$\omega_1 = \omega_0, \;\; \omega_2 = \omega_3 = 0 \tag{8.7.19}$$

That is, a projectile can rotate with constant speed about a central principal inertia axis. (We will examine the stability of such rotation in Chapter 13.)

Finally, observe that if a projectile B is rotating about a central principal axis and a point Q of B is not on the central principal axis, then Q will move on a circle whose center moves on a parabola. Moreover, a projectile always rotates about its mass center, which in turn has planar motion on a parabola.

8.8 A Rotating Circular Disk

For another illustration of the effects of inertia forces and inertia torques, consider the circular disk D with radius r rotating in a vertical plane as depicted in Figure 8.8.1. Let D be supported by frictionless bearings at its center O.

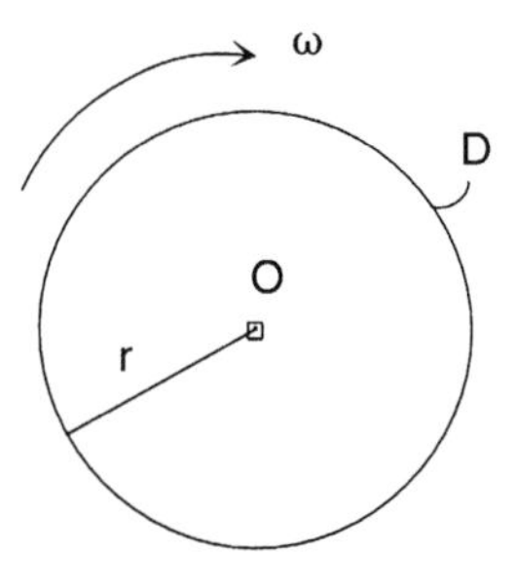

FIGURE 8.8.1
A rotating circular disk.

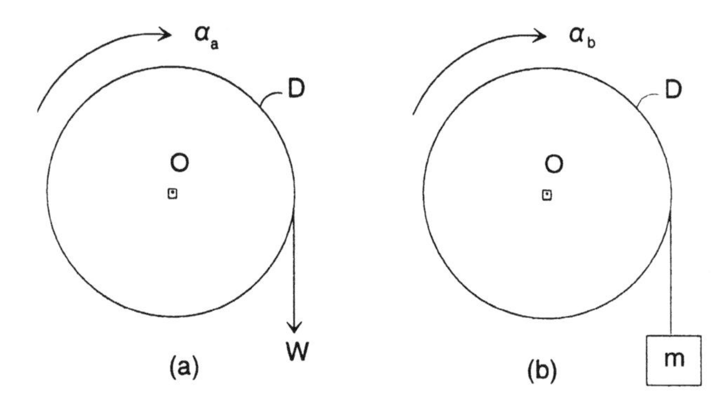

FIGURE 8.8.2
Two loading conditions on a circular disk.

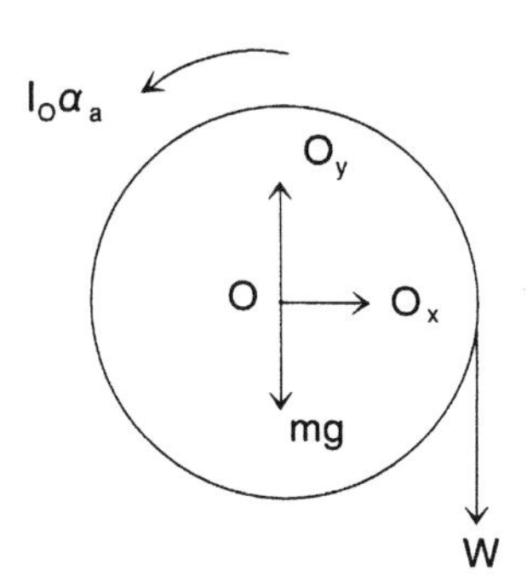

FIGURE 8.8.3
Free-body diagram of disk in Figure 8.8.2a.

Consider two loading conditions on D: First, let D be loaded by a force W applied to the rim of D as in Figure 8.8.2a. Next, let D be loaded by a weight having mass m attached by a cable to the rim of D as in Figure 8.8.2b. Let m have the value W/g (that is, the mass has weight W).

Consider first the loading of Figure 8.8.2a. The force W will cause a clockwise angular acceleration α_a as viewed in Figure 8.8.2a. This angular acceleration will in turn induce a counterclockwise inertia torque component when the equivalent inertia force is passed through O. A free-body diagram is shown in Figure 8.8.3, where I_O is the axial moment of inertia of D, M is the mass of D, and O_x and O_y are horizontal and vertical bearing reaction components. By adding forces horizontally and vertically, by setting the results equal to zero, and by setting moments about O equal to zero, we obtain:

$$O_x = 0 \tag{8.8.1}$$

$$O_y - mg - W = 0 \tag{8.8.2}$$

and

$$I_O\alpha_a - W_r = 0 \tag{8.8.3}$$

Next, for the loading of Figure 8.8.2b, the weight will create a tension T in the attachment cable which in turn will induce a clockwise angular acceleration α_b of D and a resulting counterclockwise inertia torque. Free-body diagrams for D and the attached weight are shown in Figure 8.8.4, where the term $r\alpha_b$ is the magnitude of the acceleration of the weight. By setting the resultant forces and moments about O equal to zero, we obtain:

$$O_x = 0 \tag{8.8.4}$$

$$O_y - Mg - T = 0 \tag{8.8.5}$$

$$I_O\alpha_b - Tr - 0 \tag{8.8.6}$$

$$mr\alpha_b + T - W = 0 \tag{8.8.7}$$

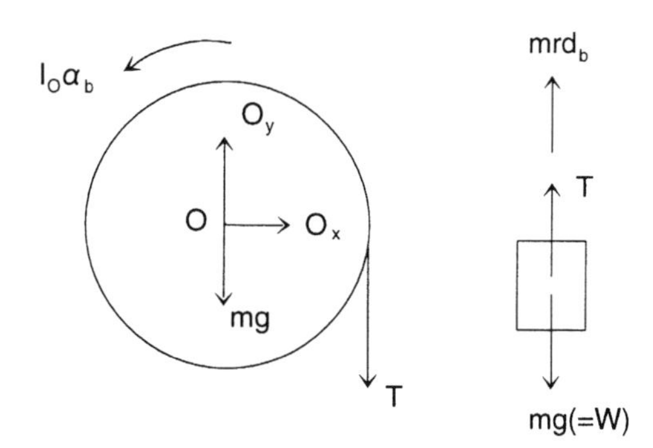

FIGURE 8.8.4
Free body diagrams of the disk and weight for the loading of Figure 8.8.2b.

By eliminating T from these last two expressions and solving for α_b, we obtain:

$$\alpha_b = Wr/\left(I_O + mr^2\right) \tag{8.8.8}$$

From Eq. (8.8.3), α_a is:

$$\alpha_a = Wr/I_O \tag{8.8.9}$$

By comparing Eqs. (8.8.8) and (8.8.9), we see the effect of the inertia of the weight in reducing the angular acceleration of the disk.

8.9 The Rod Pendulum

For another illustration of the use of Eqs. (8.6.10), (8.6.11), and (8.6.12), consider a rod of length ℓ supported by a frictionless hinge at one end and rotating in a vertical plane as shown in Figure 8.9.1. Because the rod rotates in a vertical plane about a fixed horizontal axis, the angular velocity and angular acceleration of the rod are simply:

$$\boldsymbol{\omega} = \dot{\theta}\mathbf{n}_z \quad \text{and} \quad \boldsymbol{\alpha} = \ddot{\theta}\mathbf{n}_z \tag{8.9.1}$$

where θ is the inclination angle (see Figure 8.9.1), and $\mathbf{n}_z$ is a unit vector parallel to the axis of rotation and perpendicular to the unit vectors $\mathbf{n}_x$, $\mathbf{n}_y$, $\mathbf{n}_r$, and $\mathbf{n}_\theta$ as shown in Figure 8.9.1.

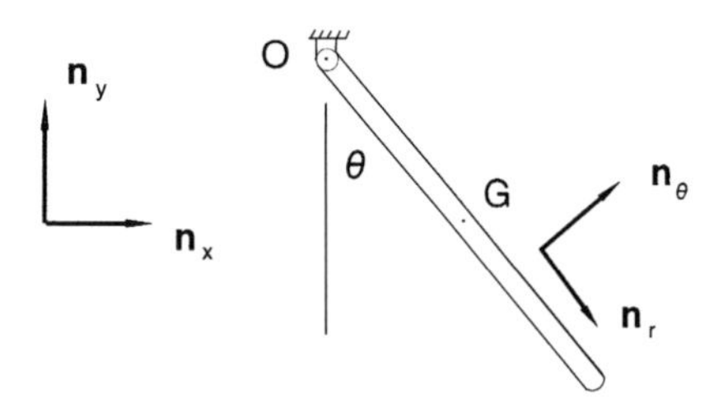

FIGURE 8.9.1
The rod pendulum.

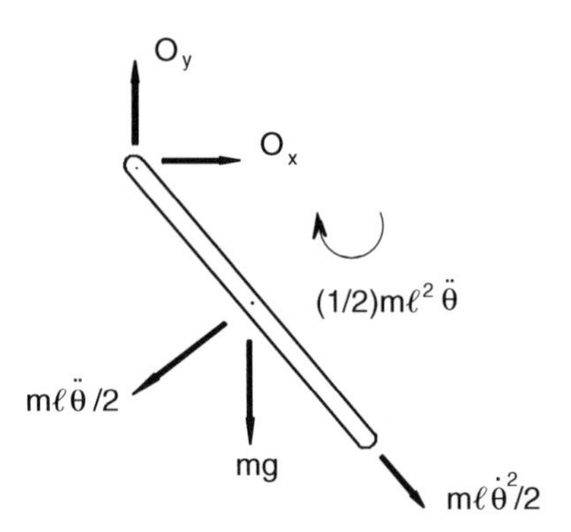

FIGURE 8.9.2
Free-body diagram of the rod pendulum.

The mass center G moves in a circle with radius $\ell/2$. The acceleration of G is:

$$\mathbf{a}^G = (\ell/2)\ddot{\theta}\mathbf{n}_\theta - (\ell/2)\dot{\theta}^2\mathbf{n}_r \tag{8.9.2}$$

Due to the symmetry of the rod, $\mathbf{n}_r$, $\mathbf{n}_\theta$, and $\mathbf{n}_z$ are principal unit vectors of inertia (see Section 7.9). The central inertia dyadic of the rod is:

$$\mathbf{I}_G = (1/12)m\ell^2\mathbf{n}_r\mathbf{n}_r + (1/12)m\ell^2 n_\theta n_\theta \tag{8.9.3}$$

where m is the mass of the rod.

Using Eqs. (8.6.5) and (8.6.6), the inertia force system on the rod is equivalent to a single force $\mathbf{F}^*$ passing through G together with a couple with torque $\mathbf{T}^*$, where $\mathbf{F}^*$ and $\mathbf{T}^*$ are:

$$\mathbf{F}^* = -m(\ell/2)\ddot{\theta}\mathbf{n}_\theta + m(\ell/2)\dot{\theta}^2\mathbf{n}_r \quad \text{and} \quad \mathbf{T}^* = -(1/12)m\ell^2\ddot{\theta}\mathbf{n}_z \tag{8.9.4}$$

Consider a free-body diagram of the rod as in Figure 8.9.2 where O_x and O_y are horizontal and vertical components of the pin reaction force: using d'Alembert's principle we can set the sum of the moments of the forces about the pinned end equal to zero. This sum leads to:

$$m(\ell/2)\ddot{\theta}(\ell/2) + mg(\ell/2)\sin\theta + (1/12)m\ell^2\ddot{\theta} = 0$$

or

$$(m\ell^2/3)\ddot{\theta} + (mg\ell/2)\sin\theta = 0 \tag{8.9.5}$$

Observe that the coefficient of $\ddot{\theta}$ in Eq. (8.9.5), $m\ell^2/3$, may be recognized as I_O, the moment of inertia of the rod about O for an axis perpendicular to the rod. That is, from the parallel axis theorem (see Section 7.6), we have:

$$I_O = I_G + m(\ell/2)^2$$

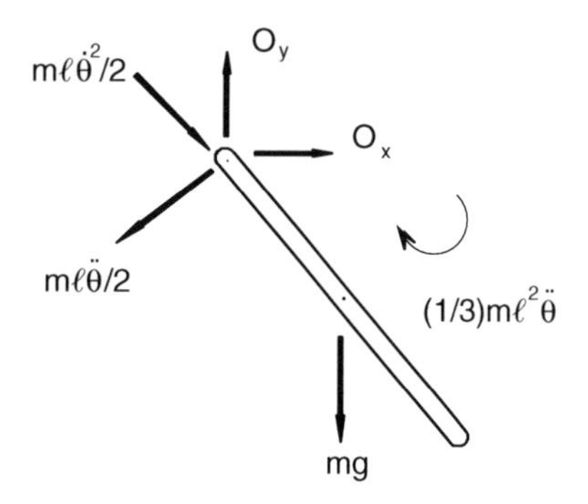

FIGURE 8.9.3
A second free-body diagram of the rod pendulum.

or

$$I_O = (1/12)m\ell^2 + (1/4)m\ell^2 = (1/3)m\ell^2 \tag{8.9.6}$$

To see the significance of this observation, suppose that the inertia force system is replaced by a single force $\mathbf{F}_O^*$ passing through O (as opposed to G) together with a couple with torque $\mathbf{T}_O^*$. Then, from Eq. (8.9.4) and the definition of equivalent force systems (see Section 6.5), we find $\mathbf{F}_O^*$ and $\mathbf{T}_O^*$ to be:

$$\mathbf{F}_O^* = \mathbf{F}^* = -m(\ell/2)\ddot{\theta}\mathbf{n}_\theta + m(\ell/2)\dot{\theta}\mathbf{n}_r \tag{8.9.7}$$

and

$$\begin{aligned} \mathbf{T}_O^* &= T^* + (\ell/2)\mathbf{n}_r \times \mathbf{F}^* \\ &= -(m\ell^2\ddot{\theta}/12)\mathbf{n}_z + (\ell/2)\mathbf{n}_r \times \left[(-m\ell\ddot{\theta}/2)\mathbf{n}_\theta + (m\ell\dot{\theta}^2/2)\mathbf{n}_r\right] \\ &= (m\ell^2\ddot{\theta}/3)\mathbf{n}_z \end{aligned} \tag{8.9.8}$$

where we have used Eq. (6.3.6).

Using this equivalent force system we can construct the free-body diagram of Figure 8.9.3. By setting the sum of the moments about end O equal to zero, we obtain the governing equation:

$$(m\ell^2/3)\ddot{\theta} + (mg\ell/2)\sin\theta = 0 \tag{8.9.9}$$

This is identical to Eq. (8.9.5), but its derivation is more direct. By dividing by the coefficient of $\ddot{\theta}$, Eq. (8.9.9) takes the form:

$$\ddot{\theta} + (3g/2\ell)\sin\theta = 0 \tag{8.9.10}$$

Equation (8.9.10) is seen to be similar to Eq. (8.4.4), the simple pendulum equation. The difference is in the coefficient of $\sin\theta$. We will explore the significance of this in Chapter 12.

8.10 Double-Rod Pendulum

As an extension of the foregoing system, consider the double-rod pendulum shown in Figure 8.10.1. It consists of two identical pin-connected rods moving in a vertical plane. Let the rods have length ℓ and mass m, and let the connecting and support pins be frictionless. Let the orientation of the rods be defined by the angles θ_1 and θ_2 shown in the figure.

Because the system has two bodies it might be called a *multibody* system. Also, because two parameters, θ_1 and θ_2, are needed to define the position and orientation of the rods, the system is said to have two degrees of freedom. Thus, two equations will govern the motion of the system.

To obtain these governing equations we can proceed as in the previous examples and examine free-body diagrams of the rods. To this end, let the rods be labeled as B_1 and B_2 and let their mass centers be G_1 and G_2 as in Figure 8.10.2. Let the pins be at O and Q, and let unit vectors $\mathbf{n}_1$, $\mathbf{n}_2$, $\mathbf{n}_{11}$, $\mathbf{n}_{12}$, $\mathbf{n}_{22}$, and $\mathbf{n}_3$ be introduced to define global directions and directions along and perpendicular to the rods as in Figure 8.10.2.

The angular velocities and angular accelerations of the rods in the inertia frame R (of the fixed support) are then:

$$\boldsymbol{\omega}_1 = \dot{\theta}_1\mathbf{n}_3, \quad \boldsymbol{\alpha}_1 = \ddot{\theta}_1\mathbf{n}_3, \quad \boldsymbol{\omega}_2 = \dot{\theta}_2\mathbf{n}_3, \quad \boldsymbol{\alpha}_2 = \ddot{\theta}_2\mathbf{n}_3 \tag{8.10.1}$$

The mass center velocities and acceleration are:

$$\mathbf{v}^{G_1} = (\ell/2)\dot{\theta}_1\mathbf{n}_{11}, \quad \mathbf{a}^{G_1} = (\ell/2)\ddot{\theta}_1\mathbf{n}_{11} + (\ell/2)\dot{\theta}_1^2\mathbf{n}_{12} \tag{8.10.2}$$

and

$$\begin{aligned} \mathbf{v}^{G_2} &= \ell\dot{\theta}_1\mathbf{n}_{11} + (\ell/2)\dot{\theta}_2\mathbf{n}_{21}, \\ \mathbf{a}^{G_2} &= \ell\ddot{\theta}_1\mathbf{n}_{11} + \ell\dot{\theta}_1^2\mathbf{n}_{12} + (\ell/2)\ddot{\theta}_2\mathbf{n}_{21} + (\ell/2)\dot{\theta}_2^2\mathbf{n}_{22} \end{aligned} \tag{8.10.3}$$

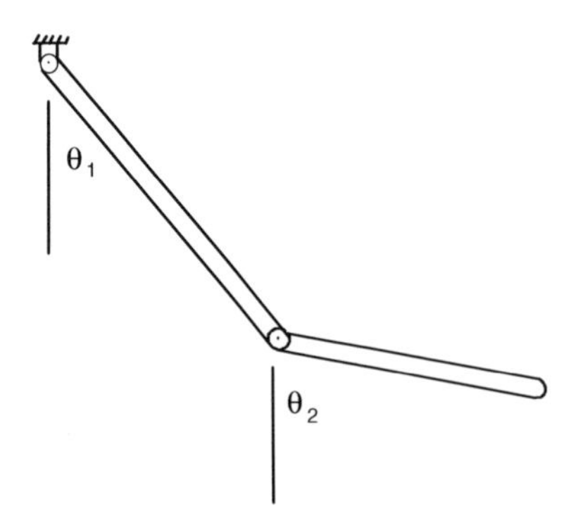

FIGURE 8.10.1
Double-rod pendulum.

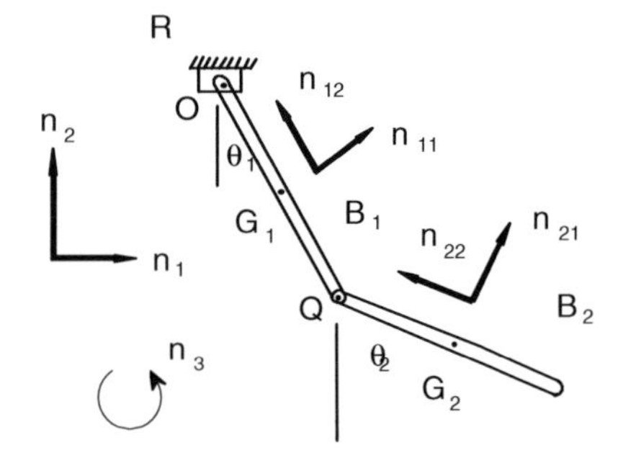

FIGURE 8.10.2
Unit vectors and notation for double-rod pendulum.

Observe that the unit vectors of the rods may be expressed in terms of the horizontal and vertical unit vectors and in terms of each other as:

$$\begin{aligned}\mathbf{n}_{11} &= c_1\mathbf{n}_1 + s_1\mathbf{n}_2, \ \mathbf{n}_{21} = c_2\mathbf{n}_1 + s_2\mathbf{n}_2 \\ \mathbf{n}_{12} &= -s_1\mathbf{n}_1 + c_1\mathbf{n}_2, \ \mathbf{n}_{22} = -s_2\mathbf{n}_1 + c_2\mathbf{n}_2\end{aligned} \tag{8.10.4}$$

and

$$\begin{aligned}n_1 &= c_1\mathbf{n}_{11} - s_1\mathbf{n}_{12}, \ n_1 = c_2\mathbf{n}_{21} - s_2\mathbf{n}_{22} \\ n_2 &= -s_1\mathbf{n}_{11} + c_1\mathbf{n}_{12}, \ n_2 = s_2\mathbf{n}_{21} + c_2\mathbf{n}_{22}\end{aligned} \tag{8.10.5}$$

and

$$\begin{aligned}\mathbf{n}_{11} &= c_{2-1}\mathbf{n}_{21} - s_{2-1}\mathbf{n}_{22}, \ \mathbf{n}_{21} = c_{2-1}\mathbf{n}_{11} + s_{2-1}\mathbf{n}_{12} \\ \mathbf{n}_{12} &= s_{2-1}\mathbf{n}_{21} + c_{2-1}\mathbf{n}_{22}, \ \mathbf{n}_{22} = -s_{2-1}\mathbf{n}_{11} + c_{2-1}\mathbf{n}_{22}\end{aligned} \tag{8.10.6}$$

where s_i and c_i are abbreviations for $\sin\theta_i$ and $\cos\theta_i$ ($i = 1, 2$), respectively, and s_{i-j} and c_{i-j} are abbreviations for $\sin(\theta_i - \theta_j)$ and $\cos(\theta_i - \theta_j)$, respectively.

To examine the forces on the rods and to invoke d'Alembert's principle it is helpful to consider a free-body diagram of the system of both rods and a free-body diagram of the second rod (B_2). These diagrams are shown in Figures 8.10.3 and 8.10.4 where O_1, O_2 and Q_1, Q_2 are horizontal and vertical components of the pin reaction forces. From Eqs. (8.6.5) and (8.6.6), the inertia forces and torques are:

$$\mathbf{F}_1^* = -m\mathbf{a}^{G_1}, \ \mathbf{F}_2^* = -m\mathbf{a}^{G_2} \tag{8.10.7}$$

and

$$\mathbf{T}_1^* = -(m\ell^2/12)\ddot{\theta}_1\mathbf{n}_3, \ \mathbf{T}_2^* = -(m\ell^2/12)\ddot{\theta}_2\mathbf{n}_3 \tag{8.10.8}$$

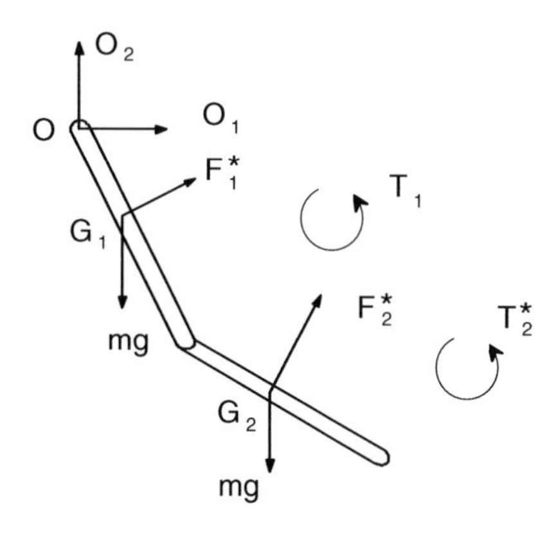

FIGURE 8.10.3
Free-body diagram of both rods.

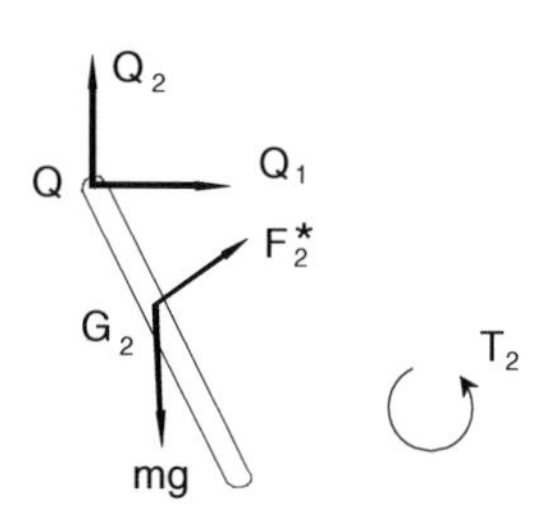

FIGURE 8.10.4
Free-body diagram of rod B_2.

Setting moments of these force systems about O and Q equal to zero leads to the equations:

$$\begin{aligned}&-(\ell/2)\mathbf{n}_{12}\times(-mg\mathbf{n}_2)-(\ell/2)\mathbf{n}_{12}\times\mathbf{F}_1^*+\mathbf{T}_1^*\\&-\left[\ell\mathbf{n}_{12}+(\ell/2)\mathbf{n}_{22}\right]\times(-mg\mathbf{n}_2)\\&-\left[\ell\mathbf{n}_{12}+(\ell/2)\mathbf{n}_{22}\right]\times\mathbf{F}_2^*+\mathbf{T}_2^*=0\end{aligned}\tag{8.10.9}$$

and

$$-(\ell/2)\mathbf{n}_{22}\times(-mg)\mathbf{n}_2-(\ell/2)\mathbf{n}_{22}\times\mathbf{F}_2^*+\mathbf{T}_2^*=0\tag{8.10.10}$$

By substituting from Eq. (8.10.4) to (8.10.8), these equations become (after simplification):

$$\begin{aligned}&(4/3)\ddot{\theta}_1+(1/2)\ddot{\theta}_1\cos(\theta_2-\theta_1)+(1/3)\ddot{\theta}_2+(1/2)\ddot{\theta}_2\cos(\theta_2-\theta_1)\\&-(1/2)\dot{\theta}_2^2\sin(\theta_2-\theta_1)+(1/2)\dot{\theta}_1^2\sin(\theta_2-\theta_1)+(3g/2\ell)\sin\theta_1\\&+(g/2\ell)\sin\theta_2=0\end{aligned}\tag{8.10.11}$$

and

$$(1/2)\cos(\theta_2-\theta_1)\ddot{\theta}_1+(1/3)\ddot{\theta}_2-(1/2)\dot{\theta}_1^2\sin(\theta_1-\theta_2)+(g/2\ell)\sin\theta_2=0\tag{8.10.12}$$

Observe that all of the terms of Eq. (8.10.12) are also contained in Eq. (8.10.11). Therefore, by eliminating these terms from Eq. (8.10.11), as their sum is zero, Eq. (8.10.11) may be written as:

$$(4/3)\ddot{\theta}_1+(1/2)\ddot{\theta}_2\cos(\theta_2-\theta_1)-(1/2)\dot{\theta}_2^2\sin(\theta_2-\theta_1)+(3g/2\ell)\sin\theta_1=0\tag{8.10.13}$$

Equations (8.10.12) and (8.10.13) are the desired governing equations for the two-rod system. They form a system of coupled ordinary differential equations. We will explore the solutions of such systems in Chapter 12.

8.11 The Triple-Rod and *N*-Rod Pendulums

As a further extension of the pendulum problems consider the three-rod and N-rod systems shown in Figures 8.11.1 and 8.11.2. In principle, these systems may be studied by following the same procedures as in the foregoing problems: that is, we can study the kinematics and from that develop expressions for the inertia forces. Then, by considering the applied forces, we can construct free-body diagrams. Finally, by using d'Alembert's principle we can assert that the systems of applied and inertia forces are zero systems and thus obtain the governing equations of motion.

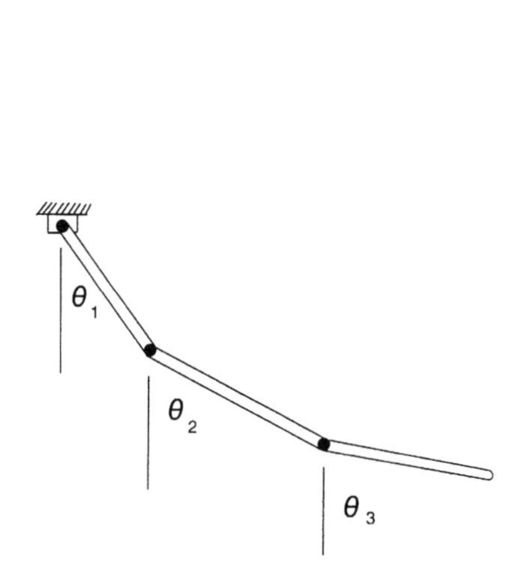

FIGURE 8.11.1
A triple-rod pendulum.

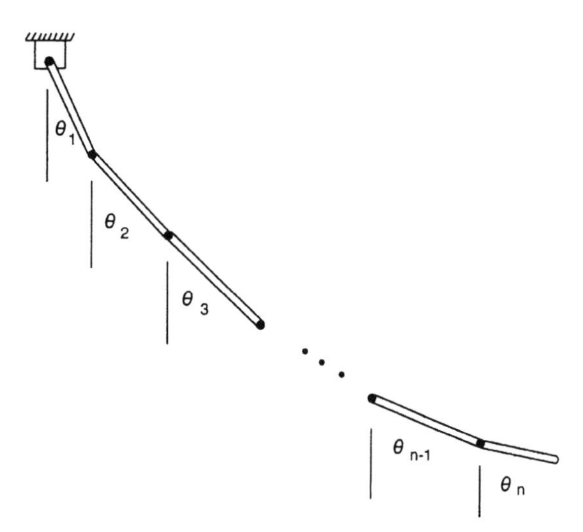

FIGURE 8.11.2
N-rod pendulum.

A little reflection, however, might suggest that, while such procedures are possible, they require extensive analyses. A comparison of the difference in the analyses of the single-rod and double-rod pendulums shows a dramatic increase in the analysis required for the double-rod system. For the triple-rod system of Figure 8.11.1, the analysis required is even more extensive. Without going into the details, we will simply present here the results that can be obtained from the analysis. As before, we assume the system is moving in a vertical plane, that the connecting and support pins are frictionless, and that the rods are identical, each having length ℓ and mass m. After simplification, the governing equations become:

$$\begin{aligned}&(7/3)\ddot{\theta}_1+(3/2)\ddot{\theta}_2\cos(\theta_1-\theta_2)+(1/2)\ddot{\theta}_3\cos(\theta_1-\theta_3)+(3/2)\dot{\theta}_2^2\sin(\theta_1-\theta_2)\\&\quad+(1/2)\dot{\theta}_3^2\sin(\theta_1-\theta_3)+(5/2)(g/\ell)\sin\theta_1=0\end{aligned} \tag{8.11.1}$$

$$\begin{aligned}&(3/2)\ddot{\theta}_1\cos(\theta_2-\theta_1)+(4/3)\ddot{\theta}_2+(1/2)\ddot{\theta}_3\cos(\theta_2-\theta_3)\\&\quad+(3/2)\dot{\theta}_1^2\sin(\theta_2-\theta_1)+(1/2)\dot{\theta}_3^2\sin(\theta_2-\theta_3)+(3/2)(g/\ell)\sin\theta_2=0\end{aligned} \tag{8.11.2}$$

$$\begin{aligned}&(1/2)\ddot{\theta}_1\cos(\theta_3-\theta_1)+(1/2)\ddot{\theta}_2\cos(\theta_3-\theta_2)+(1/3)\ddot{\theta}_3\\&\quad+(1/2)\dot{\theta}_1^2\sin(\theta_3-\theta_1)+(1/2)\dot{\theta}_2^2\sin(\theta_3-\theta_2)+(1/2)(g/\ell)\sin\theta_3=0\end{aligned} \tag{8.11.3}$$

By examining the patterns in the coefficients of these equations, it can be shown that the governing equations for the N-rod pendulum may be written in the form:

$$\sum_{j=1}^{N}\left[m_{rj}\ddot{\theta}_j+n_{rj}\dot{\theta}_j^2+(g/\ell)k_{rj}\sin\theta_j\right]=0\quad(r=1,\ldots,N) \tag{8.11.4}$$

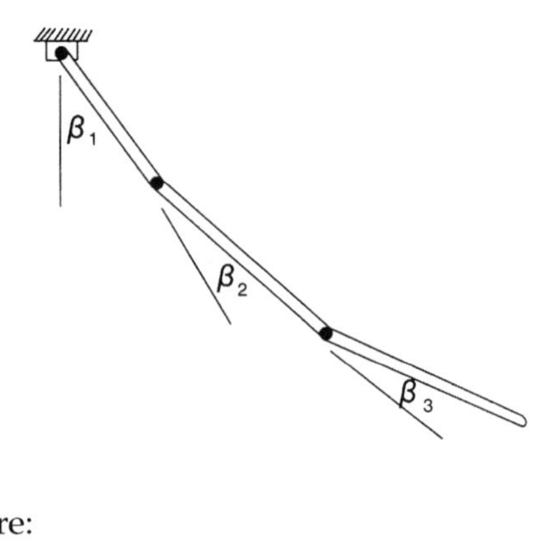

FIGURE 8.11.3
Triple-rod pendulum with relative orientation angles.

where the coefficients m_{rj}, n_{rj}, and k_{rj} are:

$$m_{rj} = [(1/2) + N - q]\cos(\theta_r - \theta_j) \quad r \neq j \text{ and } q \text{ is the largest of } r \text{ and } j \tag{8.11.5}$$

$$m_{rr} = (1/3) - N - r \quad (\text{no sum on } r) \tag{8.11.6}$$

$$n_{rj} = [(1/2) + N - q]\sin(\theta_r - \theta_j), \quad q \text{ is the largest of } r \text{ and } j \tag{8.11.7}$$

and

$$k_{rj} = [(1/2) + N - j]\delta_{rj} \tag{8.11.8}$$

where as before δ_{rj} is Kronecker's delta function having the values:

$$\delta_{rj} = \begin{cases} 0 & r \neq j \\ 1 & r = j \end{cases} \tag{8.11.9}$$

If the analysis is as detailed as would appear from an inspection of Eqs. (8.11.1) to (8.11.8) one might wonder how the results are obtained and to what extent they can be relied upon. It happens that the expressions of Eqs. (8.11.1) to (8.11.8) were not obtained through d'Alembert's principle but instead through using Lagrange's equations. (We will explore the development and use of Lagrange's equations in Chapter 11.)

Just as the choice of methods for obtaining the governing equations of motion can dramatically affect the amount of analysis required, so, too, can the choice of geometrical parameters of the system have a profound effect upon the analysis. To illustrate this last point, consider the triple-rod pendulum where the configuration of the rods is defined by the relative angles β_1, β_2, and β_3 as in Figure 8.11.3. By comparing Figures 8.11.1 and 8.11.3, we see that the relative angles (angles between adjoining rods) and the absolute angles (angles of the rods in space) are related by the equations:

$$\theta_1 = \beta_1, \quad \theta_2 = \beta_1 + \beta_2, \quad \theta_3 = \beta_1 + \beta_2 + \beta_3 \tag{8.11.10}$$

and

$$\beta_1 = \theta_1, \quad \beta_2 = \theta_2 - \theta_1, \quad \beta_3 = \theta_3 - \theta_2 \tag{8.11.11}$$

The governing equations are then:

$$\begin{aligned}
&\left[4 + 3\cos\beta_2 + \cos\beta_3 + \cos(\beta_2 + \beta_3)\right]\ddot{\beta}_1 + \left[(5/3) + (3/2)\cos\beta_2 + \cos\beta_3\right.\\
&\quad \left. + (1/2)\cos(\beta_2 + \beta_3)\right]\ddot{\beta}_2 + \left[(1/3) + (1/2)\cos\beta_3 + (1+2)\cos(\beta_2 + \beta_3)\right]\ddot{\beta}_3\\
&\quad - (3/2)\left(\dot{\beta}_1 + \dot{\beta}_2\right)^2 \sin\beta_2 + (3/2)\dot{\beta}_1^2 \sin\beta_2 + (1/2)\dot{\beta}_1^2 \sin(\beta_2 + \beta_3)\\
&\quad + (1/2)\left(\dot{\beta}_1 + \dot{\beta}_2\right)^2 \sin\beta_3 - (1/2)\left(\dot{\beta}_1 + \dot{\beta}_2 + \dot{\beta}_3\right)^2 \sin\beta_3 - (1/2)\left(\dot{\beta}_1 + \dot{\beta}_2 + \dot{\beta}_3\right)^2 \sin(\beta_2 + \beta_3)\\
&\quad + (5/2)(g/\ell)\sin\beta_1 + (3/2)(g/\ell)\sin(\beta_1 + \beta_2) + (1/2)(g/\ell)\sin(\beta_1 + \beta_2 + \beta_3) = 0
\end{aligned} \tag{8.11.12}$$

$$\begin{aligned}
&\left[(5/3) + (3/2)\cos\beta_2 + \cos\beta_3 + (1/2)\cos(\beta_2 + \beta_3)\right]\ddot{\beta}_1 + \left[(5/3) + \cos\beta_3\right]\ddot{\beta}_2\\
&\quad + \left[(1/3) + (1/2)\cos\beta_3\right]\ddot{\beta}_3 + (3/2)\dot{\beta}_1^2 \sin\beta_2 + (1/2)\dot{\beta}_1^2 \sin(\beta_2 + \beta_3)\\
&\quad + (1/2)\left(\dot{\beta}_1 + \dot{\beta}_2\right)^2 \sin\beta_3 - (1/2)\left(\dot{\beta}_1 + \dot{\beta}_2 + \dot{\beta}_3\right)^2 \sin\beta_3 + (3/2)(g/\ell)\sin(\beta_1 + \beta_2)\\
&\quad + (1/2)(g/\ell)\sin(\beta_1 + \beta_2 + \beta_3) = 0
\end{aligned} \tag{8.11.13}$$

and

$$\begin{aligned}
&\left[(1/3) + (1/2)\cos\beta_3 + (1/2)\cos(\beta_2 + \beta_3)\right]\ddot{\beta}_1 + \left[(1/3) + (1/2)\cos\beta_3\right]\ddot{\beta}_2\\
&\quad + (1/3)\ddot{\beta}_3 + (1/2)\dot{\beta}_1^2 \sin(\beta_2 + \beta_3) + (1/2)\left(\dot{\beta}_1 + \dot{\beta}_2\right)^2 \sin\beta_3\\
&\quad + (1/2)(g/\ell)\sin(\beta_1 + \beta_2 + \beta_3) = 0
\end{aligned} \tag{8.11.14}$$

Although these equations (developed using Kane's equations; see Chapter 11), are not necessarily in their simplest form, they are nevertheless far more detailed than Eqs. (8.11.1), (8.11.2), and (8.11.3).

8.12 A Rotating Pinned Rod

For an example illustrating nonplanar motion, consider the rotating pinned rod B depicted in Figure 8.12.1, where S is a shaft rotating with an angular speed Ω about a vertical axis. Let $\mathbf{k}$ be a unit vector parallel to the axis of S as shown. Let the radius of S be small. Let B be attached to S at one of its ends O by a frictionless pin whose axis is fixed on a radial

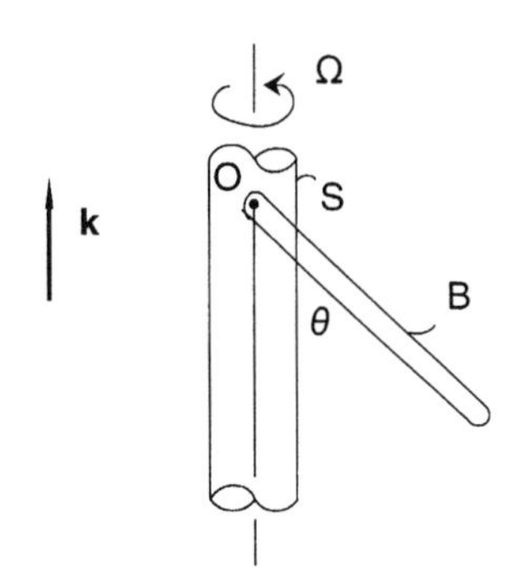

FIGURE 8.12.1
Rotating, pinned rod.

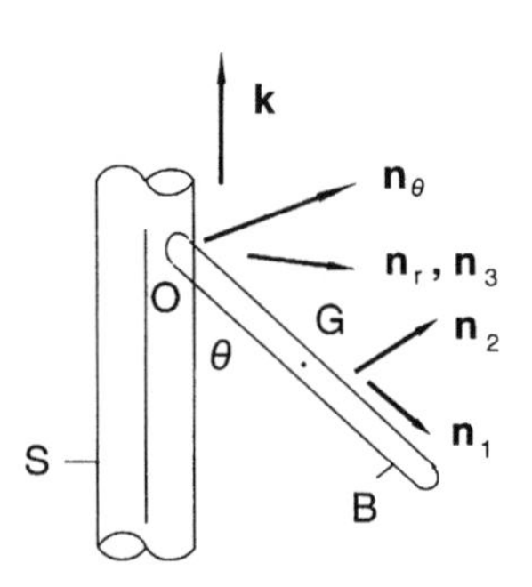

FIGURE 8.12.2
Unit vectors of the shaft and rod.

line of S. Let B have length ℓ and mass m, and let the orientation of B be defined by the angle θ as shown in Figure 8.12.1.

We can obtain the equation governing θ and, hence, the motion of B by proceeding as in the previous examples. To this end, consider first the kinematics of B: Let unit vectors $\mathbf{n}_1$, $\mathbf{n}_2$, $\mathbf{n}_3$, $\mathbf{n}_r$, and $\mathbf{n}_\theta$ be introduced as in Figure 8.12.2, where $\mathbf{n}_r$ is parallel to the pin axis of B and radial line of S; $\mathbf{n}_\theta$ is perpendicular to $\mathbf{k}$ and $\mathbf{n}_r$; and $\mathbf{n}_1$, $\mathbf{n}_2$, and $\mathbf{n}_3$ are mutually perpendicular dextral unit vectors, with $\mathbf{n}_3$ being parallel to $\mathbf{n}_r$ and $\mathbf{n}_1$ being along B, as shown. Then, these unit vectors are related to each other by the expressions:

$$\begin{aligned} \mathbf{n}_1 &= -\cos\theta\mathbf{k} + \sin\theta\mathbf{n}_\theta \\ \mathbf{n}_2 &= \sin\theta\mathbf{k} + \cos\theta\mathbf{n}_\theta \\ \mathbf{n}_3 &= \mathbf{n}_r \end{aligned} \tag{8.12.1}$$

and

$$\begin{aligned} \mathbf{n}_r &= \mathbf{n}_3 \\ \mathbf{n}_\theta &= \sin\theta\mathbf{n}_1 + \cos\theta\mathbf{n}_2 \\ \mathbf{k} &= -\cos\theta\mathbf{n}_1 + \sin\theta\mathbf{n}_2 \end{aligned} \tag{8.12.2}$$

The angular velocity $\boldsymbol{\omega}$ of B in the inertia frame R in which S is rotating may be obtained using the addition theorem for angular velocity (see Section 4.7). That is,

$$\boldsymbol{\omega} = \theta\mathbf{n}_r + \Omega\mathbf{k} \tag{8.12.3}$$

Then, by differentiating, we obtain the angular acceleration of B as:

$$\boldsymbol{\alpha} = \ddot{\theta}\mathbf{n}_r + \dot{\theta}\Omega\mathbf{n}_\theta + \dot{\Omega}\mathbf{k} \tag{8.12.4}$$

By using Eqs. (8.12.2) $\boldsymbol{\omega}$ and $\boldsymbol{\alpha}$ may be expressed in terms of $\mathbf{n}_1$, $\mathbf{n}_2$, and $\mathbf{n}_3$ as:

$$\boldsymbol{\omega} = -\Omega\cos\theta\mathbf{n}_1 + \Omega\sin\theta\mathbf{n}_2 + \dot{\theta}\mathbf{n}_3 \tag{8.12.5}$$

and

$$\boldsymbol{\alpha} = \left(\Omega\dot{\theta}\sin\theta - \Omega\dot{\theta}\cos\theta\right)\mathbf{n}_1 + \left(\Omega\dot{\theta}\cos\theta + \dot{\Omega}\sin\theta\right)\mathbf{n}_2 + \ddot{\theta}\mathbf{n}_3 \tag{8.12.6}$$

Assuming the shaft radius to be small, the velocity **v** and acceleration **a** of the mass center G of B in the inertia frame R may be obtained from the expressions (see Eqs. (4.9.4) and (4.9.6)):

$$\mathbf{v} = \boldsymbol{\omega} \times (\ell/2)\mathbf{n}_1 \tag{8.12.7}$$

and

$$\mathbf{a} = \boldsymbol{\alpha} \times (\ell/2)\mathbf{n}_1 + \boldsymbol{\omega} \times \left[\boldsymbol{\omega} \times (\ell/2)\mathbf{n}_1\right] \tag{8.12.8}$$

By substituting from Eqs. (8.12.5) and (8.12.6), **v** and **a** become:

$$\mathbf{v} = (\ell/2)\dot{\theta}\mathbf{n}_2 - (\ell/2)\Omega s_\theta \mathbf{n}_3 \tag{8.12.9}$$

and

$$\begin{aligned}\mathbf{a} = &\left[-(\ell/2)\Omega^2\sin^2\theta - (\ell/2)\dot{\theta}^2\right]\mathbf{n}_1 + \left[(\ell/2)\ddot{\theta} - (\ell/2)\Omega^2\sin\theta\cos\theta\right]\mathbf{n}_2 \\ &+ \left[-(\ell/2)\dot{\theta}\Omega\cos\theta - (\ell/2)\dot{\Omega}\sin\theta - (\ell/2)\Omega\dot{\theta}\cos\theta\right]\mathbf{n}_3\end{aligned} \tag{8.12.10}$$

Next, consider a free-body diagram of B as in Figure 8.12.3, where $\mathbf{F}^*$ and $\mathbf{T}^*$ are the force and couple torque, respectively, of an equivalent inertia force system, **w** is the weight force, **O** is the pin reaction force, and **N** is the torque of the pin reaction couple. Because the pin is frictionless, **N** has zero component in the direction of the pin axis $\mathbf{n}_3$. To eliminate **O** and **N** from the analysis, we can use d'Alembert's principle and set moments about O in the $\mathbf{n}_3$ direction equal to zero:

$$\left[(\ell/2)\mathbf{n}_1 \times \mathbf{w} + (\ell/2)\mathbf{n}_1 \times \mathbf{F}^* + \mathbf{T}^*\right] \cdot \mathbf{n}_3 = 0 \tag{8.12.11}$$

The weight force **w** is simply $-mg\mathbf{k}$ and, as before, the inertia force $\mathbf{F}^*$ is $-m\mathbf{a}$. Then, using Eq. (8.12.2), Eq. (8.12.11) may be expressed as:

$$-mg(\ell/2)\sin\theta - m(\ell/2)a_2 + T_3 = 0 \tag{8.12.12}$$

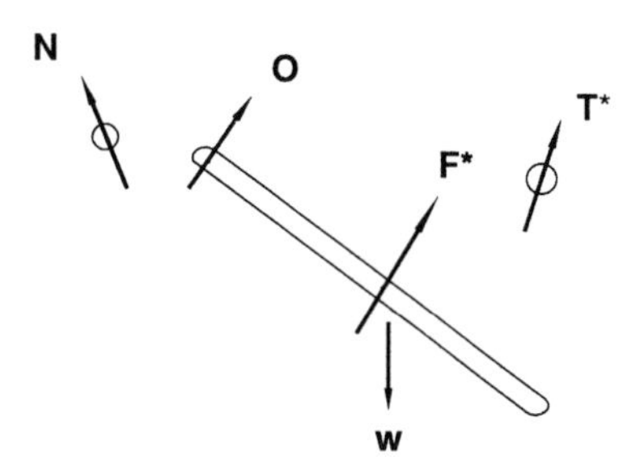

FIGURE 8.12.3
A free-body diagram of rod *B*.

where a_2 is the $\mathbf{n}_2$ component of $\mathbf{a}$ and T_3 is the $\mathbf{n}_3$ component of $\mathbf{T}^*$; a_2 may be obtained directly from Eq. (8.12.8), and, because $\mathbf{n}_1$, $\mathbf{n}_2$, and $\mathbf{n}_3$ are parallel to principal inertia directions of B for G, T_3 is given by Eq. (8.6.12) as:

$$T_3 = -\alpha_3 I_{33} + \omega_1 \omega_2 (I_{11} - I_{22}) \tag{8.12.13}$$

where α_3, ω_1, and ω_2 can be obtained directly from Eqs. (8.12.5) and (8.12.6) and where from the inertia properties of a slender rod, I_{11}, I_{22}, and I_{33} are:

$$I_{11} = 0, \quad I_{22} = I_{33} = m\ell^2/12 \tag{8.12.14}$$

Finally, by substituting from Eqs. (8.12.5), (8.12.6), (8.12.10), (8.12.13), and (8.12.14) into (8.12.12), we obtain:

$$-mg(\ell/2)\sin\theta - m(\ell/2)\left[(\ell/2)\ddot{\theta} - (\ell/2)\Omega^2 \sin\theta\cos\theta\right] - (m\ell^2/12)\ddot{\theta}$$
$$+(m\ell^2/12)\Omega^2 \sin\theta\cos\theta = 0$$

or

$$\ddot{\theta} + (3g/2\ell)\sin\theta - \Omega^2 \sin\theta\cos\theta = 0 \tag{8.12.15}$$

Observe that $\dot{\Omega}$ does not appear in this governing equation. This means that the angular speed change of S does not affect the movement of B, except indirectly through Ω. Observe further that, if Ω is constant, B has two equilibrium positions, that is, positions of constant θ:

$$\theta = 0 \quad \text{and} \quad \theta = \cos^{-1}(3g/2\ell\Omega^2) \tag{8.12.16}$$

This last result shows that the inertia torque is important even when a body is moving at a uniform rate. Indeed, if the body is not moving with uniform planar motion, its inertia torque is not generally zero. Consider again the expression for T_3 of Eq. (8.12.13):

$$T_3 = -\alpha_3 I_{33} + \omega_1 \omega_2 (I_{11} - I_{22}) \tag{8.12.17}$$

Observe that even if α_3 is zero, T_3 is not zero unless the second term is also zero.

The erroneous neglect of the inertia torque can of course produce incorrect results. For example, suppose that in the current problem, where we are seeking the equilibrium position with Ω constant, we construct an incorrect free-body diagram, as in Figure 8.12.4. By setting moments about O equal to zero, we would obtain:

$$m(\ell/2)^2 \Omega^2 \sin\theta\cos\theta - mg(\ell/2)\sin\theta = 0 \tag{8.12.18}$$

or

$$\cos\theta = 2g/\ell\Omega^2 \quad \text{or} \quad \theta = \cos^{-1}(2g/\ell\Omega^2) \tag{8.12.19}$$

The incorrectness of this result is seen by comparison with Eq. (8.12.16).

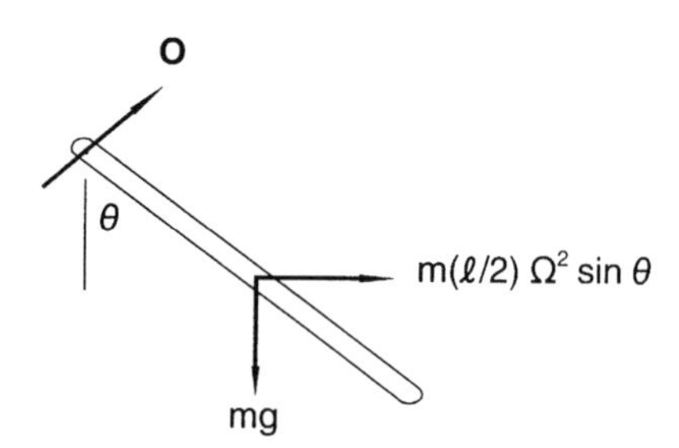

FIGURE 8.12.4
An incorrect free-body diagram of the rotating rod.

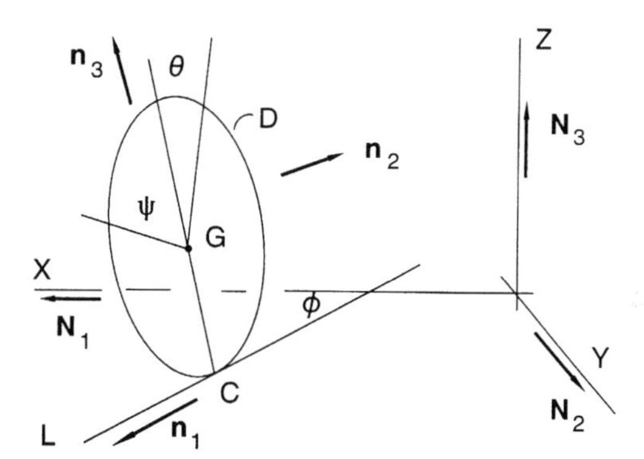

FIGURE 8.13.1
A circular disk rolling on a horizontal surface *S*.

8.13 The Rolling Circular Disk

For another illustration of a body with nonplanar motion, consider again the rolling circular disk (or rolling coin) as discussed in Section 4.12. That is, let *D* be a circular disk with mass *m* and radius *r* rolling on a "perfectly rough" flat horizontal surface *S* as in Figure 8.13.1. Let the orientation of *D* be defined by the roll, lean, and turning angles ψ, θ, and ϕ, as shown. Let *G* be the mass center of *D*, and let *C* be the contact point of *D* with *S*.

When we studied the kinematics of *D* in Section 4.12, we discovered that the requirement that *D* rolls on *S* led to the following expressions for the velocity and acceleration of *G* and the angular velocity and angular acceleration of *D* in an inertia frame *R* (in which *S* is fixed) (see Eqs. (4.12.2), (4.12.3), (4.12.7), and (4.12.8)):

$$\mathbf{v} = r\left(\dot{\psi} + \dot{\phi}\sin\theta\right)\mathbf{n}_1 - r\dot{\theta}\mathbf{n}_2 \tag{8.13.1}$$

$$\begin{aligned}\mathbf{a} &= r\left(\ddot{\psi} + \ddot{\phi}\sin\theta + 2\dot{\phi}\dot{\theta}\cos\theta\right)\mathbf{n}_1 + r\left(-\ddot{\theta} + \dot{\psi}\dot{\phi}\cos\theta + \dot{\phi}^2\sin\theta\cos\theta\right)\mathbf{n}_2 \\ &\quad + r\left(-\dot{\psi}\dot{\phi}\sin\theta - \dot{\phi}^2\sin^2\theta - \dot{\theta}^2\right)\mathbf{n}_3\end{aligned} \tag{8.13.2}$$

$$\boldsymbol{\omega} = \dot{\theta}\mathbf{n}_1 + \left(\dot{\psi} + \dot{\phi}\sin\theta\right)\mathbf{n}_2 + \dot{\phi}\cos\theta\mathbf{n}_3 \tag{8.13.3}$$

$$\begin{aligned}\boldsymbol{\alpha} = \left(\ddot{\theta} - \dot{\psi}\dot{\phi}\cos\theta\right)\mathbf{n}_1 + \left(\ddot{\psi} + \dot{\phi}\sin\theta + \dot{\theta}\dot{\phi}\cos\theta\right)\mathbf{n}_2 \\ + \left(\ddot{\phi}\cos\theta - \dot{\phi}\dot{\theta}\sin\theta + \dot{\psi}\dot{\theta}\right)\mathbf{n}_3\end{aligned} \tag{8.13.4}$$

where $\mathbf{n}_1$, $\mathbf{n}_2$, and $\mathbf{n}_3$ are the unit vectors shown in Figure 8.13.1.

As in the foregoing example, we can obtain the governing equations of motion of D using a free-body diagram as in Figure 8.13.2. As before, $\mathbf{F}^*$ and $\mathbf{T}^*$ represent the force and torque of the couple of an equivalent inertia force system, $\mathbf{w}$ is the weight force, and $\mathbf{C}$ is the contact force exerted by S on D.

By setting moments of the force system about C equal to zero, we have:

$$r\mathbf{n}_3 \times \mathbf{w} + r\mathbf{n}_3 \times \mathbf{F}^* + \mathbf{T}^* = 0 \tag{8.13.5}$$

The forces $\mathbf{w}$ and $\mathbf{F}$ may be expressed as:

$$\mathbf{w} = -mg\mathbf{N}_3 = -mg\left(\sin\theta\mathbf{n}_2 + \cos\theta\mathbf{n}_3\right) \tag{8.13.6}$$

and

$$\mathbf{F}^* = -m\mathbf{a} = -m\left(a_1\mathbf{n}_1 + a_2\mathbf{n}_2 + a_3\mathbf{n}_3\right) \tag{8.13.7}$$

where $\mathbf{N}_3$ is the vertical unit vector and where a_1, a_2, and a_3 are the $\mathbf{n}_1$, $\mathbf{n}_2$, and $\mathbf{n}_3$ components, respectively, of $\mathbf{a}$ in Eq. (8.13.2). Similarly $\mathbf{T}^*$ may be written as:

$$\mathbf{T}^* = T_1\mathbf{n}_1 + T_2\mathbf{n}_2 + T_3\mathbf{n}_3 \tag{8.13.8}$$

where as before T_1, T_2, and T_3 are given by:

$$T_1 = -\alpha_1 I_{11} + \omega_2\omega_3\left(I_{22} - I_{33}\right) \tag{8.13.9}$$

$$T_2 = -\alpha_2 I_{22} + \omega_3\omega_1\left(I_{33} - I_{11}\right) \tag{8.13.10}$$

$$T_3 = -\alpha_3 I_{33} + \omega_1\omega_2\left(I_{11} - I_{22}\right) \tag{8.13.11}$$

where ω_i and α_i ($i = 1, 2, 3$) are the $\mathbf{n}_i$ components of $\boldsymbol{\omega}$ and $\boldsymbol{\alpha}$ in Eqs. (8.13.3) and (8.13.4).

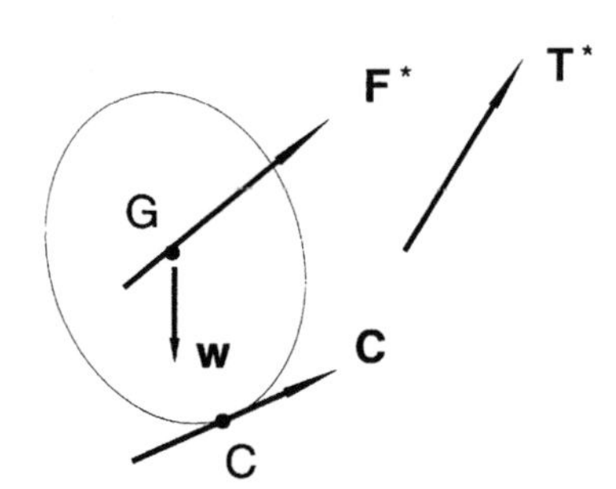

FIGURE 8.13.2
A free-body diagram of D.

For a uniform, thin circular disk, the moments of inertia are:

$$I_{11} = I_{33} = mr^2/4, \quad I_{22} = mr^2/2 \tag{8.13.12}$$

By substituting from Eqs. (8.13.6), (8.13.7), (8.13.8), and (8.13.12) into (8.13.5), we obtain:

$$mgr\sin\theta + mra_2 + T_1 = 0 \tag{8.13.13}$$

$$-mra_1 + T_2 = 0 \tag{8.13.14}$$

$$T_3 = 0 \tag{8.13.15}$$

By substituting from Eqs. (8.13.2) to (8.13.4) and (8.13.9) to (8.13.11), Eqs. (8.13.13), (8.13.14), and (8.13.15) become (after simplification):

$$(4g/r)\sin\theta - 5\ddot{\theta} + 6\dot{\psi}\dot{\phi}\cos\theta + 5\dot{\phi}^2\sin\theta\cos\theta = 0 \tag{8.13.16}$$

$$3\ddot{\psi} + 3\ddot{\phi}\sin\theta + 5\dot{\phi}\dot{\theta}\cos\theta = 0 \tag{8.13.17}$$

$$\ddot{\phi}\cos\theta + 2\dot{\psi}\dot{\theta} = 0 \tag{8.13.18}$$

Equations (8.13.16), (8.13.17), and (8.13.18) form a set of coupled nonlinear ordinary differential equations. As such we cannot obtain the general solution in closed form; however, we can obtain solutions for a few special cases. For example, if D is at rest on its rim (static, vertical equilibrium), we have (see Figure 8.13.1):

$$\theta = 0, \quad \phi = \text{constant}, \quad \psi = \text{constant} \tag{8.13.19}$$

By inspection, we see that these values of θ, ϕ, and ψ solve Eqs. (8.15.16), (8.13.17), and (8.13.18).

Next, if D is rolling in a straight line with constant speed, we have:

$$\theta = 0, \quad \phi = \text{constant}, \quad \dot{\psi} = \text{constant} \tag{8.13.20}$$

Again, by inspection, these values of θ, ϕ, and ψ are seen to be solutions of Eqs. (8.13.16), (8.13.17), and (8.13.18).

As an extension of this case, consider a disk rolling at a uniform rate on a circle. In this case, we have:

$$\theta = \theta_O, \quad \dot{\phi} = \dot{\phi}_O, \quad \dot{\psi} = \dot{\psi}_O \tag{8.13.21}$$

where θ_O, $\dot{\phi}_O$, and ψ_O are constants. Then, by inspection we see that Eqs. (8.13.17) and (8.13.18) are identically satisfied. Equation (8.13.16) becomes:

$$(4g/r)\sin\theta_O + 6\dot{\psi}_O\dot{\phi}_O\cos\theta_O + 5\dot{\phi}_O^2\sin\theta_O\cos\theta_O = 0 \tag{8.13.22}$$

Thus, if we are given two of the θ_O, $\dot{\phi}_O$, and ψ_O, we can determine the third parameter.

Finally, if D is spinning (or pivoting) at a constant rate in a vertical position, we have:

$$\theta = 0, \quad \dot{\phi} = \text{constant}, \quad \psi = \text{constant} \tag{8.13.23}$$

These values of θ_O, $\dot{\phi}_O$, and ψ_O are also seen to be solutions of Eqs. (8.13.16), (8.13.17), and (8.13.18).

In Chapter 13, we will investigate the stability of these special case solutions.

8.14 Closure

Newton's laws, together with their modification to form d'Alembert's principle, are undoubtedly the most widely used of all dynamics principles — particularly in the analysis of elementary problems. The reason for this widespread use and popularity is that Newton's laws and d'Alembert's principle are always applicable and, in principle, they always produce the governing equations of motion. For many mechanical systems, however, Newton's laws and d'Alembert's principle are not especially convenient. Their successful and efficient application often requires insight not available to many analysts. In the following chapters, we will consider other dynamics principles which, while being more specialized, are more efficient and easier to use.

References

8.1. Levinson, D. A., paper presented at the Seminar on Dynamics, University of Cincinnati, 1990.
8.2. Newton, I., *Philosophiae Naturalis Principia Mathematics*, First Ed., London, 1687.
8.3. Temple, R., *The Genius of China*, Simon & Schuster, New York, 1987, p. 161.
8.4. d'Alembert, J. L., *Traite' de Dynamique*, Paris, 1743.
8.5. Hamilton, W. R., Second essay on a general method of dynamics, *Philosophical Transactions of the Royal Society of London*, 1835.
8.6. Gibbs, J. W., On the fundamental formulae of dynamics, *Am. J. Math.*, Vol. II, pp. 49–64, 1879.
8.7. Appell, P., Sur une Forme Generale Eqs. de Dynamique, *Journal fur die Reine und Angewandte Mathematic*, Vol. 121, pp. 310–319, 1900.
8.8. Whittaker, E. T., *A Treatise on the Analytical Dynamics of Particles and Rigid Bodies*, Cambridge University Press, London, 1937.
8.9. Brand, L., *Vectorial Mechanics*, Wiley, New York, 1947.
8.10. Hamel, G., *Theoretische Mechanik*, Springer-Verlag, Berlin, 1949.
8.11. Halfman, R. L., *Dynamics*, Addison-Wesley, Reading, MA, 1959.
8.12. Housner, G. W., and Hudson, D. E., *Applied Mechanics: Dynamics*, D. van Nostrand, Princeton, NJ, 1959.
8.13. McCaskey, S. W., *An Introduction to Advanced Dynamics*, Addison-Wesley, Reading, MA, 1959.
8.14. Yeh, H., and Abrams, J. I., *Principles of Mechanics of Solids and Fluids*, Vol. 1, McGraw-Hill, New York, 1960.
8.15. Kane, T. R., *Analytical Elements of Mechanics*, Vol. 1, Academic Press, New York, 1961.
8.16. Kane, T. R., Dynamics of nonholonomic systems, *J. Appl. Mech.*, 28, 574–578, 1961.
8.17. Greenwood, D. T., *Principles of Dynamics*, Prentice Hall, Englewood Cliffs, NJ, 1965.
8.18. Kane, T. R., *Dynamics*, Holt, Rinehart & Winston, New York, 1968.
8.19. Meirovitch, L., *Methods of Analytical Dynamics*, McGraw-Hill, New York, 1970.
8.20. Tuma, J. J., *Dynamics*, Quantum Publishers, New York, 1974.

8.21. Pars, L. A., *A Treatise on Analytical Dynamics*, Ox Bow Press, Wood Bridge, CT, 1979.
8.22. Goldstein, H., *Classical Mechanics*, Addison-Wesley, Reading, MA, 1980.
8.23. Kane, T. R., and Levinson, D. A., Formulations of equations of motion for complex spacecraft, *J. Guidance Control*, 3(2), 99–112, 1980.
8.24. Beer, F. P., and Johnston, Jr., E. R., *Vector Mechanics for Engineers*, McGraw-Hill, New York, 1984.
8.25. D'Souza, A. F., and Garg, V. K., *Advanced Dynamics Modeling and Analysis*, Prentice Hall, Englewood Cliffs, NJ, 1984.
8.26. Torby, B. J., *Advanced Dynamics for Engineers*, Holt, Rinehart & Winston, New York, 1984.
8.27. T.R. Kane, T. R., and Levinson, D. A., *Dynamics: Theory and Applications*, McGraw-Hill, New York, 1985.
8.28. Meriam, J. L., and Kraige, L. G., *Engineering Mechanics*, Wiley, New York, 1986.
8.29. Marion, J. B., and Thornton, S. T., *Classical Dynamics of Particles and Systems*, Harcourt, Brace & Jovanovich, San Diego, CA, 1988.
8.30. Huston, R. L., *Multibody Dynamics*, Butterworth-Heinemann, Boston, MA, 1990.
8.31. Haug, E. J., *Intermediate Dynamics*, Prentice Hall, Englewood Cliffs, NJ, 1992.
8.32. Desloge, E. A., *Classical Mechanics*, Wiley, New York.

Problems

Section 8.4 The Simple Pendulum

P8.4.1: A simple pendulum with length ℓ of 3 ft has a speed V of 5 ft/sec when the pendulum bob P is in its lowest position as shown in Figure P8.4.1. Find the tension T in the support cable.

FIGURE P8.4.1
A simple pendulum.

FIGURE P8.4.3
An inclined simple pendulum.

P8.4.2: Repeat Problem P8.4.1 if the pendulum length ℓ is 2 m and the speed V of the bob P is 4 m/sec.

P8.4.3: See Problem P8.4.1. Find the tension T in the support cable if the bob P has a speed of 2 ft/sec when the pendulum inclination angle θ is 30°, as in Figure P8.4.3.

P8.4.4: See Problem P8.4.3. Find the horizontal and vertical components of the support reactions O_x and O_y in the configuration of Figure P8.4.3.

P8.4.5: Consider Eq. (8.4.4) governing the motion of a simple pendulum:

$$\ddot{\theta} = (g/\ell)\sin\theta = 0$$

Show that this equation can be integrated by multiplying by $\dot{\theta}$; that is,

$$\dot{\theta}\ddot{\theta} + (g/\ell)\dot{\theta}\sin\theta = 0$$

leads to:

$$\frac{d}{dt}\left(\dot{\theta}^2/2\right) + \frac{d}{dt}(-\cos\theta) = 0$$

so that

$$\dot{\theta}^2/2 - \cos\theta = C$$

where C is a constant.

P8.4.6: See Problem P8.4.5. Suppose a simple pendulum of length 3 ft is displaced through an angle of 60° and released from rest. Find the speed of the bob when θ is zero, the lowest position.

P8.4.7: Consider the small-amplitude oscillations of a simple pendulum (see Eq. (8.4.5)). Show that the general solution of the governing equation may be written in the forms:

$$\theta = A\cos\omega t + B\sin\omega t = C\cos(\omega t + \phi)$$

where ω is $\sqrt{g/\ell}$ and A, B, C, and ϕ are constants (where C is the amplitude; ϕ, the phase angle; and ω, the circular frequency).

P8.4.8 See Problem P8.4.7. Express the amplitude C and phase in terms of the constants A and B.

Section 8.5 A Smooth Particle Moving Inside a Vertical Rotating Tube

P8.5.1: A 6-oz smooth particle P moves inside a 2.5-ft-radius tube, which is rotating about a vertical diameter at a uniform rate Ω of 3 ft/sec as depicted in Figure P8.5.1. Determine the $\mathbf{n}_1$, $\mathbf{n}_2$, and $\mathbf{n}_3$ components of the inertia force $\mathbf{F}^*$ on P for an instant when the angle θ locating P is 60° and P is moving at a uniform speed v of 3 ft/sec clockwise relative to the tube.

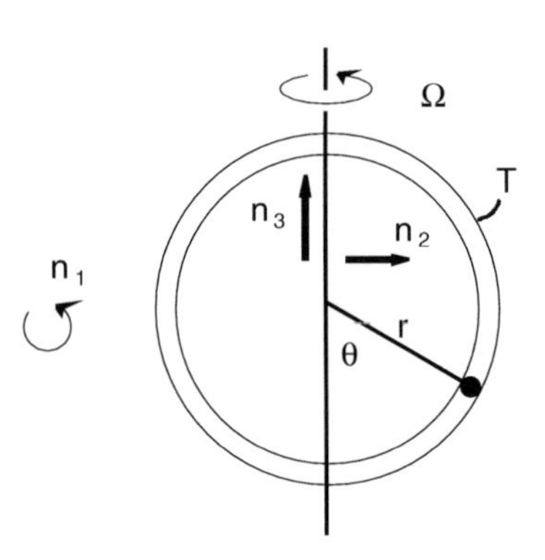

FIGURE P8.5.1
A smooth particle P inside a rotating tube T.

P8.5.2: Repeat Problem P8.5.1 for a tube radius r of 0.5 m, a particle mass m of 0.1 kg, and a speed v relative to the tube of 1 m/sec.

P8.5.3: Repeat Problems P8.5.1 and P8.5.2 if the rotation rate Ω of the tube is increasing at 7 rad/sec^2 and the speed v of P relative to the tube is increasing at 10 ft/sec^2 for P8.5.1 and 3 m/sec^2 for P8.5.2.

Section 8.7 Projectile Motion

P8.7.1: An object P is thrown vertically from the ground with a speed v_0. Using Eq. (8.7.11) show that the maximum height h reached by P is $v_0^2/2g$.

P8.7.2: See Problem P8.7.1. An object P is dropped from rest at a height h above the ground. Show that the speed v at which P strikes the ground is $\sqrt{2gh}$.

P8.7.3: See Problem P8.7.1. An object P is thrown vertically from the ground with a speed of 15 ft/sec. Find the maximum height h reached by P.

P8.7.4: See Problem P8.7.2. An object P is dropped from rest from a height of 30 m. Find the speed of impact of P with the ground.

P8.7.5: A projectile is launched from a horizontal surface at a speed v of 50 ft/sec at an angle θ of 30° relative to the horizontal as in Figure P8.7.5. Determine the range d where the projectile impacts the horizontal surface and the maximum height h reached by the projectile.

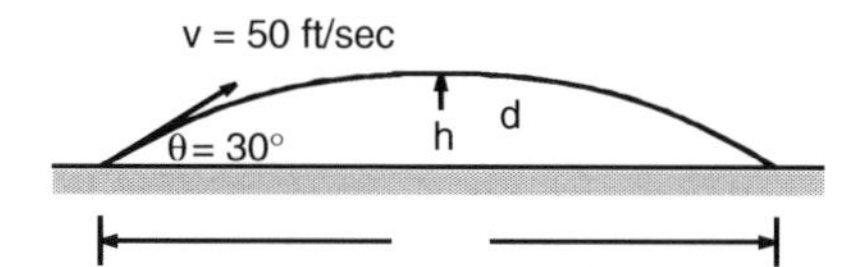

FIGURE P8.7.5
A projectile launched on a horizontal plane.

P8.7.6: See Problem P8.7.5. Repeat Problem P8.7.5 for launch angles θ of (a) 35°, (b) 40°, (c) 45°, (d) 50°, (e) 55°, and (f) 60°. Tabulate and graph the results for d and h as a function of θ. Verify that the range d is a maximum for $\theta = 45°$.

P8.7.7: See Problem P8.7.5. Repeat Problem P8.7.5 if the projectile is launched from a point 8 ft above the horizontal surface as in Figure P8.7.7.

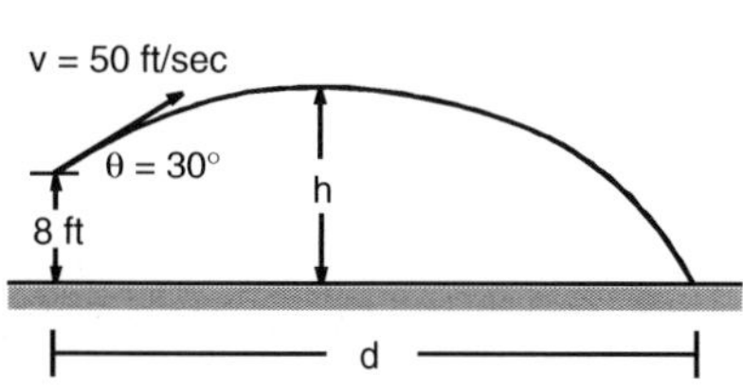

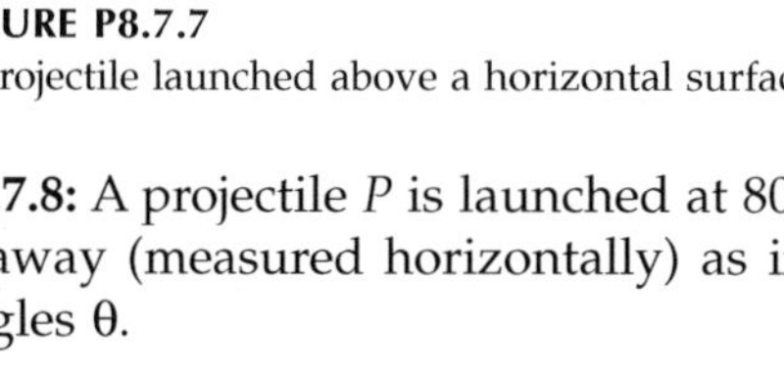

FIGURE P8.7.7
A projectile launched above a horizontal surface.

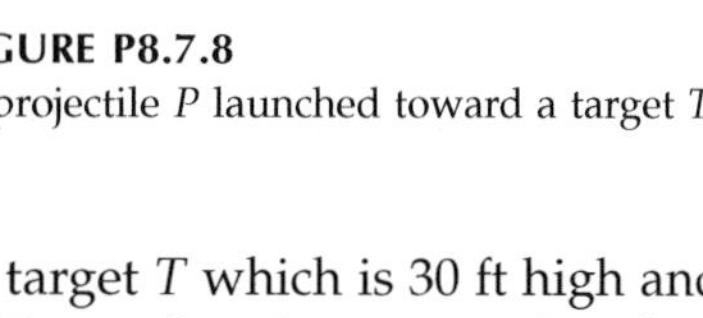

FIGURE P8.7.8
A projectile P launched toward a target T.

P8.7.8: A projectile P is launched at 80 ft/sec toward a target T which is 30 ft high and 40 ft away (measured horizontally) as in Figure P8.7.8. Determine the appropriate launch angles θ.

P8.7.9: A basketball player wants to shoot a free throw with the minimum possible speed (thus presumably using the minimum energy and therefore having the greatest control). If the foul line is 13.75 ft away from the center of the basket (measured horizontally), if the basket is 10 ft above the floor, and if the player launches the ball 7.5 ft above the floor as in Figure P8.7.9, what is the optimal angle θ and the minimum speed v?

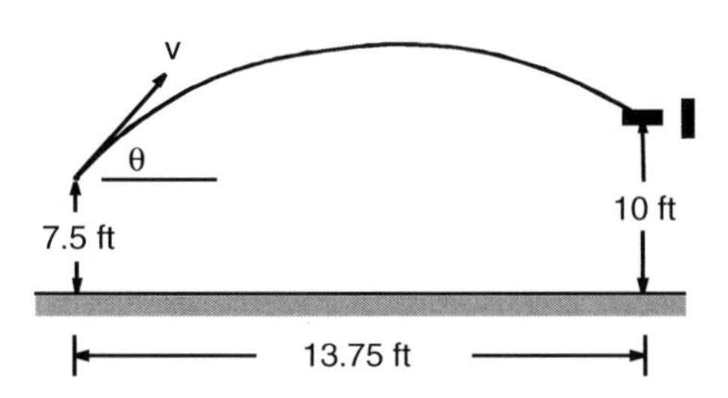

FIGURE P8.7.9
A basketball free throw.

P8.7.10: During the kickoff in a football game, the ball is kicked end-over-end, rotating 4 revolutions per second. If the ball is kicked 60 yards at a 45° launch angle, determine the maximum and minimum horizontal components of the velocity of the ends of the ball. (The ball length is 10 inches.)

Section 8.8 A Rotating Circular Disk

P8.8.1: A thin, circular disk/pulley weighing 35 lb and with radius 15 in. supports (a) a 10-lb force, and (b) a 10-lb weight, as in Figure P8.8.1. In each case, determine the angular acceleration of the disk and the cable tension.

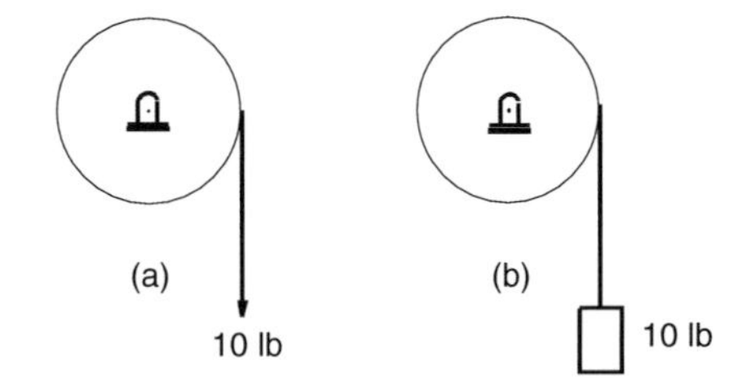

FIGURE P8.8.1
A disk supporting (a) a force, and (b) a weight.

P8.8.2: A differential pulley has an inner radius of 8 cm and an outer radius of 16 cm. Cables wrapped around the rim and the inner hub support masses of 5 kg and 3 kg, respectively, as represented in Figure P8.8.2. Let the pulley be modeled as a thin, circular disk with mass of 4 kg. If the system is released from rest in the configuration shown, determine the angular acceleration of the disk and the tensions in each cable.

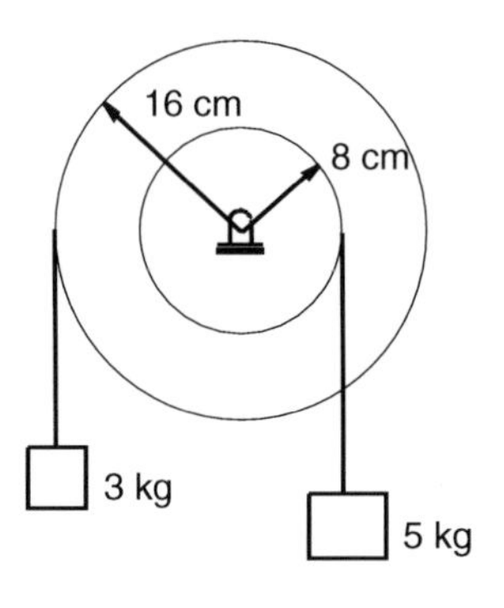

FIGURE P8.8.2
Differential pulley supporting different masses.

P8.8.3: See Problem P8.8.2. Repeat Problem P8.8.2 using pulley radii of 5 in. and 10 in., outer mass weight of 6 lb and inner mass weight of 7 lb, and a pulley weight of 4 lb.

P8.8.4: A 4-kg circular disk D with 15-cm radius is mismounted on a shaft so that its center O is 5 mm away from the shaft axis A, as in Figure P8.8.4. If the shaft rotates at 300 rpm, determine the magnitude of the resultant inertia force on D.

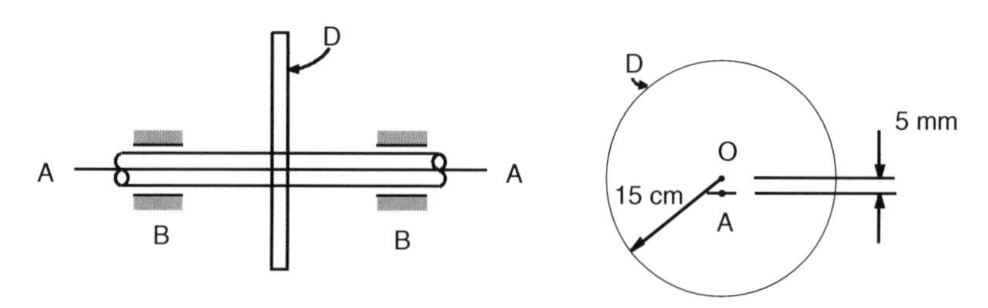

FIGURE P8.8.4
A mismounted circular disk.

P8.8.5: A thin, circular disk D with mass m and radius r is mismounted on a shaft S so that its axis makes a small angle θ with the shaft axis as in Figure P8.8.5. Using Eqs. (8.6.10), (8.6.11), and (8.6.12) determine the $\mathbf{n}_i$ and $\mathbf{N}_i$ ($i = 1, 2, 3$) components of the inertia torque exerted on D if S rotates at an angular speed Ω. ($\mathbf{N}_3$ is parallel to the axis of S and $\mathbf{n}_3$ is parallel to the axis of D.)

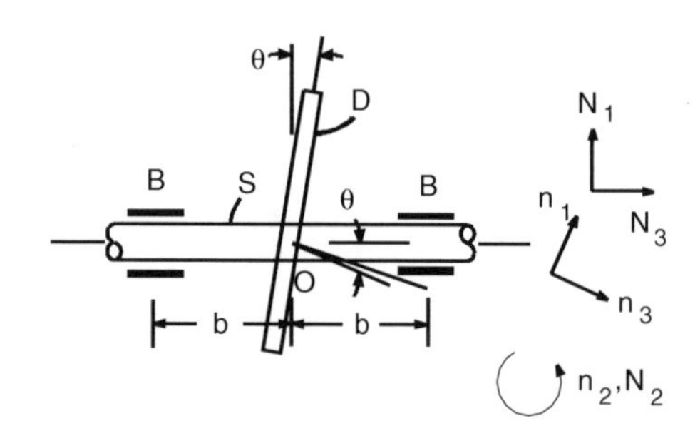

FIGURE P8.8.5
A mismounted circular disk.

P8.8.6: See Problem P8.8.5. Evaluate the torque components for the following data: $m = 4$ kg, $r = 15$ cm, $\theta = 5°$, and $\Omega = 300$ rpm. Also, if bearings B are supporting S on both sides of D as in Figure P8.8.5, find the forces on the bearings due to the misalignment angle if the distance b of the bearings from the center O of D is 20 in.

Section 8.9 The Rod Pendulum

P8.9.1: A 1-m-long rod B with mass 1 kg is supported at one end O by a frictionless pin and held in a horizontal position as in Figure P8.9.1. If the rod is suddenly released, find the angular acceleration α of the rod and the horizontal and vertical components of the reaction force (O_x and O_y) at O at the instant of release.

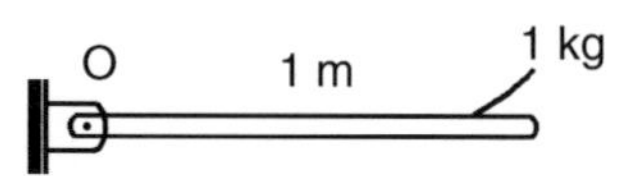

FIGURE P8.9.1
A rod held in a horizontal position before release from rest.

P8.9.2: See Problem P8.9.1. Repeat Problem P8.9.1 if B is held at a 30° inclination just before it is released, as represented in Figure P8.9.1.

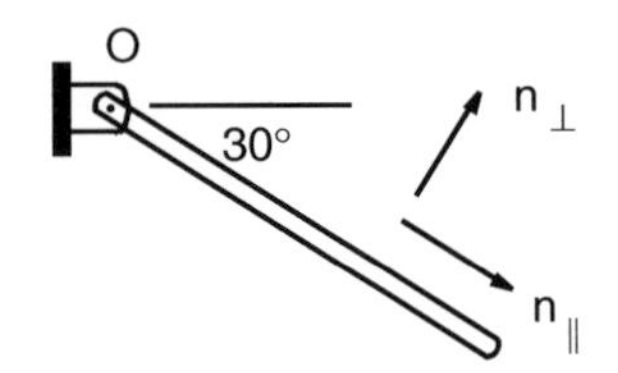

FIGURE P8.9.2
A rod held at 30° before release from rest.

P8.9.3: See Problem P8.9.2. Let $\mathbf{n}_{||}$ and $\mathbf{n}_{\perp}$ be unit vectors parallel and perpendicular to B as in Figure P8.9.2. Find the $\mathbf{n}_{||}$ and $\mathbf{n}_{\perp}$ components of the reaction force at O. Specifically, observe that $O_{||}$ is not zero. Compare these results with similar results for a simple pendulum.

P8.9.4: Consider a pendulum consisting of a rod B with length ℓ and mass m and a thin, circular disk D with radius r and mass m, as in Figure P8.9.4. Let B and D be rigidly attached to each other. Find the governing equation of motion, analogous to Eq. (8.9.10).

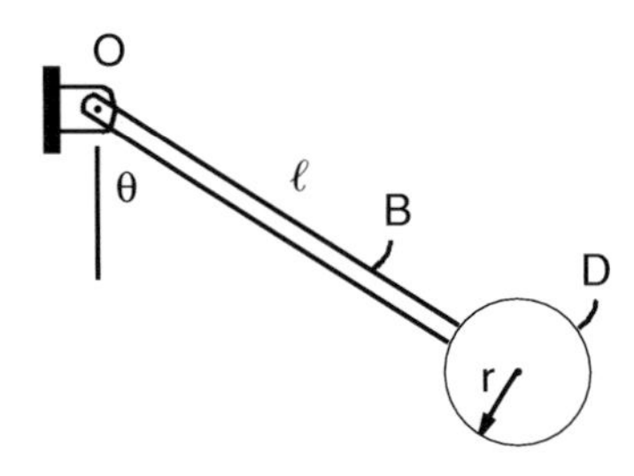

FIGURE P8.9.4
A rod/disk pendulum.

Section 8.10 Double-Rod Pendulum

P8.10.1: Consider a double-rod pendulum as in Section 8.10, with a concentrated mass m at the lower end of the second bar. Develop governing equations analogous to Eqs. (8.10.12) and (8.10.13).

Section 8.12 A Rotating Pinned Rod

P 8.12.1: Consider a rotating shaft S with a rod B affixed (that is, welded or otherwise bonded) to the surface of S with B inclined at an angle θ to the axis of S as in Figure P8.12.1. Let $\mathbf{n}_1$, $\mathbf{n}_2$, and $\mathbf{n}_3$ be mutually perpendicular unit vectors parallel to principal axes of B such that $\mathbf{n}_1$ is along B, $\mathbf{n}_3$ is perpendicular to B and parallel to a radial line of S which passes through the attachment O of B to S, and $\mathbf{n}_2$ is perpendicular to $\mathbf{n}_1$ and $\mathbf{n}_3$. Finally,

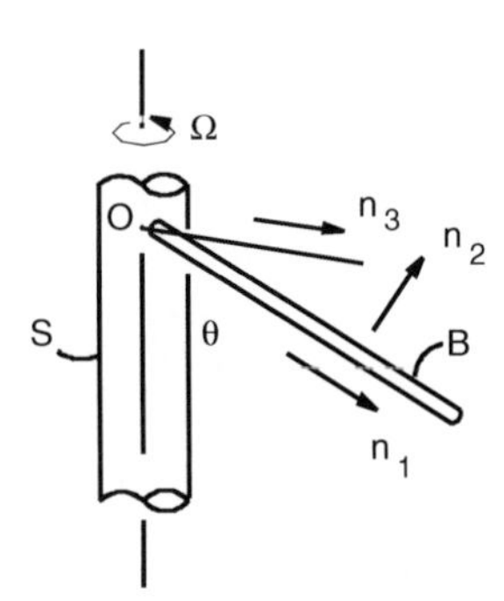

FIGURE P8.12.1
A rotating shaft S and a rod B bonded to the surface of S.

let S have radius r and angular speed Ω (a constant), and let B have length ℓ and mass m. Let the inertia forces on B be replaced by a single force $\mathbf{F}^*$ passing through the mass center G of B together with a couple with torque $\mathbf{T}^*$. Determine the $\mathbf{n}_1$, $\mathbf{n}_2$, and $\mathbf{n}_3$ components of $\mathbf{F}^*$ and $\mathbf{T}^*$.

P8.12.2: See Problem P8.12.1. Determine the moments of all forces on B (including gravity and inertia forces) about the attachment point O.

P8.12.3: See Problems P8.12.1 and P8.12.2. Repeat Problems P8.12.1 and P8.12.2 for angular speed Ω that is not constant but instead has a derivative $\dot{\Omega}$.

P8.12.4: Solve Problems P8.12.1 and P8.12.2 for the following data: $\Omega = 300$ rpm, $\ell = 3$ ft, $\theta = 30°$, $r = 2$ in., and $m = 0.5$ slug.

P8.12.5: Solve Problems P8.12.1 and P8.12.2 for the following data: $\Omega = 300$ rpm, $\ell = 1$ m, $\theta = 30°$, $r = 5$ cm, and $m = 8$ kg.

Section 8.13 The Rolling Circular Disk

P8.13.1: Consider the case of a circular disk rolling on a circle at a constant speed. Find an expression for the radius ρ of the circle on which the center G moves. Express ρ in terms of r, θ_O, $\dot{\phi}_O$, ψ_O.

9

Principles of Impulse and Momentum

9.1 Introduction

Impact is a common phenomenon in machine dynamics. Collisions occur repeatedly between machine elements such as gear teeth, cams and followers, clutch plates, brake pads, chain links, and gripper jaws. The principles of impulse and momentum are ideally suited for the study of such collisions. These collisions produce impact phenomena where large forces occur over a short time interval.

As with d'Alembert's principle, the principles of impulse and momentum may be developed directly from Newton's laws. For impact phenomena, these principles of impulse and momentum are more convenient to use than either Newton's laws or d'Alembert's principle in that accelerations do not need to be computed. In this chapter, we will examine these principles and their applications.

From a strictly theoretical perspective, impact phenomena and the resulting contact stresses and deformations are difficult topics. From a global perspective, however, the overall behavior of colliding particles or bodies is relatively easy to study using the principles of impulse and momentum, and the results agree relatively well with observed phenomena. Therefore, our emphasis will be placed upon understanding these principles and their underlying assumptions from a global perspective.

9.2 Impulse

When bodies collide the impact time is usually very short — one tenth of a second or less. The forces, however, may be quite large. Such forces are conveniently represented by impulses. Specifically, if a force **F** is applied between bodies over a time t^*, the linear impulse of **F** is defined as:

$$\mathbf{I} \overset{D}{=} \int_{O}^{t^*} \mathbf{F}\,dt \tag{9.2.1}$$

Typically, the magnitude of the impact force will have the form shown in Figure 9.2.1. It is often triangular. The magnitude of the impulse is then the average of the impact force magnitude multiplied by the impact time t^*. The direction of the impulse is the same as that of the force. The dimensions of an impulse are thus force–time.

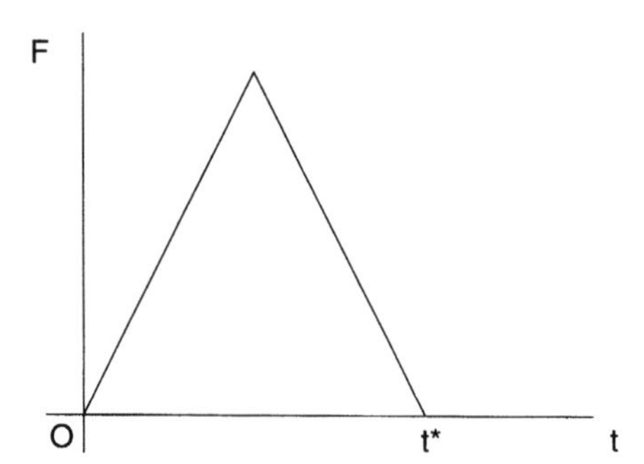

FIGURE 9.2.1
Time profile of the magnitude of a typical impact force.

On occasion, colliding bodies will be rotating so that the impact creates a force system that may be represented by a single force **F** passing through a point Q in the impact force region together with a couple with torque $\mathbf{M}_Q$. The torque can then produce an angular impulse $\mathbf{J}_Q$ which, analogous to Eq. (9.2.1), is defined as:

$$\mathbf{J}_Q \overset{D}{=} \int_O^t \mathbf{M}_Q \, dt \tag{9.2.2}$$

Finally, if an impact occurs during a time interval (t_1, t_2) the impulse system may be represented by the linear impulse **I** passing through a point Q together with an angular impulse $\mathbf{J}_Q$, where **I** and $\mathbf{J}_Q$ are:

$$\mathbf{I} = \int_{t_1}^{t_2} \mathbf{F} \, dt \quad \text{and} \quad \mathbf{J}_Q = \int_{t_1}^{t_2} \mathbf{M}_Q \, dt \tag{9.2.3}$$

where as before, $\mathbf{M}_Q$ is the moment of the impact forces about Q.

9.3 Linear Momentum

Consider a particle P with mass m and velocity **v** in a reference frame R as in Figure 9.3.1. The linear momentum $\mathbf{L}^P$ of P in R is defined as:

$$\mathbf{L}^P \overset{D}{=} m\mathbf{v} \tag{9.3.1}$$

Observe that $\mathbf{L}^P$ has the dimensions of mass–velocity or mass–length per unit time.

Next, consider a set S of N particles P_i $(i = 1,\ldots, N)$ having masses m_i and velocities $\mathbf{v}^{P_i}$ in a reference frame R as in Figure 9.3.2. Then, the linear momentum $\mathbf{L}^S$ of S in R is defined as:

$$\mathbf{L}^S \triangleq \sum_{i=1}^{N} \mathbf{L}^{P_i} = \sum_{i=1}^{N} m_i \mathbf{v}^{P_i} \tag{9.3.2}$$

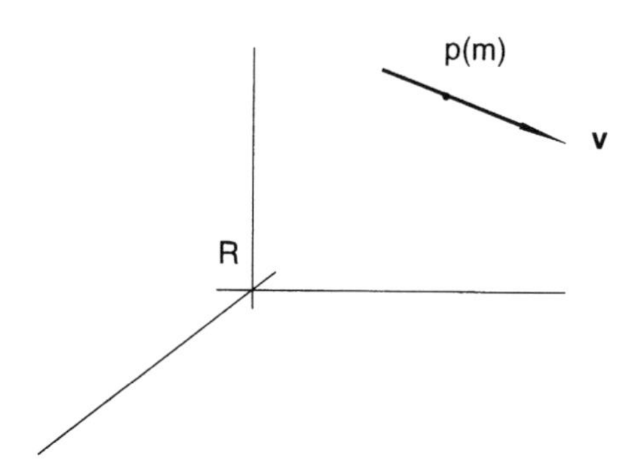

FIGURE 9.3.1
A particle P with mass m and velocity $\mathbf{v}$ in reference frame R.

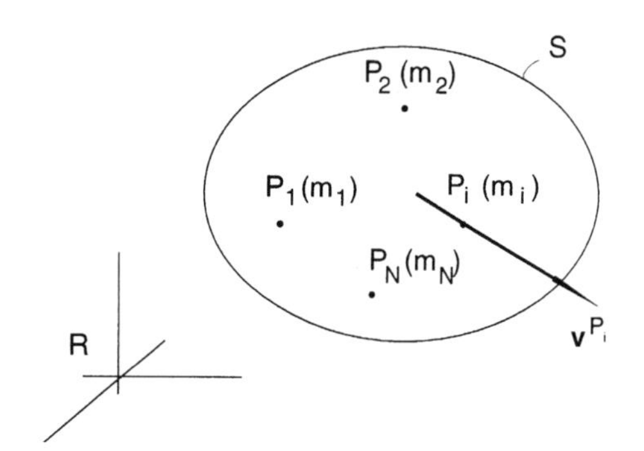

FIGURE 9.3.2
A set S of moving particles.

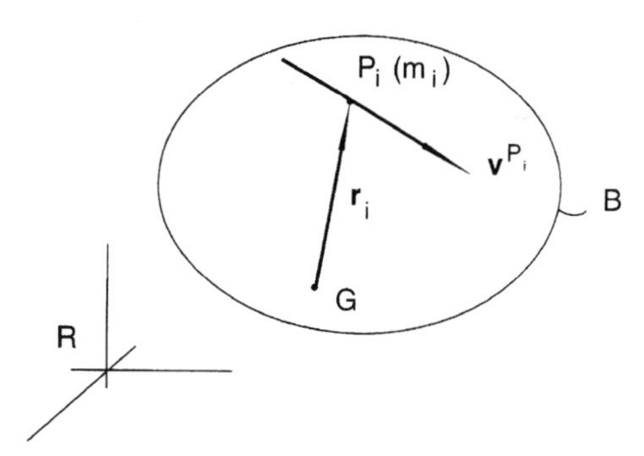

FIGURE 9.3.3
A rigid body B with mass center G moving in a reference frame R.

Finally, consider a rigid body B with mass m and mass center G moving in a reference frame R as in Figure 9.3.3. Consider B to be composed of N particles P_i ($i = 1,\ldots, N$) with masses m_i. Then, from Eq. (9.3.2), the linear momentum $\mathbf{L}^B$ of B in R is:

$$\mathbf{L}^B = \sum_{i=1}^{N} m_i \mathbf{v}^{P_i} \tag{9.3.3}$$

Because P_i and G are both fixed in B, their velocities in R are related by the expression (see Eq. (4.9.4)]:

$$\mathbf{v}^{P_i} - \mathbf{v}^G + \boldsymbol{\omega}^B \times \mathbf{r}_i \tag{9.3.4}$$

where $\boldsymbol{\omega}^B$ is the angular velocity of B in R and where $\mathbf{r}_i$ locates P_i relative to G as in Figure 9.3.3. By substituting from Eq. (9.3.4) into (9.3.3) $\mathbf{L}^B$ becomes:

$$\begin{aligned}\mathbf{L}^B &= \sum_{i=1}^{N} m_i\left(\mathbf{v}^G + \boldsymbol{\omega}^B \times \mathbf{r}_i\right) = \sum_{i=1}^{N} m_i \mathbf{v}^G + \sum_{i=1}^{N} m_i \boldsymbol{\omega}^B \times \mathbf{r}_i \\ &= \left(\sum_{i=1}^{N} m_i\right)\mathbf{v}^G + \boldsymbol{\omega}^B \times \sum_{i=1}^{N} m_i \mathbf{r}_i = M\mathbf{v}^G\end{aligned} \tag{9.3.5}$$

where the final expression is obtained by recognizing M as the total mass of the particles of B and by recalling that G is the mass center of B so that:

$$\sum_{i=1}^{N} m_i \mathbf{r}_i = 0 \tag{9.3.6}$$

(See Section 6.8.)

Equation (9.3.5) shows that for a rigid body the computation of the linear momentum is in essence as simple as the computation of linear momentum for a single particle as in Eq. (9.3.1).

9.4 Angular Momentum

Consider again a particle P having mass m and velocity $\mathbf{v}$ in a reference frame R as in Figure 9.4.1. Let Q be an arbitrary reference point. Then, the angular momentum $\mathbf{A}_Q$ of P about Q in R is defined as:

$$\mathbf{A}_Q \stackrel{D}{=} \mathbf{p} \times m\mathbf{v} \tag{9.4.1}$$

where $\mathbf{p}$ locates P relative to Q. Observe that $\mathbf{A}_Q$ has the units mass–length–velocity, or mass–(length)^2 per unit time.

Angular momentum is sometimes called *moment of momentum*. Indeed, if we recognize $m\mathbf{v}$ as the linear momentum $\mathbf{L}$ of P, we can write Eq. (9.4.1) in the form:

$$\mathbf{A}_Q = \mathbf{p} \times \mathbf{L} \tag{9.4.2}$$

Consider next a set S of N particles P_i ($i = 1,\ldots, N$) having masses m_i and velocities $\mathbf{v}^{P_i}$ in R as in Figure 9.4.2. Again, let Q be an arbitrary reference point. Then, the angular momentum of S about Q in R is defined as:

$$\mathbf{A}_{S/Q} = \sum_{i=1}^{N} \mathbf{p}_i \times m_i \mathbf{v}^{P_i} \tag{9.4.3}$$

where $\mathbf{p}_i$ locates P_i relative to Q.

Finally, consider a rigid body B with mass m and mass center G moving in a reference frame R as in Figure 9.4.3. Consider B to be composed of N particles P_i with masses m_i

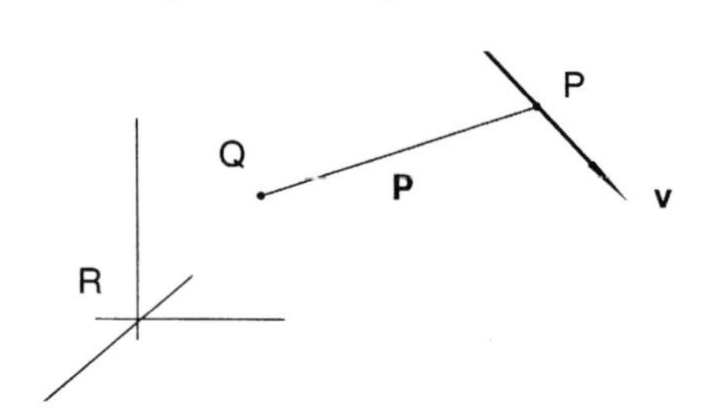

FIGURE 9.4.1
A particle P with mass m, velocity $\mathbf{v}$, and reference point Q.

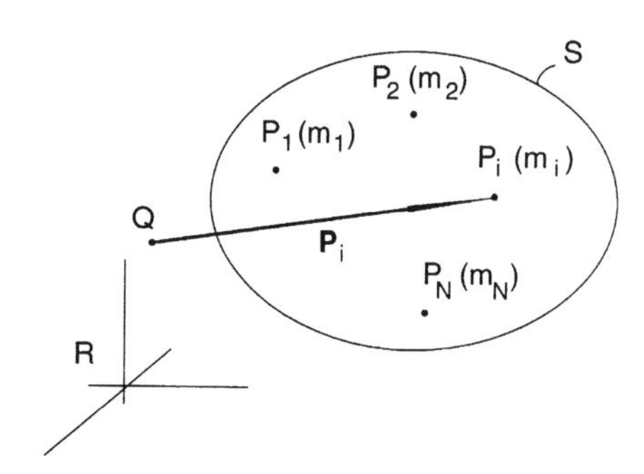

FIGURE 9.4.2
A set S of moving particles.

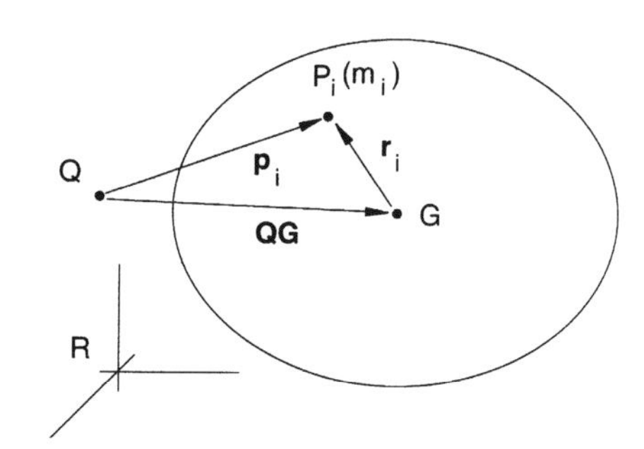

FIGURE 9.4.3
A rigid body B with mass m, mass center G, and reference point Q.

($i = 1,\ldots, N$). Again, let Q be an arbitrary reference point. Then, as in Eq. (9.4.3), the angular momentum of B about Q in R is defined as:

$$\mathbf{A}_{B/Q} = \sum_{i=1}^{N} \mathbf{p}_i \times m_i \mathbf{v}^{P_i} \tag{9.4.4}$$

where $\mathbf{p}_i$ locates P_i relative to Q.

Observe that because P_i and mass center G are both fixed in B their velocities in R are related by the expression (see Eq. (4.9.4)):

$$\mathbf{v}^{P_i} = \mathbf{v}^G + \boldsymbol{\omega} \times \mathbf{r}_i \tag{9.4.5}$$

where $\boldsymbol{\omega}$ is the angular velocity of B in R and where $\mathbf{r}_i$ locates P_i relative to G (see Figure 9.4.3). Observe further from Figure 9.4.3 that $\mathbf{p}_i$ and $\mathbf{r}_i$ are related by the expression:

$$\mathbf{p}_i = \mathbf{QG} + r_i \tag{9.4.6}$$

By substituting from Eqs. (9.4.5) and (9.4.6) into (9.4.4) $\mathbf{A}^{B/Q}$ becomes:

$$\begin{aligned}
\mathbf{A}^{B/Q} &= \sum_{i=1}^{N} m_i(\mathbf{QG} + \mathbf{r}_i) \times (\mathbf{v}^G + \boldsymbol{\omega} \times \mathbf{r}_i) \\
&= \sum_{i=1}^{N} m_i \mathbf{QG} \times \mathbf{v}^G + \sum_{i=1}^{N} m_i \mathbf{QG} \times (\boldsymbol{\omega} \times \mathbf{r}_i) \\
&\quad + \sum_{i=1}^{N} m_i \mathbf{r}_i \times \mathbf{v}^G + \sum_{i=1}^{N} m_i \mathbf{r}_i \times (\boldsymbol{\omega} \times \mathbf{r}_i) \\
&= \left(\sum_{i=1}^{N} m_i\right) \mathbf{QG} \times \mathbf{v}^G + \mathbf{QG} \times \left[\boldsymbol{\omega} \times \left(\sum_{i=1}^{N} m_i \mathbf{r}_i\right)\right] \\
&\quad + \left(\sum_{i=1}^{N} m_i \mathbf{r}_i\right) \times \mathbf{v}^G + \sum_{i=1}^{N} m_i \mathbf{r}_i \times (\boldsymbol{\omega} \times \mathbf{r}_i) \\
&= M\mathbf{QG} \times \mathbf{v}^G + 0 + 0 + \sum_{i=1}^{N} m_i \mathbf{r}_i \times (\boldsymbol{\omega} \times \mathbf{r}_i)
\end{aligned} \tag{9.4.7}$$

where the middle two terms in the last line are zero because G is the mass center of B (see Section 6.8). The first term may be recognized as the moment of the linear momentum of a particle with mass M at G about Q. That is,

$$M\mathbf{QG} \times \mathbf{v}^G = \mathbf{QG} \times M\mathbf{v}^G \overset{D}{=} QG \times \mathbf{L}^G \overset{D}{=} \mathbf{A}_{G/Q} \tag{9.4.8}$$

The last term of Eq. (9.4.7) may be expressed in terms of the central inertia dyadic (see Sections 7.4 and 7.6) as follows: Let $\mathbf{n}_\omega$ be a unit vector with the same direction as $\boldsymbol{\omega}$. Then, $\boldsymbol{\omega}$ may be expressed as:

$$\boldsymbol{\omega} = \omega \mathbf{n}_\omega \tag{9.4.9}$$

where ω is the magnitude of $\boldsymbol{\omega}$. Hence, we have:

$$\sum_{i=1}^{N} m_i \mathbf{r}_i \times (\boldsymbol{\omega} \times \mathbf{r}_i) = \sum_{i=1}^{N} m_i \mathbf{r}_i \times (\omega \mathbf{n}_\omega \times \mathbf{r}_i) = \omega \sum_{i=1}^{N} m_i \mathbf{r}_i \times (\mathbf{n}_\omega \times \mathbf{r}_i) \tag{9.4.10}$$

From Eqs. (7.2.2) and (7.4.11), this last expression in turn may be expressed in terms of the inertia dyadic as:

$$\begin{aligned} \omega \sum_{i=1}^{N} m_i \mathbf{r}_i \times (\mathbf{n}_\omega \times \mathbf{r}_i) &= \omega \mathbf{I}_\omega^{B/G} = \omega \mathbf{I}^{B/G} \cdot \mathbf{n}_\omega \\ &= \mathbf{I}^{B/G} \cdot (\omega \mathbf{n}_\omega) = \mathbf{I}^{B/G} \cdot \boldsymbol{\omega} \end{aligned} \tag{9.4.11}$$

We can obtain yet a different form of the last term of Eq. (9.4.7) by recognizing that the parenthetical term $(\boldsymbol{\omega} \times \mathbf{r}_i)$ may be identified as $\mathbf{v}^{P_i/G}$ or as $\mathbf{v}^{P_i} - \mathbf{v}^G$ (see Eq. (4.9.4)). Hence, we have:

$$\begin{aligned} \sum_{i=1}^{N} m_i \mathbf{r}_i \times (\boldsymbol{\omega} \times \mathbf{r}_i) &= \sum_{i=1}^{N} m_i \mathbf{r}_i \times (\mathbf{v}^{P_i} - \mathbf{v}^G) \\ &= \sum_{i=1}^{N} m_i \mathbf{r}_i \times \mathbf{v}^{P_i} - \sum_{i=1}^{N} m_i \mathbf{r}_i \times \mathbf{v}^G \\ &= \mathbf{A}_{B/G} - 0 \end{aligned} \tag{9.4.12}$$

where the last equality is obtained from Eq. (9.4.4) and by recalling that G is the mass center of B. Then, by comparing the results in Eqs. (9.4.11) and (9.4.12), we have:

$$\sum_{i=1}^{N} m_i \mathbf{r}_i \times (\boldsymbol{\omega} \times \mathbf{r}_i) = \mathbf{I}^{B/G} \cdot \boldsymbol{\omega} = \mathbf{A}_{B/G} \tag{9.4.13}$$

Finally, by substituting from Eqs. (9.4.8) and (9.4.13) into Eq. (9.4.7) we obtain the addition theorem for angular momentum for a rigid body:

$$\mathbf{A}_{B/Q} = \mathbf{A}_{G/Q} + \mathbf{A}_{B/G} \tag{9.4.14}$$

9.5 Principle of Linear Impulse and Momentum

Consider again a particle P with mass m moving in an inertial reference frame R as in Figure 9.5.1. Let P be acted upon by a force **F** as shown. Then, by Newton's law (see Eq. (8.3.1)), **F** is related to the acceleration **a** of P in R by the expression:

$$\mathbf{F} = m\mathbf{a} \tag{9.5.1}$$

Recalling that the acceleration is the derivative of the velocity we can use the definition of linear impulse of Eqs. (9.2.1) and (9.2.3) to integrate Eq. (9.5.1). That is,

$$\begin{aligned}\mathbf{I} &= \int_{t_1}^{t_2} \mathbf{F}dt = \int_{t_1}^{t_2} m\mathbf{a}dt = \int_{t_1}^{t_2} m(d\mathbf{v}/dt)dt \\ &= m\mathbf{v}(t_2) - m\mathbf{v}(t_1) = \mathbf{L}(t_2) - \mathbf{L}(t_1)\end{aligned} \tag{9.5.2}$$

or

$$\mathbf{I} = \Delta\mathbf{L} \tag{9.5.3}$$

where **I** represents the impulse applied between t_1 and t_2.

Equation (9.5.3) states that the linear impulse is equal to the change in linear momentum. This verbal statement is often called the *principle of linear momentum.*

This principle is readily extended to systems of particles and to rigid bodies. Consider first the system S of N particles P_i with masses m_i ($i = 1,\ldots, N$) and moving in an inertial

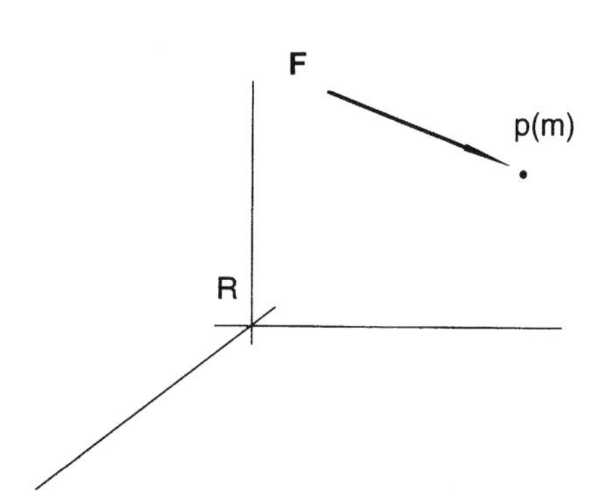

FIGURE 9.5.1
A particle P with mass m moving in an inertial reference frame.

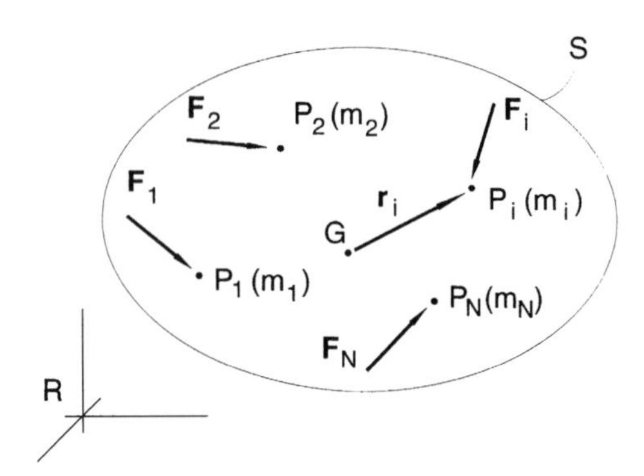

FIGURE 9.5.2
A set of particles moving in an inertial reference frame R.

frame R as in Figure 9.5.2. Let the particles be acted upon by forces $\mathbf{F}_i$ ($i = 1,\ldots, N$) as shown. Then, from Newton's law, we have for each particle:

$$\mathbf{F}_i = m_i \mathbf{a}_i \quad (i = 1,\ldots,N) \; (\text{no sum}) \tag{9.5.4}$$

where $\mathbf{a}_i$ is the acceleration of P_i in R.

Let G be the mass center of S. Then,

$$\mathbf{a}_i = \mathbf{a}_G + d^2 \mathbf{r}_i / dt^2 \tag{9.5.5}$$

where $\mathbf{r}_i$ locates P_i relative to G as in Figure 9.5.2.

Let the system of forces $\mathbf{F}_i$ be represented by an equivalent force system (see Section 6.5) consisting of a single force $\mathbf{F}$ passing through G together with a couple with torque $\mathbf{T}$. Then, $\mathbf{F}$ and $\mathbf{T}$ are:

$$\mathbf{F} = \sum_{i=1}^{N} \mathbf{F}_i \quad \text{and} \quad \mathbf{T} = \sum_{i=1}^{N} \mathbf{r}_i \times \mathbf{F}_i \tag{9.5.6}$$

Hence, from Eqs. (9.5.4) and (9.5.5), we have:

$$\begin{aligned} \mathbf{F} &= \sum_{i=1}^{N} m_i \mathbf{a}_i = \sum_{i=1}^{N} m_i \left(\mathbf{a}_G + d^2 \mathbf{r}_i / dt^2 \right) \\ &= \left(\sum_{i=1}^{N} m_i \right) \mathbf{a}_G + \frac{d^2}{dt^2} \left(\sum_{i=1}^{N} m_i \mathbf{r}_i \right) = M \mathbf{a}_G \end{aligned} \tag{9.5.7}$$

where M is the total mass of S and where the operations of summation and integration may be interchanged because N is finite. (The sum

$$\sum_{i=1}^{N} m_i \mathbf{r}_i$$

is zero because G is the mass center.)

Finally, by integrating in Eq. (9.5.7) we have:

$$\begin{aligned}\mathbf{I} &= \int_{t_1}^{t_2} \mathbf{F}\,dt = \int_{t_1}^{t_2} M\mathbf{a}_G dt = \int_{t_1}^{t_2} M\left(d\mathbf{v}_G/dt\right)dt \\ &= M\mathbf{v}_G(t_2) - M\mathbf{v}_G(t_1) = \mathbf{L}(t_2) - L(t_1)\end{aligned} \tag{9.5.8}$$

or

$$\mathbf{I} = \Delta\mathbf{L} \tag{9.5.9}$$

where **L** represents the linear momentum of S as in Eq. (9.3.2).

We can develop a similar analysis for a rigid body B as in Figure 9.5.3 where, as before, we consider B to be composed of a set of N particles P_i having masses m_i ($i = 1,\ldots, N$). Let G be the mass center of B and let R be an inertial reference frame in which B moves. Let $\mathbf{r}_i$ locate P_i relative to G. Then, because P_i and G are both fixed in B, their accelerations are related by the expression (see Eq. (4.9.6)):

$$\mathbf{a}_i = \mathbf{a}_G + \boldsymbol{\alpha} \times \mathbf{r}_i + \boldsymbol{\omega} \times \left(\boldsymbol{\omega} \times \mathbf{r}_i\right) \tag{9.5.10}$$

where as before $\mathbf{a}_i$ represents the acceleration of P_i in R and where $\boldsymbol{\alpha}$ and $\boldsymbol{\omega}$ are the angular acceleration and angular velocity of B in R.

Let P_i be acted upon by a force $\mathbf{F}_i$ as shown in Figure 9.5.3. Let the set of forces $\mathbf{F}_i$ ($i = 1,\ldots, N$) be represented by an equivalent force system consisting of a single force **F** passing through G together with a couple with torque **T**. Then, **F** and **T** are:

$$\mathbf{F} = \sum_{i=1}^{N} \mathbf{F}_i \quad \text{and} \quad \mathbf{T} = \sum_{i=1}^{N} \mathbf{r}_i \times \mathbf{F}_i \tag{9.5.11}$$

Again, from Newton's law we have:

$$\mathbf{F}_i = m_i \mathbf{a}_i \quad (\text{no sum}) \tag{9.5.12}$$

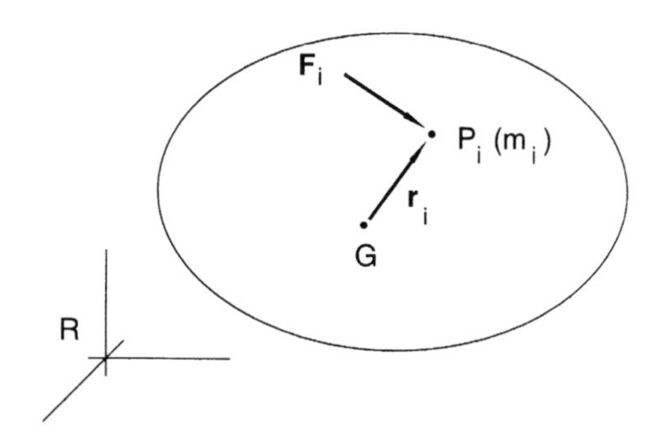

FIGURE 9.5.3
A rigid body B moving in an inertia reference frame R.

By substituting from Eqs. (9.5.10) and (9.5.12) into Eq. (9.5.11) we have:

$$\mathbf{F} = M\mathbf{a}_G \tag{9.5.13}$$

where M is the mass of B.

Finally, by integrating in Eq. (9.5.13) we obtain (as in Eqs. (9.5.8) and (9.5.9)):

$$\mathbf{I} = \Delta\mathbf{L} \tag{9.5.14}$$

where now $\mathbf{L}$ represents the linear momentum of B.

Observe the identical formats of Eqs. (9.5.3), (9.5.9), and (9.5.14) for a single particle, a set of particles, and a rigid body.

9.6 Principle of Angular Impulse and Momentum

We can develop expressions analogous to Eqs. (9.5.3), (9.5.9), and (9.5.14) for angular impulse and angular momentum. The development here, however, has the added feature of involving a reference point (or object point). Because angular momentum is always computed relative to a point, the choice of that point may affect the form of the relation between angular impulse and angular momentum.

Consider again a particle P with mass m moving in an inertial reference frame R as in Figure 9.6.1. Let P be acted upon by a force $\mathbf{F}$ as shown. Let Q be an arbitrarily chosen reference point. Consider a free-body diagram of P as in Figure 9.6.2, where $\mathbf{F}^*$ is the inertia force on P given by (see Eq. (8.3.2)):

$$\mathbf{F}^* = -m\mathbf{a}^P = -md\mathbf{v}^P/dt \tag{9.6.1}$$

where $\mathbf{v}^P$ and $\mathbf{a}^P$ are the velocity and acceleration of P in R.

From d'Alembert's principle, we have:

$$\mathbf{F} + \mathbf{F}^* = 0 \quad \text{or} \quad \mathbf{F} = md\mathbf{v}^P/dt \tag{9.6.2}$$

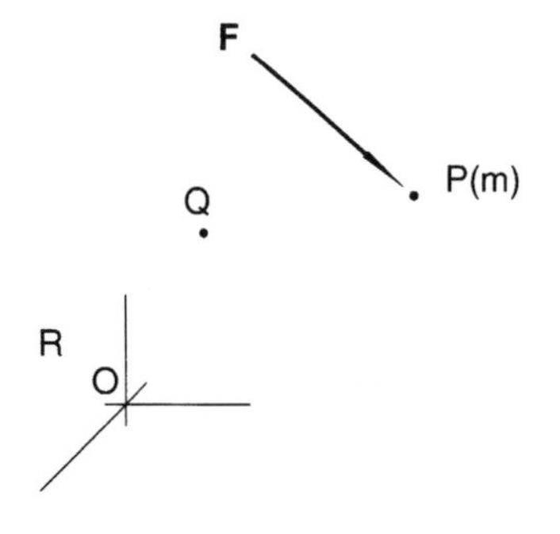

FIGURE 9.6.1
A particle P moving in an inertial reference frame R with applied force $\mathbf{F}$ and reference point Q.

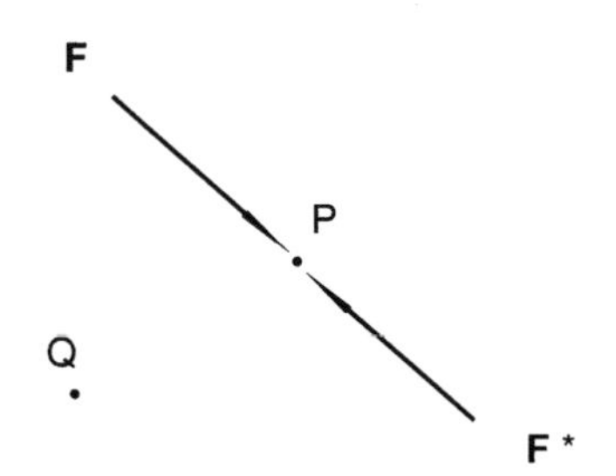

FIGURE 9.6.2
Free-body diagram of P.

By setting moments about Q equal to zero we have:

$$\mathbf{QP} \times \mathbf{F} + \mathbf{QP} \times \mathbf{F}^* = 0 \quad \text{or} \quad \mathbf{QP} \times \mathbf{F} = \mathbf{QP} \times m d\mathbf{v}^P/dt \tag{9.6.3}$$

Consider the final term in Eq. (9.6.3). By the product rule for differentiation we have:

$$\begin{aligned} \mathbf{QP} \times m d\mathbf{v}^P/dt &= d\left(\mathbf{QP} \times m d\mathbf{v}^P\right)/dt - \left(d\mathbf{QP}/dt\right) \times m\mathbf{v}^P \\ &= d\mathbf{A}_Q/dt - \mathbf{v}^{P/Q} \times m\mathbf{v}^P \\ &= d\mathbf{A}_Q/dt - \left(\mathbf{v}^P - \mathbf{v}^Q\right) \times m\mathbf{v}^P \\ &= d\mathbf{A}_Q/dt + \mathbf{v}^Q \times m\mathbf{v}^P \end{aligned} \tag{9.6.4}$$

Observe that $\mathbf{QP} \times \mathbf{F}$ is the moment of $\mathbf{F}$ about Q. Hence, by substituting from Eq. (9.6.4) into (9.6.3) we obtain:

$$\mathbf{M}_Q = d\mathbf{A}_Q/dt + \mathbf{v}^Q \times m\mathbf{v}^P \tag{9.6.5}$$

If the velocity of Q is zero (that is, if Q is fixed in R), then we have:

$$\mathbf{M}_Q = d\mathbf{A}_Q/dt \tag{9.6.6}$$

By integrating over the time interval in which $\mathbf{F}$ is applied, we have:

$$\int_{t_1}^{t_2} \mathbf{M}_Q dt = \int_{t_1}^{t_2} \left(d\mathbf{A}_Q/dt\right) dt = \mathbf{A}_Q(t_2) - \mathbf{A}_Q(t_1) \tag{9.6.7}$$

or

$$\mathbf{J}_Q = \Delta\mathbf{A}_Q \tag{9.6.8}$$

where $\mathbf{J}_Q$ is the angular impulse of $\mathbf{F}$ about Q during the time interval (t_1, t_2). That is, the angular impulse about a point Q fixed in an inertial reference frame is equal to the change in angular momentum about Q. This verbal statement of Eq. (9.6.8) is called the *principle of angular momentum*. Note, however, it is valid only if Q is fixed in the inertial reference frame.

Consider next a set S of N particles P_i with masses m_i ($i = 1,\ldots, N$) moving in an inertial reference frame R as in Figure 9.6.3. Let the particles be acted upon by forces $\mathbf{F}_i$ ($i = 1,\ldots, N$) as shown. Consider a free-body diagram of a typical particle P_i as in Figure 9.6.4 where $\mathbf{F}_i^*$ is the inertia force on P_i given by:

$$\mathbf{F}_i^* = -m_i\mathbf{a}^{P_i} = -m d\mathbf{v}^{P_i}/dt \quad (\text{no sum on } i) \tag{9.6.9}$$

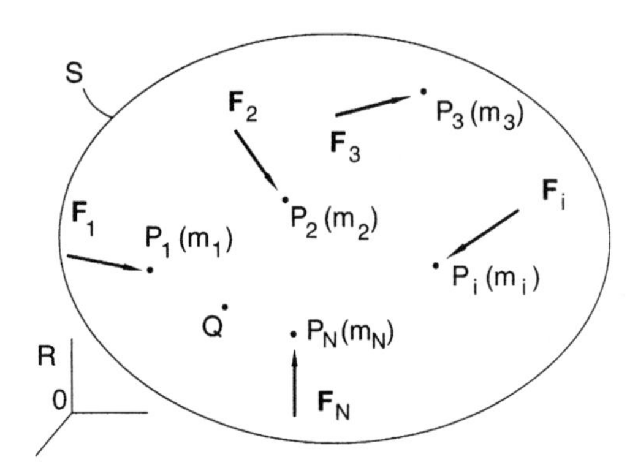

FIGURE 9.6.3
A set S of particles moving in an inertial reference frame R with reference point Q.

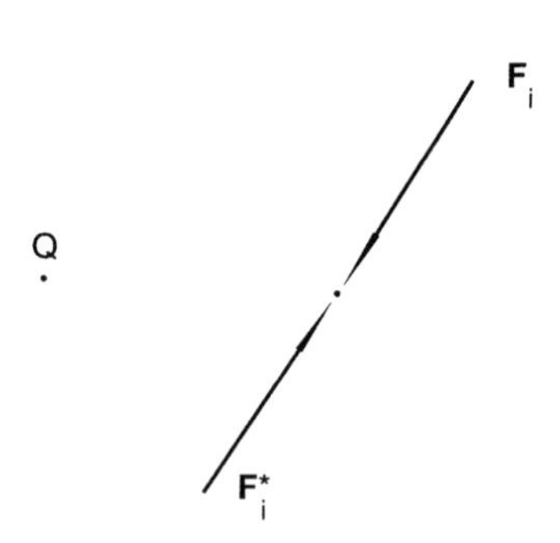

FIGURE 9.6.4
Free-body diagram of typical particle P_i.

where $\mathbf{v}^{P_i}$ and $\mathbf{a}^{P_i}$ are the velocity and acceleration of P_i in R. Then, from d'Alembert's principle, we have:

$$\mathbf{F}_i + \mathbf{F}_i^* = 0 \quad \text{or} \quad \mathbf{F}_i = m_i d\mathbf{v}^{P_i}/dt \tag{9.6.10}$$

By setting moments about Q equal to zero, we have:

$$\mathbf{QP}_i \times \mathbf{F}_i + \mathbf{QP}_i \times \mathbf{F}_i^* = 0 \quad \text{or} \quad \mathbf{QP}_i \times \mathbf{F}_i = \mathbf{QP}_i \times m_i d\mathbf{v}^{P_i}/dt \tag{9.6.11}$$

Consider the final term in Eq. (9.6.11). By the product rule for differentiation, we have (see Eq. (9.6.4)):

$$\begin{aligned} \mathbf{QP}_i \times m_i d\mathbf{v}^{P_i}/dt &= d\left(\mathbf{QP}_i \times m_i \mathbf{v}^{P_i}\right)/dt - d\mathbf{QP}_i/dt \times m_i \mathbf{v}^{P_i} \\ &= d\mathbf{A}_Q^{P_i}/dt + \mathbf{v}^Q \times m_i \mathbf{v}^{P_i} \end{aligned} \tag{9.6.12}$$

By substituting into Eq. (9.6.11) we have:

$$\mathbf{M}_{F_i/Q} = d\mathbf{A}_{P_i/Q}/dt + \mathbf{v}^Q \times m_i \mathbf{v}^{P_i} \tag{9.6.13}$$

By adding the effects from all the particles, we have:

$$\mathbf{M}_Q = d\mathbf{A}_{S/Q}/dt + \mathbf{v}^Q \times \sum_{i=1}^{N} m_i \mathbf{v}^{P_i} \tag{9.6.14}$$

If the velocity of Q is zero (that is, if Q is fixed in R), then we have:

$$\mathbf{M}_Q = d\mathbf{A}_{S/Q}/dt \tag{9.6.15}$$

By integrating over the time interval in which the forces are applied, we have:

$$\int_{t_1}^{t_2} \mathbf{M}_Q dt = \int_{t_1}^{t_2} \left(d\mathbf{A}_{S/Q}/dt \right) dt = \mathbf{A}_{S/Q}(t_2) - \mathbf{A}_{S/Q}(t_1)$$

or

$$\mathbf{J}_Q = \Delta \mathbf{A}_{S/Q} \tag{9.6.16}$$

where here $\mathbf{J}_Q$ represents the sum of the angular impulses of the applied forces about Q during the time interval (t_1, t_2). Hence, as with a single particle, the angular impulse about a point Q fixed in an inertial reference frame is equal to the change in angular momentum of the set of particles about Q.

Finally, consider a rigid body B moving in an inertial reference frame R as in Figure 9.6.5. Let G be the mass center of B, let Q be a reference point, and let O be the origin of R. Then, from Eqs. (9.4.8), (9.4.12), and (9.4.13), the angular momenta of B about O, Q, and G are:

$$\mathbf{A}_{B/O} = \mathbf{A}_{B/G} + \mathbf{A}_{G/O} = \mathbf{I}^{B/G} \cdot \boldsymbol{\omega} + \mathbf{P}_G \times m\mathbf{v}^G \tag{9.6.17}$$

$$\mathbf{A}_{B/Q} = \mathbf{A}_{B/G} + \mathbf{A}_{G/Q} = \mathbf{I}^{B/G} \cdot \boldsymbol{\omega} + \mathbf{QG} \times m\mathbf{v}^G \tag{9.6.18}$$

$$\mathbf{A}_{B/G} = \mathbf{I}^{B/G} \cdot \boldsymbol{\omega} \tag{9.6.19}$$

where $\mathbf{P}_G$ and $\mathbf{QG}$ locate G relative to O and Q as in Figure 9.6.5, and where as before $\boldsymbol{\omega}$ is the angular velocity of B in R, $\mathbf{I}^{B/G}$ is the central inertia dyadic of B, and m is the mass of B. Consider the derivatives of these momenta. For $\mathbf{A}_{B/O}$ we have:

$$\begin{aligned} {}^R d\mathbf{A}_{B/O}/dt &= {}^R d\left(\mathbf{I}^{B/G} \cdot \boldsymbol{\omega}\right)/dt + {}^R d\left(\mathbf{P}_G \times m\mathbf{v}^G\right)/dt \\ &= {}^B d\left(\mathbf{I}^{B/G} \cdot \boldsymbol{\omega}\right)/dt + \boldsymbol{\omega} \times \left(\mathbf{I}^{B/G} \cdot \boldsymbol{\omega}\right) \\ &\quad + \mathbf{v}_G \times m\mathbf{v}_G + \mathbf{P}_G \times m\mathbf{a}^G \end{aligned} \tag{9.6.20}$$

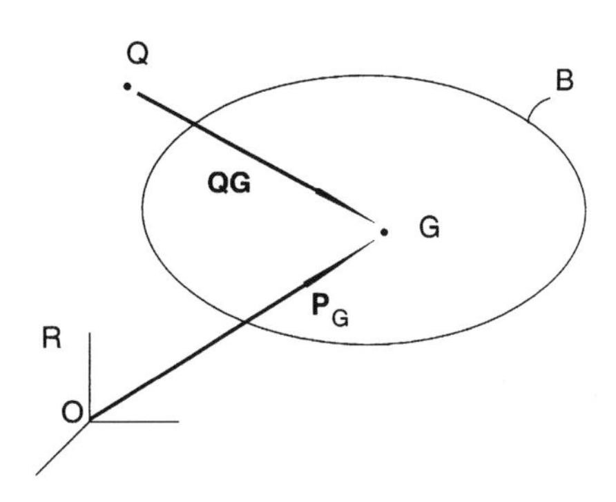

FIGURE 9.6.5
A rigid body B moving in an inertial reference frame R and a reference point Q.

where we have used Eq. (4.6.6). If $\mathbf{I}^{B/G}$ is expressed in terms of unit vectors fixed in B, its components relative to these vectors are constant. Hence, ${}^B d\mathbf{I}^{B/G}/dt$ is zero. Also, the third term is zero. Therefore, ${}^R d\mathbf{A}_{B/O}/dt$ becomes:

$$
{}^R d\mathbf{A}_{B/O}/dt = \mathbf{I}^{B/G} \cdot \boldsymbol{\alpha} + \boldsymbol{\omega} \times \left(\mathbf{I}^{B/G} \cdot \boldsymbol{\omega}\right) + \mathbf{P}_G \times m\mathbf{a}^G \tag{9.6.21}
$$

where $\boldsymbol{\alpha}$ is the angular acceleration of B in R. (Note from Eq. (4.6.6) that ${}^R d\boldsymbol{\omega}/dt = {}^B d\boldsymbol{\omega}/dt = \boldsymbol{\alpha}$.)

Next, for $\mathbf{A}_{B/Q}$ we have:

$$
\begin{aligned}
{}^R d\mathbf{A}_{B/Q}/dt &= {}^R d\left(\mathbf{I}^{B/G} \cdot \boldsymbol{\omega}\right)/dt + {}^R d\left(\mathbf{QG} \times m\mathbf{v}^G\right)/dt \\
&= {}^B d\left(\mathbf{I}^{B/G} \cdot \boldsymbol{\omega}\right)/dt + \boldsymbol{\omega} \times \left(\mathbf{I}^{B/G} \cdot \boldsymbol{\omega}\right) \\
&\quad + \mathbf{v}^{G/Q} \times m\mathbf{v}^G + \mathbf{QG} \times m\mathbf{a}^G \\
&= \mathbf{I}^{B/G} \cdot \boldsymbol{\alpha} + \boldsymbol{\omega} \times \left(\mathbf{I}^{B/G} \cdot \boldsymbol{\omega}\right) - \mathbf{v}^Q \times m\mathbf{v}^G \\
&\quad + \mathbf{QG} \times m\mathbf{a}^G
\end{aligned} \tag{9.6.22}
$$

Finally, for $\mathbf{A}_{B/G}$ we have:

$$
\begin{aligned}
{}^R d\mathbf{A}_{B/G}/dt &= {}^R d\left(\mathbf{I}^{B/G} \cdot \boldsymbol{\omega}\right)/dt = {}^B d\mathbf{I}^{B/G}/dt + \boldsymbol{\omega} \times \left(\mathbf{I}^{B/G} \cdot \boldsymbol{\omega}\right) \\
&= \mathbf{I}^{B/G} \cdot \boldsymbol{\alpha} + \boldsymbol{\omega} \times \left(\mathbf{I}^{B/G} \cdot \boldsymbol{\omega}\right)
\end{aligned} \tag{9.6.23}
$$

Let B be subjected to a system of applied forces that may be represented by a single force $\mathbf{F}$ passing through G together with a couple with torque $\mathbf{M}_G$. Similarly, let the inertia forces on B be represented by a single force $\mathbf{F}^*$ passing through G together with a couple with torque $\mathbf{T}^*$. Then, a free-body diagram of B may be constructed as in Figure 9.6.6. Recall from Eqs. (8.6.5) and (8.6.6) that $\mathbf{F}^*$ and $\mathbf{T}^*$ may be expressed as:

$$
\mathbf{F}^* = -m\mathbf{a}^G \quad \text{and} \quad \mathbf{T}^* = -\mathbf{I}^{B/G} \cdot \boldsymbol{\alpha} - \boldsymbol{\omega} \times \left(\mathbf{I}^{B/G} \cdot \boldsymbol{\omega}\right) \tag{9.6.24}
$$

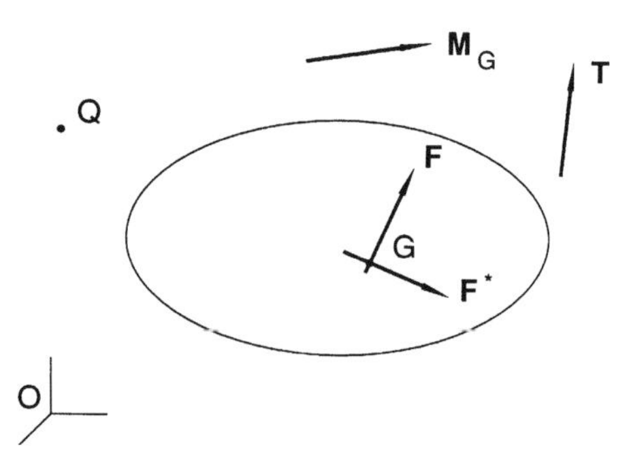

FIGURE 9.6.6
A free-body diagram of B.

From d'Alembert's principle, the entire force system of a free-body diagram is a zero system. Thus, the moment of the force system about any and all points is zero. Hence, consider setting moments about O, Q, and G equal to zero: for O we have:

$$\mathbf{P}_G \times \mathbf{F} + \mathbf{P}_G \times \mathbf{F}^* + \mathbf{M}_G + \mathbf{T}^* = 0 \tag{9.6.25}$$

By substituting for $\mathbf{F}^*$ and $\mathbf{T}^*$ from Eq. (9.6.24) we have:

$$\mathbf{M}_O = \mathbf{P}_G \times m\mathbf{a}^G + \mathbf{I}^{B/G} \cdot \boldsymbol{\alpha} + \boldsymbol{\omega} \times \left(\mathbf{I}^{B/G} \cdot \boldsymbol{\omega}\right) \tag{9.6.26}$$

where $\mathbf{M}_O$ is defined as $\mathbf{M}_G + P \times \mathbf{F}$ and is identified as the moment of the applied forces about O. By comparing Eqs. (9.6.21) and (9.6.26) we have:

$$\mathbf{M}_O = {}^R d\mathbf{A}_{B/O} / dt \tag{9.6.27}$$

Next, by setting moments about Q equal to zero we obtain:

$$\mathbf{QG} \times \mathbf{F} + \mathbf{M}_G + \mathbf{QG} \times \mathbf{F}^* + \mathbf{T}^* = 0 \tag{9.6.28}$$

Then, by substituting from Eq. (9.6.24), we have:

$$\mathbf{M}_Q = \mathbf{QG} \times m\mathbf{a}^G + \mathbf{I}^{B/G} \cdot \boldsymbol{\alpha} + \boldsymbol{\omega} \times \left(\mathbf{I}^{B/G} \cdot \boldsymbol{\omega}\right) \tag{9.6.29}$$

where $\mathbf{M}_Q$ is defined as $\mathbf{QG} \times \mathbf{F} + \mathbf{M}_G$ and is identified as the moment of the applied forces about Q. By comparing Eqs. (9.6.22) and (9.6.29), we have:

$$\mathbf{M}_Q = {}^R d\mathbf{A}_{B/Q} / dt + \mathbf{v}^Q \times m\mathbf{v}^G \tag{9.6.30}$$

Finally, by setting moments about G equal to zero we have:

$$\mathbf{M}_G + \mathbf{T}^* = 0 \tag{9.6.31}$$

Hence, by substituting from Eq. (9.6.8) we have:

$$\mathbf{M}_G + \mathbf{I}^{B/G} \cdot \boldsymbol{\alpha} + \boldsymbol{\omega} \times \left(\mathbf{I}^{B/G} \cdot \boldsymbol{\omega}\right) \tag{9.6.32}$$

By comparing Eqs. (9.6.23) and (9.6.32), we obtain:

$$\mathbf{M}_G = {}^R d\mathbf{A}_{B/G} / dt \tag{9.6.33}$$

By inspecting Eqs. (9.6.27), (9.6.30), and (9.6.33), we see that if the object point is either fixed in the inertia frame R or is the mass center of the body, then the moment of the applied forces about the object point is equal to the derivative of the angular momentum about the object point.

By integrating Eqs. (9.6.27) and (9.6.33) over the time interval when the forces are applied, we obtain:

$$\int_{t_1}^{t_2} \mathbf{M}_O dt = \int_{t_1}^{t_2} \left({}^R d\mathbf{A}_{B/O}/dt\right) dt = \mathbf{A}_{B/O}(t_2) - \mathbf{A}_{B/O}(t_1) \tag{9.6.34}$$

and

$$\int_{t_1}^{t_2} \mathbf{M}_G dt = \int_{t_1}^{t_2} \left({}^R d\mathbf{A}_{B/G}/dt\right) dt = \mathbf{A}_{B/G}(t_2) - \mathbf{A}_{B/G}(t_1) \tag{9.6.35}$$

or

$$\mathbf{J}_O = \Delta \mathbf{A}_{B/O} \tag{9.6.36}$$

and

$$\mathbf{J}_G = \Delta \mathbf{A}_{B/G} \tag{9.6.37}$$

where $\mathbf{J}_O$ and $\mathbf{J}_G$ are the angular impulses of the applied forces about O and G.

Eqs. (9.6.36) and (9.6.37) are, of course, analogous to Eqs. (9.6.8) and (9.6.16) for a single particle and for a set of particles. They all state that the angular impulse is equal to the change in angular momentum when the object point is fixed in an inertial reference frame. For a rigid body, the object point may also be the mass center. Other object points do not produce the simple relation between angular impulse and angular momentum as in Eqs. (9.6.8), (9.6.16), (9.6.36), and (9.6.37).

9.7 Conservation of Momentum Principles

Before looking at examples illustrating the impulse momentum principles, it is helpful to consider also the conservation of momentum principles. Simply stated, these principles assert that if the impulse is zero, the momentum is unchanged — that is, the momentum is conserved.

Referring to Eq. (9.5.3), if the impulse **I** on a particle is zero we have:

$$\Delta \mathbf{L} = 0 \quad \text{or} \quad \mathbf{L}(t_2) = \mathbf{L}(t_1) \tag{9.7.1}$$

Specifically, the linear momentum is the same at the beginning and end of the impulse time interval. Moreover, because the impulse is zero the time interval is arbitrary. Hence, t_1 and t_2 are arbitrary, and the linear momentum is unchanged throughout all time intervals. That is, the linear momentum is constant, or *conserved*. Expressions analogous to Eq. (9.7.1) may also be obtained for sets of particles and for a rigid body using Eqs. (9.5.9) and (9.5.14).

Similarly, from Eq. (9.6.8), if the angular impulse $\mathbf{J}_Q$ on a particle is zero, we have:

$$\Delta \mathbf{A}_Q = 0 \quad \text{or} \quad \mathbf{A}_Q(t_2) = \mathbf{A}_Q(t_1) \tag{9.7.2}$$

Therefore, if the angular impulse is zero, the angular momentum is the same at the beginning and end of the impulse time interval and thus throughout the time interval; hence, the angular momentum is constant, or conserved. Similar statements can be made for sets of particles and for a rigid body using Eqs. (9.6.16), (9.6.36), and (9.6.37).

Observe that because linear and angular impulses are vectors, they may be nonzero but still have zero projection in some directions. In such cases, the projection of the momenta along those directions is conserved. For example, if $\mathbf{n}$ is a unit vector such that:

$$\mathbf{I} \cdot \mathbf{n} = 0 \tag{9.7.3}$$

then, from Eq. (9.5.3), we have:

$$(\Delta \mathbf{L}) \cdot \mathbf{n} = 0 \quad \text{or} \quad \mathbf{L}(t_2) \cdot \mathbf{n} = \mathbf{L}(t_1) \cdot \mathbf{n} \tag{9.7.4}$$

That is, the linear momentum is conserved in the $\mathbf{n}$ direction. Similar results occur for angular momenta.

9.8 Examples

Example 9.8.1: A Sliding Collar

Consider a sliding block (or collar) of mass m moving on a smooth rod as in Figure 9.8.1. Let a force $\mathbf{F}$ be exerted on B as shown and let the magnitude of $\mathbf{F}$ have a time profile as in Figure 9.8.2. From Eq. (9.2.1), the impulse of $\mathbf{F}$ is:

$$\mathbf{I} = \int_0^t F dt \mathbf{n} = \int_0^{\hat{t}} F dt \mathbf{n} = \left(F_{\max} \hat{t}/2\right)\mathbf{n} \tag{9.8.1}$$

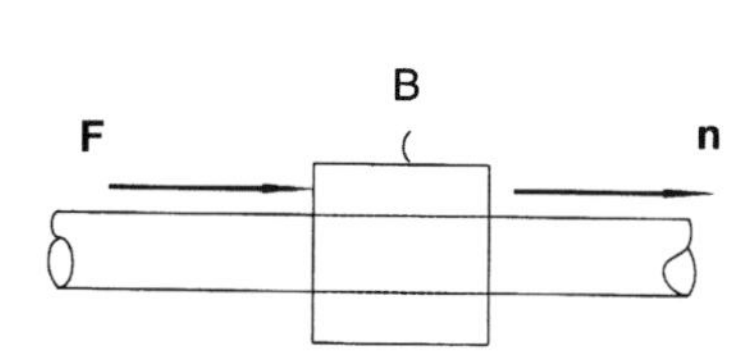

FIGURE 9.8.1
A slider block and an applied force **F**.

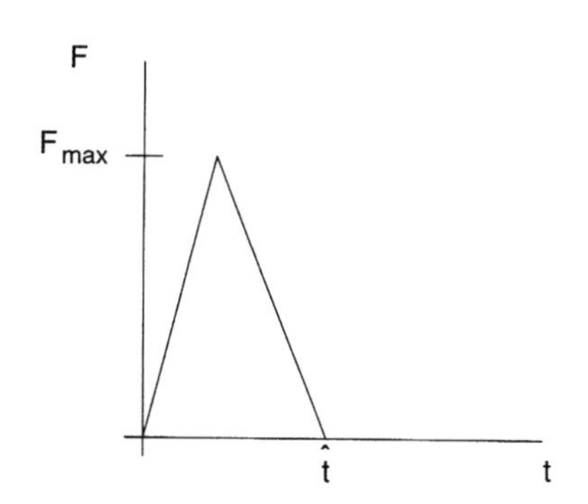

FIGURE 9.8.2
Force profile.

where **n** is a unit vector parallel to the support as in Figure 9.8.1. Let the velocity of B before and after the application of **F** be $V_O\mathbf{n}$ and $V\mathbf{n}$, respectively. Then, from the linear impulse–momentum principle, we have:

$$\mathbf{I} = \Delta\mathbf{L} \quad \text{or} \quad F_{\max}\left(\hat{t}/2\right)\mathbf{n} = mV\mathbf{n} - mV_O\mathbf{n} \tag{9.8.2}$$

or

$$V = V_O + F_{\max}\hat{t}/2m \tag{9.8.3}$$

Example 9.8.2: A Braked Flywheel

As a second example, consider a flywheel W rotating about its axis with an angular speed as in Figure 9.8.3. Let W be supported by a shaft S, and let S be subjected to a brake that exerts a moment **M** about the axis of S and W and whose magnitude M is depicted in Figure 9.8.4. Specifically, let $M(t)$ be sinusoidal such that:

$$M(t) = \begin{cases} M_{\max}\left[1 - \cos\left(2\pi t/T\right)\right] & 0 \le t \le T \\ 0 & t \ge T \end{cases} \tag{9.8.4}$$

Then, from Eq. (9.2.2), the angular impulse about O, the center of W, is:

$$\begin{aligned} \mathbf{J}_O &= \int_0^T M(t)\,dt\mathbf{k} = \int_0^T M_{\max}\left[1 - \cos\left(2\pi t/T\right)\right]\mathbf{k} \\ &= M_{\max}\left[t - \left(T/2\pi\right)/\sin\left(2\pi t/T\right)\right]\Big|_0^T \mathbf{k} = M_{\max}T\mathbf{k} \end{aligned} \tag{9.8.5}$$

where **k** is a unit vector parallel to the shaft as in Figure 9.8.3.

From Eq. (9.4.13) the angular momentum of W about O before and after braking is:

$$\mathbf{A}_O(0) = I_O\omega_O\mathbf{k} \quad \text{and} \quad \mathbf{A}_O(t) = I_O\omega\mathbf{k} \tag{9.8.6}$$

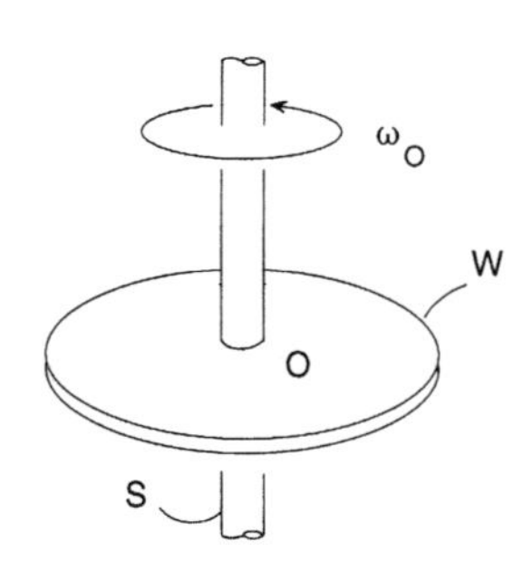

FIGURE 9.8.3
A spinning flywheel.

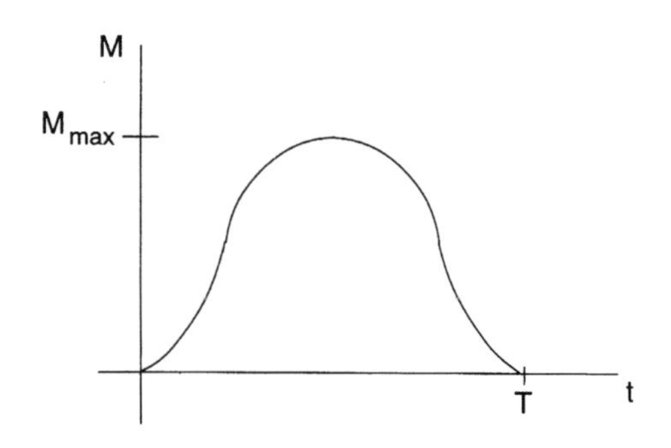

FIGURE 9.8.4
Profile of braking moment magnitude.

where I_O is the axial moment of inertia of W (the moment of inertia of W about O for the axial direction). Then from the angular impulse–momentum principle the angular speed of W after braking is determined by the expressions:

$$\mathbf{J}_O = \Delta \mathbf{A}_O \quad \text{or} \quad M_{\max} T\mathbf{k} = I_O \omega k - I_O \omega_O \mathbf{k} \tag{9.8.7}$$

or

$$\omega_O = M_{\max} T / I_O \tag{9.8.8}$$

Example 9.8.3: A Struck Pinned Bar (Center of Percussion)

Next, consider a thin horizontal bar B with length ℓ and mass m and pinned at one end O about a vertical axis as in Figure 9.8.5. Let B be struck at a point along its length as shown. If B is initially at rest it will begin to rotate about O after it is struck.

Consider a free-body diagram of B showing the impulses and momentum changes of B as in Figure 9.8.6 where P is the impact force magnitude and O_x and O_y represent components of the pin reaction force along and perpendicular to B, and where G is the mass center of B. The linear impulse change of B and the angular impulse change of B about G are:

$$\Delta \mathbf{L} = mv\mathbf{n}_y \quad \text{and} \quad \Delta \mathbf{A}_G = I_G \omega \mathbf{n}_z \tag{9.8.9}$$

where $\mathbf{n}_x$, $\mathbf{n}_y$, and $\mathbf{n}_z$ are mutually perpendicular unit vectors as in Figure 9.8.6, v is the speed of G just after impact, and ω is the angular speed of B just after impact. I_G is the central moment of inertia of B about an axis normal to its length, given by:

$$I_G = (1/12) m\ell^2 \tag{9.8.10}$$

In the free-body diagram of Figure 9.8.6, inertia forces are not shown; instead, momenta changes are given. This allows us to use the sketch to employ the impulse–momentum principles. Specifically, consider the linear impulse–momentum sums in the $\mathbf{n}_x$ and $\mathbf{n}_y$ directions:

$$\int_0^t O_x dt = 0 \tag{9.8.11}$$

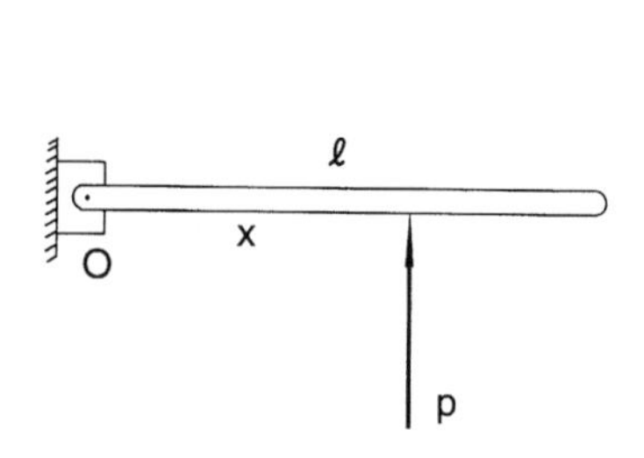

FIGURE 9.8.5
A pinned bar struck along its length.

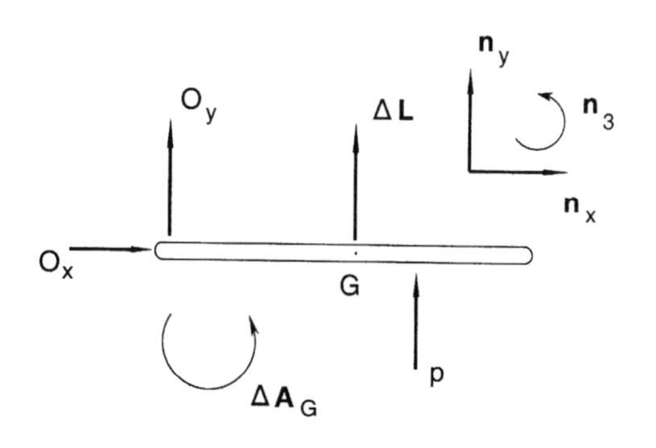

FIGURE 9.8.6
A free-body diagram of bar B.

and

$$\int_0^t O_y dt + \int_0^t P dt = mv \tag{9.8.12}$$

Similarly, the angular impulse–momentum principle about G for $\mathbf{n}_z$ leads to:

$$\int_0^t \mathbf{P}\left(x - \frac{\ell}{2}\right) dt - \int_0^t \frac{\ell}{2} O_y dt = \frac{1}{12} m\ell^2 \omega$$

or

$$\left(x - \frac{\ell}{2}\right)\int_0^t \mathbf{P}dt - \frac{\ell}{2}\int_0^t O_y dt = \frac{1}{12} m\ell^2 \omega^2 \tag{9.8.13}$$

By eliminating $\int_0^t \mathbf{P}dt$ between Eqs. (9.8.12) and (9.8.13) we have:

$$\left(x - \frac{\ell}{2}\right)\left(mv - \int_0^t O_y dt\right) - \frac{\ell}{2}\int_0^t O_y dt = \frac{1}{12} m\ell^2 \omega \tag{9.8.14}$$

Because B is pinned at O, G moves in a circle about O with radius $\ell/2$. Then, v and ω are related by:

$$v = (\ell/2)\omega \tag{9.8.15}$$

By solving Eq. (9.8.14) for $\int_0^t O_y dt$ and by using Eq. (9.8.15) we have:

$$\int_0^t O_y dt = \left(\frac{2\ell/3 - x}{\ell + x}\right) mv \tag{9.8.16}$$

Observe that if B is struck at a point such that x is $2\ell/3$ the pin reaction is zero. That is, if the impact force is applied at a point 2/3 along the bar length, no pin reaction is generated. This means that even if the bar is not pinned it will initially move such that its end away from the impact has zero velocity. This point of application of the impact force is called the *center of percussion*.

Example 9.8.4: A Pinned Double Bar Struck at One End

As an extension of the foregoing example consider two identical, pin-connected bars each having length ℓ and mass m and resting on a smooth horizontal surface as depicted in Figure 9.8.7. Let the end O of one of the bars be struck with the striking force being perpendicular to the bars and having a magnitude P as shown. The objective is to determine the motion of the bars just after the impact.

Consider a free-body diagram of the bars and also a free-body diagram of the left, or unstruck, bar as in Figures 9.8.8 and 9.8.9, where as in the foregoing example the inertia

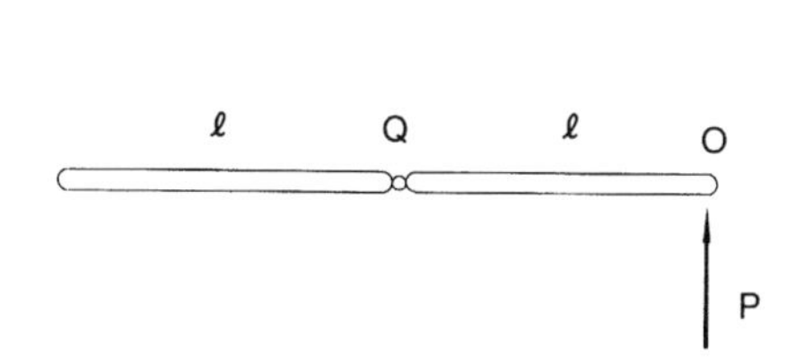

FIGURE 9.8.7
Pin-connected double bar struck at one end.

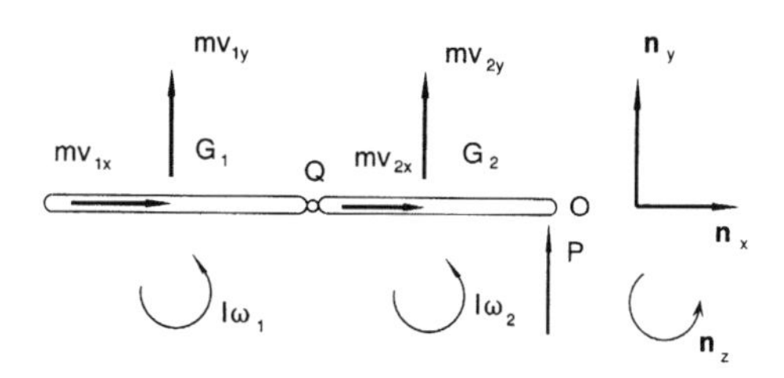

FIGURE 9.8.8
Free-body diagram of struck double bar.

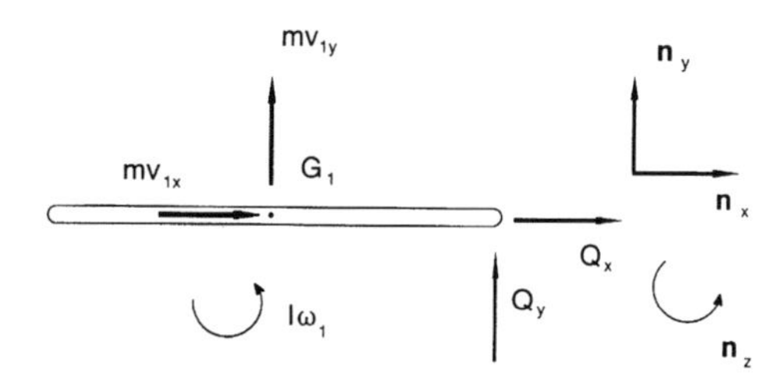

FIGURE 9.8.9
Free-body diagram of left bar.

forces are not shown. Instead, the changes in momenta of the bars are depicted where v_{1x}, v_{1y} and v_{2x}, v_{2y} are the velocity components of the mass centers, G_1 and G_2, of the bars in the directions shown; ω_1 and ω_2 are the angular speeds of bars just after impact; and $\mathbf{n}_x$, $\mathbf{n}_y$, and $\mathbf{n}_z$ are mutually perpendicular unit vectors in the directions shown. In Figure 9.8.9, Q_x and Q_y are the components of the reaction force transmitted across the pin Q when the bars are struck.

From these free-body diagrams and from the impulse–momentum principles, we obtain the following relations:

$$
\begin{aligned}
&mv_{1x} + mv_{2x} = 0 \\
&\int_0^t P dt = mv_{1y} + mv_{2y} \\
&\ell \int_0^t P dt = (\ell/2) mv_{2y} + I\omega_2 - (\ell/2) mv_{1y} + I\omega_1
\end{aligned}
\tag{9.8.17}
$$

and

$$
\begin{aligned}
&\int_0^t Q_x dt = mv_{1x} \\
&\int_0^t Q_y dt = mv_{1y} \\
&(\ell/2) \int_0^t Q_y dt = I\omega_1
\end{aligned}
\tag{9.8.18}
$$

where the third and sixth equations are obtained by evaluating moments about Q and G_1, and where I is the central moment of inertia: $(1/12)m\ell^2$. Observe from the kinematics of the bars that we can obtain the expressions:

$$v_{1x} = v_{2x}, \ v_{2y} = v_0 - (\ell/2)\omega_2, \ v_Q = v_0 - \ell\omega_2, \ v_{1y} = v_Q - (\ell/2)\omega_1 \tag{9.8.19}$$

Hence, from the first expressions of Eqs. (9.8.17), (9.8.18), and (9.8.19), we find:

$$v_{1x} = v_{2x} = 0, \ Q_x = 0 \tag{9.8.20}$$

Also, by eliminating the linear impulses $\int_0^t \mathbf{P}dt$ and $\int_0^t Q_y dt$ from Eqs. (9.8.17) and (9.8.18), respectively, we obtain:

$$\ell m v_{1y} + \ell m v_{2y} = (\ell/2)m v_{2y} + (1/12)m\ell^2\omega_2 - (\ell/2)m v_{1y} + (1/12)m\ell^2\omega_1 \tag{9.8.21}$$

and

$$(\ell/2)m v_{1y} = (1/12)m\ell^2\omega_1 \tag{9.8.22}$$

Then, by substituting for v_{1y} and v_{2y} from Eqs. (9.8.19), we have:

$$\begin{aligned} v_0 - \ell\omega_2 - (\ell/2)\omega_1 + v_0 - (\ell/2)\omega_2 &= (1/2)[v_0 - (\ell/2)\omega_2] \\ +(1/12)\ell\omega_2 - (1/2)[v_0 - \ell\omega_2 - (\ell/2)\omega_1] &+ (1/12)\ell\omega_1 \end{aligned} \tag{9.8.23}$$

and

$$(1/2)[v_0 - \ell\omega_2 - (\ell/2)\omega_1] = (1/12)\ell\omega_1 \tag{9.8.24}$$

or, after simplification:

$$v_0 = (5/12)\ell\omega_1 + (11/12)\ell\omega_2 \tag{9.8.25}$$

and

$$v_0 = (2/3)\ell\omega_1 + \ell\omega_2 \tag{9.8.26}$$

Finally, by eliminating v_0, we obtain:

$$\omega_2 - 3\omega_1 \quad \text{or} \quad \omega_1 = -(1/3)\omega_2 \tag{9.8.27}$$

Observe that the bars rotate in opposite directions immediately after the impact and that the left bar rotates clockwise.

9.9 Additional Examples: Conservation of Momentum

The most useful impulse–momentum principles are those related to the conservation of momentum, where external impulse to a system is absent. In this section we will consider a few simple examples to illustrate these principles.

Example 9.9.1: Colliding Collinear Blocks

First, consider two blocks A and B with masses m_A and m_B moving along the same line as in Figure 9.9.1. Let the blocks be moving to the right as shown, and let the speed of A be greater than the speed of B ($v_A > v_B$) so that collision occurs. Let the collision be "plastic" so that the blocks move together as a unit after collision. Given the masses and speeds of the blocks before collision, the objective is to determine their common speed after collision.

If we consider the two blocks as a system, we observe that there are no external impulses on the system. This means that the overall momentum of the system is unchanged by the collision. Thus, we have:

$$m_A v_A + m_B v_B = (m_A + m_B)v \tag{9.9.1}$$

where v represents the common speed of the blocks after collision. Solving for v we obtain:

$$v = (m_A v_A + m_B v_B)/(m_A + m_B) \tag{9.9.2}$$

Observe in Eq. (9.9.2) that if the masses are equal, v is the average of the speeds of the blocks prior to collision.

Example 9.9.2: Engaging Coaxial Disks

As a second example of momentum conservation, consider two disks A and B rotating freely on a common shaft as depicted in Figure 9.9.2. Let the disks have axial moments of inertia I_A and I_B and angular speeds ω_A and ω_B as shown. Let the disks be brought together by sliding them toward each other along their common shaft. Let the disks engage one another and then rotate together with the common angular velocity ω. The objective is to determine ω.

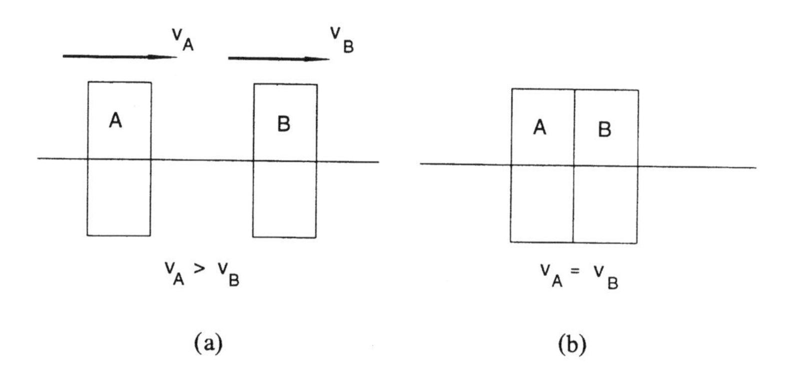

FIGURE 9.9.1
Colliding collinear blocks (plastic collision).

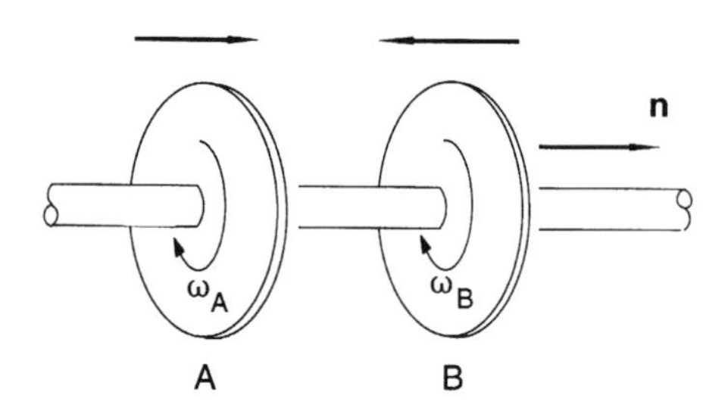

FIGURE 9.9.2
Meshing coaxial disks.

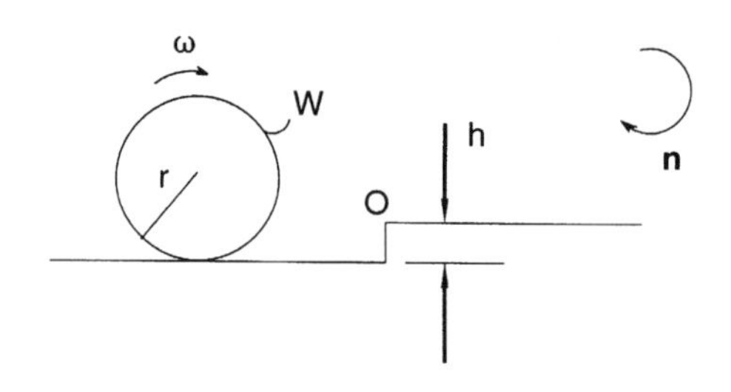

FIGURE 9.9.3
Wheel rolling over a step.

Because the disks can rotate freely on their common shaft there is no axial momentum applied to the disks; thus, there is no angular impulse about the shaft axis. This in turn means that the angular momentum of the disks about the shaft axis is conserved.

Before meshing, the axial angular momentum of the disks is:

$$A_{before}\mathbf{n} = I_A\omega_A\mathbf{n} + I_B\omega_B\mathbf{n} \tag{9.9.3}$$

where $\mathbf{n}$ is a unit vector parallel to the shaft axis. After meshing, the angular momentum of the disks is:

$$A_{after}\mathbf{n} = (I_A + I_B)\omega\mathbf{n} \tag{9.9.4}$$

By equating these expressions we have:

$$\omega = (I_A\omega_A + I_B\omega_B)/(I_A + I_B) \tag{9.9.5}$$

Observe the similarity between Eqs. (9.9.2) and (9.9.5). These expressions are useful in clutch design and in the design of slider mechanisms.

Example 9.9.2: Wheel Rolling Over a Step

As a third example of momentum conservation consider a wheel or disk W rolling in a straight line and encountering a small step as in Figure 9.9.3. Let the angular speed of W before striking the step be ω. The objective is to determine the angular speed ω of W just after impact with the step.

Let W have radius r and mass m, let its axial moment of inertia be I, and let the step height be h. Then, the angular momentum of W about the step corner O before and after impact is conserved. Before impact, the angular momentum of W about O is:

$$\mathbf{A}_{before} = \mathbf{A}_{W/G} + \mathbf{A}_{G/O} = \left[I\omega + mv(r-h)\right]\mathbf{n} \tag{9.9.6}$$

where v is the speed of the disk center before impact, and $\mathbf{n}$ is an axial unit vector as shown in Figure 9.9.3. Because W is rolling, v is simply $r\omega$. After impact with the step, W is rotating about O. Its angular momentum after impact is thus:

$$\mathbf{A}_{after} = \mathbf{A}_{W/G} + \mathbf{A}_{G/O} = \left[I\hat{\omega} + mr^2\hat{\omega}\right]\mathbf{n} \tag{9.9.7}$$

Equating the expressions of Eqs. (9.9.6) and (9.9.7), we find $\hat{\omega}$ to be:

$$\hat{\omega} = \{[I + mr(r-h)]/(I + mr^2)\}\omega \tag{9.9.8}$$

9.10 Impact: Coefficient of Restitution

The principles of impulse and momentum are ideally suited to treat problems involving impact, where large forces are exerted but only for a short time. Although a detailed analysis of the contact forces, the stresses, and the deformations of colliding bodies requires extensive analysis beyond our scope here, the principles of impulse and momentum can be used to obtain a global description of the phenomenon. That is, by making a couple of simplifying assumptions we can develop a direct and manageable procedure for studying impact.

Our first assumption is that the time interval during which impact occurs is so short that the positions and orientations of the colliding bodies do not change significantly during the impact time interval, although the velocities may have significant incremental changes. The second assumption is that after the bodies come together and collide they then generally rebound and separate again. The speed of separation is assumed to be a fraction of the speed of approach.

To develop these concepts, consider again the principle of linear impulse and momentum. Specifically, consider a particle P of mass m initially at rest but free to move along the X-axis as in Figure 9.10.1. Let P be subjected to a large force **F** acting for a short time t^*. Then, from the linear impulse–momentum principle (Eq. (9.5.3)), we have:

$$I = \int_0^{t^*} F dt = mv^* \tag{9.10.1}$$

where v^* is the speed of P at time t^*. If t^* is sufficiently small, the impulse I will be finite even though F may be very large. Let the speed v of P be expressed as:

$$v = dx/dt \tag{9.10.2}$$

where x is the displacement of P along the X-axis of Figure 9.10.1. Then, at time t^* the displacement is:

$$x^* = \int_0^{t^*} v dt < v^* t^* \tag{9.10.3}$$

where the inequality follows from Eq. (9.10.1).

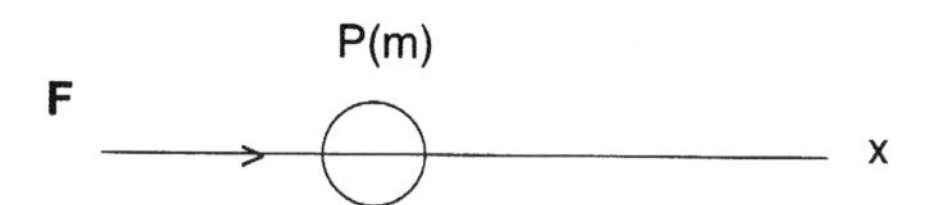

FIGURE 9.10.1
A large, short duration force applied to a particle P.

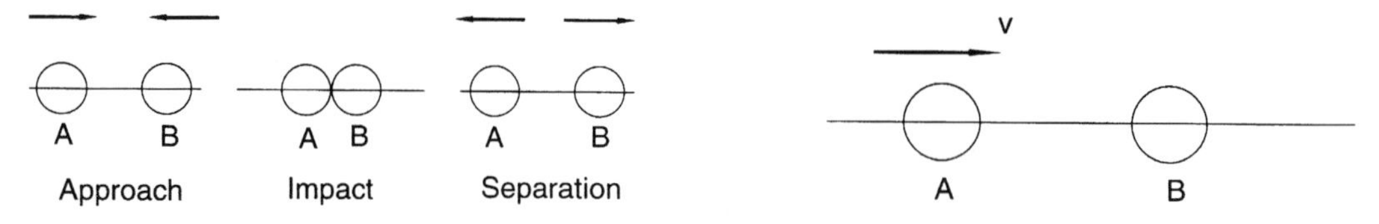

FIGURE 9.10.2
Colliding particles.

FIGURE 9.10.3
Colliding particles.

Because v^* is finite, x^* may be made arbitrarily small by making t^* small. That is, for very short impact time intervals, the displacement of P remains essentially unchanged.

Next, consider two particles A and B moving toward each other along a common line as in Figure 9.10.2. Let the relative speed of approach be v. Let the particles collide, rebound, and separate. Let the relative speed of separation be $\hat{v}$. Then, from our assumption about the impact, we have:

$$\hat{v} = ev \tag{9.10.4}$$

where e, called the *coefficient of restitution*, has values between zero and one ($0 \le e \le 1$). When e is zero, the separation speed $\hat{v}$ is zero, and the particles stay together after impact. Such collisions are said to be *plastic*. When e is one, the separation speed is the same as the approach speed (that is, $\hat{v} = v$). This is called an *elastic* collision.

Equation (9.10.4) may be used with the principle of conservation of momentum to study a variety of collision configurations. To illustrate this usage, consider first two colliding particles A and B, each having the same mass m. Let B initially be at rest, and let A approach B with speed v as in Figure 9.10.3. Then, from Eq. (9.10.4), the separation speed $\hat{v}$ of the particles is ev. That is, after collision,

$$\hat{v}_A - \hat{v}_B = ev \tag{9.10.5}$$

Also, because the impact generates only internal forces between the particles, the overall momentum of the system remains unchanged. That is,

$$mv = m\hat{v}_A + m\hat{v}_B \tag{9.10.6}$$

Solving Eqs. (9.10.5) and (9.10.6) for $\hat{v}_A$ and $\hat{v}_B$ we obtain:

$$\hat{v}_B = \left(\frac{1+e}{2}\right)v \quad \text{and} \quad \hat{v}_A = \left(\frac{1-e}{2}\right)v \tag{9.10.7}$$

Observe in Eq. (9.10.7) that if e is zero (plastic collision) $\hat{v}_A$ and $\hat{v}_B$ are equal with value $v/2$. If e is one, (elastic collision) then $\hat{v}_A$ is zero and $\hat{v}_B$ is v. In this latter case, the momentum of A is said to be transferred to B.

As a generalization of this example, consider two particles A and B with masses m_A and m_B and with each particle moving on the same line before impact. Let the pre-impact speeds be v_A and v_B. Let the post-impact speeds be $\hat{v}_A$ and $\hat{v}_B$ (see Figure 9.10.4). As before, we can determine $\hat{v}_A$ and $\hat{v}_B$ using Eq. (9.10.4) and the principle of conservation of linear momentum. From Eq. 9.10.4, we have:

$$\hat{v}_B - \hat{v}_A = e(v_A - v_B) \tag{9.10.8}$$

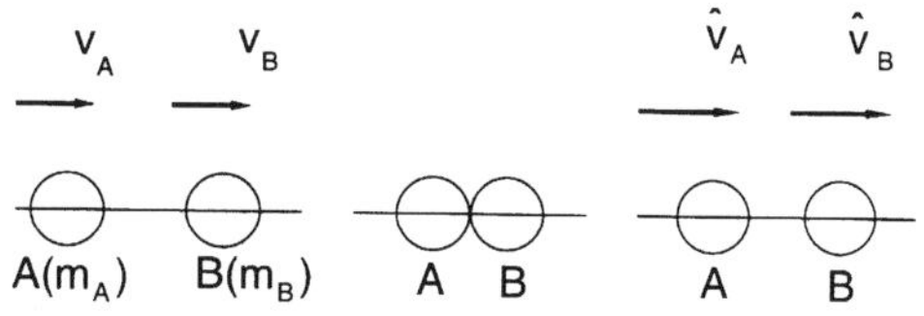

FIGURE 9.10.4
Colliding particles with different masses.

From the momentum conservation principle (see Eq. (9.7.1)), we have:

$$m_A v_A + m_B v_B = m_A \hat{v}_A + m_B \hat{v}_B \tag{9.10.9}$$

Then, by solving Eqs. (9.10.8) and (9.10.9) for $\hat{v}_A$ and $\hat{v}_B$ we obtain:

$$\hat{v}_A = (1/M)\left[(m_A - e m_B)v_A + (1+e)m_B v_B\right] \tag{9.10.10}$$

$$\hat{v}_B = (1/M)\left[m_A(1+e)v_A + (m_B - e m_A)v_B\right] \tag{9.10.11}$$

where M is defined as:

$$M = m_A + m_B \tag{9.10.12}$$

Consider three special cases: (1) $e = 0$, (2) $e = 1$, and (3) $m_A = m_B$.

Case 1: e = 0 (Plastic Collision)
In this case, Eqs. (9.10.10) and (9.10.11) simplify to the expressions:

$$\hat{v}_A = (m_A v_A + m_B v_B)/(m_A + m_B) \tag{9.10.13}$$

and

$$\hat{v}_B = (m_A v_A + m_B v_B)/(m_A + m_B) \tag{9.10.14}$$

Observe that $\hat{v}_A = \hat{v}_B$, as expected with a plastic collision.

Case 2: e = 1 (Elastic Collision)
In this case, Eqs. (9.10.10) and (9.10.11) become:

$$\hat{v}_A = \left[(m_A - m_B)v_A + 2m_B v_B\right]/(m_A + m_B) \tag{9.10.15}$$

$$\hat{v}_B = \left[2m_A v_A + (m_B - m_A)v_B\right]/(m_A + m_B) \tag{9.10.16}$$

Case 3: $m_A = m_B$
In this case, Eqs. (9.10.10) and (9.10.11) become:

$$\hat{v}_A = (1/2)\left[(1-e)v_A + (1+e)v_B\right] \tag{9.10.17}$$

$$\hat{v}_B = (1/2)[(1+e)v_A + (1-e)v_B] \tag{9.10.18}$$

Suppose further in this case that $e = 1$ (elastic collision). Then,

$$\hat{v}_A = v_B \quad \text{and} \quad \hat{v}_B = v_A \tag{9.10.19}$$

If, instead, $e = 0$ (plastic collision), then we have:

$$\hat{v}_A = (1/2)(v_A + v_B) \quad \text{and} \quad \hat{v}_B = (1/2)(v_A + v_B) \tag{9.10.20}$$

Finally, if v_B is zero in Eqs. (9.10.17) and (9.10.18), then $\hat{v}_A$ and $\hat{v}_B$ are:

$$\hat{v}_A = \left(\frac{1-e}{2}\right)v_A \quad \text{and} \quad \hat{v}_B = \left(\frac{1+e}{2}\right)v_A \tag{9.10.21}$$

These results are seen to be consistent with those of Eq. (9.10.7).

For still another specialization of Eqs. (9.10.10) and (9.10.11), let the mass of A greatly exceed that of B, let the collision be elastic, and let B be initially at rest. That is,

$$m_A >> m_B, \quad e = 1, \quad v_B = 0 \tag{9.10.22}$$

Then, $\hat{v}_A$ and $\hat{v}_B$ become:

$$\hat{v}_A = v_A \quad \text{and} \quad v_B = 2v_A \tag{9.10.23}$$

Observe that the struck particle B attains a speed twice as large as that of the striking particle A.

9.11 Oblique Impact

When colliding particles are moving along the same straight line (as in the foregoing examples) the collision is called *direct impact*. Generally, however, when particles (or bodies) collide they are not moving on the same line. Instead, they are usually moving on distinct curves that intersect. Such collisions are called *oblique impacts*. It happens that oblique impacts may be studied using the same principles we used with direct impacts. That is, oblique impact is treated as a direct impact in the direction normal to the plane of contact of the bodies. In directions parallel to the plane of contact, the momenta of the individual bodies are conserved.

To develop the analysis consider the oblique impact depicted in Figure 9.11.1, where A and B are particles moving on intersecting curves. Consider also the closer view of the impact provided in Figure 9.11.2. Let the particles be considered to be "small" bodies. That is, we will neglect rotational or angular momentum effects in the analysis. In Figure 9.11.2, let T represent a plane tangent to the contacting surfaces at the point of

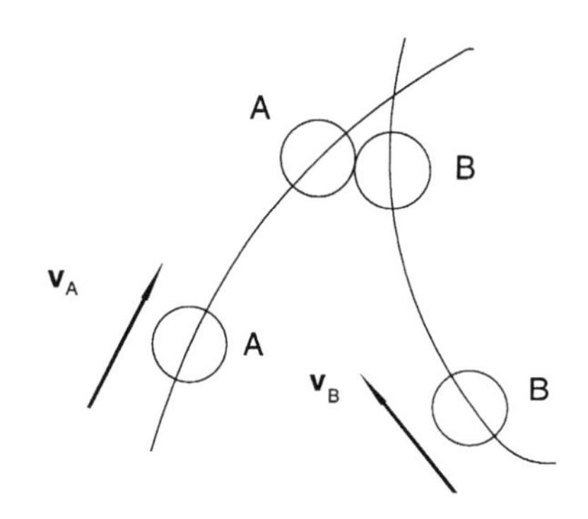

FIGURE 9.11.1
Colliding particles with curvilinear motion.

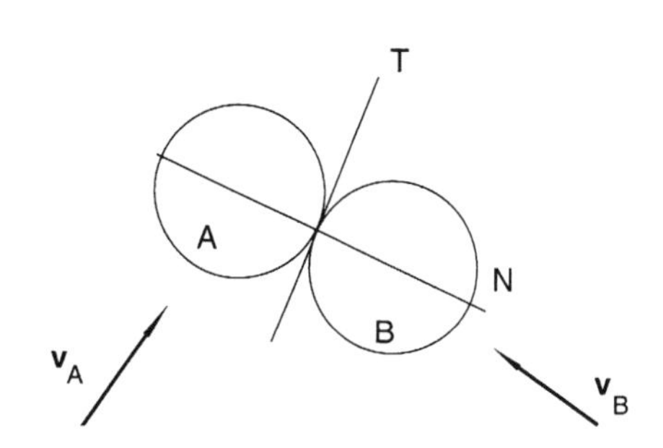

FIGURE 9.11.2
Close view of oblique impact.

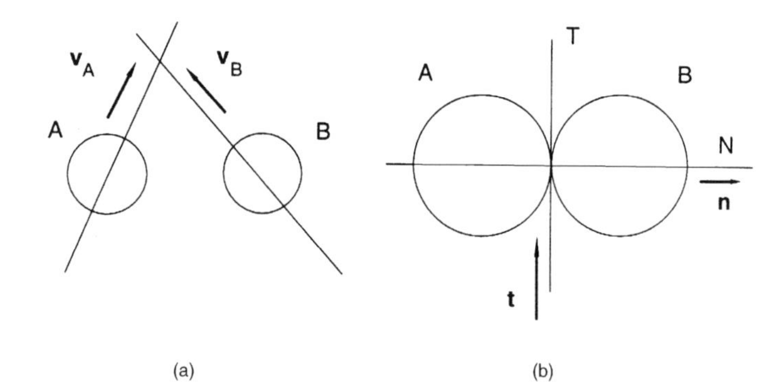

FIGURE 9.11.3
Colliding spheres.

contact, and let N be a line normal to T and passing through the contact point. We will consider the collision to be direct impact in the N direction. That is, the linear momentum of the system (A and B together) in the N direction is conserved. Also, in the N direction the particles collide and rebound according to the impact rule of Eq. (9.10.4). In directions parallel to the tangent plane, the momenta of the individual particles are conserved.

We can illustrate these ideas by considering the collision of two spheres (such as billiard balls) as in Figure 9.11.3. Let the balls A and B be moving in the same plane on intersecting lines. Let their velocities just before impact be $\mathbf{v}_A$ and $\mathbf{v}_B$ as in Figure 9.11.3a. Let the collision be modeled as in Figure 9.11.3b, where $\mathbf{n}$ and $\mathbf{t}$ are unit vectors normal and tangent to the plane of impact. Let $\mathbf{v}_A$ and $\mathbf{v}_B$ be expressed in terms of $\mathbf{n}$ and $\mathbf{t}$ as:

$$\mathbf{v}_A = v_n^A \mathbf{n} + v_t^A \mathbf{t} \quad \text{and} \quad \mathbf{v}_B = v_n^B \mathbf{n} + v_t^B \mathbf{t} \tag{9.11.1}$$

Similarly, let the velocities of A and B after impact be $\hat{\mathbf{v}}_A$ and $\hat{\mathbf{v}}_B$, and let $\hat{\mathbf{v}}_A$ and $\hat{\mathbf{v}}_B$ be expressed as:

$$\hat{\mathbf{v}}_A = \hat{v}_n^A \mathbf{n} + \hat{v}_t^A \mathbf{t} \quad \text{and} \quad \hat{\mathbf{v}}_B = \hat{v}_n^B \mathbf{n} + \hat{v}_t^B \mathbf{t} \tag{9.11.2}$$

Let the masses of A and B be m_A and m_B. Then, conservation of linear momentum of the system of both particles in the N direction produces the equations:

$$m_A v_n^A + m_B v_n^B = m_A \hat{v}_n^A + m_B \hat{v}_n^B \tag{9.11.3}$$

In the T direction, the momenta of each particle are conserved, producing the equations:

$$m_A v_t^A = m_A \hat{v}_t^A \tag{9.11.4}$$

and

$$m_B v_t^B = m_B \hat{v}_t^B \tag{9.11.5}$$

Finally, the impact–rebound condition of Eq. (9.10.4) in the N direction produces the equation:

$$\hat{v}_n^A - \hat{v}_n^A = -e\left(v_n^A - v_n^B\right) \tag{9.11.6}$$

Given $\mathbf{v}_A$ and $\mathbf{v}_B$, Eqs. (9.11.3) to (9.11.6) are four equations for the four post-impact velocity components: $\hat{v}_n^A$, $\hat{v}_t^A$, $\hat{v}_n^B$, and $\hat{v}_t^B$. Because Eqs. (9.11.4) and (9.11.5) are uncoupled, we can readily solve for these post-impact components. The results are:

$$\begin{aligned}
\hat{v}_t^A &= v_t^A \\
\hat{v}_t^B &= v_t^B \\
\hat{v}_n^A &= (1/M)\left[m_a v_n^A + m_B v_n^B - e m_B v_n^A + e m_b v_n^B\right] \\
\hat{v}_n^B &= (1/M)\left[e m_A v_n^A - e m_A v_n^B + m_A v_n^A + m_B v_n^B\right]
\end{aligned} \tag{9.11.7}$$

where M is defined as:

$$\mathbf{M} = \mathbf{m}_A + \mathbf{m}_B \tag{9.11.8}$$

For a numerical illustration of the application of these equations, consider the collision of two billiard balls moving prior to impact as in Figure 9.11.4. Let the balls have identical masses and radii, and let the collision be nearly elastic with a restitution coefficient of 0.95. From Figure 9.11.4a, the pre-impact speed components are:

$$v_n^A = 14.14\,\text{cm/sec}, v_t^A = 14.14\,\text{cm/sec}, v_n^B = -12.99\,\text{cm/sec}, v_t^B = 7.5\,\text{cm/sec} \tag{9.11.9}$$

Then, from Eq. (9.11.7), the post-impact speed components are:

$$\hat{v}_n^A = -12.31\,\text{cm/sec}, \hat{v}_t^A = 14.14\,\text{cm/sec}, \hat{v}_n^B = 1.12\,\text{cm/sec}, \hat{v}_t^B = 7.5\,\text{cm/sec} \tag{9.11.10}$$

The post-impact motion of the balls may then be depicted as in Figure 9.11.5.

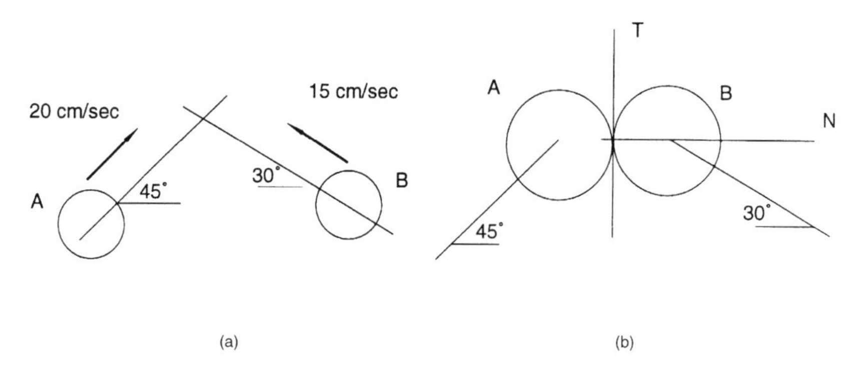

FIGURE 9.11.4
Colliding billiard balls.

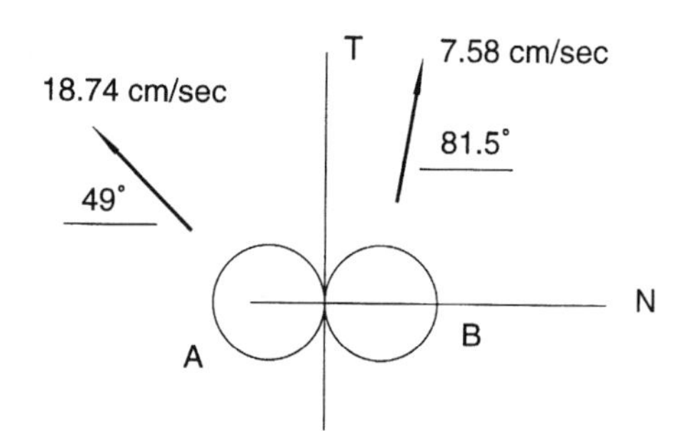

FIGURE 9.11.5
Post impact movement of colliding billiard balls of Figure 9.11.4.

9.12 Seizure of a Spinning, Diagonally Supported, Square Plate

As a final example illustrating impulse momentum concepts, consider a square plate with side length a and mass m suspended by a thin wire along a diagonal as in Figure 9.12.1. Let the plate be spinning with an angular speed $\mathbf{\Omega}$ as shown. Next, let the plate be suddenly arrested along an edge so that it is constrained to rotate about that edge as in Figure 9.12.2. Assuming that the arrested edge forms a frictionless hinge, a question arising is what is the ratio of the rotation speed $\hat{\mathbf{\Omega}}$ about the hinge relative to the rotation speed $\mathbf{\Omega}$ about the wire?

To answer this question, consider that the angular momentum of the plate about the seized edge, before and after seizure, is conserved. Let the seized edge be OQ. Then, from Figure 9.12.3, we see that the angular momentum of the plate about OQ before seizure is:

$$\mathbf{A}_{OQ} = (\mathbf{A}_O \cdot \mathbf{n})\mathbf{n} = (\mathbf{I}\mathbf{\Omega}\mathbf{k} \cdot \mathbf{n})\mathbf{n} = \left(\sqrt{2}/2\right)\mathbf{I}\mathbf{\Omega}\mathbf{n} \tag{9.12.1}$$

where $\mathbf{n}$ and $\mathbf{k}$ are unit vectors along the diagonal and edge of the plate as in Figure 9.12.3, and where I is the moment of inertia of the plate about its diagonal. From the table of inertia properties, we see that I is $(1/12)ma^2$. Hence, $\mathbf{A}_{OQ}$ before arresting is:

$$\mathbf{A}_{OQ} = \left(\sqrt{2}/24\right)ma^2\mathbf{n} \tag{9.12.2}$$

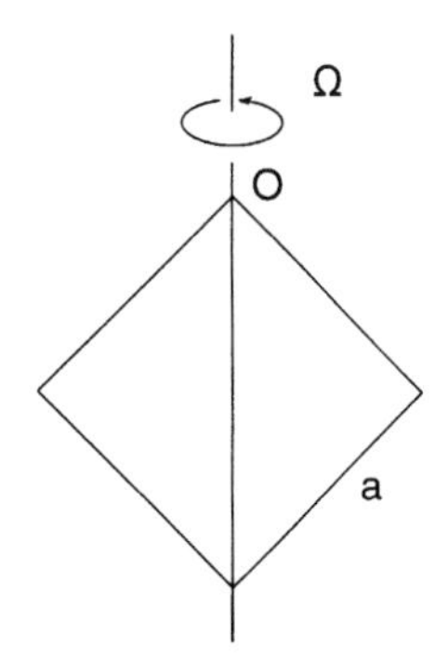

FIGURE 9.12.1
Suspended spinning square plate.

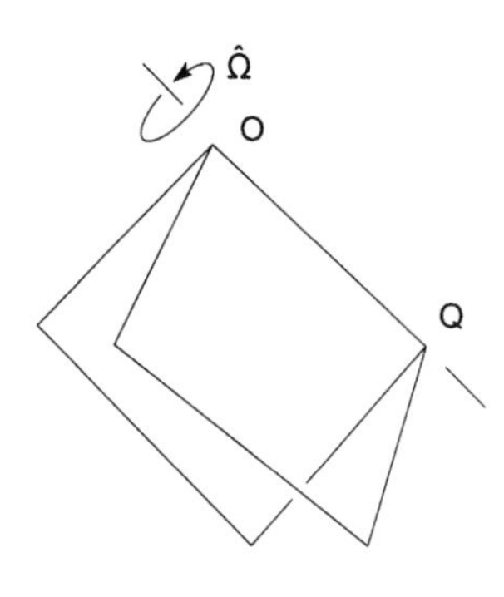

FIGURE 9.12.2
Arrested plate spinning about an edge.

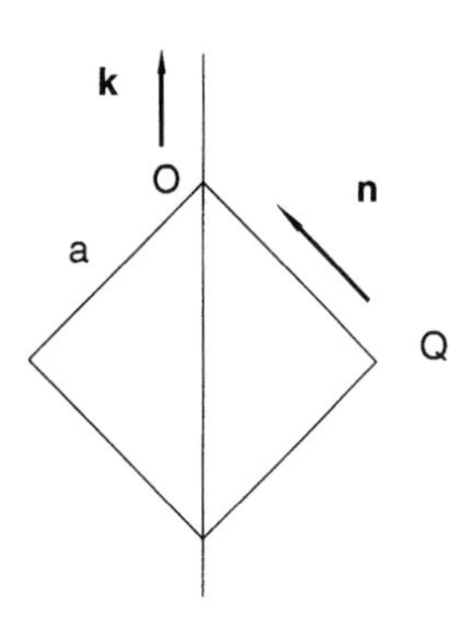

FIGURE 9.12.3
Plate geometry.

The angular momentum of the plate about the seized edge after seizure is:

$$\hat{\mathbf{A}}_{OQ} = I_{OQ}\hat{\Omega}\mathbf{n} \tag{9.12.3}$$

where I_{OQ} is the moment of inertia of the plate about edge OQ. From the parallel axis theorem (Eq. (7.6.6)), I_{OQ} is seen to be:

$$I_{OQ} = I + ma^2/4 = (1/12)ma^2 + (1/4)ma^2 = (1/3)ma^2 \tag{9.12.4}$$

(Note that due to symmetry the moment of inertia of a square plate about a diagonal is the same as the central moment of inertia about a line parallel to an edge.) Hence, $\hat{\mathbf{A}}_{OQ}$ becomes:

$$\hat{\mathbf{A}}_{OQ} = (1/3)ma^2\hat{\Omega}\mathbf{n} \tag{9.12.5}$$

By equating $\mathbf{A}_{OQ}$ and $\hat{\mathbf{A}}_{OQ}$ we obtain the result:

$$\hat{\boldsymbol{\Omega}} = \left(\sqrt{2}/8\right)\boldsymbol{\Omega} \quad \text{or} \quad \hat{\boldsymbol{\Omega}}/\boldsymbol{\Omega} = \sqrt{2}/8 \tag{9.12.6}$$

9.13 Closure

The principles of impulse and momentum are best applied with systems experiencing impact and collision. The impact forces are usually large but the time of application is short, thus producing a finite impulse. During this short impact time, the system configuration does not change significantly. The system velocities, however, may change substantially. Eqs. (9.5.14) and (9.6.37) are expressions typical of those which may be used for calculating these velocity changes.

The impulse–momentum principles are especially useful if an impact is contained within the system, as with a collision. With such events, the overall momentum of the system is conserved or unchanged. The analyst then need only be concerned with velocities, thus avoiding the calculation of accelerations and the evaluation of force systems as required with Newton's law and d'Alembert's principle. In the next chapter we will consider energy methods that also avoid the calculation of accelerations.

Problems

Section 9.2 Impulse

P9.2.1: A triangular impulsive force **F** has a profile with magnitude as in Figure P9.2.1. If **F** is in the direction of a unit vector **n**, determine the resulting impulse **I**.

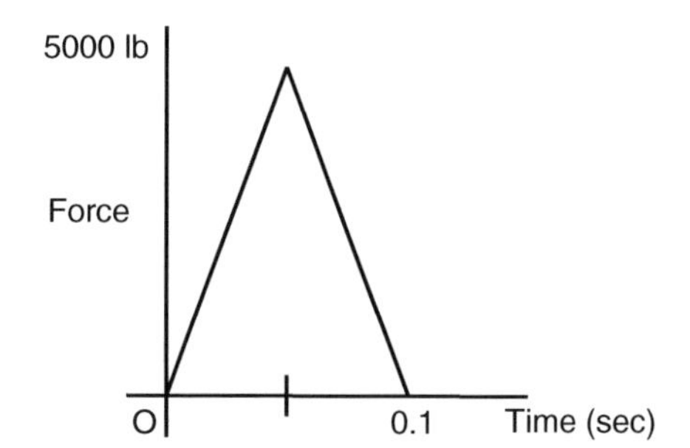

FIGURE P9.2.1
Impulsive force magnitude.

P9.2.2: A 3000-lb automobile collides with a fixed barrier producing a deceleration of the car as in Figure P9.2.2. Determine the magnitude of the impulse.

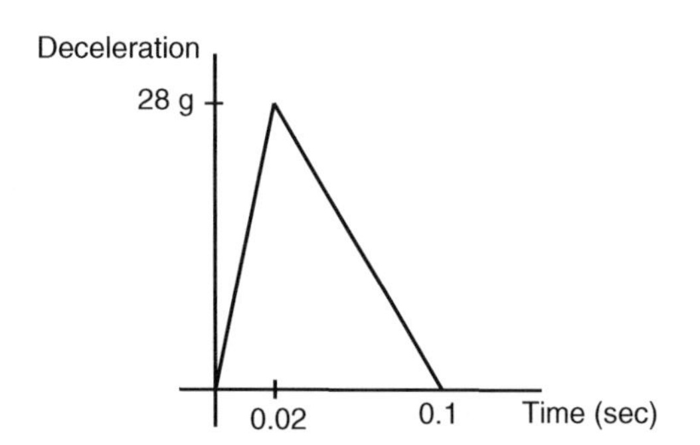

FIGURE P9.2.2
Deceleration profile.

P9.2.3: A brake for a rotor creates a moment about the axis of the rotor (or parallel to the axis of the rotor) with the magnitude of the moment depicted in Figure P9.2.3. Let **n** be a unit vector parallel to the axis of the rotor and in the direction of rotation of the rotor (according to the right-hand rule). Determine the angular impulse **J** of the braking moment.

Section 9.3 Linear Momentums

P9.3.1: A 15-lb particle *P* has a velocity $6\mathbf{n}_1 - 3\mathbf{n}_2 + 8\mathbf{n}_3$ ft/sec, where $\mathbf{n}_1$, $\mathbf{n}_2$, and $\mathbf{n}_3$ are mutually perpendicular unit vectors. Find the linear momentum of *P*. Express the result in both English and international (SI) units.

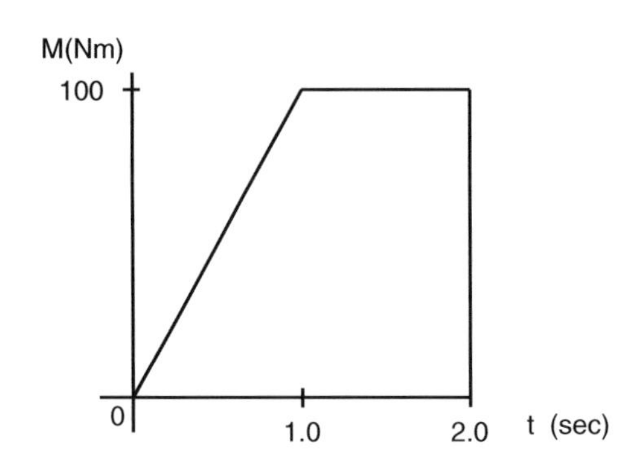

FIGURE P9.2.3
Braking moment magnitude.

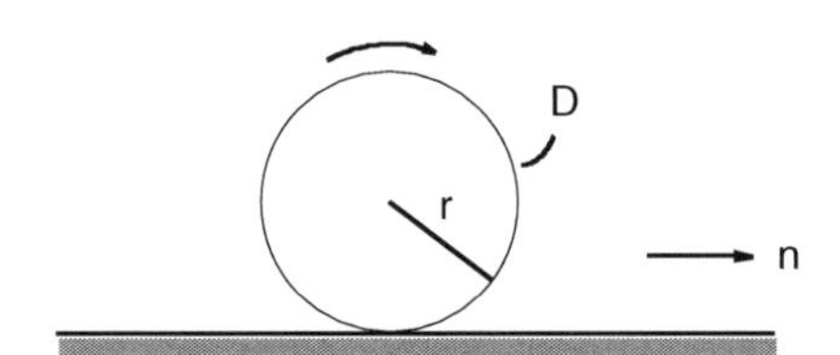

FIGURE P9.3.2
A disk *D* rolling in a straight line.

P9.3.2: A 4-kg circular disk *D* with radius *r* or 0.5 m rolls on a straight line with an angular speed of ω rad/sec (Figure P9.3.2). Find the linear momentum of *D*.

P9.3.3: A 3200-lb automobile *A* traveling at 35 mph collides with the rear of a stopped automobile *B* weighing 2800 lb. Following the collision, *A* and *B* move together as a unit. If the momenta of the automobiles are conserved during the collision (that is, the momentum of *A* just before impact is equal to the combined momenta of *A* and *B* just after impact), find the speed of the vehicles just after impact.

Section 9.4 Angular Momentum

P9.4.1: A 2-kg particle *P* moving in a Cartesian reference frame *R* has velocity **v** given by:

$$\mathbf{v} = 6\mathbf{n}_x - 3\mathbf{n}_y + 4\mathbf{n}_z \text{ m/sec}$$

where $\mathbf{n}_x$, $\mathbf{n}_y$, and $\mathbf{n}_z$ are unit vectors parallel to the *X*, *Y*, and *Z* coordinate axes of *R*. If the coordinates of *P* are (–1, 5, 4) (in meters), determine the angular momentum of *P* about the origin *O*.

P9.4.2: Repeat Problem P9.4.1 if *P* weighs 2 lb and if the units of the coordinates and velocity of *P* are in feet instead of meters.

P9.4.3: Repeat Problems P9.4.1 and P9.4.2 if the reference point instead of the origin is *Q* with coordinates (2, –1, 3).

P9.4.4: A set *S* of five particles P_i ($i = 1,\ldots, 5$) is moving in an inertial reference frame R. Let $\mathbf{n}_x$, $\mathbf{n}_y$, and $\mathbf{n}_z$ be mutually perpendicular unit vectors parallel to Cartesian axes *X*, *Y*, and *Z* fixed in *R*. Let the P_i have masses (m_i), coordinates (x_i, y_i, z_i), and velocity components (v_{ix}, v_{iy}, v_{iz} for $\mathbf{n}_x$, $\mathbf{n}_y$, $\mathbf{n}_z$) as listed in Table P9.4.4, where the masses are in kilograms, the coordinates are in meters, and the velocity components are in meters per second. Find the angular momentum of *S* about the origin O.

P9.4.5: See Problem P9.4.4. Let *G* be the mass center of *S*. Find the coordinates of *G*.

TABLE P9.4.4

Kinematic and Inertial Properties of Particles P_i

i	m_i	x_i	y_i	z_i	v_{ix}	v_{iy}	v_{iz}
1	2	1	–2	4	6	–2	3
2	1	8	3	–2	–3	5	2
3	3	–9	4	5	6	–4	8
4	4	–4	–7	6	–5	1	4
5	6	5	6	–4	3	6	–1

P9.4.6: See Problem P9.4.4. Show that the particles P_i of S, with velocity components as listed in Table P9.4.4, are *not* fixed on a rigid body B. (Hint: Select any two particles, say P_1 and P_2, and show that $\mathbf{V}^{P_1/P_2} \cdot \mathbf{P_1P_2} \neq 0$.)

P9.4.7: See Problems P9.4.4 to P9.4.6. Suppose the particles of Problem P9.4.4 are fixed relative to one another so that they form a rigid body B. Then, in view of Problem P9.4.6, the velocity components listed in Table P9.4.4 are no longer valid but instead are unknown, as represented in Table P9.4.7. Let G be the mass center of S (see Problem P9.4.5) and let G have velocity $\mathbf{V}^G$ given by:

$$\mathbf{V}^G = 5\mathbf{n}_x - 3\mathbf{n}_y + 7\mathbf{n}_z \text{ m/s}$$

Let B have angular velocity ω given by:

$$\boldsymbol{\omega} = 3\mathbf{n}_x - 2\mathbf{n}_y + 4\mathbf{n}_z \text{ rad/sec}$$

Find (a) the velocities of P_i ($i = 1,\ldots,5$) relative to G, and (b) the velocities of P_i relative to O.

TABLE P9.4.7

Position and Inertial Properties of Particles P_i

i	m_i	x_i	y_i	z_i	v_{ix} v_{iy} v_{iz}
1	2	1	–2	4	Unknown (to be determined)
2	1	8	3	–2	Unknown (to be determined)
3	3	–9	4	5	Unknown (to be determined)
4	4	–4	7	6	Unknown (to be determined)
5	6	5	6	–4	Unknown (to be determined)

P9.4.8: See Problems P9.4.4 to P9.4.7. Use the conditions and results of Problem P9.4.7 to find the angular momentum of S relative to O, $\mathbf{A}_{S/O}$; and the angular momentum of S relative to G, $\mathbf{A}_{S/G}$.

P9.4.9: See Problems P9.4.4 to P9.4.8. Let G have a mass of 16 kg. Find the angular momentum of G relative to O, $\mathbf{A}_{G/O}$.

P9.4.10: See Problems P9.4.8 and P9.4.9. Show that:

$$\mathbf{A}_{S/O} = \mathbf{A}_{S/G} + \mathbf{A}_{G/O}$$

(See Eq. (9.4.14).)

P9.4.11: A wheel W modeled as a thin circular disk with radius r and mass m rolls on a straight line as in Figure P9.4.11. If the center G has a speed V, find the angular momentum of (a) W about G, and (b) W about C.

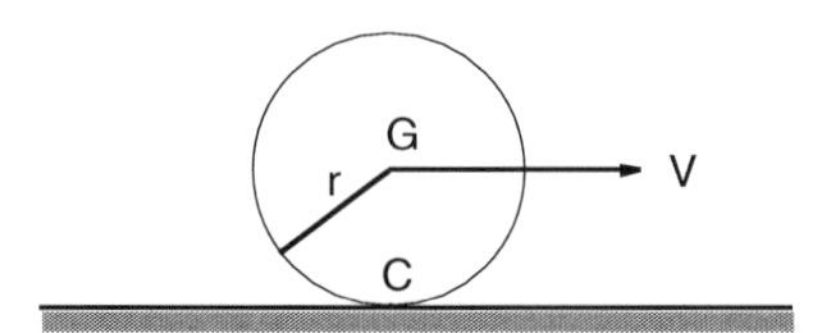

FIGURE P9.4.11
Wheel rolling in a straight line.

P9.4.12: Repeat Problem P9.4.11 if the wheel is not a disk but instead is characterized by an axial radius of gyration k. Let the radius and mass still be r and m.

P9.4.13: A wheel W modeled as a thin, circular disk with radius r and mass m rolls on a circle with radius R. Let W remain vertical during the motion so that its center G also moves on a circle with radius R. Determine the angular momentum of (a) W about G, and (b) W about its contact point C. Express the results in terms of an axial unit vector **i** and a vertical unit vector **k**.

P9.4.14: Repeat Problem P9.4.13 if the wheel is not a disk but instead is characterized by an axial radius of gyration k_x and a diametral radius of gyration k_z. Let the radius and mass still be r and m.

Section 9.5 Principle of Linear Impulse and Momentum

P9.5.1: A car traveling at 30 mph is brought to a skid stop on a level asphalt roadway with all four wheels locked. If the car weighs 3000 lb and if the coefficient of friction between the tires and the roadway is 0.75, determine the time t required for the car to skid to a stop.

P9.5.2: A 2850-lb test automobile is driven into a fixed, rigid barrier at 35 mph. Determine the impulse I exerted on the car by the barrier, assuming negligible rebound velocity.

P9.5.3: See Problem P9.5.2. Suppose the automobile in Problem P9.5.2 is stopped in 65 msec (0.065 sec). Determine the average impulsive force **F**.

P9.5.4: A 3-kg body B has a velocity $6\mathbf{n}_x - 3\mathbf{n}_y + 4\mathbf{n}_z$ m/sec in a Cartesian reference frame R where $\mathbf{n}_x$, $\mathbf{n}_y$, and $\mathbf{n}_z$ are unit vectors parallel to the X-, Y-, and Z-axes. A force **F**, expressed as $-18t\mathbf{n}_x + 9(t^2 - t)\mathbf{n}_y - 4t^3\mathbf{n}_z N$, is applied to B. Determine the velocity of B when time t is (a) 0.5 sec, (b) 1.0 sec, and (c) 1.5 sec.

Section 9.6 Principle of Angular Impulse and Momentum

P9.6.1: A drum brake acting on a rotor creates a moment M on the rotor as in Figure P9.6.1. Let the rotor have an axial moment of inertia of 4 kg m^2 and an angular speed of 420 rpm. Determine the speed reduction after (a) 0.05 sec, (b) 0.1 sec, and (c) 02 sec. Determine the time to bring the rotor to rest.

P9.6.2: A 20-lb body B has principal radii of gyration given by $k_x = 8$ in., $k_y = 4$ in., and $k_z = 4$ in. Let B have an angular velocity of $20n_x - 30n_y + 40n_z$ rad/sec, where n_x, n_y, and n_z are principal unit vectors of B for its mass center G. Let a moment be applied to B about G given by $t\mathbf{n}_x - 3t^3\mathbf{n}_y - 4(t - t^2)\mathbf{n}_z$ ft lb. Determine the angular velocity of B when t is (a) 0.5 sec, (b) 1.0 sec, and (c) 1.5 sec.

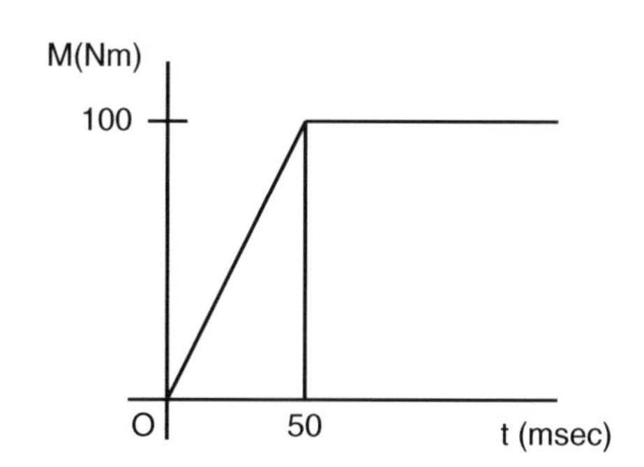

FIGURE P9.6.1
Moment applied to a rotor.

Sections 9.7, 9.8, and 9.9 Conservation of Momentum Principles and Examples

P9.7.1: A 3220-lb automobile traveling at 35 mph strikes the rear of a stopped 2800-lb car. Assuming that the vehicles move together after impact, find their common speed.

P9.7.2: A 3000-lb pickup truck traveling north at 55 mph collides head on with a 2500-lb southbound automobile traveling at 65 mph. Assuming the vehicles move together after impact, find their common speed and direction of movement.

P9.7.3: A 2900-lb eastbound automobile *A* collides at an intersection with a 3300-lb northbound automobile *B*, as shown in Figure P9.7.3. If after the collision the vehicles move together at 25 mph 30 degrees northeast, what were their speeds just prior to impact?

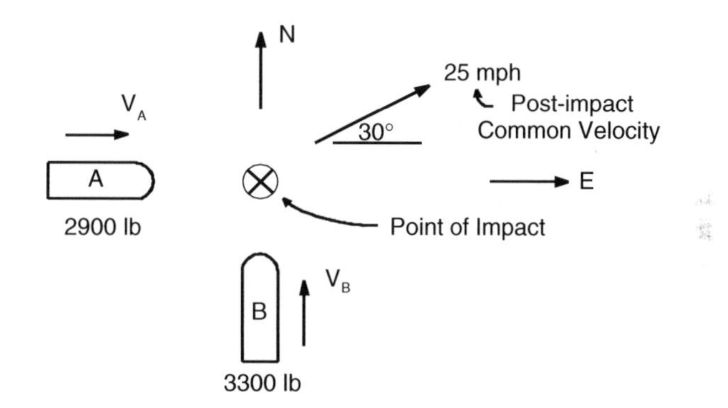

FIGURE P9.7.3
Intersection collision.

P9.7.4: See Problem P9.7.3. Suppose the pre-impact speeds of automobiles *A* and *B* are 30 mph and 25 mph, respectively. Find the common post-impact speed and direction.

P9.7.5: A 3500-lb automobile *A* traveling at 30 mph strikes a stopped automobile *B* in the right rear wheel, as shown in Figure P9.7.5. Let *B* weigh 2800 lb and let the mass center

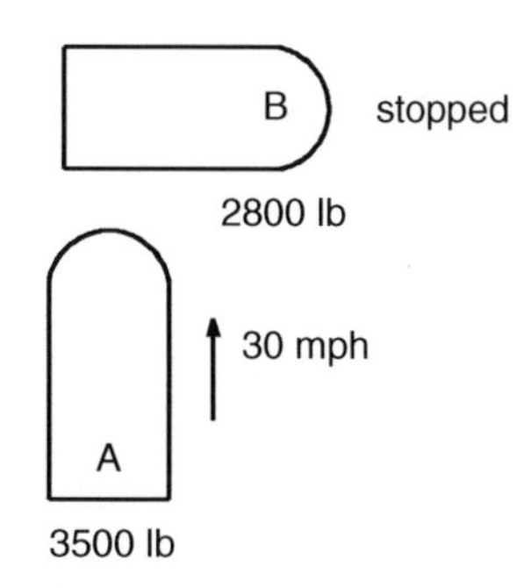

FIGURE P9.7.5
Side impact to the rear of a stopped vehicle.

of B be 64 in. in front of the rear wheels. Finally, let the central moment of inertia of B about a vertical axis be 1775 slug ft^2. Determine the angular velocity of B immediately after impact. (Assume that A remains in contact with the right rear wheel of B for a brief time after impact.)

P9.7.6: A body B modeled as a 36-in.-long bar PQ is moving to the right at 44 ft/sec as in Figure P9.7.6a. Suppose end Q of the bar encounters and strikes a small ledge as in Figure 9.7.6b and begins to rotate about the ledge corner O. Determine (a) the angular speed of B, (b) the velocity of P, and (c) the velocity of mass center G of B, immediately after impact.

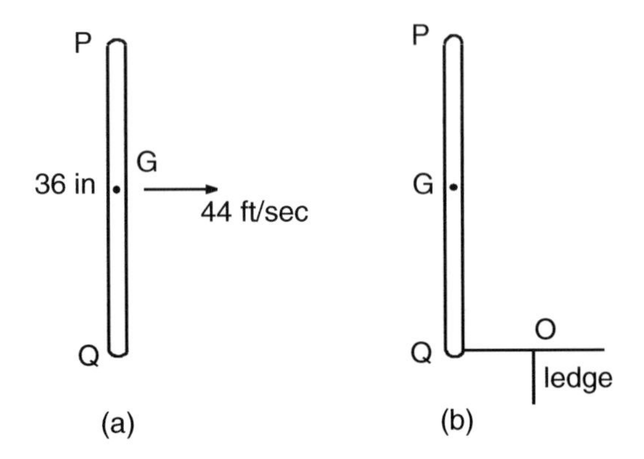

FIGURE P9.7.6
A moving bar striking a ledge at an end of the bar.

P9.7.7: A ball B weighing 5 lb is held inside a rotating tube T by an 8-in. string as depicted in Figure P9.7.7. In this position, B has a speed of 15 ft/sec. Let T have a smooth interior surface, let T be light, and let T be free to rotate about a vertical axis as shown. Let the restraining string be cut, allowing B to slide 19 in. to the end A of T. Determine the angular velocities of T before and after the string is cut.

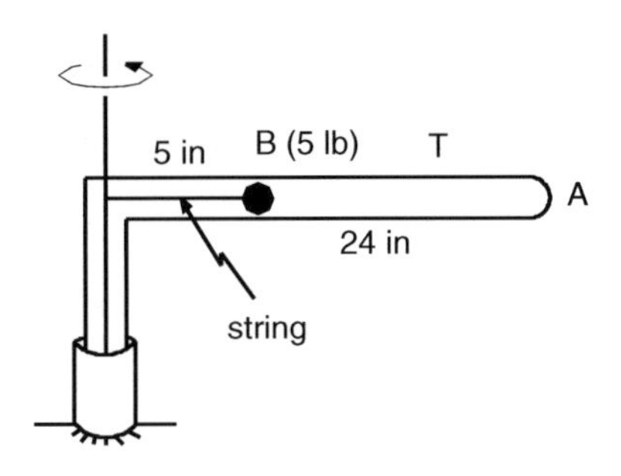

FIGURE P9.7.7
A ball inside a light, smooth rotating tube.

P9.7.8: Repeat Problem P9.7.7 if B weighs 8 lb.

P9.7.9: A bullet of mass m is shot into a vertical hanging bar with length ℓ and mass M at a distance h below a pin support O of the bar (Figure P9.7.9). If the speed of the bullet just before impact is V, determine the angular speed of the bar just after impact. Assume that the pin is frictionless and that m is small compared with M.

P9.7.10: Repeat Problem P9.7.9 if $m = 28$ g, $M = 5.5$ kg, $V = 700$ m/sec, $\ell = 1$ m, and h = 0.8 m.

P9.7.11: See Problem P9.7.6. A plate $ABCD$ with sides 3 m and 4 m is moving to the right at 10 m/sec, as represented in Figure P9.7.11. Corner D of the plate encounters and strikes a fixed ledge and begins to rotate about corner O of the ledge. Determine the $\mathbf{n}_1$, $\mathbf{n}_2$, and $\mathbf{n}_3$ components of the angular velocity of the plate just after impact. (As before, the $\mathbf{n}_i$ are mutually perpendicular unit vectors.)

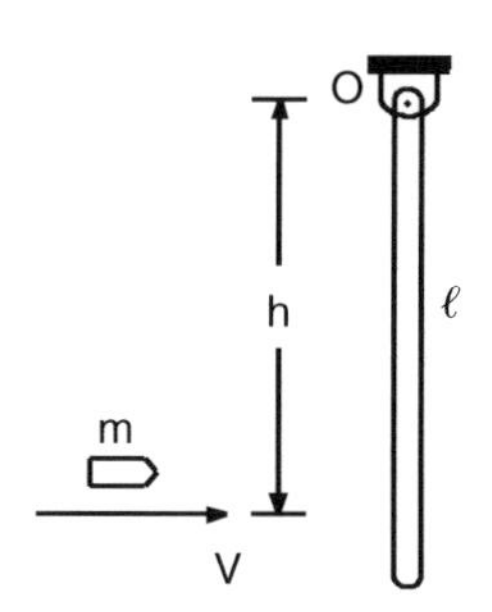

FIGURE P9.7.9
A bullet fired into a hanging bar.

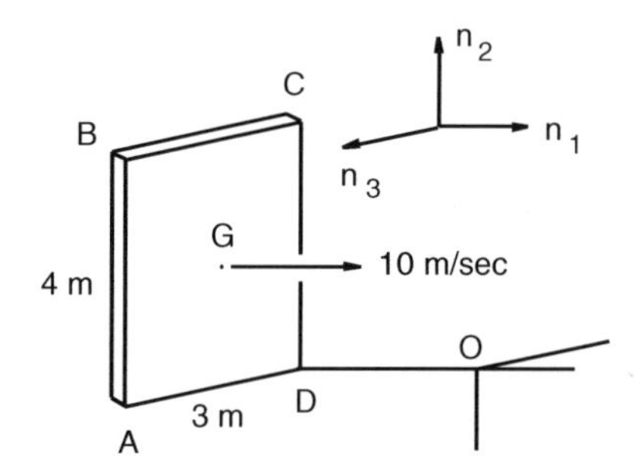

FIGURE P9.7.11
A rectangular plate striking a fixed ledge.

P9.7.12: See Problem P9.7.11. Determine the velocities of points A, B, and C and the mass center G of the plate just after impact.

P9.7.13: See Problem P9.7.10. In Problem 9.7.10 find the distance h so that the bullet strikes the bar at the center of percussion so that there is no horizontal reaction at the pin at O. Repeat Problem 9.7.10 using this value of h instead of 0.8 m.

P9.7.14: Consider again the pin-connected double bar of Figure 9.8.7 and shown again in Figure P9.7.14. Let the bars each have length 1.0 m and mass 0.5 kg. Let the impulse of the force P be 150 Nm. Determine the velocity of end O and the angular velocities of the bars just after impact.

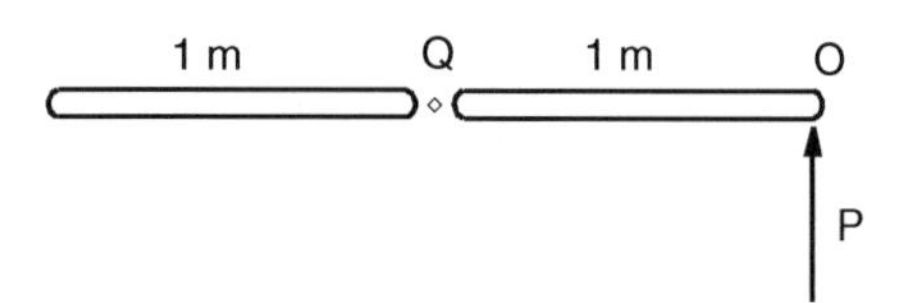

FIGURE P9.7.14
Pin-connected bars struck at one end.

P9.7.15: Repeat the analysis of Example 9.8.4 if the bars are no longer identical but instead have lengths ℓ_1 and ℓ_2 and masses m_1 and m_2 where the masses are proportional to the lengths.

P9.7.16: An automobile wheel and tire W weighing 50 lb with diameter 24 in. and axial radius of gyration 7 in. is rolling at 15 mph toward a 6-in. curb. Find the angular velocity of W immediately after impact with the curb.

Section 9.10 Impact: Coefficient of Restitution

P9.10.1: A car A traveling at 20 mph strikes the rear of a car B which is at rest as in Figure P9.10.1. Let the weights of A and B be 3220 lb and 2800 lb, respectively. Determine the speeds of A and B immediately after collision under the following assumptions.

a. A perfectly plastic collision: $e = 0$
b. A semi-elastic collision: $e = 0.5$
c. A perfectly elastic collision: $e = 1.0$

where e is the coefficient of restitution.

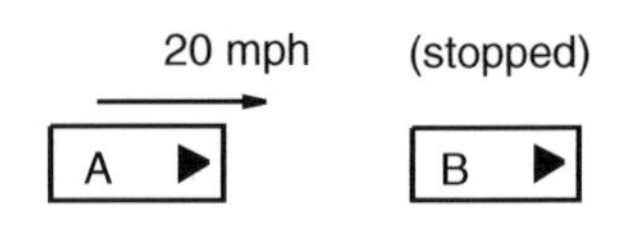

FIGURE P9.10.1
A rear-end collision.

P9.10.2: Repeat Problem P9.10.1 if B is moving to the right at 10 mph just before impact.

P9.10.3: See Problems 9.10.1 and 9.10.2. Repeat Problem P9.10.1 if B is moving to the left at 10 mph (that is, B is backing).

P9.10.4: A basketball B is dropped from a height h onto a fixed horizontal surface S. It is known that the speed v of B just before it strikes S is $\sqrt{2gh}$ (g is gravity acceleration). Also it is known that if B is projected upward from S, after bouncing, with a speed $\hat{v}$ it will reach a height $\hat{h}$ given by $\hat{v}^2/2g$. Let e be the coefficient of restitution between B and S. Let h be 7 ft and e be 0.9. Find v, $\hat{v}$ and $\hat{h}$.

P9.10.5: See Problem P9.10.4. If h is 7 ft and $\hat{h}$ is 6.5 ft, what is e?

Section 9.11 Oblique Impact

P9.11.1: Two spheres, A and B, collide as in Figure P9.11.1. Let their masses be $m_A = 0.5$ kg and $m_B = -.75$ kg. Let the collision be nearly elastic so that the coefficient of restitution e is 0.9. Determine the velocities (speeds and directions) of A and B immediately after impact.

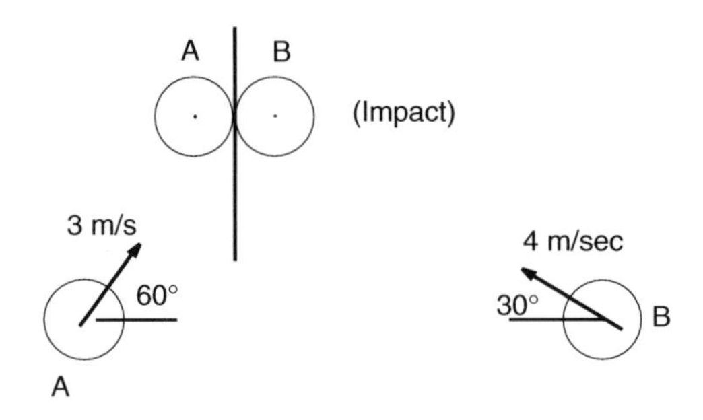

FIGURE P9.11.1
Colliding spheres.

P9.11.2: Repeat Problem P9.11.1 if $m_A = 0.75$ kg, $m_B = 0.5$ kg, and $e = 0.85$.

P9.11.3: A test car A collides with a fixed, angled barrier as in Figure P9.11.3. Let the barrier surface S be smooth and rigid so that it does not resist motion in its tangential direction. Determine the velocity (speed and direction) of A immediately after impact if the coefficient of restitution between A and C is (a) 0 (plastic), (b) 0.75 (semi-elastic), and (c) 1.0 (elastic).

P9.11.4: Repeat Problem P9.11.3 if the barrier angle is at 45°.

P9.11.5: Two cars A and B have an intersection collision at 45° to their pre-impact direction of travel as in Figure P9.11.4. Determine the post-impact velocity of the cars for the following conditions: $m_A = 100$ slug, $m_B = 80$ slug, $V_A = 30$ mph, $V_B = 25$ mph, and the coefficient of restitution $e = 0.75$. Assume the colliding surfaces are smooth. (Express the results in terms of the unit vectors $\mathbf{n}_n$ and $\mathbf{n}_e$ of Figure P9.11.4.)

P9.11.6: Repeat Problem P9.11.4 if the collision is (a) perfectly plastic ($e = 0$), and (b) perfectly elastic ($e = 1$).

P9.11.7: Repeat Problems P9.11.4 and P9.11.5 if $m_A = 150$ kg, $m_B = 1200$ kg, $V_A = 45$ km/hr, and $V_B = 37$ km/hr.

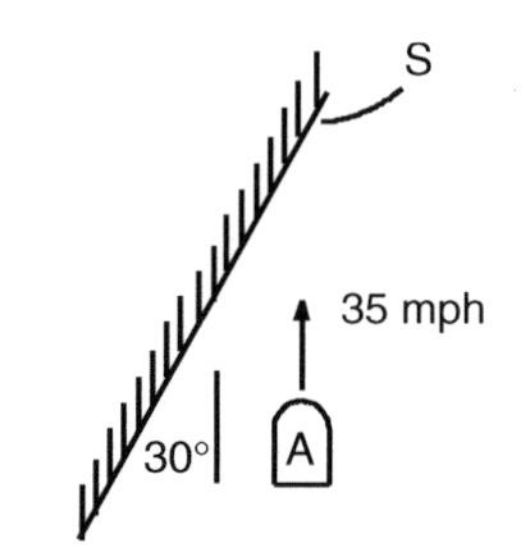

FIGURE P9.11.3
A test car colliding with a fixed rigid barrier.

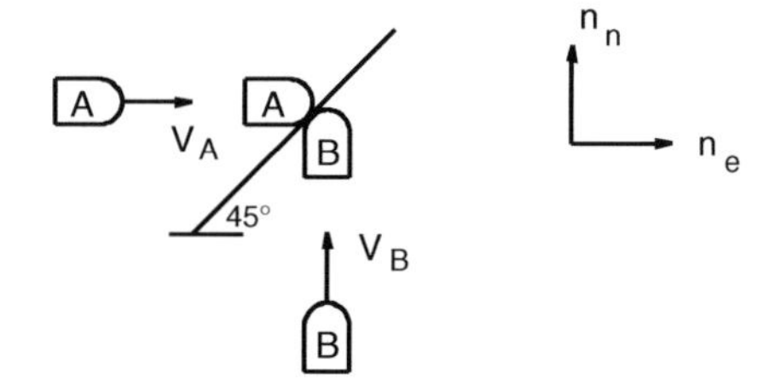

FIGURE P9.11.4
Intersection collision of cars.

10

Introduction to Energy Methods

10.1 Introduction

In this chapter, we consider energy methods with a focus on the work–energy principle. Energy methods are very convenient for a broad class of systems — particularly those with relatively simple geometrics and those for which limited information is desired. Energy methods, like impulse–momentum principles, are formulated in terms of velocities, thus avoiding the computation of accelerations as is required with Newton's laws and d'Alembert's principle. But, unlike the impulse–momentum principles, energy methods are formulated in terms of scalars. By thus avoiding vector operations, energy methods generally involve simpler analyses. The information gained, however, may be somewhat limited because often only one equation is obtained with the work–energy method.

We begin our study with a brief discussion of "work" and its computation. We then discuss power and kinetic energy and their relation to work. The balance of the chapter is then devoted to examples illustrating applications and combined use with impulse–momentum principles. In the next chapter we will discuss more advanced energy methods and the concepts of generalized mechanics.

10.2 Work

Intuitively, *work* is related to expended effort or expenditure of energy. In elementary physics, work is defined as the product of a force (effort) and the displacement (movement) of an object to which the force is applied. To develop these concepts, let P be a particle and let $\mathbf{F}$ be a force applied to P as represented in Figure 10.2.1. Let P move through a distance d in the direction of $\mathbf{F}$ as shown. Then, the work W done by $\mathbf{F}$ is defined as:

$$\mathbf{W} \stackrel{D}{=} |\mathbf{F}| d \tag{10.2.1}$$

Generally when a force is applied on a particle (or object) the force does not remain constant during the movement of the particle (or object). Both the magnitude and the direction of the force may change. Also, the particle (or object) will in general not move on a straight line.

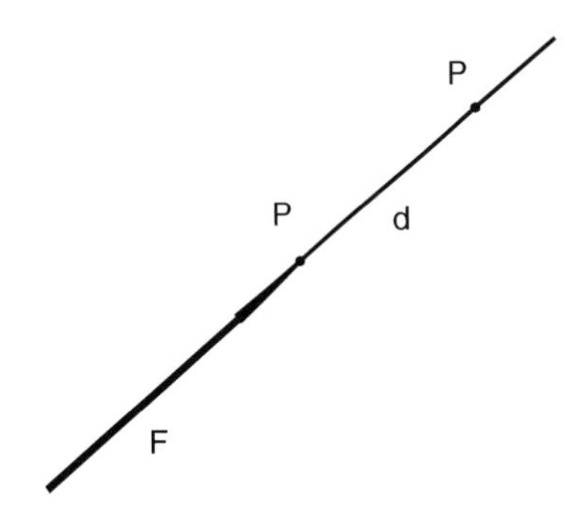

FIGURE 10.2.1
A force **F** moving a particle *P*.

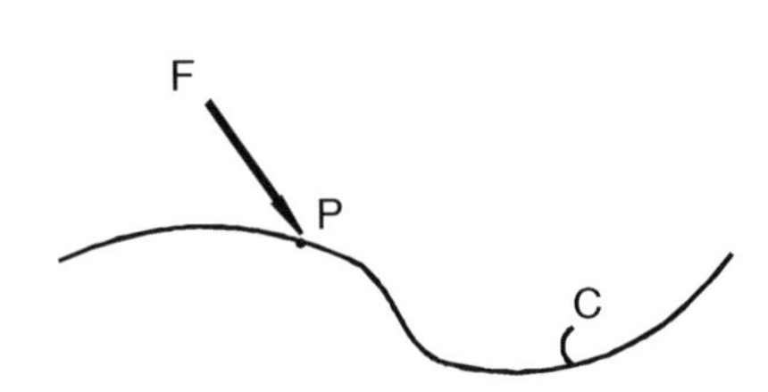

FIGURE 10.2.2.
A force **F** applied to a particle *P* moving on a curve *C*.

In view of these observations, it is necessary to generalize the definition of Eq. (10.2.1): specifically, let a force **F** be applied to a particle *P* which moves along a curve *C* as in Figure 10.2.2. Then, the work *W* done by **F** as *P* moves along *C* is defined as:

$$\mathbf{W} \overset{D}{=} \int_0^{\delta} \mathbf{F} \cdot \mathbf{ds} \tag{10.2.2}$$

where **ds** is a differential arc length vector tangent to *C* at the position of *P* and where δ is the distance *P* moves along *C* under the action of **F**.

From this generalized definition of work we see that if **F** is always directed perpendicular to the movement of *P*, the work is zero. Also, we see that if *P* moves in a direction opposed to **F**, the work is negative.

As an example, consider a particle *P* moving on a curve *C* defined by the parametric equations:

$$x = r\cos t, \quad y = r\sin t, \quad z = t \tag{10.2.3}$$

where *C* may be recognized as a circular helix as depicted in Figure 10.2.3. Let **F** be a force defined as:

$$\mathbf{F} = t\mathbf{n}_x + t^2\mathbf{n}_y + t^3\mathbf{n}_z \tag{10.2.4}$$

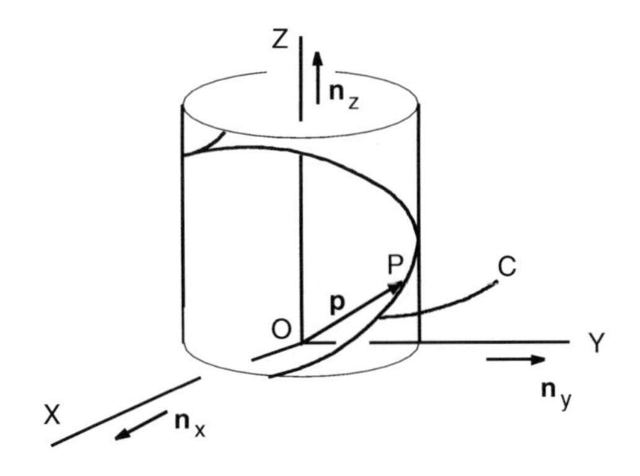

FIGURE 10.2.3
A particle moving on a circular helix.

where $\mathbf{n}_x$, $\mathbf{n}_y$, and $\mathbf{n}_z$ are unit vectors parallel to the coordinate axes as shown. We can determine the work done by **F** on P as follows: knowing **F**, we need an expression for **ds**, the differential arc length along C. Because the direction of **ds** is tangent to C, **ds** may be expressed as:

$$\mathbf{ds} = \boldsymbol{\tau} ds \tag{10.2.5}$$

where $\boldsymbol{\tau}$ is a unit vector tangent to C at the position of P and ds is the differential arc length.

We can obtain expressions for both $\boldsymbol{\tau}$ and **ds** from Eq. (10.2.3), the defining equations of C. Recall that the velocity of a particle is tangent to its path (or curve) of travel. Hence, by considering the parameter t in Eq. (10.2.3) as time, the velocity of P is:

$$\mathbf{v} = \dot{x}\mathbf{n}_x + \dot{y}\mathbf{n}_y + \dot{z}\mathbf{n}_z = -r\sin t\,\mathbf{n}_x = r\cos t\,\mathbf{n}_y + \mathbf{n}_z \tag{10.2.6}$$

Then, $\boldsymbol{\tau}$ becomes:

$$\boldsymbol{\tau} = \mathbf{v}/|\mathbf{v}| = \left(-r\sin t\,\mathbf{n}_x + r\cos t\,\mathbf{n}_y + \mathbf{n}_z\right)/\left(1+r^2\right)^{1/2} \tag{10.2.7}$$

The differential arc length is:

$$\mathbf{ds} = \left(\dot{x}^2 + \dot{y}^2 + \dot{z}^2\right)^{1/2} dt = \left(1+r^2\right)^{1/2} dt \tag{10.2.8}$$

By substituting from Eqs. (10.2.4), (10.2.7), and (10.2.8) into (10.2.2), we find the work of **F** to be:

$$W = \int_0^{t*} \mathbf{F}\cdot\mathbf{ds} = \int_0^{t*} \left(-rt\sin t + rt^2\cos t + t^3\right)^{1/2} dt \tag{10.2.9}$$

where t^* is the value of t locating the ending position of P. For example, if t^* is π, W becomes:

$$\begin{aligned} W &= \left\{-r\left[\sin t - t\cos t\right] + r\left[2t\cos t + \left(t^2-2\right)\sin t\right] + t^4/4\right\}\Big|_0^{\pi} \\ &= -3\pi r \end{aligned} \tag{10.2.10}$$

By comparing Eqs. (10.2.5) and (10.2.7), we see that **ds** may be written as:

$$\mathbf{ds} = v ds/|v| \tag{10.2.11}$$

By recognizing $|\mathbf{v}|$ as ds/dt, **ds** becomes:

$$\mathbf{ds} = v dt \tag{10.2.12}$$

Hence, the definition of work of Eq. (10.2.2) takes the form:

$$W = \int_0^{t*} \mathbf{F} \cdot \mathbf{v}\, dt \tag{10.2.13}$$

where t^* is the time of action of **F**. The integrand of Eq. (10.2.13), **F** • **v**, is often called the *power* of **F** (see Section 10.4).

As a second example, consider the work done by gravity on a simple pendulum as it falls from a horizontal position to the vertical equilibrium position as in Figure 10.2.4. Let the pendulum mass be m and let its length be ℓ as shown. The gravity (or weight) force is, then,

$$\mathbf{w} = mg\mathbf{k} \tag{10.2.14}$$

where **k** is a vertically downward directed unit vector as shown in Figure 10.2.4.

To apply Eq. (10.2.2), consider that the differential arc vector may be expressed as:

$$\mathbf{ds} = \ell d\phi \mathbf{n}_\phi \tag{10.2.15}$$

where ϕ measures the angle of the pendulum to the horizontal and $\mathbf{n}_\phi$ is a unit vector tangent to the circular arc of the pendulum as shown in Figure 10.2.4. Hence, the work of the weight force is:

$$\begin{aligned} W &= \int_{\phi=0}^{\phi=\pi/2} \mathbf{w} \cdot \mathbf{ds} \int_0^{\pi/2} mg\mathbf{k} \cdot \ell \mathbf{n}_\phi d\phi \\ &= mg\ell \int_0^{\pi/2} \cos\phi\, d\phi = mg\ell \sin\phi \Big|_0^{\pi/2} = mg\ell \end{aligned} \tag{10.2.16}$$

Observe from the next to last expression of Eq. (10.2.16) that the work done by gravity as the pendulum falls through an arbitrary angle ϕ is:

$$W = mg\ell \sin\theta \tag{10.2.17}$$

The distance $\ell \sin\phi$ may be recognized as the vertical drop h of the pendulum; hence, the gravitational work is:

$$W = mgh \tag{10.2.18}$$

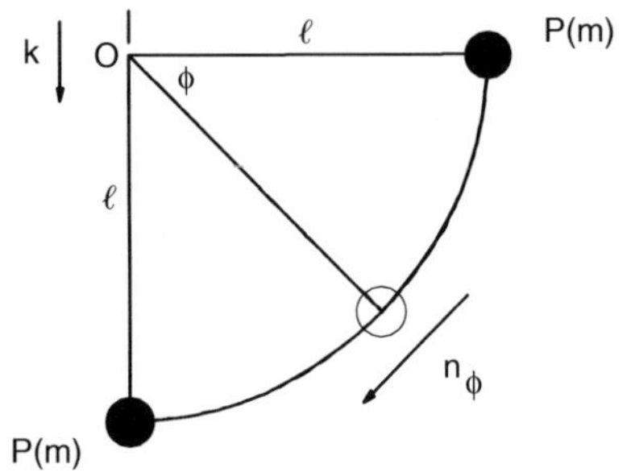

FIGURE 10.2.4
A falling pendulum.

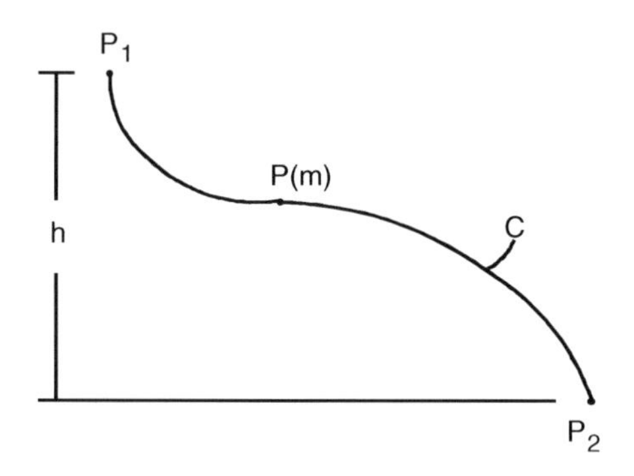

FIGURE 10.2.5
A particle P moving from P_1 to P_2 under the action of gravity.

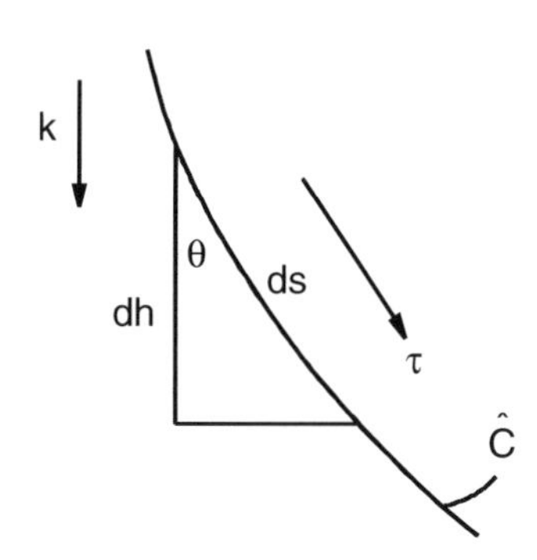

FIGURE 10.2.6
A differential arc of an arbitrary curve $\hat{C}$.

It happens that Eq. (10.2.18) is a valid expression for the work of the gravity force regardless of the path of descent of the particle P. In the pendulum example, the particle path was a circular arc. However, if P had moved on some other path (say, $\hat{C}$, as in Figure 10.2.5), the work done by gravity would still be mgh, where h is the change in elevation of P as it descends from P_1 to P_2 along $\hat{C}$. To see this, consider a differential arc of $\hat{C}$ as in Figure 10.2.6 where $\boldsymbol{\tau}$ is a unit vector tangent to $\hat{C}$ and $\mathbf{k}$ is a vertically downward directed unit vector. Then, the integrand of Eq. (10.2.2) becomes:

$$\mathbf{w} \cdot \mathbf{ds} = mg\,\mathbf{k} \cdot \boldsymbol{\tau} ds = mg \cos\theta\, ds = mg\, dh \tag{10.2.19}$$

where θ is the angle between $\hat{C}$ and the vertical as in Figure 10.2.6 and dh is the differential elevation change as shown. Hence, integrating the work done by gravity is:

$$W = \int mgdh = mgh \tag{10.2.20}$$

For a third example, consider the work done by a force in stretching or compressing a linear spring. Specifically, consider a spring as depicted in Figure 10.2.7. Let F be the magnitude of the stretching force $\mathbf{F}$. Let the natural length of the spring be ℓ, and let its stretched length, under the action of $\mathbf{F}$, be $\ell + \delta$. Then, from Eq. (10.2.2), the work is:

$$W = \int_0^\delta F dx = \int_0^\delta kx\, dx = (1/2)k\delta^2 \tag{10.2.21}$$

where k is the spring modulus, and x is the end displacement along the axis of the spring.

Similarly, if the spring is compressed, the work done by the compressing force is $(1/2)k\delta^2$, where now δ is a measure of the shortening of the spring.

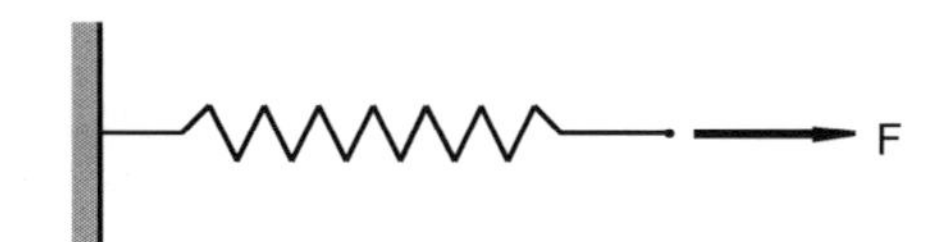

FIGURE 10.2.7
A linear spring stretched by a force **F**.

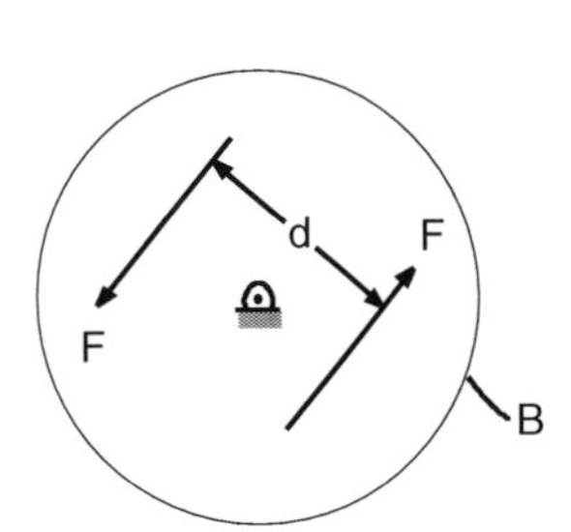

FIGURE 10.3.1
A couple applied to a flywheel.

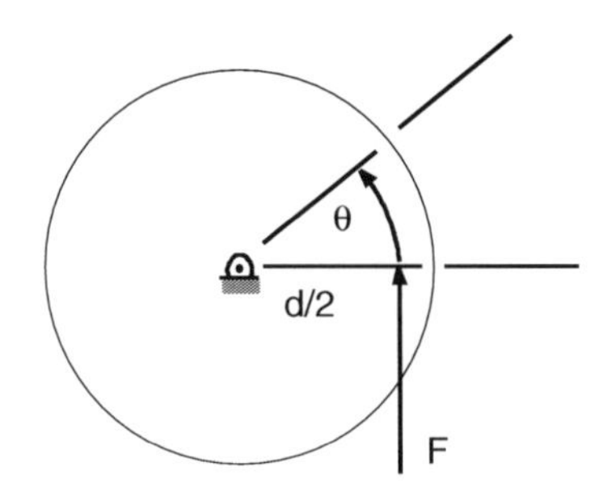

FIGURE 10.3.2
Displacement at the point of application of the force.

10.3 Work Done by a Couple

Suppose a couple is applied to a body B, causing B to rotate. Then, from our intuitive understanding of work, as an exertion creating a movement, we expect that work has been performed by the couple. To develop and quantify this, consider first a flywheel B supported by a frictionless pin at its center. Let a couple C consisting of two equal magnitude, but oppositely directed, forces be applied to B as in Figure 10.3.1. If B is initially at rest, the application of the couple will cause B to rotate about its pin. Specifically, the points of application of the forces will experience displacements with the forces, thus doing work.

Let the forces of C change orientation as B rotates (that is, let C "rotate" with B). Consider the work of one of these forces: from Figure 10.3.2, we see that the displacement at the point of application of the force is $(d/2)\theta$ where d is the distance between the couple forces and θ is the rotation angle. Then, from our definition of work of Eq. (10.2.2) we see that the work done by one of the couple forces is:

$$W = |\mathbf{F}|(d/2)\theta \tag{10.3.1}$$

Hence, the work done by the couple is:

$$W = |\mathbf{F}|d\theta = T\theta \tag{10.3.2}$$

where T is the torque of the couple (see Section 6.4).

Consider now a generalization of this example where a couple C is applied to a body B with B instantaneously rotating through a differential angle $d\theta$ about an axis parallel to a unit vector $\mathbf{n}$. Then, from Eq. (10.3.2), we see that the work done by C is:

$$W = \int_0^{\theta^*} \mathbf{T}\cdot\mathbf{n}\,d\theta \tag{10.3.3}$$

where θ^* is the total angle of rotation while C is being applied, and $\mathbf{T}$ is the torque of C.

A question that arises, however, is how do we determine **n**, the unit vector parallel to the axis of rotation of B? To answer this question, recall the example of the previous section where a particle P was pushed along a circular helix. In that example, we obtained the direction of motion of P from the velocity vector. In like manner, the instantaneous axis of rotation of body B is parallel to the angular velocity $\boldsymbol{\omega}$ of B. Then, in Eq. (10.3.3), **n** is given by:

$$\mathbf{n} = \boldsymbol{\omega}/|\boldsymbol{\omega}| \tag{10.3.4}$$

Because angular velocity is a measure of the rate of change of orientation, we see that $|\boldsymbol{\omega}|$ may be identified with $d\theta/dt$. Hence, we may also express the work of Eq. (10.3.3) in the form:

$$W = \int_0^{t^*} \mathbf{T} \cdot \boldsymbol{\omega} dt \tag{10.3.5}$$

where t^* is the time of action of **T**. Observe the similarity of Eqs. (10.2.13) and (10.3.5).

10.4 Power

Power is defined as the *rate* at which work is done. That is,

$$P = dW/dt \tag{10.4.1}$$

For a force **F** doing work on a particle or body, the power of the force may be obtained by differentiating in Eq. (10.2.13):

$$P = \mathbf{F} \cdot \mathbf{v} \tag{10.4.2}$$

where **v** is the velocity of the point of application of **F**.

Similarly, for a couple doing work on a body, the power developed by the couple may be obtained by differentiating in Eq. (10.3.5):

$$P = \mathbf{T} \cdot \boldsymbol{\omega} \tag{10.4.3}$$

where, as before, **T** is the torque of the couple and $\boldsymbol{\omega}$ is the angular velocity of the body.

10.5 Kinetic Energy

Kinetic energy is probably the most familiar and most widely used of all the energy functions. Kinetic energy is sometimes described as energy due to motion.

If P is a particle with mass m, the kinetic energy K of P is defined as:

$$K \stackrel{D}{=} \frac{1}{2} m \mathbf{v}^2 \tag{10.5.1}$$

where $\mathbf{v}$ is the velocity of P in an inertial reference frame R. The factor 1/2 is introduced for convenience in relating kinetic energy to work and to other energy functions.

If S is a set of particles, the kinetic energy of S is defined as the sum of the kinetic energies of the individual particles. Specifically, if S contains N particles P_i with masses m_i and velocities $\mathbf{v}_i$ ($i = 1,\ldots, N$) in inertial frame R, the kinetic energy of S is defined as:

$$K \stackrel{D}{=} \sum_{i=1}^{N} \frac{1}{2} m_i \mathbf{v}_i^2 \tag{10.5.2}$$

For a rigid body B we can define the kinetic energy of B as the sum of the kinetic energies of the particles making up B. To see this, consider a depiction of B as in Figure 10.5.1. Let G be the mass center of B (see Section 6.8). Then, from Eq. (4.9.4), the velocity $\mathbf{v}_i$ of typical particle P_i of B in an inertial reference frame R is:

$$\mathbf{v}_i = \mathbf{v}_G + \boldsymbol{\omega} \times \mathbf{r}_i \tag{10.5.3}$$

where $\mathbf{v}_G$ is the velocity of G in R, $\boldsymbol{\omega}$ is the angular velocity of B in R, and $\mathbf{r}_i$ locates P_i relative to G. Then, from Eq. (10.5.2), the kinetic energy of B is:

$$K = \sum_{i=1}^{N} \frac{1}{2} m_i \mathbf{v}_i^2 = \sum_{i=1}^{N} \frac{1}{2} m_i \left(\mathbf{v}_G + \boldsymbol{\omega} \times \mathbf{r}_i\right)^2 \tag{10.5.4}$$

By expanding the terms in Eq. (10.5.4), the kinetic energy becomes:

$$\begin{aligned} K &= \frac{1}{2} \sum_{i=1}^{N} m_i \mathbf{v}_G^2 + \sum_{i=1}^{N} m_i \mathbf{v}_G \cdot \boldsymbol{\omega} \times \mathbf{r}_i + \frac{1}{2} \sum_{i=1}^{N} m_i \left(\boldsymbol{\omega} \times \mathbf{r}_i\right)^2 \\ &= \frac{1}{2}\left(\sum_{i=1}^{N} m_i\right) \mathbf{v}_G^2 + \mathbf{v}_G \cdot \boldsymbol{\omega} \times \left(\sum_{i=1}^{N} m_i \mathbf{r}_i\right) + \frac{1}{2} \sum_{i=1}^{N} m_i \left(\boldsymbol{\omega} \times \mathbf{r}_i\right)^2 \\ &= \frac{1}{2} M \mathbf{v}_G^2 + 0 + \frac{1}{2} \sum_{i=1}^{N} m_i \left(\boldsymbol{\omega} \times \mathbf{r}_i\right)^2 \end{aligned} \tag{10.5.5}$$

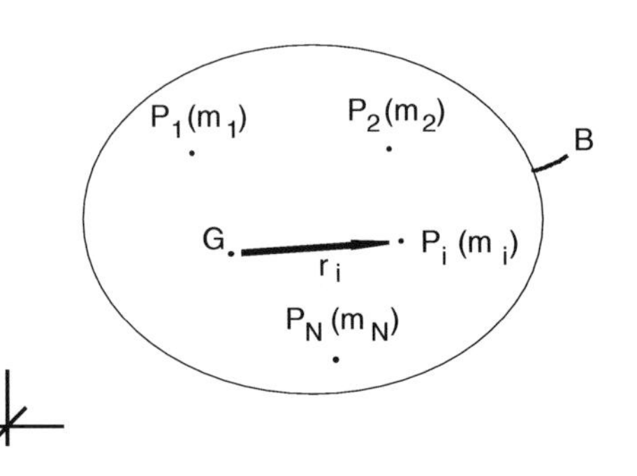

FIGURE 10.5.1
A rigid body composed.

where M is the mass of B (the total mass of all the particles composing B), and the middle term is zero because G is the mass center of B.

Consider the last term of Eq. (10.5.5). By expanding and rearranging the terms we have:

$$
\begin{aligned}
\frac{1}{2}\sum_{i=1}^{N} m_i(\boldsymbol{\omega}\times\mathbf{r}_i)^2 &= \frac{1}{2}\sum_{i=1}^{N} m_i(\boldsymbol{\omega}\times\mathbf{r}_i)\cdot(\boldsymbol{\omega}\times\mathbf{r}_i) = \frac{1}{2}\sum_{i=1}^{N} m_i\,\boldsymbol{\omega}\cdot[\mathbf{r}_i\times(\boldsymbol{\omega}\times\mathbf{r}_i)] \\
&= \frac{1}{2}\boldsymbol{\omega}\cdot\sum_{i=1}^{N} m_i\mathbf{r}_i\times(\boldsymbol{\omega}\times\mathbf{r}_i) = \frac{1}{2}\boldsymbol{\omega}\cdot\sum_{i=1}^{N} m_i[\mathbf{r}_i\times(\omega\mathbf{n}_\omega\times\mathbf{r}_i)] \\
&= \frac{1}{2}\boldsymbol{\omega}\cdot\mathbf{I}_\omega^{B/G}\,\omega = \frac{1}{2}\boldsymbol{\omega}\cdot\mathbf{I}^{B/G}\cdot\mathbf{n}_\omega\,\omega \\
&= \frac{1}{2}\boldsymbol{\omega}\cdot\mathbf{I}^{B/G}\cdot\boldsymbol{\omega}
\end{aligned}
\tag{10.5.6}
$$

where ω is the magnitude of $\boldsymbol{\omega}$, $\mathbf{n}_\omega$ is a unit vector parallel to $\boldsymbol{\omega}$, $\mathbf{I}_\omega^{B/G}$ is the second moment vector of B for G for the direction of $\mathbf{n}_\omega$, and $I^{B/G}$ is the central inertia dyadic of B (see Sections 7.4 and 7.6.).

By substituting from Eq. (10.5.6) into (10.5.5), the kinetic energy of B becomes:

$$K = \frac{1}{2}\mathbf{v}_G^2 + \frac{1}{2}\boldsymbol{\omega}\cdot\mathbf{I}^{B/G}\cdot\boldsymbol{\omega} \tag{10.5.7}$$

Finally, suppose that $\mathbf{n}_1$, $\mathbf{n}_2$, and $\mathbf{n}_3$ are mutually perpendicular principal unit vectors of B for G (see Section 7.7), then K may be expressed as:

$$K = \frac{1}{2}M\mathbf{v}_G^2 + \frac{1}{2}\left(\mathbf{I}_{11}\,\omega_1^2 + \mathbf{I}_{22}\,\omega_2^2 + \mathbf{I}_{33}\,\omega_3^2\right) \tag{10.5.8}$$

where I_{11}, I_{22}, and I_{33} are the corresponding principal moments of inertia of B for B, and the ω_i $(i = 1, 2, 3)$ are the $\mathbf{n}_i$ components of $\boldsymbol{\omega}$. Observe that in all the expressions for kinetic energy no accelerations appear, only velocities.

10.6 Work–Energy Principles

We can obtain a relationship between work and kinetic energy by integrating equations obtained from Newton's laws. To this end, consider first a particle P with mass m subjected to a force $\mathbf{F}$ as in Figure 10.6.1. Then, from Newton's laws (see Section 8.2), we have:

$$\mathbf{F} = m\mathbf{a} \tag{10.6.1}$$

where $\mathbf{a}$ is the acceleration of P in an inertial frame R. If we project the terms of Eq. (10.6.1) along the velocity $\mathbf{v}$ of P in R we have:

$$\mathbf{F}\cdot\mathbf{v} = m\mathbf{a}\cdot\mathbf{v} \tag{10.6.2}$$

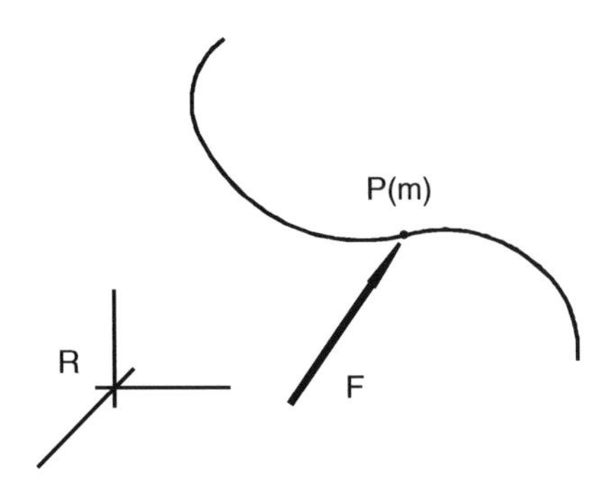

FIGURE 10.6.1
A force applied to a particle.

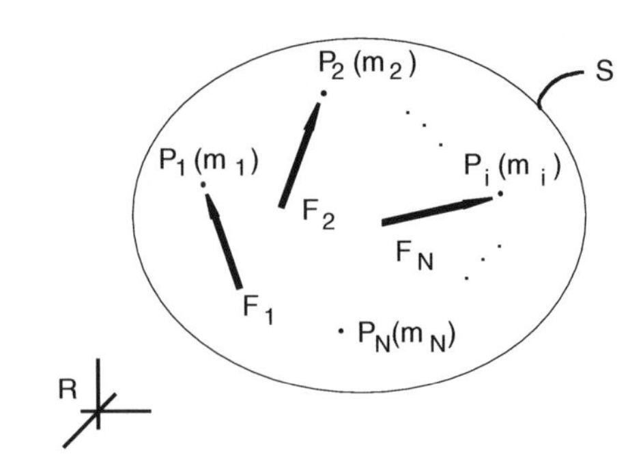

FIGURE 10.6.2
A set of particles subjected to forces.

The left side of Eq. (10.6.2) may be recognized as the power of **F** (see Eq. (10.4.2)). The right side of Eq. (10.6.2) may be expressed in terms of a derivative of the square of the velocity of P. That is,

$$m\mathbf{a}\cdot\mathbf{v} = md\left(\mathbf{v}^2/2\right)/dt = \frac{d}{dt}\left(\frac{1}{2}m\mathbf{v}^2\right) \tag{10.6.3}$$

From Eqs. (10.4.1) and (10.4.2) we recognize the left side of Eq. (10.6.2) as the derivative of the work W of **F**. In like manner, from Eq. (10.5.1), we recognize the right side of Eq. (10.6.3) as the derivative of the kinetic energy K of P. Hence, we have:

$$\frac{dW}{dt} = \frac{dK}{dt} \tag{10.6.4}$$

or

$$W = K_2 - K_1 = \Delta K \tag{10.6.5}$$

where K_2 and K_1 represent the kinetic energy of P at the beginning and end of the time interval during which **F** is applied to P.

Next, consider a set S of particles P_i $(i = 1,\ldots, N)$ subjected to forces $\mathbf{F}_i$ as in Figure 10.6.2. Then, for a typical particle P_i, Newton's laws become:

$$\mathbf{F}_i = m_i\mathbf{a}_i \quad \text{(no sum on } i\text{)} \tag{10.6.6}$$

where $\mathbf{a}_i$ is the acceleration of P_i in inertia frame R. By multiplying the terms of Eq. (10.6.6) by $\mathbf{v}_i$, the velocity of P_i in R, we obtain:

$$\mathbf{F}_i\cdot\mathbf{v}_i = m_i\mathbf{a}_i\cdot\mathbf{v}_i = \frac{d}{dt}\left(\frac{1}{2}m_i\mathbf{v}_i^2\right) = \frac{dK_i}{dt} \tag{10.6.7}$$

where K_i is the kinetic energy of P_i.

We can recognize the left side of Eq. (10.6.7) as the derivative of the work W_i of $\mathbf{F}_i$. Hence, Eq. (10.6.7) becomes:

$$\frac{dW_i}{dt} = \frac{dK_i}{dt} \tag{10.6.8}$$

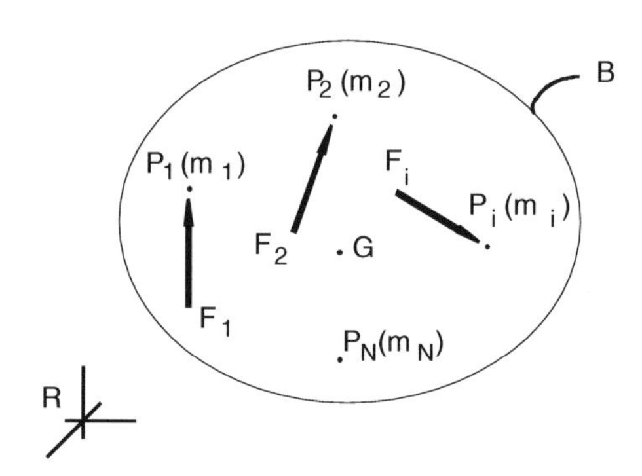

FIGURE 10.6.3
A rigid body B subjected to forces.

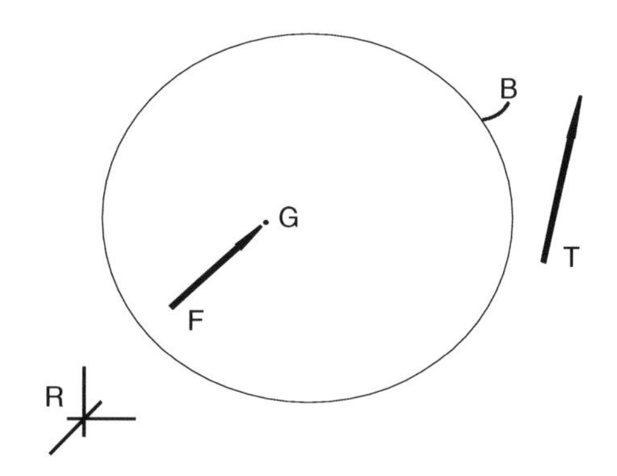

FIGURE 10.6.4
Equivalent force system on rigid body B.

By adding together the terms of similar equations for each of the particles, we have:

$$\frac{dW}{dt} = \frac{dK}{dt} \tag{10.6.9}$$

where W is the work on S from all the forces, and K is the kinetic energy of S. By integrating Eq. (10.6.9), we have:

$$W = K_2 - K_1 = \Delta K \tag{10.6.10}$$

where, as in Eq. (10.6.5), K_2 and K_1 represent the kinetic energy of S at the beginning and end of the time interval that the forces are applied to S.

Finally, consider a rigid body B acted upon by a system of forces as in Figure (10.6.3). As before, consider B to be composed of fixed particles P_i with masses m_i as shown. Let the force system applied to B be replaced by an equivalent force system (see Section 6.5) consisting of a single force $\mathbf{F}$ passing through the mass center G of B together with a couple with torque $\mathbf{T}$ as in Figure 10.6.4.

From d'Alembert's principle (see Section 8.3) we can represent the inertia force system on B by a force $\mathbf{F}^*$ passing through G together with a couple with torque $\mathbf{T}^*$ where $\mathbf{F}^*$ and $\mathbf{T}^*$ are given by (see Eqs. (8.6.5) and (8.6.6)):

$$\mathbf{F}^* = -M\mathbf{a}_G \quad \text{and} \quad \mathbf{T}^* = -\mathbf{I}\cdot\boldsymbol{\alpha} - \boldsymbol{\omega}\times(\mathbf{I}\cdot\boldsymbol{\omega}) \tag{10.6.11}$$

where, as before, M is the mass of B, $\boldsymbol{I}$ is the central inertia dyadic of B, $\mathbf{a}_G$ is the acceleration of G in inertia frame R, and $\boldsymbol{\omega}$ and $\boldsymbol{\alpha}$ are the angular velocity and angular acceleration of B in R. Then, from Newton's laws and d'Alembert's principle, we have:

$$\mathbf{F} = M\mathbf{a}_G \tag{10.6.12}$$

and

$$\mathbf{T} = \mathbf{I}\cdot\boldsymbol{\alpha} + \boldsymbol{\omega}\times(\mathbf{I}\cdot\boldsymbol{\omega}) \tag{10.6.13}$$

If we multiply the terms of Eq. (10.6.12) by $\mathbf{v}_G$, the velocity of G in R, we have:

$$\mathbf{F}\cdot\mathbf{v}_G = M\mathbf{a}_G\cdot\mathbf{v}_G = \frac{d}{dt}\left(\frac{1}{2}M\mathbf{v}_G^2\right) \tag{10.6.14}$$

Similarly, if we multiply the terms of Eq. (10.6.13) by $\boldsymbol{\omega}$ we have:

$$\mathbf{T}\cdot\boldsymbol{\omega} = (\mathbf{I}\cdot\boldsymbol{\alpha})\cdot\boldsymbol{\omega} + \overset{0}{[\boldsymbol{\omega}\times(\mathbf{I}\cdot\boldsymbol{\omega})]\cdot\boldsymbol{\omega}} = \frac{d}{dt}\left(\frac{1}{2}\boldsymbol{\omega}\cdot\mathbf{I}\cdot\boldsymbol{\omega}\right) \tag{10.6.15}$$

By adding the terms of Eqs. (10.6.14) and (10.6.15), we obtain:

$$\mathbf{F}\cdot\mathbf{v}_G + \mathbf{T}\cdot\boldsymbol{\omega} = \frac{d}{dt}\left(\frac{1}{2}M\mathbf{v}_G^2\right) + \frac{d}{dt}\left(\frac{1}{2}\boldsymbol{\omega}\cdot\mathbf{I}\cdot\boldsymbol{\omega}\right) \tag{10.6.16}$$

From Eqs. (10.4.1), (10.4.2), and (10.4.3) we can recognize the left side of Eq. (10.6.16) as the derivative of the work W of the force system acting on B. Also, from Eq. (10.5.7) we recognize the right side of Eq. (10.6.16) as the derivative of the kinetic energy K of B. Hence, Eq. (10.6.16) takes the form:

$$\frac{dW}{dt} = \frac{dK}{dt} \tag{10.6.17}$$

Then, by integrating, we have:

$$W = K_2 - K_1 = \Delta K \tag{10.6.18}$$

where, as in Eqs. (10.6.5) and (10.6.10), K_2 and K_1 represent the kinetic energy of B at the beginning and end of the time interval that forces are acting on B.

Equations (10.6.5), (10.6.10), and (10.6.18) are expressions of the principle of work and kinetic energy for a particle, a set of particles, and a rigid body, respectively. Simply stated, the work done is equal to the change in kinetic energy.

In the remaining sections of this chapter we will consider several examples illustrating application of this principle. We will also consider combined application of this principle with the impulse–momentum principles of Chapter 9.

10.7 Elementary Example: A Falling Object

Consider first the simple case of a particle P with mass m released from rest at distance h above a horizontal surface S as in Figure 10.7.1. The objective is to determine the speed v of P when it reaches S.

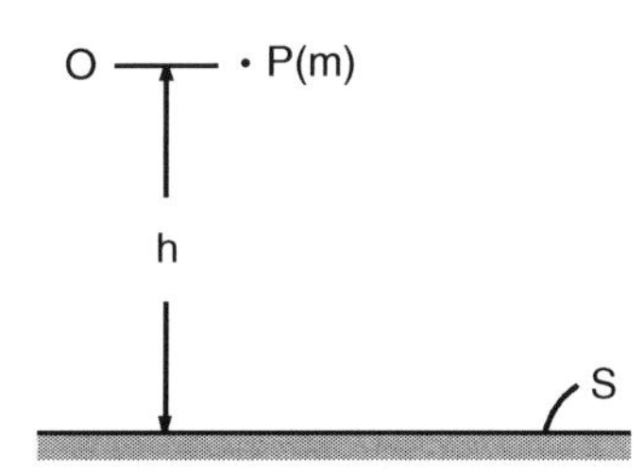

FIGURE 10.7.1
A particle released from rest in a gravitational field.

Because gravity is the only force applied to *P* and because the path of movement of *P* is parallel to the weight force through a distance *h*, the work done *W* is simply:

$$W = mgh \tag{10.7.1}$$

Because *P* is released from rest, its initial kinetic energy is zero. When *P* reaches *S*, its kinetic energy may be expressed as:

$$K = \frac{1}{2}mv^2 \tag{10.7.2}$$

Then, from the work–energy principle of Eq. (10.6.5), we have:

$$W = \Delta K \quad \text{or} \quad mgh = \frac{1}{2}mv^2 \tag{10.7.3}$$

Solving for *v*, we obtain the familiar result:

$$v = \sqrt{2gh} \tag{10.7.4}$$

10.8 Elementary Example: The Simple Pendulum

Consider next the simple pendulum depicted in Figure 10.8.1 where the mass *m* of the pendulum is concentrated in the bob *P* which is supported by a pinned, massless rod of length ℓ as shown. Let θ measure the inclination of the pendulum to the vertical. Suppose the pendulum is held in a horizontal position and released from rest. The objective is to determine the speed *v* of *P* as it passes through the equilibrium position $\theta = 0$. If we consider a free-body diagram of *P* as in Figure 10.8.2, we see that of the two forces applied to *P* the tension of the connecting rod does no work on *P* because its direction is perpendicular to the movement of *P*. From Eq. (10.2.18), we see that the work *W* done by the weight force is:

$$W = mg\ell \tag{10.8.1}$$

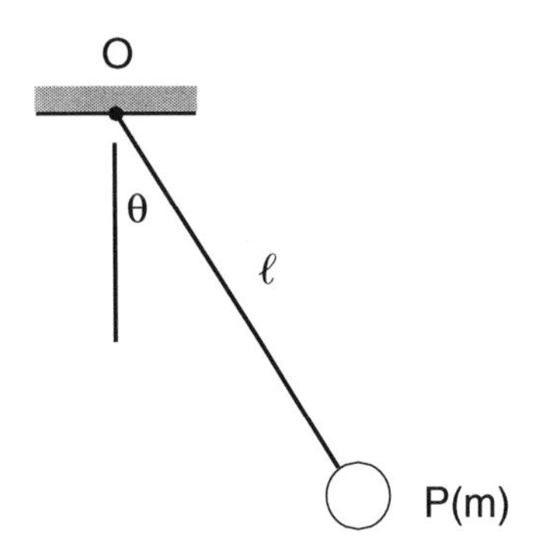

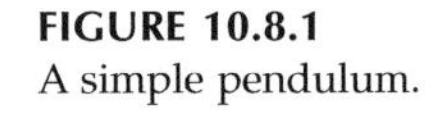

FIGURE 10.8.1
A simple pendulum.

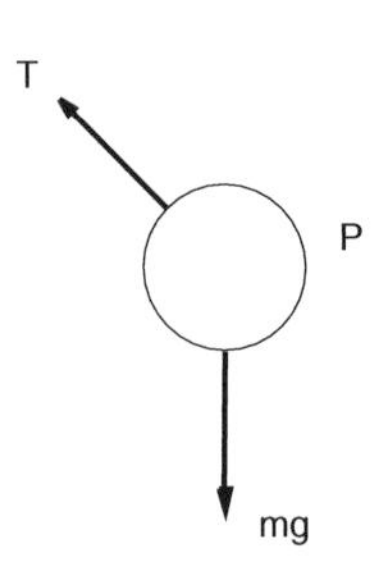

FIGURE 10.8.2
Free body diagram of the pendulum bob.

Because the pendulum is released from rest its initial kinetic energy is zero. Its kinetic energy K as it passes through the equilibrium position may be expressed as:

$$K = \frac{1}{2} m v^2 = \frac{1}{2} m \left(\ell \dot{\theta} \right)^2 \tag{10.8.2}$$

From the work–energy principle of Eq. (10.6.5), we then have:

$$W = \Delta K \quad \text{or} \quad mg\ell = \frac{1}{2} m v^2 = \frac{1}{2} m \ell \dot{\theta}^2 \tag{10.8.3}$$

or

$$v = \sqrt{mg\ell} \tag{10.8.4}$$

Observe that this speed is the same as that of an object freely falling through a distance ℓ (see Eq. (10.7.4)), even though the direction of the velocity is different.

As a generalization of this example, consider a pendulum released from rest at an angle θ_i with the objective of determining the speed of P when it falls to an angle θ_f, as in Figure 10.8.3. The work ${}_iW_f$ done by gravity as the pendulum falls from θ_i to θ_f is:

$${}_iW_f = mg\Delta h = mg\ell \left(\cos\theta_f - \cos\theta_i \right) \tag{10.8.5}$$

where Δh is the change in elevation of P as the pendulum falls (see Eq. (10.2.20)).

The kinetic energies K_i and K_f of P when θ is θ_i and θ_f are:

$$K_i = 1/2\, m \left(\ell \dot{\theta}_i \right)^2 = 0 \quad \text{and} \quad K_f = \frac{1}{2} m \left(\ell \dot{\theta}_f \right)^2 \tag{10.8.6}$$

where $\dot{\theta}_i$ and $\dot{\theta}_f$ are the values of $\dot{\theta}$ when θ is θ_i and θ_f, respectively.

The work–energy principle then leads to:

$${}_iW_f = \Delta K = K_f - K_i \tag{10.8.7}$$

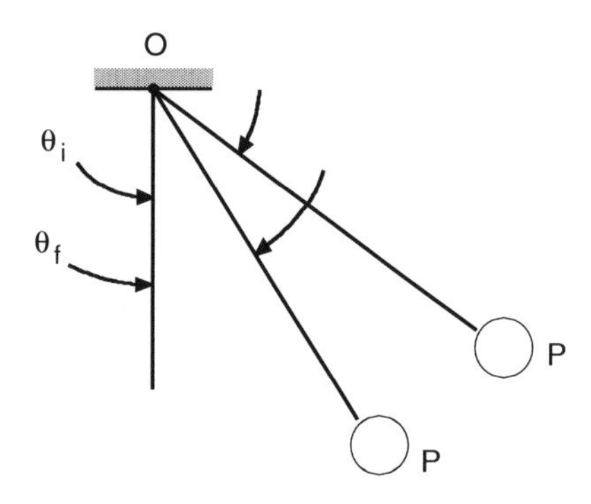

FIGURE 10.8.3
A simple pendulum released from rest and falling through angle $\theta_i - \theta_f$.

or

$$mg\ell\left(\cos\theta_f - \cos\theta_i\right) = \frac{1}{2}m\left(\ell\dot{\theta}_f\right)^2 \tag{10.8.8}$$

Solving for $\dot{\theta}_f$ we have:

$$\dot{\theta}_f = \sqrt{2g/\ell}\left(\cos\theta_f - \cos\theta_i\right)^{1/2} \tag{10.8.9}$$

The result of Eq. (10.8.9) could also have been obtained by integrating the governing differential equations of motion obtained in Chapter 8. Recall from Eq. (8.4.4) that for a simple pendulum the governing equation is:

$$\ddot{\theta} + (g/\ell)\sin\theta = 0 \tag{10.8.10}$$

Then, by multiplying both sides of this equation by $\dot{\theta}$, we have:

$$\dot{\theta}\ddot{\theta} + (g/\ell)\sin\theta\dot{\theta} = 0 \tag{10.8.11}$$

Because $\dot{\theta}\ddot{\theta}$ may be recognized as being $(1/2)d\dot{\theta}^2/dt$, we can integrate the equation and obtain:

$$(1/2)\dot{\theta}^2 - (g/\ell)\cos\theta = \text{ constant} \tag{10.8.12}$$

Because $\dot{\theta}$ is zero when θ is θ_i, the constant is $-(g/\ell)\cos\theta_i$. Therefore, we have:

$$\dot{\theta}^2 = (2g/\ell)\left(\cos\theta - \cos\theta_i\right) \tag{10.8.13}$$

When θ is θ_f, Eqs. (10.8.9) and (10.8.13) are seen to be equivalent.

The work–energy principle may also be used to determine the pendulum rise angle when the speed at the equilibrium position ($\theta = 0$) is known. Specifically, suppose the angular speed of the pendulum when θ is zero is $\dot{\theta}_o$. Then, the work W done on the pendulum as it rises to an angle θ_f is:

$$W = mg\Delta h = -mg\ell\left(1 - \cos\theta_f\right) \tag{10.8.14}$$

where the negative sign occurs because the upward movement of the pendulum is opposite to the direction of gravity, producing negative work. The work–energy principle, then, is:

$$ {}_0W_f = \Delta K = K_f - K_0 \tag{10.8.15} $$

or

$$ -mg\ell\left(1-\cos\theta_f\right) = (1/2)m\left(\ell\dot{\theta}_f\right)^2 - (1/2)m\left(\ell\dot{\theta}_0\right)^2 \tag{10.8.16} $$

or

$$ \dot{\theta}_f^2 = \dot{\theta}_0^2 - 2g\ell\left(1-\cos\theta_f\right) \tag{10.8.17} $$

If the pendulum is to rise all the way to the vertical equilibrium position ($\theta = \pi$), we have:

$$ \dot{\theta}_f^2 = \dot{\theta}_0^2 - 4g\ell \tag{10.8.18} $$

If $\dot{\theta}_0^2$ is exactly $4g\ell$, the pendulum will rise to the vertical equilibrium position and come to rest at that position. If $\dot{\theta}_0^2$ exceeds $4g\ell$, the pendulum will rise to the vertical position and rotate through it with an angular speed given by Eq. (10.8.18) (sometimes called the *rotating pendulum*).

10.9 Elementary Example — A Mass–Spring System

For a third fundamental example, consider the mass–spring system depicted in Figure 10.9.1. It consists of a block B with mass m and a linear spring, with modulus k, moving without friction or damping in a horizontal direction. Let x measure the displacement of B away from equilibrium.

Suppose B is displaced to the right (positive x displacement with the spring in tension) a distance δ away from equilibrium. Let B be released from rest in this position. Questions arising then are what is the speed v of B as it returns to the equilibrium position ($x = 0$), and how far to the left of the equilibrium position does B go?

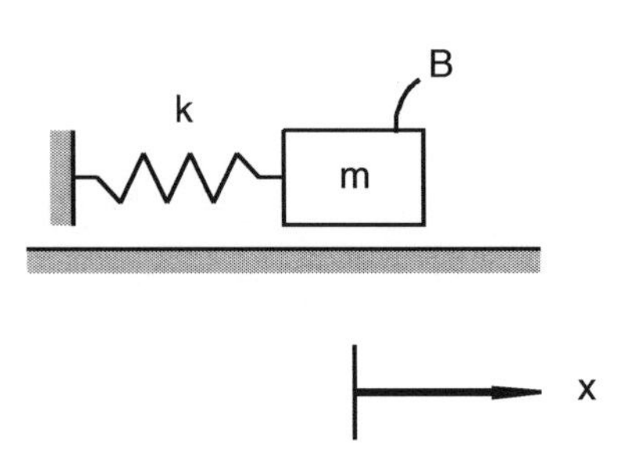

FIGURE 10.9.1
Ideal mass–spring system.

To answer these questions using the work–energy principle, recall from Eq. (10.2.22) that when a linear spring is stretched (or compressed) a distance δ, the corresponding work W done by the stretching (or compressing) force is $(1/2)k\delta^2$. Because the force exerted on the spring is equal, but oppositely directed, to the force exerted on B, the work done on B as the spring is relaxed is also $(1/2)k\delta^2$. (That is, the work on B is positive because the force of the spring on B is in the same direction as the movement of B.)

Because B is released from rest, its initial kinetic energy is zero. The kinetic energy at the equilibrium position is:

$$K = (1/2)mv^2 = (1/2)m\dot{x}^2 \tag{10.9.1}$$

Then, from the work–energy principle, we have:

$$W = \Delta K \quad \text{or} \quad (1/2)k\delta^2 = (1/2)mv^2 - 0 \tag{10.9.2}$$

or

$$v = \dot{x} = \sqrt{k/m}\,\delta \tag{10.9.3}$$

where the minus sign is selected because B is moving to the left.

Next, as B continues to move to the left past the equilibrium position, the spring force will be directed opposite to the movement of B. Therefore, the work W done on B as B moves to the left a distance d from the equilibrium position is:

$$W = -(1/2)kd^2 \tag{10.9.4}$$

When B moves to its leftmost position, its kinetic energy is zero. From Eqs. (10.9.1) and (10.9.2), the kinetic energy of B at the equilibrium position is:

$$K = (1/2)mv^2 = (1/2)k\delta^2 \tag{10.9.5}$$

The work–energy principle then produces:

$$W = \Delta K \quad \text{or} \quad -(1/2)kd^2 = 0 - (1/2)k\delta^2 \tag{10.9.6}$$

or

$$d = \delta \tag{10.9.7}$$

That is, the block moves to the left by precisely the same amount as it was originally displaced to the right.

The usual explanation of this phenomenon is that when the spring is stretched (or compressed) the work done by the stretching (or compressing) force stores energy (*potential energy*) in the spring. This stored energy in the spring is derived from the kinetic energy of the block. Then, as the spring is relaxing, its potential energy is transferred back to kinetic energy of the block. There is thus a periodic transfer of energy between the spring

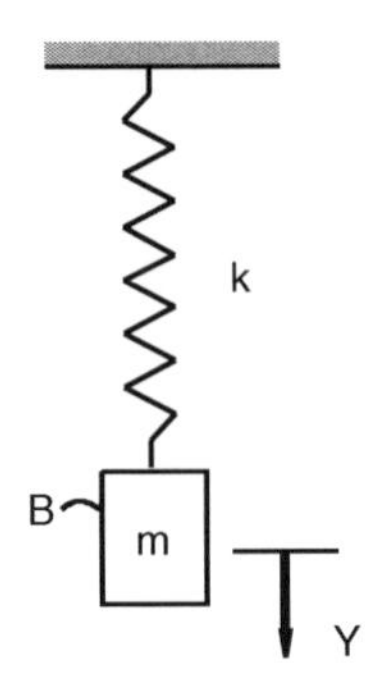

FIGURE 10.9.2
A vertical mass–spring system.

and the block, with the sum of the potential energy of the spring and the kinetic energy of the block being constant. (We will discuss potential energy in the next chapter.)

As another example of work–energy transfer, consider the mass–spring system arranged vertically as in Figure 10.9.2. Suppose B is held in a position where the spring is unstretched. If B is then released from rest from this position, it will fall and stretch the spring and eventually come to rest at an extreme downward position. Questions arising then include how far does B fall and what is the spring force when B reaches this maximum downward displacement? To answer these questions, consider that as B falls, the weight (or gravity) force on B is in the direction of the movement of B, whereas the spring force on B is opposite to the movement of B. Because B is at rest at both the beginning and the end of the movement, there is no change in the kinetic energy of B. The net work on B is then zero. That is,

$$W = 0 = mgd - (1/2)kd^2 \tag{10.9.8}$$

where d is the distance B moves downward. By solving for d we obtain:

$$d = 2mg/k \tag{10.9.9}$$

The spring force in this extended position is, then,

$$F = kd = 2mg \tag{10.9.10}$$

The result of Eq. (10.9.10) shows that a suddenly applied weight load on a spring creates a force twice that of the weight. This means that if a weight is suddenly placed on a machine or structure the force generated is twice that required to support the weight in a static equilibrium configuration.

10.10 Skidding Vehicle Speeds: Accident Reconstruction Analysis

The work–energy principle is especially useful in determining speeds of accident vehicles by using measurements of skid-mark data. Indeed, the work–energy principle together with the conservation of momentum principles are the primary methods used by accident reconstructionists when attempting to determine vehicle speeds at various stages of an accident.

To illustrate the procedure, suppose an automobile leaves skid marks from all four wheels in coming to an emergency stop. Given the length d of the skid marks, the objective is to determine the vehicle speed when the marks first began.

Skid marks are created by abrading and degrading tires sliding on a roadway surface. The tire degradation is due to friction forces and heat abrading the rubber. The friction forces are proportional to the normal (perpendicular to the roadway surface) forces on the tires and to the coefficient of friction μ. The friction coefficient, ranging in value from 0 to 1.0, is a measure of the relative slipperiness between the tires and the roadway pavement. If F and N are equivalent friction and normal forces (see Section 6.5), they are related by the expression:

$$F = \mu N \tag{10.10.1}$$

If an automobile is sliding on a flat, level (horizontal) roadway, a free-body diagram of the vehicle shows that the normal force N is equal to the vehicle weight w. Then, as the vehicle slides a distance d, the work W done by the friction force (acting opposite to the direction of the sliding) is:

$$W = -Fd = -\mu Nd = -\mu wd = -\mu mgd \tag{10.10.2}$$

where m is the mass of the automobile.

Let v be the desired speed of the automobile when the skid marks first appear. Then, the kinetic energy K_i of the vehicle at that point is:

$$K_i = (1/2)mv^2 \tag{10.10.3}$$

Because the kinetic energy K_f at the end of the skid marks is zero (the vehicle is then stopped), the work energy principle produces:

$$W = \Delta K = K_f - K_i \quad \text{or} \quad -\mu mgd = 0 - (1/2)mv^2 \tag{10.10.4}$$

or

$$v = \sqrt{2\mu gd} \tag{10.10.5}$$

Observe that the calculated speed is independent of the automobile mass.

To illustrate how the work–energy principle may be used in conjunction with the momentum conservation principles, suppose an automobile with mass m_1 slides a distance d_1 before colliding with a stopped automobile having mass m_2. Suppose further that the two vehicles then slide together (a *plastic* collision) for a distance d_2 before coming to rest. The questions arising then are what were the speeds of the vehicles just before and just after impact and what was the speed of the first vehicle when it first began to slide?

To answer these questions, consider first from Eq. (10.10.5) that the speed v_a of the vehicles just after impact is:

$$v_a = \sqrt{2\mu gd_2} \tag{10.10.6}$$

Next, during impact, the momentum is conserved. That is,

$$m_1 v_b = (m_1 + m_2) v_a \tag{10.10.7}$$

where v_b is the speed of the first vehicle just before impact (see Eq. (9.7.1)).

From Eq. (10.10.2), the work W done by friction forces on the first vehicle as it slides a distance d_1 before the collision is:

$$W = -\mu m_1 g d_1 \tag{10.10.8}$$

If v_0 is the speed of the first vehicle when skidding begins, the change in kinetic energy of the vehicle from the beginning of skidding until the collision is:

$$\Delta K = (1/2) m_1 v_b^2 - (1/2) m_1 v_0^2 \tag{10.10.9}$$

The work–energy principle then gives:

$$W = \Delta K \quad \text{or} \quad -\mu m_1 g d_1 = (1/2) m_1 v_b^2 - (1/2) m_1 v_0^2 \tag{10.10.10}$$

Finally, using Eqs. (10.10.6) and (10.10.7), we can solve Eq. (10.10.10) for v_0^2:

$$\begin{aligned} v_0^2 &= v_b^2 + 2\mu g d_1 = \left[1 + (m_2/m_1)\right]^2 v_a^2 + 2\mu g d_1 \\ &= \left[1 + (m_2/m_1)\right]^2 2\mu g d_2 + 2\mu g d_1 \end{aligned} \tag{10.10.11}$$

Observe from Eq. (10.10.7) that the speed v_a of the first vehicle just after the collision with the second vehicle is reduced by the factor $[m_1/(m_1 + m_2)]$. That is, the change in speed Δv is:

$$\Delta v = v_a - v_b = -\left[m_2/(m_1 + m_2)\right] v_b \tag{10.10.12}$$

Observe further that because the velocity changes during the impact the kinetic energy also changes. That is, even though the momentum is conserved, the kinetic energy is not conserved. Indeed, the change in kinetic energy ΔK just before and just after the impact is:

$$\Delta K = (1/2) m_1 v_a^2 - (1/2) m_1 v_b^2 = -(1/2) m_1 \left[\frac{2m_1^2 + 2m_1 m_2 + m_3^2}{(m_1 + m_2)^2}\right] v_b^2 \tag{10.10.13}$$

In actual accidents, the vehicles do not usually leave uniform skid marks from all four wheels. Also, collisions are not usually perfectly plastic nor do the vehicles always move in a straight line on a level surface. However, with minor modifications, the work–energy and momentum conservation principles may still be used to obtain reasonable estimates of vehicle speeds at various stages of an accident. The details of these modifications and the corresponding application of the principles are beyond the scope of this text; the reader is referred to the references for further information.

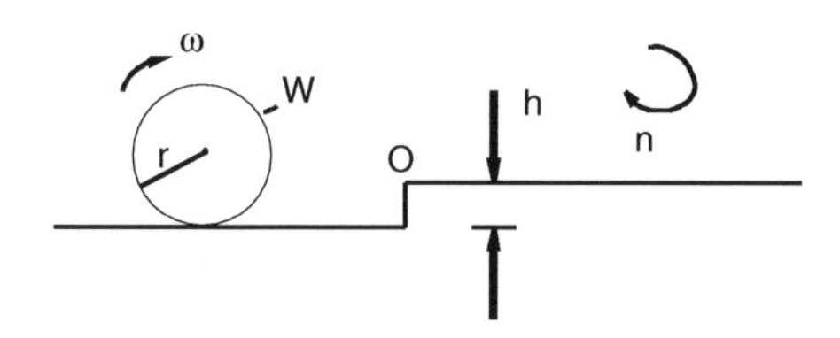

FIGURE 10.11.1
Wheel rolling over a step.

10.11 A Wheel Rolling Over a Step

For a second example illustrating the tandem use of the work–energy principle and the conservation of momentum principle, consider again the case of the wheel rolling over a step as in Figure 10.11.1 (recall that we considered this problem in Section 9.9). Let the wheel W have a radius r, mass m, and axial moment of inertia I. Suppose we are interested in knowing the speed v of the wheel center required for the wheel to roll over the step.

Recall in Section 9.9 that when the wheel encounters the step its angular momentum about the corner (or nose) O of the step is conserved. By using the conservation of angular momentum principle we found that the angular speed $\hat{\omega}$ of W just after impact is (see Eq. (9.9.8)):

$$\hat{\omega} = \left\{\left[I + mr(r-h)\right] / \left(I + mr^2\right)\right\}\omega \tag{10.11.1}$$

where ω is the angular speed of W before impact and h is the height of the step.

After impact, W rotates about the nose O of the step. For W to roll over the step it must have enough kinetic energy after impact to overcome the negative work of gravity as it rises up over the step. The work W_g of gravity as W rolls completely up the step is:

$$W_g = -mgh \tag{10.11.2}$$

If W just rolls over the step (that is, if W expends all its kinetic energy after impact in rolling over the step), it will come to rest at the top of the step and its kinetic energy K_f at that point will be zero. The kinetic energy K_i just after impact is:

$$K_i = (1/2)I\hat{\omega}^2 + (1/2)mr^2\hat{\omega}^2 = (1/2)\left(I + mr^2\right)\hat{\omega}^2 \tag{10.11.3}$$

The work–energy principle then leads to:

$$W_g = K_f - K_i \quad \text{or} \quad -mgh = 0 - (1/2)\left(I + mr^2\right)\hat{\omega}^2 \tag{10.11.4}$$

Solving $\hat{\omega}$ we have:

$$\hat{\omega} = \left[2mgh / \left(I + mr^2\right)\right]^{1/2} \tag{10.11.5}$$

Hence, from Eq. (10.11.1), the speed v of W just before impact for step rollover is:

$$v = r\omega = r\frac{\left[\left(I + mr^2\right)\left(2mgh\right)\right]^{1/2}}{I + mr\left(r - h\right)} \tag{10.11.6}$$

Suppose that W is a uniform circular disk with I then being $(1/2)mr^2$. In this case, v becomes:

$$v = r\sqrt{3gh}/\left[\left(3/2\right)r - h\right] \tag{10.11.7}$$

Observe that the disk could (at least, theoretically) roll over a step whose height h is greater than r. From a practical perspective, however, the disk will only encounter the nose of the step if h is less than r.

Observe also that, as with the colliding vehicles, the energy is not conserved when the wheel impacts the step. From Eq. (10.11.1), the energy loss L is seen to be:

$$\begin{aligned} L &= \left(1/2\right)I\hat{\omega}^2 + \left(1/2\right)mr^2\hat{\omega}^2 - \left(1/2\right)I\omega^2 - \left(1/2\right)mr^2\omega^2 \\ &= -\left(1/2\right)\left[\frac{2Imrh + 2m^2r^3h - m^2r^2h^2}{I + mr^2}\right] \end{aligned} \tag{10.11.8}$$

For a uniform circular disk, this becomes:

$$L = -mh\left[r - \left(h/3\right)\right] \tag{10.11.9}$$

10.12 The Spinning Diagonally Supported Square Plate

For a third example of combined application of the work–energy and momentum conservation principles, consider again the spinning square plate supported by a thin wire along a diagonal as in Figure 10.12.1 (recall that we studied this problem in Section 9.12). As before, let the plate have side length a, mass m, and initial angular speed Ω. Let the plate be arrested along an edge so that it rotates about that edge as in Figure 10.12.2.

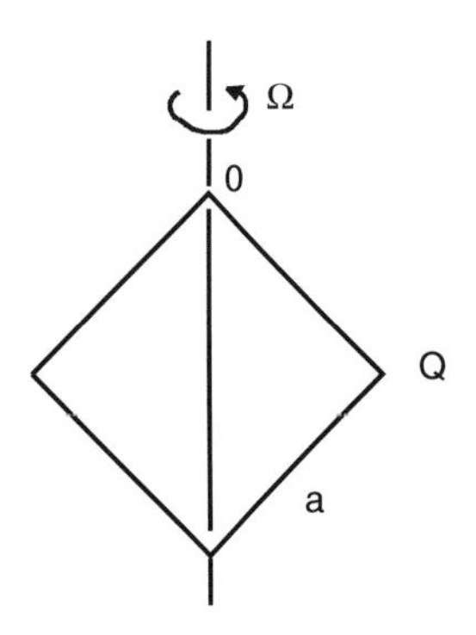

FIGURE 10.12.1
Suspended spinning square plate.

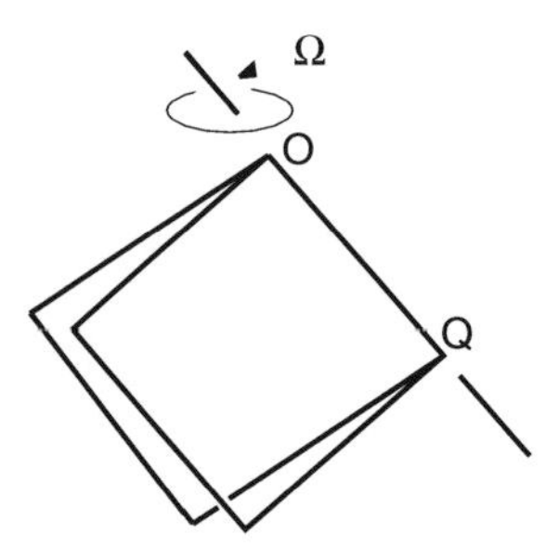

FIGURE 10.12.2
Arrested plate spinning about an edge.

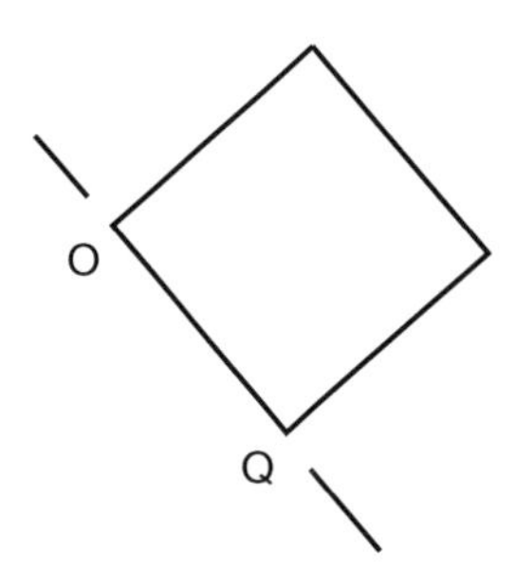

FIGURE 10.12.3
Upward resting configuration of the plate.

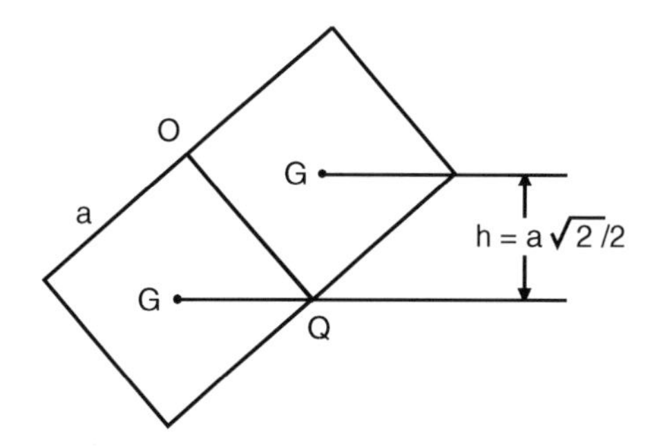

FIGURE 10.12.4
Plate mass center elevation change.

Recall that in Section 9.12 we discovered through the conservation of angular momentum principle that the post-seizure rotation speed $\hat{\Omega}$ is related to the pre-seizure speed Ω by the expression:

$$\hat{\Omega} = \left(\sqrt{2}/8\right)\Omega \tag{10.12.1}$$

In view of the result of Eq. (10.12.1), a question that may be posed is what should the pre-seizure speed be so that after seizure the plate rotates through exactly 180°, coming to rest in the upward configuration shown in Figure 10.12.3?

We can answer this question using the work–energy principle. Because gravity is the only force doing work on the plate, the work W is simply the plate weight multiplied by the mass center elevation change (see Figure 10.12.4):

$$W = -mgh = -mga\sqrt{2}/2 \tag{10.12.2}$$

where the negative sign occurs because the mass center elevation movement is opposite to the direction of gravity.

The final kinetic energy K_f is zero because the plate comes to rest. The initial kinetic energy K_i just after the plate edge is seized, is:

$$K_i = (1/2)m\left[(a/2)\hat{\Omega}\right]^2 + (1/2)\left[(1/12)ma^2\hat{\Omega}^2\right] = ma^2\hat{\Omega}^2/6 \tag{10.12.3}$$

The work–energy principle then produces:

$$W = K_f - K_i \quad \text{or} \quad -mga\sqrt{2}/2 = 0 - ma^2\hat{\Omega}^2/6 \tag{10.12.4}$$

Solving for $\hat{\Omega}$, we obtain:

$$\hat{\Omega} = \left[3\sqrt{2}\,g/a\right]^{1/2} \tag{10.12.5}$$

Then, from Eq. (10.12.1), Ω, the pre-seizure angular speed, is:

$$\Omega = \left(8/\sqrt{2}\right)\hat{\Omega} = \left[96\sqrt{2}\,g/a\right]^{1/2} \tag{10.12.6}$$

10.13 Closure

The work–energy principle is probably the most widely used of all the principles of dynamics. The primary advantage of the work–energy principle is that it only requires knowledge of velocities and not accelerations. Also, calculation of the work done is often accomplished by inspection of the system configuration.

The major disadvantage of the work–energy principle is that only a single equation is obtained. Hence, if there are several unknowns with a given mechanical system, at most one of these can be obtained using the work–energy principle. This in turn means that the principle is most advantageous for relatively simple mechanical systems. However, the utility of the principle may often be enhanced by using it in tandem with other dynamics principles — particularly impulse–momentum principles.

In the next two chapters we will consider more general energy methods. We will consider the procedures of generalized dynamics, Lagrange's equations, and Kane's equations. These procedures, while not as simple as those of the work–energy principle, have the advantage of still being computationally efficient and of producing the same number of equations as there are degrees of freedom of a system.

References (Accident Reconstruction)

10.1. Baker, J. S., *Traffic Accident Investigation Manual*, The Traffic Institute, Northwestern University, Evanston, IL, 1975.
10.2. Backaitis, S. H., Ed., *Reconstruction of Motor Vehicle Accidents: A Technical Compendium*, Publication PT-34, Society of Automotive Engineers (SAE), Warrendale, PA, 1989.
10.3. Platt, F. N., *The Traffic Accident Handbook*, Hanrow Press, Columbia, MD, 1983.
10.4. Moffatt, E. A., and Moffatt, C. A., Eds., *Highway Collision Reconstruction*, American Society of Mechanical Engineers, New York, 1980.
10.5. Gardner, J. D., and Moffatt, E. A., Eds., *Highway Truck Collision Analysis*, American Society of Mechanical Engineers, New York, 1982.
10.6. Adler, U., Ed., *Automotive Handbook*, Robert Bosch, Stuttgart, Germany, 1986.
10.7. Collins, J. C., *Accident Reconstruction*, Charles C Thomas, Springfield, IL, 1979.
10.8. Limpert, R., *Motor Vehicle Accident Reconstruction and Cause Analysis*, The Michie Company, Low Publishers, Charlottesville, VA, 1978.
10.9. Noon, R., *Introduction to Forensic Engineering*, CRC Press, Boca Raton, FL, 1992.

Problems

Sections 10.2 and 10.3 Work

P10.2.1: *A* particle *P* moves on a curve *C* defined by the parametric equations:

$$x = t\,, \quad y = t^2\,, \quad z = t^3$$

where x, y, and z are coordinates, measured in meters, relative to an X, Y, Z Cartesian system. Acting on P is a force $\mathbf{F}$ given by:

$$\mathbf{F} = 4\mathbf{n}_x - 6\mathbf{n}_y + 8\mathbf{n}_z$$

where $\mathbf{F}$ is measured in Newtons, and $\mathbf{n}_x$, $\mathbf{n}_y$, and $\mathbf{n}_z$ are unit vectors parallel to the X-, Y-, and Z-axes. Compute the work done by $\mathbf{F}$ in moving P from (0, 0, 0) to (2, 4, 8).

P10.2.2: The magnitude and direction of a force $\mathbf{F}$ acting on a particle P depend upon the coordinate position (x, y, z) of $\mathbf{F}$ (and P) in an X, Y, Z coordinate space as:

$$\mathbf{F} = 2xy^2z^2\,\mathbf{n}_x + 2x^2yz^2\mathbf{n}_y + 2x^2y^2z\mathbf{n}_z \;\; N$$

where $\mathbf{n}_x$, $\mathbf{n}_y$, and $\mathbf{n}_z$ are unit vectors parallel to X, Y, and Z. Suppose P moves from the origin O to a point C (1, 2, 3) along two different paths as in Figure P10.2.2: (1) along the line segment OC, and (2) along the rectangular segments OA, AB, BC. Calculate the work done by $\mathbf{F}$ on P in each case. (Assume that the coordinates are measured in meters.)

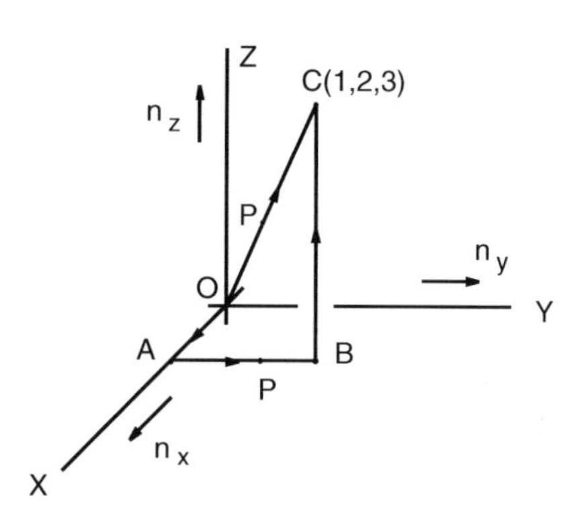

FIGURE P10.2.2
A particle P moving from O to C along two different paths.

P10.2.3: See Problem P10.2.2. Suppose a force $\mathbf{F}$ acting on a particle P depends upon the position of $\mathbf{F}$ (and P) in an X, Y, Z space as:

$$\mathbf{F} = \nabla\phi \overset{D}{=} \frac{\partial\phi}{\partial x}\mathbf{n}_x + \frac{\partial\phi}{\partial y}\mathbf{n}_y + \frac{\partial\phi}{\partial z}\mathbf{n}_z$$

Show that the work done by $\mathbf{F}$ on P as P moves from P_1 (x_1, y_1, z_1) to P_2 (x_2, y_2, z_2) is simply ϕ (x_2, y_2, z_2) – ϕ (x_1, y_1, z_1). *Comment:* When a force $\mathbf{F}$ can be represented in the form $\mathbf{F} = \nabla\phi$, $\mathbf{F}$ is said to be conservative.

P10.2.4: See Problems P10.2.2 and P10.2.3. Show that the force $\mathbf{F}$ of Problem P10.2.2 is conservative. Determine the function ϕ. Using the result of Problem P10.2.3, check the result of P10.2.2.

P10.2.5: *A* horizontal force $\mathbf{F}$ pushes a 50-lb cart C up a hill H which is modeled as a sinusoidal curve with amplitude of 7 ft and half-period of 27 ft as shown in Figure P10.2.5. Assuming that there is no frictional resistance between C and H and that $\mathbf{F}$ remains directed horizontally, find the work done by $\mathbf{F}$.

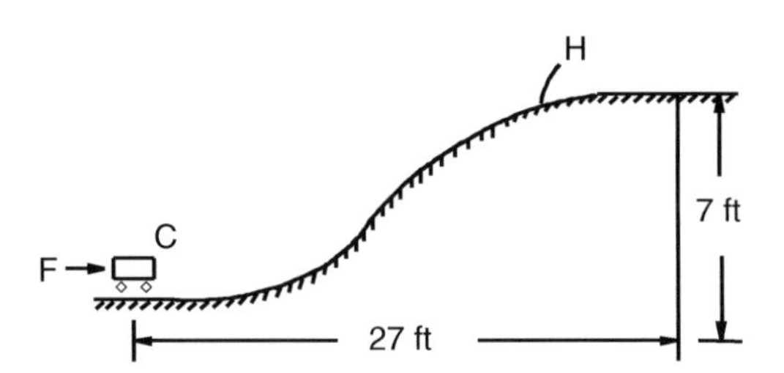

FIGURE P10.2.5
A force **F** pushing a cart up a hill.

P10.2.6: *A* force pushes a block *B* along a smooth horizontal slot from a position *O* to a position *Q* as in Figure P10.2.6. The movement of *B* is resisted by a linear spring with a natural (unstretched) length of 6 in. and modulus 12 lb/in. Determine the work done by **F**.

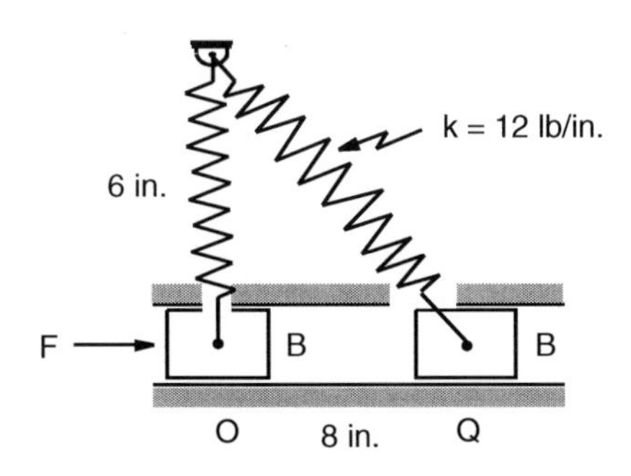

FIGURE P10.2.6
A block sliding in a slot.

P10.2.7: Repeat Problem P10.2.6 if the natural length of the spring is 4 in.

P10.2.8: A motorist, in making a turn with a 15-in.-diameter steering wheel, exerts a force of 8 lb with each hand tangent to the wheel as in Figure P10.2.8. If the wheel is turned through an angle of 150°, determine the work done by the motorist.

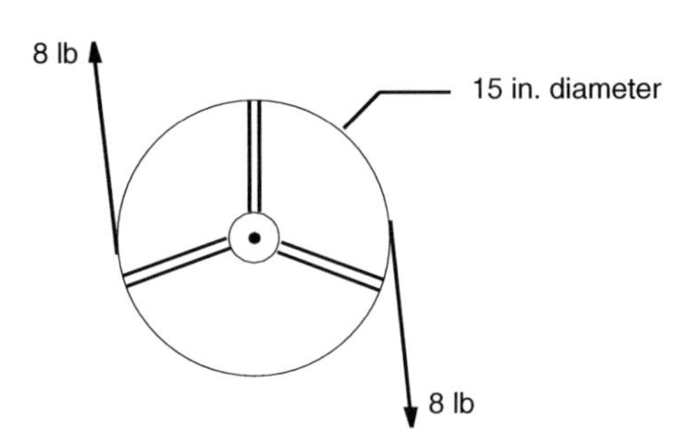

FIGURE P10.2.8
Forces on a steering wheel.

Section 10.4 Power

P10.4.1: An automobile with a 110-hp engine is traveling at 35 mph. If there are no frictional losses in the transmission or drive train, what is the tractive force exerted by the drive wheels?

P10.4.2: See Problem P10.4.1. Suppose the automobile of Problem P10.4.1 has 26-in.-diameter drive wheels, a transmission gear ratio of 8 to 1, and a drive axle gear ratio of 4.5 to 1. What is the engine speed (in rpm)? What is the torque (in ft·lb) of the engine crank shaft?

Section 10.5 Kinetic Energy

P10.5.1: An 2800-lb automobile starting from a stop accelerates at the rate of 3 mph per second. Find the kinetic energy of the vehicle after it has traveled 100 yards.

P10.5.2: A double pendulum consists of two particles P_1 and P_2 having masses m_1 and m_2 supported by light cables with lengths ℓ_1 and ℓ_2, making angles θ_1 and θ_2 with the vertical as in Figure P10.5.2. Find an expression for the kinetic energy of this system. Express the results in terms of m_1, m_2, ℓ_1, ℓ_2, θ_1, θ_2, $\dot{\theta}_1$, and $\dot{\theta}_2$.

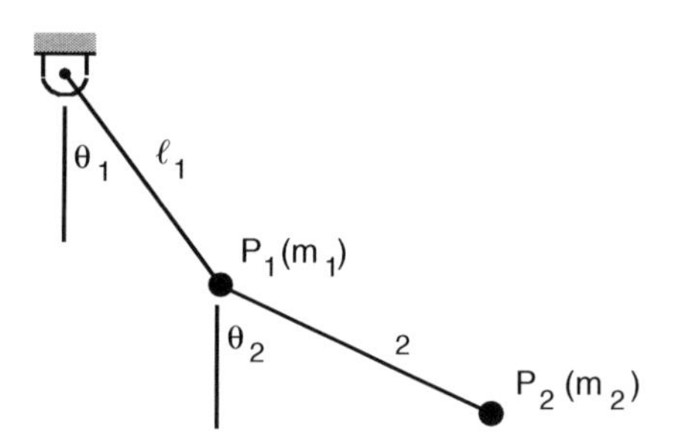

FIGURE P10.5.2
A double pendulum.

P10.5.3: Determine the kinetic energy of the rod pendulum of Figure P10.5.3. Let the rod have length ℓ and mass m. Express the result in terms of m, ℓ, θ, and $\dot{\theta}$.

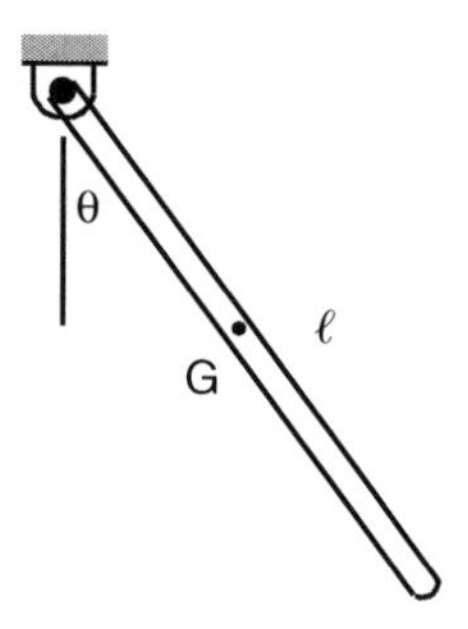

FIGURE P10.5.3
A rod pendulum.

P10.5.4: See Problem 10.5.3. Consider the double rod pendulum as in Figure 10.5.4. Let each rod have length ℓ and mass m. Determine the kinetic energy of the system. Express the results in terms of m, ℓ, θ_1, θ_2, $\dot{\theta}_1$, and $\dot{\theta}_2$.

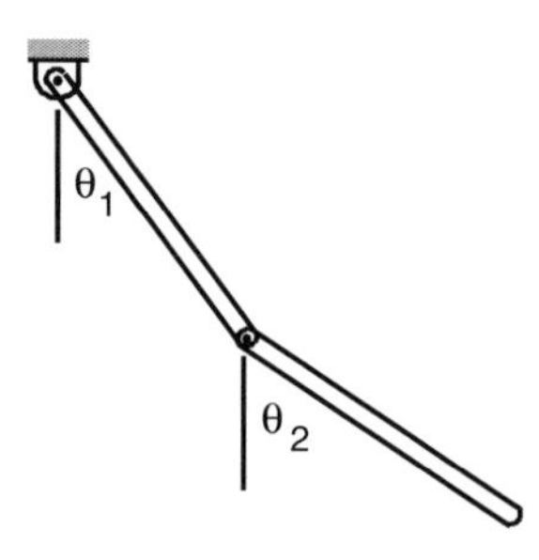

FIGURE P10.5.4
Double-rod pendulum.

P10.5.5: See Problems P10.5.3 and P10.5.4. Extend the results of Problems P10.5.3 and P10.5.4 to the triple-rod pendulum of Figure P10.5.5.

P10.5.6: An automobile with 25-in.-diameter wheels is traveling at 30 mph when the operator suddenly swerves to the left, causing the vehicle to spin out and rotate at the rate of 180°/sec. If the wheels each weigh 62 lb and if their axial and diametral radii of

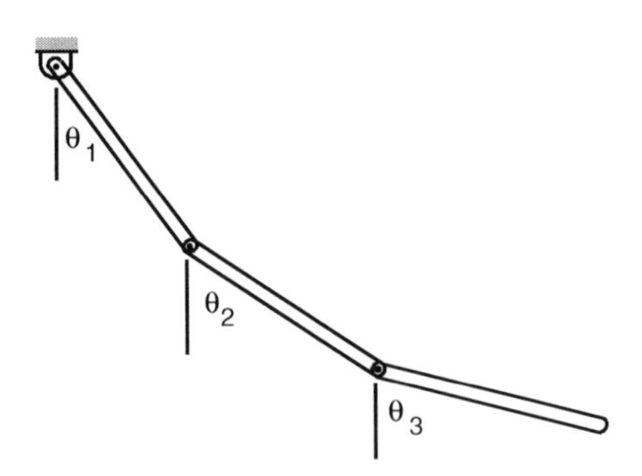

FIGURE P10.5.5
Triple-rod pendulum.

gyration are 10 and 7 in., respectively, determine the kinetic energy of one of the rear wheels for the spinning vehicle.

Sections 10.6 to 10.9: Work–Energy Principles and Applications

P10.6.1: A ball is thrown vertically upward with a speed of 12 m/sec. Determine the maximum height h reached by the ball.

P10.6.2: An object is dropped from a window which is 45 ft above the ground. What is the speed of the object when it strikes the ground?

P10.6.3: See Problem P10.6.1. What is the speed of the ball when it is 3 m above the thrower.

P10.6.4: A water faucet is dripping slowly. When a drop has fallen 1 ft, a second drop appears. What are the speeds of the drops when the first drop has fallen 3 ft? What is the separation between the drops at that time?

P10.6.5: A simple pendulum consists of a light string of length 3 ft and a concentrated mass P weighing 5 lb at the end as in Figure P10.6.5. Suppose the pendulum is displaced through an angle θ of 60° and released from rest. What is the speed of P when θ is (a) 45°, (b) 30°, and (c) 0°?

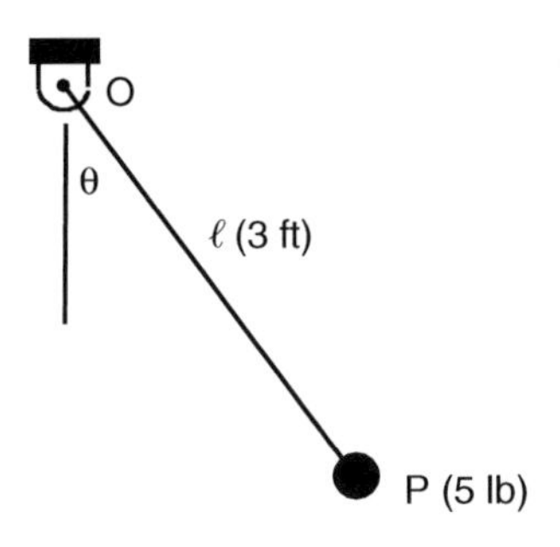

FIGURE P10.6.5
A simple pendulum.

P10.6.6: See Problem P10.6.5. Suppose P has a speed of 7.5 ft/sec when θ is 0. What is the maximum angle reached by P?

P10.6.7: See Problems 10.6.5 and P10.6.6. What is the minimum speed v of P when θ is 0 so that the pendulum will make a complete loop without the string becoming slack even in the topmost position ($\theta = 180°$)?

P10.6.8: Repeat Problems P10.6.5 to 10.6.7 if the mass of P is 10 lb instead of 5 lb.

P10.6.9: Repeat Problems P10.6.5 to 10.6.7 if the length ℓ of the pendulum is 4 ft.

P10.6.10: Repeat Problems P10.6.5 and P10.6.6 if the simple pendulum of Figure P10.6.5 is replaced by a rod pendulum of length 3 ft and weight 5 lb as in Figure P10.6.10, with P being the end point of the rod.

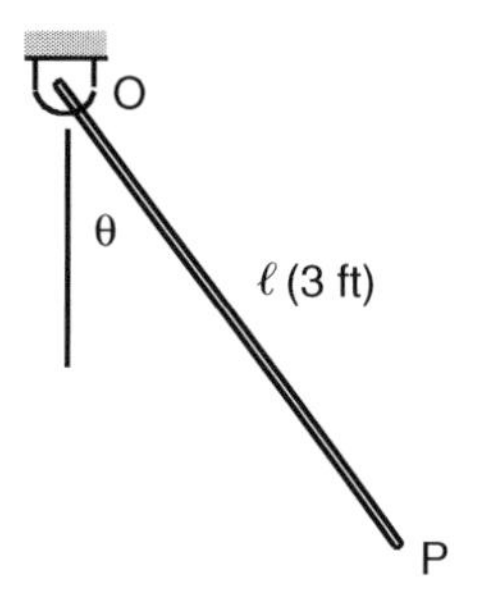

FIGURE P10.6.10
A rod pendulum.

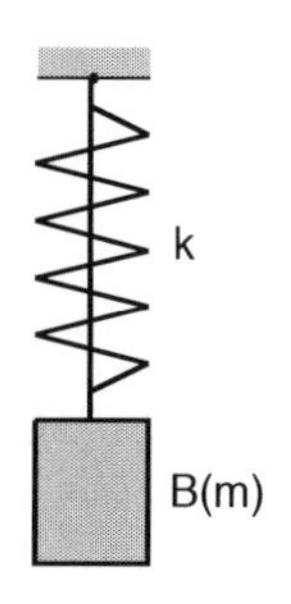

FIGURE P10.6.11
A vertical mass–spring system.

P10.6.11: A block B with mass m is attached to a vertical linear spring with modulus k as in Figure P10.6.11. Suppose B is held in a position where the spring is neither stretched nor compressed and is then released from rest. Find:

a. The maximum downward displacement of B
b. The maximum force in the spring
c. The maximum speed of B
d. The position where the maximum speed of B occurs

P10.6.12: Solve Problem P10.6.11 for $k = 7$ lb/in. and $m = 0.25$ slug.

P10.6.13: Solve Problem P10.6.11 for $k = 12$ N/cm and $m = 2$ kg.

P10.6.14: A 5-lb block B sliding in a smooth vertical slot is attached to a linear spring with modulus k of 4 lb/in. as in Figure P10.6.14. Let the natural length ℓ of the spring be 8 in. Let the displacement y of B be measured downward from O, opposite the spring anchor Q as shown. Find the speed of B when (a) $y = 0$, and (b) $y = 5$ in.

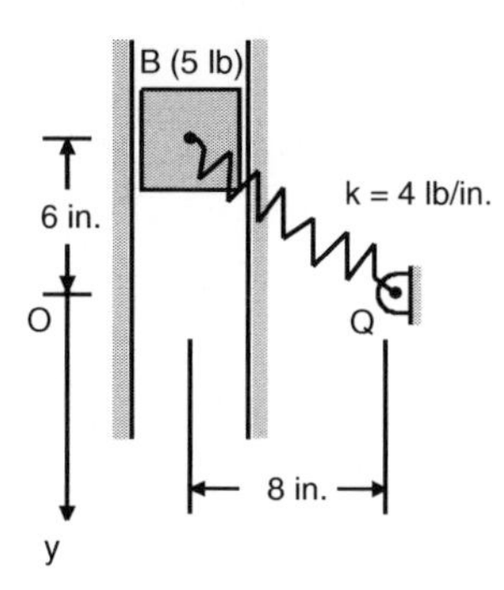

FIGURE P10.6.14
A spring connected block in a smooth vertical slot.

P10.6.15: Solve Problem P10.6.14 if the natural length of the spring is 6 in.

P10.6.16: A 10-kg block B is placed at the top of an incline which has a smooth surface as represented in Figure P10.6.16. If B is released from rest and slides down the incline, what will its speed be when it reaches the bottom of the incline?

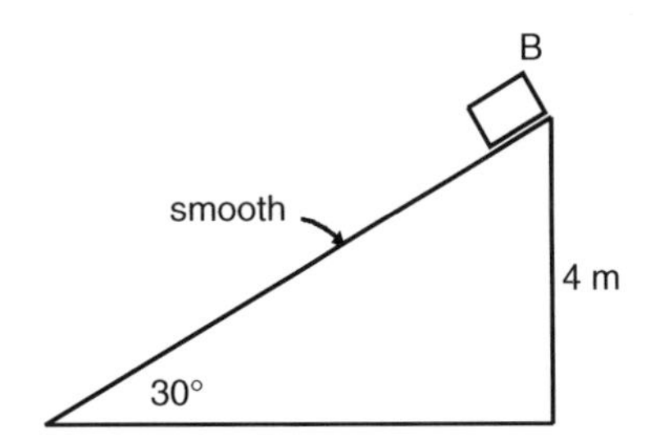

FIGURE P10.6.16
A block sliding down a smooth inclined surface.

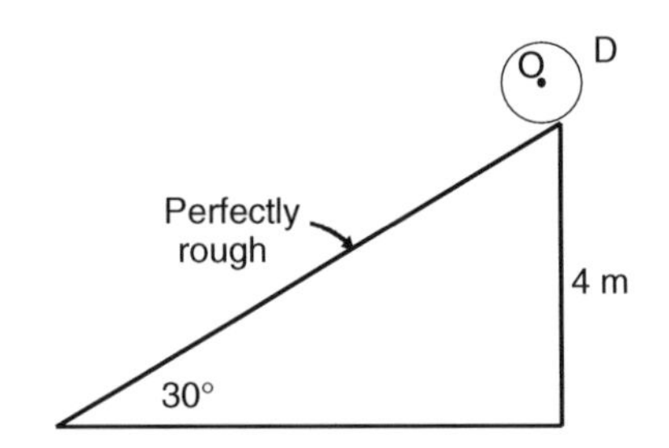

FIGURE P10.6.17
A circular disk *D* rolling down an incline plane.

P10.6.17: See Problem P10.6.16. A 10-kg circular disk *D* with 0.25-m diameter is placed at the top of an incline that has a perfectly rough surface as represented in Figure P10.6.17. If *D* is released from rest and rolls down the incline, what will be the speed of the center *O* of *D* when it reaches the bottom of the incline? Compare the result with that of Problem P10.6.16.

P10.6.18: A solid half-cylinder *C*, with radius *r* and mass *m*, is placed on a horizontal surface *S* and held with its flat side vertical as represented in Figure P10.6.18. Let *C* be released from rest and let *S* be perfectly rough so that *C* rolls on *S*. Determine the angular speed of *C* when its mass center *G* is in the lowest most position.

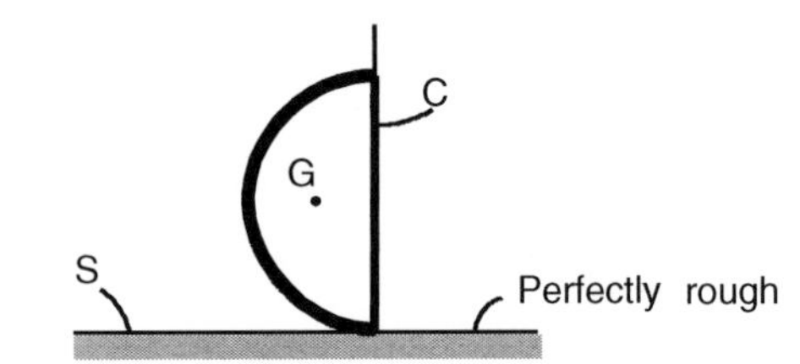

FIGURE P10.6.18
A half-cylinder (end view) on a horizontal surface.

P10.6.19: Repeat Problem P10.6.18 if *S* is smooth instead of rough.

P10.6.20: A 7-kg, 1.5-m-long rod *AB* has its end pinned to light blocks that are free to move in frictionless horizontal and vertical slots as represented in Figure P10.6.20. If the pins are also frictionless, and if *AB* is released from rest in the position shown, determine the speed of the mass center *G* and the angular speed of *AB* when *AB* falls to a horizontal position and when *AB* has fallen to a vertical position. Assume that the guide blocks at *A* and *B* remain in their vertical and horizontal slots, respectively, throughout the motion.

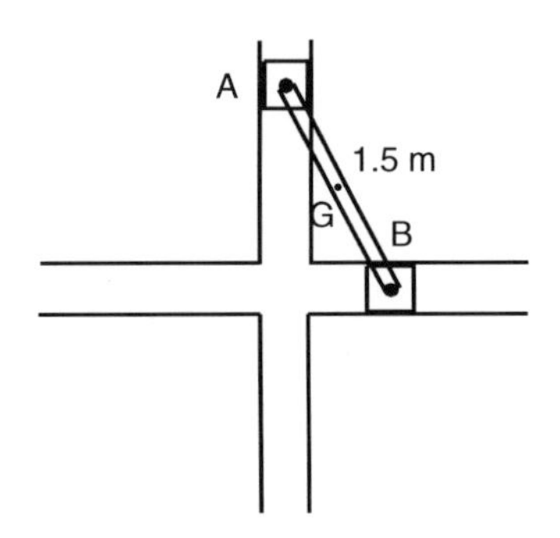

FIGURE P10.6.20
A rod with ends moving in frictionless guide slots.

P10.6.21: Repeat Problem P10.6.20 if the guide blocks at *A* and *B* are no longer light but instead have masses of 1 kg each.

P10.6.22: A 300-lb flywheel with radius of gyration of 15 in. is rotating at 3000 rpm. A bearing failure causes a small friction moment of 2 ft·lb which in turn causes the flywheel to slow and eventually stop. How many turns does the flywheel make before coming to a stop?

Section 10.10 Motor Vehicle Accident Reconstruction

P10.10.1: All four wheels of a car leave 75-ft-long skid marks in coming to a stop on a level roadway. If the coefficient of friction between the tires and the road surface is 0.75, determine the speed of the car at the beginning of the skid marks.

P10.10.2: The front wheels of a car leave 70 ft of skid marks and its rear wheels leave 50 ft of skid marks in coming to a stop on a level roadway. If the coefficient of friction between the tires and the roadway is 0.80, and if 60% of the vehicle weight is on the front wheels, determine the speed of the car at the beginning of the skid marks.

P10.10.3: A car slides to a stop leaving skid marks of length s (measured in feet) on a level roadway. If the coefficient of friction between the tires and the roadway is μ, show that the speed v (in miles per hour) of the car at the beginning of the skid marks is given by the simple expression:

$$v = \sqrt{30\mu s}$$

P10.10.4: See Problem P10.10.3. Suppose that the roadway, instead of being level, has a slight down slope in the direction of travel as represented in Figure P10.10.4. If the down slope angle is θ (measured in radians) as shown, show that the speed formula of Problem P10.10.3 should be modified to:

$$v = \sqrt{30(\mu - \theta)s}$$

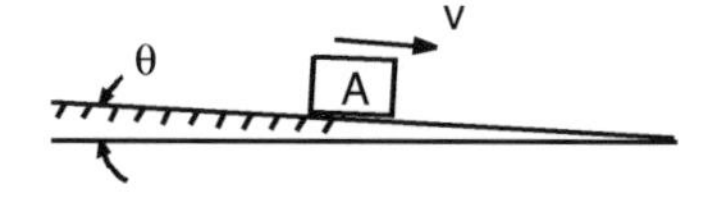

FIGURE P10.10.4
An automobile A skidding to a stop on a downslope.

P10.10.5: An automobile leaves 50 ft of skid marks before striking a pole at 20 mph. Find the speed of the vehicle at the beginning of the skid marks if the roadway is level and if the coefficient of friction between the tires and the roadway is 0.65. Assume the vehicle stops upon hitting the pole.

P10.10.6: A pickup truck leaves 30 ft of skid marks before colliding with the rear of a stopped automobile. Following the collision, the two vehicles slide together (a *plastic* collision) for 25 ft. Let the coefficient of friction between the pickup truck tires and the roadway be 0.7; after the collision, for the sliding vehicles together, let the coefficient of friction with the roadway be 0.5. Let the weights of the pickup truck and automobile be 3500 lb and 2800 lb, respectively. Find the speed of the pickup truck at the beginning of the skid marks.

P10.10.7: Repeat Problem P10.10.6 if just before collision, instead of being stopped, the automobile is moving at 10 mph in the same direction as the pickup truck.

P10.10.8: Repeat Problem P10.10.6 if just before collision, instead of being stopped, the automobile is moving toward the pickup truck at 10 mph (that is, a head-on collision).

Sections 10.11 and 10.12 Work, Energy, and Impact

P10.11.1: *A* 60-gauge bullet is fired into a 20-kg block suspended by a 7-m cable as depicted in Figure P10.11.1. The impact causes the block pendulum to swing through an angle of 15°. Determine the speed v of the bullet.

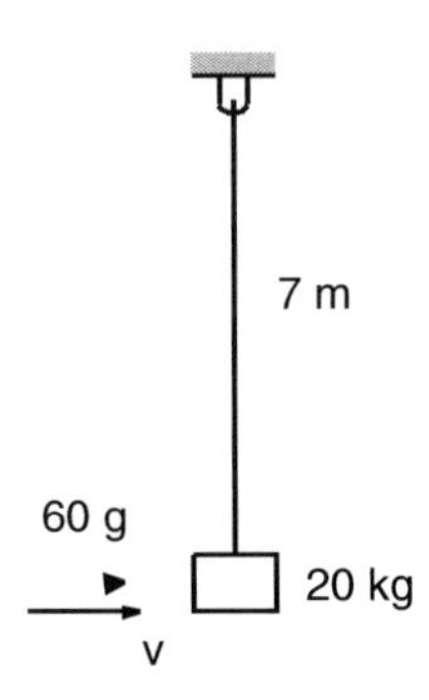

FIGURE P10.11.1
A bullet fired into a block.

P10.11.2: *A* 14-in. diameter wheel W_1 is rotating at 350 rpm when it is brought into contact with a 10-in.-diameter wheel W_2, which is initially at rest, as represented in Figure P10.11.2. After the wheels come into contact, they roll together without slipping. Although W_1 is free to rotate, with negligible friction, W_2 is subjected to a frictional moment of 2 ft·lb in its bearings. Let the weights of W_1 and W_2 be 28 lb and 20 lb, respectively. Let the radii of gyration of W_1 and W_2 be 5 in. and 3.5 in., respectively. Determine the number of revolutions N_1 and N_2 turned by each wheel after the meshing contact until they come to rest.

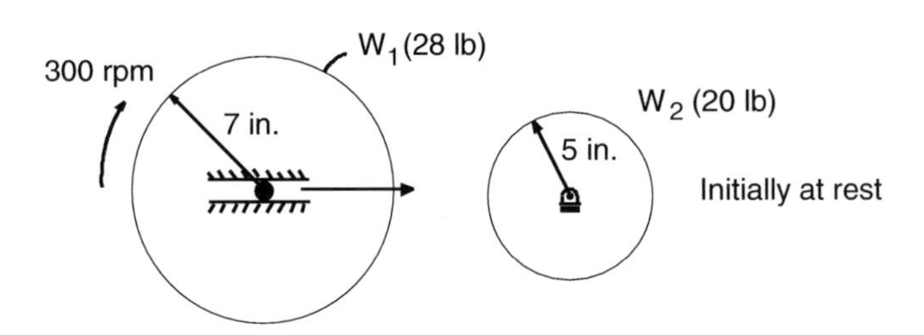

FIGURE P10.11.2
Meshing wheels.

P10.11.3: A 1-m rod B with a mass of 1 kg is hanging vertically and supported by a frictionless pin at O as in Figure P10.11.3. *A* particle P with a mass of 0.25 kg moving horizontally with speed v collides with the rod as also indicated in Figure P10.11.3. If the collision is perfectly plastic (with coefficient of restitution $e = 0$), determine v so that B completes exactly one half of a revolution and then comes to rest in a vertically up position. Let the point of impact x be (a) 0.5 m; (b) 0.667 m; and (c) 1.0 m.

P10.11.4: Repeat Problem P10.11.3 for a perfectly elastic collision (coefficient of restitution $e = 1$).

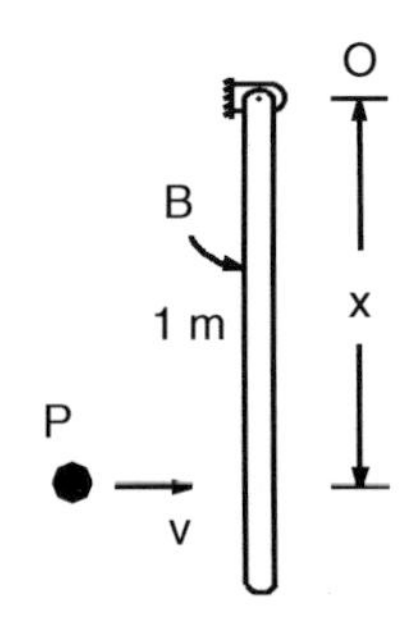

FIGURE P10.11.3
A particle P colliding with a vertical pin supported rod initially at rest.

11

Generalized Dynamics: Kinematics and Kinetics

11.1 Introduction

Recall in the analysis of elementary statics problems we discover, after gaining experience, that by making insightful choices about force directions and moment points, we can greatly simplify the analysis. Indeed, with sufficient insight, we discover that we can often obtain precisely the same number of equations as there are unknowns in the problem statement. Moreover, these equations are often uncoupled, thus producing answers with little further analysis.

In Chapter 10, we found that the work–energy principle, like clever statics solution procedures, can often produce simple and direct solutions to dynamics problems. We also found, however, that while the work–energy principle is simple and direct, it is also quite restricted in its range of application. The work–energy principle leads to a single scalar equation, thus enabling the determination of a single unknown. Hence, if two or more unknowns are to be found, the work–energy principle is inadequate and is restricted to relatively simple problems.

The objective of *generalized dynamics* is to extend the relatively simple analysis of the work–energy principle to complex dynamics problems having a number of unknowns. The intention is to equip the analyst with the means of determining unknowns with a minimal effort — as with insightful solutions of statics problems.

In this chapter, we will introduce and discuss the elementary procedures of generalized dynamics. These include the concepts of generalized coordinates, partial velocities and partial angular velocities, generalized forces, and potential energy. In Chapter 12, we will use these concepts to obtain equations of motion using Kane's equations and Lagrange's equations.

11.2 Coordinates, Constraints, and Degrees of Freedom

In the context of generalized dynamics, a *coordinate* (or *generalized coordinate*) is a parameter used to define the configuration of a mechanical system. Consider, for example, a particle P moving on a straight line L as in Figure 11.2.1. Let x locate P relative to a fixed point O on L. Specifically, let x be the distance between O and P. Then, x is said to be a *coordinate* of P.

Next, consider the simple pendulum of Figure 11.2.2. In this case, the configuration of the system and, as a consequence, the location of the bob P are determined by the angle θ. Thus, θ is a coordinate of the system.

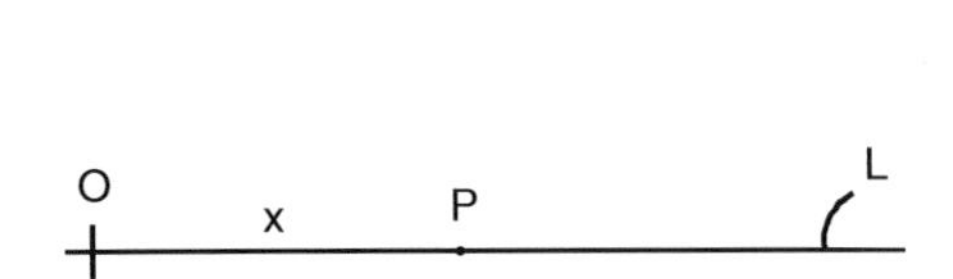

Figure 11.2.1
A particle moving on a straight line with coordinate x.

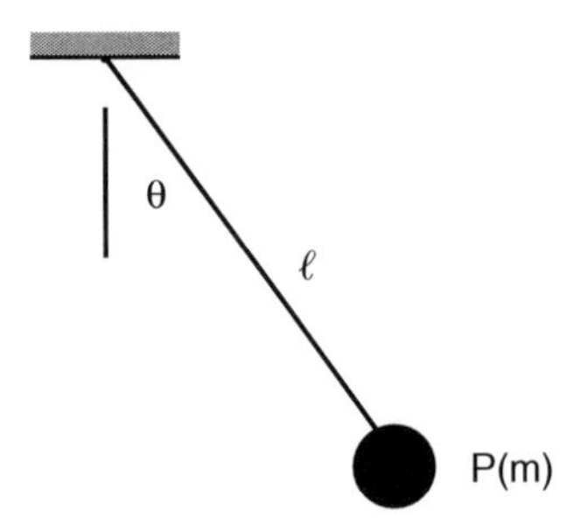

FIGURE 11.2.2
A simple pendulum with coordinate θ.

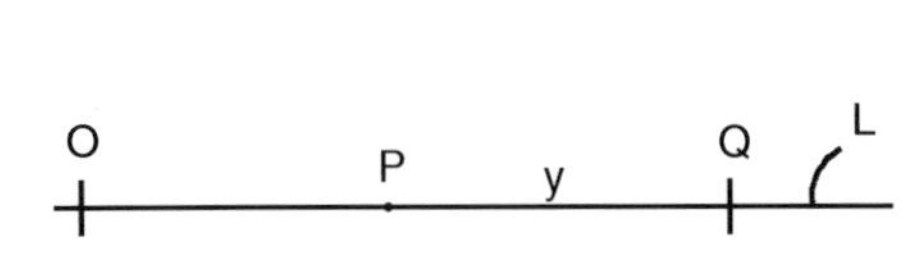

FIGURE 11.2.3
A particle moving on a single line with coordinate y.

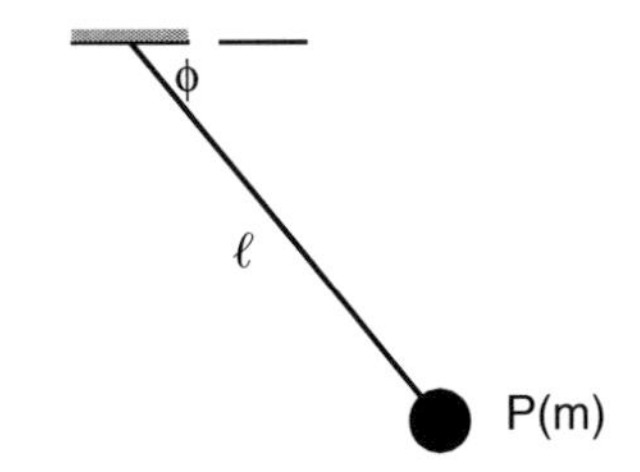

FIGURE 11.2.4
Simple pendulum with coordinate ϕ.

System coordinates are not unique. For the systems of Figures 11.2.1 and 11.2.2 we could also define the configurations by the coordinates y and ϕ as in Figures 11.2.3 and 11.2.4. (In Figure 11.2.3, Q, like O, is fixed on L.)

As a mechanical system moves and its configuration changes, the values of the coordinates change. This means that the coordinates are functions of time t. In a dynamical analysis of the system, the coordinates become the dependent variables in the governing differential equations of the system. From this perspective, constant geometrical parameters, such as the pendulum length ℓ in Figure 11.2.2, are not coordinates.

The minimum number of coordinates needed to define a system's configuration is the number of degrees of freedom of the system. Suppose, for example, that a particle P moves in the X–Y plane as in Figure 11.2.5. Then, (x, y) or, alternatively, (r, θ) are coordinates of P. Because P has two coordinates defining its position, P is said to have two degrees of freedom.

A restriction on the movement of a mechanical system is said to be a *constraint*. For example, in Figure 11.2.5, suppose P is restricted to move only in the X–Y plane. This restriction is then a constraint that can be expressed in the three-dimensional X, Y, Z space as:

$$z = 0 \tag{11.2.1}$$

Expressions describing movement restrictions, such as Eq. (11.2.1) are called *constraint equations*. A mechanical system may have any number of constraint equations, often more than the number of degrees of freedom. For example, the particle of Figure 11.2.1, restricted to move on the straight line, has two constraint equations in the three-dimensional X, Y, Z space. That is,

$$y = 0 \quad \text{and} \quad z = 0 \tag{11.2.2}$$

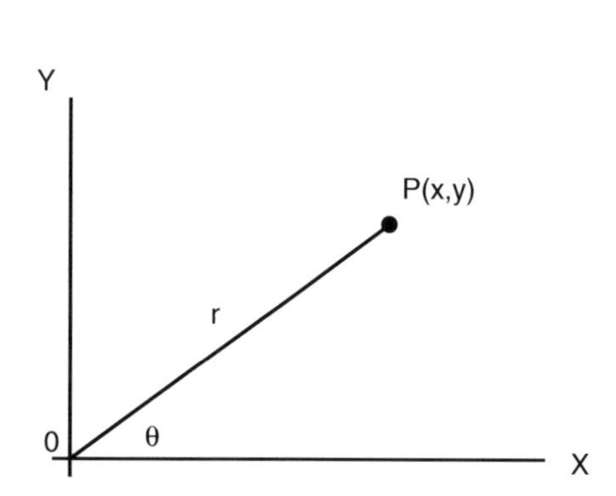

FIGURE 11.2.5
A particle P moving in a plane.

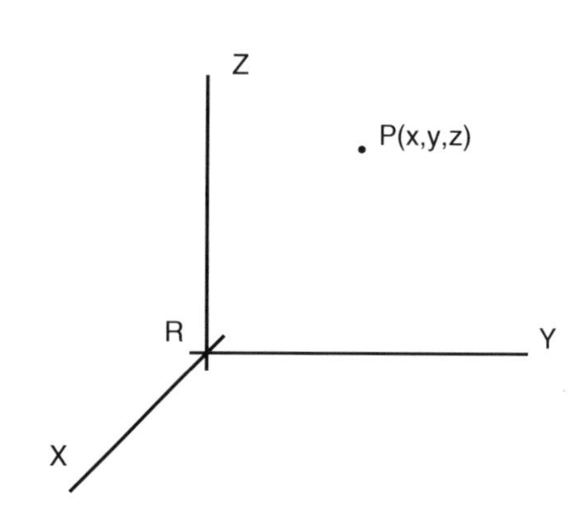

FIGURE 11.2.6
A particle moving in a Cartesian reference frame.

The number of degrees of freedom of a mechanical system is the number of coordinates of the system if it were unrestricted minus the number of constraint equations. For example, if a particle P moves relative to a Cartesian reference frame R as in Figure 11.2.6, then it has, if unrestricted, three degrees of freedom. If, however, P is restricted to move in a plane (say, a plane parallel to the X–Z plane), then P is *constrained*, and its constraint may be described by a single constraint equation of the form:

$$y = \text{constant} \tag{11.2.3}$$

Hence, in this case there are three minus one, or two, degrees of freedom.

To further illustrate these concepts, consider the system of two particles P_1 and P_2 at opposite ends of a light rod (a "dumbbell") as in Figure 11.2.7. In a three-dimensional space, the two particles with unrestricted motion require six coordinates to specify their positions. Let these coordinates be (x_1, y_1, z_1) and (x_2, y_2, z_2) as shown in Figure 11.2.7. Now, for the particles to remain at opposite ends of the rod, the distance between them must be maintained at the constant value ℓ, the rod length. That is,

$$(x_1 - x_2)^2 + (y_1 - y_2)^2 + (z_1 - z_2)^2 = \ell^2 \tag{11.2.4}$$

This is a single constraint equation; thus, the system has a net of five degrees of freedom.

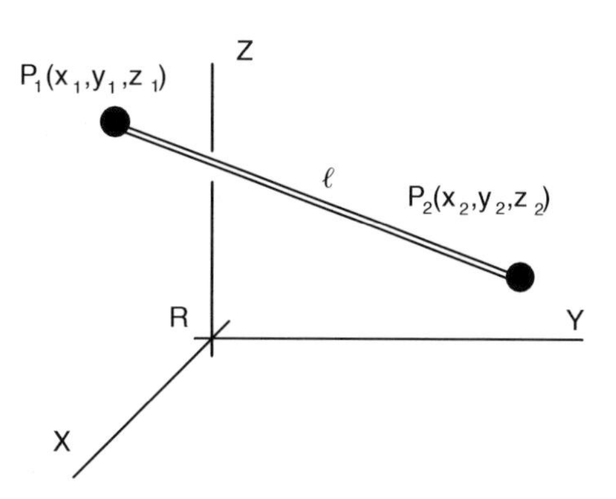

FIGURE 11.2.7
Particles at opposite ends of a light rod.

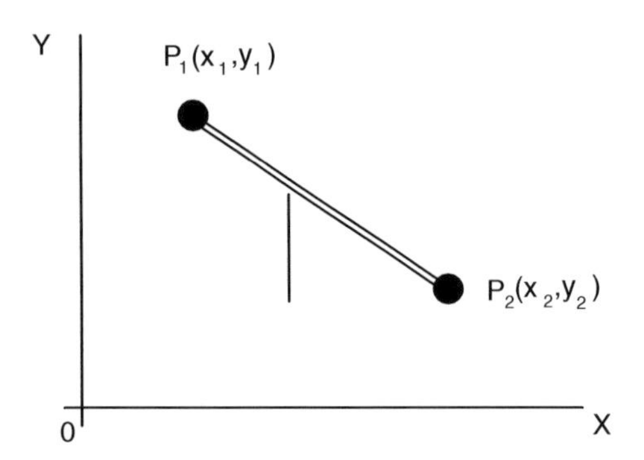

FIGURE 11.2.8
A dumbbell moving in the X–Y plane.

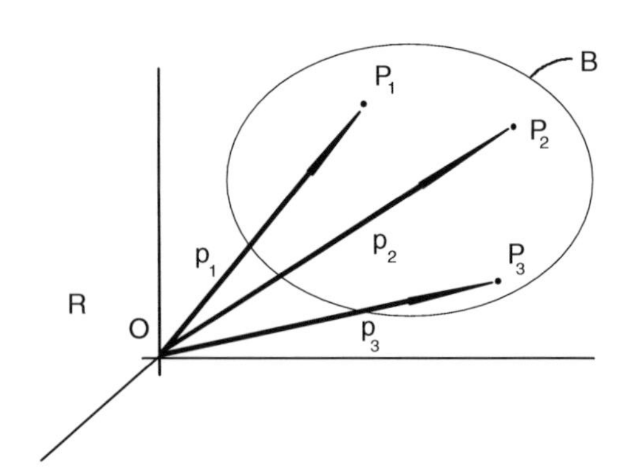

FIGURE 11.2.9
A rigid body B moving in a reference frame R.

If the movement of the dumbbell system is further restricted to the X–Y plane, additional constraints occur, as represented by the equations:

$$z_1 = 0 \text{ and } z_2 = 0 \tag{11.2.5}$$

These expressions together with Eq. (11.2.4) then form three constraint equations, leaving the system with six minus three, or three, degrees of freedom. These degrees of freedom might be represented by either the parameters (x_1, y_1, θ) or (x_2, y_2, θ) as shown in Figure 11.2.8.

As a final illustration of these ideas consider a rigid body B composed of N particles P_i ($i = 1,\ldots, N$) moving in a reference frame R as in Figure 11.2.9. The rigidity of B requires that the distances between the respective particles are maintained at constant values. Suppose, for example, that P_1, P_2, and P_3 are noncollinear points. Let $\mathbf{p}_1$, $\mathbf{p}_2$, and $\mathbf{p}_3$ locate P_1, P_2, and P_3 relative to O in R. Then, the respective distances between these particles are maintained by the equations:

$$(\mathbf{p}_1 - \mathbf{p}_2)^2 = d_1^2,\ (\mathbf{p}_2 - \mathbf{p}_3)^2 = d_2^2,\ (\mathbf{p}_3 - \mathbf{p}_1)^2 = d_3^2 \tag{11.2.6}$$

where the distances d_1, d_2, and d_3 are constants.

P_1, P_2, and P_3 thus form a rigid triangle. The other particles of B are then maintained in fixed positions relative to the triangle of P_1, P_2, and P_3 by the expressions:

$$(\mathbf{p}_i - \mathbf{p}_1)^2 = d_i^2,\ (\mathbf{p}_i - \mathbf{p}_2)^2 = f_i^2,\ (\mathbf{p}_i - \mathbf{p}_3)^2 = g_i^2\ (i = 4,\ldots,N) \tag{11.2.7}$$

Equations (11.2.6) and (11.2.7) form $3 + 3(N - 3)$ or $3N - 6$ constraint equations. If the particles of B are unrestricted in their movement, $3N$ coordinates would be required to specify their position in R. Hence, the number of degrees of freedom n of the rigid body is:

$$n = 3N - (3N - 6) = 6 \tag{11.2.8}$$

11.3 Holonomic and Nonholonomic Constraints

The constraint equations of the examples of the foregoing section might be described as being *positional* or *geometrical*. That is, they involve the relative positions of, or distances between, points of a mechanical system. They are developed from position vectors of the system. They do not involve velocities, accelerations, or derivatives of system coordinates. Such positional or geometrical constraints are said to be *holonomic*, and the associated mechanical system is said to be a *holonomic system*. If, however, a system has constraint equations that involve velocities, accelerations, or derivatives of system coordinates, the constraint equations are said to be *nonholonomic*, or *kinematic*, and the mechanical system is said to be a *nonholonomic system*.

Specifically, suppose a constraint equation has the form:

$$f(q_r, t) = 0 \tag{11.3.1}$$

where the q_r ($r = 1,\ldots, n$) are coordinates of the system (n is the number of degrees of freedom of the unrestrained system). Such a constraint is said to be *holonomic* (or *geometric*). Alternatively, suppose a constraint equation has the form:

$$f(q_r, \dot{q}_r, \ddot{q}_r, \ldots, t) = 0 \tag{11.3.2}$$

Such a constraint is said to be *nonholonomic* (or *kinematic*).

If a nonholonomic constraint equation does not involve second-order or higher-order derivatives of the coordinates, and if it is a linear function of the first derivatives, then it is said to be a *simple nonholonomic constraint*.

As might be expected, holonomic systems are easier to study and analyze than nonholonomic systems. The reason is that holonomic systems produce algebraic constraint equations, whereas nonholonomic systems produce differential constraint equations. Indeed, expressions of the form of Eq. (11.3.2) are generally nonintegrable and as a consequence cannot be solved in terms of elementary functions. Fortunately, the vast majority of mechanical systems of interest and importance in machine dynamics can be modeled as holonomic systems.

To illustrate these concepts, consider again the rolling circular disk (or "rolling coin") discussed in Sections 4.12 and 8.13 and as shown in Figure 11.3.1. Recall from Eqs. (4.12.3) and (8.13.1) that the velocity $\mathbf{v}^G$ of the mass center G of the disk D in the inertia reference frame R was found to be:

$$\mathbf{v}^G = r(\dot{\psi} + \dot{\phi}\sin\theta)\mathbf{n}_1 - r\dot{\theta}\mathbf{n}_2 \tag{11.3.3}$$

where, as before, θ, ϕ, and ψ are the angles shown in Figure 11.3.1. Also, from Eqs. (4.12.2) and (8.13.3) the angular velocity $\boldsymbol{\omega}$ of D in R is:

$$\boldsymbol{\omega} = \dot{\theta}\mathbf{n}_1 + (\dot{\psi} + \dot{\phi}\sin\theta)\mathbf{n}_2 - \dot{\phi}\cos\theta\,\mathbf{n}_3 \tag{11.3.4}$$

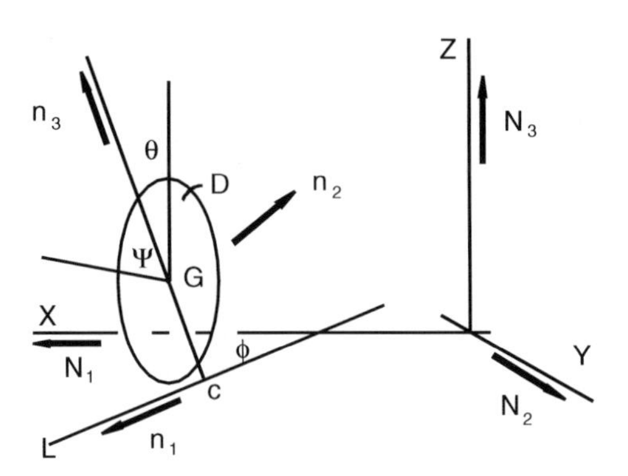

FIGURE 11.3.1
A circular disk rolling on a horizontal surface *S*.

Using these expressions, we can obtain the velocity $\mathbf{v}^C$, the contact point *C* of *D*, as:

$$\mathbf{v}^C = \mathbf{v}^G + \boldsymbol{\omega} \times (-r\mathbf{n}_3) \tag{11.3.5}$$

or

$$\begin{aligned} \mathbf{v}^C &= r(\dot{\psi} + \dot{\phi}\sin\theta)\mathbf{n}_1 - r\dot{\theta}\mathbf{n}_2 + r\dot{\theta}\mathbf{n}_2 \\ &\quad - r(\dot{\psi} + \dot{\phi}\sin\theta)\mathbf{n}_1 = 0 \end{aligned} \tag{11.3.6}$$

Recall that the condition that the contact point has zero velocity is a condition of rolling. This is a velocity or kinematic constraint and thus, in the foregoing terminology, a *nonholonomic* constraint.

Equation (11.3.6), being a vector equation, is equivalent to three scalar equations. Because an unrestrained rigid body has six degrees of freedom, the three scalar constraint equations reduce the number of degrees of freedom to three. To explore this further let (x, y, z) be the *X*, *Y*, *Z* coordinates of *G*. Then, if *D* is unrestrained with six degrees of freedom, these degrees of freedom could be defined by the coordinates $(x, y, z, \theta, \phi, \psi)$. If now *D* is constrained to be rolling on the horizontal surface *S* as by Eq. (11.3.6), we can obtain three constraint equations relating these six variables: specifically, the velocity of *G* may be expressed in terms of $\dot{x}$, $\dot{y}$, and $\dot{z}$ as:

$$\mathbf{v}^G = \dot{x}\mathbf{N}_1 + \dot{y}\mathbf{N}_2 + \dot{z}\mathbf{N}_3 \tag{11.3.7}$$

where $\mathbf{N}_1$, $\mathbf{N}_2$ and $\mathbf{N}_3$ are unit vectors parallel to the *X*, *Y*, and *Z* axes as in Figure 11.3.1. By equating the expression for $\mathbf{v}^G$ of Eq. (11.3.3) with that of Eq. (11.3.7), we obtain the desired constraint equations:

$$\dot{x} = r\left[(\dot{\psi} + \phi\ \sin\theta)\cos\phi + \dot{\theta}\cos\theta\sin\phi\right] \tag{11.3.8}$$

$$\dot{y} = r\left[(\dot{\psi} + \dot{\phi}\sin\theta)\sin\phi - \dot{\theta}\cos\theta\cos\phi\right] \tag{11.3.9}$$

and

$$\dot{z} = -r\dot{\theta}\sin\theta \tag{11.3.10}$$

Observe that Eqs. (11.3.8), (11.3.9), and (11.3.10) constitute three constraint equations. Hence, the disk D has three degrees of freedom, as expected. Observe next that the third equation, Eq. (11.3.10), is integrable in terms of elementary functions. That is,

$$z = r\cos\theta \tag{11.3.11}$$

From Figure 11.3.1, we see that this expression is simply a requirement that D remains in contact with the surface S — a geometric, or holonomic, constraint. Finally, observe that Eqs. (11.3.8) and (11.3.9) are not integrable in terms of elementary functions. Instead, they are coupled nonlinear differential equations requiring numerical solution. These expressions arise from the requirement that C must have zero velocity in S — a kinematic, or nonholonomic, constraint.

In the following and subsequent sections we will see that simple analyses using energy functions (that is, the use of kinetic energy and Lagrange's equations) are precluded with nonholonomic systems. With Kane's equations, however, the analysis is essentially the same for both holonomic and nonholonomic systems. Nevertheless, our focus will be on holonomic systems because, as noted earlier, the vast majority of mechanical systems of interest are holonomic systems.

11.4 Vector Functions, Partial Velocity, and Partial Angular Velocity

Consider a holonomic mechanical system S having n degrees of freedom represented by the coordinates q_r ($r = 1,\ldots, n$). Let P be a typical point of S. Then, a position vector $\mathbf{p}$ locating P relative to a fixed point O in an inertia frame R will, in general, be a function of the q_r and time t. That is,

$$\mathbf{p} = \mathbf{p}(q_r, t) \tag{11.4.1}$$

Then, from the chain rule for differentiation, the velocity of P in R is:

$$\mathbf{v} = \frac{d\mathbf{p}}{dt} = \frac{\partial \mathbf{p}}{\partial t} + \sum_{r=1}^{n} \frac{\partial \mathbf{p}}{\partial q_r}\dot{q}_r \tag{11.4.2}$$

The terms $\partial\mathbf{p}/\partial t$ and $\partial\mathbf{p}/\partial q_r$ ($r = 1,\ldots, n$) are called *partial velocity vectors*. These vectors are fundamental vectors in the development of generalized dynamics theories and procedures. Indeed, they may be viewed as base vectors in the generalized space of motion of a mechanical system. Because of their fundamental nature and widespread use, they are given a separate notation defined as:

$$\mathbf{v}_t \stackrel{D}{=} \partial\mathbf{p}/\partial t \quad \text{and} \quad \mathbf{v}_{\dot{q}_r} \stackrel{D}{=} \partial\mathbf{p}/\partial q_r \tag{11.4.3}$$

Then, from Eq. (11.4.2), we have:

$$\mathbf{v}_{\dot{q}_r} = \partial\mathbf{v}/\partial\dot{q}_r \tag{11.4.4}$$

Thus, Eq. (11.4.2) may be expressed in the form:

$$\mathbf{v} = \mathbf{v}_t + \sum_{r=1}^{n} \mathbf{v}_{\dot{q}_r} \dot{q}_r \tag{11.4.5}$$

From Eq. (11.4.2), we see that the partial rate of change of the velocity $\mathbf{v}$ with respect to q_s is:

$$\begin{aligned}\frac{\partial \mathbf{v}}{\partial q_s} &= \frac{\partial}{\partial q_s}\left(\frac{\partial \mathbf{p}}{\partial t} + \sum_{r=1}^{n} \frac{\partial \mathbf{p}}{\partial q_r}\dot{q}_r\right) = \frac{\partial}{\partial t}\left(\frac{\partial \mathbf{p}}{\partial q_s}\right) + \sum_{r=1}^{n} \frac{\partial}{\partial q_r}\left(\frac{\partial \mathbf{p}}{\partial q_s}\right)\dot{q}_r \\ &= \frac{d}{dt}\left(\frac{\partial \mathbf{p}}{\partial q_s}\right) = \frac{d\mathbf{v}_{\dot{q}_s}}{dt}\end{aligned} \tag{11.4.6}$$

Because the partial velocity vectors are analogous to base vectors or to unit vectors, it is useful to compute the projections of forces along these vectors. These force projections are called *generalized forces*. For example, the projections of inertia forces along partial velocity vectors are called *generalized inertia forces*.

To further develop and illustrate these concepts, recall that inertia forces are proportional to accelerations; hence, it is helpful to consider the projection of the acceleration $\mathbf{a}$ of a particle P along the partial velocities $\mathbf{v}_{\dot{q}_r}$. Thus, we have:

$$\begin{aligned}\mathbf{a} \cdot \mathbf{v}_{\dot{q}_r} &= \frac{d\mathbf{v}}{dt} \cdot \mathbf{v}_{\dot{q}_r} = \frac{d}{dt}\left(\mathbf{v} \cdot \mathbf{v}_{\dot{q}_r}\right) - \mathbf{v} \cdot \frac{d}{dt}\left(\mathbf{v}_{\dot{q}_r}\right) \\ &= \frac{1}{2}\frac{d}{dt}\left(\frac{\partial \mathbf{v}^2}{\partial \dot{q}_r}\right) - \mathbf{v} \cdot \frac{\partial \mathbf{v}}{\partial q_r} \\ &= \frac{1}{2}\frac{d}{dt}\left(\frac{\partial \mathbf{v}^2}{\partial \dot{q}_r}\right) - \frac{1}{2}\frac{\partial \mathbf{v}^2}{\partial q_r}\end{aligned} \tag{11.4.7}$$

Observe in Eq. (11.4.7) that the common factor $(1/2)\mathbf{v}^2$ in each term is proportional to kinetic energy. Indeed, Eq. (11.4.7) forms a basis for the development of Lagrange's equations that use kinetic energy. We will use it for this purpose in subsequent sections.

Consider next a rigid body B. Recall that in Section 4.5 we saw that, if $\mathbf{c}$ is a vector fixed in B, the derivative of $\mathbf{c}$ in a reference frame R is ((see Eq. (4.5.2)):

$$\frac{{}^R d\mathbf{c}}{dt} = \boldsymbol{\omega} \times \mathbf{c} \tag{11.4.8}$$

where $\boldsymbol{\omega}$ is the angular velocity of B in R defined as (see Eq. (4.5.1)):

$$\boldsymbol{\omega} = \left(\frac{{}^R d\mathbf{n}_2}{dt} \cdot \mathbf{n}_3\right)\mathbf{n}_1 + \left(\frac{{}^R d\mathbf{n}_3}{dt} \cdot \mathbf{n}_1\right)\mathbf{n}_2 + \left(\frac{{}^R d\mathbf{n}_1}{dt} \cdot \mathbf{n}_2\right)\mathbf{n}_3 \tag{11.4.9}$$

where $\mathbf{n}_1$, $\mathbf{n}_2$, and $\mathbf{n}_3$ are mutually perpendicular unit vectors fixed in B.

Let B be a part of a mechanical system S which has n degrees of freedom represented by n generalized coordinates q_r ($r = 1,\ldots, n$). Specifically, let the movement of B in an inertial reference frame R be determined by the coordinates q_r and time t. Then, if $\mathbf{c}$ is fixed in B, the derivative of $\mathbf{c}$ in R is:

$$\frac{{}^R d\mathbf{c}}{dt} = \frac{\partial \mathbf{c}}{\partial t} + \sum_{r=1}^{n} \frac{\partial \mathbf{c}}{\partial q_r}\dot{q}_r \tag{11.4.10}$$

By following the exact same procedure as in Section 4.5 in the development of Eq. (11.4.8) we can readily show that the partial derivatives $\partial\mathbf{c}/\partial t$ and $\partial\mathbf{c}/\partial q_r$ may be expressed as:

$$\partial\mathbf{c}/\partial t = \boldsymbol{\omega}_t \times \mathbf{c} \quad \text{and} \quad \partial\mathbf{c}/\partial q_r = \boldsymbol{\omega}_{\dot{q}_r} \times \mathbf{c} \tag{11.4.11}$$

where $\boldsymbol{\omega}_t$ and $\boldsymbol{\omega}_{\dot{q}_r}$ are defined as:

$$\boldsymbol{\omega}_t = \left(\frac{\partial \mathbf{n}_2}{\partial t}\cdot\mathbf{n}_3\right)\mathbf{n}_1 + \left(\frac{\partial \mathbf{n}_3}{\partial t}\cdot\mathbf{n}_1\right)\mathbf{n}_2 + \left(\frac{\partial \mathbf{n}_1}{\partial t}\cdot\mathbf{n}_2\right)\mathbf{n}_3 \tag{11.4.12}$$

$$\boldsymbol{\omega}_{\dot{q}_r} = \left(\frac{\partial \mathbf{n}_2}{\partial q_r}\cdot\mathbf{n}_3\right)\mathbf{n}_1 + \left(\frac{\partial \mathbf{n}_3}{\partial q_r}\cdot\mathbf{n}_1\right)\mathbf{n}_2 + \left(\frac{\partial \mathbf{n}_1}{\partial q_r}\cdot\mathbf{n}_2\right)\mathbf{n}_3 \tag{11.4.13}$$

where, as in Eq. (11.4.9) $\mathbf{n}_1$, $\mathbf{n}_2$, and $\mathbf{n}_3$ are mutually perpendicular unit vectors fixed in B, and the partial derivatives are all computed in R. Then, by substituting from Eq. (11.4.11) into (11.4.10), ${}^R d\mathbf{c}/dt$ becomes:

$$\frac{{}^R d\mathbf{c}}{dt} = \boldsymbol{\omega}_t \times \mathbf{c} + \sum_{r=1}^{n}\left(\boldsymbol{\omega}_{\dot{q}_r} \times \mathbf{c}\right)\dot{q}_r = \left(\boldsymbol{\omega}_t + \sum_{r=1}^{n} \omega_{\dot{q}_r}\dot{q}_r\right)\times \mathbf{c} \tag{11.4.14}$$

Recalling that $\mathbf{c}$ is an arbitrary vector fixed in B and by comparing Eqs. (11.4.8) and (11.4.14) we see that the angular velocity of B in R may be expressed as:

$$\boldsymbol{\omega} = \boldsymbol{\omega}_t + \sum_{r=1}^{n} \boldsymbol{\omega}_{\dot{q}_r}\dot{q}_r \tag{11.4.15}$$

The vectors ω_t and $\boldsymbol{\omega}_{\dot{q}_r}$ are called *partial angular velocity vectors* of B in R. As with the partial velocity vectors of Eq. (11.4.3), the partial angular velocity vectors may be thought of as being base vectors for the movement of B in the n-dimensional space of the q_r.

Partial velocity and partial angular velocity vectors are remarkably easy to evaluate: from Eqs. (11.4.5) and (11.4.15) we see that the partial velocity and partial angular velocity vectors $\mathbf{v}_{\dot{q}_r}$ and $\boldsymbol{\omega}_{\dot{q}_r}$ are simply the coefficients of the $\dot{q}_r$. Specifically, we see from Eq. (11.4.15) that analogous to Eq. (11.4.4):

$$\boldsymbol{\omega}_{\dot{q}_r} = \partial\boldsymbol{\omega}/\partial\dot{q}_r \tag{11.4.16}$$

To illustrate the ease of evaluation of these vectors, consider first the motion of a particle P in a three-dimensional inertia frame R as in Figure 11.4.1. Let $\mathbf{p}$ locate P relative to a

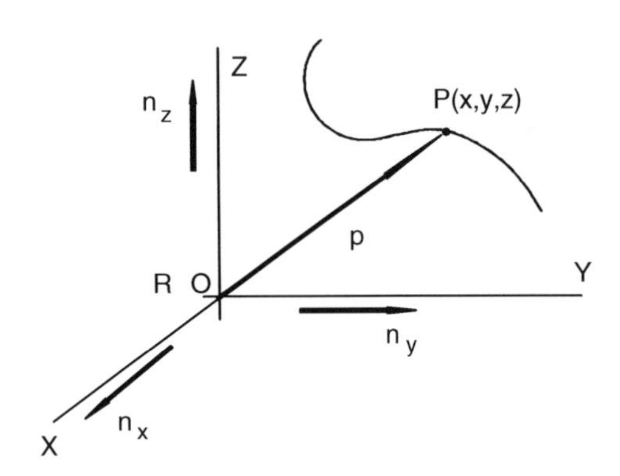

FIGURE 11.4.1
A particle P moving in an inertial frame R.

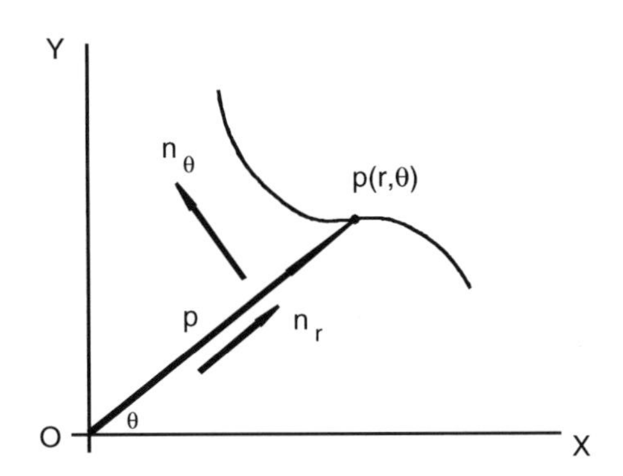

FIGURE 11.4.2
Movement of a particle P in a plane.

fixed point O in R. Let the coordinates of P be (x, y, z), the X, Y, Z Cartesian coordinates of a point at the same position as P. Then the position vector $\mathbf{p}$ may be expressed as:

$$\mathbf{p} = x\mathbf{n}_x + y\mathbf{n}_y + z\mathbf{n}_z \tag{11.4.17}$$

where $\mathbf{n}_x$, $\mathbf{n}_y$, and $\mathbf{n}_z$ are constant unit vectors parallel to the X-, Y-, and Z-axes as in Figure 11.4.1. The velocity of P in R is then:

$$\mathbf{v} = \dot{x}\mathbf{n}_x + \dot{y}\mathbf{n}_y + \dot{z}\mathbf{n}_z \tag{11.4.18}$$

From Eqs. (11.4.4) and (11.4.5), the partial velocities of P relative to x, y, and z are:

$$\mathbf{v}_{\dot{x}} = \mathbf{n}_x\,, \quad \mathbf{v}_{\dot{y}} = \mathbf{n}_y\,, \quad \mathbf{v}_{\dot{z}} = \mathbf{n}_z \tag{11.4.19}$$

Observe in this case that the partial velocity vectors are simply the unit vectors normally used in analyses with Cartesian reference frames.

As a second example, consider the motion of a particle P in a plane as in Figure 11.4.2. In this case, let the coordinates of P be the polar coordinates (r, θ) as shown. Then, the position vector $\mathbf{p}$ locating P relative to the origin O is:

$$\mathbf{p} = r\mathbf{n}_r \tag{11.4.20}$$

where $\mathbf{n}_r$ (and $\mathbf{n}_\theta$) are the radial (and transverse) unit vectors shown in Figure 11.4.2. The velocity of P is then (see Eq. (3.8.3)):

$$\mathbf{v} = \dot{r}\mathbf{n}_r + r\dot{\theta}\mathbf{n}_\theta \tag{11.4.21}$$

Hence, by comparison with Eq. (11.4.5), we see that the partial velocities of P relative to r and θ are:

$$\mathbf{v}_{\dot{r}} = \mathbf{n}_r \text{ and } \mathbf{v}_{\dot{\theta}} = r\mathbf{n}_\theta \tag{11.4.22}$$

Observe in this case the partial velocity $\mathbf{v}_{\dot{\theta}}$ is not a unit vector, neither is it dimensionless. Instead, $\mathbf{v}_{\dot{\theta}}$ has the unit of length, as a consequence of the differentiation with respect to $\dot{\theta}$.

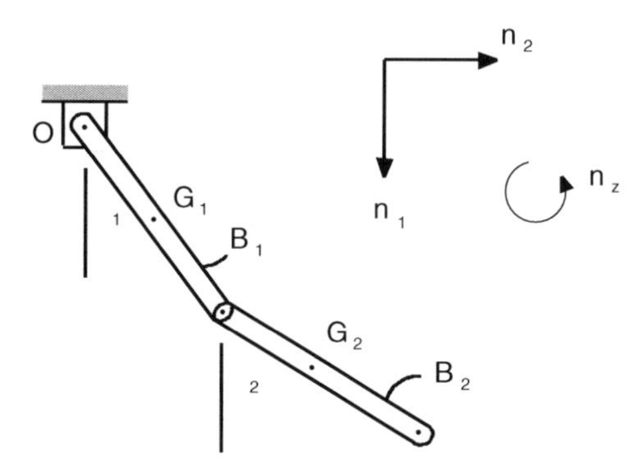

FIGURE 11.4.3
A double rod pendulum.

As a third example, consider the double rod pendulum of Figure 11.4.3 consisting of two identical rods B_1 and B_2 having lengths ℓ and mass centers G_1 and G_2. If the system is restricted to move in a plane, then the system has two degrees of freedom represented by the parameters θ_1 and θ_2 as shown. The velocities of G_1 and G_2 and the angular velocities of B_1 and B_2 are readily seen to be:

$$\begin{aligned}
\mathbf{v}^{G_1} &= (\ell/2)\dot{\theta}_1(\cos\theta_1\,\mathbf{n}_2 - \sin\theta_1\,\mathbf{n}_1) \\
\mathbf{v}^{G_2} &= \ell\dot{\theta}_1(\cos\theta_1\,\mathbf{n}_2 - \sin\theta_1\,\mathbf{n}_1) + (\ell/2)\dot{\theta}_2(\cos\theta_2\,\mathbf{n}_2 - \sin\theta_2\,\mathbf{n}_1) \\
\boldsymbol{\omega}^{B_1} &= \dot{\theta}_1\,un_3 \quad \text{and} \quad \boldsymbol{\omega}^{B_2} = \dot{\theta}_2\,\mathbf{n}_3
\end{aligned} \tag{11.4.23}$$

where $\mathbf{n}_1$, $\mathbf{n}_2$, and $\mathbf{n}_3$ are the unit vectors of Figure 11.4.3. The partial velocities of G_1 and G_2 and the partial angular velocities of B_1 and B_2 are then:

$$\begin{aligned}
&\mathbf{v}_{\dot{\theta}_1}^{G_1} = (\ell/2)(\cos\theta_1\,\mathbf{n}_2 - \sin\theta_1\,\mathbf{n}_1), \quad \mathbf{v}_{\dot{\theta}_2}^{G_1} = 0 \\
&\mathbf{v}_{\dot{\theta}_1}^{G_2} = \ell(\cos\theta_1\,\mathbf{n}_2 - \sin\theta_1\,\mathbf{n}_1), \quad \mathbf{v}_{\dot{\theta}_2}^{G_2} = (\ell/2)(\cos\theta_2\,\mathbf{n}_2 - \sin\theta_2\,\mathbf{n}_1) \\
&\boldsymbol{\omega}_{\dot{\theta}_1}^{B_1} = \mathbf{n}_3, \quad \boldsymbol{\omega}_{\dot{\theta}_2}^{B_1} = 0, \quad \boldsymbol{\omega}_{\dot{\theta}_1}^{B_2} = 0, \quad \boldsymbol{\omega}_{\dot{\theta}_2}^{B_2} = \mathbf{n}_3
\end{aligned} \tag{11.4.24}$$

11.5 Generalized Forces: Applied (Active) Forces

With the partial velocity vectors regarded as base vectors (or *direction* vectors) in the generalized motion space (n-dimensional space of the coordinates), it is useful to project forces along these vectors. Such projections are called *generalized forces*. To illustrate this concept, let P be a point of a body or a particle of a mechanical system S. Let S have n degrees of freedom represented by coordinates q_r ($r = 1,\ldots, n$). Let $\mathbf{F}$ be a force applied to P as in Figure 11.5.1. Then, the generalized force on P corresponding to the coordinates q_r ($r = 1,\ldots, n$) is defined as:

$$\mathbf{F}_r \stackrel{D}{=} \mathbf{F}\cdot\mathbf{v}_{\dot{q}_r} \quad (r = 1,\ldots,n) \tag{11.5.1}$$

where $\mathbf{v}_{\dot{q}_r}$ is the partial velocity of P relative to q_r in R.

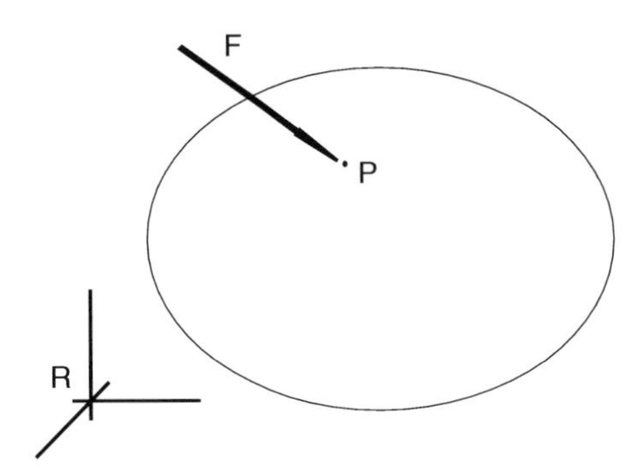

FIGURE 11.5.1
A force **F** applied at a point P of a mechanical system S.

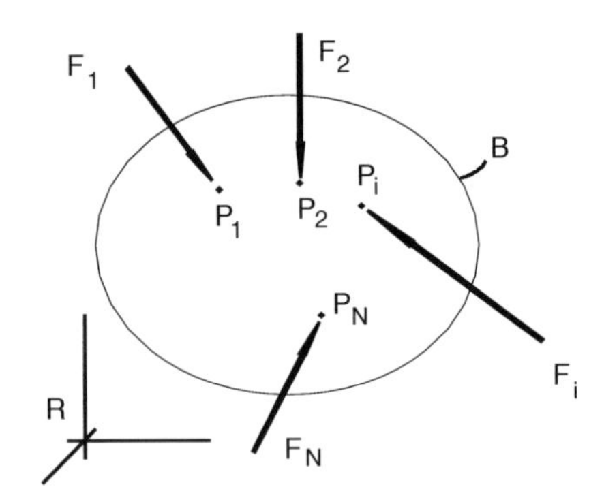

FIGURE 11.5.2
A rigid body B with applied forces $\mathbf{F}_i$ at points P_i.

Observe in Eq. (11.5.1) that the number of generalized forces is the same as the degrees of freedom of the system. Observe further that if the force **F** is perpendicular to the partial velocity $\mathbf{v}_{\dot{q}_r}$ then the generalized force F_r is zero. Also, if $\mathbf{v}_{\dot{q}_r}$ is zero, F_r is zero. Finally, observe that the dimensions and units of generalized forces are not necessarily the same as those of forces. Indeed, the dimensions and units of generalized forces depend directly upon the dimensions and units of the partial velocity vectors which in turn depend upon the choice of the coordinate q_r.

Consider next a set of forces $\mathbf{F}_i$ ($i = 1,\ldots, N$) applied at points P_i of the system S. The generalized forces for this set of forces are then obtained by adding the generalized forces from the individual forces. That is,

$$F_r = \sum_{i=1}^{N} \mathbf{F}_i \cdot \mathbf{v}_{\dot{q}_r}^{P_i} \quad (r = 1,\ldots,n) \tag{11.5.2}$$

where, as before, n is the number of degrees of freedom, represented by the coordinates q_r.

Finally, consider a rigid body B moving in an inertial frame R as in Figure 11.5.2. Let forces F_i ($i = 1,\ldots, N$) be applied at points P_i of B as shown. Then, from Eq. 11.5.2, the generalized forces are:

$$F_r = \sum_{i=1}^{N} \mathbf{F}_i \cdot \mathbf{v}_{\dot{q}_r}^{P_i} \quad (r = 1,\ldots,n) \tag{11.5.3}$$

where as before n is the number of degrees of freedom represented by the coordinates q_r. Because in this case we have a rigid body we can obtain a different form of Eq. (11.5.3) by using the rigidity of the body. Specifically, let Q be an arbitrary point of B. Then, from Eq. (4.94), we can express the velocity of P_i in terms of Q as:

$$\mathbf{v}^{P_i} = \mathbf{v}^{Q} + \boldsymbol{\omega} \times \mathbf{r}_i \tag{11.5.4}$$

where $\boldsymbol{\omega}$ is the angular velocity of B in R, and $\mathbf{r}_i$ locates P_i relative to Q as in Figure 11.5.3. By using Eqs. (11.4.5) and (11.4.14), we can differentiate Eq. (11.5.4) with respect to the $\dot{q}_r$ ($r = 1,\ldots, n$) leading to the expression:

$$\mathbf{v}_{\dot{q}_r}^{P_i} = \mathbf{v}_{\dot{q}_r}^{Q} + \boldsymbol{\omega}_{\dot{q}_r} \times r_i \quad (r = 1,\ldots,n) \tag{11.5.5}$$

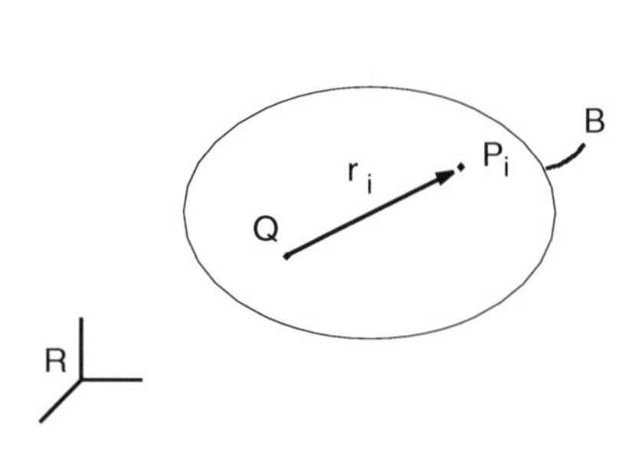

FIGURE 11.5.3
Typical point P_i and arbitrary reference point Q of R.

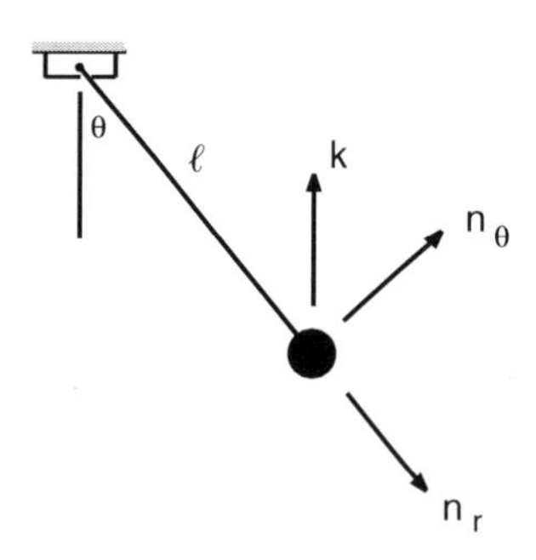

FIGURE 11.5.4
A simple pendulum.

Then, by substituting into Eq. (11.5.3), F_r becomes:

$$\begin{aligned} F_r &= \sum_{i=1}^{N} \mathbf{F}_i \cdot \left(\mathbf{v}^Q_{\dot{q}_r} + \boldsymbol{\omega}_{\dot{q}_r} \times r_i\right) \\ &= \sum_{i=1}^{N} \mathbf{F}_i \cdot \mathbf{v}^Q_{\dot{q}_r} + \sum_{i=1}^{N} \mathbf{F}_i \cdot \left(\boldsymbol{\omega}_{\dot{q}_r} \times \mathbf{r}_i\right) \\ &= \mathbf{v}^Q_{\dot{q}_r} \cdot \left(\sum_{i=1}^{N} F_i\right) + \boldsymbol{\omega}_{\dot{q}_r} \cdot \left(\sum_{i=1}^{N} \mathbf{r}_i \times \mathbf{F}_i\right) \end{aligned} \tag{11.5.6}$$

where the form of the last term is obtained using the properties of triple scalar products. We can rewrite Eq. (11.5.6) in the form:

$$F_r = \mathbf{v}^Q_{\dot{q}_r} \cdot \mathbf{F} + \boldsymbol{\omega}_{\dot{q}_r} \cdot \mathbf{T} \tag{11.5.7}$$

where **F** and **T** are defined by comparing Eqs. (11.5.6) and (11.5.7). (Observe and recall from Section 6.5 that **F** and **T** are the resultant and couple torque of a force system equivalent to the set of forces $\mathbf{F}_i$ $[i = 1,\ldots, N]$.)

To illustrate the computation of generalized forces, consider first the simple pendulum of Figure 11.5.4. Moving in a vertical plane, the pendulum has only one degree of freedom represented by the angle θ. The velocity and partial velocity of the pendulum bob P are then:

$$\mathbf{v} = \ell\dot{\theta}\mathbf{n}_\theta \quad \text{and} \quad \mathbf{v}_{\dot{\theta}} = \ell\mathbf{n}_\theta \tag{11.5.8}$$

where ℓ is the length of the connecting string or massless connecting rod.

Consider a free-body diagram showing the applied forces on P as in Figure 11.5.5, where T is the tension in the string and m is the mass of P. From Eq. (11.5.1), the generalized force F_θ on P is then:

$$F_\theta = -T\mathbf{n}_r \cdot \mathbf{v}_{\dot{\theta}} - mg\mathbf{k} \cdot \mathbf{v}_{\dot{\theta}} = 0 - mg\ell \sin\theta \tag{11.5.9}$$

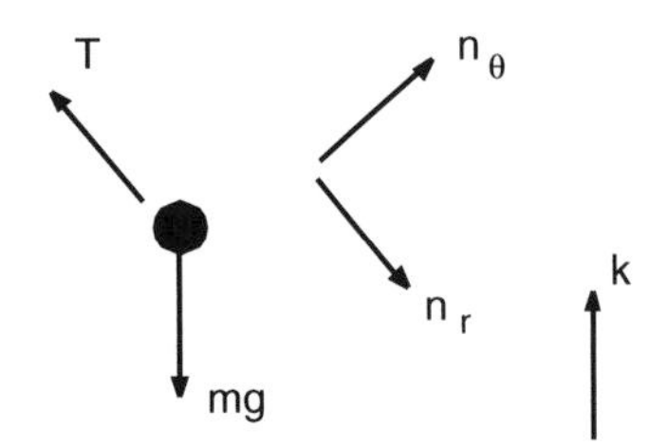

FIGURE 11.5.5
A free-body diagram of the pendulum bob.

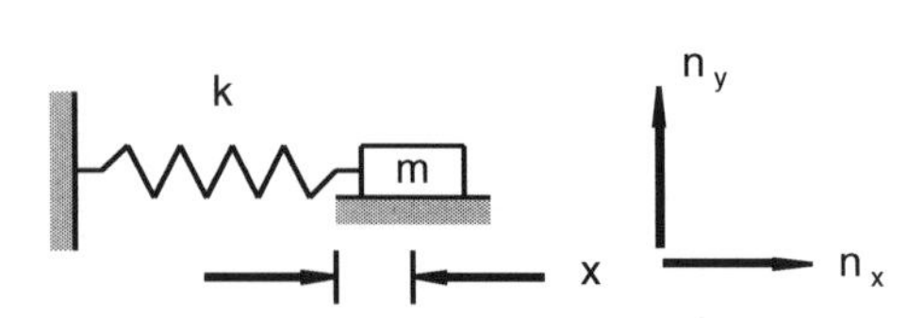

FIGURE 11.5.6
A simple mass–spring system.

Observe in Eq. (11.5.9) that the constraint tension force T does not contribute to the generalized force. Observe further that the dimensions of F_θ are force–length.

Consider next the simple mass–spring system of Figure 11.5.6, where the mass m sliding on a smooth surface is attached to a linear spring with modulus k. This system also has only one degree of freedom represented by the displacement x of the mass away from its static equilibrium position. The velocity and partial velocity of the mass are then:

$$\mathbf{v} = \dot{x}\mathbf{n}_x \quad \text{and} \quad \mathbf{v}_{\dot{x}} = \mathbf{n}_x \tag{11.5.10}$$

Consider a free-body diagram showing the applied forces on the mass as in Figure 11.5.7 where N is the equivalent normal force of the surface. The generalized force on the mass is then:

$$\begin{aligned} F_x &= -kx\,\mathbf{n}_x \cdot \mathbf{v}_{\dot{x}} + N\mathbf{n}_y \cdot \mathbf{v}\dot{x} - mg\mathbf{n}_y \cdot \mathbf{v}_{\dot{x}} \\ &= -kx + 0 - 0 \end{aligned} \tag{11.5.11}$$

Observe that the normal force N and the weight force mg do not contribute to the generalized force.

Finally, consider the rod pendulum of Figure 11.5.8 consisting of a rod B of length ℓ and mass m pinned at one end and moving in a vertical plane. Let B also be attached to a linear torsion spring with modulus k. Let the spring be a "restoring spring" such that

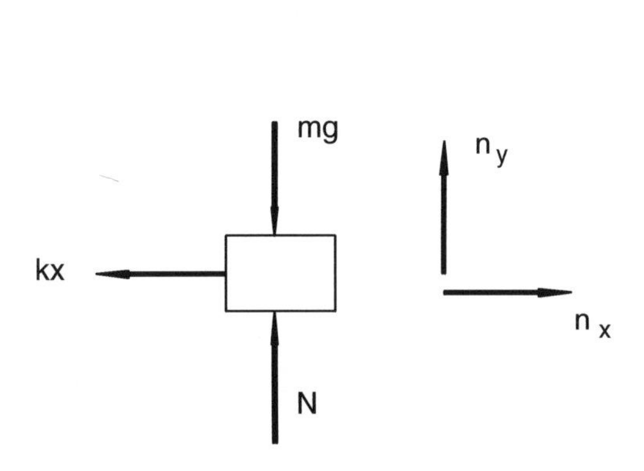

FIGURE 11.5.7
A free-body diagram of the mass.

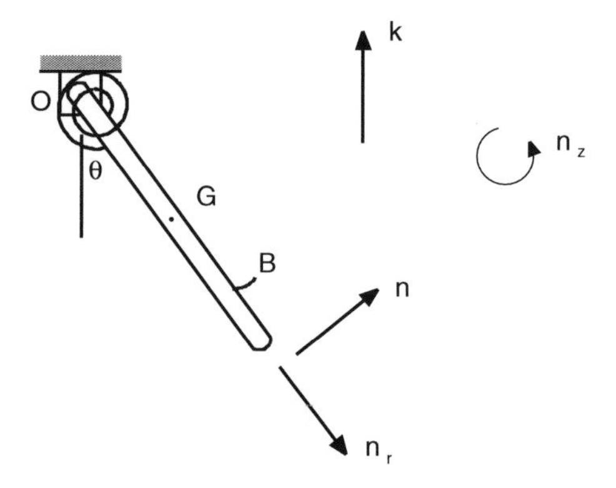

FIGURE 11.5.8
A rod pendulum with linear torsion restoring spring.

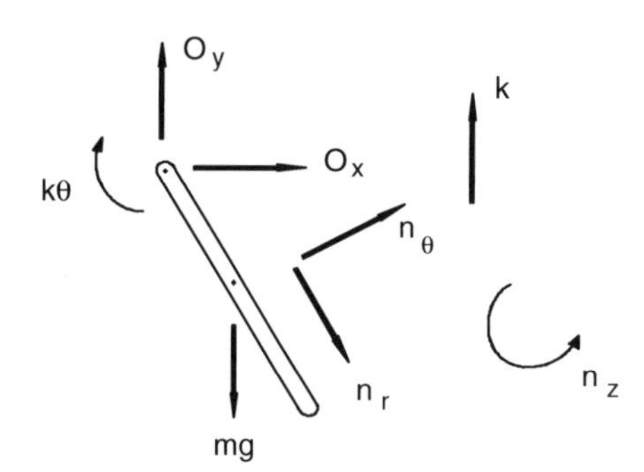

FIGURE 11.5.9
A free-body diagram of the pendulum rod.

the moment exerted on B is zero when B is in the vertical static equilibrium position. As with the two previous examples, this system also has only one degree of freedom, represented by the angle θ. Let G be the mass center of B. Then, the velocity and partial velocity of G are:

$$\mathbf{v}^G = (\ell/2)\dot{\theta}\mathbf{n}_\theta \quad \text{and} \quad \mathbf{v}^G_{\dot{\theta}} = (\ell/2)\mathbf{n}_\theta \tag{11.5.12}$$

The angular velocity and partial angular velocity of B are:

$$\boldsymbol{\omega} = \dot{\theta}\mathbf{n}_z \quad \text{and} \quad \boldsymbol{\omega}_{\dot{\theta}} = \mathbf{n}_z \tag{11.5.13}$$

Consider a free-body diagram showing the applied forces on B as in Figure 11.5.9 where O_x and O_y represent horizontal and vertical components of the pin reaction forces. From Eq. (11.5.7), the generalized force is:

$$\begin{aligned} F_\theta &= -mg\mathbf{k}\cdot\mathbf{v}^G_{\dot{\theta}} + \left(O_x\mathbf{n}_x + O_y\mathbf{n}_y\right)\cdot v^O_{\dot{\theta}} - k\theta\mathbf{n}_z\cdot\boldsymbol{\omega}_{\dot{\theta}} \\ &= -mg(\ell/2)\sin\theta + 0 - k\theta \end{aligned} \tag{11.5.14}$$

Observe that because the pin O has zero velocity and as a consequence, zero partial velocity, the pin reaction forces do not contribute to the generalized force.

11.6 Generalized Forces: Gravity and Spring Forces

The simple examples of the foregoing section demonstrate the ease of determining generalized forces. Indeed, all that is required is to project the forces and moments along the partial velocity and partial angular velocity vectors. In this section, we will see that it is possible to obtain general expressions for the contributions to generalized forces from gravity and spring forces. That is, for gravity and spring forces we will see that we can obtain their contributions to the generalized forces without computing the projections onto the partial velocity and partial angular velocity vectors.

To this end, consider first a particle P of a mechanical system S where S has n degrees of freedom in an inertial frame R represented by the coordinates q_r ($r = 1,\ldots, n$). Let P have

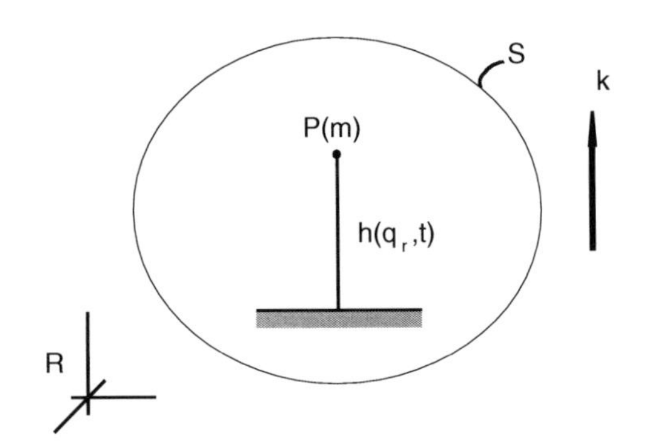

FIGURE 11.6.1
Elevation of a particle P of a mechanical system S.

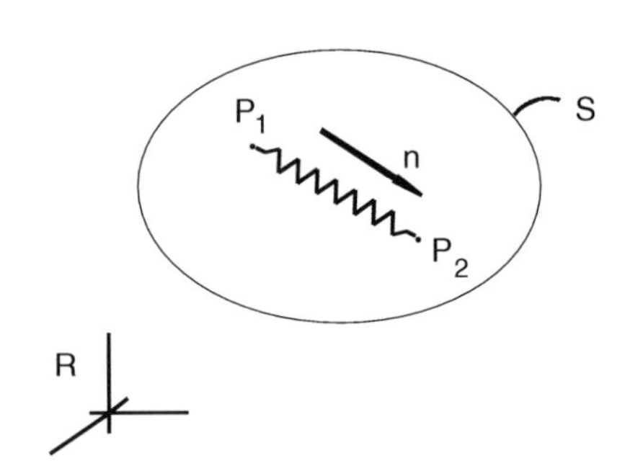

FIGURE 11.6.2
A spring within a mechanical system S.

mass m. Then, from Eq. (11.5.1), the contribution $\hat{\mathbf{F}}_r$ of the gravitational (or weight) force on P to the generalized force F_r on S, for the coordinate q_r, may be expressed as:

$$\hat{\mathbf{F}}_r = -mg\mathbf{k} \cdot \mathbf{v}^P_{\dot{q}_r} \quad (r = 1, \ldots, n) \tag{11.6.1}$$

where $\mathbf{k}$ is a vertical unit vector.

Let h measure the elevation of P above an arbitrary, but fixed, reference level of R as illustrated in Figure 11.6.1. Then, in general, h is a function of the q_r and time t. From Eq. (11.4.4), we see that the partial velocity $\mathbf{v}^P_{\dot{q}_r}$ may be written as:

$$\mathbf{v}^P_{\dot{q}_r} = \partial \mathbf{v}^P / \partial \dot{q}_r \tag{11.6.2}$$

Hence, because $\mathbf{k}$ is a constant unit vector, we can write Eq. (11.6.1) in the form:

$$\hat{\mathbf{F}}_r = -mg\mathbf{k} \cdot \partial \mathbf{v}^P / \partial \dot{q}_r = -mg\, \partial\left(\mathbf{v}^P \cdot \mathbf{k}\right) / \partial \dot{q}_r \tag{11.6.3}$$

However, $\mathbf{v}^P \cdot \mathbf{k}$ is simply the vertical projection of the velocity of P in R which we can recognize as dh/dt. That is,

$$\mathbf{v}^P \cdot \mathbf{k} = \frac{dh}{dt} = \frac{\partial h}{\partial t} + \sum_{r=1}^{n} \frac{\partial h}{\partial q_r} \dot{q}_r \tag{11.6.4}$$

By substituting from Eq. (11.6.4) into (11.6.3) we see that $\hat{\mathbf{F}}_r$ becomes simply:

$$\hat{\mathbf{F}}_r = -mg\, \partial h / \partial q_r \quad (r = 1, \ldots, n) \tag{11.6.5}$$

Next, regarding spring forces, let P_1 and P_2 be points at opposite ends of a spring that is part of the mechanical system S, as depicted in Figure 11.6.2. Let $\mathbf{n}$ be a unit vector parallel to the axis of the spring as shown. Let x represent the elongation of the spring, and let the magnitude of the resulting spring tension force be $f(x)$. If the spring is linear, $f(x)$ is simply kx, where k is the spring modulus. (If x is negative, representing a shortening of the spring, the resulting spring force is compressive.) Specifically, if the spring is elongated, it will exert equal but oppositely directed forces on P_1 and P_2 as $f(x)\mathbf{n}$ and $-f(x)\mathbf{n}$.

The contribution $\hat{\mathbf{F}}_r$ of the spring force to the generalized force F_r on S is then:

$$\begin{aligned}\hat{\mathbf{F}}_r &= f(x)\mathbf{n}\cdot\mathbf{v}^{P_1}_{\dot{q}_r} - f(x)\mathbf{n}\cdot\mathbf{v}^{P_2}_{\dot{q}_r}\\ &= f(x)\mathbf{n}\cdot\left(\partial\mathbf{v}^{P_1}_{\dot{q}_r} - \mathbf{v}^{P_2}_{\dot{q}_r}\right)\\ &= f(x)\mathbf{n}\cdot\left(\partial\mathbf{v}^{P_1}/\partial\dot{q}_r - \partial\mathbf{v}^{P_2}/\partial\dot{q}_r\right)\\ &= f(x)\mathbf{n}\cdot\frac{\partial}{\partial\dot{q}_r}\left(\mathbf{v}^{P_1} - \mathbf{v}^{P_2}\right) \quad (r=1,\ldots,n)\end{aligned} \tag{11.6.6}$$

Observe that the direction of the unit vector $\mathbf{n}$ will be a function of the coordinates q_r but not the coordinate derivative $\dot{q}_r$. Hence, in Eq. (11.6.6) $\mathbf{n}$ may be taken inside the partial derivative with respect to the $\dot{q}_r$. $\hat{\mathbf{F}}_r$ may thus be written as:

$$\begin{aligned}\hat{\mathbf{F}}_r &= f(x)\frac{\partial}{\partial\dot{q}_r}\left[\mathbf{n}\cdot\left(\mathbf{v}^{P_1} - \mathbf{v}^{P_2}\right)\right]\\ &= f(x)\frac{\partial}{\partial\dot{q}_r}\left[\mathbf{n}\cdot\left(-\frac{dx}{dt}\mathbf{n}\right)\right]\\ &= -f(x)\frac{\partial}{\partial\dot{q}_r}\left(\frac{dx}{dt}\right)\\ &= -f(x)\frac{\partial}{\partial\dot{q}_r}\left[\frac{\partial x}{\partial t} + \sum_{s=1}^{n}\frac{\partial x}{\partial q_s}\dot{q}_s\right]\end{aligned}$$

or

$$\hat{\mathbf{F}}_r = -f(x)\frac{\partial x}{\partial q_r} \tag{11.6.7}$$

11.7 Example: Spring-Supported Particles in a Rotating Tube

For an example illustrating these concepts, consider a cylindrical tube T with mass M and length L. Let T be pinned at one of its ends O, and let T rotate in a vertical plane with the angle of rotation being as in Figure 11.7.1. Let T contain three identical spring-supported particles (or small spheres or balls) each having mass m as in Figure 11.7.2. Let the four connecting springs also be identical. Let the springs be linear with each having modulus k and natural length ℓ. Let the interior of T be smooth so that the particles can move freely along the axis of T. Let the movement of the particles within T be defined by the coordinates x_1, x_2, and x_3 shown in Figure 11.7.3. These coordinates measure the displacements of the particles away from their static equilibrium position where the tube is horizontal.

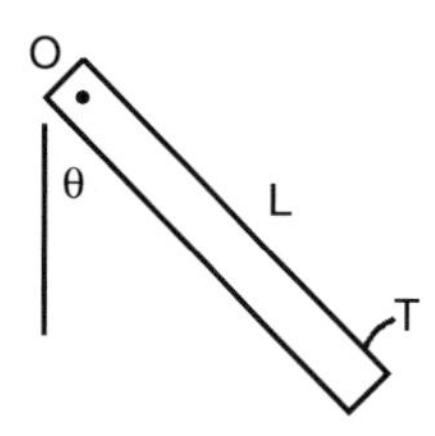

FIGURE 11.7.1
A rotating cylindrical tube.

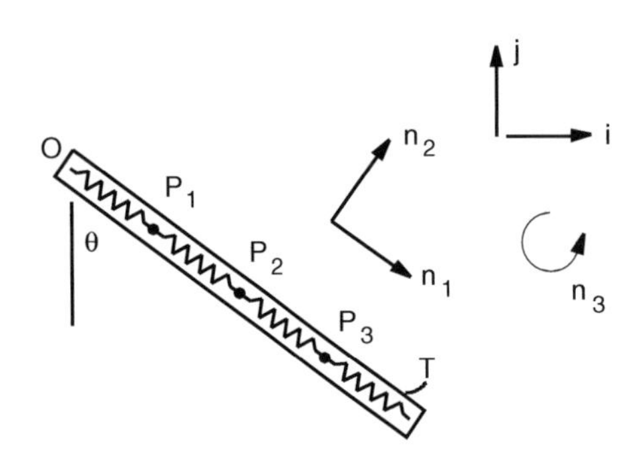

FIGURE 11.7.2
A rotating tube containing spring-connected particles.

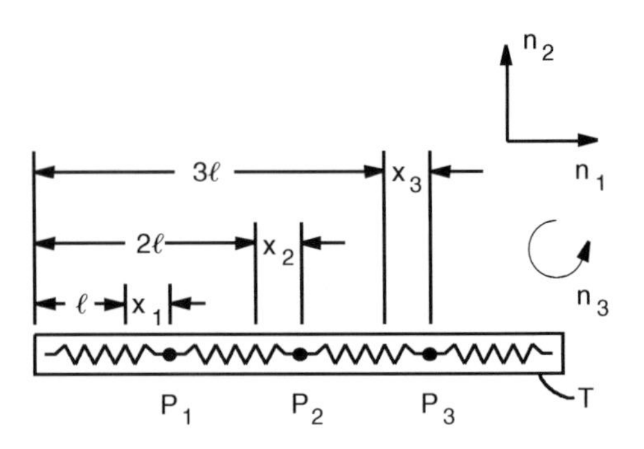

FIGURE 11.7.3
Coordinates of the particles within the tube.

This system has four degrees of freedom described by the coordinates x_1, x_2, x_3, and θ. Our objective is to determine the generalized forces applied to the system corresponding to these coordinates. (This example is a modified and extended version of a similar example appearing in References 11.1 and 11.2.)

To determine the generalized forces, it is helpful to first determine the partial velocities of the particles, the partial velocities of the mass center G of T and the partial angular velocities of T. By using the procedures of Chapter 4, we find the angular velocity of T, its mass center velocity, and the particle velocities to be:

$$\boldsymbol{\omega}^T = \dot{\theta}\mathbf{n}_3 \,, \quad \mathbf{v}^G = (\ell/2)\dot{\theta}\,\mathbf{n}_2 \tag{11.7.1}$$

and

$$\mathbf{v}^{P_1} = \dot{x}_1\mathbf{n}_1 + (\ell + x_1)\dot{\theta}\,\mathbf{n}_2 \tag{11.7.2}$$

$$\mathbf{v}^{P_2} = \dot{x}_2\mathbf{n}_1 + (2\ell + x_2)\dot{\theta}\,\mathbf{n}_2 \tag{11.7.3}$$

$$\mathbf{v}^{P_3} = \dot{x}_3\mathbf{n}_1 + (3\ell + x_3)\dot{\theta}\,\mathbf{n}_2 \tag{11.7.4}$$

where $\mathbf{n}_1$, $\mathbf{n}_2$, and $\mathbf{n}_3$ are the unit vectors of Figures 11.7.2 and 11.7.3. From Eqs. (11.4.4) and (11.4.15) we obtain the partial velocities and partial angular velocities:

$$\boldsymbol{\omega}^T_{\dot{x}_1} = \boldsymbol{\omega}^T_{\dot{x}_2} = \boldsymbol{\omega}^T_{\dot{x}_3} = 0 \,, \quad \boldsymbol{\omega}^T_{\dot{\theta}} = \mathbf{n}_3 \tag{11.7.5}$$

and

$$\begin{aligned}
&\mathbf{v}^{P_1}_{\dot{x}_1} = \mathbf{n}_1 \,, \ \mathbf{v}^{P_1}_{\dot{x}_2} = 0 \,, \ \mathbf{v}^{P_1}_{\dot{x}_3} = 0 \,, \ \mathbf{v}^{P_1}_{\dot{\theta}} = (\ell + x_1)\mathbf{n}_2 \\
&\mathbf{v}^{P_2}_{\dot{x}_1} = 0 \,, \ \mathbf{v}^{P_2}_{\dot{x}_2} = \mathbf{n}_1 \,, \ \mathbf{v}^{P_2}_{\dot{x}_3} = 0 \,, \ \mathbf{v}^{P_3}_{\dot{\theta}} = (2\ell + x_2)\mathbf{n}_2 \\
&\mathbf{v}^{P_3}_{\dot{x}_1} = 0 \,, \ \mathbf{v}^{P_3}_{\dot{x}_2} = 0 \,, \ \mathbf{v}^{P_3}_{\dot{x}_3} = \mathbf{n}_1 \,, \ \mathbf{v}^{P_3}_{\dot{\theta}} = (3\ell + x_3)\mathbf{n}_2
\end{aligned} \tag{11.7.6}$$

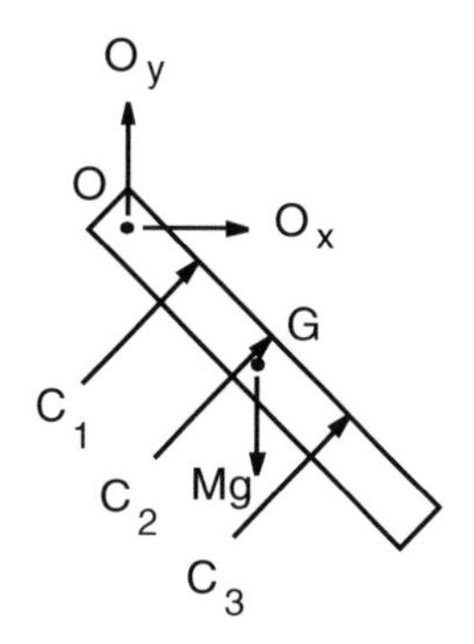

FIGURE 11.7.4
Free-body diagram of the applied forces on the tube T.

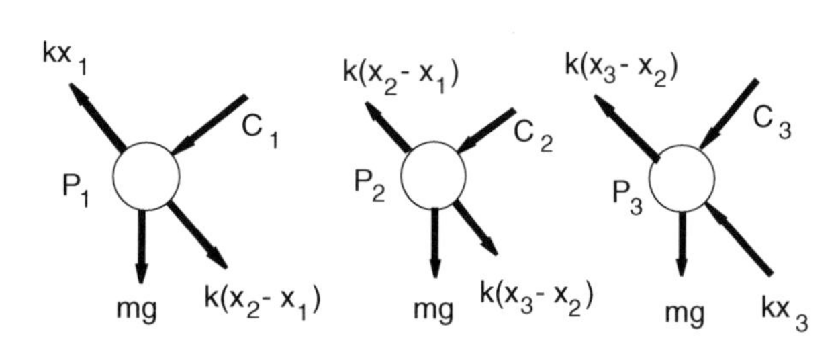

FIGURE 11.7.5
Free-body diagrams of the applied forces on the particles.

and

$$\mathbf{v}^G_{\dot{x}_1} = 0\,,\ \mathbf{v}^G_{\dot{x}_2} = 0\,,\ \mathbf{v}^G_{\dot{x}_3} = 0\,,\ \mathbf{v}^G_{\dot{\theta}} = (L/2)\mathbf{n}_2 \tag{11.7.7}$$

Consider next the forces acting on the particles and on the tube T. Figures 11.7.4 and 11.7.5 contain free-body diagrams showing applied forces on T and the particles. In these figures C_1, C_2, and C_3 represent contact forces between the particles and T applied at points Q_1, Q_2, and Q_3 of T. O_x and O_y are components of the pin reaction forces at O on T.

With forces applied at O and at Q_1, Q_2, and Q_3, it is necessary to also determine the partial velocities of O, Q_1, Q_2, and Q_3. Because O is fixed in our inertial frame we have:

$$\mathbf{v}^O = 0 \tag{11.7.8}$$

and then:

$$\mathbf{v}^O_{\dot{x}_1} = \mathbf{v}^O_{\dot{x}_2} = \mathbf{v}^O_{\dot{x}_3} = \mathbf{v}^O_{\dot{\theta}} = 0 \tag{11.7.9}$$

Points Q_1, Q_2, and Q_3 are points of T at the positions of P_1, P_2, and P_3. Their velocities are:

$$\mathbf{v}^{Q_1} = (\ell + x_1)\dot{\theta}\mathbf{n}_2 \tag{11.7.10}$$

$$\mathbf{v}^{Q_2} = (2\ell + x_2)\dot{\theta}\mathbf{n}_2 \tag{11.7.11}$$

$$\mathbf{v}^{Q_3} = (3\ell + x_3)\dot{\theta}\mathbf{n}_3 \tag{11.7.12}$$

Hence, the partial velocities of Q_1, Q_2, and Q_3 are:

$$\begin{aligned}
&\mathbf{v}^{Q_1}_{\dot{x}_1} = 0\,,\ \mathbf{v}^{Q_1}_{\dot{x}_2} = 0\,,\ \mathbf{v}^{Q_1}_{\dot{x}_3} = 0\,,\ \mathbf{v}^{Q_1}_{\dot{\theta}} = (\ell + x_1)\mathbf{n}_2 \\
&\mathbf{v}^{Q_2}_{\dot{x}_1} = 0\,,\ \mathbf{v}^{Q_2}_{\dot{x}_2} = 0\,,\ \mathbf{v}^{Q_2}_{\dot{x}_3} = 0\,,\ \mathbf{v}^{Q_2}_{\dot{\theta}} = (2\ell + x_2)\mathbf{n}_2 \\
&\mathbf{v}^{Q_3}_{\dot{x}_1} = 0\,,\ \mathbf{v}^{Q_3}_{\dot{x}_2} = 0\,,\ \mathbf{v}^{Q_3}_{\dot{x}_3} = 0\,,\ \mathbf{v}^{Q_3}_{\dot{\theta}} = (3\ell + x_3)\mathbf{n}_2
\end{aligned} \tag{11.7.13}$$

By projecting the forces applied to T and the particles along the partial velocities at the points of application, we obtain the generalized forces as:

$$\begin{aligned}
\mathbf{F}_{x_1} &= (O_x\mathbf{i}+O_y\mathbf{j})\cdot\mathbf{v}^O_{\dot{x}_1} - Mg\mathbf{j}\cdot\mathbf{v}^G_{\dot{x}_1} + C_1\mathbf{n}_2\cdot\mathbf{v}^{Q_1}_{\dot{x}_1} + C_2\mathbf{n}_2\cdot\mathbf{v}^{Q_2}_{\dot{x}_1} \\
&\quad + C_3\mathbf{n}_2\cdot\mathbf{v}^{Q_3}_{\dot{x}_1} + \left[-mg\mathbf{j} + k(x_2-x_1)\mathbf{n}_1 - kx_1\mathbf{n}_1 - C_1\mathbf{n}_2\right]\cdot\mathbf{v}^{P_1}_{\dot{x}_1} \\
&\quad + \left[-mg\mathbf{j} + k(x_3-x_2)\mathbf{n}_1 - k(x_2-x_1)\mathbf{n}_1 - C_2\mathbf{n}_2\right]\cdot\mathbf{v}^{P_2}_{\dot{x}_1} \\
&\quad + \left[-mg\mathbf{j} - kx_3\mathbf{n}_1 - k(x_3-x_2)\mathbf{n}_1 - C_3\mathbf{n}_2\right]\cdot\mathbf{v}^{P_3}_{\dot{x}_1} \\
&= 0 - 0 + 0 + 0 + 0 + mg\cos\theta + kx_2 - 2kx_1 + 0 + 0
\end{aligned}$$

or

$$\mathbf{F}_{x_1} = mg\cos\theta + kx_2 - 2kx_1 \tag{11.7.14}$$

and

$$\begin{aligned}
\mathbf{F}_{x_2} &= (O_x\mathbf{i}+O_y\mathbf{j})\cdot\mathbf{v}^O_{\dot{x}_2} - Mg\mathbf{j}\cdot\mathbf{v}^G_{\dot{x}_2} + C_1\mathbf{n}_2\cdot\mathbf{v}^{Q_1}_{\dot{x}_2} + C_2\mathbf{n}_2\cdot\mathbf{v}^{Q_2}_{\dot{x}_2} \\
&\quad + C_3\mathbf{n}_2\cdot\mathbf{v}^{Q_3}_{\dot{x}_2} + \left[-mg\mathbf{j} + k(x_2-x_1)\mathbf{n}_1 - kx_1\mathbf{n}_1 - C_1\mathbf{n}_2\right]\cdot\mathbf{v}^{P_1}_{\dot{x}_2} \\
&\quad + \left[-mg\mathbf{j} + k(x_3-x_2)\mathbf{n}_1 - k(x_2-x_1)\mathbf{n}_1 - C_2\mathbf{n}_2\right]\cdot\mathbf{v}^{P_2}_{\dot{x}_2} \\
&\quad + \left[-mg\mathbf{j} - kx_3\mathbf{n}_1 - k(x_3-x_2)\mathbf{n}_1 - C_3\mathbf{n}_2\right]\cdot\mathbf{v}^{P_3}_{\dot{x}_2} \\
&= 0 - 0 + 0 + 0 + 0 + mg\cos\theta + kx_3 + kx_1 - 2kx_2 + 0
\end{aligned}$$

or

$$\mathbf{F}x_2 = mg\cos\theta + kx_3 + kx_1 - 2kx_2 \tag{11.7.15}$$

and

$$\begin{aligned}
\mathbf{F}x_3 &= (O_x\mathbf{i}+O_y\mathbf{j})\cdot\mathbf{v}^O_{\dot{x}_3} - Mg\mathbf{j}\cdot\mathbf{v}^G_{\dot{x}_3} + C_1\mathbf{n}_2\cdot\mathbf{v}^{Q_1}_{\dot{x}_3} + C_2\mathbf{n}_2\cdot\mathbf{v}^{Q_2}_{\dot{x}_3} \\
&\quad + C_3\mathbf{n}_2\cdot\mathbf{v}^{Q_3}_{\dot{x}_3} + \left[-mg\mathbf{j} + k(x_2-x_1)\mathbf{n}_1 - kx_1\mathbf{n}_1 - C_1\mathbf{n}_2\right]\cdot\mathbf{v}^{P_1}_{\dot{x}_3} \\
&\quad + \left[-mg\mathbf{j} + k(x_3-x_2)\mathbf{n}_1 - k(x_2-x_1)\mathbf{n}_1 - C_2\mathbf{n}_2\right]\cdot\mathbf{v}^{P_2}_{\dot{x}_3} \\
&\quad + \left[-mg\mathbf{j} - kx_3\mathbf{n}_1 - k(x_3-x_2)\mathbf{n}_1 - C_3\mathbf{n}_2\right]\cdot\mathbf{v}^{P_3}_{\dot{x}_3} \\
&= 0 - 0 + 0 + 0 + 0 + 0 + 0 + mg\cos\theta - 2kx_3 + kx_2
\end{aligned}$$

or

$$\mathbf{F}_{x_3} = mg\cos\theta - 2kx_3 + kx_2 \tag{11.7.16}$$

and

$$\begin{aligned}
\mathbf{F}_\theta &= \left(O_x\mathbf{i}+O_y\mathbf{j}\right)\cdot\mathbf{v}_{\dot\theta}^{O} - Mg\mathbf{j}\cdot\mathbf{v}_{\dot\theta}^{G} + C_1\mathbf{n}_2\cdot\mathbf{v}_{\dot\theta}^{Q_1} + C_2\mathbf{n}_2\cdot\mathbf{v}_{\dot\theta}^{Q_2} \\
&\quad + C_3\mathbf{n}_2\cdot\mathbf{v}_{\dot\theta}^{Q_3} + \left[-mg\mathbf{j} + k(x_2-x_1)\mathbf{n}_1 - kx_1\mathbf{n}_1 - C_1\mathbf{n}_2\right]\cdot\mathbf{v}_{\dot\theta}^{P_1} \\
&\quad + \left[-mg\mathbf{j} + k(x_3-x_2)\mathbf{n}_1 - k(x_2-x_1)\mathbf{n}_1 - C_2\mathbf{n}_2\right]\cdot\mathbf{v}_{\dot\theta}^{P_2} \\
&\quad + \left[-mg\mathbf{j} - kx_3\mathbf{n}_1 - k(x_3-x_2)\mathbf{n}_1 - C_3\mathbf{n}_2\right]\cdot\mathbf{v}_{\dot\theta}^{P_3} \\
&= 0 - Mg(L/2)\sin\theta + C_1(\ell+x_1) + C_2(2\ell+x_2) + C_3(3\ell+x_3) \\
&\quad - mg(\ell+x_1)\sin\theta - C_1(\ell+x_1) - mg(2\ell+x_2)\sin\theta - C_2(2\ell+x_2) \\
&\quad - mg(3\ell+x_3)\sin\theta - C_3(3\ell+x_3)
\end{aligned}$$

or

$$\mathbf{F}_\theta = -Mg(L/2)\sin\theta - mg(6\ell + x_1 + x_2 + x_3)\sin\theta \tag{11.7.17}$$

Observe that neither the pin forces nor the contact forces exerted across the smooth surface of T appear in the expressions for the generalized forces. The pin forces are absent because they are exerted at the point O, which has zero velocity and zero partial velocities. The contact forces across the smooth surface of T do not appear because they are perpendicular to the partial velocities of their points of application. In the following section, we will explore and identify conditions that will always lead to the elimination of force components from the generalized forces.

Finally, observe that the force components appearing in the generalized forces are from gravitational and spring forces. This means that we could have obtained the generalized forces using Eqs. (11.6.5) and (11.6.7). Specifically, the contribution $\hat{\mathbf{F}}_r$ to the generalized forces from the gravitational forces are:

$$\hat{\mathbf{F}}_r = -m_i g\,\partial h_i/\partial q_r \tag{11.7.18}$$

where m_i is the mass of particle P_i and h_i is the elevation of P_i above an arbitrary reference level. If we take the reference level to be the elevation of the pin O, we find the h_i (i = 1, 2, 3) to be (see Figure 11.7.6):

$$\begin{aligned}
h_1 &= -(\ell + x_1)\cos\theta \\
h_2 &= -(2\ell + x_2)\cos\theta \\
h_3 &= -(3\ell + x_3)\cos\theta
\end{aligned} \tag{11.7.19}$$

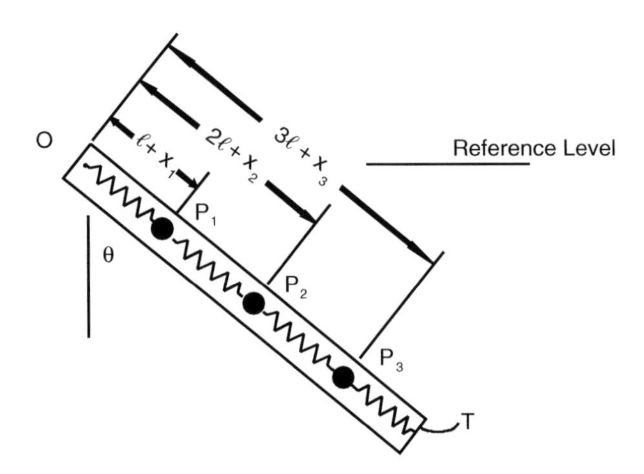

FIGURE 11.7.6
Position of tube particles relative to the reference level.

Similarly, for the mass center G of T the elevation H relative to O is:

$$H = -(L/2)\cos\theta \tag{11.7.20}$$

Then, from Eq. (11.7.18) we have:

$$\hat{\mathbf{F}}_{x_1} = \hat{\mathbf{F}}_{x_2} = \hat{\mathbf{F}}_{x_3} = mg\cos\theta \tag{11.7.21}$$

and

$$\hat{\mathbf{F}}_\theta = -mg(\ell + x_1)\sin\theta - mg(2\ell + x_2)\sin\theta - mg(3\ell + x_3)\sin\theta - Mg(L/2)\sin\theta$$

or

$$\hat{\mathbf{F}}_\theta = -mg\,(6\ell + x_1 + x_2 + x_3)\sin\theta - Mg(L/2)\sin\theta \tag{11.7.22}$$

Comparing Eqs. (11.7.21) and (11.7.22) with Eqs. (11.7.14) to (11.7.17), we see that the results are consistent.

Next, for the springs, we see from Eq. (11.6.7) that the spring force contributions $\hat{\mathbf{F}}_r$ are given by:

$$\hat{\mathbf{F}}_r = -kx\partial x/\partial q_r \tag{11.7.23}$$

where x is the spring elongation. From Figure (11.7.3), we see that the spring elongations are x_1, $x_2 - x_1$, $x_3 - x_2$, and $-x_3$. Hence, the $\hat{\mathbf{F}}_r$ are:

$$\begin{aligned}
\hat{\mathbf{F}}_{x_1} &= -kx_1 + k(x_2 - x_1) = -2kx_1 + kx_2 \\
\hat{\mathbf{F}}_{x_2} &= -k(x_2 - x_1) + k(x_3 - x_2) = kx_1 + kx_3 - 2kx_2 \\
\hat{\mathbf{F}}_{x_3} &= -k(x_3 - x_2) - kx_3 = -2kx_3 + kx_2 \\
\hat{\mathbf{F}}_\theta &= 0
\end{aligned} \tag{11.7.24}$$

Comparing Eq. (11.7.24) with Eqs. (11.7.14) to (11.7.17), we see that the results are consistent.

11.8 Forces That Do Not Contribute to the Generalized Forces

In the example of the foregoing section we observed that some of the forces exerted on the system did not contribute to the generalized forces. These noncontributing forces were exerted at the pin and across the smooth surface in the interior of the tube T. If it had been known in advance that these forces would not contribute to the generalized forces, then we could have saved the computation labor of evaluating their contribution — only to find it to be zero. That is, if we could identify noncontributing forces at the onset of the analysis, we could ignore those forces in the remainder of the analysis.

It happens that it is possible to identify classes of forces that will not contribute to the generalized forces. These are forces exerted at fixed pins, forces exerted at points of zero velocity, forces exerted across smooth surfaces interior to a mechanical system, and forces exerted across smooth surfaces whose motion is prescribed.

To see this consider first forces exerted at fixed pins and at points with a velocity of zero. Let O be a point at the pin center, or a point with zero velocity, and let $\mathbf{F}$ be a force applied at O (see Figure 11.8.1). Let O be a point of a mechanical system S having n degrees of freedom described by coordinates q_r $(r = 1,\ldots, n)$. Then, the velocity of O in the reference inertial frame and the associated partial velocities of O are zero. That is,

$$\mathbf{v}^O = 0 \ \text{ and } \ \mathbf{v}^O_{\dot{q}_r} = 0 \ \left(r = 1,\ldots,n\right) \tag{11.8.1}$$

Then, from Eq. (11.5.1), the contribution F_r to the generalized force F_r is:

$$\hat{\mathbf{F}}_r = \mathbf{F}\cdot\mathbf{v}^O_{\dot{q}_r} = 0 \quad \left(r = 1,\ldots n\right) \tag{11.8.2}$$

In addition to fixed pins, other points having zero velocity are contact points of bodies rolling on fixed surfaces (such as in our rolling disk example).

Consider next forces exerted across smooth surfaces in the interior of a mechanical system. Specifically, let S_1 and S_2 be contacting smooth surfaces of a mechanical system S having n degrees of freedom characterized by the coordinates q_r $(r = 1,\ldots, n)$ as depicted in Figure 11.8.2. Because the surfaces are smooth, forces transmitted across the surfaces will be normal to the surfaces. Let the points of contact of S_1 and S_2 be C_1 and C_2, and let the forces exerted by S_1 on S_2 be represented by $\mathbf{F}_2$ and the forces exerted by S_2 on S_1 be represented by $\mathbf{F}_1$, as in Figure 11.8.2.

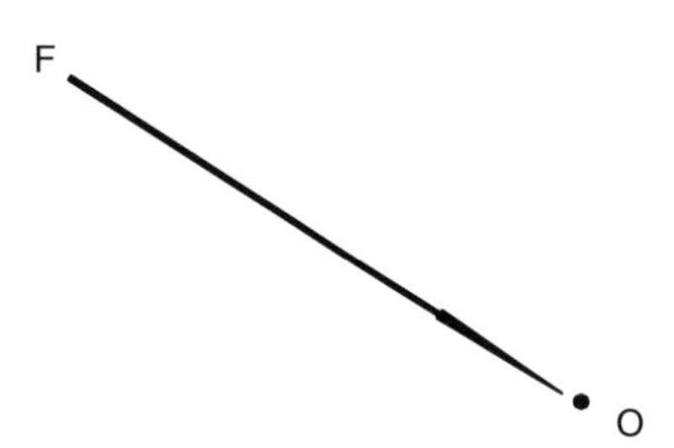

FIGURE 11.8.1
A force $\mathbf{F}$ applied at a fixed point O.

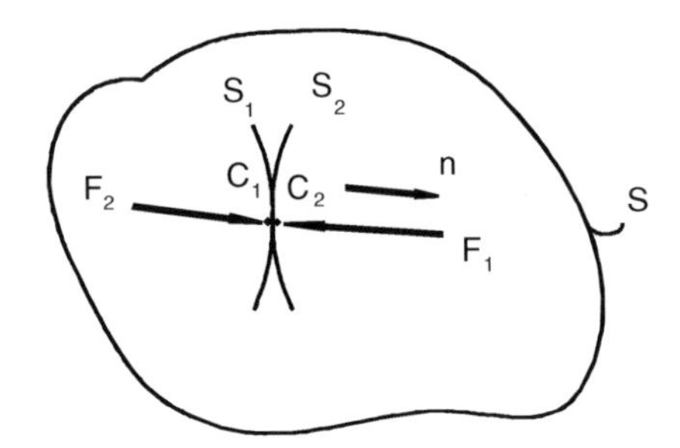

FIGURE 11.8.2
Forces transmitted across smooth contacting surfaces of a mechanical system.

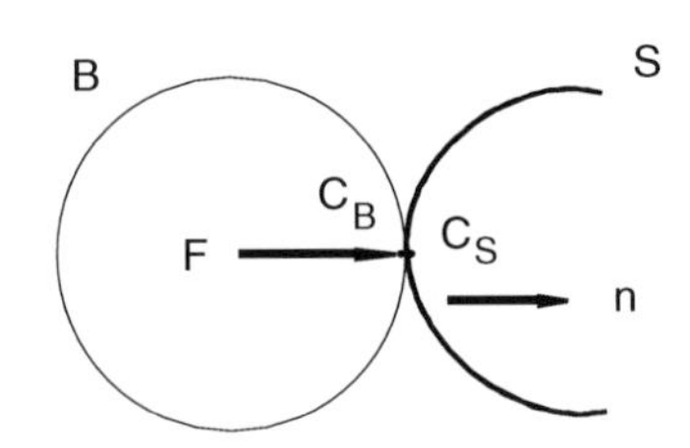

FIGURE 11.8.3
A force exerted on a mechanical system by a smooth body with specified motion.

Because $\mathbf{F}_1$ and $\mathbf{F}_2$ are both normal to the contacting surfaces at the point of contact, they may be expressed as:

$$\mathbf{F}_1 = -\mathbf{F}_1\mathbf{n} \quad \text{and} \quad \mathbf{F}_2 = \mathbf{F}_2\mathbf{n} \tag{11.8.3}$$

where n is a unit vector normal to the contacting surfaces at the point of contact. Then, by the law of action and reaction (see Reference 11.3), we have:

$$\mathbf{F}_1 = -\mathbf{F}_2 \quad \text{or} \quad \mathbf{F}_1 + \mathbf{F}_2 = 0 \quad \text{or} \quad \mathbf{F}_1 + \mathbf{F}_2 = 0 \tag{11.8.4}$$

Because S_1 and S_2 are in contact at the instant of interest, the relative velocities of the contact points C_1 and C_2 in the normal direction are zero. That is,

$$\left(\mathbf{v}^{C_1} - \mathbf{v}^{C_2}\right)\cdot\mathbf{n} = 0 \quad \text{or} \quad \mathbf{v}^{C_1}\cdot\mathbf{n} = \mathbf{v}^{C_2}\cdot\mathbf{n} \tag{11.8.5}$$

Then by differentiating with respect to $\dot{q}_r$ we have:

$$\left(\mathbf{v}^{C_1}_{\dot{q}_r} - \mathbf{v}^{C_2}_{\dot{q}_r}\right)\cdot\mathbf{n} = 0 \quad \text{or} \quad \mathbf{v}^{C_1}_{\dot{q}_r}\cdot\mathbf{n} = {}^{C_2}_{\dot{q}_r}\cdot\mathbf{n} \tag{11.8.6}$$

(Observe that n is not a function of the $\dot{q}_r$)

The contributions $\hat{\mathbf{F}}_r$ of $\mathbf{F}_1$ and $\mathbf{F}_2$ to the generalized forces $\mathbf{F}_r$ are then:

$$\hat{\mathbf{F}}_r = \mathbf{F}_1\cdot\mathbf{v}^{C_1}_{\dot{q}_r} + \mathbf{F}_2\cdot\mathbf{v}^{C_2}_{\dot{q}_r} \tag{11.8.7}$$

Then by using Eqs. (11.8.4) and (11.3.6), we have:

$$\hat{F}_r = \mathbf{F}_1\cdot\left(\mathbf{v}^{C_1}_{\dot{q}_r} - \mathbf{v}^{C_2}_{\dot{q}_r}\right) = F_1\mathbf{n}\cdot\left(\mathbf{v}^{C_1}_{\dot{q}_r} - \mathbf{v}^{C_2}_{\dot{q}_r}\right) = 0 \tag{11.8.8}$$

Finally, consider the forces exerted on a mechanical system S by a body B which has a smooth surface and whose motion is specified (that is, known or given) in an inertial frame R (see Figure 11.8.3). As before, let S have n degrees of freedom with coordinates

q_r ($r = 1,\ldots, n$). Let the points of contact between B and S be C_B and C_S, and let n be a unit vector normal to the contacting surfaces at the point of contact.

Let **F** represent the forces exerted by B on S at C_B. Then, because the surface of B is smooth, **F** may be expressed as:

$$\mathbf{F} = F\mathbf{n} \tag{11.8.9}$$

Because B and S are in contact at C_B and C_S at the instant of interest, we have:

$$\left(\mathbf{v}^{C_B} - \mathbf{v}^{C_S}\right)\cdot\mathbf{n} = 0 \quad \text{or} \quad \mathbf{v}^{C_B}\cdot\mathbf{n} = \mathbf{v}^{C_S}\cdot\mathbf{n} \tag{11.8.10}$$

Then, by differentiating with respect to $\dot{q}_r$, we have:

$$\mathbf{v}_{\dot{q}_r}^{C_B}\cdot\mathbf{n} = \mathbf{v}_{\dot{q}_r}^{C_S}\cdot\mathbf{n} \tag{11.8.11}$$

where, as before, n is not a function of the $\dot{q}_r$. Because the motion of B is specified, the velocity of C_B is also independent of the q_r. Hence, we have:

$$\mathbf{v}_{\dot{q}_r}^{C_B} = 0 \tag{11.8.12}$$

Thus, we also have:

$$\mathbf{v}^{C_S}\cdot\mathbf{n} = 0 \tag{11.8.13}$$

The contributions $\hat{F}_r$ of **F** to the generalized forces F_r are then:

$$\hat{F}_r = \mathbf{F}\cdot\mathbf{v}_{\dot{q}_r}^{C_S} = F\mathbf{n}\cdot\mathbf{v}_{\dot{q}_r}^{C_S} = 0 \tag{11.8.14}$$

Forces that do not contribute to the generalized forces are sometimes called *nonworking* because they are analogous to forces that do no work (see Section 10.2). A principal advantage of the procedures of generalized mechanics is that nonworking forces (which usually are of little interest) do not enter the analysis and thus can be ignored at the onset.

11.9 Generalized Forces: Inertia (Passive) Forces

As with applied forces, we can also introduce and define generalized inertia forces. Specifically, a generalized inertia force is defined as the projection of an inertia force along a partial velocity vector. Consider, for example, a particle P, having mass m and moving in an inertial reference frame R as in Figure 11.9.1. From Eq. (8.2.5), we recall that the inertia force $\mathbf{F}^*$ on P is:

$$\mathbf{F}^* = -m\mathbf{a} \tag{11.9.1}$$

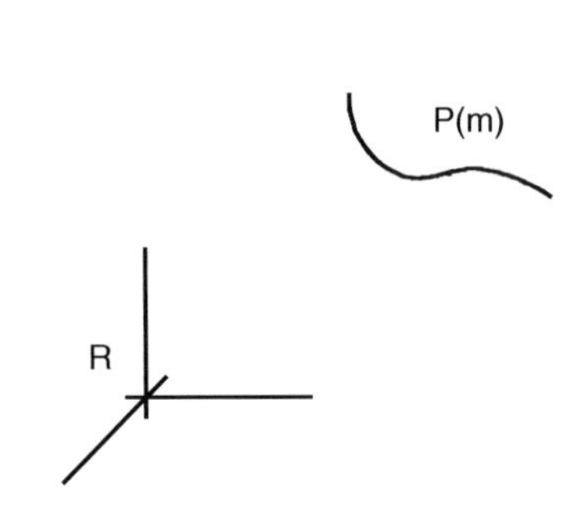

FIGURE 11.9.1
A particle P moving in an inertial reference frame R.

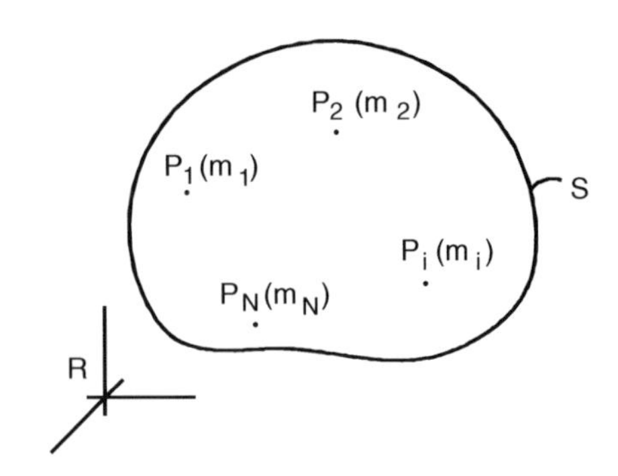

FIGURE 11.9.2
A set S of N particles moving in an inertial reference frame R.

where **a** is the acceleration of P in R. If P is part of a mechanical system S having n degrees of freedom represented by coordinates q_r $(r = 1,\ldots, n)$, then the generalized inertia forces F_r^* associated with these coordinates are defined as:

$$F_r^* \stackrel{D}{=} \mathbf{F}^* \cdot \mathbf{v}_{\dot{q}_r} = -m\mathbf{a} \cdot \mathbf{v}_{\dot{q}_r} \quad (r = 1,\ldots,n) \tag{11.9.2}$$

Consider next a set S of N particles P_i $(i = 1,\ldots, N)$ having masses m_i and moving in an inertial reference frame R as depicted in Figure 11.9.2. Let S have n degrees of freedom with coordinates q_r $(r = 1,\ldots, n)$. Then, the generalized inertia forces F_r^* on S are defined as:

$$F_r^* \stackrel{D}{=} \sum_{i=1}^{N} \mathbf{F}_i^* \cdot \mathbf{v}_{\dot{q}_r}^{P_i} = -\sum_{i=1}^{N} m_i \mathbf{a}_i \cdot \mathbf{v}_{\dot{q}_r}^{P_i} \quad (r = 1,\ldots,n) \tag{11.9.3}$$

where, as before, F_i^* is the inertia force on P_i and $\mathbf{a}_i$ is the acceleration of P_i in R.

Finally, consider a rigid body B that is part of a mechanical system S. As before, let S have n degrees of freedom represented by coordinates q_r $(r = 1,\ldots, n)$. Also, as we have done before, let B be considered to be made up of particles P_i $(i = 1,\ldots, N)$ having masses m_i as in Figure 11.9.3 where G is the mass center of B and R is an inertial reference frame. Then, based upon the definitions of Eqs. (11.9.2) and (11.9.3), the generalized inertia forces F_r^* on B are:

$$F_r^* = \sum_{i=1}^{N} \mathbf{F}_{i^*} \cdot \mathbf{v}_{\dot{q}_r}^{P_i} = -\sum_{i=1}^{N} m_i \mathbf{a}_i \cdot \mathbf{v}_{\dot{q}_r}^{P_i} \quad (r = 1,\ldots,n) \tag{11.9.4}$$

where, as before, $\mathbf{F}_i^*$ is the inertia force on P_i and $\mathbf{a}_i$ is the acceleration of P_i in R.

We can simplify Eq. (11.9.4) by taking advantage of the rigidity of B. Specifically, we can represent the system of inertia forces on B by a single force $\mathbf{F}^*$ passing through the mass center G together with a couple with torque $\mathbf{T}^*$, where from Eqs. (8.6.5) and (8.6.6) $\mathbf{F}^*$ and $\mathbf{T}^*$ are:

$$\mathbf{F}^* = -\mathbf{M}\mathbf{a}^G \quad \text{and} \quad \mathbf{T}^* = -\mathbf{I} \cdot \boldsymbol{\alpha} - \boldsymbol{\omega} \times (\mathbf{I} \cdot \boldsymbol{\omega}) \tag{11.9.5}$$

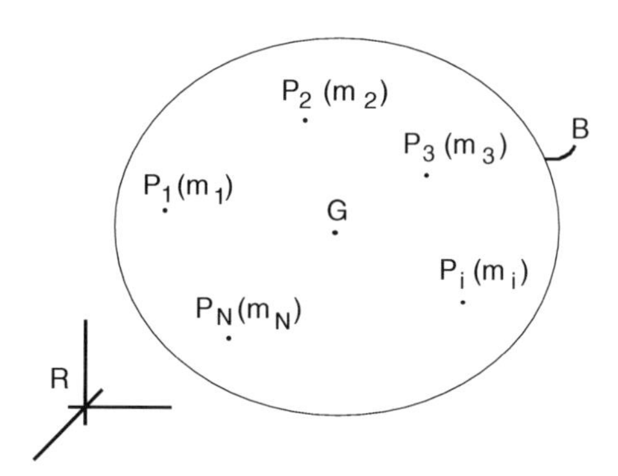

FIGURE 11.9.3
A rigid body B, modeled as a set of particles, moving in an inertial reference frame R.

where M is the mass of B, $\mathbf{I}$ is the central inertia dyadic of B, $\mathbf{a}^G$ is the acceleration of G in R, and $\boldsymbol{\omega}$ and $\boldsymbol{\alpha}$ are the angular velocity and angular acceleration of B in R. Then, by following the procedures of Section 11.5 leading to Eq. (11.5.7), we can express the generalized inertia forces on B as:

$$F_r^* = \mathbf{v}_{\dot{q}_r}^G \cdot \mathbf{F}^* + \boldsymbol{\omega}_{\dot{q}_r} \cdot \mathbf{T}^* \quad (r = 1, \ldots, n) \tag{11.9.6}$$

As noted earlier, inertia forces are sometimes called *passive forces*. In this context, applied forces (such as gravity and contact forces) are sometimes called *active forces*; hence, the generalized forces of the foregoing section are often called *generalized active forces*.

11.10 Examples

We can illustrate the concept of generalized inertia forces with a few elementary examples.

Example 11.10.1: A Simple Pendulum

Consider first the simple pendulum of Figure 11.10.1. Recall that the pendulum bob P moves in a circle with radius ℓ and that the velocity and acceleration of P in an inertial reference frame R may be expressed as:

$$\mathbf{v}^P = \ell\dot{\theta}\mathbf{n}_\theta \tag{11.10.1}$$

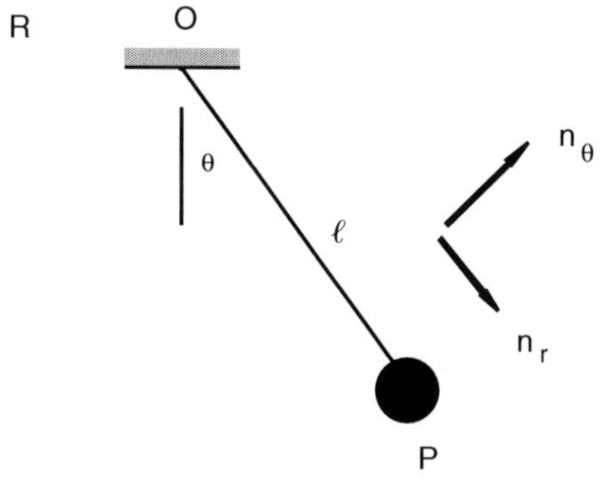

FIGURE 11.10.1
The simple pendulum.

and

$$\mathbf{a}^P = -\ell\dot{\theta}^2\mathbf{n}_r + \ell\ddot{\theta}\mathbf{n}_\theta \tag{11.10.2}$$

where ℓ is the pendulum length, θ is the orientation angle, and $\mathbf{n}_r$ and $\mathbf{n}_\theta$ are the radial and transverse unit vectors, respectively, shown in Figure 11.10.1. The inertia force $\mathbf{F}^*$ on P is, then,

$$\mathbf{F}^* = -m\mathbf{a}^P = -m\ell\dot{\theta}^2\mathbf{n}_r - m\ell\ddot{\theta}\mathbf{n}_\theta \tag{11.10.3}$$

where m is the mass of P.

From Eqs. (11.4.4) and (11.10.1), the partial velocity $\mathbf{v}_\theta$ of P with respect to $\dot{\theta}$ is:

$$\mathbf{v}_{\dot{\theta}} = \ell\mathbf{n}_\theta \tag{11.10.4}$$

From Eq. (11.9.2), the generalized inertia force F_θ^* is:

$$F_\theta^* = \mathbf{F}^* \cdot \mathbf{v}_{\dot{\theta}} = -m\ell^2\ddot{\theta} \tag{11.10.5}$$

Example 11.10.2: Rod Pendulum

Consider next the rod pendulum consisting of a uniform rod B having mass m, length ℓ, and mass center G. Let B be pinned at one end so that it is free to move in a vertical plane as described by the angle θ as in Figure 11.10.2. The velocity and acceleration of G and the angular velocity and angular acceleration of B itself in an inertial reference frame R are:

$$\mathbf{v}^G = (\ell/2)\dot{\theta}\mathbf{n}_\theta \quad \text{and} \quad \mathbf{a}^G = (\ell/2)\ddot{\theta}\mathbf{n}_\theta - (\ell/2)\dot{\theta}\mathbf{n}_r \tag{11.10.6}$$

and

$$\boldsymbol{\omega} = \dot{\theta}\mathbf{n}_z \quad \text{and} \quad \boldsymbol{\alpha} = \ddot{\theta}\mathbf{n}_z \tag{11.10.7}$$

where $\mathbf{n}_r$, $\mathbf{n}_\theta$, and $\mathbf{n}_z$ are mutually perpendicular unit vectors as shown in Figure 11.10.2.

Let the inertia force system on B be represented by an equivalent force system consisting of a single force $\mathbf{F}^*$ passing through G together with a couple with torque $\mathbf{T}$. Then, $\mathbf{F}^*$ and $\mathbf{T}^*$ are:

$$\mathbf{F}^* = -m\mathbf{a}^G = -m(\ell/2)\ddot{\theta}\mathbf{n}_\theta + m(\ell/2)\dot{\theta}\mathbf{n}_r \tag{11.10.8}$$

and

$$\mathbf{T}^* = -m(\ell^2/12)\ddot{\theta}\mathbf{n}_z \tag{11.10.9}$$

The partial velocity of G and the partial angular velocity of B with respect to θ are:

$$\mathbf{v}_{\dot{\theta}}^G = (\ell/2)\mathbf{n}_\theta \quad \text{and} \quad \boldsymbol{\omega}_{\dot{\theta}} = \mathbf{n}_z \tag{11.10.10}$$

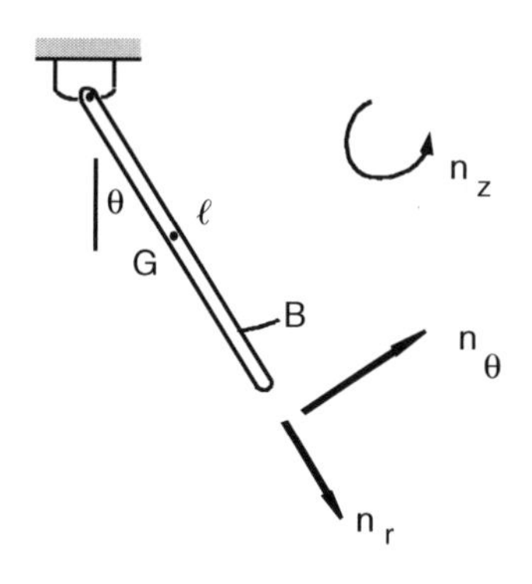

FIGURE 11.10.2
A rod pendulum.

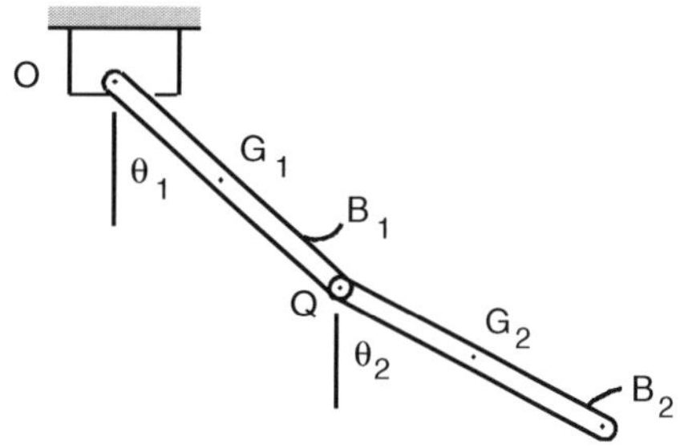

FIGURE 11.10.3
A double-rod pendulum.

From Eq. (11.9.6) we find the generalized inertia force to be:

$$F_\theta^* = \mathbf{v}_{\dot{\theta}}^G \cdot \mathbf{F}^* + \boldsymbol{\omega}_{\dot{\theta}} \cdot \mathbf{T}^*$$

$$= -m\left(\ell^2/4\right)\ddot{\theta} - m\left(\ell^2/12\right)\ddot{\theta}$$

or

$$F_\theta^* = -m\left(\ell^2/3\right)\ddot{\theta} \tag{11.10.11}$$

Example 11.10.3: Double-Rod Pendulum

As an extension of the foregoing example, consider the double-rod pendulum as in Figure 11.10.3. Let the rods, B_1 and B_2, be identical, each having mass m and length ℓ. Let the rods be pinned together at Q and supported at O by frictionless pins such that the system is free to move in a vertical plane as depicted in Figure 11.10.3. The system then has two degrees of freedom as represented by the angles θ_1 and θ_2 as shown.

The velocities and accelerations of the mass centers G_1 and G_2 and the angular velocities and angular accelerations of the rods themselves are:

$$\mathbf{v}^{G_1} = (\ell/2)\dot{\theta}_1\mathbf{n}_{1\theta}, \quad \mathbf{v}^{G_2} = \ell\dot{\theta}_1\mathbf{n}_{1\theta} + (\ell/2)\dot{\theta}_2\mathbf{n}_{2\theta}$$

$$\mathbf{a}_1^G = (\ell/2)\ddot{\theta}_1\mathbf{n}_{1\theta} - (\ell/2)\dot{\theta}_1^2\mathbf{n}_{1r} \tag{11.10.12}$$

$$\mathbf{a}^{G_2} = \ell\ddot{\theta}_1\mathbf{n}_{1\theta} - \ell\dot{\theta}_1^2\mathbf{n}_{1r} + (\ell/2)\ddot{\theta}_2\mathbf{n}_{2\theta} - (\ell/2)\dot{\theta}_2^2\mathbf{n}_{2r}$$

$$\boldsymbol{\omega}^{B_1} = \dot{\theta}_1\mathbf{n}_3, \quad \boldsymbol{\omega}^{B_2} = \dot{\theta}_2\mathbf{n}_3$$

$$\boldsymbol{\alpha}^{B_1} = \ddot{\theta}_1\mathbf{n}_3, \quad \boldsymbol{\alpha}^{B_2} = \ddot{\theta}_2\mathbf{n}_3$$

where the unit vectors are shown in Figure 11.10.4. The unit vectors are related to one another by the expressions:

$$\mathbf{n}_{1r} = \cos(\theta_2 - \theta_1)\mathbf{n}_{2r} - \sin(\theta_2 - \theta_1)\mathbf{n}_{2\theta} = \cos\theta_1\mathbf{n}_1 + \sin\theta_1\mathbf{n}_2$$
$$\mathbf{n}_{1\theta} = \sin(\theta_2 - \theta_1)\mathbf{n}_{2r} + \cos(\theta_2 - \theta_1)\mathbf{n}_{2\theta} = -\sin\theta_1\mathbf{n}_1 + \cos\theta_1\mathbf{n}_2 \tag{11.10.13}$$

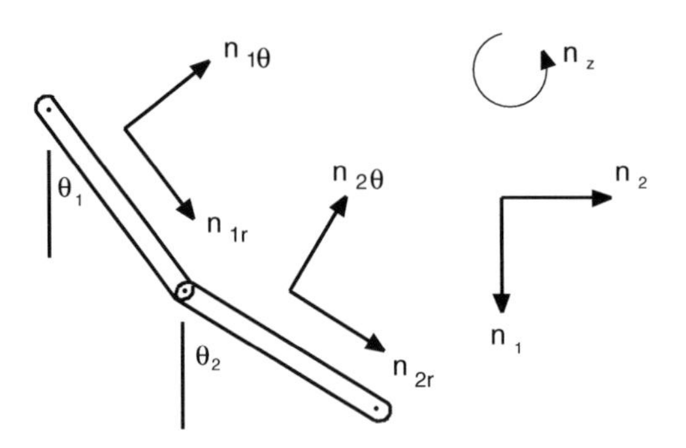

FIGURE 11.10.4
Unit vector geometry for the double-rod pendulum.

$$\begin{aligned}\mathbf{n}_{2r} &= \cos(\theta_2 - \theta_1)\mathbf{n}_{1r} + \sin(\theta_2 - \theta_1)\mathbf{n}_{1\theta} = \cos\theta_2\mathbf{n}_1 + \sin\theta_2\mathbf{n}_2 \\ \mathbf{n}_{2\theta} &= \sin(\theta_2 - \theta_1)\mathbf{n}_{1r} + \cos(\theta_2 - \theta_1)\mathbf{n}_{1\theta} = -\sin\theta_2\mathbf{n}_1 + \cos\theta_2\mathbf{n}_2\end{aligned} \tag{11.10.14}$$

Let the inertia forces on B_1 and B_2 be represented by forces $\mathbf{F}_1^*$ and $\mathbf{F}_2^*$ passing through G_1 and G_2 together with couples having torques $\mathbf{T}_1^*$ and $\mathbf{T}_2^*$. Then $\mathbf{F}_1^*$, $\mathbf{F}_2^*$, $\mathbf{T}_1^*$, and $\mathbf{T}_2^*$ are:

$$F_1^* = -m\mathbf{a}^{G_r} = m(\ell/2)\dot{\theta}_1^2\mathbf{n}_{1r} - m(\ell/2)\ddot{\theta}_1\mathbf{n}_{1\theta}$$

or

$$\mathbf{F}_1^* = m(\ell/2)\left(\dot{\theta}_1^2\cos\theta_1 + \ddot{\theta}_1\sin\theta_1\right)\mathbf{n}_1 + m(\ell/2)\left(\dot{\theta}_1^2\sin\theta_1 - \ddot{\theta}_1\cos\theta_1\right)\mathbf{n}_2 \tag{11.10.15}$$

$$\mathbf{F}_2^* = -m\mathbf{a}^{G_2} = m\ell\dot{\theta}_1^2\mathbf{n}_{1r} - m\ell\ddot{\theta}_1\mathbf{n}_{1\theta} + m(\ell/2)\dot{\theta}_2^2\mathbf{n}_{2r} - m(\ell/2)\ddot{\theta}_2\mathbf{n}_{2\theta}$$

or

$$\begin{aligned}\mathbf{F}_2^* &= m\ell\left[\ddot{\theta}_1\sin\theta_1 + \theta_1^2\cos\theta_1 + (\ddot{\theta}_2/2)\sin\theta_2 + (\dot{\theta}_2^2/2)\cos\theta_2\right]\mathbf{n}_1 \\ &\quad + m\ell\left[-\ddot{\theta}_1\cos\theta_1 + \dot{\theta}_1^2\sin\theta_1 - (\ddot{\theta}_2/2)\cos\theta_2 + (\dot{\theta}_2^2/2)\sin\theta_2\right]\mathbf{n}_2\end{aligned} \tag{11.10.16}$$

$$\mathbf{T}_1^* = -\left(m\ell^2/12\right)\ddot{\theta}_1\mathbf{n}_3 \tag{11.10.17}$$

and

$$\mathbf{T}_2^* = -\left(m\ell^2/12\right)\ddot{\theta}_2\mathbf{n}_3 \tag{11.10.18}$$

From Eq. (11.10.12) the partial velocities of G_1 and G_2 and the partial angular velocities of B_1 and B_2 are (see also Eqs. (11.4.23)):

$$\begin{aligned}
\mathbf{v}^{G_1}_{\dot{\theta}_1} &= (\ell/2)\,\mathbf{n}_{1\theta} = -(\ell/2)\sin\theta_1\,\mathbf{n}_1 + (\ell/2)\cos\theta_1\,\mathbf{n}_2 \\
\mathbf{v}^{G_2}_{\dot{\theta}_1} &= \ell\,\mathbf{n}_{1\theta} = -\ell\sin\theta_1\,\mathbf{n}_1 + \ell\cos\theta_1\,\mathbf{n}_2 \\
\mathbf{v}^{G_1}_{\dot{\theta}_2} &= 0 \\
\mathbf{v}^{G_2}_{\dot{\theta}_2} &= (\ell/2)\,\mathbf{n}_{2\theta} = -(\ell/2)\sin\theta_2\,\mathbf{n}_1 + (\ell/2)\cos\theta_2\,\mathbf{n}_2
\end{aligned} \tag{11.10.19}$$

and

$$\boldsymbol{\omega}^{B_1}_{\dot{\theta}_1} = \mathbf{n}_3\,,\quad \boldsymbol{\omega}^{B_2}_{\dot{\theta}_1} = 0\,,\quad \boldsymbol{\omega}^{B_1}_{\dot{\theta}_2} = 0\,,\quad \boldsymbol{\omega}^{B_2}_{\dot{\theta}_2} = \mathbf{n}_3 \tag{11.10.20}$$

Finally, from Eq. (11.9.6) the generalized inertia forces are:

$$\mathbf{F}^*_{\theta_1} = -(4/3)m\ell^2\ddot{\theta}_1 - (1/2)m\ell^2\ddot{\theta}_2\cos(\theta_2-\theta_1) + (1/2)m\ell^2\dot{\theta}_2^2\sin(\theta_2-\theta_1) \tag{11.10.21}$$

and

$$\mathbf{F}_{\theta^*_2} = -(1/3)m\ell^2\ddot{\theta}_2 - (1/2)m\ell^2\ddot{\theta}_1\cos(\theta_2-\theta_1) - (1/2)m\ell^2\dot{\theta}_1^2\sin(\theta_2-\theta_1) \tag{11.10.22}$$

Observe how routine the computation is; the principal difficulty is the detail. We will discuss this later.

Example 11.10.4: Spring-Supported Particles in a Rotating Tube

Consider again the system of Section 11.7 consisting of a cylindrical tube T containing three spring-supported particles P_1, P_2, and P_3 (or small spheres) as in Figure 11.10.5. As before, T has mass M and length L and it rotates in a vertical plane with the angle of rotation being θ as shown. The particles each have mass m, and their positions within T are defined by the coordinates x_1, x_2, and x_3 as in Figure 11.10.6, where ℓ is the natural length of each of the springs.

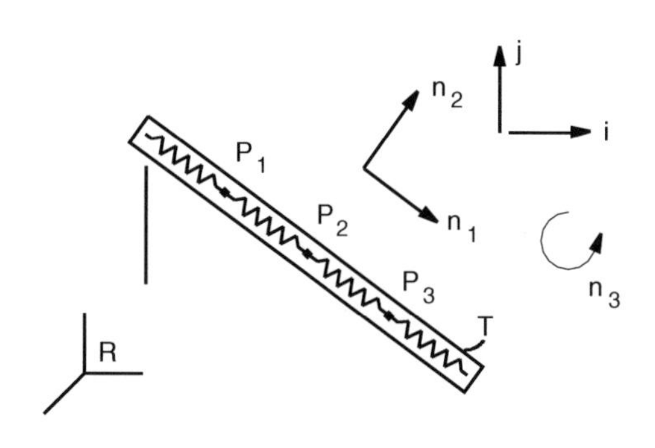

FIGURE 11.10.5
A rotating tube containing spring-supported particles.

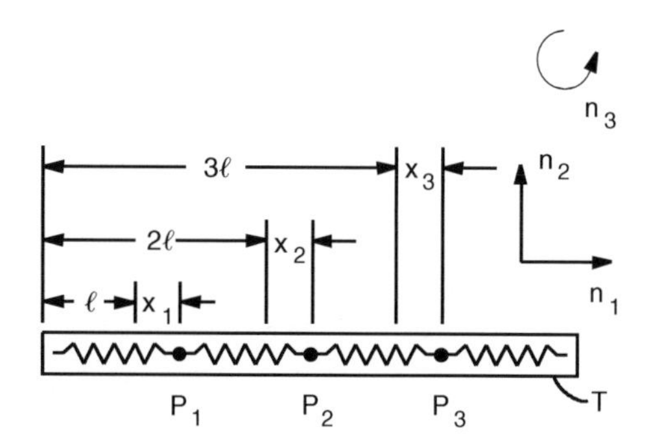

FIGURE 11.10.6
Coordinates of particles within the tube.

The velocities and accelerations of P_1, P_2, and P_3 in the fixed or inertial frame R are (see Eqs. (11.7.2), (11.7.3), and (11.7.4)):

$$\mathbf{v}^{P_1} = \dot{x}_1\mathbf{n}_1 + (\ell + x_1)\dot{\theta}\mathbf{n}_2\,, \quad \mathbf{a}^{P_1} = \left[\ddot{x}_1 - (\ell + x_1)\dot{\theta}^2\right]\mathbf{n}_1 + \left[(\ell + x_1)\ddot{\theta} + 2\dot{\theta}\dot{x}_1\right]\mathbf{n}_2 \tag{11.10.23}$$

$$\mathbf{v}^{P_2} = \dot{x}_2\mathbf{n}_1 + (2\ell + x_2)\dot{\theta}\mathbf{n}_2\,, \quad \mathbf{a}^{P_2} = \left[\ddot{x}_2 - (2\ell + x_2)\dot{\theta}^2\right]\mathbf{n}_1 + \left[(2\ell + x_2)\ddot{\theta} + 2\dot{\theta}\dot{x}_2\right]\mathbf{n}_2 \tag{11.10.24}$$

$$\mathbf{v}^{P_3} = \dot{x}_3\mathbf{n}_1 + (3\ell + x_3)\dot{\theta}\mathbf{n}_2\,, \quad \mathbf{a}^{P_3} = \left[\ddot{x}_3 - (3\ell + x_3)\dot{\theta}^2\right]\mathbf{n}_1 + \left[(3\ell + x_3)\ddot{\theta} + 2\dot{\theta}\dot{x}_3\right]\mathbf{n}_2 \tag{11.10.25}$$

where $\mathbf{n}_1$, $\mathbf{n}_2$, and $\mathbf{n}_3$ are the unit vectors shown in Figures 11.10.5 and 11.10.6. Let G be the mass center of T. Then, from Eq. (11.7.1), the velocity and acceleration of G in R are:

$$\mathbf{v}^G = (L/2)\dot{\theta}\mathbf{n}_2\,, \quad \mathbf{a}^G = -(L/2)\dot{\theta}^2\,\mathbf{n}_1 + (L/2)\,\ddot{\theta}\mathbf{n}_2 \tag{11.10.26}$$

Finally, the angular velocity and the angular acceleration of T in R are:

$$\boldsymbol{\omega} = \dot{\theta}\mathbf{n}_3 \quad \text{and} \quad \boldsymbol{\alpha} = \ddot{\theta}\mathbf{n}_3 \tag{11.10.27}$$

As noted in Section 11.7, the system has four degrees of freedom represented by the coordinates x_1, x_2, x_3, and θ. The corresponding partial velocities and partial angular velocities are recorded in Eqs. (11.7.5), (11.7.6), and (11.7.7) as:

$$\begin{aligned}
&\mathbf{v}^{P_1}_{\dot{x}_1} = \mathbf{n}_1\,, \quad \mathbf{v}^{P_1}_{\dot{x}_2} = 0\,, \quad \mathbf{v}^{P_1}_{\dot{x}_3} = 0\,, \quad \mathbf{v}^{P_1}_{\dot{\theta}} = (\ell + x_1)\mathbf{n}_2 \\
&\mathbf{v}^{P_2}_{\dot{x}_1} = 0\,, \quad \mathbf{v}^{P_2}_{\dot{x}_2} = \mathbf{n}_1\,, \quad \mathbf{v}^{P_2}_{\dot{x}_3} = 0\,, \quad \mathbf{v}^{P_2}_{\dot{\theta}} = (2\ell + x_2)\mathbf{n}_2 \\
&\mathbf{v}^{P_3}_{\dot{x}_1} = 0\,, \quad \mathbf{v}^{P_3}_{\dot{x}_2} = 0\,, \quad \mathbf{v}^{P_3}_{\dot{x}_3} = \mathbf{n}_1\,, \quad \mathbf{v}^{P_3}_{\dot{\theta}} = (3\ell + x_3)\mathbf{n}_2 \\
&\mathbf{v}^{G}_{\dot{x}_1} = 0\,, \quad \mathbf{v}^{G}_{\dot{x}_2} = 0\,, \quad \mathbf{v}^{G}_{\dot{x}_3} = 0\,, \quad \mathbf{v}^{G}_{\dot{\theta}} = (L/2)\mathbf{n}_2 \\
&\boldsymbol{\omega}_{\dot{x}_1} = 0\,, \quad \boldsymbol{\omega}_{\dot{x}_2} = 0\,, \quad \boldsymbol{\omega}_{\dot{x}_3} = 0\,, \quad \boldsymbol{\omega}_{\dot{\theta}} = \mathbf{n}_3
\end{aligned} \tag{11.10.28}$$

From Eqs. (11.10.23) to (11.10.27), the inertia forces on the particles and the tube may be represented by:

$$\begin{aligned}
&\mathbf{F}^*_{P_1} = -m\mathbf{a}^{P_1} = -m\left[\ddot{x}_1 - (\ell + x_1)\dot{\theta}^2\right]\mathbf{n}_1 - m\left[(\ell + x_1)\,\ddot{\theta} + 2\dot{\theta}\dot{x}_1\right]\mathbf{n}_2 \\
&\mathbf{F}^*_{P_2} = -m\mathbf{a}^{P_2} = -m\left[\ddot{x}_2 - (2\ell + x_2)\dot{\theta}^2\right]\mathbf{n}_1 - m\left[(2\ell + x_2)\,\ddot{\theta} + 2\dot{\theta}\dot{x}_2\right]\mathbf{n}_2 \\
&\mathbf{F}^*_{P_3} = -m\mathbf{a}^{P_3} = -m\left[\ddot{x}_3 - (3\ell + x_3)\dot{\theta}^2\right]\mathbf{n}_1 - m\left[(3\ell + x_3)\,\ddot{\theta} + 2\dot{\theta}\dot{x}_3\right]\mathbf{n}_2 \\
&\mathbf{F}^*_T = -M\mathbf{a}^G = M(L/2)\dot{\theta}^2\,\mathbf{n}_1 - M(L/2)\ddot{\theta}\mathbf{n}_2 \\
&\mathbf{T}^*_T = -\mathbf{I}_G\cdot\boldsymbol{\alpha} - \boldsymbol{\omega}\times(\mathbf{I}_G\cdot\boldsymbol{\omega}) = -(ML^2/12)\,\ddot{\theta}\mathbf{n}_3
\end{aligned} \tag{11.10.29}$$

Finally, from Eq. (11.9.6), the generalized inertia forces are:

$$\begin{aligned} \mathbf{F}^*_{x_1} &= \mathbf{v}^{P_1}_{\dot{x}_1} \cdot \mathbf{F}^*_{P_1} + \mathbf{v}^{P_2}_{\dot{x}_1} \cdot \mathbf{F}^*_{P_2} + \mathbf{v}^{P_3}_{\dot{x}_1} \cdot \mathbf{F}^*_{P_3} + \mathbf{v}^*_{\dot{x}_1} \cdot \mathbf{F}^*_T + \boldsymbol{\omega}_{\dot{x}_1} \cdot \mathbf{T}^*_T \\ &= -m\left[\ddot{x}_1 - (\ell + x_1)\dot{\theta}^2\right] \end{aligned} \tag{11.10.30}$$

$$\begin{aligned} \mathbf{F}^*_{x_2} &= \mathbf{v}^{P_1}_{\dot{x}_2} \cdot \mathbf{F}^*_{P_1} + v^{P_2}_{\dot{x}_2} \cdot \mathbf{F}^*_{P_2} + \mathbf{v}^{P_3}_{\dot{x}_2} \cdot \mathbf{F}^*_{P_3} + \mathbf{v}^G_{\dot{x}_2} \cdot \mathbf{F}^*_T + \boldsymbol{\omega}_{\dot{x}_2} \cdot \mathbf{T}^*_T \\ &= -m\left[\ddot{x}_2 - (2\ell + x_2)\dot{\theta}^2\right] \end{aligned} \tag{11.10.31}$$

$$\begin{aligned} \mathbf{F}^*_{x_3} &= \mathbf{v}^{P_1}_{\dot{x}_3} \cdot \mathbf{F}^*_{P_1} + v^{P_2}_{\dot{x}_3} \cdot \mathbf{F}^*_{P_2} + \mathbf{v}^{P_3}_{\dot{x}_3} \cdot \mathbf{F}^*_{P_3} + \mathbf{v}^G_{\dot{x}_3} \cdot \mathbf{F}^*_T + \boldsymbol{\omega}_{\dot{x}_3} \cdot \mathbf{T}^*_T \\ &= -m\left[\ddot{x}_3 - (3\ell + x_3)\dot{\theta}^2\right] \end{aligned} \tag{11.10.32}$$

$$\begin{aligned} F^*_\theta &= \mathbf{v}^{P_1}_{\dot{\theta}} \cdot \mathbf{F}^*_{P_1} + \mathbf{v}^{P_2}_{\dot{\theta}} \cdot \mathbf{F}^*_{P_2} + \mathbf{v}^{P_3}_{\dot{\theta}} \cdot \mathbf{F}^*_{P_3} + \mathbf{v}^G_{\dot{\theta}} \cdot \mathbf{F}^*_T + \boldsymbol{\omega}_{\dot{\theta}} \cdot \mathbf{T}^*_T \\ &= -m(\ell + \dot{x}_1)\left[(\ell + x_1)\ddot{\theta} + 2\dot{\theta}\dot{x}_1\right] - m(2\ell + x_2)\left[(2\ell + x_2)\ddot{\theta} + 2\dot{\theta}\dot{x}_2\right] \\ &\quad - m(3\ell + x_3)\left[(3\ell + x_3)\ddot{\theta} + 2\dot{\theta}\dot{x}_3\right] - M(L/2)^2\ddot{\theta} - M(L^2/12)\ddot{\theta} \end{aligned}$$

or

$$\begin{aligned} F^*_\theta &= -m\left[(\ell + x_1)^2 + (2\ell + x_2)^2 + (3\ell + x_3)^2\right]\ddot{\theta} - M(L^2/3)\ddot{\theta} \\ &\quad - 2m\left[(\ell + x_1)\dot{x}_1 + (2\ell + x_2)\dot{x}_2 + (3\ell + x_3)\dot{x}_3\right]\dot{\theta} \end{aligned} \tag{11.10.33}$$

Example 11.10.5: Rolling Circular Disk

As a final example consider again the rolling circular disk D with mass m and radius r of Figure 11.10.7. (We first considered this system in Section 4.12 and later in Sections 8.13 and 11.3.) As we observed in Section 11.3, this is a nonholonomic system having three

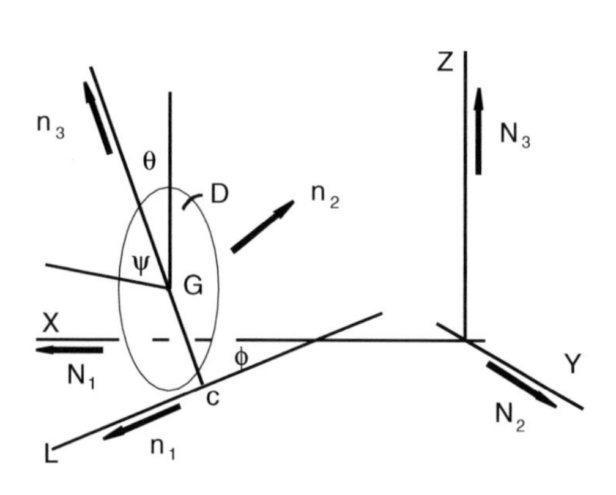

FIGURE 11.10.7
A circular disk rolling on a horizontal surface.

degrees of freedom. Recall that if we introduce six parameters (say, x, y, z, θ, ϕ, and ψ) to define the position and orientation of D we find that the conditions of rolling lead to the constraint equations (see Eq. (11.3.8), (11.3.9), and (11.3.10)):

$$\dot{x} = r\left[\left(\dot{\psi} + \phi\sin\theta\right)\cos\phi + \dot{\theta}\cos\theta\sin\theta\right] \tag{11.10.34}$$

$$\dot{y} = r\left[\left(\dot{\psi} + \phi\sin\theta\right)\sin\phi - \dot{\theta}\cos\theta\cos\phi\right] \tag{11.10.35}$$

$$\dot{z} = -r\dot{\theta}\sin\theta \tag{11.10.36}$$

where x, y, and z are the Cartesian coordinates of G and θ, ϕ, and ψ are the orientation angles of D as in Figure 11.10.7.

The last of these equations is integrable leading to the expression:

$$z = r\cos\theta \tag{11.10.37}$$

This expression simply means that D must remain in contact with the surface S. It is therefore a geometric (or holonomic) constraint. Equations (11.10.34) and (11.10.35), however, are not integrable in terms of elementary functions. These equations are kinematic (or nonholonomic) constraints. They ensure that the instantaneous velocity of the contact point C, relative to S, is zero.

To obtain the generalized inertia forces for this nonholonomic system we may simply select three of the six parameters as our independent variables. Then the partial velocities and partial angular velocities may be determined from the coefficients of the derivatives of these three variables in the expressions for the velocities and angular velocity. That is, observe that the coordinate derivatives are linearly related in Eqs. (11.10.34), (11.10.35), and (11.10.36). This means that we can readily solve for the nonselected coordinate derivatives in terms of the selected coordinate derivatives. To illustrate this, suppose we want to describe the movement of D in terms of the orientation angles θ, ϕ, and ψ. We may express the velocity of the mass center G as:

$$\mathbf{v}^G = \dot{x}\mathbf{N}_1 + \dot{y}\mathbf{N}_2 + \dot{z}\mathbf{N}_3 \tag{11.10.38}$$

where $\mathbf{N}_1$, $\mathbf{N}_2$, and $\mathbf{N}_3$ are unit vectors parallel to the X-, Y-, and Z-axes as in Figure 11.10.7. Then, from Eqs. (11.10.34), (11.10.35), and (11.10.36), we may express $\mathbf{v}_G$ as:

$$\begin{aligned} \mathbf{v}_G = {} & r\left[\left(\dot{\psi} + \dot{\phi}\sin\theta\right)\cos\phi + \dot{\theta}\cos\theta\sin\phi\right]\mathbf{N}_1 \\ & + r\left[\left(\dot{\psi} + \dot{\phi}\sin\theta\right)\sin\phi - \dot{\theta}\cos\theta\cos\phi\right]\mathbf{N}_2 \\ & - r\dot{\theta}\sin\theta\mathbf{N}_3 \end{aligned} \tag{11.10.39}$$

From Figure 11.10.7, we see that $\mathbf{N}_1$, $\mathbf{N}_2$, and $\mathbf{N}_3$ may be expressed in terms of the unit vectors $\mathbf{n}_1$, $\mathbf{n}_2$, and $\mathbf{n}_3$ as:

$$\begin{aligned} \mathbf{N}_1 &= \cos\phi\,\mathbf{n}_1 - \sin\phi\cos\theta\,\mathbf{n}_1 + \sin\theta\sin\phi\,\mathbf{n}_3 \\ \mathbf{N}_2 &= \sin\phi\,\mathbf{n}_1 + \cos\phi\cos\theta\,\mathbf{n}_2 - \cos\phi\sin\theta\,\mathbf{n}_3 \\ \mathbf{N}_3 &= \sin\theta\,\mathbf{n}_2 + \cos\theta\,\mathbf{n}_3 \end{aligned} \tag{11.10.40}$$

Hence, by substituting into Eq. (11.10.39) $\mathbf{v}_G$ becomes:

$$\mathbf{v}_G = r\left(\dot{\psi} + \dot{\phi}\,\sin\theta\right)\mathbf{n}_1 - r\dot{\theta}\mathbf{n}_2 \tag{11.10.41}$$

Also, the angular velocity $\boldsymbol{\omega}$ of D relative to S may be expressed as (see Eq. (4.12.2)):

$$\boldsymbol{\omega} = \dot{\theta}\mathbf{n}_1 + \left(\dot{\psi} + \dot{\phi}\sin\theta\right)\mathbf{n}_2 + \dot{\phi}\cos\theta\,\mathbf{n}_3 \tag{11.10.42}$$

Observe that the result of Eq. (11.10.41) could have been obtained directly as $\boldsymbol{\omega} \times r\mathbf{n}^3$, the expression for velocities of points of rolling bodies (see Section 4.11, Eq. (4.11.5)).

The partial velocities of G and the partial angular velocities of D with respect to θ, ϕ, and ψ are then:

$$\mathbf{v}^G_{\dot{\theta}} = -r\mathbf{n}_2\,,\quad \mathbf{v}^G_{\dot{\phi}} = r\,\sin\theta\,\mathbf{n}_1\,,\quad \mathbf{v}^G_{\dot{\psi}} = r\mathbf{n}_1 \tag{11.10.43}$$

and

$$\boldsymbol{\omega}_{\dot{\theta}} = \mathbf{n}_1\,,\quad \boldsymbol{\omega}_{\dot{\phi}} = \sin\theta\,\mathbf{n}_2 + \cos\theta\,\mathbf{n}_3\,,\quad \boldsymbol{\omega}_{\dot{\psi}} = \mathbf{n}_2 \tag{11.10.44}$$

The inertia force system on D may be represented by a single force $\mathbf{F}^*$ passing through G together with a couple with torque $\mathbf{T}^*$ where $\mathbf{F}^*$ and $\mathbf{T}^*$ are (see Section 8.13, Eqs. (8.13.7) to (8.13.12)):

$$\mathbf{F}^* = -m\mathbf{a}^G = -m\left(a_1\mathbf{n}_1 + a_2\mathbf{n}_2 + a_3\mathbf{n}_3\right) \tag{11.10.45}$$

and

$$\mathbf{T}^* = T_1\mathbf{n}_1 + T_2\mathbf{n}_2 + T_3\mathbf{n}_3 \tag{11.10.46}$$

where

$$\begin{aligned} T_1 &= -\alpha_1 \mathrm{I}_{11} + \omega_2\omega_3\left(\mathrm{I}_{22} - \mathrm{I}_{33}\right) \\ T_2 &= -\alpha_2 \mathrm{I}_{22} + \omega_3\omega_1\left(\mathrm{I}_{33} - \mathrm{I}_{11}\right) \\ T_3 &= -\alpha_3 \mathrm{I}_{33} + \omega_1\omega_2\left(\mathrm{I}_{11} - \mathrm{I}_{22}\right) \end{aligned} \tag{11.10.47}$$

where $\mathbf{a}^G$ is the acceleration of G relative to the fixed surface S (the inertia frame); where ω_i and α_i ($i = 1, 2, 3$) are the $\mathbf{n}_i$ components of $\boldsymbol{\omega}$ and the angular acceleration $\boldsymbol{\alpha}$ of D relative to S; and, finally, where:

$$\mathrm{I}_{11} = \mathrm{I}_{33} = mr^2/4 \,, \quad \mathrm{I}_{22} = mr^2/2 \tag{11.10.48}$$

From Eqs. (4.12.7) and (4.12.8), $\mathbf{a}^G$ and $\boldsymbol{\alpha}$ are:

$$\begin{aligned} \mathbf{a}^G &= r\left(\ddot{\psi} + \ddot{\phi}\sin\theta + 2\dot{\phi}\dot{\theta}\cos\theta\right)\mathbf{n}_1 + r\left(-\ddot{\theta} + \dot{\psi}\dot{\phi}\cos\theta + \dot{\phi}^2\sin\theta\cos\theta\right) \\ &\quad + r\left(-\dot{\psi}\dot{\phi}\sin\theta - \dot{\phi}^2\sin^2\theta - \dot{\theta}^2\right)\mathbf{n}_3 \end{aligned} \tag{11.10.49}$$

and

$$\begin{aligned} \boldsymbol{\alpha} &= \left(\ddot{\theta} - \dot{\psi}\dot{\phi}\cos\theta\right)\mathbf{n}_1 + \left(\ddot{\psi} + \ddot{\phi}\sin\theta + \dot{\theta}\dot{\phi}\cos\theta\right)\mathbf{n}_2 \\ &\quad + \left(\ddot{\phi}\cos\theta - \dot{\phi}\dot{\theta}\sin\theta + \dot{\psi}\dot{\theta}\right)\mathbf{n}_3 \end{aligned} \tag{11.10.50}$$

Hence, $\mathbf{F}^*$ and $\mathbf{T}^*$ may be written as:

$$\begin{aligned} \mathbf{F}^* &= -mr\Big[\left(\ddot{\psi} + \ddot{\phi}\sin\theta + 2\dot{\phi}\dot{\theta}\cos\theta\right)\mathbf{n}_1 + \Big(-\ddot{\theta} + \dot{\psi}\dot{\phi}\cos\theta \\ &\quad + \dot{\phi}^2\sin\theta\cos\theta\Big)\mathbf{n}_2 + \left(-\dot{\psi}\dot{\phi}\sin\theta - \dot{\phi}^2\sin^2\theta - \dot{\theta}^2\right)\mathbf{n}_3\Big] \end{aligned} \tag{11.10.51}$$

and

$$\begin{aligned} \mathbf{T}^* &= -\left(mr^2/4\right)\Big[\left(\ddot{\theta} - 2\dot{\psi}\dot{\phi}\cos\theta - \dot{\phi}^2\sin\theta\cos\theta\right)\mathbf{n}_1 \\ &\quad + \left(2\ddot{\psi} + 2\ddot{\phi}\sin\theta + 2\dot{\theta}\dot{\phi}\cos\theta\right)\mathbf{n}_2 \\ &\quad + \left(\ddot{\phi}\cos\theta + 2\dot{\psi}\dot{\theta}\right)\mathbf{n}_3\Big] \end{aligned} \tag{11.10.52}$$

Finally, by using Eq. (11.9.6), the generalized inertia forces become:

$$\begin{aligned} F_\theta^* &= \mathbf{v}_{\dot{\theta}}^G \cdot \mathbf{F}^* + \boldsymbol{\omega}_{\dot{\theta}} \cdot \mathbf{T}^* \\ &= mr^2\left(-\ddot{\theta} + \dot{\psi}\dot{\theta}\cos\theta + \dot{\phi}^2\sin\theta\cos\theta\right) \\ &\quad -\left(mr^2/4\right)\left(\ddot{\theta} - 2\dot{\psi}\dot{\phi}\cos\theta - \dot{\phi}^2\sin\theta\cos\theta\right) \\ &= mr^2\left[-(5/4)\,\ddot{\theta} + (3/2)\dot{\psi}\dot{\phi}\cos\theta + (5/4)\dot{\phi}^2\sin\theta\cos\theta\right] \end{aligned} \tag{11.10.53}$$

$$
\begin{aligned}
F_\phi^* &= \mathbf{v}_{\dot{\phi}}^G \cdot \mathbf{F}^* + \boldsymbol{\omega}_{\dot{\phi}} \cdot \mathbf{T}^* \\
&= -mr^2 \sin\theta\left(\ddot{\psi} + \ddot{\phi}\sin\theta + 2\dot{\phi}\dot{\theta}\cos\theta\right) \\
&\quad -\left(mr^2/4\right) \sin\theta\left(2\ddot{\psi} + 2\ddot{\phi}\sin\theta + 2\dot{\theta}\dot{\phi}\cos\theta\right) \\
&\quad -\left(mr^2/4\right)\cos\theta\left(\ddot{\phi}\cos\theta + 2\dot{\psi}\dot{\theta}\right) \\
&= mr^2\left[-(3/2)\ddot{\psi}\sin\theta - (3/2)\ddot{\phi}\sin^2\theta - (5/2)\dot{\phi}\dot{\theta}\sin\theta\cos\theta\right. \\
&\quad \left. -(1/4)\ddot{\phi}\cos^2\theta + (1/2)\dot{\psi}\dot{\theta}\cos\theta\right]
\end{aligned}
\tag{11.10.54}
$$

and

$$
\begin{aligned}
F_\psi^* &= -mr^2\left(\ddot{\psi} + \ddot{\phi}\sin\theta + 2\dot{\phi}\dot{\theta}\cos\theta\right) \\
&\quad -\left(mr^2/4\right)\left(2\ddot{\psi} + 2\ddot{\phi}\sin\theta + 2\dot{\theta}\dot{\phi}\cos\theta\right) \\
&= -mr^2\left[(3/2)\ddot{\psi} + (3/2)\ddot{\phi}\sin\theta + (5/2)\dot{\theta}\dot{\phi}\cos\theta\right]
\end{aligned}
\tag{11.10.55}
$$

As an aside, we can readily develop the generalized applied (or active) forces for this system. Indeed, the only applied forces are gravity and contact forces, and of these only the gravity (or weight) forces contribute to the generalized forces. (The contact forces do not contribute because they are applied at a point of zero velocity.)

The weight forces may be represented by a single vertical force **W** passing through G given by:

$$
\mathbf{W} = -mg\,\mathbf{N}_3 = -mg\left(\sin\theta\,\mathbf{n}_2 + \cos\theta\,\mathbf{n}_3\right) \tag{11.10.56}
$$

Then, from Eq. (11.10.43), the generalized forces are:

$$
F_\theta = mg\sin\theta\,, \quad F_\phi = 0, \quad F_x = 0 \tag{11.10.57}
$$

11.11 Potential Energy

In elementary mechanics, potential energy is often defined as the "ability to do work." While this is an intuitively satisfying concept it requires further development to be computationally useful. To this end, we will define potential energy to be a scalar function of the generalized coordinates which when differentiated with respect to one of the coordinates produces the negative of the generalized force for that coordinate. Specifically, we define potential energy $P(q_r)$ as the function such that:

$$
F_r \stackrel{D}{=} -\partial \mathbf{P}/\partial q_r \quad (r = 1,\ldots,n) \tag{11.11.1}
$$

where as before, n is the number of degrees of freedom of the system. (The minus sign is chosen so that P is positive in the usual physical applications.)

To illustrate the consistency of this definition with the intuitive concept, consider a particle Q having a mass m in a gravitational field. Let Q be at an elevation h above a fixed level surface S as in Figure 11.11.1. Then, if Q is released from rest in this position, the work w done by gravity as Q falls to S is:

$$w = mgh \tag{11.11.2}$$

From a different perspective, if h is viewed as a generalized coordinate, the velocity $\mathbf{v}$ of Q and, consequently, the partial velocity $\mathbf{v}_h$ of Q (relative to h) are:

$$\mathbf{v} = \dot{h}\mathbf{k} \quad \text{and} \quad v_h = \mathbf{k} \tag{11.11.3}$$

where $\mathbf{k}$ is the vertical unit vector as in Figure 11.11.1. The generalized force due to gravity is then:

$$F_h = -mg\mathbf{k} \cdot \mathbf{v}_h = -mg \tag{11.11.4}$$

Let P be a potential energy defined as:

$$\mathbf{P} = w = mgh \tag{11.11.5}$$

Then, from Eq. (11.11.1), F_h is:

$$F_h = -\partial \mathbf{P}/\partial h = -mg \tag{11.11.6}$$

which is consistent with the results of Eq. (11.11.4).

As a second illustration, consider a linear spring as in Figure 11.11.2. Let the spring have modulus k and natural length ℓ. Let the spring be supported at one end, O, and let its other end, Q, be subjected to a force with magnitude F producing a displacement x of Q, as depicted in Figure 11.11.2. The movement of Q has one degree of freedom represented by the parameter x. The velocity $\mathbf{v}$ and partial velocity $\mathbf{v}_{\dot{x}}$ of Q are then:

$$\mathbf{v} = \dot{x}\mathbf{n} \quad \text{and} \quad \mathbf{v}_{\dot{x}} = \mathbf{n} \tag{11.11.7}$$

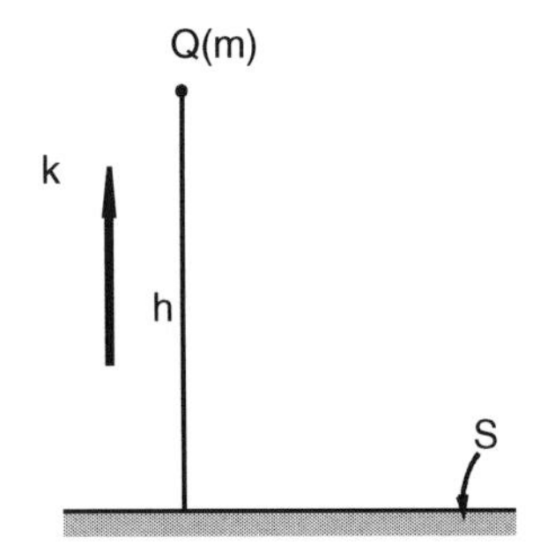

FIGURE 11.11.1
A particle Q above a level surface S.

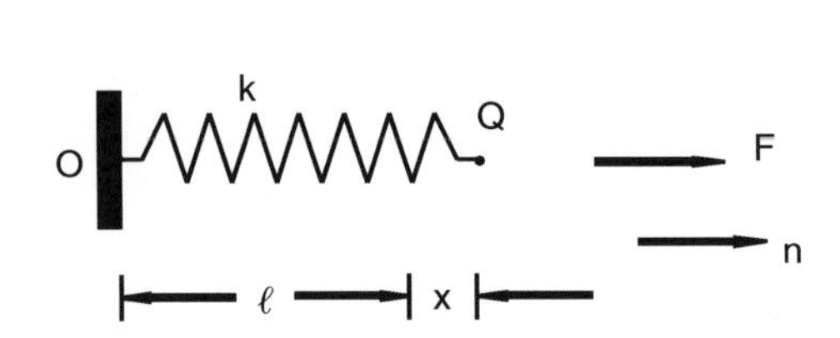

FIGURE 11.11.2
A force applied to a linear spring.

The force S exerted on Q by the spring is:

$$\mathbf{S} = -F\mathbf{n} = -kx\,\mathbf{n} \tag{11.11.8}$$

The generalized force F_x relative to x is then:

$$F_x = \mathbf{S} \cdot \mathbf{v}_{\dot{x}} = -kx \tag{11.11.9}$$

Let a potential energy function P be defined as:

$$\mathbf{P} = (1/2)kx^2 \tag{11.11.10}$$

Then, from Eq. (11.11.1), we have:

$$-\partial\mathbf{P}/\partial x = -kx = F_x \tag{11.11.11}$$

which is consistent with Eq. (11.11.9).

It happens that Eqs. (11.11.5) and (11.11.10) are potential energy functions for gravity and spring forces in general. Consider first Eq. (11.11.5) for gravity forces. Suppose Q is a particle with mass m. Let Q be a part of a mechanical system S having n degrees of freedom represented by the coordinates q_r ($r = 1,\ldots, n$). Recall from Eq. (11.6.5) that the contribution $\hat{F}_r$ of the weight force on Q to the generalized force F_r for the coordinate q_r is:

$$\hat{F}_r = -mg\,\partial h/\partial q_r \tag{11.11.12}$$

where h is the elevation of Q above a reference level as in Figure 11.11.3.

From Eq. (11.11.2), if a potential energy function P is given by mgh, Eq. (11.11.1) gives the contribution $\hat{F}_r$ to the generalized force of q_r for the weight force as:

$$\hat{F}_r = -\partial\mathbf{P}/\partial q_r = -mg\,\partial h/\partial q_r \tag{11.11.13}$$

which is consistent with Eq. (11.11.12).

Consider next Eq. (11.11.10) for spring forces. Suppose Q is a point at the end of a spring which is part of a mechanical system S as depicted in Figure 11.11.4. Let S have n degrees

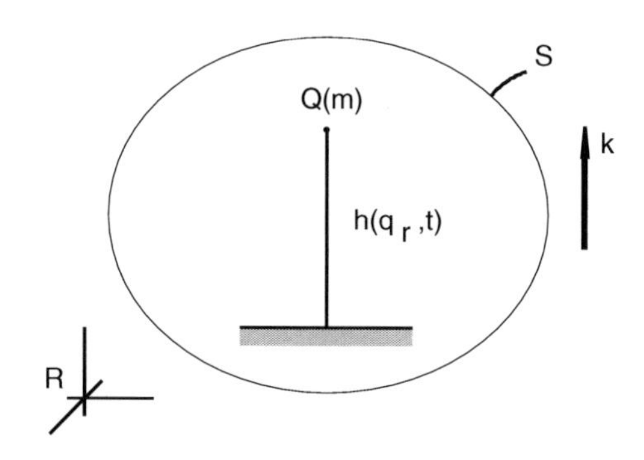

FIGURE 11.11.3
Elevation of a particle Q of a mechanical system S.

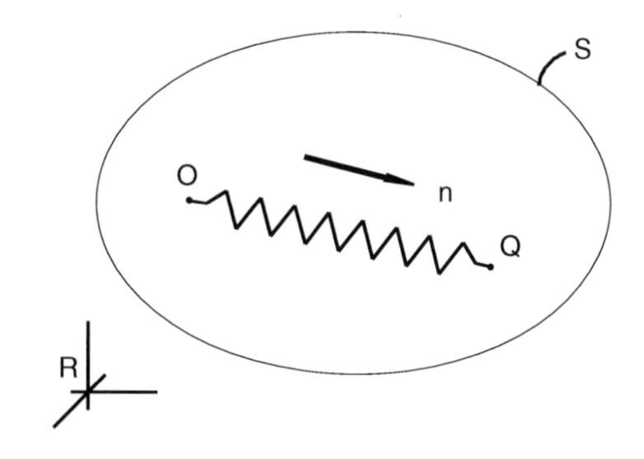

FIGURE 11.11.4
A spring within a mechanical system S.

of freedom represented by the coordinates q_r ($r = 1,\dots, n$). Then, from Eq. (11.6.7), we recall that the contribution $\hat{F}_r$ of the spring force to the generalized force on S, for the coordinate q_r, is:

$$\hat{F}_r = -f(x)\frac{\partial x}{\partial q_r} \tag{11.11.14}$$

where $f(x)$ is the magnitude of the spring force due to a spring extension, or compression, x. For a linear spring $f(x)$ is simply kx; hence, for a linear spring, $\hat{F}_r$ is:

$$\hat{F}_r = -kx\frac{\partial x}{\partial q_r} \tag{11.11.15}$$

From Eq. (11.11.10), if we let the potential energy function P be $(1/2)kx^2$, we obtain $\hat{F}_r$ from Eq. (11.11.1) as:

$$\hat{F}_r = -\partial \mathbf{P}/\partial q_r = -\frac{\partial}{\partial q_r}\left[(1/2)kx^2\right] = -kx\frac{\partial x}{\partial q_r} \tag{11.11.16}$$

which is consistent with Eq. (11.11.15).

Alternatively, for a nonlinear spring, we may let the potential energy function have the form:

$$\mathbf{P} = \int_0^x f(\xi)\,d\xi \tag{11.11.17}$$

Then, from Eq. (11.11.1), $\hat{F}_r$ becomes

$$\hat{\mathbf{F}}_r = -\partial \mathbf{P}/\partial q_r = \frac{\partial}{\partial q_r}\left[\int_0^x f(\xi)\,d\xi\right] = -f(x)\frac{\partial x}{\partial q_r} \tag{11.11.18}$$

which is consistent with Eq. (11.11.14).

The definition of Eq. (11.11.1) and these simple examples show that if a potential energy function is known we can readily obtain the generalized applied (or active) forces. Indeed, the examples demonstrate the utility of potential energy for finding generalized forces. Moreover, for gravity and spring forces, Eqs. (11.11.5), (11.11.10), and (11.11.17) provide expressions for potential energy functions. This, however, raises a question about other forces; that is, what are the potential energy functions for forces other than gravitational or spring forces? The answer is found by considering again Eq. (11.11.1): If

$$\frac{\partial x}{\partial q_r} = -\mathbf{F}_r \quad \text{then} \quad \mathbf{P} = -\int F_r\,dq_r \tag{11.11.19}$$

Unfortunately, it is not always possible to perform the integration indicated in Eq. (11.11.19). Indeed, the integral represents an anti-partial differentiation with respect to

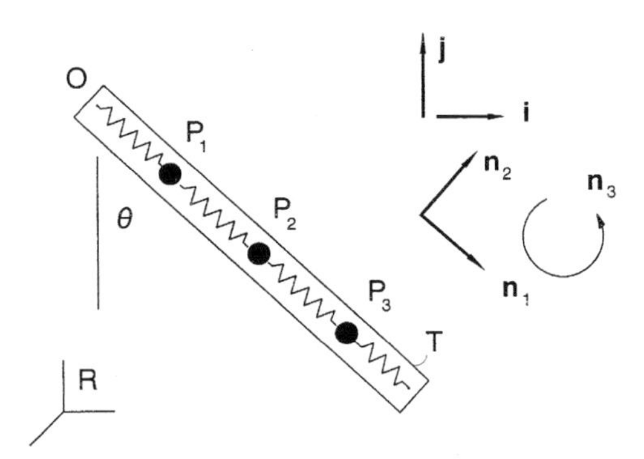

FIGURE 11.11.5
A rotating tube containing spring-connected particles.

each of the q_r. Also, if the F_r involve derivatives of the $\dot{q}_r$ (such as with "dissipation forces"), the integrals will not exist, because P is to depend only on the q_r.

Further, observe that if knowledge of the generalized forces is needed to obtain a potential energy function which in turn is to be used to obtain the generalized forces, little progress has been made. Therefore, in the solution of practical problems as in machine dynamics, potential energy is useful primarily for gravity and spring forces. Finally, it should be noted that potential energy is not a unique function. Indeed, from Eq. (11.11.1), we see that the addition of a constant to any valid potential energy function P also produces a valid potential energy function.

It may be helpful to consider another illustration. Consider again the system of the rotating tube containing three spring-supported particles as in Figures 11.11.5 and 11.11.6. We discussed this system in Section 11.7, and we will use the same notation here without repeating the description (see Section 11.7 for the details).

Recall from Eqs. (11.7.14) to (11.7.17) that the generalized forces for the coordinates x_1, x_2, x_3, and θ are:

$$\mathbf{F}_{x_1} = mg\cos\theta + kx_2 - 2kx_1 \tag{11.11.20}$$

$$\mathbf{F}_{x_2} = mg\cos\theta + kx_3 + kx_1 - 2kx_2 \tag{11.11.21}$$

$$\mathbf{F}_{x_3} = mg\cos\theta - 2kx_3 + kx_2 \tag{11.11.22}$$

$$\mathbf{F}_{\theta} = -Mg(L/2)\sin\theta - mg(6\ell + x_1 + x_2 + x_3)\sin\theta \tag{11.11.23}$$

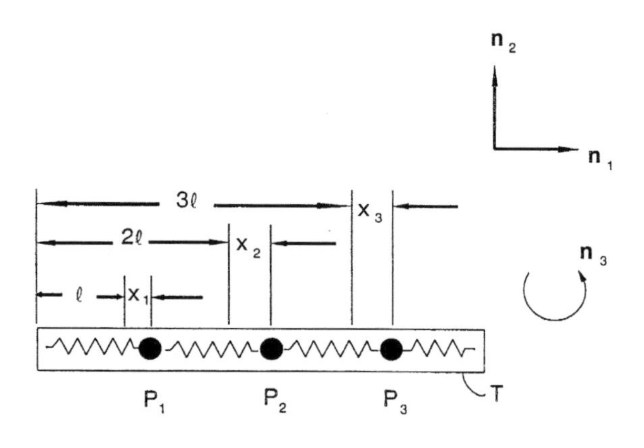

FIGURE 11.11.6
Coordinates of the particles within the tube.

In the development of these generalized forces we discovered that the only forces contributing to these forces were gravity and spring forces. The contact forces between the smooth tube and the particles did not contribute to the generalized forces; hence, we should be able to form a potential energy function for the system. To this end, using Eqs. (11.11.5) and (11.11.10) we obtain the potential energy function:

$$\begin{aligned}\mathbf{P} &= -mg(\ell + x_1)\cos\theta - mg(2\ell + x_2)\cos\theta - mg(3\ell + x_3)\cos\theta \\ &\quad - Mg(L/2)\cos\theta + (1/2)kx_1^2 + (1/2)\,k(x_2 - x_1)^2 + (1/2)k(x_3 - x_2)^2\end{aligned} \tag{11.11.24}$$

where the reference level for the gravitational forces is taken through the support pin O. From Eq. (11.11.1) we can immediately determine the generalized forces. That is,

$$\mathbf{F}_{x_1} = -\partial\mathbf{P}/\partial x_1,\ \mathbf{F}_{x_2} = -\partial\mathbf{P}/\partial x_2,\ F_{x_3} = -\partial\mathbf{P}/\partial x_3,\ \mathbf{F}_\theta = -\partial\mathbf{P}/\partial\theta \tag{11.11.25}$$

By substituting from Eq. (11.11.24) into (11.11.25), we obtain results identical to those of Eqs. (11.11.20) to (11.11.23).

11.12 Use of Kinetic Energy to Obtain Generalized Inertia Forces

Just as potential energy can be used to obtain generalized applied (active) forces, kinetic energy can be used to obtain generalized inertia (passive) forces. In each case, the forces are obtained through differentiation of the energy functions. In this section, we will establish the procedures for using kinetic energy to obtain the generalized inertia forces.

To begin the analysis, recall Eq. (11.4.7) concerning the projection of the acceleration of a particle on the partial velocity vectors:

$$\mathbf{a}\cdot\mathbf{v}_{\dot{q}_r} = \frac{1}{2}\frac{d}{dt}\left(\frac{\partial\mathbf{v}^2}{\partial\dot{q}_r}\right) - \frac{1}{2}\left(\frac{\partial\mathbf{v}^2}{\partial q_r}\right) \tag{11.12.1}$$

This expression, which is valid only for holonomic systems (see Section 11.3), provides a means for relating the generalized inertia forces and kinetic energy. To this end, let P be a particle with mass m of a holonomic mechanical system S having n degrees of freedom represented by the coordinates q_r ($r = 1,\ldots, n$). Then, the inertia force $\mathbf{F}^*$ on P is:

$$\mathbf{F}^* = -m\mathbf{a} \tag{11.12.2}$$

where $\mathbf{a}$ is the acceleration of P in an inertial reference frame R. Recall further that the kinetic energy K of P is:

$$K = \frac{1}{2}\mathbf{v}^2 \tag{11.12.3}$$

By multiplying Eq. (11.12.1) by the mass m of P, we have:

$$m\mathbf{a}\cdot\mathbf{v}_{\dot{q}_r} = \frac{d}{dt}\left[\frac{\partial}{\partial \dot{q}_r}\left(\frac{1}{2}m\mathbf{v}^2\right)\right] - \frac{\partial}{\partial q_r}\left(\frac{1}{2}\mathbf{v}^2\right) \tag{11.12.4}$$

or by using Eqs. (11.12.2) and (11.12.3) we have:

$$\mathbf{F}^*_{q_r} = -\frac{d}{dt}\left(\frac{\partial K}{\partial \dot{q}_r}\right) + \frac{\partial K}{\partial q_r} \tag{11.12.5}$$

Consider next a set of particles P_i ($i = 1,\ldots, N$) as parts of a mechanical system S having n degrees of freedom. Then, by superposing (or adding together) equations as Eq. (11.12.5) for each of the particles, we obtain an expression identical in form to Eq. (11.12.5) and valid for the set of particles. Finally, if the set of particles is a rigid body, Eq. (11.12.5) also holds.

To illustrate the use of Eq. (11.12.5), consider again the elementary examples of the foregoing section.

Example 11.12.1: Simple Pendulum

Consider first the simple pendulum as in Figure 11.12.1, where we are using the same notation as before. Recall that this system has one degree of freedom represented by the angle and that the velocity and partial velocity of the bob P are (see Eqs. (11.10.1) and (11.10.4)):

$$\mathbf{v}^P = \ell\dot{\theta}\mathbf{n}_\theta \quad \text{and} \quad \mathbf{v}^P_{\dot{\theta}} = \ell\mathbf{n}_\theta \tag{11.12.6}$$

The kinetic energy of P is then:

$$K = \frac{1}{2}m\left(\mathbf{v}^P\right)^2 = (1/2)m\ell^2\dot{\theta}^2 \tag{11.12.7}$$

Then, by using Eq. (11.12.5), the generalized inertia force $\mathbf{F}^*_\theta$ is:

$$\mathbf{F}^*_\theta = -\frac{d}{dt}\left[\frac{\partial}{\partial \theta}\left(\frac{1}{2}m\ell^2\dot{\theta}^2\right)\right] + \frac{\partial}{\partial \theta}\left(\frac{1}{2}m\ell^2\theta^2\right) = -m\ell^2\,\ddot{\theta} \tag{11.12.8}$$

This result is identical with that of Eq. (11.10.5).

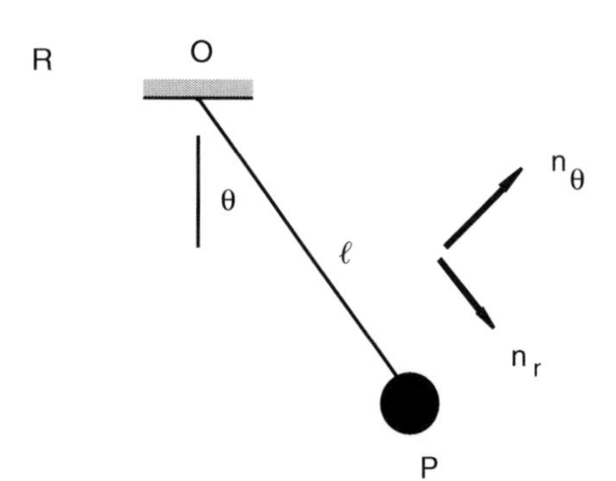

FIGURE 11.12.1
The simple pendulum.

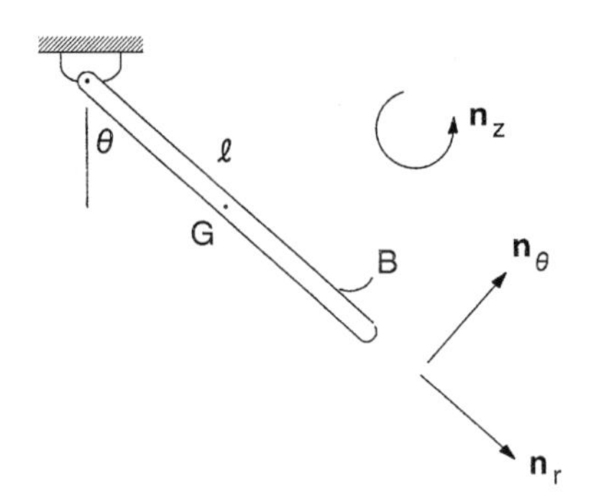

FIGURE 11.12.2
The rod pendulum.

Example 11.12.2: Rod Pendulum

Consider next the rod pendulum of Figure 11.12.2, using the same notation as before. Recall that the mass center velocity v_G and the angular velocity of the rod are (see Eqs. (11.10.6) and (11.10.7)):

$$\mathbf{v}_G = (\ell/2)\dot{\theta}\mathbf{n}_\theta \quad \text{and} \quad \boldsymbol{\omega} = \dot{\theta}\mathbf{n}_z \tag{11.12.9}$$

Hence, the kinetic energy K of the rod is:

$$\begin{aligned} K &= (1/2)m\mathbf{v}_G^2 + (1/2)\mathbf{I}_G\boldsymbol{\omega}^2 \\ &= (1/2)m(\ell/2)^2\dot{\theta}^2 + (1/2)(1/12)m\ell^2\dot{\theta}^2 \\ &= (1/6)m\ell^2\dot{\theta}^2 \end{aligned} \tag{11.12.10}$$

Then, from Eq. (11.12.5), the generalized inertia force $\mathbf{F}_\theta^*$ is:

$$\mathbf{F}_\theta^* = -(1/3)m\ell^2\ddot{\theta} \tag{11.12.11}$$

This result is the same as in Eq. (11.10.11).

Example 11.12.3: Double-Rod Pendulum

To extend this last example, consider again the double-rod pendulum of Figure 11.12.3. Unlike the two above examples, this system has two degrees of freedom, as represented by the angles θ_1 and θ_2.

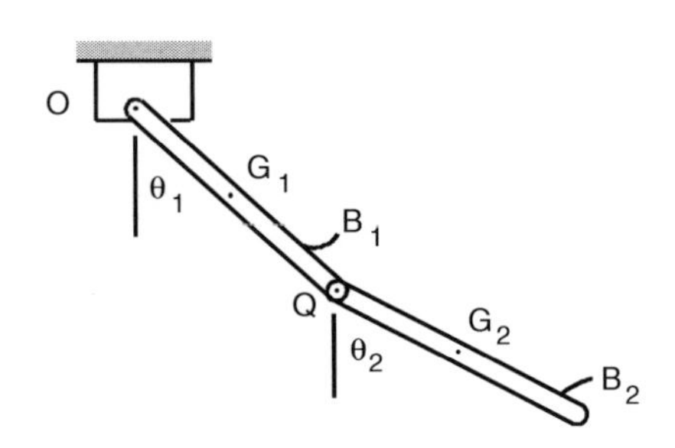

FIGURE 11.12.3
A double-rod pendulum.

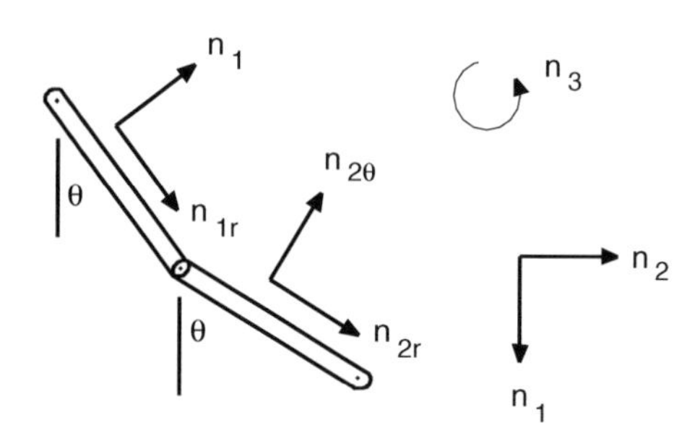

FIGURE 11.12.4
Unit vector geometry for the double-rod pendulum.

In our earlier analysis of this system we found that the mass center velocities and the angular velocities of the rods were (see Eqs. (11.10.12)):

$$\mathbf{v}^{G_1} = (\ell/2)\dot{\theta}_1 \mathbf{n}_{1\theta} \quad , \quad \mathbf{v}^{G_2} = \ell\dot{\theta}_1 \mathbf{n}_{1\theta} + (\ell/2)\dot{\theta}_2 \mathbf{n}_{2\theta}$$
$$\boldsymbol{\omega}^{B_1} = \dot{\theta}_1 \mathbf{n}_3 \quad , \quad \boldsymbol{\omega}^{B_2} = \dot{\theta}_2 \mathbf{n}_3 \tag{11.12.12}$$

where, as before, the unit vectors are as shown in Figure 11.12.4.

The kinetic energy of the system is then:

$$\begin{aligned}
\mathbf{K} &= (1/2)m(\mathbf{v}^{G_1})^2 + (1/2)\mathbf{I}_{G_1}(\boldsymbol{\omega}^{B_1})^2 \\
&\quad + (1/2)m(\mathbf{v}^{G_2})^2 + (1/2)\mathbf{I}_{G_2}(\boldsymbol{\omega}^{B_2})^2 \\
&= (1/2)m(\ell^2/4)\dot{\theta}_1^2 + (1/2)(1/12)m\ell^2\dot{\theta}\ell_1^2 \\
&\quad + (1/2)m\left[\ell^2\dot{\theta}_1^2 + \ell^2\dot{\theta}_1\dot{\theta}_2\cos(\theta_2 - \theta_1) + (\ell^2/4)\dot{\theta}_2^2\right. \\
&\quad + (1/2)m(1/12)m\ell^2\dot{\theta}_2^2 \\
&= (2/3)m\ell^2\dot{\theta}_1^2 + (1/2)m\ell^2\dot{\theta}_1\dot{\theta}_2\cos(\theta_2 - \theta_1) \\
&\quad + (1/6)m\ell^2\dot{\theta}_2^2
\end{aligned} \tag{11.12.13}$$

Then, from Eq. (11.12.5), the generalized inertia forces $\mathbf{F}_{\theta_1}^*$ and $\mathbf{F}_{\theta_2}^*$ are:

$$\mathbf{F}_{\theta_1}^* = -(4/3)m\ell^2\ddot{\theta}_1 - (1/2)m\ell^2\ddot{\theta}_2\cos(\theta_2 - \theta_1) + (1/2)m\ell^2\dot{\theta}_2^2\sin(\theta_2 - \theta_1) \tag{11.12.14}$$

and

$$\mathbf{F}_{\theta_2}^* = -(1/3)m\ell^2\ddot{\theta}_2 - (1/2)m\ell^2\ddot{\theta}_1\cos(\theta_2 - \theta_1) - (1/2)m\ell^2\dot{\theta}_1^2\sin(\theta_2 - \theta_1) \tag{11.12.15}$$

These results are identical with the Eqs. (11.10.21) and (11.10.22).

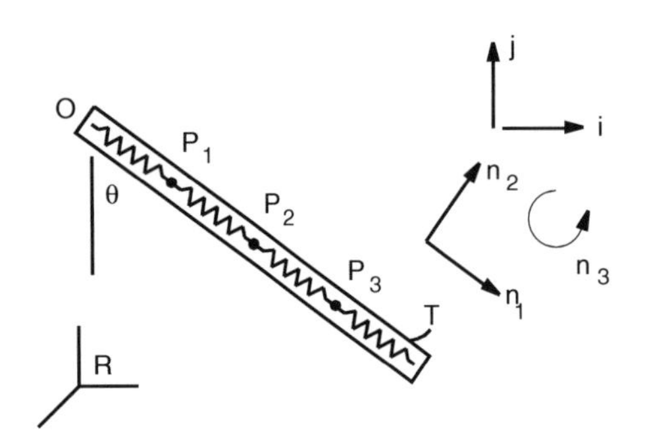

FIGURE 11.12.5
Rotating tube containing spring-supported particles.

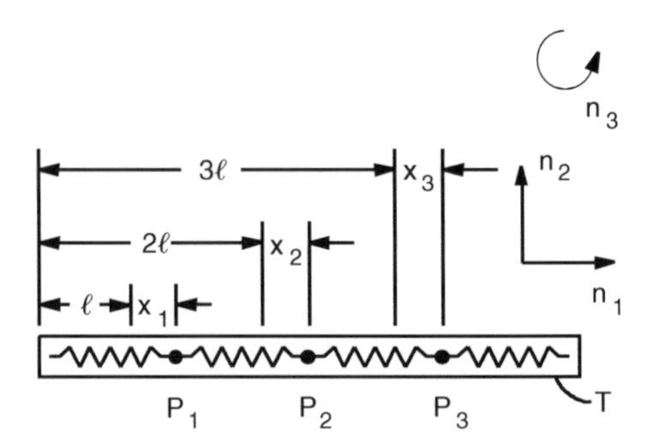

FIGURE 11.12.6
Coordinates of particles in the tube.

Example 11.12.4: Spring-Supported Particles in a Rotating Tube

For another example illustrating the use of kinetic energy to obtain generalized inertia forces, consider again the system of spring-supported particles in a rotating tube as in Figure 11.12.5. (We considered this system in Sections 11.7 and 11.11.) Recall that this system has four degrees of freedom represented by the coordinates x_1, x_2, x_3, and θ, as shown in Figures 11.12.5 and 11.12.6.

The velocities of the particles, the tube mass center velocity, and the angular velocity of the tube in the inertia frame R are (see Eqs. (11.10.23) through (11.10.27)):

$$\begin{aligned}
\mathbf{v}^{P_1} &= \dot{x}_1\mathbf{n}_1 + (\ell + x_1)\dot{\theta}\mathbf{n}_2 \\
\mathbf{v}^{P_2} &= \dot{x}_2\mathbf{n}_1 + (2\ell + x_2)\dot{\theta}\mathbf{n}_2 \\
\mathbf{v}^{P_3} &= \dot{x}_3\mathbf{n}_1 + (3\ell + x_3)\dot{\theta}\mathbf{n}_3 \\
\mathbf{v}^{G} &= (L/2)\dot{\theta}\mathbf{n}_2 \\
\boldsymbol{\omega} &= \dot{\theta}\mathbf{n}_3
\end{aligned} \tag{11.12.16}$$

where the notation is the same as we used in Section 11.7 and 11.11.

The kinetic energy of the system is then:

$$\begin{aligned}
\mathbf{K} &= \frac{1}{2}m\left(\mathbf{v}^{P_1}\right)^2 + \frac{1}{2}m\left(\mathbf{v}^{P_2}\right)^2 + \frac{1}{2}m\left(\mathbf{v}^{P_3}\right)^2 \\
&\quad + \frac{1}{2}M\left(\mathbf{v}^{G}\right)^2 + \frac{1}{2}\mathbf{I}(\boldsymbol{\omega})^2 \\
&= (1/2)m\left[\dot{x}_1^2 + (\ell + x_1)^2\dot{\theta}^2 + \dot{x}_2^2 + (2\ell + x_2)^2\dot{\theta}^2 + \dot{x}_3^2\right. \\
&\quad \left. + (3\ell + x_3)^2\dot{\theta}^2\right] + (1/2)M(L/2)^2\dot{\theta}^2 + (1/2)(1/12)ML^2\dot{\theta}^2
\end{aligned} \tag{11.12.17}$$

Using Eq. (11.12.5), the generalized inertia forces are then:

$$\mathbf{F}^*_{x_1} = -m\ddot{x}_1 + m(\ell + x_1)\dot{\theta}^2 \tag{11.12.18}$$

$$\mathbf{F}^*_{x_2} = -m\ddot{x}_2 + m(2\ell + x_2)\dot{\theta}^2 \tag{11.12.19}$$

$$\mathbf{F}^*_{x_3} = -m\ddot{x}_3 + m(3\ell + x_3)\dot{\theta}^2 \tag{11.12.20}$$

$$\begin{aligned}\mathbf{F}^*_{\theta} = &-m\left[(\ell + x_1)^2 + (2\ell + x_2)^2 + (3\ell + x_3)^2\right]\ddot{\theta} \\ &-2m\left[(\ell + x_1)\dot{x}_1 + (2\ell + x_2)\dot{x}_2 + (3\ell + x_3)\dot{x}_3\right]\dot{\theta} \\ &-M(L^2/3)\,\ddot{\theta}\end{aligned} \tag{11.12.21}$$

These results are the same as those of Eqs. (11.10.30) to (11.10.33).

Example 11.12.5: Rolling Circular Disk

For an example where Eq. (11.12.5) cannot be used to determine generalized inertia forces, consider again the nonholonomic system consisting of the rolling circular disk on the flat horizontal surface as shown in Figure 11.12.7. (Recall that Eq. (11.12.5) was developed for holonomic systems [that is, systems with integrable or nonkinematic constraint equations] and, as such, it is not applicable for nonholonomic systems [that is, systems with kinematic or nonintegrable constraint equations].)

Due to the rolling constraint, the disk has three degrees of freedom instead of six, as would be the case if the disk were unrestrained. Recall from Eqs. (11.10.34), (11.10.35), and (11.10.36) that the condition of rolling (zero contact point velocity) produces the constraint equations:

$$\dot{x} = r\left[(\dot{\psi} + \dot{\phi}\sin\theta)\cos\phi + \dot{\theta}\cos\theta\sin\phi\right] \tag{11.12.22}$$

$$\dot{y} = r\left[(\dot{\psi} + \dot{\phi}\ \sin\theta)\sin\phi - \dot{\theta}\cos\theta\cos\phi\right] \tag{11.12.23}$$

$$\dot{z} = -r\dot{\theta}\sin\theta \tag{11.12.24}$$

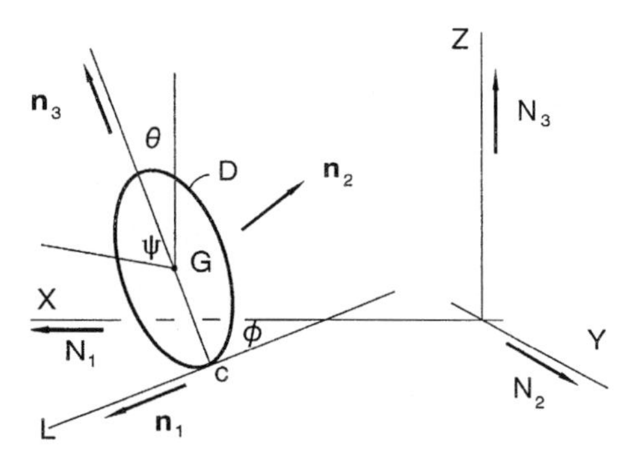

FIGURE 11.12.7
Circular disk rolling on a horizontal surface *S*.

The first two of these equations are nonintegrable in terms of elementary functions; therefore, the system is nonholonomic.

With three constraint equations, the disk has three degrees of freedom. These may conveniently be represented by the angles θ, ϕ, and ψ, and in terms of these angles the mass center velocity and the disk angular velocity are (see Eqs. (11.10.41) and (11.10.42)):

$$\mathbf{v}_G = r\left(\dot{\psi} + \dot{\phi}\sin\theta\right)\mathbf{n}_1 - r\dot{\theta}\mathbf{n}_2 \tag{11.12.25}$$

and

$$\boldsymbol{\omega} = \dot{\theta}\mathbf{n}_1 + \left(\dot{\psi} + \dot{\phi}\sin\theta\right)\mathbf{n}_2 + \dot{\phi}\,\cos\theta\,\mathbf{n}_3 \tag{11.12.26}$$

Hence, the kinetic energy of the disk is:

$$\begin{aligned} K &= (1/2)m\left(\mathbf{v}_G\right)^2 + \frac{1}{2}\boldsymbol{\omega}\cdot\mathbf{I}\cdot\boldsymbol{\omega} \\ &= \frac{1}{2}m\left[r^2\left(\dot{\psi} + \dot{\phi}\sin\theta\right)^2 + r^2\dot{\theta}^2\right] \\ &+ \frac{1}{2}\left[\mathbf{I}_{11}\dot{\theta}^2 + \mathbf{I}_{22}\left(\dot{\psi} + \dot{\phi}\sin\theta\right)^2 + \mathbf{I}_{33}\left(\dot{\phi}\cos\theta\right)^2\right] \end{aligned} \tag{11.12.27}$$

where the moments of inertia I_{11}, I_{22}, and I_{33} are:

$$\mathbf{I}_{11} = \mathbf{I}_{33} = mr^2/4\,, \quad \mathbf{I}_{22} = mr^2/2 \tag{11.12.28}$$

Assuming (erroneously) that Eq. (11.12.5) can be used to determine the generalized inertia forces, we have:

$$\begin{aligned} \mathbf{F}_\theta^* &= -\frac{d}{dt}\left(\frac{\partial K}{\partial \dot{\theta}}\right) + \frac{\partial K}{\partial \theta} \\ &= -mr^2\left[(5/4)\,\ddot{\theta} - (3/2)\,\dot{\psi}\,\dot{\phi}\cos\theta - (5/4)\dot{\phi}^2\sin\theta\cos\theta\right] \end{aligned} \tag{11.12.29}$$

$$\begin{aligned} \mathbf{F}_\theta^* = -mr^2&\left[(3/2)\,\ddot{\psi}\sin\theta + (3/2)\ddot{\phi}\sin^2\theta + (1/4)\ddot{\phi}\cos^2\theta\right. \\ &\left.+ (3/2)\,\dot{\psi}\dot{\theta}\,\cos\theta + (11/4)\dot{\theta}\dot{\phi}\sin\theta\cos\theta\right] \end{aligned} \tag{11.12.30}$$

$$\mathbf{F}_\psi^* = -mr^2\left[(3/2)\left(\ddot{\psi} + \ddot{\phi}\sin\theta + \dot{\phi}\dot{\theta}\cos\theta\right)\right. \tag{11.12.31}$$

In Section 11.10, in Eqs. (11.10.53), (11.10.54), and (11.10.55), we found $\mathbf{F}_\theta^*$, $\mathbf{F}_\phi^*$, and $\mathbf{F}_\psi^*$ to be:

$$\mathbf{F}_\theta^* = -mr^2\left[(5/4)\ddot{\theta} - (3/2)\,\dot{\psi}\,\dot{\phi}\,\cos\theta - (5/4)\dot{\phi}^2\sin\theta\cos\theta\right] \tag{11.12.32}$$

$$\mathbf{F}_{\phi}^{*} = -mr^{2}\Big[(3/2)\ddot{\psi}\sin\theta + (3/2)\ddot{\phi}\sin^{2}\theta + (5/2)\dot{\phi}\dot{\theta}\sin\theta\cos\theta$$
$$+(1/4)\ddot{\phi}\cos\theta - (1/2)\dot{\psi}\dot{\theta}\cos\theta\Big] \tag{11.12.33}$$

and

$$\mathbf{F}_{\psi}^{*} = -mr^{2}\Big[(3/2)\ddot{\psi} + (3/2)\ddot{\phi}\sin\theta + (5/2)\,\dot{\theta}\dot{\phi}\cos\theta\Big] \tag{11.12.34}$$

While Eqs. (11.12.29), (11.12.30), and (11.12.31) are similar to Eqs. (11.12.32), (11.12.33), and (11.12.34), they are not identical; therefore, Eq. (11.12.5) is not valid for nonholonomic systems. It happens, however, that Eq. (11.12.5) may be modified and expanded to also accommodate nonholonomic systems. Although the details of this expansion are beyond the scope of this text, the interested reader is referred to References 11.2 and 11.4.

11.13 Closure

Our objective in this chapter has been to introduce the principles and procedures of generalized dynamics. Our intention was to obtain a working knowledge of the elementary procedures. More advanced procedures, such as those applicable with large multibody systems and, to some extent, those concerned with nonholonomically constrained systems, have not been discussed because they are beyond our scope at this time. The interested reader may want to refer to References 11.2 and 11.4 and later chapters for a discussion of these topics.

The examples are intended to demonstrate the principal advantages of the generalized procedures over the procedures used in elementary mechanics, including: (1) non-working constraint forces do not contribute to the generalized forces so these forces may be simply ignored in the analysis; and (2) for holonomic systems (which include the vast majority of systems of interest in machine dynamics), generalized inertia forces may be computed from kinetic energy functions. This in turn means that vector acceleration need not be computed, thus saving considerable analysis effort.

In addition to these advantages, if a system possesses a potential energy function, the generalized active (or applied) forces may be obtained by a single derivative of the potential energy function. Moreover, for gravity and spring forces (which are prevalent in machine dynamics), the potential energy functions may often be directly obtained from Eqs. (11.11.5) and (11.11.10). In the next chapter, we will consider the application of these procedures in obtaining equations of motion.

References

11.1. Kane, T. R., *Dynamics*, Holt, Rinehart & Winston, New York, 1968, p. 78.
11.2. Kane, T. R., and Levinson, D. A., *Dynamics: Theory and Applications*, McGraw-Hill, New York, 1985, p. 100.

11.3. Kane, T. R., *Analytical Elements of Mechanics*, Vol. 1, Academic Press, New York, 1959, p. 128.
11.4. Huston, R. L., and Passerello, C. E., Another look at nonholonomic systems, *J. Appl. Mech.*, 40, 101–104, 1973.
11.5. Huston, R. L., *Multibody Dynamics*, Butterworth-Heinemann, Stoneham, MA, 1990.
11.6 Huston, R. L., and Liu, C. Q., *Formulas for Dynamic Analysis*, Marcel Dekker, New York, 2001.

Problems

Section 11.2 Coordinates, Constraints, and Degrees of Freedom

P11.2.1: Determine the number of degrees of freedom of the following systems:

a. Pair of eyeglasses
b. Pair of pliers or pair of scissors
c. Child's tricycle rolling on a flat horizontal surface (let the tricycle be modeled by a frame, two rear wheels, and a front steering wheel)
d. Human arm model consisting of three rigid bodies representing the upper arm, the lower arm, and the hand (with spherical joints at the wrist and shoulder and a hinge joint at the elbow)
e. Pencil writing on a sheet of paper
f. Eraser, erasing a chalk board

P11.2.2: An insect A is crawling on the surface of a sphere B with radius r as represented in Figure P11.2.2. Let the coordinates of A relative to a Cartesian X, Y, Z system with origin O at the center of the sphere be (x, y, z). What are the constraints and degrees of freedom?

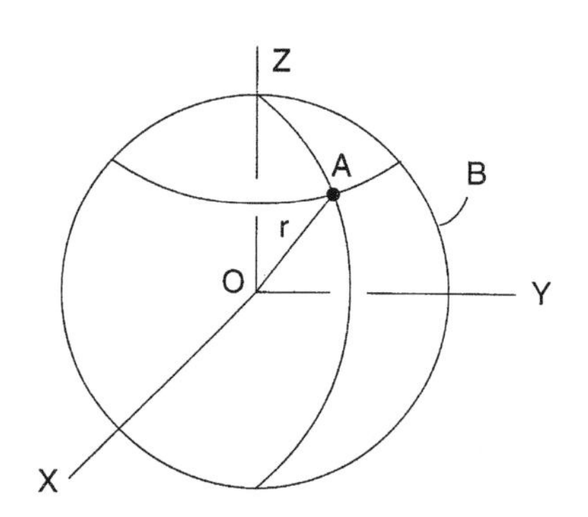

FIGURE P11.2.2
An insect crawling on a sphere.

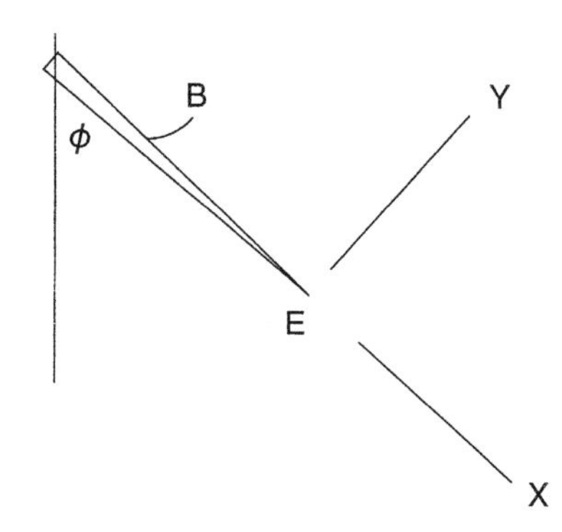

FIGURE P11.2.3
A knife blade cutting a soft medium.

P11.2.3: A knife blade B is being held such that its cutting edge E remains horizontal (Figure P11.2.3). Let B be used to cut a soft medium such as a soft cheese. Let the cheese restrict B so that B has no velocity perpendicular to its plane. How many degrees of freedom does B have? What are the constraint equations?

Section 11.4 Vector Functions, Partial Velocity, and Partial Angular Velocity

P11.4.1: Consider again the double-rod pendulum of Figure 11.4.3 and as shown again in Figure P11.4.1. As before, let each rod be identical with length ℓ. Let the rod orientations be defined by the "relative" orientation angles β_1 and β_2 as in Figure P11.4.1. Find the velocities of mass centers G_1 and G_2, and the angular velocities of B_1 and B_2. Find the partial velocities of G_1 and G_2 for β_1 and β_2 and the partial angular velocities of B_1 and B_2 for β_1 and β_2. Compare the results with those of Eqs. (11.4.23) and (11.4.24).

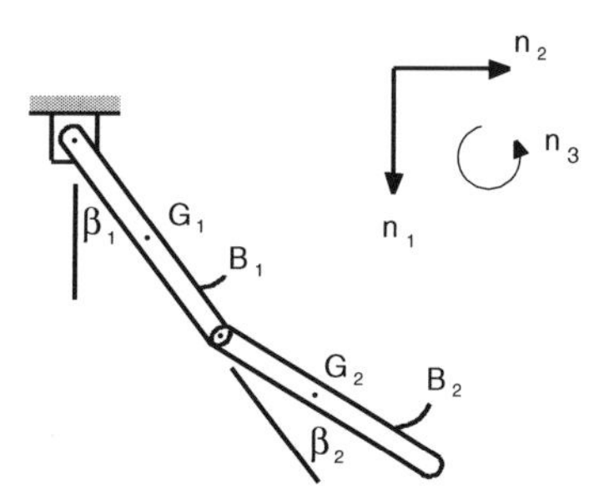

FIGURE P11.4.1
A double-rod pendulum with relative orientation angles.

P11.4.2: A rotating rod B has two collars C_1 and C_2 which can slide relative to B as indicated in Figure P11.4.2. Let the orientation of B be and let x_1 locate C_1 relative to the pin O and let x_2 locate C_2 relative to C_1. Find the partial velocities of C_1 and C_2 with respect to θ, x_1, and x_2.

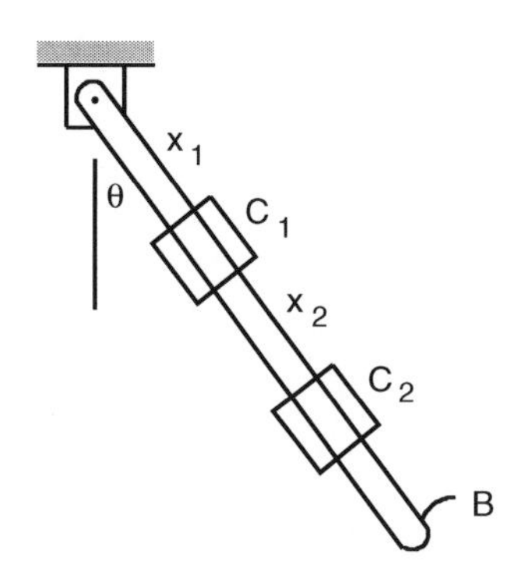

FIGURE P11.4.2
A rotating rod with sliding collars.

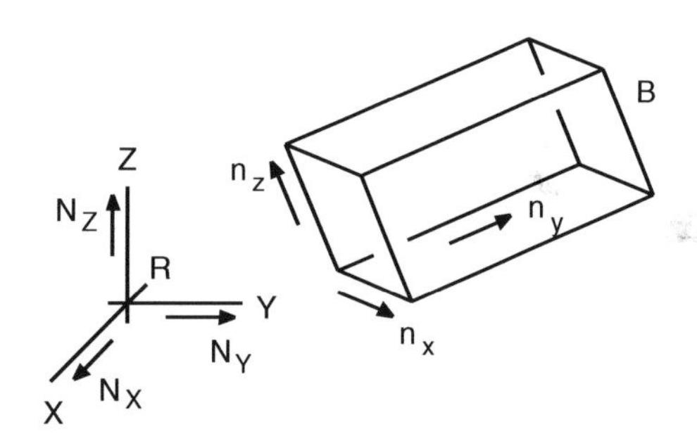

FIGURE P11.4.3
A box B moving in a reference frame R.

P11.4.3: A box B moving in a reference frame R is oriented in R by dextral (Bryan) orientation angles and defined as follows: let unit vectors $\mathbf{n}_x$, $\mathbf{n}_y$, and $\mathbf{n}_z$ fixed in B be aligned with unit vectors $\mathbf{N}_X$, $\mathbf{N}_Y$, and $\mathbf{N}_Z$, respectively, fixed in R as depicted in Figure P11.4.3. Then, let B be successively rotated in R about axes parallel to $\mathbf{n}_x$, $\mathbf{n}_y$, and $\mathbf{n}_z$ through angles α, β, and γ, bringing B into a general orientation in R. Show that the angular velocity of B in R may then be expressed as:

$$
\begin{aligned}
{}^R\boldsymbol{\omega}^B = &\left(\dot{\alpha}c_\beta c_\gamma + \dot{\beta}s_\gamma\right)\mathbf{n}_x + \left(\dot{\beta}c_\gamma - \dot{\alpha}c_\beta s_\gamma\right)\mathbf{n}_y \\
&+ \left(\dot{\alpha}s_\beta + \dot{\gamma}\right)\mathbf{n}_z
\end{aligned}
$$

where s and c are abbreviations for sine and cosine (see Section 4.7).

P11.4.4: See Problem P11.4.3. Express ${}^R\boldsymbol{\omega}^B$ in terms of the unit vectors $\mathbf{N}_X$, $\mathbf{N}_Y$, and $\mathbf{N}_Z$.

P11.4.5: See Problems 11.4.3 and P11.4.4. Find the partial angular velocities of B in R for α, β, and γ. (Express the results in terms of both $\mathbf{n}_x$, $\mathbf{n}_y$, and $\mathbf{n}_z$ and $\mathbf{N}_X$, $\mathbf{N}_Y$, and $\mathbf{N}_Z$.

P11.4.6: See Problems P11.4.3, P11.4.4, and P11.4.5. Let O be the origin of a Cartesian axis system X, Y, Z fixed in R, and let Q be a vertex of B as in Figure P11.4.6. Let the velocity of Q in R be expressed alternatively as:

$${}^R\mathbf{V}^Q = \dot{X}\mathbf{N}_X + \dot{Y}\mathbf{N}_Y + \dot{Z}\mathbf{N}_Z$$

and

$${}^R\mathbf{V}^Q = \dot{x}\mathbf{n}_x + \dot{y}\mathbf{n}_y + \dot{z}\mathbf{n}_z$$

a. Express $\dot{x}$, $\dot{y}$, and $\dot{z}$ in terms of $\dot{X}$, $\dot{Y}$, and $\dot{Z}$.
b. Express $\dot{X}$, $\dot{Y}$, and $\dot{Z}$ in terms of $\dot{x}$, $\dot{y}$, and $\dot{z}$.

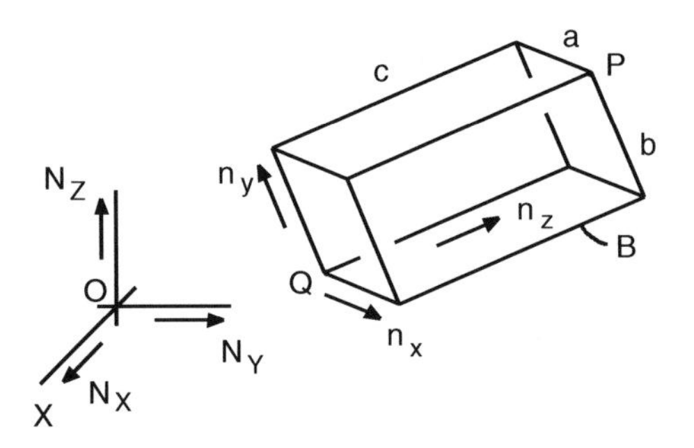

FIGURE P11.4.6
A box B moving in a reference frame R.

P11.4.7: See Problem P11.4.6. Find the partial velocities of Q in R for the coordinates x, y, z, X, Y, and Z. Express the results in terms of both $\mathbf{n}_x$, $\mathbf{n}_y$, and $\mathbf{n}_z$ and $\mathbf{N}_X$, $\mathbf{N}_Y$, and $\mathbf{N}_Z$.

P11.4.8: See Problem 11.4.6. Let the sides of the box B have lengths a, b, and c, as represented in Figure P11.4.6, and let P be a vertex of B as shown. Find the velocity of P in R. Express the results in terms of both $\mathbf{n}_x$, $\mathbf{n}_y$, and $\mathbf{n}_z$ and $\mathbf{N}_X$, $\mathbf{N}_Y$, and $\mathbf{N}_Z$. Use either $\dot{x}$, $\dot{y}$, and $\dot{z}$ or $\dot{X}$, $\dot{Y}$, and $\dot{Z}$ for convenience.

P11.4.9: See Problems P11.4.3 to P11.4.8. Find the partial velocities of P in R for the coordinates x, y, z, X, Y, Z, α, β, and γ. Express the results in terms of both $\mathbf{n}_x$, $\mathbf{n}_y$, and $\mathbf{n}_z$ and $\mathbf{N}_X$, $\mathbf{N}_Y$, and $\mathbf{N}_Z$.

Section 11.5 Generalized Forces: Applied (Active) Forces

P11.5.1: A simple pendulum with length ℓ of 0.5 m and bob P mass of 2 kg is subjected to a constant horizontal force $\mathbf{F}$ of 5 N as represented in Figure P11.5.1. Determine the generalized active force F_θ on P for the orientation angle θ. (Include the contributions of $\mathbf{F}$, gravity, and the cable tension.)

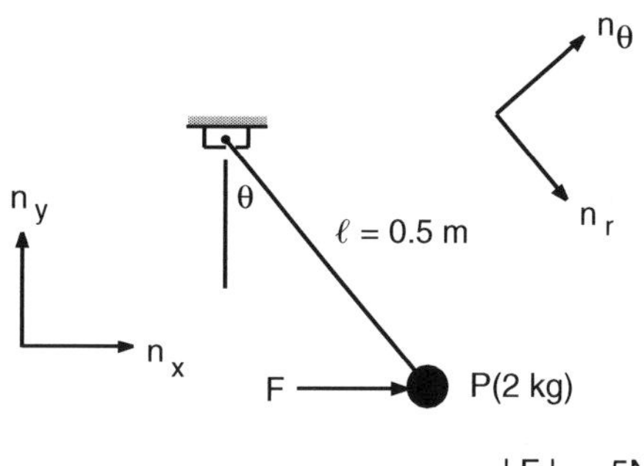

FIGURE P11.5.1
A simple pendulum subjected to a horizontal force.

P11.5.2: Repeat Problem P11.5.1 for a force **F** given as follows:

a. $\mathbf{F} = 4\mathbf{n}_y$ N
b. $\mathbf{F} = 3\mathbf{n}_x + 4\mathbf{n}_y$ N
c. $\mathbf{F} = 5\mathbf{n}_r$ N
d. $\mathbf{F} = 5\mathbf{n}_\theta$ N

P11.5.3: A double-rod pendulum consists of two identical pin-connected rods each having mass m and length ℓ as depicted in Figure P11.5.3. Let there be a particle P with mass M at the end of the lower rod. Assuming the pin connections are frictionless, find the generalized active forces for the orientation angles θ_1 and θ_2 shown in Figure P11.5.3.

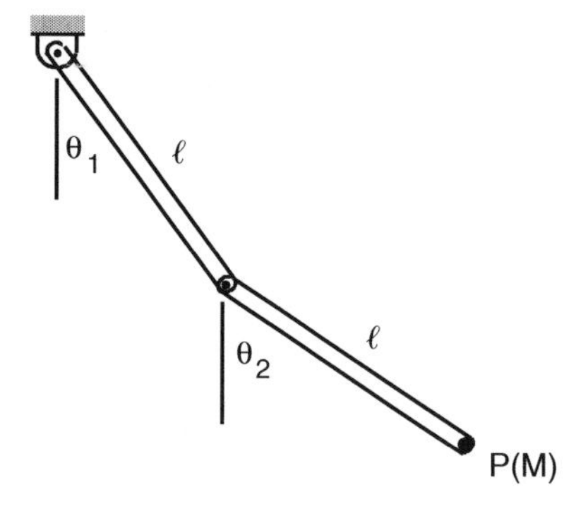

FIGURE P11.5.3
A double-rod pendulum with a concentrated end mass.

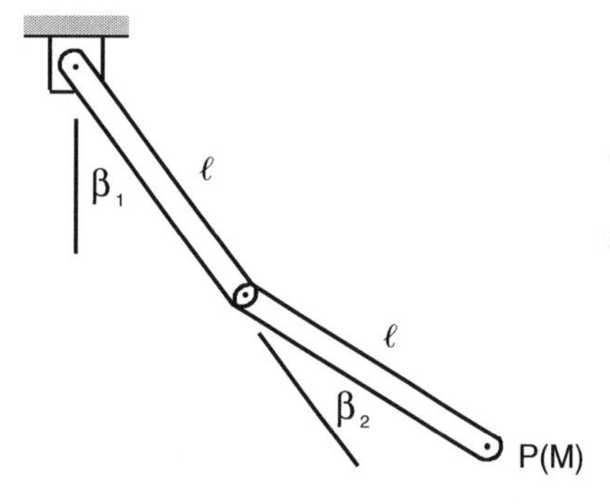

FIGURE P11.5.4
A double-rod pendulum with a concentrated end mass and relative orientation angles.

P11.5.4: Repeat Problem P11.5.3 for the "relative" orientation angles β_1 and β_2 shown in Figure P11.5.4.

P11.5.5: See Problem P11.5.3. A triple-rod pendulum consists of three pin-connected rods each having mass m and length ℓ as depicted in Figure P11.5.5. Let there be a particle P with mass M at the end of the lower rod. Assuming the pin connections are frictionless, find the generalized active forces for the orientation angles θ_1, θ_2, and θ_3 shown in Figure P11.5.3.

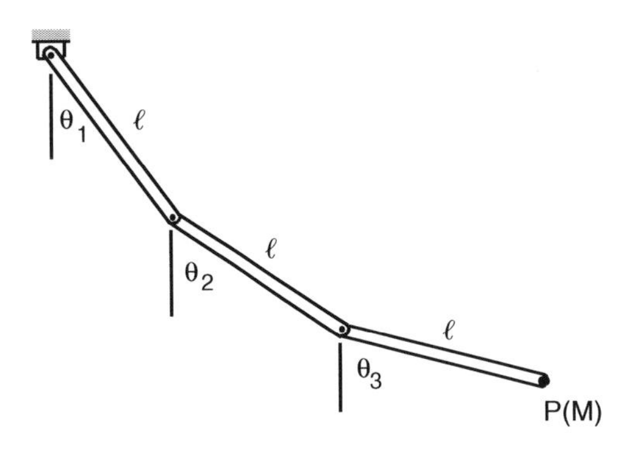

FIGURE P11.5.5
A triple-rod pendulum with a concentrated end mass.

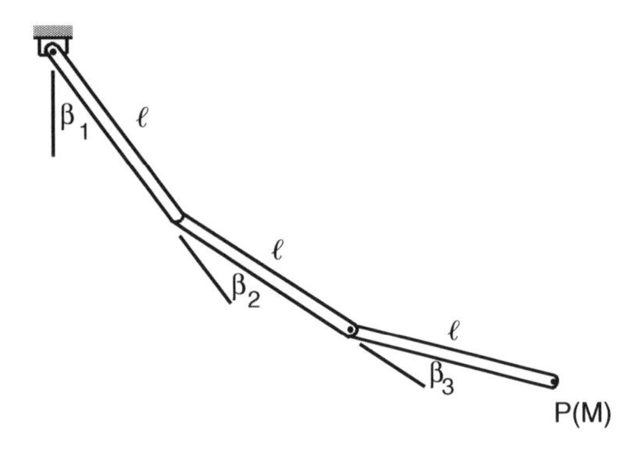

FIGURE P11.5.6
A triple-rod pendulum with concentrated end mass and relative orientation angles.

P11.5.6: Repeat Problem P11.5.5 for the relative orientation angles β_1, β_2, and β_3 shown in Figure P11.5.6.

P11.5.7: See Problem P11.5.5. Let there be linear torsion springs at the pin joints of the triple-rod pendulum. Let these springs have moduli k_1, k_2, and k_3, and let the resulting moments generated by the springs be parallel to the pin axes and proportional to the relative angles β_1, β_2, and β_3 between the rods as shown in Figure P11.5.6. Find the contribution of the spring moments to the generalized active forces for the angles θ_1, θ_2, and θ_3.

P11.5.8: Repeat Problem P11.5.7 for the relative orientation angles β_1, β_2, and β_3.

P11.5.9: An elongated box kite K, depicted in Figure P11.5.9A, is suddenly subjected to turbulent wind gusts creating forces on K as represented in Figure P11.5.9B. Let the wind forces be modeled by forces $\mathbf{F}_1, \mathbf{F}_2, \ldots, \mathbf{F}_8$, whose directions are shown in Figure P11.5.9B and whose magnitudes are:

$$\mathbf{F}_1 = 5 \text{ lb}, \quad \mathbf{F}_2 = 6 \text{ lb}, \quad \mathbf{F}_3 = 6 \text{ lb}, \quad \mathbf{F}_4 = 7 \text{ lb},$$
$$\mathbf{F}_5 = 8 \text{ lb}, \quad \mathbf{F}_6 = 10 \text{ lb}, \quad \mathbf{F}_7 = 13 \text{ lb}, \quad \mathbf{F}_8 = 8 \text{ lb}$$

Let the orientation of K be defined by dextral orientation angles α, β, and γ, and let the angular velocity $\boldsymbol{\omega}$ of K be expressed as (see Problem P11.4.3):

$$\boldsymbol{\omega} = \left(\dot{\alpha} c_\beta c_\gamma + \dot{\beta} s_\gamma\right)\mathbf{n}_1 + \left(\dot{\beta} c_\gamma - \dot{\alpha} c_\beta s_\gamma\right)\mathbf{n}_2 + \left(\dot{\alpha} s_\beta + \dot{\gamma}\right)\mathbf{n}_3$$

where $\mathbf{n}_1$, $\mathbf{n}_2$, and $\mathbf{n}_3$ are unit vectors parallel to the edges of K as shown in Figure P11.5.9B. Finally, let the velocity $\mathbf{v}$ of the tether attachment point O of K be:

$$\mathbf{v} = 3\mathbf{n}_1 - 2\mathbf{n}_2 + 5\mathbf{n}_3 \quad \text{ft/sec}$$

Find the generalized forces F_α, F_β, and F_γ due to the wind forces $\mathbf{F}_1, \mathbf{F}_2, \ldots, \mathbf{F}_8$.

P11.5.10: Repeat Problem P11.5.9 by first replacing $\mathbf{F}_1, \mathbf{F}_2, \ldots, \mathbf{F}_8$ by a single force $\mathbf{F}$ passing through O together with a couple with torque $\mathbf{T}$; then, use Eq. (11.5.7) to determine F_α, F_β, and F_γ.

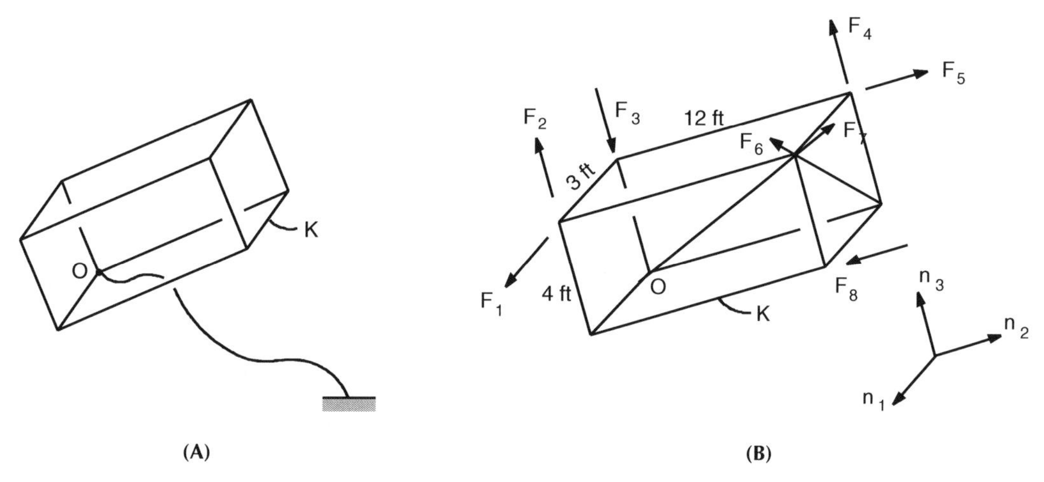

FIGURE P11.5.9
(A) An elongated box kite. (B) A modeling and representation of turbulent wind forces on a box kite.

Section 11.6 Generalized Forces: Gravity and Spring Forces

P11.6.1: Consider again the triple-rod pendulum with a concentrated end mass of Problem P11.5.5 and as shown again in Figure P11.6.1. As before, let the rods be identical with each having length ℓ and mass m. Let G_1, G_2, and G_3 be the mass centers of the rods. Let the orientations of the rods be defined by the angles θ_1, θ_2, and θ_3 shown in Figure P11.6.1. Finally, let L be a reference level passing through the upper support pin at O.

a. Let h_1, h_2, h_3, and h_p measure the elevation of G_1, G_2, G_3, and P above L. (Observe that h_1, h_2, h_3, and h_p are negative in the configuration shown in Figure P11.6.1.) Determine h_1, h_2, h_3, and h_p in terms of θ_1, θ_2, and θ_3.
b. Use Eq. (11.6.5) to determine the contribution of the weight forces of the rods and the weight of P to the generalized active forces F_{θ_1}, F_{θ_2}, and F_{θ_3}.

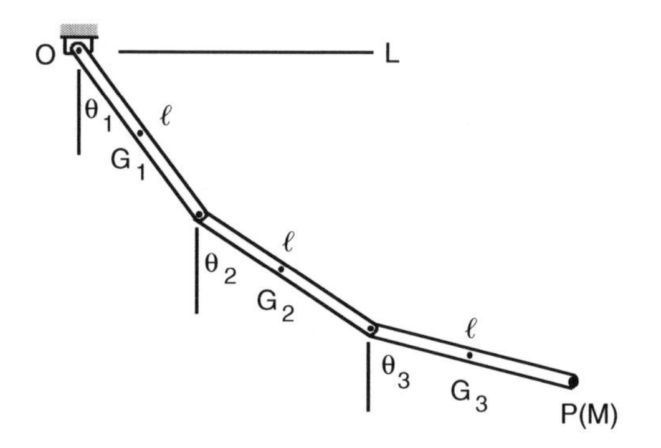

FIGURE P11.6.1
A triple-rod pendulum with concentrated end mass.

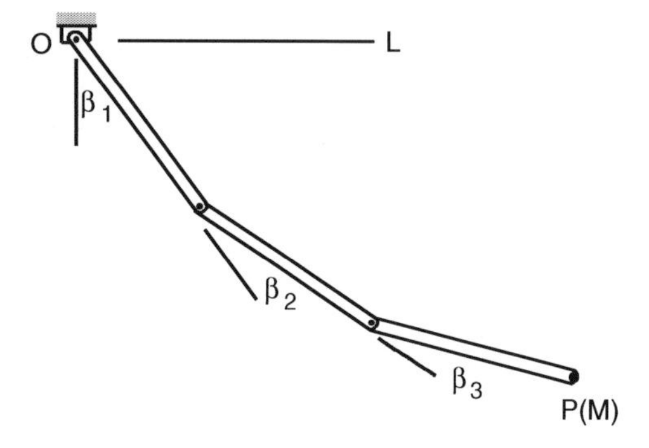

FIGURE P11.6.2
A triple-rod pendulum with concentrated end mass with relative orientation angles.

P11.6.2: Repeat Problem P11.6.1 if the orientations of the rods are defined by the relative orientation angles β_1, β_2, and β_3, as shown in Figure P11.6.2. Specifically, use Eq. (11.6.5) to determine the generalized active forces F_{β_1}, F_{β_2}, and F_{β_3}.

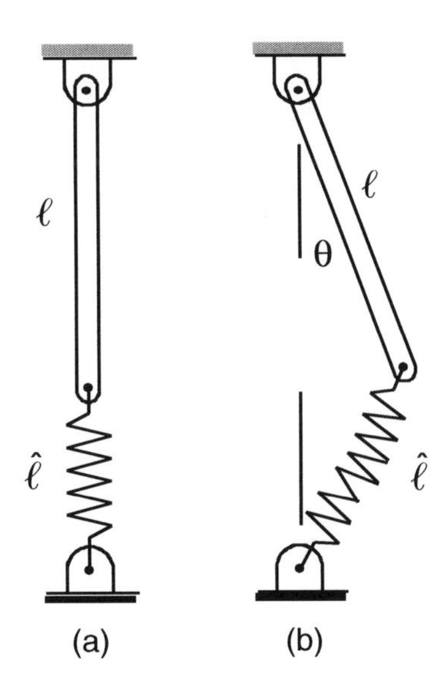

FIGURE P11.6.3
A rod pendulum with a spring attachment.

P11.6.3: A rod pendulum's movement is restricted by a linear spring attached to its end as in Figure P11.6.3. Let the rod length be ℓ and the spring length be $\hat{\ell}$, and let the spring be unstretched in the equilibrium position (Figure 11.6.3a). If the pendulum is restricted to moving in a plane, the pendulum–spring system has only one degree of freedom, which may be represented by the angle θ.

a. Determine an expression for the spring extension x as a function of θ.
b. Use Eq. (11.6.7) to determine the contribution of the spring force to the generalized active force for θ. (Let k be the spring modulus so that $f(x) = kx$.)

P11.6.4: Develop an expression, analogous to Eq. (11.6.7), which is applicable for torsion springs.

P11.6.5: Repeat Problems P11.5.7 and P11.5.8 by using the expression developed in Problem P11.6.4.

P11.6.6: A particle P with mass m slides inside a smooth surfaced tube T, with radius r, which rotates at constant speed Ω about a vertical diameter as represented in Figure P11.6.6. Let the position of P be located by the angle θ as shown. Determine the generalized active force F_θ.

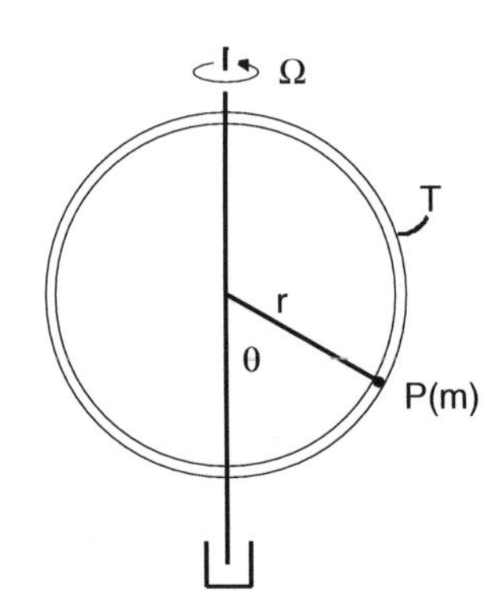

FIGURE P11.6.6
A particle moving inside a smooth-surfaced tube.

P11.6.7: A right circular cone C with altitude h and half-central angle α rolls on a plane inclined at an angle β as shown in Figure P11.6.7. Let the position of C be located by the angle ϕ between a contacting element of C with the plane and a line fixed in the plane. Let the element length of C be ℓ, and let the base radius be r. Let O be the apex of C, and let G be its mass center. Let ψ measure the "roll" of C as shown in Figure P11.6.7.

a. Express r and ℓ in terms of h and α.
b. Observe that while C rolls and oscillates on the inclined plane its apex O does not move relative to the plane.
c. Find the distance x between O and G.
d. Find a relationship between ψ and ϕ.
e. Show that C has only one degree of freedom. If that degree of freedom is represented by the angle ϕ, determine the generalized active force F_θ.

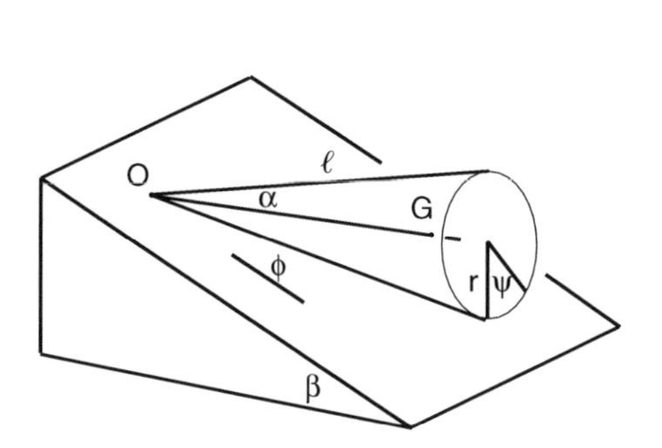

FIGURE P11.6.7
A cone rolling on an inclined plane. (Reprinted from Huston, R. L., and Liu, C. Q., *Formulas for Dynamic Analysis*, p. 517, by courtesy of Marcel Dekker, Inc., 2001.

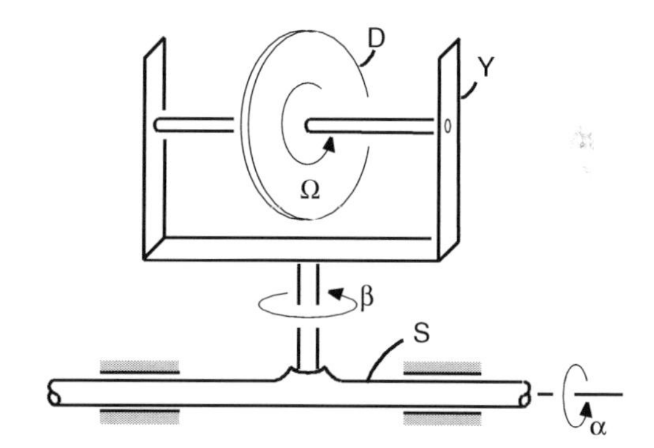

FIGURE P11.6.8
A disk spinning in a free-turning yoke and supported by a shaft S.

P11.6.8: A heavy disk D rotates with a constant angular speed Ω in a light yoke Y which in turn rotates relative to a light horizontal shaft S as depicted in Figure P11.6.8. Let D have mass m and radius r. Let the rotation of Y relative to S be measured by the angle β, and the rotation of S relative to its bearings be measured by the angle α. Let all the bearings be frictionless. Observe that the system has two degrees of freedom, which may be represented by the angles α and β. Determine the generalized forces F_α and F_β. *Suggestion:* introduce unit vector sets in the fixed frame, in S, in Y, and in D, and then develop the kinematics of the system. Identify the contributing forces to the generalized forces. Then, from partial velocity vectors, determine the generalized active forces.

Section 11.9 Generalized Forces: Inertia (Passive) Forces

P11.9.1: A pendulum consists of a rod with length ℓ and mass m supported by a frictionless pin at one end and rigidly attached to a circular disk with radius r and mass M (as in a clock) as depicted in Figure P11.9.1. This system has one degree of freedom represented by the angle θ. Determine the generalized inertia force F_θ^*.

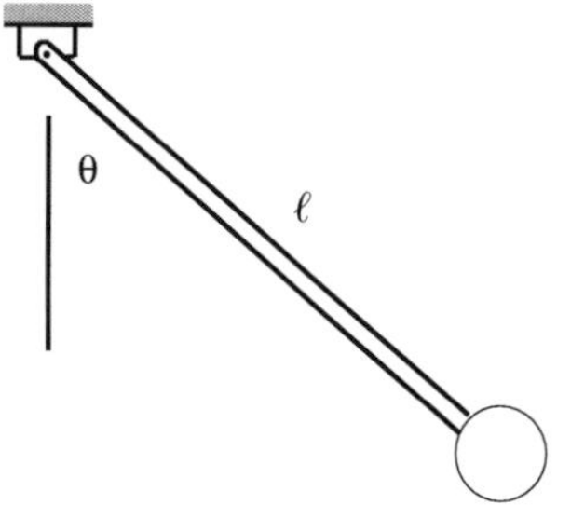

FIGURE P11.9.1
A rod/disk pendulum.

P11.9.2: A rod B with length ℓ and mass m is pinned to a vertical rotating shaft S as depicted in Figure P11.9.2. Let the pin axis be along a radial line of S. Let θ define the angle between B and S. Let the rotation speed of S be specified as Ω. This system has one degree of freedom represented by the angle θ. Determine the generalized inertia force F_θ^*, assuming the shaft radius r is small.

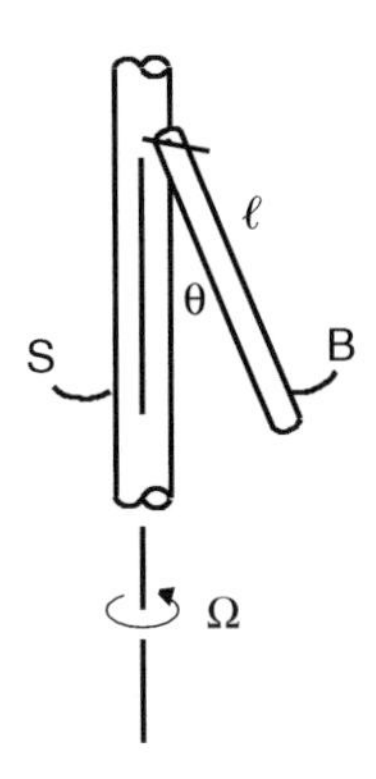

FIGURE P11.9.2
A rod B pinned to a rotating shaft S.

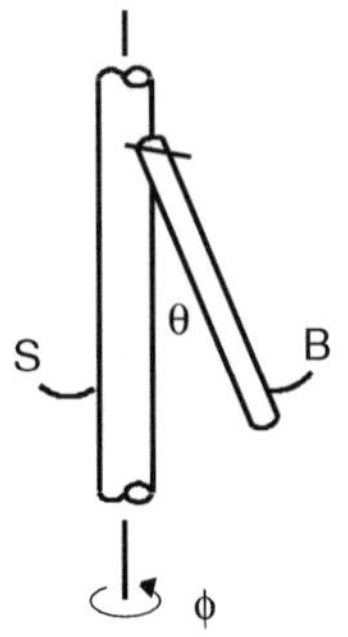

FIGURE P11.9.4
A rod B pinned to a rotating shaft S.

P11.9.3: Repeat Problem P11.9.2 by not assuming a small shaft radius; that is, include the effect of the shaft radius in the analysis.

P11.9.4: See Problem P11.9.2. Suppose the rotation of S is not specified but instead is free, or arbitrary, and defined by the angle ϕ as in Figure P11.9.4. This system now has two degrees of freedom, represented by the angles θ and ϕ. Determine the generalized inertia forces F_θ^* and F_ϕ^*, assuming the shaft radius r is small.

P11.9.5: Repeat Problem P11.9.4 by not assuming the shaft radius is small; that is, include the effect of the shaft radius in the analysis and in this regard, let the mass of S be M.

P11.9.6: See Problem P11.6.6. Consider again the rotating tube with a particle P of Problem P11.6.6 and as shown again in Figure P11.9.6. Letting the rotating speed Ω be constant as before, and, letting the mass of P be m, determine the generalized inertia force for the angle θ, F_θ^*.

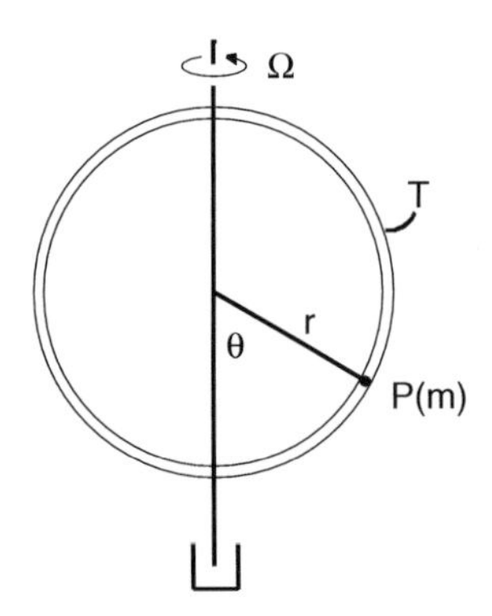

FIGURE P11.9.6
A particle moving inside a smooth-surfaced tube.

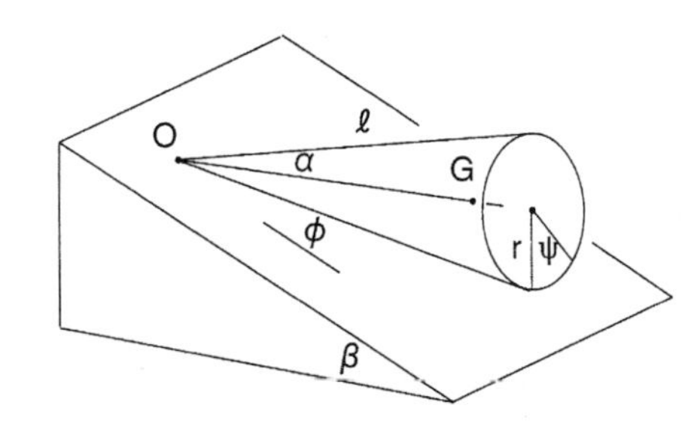

FIGURE P11.9.7
A cone rolling on an inclined plane.

P11.9.7: See Problem P11.6.7. Consider again the cone rolling on an inclined plane of Problem P11.6.7 and as shown again in Figure P11.9.7. As determined in Problem P11.6.7, this system has one degree of freedom, which may be represented by the angle ϕ. Determine the generalized inertia force F_θ^*.

P11.9.8: See Problem P11.6.8. Consider again the rotating disk in the shaft-supported yoke of Problem P11.6.8 and as shown again in Figure P11.9.8. With the angular speed of the disk being specified, the system has two degrees of freedom, represented by the angles α and β. Following the solution outline of Problem P11.6.8, determine the generalized inertia forces F_α^* and F_β^*.

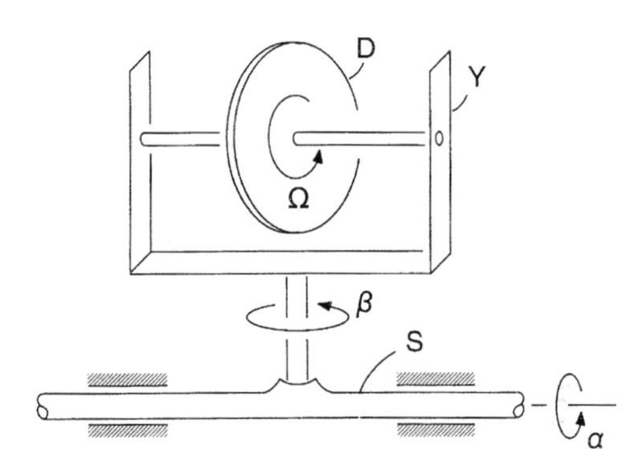

FIGURE P11.9.8
A disk spinning in a free-turning yoke and supported by a shaft S.

Section 11.11 Potential Energy

P11.11.1: See Problem P11.5.3. Consider again the double-rod pendulum with a concentrated end mass of Problem P11.5.3 and as shown again in Figure P11.11.1. Determine the potential energy of the system using a horizontal line through the pin support O as the reference.

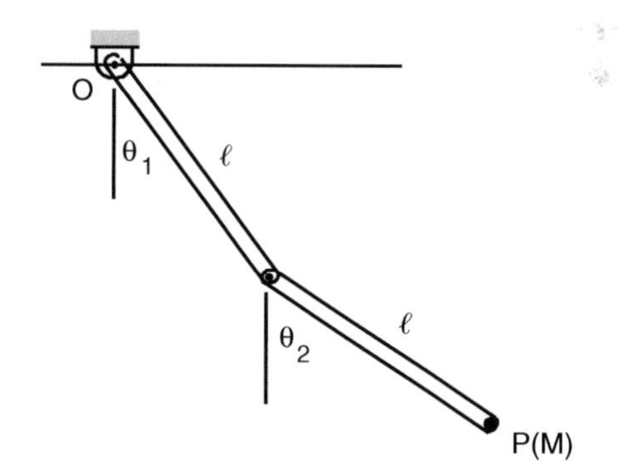

FIGURE P11.11.1
A double-rod pendulum with a concentrated end mass.

P11.11.2: Repeat Problem P11.11.1 using a horizontal line that is 2ℓ below the pin support at O as the reference for the potential energy.

P11.11.3: Use the results of Problems P11.11.1 and P11.11.2 and Eq. (11.11.1) to evaluate the generalized active forces F_{θ_1} and F_{θ_2}. Compare the results and analysis effort with those of Problem P11.5.3.

P11.11.4: See Problem P11.5.4. Consider still further the double-rod pendulum with a concentrated end mass. Let the orientations of the rods be defined by the relative orientation angles β_1 and β_2, as shown in Figure P11.11.4. Repeat Problems P11.11.1 and P11.11.2 finding the potential energies for the relative orientation angles.

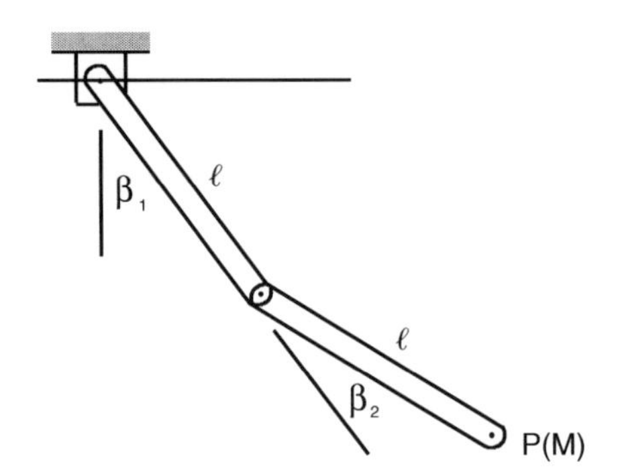

FIGURE P11.11.4
A double-rod pendulum with a concentrated end mass and relative orientation angles.

P11.11.5: Use the results of Problem P11.11.4 and Eq. (11.11.1) to evaluate the generalized active forces F_{β_1} and F_{β_2}. Compare the results and analysis effort with those of Problem P11.5.4.

P11.11.6: See Problem P11.5.5. Consider again the triple-rod pendulum with concentrated end mass of Problem P11.5.5 and as shown again in Figure P11.11.6. Determine the potential energy of the system using a horizontal line through the pin support O as the reference.

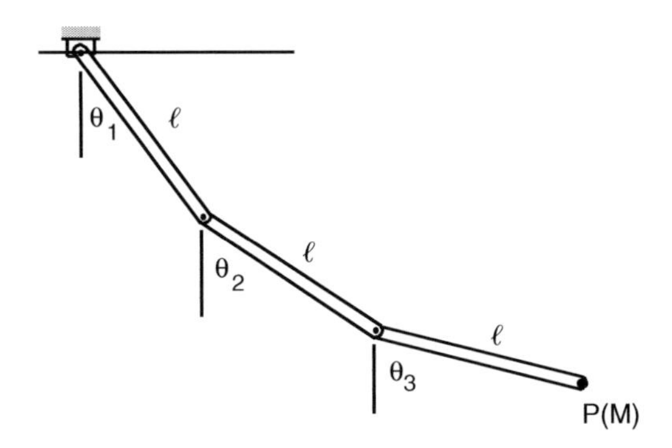

FIGURE P11.11.6
A triple-rod pendulum with a concentrated end mass.

P11.11.7: Repeat Problem P11.11.6 using a horizontal line that is 3ℓ below the pin support of O as the reference for the potential energy.

P11.11.8: Use the results of Problems P11.11.6 and P11.11.7 and Eq. (11.11.1) to evaluate the generalized active forces and F_{θ_1}, F_{θ_2}, and F_{θ_3}. Compare the results and analysis effort with those of Problem P11.5.5.

P11.11.9: See Problem P11.5.6. Consider still further the triple-rod pendulum with a concentrated end mass. Let the orientation of the rods be defined by the relative orientation angles β_1, β_2, and β_3, as shown in Figure P11.11.9. Repeat Problems P11.11.6 and P11.11.7 finding the potential energies for the relative orientation angles.

P11.11.10: Use the results of Problem P11.11.9 and Eq. (11.11.1) to evaluate the generalized active forces F_{β_1}, F_{β_2}, and F_{β_3}. Compare the results and analysis effort with those of Problem P11.5.6.

P11.11.11: See Problem P11.6.3. Consider again the rod pendulum with spring attachment of Problem P11.6.3 and as shown again in Figure P11.11.11. Use Eq. (11.11.10) to determine the potential energy of the spring. Then, use Eq. (11.11.1) to determine the contribution of the spring force to the generalized active force F_θ. Compare the results and analysis effort with those of Problem P11.6.3.

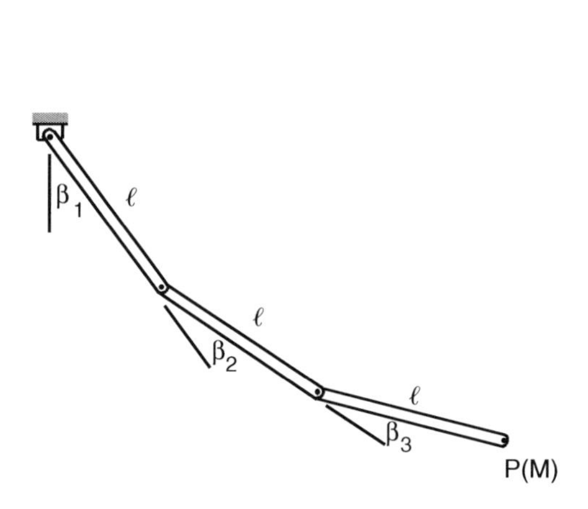

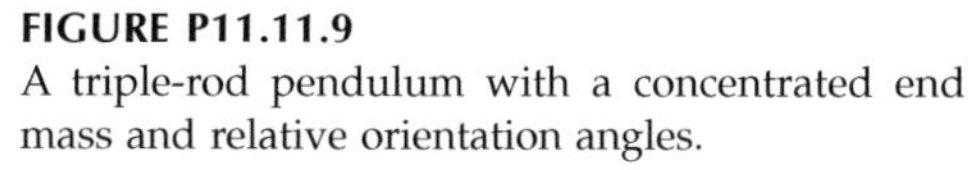
FIGURE P11.11.9
A triple-rod pendulum with a concentrated end mass and relative orientation angles.

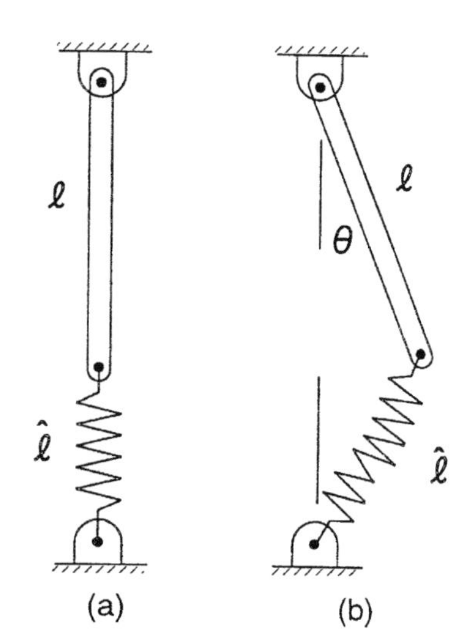

FIGURE P11.11.11

Section 11.12 Use of Kinetic Energy To Obtain Generalized Inertia Forces

P11.12.1: See Problem P11.9.1. Consider again the rod pendulum with the attached circular disk shown in Figure P11.12.1. As before, let the rod have length ℓ and mass m, and let the disk have radius r and mass M. Use Eq. (11.12.5) to determine the generalized inertia force F_θ^* for the orientation angle θ. Compare the results and analysis effort with those of Problem P11.9.1.

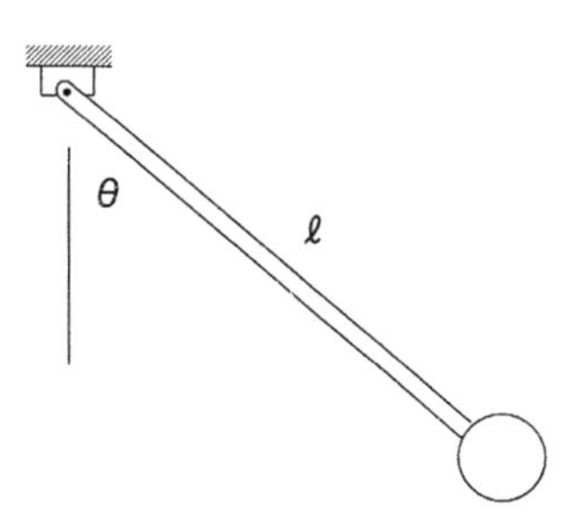

FIGURE P11.12.1
A rod/disk pendulum.

P11.12.2: See Problems P11.9.2 to P11.9.5. Consider again the rod pinned to the vertically rotating shaft shown in Figure P11.12.2. As before, let the rod have length ℓ and mass m, and let the shaft have radius r and mass M. Use Eq. (11.12.5) to determine the generalized inertia forces F_θ^* and F_ϕ^*. Compare the results and analysis effort with those of Problem P11.9.5.

P11.12.3: See Problem P11.9.6. Consider again the rotating tube with a particle P of Problem P11.6.6 and P11.9.6 and as shown again in Figure P11.12.3. Use Eq. (11.12.5) to determine the generalized inertia force F_θ^*.

P11.12.4: See Problem P11.9.7. Consider again the cone rolling on an inclined plane of Problems P11.6.7 and P11.9.7 and as shown again in Figure P11.12.4. Determine the generalized inertia force F_θ^*.

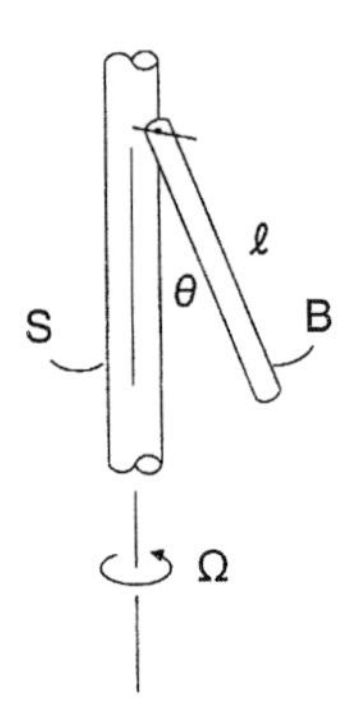

FIGURE P11.12.2
A rod pinned to a rotating shaft.

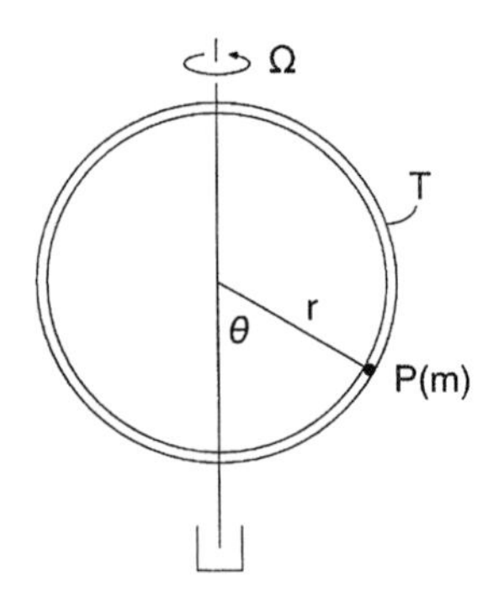

FIGURE P11.12.3
A particle moving inside a smooth-surfaced tube.

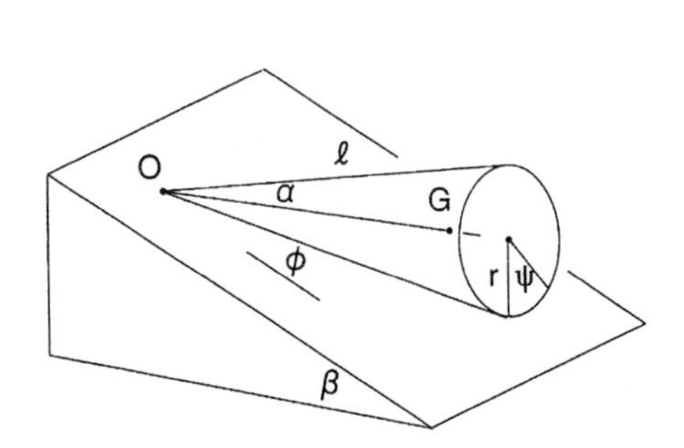

FIGURE P11.12.4
A cone rolling on an inclined plane.

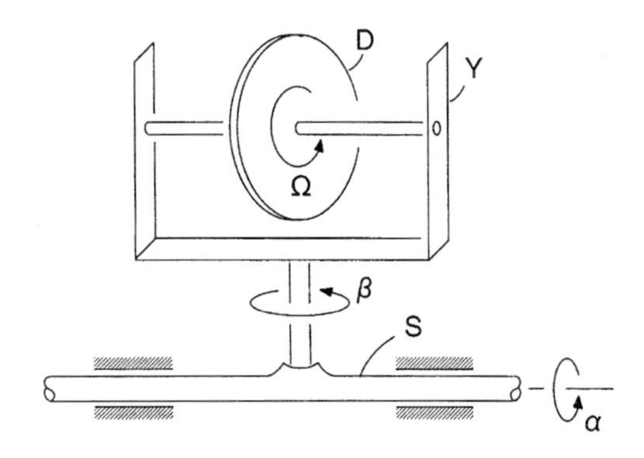

FIGURE P11.12.5
A disk spinning in a free-turning yoke and suported by a shaft *S*.

P11.12.5: See Problem P11.9.8. Consider again the rotating disk in the shaft-supported yoke of Problems P11.6.8 and P11.9.8 and as shown again in Figure P11.12.5. Determine the generalized inertia forces F_α^* and F_β^*.

12

Generalized Dynamics: Kane's Equations and Lagrange's Equations

12.1 Introduction

In this chapter, we apply the procedures developed in Chapter 11 to obtain governing dynamical equations of motion. We will consider two approaches: the use of Kane's equations and the use of Lagrange's equations. We will discover that these approaches are related. Indeed, we will use Kane's equations to develop Lagrange's equations. We will illustrate and compare the use of Kane's and Lagrange's equations by obtaining equations of motion for the various examples considered in Chapter 11. In addition, we will also consider the triple-bar pendulum and the N-bar pendulum (an N-bar pendulum may be considered as a model of a chain or cable).

Using either Kane's equations or Lagrange's equations to obtain equations of motion has distinct advantages over using Newton's laws, over the method of impulse and momentum, and over the work–energy principle. These advantages include the automatic elimination of nonworking internal constraint forces from the analysis. In addition, with Kane's equations and Lagrange's equations, the exact same number of equations are obtained as there are degrees of freedom. Moreover, these equations are obtained without needing to make insightful or clever choices of summation directions or of reference points to take moments about. Lagrange's equations have the additional advantage of not requiring computation of accelerations.

The advantages of Kane's and Lagrange's equations, however, do not come without corresponding disadvantages. Indeed, if the constraint forces are eliminated, then they are not determined and remain unknown in the analysis. On occasion, such forces may be of interest, particularly in machine design. Also, with Lagrange's equations we cannot study nonholonomic systems (at least, not without a modification of the equations). Nevertheless, on balance, the advantages of Kane's and Lagrange's equations outweigh the disadvantages for a large class of systems of importance in machine dynamics.

12.2 Kane's Equations

Kane's equations provide an elegant formulation of the dynamical equations of motion. Kane's equations simply state that the sum of the generalized forces (both applied and inertia forces), for each generalized coordinate, is zero. That is, for a mechanical system

S having n degrees of freedom, represented by generalized coordinates q_r, Kane's equations state that:

$$F_r + F_r^* = 0 \tag{12.2.1}$$

Professor Thomas R. Kane first published Eq. (12.2.1) in 1961 [12.1]. The equations were intended primarily to be a means for studying nonholonomic systems. The objective was to obtain a "Lagrangian" formulation that would automatically eliminate nonworking constraint forces from the analysis. As such, Kane's equations were originally called *Lagrange's form of d'Alembert's principle* [12.2, 12.3].

Some writers suggest that Kane's equations are simply a reformulation of principles developed earlier by Appell and Jourdain [12.9]. It appears, however, that those principles and their resulting equations are not as simple nor as intuitive as the expressions of Eq. (12.2.1).

With the advent of high-speed digital computers and the corresponding development of procedures in computational mechanics, Kane's equations have found application in areas far beyond those envisioned in 1961. Indeed, Kane's equations are currently the equations of choice for automated (numerical) formulation of the governing equations of motion for large mechanical systems [12.5, 12.6].

Intuitively, Kane's equations may be interpreted as follows: if the partial velocity vectors define the directions of motion of a mechanical system, then Kane's equations represent a projection of the applied and inertia forces along those directions.

From this perspective, Kane's equations may be developed from d'Alembert's principle (see Section 8.3) by simply projecting the forces along the partial velocity vectors $\mathbf{v}_{\dot{q}_r}$. Specifically, let a mechanical system S be regarded as a set of N particles P_i ($i = 1,\ldots, N$) having masses m_i. Let $\mathbf{F}$ represent the resultant of the applied forces on S, and let $\mathbf{F}^*$ represent the resultant of the inertia forces on S. That is,

$$\mathbf{F} = \sum_{i=1}^{N} \mathbf{F}_i \quad \text{and} \quad \mathbf{F}^* = \sum_{i=1}^{N} \left(-m_i \mathbf{a}^{P_i}\right) \tag{12.2.2}$$

where $\mathbf{F}_i$ represents the applied forces on P_i, and $\mathbf{a}^{P_i}$ is the acceleration of P_i in an inertial frame R. Then, d'Alembert's principle states that [see Eq. (8.3.3)]:

$$\mathbf{F} + \mathbf{F}^* = 0 \tag{12.2.3}$$

Then, by taking the scalar product (projection) of Eq. (12.2.3) with the partial velocity vectors $\mathbf{v}_{\dot{q}_r}$ ($r = 1,\ldots, n$), we have:

$$\mathbf{F} \cdot \mathbf{v}_{\dot{q}_r} + \mathbf{F}^* \cdot \mathbf{v}_{\dot{q}_r} = 0 \quad \text{or} \quad F_r + F_r^* = 0 \quad (r = 1,\ldots,n) \tag{12.2.4}$$

(See Eqs. (11.5.1) and (11.9.2).)

Because d'Alembert's principle may be obtained from Newton's laws, as discussed in Section 8.3, we may consider Kane's equations to be derived from Newton's laws. But, unlike Newton's laws or d'Alembert's principle, Kane's equations provide for the automatic elimination of nonworking constraint forces. Also, Kane's equations produce exactly the same number of equations as the degrees of freedom. That is, analysts need not be clever in their choice of points to take moments about or in the choice of directions to add forces. Finally, as noted before, Kane's equations are applicable with both holonomic and nonholonomic systems. We can illustrate the use of Kane's equations with the examples of the previous chapter.

Example 12.2.1: The Simple Pendulum

Consider again the simple pendulum as in Figure 12.2.1 (we considered this system in Sections 11.5, 11.10, and 11.12). This system has one degree of freedom represented by the angle θ. In Section 11.5, Eq. (11.5.9), we found that the generalized active force F_θ was:

$$F_\theta = -mg\ell \sin\theta \tag{12.2.5}$$

In Sections 11.10 and 11.12, Eqs. (11.10.5) and (11.12.8), we found the generalized inertia force F_θ^* to be:

$$F_\theta^* = -m\ell^2 \ddot{\theta} \tag{12.2.6}$$

Hence, from Kane's equations (Eq. (12.2.1)), we have the dynamical equation of motion:

$$F_\theta + F_\theta^* = 0 \tag{12.2.7}$$

or

$$-mg\ell \sin\theta - m\ell^2 \ddot{\theta} = 0 \tag{12.2.8}$$

or

$$\ddot{\theta} + (g/\ell)\sin\theta = 0 \tag{12.2.9}$$

Equation (12.2.9) is identical to Eq. (8.4.4) obtained using d'Alembert's principle.

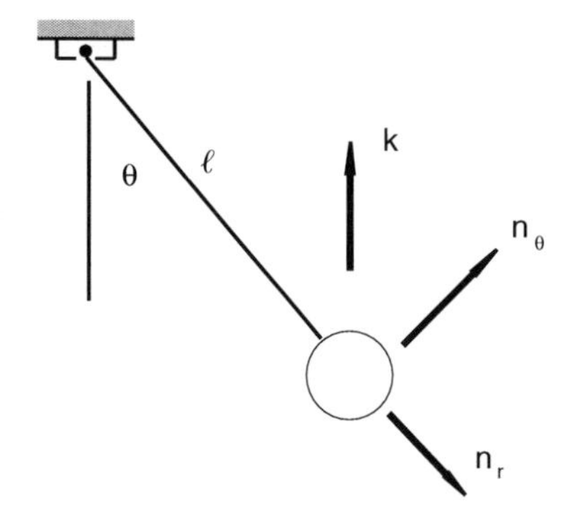

FIGURE 12.2.1
A simple pendulum.

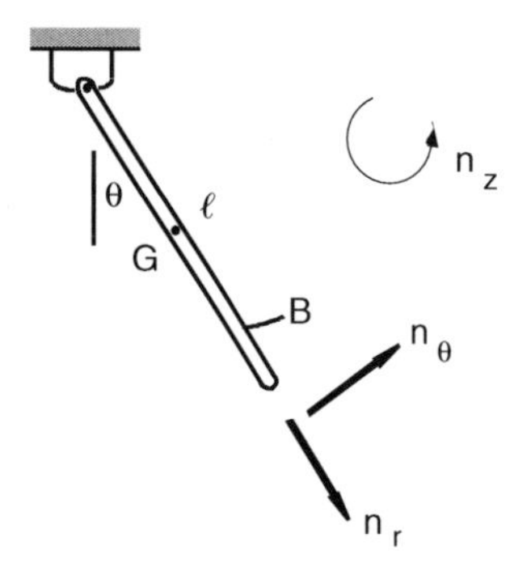

FIGURE 12.2.2
A rod pendulum.

Example 12.2.2: The Rod Pendulum

Consider next the rod pendulum of Figure 12.2.2. (We also considered this system in Sections 11.5, 11.10, and 11.12.) As with the simple pendulum, this system also has one degree of freedom represented by the angle θ. In Section 11.5, Eq. (11.5.14), we found that the generalized active force F_θ was (neglecting the torsion spring):

$$F_\theta = -mg(\ell/2)\sin\theta \tag{12.2.10}$$

In Sections 11.10 and 11.12, Eqs. (11.10.11) and (11.12.11), we found the generalized inertia force F_θ^* to be:

$$F_\theta^* = -m(\ell^2/3)\,\ddot{\theta} \tag{12.2.11}$$

Hence, from Kane's equations [Eq. (12.2.1)], we have:

$$F_\theta + F_\theta^* = 0 \tag{12.2.12}$$

or

$$-mg(\ell/2)\sin\theta - m(\ell^2/3)\,\ddot{\theta} = 0 \tag{12.2.13}$$

or

$$\ddot{\theta} + (3g/2\ell)\sin\theta \tag{12.2.14}$$

Equation (12.2.14) is seen to be equivalent to Eq. (8.9.9), obtained using d'Alembert's principle.

Example 12.2.3: Double-Rod Pendulum

As an extension of the rod pendulum, consider again the double-rod pendulum of Figure 12.2.3 (we previously considered this system in Sections 11.10 and 11.12). This system has two degrees of freedom as represented by the angles θ_1 and θ_2. In Section 11.10, Eq. (11.10.19), we found that the partial velocities of the mass centers G_1 and G_2 are:

$$\begin{aligned}
\mathbf{v}_{\dot{\theta}_1}^{G_1} &= -(\ell/2)\sin\theta_1\,\mathbf{n}_1 + (\ell/2)\cos\theta_1\mathbf{n}_2 \\
\mathbf{v}_{\dot{\theta}_1}^{G_2} &= -\ell\sin\theta_1\mathbf{n}_1 + \ell\cos\theta_1\mathbf{n}_2 \\
\mathbf{v}_{\dot{\theta}_1}^{G_2} &= 0 \\
\mathbf{v}_{\dot{\theta}_2}^{G_1} &= -(\ell/2)\sin\theta_2\,\mathbf{n}_1 + (\ell/2)\cos\theta_2\mathbf{n}_2
\end{aligned} \tag{12.2.15}$$

The only applied forces contributing to the generalized active forces are the weight, or gravity, forces. These may be represented by single vertical forces $mg\mathbf{n}_1$ passing through G_1 and G_2. Hence, the generalized active forces are:

$$F_{\theta_1} = \mathbf{v}_{\dot{\theta}_1}^{G_1}\cdot(mg\mathbf{n}_1) + \mathbf{v}_{\dot{\theta}_1}^{G_2}\cdot(mg\mathbf{n}_1)$$

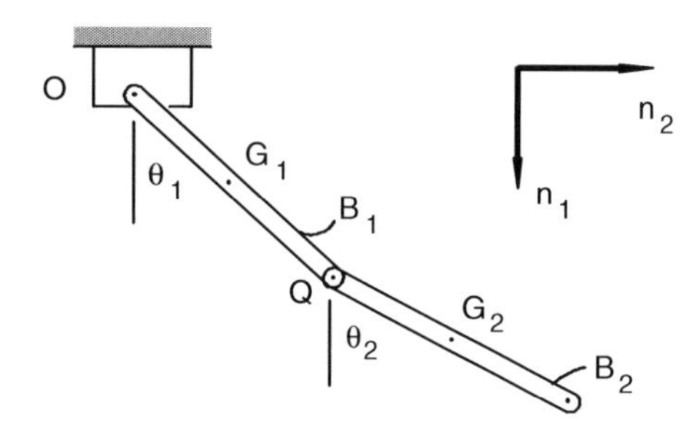

FIGURE 12.2.3
Double-rod pendulum.

or

$$F_{\theta_1} = -(3mg\,\ell/2)\sin\theta_1 \tag{12.2.16}$$

and

$$F_{\theta_2} = \mathbf{v}_{\dot{\theta}_2}^{G_1} \cdot (mg\mathbf{n}_1) + \mathbf{v}_{\dot{\theta}_2}^{G_2} \cdot (mg\mathbf{n}_1)$$

or

$$F_{\theta_2} = -(mg\ell/2)\sin\theta_2 \tag{12.2.17}$$

In Sections 11.10 and 11.12, Eqs. (11.10.21), (11.10.22), (11.12.14), and (11.12.15), we found the generalized inertia forces $F_{\theta_1}^*$ and $F_{\theta_2}^*$ to be:

$$F_{\theta_1}^* = -(4/3)m\ell^2\ddot{\theta}_1 - (1/2)m\ell^2\ddot{\theta}_2\cos(\theta_2 - \theta_1) + (1/2)m\ell^2\dot{\theta}_2^2\sin(\theta_2 - \theta_1) \tag{12.2.18}$$

and

$$F_{\theta_2}^* = -(1/3)m\ell^2\ddot{\theta}_2 - (1/2)m\ell^2\ddot{\theta}_1\cos(\theta_2 - \theta_1) - (1/2)m\ell^2\dot{\theta}_1^2\sin(\theta_2 - \theta_1) \tag{12.2.19}$$

Kane's equations then produce the governing equations:

$$F_{\theta_1} + \mathbf{F}_{\theta_1}^* = 0 \tag{12.2.20}$$

and

$$F_{\theta_2} + \mathbf{F}_{\theta_2}^* = 0 \tag{12.2.21}$$

or

$$(4/3)\,\ddot{\theta}_1 + (1/2)\ddot{\theta}_2\cos(\theta_2 - \theta_1) - (1/2)\dot{\theta}_2^2\sin(\theta_2 - \theta_1) + (3g/2\ell)\,\sin\theta_1 = 0 \tag{12.2.22}$$

and

$$(1/2)\cos(\theta_2-\theta_1)\ddot{\theta}_1+(1/3)\ddot{\theta}_2-(1/2)\dot{\theta}_1^2\sin(\theta_1-\theta_2)+(g/2\ell)\sin\theta_2=0 \tag{12.2.23}$$

Equations (12.2.22) and (12.2.23) are identical to Eqs. (8.10.12) and (8.10.13) obtained using d'Alembert's principle. By reviewing the analysis of Section 8.10, we see that the final form of the governing equations is obtained more directly (that is, without subsequent simplification) using Kane's equations.

Example 12.2.4: Spring-Supported Particles in a Rotating Tube

Consider next the system of spring-supported particles in a rotating tube that we considered in Sections 11.7, 11.10, 11.11, and 11.12 and as shown in Figure 12.2.4. This system has four degrees of freedom represented by the angle θ and the coordinates x_1, x_2, and x_3 as in Figure 12.2.5.

In Section 11.7 we found the generalized active forces to be (see Eqs. (11.7.14) to (11.7.17)):

$$F_{x_1}=mg\cos\theta+kx_2-2kx_1 \tag{12.2.24}$$

$$F_{x_2}=mg\cos\theta+kx_3+kx_1-2kx_2 \tag{12.2.25}$$

$$F_{x_3}=mg\cos\theta-2kx_3+kx_2 \tag{12.2.26}$$

and

$$F_\theta=-Mg(L/2)\sin\theta-mg(6\ell+x_1+x_2+x_3)\sin\theta \tag{12.2.27}$$

where, as before, m is the mass of each particle, M is the mass of the tube T, k is the modulus of each spring, L is the length of T, and ℓ is the natural spring length.

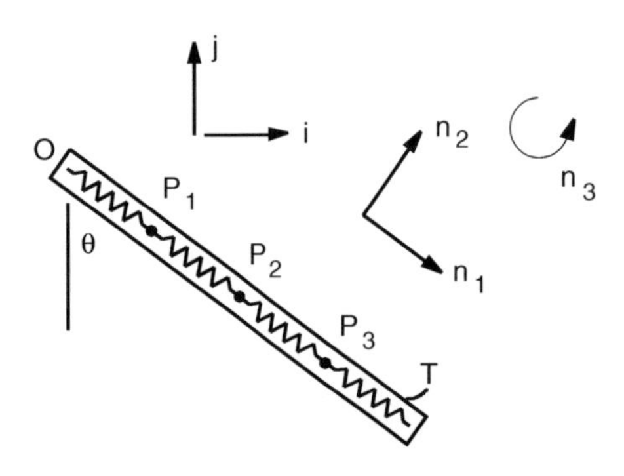

FIGURE 12.2.4
A rotating tube containing spring-connected particles.

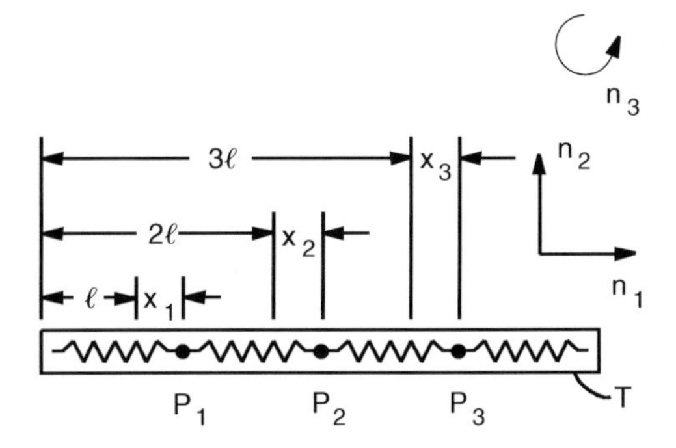

FIGURE 12.2.5
Coordinates of the particles within the tube.

In Section 11.10, we found the generalized inertia forces to be (see Eqs. (11.10.30) to (11.10.33)):

$$F_{x_1}^* = -m\left[\ddot{x}_1 - (\ell + x_1)\dot{\theta}^2\right] \tag{12.2.28}$$

$$F_{x_2}^* = -m\left[\ddot{x}_2 - (2\ell + x_2)\dot{\theta}^2\right] \tag{12.2.29}$$

$$F_{x_3}^* = -m\left[\ddot{x}_3 - (3\ell + x_3)\dot{\theta}^2\right] \tag{12.2.30}$$

and

$$\begin{aligned} F_\theta^* = &-m\left[(\ell + x_1)^2 + (2\ell + x_2)^2 + (3\ell + x_3)^2\right]\ddot{\theta} - M(L^2/3)\,\ddot{\theta} \\ &-2m\left[(\ell + x_1)\dot{x}_1 + (2\ell + x_2)\dot{x}_2 + (3\ell + x_3)\dot{x}_3\right]\dot{\theta} \end{aligned} \tag{12.2.31}$$

Hence, from Kane's equations, the governing dynamical equations are:

$$mg\cos\theta + kx_2 - 2kx_1 - m\left[\ddot{x}_1 - (\ell + x_1)\dot{\theta}^2\right] = 0 \tag{12.2.32}$$

$$mg\cos\theta + kx_3 + kx_1 - m\left[\ddot{x}_2 - (2\ell + x_2)\dot{\theta}^2\right] = 0 \tag{12.2.33}$$

$$mg\cos\theta - 2kx_3 + kx_2 - m\left[\ddot{x}_3 - (3\ell + x_3)\dot{\theta}^2\right] = 0 \tag{12.2.34}$$

and

$$\begin{aligned} &-Mg(L/2)\sin\theta - mg(6\ell + x_1 + x_2 + x_3)\sin\theta \\ &-m\left[(\ell + x_1)^2 + (2\ell + x_2)^2 + (3\ell + x_3)^2\right]\ddot{\theta} - M(L^2/3)\ddot{\theta} \\ &-2m\left[(\ell + x_1)\dot{x}_1 + (2\ell + x_2)\dot{x}_2 + (3\ell + x_3)\dot{x}_3\right]\dot{\theta} = 0 \end{aligned} \tag{12.2.35}$$

Example 12.2.5: Rolling Circular Disk

As a final example, consider again the rolling circular disk of Figure 12.2.6 (we considered this system in Sections 4.12, 8.13, 11.3, 11.10, and 11.12). This is a nonholonomic system with three degrees of freedom represented by the angles as shown.

The applied forces acting on the disk *D* may be represented by a weight force $-mg\mathbf{N}_3$ passing through the mass center *G* together with a contact force **C** passing through the contact point *C*. Because the contact point has zero velocity (the rolling condition), **C** does

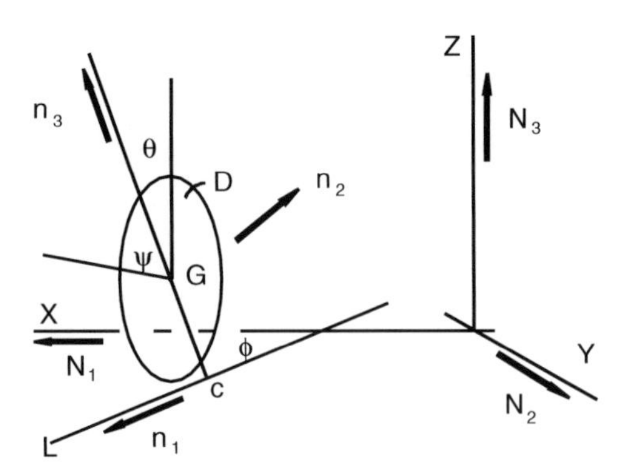

FIGURE 12.2.6
A rolling circular disk.

not contribute to the generalized active forces. Regarding the weight force, we found in Section 11.10 that the partial velocities of G relative to θ, ϕ, and ψ are (see Eq. (11.10.43)):

$$\mathbf{v}^G_{\dot\theta} = -r\mathbf{n}_2\ ,\quad \mathbf{v}^G_{\dot\phi} = r\sin\theta\mathbf{n}_1\ ,\quad \mathbf{v}^G_{\dot\psi} = r\mathbf{n}_1 \tag{12.2.36}$$

Hence, the generalized active forces are:

$$F_\theta = (-mg\mathbf{N}_3)\cdot(-r\mathbf{n}_2) = mgr\sin\theta \tag{12.2.37}$$

$$F_\phi = (-mg\mathbf{N}_3)\cdot(-r\sin\theta\,\mathbf{n}_1) \tag{12.2.38}$$

and

$$F_\psi = (-mg\mathbf{N}_3)\cdot(r\mathbf{n}_1) = 0 \tag{12.2.39}$$

Also in Section 11.10 we found that the generalized inertia forces for the angles are (see Eqs. (11.10.53), (11.10.54), and (11.10.55)):

$$F^*_\theta = mr^2\left[-(5/4)\,\ddot\theta + (3/2)\dot\psi\,\dot\phi\cos\theta + (5/4)\dot\phi^2\sin\theta\cos\theta\right] \tag{12.2.40}$$

$$\begin{aligned} F^*_\phi = mr^2\big[&-(3/2)\ddot\psi\sin\theta - (3/2)\ddot\phi\sin^2\theta - (5/2)\dot\phi\dot\theta\sin\theta\cos\theta \\ &-(1/4)\ddot\phi\cos^2\theta + (1/2)\dot\psi\dot\theta\cos\theta\big] \end{aligned} \tag{12.2.41}$$

and

$$F^*_\psi = -mr^2\left[(3/2)\ddot\psi + (3/2)\ddot\phi\sin\theta + (5/2)\dot\theta\dot\phi\cos\theta\right] \tag{12.2.42}$$

Kane's equations then lead to the expressions:

$$F_\theta + F^*_\theta = 0\ ,\quad F_\phi + F^*_\phi = 0\ ,\quad F_\psi + F^*_\psi = 0 \tag{12.2.43}$$

or

$$mgr\sin\theta + mr^2\left[-(5/4)\,\ddot{\theta} + (3/2)\,\dot{\psi}\,\dot{\phi}\cos\theta + (5/4)\dot{\phi}\sin\theta\cos\theta = 0\right. \tag{12.2.44}$$

$$\begin{aligned}0 + mr^2\Big[&-(3/2)\,\ddot{\psi}\sin\theta - (3/2)\ddot{\phi}\sin^2\theta - (5/2)\dot{\phi}\dot{\theta}\,\sin\theta\cos\theta \\ &-(1/4)\ddot{\phi}\cos^2\theta + (1/2)\,\dot{\psi}\,\dot{\theta}\cos\theta\Big] = 0\end{aligned} \tag{12.2.45}$$

$$0 - mr^2\left[(3/2)\,\ddot{\psi} + (3/2)\ddot{\phi}\sin\theta + (5/2)\dot{\theta}\dot{\phi}\cos\theta\right] = 0 \tag{12.2.46}$$

After simplification, these equations may be written in the forms:

$$(4g/r)\sin\theta - 5\ddot{\theta} + 6\,\dot{\psi}\,\dot{\phi}\cos\theta + 5\dot{\phi}^2\,\sin\theta\cos\theta = 0 \tag{12.2.47}$$

$$\ddot{\phi}\cos\theta + 2\,\dot{\psi}\,\dot{\theta} = 0 \tag{12.2.48}$$

$$3\,\ddot{\psi} + 3\ddot{\phi}\sin\theta + 5\dot{\phi}\dot{\theta}\cos\theta = 0 \tag{12.2.49}$$

(Note that in this simplification process the terms of Eqs. (12.2.46) were recognized as the first three terms in Eq. (12.2.45).)

Equations (12.2.47), (12.2.48), and (12.2.49) are identical to Eqs. (8.13.16), (8.13.17), and (8.13.18), obtained using d'Alembert's principle.

12.3 Lagrange's Equations

Next to Newton's laws, Lagrange's equations are probably the most widely used equations for studying systems with several degrees of freedom. Lagrange's equations can be obtained directly from Kane's equations. Unlike Kane's equations, however, Lagrange's equations are primarily restricted to holonomic systems.

Consider a holonomic mechanical system S with n degrees of freedom represented by the coordinates q_r ($r = 1,\ldots, n$). Then, the generalized inertia forces F_r^* may be expressed in terms of the kinetic energy K of S as [see Eq. 11.12.5)]:

$$F_r^* = -\frac{d}{dt}\left(\frac{\partial K}{\partial \dot{q}_r}\right) + \frac{\partial K}{\partial q_r} \quad (r = 1,\ldots,n) \tag{12.3.1}$$

Kane's equations state that (see Eq. (12.2.1)):

$$F_r + F_r^* = 0 \quad (r = 1,\ldots,n) \tag{12.3.2}$$

By substituting from Eqs. (12.3.1) into (12.3.2) we have:

$$\frac{d}{dt}\left(\frac{\partial K}{\partial \dot{q}_r}\right) - \frac{\partial K}{\partial q_r} = F_r \quad (r = 1, \ldots, n) \tag{12.3.3}$$

Next, if the system S is such that a potential energy function P exists, the generalized active forces F_r may be expressed as (see Eq. (11.11.1)):

$$F_1 = -\frac{\partial \mathbf{P}}{\partial q_r} \quad (r = 1, \ldots, n) \tag{12.3.4}$$

By substituting into Eq. (12.3.3), we have:

$$\frac{d}{dt}\left(\frac{\partial L}{\partial \dot{q}_r}\right) - \frac{\partial L}{\partial q_r} = 0 \quad (r = 1, \ldots, n) \tag{12.3.5}$$

where L is defined as:

$$L \overset{D}{=} K - \mathbf{P} \tag{12.3.6}$$

and is called the *Lagrangian*.

Equations (12.3.3) and (12.3.5) are the common forms of Lagrange's equations. The principle advantage of these equations is that vector accelerations need not be computed; that is, the inertia forces are developed independently by differentiation of the kinetic energy function. Another advantage of Lagrange's equations is that, like Kane's equations, nonworking forces are automatically eliminated from the equations through use of generalized active forces. As such, nonworking forces may be neglected at the onset of an analysis. Finally, as with Kane's equations, Lagrange's equations produce exactly the same number of equations as the degrees of freedom.

Disadvantages of Lagrange's equations are that they are not readily applicable with nonholonomic systems and the scalar differentiations in Eqs. (12.3.3) and (12.3.5) can be tedious and burdensome for large systems. (References 12.7 and 12.8 discuss extending Lagrange's equations to nonholonomic systems.)

To illustrate the use of Lagrange's equations we can again consider the examples of the foregoing sections.

Example 12.3.1: The Simple Pendulum

Consider first the simple pendulum of Figure 12.3.1. We considered this system in several previous sections including Sections 11.5, 11.10. 11.12, and 12.2. The system has one degree of freedom represented by the angle θ. In Section 11.12, we found the kinetic energy K of the pendulum to be (see Eq. (11.12.7)):

$$K = 1/2\, m\ell^2 \dot{\theta}^2 \tag{12.3.7}$$

The only forces doing work on the pendulum are gravitational. In Section 11.11, we found (see Eqs. (11.11.5)) that the potential energy of the gravitational forces on the pendulum bob may be expressed as:

$$P = mgh \tag{12.3.8}$$

where h is the elevation of the bob above an arbitrary but fixed reference level. If we take the reference level through the support O, we find h to be:

$$h = -\ell \cos\theta \tag{12.3.9}$$

The Lagrangian L of the pendulum is then

$$L = K - P = (1/2) m\ell^2 \dot{\theta}^2 + mg\ell \cos\theta \tag{12.3.10}$$

Substituting L into Lagrange's equations, Eq. (12.3.5), then produces the expression:

$$m\ell^2 \ddot{\theta} + mg\ell \sin\theta = 0 \tag{12.3.11}$$

or

$$\ddot{\theta} + (g/\ell)\sin\theta = 0 \tag{12.3.12}$$

Equation (12.3.12) is seen to be identical to Eqs. (8.4.5) and (12.2.9) obtained using d'Alembert's principle and Kane's equations.

Example 12.3.2: The Rod Pendulum

Consider next the rod pendulum of Figure 12.3.2 (we previously considered this system in Sections 11.5, 11.10, 11.12, and 12.2). As with the simple pendulum, this system has one degree of freedom represented by the angle θ.

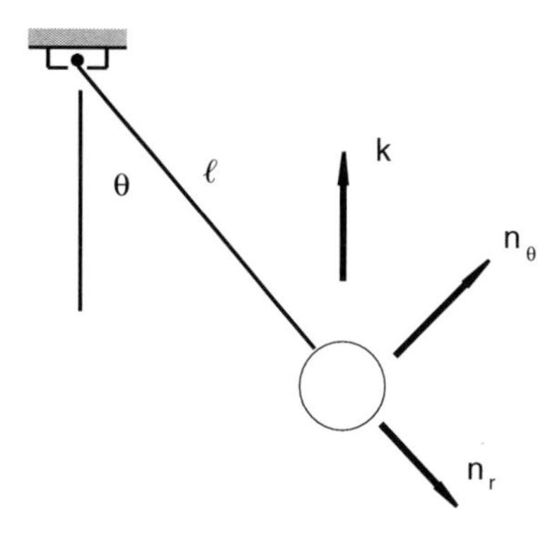

FIGURE 12.3.1
A simple pendulum.

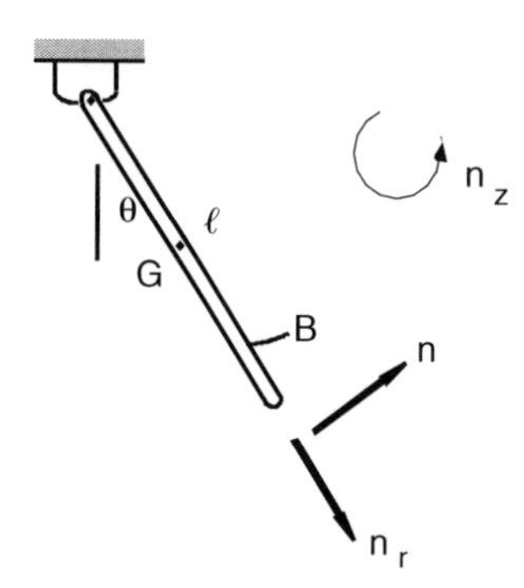

FIGURE 12.3.2
A rod pendulum.

In Section 11.12, we found the kinetic energy K of the pendulum to be (see Eq. (11.12.10)):

$$K = (1/6)m\ell^2 \dot{\theta}^2 \tag{12.3.13}$$

As with the simple pendulum, the only forces doing work on the rod pendulum are gravitational forces. In Section 11.11, we found (see Eq. (11.11.5)) that the potential energy P of the gravitation forces on the rod may be expressed as:

$$P = mgh \tag{12.3.14}$$

where here h is the elevation of the rod mass center G above an arbitrary but fixed reference level. If, as before, we take the reference level through the support O, we find h to be:

$$h = -(\ell/2)\cos\theta \tag{12.3.15}$$

The Lagrangian L of the rod pendulum is then

$$L = K - P = (1/6)m\ell^2 \dot{\theta}^2 + (mg\ell/2)\cos\theta \tag{12.3.16}$$

Substituting L into Lagrange's equations, Eq. (12.3.5), thus produces the expression:

$$(1/3)m\ell^2 \ddot{\theta} + (mg\ell/2)\sin\theta = 0 \tag{12.3.17}$$

or

$$\ddot{\theta} + (3g/2\ell)\sin\theta = 0 \tag{12.3.18}$$

Equation (12.3.18) is identical to Eqs. (8.9.10) and (12.2.14) obtained using d'Alembert's principle and Kane's equations.

Example 12.3.3: Double Rod Pendulum

As an extension of this previous system, consider again the double-rod pendulum as in Figure 12.3.3 (we previously considered the double rod pendulum in Sections 11.10, 11.12, and 12.2). The system has two degrees of freedom represented by the angles θ_1 and θ_2.

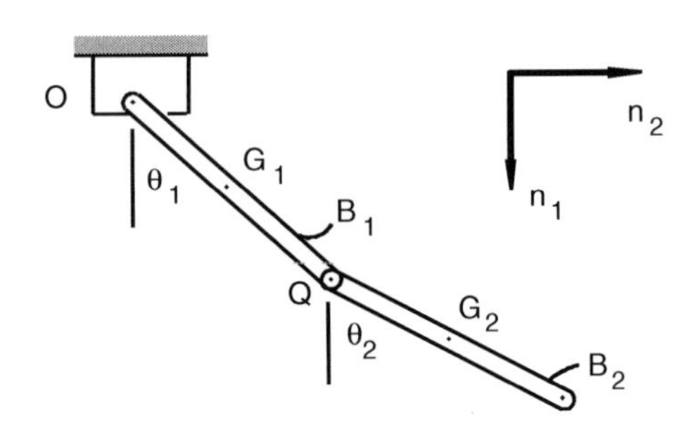

FIGURE 12.3.3
Double-rod pendulum.

In Section 11.12, we found the kinetic energy of the system to be (see Eq. (11.12.13)):

$$K = (2/3)m\ell^2\dot{\theta}_1^2 + (1/2)m\ell^2\dot{\theta}_1\dot{\theta}_2\cos(\theta_2 - \theta_1) + (1/6)m\ell^2\dot{\theta}_2^2 \tag{12.3.19}$$

As with the single-rod pendulum, gravitational forces are the only forces doing work on the system. In Section 11.11, we found the potential energy P of the gravitational forces to be (see Eq. (11.11.5)):

$$P = mgh_1 + mgh_2 \tag{12.3.20}$$

where h_1 and h_2 are the elevations of the rod mass centers G_1 and G_2 above an arbitrary but fixed reference level. If, as before, we take the reference level through the support O we find h_1 and h_2 to be (see Figure 12.3.3):

$$h_1 = -(\ell/2)\cos\theta_1 \quad \text{and} \quad h_2 = -\ell\cos\theta_1 - (\ell/2)\cos\theta_2 \tag{12.3.21}$$

where, as before, each rod has length ℓ and mass m.

Using Eqs. (12.3.19) and (12.3.20), the Lagrangian L becomes:

$$\begin{aligned} L = K - P = (2/3)m\ell^2\dot{\theta}_1^2 + (1/2)m\ell^2\dot{\theta}_1\dot{\theta}_2\cos(\theta_2 - \theta_1) \\ +(1/6)m\ell^2\dot{\theta}_2^2 = (mg\ell/2)\cos\theta_1 \\ +mg\ell\cos\theta_1 + (mg\ell/2)\cos\theta_2 \end{aligned} \tag{12.3.22}$$

Substituting L into Lagrange's equations, Eq. (12.3.5) then produces the expressions:

$$(4/3)\ddot{\theta}_1 + (1/2)\ddot{\theta}_2\cos(\theta_2 - \theta_1) - (1/2)\dot{\theta}_2^2\sin(\theta_2 - \theta_1) + (3g/2\ell)\sin\theta_1 = 0 \tag{12.3.23}$$

$$(1/2)\cos(\theta_2 - \theta_1)\ddot{\theta}_1 + (1/3)\ddot{\theta}_2 - (1/2)\dot{\theta}_1^2\sin(\theta_1 - \theta_2) + (g/2\ell)\sin\theta_2 = 0 \tag{12.3.24}$$

Eqs. (12.3.23) and (12.3.24) are identical to Eqs. (8.10.12) and (8.10.13) obtained using d'Alembert's principle, and Eqs. (12.2.22) and (12.2.23) using Kane's equations. Comparing the three approaches it should be clear that Lagrange's equations produce the governing equations with the least amount of analysis effort.

Example 12.3.4: Spring-Supported Particles in a Rotating Tube

Finally, consider the system of spring-supported particles in a rotating tube as in Figure 12.3.4. We considered this system in Sections 11.7, 11.10, 11.11, 11.12, and 12.12. Recall that the system has four degrees of freedom represented by the angle θ and the coordinates x_1, x_2, and x_3 as in Figure 12.3.5.

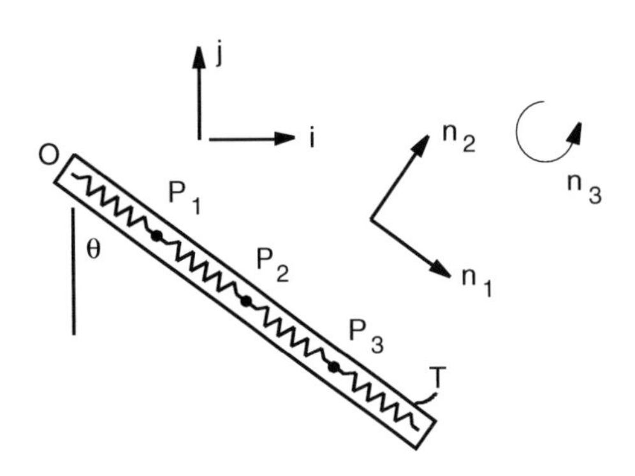

FIGURE 12.3.4
Spring-supported particles in a rotating tube.

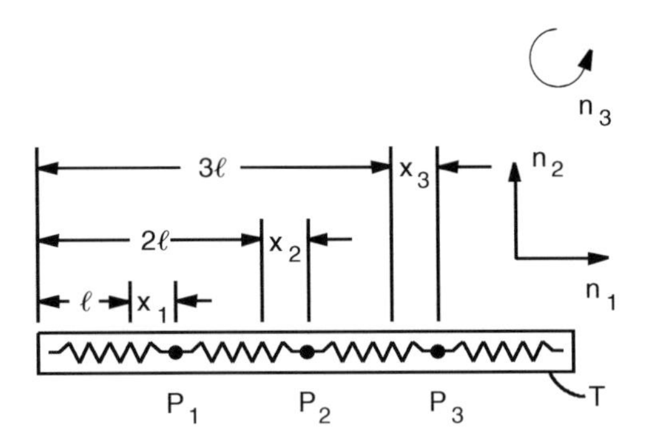

FIGURE 12.3.5
Coordinates of the particles within the tube.

In Section 11.12 (Eq. (11.12.17)), we found the kinetic energy of the system to be:

$$K = (1/2)m\left[\dot{x}_1^2 + (\ell + x_1)^2 \dot{\theta}^2 + \dot{x}_2^2 + (2\ell + x_2)^2 \dot{\theta}^2 + \dot{x}_3^2 + (3\ell + x_3)^2 \dot{\theta}^2\right] + (1/6)ML^2 \dot{\theta}^2 \tag{12.3.25}$$

where, as before, m is the mass of each particle, M is the mass of the tube T, ℓ is the natural length of the springs, and L is the length of the tube T.

The forces doing work in this system are the gravity and spring forces. In Section 11.11 (Eq. (11.11.24)), we found the potential energy of the system to be:

$$P = -mg(\ell + x_1)\cos\theta - mg(2\ell + x_2)\cos\theta - mg(3\ell + x_3)\cos\theta - Mg(L/2)\cos\theta + (1/2)kx_1^2 + (1/2)k(x_2 - x_1)^2 + (1/2)k(x_3 - x_2)^2 \tag{12.3.26}$$

At this point we could use Eq. (12.3.6) to form the Lagrangian and then use Lagrange's equations in the form of Eq. (12.3.5) to obtain the equations of motion. For this system, however, it might be simpler to use Lagrange's equations in the form of Eq. (12.3.3), employing generalized active forces and thus avoiding the differentiation of the potential energy function. In Section 11.7 (Eqs. (11.7.14) to (11.7.17)), we found the generalized active forces to be:

$$F_{x_1} = mg\cos\theta + kx_2 - 2kx_1 \tag{12.3.27}$$

$$F_{x_2} = mg\cos\theta + kx_3 + kx_1 - 2kx_2 \tag{12.3.28}$$

$$F_{x_3} = mg\cos\theta - 2kx_3 + kx_2 \tag{12.3.29}$$

$$F_\theta = -Mg(L/2)\sin\theta - mg(6\ell + x_1 + x_2 + x_3)\sin\theta \tag{12.3.30}$$

Lagrange's equations are then (Eq. (12.3.3)):

$$\frac{d}{dt}\left(\frac{\partial K}{\partial \dot{q}_r}\right) - \frac{\partial K}{\partial q_r} = F_r \quad r = x_1, x_2, x_3, \theta \tag{12.3.31}$$

By substituting from Eqs. (12.3.15), (12.3.17), (12.3.18), (12.3.19), and (12.3.20), the governing equations become:

$$m\ddot{x}_1 - m(\ell + x_1)\dot{\theta}^2 = mg\cos\theta + kx_2 - 2kx_1 \tag{12.3.32}$$

$$m\ddot{x}_2 - m(2\ell + x_2)\dot{\theta}^2 = mg\cos\theta + kx_3 + kx_1 \tag{12.3.33}$$

$$m\ddot{x}_3 - m(3\ell + x_3)\dot{\theta}^2 = mg\cos\theta - 2kx_3 + kx_2 \tag{12.3.34}$$

and

$$\begin{gathered} m\left[(\ell + x_1)^2 + (2\ell + x_2)^2 + (3\ell + x_3)^2\right]\ddot{\theta} + (1/3)ML^2\ddot{\theta} \\ + 2m\left[(\ell + x_1)\dot{x}_1\dot{\theta} + (2\ell + x_2)\dot{x}_2\dot{\theta} + (3\ell + x_3)\dot{x}_3\dot{\theta}\right] \\ = -Mg(L/2)\sin\theta - mg(6\ell + x_1 + x_2 + x_3)\sin\theta \end{gathered} \tag{12.3.35}$$

Equations (12.3.22), (12.3.23), (12.3.24), and (12.3.25) are seen to be identical to Eqs. (12.2.32), (12.2.33), (12.2.34), and (12.2.35) obtained using Kane's equations.

12.4 The Triple-Rod Pendulum

As a further illustration of the use of Lagrange's equations, consider the triple-rod pendulum depicted in Figure 12.4.1. We initially considered this system in Section 8.11. As before, let the three rods be identical, each having length ℓ and mass m. Let the rods be connected by frictionless pins. As a generalization of the system of Section 8.11, however, let there be moments M_1, M_2, and M_3 on and between the rods as in Figure 12.4.1. Also, let there be a point mass Q (with mass M) at the end of the third rod as shown. This system has three degrees of freedom as represented by the angles θ_1, θ_2, and θ_3.

To develop the kinematics it is helpful to introduce the unit vectors and points shown in Figure 12.4.2, where G_1, G_2, and G_3 are the mass centers of the rods, O, Q_1, and Q_2 are

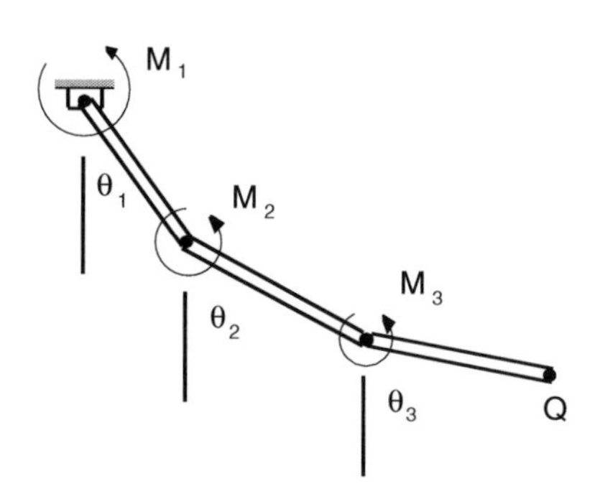

FIGURE 12.4.1
A triple-rod pendulum.

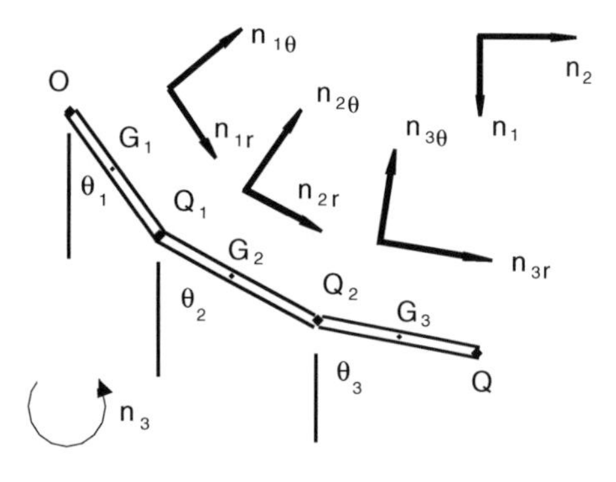

FIGURE 12.4.2
Unit vector geometry.

points on the pin axes, and B_1, B_2, and B_3 are the rods themselves. Using this notation, the angular velocities of the rods are:

$$\boldsymbol{\omega}^{B_1} = \dot{\theta}_1 \mathbf{n}_3 , \quad \boldsymbol{\omega}^{B_2} = \dot{\theta}_2 \mathbf{n}_3 , \quad \boldsymbol{\omega}^{B_3} = \dot{\theta}_3 \mathbf{n}_3 \tag{12.4.1}$$

Also with this notation, the mass center and pin velocities are:

$$\begin{aligned} \mathbf{v}^{G_1} &= (\ell/2)\dot{\theta}_1 \mathbf{n}_{1\theta} \\ \mathbf{v}^{G_2} &= \ell\dot{\theta}_1 \mathbf{n}_{1\theta} + (\ell/2)\dot{\theta}_2 \mathbf{n}_{2\theta} \\ \mathbf{v}^{G_3} &= \ell\dot{\theta}_1 \mathbf{n}_{1\theta} + \ell\dot{\theta}_2 \mathbf{n}_{2\theta} + (\ell/2)\dot{\theta}_3 \mathbf{n}_{3\theta} \end{aligned} \tag{12.4.2}$$

and

$$\begin{aligned} \mathbf{v}^{Q_1} &= \ell\dot{\theta}_1 \mathbf{n}_{1\theta} \\ \mathbf{v}^{Q_2} &= \ell\dot{\theta}_1 \mathbf{n}_{1\theta} + \ell\dot{\theta}_2 \mathbf{n}_{2\theta} \\ \mathbf{v}^{Q_3} &= \ell\dot{\theta}_1 \mathbf{n}_{1\theta} + \ell\dot{\theta}_2 \mathbf{n}_{2\theta} + \ell\dot{\theta}_3 \mathbf{n}_{3\theta} \end{aligned} \tag{12.4.3}$$

The partial angular velocities of the rods are then

$$\begin{array}{lll} \boldsymbol{\omega}^{B_1}_{\dot{\theta}_1} = \mathbf{n}_3 , & \boldsymbol{\omega}^{B_1}_{\dot{\theta}_2} = 0 , & \boldsymbol{\omega}^{B_1}_{\dot{\theta}_3} = 0 \\ \boldsymbol{\omega}^{B_2}_{\dot{\theta}_1} = 0 , & \boldsymbol{\omega}^{B_2}_{\dot{\theta}_2} = \mathbf{n}_3 , & \boldsymbol{\omega}^{B_2}_{\dot{\theta}_3} = 0 \\ \boldsymbol{\omega}^{B_3}_{\dot{\theta}_1} = 0 , & \boldsymbol{\omega}^{B_3}_{\dot{\theta}_2} = 0 , & \boldsymbol{\omega}^{B_3}_{\dot{\theta}_3} = \mathbf{n}_3 \end{array} \tag{12.4.4}$$

Similarly, the partial velocities of the mass centers and of point Q are:

$$\begin{array}{lll} \mathbf{v}^{G_1}_{\dot{\theta}_1} = (\ell/2)\mathbf{n}_{1\theta} , & \mathbf{v}^{G_1}_{\dot{\theta}_2} = 0 , & \mathbf{v}^{G_1}_{\dot{\theta}_3} = 0 \\ \mathbf{v}^{G_2}_{\dot{\theta}_1} = \ell\mathbf{n}_{1\theta} , & \mathbf{v}^{G_2}_{\dot{\theta}_2} = (\ell/2)\mathbf{n}_{2\theta} , & \mathbf{v}^{G_2}_{\dot{\theta}_3} = 0 \\ \mathbf{v}^{G_3}_{\dot{\theta}_1} = \ell\mathbf{n}_{1\theta} , & \mathbf{v}^{G_3}_{\dot{\theta}_2} = \ell\mathbf{n}_{2\theta} , & \mathbf{v}^{G_3}_{\dot{\theta}_2} = (\ell/2)\mathbf{n}_{3\theta} \end{array} \tag{12.4.5}$$

and

$$\mathbf{v}^{Q}_{\dot{\theta}_1} = \ell\mathbf{n}_{1\theta} , \quad \mathbf{v}^{Q}_{\dot{\theta}_2} = \ell\mathbf{n}_{2\theta} , \quad \mathbf{v}^{Q}_{\dot{\theta}_3} = \ell\mathbf{n}_{3\theta} \tag{12.4.6}$$

The applied forces that contribute to the generalized forces are weight forces $mg\mathbf{n}_1$ through the mass centers, the weight force $Mg\mathbf{n}_1$ through Q, and the moments at the pin joints. Hence, the generalized forces become:

$$\begin{aligned} F_{\theta_1} = {} & mg\mathbf{n}_1 \cdot \mathbf{v}^{G_1}_{\dot{\theta}_1} + mg\mathbf{n}_1 \cdot \mathbf{v}^{G_2}_{\dot{\theta}_1} + mg\mathbf{n}_1 \cdot \mathbf{v}^{G_3}_{\dot{\theta}_1} \\ & + Mg\mathbf{n}_1 \cdot \mathbf{v}^{Q}_{\dot{\theta}_1} + (M_1 - M_2)\mathbf{n}_3 \cdot \boldsymbol{\omega}^{B_1}_{\dot{\theta}_1} \\ & + (M_2 - M_3)\mathbf{n}_3 \cdot \boldsymbol{\omega}^{B_2}_{\dot{\theta}_1} + M_3\mathbf{n}_3 \cdot \boldsymbol{\omega}^{B_3}_{\dot{\theta}_1} \end{aligned} \tag{12.4.7}$$

By substituting from Eqs. (12.4.4), (12.4.5), and (12.4.6), F_{θ_1} becomes:

$$F_{\theta_1} = -mg(\ell/2)\sin\theta_1 - mg\ell\sin\theta_1 - mg\ell\sin\theta_1 + M_1 - M_2 - Mg\ell\sin\theta_1$$

or

$$F_{\theta_1} = -(5/2)mg\ell\sin\theta_1 + M_1 - M_2 - Mg\ell\sin\theta_1 \tag{12.4.8}$$

Similarly, F_{θ_2} and F_{θ_3} are:

$$F_{\theta_2} = -(3/2)mg\ell\sin\theta_2 + M_2 - M_3 - Mg\ell\sin\theta_2 \tag{12.4.9}$$

and

$$F_{\theta_3} = -(1/2)mg\ell\sin\theta_3 + M_3 - Mg\ell\sin\theta_3 \tag{12.4.10}$$

The kinetic energy K of the system may be expressed as:

$$\begin{aligned} K = {} & (1/2)m(\mathbf{v}^{G_1})^2 + (1/2)\mathbf{I}(\boldsymbol{\omega}^{B_1})^2 + (1/2)m(\mathbf{v}^{G_2})^2 + (1/2)\mathbf{I}(\boldsymbol{\omega}^{B_2})^2 \\ & + (1/2)m(\mathbf{v}^{G_3})^2 + (1/2)\mathbf{I}(\boldsymbol{\omega}^{B_3})^2 + (1/2)\mathbf{M}(\mathbf{v}^{Q})^2 \end{aligned} \tag{12.4.11}$$

where I is the central moment of inertia of a rod about an axis normal to the rod and given by:

$$\mathrm{I} = (1/2)m\ell^2 \tag{12.4.12}$$

Using Eqs. (12.4.1) and (12.4.2) K becomes (after simplification):

$$\begin{aligned} K = {} & (1/2)m\ell^2\big[(7/3)\dot\theta_1^2 + (4/3)\dot\theta_2^2 + (1/3)\dot\theta_3^2 + 3\dot\theta_1\dot\theta_2\cos(\theta_2-\theta_1) \\ & + \dot\theta_2\dot\theta_3\cos(\theta_3-\theta_2) + \dot\theta_1\dot\theta_3\cos(\theta_3-\theta_1)\big] \\ & + (1/2)M\ell^2\big[\dot\theta_1^2 + \dot\theta_2^2 + \theta_3^2 + 2\dot\theta_1\dot\theta_3\cos(\theta_2-\theta_1) \\ & + 2\dot\theta_2\dot\theta_3\cos(\theta_3-\theta_2) + 2\dot\theta_3\dot\theta_1\cos(\theta_3-\theta_1)\big] \end{aligned} \tag{12.4.13}$$

By differentiating in Eq. (12.4.13), we obtain the following terms, useful in Lagrange's equations:

$$\begin{aligned} \partial K/\partial\theta_1 = {} & (1/2)m\ell^2\big[3\dot\theta_1\dot\theta_2\sin(\theta_2-\theta_1) - \dot\theta_1\dot\theta_3\sin(\theta_1-\theta_3)\big] \\ & + (1/2)M\ell^2\big[2\dot\theta_1\theta_2\sin(\theta_2-\theta_1) - 2\dot\theta_3\theta_1\sin(\theta_1-\theta_3)\big] \end{aligned} \tag{12.4.14}$$

$$
\begin{aligned}
\partial K/\partial\theta_2 &= (1/2)m\ell^2\left[-3\dot{\theta}_1\dot{\theta}_2\sin(\theta_2-\theta_1)+\dot{\theta}_2\,\dot{\theta}_3\sin(\theta_3-\theta_2)\right] \\
&\quad +(1/2)M\ell^2\left[-2\dot{\theta}_1\,\theta_2\sin(\theta_2-\theta_1)+2\dot{\theta}_2\,\theta_3\sin(\theta_3-\theta_2)\right]
\end{aligned}
\tag{12.4.15}
$$

$$
\begin{aligned}
\partial K/\partial\theta_3 &= (1/2)m\ell^2\left[-\dot{\theta}_2\dot{\theta}_3\sin(\theta_3-\theta_2)+\dot{\theta}_1\,\dot{\theta}_3\sin(\theta_2-\theta_3)\right] \\
&\quad +(1/2)M\ell^2\left[-2\dot{\theta}_2\,\theta_3\sin(\theta_3-\theta_2)+2\dot{\theta}_3\,\dot{\theta}_1\sin(\theta_1-\theta_3)\right]
\end{aligned}
\tag{12.4.16}
$$

$$
\begin{aligned}
\partial K/\partial\dot{\theta}_1 &= (1/2)m\ell^2\left[(14/3)\dot{\theta}_1+3\dot{\theta}_2\cos(\theta_2-\theta_1)+\dot{\theta}_3\cos(\theta_1-\theta_3)\right] \\
&\quad +(1/2)M\ell^2\left[2\dot{\theta}_1+2\dot{\theta}_2\cos(\theta_2-\theta_1)+2\dot{\theta}_3\cos(\theta_1-\theta_3)\right]
\end{aligned}
\tag{12.4.17}
$$

$$
\begin{aligned}
\partial K/\partial\dot{\theta}_2 &= (1/2)m\ell^2\left[(8/3)\dot{\theta}_2+3\dot{\theta}_1\cos(\theta_2-\theta_1)+\dot{\theta}_3\cos(\theta_3-\theta_2)\right] \\
&\quad +(1/2)M\ell^2\left[2\dot{\theta}_2+2\dot{\theta}_1\cos(\theta_2-\theta_1)+2\dot{\theta}_3\cos(\theta_3-\theta_2)\right]
\end{aligned}
\tag{12.4.18}
$$

$$
\begin{aligned}
\partial K/\partial\dot{\theta}_3 &= (1/2)m\ell^2\left[(2/3)\dot{\theta}_3+\dot{\theta}_2\cos(\theta_3-\theta_2)+\dot{\theta}_1\cos(\theta_1-\theta_3)\right] \\
&\quad +(1/2)M\ell^2\left[2\dot{\theta}_3+2\dot{\theta}_2\cos(\theta_3-\theta_2)+2\dot{\theta}_1\cos(\theta_1-\theta_3)\right]
\end{aligned}
\tag{12.4.19}
$$

By substituting from Eqs. (12.4.8), (12.4.9), and (12.4.10) and Eqs. (12.4.14) through (12.4.19) into Lagrange's equations, Eq. (12.3.3), we obtain the governing equations:

$$
\begin{aligned}
&(7/3)\,\ddot{\theta}_1+(3/2)\,\ddot{\theta}_2\cos(\theta_1-\theta_2)+(1/2)\,\ddot{\theta}_3\cos(\theta_1-\theta_3)+(3/2)\dot{\theta}_2^2\sin(\theta_1-\theta_2) \\
&\quad +(1/2)\dot{\theta}_3^2\sin(\theta_1-\theta_3)+(M/m)\left[\ddot{\theta}_1+\ddot{\theta}_2\cos(\theta_1-\theta_2)+\ddot{\theta}_3\cos(\theta_1-\theta_3)\right. \\
&\quad \left.+\dot{\theta}_2^2\sin(\theta_1-\theta_2)+\dot{\theta}_3^2\sin(\theta_1-\theta_3)\right]+(5/2)(g/\ell)\sin\theta_2-M_2+M_3=0
\end{aligned}
\tag{12.4.20}
$$

$$
\begin{aligned}
&(4/3)\,\ddot{\theta}_2+(3/2)\,\ddot{\theta}_1\cos(\theta_2-\theta_1)+(1/2)\,\ddot{\theta}_3\cos(\theta_2-\theta_3)+(3/2)\dot{\theta}_2^1\sin(\theta_2-\theta_1) \\
&\quad +(1/2)\dot{\theta}_3^2\sin(\theta_2-\theta_3)+(M/m)\left[\ddot{\theta}_2+\ddot{\theta}_1\cos(\theta_2-\theta_1)+\ddot{\theta}_3\cos(\theta_2-\theta_3)\right. \\
&\quad \left.+\dot{\theta}_1^2\sin(\theta_2-\theta_1)+\dot{\theta}_3^2\sin(\theta_2-\theta_3)\right]+(3/2)(g/\ell)\sin\theta_2-M_2+M_3=0
\end{aligned}
\tag{12.4.21}
$$

$$
\begin{aligned}
&(1/3)\,\ddot{\theta}_3+(1/2)\,\ddot{\theta}_1\cos(\theta_3-\theta_1)+(1/2)\,\ddot{\theta}_2\cos(\theta_3-\theta_2)+(1/2)\dot{\theta}_1^2\sin(\theta_3-\theta_1) \\
&\quad +(1/2)\dot{\theta}_2^2\sin(\theta_3-\theta_2)+(M/m)\left[\ddot{\theta}_3+\ddot{\theta}_1\cos(\theta_3-\theta_1)+\ddot{\theta}_2\cos(\theta_3-\theta_2)\right. \\
&\quad \left.+\dot{\theta}_1^2\sin(\theta_3-\theta_1)+\dot{\theta}_2^2\sin(\theta_3-\theta_2)\right]+(1/2)(g/\ell)\sin\theta_3-M_3=0
\end{aligned}
\tag{12.4.22}
$$

If we let the joint moments (M_1, M_2, and M_3) be zero and if we let the point mass M also be zero in Eqs. (12.4.20), (12.4.21), and (12.4.22), we see that the equations are identical to Eqs. (8.11.1), (8.11.2), and (8.11.3) reportedly developed using d'Alembert's principle. Although the details of that development were not presented, the use of Lagrange's equations has the clear advantage of providing the simpler analysis. As noted before, the simplicity and efficiency of the Lagrangian analysis stem from the avoidance of the evaluation of accelerations and from the automatic elimination of nonworking constraint forces. In the following section, we outline the extension of this example to include N rods.

12.5 The *N*-Rod Pendulum

Consider a pendulum system composed of N identical pin-connected rods with a point mass Q at the end as in Figure 12.5.1 (we considered this system [without the end mass] in Section 8.11). As before, let each rod have mass m and length ℓ. Also, let us restrict our analysis to motion in the vertical plane. The system then has N degrees of freedom represented by the angles θ_i ($i = 1,\dots, n$) as shown in Figure 12.5.1. This system is useful for modeling the dynamic behavior of chains and cables.

We can study this system by generalizing our analysis for the triple-rod pendulum. Indeed, by examining Eqs. (12.4.1) through (12.4.10), we see patterns that can readily be generalized. To this end, consider a typical rod B_i as in Figure 12.5.2. Then, from Eq. (12.4.1), the angular velocity of B_i in the fixed inertia frame may be expressed as:

$$\boldsymbol{\omega}^{B_1} = \dot{\theta}_i \mathbf{n}_3 \tag{12.5.1}$$

where $\mathbf{n}_3$ is a unit vector normal to the plane of motion as in Figure 12.5.2.

Next, from Eq. (12.4.2), the velocity of the mass center G_i may be expressed as:

$$\mathbf{v}^{G_i} = \ell\dot{\theta}_1\mathbf{n}_{1\theta} + \ell\dot{\theta}_2\mathbf{n}_{2\theta} + \dots + \ell\dot{\theta}_{i-1}\mathbf{n}_{(i-1)\theta} + (\ell/2)\dot{\theta}_i\mathbf{n}_{i\theta} \tag{12.5.2}$$

where as before, and as in Figure 12.5.2, the $\mathbf{n}_{i\theta}$ are unit vectors normal to the rods.

Similarly, the velocity of the point mass Q is:

$$\mathbf{v}^{Q} = \ell\dot{\theta}_1\mathbf{n}_{1\theta} + \ell\dot{\theta}_2\mathbf{n}_{2\theta} + \dots + \ell\dot{\theta}_N\mathbf{n}_{N\theta} \tag{12.5.3}$$

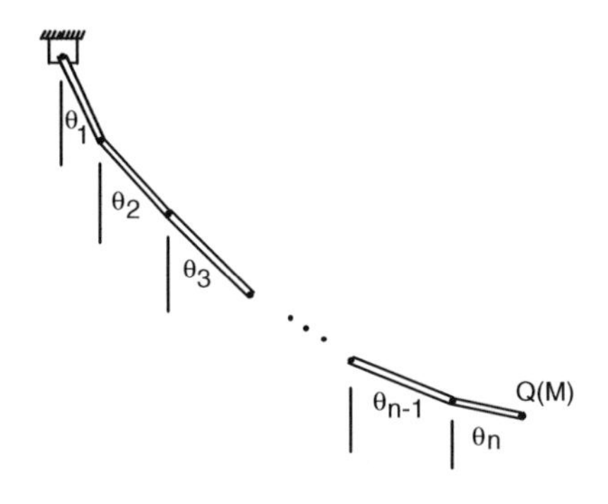

FIGURE 12.5.1
N-rod pendulum with endpoint mass.

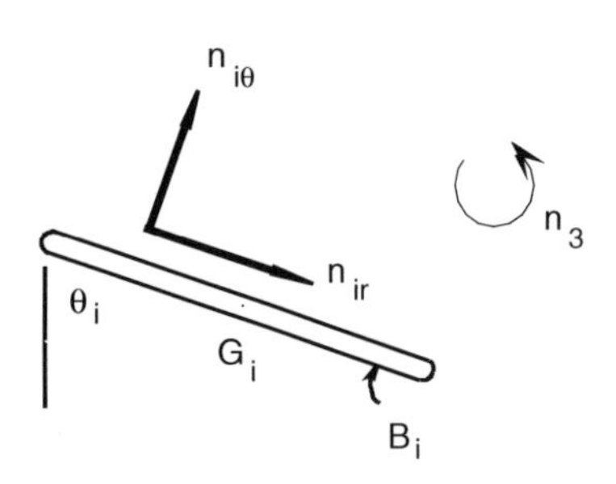

FIGURE 12.5.2
A typical rod of the pendulum.

From Eq. (12.4.4) we see that the partial angular velocities of B_i are:

$$\boldsymbol{\omega}_{\dot{\theta}_j}^{B_i} = \begin{cases} 0 & i \neq j \\ \mathbf{n}_3 & i = j \end{cases} \tag{12.5.4}$$

Similarly, from Eq. (12.4.5), we see that the partial velocities of G_i are:

$$\boldsymbol{\omega}_{\dot{\theta}_j}^{G_i} = \begin{cases} \ell \mathbf{n}_{j\theta} & j < i \\ (\ell/2)\mathbf{n}_{j\theta} & j = i \\ 0 & j > i \end{cases} \tag{12.5.5}$$

From an examination of Eqs. (12.4.8), (12.4.9), and (12.4.10), the generalized forces are:

$$F_{\theta_i} = -[N - i + (1/2)]mg\ell \sin\theta_i - Mg\ell \sin\theta_i \tag{12.5.6}$$

where M is the mass of Q.

From a generalization of Eq. (12.4.11) the kinetic energy K of the N-rod system is:

$$\begin{aligned} K = {} & (1/2)m\left(\mathbf{v}^{G_1}\right)^2 + (1/2)m\left(\mathbf{v}^{G_2}\right)^2 + \ldots + (1/2)m\left(\mathbf{v}^{G_N}\right)^2 \\ & + (1/2)I\left(\boldsymbol{\omega}^{B_1}\right)^2 + (1/2)I\left(\boldsymbol{\omega}^{B_2}\right)^2 + \ldots + (1/2)I\left(\boldsymbol{\omega}^{B_N}\right)^2 \\ & + (1/2)M\left(\mathbf{v}^{Q}\right)^2 \end{aligned} \tag{12.5.7}$$

where as before I is the central moment of inertia of a rod about an axis parallel to $\mathbf{n}_3$ and is given by:

$$I = (1/12)m\ell^2 \tag{12.5.8}$$

Then, by substituting into Eq. (12.5.7) from Eqs. (12.5.1), (12.5.2), and (12.5.3), K becomes:

$$K = (1/2)m\ell^2 \sum_{i=1}^{N} \sum_{j=1}^{N} m_{ij} \dot{\theta}_i \dot{\theta}_j \tag{12.5.9}$$

where the coefficients m_{ij} are:

$$m_{ij} = (1/2)[1 + 2(N - p) + 2(M/m)]\cos(\theta_j - \theta_i) \tag{12.5.10}$$

$$i \neq j \text{ and } p \text{ is the larger of } i \text{ and } j$$

and

$$m_{ii} = [N - i + (1/3) + (M/m)] \tag{12.5.11}$$

Finally, substituting from Eqs. (12.5.6) and (12.5.9) into Lagrange's equations in the form of Eq. (12.3.3), we obtain the governing dynamical equations of the system:

$$\sum_{j=1}^{N}\left[m_{rj}\ddot{\theta}_j + n_{rj}\dot{\theta}_j^2 + (g/\ell)k_{rj}\right] = 0 \tag{12.5.12}$$

where the coefficients m_{rj}, n_{rj}, and k_{rj} are:

$$m_{rj} = (1/2)\left[1 + 2(N-p) + 2(M/m)\right]\cos(\theta_j - \theta_r)$$
$$r \neq j \text{ and } p \text{ is the larger of } i \text{ and } j \tag{12.5.13}$$

$$m_{rr} = \left[N - r + (1/3) + (M/m)\right] \tag{12.5.14}$$

$$n_{rj} = -(1/2)\left[1 + 2(N-p) + 2(M/m)\right]\sin(\theta_j - \theta_r)$$
$$r \neq j \text{ and } p \text{ is the larger of } i \text{ and } j \tag{12.5.15}$$

$$n_{rr} = 0 \tag{12.5.16}$$

$$k_{rj} = 0 \qquad r \neq j \tag{12.5.17}$$

$$k_{rr} = \left[N - r + (1/2) + (M/m)\right]\sin\theta_r \tag{12.5.18}$$

12.6 Closure

The computational and analytical advantages of Kane's equations and Lagrange's equations are illustrated by the examples. In each case, the effort required to obtain the governing dynamical equations is significantly less than that with d'Alembert's principle or Newton's laws. As noted earlier, the reason for the reduction in effort is that non-working constraint forces are automatically eliminated from the analysis with Kane's and Lagrange's equations; hence, an analyst can ignore such forces at the onset. Also, with Kane's and Lagrange's equations, the exact same number of governing equations are obtained as the degrees of freedom. Finally, Lagrange's equations offer the additional advantage of using energy functions, which makes the computation of vector acceleration unnecessary. The disadvantages of Lagrange's equations are that they are not applicable with nonholonomic systems, and the differentiation of the energy functions may be tedious and even unwieldy for large systems.

In the following chapters we will consider applications of these principles in vibrations, stability, balancing, and in the study of mechanical components such as gears and cams.

References

12.1. Kane, T. R., Dynamics of nonholonomic systems, *J. Appl. Mech.*, 28, 574, 1961.
12.2. Kane, T. R., *Dynamics*, Holt, Rinehart & Winston, New York, 1968, p. 177.
12.3. Huston, R. L., and Passerello, C. E., On Lagrange's form of d'Alembert's principle, *Matrix Tensor Q*, 23, 109, 1973.
12.4. Papastavridis, J. G., On the nonlinear Appell's equations and the determination of generalized reaction forces, *Int. J. Eng. Sci.*, 26(6), 609, 1988.
12.5. Huston, R. L., Multibody dynamics: modeling and analysis methods [feature article], *Appl. Mech. Rev.*, 44(3), 109, 1991.
12.6. Huston, R. L., Multibody dynamics formulations via Kane's equations, in *Mechanics and Control of Large Flexible Structures*, J. L. Jenkins, Ed., Vol. 129 of *Progress in Aeronautics and Astronautics*, American Institute of Aeronautics and Astronautics (AIAA), 1990, p. 71.
12.7. Huston, R. L., and Passerello, C. E., Another look at nonholonomic systems, *J. Appl. Mech.*, 40, 101, 1973.
12.8. Kane, T. R., and Levinson, D. A., *Dynamics, Theory, and Applications*, McGraw-Hill, New York, 1985, p. 100.

Problems

Section 12.2 Kane's Equations

P12.2.1: Consider the rotating tube T, with a smooth interior surface, containing a particle P with mass m, and rotating about a vertical diameter as in Problems P11.6.6 and P11.9.6 and as shown again in Figure P12.2.1. As before, let the radius of T be r, let the angular speed of T be Ω, and let P be located by the angle θ as shown in Figure P12.2.1. This system has one degree of freedom, which may be represented by θ. Use Kane's equations, Eq. (12.2.1), to determine the governing dynamical equation.

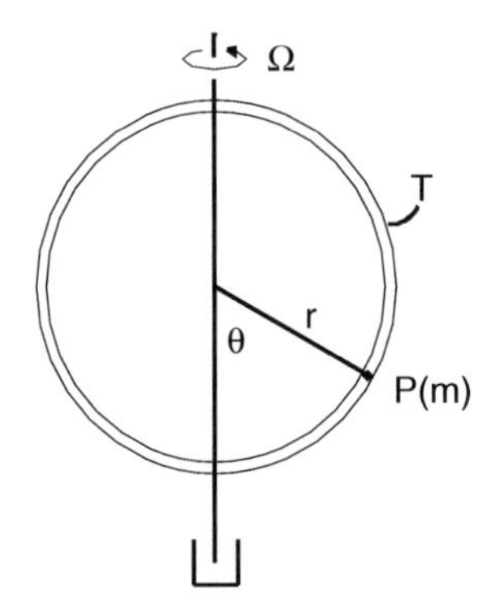

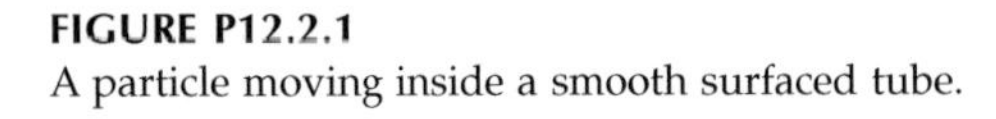
FIGURE P12.2.1
A particle moving inside a smooth surfaced tube.

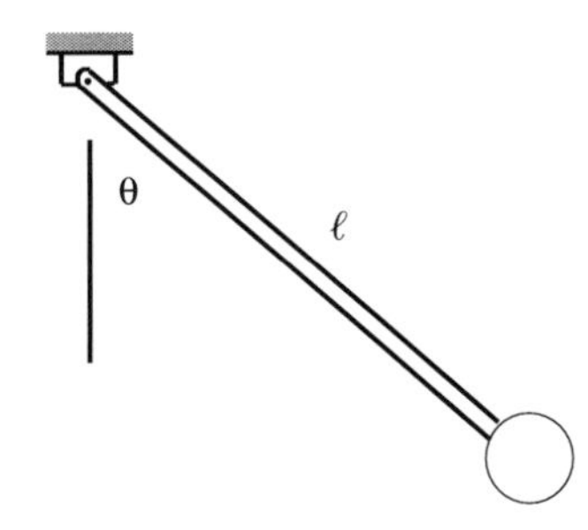

FIGURE P12.2.2
A rod/disk pendulum.

P12.2.2: Consider the pendulum consisting of a rod with length ℓ and mass m attached to a circular disk with radius r and mass M and supported by a frictionless pin as in Problems P11.9.1 and P11.12.1 and as shown again in Figure P12.2.2. This system has one degree of freedom represented by the angle θ. Use Kane's equations to determine the governing dynamical equation.

P12.2.3: Consider the rod pinned to the vertically rotating shaft as in Problems P11.9.2 and as shown again in Figure P12.2.3. If the shaft *S* has a specified angular speed Ω, the system has only one degree of freedom: the angle θ between the rod *B* and *S*. Use Kane's equations to determine the governing dynamical equation where *B* has mass *m* and length ℓ. Assume the radius of *S* is small.

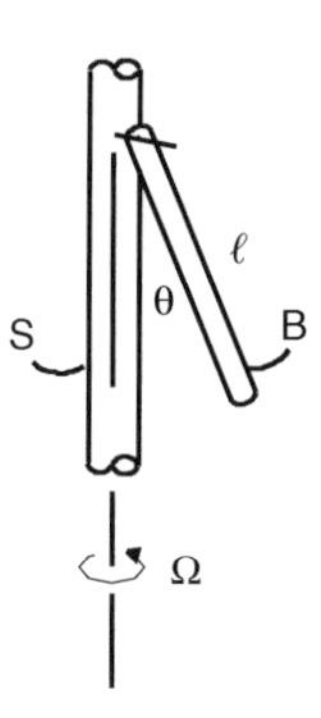

FIGURE P12.2.3
A rod *B* pinned to a rotating shaft *S*

P12.2.4: Repeat Problem P12.2.3 by including the effect of the radius *r* of the shaft *S*. Let the mass of *S* be *M*.

P12.2.5: See Problems P11.9.4 and P12.2.3. Suppose the rotation of *S* is not specified but instead is free, or arbitrary, and defined by the angle ϕ as in Problem P11.9.4 and as represented in Figure P12.2.5. This system now has two degrees of freedom represented by the angles θ and ϕ. Use Kane's equations to determine the governing dynamical equations, assuming the shaft radius *r* is small.

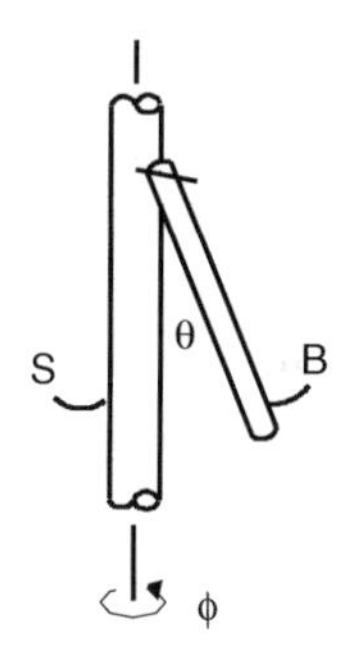

FIGURE P12.2.5
A rod *B* pinned to a rotating shaft *S*.

P12.2.6: Repeat Problem P12.2.5 by including the effect of the shaft radius *r* and the shaft mass *M*.

P12.2.7: Consider a generalization of the double-rod pendulum where the rods have unequal lengths and unequal masses as in Figure P12.2.7. Let the rod lengths be ℓ_1 and ℓ_2, and let their masses be m_1 and m_2. Let the rod orientations be defined by the angles θ_1 and θ_2, as shown. Assuming frictionless pins, determine the equations of motion by using Kane's equations.

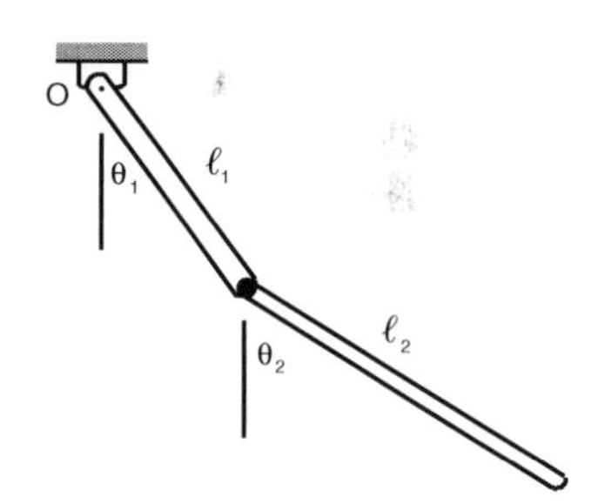

FIGURE P12.2.7
A double-rod pendulum with unequal rod lengths and masses.

P12.2.8: See Problem P12.2.7. Suppose an actuator (or motor) is exerting a moment M_1 at support *O* on the upper bar and suppose further that an actuator at the pin connection between the rods is exerting a moment M_2 on the lower rod by the upper rod (and hence a moment $-M_2$ on the upper rod by the lower rod). Finally, let there be a concentrated mass *M* at the lower end *Q* of the second rod, as represented in Figure P12.2.8. Use Kane's equations to determine the equations of motion of this system.

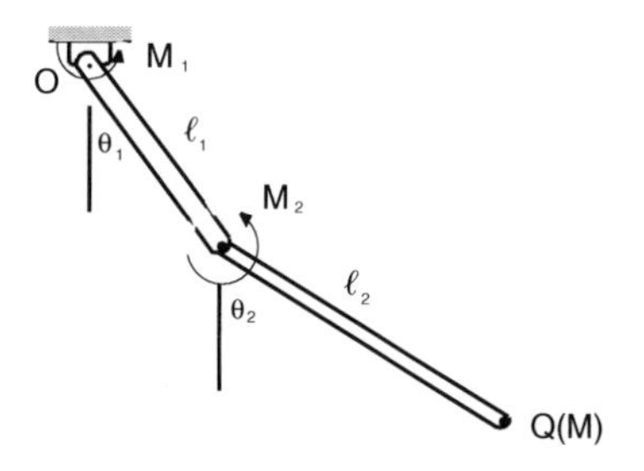

FIGURE P12.2.8
A double-rod pendulum with unequal rods, joint moments, and a concentrated end mass.

P12.2.9: Repeat Problems P12.2.7 and P12.2.8 using the relative orientation angles β_1 and β_2, as shown in Figure P12.2.9, to define the orientations of the rods.

P12.2.10: See Problems P11.6.8 and P11.9.8. Consider again the heavy rotating disk *D* supported by a light yoke *Y* which in turn can rotate relative to a light horizontal shaft *S* which is supported by frictionless bearings as depicted in Figure P12.2.10. Let *D* have mass *m* and radius *r*. Let angular speed Ω of *D* in *Y* be constant. Let the rotation of *Y* relative to *S* be

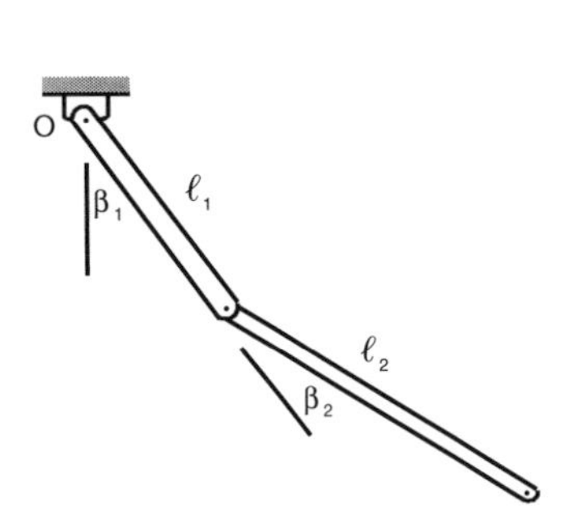

FIGURE P12.2.9
Double, unequal-rod pendulum with relative orientation angles.

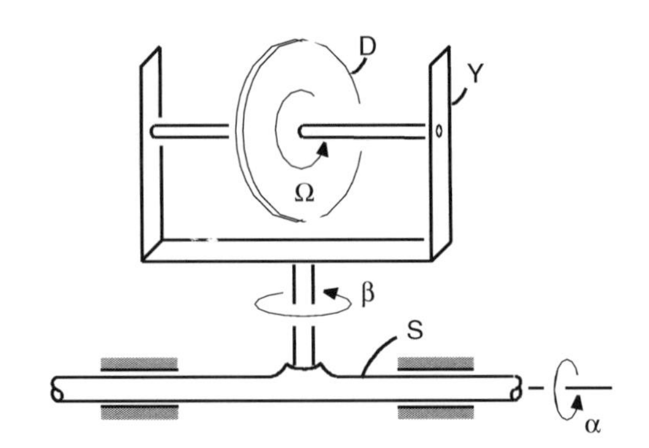

FIGURE P12.2.10
A disk spinning in a free-turning yoke and supported by a shaft *S*.

measured by the angle β and the rotation of *S* in its bearings be measured by the angle α as shown. Recall that this system has two degrees of freedom, which may be represented by the angles α and β. By following the procedures outlined in Problems P11.6.8 and P11.9.8, use Kane's equations to determine the governing equations of motion.

P12.2.11: See Problems P11.6.7 and P11.9.7. Consider again the right circular cone C with altitude h and half-central angle rolling on an inclined plane as in Figure P12.2.11. As before let the incline angle be β and let the position of C be determined by the angle ϕ between the contacting element of C and the plane and a line fixed in the plane as shown. Let ℓ be the element length of C, and let r be the base radius of C. Let O be the apex of C, and let G be the mass center of C. Finally, let ψ measure the roll of C, as shown in Figure P12.2.11. This system has one degree of freedom (see Problem P12.2.11). Use Kane's equations to determine the equations of motion.

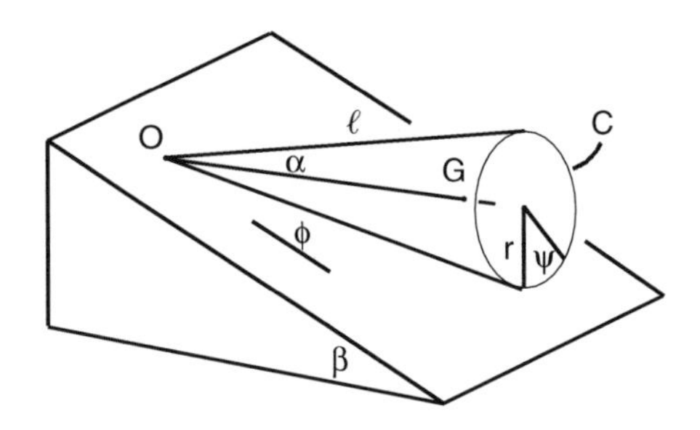

FIGURE P12.2.11
A cone rolling on an inclined plane.

Section 12.3 Lagrange's Equations

P12.3.1 to P12.3.10: Repeat Problems P12.2.1 to P12.2.10 by using Lagrange's equations to obtain the equations of motion. Compare the analysis effort with that of using Kane's equations.

13

Introduction to Vibrations

13.1 Introduction

Vibration is sometimes defined as periodic (repeating) oscillatory movement. It occurs in virtually all mechanical systems. Vibration produces noise, unwanted wear, and even catastrophic failure. On the other hand, for many systems vibration is essential for the proper functioning of the systems. Therefore, analysis and control of vibration are principal problems of mechanical design.

In this chapter we will develop a brief and elementary introduction to mechanical vibration. It is only an introduction and is not intended to replace a course or a more intense study. The reader is referred to the references, which provide a partial listing of the many books devoted to the subject.

We begin with a brief review of solutions to second-order ordinary differential equations. We then consider single and multiple degree of freedom systems. We conclude with a brief discussion of nonlinear vibrations.

13.2 Solutions of Second-Order Differential Equations

Vibration phenomena are often modeled by second-order ordinary differential equations. Solutions of these equations provide a representation of the movement of vibrating systems; therefore, to begin our study, it is helpful to review the solution procedures of second-order ordinary differential equations. The reader is encouraged to also independently review these procedures. References 13.1 to 13.7 provide a sampling of the many texts available on the subject.

We will consider first the so-called *linear oscillator* equation:

$$\ddot{x} + \omega^2 x = 0 \tag{13.2.1}$$

where, as before, the overdot represents differentiation with respect to time, and ω is a constant.

In Eq. (13.2.1) the time t is the independent variable and x is the dependent variable to be determined. It is readily verified that the solution of Eq. (13.2.1) may be expressed in the form:

$$x = A\cos\omega t + B\sin\omega t \tag{13.2.2}$$

where A and B are constants that may be evaluated by auxiliary conditions (*initial conditions* or *boundary conditions*). (We can verify that the expression of Eq. (13.2.2) is indeed a solution of Eq. (13.2.1) by direct substitution. It is in fact a general solution, as $\cos\omega t$ and $\sin\omega t$ are independent functions and there are two arbitrary constants, A and B.)

Through use of trigonometric identities, we can express Eq. (13.2.2) in the form:

$$x = \hat{A}\cos(\omega t + \phi) \tag{13.2.3}$$

where $\hat{A}$ and ϕ are constants. To develop this, recall the identity:

$$\cos(\alpha + \beta) \equiv \cos\alpha\cos\beta - \sin\alpha\sin\beta \tag{13.2.4}$$

Then, by thus expanding the expression of Eq. (13.2.3) we have:

$$\hat{A}\cos(\omega t + \phi) = \hat{A}\cos\omega t\cos\phi - \hat{A}\sin\omega t\sin\phi \tag{13.2.5}$$

By comparing this with Eq. (13.2.2), we have:

$$\begin{aligned} A &= \hat{A}\cos\phi \qquad & B &= \hat{A}\sin\phi \\ \hat{A}^2 &= A^2 + B^2 \qquad & \tan\phi &= -B/A \end{aligned} \tag{13.2.6}$$

In the expression $\hat{A}\cos(\omega t + \phi)$ of Eq. (13.2.3), $\hat{A}$ is the *amplitude*, ω is the *circular frequency*, and ϕ is the *phase*.

Recall the form of the cosine function as depicted in Figure 13.2.1. Observe that the function is periodic with period 2π. By comparing Eq. (13.2.3) with Figure 13.2.1, we see that the displacement of the linear oscillator changes periodically between the extremes of $\hat{A}$ and $-\hat{A}$. Also, the period T of the oscillation is determined by:

$$\omega T = 2\pi \quad \text{or} \quad T = 2\pi/\omega \tag{13.2.7}$$

The *frequency* is the rate at which the oscillation occurs; that is, the frequency f is the reciprocal of the period:

$$f = 1/T = \omega/2\pi \quad \text{or} \quad \omega = 2\pi f \tag{13.2.8}$$

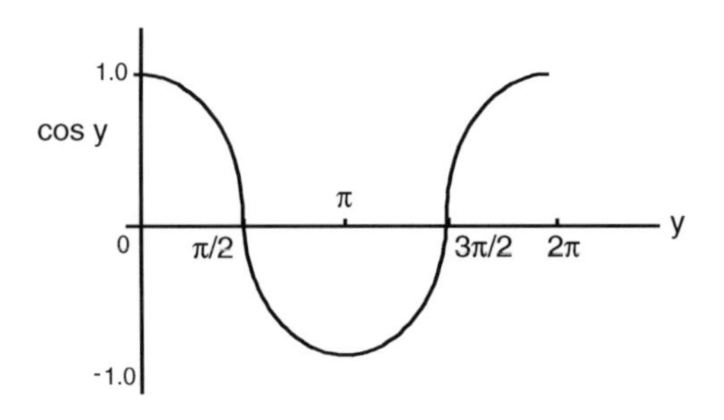

FIGURE 13.2.1
Cosine function.

As noted earlier, the constants in Eqs. (13.2.2) and (13.2.3) are to be evaluated by auxiliary conditions. These auxiliary conditions are generally initial conditions or boundary conditions. Initial conditions (where $t = 0$) might be expressed as:

$$x(0) = x_0 \quad \text{and} \quad \dot{x}(0) = \dot{x}_0 \tag{13.2.9}$$

Then, by requiring the solution of Eq. (13.2.2) to meet these conditions, we have:

$$A = x_0 \quad \text{and} \quad B = \dot{x}_0/\omega \tag{13.2.10}$$

and thus,

$$x = x_0 \cos \omega t + (\dot{x}_0/\omega) \sin \omega t \tag{13.2.11}$$

Boundary conditions might be expressed as:

$$x(0) = 0 \quad \text{and} \quad x(\ell) = 0 \tag{13.2.12}$$

Then, by requiring the solution of Eq. (13.2.2) to meet these conditions, we see that the constants A and B must satisfy:

$$A = 0 \quad \text{and} \quad B \sin \omega \ell = 0 \tag{13.2.13}$$

The second expression is satisfied by either:

$$B = 0 \quad \text{or} \quad \sin \omega \ell = 0 \tag{13.2.14}$$

If $B = 0$, we have the trivial solution $x = 0$. Alternatively, if $\sin\omega\ell = 0$, $\omega\ell$ must be an integer multiple of π. That is,

$$\omega \ell = n\pi \quad \text{or} \quad \omega = \omega_n = n\pi/\ell \tag{13.2.15}$$

Thus, there is a nontrivial solution only for selected values of ω — that is, for $\omega = n\pi/\ell$. The solution for the displacement then takes the multiple forms:

$$x = x_n = B_n \sin n\pi t/\ell \qquad n = 1, 2, \ldots \tag{13.2.16}$$

Then, by superposition, we have:

$$x = \sum_{n=1}^{\infty} B_n \sin n\pi t/\ell \tag{13.2.17}$$

where the constants B_n are to be determined from other conditions of the specific system being studied. (Eq. (13.2.17) is a Fourier series representation of the solution.)

By comparing the solution procedures between the initial and boundary value problems, we see that the solution of the initial value problem is simpler and more direct than that of the boundary value problem. Therefore, for simplicity, in the sequel we will consider primarily initial value problems.

Consider next the damped linear oscillator equation:

$$m\ddot{x} + c\dot{x} + kx = 0 \tag{13.2.18}$$

where m, c, and k are constants. This is the classical second-order homogeneous linear ordinary differential equation with constant coefficients. From References 13.9 to 13.13, we see that the solution depends upon the relative magnitudes of m, c, and k. If the product $4km$ is less than c^2, we can write the solution in the form:

$$x = e^{-\mu t}\left[A\cos\omega t + B\sin\omega t\right] \tag{13.2.19}$$

where μ and ω are defined as:

$$\mu \overset{D}{=} c/2m \quad \text{and} \quad \omega \overset{D}{=} \left[\frac{k}{m} - \frac{c^2}{4m^2}\right]^{1/2} \tag{13.2.20}$$

If, in Eq. (13.2.18), the product c^2 exceeds $4km$, the solution takes the form:

$$x = Ae^{-(\mu+v)t} + Be^{-(\mu-v)t} \tag{13.2.21}$$

where μ and ν are defined as:

$$\mu \overset{D}{=} c/2m \quad \text{and} \quad \nu = \left[\frac{c^2}{4m^2} - \frac{k}{m}\right]^{1/2} \tag{13.2.22}$$

Finally, if in Eq. (13.2.18), the product $4km$ is exactly equal to c^2, the solution takes the form:

$$x = e^{-\mu t}(A + Bt) \tag{13.2.23}$$

where μ is:

$$\mu = c/2m = \sqrt{k/m} \tag{13.2.24}$$

Next, consider the forced linear oscillator described by the equation (with $4km > c^2$):

$$m\ddot{x} + c\dot{x} + kx = f(t) \tag{13.2.25}$$

where $f(t)$ is the *forcing function*. The forcing function typically has the form:

$$f(t) = F\cos pt \tag{13.2.26}$$

Equation (13.2.25) then becomes:

$$m\ddot{x} + c\dot{x} + kx = F\cos pt \tag{13.2.27}$$

From References 13.9 to 13.13, we see that the general solution may be expressed as:

$$x = x_h + x_p \tag{13.2.28}$$

where x_h is the general solution of the homogeneous equation (right side equal to zero) as in Eqs. (13.2.18) and (13.2.19), and where x_p is *any* solution of the nonhomogeneous equation (right side equal to $F\cos pt$, as in Eq. (13.2.27)). x_p is commonly called the *particular solution*. From Eq. (13.2.19). we see that x_h is:

$$x_h = e^{-\mu t}\left[A\cos\omega t + B\sin\omega t\right] \tag{13.2.29}$$

where, as before, μ and ω are defined by Eq. (13.2.20). Also, from the references, we see that x_p may be expressed as:

$$x_p = (F/\Delta)\left[(k - mp^2)\cos pt + cp\sin pt\right] \tag{13.2.30}$$

where Δ is defined as:

$$\Delta \stackrel{D}{=} (k - mp^2)^2 + c^2p^2 \tag{13.2.31}$$

(The validity of Eq. (13.2.26) may be verified by direct substitution into Eq. (13.2.23).)

Finally, if c is zero in Eq. (13.2.27), we have the forced undamped linear oscillator equation:

$$m\ddot{x} + kx = F\cos pt \tag{13.2.32}$$

From Eqs. (13.2.28), (13.2.29), and (13.2.30), we see that the solution is:

$$x = x_h + x_p \tag{13.2.33}$$

where x_h and x_p are:

$$x_h = A\cos\omega t + B\sin\omega t \tag{13.2.34}$$

and

$$x_p = \left[F/(k - mp^2)\right]\cos pt \tag{13.2.35}$$

where from Eq. (13.2.20) ω is defined as:

$$\omega \stackrel{D}{=} \sqrt{k/m} \tag{13.2.36}$$

In the following section, we will examine the undamped linear oscillator in more detail.

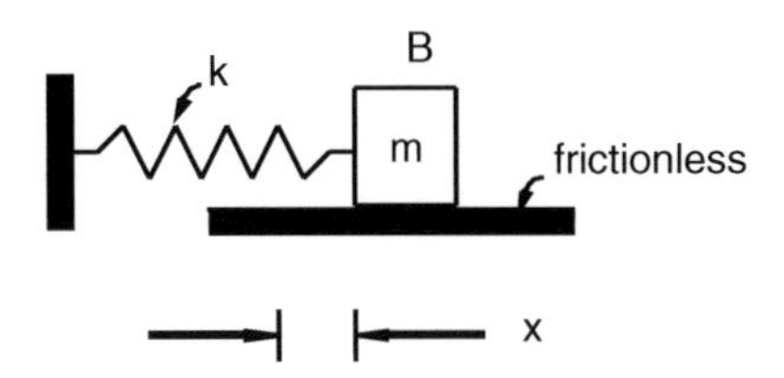

FIGURE 13.3.1
An undamped mass–spring system.

13.3 The Undamped Linear Oscillator

Consider the undamped linear oscillator consisting of the mass–spring system, which we considered in Chapter 11, Section 11.5, and as shown in Figure 13.3.1, where m is the mass of a block B sliding on a smooth (frictionless) horizontal surface, k is the modulus of a linear supporting spring, and x measures the displacement of B away from its equilibrium configuration. The system is said to be *undamped* because B moves on a frictionless surface and the total energy of the mass–spring system is unchanged during the motion. Using any of the principles of dynamics, we find the equation of motion to be:

$$m\ddot{x} + kx = 0 \tag{13.3.1}$$

or

$$\ddot{x} + \omega^2 x = 0 \tag{13.3.2}$$

where:

$$\omega^2 = k/m \tag{13.3.3}$$

From Eq. (13.2.2), we see that the solution of Eq. (13.3.2) is:

$$x = A\cos\omega t + B\sin\omega t \tag{13.3.4}$$

where, as we noted, A and B are constants to be evaluated from auxiliary conditions such as initial conditions for the mass–spring system. For example, suppose that at time $t = 0$ the displacement and displacement rate are:

$$x = x_0 \quad \text{and} \quad \dot{x} = \dot{x}_0 \tag{13.3.5}$$

Then, from Eq. (13.3.4), we have:

$$A = x_0 \quad \text{and} \quad B = \dot{x}_0/\omega \tag{13.3.6}$$

Hence, the solution becomes:

$$x = x_0\cos\omega t + (\dot{x}_0/\omega)\sin\omega t \tag{13.3.7}$$

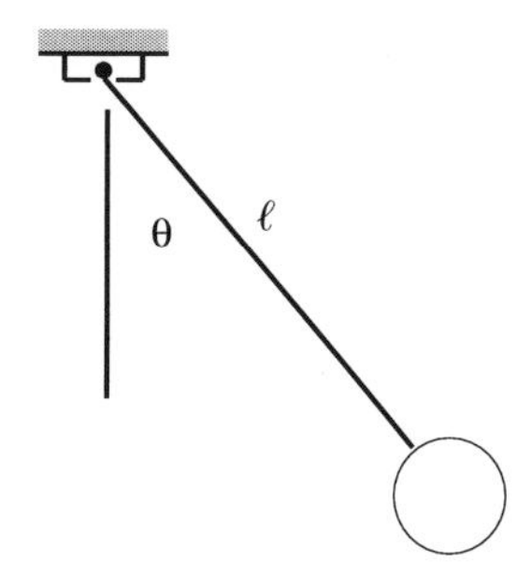

FIGURE 13.3.2
The simple pendulum.

If B is at rest when $t = 0$, (that is, $\dot{x}_0 = 0$), then the displacement x becomes:

$$x = x_0 \cos \omega t \tag{13.3.8}$$

In this context, the initial displacement x_0 is the amplitude and ω is the circular frequency. That is,

$$\omega = 2\pi f = \sqrt{k/m} \tag{13.3.9}$$

or

$$f = (1/2\pi)\sqrt{k/m} \quad \text{and} \quad T = 2\pi\sqrt{m/k} \tag{13.3.10}$$

where f is the frequency (or *natural frequency*) and T is the period.

Consider next the simple pendulum as in Figure 13.3.2. From Eq. (12.2.9), we see that the equation of motion is:

$$\ddot{\theta} + (g/\ell)\sin\theta \tag{13.3.11}$$

where as before, θ measures the displacement away from equilibrium, g is the gravity constant, and ℓ is the pendulum length. Observe that if θ is small, as is the case with most pendulums, we can approximate $\sin\theta$ by the first few terms of a Taylor series expansion of $\sin\theta$ about the equilibrium position $\theta = 0$. That is,

$$\sin\theta = \theta - \theta^3/3! + \theta^5/5! - \ldots \tag{13.3.12}$$

Hence, for small θ, we have:

$$\sin\theta \approx \theta \tag{13.3.13}$$

The governing equation. Eq. (13.3.11), then becomes:

$$\ddot{\theta} + (g/\ell)\theta = 0 \tag{13.3.14}$$

Equation (13.3.14) is identical in form to Eq. (13.3.2); therefore, the solution will have the form of Eq. (13.3.7). That is,

$$\theta = \theta_0 \cos \omega t + \left(\dot{\theta}_0/\omega\right) \sin \omega t \tag{13.3.15}$$

where θ_0 and $\dot{\theta}_0$ are the initial values of θ and $\dot{\theta}$, respectively, and where ω is:

$$\omega = \sqrt{g/\ell} \tag{13.3.16}$$

Suppose, for example, that the pendulum is released from rest at an angle θ_0. The subsequent motion of the pendulum is:

$$\theta = \theta_0 \cos \sqrt{g/\ell}\, t \tag{13.3.17}$$

Observe in Eq. (13.3.17) that the amplitude of the periodic motion is θ_0 and that the frequency and period are:

$$f = (1/2\pi)\sqrt{g/\ell} \quad \text{and} \quad T = 2\pi\sqrt{\ell/g} \tag{13.3.18}$$

Finally, observe that the frequency and period of the pendulum depend only upon the length ℓ of the pendulum. That is, the pendulum movement is independent of the mass m. (This is the reason why pendulums have been used extensively in clocks and timing devices.)

13.4 Forced Vibration of an Undamped Oscillator

Consider again the undamped mass–spring system of the foregoing section and as depicted again in Figure 13.4.1. This time, let the mass B be subjected to a time-varying force $F(t)$ as shown. Then, it is readily seen by using any of the principles of dynamics discussed earlier, that the governing equation of motion for this system is:

$$m\ddot{x} + kx = F(t) \tag{13.4.1}$$

Suppose $F(t)$ is itself a periodic function such as:

$$F(t) = F_0 \cos pt \tag{13.4.2}$$

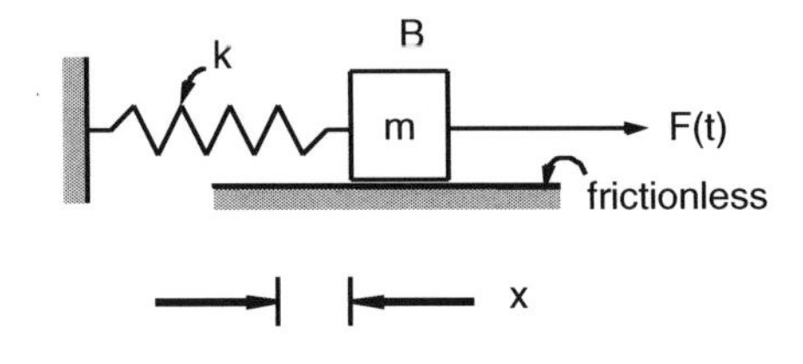

FIGURE 13.4.1
An undamped forced mass–spring system.

where F_0 and p are constants. The governing equation then becomes:

$$m\ddot{x} + kx = A\cos pt \tag{13.4.3}$$

From Eqs. (13.2.32) through (13.2.35) we see that the general solution of Eq. (13.4.3) is:

$$x = A\cos\omega t + B\sin\omega t + \left[F_0/\left(k - mp^2\right)\right]\cos pt \tag{13.4.4}$$

where as before A and B are constants to be determined from auxiliary conditions (such as initial conditions) and where ω is defined as:

$$\omega \overset{D}{=} \sqrt{k/m} \tag{13.4.5}$$

Consider the last term, $[F_0/(k - mp^2)]\cos pt$ of Eq. (13.4.4). Observe that if p^2 has values nearly equal to k/m (that is, if p is nearly equal to ω) the denominator becomes very small, producing large-amplitude oscillation. Indeed, if p is equal to ω, the oscillation amplitude becomes unbounded. This means that by stimulating the mass B with a periodic force having a frequency nearly equal to ω (that is, $\sqrt{k/m}$), the amplitude of the resulting oscillation becomes unbounded.

The quantity $\sqrt{k/m}$ is called the *natural frequency* of the system. When the frequency of the loading function is equal to the natural frequency, giving rise to a large-amplitude response, we have the phenomenon commonly referred to as *resonance*.

13.5 Damped Linear Oscillator

Consider next the damped linear oscillator as depicted in Figure 13.5.1. This is the same system we considered in the previous sections, but here the movement of the mass B is restricted by a "damper" in the form of a dashpot. For simplicity of illustration, we will assume *viscous damping*, where the force exerted by the dashpot on B is proportional to the speed of B and directed opposite to the motion of B with c being the constant of proportionality. That is, the damping force $\mathbf{F}_D$ on B is:

$$\mathbf{F}_D = -c\dot{x}\mathbf{n} \tag{13.5.1}$$

where $\mathbf{n}$ is a unit vector in the positive X direction as shown in Figure 13.5.1.

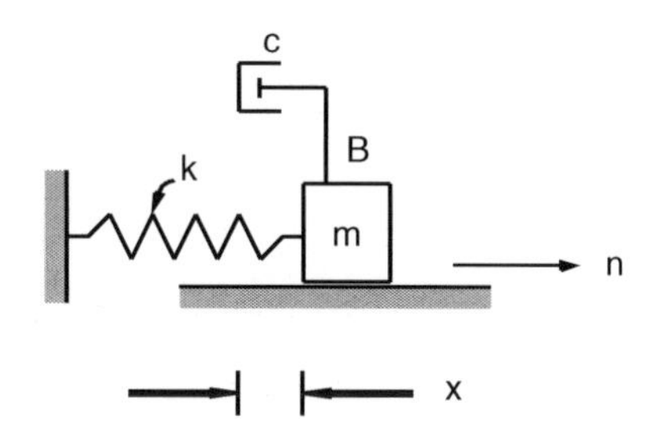

FIGURE 13.5.1
A damped mass–spring system.

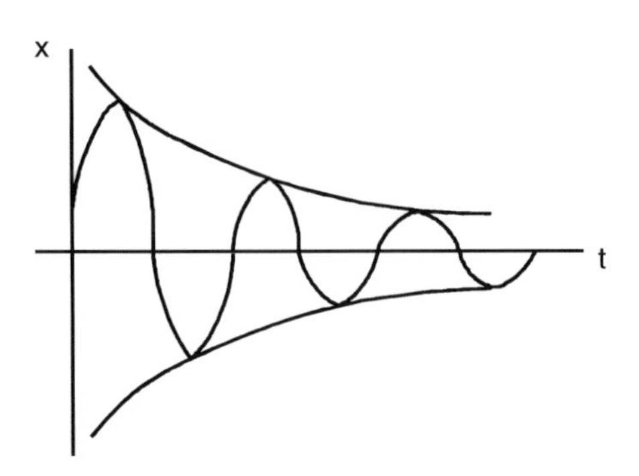

FIGURE 13.5.2
Underdamped oscillation.

By considering a free-body diagram of B and by using the principles of dynamics (for example, Newton's laws or d'Alembert's principle), we readily see that the governing equation of motion is:

$$m\ddot{x} + c\dot{x} + kx = 0 \tag{13.5.2}$$

where as before, m is the mass of B and k is the linear spring modulus.

From Eqs. (13.2.18), (13.2.19), and (13.2.20), we see that the solution of Eq. (13.5.2) may be written in the form:

$$x = e^{-\mu t}\left[A\cos\omega t + B\sin\omega t\right] \tag{13.5.3}$$

where μ and ω are defined as:

$$\mu \overset{D}{=} c/2m \quad \text{and} \quad \omega \overset{D}{=} \left[\frac{k}{m} - \frac{c^2}{4m^2}\right]^{1/2} \tag{13.5.4}$$

and where, as before, A and B are constants to be evaluated from auxiliary conditions, such as initial conditions, on the system.

Observe in Eq. (13.5.3) that if the damping constant c is small we have an oscillating system where the amplitude of the oscillation slowly decreases, as in Figure 13.5.2. This phenomenon is called *underdamped vibration*.

Next, observe in Eqs. (13.5.3) and (13.5.4) that if the damping constant c is such that ω is zero, there will be no oscillation. That is, if

$$c = 2\sqrt{km} \tag{13.5.5}$$

then

$$\omega = 0 \tag{13.5.6}$$

and

$$x = e^{-\mu t}(A + Bt) \tag{13.5.7}$$

where μ is:

$$\mu = \sqrt{k/m} \tag{13.5.8}$$

In this case, the block moves to its static equilibrium position without oscillation. This phenomenon is called *critical damping*.

Finally, suppose the damping constant is larger than critical damping, that is, larger than $2\sqrt{km}$. Then, from Eq. (13.5.4), we see that ω is imaginary and that the solution for the displacement x of B becomes:

$$x = Ae^{-(\mu+\nu)t} + Be^{-(\mu-\nu)t} \tag{13.5.9}$$

where μ and ν are:

$$\mu = c/2m \quad \text{and} \quad \nu = \left[\frac{c^2}{4m^2} - \frac{k}{m}\right]^{1/2} \tag{13.5.10}$$

Observe that μ is always larger than ν and we have a relatively rapidly decaying motion of B. This phenomenon is called *overdamping*.

13.6 Forced Vibration of a Damped Linear Oscillator

Consider next the forced vibration of a damped linear oscillator as depicted in Figure 13.6.1. From the principles of dynamics, we readily find the governing equation of motion to be:

$$m\ddot{x} + c\dot{x} + kx = F(t) \tag{13.6.1}$$

where as before m is the mass of the block B, c is the viscous damping coefficient, k is the linear spring constant, and $F(t)$ is the forcing function. Suppose that as in Section 13.4 $F(t)$ has the periodic form:

$$F(t) = F_0 \cos pt \tag{13.6.2}$$

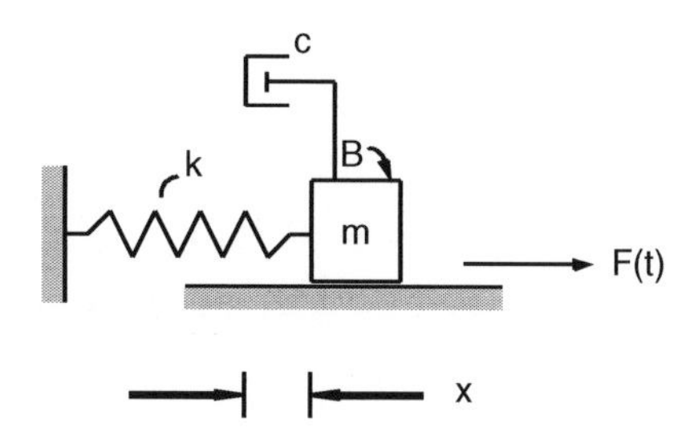

FIGURE 13.6.1
Forced motion of a damped mass–spring system.

Equation (13.6.1) then takes the form:

$$m\ddot{x} + c\dot{x} + kx = F_0 \cos pt \tag{13.6.3}$$

From Eqs. (13.2.27) through (13.2.31), we see that the solution of Eq. (13.6.3) may be written as:

$$x = x_h + x_p \tag{13.6.4}$$

where x_h and x_p are:

$$x_h = e^{-\mu t}\left[A\cos\omega t + B\sin\omega t\right] \tag{13.6.5}$$

and

$$x_p = \left(F_0/\Delta\right)\left[\left(k - mp^2\right)\cos pt + cp\sin pt\right] \tag{13.6.6}$$

where μ, ω, and Δ are defined by:

$$\begin{gathered} \mu \overset{D}{=} c/2m, \quad \omega \overset{D}{=} \left[\frac{k}{m} - \frac{c^2}{4m^2}\right]^{1/2} \\ \Delta \overset{D}{=} \left(k - mp^2\right)^2 + c^2p^2 \end{gathered} \tag{13.6.7}$$

where we assume that $c^2 < 4km$. As before, the constants A and B in Eq. (13.6.5) are to be evaluated by auxiliary (initial) conditions on the system.

Observe that the last term of Eq. (13.6.6) can become quite large if the damping coefficient c is small and if the frequency p of the forcing function is nearly equal to the natural frequency $\sqrt{k/m}$ of the undamped system. Note, however, that unlike the undamped system, the presence of damping assures that the amplitude of the oscillation remains finite. Thus, we see that damping (or effects of friction and viscosity) can have a beneficial effect in preventing harmful or unbounded vibration of a mechanical system.

The phenomenon of damping in physical systems, however, is generally more complex than our relatively simple model of Eq. (13.5.1). Indeed, damping is generally a nonlinear phenomenon that varies from system to system and is generally not well understood. Theoretical and experimental research on damping is currently a major interest of vibration analysts.

13.7 Systems with Several Degrees of Freedom

We consider next mechanical systems where more than one body can oscillate. An example of such a system might be a double mass–spring system as in Figure 13.7.1. This system has two degrees of freedom as represented by the displacements x_1 and x_2 of the masses. Accordingly, we expect to obtain two governing differential equations that must be solved

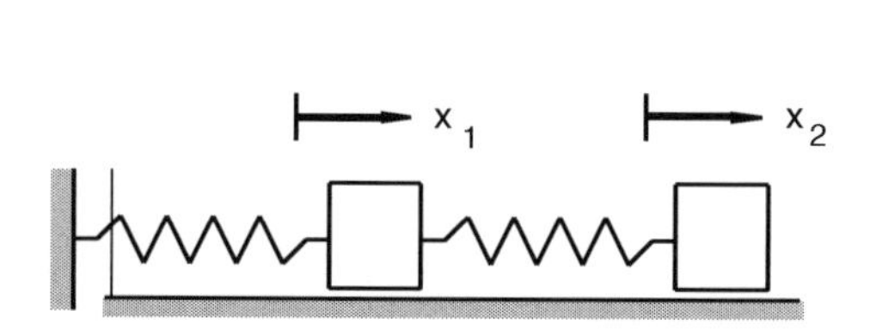

FIGURE 13.7.1
A double mass–spring system.

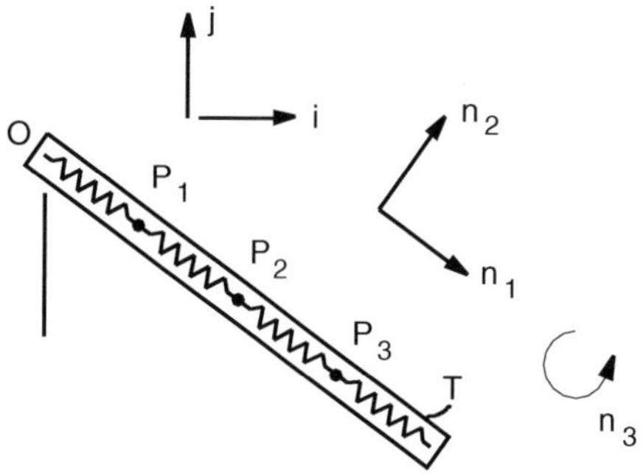

FIGURE 13.7.2
Spring-supported particles in a rotating tube.

simultaneously. Similarly, systems with three or more degrees of freedom will have three or more governing differential equations to be solved simultaneously.

To illustrate a procedure for studying such systems, consider again the system of spring-connected smooth particles (or balls) in the rotating tube as in Figure 13.7.2. As before, let each particle have mass m and let the connecting springs be linear with natural length ℓ and modulus k.

To simplify our analysis, let θ be fixed at 90° so that the particles move in a fixed horizontal tube with their position defined by the coordinates x_1, x_2, and x_3 as in Figure 13.7.3. From Eqs. (12.2.27), (12.2.28), and (12.2.29), we see that, with θ fixed at 90°, the equations of motion may be written as:

$$m\ddot{x}_1 + 2kx_1 - kx_2 = 0 \tag{13.7.1}$$

$$m\ddot{x}_2 - kx_1 + 2kx_2 - kx_3 = 0 \tag{13.7.2}$$

$$m\ddot{x}_3 - kx_2 + 2kx_3 = 0 \tag{13.7.3}$$

Equations (13.7.1), (13.7.2), and (13.7.3) may be written in the matrix form:

$$M\ddot{x} + Kx = 0 \tag{13.7.4}$$

where the matrices M and K are:

$$M = \begin{bmatrix} m & 0 & 0 \\ 0 & m & 0 \\ 0 & 0 & m \end{bmatrix} \quad \text{and} \quad K = \begin{bmatrix} 2k & -k & 0 \\ -k & 2k & -k \\ 0 & -k & 2k \end{bmatrix} \tag{13.7.5}$$

and x is the array:

$$x = \begin{bmatrix} x_1 \\ x_2 \\ x_3 \end{bmatrix} \tag{13.7.6}$$

To solve Eq. (13.7.4), let x have the matrix form:

$$x = Ae^{i\omega t} \tag{13.7.7}$$

where ω is a frequency to be determined and A is the array:

$$A = \begin{bmatrix} A_1 \\ A_2 \\ A_3 \end{bmatrix} \tag{13.7.8}$$

Equivalently, the coordinates of x may be written as:

$$x_1 = A_1 e^{i\omega t}, \quad x_2 = A_2 e^{i\omega t}, \quad x_3 = A_3 e^{i\omega t} \tag{13.7.9}$$

By substituting from Eq. (13.7.7) into (13.7.4) we have:

$$-\omega^2 MA + KA = 0 \tag{13.7.10}$$

or

$$\left(-\omega^2 M + K\right)A = 0 \tag{13.7.11}$$

That is,

$$\begin{bmatrix} \left(-m\omega^2 + 2k\right) & -k & 0 \\ -k & \left(-m\omega^2 + 2k\right) & -k \\ 0 & -k & \left(-m\omega^2 + 2k\right) \end{bmatrix} \begin{bmatrix} A_1 \\ A_2 \\ A_{33} \end{bmatrix} = 0 \tag{13.7.12}$$

Equation (13.7.12) is equivalent to the scalar equations:

$$\left(-m\omega^2 + 2k\right)A_1 - kA_2 = 0 \tag{13.7.13}$$

$$-kA_1 + \left(-m\omega^2 + 2k\right)A_2 - kA_3 = 0 \tag{13.7.14}$$

$$-kA_2 + \left(-m\omega^2 + 2k\right) = 0 \tag{13.7.15}$$

As with the eigenvalue problem we encountered in Sections 7.7, 7.8, and 7.9 in studying inertia, we have a system of simultaneous linear algebra equations for the amplitudes A_1, A_2, and A_3. The system is homogeneous in that the right sides are zero. Thus, we have a nontrivial (or nonzero) solution only if the determinant of the coefficients is zero. Hence, we have:

$$\det\begin{bmatrix} \left(-m\omega^2 + 2k\right) & -k & 0 \\ -k & \left(-m\omega^2 + 2k\right) & -k \\ 0 & -k & \left(-m\omega^2 + 2k\right) \end{bmatrix} = 0 \tag{13.7.16}$$

or

$$\left[-\omega^2+2(k/m)\right]\left\{\left[-\omega^2+2(k/m)\right]^2-(k/m)^2\right\}-(k/m)^2\left[-\omega^2+2(k/m)\right]=0 \tag{13.7.17}$$

Solving for ω^2 we obtain:

$$\omega_1^2=\left(2-\sqrt{2}\right)(k/m)\,,\quad \omega_2^2=2(k/m)\,,\quad \omega_3^2=\left(2+\sqrt{2}\right)(k/m) \tag{13.7.18}$$

Observe that instead of obtaining one solution we have three solutions. That is, there are three positive frequencies (three *natural frequencies*) that make the determinant of Eq. (13.7.16) equal to zero and thus allow a nonzero solution of Eqs. (13.7.13), (13.7.14), and (13.7.15) to occur. This in turn means that we can expect to find three sets of amplitudes solving Eqs. (13.7.13), (13.7.14), and (13.7.15), one set for each frequency ω_i (i = 1, 2, 3).

To find these amplitude solutions, we can select one of the ω_i (say, the smallest of the ω_i), substitute it into Eqs. (13.7.13) to (13.7.15), and thus solve for the amplitudes A_i (i = 1, 2, 3). We can then repeat the process for the other two values of ω_i. Notice, however, that although there are three equations for the three amplitudes, the equations are not independent in view of Eq. (13.7.16). That is, there are at most two independent equations, thus we need another equation to obtain a unique solution for the amplitudes. Such an equation can be obtained by arbitrarily specifying the magnitude of one or more of the amplitudes. For example, we could "normalize" the amplitudes such that

$$A_1^2+A_2^2+A_3^2=1 \tag{13.7.19}$$

Then, Eqs. (13.7.13), (13.7.14), (13.7.15), and (13.7.19) are equivalent to an independent set of three equations, enabling us to determine unique values of the amplitudes.

To this end, let us select the smallest of the ω_i (ω_1) and substitute it (that is, $\{(2-\sqrt{2})(k/m)\}^{1/2}$) into Eqs. (13.7.13), (13.7.14), and (13.7.15). This produces the equations:

$$\sqrt{2}\,A_1-A_2=0 \tag{13.7.20}$$

$$-A_1+\sqrt{2}\,A_2-A_3=0 \tag{13.7.21}$$

$$-A_2+\sqrt{2}\,A_3=0 \tag{13.7.22}$$

Observe that these equations are dependent. (If we multiply the first and third equations by $\sqrt{2}$ and add them we obtain a multiple of the second equation.) Thus, by selecting two (say, the first two) of these equations and by combining them with Eq. (13.7.19), we obtain upon solving the expressions:

$$A_1=1/2\,,\quad A_2=\sqrt{2}/2\,,\quad A_3=1/2 \tag{13.7.23}$$

Next, let ω be ω_2 (that is, $\sqrt{2k/m}$). By substituting into Eqs. (13.7.13), (13.7.14), and (13.7.15), we obtain the equations:

$$A_2 = 0 \tag{13.7.24}$$

$$A_1 = -A_3 \tag{13.7.25}$$

$$A_2 = 0 \tag{13.7.26}$$

These equations are also dependent (the first and third are the same). By using the first two of these together with Eq. (13.7.19) and solving for A_1, A_2, and A_3, we have:

$$A_1 = \sqrt{2}/2\,,\quad A_2 = 0\,,\quad A_3 = -\sqrt{2}/2 \tag{13.7.27}$$

Finally, let ω be ω_3 (that is, $\{(2 + \sqrt{2})\,(k/m)\}^{1/2}$). Substituting into Eqs. (13.7.13), (13.7.14), and (13.7.15), we obtain the equations:

$$\sqrt{2}\,A_1 + A_2 = 0 \tag{13.7.28}$$

$$A_1 + \sqrt{2}\,A_2 + A_3 = 0 \tag{13.7.29}$$

$$A_2 + \sqrt{2}\,A_3 = 0 \tag{13.7.30}$$

Observe that these equations are also dependent. (If we multiply the first and third by $\sqrt{2}$ and add them, we obtain a multiple of the second equation.) Using the first two of these with Eq. (13.7.19) then produces:

$$A_1 = -1/2\,,\quad A_2 = \sqrt{2}/2\,,\quad A_3 = -1/2 \tag{13.7.31}$$

Equations (13.7.23), (13.7.27), and (13.7.26) represent the amplitude solutions corresponding to the three frequencies of Eq. (13.7.18). Figures 13.7.4, 13.7.5, and 13.7.6 provide a pictorial representation of these solutions depicting the movements of the particles of the tube. We will discuss these solutions in greater detail in the following section.

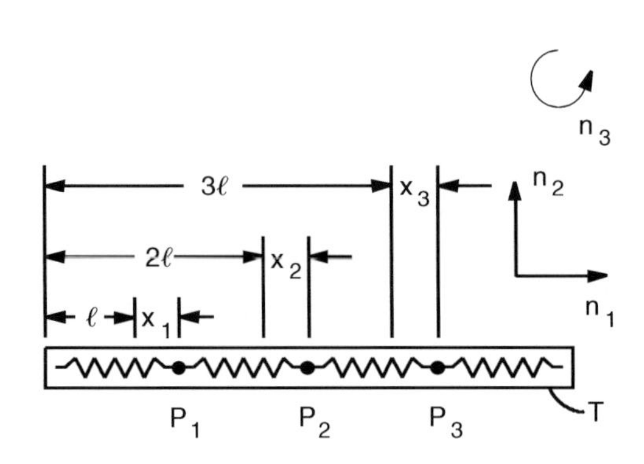

FIGURE 13.7.3
Coordinates of particles in the fixed horizontal tube.

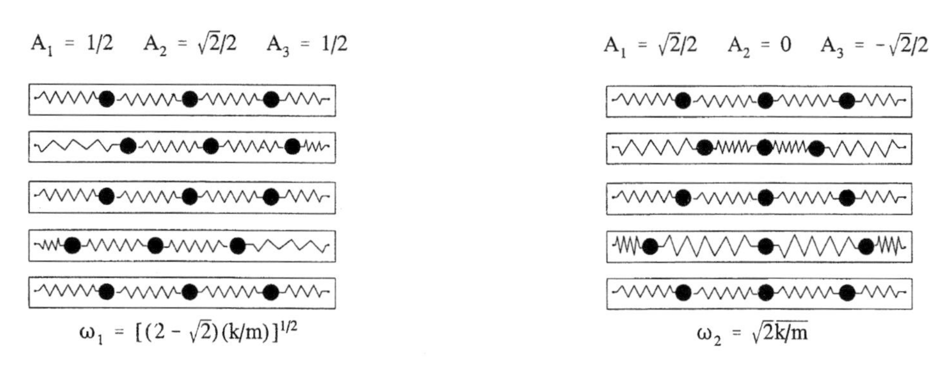

FIGURE 13.7.4
Particle movement at lowest frequency ω_1.

FIGURE 13.7.5
Particle movement at intermediate frequency ω_2.

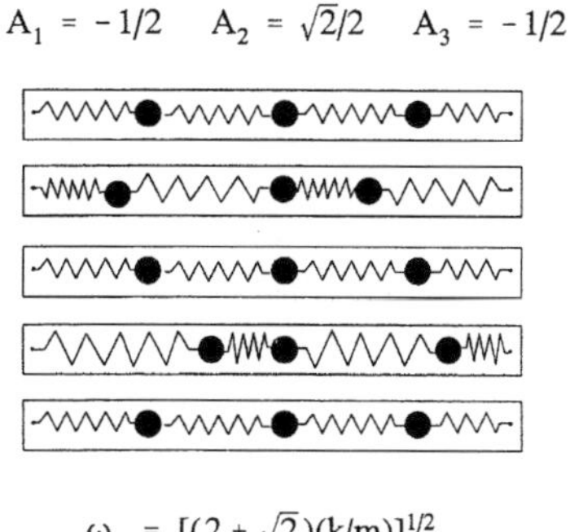

FIGURE 13.7.6
Particle movement at highest frequency ω_3.

13.8 Analysis and Discussion of Three-Particle Movement: Modes of Vibration

In reviewing the foregoing analysis of the three-particle system we see a striking similarity to the analysis for the eigenvalue problem of Sections 7.7, 7.8, and 7.9. Indeed, a comparison of Eqs. (13.7.13), (13.7.14) and (13.7.15) with Eq. (7.7.10), shows that they are in essence the same problem. This means that analyses similar to those of Sections 7.7, 7.8, and 7.9 (such as determination of eigenvalues, eigenvectors, orthogonality, etc.) could also be conducted for the three-particle vibration problem. In our relatively brief introduction to vibrations, however, it is not our intention to develop such detail. Instead, we plan to simply introduce the concept of vibration modes (analogous to eigenvectors).

To this end, consider again the governing differential equations for the spring-supported particles (see Eqs. (13.7.1), (13.7.2), and (13.7.3)):

$$m\ddot{x}_1 + 2kx_1 - kx_2 = 0 \tag{13.8.1}$$

$$m\ddot{x}_2 - kx_1 + 2kx_2 - kx_3 = 0 \tag{13.8.2}$$

$$m\ddot{x}_3 - kx_2 + 2kx_3 = 0 \tag{13.8.3}$$

TABLE 13.8.1

Modes of Vibration of Spring-Supported Particles

Mode	Frequency	Normalized Amplitudes		
1	$\left[\left(2-\sqrt{2}\right)(k/m)\right]^{1/2}$	$A_1 = 1/2,$	$A_2 = \sqrt{2}/2,$	$A_3 = 1/2$
2	$[2k/m]^{1/2}$	$A_1 = \sqrt{2}/2,$	$A_2 = 0,$	$A_3 = -\sqrt{2}/2$
3	$\left[\left(2+\sqrt{2}\right)(k/m)\right]^{1/2}$	$A_1 = -1/2,$	$A_2 = \sqrt{2}/2,$	$A_3 = -1/2$

Recall that we found not one but *three* nontrivial solutions to these equations. Each solution had its own frequency, which means that the system can vibrate in three ways, or in three "modes," as depicted in Figures 13.7.4, 13.7.5, and 13.7.6. These are called the *natural modes of vibration* of the system.

To discuss this further, consider the amplitudes of these vibration modes as in Eqs. (13.7.23), (13.7.27), and (13.7.31) and as listed in Table 13.8.1. In view of the amplitude ratios, let new variables ξ_1, ξ_2, and ξ_3 be introduced as:

$$\xi_1 \stackrel{D}{=} (1/2)x_1 + \left(\sqrt{2}/2\right)x_2 + (1/2)x_3 \tag{13.8.4}$$

$$\xi_2 \stackrel{D}{=} \left(\sqrt{2}/2\right)x_1 + 0x_2 - \left(\sqrt{2}/2\right)x_3 \tag{13.8.5}$$

$$\xi_3 \stackrel{D}{=} -(1/2)x_1 + \left(\sqrt{2}/2\right)x_2 - (1/2)x_3 \tag{13.8.6}$$

Then, by differentiating, we have:

$$\ddot{\xi}_1 \stackrel{D}{=} -(1/2)\ddot{x}_1 + \left(\sqrt{2}/2\right)\ddot{x}_2 + (1/2)\ddot{x}_3 \tag{13.8.7}$$

$$\ddot{\xi}_2 \stackrel{D}{=} \left(\sqrt{2}/2\right)\ddot{x}_1 - \left(\sqrt{2}/2\right)\ddot{x}_3 \tag{13.8.8}$$

$$\ddot{\xi}_3 \stackrel{D}{=} -(1/2)\ddot{x}_1 + \left(\sqrt{2}/2\right)\ddot{x}_2 + (1/2)\ddot{x}_3 \tag{13.8.9}$$

Consider first $\ddot{\xi}_1$. By substituting from Eqs. (13.8.1), (13.8.2), and (13.8.3), we have:

$$\begin{aligned}
\ddot{\xi}_1 &= (1/2)(k/m)(-2x_1 + x_2) + \left(\sqrt{2}/2\right)(k/m)(x_1 - 2x_2 + x_3) \\
&\quad + (1/2)(k/m)(x_2 - 2x_3) \\
&= (k/m)\left[\left(-2 + \sqrt{2}/2\right)\right]x_1 + \left[\left(1 - \sqrt{2}\right)\right]x_2 + \left[\left(\sqrt{2} - 2\right)/2\right]x_3 \\
&= (k/m)\left(-2 + \sqrt{2}\right)\left[(1/2)x_1 + \left(\sqrt{2}/2\right)x_2 + (1/2)x_3\right] \\
&= (k/m)\left(-2 + \sqrt{2}\right)\xi_1
\end{aligned}$$

or

$$\ddot{\xi}_1 + \left(2 - \sqrt{2}\right)(k/m)\xi_1 = 0 \tag{13.8.10}$$

or

$$\ddot{\xi}_1 + \omega_1^2 \xi_1 = 0 \tag{13.8.11}$$

Similarly, by considering $\ddot{\xi}_2$ and $\ddot{\xi}_3$ we have:

$$\ddot{\xi}_2 + \omega_2^2 \xi_2 = 0 \tag{13.8.12}$$

and

$$\ddot{\xi}_3 + \omega_3^2 \xi_3 = 0 \tag{13.8.13}$$

We can recognize Eqs. (13.8.11), (13.8.12), and (13.8.13) as being in the form of the undamped linear oscillator equation, Eq. (13.2.1). The solutions of the equations may thus be expressed as:

$$\xi_1 = A_1 \cos\omega_1 t + B_1 \sin\omega_1 t \tag{13.8.14}$$

$$\xi_2 = A_2 \cos\omega_2 t + B_2 \sin\omega_2 t \tag{13.8.15}$$

$$\xi_3 = A_3 \cos\omega_3 t + B_3 \sin\omega_3 t \tag{13.8.16}$$

where, as before, the constants A_i and B_i (i = 1, 2, 3) are to be evaluated from auxiliary conditions.

Equations (13.8.14), (13.8.15), and (13.8.16) show that. given suitable auxiliary conditions. the three-particle system will have sinusoidal oscillation, with the "shape," or mode, of the oscillation being in the form of ξ_1, ξ_2, or ξ_3 as defined by Eqs. (13.8.4), (13.8.5), and (13.8.6).

This, however, raises a question: suppose the auxiliary conditions (say, the initial conditions) are such that they do not correspond to a shape of one of the modes ξ_1, ξ_2, or ξ_3; what, then, is the subsequent motion of the system? The answer is that the subsequent motion may be expressed as a linear combination of the ξ_1, ξ_2, and ξ_3. That is, the ξ_1, ξ_2, and ξ_3 may be regarded as "base vectors" in a three-dimensional space (for the three degrees of freedom of the system).

To see this, consider again the solutions of Eqs. (13.8 .11), (13.8.12), and (13.8.13) in the form of Eqs. (13.8.14), (13.8.15), and (13.8.16). The derivatives of ξ_1, ξ_2, and ξ_3 may be expressed as:

$$\dot{\xi}_1 = -A_1\omega_1 \sin\omega_1 t + B_1\omega_1 \cos\omega_1 t \tag{13.8.17}$$

$$\dot{\xi}_2 = -A_2\omega_2 \sin\omega_2 t + B_2\omega_2 \cos\omega_2 t \tag{13.8.18}$$

$$\dot{\xi}_3 = -A_3\omega_3 \sin\omega_3 t + B_3 \omega_3 \cos\omega_3 t \tag{13.8.19}$$

Next, suppose that at time $t = 0$, the values of x_1, x_2, and x_3 and their derivatives are x_{10}, x_{20}, x_{30}, $\dot{x}_{10}$, $\dot{x}_{20}$, $\dot{x}_{30}$, respectively. Then, from Eqs. (13.8.4), (13.8.5), and (13.8.6), the initial values of the ξ_1, ξ_2, and ξ_3 and of the $\dot{\xi}_1$, $\dot{\xi}_2$, and $\dot{\xi}_3$ are determined. This in turn enables us to use Eqs. (13.8.14) through (13.8.19) to determine the constants A_i and B_i ($i = 1, 2, 3$).

Finally, if we know the ξ_1, ξ_2, and ξ_3 we can readily find the x_1, x_2, and x_3 through Eqs. (13.8.4), (13.8.5), and (13.8.6) in the matrix form:

$$\begin{bmatrix} \xi_1 \\ \xi_2 \\ \xi_3 \end{bmatrix} = \begin{bmatrix} 1/2 & \sqrt{2}/2 & 1/2 \\ \sqrt{2}/2 & 0 & -\sqrt{2}/2 \\ -1/2 & \sqrt{2}/2 & -1/2 \end{bmatrix} \begin{bmatrix} x_1 \\ x_2 \\ x_3 \end{bmatrix} \tag{13.8.20}$$

or as:

$$\xi = Sx \tag{13.8.21}$$

where S is regarded as a transformation matrix. An inspection of S shows that it is orthogonal. That is, the inverse is equal to the transpose. Hence, we have:

$$x = S^{-1}\xi = S^T\xi \tag{13.8.22}$$

or

$$\begin{bmatrix} x_1 \\ x_2 \\ x_3 \end{bmatrix} = \begin{bmatrix} 1/2 & \sqrt{2}/2 & -1/2 \\ \sqrt{2}/2 & 0 & \sqrt{2}/2 \\ 1/2 & -\sqrt{2}/2 & -1/2 \end{bmatrix} \begin{bmatrix} \xi_1 \\ \xi_2 \\ \xi_3 \end{bmatrix} \tag{13.8.23}$$

or

$$x_1 = (1/2)\xi_1 + \left(\sqrt{2}/2\right)\xi_2 - (1/2)\xi_3 \tag{13.8.24}$$

$$x_2 = \left(\sqrt{2}/2\right)\xi_1 + 0\xi_2 + \left(\sqrt{2}/2\right)\xi_3 \tag{13.8.25}$$

$$x_3 = (1/2)\xi_1 - \left(\sqrt{2}/2\right)\xi_2 - (1/2)\xi_3 \tag{13.8.26}$$

13.9 Nonlinear Vibrations

The foregoing analyses are all based upon the solutions of linear differential equations. A convenient property of linear equations is that we can superpose or add solutions.

Specifically, if x_1 and x_2 are solutions to a linear equation, then $a_1x_1 + a_2x_2$, where a_1 and a_2 are constants, is also a solution.

Linear differential equations arise from assumptions made in the modeling of mechanical systems. Typical of such assumptions are "small displacements" (as with the simple pendulum) and forces proportional to the displacement (as with linear mass–spring systems).

An assumption of small displacements means that the modeling, and hence the solution of the ensuing linear equation, becomes more and more accurate the smaller the displacement becomes. For large displacements, however, the solution becomes less accurate — that is, less representative of the behavior of the system. In this section, we will explore the extent to which large displacement affects the solution.

If the displacements are large, the use of linear differential equations to model the displacement may no longer be appropriate. Instead, we need to use a nonlinear differential equation and attempt to find a solution to it — or at least an approximate solution. Therein, however, lies the difficulty — known solutions to nonlinear differential equations are very few, thus we generally have to resort to approximation methods to obtain a solution.

We will illustrate such an approach with an analysis of large movements of a simple pendulum as in Figure 13.9.1. Recall from Eq. (8.4.4) that the governing differential equation is:

$$\ddot{\theta} + (g/\ell)\sin\theta = 0 \tag{13.9.1}$$

where, as before, θ is the displacement angle and ℓ is the pendulum length. The presence of $\sin\theta$ in Eq. (13.9.1) makes the equation nonlinear and prevents us from obtaining a solution by conventional methods; however, we can obtain what is called a *first integral* of the equation by multiplying the equation by $\dot{\theta}$ and integrating. That is,

$$\dot{\theta}\ddot{\theta} + (g/\ell)\dot{\theta}\sin\theta = 0 \tag{13.9.2}$$

Thus, we have:

$$d\left(\dot{\theta}^2/2\right)/dt - (g/\ell)d\left(\cos\theta\right)/dt = 0 \tag{13.9.3}$$

or

$$\dot{\theta}^2/2 - (g/\ell)\cos\theta = C \tag{13.9.4}$$

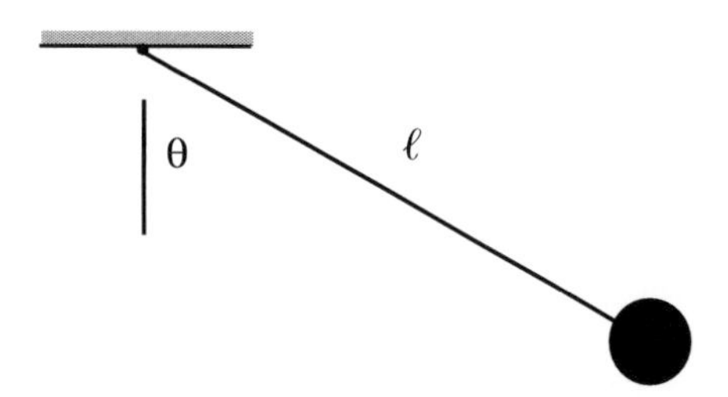

FIGURE 13.9.1
The simple pendulum with large displacement.

where C is a constant that can be evaluated by initial conditions. Suppose, for example, that the pendulum is displaced through an angle θ_0 and released from rest. Then, the initial conditions are:

$$\text{when } t = 0\,,\quad \theta = \theta_0 \text{ and } \dot{\theta} = 0 \tag{13.9.5}$$

Thus, the first integral of Eq. (13.9.4) becomes:

$$\dot{\theta}^2/2 - (g/\ell)\cos\theta = -(g/\ell)\cos\theta_0 \tag{13.9.6}$$

By solving Eq. (13.9.4) for $\dot{\theta}$ we have:

$$\dot{\theta} = \pm\left[(2g/\ell)\left(\cos\theta - \cos\theta_0\right)\right]^{1/2} = d\theta/dt \tag{13.9.7}$$

Thus, we have a nonlinear first-order differential equation to solve. By separating the variables, we have:

$$dt = \pm\sqrt{\ell/2g}\left(\cos\theta - \cos\theta_0\right)^{-1/2} d\theta \tag{13.9.8}$$

Suppose we are interested in determining the time of descent of the pendulum from the angle θ_0 to some smaller angle $\hat{\theta}$. Hence, by integration of Eq. (13.9.8) we have:

$$t = \sqrt{\ell/2g}\int_{\hat{\theta}}^{\theta_0}\left(\cos\theta - \cos\theta_0\right)^{-1/2} d\theta \tag{13.9.9}$$

where we have selected the negative sign and eliminated it by interchanging the limits on the integral.

Unfortunately, the integral of Eq. (13.9.9) cannot be evaluated in terms of elementary functions. We could, however, expand the integrand in a series (say, the binomial series) and integrate term by term. Alternatively, we can change the variables and convert the integral into the form of an *elliptic integral* (see, for example, Reference 13.14). To this end, let us introduce the constant k and parameter z as:

$$k \stackrel{D}{=} \sin(\theta_0/2)\,,\quad |k| < 1 \tag{13.9.10}$$

and

$$k\sin z \stackrel{D}{=} \sin(\theta/2) \tag{13.9.11}$$

(The motivation for such substitutions is to convert the integral of Eq. (13.9.9) into a standard form of elliptic integrals.)*

* The authors initially learned of this substitution in a course given by Professor T. R. Kane at the University of Pennsylvania.

We can express $\cos\theta$ and $\cos\theta_0$ in terms of k and z by using the identity:

$$\sin^2(\theta/2) \equiv (1/2) - (1/2)\cos\theta \tag{13.9.12}$$

That is,

$$\cos\theta = 1 - 2\sin^2(\theta/2) = 1 - 2k^2\sin^2 z \tag{13.9.13}$$

Then, in view of Eqs. (13.9.10) and (13.9.11), we have:

$$\cos\theta_0 = 1 - 2k^2 \tag{13.9.14}$$

Using these results, we can develop the integrand of Eq. (13.9.9) as:

$$\begin{aligned}
(\cos\theta - \cos\theta_0)^{-1/2} &= \left[1 - 2k^2\sin^2 z - (1 - 2k^2)\right]^{1/2} \\
&= \left[2k^2(1 - \sin^2 z)\right]^{1/2} \\
&= (2k^2\cos^2 z)^{-1/2} \\
&= 1/(\sqrt{2}\,k\cos z)
\end{aligned} \tag{13.9.15}$$

Consider next the differential $d\theta$. From Eq. (13.9.11), we have:

$$\begin{aligned}
\frac{d}{d\theta}\sin(\theta/2) &= (1/2)\cos(\theta/2) \\
&= \frac{d}{d\theta}(k\sin z) \\
&= k\cos z\frac{dz}{d\theta}
\end{aligned} \tag{13.9.16}$$

Observe that in view of Eq. (13.9.11), $(1/2)\cos(\theta/2)$ may be expressed as:

$$\begin{aligned}
(1/2)\cos(\theta/2) &= (1/2)\left[1 - \sin^2(\theta/2)\right]^{1/2} \\
&= (1/2)(1 - k^2\sin^2 z)^{1/2}
\end{aligned} \tag{13.9.17}$$

Then, by combining the terms of Eqs. (13.9.16) and (13.9.17) we have:

$$k\cos z\frac{dz}{d\theta} = (1/2)\cos(\theta/2) = (1/2)(1 - k^2\sin^2 z)^{1/2} \tag{13.9.18}$$

and then

$$\frac{dz}{d\theta} = (1/2)(1 - k^2 \sin^2 z)^{1/2} / k \cos z \tag{13.9.19}$$

or

$$d\theta = \frac{2k \cos z \, dz}{(1 - k^2 \sin^2 z)^{1/2}} \tag{13.9.20}$$

Regarding the limits of the integral of Eq. (13.9.9) we have from Eq. (13.9.10) and (13.9.11) that:

$$\text{when } \theta = \theta_0 , \quad z = \pi/2 , \text{ and when } \quad \theta = \hat{\theta} , \quad z = \hat{z} \tag{13.9.21}$$

Finally, by substituting from Eqs. (13.9.15), (13.9.20), and (13.9.21), the time of descent integral may be expressed as:

$$\begin{aligned} t &= \sqrt{\ell/g} \int_{\hat{z}}^{\pi/2} \frac{dz}{(1 - k^2 \sin^2 z)^{1/2}} \\ &= \sqrt{\ell/g} \int_0^{\pi/2} \frac{dz}{(1 - k^2 \sin^2 z)^{1/2}} - \sqrt{\ell/g} \int_0^{\hat{z}} \frac{dz}{(1 - k^2 \sin^2 z)^{1/2}} \\ &= \sqrt{\ell/g} \left[F(k, \pi/2) - F(k, \hat{z}) \right] \end{aligned} \tag{13.9.22}$$

where $F(k, \pi/2)$ and $F(k, \hat{z})$ are "complete" and "incomplete" elliptic integrals of the first kind [13.14].

To illustrate the use of Eq. (13.9.22), suppose we are interested in determining the period T and frequency f of the oscillation of a simple, undamped pendulum released from rest from an initial angle θ_0. For one complete cycle of oscillation, the orientation angle θ of the pendulum will first decrease from θ_0 to zero, then rise to $-\theta_0$, and then go back to zero again, finally returning again to θ. The time for each of these four phases is the same: $T/4$. Hence, from Eq. (13.9.22), the period T is:

$$T = 4\sqrt{\ell/g} \int_0^{\pi/2} \frac{dz}{(1 - k^2 \sin^2 z)^{1/2}} \tag{13.9.23}$$

The complete elliptic integral of Eq. (13.9.23) is sometimes designated by $K(k)$ [13.14].

As an example, suppose the pendulum is initially displaced 60°. Then, from Eq. (13.9.10), we see that k is 0.5. From Reference 13.14 (or a comparable book of tables), we find $K(k)$ to be 1.6858. Therefore, the period T is:

$$T = 4\sqrt{\ell/g}\,(1.6858) = 6.7432\sqrt{\ell/g} \tag{13.9.24}$$

Consequently the frequency f is:

$$f = 0.1483\sqrt{g/\ell} \tag{13.9.25}$$

Recall from Eq. (13.3.14) that the linearized form of the pendulum equation is:

$$\ddot{\theta} + (g/\ell)\theta = 0 \tag{13.9.26}$$

Then, from Eq. (13.3.18), the period and frequency for this linear representation are:

$$T = 2\pi\sqrt{\ell/g} = 6.283\sqrt{\ell/g} \tag{13.9.27}$$

and

$$f = (1/2\pi)\sqrt{g/\ell} = 0.1592\sqrt{g/\ell} \tag{13.9.28}$$

Comparing Eqs. (13.9.24) and (13.9.27), we see that the linearizing approximation of $\sin\theta$ by θ produces a shorter period and consequently a higher frequency. That is, linearization produces a higher frequency estimate than that occurring in the physical system.

13.10 The Method of Krylov and Bogoliuboff

The foregoing analysis with the nonlinear pendulum equation is illustrative of how complex the analysis can become for even a relatively simple physical system. Exact solutions in closed form for nonlinear governing differential equations are virtually impossible to obtain.

In this section we present an approximate method for finding solutions to nonlinear differential equations that are "nearly linear" — that is, where the nonlinear terms are relatively small compared to the linear terms. The method, called the *Krylov–Bogoliuboff method*, is intended for use with equations of the type:

$$\ddot{x} + \in f(x, \dot{x}) + p^2 x \tag{13.10.1}$$

where p is a constant and ε is a small parameter. The method is developed through use of an approximate Fourier series expansion of the solution. Although it is beyond our scope to present details of the derivation of the method, we will nonetheless outline the procedure itself. Readers interested in more details may consult References 13.2 and 13.15.

The method proceeds as follows: For equations that may be put in the form of Eq. (13.10.1), let a function $F(A, \psi)$ be introduced such that

$$F(A, \psi) = f(A\sin\psi, Ap\cos\psi) \tag{13.10.2}$$

where A is an amplitude to be determined and where ψ is defined as:

$$\psi \stackrel{D}{=} pt + \phi \tag{13.10.3}$$

where ϕ is a phase angle to be determined.

Next, let dA/dt and $d\phi/dt$ be given by:

$$\frac{dA}{dt} = -\frac{\epsilon}{2\pi p}\int_0^{2\pi} F\cos\psi\, d\psi = \alpha(A) \tag{13.10.4}$$

and

$$\frac{d\phi}{dt} = \frac{\epsilon}{2\pi pA}\int_0^{2\pi} F\sin\psi\, d\psi = \beta(A) \tag{13.10.5}$$

If Eqs. (13.10.4) and (13.10.5) can be solved for A and ϕ, then an approximate solution to Eq. (13.10.1) may be expressed in the form:

$$x = A\sin(pt + \phi) \tag{13.10.6}$$

with the derivative $\dot{x}$ given by:

$$\dot{x} = Ap\cos(pt + \phi) \tag{13.10.7}$$

To illustrate the procedure, consider again the nonlinear pendulum equation:

$$\ddot{\theta} + (g/\ell)\sin\theta = 0 \tag{13.10.8}$$

Suppose we approximate $\sin\theta$ as:

$$\sin\theta \approx \theta - \theta^3/6 \tag{13.10.9}$$

Then, Eq. (13.10.8) becomes:

$$\ddot{\theta} - (g/6\ell)\theta^3 + (g/\ell)\theta = 0 \tag{13.10.10}$$

Hence, by comparison with Eq. (13.10.1), we can identify ϵ, f, and p as:

$$\epsilon = g/6\ell \tag{13.10.11}$$

$$f(\theta,\dot{\theta}) = -\theta^3 \tag{13.10.12}$$

and

$$p = \sqrt{g/\ell} \tag{13.10.13}$$

Hence, from Eq. (13.11.2), $F(A, \psi)$ is:

$$F(A,\psi) = -A^3 \sin^3 \psi \tag{13.10.14}$$

Then, Eqs. (13.10.4) and (13.10.5) become:

$$\begin{aligned} \frac{dA}{dt} &= -(g/6\ell)\left(1/2\pi\sqrt{g/\ell}\int_0^{2\pi} -A^3 \sin^3 \psi \cos \psi \, d\psi\right) \\ &= \frac{\sqrt{g/\ell}}{12\pi} A^3 \frac{\sin^4 \psi}{4} \Big|_0^{2\pi} = 0 \end{aligned} \tag{13.10.15}$$

and

$$\begin{aligned} \frac{d\phi}{dt} &= -(g/6\ell)(1/2\pi\sqrt{g/\ell}\, A)\int_0^{2\pi} A^3 \sin^4 \psi \, d\psi \\ &= -\frac{\sqrt{g/\ell}\, A^2}{12\pi}\left(\frac{3}{8}\psi - \frac{\sin 2\psi}{4} + \frac{\sin 4\psi}{32}\right)\Big|_0^{\pi} \\ &= -\sqrt{g/\ell}\, A^2/16 \end{aligned} \tag{13.10.16}$$

Upon integration of Eqs. (13.10.15) and (13.10.16) we obtain:

$$A = A_0 \quad \text{(a constant)} \tag{13.10.17}$$

and

$$\phi = -(1/16)\sqrt{g/\ell}\, A_0^2 t + \phi_0 \tag{13.10.18}$$

where ϕ_0 is a constant.

Therefore, from Eq. (13.10.6), the approximate solution to Eq. (13.10.8) is:

$$\theta = A_0 \sin\left\{\sqrt{g/\ell}\left[1 - (A_0/16)\right]t - \phi_0\right\} \tag{13.10.19}$$

Comparing Eq. (13.10.19) with the linear equation solution, $A_0 \sin\omega t$, we see that ω is:

$$\omega = 2\pi/T = \sqrt{g/\ell}\left[1 - (A_0/16)\right] \tag{13.10.20}$$

where as before T is the period. Solving Eq. (13.11.20) for T we obtain:

$$T = 2\pi\sqrt{\ell/g}\Big/\left[1-\left(A_0^2/16\right)\right] \tag{13.10.21}$$

For a measure of the accuracy of the approximation, consider again the example of the foregoing section where the pendulum is displaced through 60° and released from rest. Then, A_0 is $\pi/3$, and T becomes:

$$T = 2\pi\sqrt{\ell/g}\Big/\left\{1-\left[(\pi/3)^2/16\right]\right\} = 6.7455\sqrt{\ell/g} \tag{13.10.22}$$

Recall that for the linearized equation we obtained (see Eq. (13.3.18)):

$$T = 2\pi\sqrt{\ell/g} = 6.2832\sqrt{\ell/g} \tag{13.10.23}$$

Recall also that the "exact" solution, obtained with elliptic integrals, is (see Eq. (13.9.24)):

$$T = 6.7432\sqrt{\ell/g} \tag{13.10.24}$$

13.11 Closure

Vibration analysis is based primarily upon the solution of differential equations; therefore, vibration analyses are largely dependent upon available methods for solving differential equations. This means that most analyses are confined to systems that can be modeled by linear equations. Linear models, however, may fail to adequately represent a mechanical system if displacements are large or if damping is not proportional to the velocity. Thus, most vibration analyses ultimately involve approximate numerical and experimental procedures. Research in vibration is ongoing, with an emphasis upon experimental techniques for determining natural frequencies, mode shapes, and damping characteristics. In the following chapter, we will use some of our fundamental results in examining stability of mechanical systems.

References

13.1. Meirovitch, L., *Elements of Vibration Analysis*, McGraw-Hill, New York, 1975.
13.2. Newland, D. E., *Mechanical Vibration Analysis and Computation*, Wiley, New York, 1989.
13.3. Thompson, W. T., *Theory of Vibration with Application*, Prentice Hall, Englewood Cliffs, NJ, 1988.
13.4. Tse, F. S., Morse, I. E., and Hinkle, R. T., *Mechanical Vibration Theory and Applications*, Allyn & Bacon, Boston, MA, 1978.
13.5. Steidel, R. F., *An Introduction to Mechanical Vibration*, Wiley, New York, 1989.
13.6. Weaver, W., Timoshenko, S. P., and Young, D. H., *Vibration Problems in Engineering*, John Wiley & Sons, New York, 1990.

13.7. Wowk, V., *Machinery Vibration*, McGraw-Hill, New York, 1991.
13.8. Ayres, F., *Theory and Problems of Differential Equations*, Schaum's Outline Series, McGraw-Hill, New York, 1958.
13.9. Boyce, W. E., and DiPrima, R. C., *Elementary Differential Equations*, Wiley, New York, 1965.
13.10. Coddington, E. A., *An Introduction to Ordinary Differential Equations*, Prentice Hall, Englewood Cliffs, NJ, 1961.
13.11. Golomb, M., and Shanks, M., *Elements of Ordinary Differential Equations*, McGraw-Hill, New York, 1965.
13.12. Murphy, G. M., *Ordinary Differential Equations and their Solutions*, Van Nostrand, New York, 1960.
13.13. Rainville, E. D., *Elementary Differential Equations*, Macmillan, New York, 1964.
13.14. *CRC Standard Mathematical Tables*, CRC Press, Boca Raton, FL, 1972.
13.15. Schmidt, G., and Tondl, A., *Non-Linear Vibrations*, Cambridge University Press, New York, 1986.

Problems

Section 13.2 Solutions of Second-Order Ordinary Differential Equations

P13.2.1: Show that the solution of Eq. (13.2.1) may also be expressed in the form:

$$x = \hat{B}\sin\left(\omega t + \hat{\phi}\right)$$

where $\hat{B}$ and $\hat{\phi}$ are constants. Compare this solution with the solutions given by Eqs. (13.2.2) and (13.2.3).

P13.2.2: Find the solution to the equation:

$$\ddot{x} + \lambda x = 0$$

if $\lambda = 36$ sec^{-2}, $x(0) = 0$, and $\dot{x}(0) = 3$ m/sec.

P13.2.3: Solve Problem P13.2.2 if $\lambda = -36$ sec^{-2}.

P13.2.4: A particle P moves on a straight line as represented in Figure P13.2.4. Let x measure the displacement of P away from a fixed point O as shown. Suppose x is given by the expression:

$$x = 6\cos 5t + 8\sin 5t$$

where x is measured in feet and t in seconds. Find the following:

a. Amplitude of the motion
b. Circular frequency
c. Frequency
d. Period
e. Phase

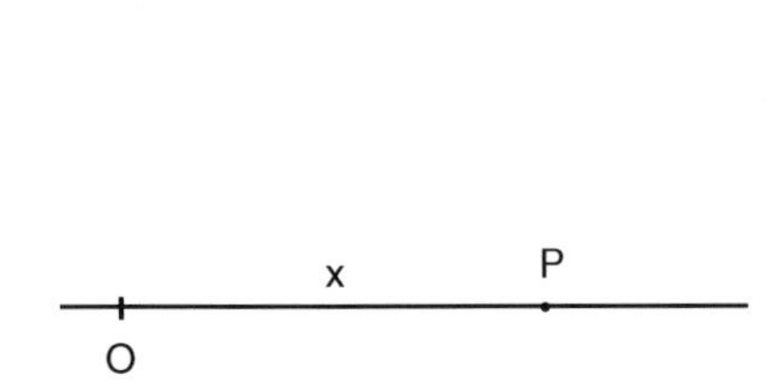

FIGURE P13.2.4
A particle P moving on a straight line.

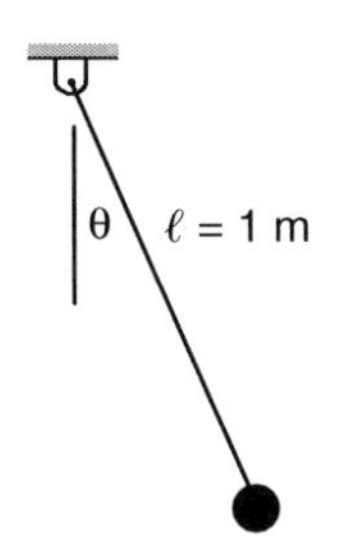

FIGURE P13.2.5
A simple pendulum.

P13.2.5: A simple pendulum with length ℓ = 1 m has its angular position measured by the angle θ as shown in Figure P13.2.5. The pendulum motion is approximated by the differential equation:

$$\ddot{\theta} + (g/\ell)\theta = 0$$

Suppose the pendulum bob speed when θ is zero is 0.5 m/sec. Determine:

a. Amplitude of the pendulum
b. Frequency
c. Period

P13.2.6: See Problem P13.2.5. Suppose the amplitude of the simple pendulum of Problem P13.2.5 is 15°. What is the speed of the pendulum bob as it passes the lowermost position?

P13.2.7: Suppose a simple pendulum has an amplitude of 10° and a speed at the lowest position of 6 in./sec. What is the length of the pendulum?

P13.2.8: Find the general solution to the following equations:

a. $3\ddot{x} + 2\dot{x} + 4x = 0$
b. $4\ddot{x} + \dot{x} + 8x = 0$
c. $7\ddot{x} - 4\dot{x} + 9x = 0$
d. $\ddot{x} + 5\dot{x} + 2x = 0$
e. $2\ddot{x} + 4\dot{x} + 2x = 0$

P13.2.9: Consider the damped linear oscillator equation:

$$m\ddot{x} + c\dot{x} + kx = 0$$

where m is the oscillator mass, c is the damping, k is the spring stiffness, and x measures the displacement. Suppose m is 0.25 slug, c is 1.25 lb·sec/ft, and k is 2 lb/ft. Determine a general expression for the displacement $x(t)$ of the oscillator.

P13.2.10: See Problem P13.2.9. Suppose that initially (t = 0) that the oscillator is displaced 4 inches and released from rest. Determine the subsequent displacement x(t).

P13.2.11: See Problem P13.2.19. What is the necessary value of the damping c so that there will be no oscillation?

P13.2.12: A damped linear oscillator has a mass of 2 kg. Suppose the displacement $x(t)$ of the oscillator is given by:

$$x(t) = e^{-1.5t}\ (A\cos 4t + B\sin 4t)$$

Determine the values of the damping c and spring stiffness k of the oscillator.

P13.2.13: See Problem P13.2.8. Find the solution to the following equations if $x(0) = 1$ and $\dot{x}(0) = -2$.

a. $3\ddot{x} + 2\dot{x} + 4x = 0$
b. $4\ddot{x} + \dot{x} + 8x = 0$
c. $7\ddot{x} - 4\dot{x} + 9x = 0$
d. $\ddot{x} + 5\dot{x} + 2x = 0$
e. $2\ddot{x} + 4\dot{x} + 2x = 0$

P13.2.14: Find the general solution of the following equations:

a. $3\ddot{x} + 6x = 7\cos 3t$
b. $4\ddot{x} + 7x = 8\cos 4t$
c. $2\ddot{x} + 5x = 6\sin 2t$
d. $2\ddot{x} + 5x = 6\sin 2t + 7\cos 3t$

P13.2.15: See Problem P13.2.14. Find the solution to the following equations if $x(0) = -1$ and $\dot{x}(0) = 2$.

a. $3\ddot{x} + 6x = 7\cos 3t$
b. $3\ddot{x} + 7x = 8\cos 4t$
c. $2\ddot{x} + 5x = 6\sin 2t$
d. $2\ddot{x} + 5x = 6\sin 2t + 7\cos 3t$

P13.2.16: Find the general solution of the following equations:

a. $3\ddot{x} + 2\dot{x} + 6x = 7\cos 3t$
b. $4\ddot{x} + 3\dot{x} + 7x = 8\cos 4t$
c. $2\ddot{x} + \dot{x} + 5x = 6\sin 2t$
d. $2\ddot{x} + \dot{x} + 5x = 6\sin 2t + 7\cos 3t$

P13.2.17: See Problem P13.2.16. Find the solution to the following equations if $\dot{x}(0) = -1$ and $x(0) = 2$.

a. $3\ddot{x} + 2\dot{x} + 6x = 7\cos 3t$
b. $4\ddot{x} + 3\dot{x} + 7x = 8\cos 4t$
c. $2\ddot{x} + \dot{x} + 5x = 6\sin 2t$
d. $2\ddot{x} + \dot{x} + 5x = 6\sin 2t + 7\cos 3t$

Section 13.3 The Undamped Linear Oscillator

P13.3.1: For the mass–spring system shown in Figure P13.3.1 let the mass have a weight of 25 lb and the spring have a modulus of 3 lb/in. Determine the natural frequency and period of the system.

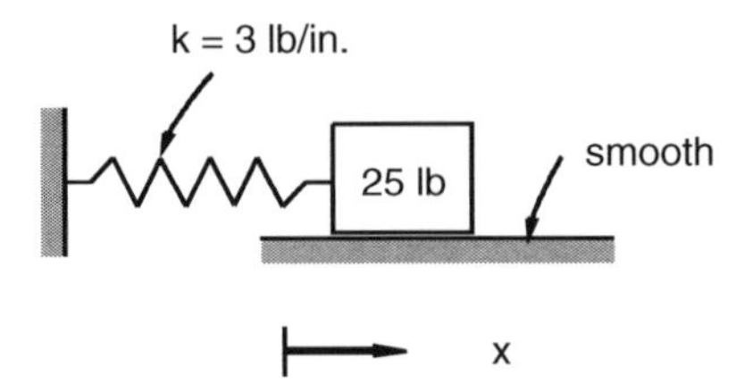

FIGURE P13.3.1
An undamped mass–spring system.

P13.3.2: See Problem P13.3.1. Suppose that at time $t = 0$ the mass is displaced 2 in. to the right and also given a speed of 5 in./sec to the right. Find and express $x(t)$ for the subsequent displacement of the mass. What is the amplitude of the oscillation?

P13.3.3: Repeat Problem P13.3.1 if the mass is 10 kg and the spring modulus is 6 N/cm.

P13.3.4: See Problems P13.3.2 and P13.3.3. Repeat Problem P13.3.2 if the mass is 10 kg, the spring modulus is 6 N/cm, the initial displacement is 5 cm to the right, and the initial speed is 12 cm/sec to the right.

P13.3.5: Consider a simple pendulum with small oscillation. If the pendulum length is 1 m, what is the period? What should be the length if the period is to be 1 second?

P13.3.6: Repeat Problem P13.3.5 if the simple pendulum is replaced by a rod pendulum.

Section 13.4 Forced Vibration of an Undamped Oscillator

P13.4.1: Consider the mass–spring system shown in Figure P13.4.1. Let the block B weigh 15 lb and let the spring modulus be 4 lb/in. Let B be subject to a force $F(t)$ as shown and given by:

$$F(t) = 12 \sin 8t \text{ lb}$$

where t is in seconds. If the system is initially at rest in its equilibrium position, determine the subsequent motion $x(t)$ of B.

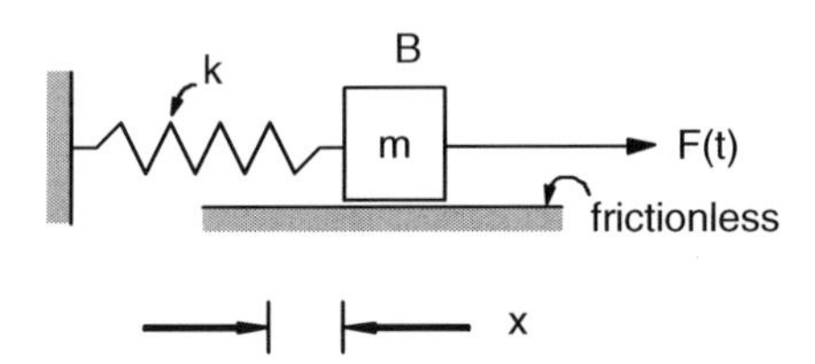

FIGURE P13.4.1
An undamped forced mass–spring system.

P13.4.2: Repeat Problem P13.4.1 if initially ($t = 0$) B is displaced to the right 3 in. and is moving to the right with a speed of 6 in./sec.

P13.4.3: Repeat Problem P13.4.1 if the block B has a mass of 10 kg, the spring modulus is 8 N/cm, and the forcing function is:

$$F(t) = 50 \sin 9t \text{ N}$$

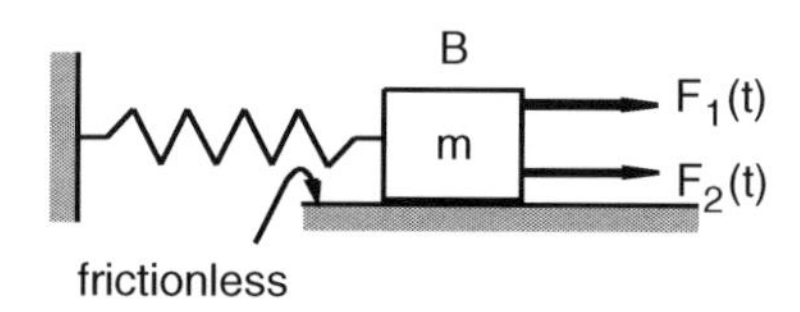

FIGURE P13.4.4
An undamped mass–spring system with two forcing functions.

P13.4.4: Suppose an undamped mass–spring system is subjected to two forcing functions as represented in Figure P13.4.4. Determine the governing equation of motion and its general solution. Let the forcing functions $F_1(t)$ and $F_2(t)$ have the forms:

$$F_1(t) = F_{10} \cos p_1 t$$

and

$$F_2(t) = F_{20} \cos p_2 t$$

P13.4.5: Solve Eq. (13.4.1) if $F(t)$ has the form:

$$F(t) = F_0 \cos(pt + \phi)$$

where F_0, p, and ϕ are constants.

Section 13.5 Damped Linear Oscillator

P13.5.1: Consider the damped mass–spring system depicted in Figure P13.5.1. Let the block B have a mass of 0.25 slug, let the spring modulus be 4 lb/ft, and let the damping coefficient c be 0.5 lb·sec/ft. Let B be displaced to the right 1.5 ft from its equilibrium position and then released from rest. Determine the subsequent motion $x(t)$ of the system.

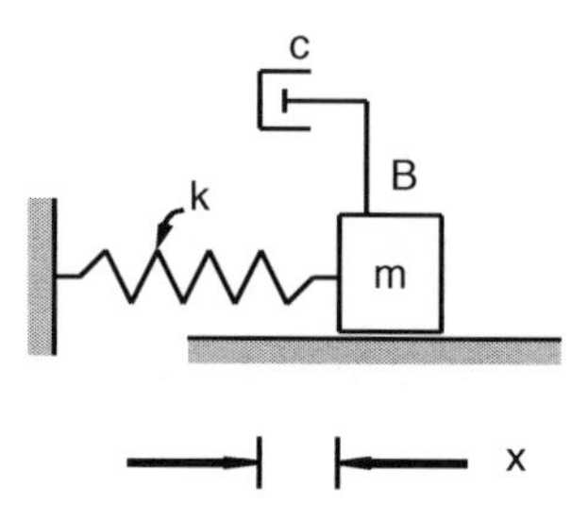

FIGURE P13.5.1
A damped mass–spring system.

P13.5.2: See Problem P13.5.1. Show that the system is underdamped (see Eq. (13.5.5)).

P13.5.3: See Problem P13.5.1. The logarithm δ of the ratio of amplitudes of successive cycles of vibration is called the *logarithmic decrement*. Compute δ for the system of Problem P13.5.1.

P13.5.4: See Problem P13.5.1. Determine the value of the damping coefficient c so that the system is critically damped.

P13.5.4: Repeat Problems P13.5.1 through P13.5.4 if m, k, and c have the values 3.5 kg, 60 N/m, and 6.5 N·sec/m, respectively.

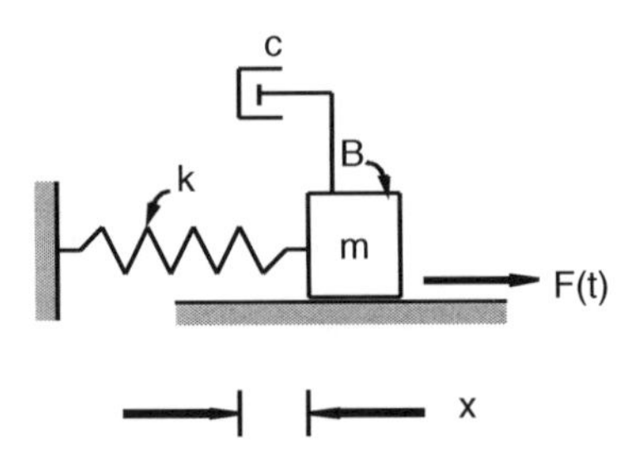

FIGURE P13.6.1
Forced motion of a damped mass–spring system.

Section 13.6 Forced Vibration of a Damped Linear Oscillator

P13.6.1: Consider the damped mass–spring system subjected to a forcing function as represented in Figure P13.6.1. Let the block B have a mass of 0.5 slug, let the spring stiffness be 8 lb/ft, let the damping coefficient be 1.0 lb·sec/ft, and let the forcing function be given by:

$$F(t) = 10 \sin 3t \text{ lb}$$

At time $t = 0$, let B be at rest in its equilibrium position $x = 0$. Determine the subsequent movement $x(t)$ of B.

P13.6.2: See Problem P13.6.1. What should be the value of the forcing function frequency so that the vibration amplitude is maximized? What is the corresponding maximum amplitude?

P13.6.3: Repeat Problem P13.6.1 if at $t = 0$, B is displaced to the right 9 in. with a speed of 3 ft/sec to the right.

P13.6.4: Repeat Problem P13.6.1 if the mass, stiffness, and damping parameters are $m = 6$ kg, $k = 100$ N/m, and $c = 10$ N·sec/m. Let the forcing function be $F(t) = 50 \sin 3t$ N.

P13.6.5: Repeat Problem P13.6.2 using the data of Problem P13.6.4.

Section 13.7 Systems with Several Degrees of Freedom

P13.7.1: Consider the mass–spring system consisting of two blocks B_1 and B_2, having masses m_1 and m_2, respectively, supported by three springs with moduli k_1, k_2, and k_3 as depicted in Figure P13.7.1. Let B_1 and B_2 move in a straight line on a frictionless horizontal surface. Let the natural lengths of the springs be ℓ_1, ℓ_2, and ℓ_3, and let the springs have their natural lengths in the static equilibrium configuration of the system. Finally, let the displacements of the blocks be measured by the coordinates x_1 and x_2 as shown. Determine the governing equations of motion of the system.

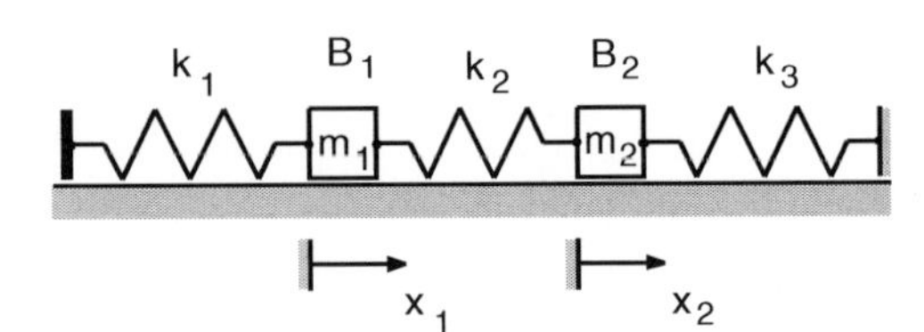

FIGURE P13.7.1
Spring-supported mass–spring system.

P13.7.2: See Problem P13.7.1. Suppose the end supports of the system are such that in the static equilibrium position the springs are stretched by a distance δ. Determine the deformations δ_1, δ_2, and δ_3 of the springs.

P13.7.3: Determine the governing equations of motion for the systems of Problem P13.7.2.

P13.7.4: Consider the following system of equations for a vibrating system with three degrees of freedom:

$$3\ddot{x}_1 + 2x_1 - x_2 = 0$$
$$3\ddot{x}_2 - x_1 + 2x_2 - x_3 = 0$$
$$3\ddot{x}_3 - x_2 + 2x_3 = 0$$

Determine the frequencies (the *natural frequencies*) of the vibration.

P13.7.5: Find the natural frequencies for the vibrating system governed by the following equations:

$$4\ddot{x}_1 + 3x_1 - 2x_2 = 0$$
$$4\ddot{x}_2 - 2x_1 + 3x_2 - 2x_3 = 0$$
$$4\ddot{x}_3 - 2x_2 + 3x_3 = 0$$

P13.7.6: Find the general solution to the following system of equations:

$$m\ddot{x}_1 + 2kx_1 - kx_2 = a$$
$$m\ddot{x}_2 - kx_1 + 2kx_2 = a$$

where k, m, and a are constants.

P13.7.7: Find the general solution to the following system of equations:

$$2\ddot{x}_1 + 3x_1 - 2x_2 = 0$$
$$3\ddot{x}_2 - 2x_1 + 5x_2 = 0$$

P13.7.8: Consider again the double-rod pendulum as discussed in Section 8.10 and as shown in Figure P13.7.8. As before, let the rods be identical, with each having length ℓ and mass m. Let the system move in a vertical plane with the rod orientation being defined by the

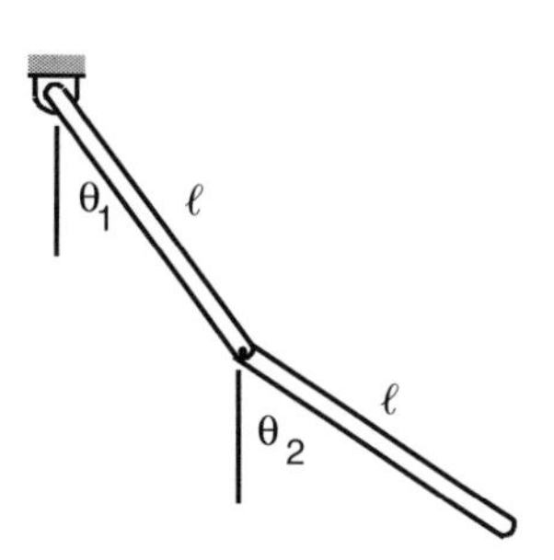

FIGURE P13.7.8
A double-rod pendulum.

angles θ_1 and θ_2 as shown. Determine the governing equation of motion assuming θ_1 and θ_2 are sufficiently small that all nonlinear terms in θ_1, θ_2, $\dot{\theta}_1$, and $\dot{\theta}_2$ may be neglected.

P13.7.9: See Problem P13.7.8. Determine the natural frequencies of vibration for the small-displacement, double-rod pendulum.

P13.7.10: Repeat Problems P13.7.8 and P13.7.9 for the triple-rod pendulum of Section 8.11 and as shown in Figure P13.7.10. As before, let the rods be identical with each having length ℓ and mass m. Let the joint pins be frictionless. Let the system move in a vertical plane with the rod orientation being defined by the angles θ_1, θ_2, and θ_3 as shown.

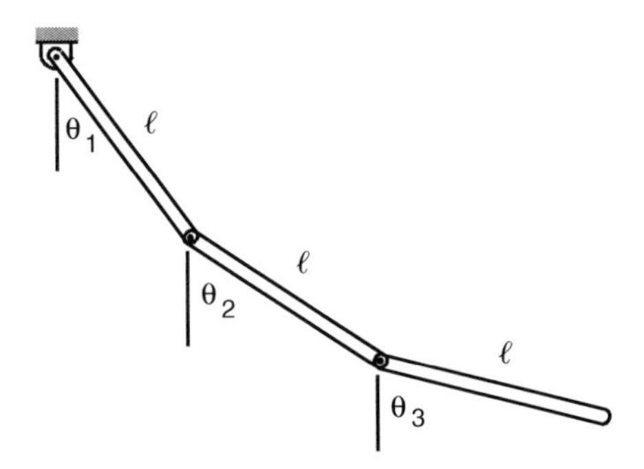

FIGURE P13.7.10
A triple-rod pendulum.

Section 13.8 Modes of Vibration

P13.8.1: Consider a vibrating system with two degrees of freedom governed by the following equations:

$$m_1\ddot{x}_1 + (k_1 + k_2)x_1 - k_2x_2 = 0$$

$$m_2\ddot{x}_2 + (k_2 + k_3)x_2 - k_2x_1 = 0$$

where m_1 and m_2 are masses; k_1, k_2, and k_3 are spring stiffness; and x_1 and x_2 are displacements. Determine the natural frequencies for this system.

P13.8.2: See Problem P13.8.1. Let m_1, m_2, k_1, k_2, and k_3 have the following values: $m_1 = 2$ kg, $m_2 = 3$ kg, $k_1 = 200$ N/m, $k_2 = 150$ N/m, and $k_3 = 250$ N/m. Determine the natural frequencies and the modes of vibration.

P13.8.3: See Problem P13.7.4. Determine the modes of vibration for the system governed by the equation:

$$3\ddot{x}_1 + 2x_1 - x_2 = 0$$

$$3\ddot{x}_2 - x_1 + 2x_2 - x_3 = 0$$

$$3\ddot{x}_3 - x_2 + 2x_3 = 0$$

P13.8.4: See Problem P13.7.5. Determine the modes of vibration for the system governed by the equations:

$$4\ddot{x}_1 + 3x_1 - 2x_2 = 0$$

$$4\ddot{x}_2 - 2x_1 + 3x_2 - 2x_3 = 0$$

$$4\ddot{x}_3 - 2x_2 + 3x_3 = 0$$

P13.8.5: See Problem P13.7.8 and P13.7.9. Determine the modes of vibration of the double-rod pendulum with small oscillation.

P13.8.6: See Problem P13.7.10. Determine the modes of vibration of the triple-rod pendulum with small oscillation.

P13.8.7: Consider a system with three degrees of freedom whose governing equations are:

$$2\ddot{x}_1 + 6x_1 - 3x_2 = 0$$

$$2\ddot{x}_2 - 3x_1 + 6x_2 - 3x_3 = 0$$

$$2\ddot{x}_3 - 3x_2 + 6x_3 = 0$$

where x_1, x_2, and x_3 are measured in feet. Determine the natural frequencies and the modes of vibration.

P13.8.8: See Problem P13.8.7. Determine expressions describing the movement of the system of Problem P13.8.7 if initially ($t = 0$) the system is at rest and x_1, x_2, and x_3 have the values:

$$x_1(0) = 0.5 \text{ ft}, \quad x_2(0) = -2.0 \text{ ft}, \quad x_3(0) = 1.0 \text{ ft}$$

P13.8.9: Repeat Problem P13.8.8 if initially x_1, x_2, and x_3 are:

$$x_1(0) = 1/2 \text{ ft}, \quad x_2(0) = \sqrt{2}/2 \text{ ft}, \quad x_3(0) = 1/2 \text{ ft}$$

P13.8.10 See Problem P13.8.7. Determine expressions describing the movement of the system of Problem P13.8.7 if initially ($t = 0$) x_1, x_2, and x_3 are zero but $\dot{x}_1$, $\dot{x}_2$, and $\dot{x}_3$ have the values:

$$\dot{x}_1 = 0.5 \text{ ft/sec}, \quad \dot{x}_2 = -1.0 \text{ ft/sec}, \quad \dot{x}_3 = 0.75 \text{ ft/sec}$$

Section 13.9 Nonlinear Vibrations

P13.9.1: Consider the rod pendulum depicted in Figure P13.9.1 where the rod has length ℓ and mass m and oscillates in a vertical plane supported by a frictionless pin. Develop equations analogous to Eqs. (13.9.24) and (13.9.25) for the large-angle oscillations of the pendulum.

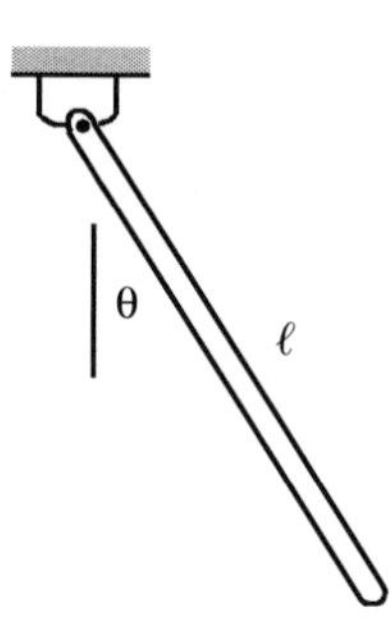

FIGURE P13.9.1
A rod pendulum.

P13.9.2: See Problem P13.9.2. Let a rod pendulum with length 1 m be displaced 45° and released from rest. Find the period and frequency.

P13.9.3: See Problem P13.9.2. Repeat Problem 13.9.2 for initial displacement angles of 60, 90, and 135°.

P13.9.4: A simple pendulum with length of 2 ft has a bob speed of 8 ft/sec in its lowest (static equilibrium) position (Figure 13.9.4). Determine the amplitude, frequency, and period.

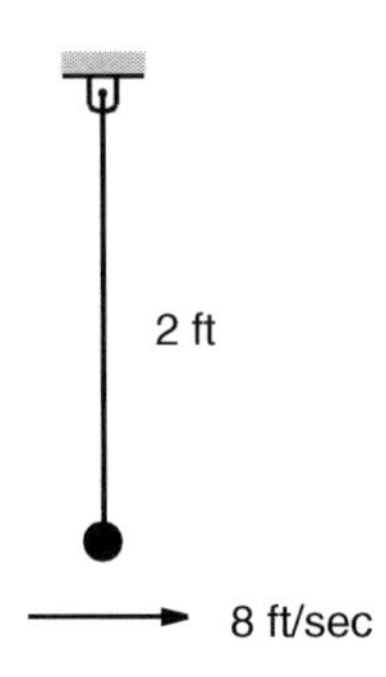

FIGURE P13.9.4
A simple pendulum.

P13.9.5: Repeat Problem P13.9.4 if the pendulum length is 0.5 m and the speed at the lowest position is 3 m/sec.

P13.9.6: The low end of a rod pendulum has a speed of 20 ft/sec as it passes through the static equilibrium position as represented in Figure P13.9.6. Determine the amplitude, frequency, and period of the pendulum for a pendulum length of 3 ft.

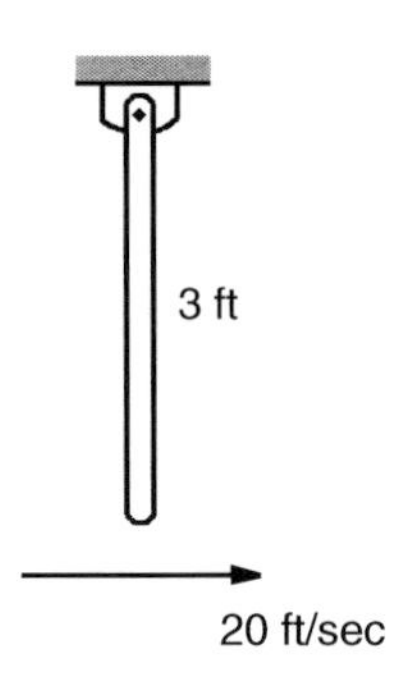

FIGURE P13.9.6
A rod pendulum.

Section 13.10 The Method of Krylov and Bogoliuboff

P13.10.1: Use the method of Krylov and Bogoliuboff to find an approximate solution to the equation:

$$\ddot{x} + \varepsilon\left(1 - x^2\right)\dot{x} + p^2 x = 0$$

where ε is a small constant parameter and p is an arbitrary constant. (This is a form of the Van de Pol equation.)

P13.10.2: Use the method of Krylov and Bogoliuboff to find an approximate solution to the equation:

$$\ddot{x} + c\dot{x} + p^2 x + dx^3 = 0$$

where c and d are small constant parameters and where p is an arbitrary constant (this is a form of the Duffing equation).

14

Stability

14.1 Introduction

Stability and balancing are among the topics of greatest interest to designers and analysts of mechanical systems with moving parts. In this chapter, we introduce and discuss the fundamentals of infinitesimal stability theory. We will consider balancing in the next chapter.

Engineers, physicists, and mathematicians have been exploring and defining concepts of stability for many years. Indeed, stability is a subject with numerous theoretical aspects, many of which are still being developed. In our discussions, however, we will simply introduce the fundamentals of infinitesimal stability theory and the associated analysis procedures. As in earlier chapters, we introduce the concepts through elementary examples. Readers interested in more theoretical and more extensive discussion may want to consult the references at the end of the chapter.

14.2 Infinitesimal Stability

The term *stability* has an intuitive meaning in everyday conversation. Stability means "firmness" or "ability to resist change." For our purposes, however, we can simply define stability as the ability to return to or maintain a closeness to an equilibrium position following a small displacement or disturbance away from that equilibrium position. The term *small displacement* is often described as a "perturbation." As before, the term *small* means that, if approximations are made based upon a quantity being "small," then those approximations become increasingly reliable the smaller the quantity becomes. Finally, the expression *equilibrium position* for a mechanical system means a constant or steady-state solution of the governing equations of motion of the system.

To illustrate these concepts, consider again the simple pendulum of Figure 14.2.1, here consisting of the concentrated mass (or bob) P connected to a pin supported by a light (massless) rod, as opposed to a string. Recall from Eq. (8.4.4) that the governing differential equation for this system is:

$$\ddot{\theta} + (g/\ell)\sin\theta = 0 \tag{14.2.1}$$

where as before θ describes the orientation of the pendulum, ℓ is its length, and g is the gravity constant.

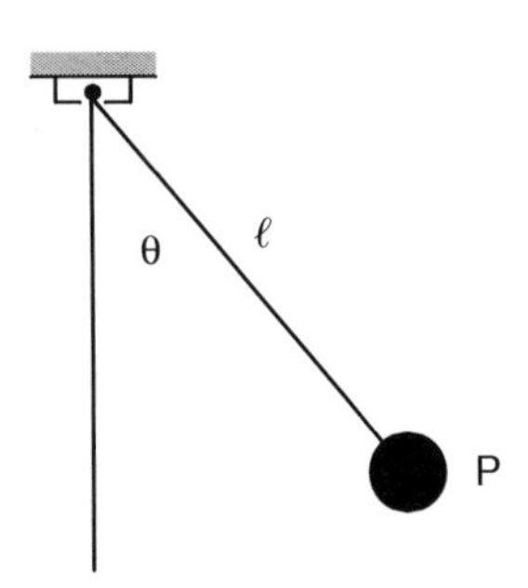

FIGURE 14.2.1
The simple pendulum.

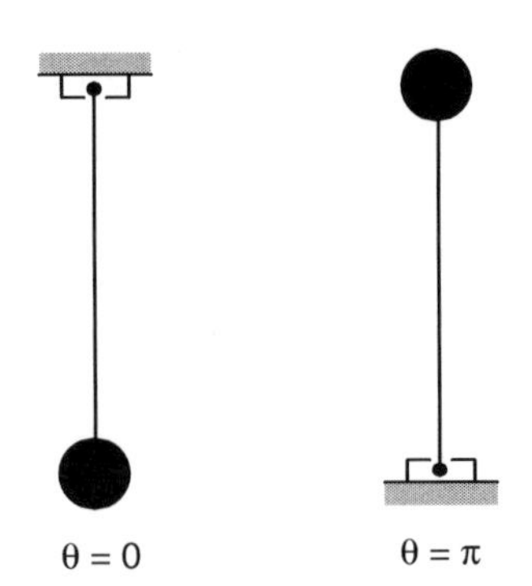

FIGURE 14.2.2
Static equilibrium position for the simple pendulum.

Observe that there are two equilibrium, or "constant," solutions to Eq. (14.2.1). That is,

$$\theta = 0 \quad \text{or} \quad \theta = \pi \quad (\text{or } 180^\circ) \tag{14.2.2}$$

These solutions represent static equilibrium positions shown graphically in Figure 14.2.2.

Intuitively, we would expect the solution $\theta = 0$ to be stable and the solution $\theta = \pi$ to be unstable. That is, if we displace (or "perturb") the pendulum a small amount away from $\theta = 0$ and then release it, we would expect the pendulum to simply oscillate about $\theta = 0$. Alternatively, if we perturb the pendulum by a small amount away from $\theta = \pi$ and then release it, we would expect the pendulum to fall increasingly further away from the equilibrium position.

We can show this mathematically through an analysis of the governing equation (Eq. (14.2.1)). Let us designate "small" quantities by a superscript star or asterisk (*). Specifically, a small displacement of the pendulum will be represented by the angle θ^*. Hence, a perturbation or disturbance away from the equilibrium position $\theta = 0$ may be expressed as:

$$\theta = 0 + \theta^* \tag{14.2.3}$$

By substituting from Eq. (14.2.3) into Eq. (14.2.1), we obtain:

$$\ddot{\theta}^* + (g/\ell)\sin\theta^* = 0 \tag{14.2.4}$$

Because θ^* is small, we can approximate and represent $\sin\theta^*$ by θ^* leading to the equation:

$$\ddot{\theta}^* + (g/\ell)\theta^* = 0 \tag{14.2.5}$$

The solution of Eq. (14.2.5) may be expressed in the form:

$$\theta^* + A\cos\sqrt{g/\ell}\,t + B\sin\sqrt{g/\ell}\,t \tag{14.2.6}$$

where A and B are constants to be determined from auxiliary initial conditions. To this end, suppose that initially the pendulum is displaced away from equilibrium through a small angle θ_0^* and released from rest. From Eq. (14.2.6) we then have:

$$A = \theta_0^*, \quad B = 0 \tag{14.2.7}$$

and then

$$\theta = \theta_0^* \cos\sqrt{g/\ell}\, t \tag{14.2.8}$$

Equation (14.2.8) shows that the movement of the pendulum never gets larger than θ_0^*, following the small disturbance (θ_0^*) away from equilibrium ($\theta = 0$). That is, the small disturbance remains small so $\theta = 0$ is a *stable* equilibrium position.

Consider next the equilibrium position $\theta = \pi$. In this case, a small disturbance or perturbation away from $\theta = \pi$ may be represented by:

$$\theta = \pi + \theta^* \tag{14.2.9}$$

By substituting from Eq. (14.2.9) into Eq. (14.2.1), we have:

$$\ddot{\theta}^* + (g/\ell)\sin(\pi + \theta^*) = 0 \tag{14.2.10}$$

Using a trigonometric identity we can express $\sin(\pi + \theta^*)$ as:

$$\sin(\pi + \theta^*) \equiv \sin\pi\cos\theta^* + \cos\pi\sin\theta^{**} = -\sin\theta^* \tag{14.2.11}$$

Then, because θ^* is small, we can approximate $-\sin\theta^*$ by $-\theta^*$; thus, Eq. (14.2.10) becomes:

$$\ddot{\theta}^* - (g/\ell)\theta^* = 0 \tag{14.2.12}$$

The solution of Eq. (14.2.12) may be written as:

$$\begin{aligned}\theta^* &= Ae^{\sqrt{g/\ell}\,t} + Be^{-\sqrt{g/\ell}\,t} \\ &= (A+B)\cosh\sqrt{g/\ell}\,t + (A-B)\sinh\sqrt{g/\ell}\,t\end{aligned} \tag{14.2.13}$$

where, as before, A and B are constants. Suppose again that initially the pendulum is displaced away from the equilibrium position ($\theta = \pi$) through a small angle θ_0^* and released from rest. From Eq. (14.2.13), we then have:

$$A = \theta_0^*/2 \quad \text{and} \quad B = \theta_0^*/2 \tag{14.2.14}$$

and then

$$\theta^* = \theta_0 \cosh\sqrt{g/\ell}\, t \tag{14.2.15}$$

Equation (14.2.15) shows that the small disturbance gets larger and larger. Thus, $\theta = \pi$ is an *unstable* equilibrium position. In the next several sections, we will use the foregoing technique to explore the stability of several other mechanical systems.

14.3 A Particle Moving in a Vertical Rotating Tube

Consider the system consisting of a particle free to move in the smooth interior of a vertical rotating tube as depicted in Figure 14.3.1 (we considered the kinematics and dynamics of this system in Section 8.5). If the angular speed Ω of the tube is prescribed, the system has one degree of freedom represented by the angle θ. From Eq. (8.5.15) we see that the equation of motion is then:

$$\ddot{\theta} - \Omega^2 \sin\theta\cos\theta + (g/r)\sin\theta = 0 \tag{14.3.1}$$

where r is the tube radius.

If the particle P has reached an equilibrium position, $\ddot{\theta}$ will be zero. The equilibrium angle will then satisfy the equation:

$$-\Omega^2 \sin\theta\cos\theta + (g/r)\sin\theta = 0 \tag{14.3.2}$$

It is readily seen that the solutions to this equation are:

$$\theta = \theta_1 = 0, \quad \theta = \theta_2 = \pi, \quad \text{and} \quad \theta = \theta_3 = \cos^{-1}(g/r\Omega^2) \tag{14.3.3}$$

Thus, there are three equilibrium positions. In the following paragraphs, we consider the stability of each of these.

Case 1: $\theta = \theta_1 = 0$

Consider first the equilibrium position $\theta = 0$. By introducing a small disturbance θ^* about $\theta = 0$, we have:

$$\theta = 0 + \theta^* \tag{14.3.4}$$

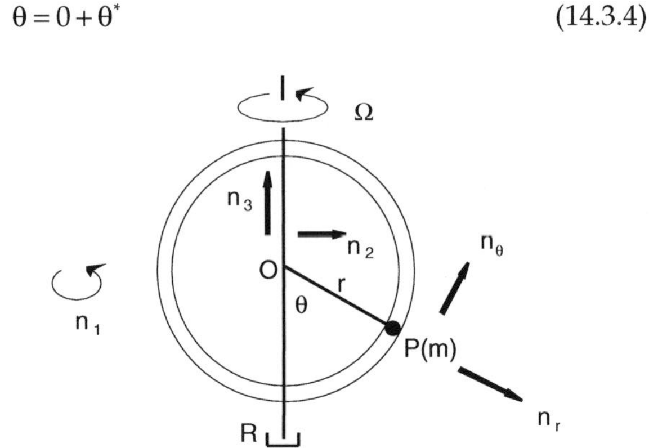

FIGURE 14.3.1
A vertical rotating tube with a smooth interior surface and containing a particle *P*.

Then, to the first order in θ^*, $\sin\theta$ and $\cos\theta$ may be approximated as:

$$\sin\theta = \sin(0 + \theta^*) = \theta^*$$

and

$$\cos\theta = \cos(0 + \theta^*) = 1 \tag{14.3.5}$$

By substituting from Eqs. (14.3.4) and (14.3.5) into (14.3.1), we obtain:

$$\ddot{\theta}^* + [(g/r) = \Omega^2]\theta^* = 0 \tag{14.3.6}$$

By referring to the solutions of Eqs. (14.2.5) and (14.2.12), we see that if $[(g/r) - \Omega^2]$ is positive, the solution of Eq. (14.3.6) may be expressed in terms of trigonometric functions and thus will be bounded and stable. Alternatively, if $[g/r - \Omega^2]$ is negative, the solution of Eq. (14.3.6) will be expressed in terms of exponential or hyperbolic functions and thus will be unbounded and unstable. Hence, the equilibrium position $\theta = 0$ is stable if:

$$g/r > \Omega^2 \quad \text{or} \quad \Omega^2 < g/r \tag{14.3.7}$$

Case 2: $\theta = \theta_2 = 0$

Consider next the equilibrium position $\theta = \pi$. A small disturbance about $\theta = \pi$ may be expressed as:

$$\theta = \pi + \theta^* \tag{14.3.8}$$

Then, $\sin\theta$ and $\cos\theta$ may be approximated to the first order in θ^* as:

$$\begin{aligned} \sin\theta &= \sin(\pi + \theta^*) = -\theta^* \\ \cos\theta &= \cos(\pi + \theta^*) = -1 \end{aligned} \tag{14.3.9}$$

By substituting from Eqs. (14.3.8) and (14.3.9) into (14.3.1) we obtain:

$$\ddot{\theta}^* - [\Omega^2 + (g/r)]\theta^* = 0 \tag{14.3.10}$$

In this case, the coefficient of θ^* is negative for all values of Ω. Therefore, the solution of Eq. (14.3.10) will involve exponential or hyperbolic functions; thus, the equilibrium position is unstable.

Case 3: $\theta = \theta_3 = \cos^{-1}(g/r\Omega^2)$

Finally, consider the equilibrium position $\theta = \cos^{-1}(g/r\Omega^2)$. Observe that this equilibrium position will not exist unless $g/r\Omega^2$ is smaller than 1. That is, Ω^2 must be greater than g/r for equilibrium at $\theta = \cos^{-1}(g/r\Omega^2)$.

If $\Omega^2 > g/r$, a small disturbance about the equilibrium position may be expressed as:

$$\theta = \theta_3 + \theta^* \tag{14.3.11}$$

Then, to the first order in θ^*, $\sin\theta$ and $\cos\theta$ and may be approximated as:

$$\sin\theta = \sin(\theta_3 + \theta^*) = \sin\theta_3 + \theta^* \cos\theta_3$$

and

$$\cos\theta = \cos(\theta_3 + \theta^*) = \cos\theta_3 - \theta^* \sin\theta_3 \tag{14.3.12}$$

By substituting from Eqs. (14.3.11) and (14.3.12) into (14.3.1) we obtain:

$$\begin{aligned} &\ddot{\theta}^* + \left[\Omega^2(\sin^2\theta_3 - \cos^2\theta_3) + (g/r)\cos\theta_3\right]\theta^* \\ &= \Omega^2 \sin\theta_3 \cos\theta_3 - (g/r)\sin\theta_3 \end{aligned} \tag{14.3.13}$$

Because the right side of Eq. (14.3.13) is a constant, the stability (or instability) of the equilibrium position is determined by the sign of the coefficient of θ^*. Specifically, the equilibrium position is stable if the term $[\Omega^2(\sin^2\theta_3 - \cos^2\theta_3) + (g/r)\cos\theta_3]$ is positive. If the term is negative, the equilibrium position is unstable.

From the definition of θ_3 in Eq. (14.3.3), we see that:

$$\cos\theta_3 = g/r\Omega^2, \quad \cos^2\theta_3 = g^2/r^2\Omega^4$$

and

$$\sin^2\theta_3 = 1 - \cos^2\theta_3 = 1 - (g^2/r^2\Omega^4) \tag{14.3.14}$$

Hence, the coefficient of θ^* in Eq. (14.3.13) becomes:

$$\begin{aligned} &\Omega^2 \sin^2\theta_3 - \Omega^2\cos^2\theta_3 + (g/r)\cos\theta_3 = \Omega^2\left[1 - (g^2/r^2\Omega^4)\right] \\ &-\Omega^2(g^2/r^2\Omega^4) + (g^2/r^2\Omega^2) \\ &= \Omega^2\left[1 - (g/r\Omega^2)^2\right] \end{aligned} \tag{14.3.15}$$

Therefore, the equilibrium position is stable if $[1 - (g/r\Omega^2)^2]$ is positive or if:

$$\Omega^2 > g/r \tag{14.3.16}$$

Comparing the inequality of Eq. (14.3.16) with that of Eq. (14.3.7) we see that they are opposite. Also, as noted earlier, we see that Eq. (14.3.16) is a necessary condition for the existence of the equilibrium position $\theta = \theta_3 = \cos^{-1}(g/r\Omega^2)$. That is, for slow tube rotation (specifically, $\Omega^2 < g/r$), there are only two equilibrium positions: $\theta = \theta_1 = 0$ and $\theta = \theta_2 = \pi$, with $\theta = \theta_1 = 0$ being stable (see Eq. (14.3.7)) and $\theta = \theta_2 = \pi$ being unstable (see Eq. (14.3.10)). For fast tube rotation (specifically, $\Omega^2 > g/r$), there are three equilibrium positions: $\theta = \theta_1 = 0$, $\theta = \theta_2 = \pi$, and $\theta = \theta_3 = \cos^{-1}(g/r\Omega^2)$, with $\theta = \theta_1 = 0$ and $\theta = \theta_2 = \pi$ being unstable (see Eqs. (14.3.7) and (14.3.10)) and $\theta = \theta_3 = \cos^{-1}(g/r\Omega^2)$ being stable (see Eq. (14.3.16)). That is, the third equilibrium position does not exist unless the tube rotation is such that $\Omega^2 > g/r$, but if it does exist, it is stable.

14.4 A Freely Rotating Body

Consider next an arbitrarily shaped body B that is thrown into the air, rotating about one of its central principal axes of inertia. Our objective is to explore the stability of that motion; that is, will the body continue to rotate about the principal inertia axis or will it be unstable, wobbling away from the axis?

To answer this question, consider a free-body diagram of B as in Figure 14.4.1, where G is the mass center of B; m is the mass of B; $\mathbf{k}$ is a vertical unit vector; $\mathbf{F}^*$ and $\mathbf{T}^*$ are the inertia force and couple torque, respectively, of a force system equivalent to the inertia forces on B; $-mg\mathbf{k}$ is equivalent to the gravitational forces on B, with g being the gravity constant; and R is an inertial reference frame in which B moves. In the free-body diagram, we have neglected air resistance; thus, the gravitational (or weight) force $-mg\mathbf{k}$ is the only applied (or active) force on B.

From Eqs. (7.12.1) and (7.12.8), we recall that the inertia force $\mathbf{F}^*$ and couple torque $\mathbf{T}^*$ may be expressed as:

$$\mathbf{F}^* = -m\mathbf{a} \quad \text{and} \quad \mathbf{T}^* = -\mathbf{I}\cdot\boldsymbol{\alpha} - \boldsymbol{\omega}\times(\mathbf{I}\cdot\boldsymbol{\omega}) \tag{14.4.1}$$

where $\mathbf{a}$ is the acceleration of G in R; $\boldsymbol{\omega}$ and $\boldsymbol{\alpha}$ are the angular velocity and angular acceleration, respectively, of B in R; and $\mathbf{I}$ is the central inertia dyadic of B (see Sections 7.4 to 7.9).

From the free-body diagram we then have:

$$-mg\mathbf{k} + \mathbf{F}^* = 0 \tag{14.4.2}$$

and

$$\mathbf{T}^* = 0 \tag{14.4.3}$$

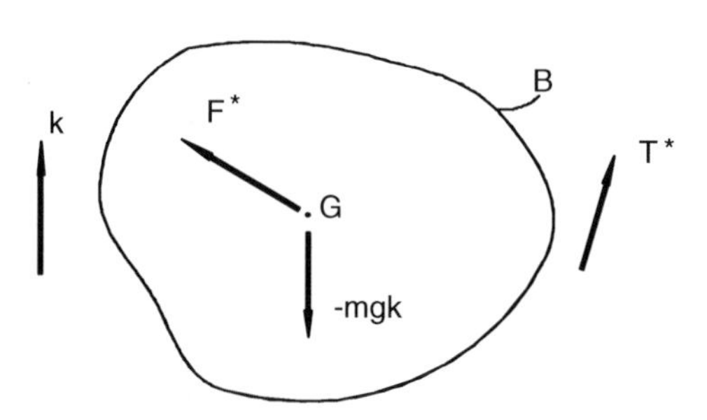

FIGURE 14.4.1
Free-body diagram of freely rotating body.

By inspection of Eqs. (14.4.1) and (14.4.2), we have:

$$\mathbf{a} = -g\mathbf{k} \tag{14.4.4}$$

Thus, G moves as a projectile particle having a parabolic path (see Section 8.7). Note further that points of B not lying on the central principal inertia axis of rotation will not have a parabolic path. That is, as B rotates it rotates about the central inertia axis.

From an inspection of Eqs. (14.4.1) and (14.4.3) we have:

$$\mathbf{I}\cdot\boldsymbol{\alpha} + \boldsymbol{\omega}\times(\mathbf{I}\cdot\boldsymbol{\omega}) = 0 \tag{14.4.5}$$

Let $\mathbf{n}_1$, $\mathbf{n}_2$, and $\mathbf{n}_3$ be mutually perpendicular unit vectors fixed in B and parallel to the central principal inertia axes of B. Let ω and α be expressed in terms of $\mathbf{n}_1$, $\mathbf{n}_2$, and $\mathbf{n}_3$ as:

$$\begin{aligned} \boldsymbol{\omega} &= \omega_1\mathbf{n}_1 + \omega_2\mathbf{n}_2 + \omega_3\mathbf{n}_3 = \omega_i\mathbf{n}_i \\ \boldsymbol{\alpha} &= \alpha_1\mathbf{n}_1 + \alpha_2\mathbf{n}_2 + \alpha_3\mathbf{n}_3 = \alpha_i\mathbf{n}_i \end{aligned} \tag{14.4.6}$$

Because the $\mathbf{n}_i$ $(i = 1, 2, 3)$ are fixed in B the α_i are derivatives of the ω_i (see Eq. (4.4.6)). That is,

$$\alpha_i = \dot{\omega}_i \tag{14.4.7}$$

By substituting from Eqs. (14.4.6) and (14.4.7) into (14.4.5), we obtain:

$$-I_{11}\dot{\omega}_1 + \omega_2\omega_3(I_{22} - I_{33}) = 0 \tag{14.4.8}$$

$$-I_{22}\dot{\omega}_2 + \omega_3\omega_1(I_{33} - I_{11}) = 0 \tag{14.4.9}$$

$$-I_{33}\dot{\omega}_3 + \omega_1\omega_2(I_{11} - I_{22}) = 0 \tag{14.4.10}$$

where I_{11}, I_{22}, and I_{33} are the central principal moments of inertia.

Equations (14.4.8), (14.4.9), and (14.4.10) form a set of three coupled nonlinear ordinary differential equations for the three ω_i $(i = 1, 2, 3)$. To use these equations to determine the stability of rotation of the body, let B be thrown into the air such that B is initially rotating about the central principal inertia axis parallel to $\mathbf{n}_1$. That is, let B be thrown into the air such that its initial angular velocity components ω_i are:

$$\omega_1 = \Omega \tag{14.4.11}$$

$$\omega_3 = 0 \tag{14.4.12}$$

$$\omega_2 = 0 \tag{14.4.13}$$

By inspection, we see that the ω_i ($i = 1, 2, 3$) of Eqs. (14.4.11), (14.4.12), and (14.4.13) are solutions of Eqs. (14.4.8), (14.4.9), and (14.4.10). To test for the stability of this solution, let small disturbances to the motion occur such that:

$$\omega_1 = \Omega + \omega_1^* \tag{14.4.14}$$

$$\omega_2 = 0 + \omega_2^* \tag{14.4.15}$$

$$\omega_3 = 0 + \omega_3^* \tag{14.4.16}$$

where as before the (*) quantities are small. Then, by substituting these expressions into Eqs. (14.4.8), (14.4.9), and (14.4.10), we have:

$$-I_{11}\dot{\omega}_1^* + \omega_2^*\omega_3^*(I_{22} - I_{33}) = 0 \tag{14.4.17}$$

$$-I_{22}\dot{\omega}_2^* + \omega_3^*(\Omega + \omega_1^*)(I_{33} - I_{11}) = 0 \tag{14.4.18}$$

$$-I_{33}\dot{\omega}_3^* + (\Omega + \omega_1^*)\omega_2^*(I_{11} - I_{22}) = 0 \tag{14.4.19}$$

By neglecting products of small quantities, these equations take the form:

$$\dot{\omega}_1^* = 0 \tag{14.4.20}$$

$$-I_{22}\dot{\omega}_2^* + \omega_3^*\Omega(I_{33} - I_{11}) = 0 \tag{14.4.21}$$

$$-I_{33}\dot{\omega}_3^* + \Omega\omega_2^*(I_{11} - I_{22}) = 0 \tag{14.4.22}$$

Equation (14.4.20) has a solution:

$$\omega_1^* = \omega_{10}^*, \quad \text{a constant} \tag{14.4.23}$$

Equations (14.4.21) and (14.4.22) may be solved by eliminating one of the variables (say, ω_3^*) between the equations. Specifically, from Eq. (14.4.21), we have:

$$\omega_3^* = \frac{I_{22}}{\Omega(I_{33} - I_{11})}\dot{\omega}_2^* \tag{14.4.24}$$

Then, by substituting into Eq. (14.4.22), we obtain:

$$\frac{-I_{33}I_{22}}{\Omega(I_{33} - I_{11})}\ddot{\omega}_2^* + \Omega(I_{11} - I_{22})\omega_2^* = 0$$

or

$$\ddot{\omega}_2^* + \Omega^2 \frac{(I_{11} - I_{22})(I_{11} - I_{33})}{I_{22} I_{33}} \omega_2^* = 0 \tag{14.4.25}$$

Inspection of Eq. (14.4.25) shows that the disturbance ω_2^* will remain small and the motion of B will be *stable* if the coefficient of ω_2^* is positive, and that this will occur if I_{11} is either a *maximum* or *minimum* movement of inertia. If I_{11} is an intermediate valued moment of inertia (that is, if $I_{33} < I_{11} < I_{22}$ or $I_{22} < I_{11} < I_{33}$), the motion will be *unstable.*

14.5 The Rolling/Pivoting Circular Disk

Consider again the rolling circular disk (or "rolling coin") as discussed earlier in Sections 4.12 and 8.13 and as shown in Figure 14.5.1. As before, D is the disk, with radius r, mass m, mass center G, contact point C, and orientation angles θ, ϕ, and ψ. Recall that the condition of rolling requires that the contact point C of D has zero velocity relative to the rolling surface. Recall further that by setting the moments of forces on D about C equal to zero, we obtained the governing equations:

$$(4g/r)\sin\theta - 5\ddot{\theta} + 6\dot{\psi}\dot{\phi}\cos\theta + 5\dot{\phi}^2 \sin\theta\cos\theta = 0 \tag{14.5.1}$$

$$3\ddot{\psi} + 3\ddot{\phi}\sin\theta + 5\dot{\phi}\dot{\theta}\cos\theta = 0 \tag{14.5.2}$$

$$\ddot{\phi}\cos\theta + 2\dot{\psi}\dot{\theta} = 0 \tag{14.5.3}$$

In the following paragraphs, we consider the stability of the motions represented by elementary solutions of Eqs. (14.5.1), (14.5.2), and (14.5.3): straight-line rolling, rolling in a circle, and pivoting.

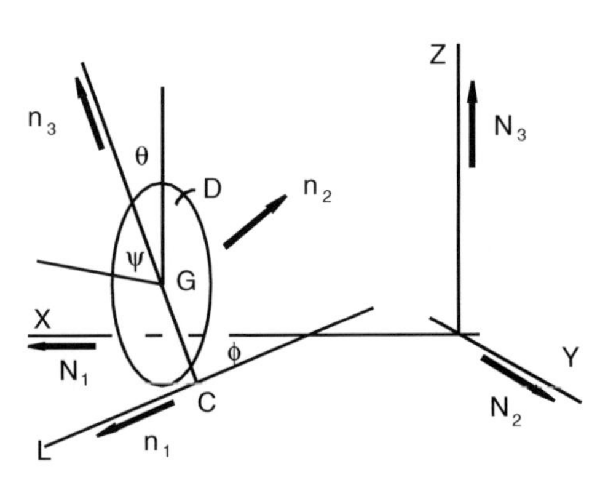

FIGURE 14.5.1
Rolling circular disk.

Case 1: Straight-Line Rolling

Recall from Eq. (8.13.20) that if D is rolling in a straight line with constant speed the angles θ, ϕ, and ψ are (see Figure 14.5.1):

$$\theta = 0, \quad \phi = \phi_0, \quad \text{a constant}, \quad \dot{\psi} = \dot{\psi}_0, \quad \text{a constant} \tag{14.5.4}$$

By inspection we readily see that Eq. (14.5.4) forms a solution of Eqs. (14.5.1), (14.5.2), and (14.5.3).

Suppose that the disk D encounters a small disturbance such that θ, ϕ, and $\dot{\psi}$ have the forms:

$$\theta = 0 + \theta^*, \quad \phi = \phi_0 + \phi^*, \quad \dot{\psi} = \dot{\psi}_0 + \dot{\psi}^* \tag{14.5.5}$$

where as before the quantities with a (*) are small. By substituting from Eq. (14.5.5) into Eqs. (14.5.1), (14.5.2), and (14.5.3), we obtain:

$$(4g/r)\theta^* - 5\ddot{\theta}^* + 6\dot{\psi}_0\dot{\phi}^* = 0 \tag{14.5.6}$$

$$\ddot{\psi}^* = 0 \tag{14.5.7}$$

$$\ddot{\phi}^* + 2\dot{\psi}_0\dot{\theta}^* = 0 \tag{14.5.8}$$

Equations (14.5.7) and (14.5.8) may be integrated, leading to:

$$\dot{\psi}^* = c_1^* \tag{14.5.9}$$

$$\dot{\phi}^* + 2\dot{\psi}_0\theta^* = c_2^* \tag{14.5.10}$$

where c_1^* and c_2^* are small constants. By substituting from Eq. (14.5.10) into Eqs. (14.5.5), eliminating ϕ^*, we obtain:

$$(4g/r)\theta^* - 5\ddot{\theta}^* + 6\dot{\psi}_0(c_2^* - 2\dot{\psi}_0\theta^*) = 0 \tag{14.5.11}$$

or

$$\ddot{\theta}^* + \left[(12/5)\dot{\psi}_0^2 - (4g/5r)\right]\theta^* = c_3^* \tag{14.5.12}$$

where c_3^* is the small constant $6\dot{\psi}_0^2 c_2^*/5$. As before, the motion is stable if the coefficient of θ^* is positive. That is, the motion is stable if:

$$\dot{\psi}_0 > \sqrt{g/3r} \tag{14.5.13}$$

Thus, if the angular speed of D exceeds $\sqrt{g/3r}$, D will remain erect and continue to roll in a straight line. If the angular speed of D is less than $\sqrt{g/3r}$, the motion is unstable. D will wobble and eventually fall.

Case 2: Rolling in a Circle

Next, suppose D is rolling in a circle with uniform speed such that θ, $\dot{\phi}$, and $\dot{\psi}$ are (see Figure 14.5.1):

$$\theta = \theta_0, \quad \dot{\phi} = \dot{\phi}_0, \quad \dot{\psi} = \dot{\psi}_0 \tag{14.5.14}$$

By inspection of the governing equations, we see that Eqs. (14.5.2) and (14.5.3) are then identically satisfied and that Eq. (14.5.1) becomes:

$$(4g/r)\sin\theta_0 + 6\dot{\psi}_0\dot{\phi}_0\cos\theta_0 + 5\dot{\phi}_0^2\sin\theta_0\cos\theta_0 = 0 \tag{14.5.15}$$

Equation (14.5.15) provides a relationship between θ_0, $\dot{\phi}_0$, and $\dot{\psi}_0$. By inspection of Figure 14.5.1 we see that if $\dot{\phi}_0$ and $\dot{\psi}_0$ are positive, then Eq. (14.5.15) requires that θ_0 be negative. That is, the disk will lean toward the interior of the circle on which it rolls.

To test the stability of this motion, let the disk encounter a small disturbance such that θ, $\dot{\phi}$, and $\dot{\psi}$ have the forms:

$$\theta = \theta_0 + \theta^*, \quad \dot{\phi} = \dot{\phi}_0 + \dot{\phi}^*, \quad \dot{\psi} = \dot{\psi}_0 + \dot{\psi}^* \tag{14.5.16}$$

where, as before, the quantities with the (*) are small. Then, $\sin\theta$ and $\cos\theta$ are:

$$\sin\theta = \sin(\theta_0 + \theta^*) = \sin\theta_0 + \theta^*\cos\theta_0 \tag{14.5.17}$$

and

$$\cos\theta = \cos(\theta_0 + \theta^*) = \cos\theta_0 - \theta^*\sin\theta_0 \tag{14.5.18}$$

By substituting from Eqs. (14.5.16), (14.5.17), and (14.5.18) into Eqs. (14.5.1), (14.5.2), and (14.5.3) and by neglecting quadratic (second) and higher order terms in the (*) terms we obtain:

$$\begin{aligned}&(4g/r)\sin\theta_0 + 6\dot{\psi}_0\dot{\phi}_0\cos\theta_0 + 5\dot{\phi}_0^2\sin\theta_0\cos\theta_0 + (4g/r)\theta^*\cos\theta_0\\&-5\ddot{\theta}^* + 6\dot{\psi}_0\dot{\phi}^*\cos\theta_0 + 6\dot{\psi}^*\dot{\phi}_0\cos\theta_0 - 6\dot{\psi}_0\dot{\phi}_0\theta^*\sin\theta_0\\&+10\dot{\phi}_0\dot{\phi}^*\sin\theta_0\cos\theta_0 + 5\dot{\phi}_0^2\theta^*\cos^2\theta_0 - 5\dot{\phi}_0^2\theta^*\sin^2\theta_0 = 0\end{aligned} \tag{14.5.19}$$

$$3\ddot{\psi}^* + 3\ddot{\phi}^*\sin\theta_0 + 5\dot{\phi}_0\dot{\theta}^*\cos\theta_0 = 0 \tag{14.5.20}$$

and

$$\ddot{\phi}^* \cos\theta_0 + 2\dot{\psi}_0\dot{\theta}^* = 0 \tag{14.5.21}$$

In view of Eq. (14.5.15), the sum of the first three terms of Eq. (14.5.19) is zero, thus they may be neglected. Also, Eqs. (14.5.20) and (14.5.21) may be integrated as:

$$3\dot{\psi}^* + 3\dot{\phi}^* \sin\theta_0 + 5\dot{\phi}_0\theta^* \cos\theta_0 = c_1^* \tag{14.5.22}$$

and

$$\dot{\phi}^* \cos\theta_0 + 2\dot{\psi}_0\theta^* = c_2^* \tag{14.5.23}$$

where c_1^* and c_2^* are constants. Solving Eq. (14.5.23) for $\dot{\phi}^*$ we have:

$$\dot{\phi}^* = \left(-2\dot{\psi}_0\theta^* + c_2^*\right)/\cos\theta_0 \tag{14.5.24}$$

Then, by substituting into Eq. (14.5.22), we have:

$$3\dot{\psi}^* + \left(-6\dot{\psi}_0\theta^* + 3c_2^*\right)\tan\theta_0 + 5\dot{\phi}_0\theta^* \cos\theta_0 = c_1^* \tag{14.5.25}$$

Finally, by solving for $\dot{\psi}^*$ and by substituting for $\dot{\phi}^*$ and $\dot{\psi}^*$ into Eq. (14.5.19) (without the first three terms) we have:

$$\ddot{\theta} + \lambda\theta^* = \kappa^* \tag{14.5.26}$$

where λ and κ^* are defined as:

$$\lambda \stackrel{D}{=} (12/5)\dot{\psi}_0^2 + (14/5)\dot{\phi}_0\dot{\psi}_0 \sin\theta_0 + \dot{\phi}_0^2 \\ -(4g/5r)\cos\theta_0 \tag{14.5.27}$$

and

$$\kappa^* \stackrel{D}{=} (6/5)\dot{\psi}_0 c_2^* + (2/5)\dot{\phi}_0 c_1^* \cos\theta_0 + (4/5)\dot{\phi}_0 c_2^* \sin\theta_0 \tag{14.5.28}$$

We recall from our previous analyses that stability will occur if the coefficient λ of θ^* in Eq. (14.5.26) is positive; as a corollary, instability will occur if λ is negative. ($\lambda = 0$ represents a neutral condition, bordering between stability and instability.) Recall also from Eq. (14.5.15) and by inspection of Figure 14.5.1, that if $\dot{\psi}_0$ and $\dot{\phi}_0$ are positive, then θ_0 must be negative. Hence, from Eq. (14.5.27) we see that λ is positive; thus, stability occurs, if:

$$(12/5)\dot{\psi}_0^2 + \dot{\phi}_0^2 > (4g/5r)\cos\theta_0 - (14/5)\dot{\phi}_0\dot{\psi}_0 \sin\theta_0 \tag{14.5.29}$$

Therefore, by making $\dot{\psi}_0$ and $\dot{\phi}_0$ sufficiently large, we can attain stability. Finally, note that in the limiting case when $\dot{\phi}_0 = 0$ and $\theta_0 = 0$, we have rolling in a straight line and then Eq. (14.5.29) becomes identical with Eq. (14.5.27).

Case 3. Pivoting

Finally, suppose that disk D is pivoting such that (see Figure 14.5.1):

$$\theta = 0, \quad \psi = \psi_0, \quad \text{a constant}, \quad \dot{\phi} = \dot{\phi}_0, \quad \text{a constant} \tag{14.5.30}$$

Then, by inspection, we readily see that Eq. (14.5.30) forms a solution of Eqs. (14.5.1), (14.5.2), and (14.5.3).

Suppose that during the pivoting D encounters a small disturbance such that θ, ψ, and $\dot{\phi}$ have the forms:

$$\theta = 0 + \theta^*, \quad \psi = \psi_0 + \psi^*, \quad \dot{\phi} = \dot{\phi}_0 + \dot{\phi}^* \tag{14.5.31}$$

where, as before, the quantities with a (*) are small. By substituting from Eq. (14.5.31) into Eqs. (14.5.1), (14.5.2), and (14.5.3) and by neglecting quadratic and higher powers of small quantities, we obtain:

$$(4g/r)\theta^* - 5\ddot{\theta}^* + 6\dot{\psi}^*\dot{\phi}_0 + 5\dot{\phi}_0^2\theta^* = 0 \tag{14.5.32}$$

$$3\ddot{\psi}^* + 5\dot{\phi}_0\dot{\theta}^* = 0 \tag{14.5.33}$$

$$\ddot{\phi}^* = 0 \tag{14.5.34}$$

Equations (14.5.33) and (14.5.34) may be integrated, leading to:

$$3\dot{\psi}^* + 5\dot{\phi}_0\theta^* = c_1^* \tag{14.5.35}$$

and

$$\dot{\phi}^* = c_2^* \tag{14.5.36}$$

where, as before, c_1^* and c_2^* are small constants. By substituting from Eq. (14.5.35) into (14.5.32), eliminating $\dot{\psi}^*$, we obtain:

$$(4g/r)\theta^* - 5\ddot{\theta}^* + 6\dot{\phi}_0\left[c_1^* - (5/3)\dot{\phi}_0\theta^*\right] + 5\dot{\phi}_0^2\theta^* = 0$$

or

$$\ddot{\theta} + \left[\dot{\phi}_0^2 - (4g/5r)\right]\theta^* = 6c_1^*/5 \tag{14.5.37}$$

As before, the motion is stable if the coefficient of θ^* is positive. That is, the pivoting motion is stable if:

$$\dot{\phi}_0 > \sqrt{4g/5r} \tag{14.5.38}$$

14.6 Pivoting Disk with a Concentrated Mass on the Rim

For a generalization of the foregoing analysis consider a pivoting circular disk D with a concentrated mass at a point Q on the rim of the disk as depicted in Figure 14.6.1. Let the mass at Q be m, and let the mass of D be M. As before, let the radius of D be r, and let the orientation of D be defined by the angles θ, ϕ, and ψ as in Figure 14.6.1.

The presence of the mass at Q makes it necessary to know the kinematics of Q. The kinematics of Q, however, are directly dependent upon the kinematics of D. Therefore, in developing the kinematics of Q it is helpful to review and summarize the kinematics of D as previously developed in Section 4.12.

In developing the kinematics and dynamics of D and Q it is helpful to introduce various unit vector sets as shown in Figure 14.6.1. As before, let $\mathbf{N}_1$, $\mathbf{N}_2$, and $\mathbf{N}_3$ be mutually perpendicular unit vectors fixed in an inertial frame R; similarly, let $\mathbf{d}_1$, $\mathbf{d}_2$, and $\mathbf{d}_3$ be unit vectors fixed in D; let $\mathbf{n}_1$, $\mathbf{n}_2$, and $\mathbf{n}_3$ be, as before, unit vectors parallel to principal inertia axes of D; and, finally, let $\hat{\mathbf{n}}_1$ and $\hat{\mathbf{n}}_2$ be mutually perpendicular unit vectors parallel and perpendicular to the surface on which D pivots as indicated in Figure 14.6.1.

To obtain relations between these unit vector sets it is helpful to use a configuration graph (see Section 4.3) as in Figure 14.6.2 where $\hat{D}$ and $\hat{R}$ are reference frames containing the $\mathbf{n}_i$ and the $\hat{\mathbf{n}}$ $(i = 1, 2, 3)$ respectively. As before, it is convenient to express the

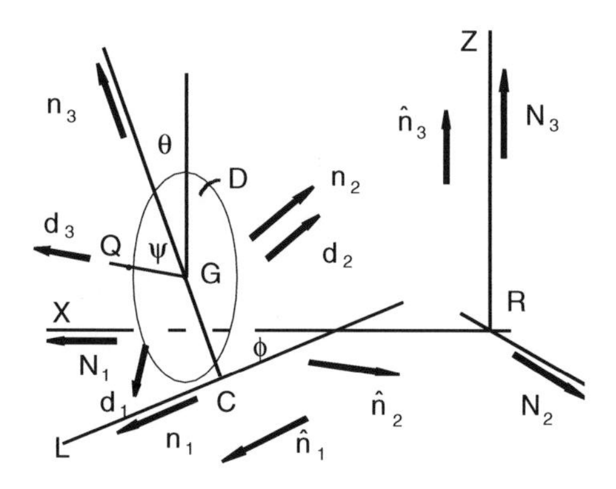

FIGURE 14.6.1
Pivoting disk with a concentrated mass on the rim.

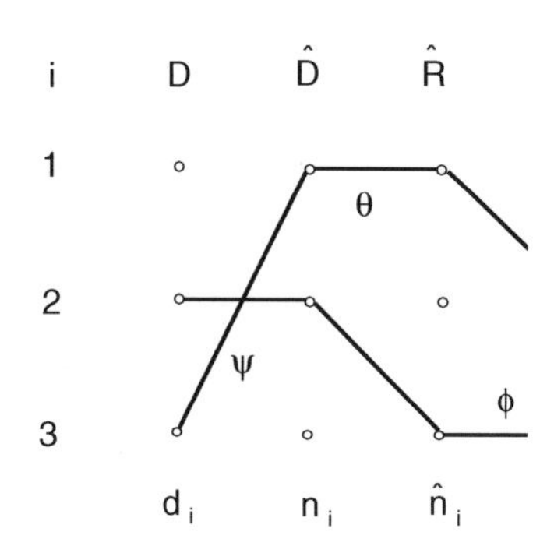

FIGURE 14.6.2
Configuration graph for the unit vector sets of Figure 14.6.1.

kinematical quantities in terms of $\mathbf{n}_1$, $\mathbf{n}_2$, and $\mathbf{n}_3$. To this end, the configuration graph may be used to obtain the relations:

$$\begin{aligned}\mathbf{d}_1 &= \cos\psi\mathbf{n}_1 - \sin\psi\mathbf{n}_3 \\ \mathbf{d}_2 &= \mathbf{n}_2 \\ \mathbf{d}_3 &= \sin\psi\mathbf{n}_1 + \cos\psi\mathbf{n}_3\end{aligned} \tag{14.6.1}$$

$$\begin{aligned}\hat{\mathbf{n}}_1 &= \mathbf{n}_1 \\ \hat{\mathbf{n}}_2 &= \cos\theta\mathbf{n}_2 - \sin\theta\mathbf{n}_3 \\ \hat{\mathbf{n}}_3 &= \sin\theta\mathbf{n}_2 + \cos\theta\mathbf{n}_3\end{aligned} \tag{14.6.2}$$

and

$$\begin{aligned}\mathbf{N}_1 &= \cos\phi\mathbf{n}_1 - \sin\phi\cos\theta\mathbf{n}_2 + \sin\phi\sin\theta\mathbf{n}_3 \\ \mathbf{N}_2 &= \sin\phi\mathbf{n}_1 + \cos\phi\cos\theta\mathbf{n}_2 - \cos\phi\sin\theta\mathbf{n}_3 \\ \mathbf{N}_3 &= \sin\theta\mathbf{n}_2 + \cos\theta\mathbf{n}_3\end{aligned} \tag{14.6.3}$$

Using the procedures of Chapter 4 and the configuration graph of Figure 14.6.2, we readily find the angular velocity of D in R to be (see Eq. (4.7.6)):

$${}^R\boldsymbol{\omega}^D = \dot{\psi}\mathbf{n}_2 + \dot{\theta}\mathbf{n}_1 + \dot{\phi}\hat{\mathbf{n}}_3 \tag{14.6.4}$$

Then, by using the third expression of Eq. (14.6.2) to express $\mathbf{n}_3$ in terms of $\mathbf{n}_2$ and $\mathbf{n}_3$, ${}^R\boldsymbol{\omega}^D$ becomes:

$${}^R\boldsymbol{\omega}^D = \dot{\theta}\mathbf{n}_1 + (\dot{\psi} + \dot{\phi}\sin\theta)\mathbf{n}_2 + \dot{\phi}\cos\theta\mathbf{n}_3 \tag{14.6.5}$$

Observe in Eq. (14.6.4) that in computing the angular acceleration of D in R we will need to compute the derivatives of the unit vectors $\mathbf{n}_1$, $\mathbf{n}_2$, and $\mathbf{n}_3$. Observe further from the configuration graph of Figure 14.6.1 that the $\mathbf{n}_i$ ($i = 1, 2, 3$) are fixed in reference frame $\hat{D}$. Then, also from the configuration graph, we see that:

$${}^R\boldsymbol{\omega}^{\hat{D}} = \dot{\theta}\mathbf{n}_1 + \dot{\phi}\sin\theta\mathbf{n}_2 + \dot{\phi}\cos\theta\mathbf{n}_3 \tag{14.6.6}$$

Hence, from Eq. (4.5.2) the derivatives of the $\mathbf{n}_i$ in R are:

$$\begin{aligned}d\mathbf{n}_1/dt &= {}^R\boldsymbol{\omega}^{\hat{D}} \times \mathbf{n}_1 = \dot{\phi}\cos\theta\mathbf{n}_2 - \dot{\phi}\sin\theta\mathbf{n}_3 \\ d\mathbf{n}_2/dt &= {}^R\boldsymbol{\omega}^{\hat{D}} \times \mathbf{n}_2 = -\dot{\phi}\cos\theta\mathbf{n}_1 + \dot{\theta}\mathbf{n}_3 \\ d\mathbf{n}_3/dt &= {}^R\boldsymbol{\omega}^{\hat{D}} \times \mathbf{n}_3 = -\dot{\theta}\mathbf{n}_2 + \dot{\phi}\sin\theta\mathbf{n}_1\end{aligned} \tag{14.6.7}$$

Using Eq. (14.6.7) we can now differentiate Eq. (14.6.5) to obtain the angular acceleration of D in R as:

$$
\begin{aligned}
{}^R\boldsymbol{\alpha}^D &= \left(\ddot{\theta} - \dot{\psi}\dot{\phi}\cos\theta\right)\mathbf{n}_1 + \left(\ddot{\psi} + \dot{\theta}\dot{\phi}\cos\theta + \ddot{\phi}\sin\theta\right)\mathbf{n}_2 \\
&\quad + \left(\dot{\psi}\dot{\theta} + \ddot{\phi}\cos\theta - \dot{\phi}\dot{\theta}\sin\theta\right)\mathbf{n}_3
\end{aligned}
\tag{14.6.8}
$$

Recall that because D is pivoting (a special case of rolling; see Section 4.11) on a surface in the X–Y plane, the velocity of the center G of D in R is (see Eq. (4.11.5)):

$$
{}^R\mathbf{V}^G = {}^R\boldsymbol{\omega}^D \times r\mathbf{n}_3 = r\left(\dot{\psi} + \dot{\phi}\sin\theta\right)\mathbf{n}_1 - r\dot{\theta}\mathbf{n}_2 \tag{14.6.9}
$$

Similarly, because Q is fixed in D, the velocity of Q in R is:

$$
{}^R\mathbf{V}^G = {}^R\boldsymbol{\omega}^D \times \left(r\mathbf{n}_3 + r\mathbf{d}_3\right) \tag{14.6.10}
$$

Then, by using Eqs. (14.6.1) and (14.6.6), we obtain:

$$
\begin{aligned}
{}^R\mathbf{V}^Q &= r\left[\dot{\psi}(1+\cos\psi) + \dot{\phi}(1+\cos\psi)\sin\theta\right]\mathbf{n}_1 \\
&\quad + r\left[\dot{\phi}\sin\psi\cos\theta - \dot{\theta}(1+\cos\psi)\right]\mathbf{n}_2 \\
&\quad - r\left[\left(\dot{\psi} + \dot{\phi}\sin\theta\right)\sin\psi\right]\mathbf{n}_3
\end{aligned}
\tag{14.6.11}
$$

Finally, by differentiating in Eqs. (14.6.9) and (14.6.11) and by using Eq. (14.6.7), we obtain the accelerations of G and Q in R to be:

$$
\begin{aligned}
{}^R\mathbf{a}^G &= r\left(\ddot{\psi} + \ddot{\phi}\sin\theta + 2\dot{\theta}\dot{\phi}\cos\theta\right)\mathbf{n}_1 \\
&\quad + r\left(-\ddot{\theta} + \dot{\phi}^2\sin\theta\cos\theta + \dot{\psi}\dot{\phi}\cos\theta\right)\mathbf{n}_2 \\
&\quad + r\left(-\dot{\psi}\dot{\phi}\sin\theta - \dot{\phi}^2\sin^2\theta - \dot{\theta}^2\right)\mathbf{n}_3
\end{aligned}
\tag{14.6.12}
$$

and

$$
\begin{aligned}
{}^R\mathbf{a}^Q &= r\left[\ddot{\psi}(1+\cos\psi) - \dot{\psi}^2\sin\psi + \ddot{\phi}(1+\cos\psi)\sin\theta\right. \\
&\quad - 2\dot{\phi}\dot{\psi}\sin\theta\sin\psi + 2\dot{\phi}\dot{\theta}(1+\cos\psi)\cos\theta \\
&\quad \left. - \dot{\phi}^2\sin\psi\right]\mathbf{n}_1 \\
&\quad + r\left[\dot{\psi}\dot{\phi}(1+\cos\psi)\cos\theta + \dot{\phi}^2(1+\cos\psi)\sin\theta\cos\theta\right. \\
&\quad - \ddot{\theta}(1+\cos\psi) + \ddot{\phi}\sin\psi\cos\theta \\
&\quad \left. + \dot{\phi}\dot{\psi}\cos\psi\cos\theta + 2\dot{\theta}\dot{\psi}\sin\psi\right]\mathbf{n}_2 \\
&\quad + \left[-\dot{\psi}\dot{\phi}\sin\theta - \dot{\phi}^2(1+\cos\psi)\sin^2\theta - 2\dot{\psi}\dot{\phi}\cos\psi\sin\theta\right. \\
&\quad \left. - \dot{\theta}^2(1+\cos\psi) - \dot{\psi}^2\cos\psi - \ddot{\psi}\sin\psi - \ddot{\phi}\sin\theta\sin\psi\right]\mathbf{n}_3
\end{aligned}
\tag{14.6.13}
$$

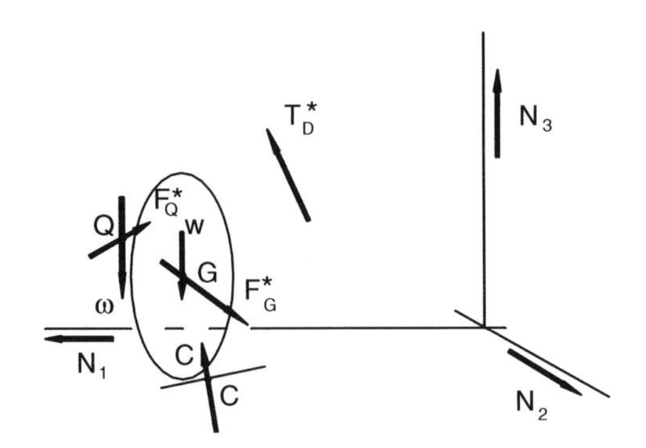

FIGURE 14.6.3
Free-body diagram of pivoting disk.

Next, consider the forces acting on D and Q. The applied (or active) forces may be represented by a contact force **C** acting through the contact point together with weight forces **w** and **W** acting through Q and G. The inertia (or passive) forces may be represented by forces $\mathbf{F}_Q^*$ and $\mathbf{F}_G^*$ and a couple with torque $\mathbf{T}_D^*$. Figure 14.6.3 contains a free-body diagram depicting these forces and the torque. Analytically, they may be expressed as:

$$\mathbf{w} = -mg\mathbf{N}_3 \tag{14.6.14}$$

$$\mathbf{W} = -Mg\mathbf{N}_3 \tag{14.6.15}$$

$$\mathbf{F}_Q^* = -m\,{}^R\mathbf{a}^Q \tag{14.6.16}$$

$$\mathbf{F}_D^* = -M\,{}^R\mathbf{a}^G \tag{14.6.17}$$

and

$$\mathbf{T}_D^* = -\mathbf{I}^D \cdot {}^R\boldsymbol{\alpha}^D - {}^R\boldsymbol{\omega}^D \times \left(\mathbf{I}^D \cdot {}^R\boldsymbol{\omega}^D\right) \tag{14.6.18}$$

where as before, I^D is the central inertia dyadic of D. (Regarding notation, the superscript [*] in Eqs. (14.6.15), (14.6.16), and (14.5.17) designates inertia forces and not "small" quantities.)

The unit vectors $\mathbf{n}_1$, $\mathbf{n}_2$, and $\mathbf{n}_3$ are parallel to principal inertia directions for D (see Figure 14.6.1). In terms of the $\mathbf{n}_i$ ($i = 1, 2, 3$), the inertia torque $\mathbf{T}_D^*$ may be expressed as (see Eqs. (8.6.10), (8.6.11), and (8.6.12)):

$$\mathbf{T}_D^* = T_1\mathbf{n}_1 + T_2\mathbf{n}_2 + T_3\mathbf{n}_3 \tag{14.6.19}$$

where T_1, T_2, and T_3 are:

$$T_1 = -\alpha_1 I_{11} + \omega_2\omega_3\left(I_{22} - I_{33}\right) \tag{14.6.20}$$

$$T_2 = -\alpha_2 I_{22} + \omega_3\omega_1\left(I_{33} - I_{11}\right) \tag{14.6.21}$$

$$T_3 = -\alpha_3 I_{33} + \omega_1\omega_2\left(I_{11} - I_{22}\right) \tag{14.6.22}$$

where the α_i and the ω_i (i = 1, 2, 3) are the $\mathbf{n}_i$ components of ${}^R\boldsymbol{\alpha}^D$ and ${}^R\boldsymbol{\omega}^D$ and where I_{11}, I_{22}, and I_{33} are:

$$I_{11} = I_{33} = Mr^2/4, \quad I_{22} = Mr^2/2 \tag{14.6.23}$$

Consider next the free-body diagram itself in Figure 14.6.2. By setting moments of the forces, about C, equal to zero we have:

$$r\mathbf{n}_3 \times \left(\mathbf{W} + \mathbf{F}_G^*\right) + \left(r\mathbf{n}_3 + r\mathbf{d}_3\right) \times \left(\mathbf{w} + \mathbf{F}_Q^*\right) + \mathbf{T}_D^* = 0 \tag{14.6.24}$$

Then, by substituting from Eqs. (14.6.13) through (14.6.18), we obtain the scalar equations:

$$\begin{aligned} &Mgr\sin\theta + Mra_2^G + mr(1+\cos\psi)a_2^Q \\ &\quad + T_1 + mgr(1+\cos\psi)\sin\theta = 0 \end{aligned} \tag{14.6.25}$$

$$\begin{aligned} &-Mra_1^G - mr(1+\cos\psi)a_1^Q + mra_3^Q\sin\psi \\ &\quad + T_2 + mgr\sin\psi\cos\theta = 0 \end{aligned} \tag{14.6.26}$$

and

$$-mra_2^Q\sin\psi + T_3 - mgr\sin\theta\sin\psi = 0 \tag{14.6.27}$$

where the a_i^G and the a_i^Q are the $\mathbf{n}_i$ (i = 1, 2, 3) components of ${}^R\mathbf{a}^G$ and ${}^R\mathbf{a}^Q$. Then, by substituting from Eqs. (14.6.12), (14.6.13), (14.6.25), (14.6.26), and (14.6.27), we have:

$$\begin{aligned} &Mgr\sin\theta + Mr^2\left(-\ddot{\theta} + \dot{\phi}^2\sin\theta\cos\theta + \dot{\psi}\dot{\phi}\cos\theta\right) \\ &-I_{11}\left(\ddot{\theta} - \dot{\psi}\dot{\phi}\cos\theta\right) + \left(I_{22} - I_{33}\right)\left(\dot{\psi} + \dot{\phi}\sin\theta\right)\dot{\phi}\cos\theta \\ &+mr^2(1+\cos\psi)\left[(1+\cos\psi)\left(\dot{\psi}\dot{\phi}\cos\theta + \dot{\phi}^2\sin\theta\cos\theta - \ddot{\theta}\right)\right. \\ &\left.+\ddot{\phi}\sin\psi\cos\theta + \dot{\phi}\dot{\psi}\cos\psi\cos\theta + 2\dot{\theta}\dot{\psi}\sin\psi\right] \\ &+mgr(1+\cos\psi)\sin\theta = 0 \end{aligned} \tag{14.6.28}$$

$$\begin{aligned} &-Mr^2\left(\ddot{\psi} + \ddot{\phi}\sin\theta + 2\dot{\theta}\dot{\phi}\cos\theta\right) - I_{22}\left(\ddot{\psi} + \dot{\theta}\dot{\phi}\cos\theta\right. \\ &\left.+\ddot{\phi}\sin\theta\right) + \left(I_{33} - I_{11}\right)\dot{\theta}\dot{\phi}\cos\theta - mr^2(1+\cos\psi)\left[\left(\ddot{\psi}\right.\right. \\ &\left.+\ddot{\phi}\sin\theta + 2\dot{\phi}\dot{\theta}\cos\theta\right)(1+\cos\psi) - \left(\dot{\psi}^2 + \dot{\phi}^2\right)\sin\psi \\ &\left.-2\dot{\phi}\dot{\psi}\sin\theta\sin\psi\right] + mr^2\sin\psi\left[\left(-\dot{\phi}^2\sin^2\theta\right.\right. \\ &\left.-\dot{\psi}\dot{\phi}\sin\theta - \dot{\theta}^2\right)(1+\cos\psi) - \dot{\psi}\dot{\phi}\cos\psi\sin\theta \\ &\left.-\dot{\psi}^2\cos\psi - \ddot{\psi}\sin\psi - \ddot{\phi}\sin\theta\sin\psi\right] \\ &+mgr\sin\psi\cos\theta = 0 \end{aligned} \tag{14.6.29}$$

and

$$\begin{aligned}
&-mr^2 \sin\psi\left[\left(\dot{\psi}\dot{\phi}\cos\theta + \dot{\phi}^2 \sin\theta\cos\theta - \ddot{\theta}\right)(1+\cos\psi)\right. \\
&\left.+\ddot{\phi}\sin\psi\cos\theta + \dot{\phi}\dot{\psi}\cos\psi\cos\theta + 2\dot{\theta}\dot{\psi}\sin\psi\right] \\
&-I_{33}\left(\dot{\psi}\dot{\theta} + \ddot{\phi}\cos\theta - \dot{\phi}\dot{\theta}\sin\theta\right) \\
&+(I_{11} - I_{22})\dot{\theta}\left(\dot{\psi} + \dot{\phi}\sin\theta\right) - mgr\sin\theta\sin\psi = 0
\end{aligned} \tag{14.6.30}$$

Equations (14.6.28), (14.6.29), and (14.6.30) are the governing equations of motion of the rim-weighted disk. Observe that if we remove the mass on the rim (that is, let $m = 0$) and if we let I_{11}, I_{22}, and I_{33} have the values as in Eq. (14.6.23), then Eqs. (14.6.28), (14.6.29), and (14.6.30) reduce to:

$$\begin{aligned}
&4(g/r)\sin\theta - 4\ddot{\theta} + 4\dot{\phi}^2\sin\theta\cos\theta + 4\dot{\psi}\dot{\phi}\cos\theta - \ddot{\theta} \\
&+\dot{\psi}\dot{\phi}\cos\theta\left(\dot{\psi} + \dot{\phi}\sin\theta\right)\dot{\phi}\cos\theta = 0
\end{aligned} \tag{14.6.31}$$

$$3\ddot{\psi} + 3\ddot{\phi}\sin\theta + 5\dot{\theta}\dot{\phi}\cos\theta \tag{14.6.32}$$

and

$$\dot{\psi}\dot{\theta} + \ddot{\phi}\cos\theta - \dot{\phi}\dot{\theta}\sin\theta + \dot{\theta}\left(\dot{\psi} + \dot{\phi}\sin\theta\right) = 0 \tag{14.6.33}$$

After simplification, these equations are seen to be identical to Eqs. (14.5.1), (14.5.2), and (14.5.3), the governing equations for the uniform rolling/pivoting circular disk.

Next, observe that two steady-state solutions of Eqs. (14.5.28), (14.6.29), and (14.6.30) occur when the disk D is pivoting with the mass at Q in either the upper-most or lower-most position. Specifically, for Q in the upper- or top-most position, we have:

$$\theta = 0, \quad \dot{\phi} = \dot{\phi}_0, \quad \psi = 0 \tag{14.6.34}$$

where $\dot{\phi}_0$ is the steady-state spin speed. Similarly, for Q in the lower- or bottom-most position, we have:

$$\theta = 0, \quad \dot{\phi} = \dot{\phi}_0, \quad \psi = \pi \tag{14.6.35}$$

In the following paragraphs we examine the stability of these two positions.

14.6.1 Rim Mass in the Uppermost Position

Let there be a small disturbance to the equilibrium position with the rim mass in its uppermost position, as defined by Eq. (14.6.34). Specifically, let θ, $\dot{\phi}$, and ψ have the forms:

$$\theta = 0 + \theta^*, \quad \dot{\phi} = \dot{\phi}_0 + \dot{\phi}^*, \quad \psi = 0 + \psi^* \tag{14.6.36}$$

where, as before, the quantities with the (*) are small. Then, Eqs. (14.6.28), (14.6.29), and (14.6.30) become (after simplification):

$$\left(Mr^2 + I_{11}4mr^2\right)\ddot{\theta}^* + \left(-Mr^2 - I_{22} - 6mr^2\right)\dot{\phi}_0\dot{\psi}^* + \left[-Mgr - 2mgr - \left(Mr^2 + I_{22} - I_{33} + 4mr^2\right)\dot{\phi}_0^2\right]\theta^* = 0 \tag{14.3.37}$$

$$\left(Mr^2 + I_{22} + 4mr^2\right)\ddot{\psi}^* + \left(2Mr^2 + I_{22} + 8mr^2\right)\dot{\phi}_0\dot{\theta}^* - \left(2mr^2\dot{\phi}_0^2 + mgr\right)\psi^* = 0 \tag{14.6.38}$$

and

$$\ddot{\phi}^* = 0 \tag{14.6.39}$$

To solve these equations, let $I_{11} = I_{33} = (1/2)I_{22} = (1/4)Mr^2$, and let Eqs. (14.6.37) and (14.6.38) be written in the forms:

$$a_1\ddot{\theta}^* + a_2\dot{\psi}^* + a_3\theta^* = 0 \tag{14.6.40}$$

and

$$b_1\ddot{\psi}^* + b_2\dot{\theta}^* + b_3\psi^* = 0 \tag{14.6.41}$$

where the coefficients a_i and b_i ($i = 1, 2, 3$) are:

$$\begin{aligned}
&a_1 = (4/5)Mr^2 + 4mr^2, \quad a_2 = -\left[(3/2)Mr^2 + 6mr^2\right]\dot{\phi}_0 \\
&a_3 = -Mgr - 2mgr - \left[(5/4)Mr^2 + 4mr^2\right]\dot{\phi}_0^2 \\
&b_1 = (3/2)Mr^2 + 4mr^2, \quad b_2 = \left[(5/2)Mr^2 + 8mr^2\right]\dot{\phi}_0 \\
&b_3 = -mr^2\left[(g/r) + 2\dot{\phi}_0^2\right]
\end{aligned} \tag{14.6.42}$$

Observe that the signs of these coefficients are:

$$\begin{aligned}
&a_1 > 0, \quad a_2 < 0, \quad a_3 < 0 \\
&b_1 > 0, \quad b_2 > 0, \quad b_3 < 0
\end{aligned} \tag{14.6.43}$$

Observe further that the a_i and b_i ($i = 1, 2, 3$) may be written in the forms:

$$\begin{aligned}
&a_1 = c_1, \quad a_2 = -c_2\dot{\phi}_0, \quad a_3 = -c_3 - c_4\dot{\phi}_0^2 \\
&b_1 = c_5, \quad b_2 = c_6\dot{\phi}_0, \quad b_3 = -c_7 - c_8\dot{\phi}_0^2
\end{aligned} \tag{14.6.44}$$

where by comparison of Eqs. (14.6.42) and (14.6.44) the c_i ($i = 1,\ldots, 8$) are seen to be:

$$
\begin{aligned}
c_1 &= (5/4)Mr^2 + 4mr^2, & c_2 &= (3/2)Mr^2 + 6mr^2 \\
c_3 &= Mgr + 2mgr, & c_4 &= (5/4)Mr^2 + 4mr^2 \\
c_5 &= (3/2)Mr^2 + 4mr^2, & c_6 &= (5/2)Mr^2 + 8mr^2 \\
c_7 &= mgr, & c_8 &= 2mr^2
\end{aligned}
\tag{14.6.45}
$$

Note that each c_i is positive.

To solve Eqs. (14.6.40) and (14.6.41), let θ^* and ψ^* have the forms:

$$\theta^* = \Theta^* e^{\lambda t} \quad \text{and} \quad \psi^* = \Psi^* e^{\lambda t} \tag{14.6.46}$$

Then, Eqs. (14.6.40) and (14.6.41) become:

$$a_1\lambda^2\Theta^* e^{\lambda t} + a_2\lambda\Psi^* e^{\lambda t} + a_3\Theta^* e^{\lambda t} = 0 \tag{14.6.47}$$

and

$$b_1\lambda^2\Psi^* e^{\lambda t} + b_2\lambda\Theta^* e^{\lambda t} + b_3\Psi^* e^{\lambda t} = 0 \tag{14.6.48}$$

or

$$\left(a_1\lambda^2 + a_3\right)\Theta^* + a_2\lambda\Psi^* = 0 \tag{14.6.49}$$

and

$$b_2\lambda\Theta^* + \left(b_1\lambda^2 + b_3\right)\Psi^* = 0 \tag{14.6.50}$$

Equations (14.6.49) and (14.6.50) are simultaneous, homogeneous, linear, algebraic equations similar to those we obtained in the eigenvalue inertia problem (see Section 7.7, Eqs. (7.7.10) and (7.7.11)). There is a nonzero solution to these equations only if the determinant of the coefficients is zero. That is,

$$\left(a_1\lambda^2 + a_3\right)\left(b_1\lambda^2 + b_3\right) - a_2b_2\lambda^2 = 0 \tag{14.6.51}$$

or

$$a_1b_1\lambda^4 + \left(a_3b_1 + b_3a_1 - a_2b_2\right)\lambda^2 + a_3b_3 = 0 \tag{14.6.52}$$

or

$$A\lambda^4 + B\lambda^2 + C = 0 \tag{14.6.53}$$

where A, B, and C are (see Eqs. (14.6.43)):

$$A = a_1 b_1 = c_1 c_5$$

$$B = a_3 b_1 + b_3 a_1 - a_2 b_2 \tag{14.6.54}$$

or

$$B = -(c_3 c_5 + c_1 c_7) + (c_2 c_6 - c_4 c_5 - c_1 c_8)\dot{\phi}_0^2 \tag{14.6.55}$$

and

$$C = a_3 b_3 = c_3 c_7 + (c_3 c_8 + c_4 c_7)\dot{\phi}_0^2 + c_4 c_8 \dot{\phi}_0^4 \tag{14.6.56}$$

Finally, by solving Eq. (14.6.53) for λ^2, we have:

$$\lambda^2 = \frac{-B \pm (B^2 - 4AC)^{1/2}}{2A} \tag{14.6.57}$$

Observe from Eq. (14.6.46) that the solutions θ^* and ψ^* for the motion following the disturbance will be bounded (that is, the motion is stable) if λ is either negative or imaginary. Expressed another way, the motion is stable if λ does not have any positive real part. Thus, from Eq. (14.6.57) we see that λ^2 will be negative, and there will be stability if:

$$B/A > 0 \quad \text{and either} \quad 4AC > B^2 \quad \text{or} \quad [B^2 - 4AC]^{1/2} < B \tag{14.6.58}$$

The first of these conditions is satisfied if A and B have the same signs. From Eq. (14.6.54) we see that A is positive because c_1 and c_5 are positive. Therefore, the first inequality in Eq. (14.6.58) is satisfied if B is positive. From Eq. (14.6.55), we see that B will be positive if:

$$c_2 c_6 - c_4 c_5 - c_1 c_8 > 0 \tag{14.6.59}$$

and

$$\dot{\phi}_0^2 > (c_3 c_5 + c_1 c_7)/(c_2 c_6 - c_4 c_5 - c_1 c_8) \tag{14.6.60}$$

Because each of the c_i ($i = 1, \ldots, 8$) is positive (see Eq. (14.6.45)), the inequality of Eq. (14.6.60) will be satisfied if the spin rate $\dot{\phi}_0$ is sufficiently large and the inequality of Eq. (14.6.59) is satisfied. That is, the first inequality of Eq. (14.6.58) will be satisfied if $\dot{\phi}_0$ is sufficiently large and if:

$$c_2 c_6 - c_4 c_5 - c_1 c_8 > 0 \tag{14.6.61}$$

From Eq. (14.6.45) we see after computation and simplification that:

$$c_2c_6 - c_4c_5 - c_1c_8 - (15/8)M^2r^4 + (27/2)mMr^2 + 24m^2r^4 \qquad (14.6.62)$$

Therefore, the inequality of Eq. (14.6.61) is inherently satisfied.

Next, the second inequality of Eq. (14.6.58) will be satisfied if AC is positive. By inspection of Eqs. (14.6.54) and (14.6.56) we see immediately that both A and C are positive, thus the second inequality of Eq. (14.6.58) is satisfied. Therefore, the motion of the pivoting disk with the rim mass in the upper-most position is stable if $\dot{\phi}_0$ is sufficiently large to satisfy Eq. (14.6.60). From Eq. (14.6.45) we see that:

$$c_3c_5 + c_1c_7 = \left[(3/2)M^2 + (33/4)Mm + 12m^2\right]gr^3 \qquad (14.6.63)$$

Hence, from Eqs. (14.6.60), (14.6.62), and (14.6.63) we have the stability criterion:

$$\dot{\phi}_0^2 > \left(\frac{4M^2 + 22Mm + 32m^2}{5M^2 + 36Mm + 64m^2}\right)(g/r) \qquad (14.6.64)$$

Finally, observe that if $m = 0$, the stability criterion reduces to:

$$\dot{\phi}_0^2 > (4g/5r) \qquad (14.6.65)$$

which is identical to Eq. (14.5.38).

14.6.2 Rim Mass in the Lowermost Position

Next, let there be a small disturbance to the equilibrium position with the rim mass in its lower-most position as defined by Eq. (14.6.35). That is, let θ, $\dot{\phi}$, and ψ have the forms:

$$\theta = 0 + \theta^*, \quad \dot{\phi} = \dot{\phi}_0 + \dot{\phi}^*, \quad \psi = \pi + \dot{\psi}^* \qquad (14.6.66)$$

where, as before, the quantities with the (*) are small. The governing equations, Eqs. (14.6.28), (14.6.29), and (14.6.30), then become (after simplification):

$$\begin{gathered}(5/4)Mr^2\ddot{\theta}^* - (3/2)Mr^2\dot{\phi}_0\dot{\psi}^* \\ -\left[Mgr + (5/4)Mr^2\dot{\phi}_0^2\right]\theta^* = 0\end{gathered} \qquad (14.6.67)$$

$$(3/2)Mr^2\ddot{\psi}^* + (5/2)Mr^2\dot{\phi}_0\dot{\theta}^* + mgr\psi^* = 0 \qquad (14.6.68)$$

and

$$\ddot{\phi}^* = 0 \qquad (14.6.69)$$

Note that in the development of these equations we have neglected all products of small quantities (those with a [*]). Also, we have approximated trigonometric functions as:

$$\sin\theta = \theta^*, \quad \cos\theta = 1, \quad \sin\psi = -\psi^*, \quad \cos\psi = -1 \tag{14.6.70}$$

The last of these makes the sum (1 + cosψ) vanish, leading to the simple forms of Eqs. (14.6.67), (14.6.68), and (14.6.69).

After further simplification, we see that Eqs. (14.6.67) and (14.6.68) may be written in the forms:

$$\hat{a}_1\ddot{\theta}^* + \hat{a}_2\dot{\psi}^* + \hat{a}_3\theta^* = 0 \tag{14.6.71}$$

and

$$\hat{b}_1\ddot{\psi}^* + \hat{b}_2\dot{\theta}^* + \hat{b}_3\psi^* = 0 \tag{14.6.72}$$

where the a_i and b_i (i = 1, 2, 3) are:

$$\hat{a}_1 = 5, \quad \hat{a}_2 = -6\dot{\phi}_0, \quad \hat{a}_3 = -(4g/r) - 5\dot{\phi}_0^2$$

and

$$\hat{b}_1 = 3, \quad \hat{b}_2 = 5\dot{\phi}_0, \quad \hat{b}_3 = (2mg/Mr) \tag{14.6.73}$$

Following the procedure of the foregoing case we can solve Eqs. (14.6.67) and (14.6.68) by letting θ^* and ψ^* have the forms:

$$\theta^* = \Theta^* e^{\lambda t} \quad \text{and} \quad \psi^* = \Psi^* e^{\lambda t} \tag{14.6.74}$$

Then, Eqs. (14.6.67) and (14.6.68) become:

$$\hat{a}_1\lambda^2\Theta^* e^{\lambda t} + \hat{a}_2\lambda\Psi^* e^{\lambda t} + \hat{a}_3\Theta^* e^{\lambda t} = 0 \tag{14.6.75}$$

and

$$\hat{b}_1\lambda^2\Psi^* e^{\lambda t} + \hat{b}_2\lambda\Theta^* e^{\lambda t} + \hat{b}_3\Psi^* e^{\lambda t} = 0 \tag{14.6.76}$$

or

$$(\hat{a}_1\lambda^2 + \hat{a}_3)\Theta^* + \hat{a}_2\lambda\Psi^* = 0 \tag{14.6.77}$$

and

$$\hat{b}_2\lambda\Theta^* + (\hat{b}_1\lambda^2 + \hat{b}_3)\Psi^* = 0 \tag{14.6.78}$$

As with Eqs. (14.6.49) and (14.6.50), Eqs. (14.6.77) and (14.6.78) are simultaneous, homogeneous, linear, algebraic equations. Thus, there is a nonzero solution to these equations only if the determinant of the coefficients is zero. That is,

$$\left(\hat{a}_1\lambda^2+\hat{a}_3\right)\left(\hat{b}_1\lambda^2+\hat{b}_3\right)-\hat{a}_2\hat{b}_2\lambda^2=0 \tag{14.6.79}$$

or

$$\hat{a}_1\hat{b}_1\lambda^4+\left(\hat{a}_3\hat{b}_1+\hat{b}_3\hat{a}_1-\hat{a}_2\hat{b}_2\right)\lambda^2+\hat{a}_3\hat{b}_3=0 \tag{14.6.80}$$

or

$$\hat{A}\lambda^4+\hat{B}\lambda^2+\hat{C}=0 \tag{14.6.81}$$

where $\hat{A}$, $\hat{B}$, and $\hat{C}$ are:

$$\hat{A}=\hat{a}_1\hat{b}_1 \tag{14.6.82}$$

$$\hat{B}=\hat{a}_3\hat{b}_1+\hat{b}_3\hat{a}_1-\hat{a}_2\hat{b}_2 \tag{14.6.83}$$

$$\hat{C}=\hat{a}_3\hat{b}_3 \tag{14.6.84}$$

Solving for λ^2 we have:

$$\lambda^2=\frac{-\hat{B}\pm\left[\hat{B}^2-4\hat{A}\hat{C}\right]^{1/2}}{2\hat{A}} \tag{14.6.85}$$

As before, the solutions θ^* and ψ^* of Eq. (14.6.74) are stable if λ does not have any positive real part. From Eq. (14.6.85), this means that stability occurs if (see Eq. (14.6.58)):

$$\hat{B}/\hat{A}>0 \quad \text{and} \quad \left[\hat{B}^2-4\hat{A}\hat{C}\right]^{1/2}<\hat{B} \tag{14.6.86}$$

From Eqs. (14.6.73), (14.6.82), (14.6.83), and (14.6.84), we see that $\hat{A}$, $\hat{B}$, and $\hat{C}$ are:

$$\hat{A}=15 \tag{14.6.87}$$

$$\hat{B}=15\dot{\phi}_0^2+\left[10(m/M)-12\right](g/r) \tag{14.6.88}$$

and

$$\hat{C}=-8(m/M)(g/r)^2-10(m/M)(g/r)\dot{\phi}_0^2 \tag{14.6.89}$$

By inspection of Eqs. (14.6.87) and (14.6.88), we see that the first condition of Eq. (14.6.86) is satisfied if $\dot{\phi}_0$ is sufficiently large. The second condition of Eq. (14.6.86), however, will not be satisfied if $\hat{A}$ is positive and $\hat{C}$ is negative independent of the magnitude or sign of $\dot{\phi}_0$. Thus, for all pivoting speeds, the disk with the rim mass in the lower position is unstable.

Finally, observe that with $m = 0$, Eq. (14.6.89) shows that $\hat{C} = 0$, thus stability can be attained by having $\hat{B} > 0$. This in turn will occur if:

$$\dot{\phi}_0^2 > (4/5)(g/r) \tag{14.6.90}$$

which is identical to Eqs. (14.6.65) and (14.5.38).

14.7 Discussion: Routh–Hurwitz Criteria

Recall that in the previous section we solved the governing equations for the disturbance motion by seeking to simultaneously solve the equations. Consider again Eqs. (14.6.50) and (14.6.41):

$$a_1\ddot{\theta}^* + a_2\dot{\psi}^* + a_3\theta^* = 0 \tag{14.7.1}$$

$$b_1\ddot{\psi}^* + b_2\dot{\theta}^* + b_3\psi^* = 0 \tag{14.7.2}$$

Instead of seeking simultaneous solutions to these equations we could have eliminated one of the dependent variables, leaving a single equation of higher order for the other variable. Specifically, suppose we solve Eq. (14.7.1) for $\dot{\psi}^*$, giving us:

$$\dot{\psi}^* = -(a_1/a_2)\ddot{\theta}^* - (a_3/a_2)\theta^* \tag{14.7.3}$$

Then, by differentiating, we have:

$$\ddot{\psi}^* = -(a_1/a_2)\dddot{\theta}^* - (a_3/a_2)\dot{\theta}^* \tag{14.7.4}$$

Also, by differentiating Eq. (14.7.2), we have:

$$b_1\dddot{\psi}^* + b_2\ddot{\theta}^* + b_3\dot{\psi}^* = 0 \tag{14.7.5}$$

Finally, by substituting from Eqs. (14.7.3) and (14.7.4), we obtain (after simplification):

$$a_1b_1\ddddot{\theta}^* + (a_3b_1 + a_1b_3 - a_2b_2)\ddot{\theta}^* + a_3b_3\theta^* = 0 \tag{14.7.6}$$

To solve Eq. (14.7.6), let θ^* have the form:

$$\theta^* = \Theta^* e^{\lambda t} \tag{14.7.7}$$

Then, we obtain:

$$a_1 b_1 \lambda^4 + (a_3 b_1 + a_1 b_3 - a_2 b_2)\lambda^2 + a_3 b_3 = 0 \tag{14.7.8}$$

Equation (14.7.8) is identical to Eq. (14.6.52), which was obtained by the simultaneous solution of Eqs. (14.6.40) and (14.6.41).

It happens, in infinitesimal stability procedures in the analysis of small disturbances of a system from equilibrium, that we can generally convert the governing simultaneous linear differential equations into a single differential equation of higher order, as Eq. (14.7.6). Then, in the solution of the higher order equation, we generally obtain a polynomial equation of the form:

$$a_0 \lambda^n + a_1 \lambda^{n-1} + \ldots + a_{n-1}\lambda + a_n = 0 \tag{14.7.9}$$

From our previous analyses we see that a solution to the governing equations in the form of θ^* in Eq. (14.7.7) is stable if (and only if) λ does not have a positive real part. Necessary and sufficient conditions such that Eq. (14.7.9) will produce no λ_i ($i = 1,\ldots, n$) with positive real parts may be determined using the Routh–Hurwitz criteria (see Reference 14.1). Specifically, let determinants Δ_i ($i = 1,\ldots, n$) be defined as:

$$\Delta_1 = a_1$$

$$\Delta_2 = \begin{vmatrix} a_1 & a_0 \\ a_3 & a_2 \end{vmatrix}$$

$$\Delta_3 = \begin{vmatrix} a_1 & a_0 & 0 \\ a_3 & a_2 & a_1 \\ a_5 & a_4 & a_3 \end{vmatrix}$$

$$\Delta_4 = \begin{vmatrix} a_1 & a_0 & 0 & 0 \\ a_3 & a_2 & a_1 & a_0 \\ a_5 & a_4 & a_3 & a_2 \\ a_7 & a_6 & a_5 & a_4 \end{vmatrix} \tag{14.7.10}$$

- - - - - - - - - - - - -

$$\Delta_n = \begin{vmatrix} a_1 & a_0 & 0 & 0 & \ldots & 0 \\ a_3 & a_2 & a_1 & a_0 & \ldots & 0 \\ a_5 & a_4 & a_3 & a_2 & \ldots & 0 \\ --- & --- & --- & --- & --- & --- \\ a_{2m-1} & a_{2m-2} & --- & --- & --- & a_m \end{vmatrix}$$

Then, the criteria, such that the solutions λ_i ($i = 1,\ldots, n$) have no positive real part, are that each of the Δ_i ($i = 1,\ldots, n$) must be positive and that each of the a_k ($k = 0,\ldots, n$) must have the same sign.

The above statements asserting that the a_i and the Δ_i need to be positive for the λ_i to have no real parts are commonly called the *Routh–Hurwitz stability criteria*. It is seen that

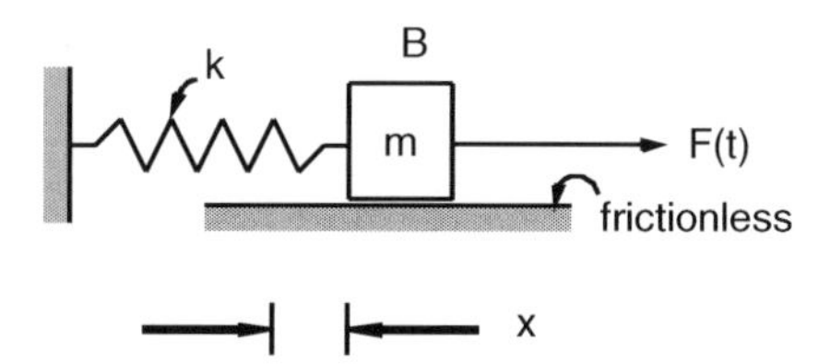

FIGURE 14.7.1
Undamped linear mass–spring oscillator.

these criteria form a stronger stability test than that used in the foregoing sections. Specifically, if the Routh–Hurwitz criteria are satisfied, the dynamical system will return to the equilibrium position after a disturbance. In the foregoing sections, however, we simply required that a small disturbance away from equilibrium remain small or bounded.

To illustrate these concepts consider a system governed by the differential equation:

$$a_0 \frac{d^2x}{dt^2} + a_1 \frac{dx}{dt} + a_2 x = 0 \tag{14.7.11}$$

Let the solution x have the form of Eq. (14.6.46). That is,

$$x = Xe^{\lambda t} \tag{14.7.12}$$

Then, we immediately obtain:

$$a_0 \lambda^2 + a_1 \lambda + a_2 = 0 \tag{14.7.13}$$

The Routh–Hurwitz criteria for stability are then:

$$a_0 > 0, \quad a_1 > 0, \quad \text{and} \quad a_2 > 0 \tag{14.7.14}$$

To interpret this result, consider again the undamped linear mass–spring oscillator depicted in Figure 14.7.1. Recall that the governing equation for this system is (see Eq. (13.3.1)):

$$m\ddot{x} + kx = 0 \tag{14.7.15}$$

where, as before, m is the mass, x is the displacement, and k is the spring modulus. Then, by seeking a solution in the form of Eq. (14.7.12), we immediately obtain:

$$m\lambda^2 + k = 0 \tag{14.7.16}$$

By comparison with Eq. (14.7.13), see that the coefficients a_0, a_1, and a_2 are:

$$a_0 = m, \quad a_1 = 0, \quad \text{and} \quad a_2 = k \tag{14.7.17}$$

The first and third expressions of Eq. (14.7.17) satisfy Eq. (14.7.14), but the second expression violates the requirement $a_1 > 0$. Thus, the equilibrium position $x = 0$ of the mass–spring oscillator is unstable according to the Routh–Hurwitz criteria. Intuitively, however, we would expect the equilibrium position to be stable because a small disturbance

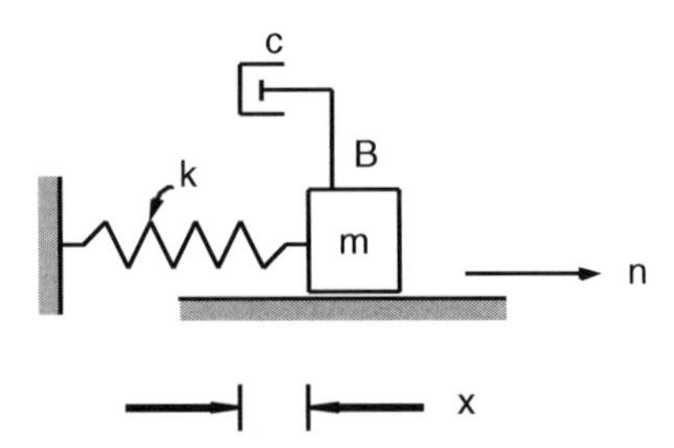

FIGURE 14.7.2
Damped linear mass–spring oscillator.

x^* away from equilibrium ($x = 0$) simply produces a bounded oscillation of the form (see Eq. (13.3.4)):

$$x^* = A^* \cos\sqrt{k/m}\,t + B^* \sin\sqrt{k/m}\,t \tag{14.7.18}$$

where A^* and B^* are small constants. But, because the amplitude of the disturbance oscillation does not diminish to zero, the system is not stable according to the Routh–Hurwitz criteria.

To discuss this further, consider the damped linear mass–spring oscillator depicted in Figure 14.7.2. Recall that the governing equation for this system is (see Eq. (13.5.2)):

$$m\ddot{x} + c\dot{x} + kx = 0 \tag{14.7.19}$$

where, as before, c is the damping coefficient. Then, by seeking a solution in the form of Eq. (14.7.6), we obtain:

$$m\lambda^2 + c\lambda + k = 0 \tag{14.7.20}$$

By comparison with Eq. (14.7.13) we see that a_0, a_1, and a_2 are:

$$a_0 = m, \quad a_1 = c, \quad \text{and} \quad a_2 = k \tag{14.7.21}$$

Because each of these coefficients is zero, we see that the Routh–Hurwitz criteria of Eq. (14.7.14) are satisfied. This means that a small disturbance x^* away from the equilibrium position ($x = 0$) will diminish to zero. Indeed, from Eq. (13.5.3) we see that the behavior of such a disturbance is governed by the equation:

$$\dot{x}^* = e^{-\mu t}\left[A^* \cos\omega t + B^* \sin\omega t\right] \tag{14.7.22}$$

where A^* and B^* are small constants and where, as before, μ and ω are (see Eqs. (13.5.3) and (13.5.4)):

$$\mu = c/2m \quad \text{and} \quad \omega = \left[\frac{k}{m} - \frac{c^2}{4m^2}\right]^{1/2} \tag{14.7.23}$$

Hence, as time t increases, the exponential function $e^{-\mu t}$ causes x^* to diminish to zero.

To further illustrate the Routh–Hurwitz criteria, consider a system governed by the equation:

$$a_0 \frac{d^3x}{dt^3} + a_1 \frac{d^2x}{dt^2} + a_2 \frac{dx}{dt} + a_3 x = 0 \tag{14.7.24}$$

In this case the criteria of Eqs. (14.7.10) become:

$$a_0 > 0, \quad a_1 > 0, \quad a_2 > 0, \quad \text{and} \quad a_3 > 0 \tag{14.7.25}$$

and

$$a_1 a_2 > a_0 a_3 \tag{14.7.26}$$

Finally, consider a system governed by the equation:

$$a_0 \frac{d^4x}{dt^4} + a_1 \frac{d^3x}{dt^3} + a_2 \frac{d^2x}{dt^2} + a_3 \frac{dx}{dt} + a_4 k = 0 \tag{14.7.27}$$

Here, the Routh–Hurwitz criteria of Eqs. (14.7.10) become:

$$a_0 > 0, \quad a_1 > 0, \quad a_2 > 0, \quad a_3 > 0, \quad \text{and} \quad a_4 > 0 \tag{14.7.28}$$

and

$$a_1 a_2 > a_0 a_3 \quad \text{and} \quad a_1 a_2 a_3 > a_0 a_3^2 + a_1^2 a_4 \tag{14.7.29}$$

14.8 Closure

This concludes our relatively brief introduction to stability and its associated computational procedures. Our discussions have been limited to infinitesimal stability. There are other less stringent stability criteria, but these are beyond the limited scope of our discussion. Those interested in these more advanced concepts and their associated procedures may want to refer to the references or to more theoretical works devoted exclusively to stability.

References

14.1. Meirovitch, L., *Methods of Analytical Dynamics*, McGraw-Hill, New York, 1970, pp. 222–224.

14.2. Di Stefano, J. J., Stubberud, A. R., and Williams, I. J., *Theory and Problems of Feedback and Control Systems*, Schaum's Outline Series, McGraw-Hill, New York, 1967, 114 ff.

14.3. Davis, S. A., *Feedback and Control Systems*, Simon & Shuster, New York, 1974, 262 ff.

14.4. Skelton, R. E., *Dynamic Systems Control: Linear Systems Analysis and Synthesis*, Wiley, New York, 1988, chap. 7.

14.5. Meirovitch, L., *Elements of Vibration Analysis*, McGraw-Hill, New York, 1975, 329 ff.
14.6. Meirovitch, L., *Dynamics and Control of Structures*, Wiley-Interscience, New York, 1990, 68 ff.

Problems

Section 14.2 Infinitesimal Stability

P14.2.1: A rod with length ℓ and mass m is supported by a frictionless pin and a torsion spring at the support as represented in Figure P14.2.1. Determine the value of the minimum spring modulus κ_{min} so that the rod is stable in a vertical equilibrium position while being supported at its lower end.

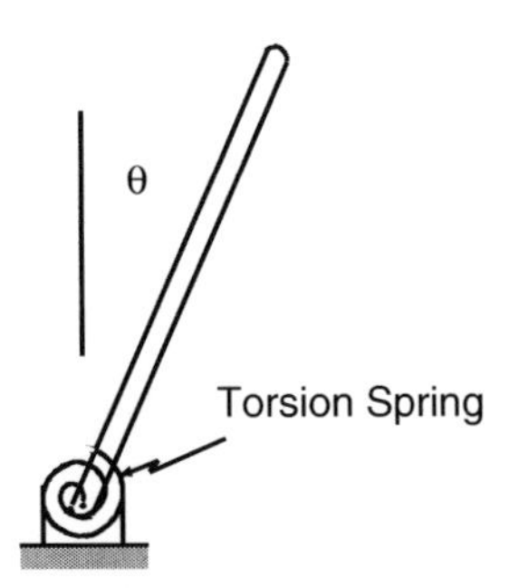

FIGURE P14.2.1
A pin-supported rod with a torsion spring at the support.

P14.2.2: See Problem P14.2.1. Let the torsion spring modulus have a value of 0.8 κ_{min}. Determine the governing equation defining the static equilibrium position θ (in addition to $\theta = 0$).

P14.2.3: See Problem P14.2.1. Let the rod length be 1 m and let the mass be 2 kg. Determine κ_{min}.

P14.2.4: See Problems P14.2.1, P14.2.2, and P14.2.3. Determine the equilibrium angle θ (aside from $\theta = 0$) for the values of ℓ and m of Problem P14.2.3.

P14.2.5: Repeat Problem P14.2.4 if the rod length ℓ is 3 ft and the rod weight is 5 lb.

Section 14.3 A Particle in a Vertical Rotating Tube

P14.3.1: Consider the particle moving inside a smooth vertical rotating tube as discussed in Section 14.3 and as depicted again in Figure P14.3.1. Determine the angular speed Ω so that there is an equilibrium position at $\theta = 30°$ if the tube radius r is 18 in.

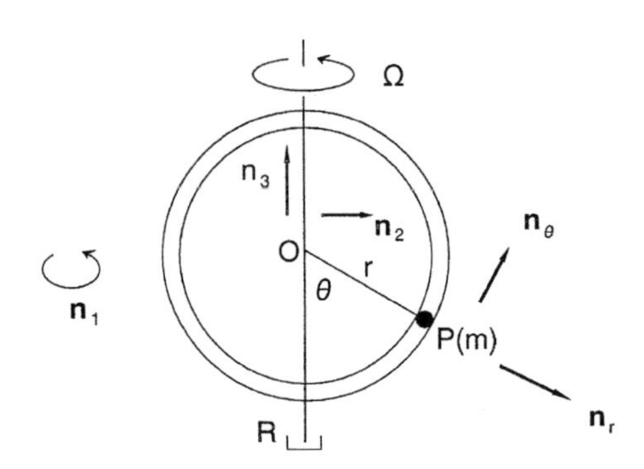

FIGURE P14.3.1
A vertical rotating tube with a smooth interior surface and containing a particle *P*.

P14.3.2: See Problem P14.3.1. If the tube radius r is 0.5 m, determine the angular speed Ω so that the particle position $\theta = 0$ is unstable.

Section 14.5 The Rolling/Pivoting Circular Disk

P14.5.1: A uniform circular disk with radius 8 in. is rolling on a straight line on a level, perfectly rough surface as represented in Figure P14.5.1. Determine the minimum speed v needed for stability of the motion.

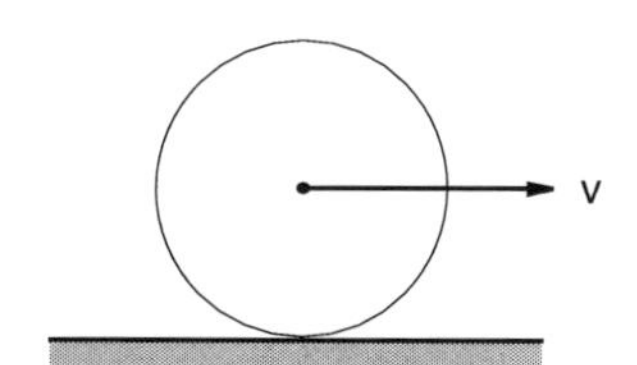

FIGURE P14.5.1
A circular disk rolling in a straight line on a perfectly rough surface.

P14.5.2: Repeat Problem P14.5.1 for a disk with radius 25 cm.

P14.5.3: A uniform circular disk with radius 10 in. is pivoting on a flat horizontal surface as represented in Figure P14.5.3. Determine the minimum angular speed Ω for stable rotation.

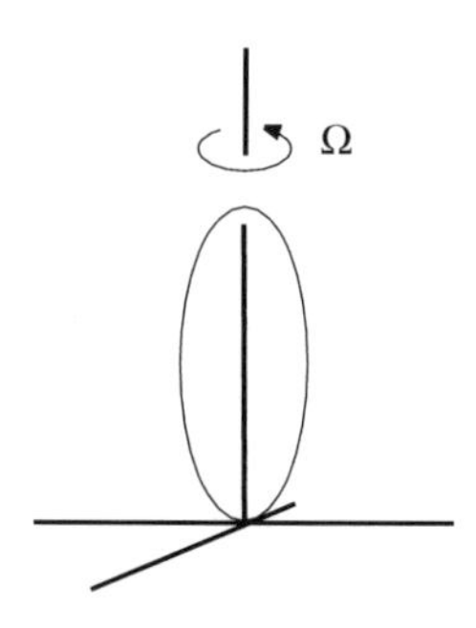

FIGURE P14.5.3
A pivoting circular disk.

P14.5.4: Repeat Problem P14.5.3 for a disk with radius 30 cm.

P14.5.5: Consider a ring or hoop with radius r rolling on a perfectly rough horizontal surface as in Figure 14.5.5. Following the procedures of Section 14.5, develop the governing dynamical equations analogous to Eqs. (14.5.1), (14.5.2), and (14.5.3).

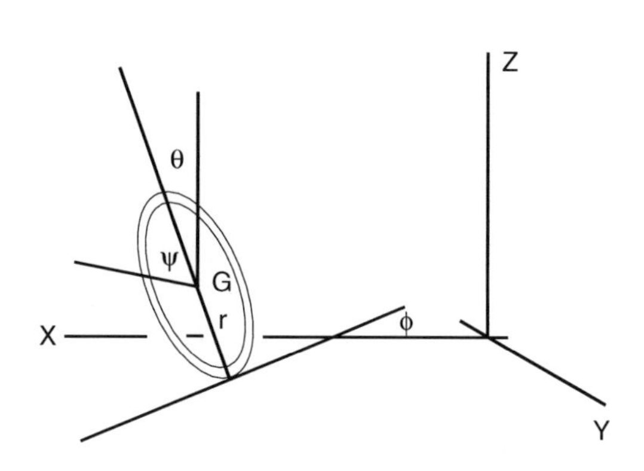

FIGURE P14.5.5
A rolling ring.

P14.5.6: A ring with radius 8 in. is rolling on a straight line on a level perfectly rough surface as represented in Figure P14.5.6. Determine the minimum speed v needed for stability of the motion. Compare the result with that of Problem P14.5.1.

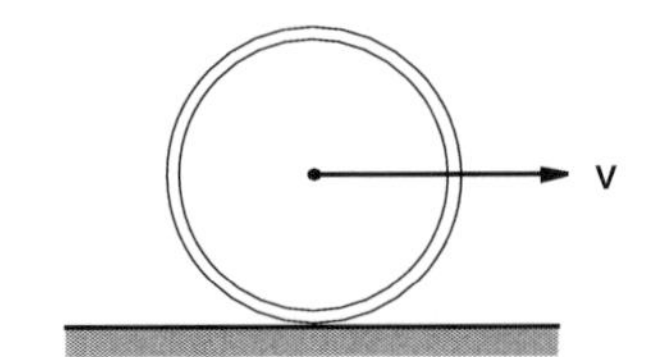

FIGURE P14.5.6
A ring rolling in a straight line on a perfectly rough surface.

P14.5.7: Repeat Problem P14.5.6 for a ring with radius 25 cm. Compare the result with that of Problem P14.5.2.

P14.5.8: A ring with radius 10 in. is pivoting on a flat horizontal surface as represented in Figure P14.5.8. Determine the minimum angular speed Ω for stable rotation. Compare the result with that of Problem P14.5.3.

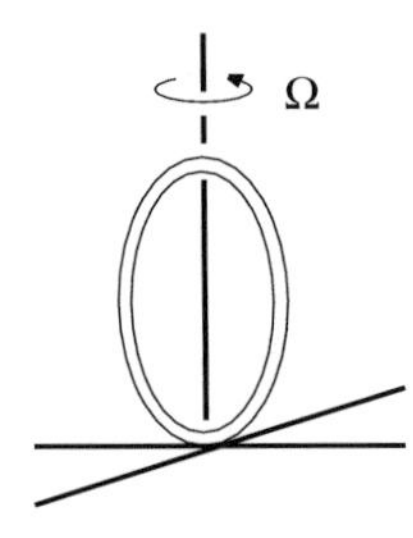

FIGURE P14.5.8
A pivoting ring.

P14.5.9: Repeat Problem P14.5.8 for a ring with radius 30 cm. Compare the result with that of Problem P14.5.4.

15

Balancing

15.1 Introduction

As with stability, balancing is a subject of great interest to engineers and to designers of mechanical systems — particularly in the design of rotating machinery. Unlike stability, however, balancing is a more applied subject, although balancing procedures can be rather detailed and technical. In what follows, we will discuss fundamental concepts of balancing. Readers interested in more detail or in more technical aspects should consult references at the end of the chapter or in an engineering library.

In classical dynamical analyses of machines and mechanical systems, the machines or systems are generally modeled as ideal bodies with perfect geometry and uniform mass distribution — and, specifically, with no imbalance. In reality, of course, it is impossible to manufacture ideal bodies. Even with modern technologies, there will be variations in geometry and variations in mass distribution. Often, these variations are insignificant and unimportant. With high-speed and high-precision machines, however, even minor variations in geometry and mass distribution can have significant and deleterious effects upon the function and life of a machine. This is especially the case with high-speed rotating bodies. A variation in geometry or in mass distribution away from the ideal is referred to as an *imbalance*.

Because imbalances often lead to instability, particularly with rotating equipment, a technology of *balancing* has arisen which is intended to counter or eliminate the imbalances. Indeed, balancing technology has become an integral part of modern manufacturing processes for a wide variety of products ranging from rotors to propellers, automobile wheels, and bowling balls. In the following sections, we will briefly explore the fundamentals of balancing and will restrict our attention to rotating bodies. Initially, we will study static balancing and then go on to dynamic balancing.

15.2 Static Balancing

Consider a rotor consisting of a cylindrical shaft together with a circular disk D mounted on S as in Figure 15.2.1. Let A–A be the axis of rotation. Ideally, with perfect geometry, we would expect the mass center G of such a system to lie on axis A–A. In a physical system, however, even with precision machining and careful fabrication, G will not lie exactly on A–A. Instead, G will generally be a small distance δ away from A–A. Figure 15.2.2 depicts an end view of the rotor showing G and a point O on the axis A–A.

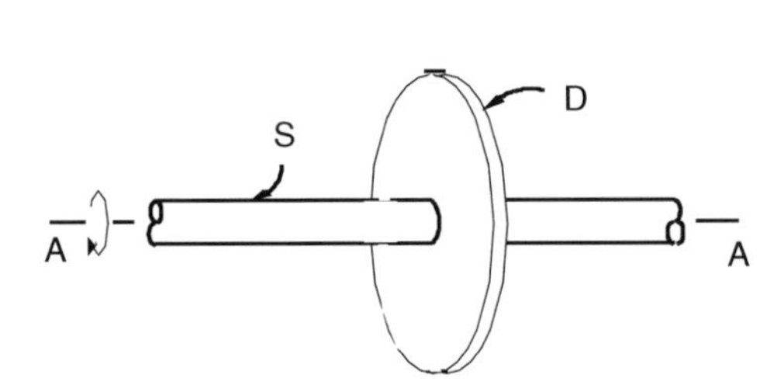

FIGURE 15.2.1
A cylindrical shaft with a mounted circular disk.

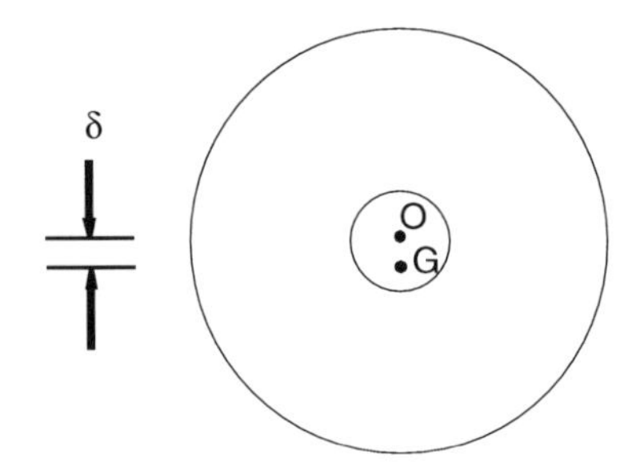

FIGURE 15.2.2
End view of rotor showing mass center offset.

To "balance" the system we need to add or remove weight to diminish δ so that G moves toward the axis A–A. Although a perfect balance cannot be obtained, we can nevertheless generally improve the balance by the small addition or removal of mass along the line passing through O and G. Specifically, suppose we add a mass m at a point P along the line passing through O and G where P is on the opposite side of O from G. Then, a theoretical balance occurs if P is a distance d_P from O such that:

$$md_p = M\delta \tag{15.2.1}$$

where M is the mass of the rotor.

In practice, it is often relatively easy to locate the line through O and G. The rotor is simply placed horizontally and supported by frictionless or nearly frictionless bearings. Then, if the rotor is given a slow rotation, it should eventually oscillate and come to rest with the mass center G directly below O.

By trial and error, a mass m may be placed along the upper part of the line through O and G such that the rotor mass center (including m) is nearly on the axis. This procedure is called *static balancing* because the rotor with the mass center on the axis can be at rest at any orientation about the axis. Note that if the rotor is not in static balance, a whirling centrifugal inertia force is created when the rotor is rotating. This force will increase linearly with the square of the angular speed, possibly reaching harmful and undesirable magnitudes. Finally, observe that an alternative to placing a mass on the upper portion of the line through O and G is to remove some mass along the lower portion of the line through O and G.

15.3 Dynamic Balancing: A Rotating Shaft

Even if a body is statically balanced, it still may not rotate freely without generating unwanted inertia forces. To see this, consider the cylindrical shaft of Figure 15.3.1 having equal point masses P_1 and P_2 on the cylindrical surface separated from each other axially and also by 180° circumferentially. With ideal geometry, the mass center of this system is at G, a point on the axis of the cylinder.

In the context of the foregoing section, the system of Figure 15.3.1 is statically balanced; however, as the cylindrical shaft rotates, the point masses generate radial inertia (*centrifugal*) forces axially separated from each other. These separated forces will have equal

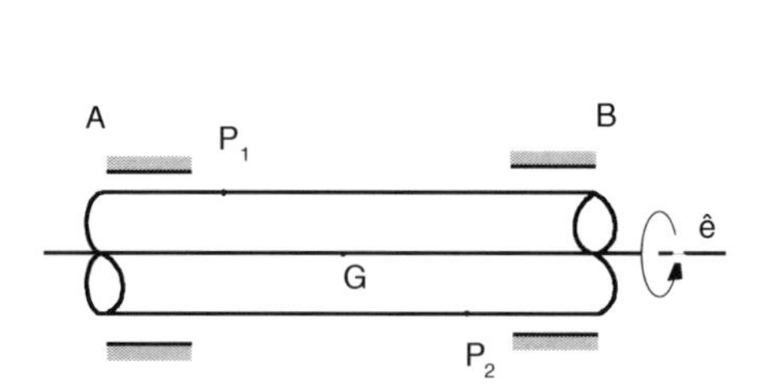

FIGURE 15.3.1
Rotating cylindrical shaft with 180° offset and separated point masses.

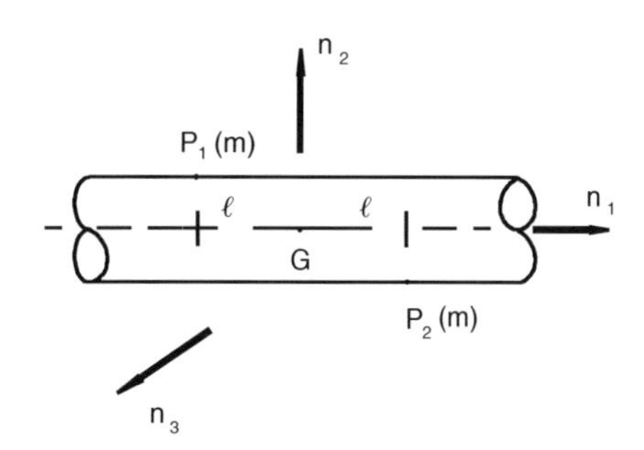

FIGURE 15.3.2
Cylindrical shaft with offset masses and mutually perpendicular unit vectors.

magnitudes and be oppositely directed; thus, they form a simple couple having a torque perpendicular to the axis of the shaft. This torque will in turn create forces at the bearings A and B, possibly leading to unwanted bearing wear. Such a system is said to be *dynamically out of balance*.

To further explore these concepts, let the cylindrical shaft have radius r, length L, and mass M. Let P_1 and P_2 each have mass m, and let these masses be separated axially by a distance 2ℓ. Finally, let $\mathbf{n}_1$, $\mathbf{n}_2$, and $\mathbf{n}_3$ be mutually perpendicular unit vectors fixed in the shaft, with $\mathbf{n}_1$ being parallel to the shaft axis; $\mathbf{n}_2$ parallel to the plane of P_1, P_2, and the shaft axis; and $\mathbf{n}_3$ normal to this plane such that $\mathbf{n}_3$ is generated by $\mathbf{n}_1 \times \mathbf{n}_2$, as shown in Figure 15.3.2.

Using the procedures of Chapter 7 we readily see that the central inertia dyadic of the shaft with the offset weights may be expressed as:

$$\begin{aligned}\mathbf{I} = &\left[mr^2 + (Mr^2/2)\right]\mathbf{n}_1\mathbf{n}_1 + 2mr\ell\mathbf{n}_1\mathbf{n}_2 + 2mr\ell\mathbf{n}_2\mathbf{n}_1 \\ &+ \left[mr^2 + (M/12)(3r^2 + \ell^2)\right]\mathbf{n}_2\mathbf{n}_2 + \left[m(r^2 + \ell^2) + (M/12)(3r^2 + \ell^2)\right]\mathbf{n}_3\mathbf{n}_3\end{aligned} \tag{15.3.1}$$

In matrix form, the $\mathbf{n}_i\ \mathbf{n}_j$ components of $\mathbf{I}$ may be expressed as:

$$\left[I_{ij}\right] = \begin{bmatrix} \left[mr^2 + (M/2)r^2\right] & 2mr\ell & 0 \\ 2mr\ell & \left[m\ell^2 + (M/12)(3r^2 + \ell^2)\right] & 0 \\ 0 & 0 & \left[M(r^2 + \ell^2) + (M/12)(3r^2 + \ell^2)\right] \end{bmatrix} \tag{15.3.2}$$

Recall from Chapters 7 and 8 (see Section 7.12 and 8.6) that if we represent the inertia forces on a body, or a mechanical system, by a single force $\mathbf{F}^*$ passing through the mass center G together with a couple with torque $\mathbf{T}^*$, then $\mathbf{F}^*$ and $\mathbf{T}^*$ are (Equation (7.12.1) and (7.12.8)):

$$\mathbf{F}^* = -\hat{M}\mathbf{a}_G \tag{15.3.3}$$

and

$$\mathbf{T}^* = -\mathbf{I}\cdot\boldsymbol{\alpha} - \boldsymbol{\omega}\times(\mathbf{I}\cdot\boldsymbol{\omega}) \tag{15.3.4}$$

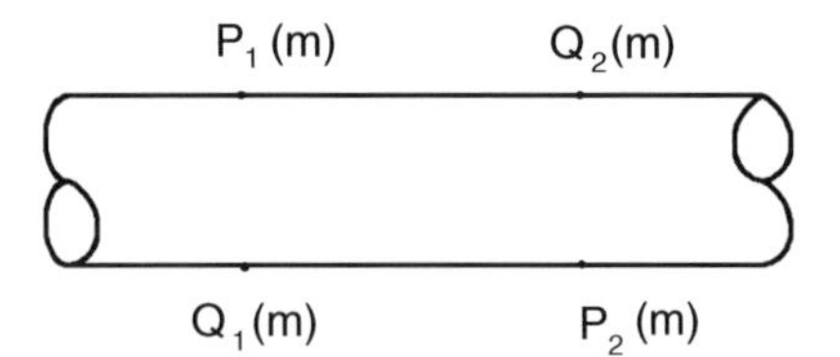

FIGURE 15.3.3
Dynamic balancing of the shaft of Figure 15.3.2.

where $\hat{M}$ is the mass of the body (or mechanical system), $\mathbf{a}_G$ is the mass center acceleration in an inertial reference frame R, $\boldsymbol{I}$ is the central inertia dyadic, $\boldsymbol{\omega}$ is the angular velocity in R, and $\boldsymbol{\alpha}$ is the angular acceleration in R. (In our case, $\hat{M}$ is $M + 2m$.)

If the offset weighted cylindrical shaft of Figure 13.3.2 is rotating at a uniform speed Ω about its axis, then $\mathbf{a}_G$, $\boldsymbol{\omega}$, and $\boldsymbol{\alpha}$ are:

$$\mathbf{a}_G = 0, \quad \boldsymbol{\omega} = \Omega \mathbf{n}_1, \quad \boldsymbol{\alpha} = 0 \tag{15.3.5}$$

Then, from Eq. (15.3.2), we see that $I \cdot \omega$ is:

$$\mathbf{I} \cdot \boldsymbol{\omega} = \mathbf{I} \cdot \Omega \mathbf{n}_1 = \left[mr^2 + (M/2) r^2 \right] \Omega \mathbf{n}_1 + 2mr\ell\Omega \mathbf{n}_2 \tag{15.3.6}$$

Hence, $\mathbf{T}^*$ becomes:

$$\mathbf{T}^* = -2mr\ell\Omega^2 \mathbf{n}_3 \tag{15.3.7}$$

(Observe that we could have predicted this result by inspection of Figure 15.3.2.)

Equation (15.3.7) shows that the offset weights of P_1 and P_2 create a moment on the shaft about $\mathbf{n}_3$ or perpendicular to the shaft axis. This moment in turn will create forces on the bearings. These forces will be perpendicular to the cylindrical shaft axis but will be continuously changing directions as the shaft rotates. Eq. (15.3.7) also shows that these bearing forces are proportional to the square of the angular speed Ω. Therefore, with high-speed machinery, we see that even small offset masses can produce significant bearing loads. Moreover, these loads can occur even if the shaft is statically balanced as in this example. This phenomenon is called *dynamic unbalance*.

With this understanding of the dynamic unbalance of the rotating cylindrical shaft, it is relatively easy to conceive of ways to eliminate the unbalance. Perhaps the simplest way is to add counterbalancing weights Q_1 and Q_2 at opposite sides of the shaft, as in Figure 15.3.3. The relatively simple geometry of this system makes the dynamic balancing procedure appear to be straightforward. While this is true in principal, the actual procedure with a physical system may involve some trial and error and the use of rather sophisticated balancing machines. Nevertheless, the concepts remain relatively simple. In the following section, we will consider the more general case of balancing a rotating body with arbitrary geometry.

15.4 Dynamic Balancing: The General Case

Consider an arbitrarily shaped body B rotating about one of its principal axes of inertia (see Section 7.7) as represented in Figure 15.4.1. Specifically, let $\mathbf{n}_1$, $\mathbf{n}_2$, and $\mathbf{n}_3$ be mutually perpendicular principal unit vectors fixed in B, and let B rotate about its first central principal axis with a constant angular speed Ω as shown, where G is the mass center of B.

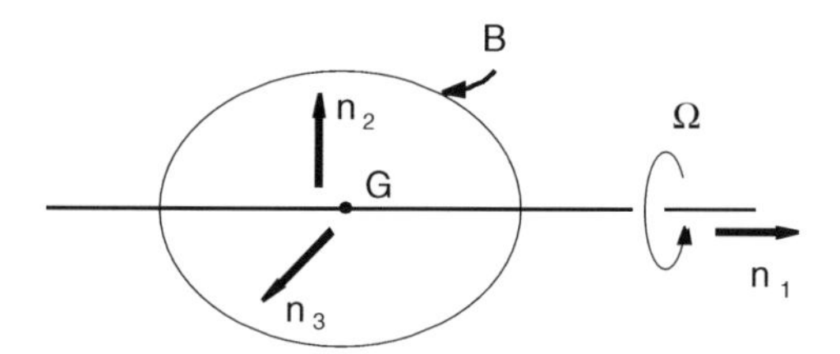

FIGURE 15.4.1
An arbitrarily shaped body rotating about its first principal axis of inertia.

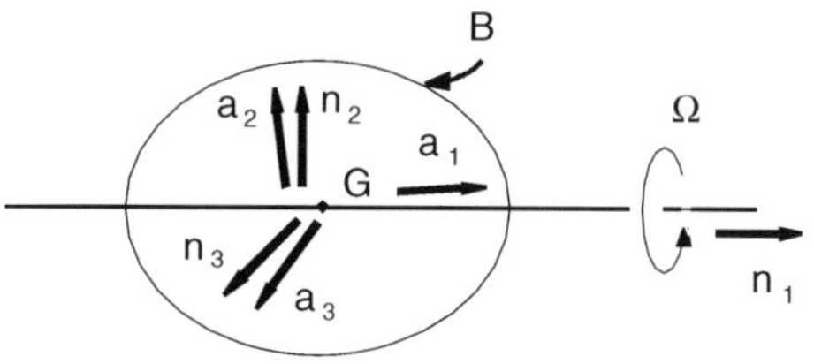

FIGURE 15.4.2
An arbitrarily shaped body rotating about a non-principal axis.

If the inertia forces on B are represented by a single force $\mathbf{F}^*$ passing through G, together with a couple with torque $\mathbf{T}^*$, then $\mathbf{F}^*$ and $\mathbf{T}^*$ may be expressed as (see Eqs. (8.6.5) and (8.6.6)):

$$\mathbf{F}^* = -m\mathbf{a}_G = 0 \tag{15.4.1}$$

and

$$\begin{aligned} \mathbf{T}^* &= -\mathbf{I}\cdot\dot{\Omega}\mathbf{n}_1 - \Omega\mathbf{n}_1 \times (\mathbf{I}\cdot\Omega\mathbf{n}_1) \\ &= 0 - \Omega^2\mathbf{n}_1 \times (\mathbf{I}\cdot\mathbf{n}_1) \end{aligned} \tag{15.4.2}$$

where m is the mass of B and $\boldsymbol{I}$ is the central inertia dyadic, which may be expressed as:

$$\mathbf{I} = I_{11}\mathbf{n}_1\mathbf{n}_1 + I_{22}\mathbf{n}_2\mathbf{n}_2 + I_{33}\mathbf{n}_3\mathbf{n}_3 \tag{15.4.3}$$

By substituting from Eq. (15.4.3) into (15.4.2), $\mathbf{T}^*$ becomes:

$$\mathbf{T}^* = -\Omega^2\mathbf{n}_1 \times I_{11}\mathbf{n}_1 = 0 \tag{15.4.4}$$

Thus, from Eqs. (15.4.1) and (15.4.4), we see that B, while rotating about its first principal axis, is both statically and dynamically balanced.

Next, suppose that the same body B is rotating at constant speed Ω about an axis other than a principal axis. Specifically, suppose that $\mathbf{n}_1$, $\mathbf{n}_2$, and $\mathbf{n}_3$ are only approximately parallel to the central principal axes and that instead $\mathbf{a}_1$, $\mathbf{a}_2$, and $\mathbf{a}_3$ are parallel to the principal axes, as depicted in Figure 15.4.2. As before, let the mass center G of B lie on the axis of rotation.

If the inertia forces on B are represented by a single force $\mathbf{F}^*$ passing through G together with a couple with torque $\mathbf{T}^*$, then $\mathbf{F}^*$ and $\mathbf{T}^*$ are (see Eqs. (15.4.1) and (15.4.2)):

$$\mathbf{F}^* = 0 \quad \text{and} \quad \mathbf{T}^* = -\Omega^2\mathbf{n}_1 \times (\mathbf{I}\cdot\mathbf{n}_1) \tag{15.4.5}$$

Because $\mathbf{n}_1$, $\mathbf{n}_2$, and $\mathbf{n}_3$ are no longer parallel to the central principal inertia axes, the inertia dyadic $\boldsymbol{I}$ no longer has the simple form of Eq. (15.4.3). Instead, $\boldsymbol{I}$ will have the general form:

$$\begin{aligned} \mathbf{I} = {} & I_{11}\mathbf{n}_1\mathbf{n}_1 + I_{12}\mathbf{n}_1\mathbf{n}_2 + I_{13}\mathbf{n}_1\mathbf{n}_3 \\ & + I_{21}\mathbf{n}_2\mathbf{n}_1 + I_{22}\mathbf{n}_2\mathbf{n}_2 + I_{23}\mathbf{n}_2\mathbf{n}_3 \\ & + I_{31}\mathbf{n}_3\mathbf{n}_1 + I_{32}\mathbf{n}_3\mathbf{n}_2 + I_{33}\mathbf{n}_3\mathbf{n}_3 \end{aligned} \tag{15.4.6}$$

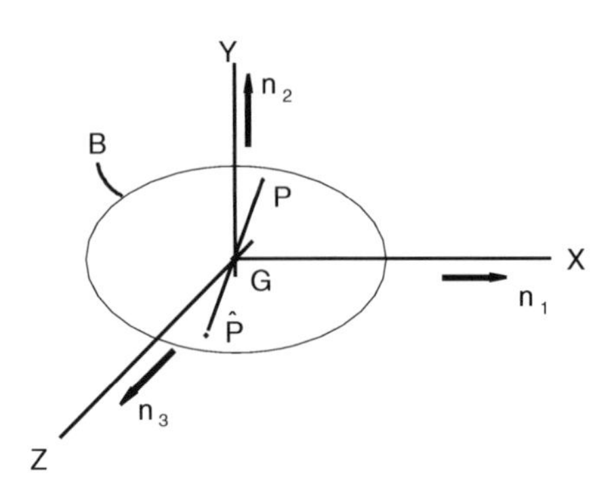

FIGURE 15.4.3
An arbitrarily shaped body B with particle weights P and $\hat{P}$.

Then, by substituting into Eq. (15.4.5), we find $\mathbf{T}^*$ to be:

$$\begin{aligned}\mathbf{T}^* &= -\Omega^2\mathbf{n}_1 \times \left(I_{11}\mathbf{n}_1 + I_{12}\mathbf{n}_2 + I_{13}\mathbf{n}_3\right) \\ &= -I_{12}\Omega^2\mathbf{n}_3 + I_{13}\Omega^2\mathbf{n}_2\end{aligned} \tag{15.4.7}$$

Observe that the product of inertia terms I_{12} and I_{13} cause the body B to have inertia force moments perpendicular to the axis of rotation and thus be out of balance. Hence, the product of inertia terms are a measure of the imbalance. Knowing this, we can bring the body into balance by judiciously adding weights that will reduce the product of inertia terms to zero.

To explore these concepts, suppose an arbitrarily shaped body B has an inertia dyadic as in Eq. (15.4.6) whose scalar components I_{ij}^B relative to $\mathbf{n}_i$ and $\mathbf{n}_j$ have the matrix form:

$$I_{ij}^B = \begin{bmatrix} I_{11}^B & I_{12}^B & I_{13}^B \\ I_{21}^B & I_{22}^B & I_{23}^B \\ I_{31}^B & I_{32}^B & I_{33}^B \end{bmatrix} \tag{15.4.8}$$

Suppose further that the product of inertia terms I_{12}, I_{23}, and I_{31} are small compared with the moments of inertia terms I_{11}, I_{22}, and I_{33}. To see how we can balance the body by adding weights, let the body and the weights be as depicted in Figure 15.4.3, where B, as before, is the body and P and $\hat{P}$ are particle weights, each having mass m.

Specifically, let X, Y, and Z be mutually perpendicular axes fixed in B with origin at the mass center G of B. Let P and $\hat{P}$ be placed at opposite ends of a line segment whose midpoint is at G (see Figure 15.4.3). Finally, let $\mathbf{n}_1$, $\mathbf{n}_2$, and $\mathbf{n}_3$ be mutually perpendicular unit vectors parallel to X, Y, and Z.

It is readily seen that the $\mathbf{n}_i$, $\mathbf{n}_j$ inertia dyadic components of P and $\hat{P}$ are (see Eq. (7.3.11)):

$$I_{ij}^P = m\begin{bmatrix} \left(y^2+z^2\right) & -xy & -xz \\ -xy & \left(x^2+z^2\right) & -yz \\ -xz & -yz & \left(x^2+y^2\right) \end{bmatrix} \tag{15.4.9}$$

and

$$I_{ij}^{\hat{P}} = m\begin{bmatrix} \left(\hat{y}^2+\hat{z}^2\right) & -\hat{x}\hat{y} & -\hat{x}\hat{z} \\ -\hat{x}\hat{y} & \left(\hat{x}^2+\hat{z}^2\right) & -\hat{y}\hat{z} \\ -\hat{x}\hat{z} & -\hat{y}\hat{z} & \left(\hat{x}^2+\hat{y}^2\right) \end{bmatrix} \tag{15.4.10}$$

where (x, y, z) and $(\hat{x}, \hat{y}, \hat{z})$ are the X, Y, Z components of P and $\hat{P}$, respectively. However, because P and $\hat{P}$ are at opposite ends of a line segment with midpoint at the origin G, we have:

$$\hat{x} = -x, \quad \hat{y} = -y, \quad \hat{z} = -z \tag{15.4.11}$$

Hence, we have:

$$I_{ij}^{P} = I_{ij}^{\hat{P}} \tag{15.4.12}$$

and then

$$I_{ij}^{P} + I_{ij}^{\hat{P}} = 2m \begin{bmatrix} (y^2 + z^2) & -xy & -xz \\ -xy & (x^2 + z^2) & -yz \\ -xz & -yz & (x^2 + y^2) \end{bmatrix} \tag{15.4.13}$$

Let $\hat{B}$ represent the body B with the added weights at P and $\hat{P}$. Then, the $\mathbf{n}_i$, $\mathbf{n}_j$ components of the inertia dyadic of $\hat{B}$ are:

$$\begin{aligned} I_{ij}^{\hat{B}} &= I_{ij}^{B} + I_{ij}^{P} + I_{ij}^{\hat{P}} \\ &= \begin{bmatrix} [I_{11}^{B} + 2m(y^2 + z^2)] & [I_{12}^{B} - 2mxy] & [I_{13}^{B} - 2mxz] \\ [I_{21}^{B} - 2mxy] & [I_{22}^{B} + 2m(x^2 + z^2)] & [I_{23}^{B} - 2myz] \\ [I_{31}^{B} - 2mxz] & [I_{32}^{B} - 2myz] & [I_{33}^{B} + 2m(x^2 + y^2)] \end{bmatrix} \end{aligned} \tag{15.4.14}$$

Therefore, the products of inertia of the weighted body $\hat{B}$ are:

$$\begin{aligned} I_{12}^{\hat{B}} &= I_{21}^{\hat{B}} = I_{12}^{B} - 2mxy \\ I_{13}^{\hat{B}} &= I_{31}^{\hat{B}} = I_{13}^{B} - 2mxz \\ I_{23}^{\hat{B}} &= I_{32}^{\hat{B}} = I_{23}^{B} = 2myz \end{aligned} \tag{15.4.15}$$

Hence, if these products of inertia are to be zero so that the weighted body is balanced both statically and dynamically, we must find m, x, y, and z such that

$$\begin{aligned} 2mxy &= I_{12}^{B} \\ 2mxz &= I_{13}^{B} \\ 2myz &= I_{23}^{B} \end{aligned} \tag{15.4.16}$$

Solving these equations for x, y, and z we obtain:

$$x = -\hat{x} = \left[I_{23}^B I_{12}^B / 2m I_{13}^B \right]^{1/2}$$
$$y = -\hat{y} = \left[I_{12}^B I_{13}^B / 2m I_{23}^B \right]^{1/2} \tag{15.4.17}$$
$$z = -\hat{z} = \left[I_{13}^B I_{23}^B / 2m I_{12}^B \right]^{1/2}$$

These results show that if the products of inertia are relatively small, B can be balanced dynamically by placing small weights at points P and $\hat{P}$ at opposite ends of a line whose midpoint is at the mass center of B. The resulting inertia dyadic components are then:

$$I_{ij}^{\hat{B}} = \begin{bmatrix} I_{11}^B + 2m\left(y^2 + z^2\right) & 0 & 0 \\ 0 & I_{22}^B + 2m\left(x^2 + z^2\right) & 0 \\ 0 & 0 & I_{33}^B + 2m\left(x^2 + y^2\right) \end{bmatrix} \tag{15.4.18}$$

Hence, the axes X, Y, and Z become principal inertia axes; thus, the weighted body is stable in rotation about each of these axes.

15.5 Application: Balancing of Reciprocating Machines

We can further illustrate the application of the foregoing ideas in the balancing of reciprocating machines such as the simple slider/crank mechanism of Figure 15.5.1. The system consists of a rotating rod AB (called the *crank*), a block C (called the *slider* or *piston*) restricted to sliding in a slot, and a connecting rod BC. We will see that, as before, we can balance the rotating crank through the simple addition of mass. The connecting rod and slider, however, are more difficult to balance, although there are steps that can be taken.

To develop these ideas, and to keep our analysis simple, let us assume that the system is symmetrical about the central plane of motion — that is, about the X–Y plane. The system then has planar motion (see Chapter 5). This means that inertia forces are parallel to the X–Y plane; therefore, the addition of balancing point masses in the plane of motion will not disrupt the stability, or balance, in the X–Z and Y–Z planes.

Expressed analytically, if the inertia forces on the crank AB and the connecting rod BC are replaced by equivalent force systems consisting of forces $\mathbf{F}_{AB}$ and $\mathbf{F}_{BC}$, passing through

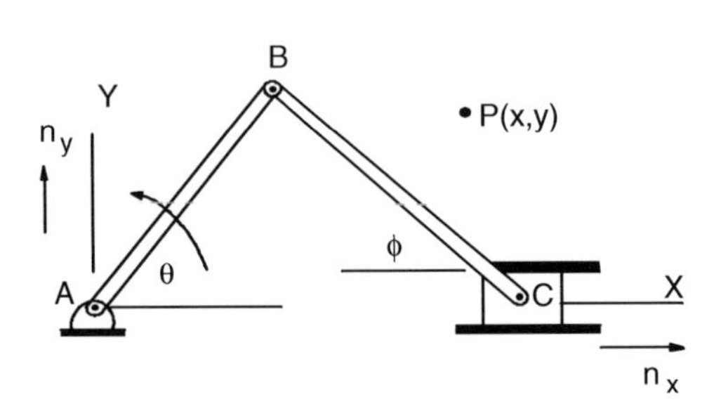

FIGURE 15.5.1
A simple slider crank mechanism.

the mass centers of the crank and connecting rod, together with couples having torques $\mathbf{T}_{AB}$ and $\mathbf{T}_{BC}$, then $\mathbf{F}_{AB}$, $\mathbf{F}_{BC}$, $\mathbf{T}_{AB}$, and $\mathbf{T}_{BC}$ may be expressed as (see Eqs. (8.6.5) and (8.6.6)):

$$\mathbf{F}_{AB} = -m_{AB}\mathbf{a}_{G_{BC}} \tag{15.5.1}$$

$$\mathbf{F}_{BC} = -m_{BC}\mathbf{a}_{G_{BC}} \tag{15.5.2}$$

$$\mathbf{T}_{AB} = -\mathbf{I}_{AB} \cdot \alpha_{AB} - \boldsymbol{\omega}_{AB} \times (\mathbf{I}_{AB} \cdot \boldsymbol{\omega}_{AB}) \tag{15.5.3}$$

and

$$\mathbf{T}_{BC} = -\mathbf{I}_{BC} \cdot \boldsymbol{\alpha}_{BC} - \boldsymbol{\omega}_{BC} \times (\mathbf{I}_{BC} \cdot \boldsymbol{\omega}_{BC}) \tag{15.5.4}$$

where m_{AB}, m_{BC}, I_{AB}, and I_{BC} are the masses and central inertia dyadics of AB and BC; $\mathbf{a}_{G_{AB}}$ and $\mathbf{a}_{G_{BC}}$ are the accelerations of the mass centers of AB and BC; and $\boldsymbol{\omega}_{AB}$, $\boldsymbol{\omega}_{BC}$, $\boldsymbol{\alpha}_{AB}$, and $\boldsymbol{\alpha}_{BC}$ may be expressed as:

$$\boldsymbol{\omega}_{AB} = \dot{\theta} n_z \quad \text{and} \quad \boldsymbol{\alpha}_{AB} = \ddot{\theta}\mathbf{n}_z \tag{15.5.5}$$

and

$$\boldsymbol{\omega}_{BC} = -\dot{\phi}\mathbf{n}_z \quad \text{and} \quad \boldsymbol{\alpha}_{BC} = -\ddot{\phi}\mathbf{n}_z \tag{15.5.6}$$

where θ and ϕ are the inclination angles of AB and BC as shown in Figure 15.5.1 and where $\mathbf{n}_z$ is a unit vector perpendicular to the plane of motion.

If the rods AB and BC are slender, the components of their central inertia dyadics, relative to principal unit vectors (parallel and perpendicular to the rods) may be expressed as (see Chapter 7 and Appendix II):

$$I_{ij}^{AB} = \begin{bmatrix} 0 & 0 & 0 \\ 0 & m_{AB}\ell_{AB}^2/12 & 0 \\ 0 & 0 & m_{AB}\ell_{AB}^2/12 \end{bmatrix} \quad \text{and} \quad I_{ij}^{BC} = \begin{bmatrix} 0 & 0 & 0 \\ 0 & m_{BC}\ell_{BC}^2/12 & 0 \\ 0 & 0 & m_{BC}\ell_{BC}^2/12 \end{bmatrix} \tag{15.5.7}$$

where ℓ_{AB} and ℓ_{BC} are the lengths of rods AB and BC. By substituting from Eqs. (15.5.5), (15.5.6), and (15.5.7) into Eqs. (15.5.3) and (15.5.4), we obtain the inertia torques in the forms:

$$\mathbf{T}_{AB} = -m_{AB}(\ell_{AB}^2/12)\ddot{\theta}\mathbf{n}_z \tag{15.5.8}$$

and

$$\mathbf{T}_{AB} = m_{BC}(\ell_{BC}^2/12)\ddot{\phi}\mathbf{n}_z \tag{15.5.9}$$

Hence, we see that the rod movements do not create any inertia moments along n_x or n_y.

Next, consider the addition of a particle P with mass m and coordinates (x, y) as in Figure 15.5.1. The components of the inertia dyadic of P relative to A relative to n_x, n_y, and n_z are (see Eq. (7.3.11)):

$$I_{ij}^P = m\begin{bmatrix} y^2 & -xy & 0 \\ -xy & x^2 & 0 \\ 0 & 0 & x^2+y^2 \end{bmatrix} \tag{15.5.10}$$

Suppose P is attached to one of the rods. Then, by inspection of Eqs. (15.5.3) to (15.5.6), we see that, with the zeroes in the third rows and columns of the inertia matrix of Eq. (15.5.10) and with both $\boldsymbol{\omega}_{AB}$ and $\boldsymbol{\omega}_{BC}$ having only components along n_z, there is no contribution to the inertia torques with components along n_x or n_y. This means that for the purposes of balancing we can focus our attention upon reducing the magnitudes of the inertia forces in the X–Y plane.

To this end, let the masses of the crank AB, the connecting rod BC, and the slider C be represented by point masses at the joints A, B, and C, and let these masses be m_A, m_B, and m_C. That is, we let the crank mass be represented by point masses at A and B such that the sum of the point masses is m_{AB} and such that the mass center remains at G_{AB}. Similarly, the connecting rod mass is distributed between joints B and C. Then, the resultant mass at B represents a contribution from both the crank and the connecting rod. This representation of the mass of the system then produces a system model consisting of three point masses at A, B, and C connected by massless rods AB and BC as depicted in Figure 15.5.2, where, for simplicity in notation, r and ℓ are used to represent the lengths of the crank and connecting rod. In considering the balancing of the inertia forces for this system, we observe that A does not move, B moves on a circle with radius r, and C oscillates along the X-axis.

If we assume further that the crank AB rotates with a constant angular speed ω ($\omega = \dot{\theta}$), then the inertia force $\mathbf{F}_B^*$ on B is directed radially outward along AB with magnitude $m_B r\omega^2$. That is,

$$\mathbf{F}_B^* = m_B r\omega^2 \mathbf{n}_r \tag{15.5.11}$$

where $\mathbf{n}_r$ is the radial unit vector (see Figure 15.5.2).

This imbalance created by the mass at B may be eliminated by adding a balancing mass to the crank on the opposite side of the rotation axis from B. Specifically, we might add a mass $\hat{m}_B$ at a point $\hat{B}$ a distance $\hat{r}$ away from A such that:

$$\hat{m}_B\hat{r} = m_B r \tag{15.5.12}$$

Figure 15.5.3 depicts such a balancing.

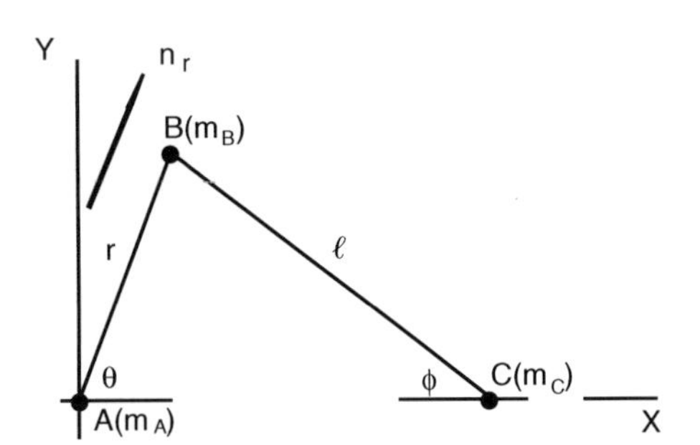

FIGURE 15.5.2
Point mass model of the system of Figure 15.5.1.

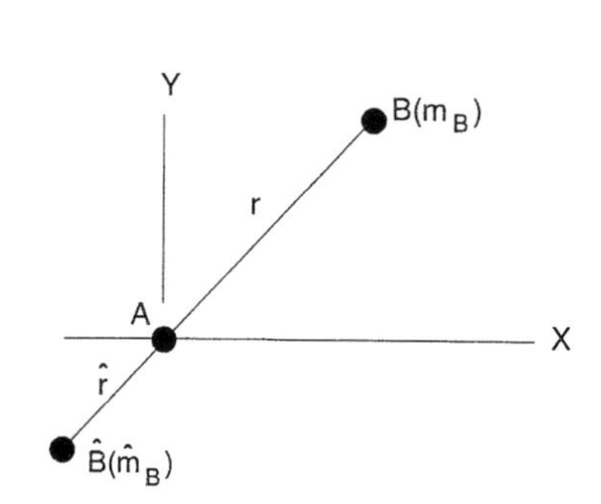

FIGURE 15.5.3
A balancing of the unbalance mass at *B*.

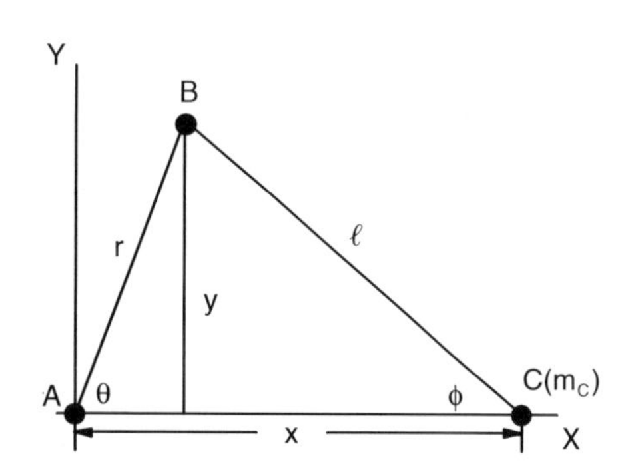

FIGURE 15.5.4
Oscillating mass at C for the model of Figure 15.5.2.

Unfortunately, balancing the oscillating mass at *C* is not as simple. The difficulty stems from the necessity to develop an oscillating force to counterbalance the oscillating inertia force of the mass at *C*. Such a force cannot be generated by simply adding mass to the system. Indeed, to generate an oscillating force, it is necessary to append some mechanism that will generate the force. Hence, for practical reasons, balancing the oscillating mass at *C* is usually accomplished only approximately.

To explore this, consider again the system of Figure 15.5.2. Let x represent the displacement of the mass at *C* relative to the crank axis at *A*. Then, from Figure 15.5.4, x is seen to be:

$$x = r\cos\theta + \ell\cos\phi \tag{15.5.13}$$

If we let y represent the displacement of *B* above the *X*-axis as in Figure 15.5.4, then y may be expressed as:

$$y = r\sin\theta = \ell\sin\phi \tag{15.5.14}$$

Then, $\sin\phi$ is:

$$\sin\phi = (r/\ell)\sin\theta \tag{15.5.15}$$

and as a consequence $\cos\phi$ is:

$$\cos\phi = \sqrt{1-\sin^2\phi} = \left[1-(r/\ell)^2\sin^2\theta\right]^{1/2} \tag{15.5.16}$$

Observe that the crank length r is usually somewhat smaller than the connecting rod length ℓ. Hence, we may approximate $\cos\phi$ by a binomial expansion of the radical. That is, we may use the expansion [15.6]:

$$\begin{aligned}(1+\varepsilon)^n &= 1 + n\varepsilon + \frac{n(n-1)}{2!}\varepsilon^2 + \frac{n(n-1)(n-2)}{3!}\varepsilon^3 \\ &\quad + \ldots + \frac{n!}{(n-p)!\,p!}\varepsilon^p + \ldots\end{aligned} \tag{15.5.17}$$

Then, by comparing Eqs. (15.5.16) and (15.5.17) we have:

$$\cos\phi = 1 - (1/2)(r/\ell)^2 \sin^2\theta - (1/8)(r/\ell)^4 \sin^4\theta - \ldots \tag{15.5.18}$$

By substituting into Eq. (15.5.13), the displacement x has the approximate form:

$$x \approx r\cos\theta + \ell - (1/2)\ell(r/\ell)^2 \sin^2\theta \tag{15.5.19}$$

Using the identity $\sin^2\theta \equiv (1/2) - (1/2)\cos 2\theta$, we then have:

$$x \approx r\cos\theta + \ell - (1/4)(r/\ell)^2 + (\ell/4)(r/\ell)^2 \cos 2\theta \tag{15.5.20}$$

Let us assume that the crank rotates at a constant angular rate ω. Then, by differentiating in Eq. (15.5.20), the acceleration of the mass C is given by the approximate expression:

$$\ddot{x} \approx -r\omega^2\left[\cos\theta + (r/\ell)\cos 2\theta\right] \tag{15.5.21}$$

Hence, the inertia force $\mathbf{F}_\theta^*$ on C is approximately:

$$\mathbf{F}_C^* \approx m_C r\omega^2\left[\cos\theta + (r/\ell)\cos 2\theta\right]\mathbf{n}_x \tag{15.5.22}$$

The first term of Eq. (15.5.22) is called the *primary unbalancing force* and the second smaller term is called the *secondary unbalancing force*.

We can counteract the primary unbalancing force by adding a balancing mass $\hat{m}$ at a point Q on the crank, at a distance h away from A, on the opposite side of B, as in Figure 15.5.5. Then, the resultant $\mathbf{F}^*$ of the inertia forces on the system due to the masses $\hat{m}$ and m_C become:

$$\begin{aligned}\mathbf{F}^* = &\left[(-\hat{m}h + m_C r)\omega^2\cos\theta + m_C(r/\ell)r\omega^2\cos 2\theta\right]\mathbf{n}_x \\ &+ \left[-\hat{m}h\omega^2\sin\theta\right]\mathbf{n}_y\end{aligned} \tag{15.5.23}$$

Then, by adjusting the magnitude of $\hat{m}h$, we can eliminate the primary unbalancing force ($m_C r\omega^2\cos\theta\mathbf{n}_x$). In so doing, however, we introduce an unbalance in the $\mathbf{n}_y$ direction. Also, the secondary unbalancing force ($m_C(r/\ell)r\omega^2\cos 2\theta\mathbf{n}_x$) remains. Therefore, as a compromise, mh is usually selected as a fraction (say, 1/2 to 2/3) of $m_C r$ [15.7]. In the following section, we will explore means for counteracting the secondary unbalancing forces.

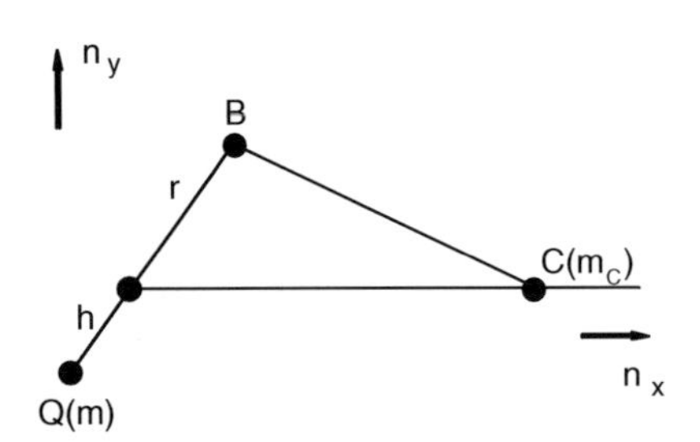

FIGURE 15.5.5
Balancing mass Q ($\hat{m}$) on the rotating crank.

15.6 Lanchester Balancing Mechanism

With the primary unbalancing force somewhat counteracted we may direct our attention to balancing the smaller secondary unbalancing force. This may be accomplished by a mechanism that develops an equal but opposite inertia force to that of the secondary unbalancing force. One such mechanism, called a *Lanchester balancing mechanism*, is depicted in Figure 15.6.1. It consists of two identical but oppositely rotating disks or gears with weights attached to their perimeters, as represented in Figure 15.6.1. If these weights each have mass m_2, then the resultant inertia force $\mathbf{F}_\ell^*$ generated as the disks rotate is:

$$\mathbf{F}_\ell^* = -2m_\ell \xi \dot{\beta}^2 \cos\beta \mathbf{n}_x \tag{15.6.1}$$

where ξ is the distance from the mass center of the weight to the disk center and β is the rotation angle as shown in Figure 15.6.1.

If this device is placed in line with the oscillating slider or piston *C*, as shown in Figure 15.6.2, and supported by the same structure that supports the crank bearing at *A*, then from Eqs. (15.5.23) and (15.6.1) the resultant inertia force $\mathbf{F}^*$ on the system may be expressed as:

$$\begin{aligned} \mathbf{F}^* = \Big[& -\hat{m}h\omega^2 \cos\theta + m_C r\omega^2 \cos\theta + m_C (r/\ell) r\omega^2 \cos 2\theta \\ & - 2m_\ell \xi \dot{\beta}^2 \cos\beta \Big] \mathbf{n}_x + \left[-\hat{m}h\omega^2 \sin\theta \right] \mathbf{n}_y \end{aligned} \tag{15.6.2}$$

If $\hat{m}h$ is adjusted to counteract the primary unbalance force ($m_C r\omega^2 \cos\theta n$), we can counteract the secondary unbalance force ($m_C(r/\ell)r\omega^2\cos 2\theta$) by adjusting m_ℓ, ξ, and β such that:

$$2m_\ell \xi \dot{\beta}^2 \cos\beta = m_C (r/\ell) r\omega^2 \cos 2\theta \tag{15.6.3}$$

Hence, we have:

$$\beta = 2\theta, \quad \dot{\beta} = 2\omega \tag{15.6.4}$$

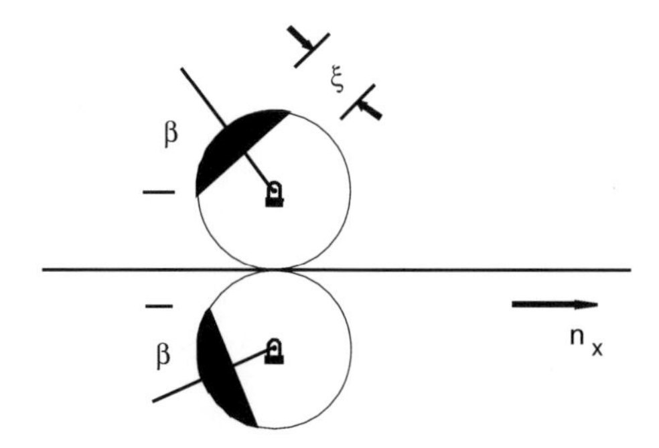

FIGURE 15.6.1
Lanchester balancing wheels.

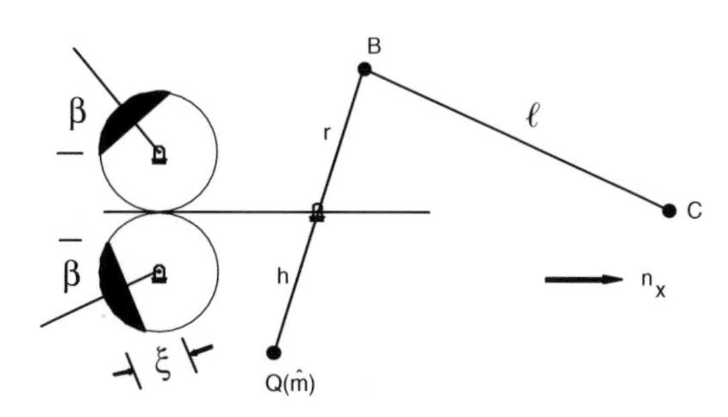

FIGURE 15.6.2
Lanchester balances attached to the slider crank.

and

$$m_\ell \xi = m_C r^2 / 8\ell \tag{15.6.5}$$

That is, the Lanchester gears are driven at twice the angular speed as the rotating crank.

15.7 Balancing of Multicylinder Engines

We can further apply the concepts of balancing and the analysis of the foregoing sections in the balance of multicylinder engines, such as automobile engines. To develop this concept, consider first the system of Figure 15.7.1 which depicts an inline multi-slider/crank system with n sliders. The sliders represent pistons moving in cylinders. The crank angles vary from one slider (or piston) to the next as in Figure 15.7.2.

Recall that the pistons, cylinders, connecting rods, and crank form the essence of internal combustion engines. In a typical cylinder, combustible gas is brought into the cylinder above the piston. The gas is then ignited, and the burning gas expands rapidly, pushing the piston out of the cylinder, thus driving the crankshaft. The crankshaft is joined to the connecting rods at various angles (as in Figure 15.7.2) to distribute the loading as the gas is ignited in the various cylinders. As we will see, the values of the various crank angles also determine the balance (or unbalance) of the engine.

To explore this, we can examine the unbalance forces of the individual cylinders and then add the forces to determine the magnitude of the unbalance forces for the entire engine. For each cylinder we can then follow the procedures of Section 15.5. That is, for a given cylinder we can assume that the crank, at the connecting rod, is rotationally balanced as before, with counterweights. Then, the unbalance of the cylinder may be approximately represented by primary and secondary unbalance inertia forces as in Eq. (15.5.22). Specifically, for typical cylinder i of Figure 15.7.1, the unbalance forces may be approximately represented by the force $\mathbf{F}_i^*$ expressed as:

$$\mathbf{F}_i^* = mr\omega^2 \left[\cos\theta_i + (r/\ell)\cos 2\theta_i\right]\mathbf{n}_y \tag{15.7.1}$$

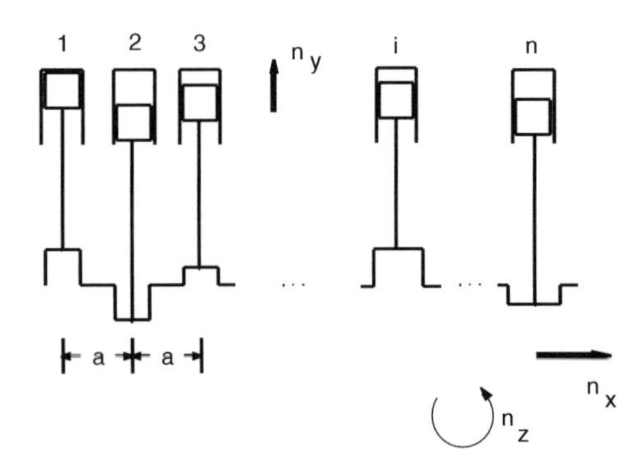

FIGURE 15.7.1
A slider/crank representation of a multicylinder engine.

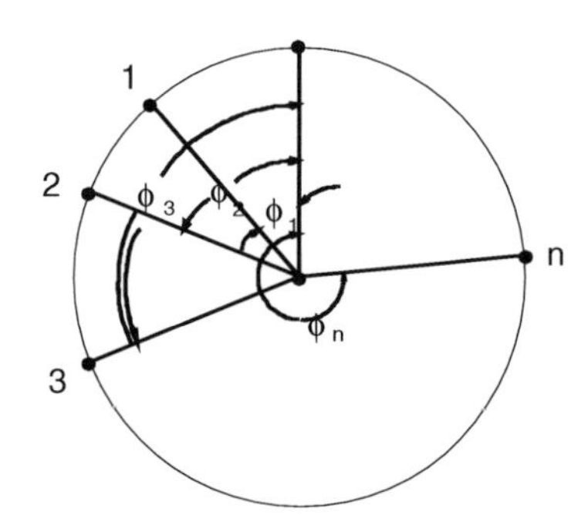

FIGURE 15.7.2
Crank angles.

where as before m is the mass concentrated at the piston, r is the crank radius, ℓ is the length of the connecting rod, ω is the angular speed of the crankshaft (assumed to be constant), and θ_i is the angular position of the crank arm. In terms of the connecting rod crank angles of Figure 15.7.2, θ_i may be expressed as:

$$\theta_i = \theta + \phi_i \tag{15.7.2}$$

where θ is the crank rotation angle. In our analysis we will assume that the n piston/cylinders are identical, with each having the same geometrical dimensions and each having the same unbalance mass m. We will also assume that the cylinders are equally spaced along the axis of the crankshaft with the distance between them being a, as in Figure 15.7.1. It happens that the unbalancing forces of the various cylinders may to some extent cancel one another; however, the unbalancing forces may also create a moment unbalance in the $\mathbf{n}_z$ direction.

From Eqs. (15.7.1) and (15.7.2), we see that the equivalent unbalance force $\mathbf{F}^*$ is:

$$\mathbf{F}^* = \sum_{i=1}^{m} \mathbf{F}_i^* = mr\omega^2 \left[\sum_{i=1}^{n} \cos(\theta + \phi_i) + (r/\ell) \sum_{i=1}^{n} \cos 2(\theta + \phi_i) \right] \mathbf{n}_y \tag{15.7.3}$$

Similarly, the moment $\mathbf{M}^*$ of the unbalancing forces about the crankshaft point at the center of the first cylinder is:

$$\mathbf{M}^* = mr\omega^2 a \left[\sum_{i=1}^{n} (i-1)\cos(\theta + \phi_i) + (r/\ell) \sum_{i=1}^{n} (i-1)\cos 2(\theta + \phi_i) \right] \mathbf{n}_z \tag{15.7.4}$$

To explore the implications of these results, we can expand the terms on the right sides of Eqs. (15.7.3) and (15.7.4) using the identity:

$$\cos(\theta + \phi_i) \equiv \cos\theta \cos\phi_i - \sin\theta \sin\phi_i \tag{15.7.5}$$

Equations (15.7.3) and (15.7.4) then become:

$$\begin{aligned} \mathbf{F}^* = mr\omega^2 \Bigg[& \cos\theta \sum_{i=1}^{n} \cos\phi_i - \sin\theta \sum_{i=1}^{n} \sin\phi_i \\ & + (r/\ell)\cos 2\theta \sum_{i=1}^{n} \cos 2\phi_i - (r/\ell)\sin 2\theta \sum_{i=1}^{n} \sin 2\phi_i \Bigg] \mathbf{n}_y \end{aligned} \tag{15.7.6}$$

and

$$\begin{aligned} \mathbf{M}^* = mr\omega^2 a \Bigg[& \cos\theta \sum_{i=1}^{n} (i-1)\cos\phi_i - \sin\theta \sum_{i=1}^{n} (i-1)\sin\phi_i \\ & + (r/\ell)\cos 2\theta \sum_{i=1}^{n} (i-1)\cos 2\phi_i - (r/\ell)\sin 2\theta \sum_{i=1}^{n} (i-1)\sin 2\phi_i \Bigg] \mathbf{n}_z \end{aligned} \tag{15.7.7}$$

From inspection of Eq. (15.7.6), we see that the primary unbalance force will be zero if:

$$\sum_{i=1}^{n} \cos\phi_i = 0 \quad \text{and} \quad \sum_{i=1}^{n} \sin\phi_i = 0 \tag{15.7.8}$$

We also see that the secondary unbalancing force will be zero if:

$$\sum_{i=1}^{n} \cos 2\phi_i = 0 \quad \text{and} \quad \sum_{i=1}^{n} \sin 2\phi_i = 0 \tag{15.7.9}$$

Similarly, from Eq. (15.7.7), we see that the primary moment unbalance will be zero if:

$$\sum_{i=1}^{n} (i-1)\cos\phi_i = 0 \quad \text{and} \quad \sum_{i=1}^{n} (i-1)\sin\phi_i = 0 \tag{15.7.10}$$

Also, the secondary moment unbalance will be zero if:

$$\sum_{i=1}^{n} (i-1)\cos 2\phi_i = 0 \quad \text{and} \quad \sum_{i=1}^{n} (i-1)\sin 2\phi_i = 0 \tag{15.7.11}$$

The challenge for the engine designer is then to determine and adjust the ϕ_i so that the eight sums of Eqs. (15.7.8 to 15.7.11) are zero, or nearly zero. In the following sections, we will explore the use of these equations in the balancing of four- and eight-cylinder engines.

15.8 Four-Stroke Cycle Engines

Four-, six-, and eight-cylinder engines are commonly used in automobiles, light trucks, and utility vehicles. Most of these engines use a *four-stroke* design, where two complete revolutions of the crankshaft are required to complete a cycle of a cylinder.

To review the terminology and to describe the strokes of a typical piston within a cylinder, consider the piston/cylinder configurations depicted in Figure 15.8.1. When the piston is farthest within the cylinder, that is, when the space between the piston and the closed cylinder end is smallest (as in Figure 15.8.1a), the piston is said to be in the top-dead-center position. Alternatively, when the piston is farthest out of the cylinder, as in Figure 15.8.1b, the piston is said to be in the bottom-dead-center position.

The movement of the piston from its top-dead-center position to its bottom-dead-center position is called a *stroke* of the engine. The movement of the piston from bottom-dead-center to top-dead-center is also called a stroke. A stroke requires a half turn of the crankshaft. Thus, one revolution of the crankshaft taking the piston from top-dead-center to bottom-dead-center and back to top-dead-center again constitutes two strokes.

When the piston is in the top-dead-center position, the volume between the piston and the cylinder is smallest, and gas within the cylinder is therefore compressed. In a four-stroke cylinder engine, compressed combustible gas is ignited when the piston is in the

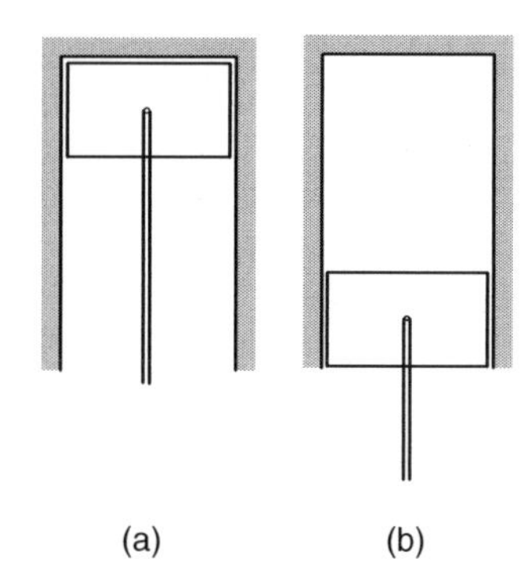

FIGURE 15.8.1
Top- and bottom-dead-center positions for a piston in a cylinder.

top-dead-center position. The ignited gas then expands, pushing the piston to its bottom-dead-center position. This movement is the *first stroke* or *power stroke* of the engine.

As the crankshaft continues to turn, the piston will move from bottom-dead-center back into the cylinder toward the top-dead-center position. During this movement, a valve in the cylinder is opened, allowing the combusted gas to be pushed out of the cylinder by the upward-moving piston. This movement is the *second stroke* or *exhaust stroke* of the engine.

When the piston reaches top-dead-center and the combusted gas is exhausted, the *exhaust valve*, which allows the gas to go out of the cylinder, is closed. At the same time, another valve (the *intake valve*) is opened, allowing fresh gas to be drawn into the cylinder by the vacuum created as the crank continues to turn, moving the piston from top-dead-center back toward the bottom-dead-center position. This movement is the *third stroke* or *intake stroke* of the engine.

Finally, as the crankshaft continues to turn, the intake valve is closed. As the piston moves again toward top-dead-center, the fresh, uncombusted gas is compressed and ready for ignition. This movement is called the *fourth stroke* or *compression stroke* of the engine. The four-stroke cycle of power, exhaust, intake, and compression is then repeated.

In a multicylinder engine, the strokes of the various cylinders are staggered so as to obtain a smooth or balanced operation. In the following sections, we will examine this staggering for four- and eight-cylinder engines.

15.9 Balancing of Four-Cylinder Engines

Consider the inline, four-stroke, four-cylinder engine as represented in Figure 15.9.1. From our discussion in Section 15.7, we discovered that the angular positioning of the connecting

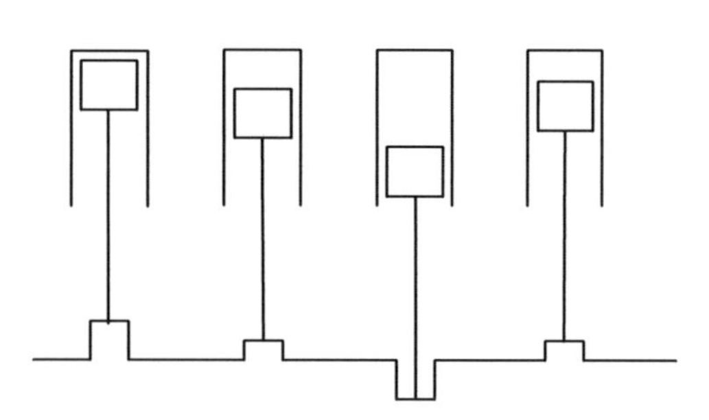

FIGURE 15.9.1
A representation of an inline four-cylinder engine.

TABLE 5.9.1

Cylinder Strokes for Various Crankshaft Angles for 180° Connecting Rod Spacing

Cylinder	Connecting Rod Crank Angle (°)	Crankshaft Rotation/Stroke Type			
		0 to 180°	180 to 360°	360 to 540°	540 to 720°
1	0	Power	Exhaust	Intake	Compression
2	180	Exhaust	Intake	Compression	Power
3	360	Intake	Compression	Power	Exhaust
4	540	Compression	Power	Exhaust	Intake

TABLE 15.9.2

Cylinder Strokes for Various Crankshaft Angles for 90° Connecting Rod Spacing

Cylinder	Connecting Rod Crank Angle (°)	Crankshaft Rotation/Stroke Type			
		0 to 180°	180 to 360°	360 to 540°	540 to 720°
1	0	Power	Exhaust	Intake	Compression
2	90	Power Exhaust	Intake	Compression	Power
3	180	Exhaust	Intake	Compression	Power
4	270	Exhaust Intake	Compression	Power	Exhaust

TABLE 15.9.3

Engine Balance: Listing of Terms of Eqs. (15.7.8) to (15.7.11) for the Connecting Rod/Crankshaft Configuration of Table 15.9.1

i	ϕ_i (°)	$\cos\phi_i$	$\sin\phi_i$	$\cos 2\phi_i$	$\sin 2\phi_i$	$(i-1)\cos\phi_i$	$(i-1)\sin\phi_i$	$(i-1)\cos 2\phi_i$	$(i-1)\sin 2\phi_i$
1	0	1	0	1	0	0	0	0	0
2	180	–1	0	1	0	–1	0	1	0
3	360	1	0	1	0	2	0	2	0
4	540	–1	0	1	0	–3	0	3	0
Totals		0	0	4	0	–2	0	6	0

rods along the crankshaft will directly affect the stability of the engine. To explore this, suppose we stagger the power strokes uniformly. Specifically, because a four-stroke cycle consumes 720°, let us space the connecting rods on the crankshaft at 180° intervals. Then, the angular position and strokes are those as listed in Table 15.9.1. With this arrangement, there is a power stroke occurring at all times during engine operation.

A disadvantage of this crankshaft arrangement, however, is that there are orientations of the crankshaft for which the pistons are all either in an uppermost, or top-dead-center (TDC), or lowermost, or bottom-dead-center (BDC), position. These orientations could interfere with the functioning of the engine by producing a locked crankshaft.

An alternative arrangement, avoiding crankshaft lock, would be to have the connecting rods spaced at 90° intervals along the crankshaft as in Table 15.9.2. This configuration, however, does not have a uniform distribution of power strokes.

The arrangement of the strokes and, in particular, the occurrence of the power strokes, is often called the *firing order* of the engine. To explore the effect of the firing order on engine balancing, consider first the sequence of Table 15.9.1. We can evaluate the engine balance by referring to the balancing conditions of Eqs. (15.7.8) to (15.7.11). That is, if we identify the index i with the cylinder number, then from Table 15.9.1 we can immediately develop the data of Table 15.9.3.

TABLE 15.9.4

Engine Balance: Listing of Terms of Eqs. (15.7.8) to (15.7.11) for the Connecting Rod/Crankshaft Configuration of Table 15.9.2

i	ϕ_i (°)	$\cos\phi_i$	$\sin\phi_i$	$\cos 2\phi_i$	$\sin 2\phi_i$	$(i-1)\cos\phi_i$	$(i-1)\sin\phi_i$	$(i-1)\cos 2\phi_i$	$(i-1)\sin 2\phi_i$
1	0	1	0	1	0	0	0	0	0
2	90	0	1	–1	0	0	1	–1	0
3	180	–1	0	1	0	–2	0	2	0
4	270	0	–1	–1	0	0	–3	–3	0
Totals		0	0	0	0	–2	–2	–2	0

TABLE 15.9.5

Cylinder Strokes for Various Crankshaft Angles for 0, 90, 270, and 180° Connecting Rod Spacing

Cylinder	Connecting Rod Crank Angle (°)	Crankshaft Rotation/Stroke Type			
		0 to 180°	180 to 360°	360 to 540°	540 to 720°
1	0	Power	Exhaust	Intake	Compression
2	90	Intake Compression	Power	Exhaust	Intake
3	270	Exhaust Intake	Compression	Power	Exhaust
4	180	Compression	Power	Exhaust	Intake

TABLE 15.9.6

Engine Balance: Listing of Terms of Eqs. (15.7.8) to (15.7.11) for the Connecting Rod/Crankshaft Configuration of Table 15.9.5

i	ϕ_i (°)	$\cos\phi_i$	$\sin\phi_i$	$\cos 2\phi_i$	$\sin 2\phi_i$	$(i-1)\cos\phi_i$	$(i-1)\sin\phi_i$	$(i-1)\cos 2\phi_i$	$(i-1)\sin 2\phi_i$
1	0	1	0	1	0	0	0	0	0
2	90	0	1	–1	0	0	1	–1	0
3	270	0	–1	–1	0	0	–2	–2	0
4	180	–1	0	1	0	–3	0	3	0
Totals		0	0	0	0	–3	–1	0	0

By comparing the totals of the columns of Table 15.9.3 with the expressions of Eq. (15.7.8) through (15.7.11) we see that, although the resultant primary unbalance force is zero, the secondary unbalance force is not zero. Also, the primary and secondary unbalance moments are nonzero. Hence, this engine configuration, even with its uniformly distributed power strokes, is unbalanced.

Consider next the arrangement of Table 15.9.2. By again identifying the cylinder number with the index i of the balancing conditions of Eqs. (15.7.8) to (15.7.11), we obtain the data of Table 15.9.4.

In this case, we see that both the resultants of both the primary and secondary inertia forces are zero (or balanced), but the resultant primary and secondary moments of these forces are not zero, which still leaves an engine unbalance. In comparing the results of Table 15.9.4 with those of Table 15.9.3, however, we see that there is a marked improvement with the stroke arrangement of Table 15.9.2 compared to that of Table 15.9.1

This raises the question as to whether we improve the balance even further by moving away from both the uniform power stroke design of Table 15.9.1 and the crankshaft symmetry design of Table 15.9.2. To answer this question, consider the crankshaft/connecting rod configuration of Table 15.9.5 (see Reference 15.8). Then, Table 15.9.6 provides a listing of data for the balancing conditions of Eqs. (15.7.8) to (15.7.11).

Here we see that both the resultant primary and secondary inertia forces are also balanced, leaving the only unbalance with the resultant primary moments. Thus, we have still further improvement in the balance.

These examples demonstrate the wide range of possibilities available to the engine designer; however, the examples are not meant to be exhaustive. Many other practical configurations are possible. The examples simply show that the crankshaft configuration can have a significant effect upon the engine balance. In the following section, we will extend these concepts and analyses to eight-cylinder engines.

15.10 Eight-Cylinder Engines: The Straight-Eight and the V-8

If we consider engines with eight cylinders the number of options for balancing increases dramatically. The analysis procedure, however, is the same as in the foregoing section. To illustrate the balancing procedure, consider an engine with eight cylinders in a line (the *straight-eight* engine) and with crank angles ϕ_i for the connecting rods arranged incrementally at 90° along the shaft. Table 15.10.1 provides the listing of the ϕ_i for the eight cylinders ($i = 1,\ldots, 8$) together with the trigonometric functions needed to test for balancing. A glance at the table immediately shows that the engine with this incremental 90° crank angle sequence has a moment unbalance.

This raises the question as to whether there are crank angle configurations where complete balancing would occur, within the approximations of our analysis. To respond to this question, consider again the four-cylinder engine of the previous section. For the crank angle configuration of Table 15.9.5, we found that the four-cylinder engine was balanced except for the primary moment. Hence, it appears that we could balance the eight-cylinder engine by considering it as two four-cylinder engines having reversed crank angle configurations. Specifically, for the four-cylinder engine of Table 15.9.5, the crank angles are 0, 90, 270, and 180°; therefore, let the first four cylinders have the crank angles 0, 90, 270, and 180°, and let the second set of four cylinders have the reverse crank angle sequence of 180, 270, 90, and 0°. Table 15.10.2 provides a listing of the crank angles ϕ; together with the trigonometric functions needed to test for balancing. As desired (and expected), the engine is balanced for both primary and secondary forces and moments.

TABLE 15.10.1

Engine Balance: Listing of Terms of Eqs. (15.7.8) to (15.7.11) for the Eight-Cylinder Uniformly Ascending Crank Angle Configuration

i	ϕ_i (°)	$\cos\phi_i$	$\sin\phi_i$	$\cos2\phi_i$	$\sin2\phi_i$	$(i-1)\cos\phi_i$	$(i-1)\sin\phi_i$	$(i-1)\cos2\phi_i$	$(i-1)\sin2\phi_i$
1	0	1	0	1	0	0	0	0	0
2	90	0	1	–1	0	0	1	–1	0
3	180	–1	0	1	0	–2	0	2	0
4	270	0	–1	–1	0	0	–3	–3	0
5	360	1	0	1	0	4	0	4	0
6	450	0	1	–1	0	0	5	–5	0
7	540	–1	0	1	0	–6	0	6	0
8	630	0	–1	–1	0	0	–7	–7	0
Totals		0	0	0	0	–4	–4	–4	0

TABLE 15.10.2

Engine Balance: Listing of Terms of Eqs. (15.7.8) to (15.7.11) for the Eight-Cylinder Crank Angle Configuration from Table 15.9.6.

i	ϕ_i (°)	$\cos\phi_i$	$\sin\phi_i$	$\cos 2\phi_i$	$\sin 2\phi_i$	$(i-1)\cos\phi_i$	$(i-1)\sin\phi_i$	$(i-1)\cos 2\phi_i$	$(i-1)\sin 2\phi_i$
1	0	1	0	1	0	0	0	0	0
2	90	0	1	–1	0	0	1	–1	0
3	270	0	–1	–1	0	0	–2	–2	0
4	180	–1	0	1	0	–3	0	3	0
5	180	–1	0	1	0	–4	0	4	0
6	270	0	–1	–1	0	0	–5	–5	0
7	90	0	1	–1	0	0	6	–6	0
8	0	1	0	1	0	7	0	7	0
Totals		0	0	0	0	0	0	0	0

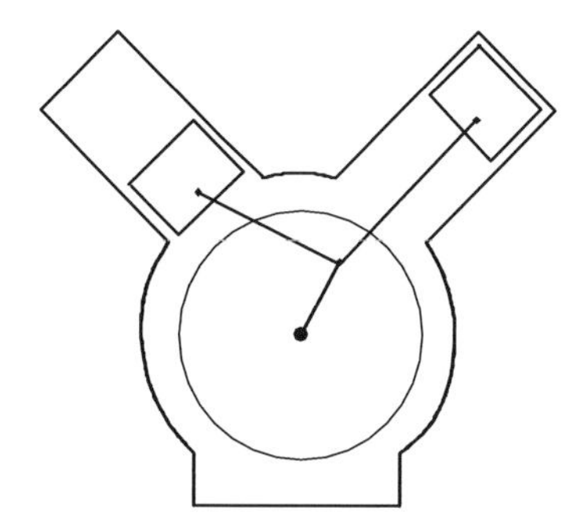

FIGURE 15.10.1
A V-type engine.

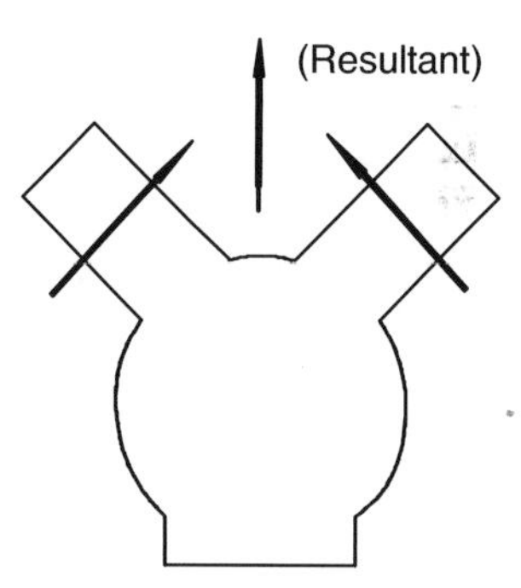

FIGURE 15.10.2
Unbalance primary moments and their resultants.

A practical difficulty with a straight-eight engine, however, is that it is often too long to conveniently fit into a vehicle engine compartment. One approach to solving this problem is to divide the engine into two parts, into a *V-type* engine as depicted in Figure 15.10.1. The two sides of the engine are called *banks*, each containing four cylinders. Because the total number of cylinders is eight, the engine configuration is commonly referred to as a *V-8*.

A disadvantage of this engine configuration, however, is that the engine is no longer in balance, as compared to the straight-eight engine: To see this, consider again the crank configuration of the straight-eight as listed in Table 15.10.2. Taken by themselves, the first four cylinders are unbalanced with an unbalanced primary moment perpendicular to the plane of the cylinders as seen in Table 15.9.6; hence, the second set of four cylinders has an unbalanced primary moment perpendicular to its plane. With the cylinder planes themselves being perpendicular, these unbalanced moments no longer cancel but instead have a vertical resultant as represented in Figure 15.10.2. This unbalance will have a tendency to cause the engine to oscillate in a yaw mode relative to the engine compartment. This yawing, however, can often be kept small by the use of motor mounts having high damping characteristics. Thus, the moment unbalance is usually an acceptable tradeoff in exchange for obtaining a more compact engine.

15.11 Closure

Our analysis shows that if a system is out of balance it can create undesirable forces at the bearings and supports. If the balance is relatively small, it can often be significantly reduced or even eliminated by judicious placing of balancing weights.

Perhaps the most widespread application of balancing principles is with the balancing of internal-combustion engines and with similar large systems. Because such systems have a number of moving parts, complete balance is generally not possible. Designers of such systems usually attempt to minimize the unbalance while at the same time making compromises or tradeoffs with other design objectives.

We saw an example of such a tradeoff in the balancing of an eight-cylinder engine: the engine could be approximately balanced if the cylinders were all in a line. This arrangement, however, creates a relatively long engine, not practical for many engine compartments. An alternative is to divide the engine into two banks of four cylinders, inclined relative to each other (the V-8 engine); however, the engine is then out of balance in yaw moments, requiring damping at the engine mounts to reduce harmful vibration.

Optimal design of large engines thus generally involves a number of issues that must be resolved for each individual machine. While there are no specific procedures for such optimal design, the procedures outlined herein, together with information available in the references, should enable designers and analysts to reach toward optimal design objectives.

References

15.1. Wilson, C. E., Sadler, J. P., and Michaels, W. J., *Kinematics and Dynamics of Machinery,* Harper & Row, New York, 1983, pp. 609–632.
15.2. Wowk, V., *Machine Vibrations,* McGraw-Hill, New York, 1991, pp. 128–134.
15.3. Paul, B., *Kinematics and Dynamics of Planar Machinery,* Prentice Hall, Englewood Cliffs, NJ, 1979, chap. 13.
15.4. Mabie, H. H., and Reinholtz, C. F., *Mechanisms and Dynamics of Machinery,* Wiley, New York, 1987, chap. 10.
15.5. Sneck, H. J., *Machine Dynamics,* Prentice Hall, Englewood Cliffs, NJ, 1991, pp. 211–227.
15.6. Swight, H. B., *Tables of Integrals and Other Mathematical Data,* Macmillan, New York, 1057, p. 1.
15.7. Shigley, J. E., and Uicker, J. J., Jr., *Theory of Machines and Mechanisms,* McGraw-Hill, New York, 1980, p. 499.
15.8. Martin, G. H., *Kinematics and Dynamics of Machines,* McGraw-Hill, New York, 1982, p. 419.
15.9. Taylor, C. F., *The Internal-Combustion Engine in Theory and Practice,* Vol. II: *Combustion, Fuels, Materials, Design,* MIT Press, Cambridge, MA, 1985, pp. 240–305.

Problems

Section 15.2 Static Balancing

P15.2.1: A 125-lb flywheel in the form of a thin circular disk with radius 1.0 ft and thickness 1.0 in. is mounted on a light (low-weight) shaft which in turn is supported by nearly

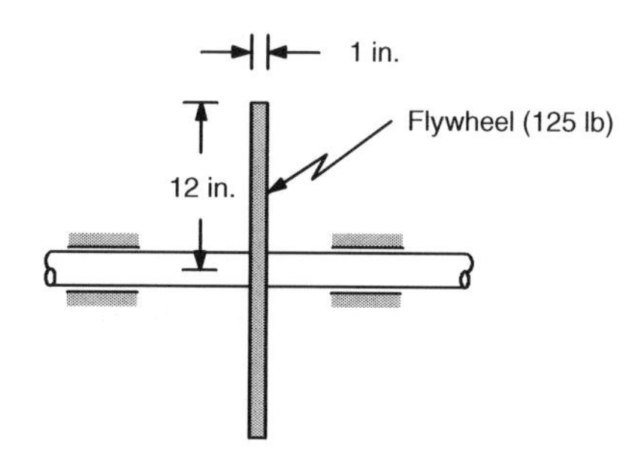

FIGURE P15.2.1
A flywheel on a light shaft in nearly frictionless bearings.

frictionless bearings as represented in Figure P15.2.1. If the flywheel is mounted off-center by 0.25 in., what weight should be placed on the flywheel rim, opposite to the off-center offset, so that the flywheel is statically balanced?

P15.2.2: Repeat Problem P15.2.1 if the flywheel mass is 50 kg, with a radius of 30 cm, a thickness of 2.5 cm, and an off-center mounting of 7 mm.

P15.2.3: See Problem P15.2.1. Suppose the flywheel shaft has frictionless bearings. What would be the period of small oscillations?

P15.2.4: Repeat Problem P15.2.3 for the data of Problem P15.2.2.

P15.2.5: See Problem P15.2.3. Suppose a slightly unbalanced disk flywheel, supported in a light shaft with frictionless bearings, is found to oscillate about a static equilibrium position with a period of 7 sec. How far is the flywheel mass center displaced from the shaft axis?

Section 15.3 Dynamic Balancing

P15.3.1: A shaft with radius r of 3 in. is rotating with angular speed Ω of 1300 rpm. Particles P_1 and P_2, each with weight w of 2 oz. each, are placed on the surface of the shaft as shown in Figure P15.3.1. If P_1 and P_2 are separated axially by a distance ℓ of 12 in., determine the magnitude of the dynamic unbalance.

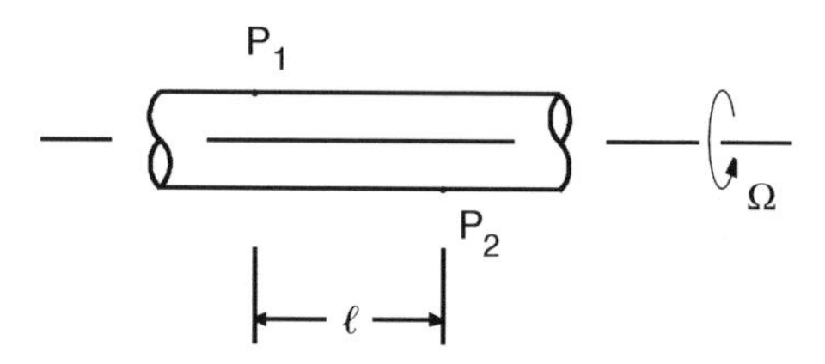

FIGURE P15.3.1
A rotating shaft with unbalance particles.

P15.3.2: Repeat Problem P15.3.1 if r, w, and ℓ have the values $r = 7$ cm, $w = 50$ g, and $\ell = 0.333$ m.

P15.3.3: See Problem P15.3.1. Suppose the dynamically unbalanced shaft is made of steel with a density of 489 lb/ft^3. Suppose further that it is proposed that the shaft be balanced by removing material by drilling short holes on opposite sides of the shaft. Discuss the feasibility of this suggestion. Specifically, if the holes are to be no more than 0.5 in. in diameter, no more than 0.5 in. deep, and separated axially by no more than 18 in., suggest a drilling procedure to balance the shaft. That is, suggest the number, size, and positioning of the holes.

P15.3.4: Repeat Problem P15.3.3 for the shaft unbalance of Problem P15.3.2 if the mass density of the shaft is 7800 kg/m^3.

Section 15.4 Dynamic Balancing: Arbitrarily Shaped Rotating Bodies

P15.4.1: Suppose $\mathbf{n}_1$, $\mathbf{n}_2$, and $\mathbf{n}_3$ are mutually perpendicular unit vectors fixed in a body B, and suppose that B is intended to be rotated with a constant speed Ω about an axis X which passes through the mass center G of B and which is parallel to $\mathbf{n}_1$. Let $\mathbf{n}_1$, $\mathbf{n}_2$, and $\mathbf{n}_3$ be nearly parallel to principal inertia directions of B for G so that the components I_{ij} of the inertia dyadic of B for G relative to $\mathbf{n}_1$, $\mathbf{n}_2$, and $\mathbf{n}_3$ are:

$$I_{ij} = \begin{bmatrix} 18 & -0.1 & 0.25 \\ -0.1 & 12 & -0.15 \\ 0.25 & -0.15 & 6 \end{bmatrix} \text{ slug ft}^2$$

Show that with this configuration and inertia dyadic that B is dynamically out of balance. Next, suppose we intend to balance B by the addition of two 12-oz. weights P and $\hat{P}$ placed opposite one another about the mass center G. Determine the coordinates of P and $\hat{P}$ relative to the X-, Y-, and Z-axes with origin at G and parallel to $\mathbf{n}_1$, $\mathbf{n}_2$, and $\mathbf{n}_3$.

P15.4.2: Repeat Problem P15.4.1 if the inertia dyadic components are:

$$I_{ij} = \begin{bmatrix} 30 & -0.2 & -0.3 \\ -0.2 & 20 & 0.25 \\ -0.3 & 0.25 & 10 \end{bmatrix} \text{ kg m}^2$$

and if the masses of P and $\hat{P}$ are each 0.5 kg.

P15.4.3: Repeat Problems P15.4.1 and P15.4.2 if B is rotating about the Z-axis instead of the X-axis.

Section 15.5 Balancing Reciprocating Machines

P15.5.1: Suppose the crank AB of a simple slider/crank mechanism (see Figures 15.5.1 and P15.5.1, below) is modeled as a rod with length of 4 in. and weight of 2 lb. At what distance $\hat{r}$ away from A should a weight of 4 lb be placed to balance AB?

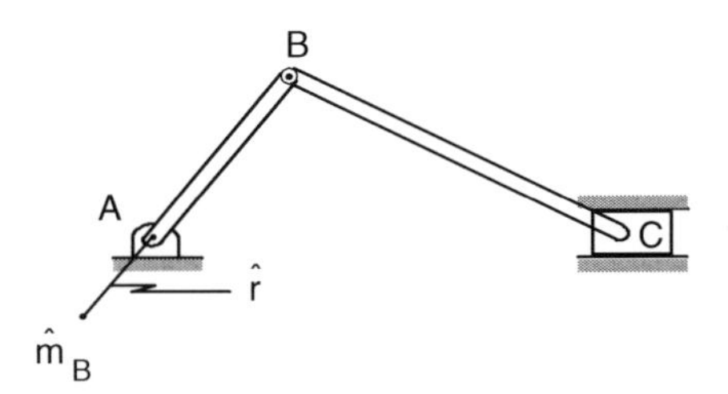

FIGURE P15.5.1
A simple slider crank mechanism.

P15.5.2: See Problem P15.5.1. Suppose $\hat{r}$ is to be 1.5 in. What should be the weight of the balancing mass $\hat{m}_B$?

P15.5.3: Repeat Problem P15.5.1 if rod AB has length 10 cm and mass 1 kg.

P15.5.4: Consider again the simple slider/crank mechanism as in Figure P15.5.4, this time with an objective of eliminating or reducing the primary unbalancing force as developed in Eq. (15.5.23). Specifically, let the length r of the crank arm be 4 in., the length ℓ of the connecting rod be 9 in., the weight of the piston C be 3.5 lb, and the angular speed w of

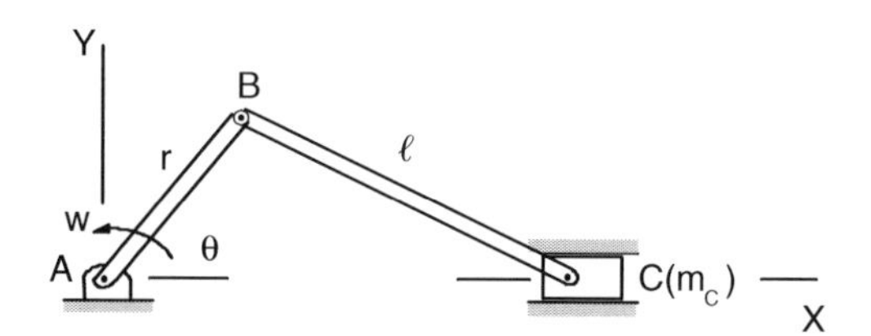

FIGURE P15.5.4
A slider crank mechanism.

the crank be 1000 rpm. Determine the weight (or mass $\hat{m}$) of the balancing weight to eliminate the primary unbalance if the distance h of the weight from the crank axis is 3 in. Also, determine the maximum secondary unbalance that remains and the maximum unbalance in the Y-direction created by the balancing weight.

P15.5.5: Repeat Problem P15.5.4 if the balancing weight compensates for only (a) 1/2 and (b) 2/3 of the primary unbalance.

P15.5.6: Repeat Problems P15.5.4 and P15.5.5 if r is 10 cm, ℓ is 25 cm, h is 5 cm, and the mass of C is 2 kg.

Section 15.6 Lanchester Balancing

P15.6.1: Verify Eqs. (15.6.1), (15.6.2), and (15.6.3).

P15.6.2: Observe that the geometric parameters of Eq. (15.6.5) are such that the Lanchester balancing mass m_ℓ need only be a fraction of the piston mass m_C. Specifically, suppose that (r/ℓ) is 0.5 and (ξ/r) is also 0.5. What, then, is the mass ratio m_ℓ/m_C?

Sections 15.7, 15.8, 15.9 Balancing Multicylinder Engines

P15.7.1: Consider a three-cylinder, four-stroke engine. Following the procedures outlined in Sections 15.7, 15.8, and 15.9, develop a firing order and angular positioning to optimize the balancing of the engine.

16

Mechanical Components: Cams

16.1 Introduction

In this chapter, we consider the design and analysis of cams and cam–follower systems. As before, we will focus upon basic and fundamental concepts. Readers interested in more details than those presented here may want to consult the references at the end of the chapter.

A cam (or *cam-pair*) is a mechanical device intended to transform one kind of motion into another kind of motion (for example, rotation into translation). In this sense, a cam is primarily a kinematic device or mechanism. The most common cam-pairs transform simple uniform (constant speed) rotation into translation (or rectilinear motion). Figure 16.1.1 shows a sketch of such a device. In such mechanisms, the actuator (or active) component (in this case, the rotating disk) is called the *cam*. The responding (or passive) component is called the *follower.*

Some cam-pairs simply convert one kind of translation into another kind of translation, as in Figure 16.1.2. Here, again, the actuator or driving component is the cam and the responding component is the follower. An analogous cam-pair that transforms simple rotation into simple rotation is a pair of meshing gears depicted schematically in Figure 16.1.3. Because gears are used so extensively, they are usually studied separately, which we will do in the next chapter. Here, again, however, the actuator gear is called the *driver* and the responding gear is called the *follower.* Also, the smaller of the gears is called the *pinion* and the larger is called the *gear.*

To some extent, the study of cams employs different procedures than those used in our earlier chapters; the analysis of cams is primarily a kinematic analysis. Although the study of forces is important in mechanism analyses, the forces generated between cam–follower pairs are generally easy to determine once the kinematics is known. The major focus of cam analysis is that of cam design. Whereas in previous chapters we were generally given a mechanical system to analyze, with cams we are generally given a desired motion and asked to determine or design a cam–follower pair to produce that motion.

In the following sections, we will consider the essential features of cam and follower design. We will focus our attention upon simple cam pairs as in Figure 16.1.1. Similar analyses of more complex cam mechanisms can be found in the various references. Before directing our attention to simple cam pairs, it may be helpful to briefly review configurations of more complex cam mechanisms. We do this in the following section and then consider simple cam–follower design in the subsequent sections.

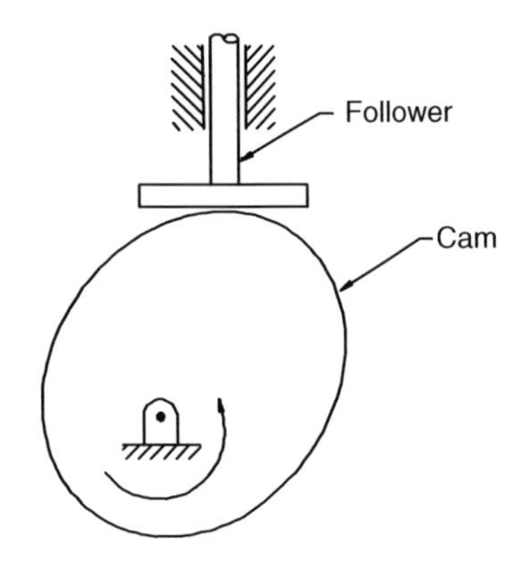

FIGURE 16.1.1
A simple cam–follower pair.

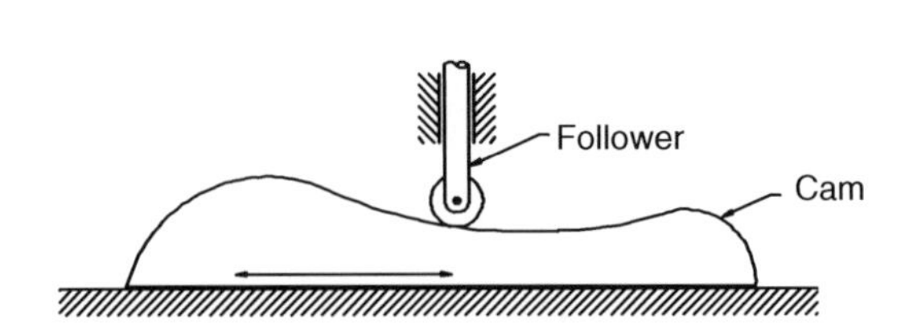

FIGURE 16.1.2
A translation to translation cam–follower pair.

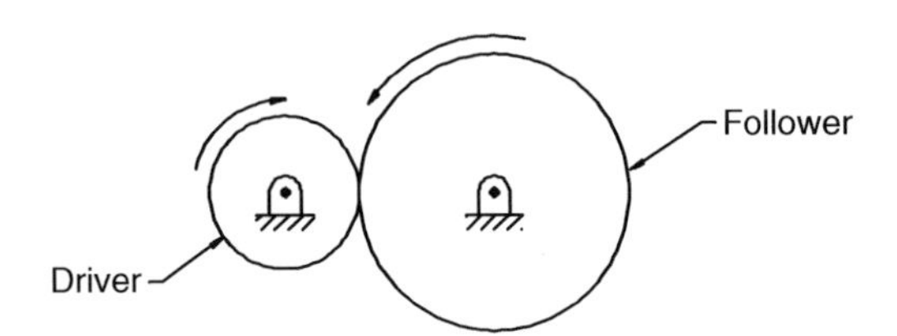

FIGURE 16.1.3
A gear pair.

16.2 A Survey of Cam Pair Types

As noted earlier, the objective of a cam–follower pair is to transform one kind of motion into another kind of motion. As such, cam-pairs are kinematic devices. When, in addition to transforming and transmitting motion, cam-pairs are used to transmit forces, they are called *transmission* devices.

There are many ways to transform one kind of motion into another kind of motion. Indeed, the variety of cam–follower pair designs is limited only by one's imagination. In addition to those depicted in Figures 16.1.1, 16.1.2, and 16.1.3, Figures 16.2.1, 16.2.2, and 16.2.3 depict other common types of cams. Figures 16.2.4 and 16.2.5 depict various common

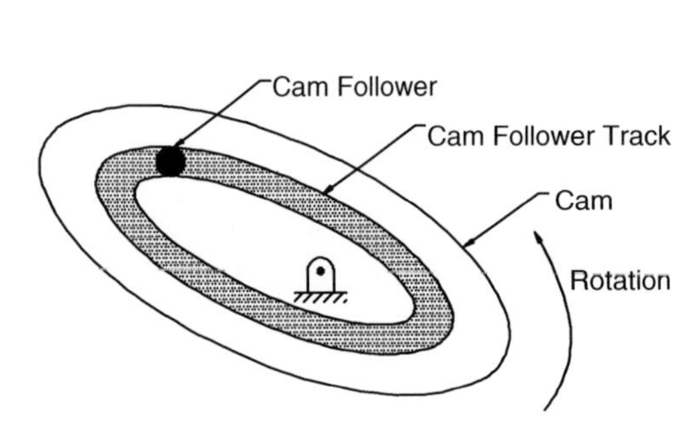

FIGURE 16.2.1
Cam–follower with cam–follower track.

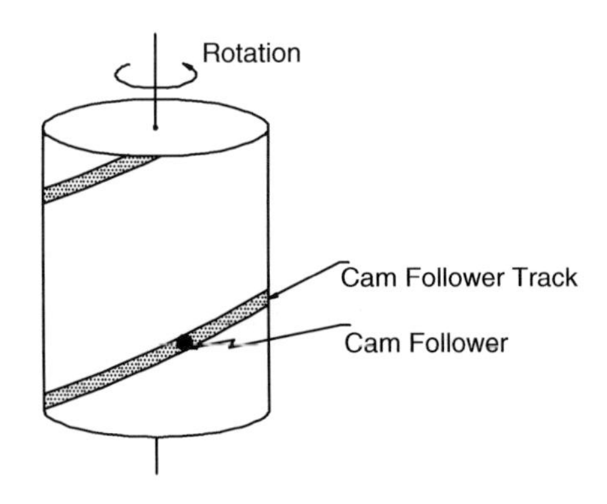

FIGURE 16.2.2
Cylindrical cam with cam–follower track.

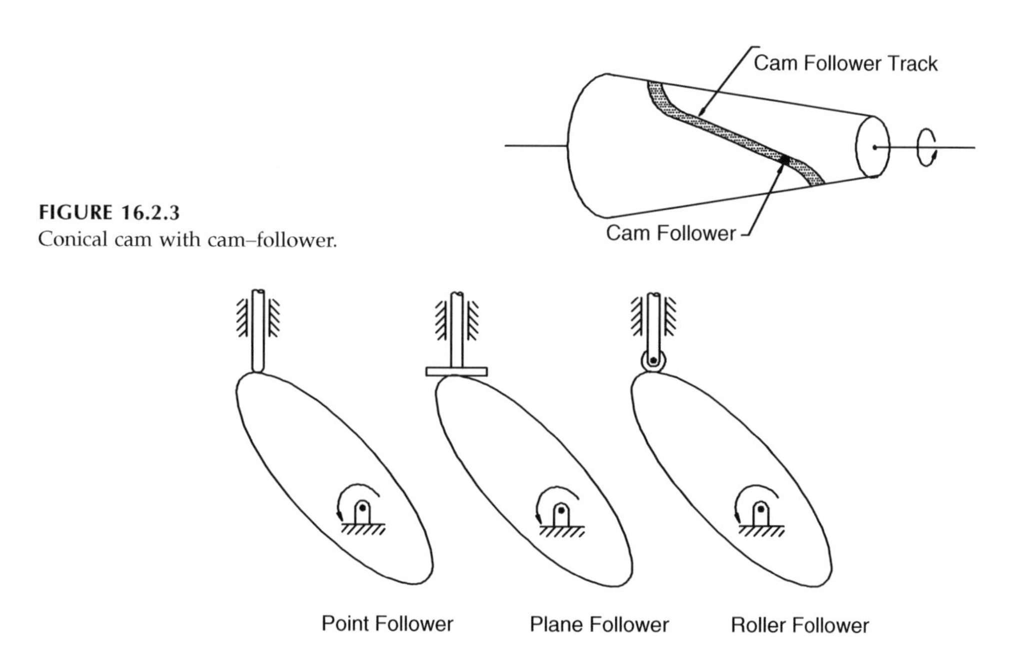

FIGURE 16.2.3
Conical cam with cam–follower.

FIGURE 16.2.4
Types of translation followers.

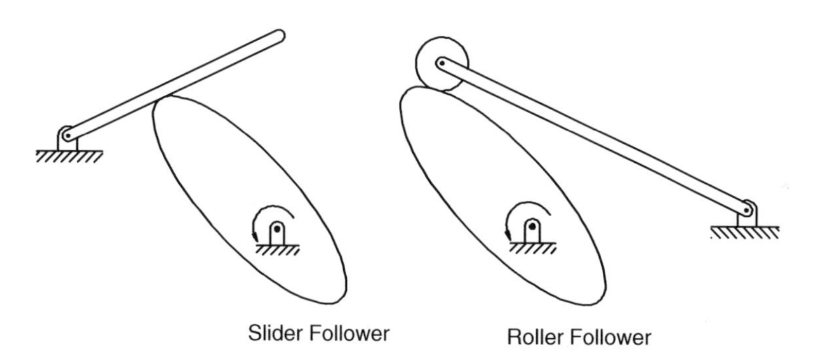

FIGURE 16.2.5
Types of rotation followers.

follower types. Figures 16.2.2 and 16.2.3 show that cam–follower pairs may be used not only for exchanging types of planar motion but also for converting planar motion into three-dimensional motion.

In the following sections, we will direct our attention to planar rotating cams (or *radial cams*) with translating followers as in Figure 16.2.4. Initially, we will consider common nomenclature and terminology for such cams.

16.3 Nomenclature and Terminology for Typical Rotating Radial Cams with Translating Followers

Consider a radial cam–follower pair as in Figure 16.3.1 where the axis of the follower shaft intersects the axis of rotation of the cam. Let the follower shaft be driven by the cam

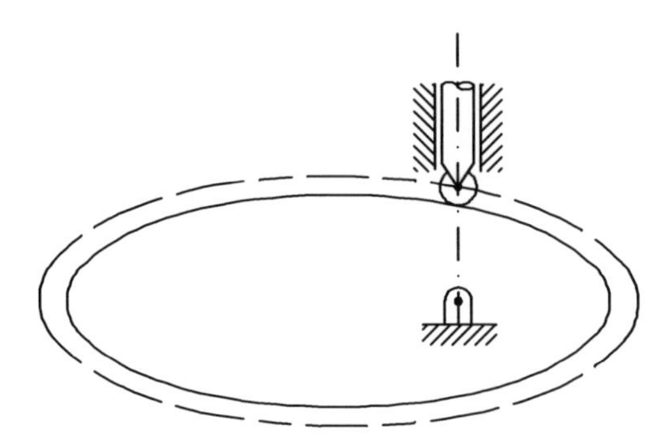

FIGURE 16.3.1
A radial cam–follower pair.

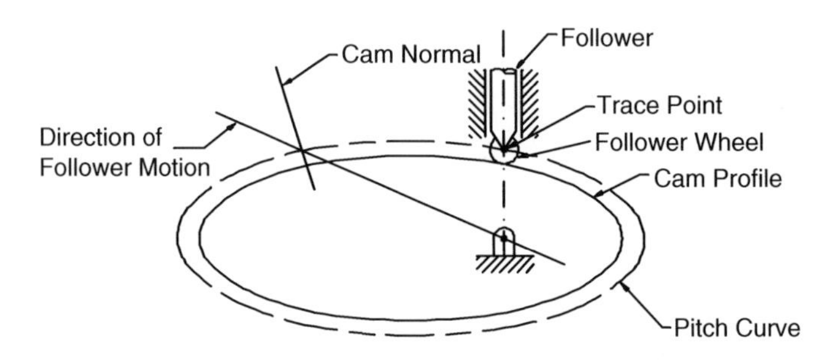

FIGURE 16.3.2
Radial cam–follower pair terminology and nomenclature.

through a follower wheel as shown. Figure 16.3.2 shows the commonly employed terminology and nomenclature of the radial cam–follower pair [16.1]. Brief descriptions of these items follow:

- Cam profile — contact boundary of the rotating cam
- Follower wheel — rolling wheel of the follower contacting the cam profile to reduce wear and to provide smooth operation
- Trace point — center of rolling follower wheel; also, point of *knife-edge* follower
- Pitch curve — path of the trace point
- C — typical location of the trace point on the pitch curve
- θ — pressure angle

Observe the pressure angle in Figure 16.3.2. It is readily seen that the pressure angle is a measure of the lateral force exerted by the cam on the follower. Large lateral forces (as opposed to longitudinal or axial forces) can lead to high system stress and eventually deleterious wear. A large pressure angle θ can create large lateral forces. Specifically, the lateral force is proportional to $\sin\theta$.

Observe further in Figure 16.3.2 that if the cam profile is circular the pressure angle is zero. In this case, no lateral forces are exerted on the follower by the cam. However, with the circular profile, no longitudinal force will be exerted on the follower, so the follower will remain stationary. A stationary follower corresponds to a *dwell* region for the cam.

These observations show that there is no follower movement without a nonzero pressure angle; hence, a cost of follower movement is the lateral force generated by the pressure angle. Thus, cam designers generally attempt to obtain the desired follower motion with the smallest possible pressure angle. This is usually accomplished by a large cam, as space permits.

Finally, observe that the follower wheel simply has the effect of enlarging the cam profile creating the *pitch curve*, with the follower motion then defined by the movement of the trace point.

16.4 Graphical Constructions: The Follower Rise Function

The central problem confronting the cam designer is to determine the cam profile that will produce a desired follower movement. To illustrate these concepts and some graphical design procedures, consider again the simple knife-edge radial follower pair of Figure 16.4.1. Let the cam profile be represented analytically as (see Figure 16.4.2):

$$r = f(\theta) \tag{16.4.1}$$

where r and θ are polar coordinates of a typical point P on the cam profile. The angle θ may also be identified with the rotation angle of the cam. Thus, given $f(\theta)$, we know r as a function of the rotation angle. Then with the knife-edge follower, we can identify r with the movement of the follower.

If the angular speed ω of the cam ($\omega = \dot{\theta}$) is constant, the speed v of the follower is:

$$v = \dot{r} = dr/dt = df/dt = \frac{df}{d\theta}\frac{d\theta}{dt} = \omega\frac{df}{d\theta} \tag{16.4.2}$$

and the acceleration a of the follower is:

$$\begin{aligned} a = \ddot{r} &= \frac{d}{dt}\left(\omega\frac{df}{d\theta}\right) = \omega\frac{d}{dt}\left(\frac{df}{d\theta}\right) \\ &= \omega\left[\frac{d}{d\theta}\left(\frac{df}{d\theta}\right)\right]\frac{d\theta}{dt} = \omega^2\frac{d^2f}{d\theta^2} \end{aligned} \tag{16.4.3}$$

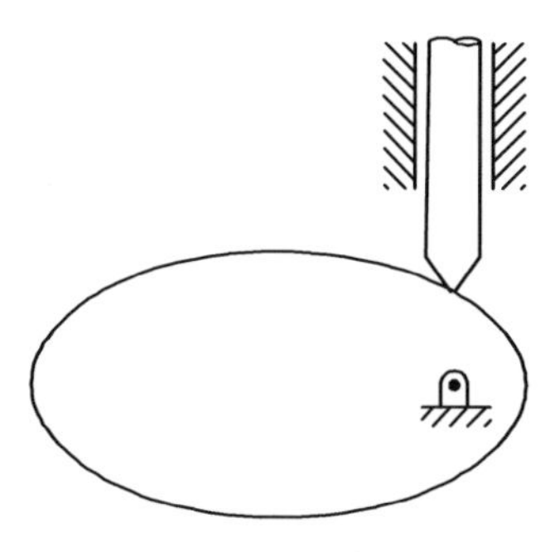

FIGURE 16.4.1
Knife-edge radial cam–follower pair.

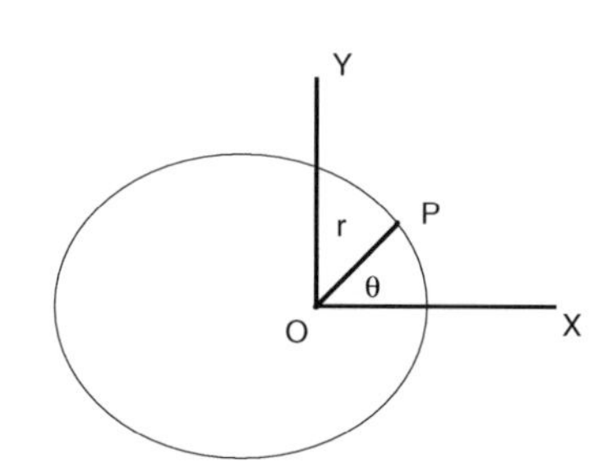

FIGURE 16.4.2
Polar coordinates of a cam profile.

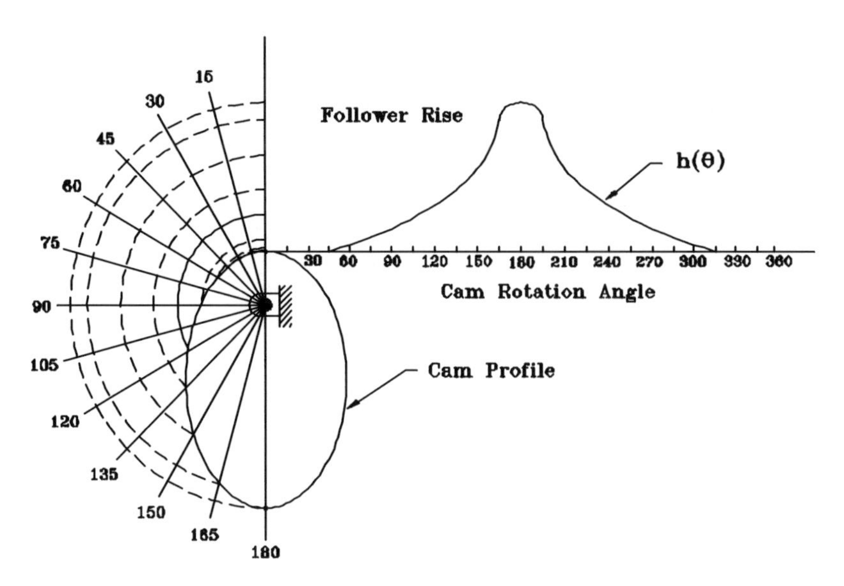

FIGURE 16.4.3
Graphical construction of follower rise profile.

Let r_{min} be the minimum value of r in Eq. (16.4.1). Let h be the difference between r and r_{min} at any cam position: That is, let

$$h = r - r_{\min} = f(\theta) - r_{\min} = h(\theta) \tag{16.4.4}$$

Then, $h(\theta)$ represents the rise (height) of the follower above its lowest position.

We can obtain a graphical representation of $h(\theta)$ by dividing the cam into equiangular segments and then by constructing circular arcs from the segment line/cam profile intersection to the vertical radial line. Then, horizontal lines from these intersections determine the $h(\theta)$ profile, as illustrated in Figure 16.4.3.

By inspecting the construction of Figure 16.4.3, we see that given the cam profile it is relatively easy — at least, in principle — to obtain the follower rise function $h(\theta)$. The inverse problem, however, is not as simple; that is, given the follower rise function $h(\theta)$, determine the driving cam profile. This is a problem commonly facing cam designers which we discuss in the following section.

16.5 Graphical Constructions: Cam Profiles

If we are given the follower rise function $f(\theta)$, we can develop the cam profile by reversing the graphical construction of the foregoing section. To illustrate this, consider a general follower rise function as in Figure 16.5.1.

Next, we can construct the cam profile by first selecting a minimum cam radius r_{min}. Then, the cam radius at any angular position is defined by Figure 16.5.1. Graphically, this may be developed by constructing a circle with radius r_{min} and then by scaling points on radial lines according to the follower rise function. To illustrate this, let this minimum radius circle (called the *base circle*) and the radial lines be constructed as in Figure 16.5.2. Next, let the follower rise be plotted on radial lines according to the angular position as

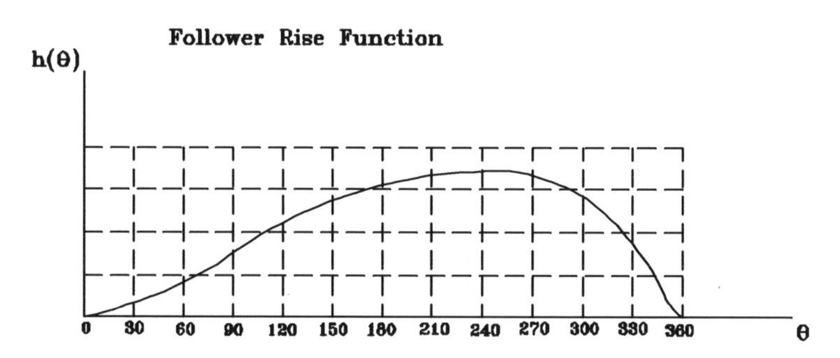

FIGURE 16.5.1
A typical follower rise function.

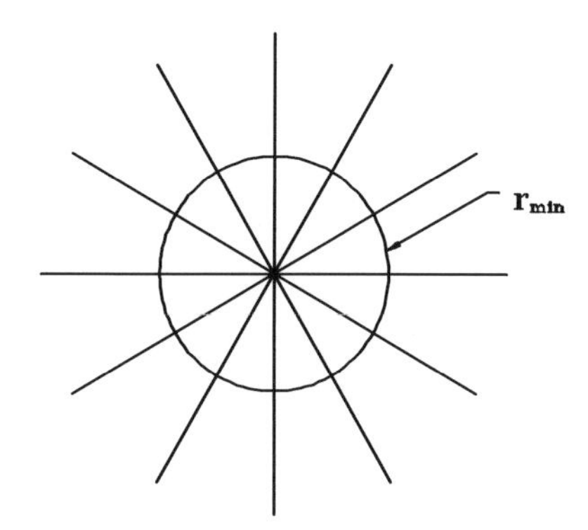

FIGURE 16.5.2
Base circle and radial lines.

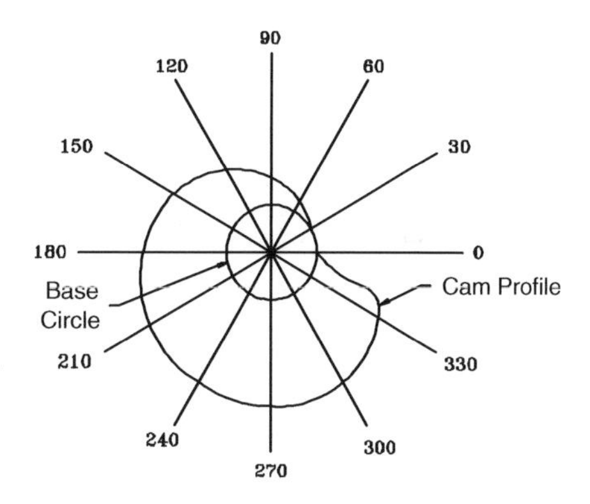

FIGURE 16.5.3
Cam profile for follower rise function of Figure 16.5.1.

in Figure 16.5.3 where, for illustrative purposes, we have used the data of Figure 16.5.1. Then, by fitting a curve through the plotted points on the radial line we obtain the cam profile as in Figure 16.5.3. The precision of this process can be improved by increasing the number of radial lines.

Although this procedure seems to be simple enough, hidden difficulties may emerge in actual construction, including the effects of differing cam–follower geometries and the more serious problem of obtaining a cam profile that may produce undesirable accelerations of the follower. We discuss these problems and their solutions in the following sections.

16.6 Graphical Construction: Effects of Cam–Follower Design

In reviewing the procedures of the two foregoing sections, we see and recall from Figure 16.4.1 that our discussions assumed a simple knife-edge radial follower. If the follower design is modified, however, the graphical procedures will generally need to be modified as well.

To illustrate these changes, consider again an elliptical cam profile for which the follower is no longer radial but instead is offset from the cam rotation axis. Also, let the contact surface between the follower and cam be an extended flat surface as in Figure 16.6.1.

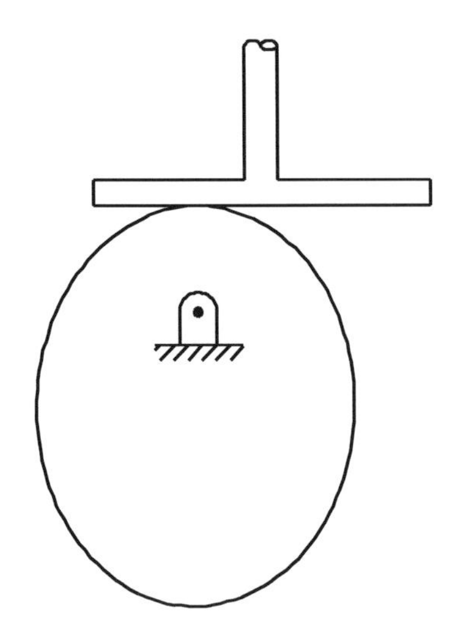

FIGURE 16.6.1
An offset flat-surface follower.

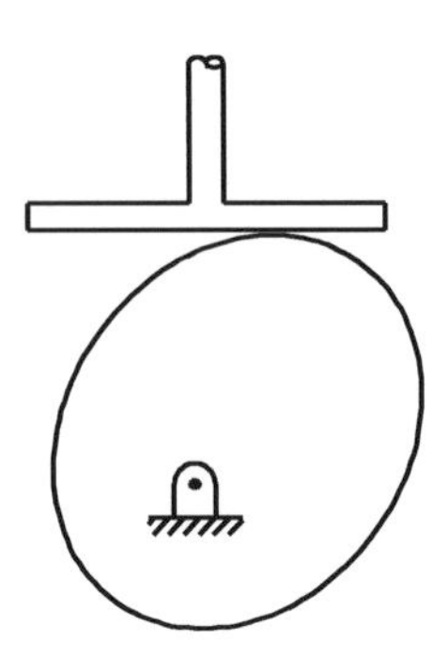

FIGURE 16.6.2
Contact between cam and follower at a typical cam orientation.

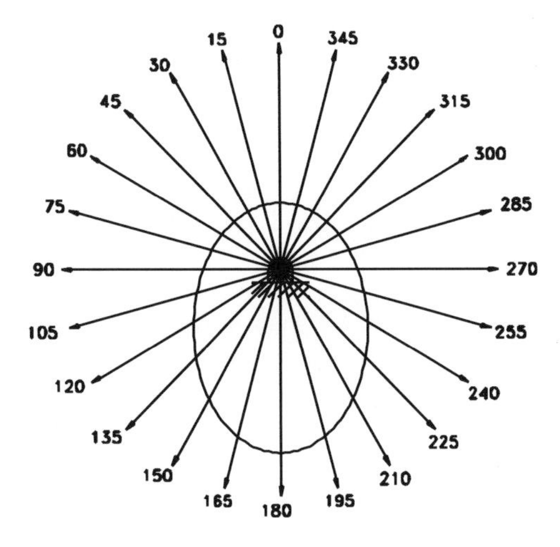

FIGURE 16.6.3
Cam profile with superposed radial lines.

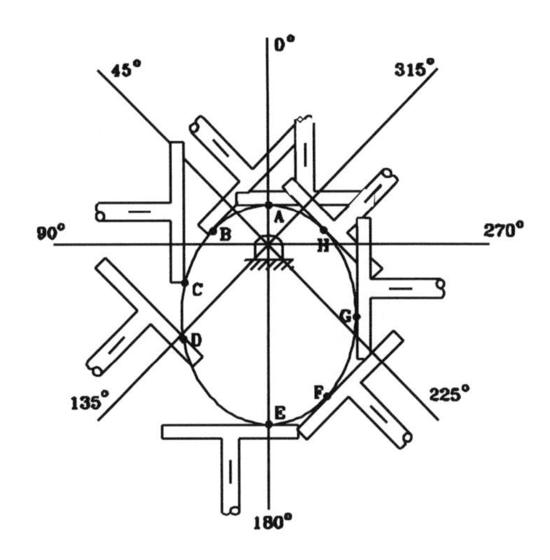

FIGURE 16.6.4
Representation of the follower at various positions about the cam profile.

If, as before, we attempt to determine the follower rise function (assuming constant cam angular speed) by considering equiangular radial lines, we see that, due to the offset, the follower axis will not coincide with the radial lines. Also, due to the flat follower surface, we see that the base of the follower at the axis of the follower will not, in general, be in contact with the cam profile (see Figure 16.6.2).

A question that arises then is how do these geometrical changes affect the graphical construction of the follower rise profile as in Figure 16.4.3? To answer this question, consider again constructing radial lines on the cams as in Figure 16.4.3 and as shown again here in Figure 16.6.3.

Next, for each radial line let a profile of the follower be superposed on the figure such that the axis of the follower is parallel to the radial line but offset from it by the offset of the follower axis and the cam axis. Let the follower profile be placed so that the flat edge

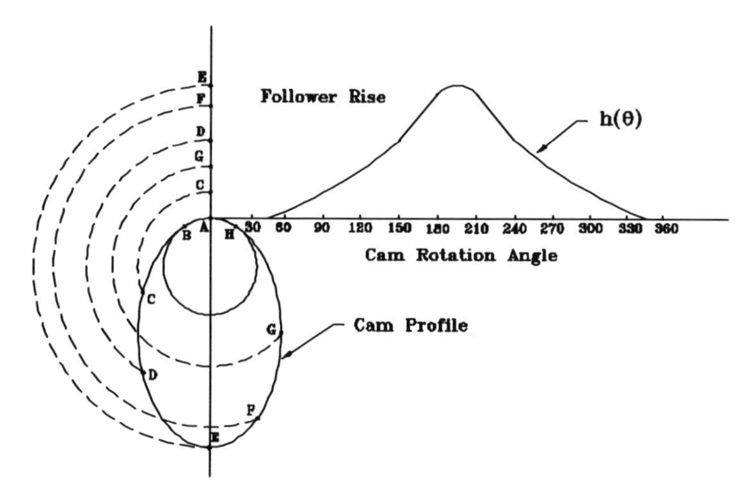

FIGURE 16.6.5
Graphical construction of follower rise profile.

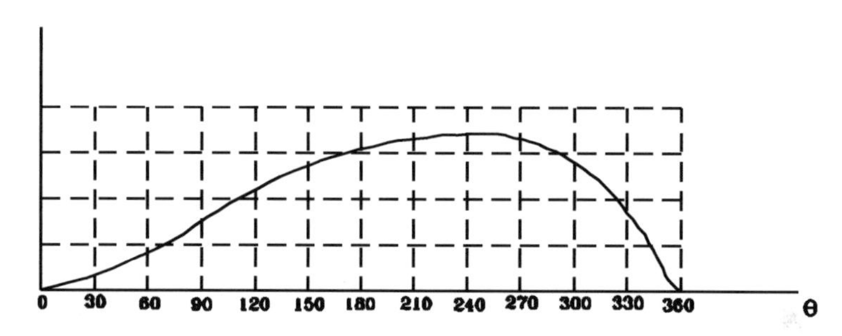

FIGURE 16.6.6
A typical follower rise function.

is tangent to the cam profile. A representation of this procedure at 45°-angle increments is shown in Figure 16.6.4.

Finally, let the points of tangency between the flat edge of the follower and the cam profile be used to develop the follower rise function using the procedure of Figure 16.4.3 and as demonstrated in Figure 16.6.5.

The inverse problem of determining the cam profile when given the follower rise function is a bit more difficult, but it may be solved by following similar procedures to those of Figures 16.6.3, 16.6.4, and 16.6.5. (As noted earlier, this problem is of greater interest to designers, whereas the problem of determining the follower rise function is of greater interest to analysts.)

To illustrate the procedure, suppose we are given a follower rise function as in Figure 16.5.1 and as shown again in Figure 16.6.6. To accommodate such a function with an offset flat surface follower as in Figure 16.6.1, we need to slightly modify the procedure of Section 16.5 in Figures 16.5.2 and 16.5.3. Specifically, instead of constructing radial lines through the cam axis as in Figure 16.5.2, we need to account for the offset of the follower axis. This means that, instead of radial lines through the cam axis, we need to construct tangential lines to a circle centered on the cam axis with radius equal to the offset and thus tangent to the follower axis.

To see this consider again the representation of the offset flat surface follower of Figure 16.6.1 and as shown in Figure 16.6.7. Let the offset between the follower and cam

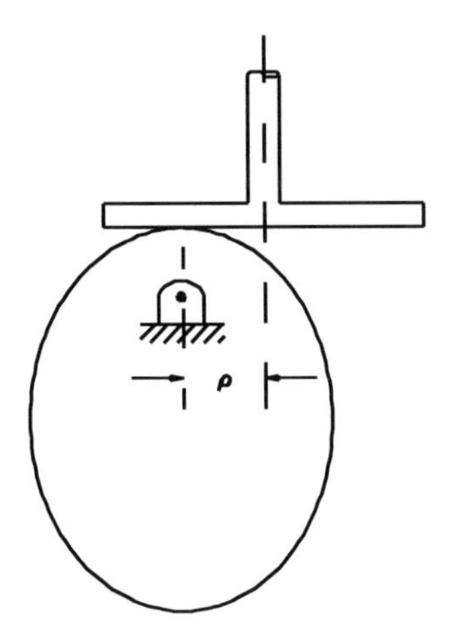

FIGURE 16.6.7
Offset flat surface follower of Figure 16.6.1.

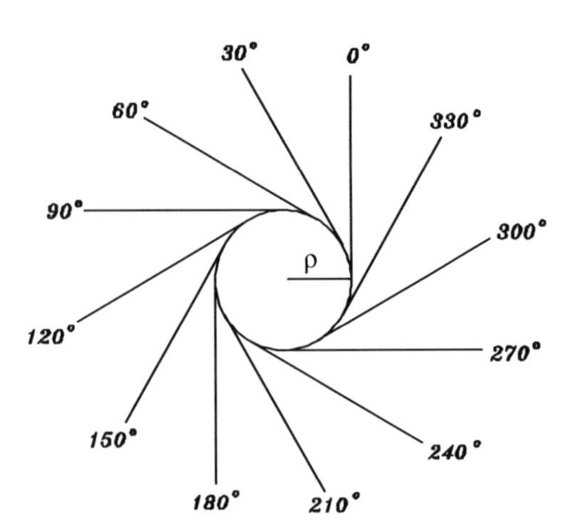

FIGURE 16.6.8
Offset axes circle and tangent line.

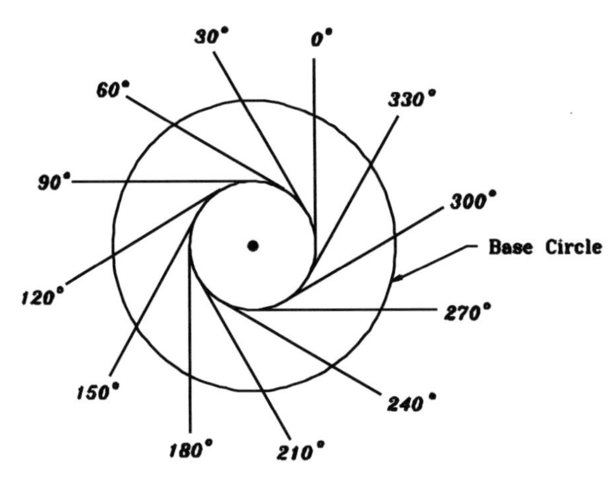

FIGURE 16.6.9
Base circle superposed over offset axis circle.

rotation axis be ρ as shown. Next, construct a circle with radius ρ and with tangent lines representing the follower axis and placed at equal interval angles around the circle as shown in Figure 16.6.8.

The third step is to select a base circle radius, as before, and to superimpose it over the offset axes circle (coincident centers) as in Figure 16.6.9. Fourth, using the follower rise function of Figure 16.6.6, we can use the curve ordinates of the various cam rotation angles as the measure of the follower displacement above the base circle, along the corresponding radial lines of Figure 16.6.9. Figure 16.6.10 shows the location of these displacement points.

Finally, by sketching the flat surface of the follower placed at these displacement points we form an *envelope* [16.2] of the cam profile. That is, the flat follower surfaces are tangent to the cam profile. Then, by sketching the curve tangent to these surfaces, we have the cam profile as shown in Figure 16.6.11.

Observe in this process that accuracy is greatly improved by increasing the number of angle intervals. Observe further that, while this process is, at least in principle, the same as that of the foregoing section (see Figure 16.5.3), the offset of the follower axis and the flat surface of the follower require additional graphical construction details. (One effect

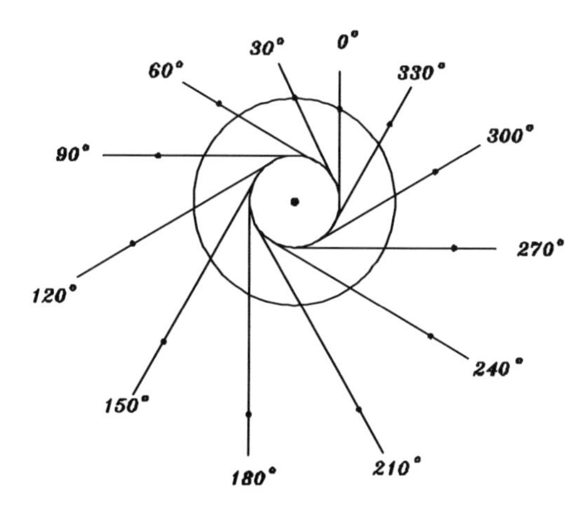

FIGURE 16.6.10
Follower displacement points relative to the base circle.

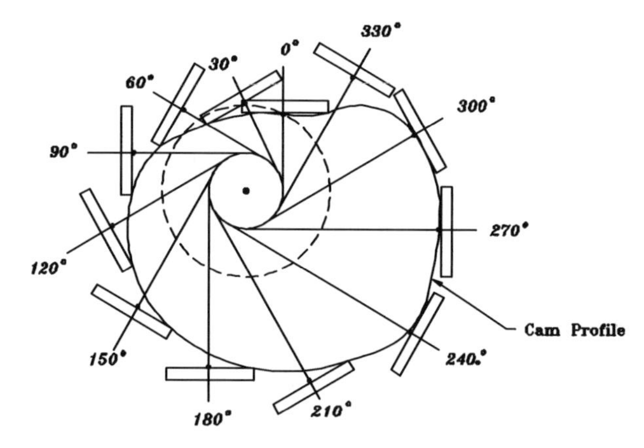

FIGURE 16.6.11
Envelope of follower flat defining the cam profile.

of this is that the cam profile can fall inside the base circle as seen in Figure 16.6.11.) Finally, observe that, although the cam profiles of Figures 16.5.3 and 16.6.11 are not identical, they are similar.

16.7 Comments on Graphical Construction of Cam Profiles

Although the foregoing constructions are conceptually relatively simple, difficulties are encountered in implementing them. First, because they are graphical constructions, their accuracy is somewhat limited — although emerging computer graphics techniques will undoubtedly provide a means for improving the accuracy. Next, analyses assuming point contact between the cam and follower (as in Sections 16.4 and 16.5) may produce cam profiles that are either impractical or incompatible with roller-tipped followers. For example, for an arbitrary follower rise function, it is possible to obtain cam profiles with cusps or concavities as in Figure 16.7.1. Finally, if the cam is rotating rapidly (as is often the

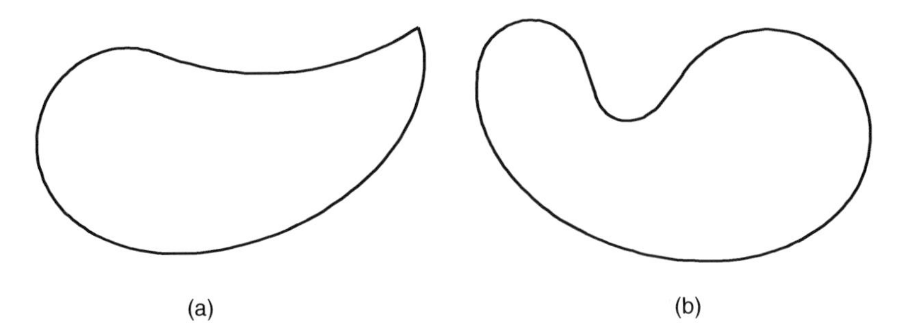

FIGURE 16.7.1
Incompatible and impractical cam profiles for a roller follower.

case with modern mechanical systems), the cam profile may create very large follower accelerations. These accelerations in turn may lead to large and potentially destructive inertia forces. In the next section, we consider analytical constructions intended to overcome these difficulties.

16.8 Analytical Construction of Cam Profiles

As with graphical cam profile construction, analytical profile construction has the same objective; that is, given the follower motion, determine the cam profile that will produce that motion. Because the follower motion may often be represented in terms of elementary functions, it follows that the cam profile may also often be represented in terms of elementary functions. When this occurs, the analytical method has distinct advantages over the graphical method. Conversely, if the follower motion cannot be represented in terms of elementary functions, the analytical approach can become unwieldy and impractical, necessitating the use of numerical procedures and approximations. Also, as we will see, the design of cam profiles with elementary functions may create undesirable inertia forces with high-speed systems.

With these possible deficiencies in mind, we outline the analytical construction in the following text. As with the graphical construction, we develop the fundamentals using the simple planar cam–follower pair, with the follower axis intersecting the rotation axis of the cam as in Figure 16.8.1. Also, we will assume the cam rotates with a constant angular speed ω_0.

In many applications, the follower is to remain stationary for a fraction of the cam rotation period. After passing through this stationary period, the follower is then generally required to rise away from this stationary position and then ultimately return back to the stationary position. Figure 16.8.2 shows such a typical follower motion profile. The stationary positions are called *periods of dwell* for the follower and, hence, also for the cam.

The principle of analytical cam profile construction is remarkably simple. The follower rise function $h(\theta)$ (as shown in Figure 16.8.2) is directly related to the cam profile function. To see this, note that, because the cam is rotating at a constant angular speed ω_0, the angular displacement θ of the cam is a linear function of time t. That is,

$$\text{If} \quad \omega_0 = d\theta/dt, \quad \text{then} \quad \theta = \omega_0 t + \theta_0 \tag{16.8.1}$$

where the initial angle θ_0 may conveniently be taken as zero without loss of generality.

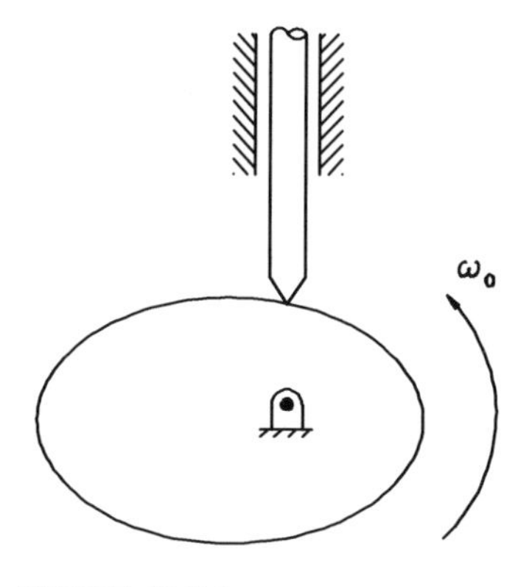

FIGURE 16.8.1
Knife-edge radial cam–follower pair.

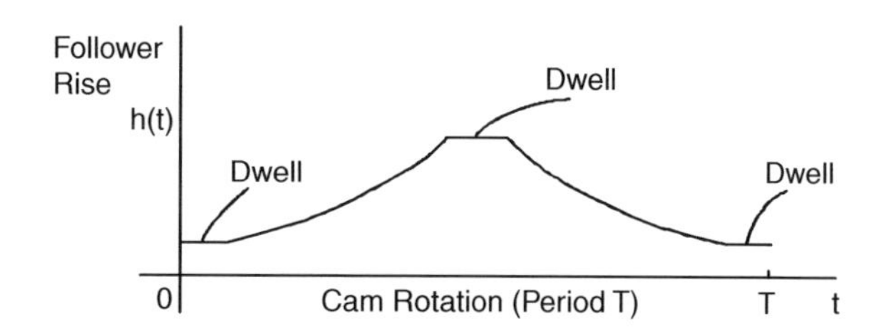

FIGURE 16.8.2
Typical follower rise function.

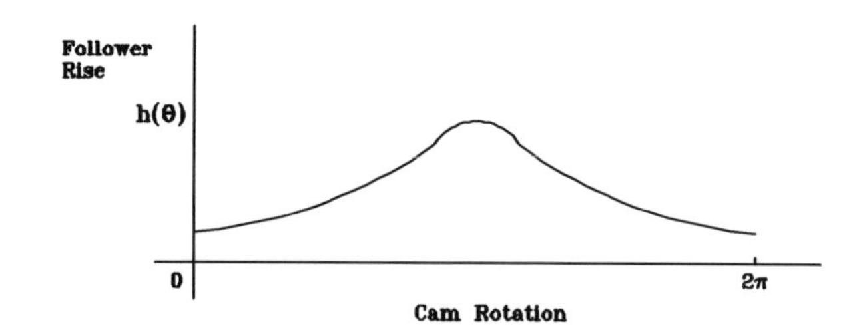

FIGURE 16.8.3
Follower displacement as a function of the cam rotation angle θ.

Next, because the follower rise h is often expressed as a function of time t, say $h(t)$, we can solve Eq. (16.8.1) for t ($t = \theta/\omega_0$) and then express the follower displacement as a function of the rotation angle θ as in Figure 16.8.3. Then, in view of the graphical constructions of the foregoing sections and in view of Figure 16.8.1, we see that the follower rise function is simply the deviation of the cam profile from its base circle. That is, if the cam profile is expressed in polar coordinates as $r(\theta)$, then $r(\theta)$ and $h(\theta)$ are related as:

$$r(\theta) = r_0 + h(\theta) \tag{16.8.2}$$

where r_0 is the base circle radius.

16.9 Dwell and Linear Rise of the Follower

To illustrate the application of Eq. (16.8.2), suppose the follower displacement is in dwell. That is, let the rise function $h\theta$ be zero as in Figure 16.9.1. Then, the cam profile $r(\theta)$ has the simple form:

$$r(\theta) = r_0 \quad \text{(a constant)} \tag{16.9.1}$$

That is, the cam profile is circular, with radius equal to the base circle, and no follower motion is generated.

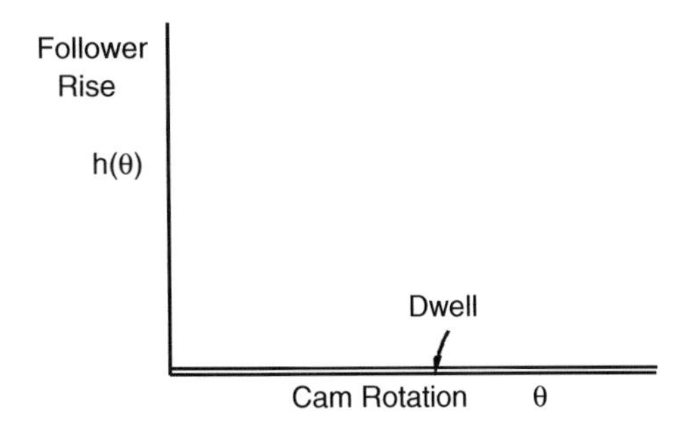

FIGURE 16.9.1
A zero or dwell follower rise function.

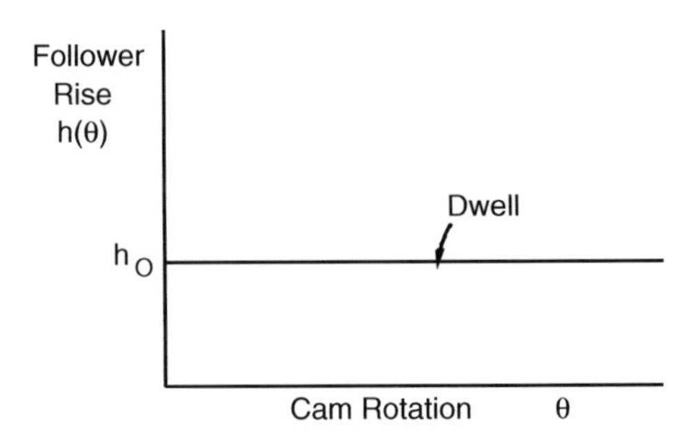

FIGURE 16.9.2
A nonzero dwell follower rise function.

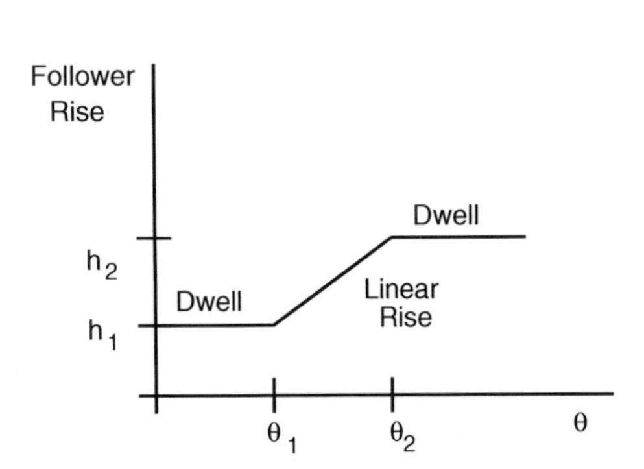

FIGURE 16.9.3
A linear rise in the follower displacement.

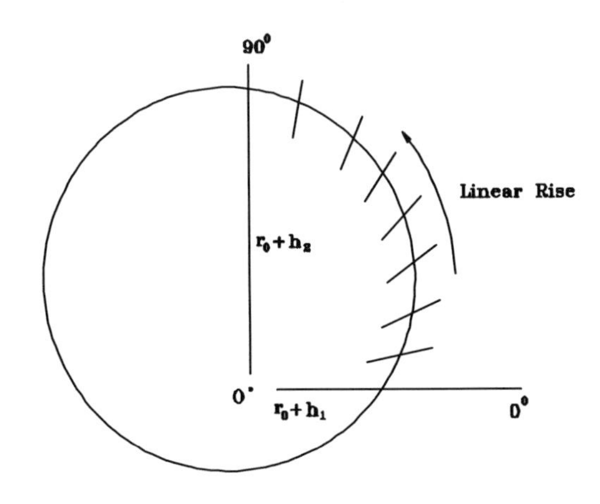

FIGURE 16.9.4
Linear rise over quarter cam angle.

Next, suppose the follower rise function is not zero but instead is still a constant (say, h_0) as in Figure 16.9.2. In this case, the follower is still in dwell, but after a displacement (or rise) of h_0. From Eq. (16.8.2) the cam profile is then:

$$r(\theta) = r_0 + h_0 \quad \text{(a constant)} \tag{16.9.2}$$

Here, again, the cam profile is circular with a radius h_0 greater than the base circle radius.

Next, suppose the rise of the follower increases linearly with time (or cam rotation) as in Figure 16.9.3. If the follower rise is to begin at $\theta = \theta_1$, then the slope of the rise is:

$$\text{slope} = \frac{h_2 - h_1}{\theta_2 - \theta_1} \stackrel{D}{=} m \tag{16.9.3}$$

The follower displacement may then be expressed in the form:

$$h(\theta) = \begin{cases} h_1 & \theta \le \theta_1 \\ m(\theta - \theta_1) & \theta_1 \le \theta \le \theta_2 \\ h_2 & \theta \ge \theta_2 \end{cases} \tag{16.9.4}$$

Hence, from Eq. (16.8.2) the cam profile radius $r(\theta)$ is:

$$r(\theta) = r_0 + h(\theta) \tag{16.9.5}$$

Using Eq. (16.9.4) produces a profile as in Figure 16.9.4, where a linearly increasing radius is shown for 90° of the cam angle.

16.10 Use of Singularity Functions

The foregoing analysis may be facilitated by the use of *singularity* or *delta* functions. To develop this idea, let us first introduce the *step function* $\delta_1(x)$ defined as (see References 16.3 to 16.6):

$$\delta_1(x) \stackrel{D}{=} \begin{cases} 0 & x < 0 \\ 1 & x > 0 \end{cases} \tag{16.10.1}$$

where $\delta_1(0)$ is left undefined (or, on occasion, defined as 1/2). Then, by shifting the independent variable, we have:

$$\delta_1(x-a) = \begin{cases} 0 & x < a \\ 1 & x > a \end{cases} \tag{16.10.2}$$

We can immediately apply Eq. (16.10.2) with the follower rise function $h(\theta)$. For example, consider the linear rise function of Eq. (16.9.4):

$$h(\theta) = \begin{cases} h_1 & \theta \le \theta_1 \\ m(\theta - \theta_1) & \theta_1 \le \theta \le \theta_2 \\ h_2 & \theta \ge \theta_2 \end{cases} \tag{16.10.3}$$

Then, by using Eq. (16.10.2), the follower rise function may be written as:

$$\begin{aligned} h(\theta) &= h_1 + m(\theta - \theta_1)\left[\delta_1(\theta - \theta_1) - \delta_1(\theta - \theta_2)\right] \\ &= h_1 + (h_2 - h_1)\left(\frac{\theta - \theta_1}{\theta_2 - \theta_1}\right)\left[\delta_1(\theta - \theta_1) - \delta_1(\theta - \theta_2)\right] \end{aligned} \tag{16.10.4}$$

By integrating, we may generalize the definition of Eq. (16.10.1) and thus introduce higher order step-type functions (or *singularity* [16.6] functions) as:

$$\delta_2(x) \stackrel{D}{=} x\delta_1(x) = \begin{cases} x & x > 0 \\ 0 & x < 0 \end{cases} \tag{16.10.5}$$

where $\delta_2(0)$ is defined as 0. Similarly, we have:

$$\delta_3(x) \stackrel{D}{=} (x^2/2)\delta_1(x) = \begin{cases} x^2/2 & x>0 \\ 0 & x<0 \end{cases}, \quad \delta_3(0) \stackrel{D}{=} 0 \tag{16.10.6}$$

$$\delta_4(x) \stackrel{D}{=} (x^3/6)\delta_1(x) = \begin{cases} x^3/6 & x>0 \\ 0 & x<0 \end{cases}, \quad \delta_4(0) \stackrel{D}{=} 0 \tag{16.10.7}$$

$$\delta_{n+1}(x) \stackrel{D}{=} (x^n/n!)\delta_1(x) = \begin{cases} x^n/n! & x>0 \\ 0 & x<0 \end{cases}, \quad \delta_{n+1}(0) \stackrel{D}{=} 0 \tag{16.10.8}$$

Observe that the definitions:

$$\delta_2(0) = \delta_3(0) = \ldots = \delta_n(0) = \delta_{n+1}(0) = 0 \tag{16.10.9}$$

allow the functions to be continuous at $x = 0$.

Graphically, these functions may be represented as in Figure 16.10.1. Observe that except for $\delta_1(x)$ all the functions are continuous at $x = 0$.

Observe further that just as $\delta_{n+1}(x)$ represents the integration (or *antiderivative*) of $\delta_n(x)$, it follows that $\delta_n(x)$ is the derivative of $\delta_{n+1}(x)$. This raises the question, however, as to the derivative of $\delta_1(x)$. From the definition of Eq. (16.10.1), we see that, if $x \neq 0$, then $d\delta_1(x)/dx$ is zero. At $x = 0$, however, the derivative is undefined, representing an infinite change in the function. Thus, we have:

$$d\delta_1(x)/dx \stackrel{D}{=} \delta_0(x) = \begin{cases} 0 & x \neq 0 \\ \infty & x = 0 \end{cases} \tag{16.10.10}$$

$\delta_0(x)$ is often referred to as *Dirac's delta function* — analogous to Kronecker's delta function of Section 2.6. The analogy is seen through the "sifting" or "substitution" property of the functions. Recall from Eq. (2.6.7) that we have:

$$\delta_{ij} v_j = v_i \tag{16.10.11}$$

Similarly, it is seen (see References 16.1 to 16.5) that:

$$\int_{-b}^{b} f(x)\delta_0(x)dx = f(0) \quad (b>0) \tag{16.10.12}$$

or

$$\int_{-b}^{b} f(x)\delta_0(x-a)dx = f(a) \quad (b>a) \tag{16.10.13}$$

Interpreting an integral as a sum, we see that Eqs. (16.10.11) and (16.10.11) are simply discrete and continuous forms of the same procedure with analogous results.

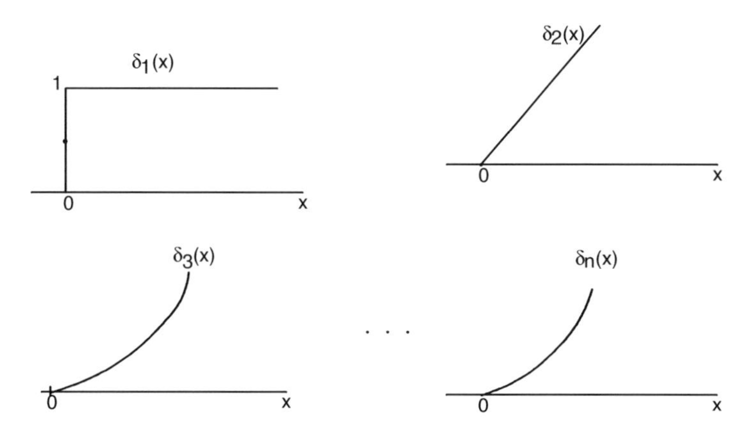

FIGURE 16.10.1
Graphical form of the singularity function.

The functions $\delta_2(x)$, $\delta_3(x)$, $\delta_4(x)$,..., $\delta_{n+1}(x)$ of Eqs. (16.10.5) through (16.10.8), as well as $\delta_0(x)$ of Eq. (16.10.10), may also be represented with the independent variable shifted as in Eq. (16.10.2). That is,

$$\delta_2(x-a)=\begin{cases} x-a & x \ge a \\ 0 & x \le a \end{cases}, \quad \delta_3(x-a)=\begin{cases} (x-a)^2/2 & x \ge a \\ 0 & x \le a \end{cases},$$
$$\delta_4(x-a)=\begin{cases} (x-a)^3/6 & x \ge a \\ 0 & x \ge a \end{cases}, \ldots, \delta_{n+1}(x-a)=\begin{cases} (x-a)^n/n! & x \ge a \\ 0 & x \le a \end{cases} \tag{16.10.14}$$

and

$$\delta_0(x-a)=\begin{cases} 0 & x \ne a \\ \infty & x = a \end{cases} \tag{16.10.15}$$

From Eq. (16.10.15), we see that $\delta_0(x - a)$ may be represented graphically as in Figure 16.10.2.

Symbolically, or by appealing to generalized function theory [16.5], we may develop derivatives of Dirac's delta function as:

$$d\delta_0(x)/dx=\delta_{-1}(x), \quad d\delta_{-1}(x)/dx=\delta_{-2}(x)$$
$$\ldots \; d\delta_{-n}(x)/dx=\delta_{-(n+1)}(x) \tag{16.10.16}$$

where $\delta_{-1}(x)$ has the form of an *impulsive doublet* function as in Figure 16.10.3. (Graphical representations of the higher order derivatives are not as readily obtained, thus the graphical representations of the higher order derivatives are not as helpful.)

There has been extensive application of these influence or interval functions in structural mechanics — particularly in beam theory (see, for example, Reference 16.6). For cam profile analysis, they may be used to conveniently define the profile, as in Eq. (16.10.4). The principal application of the functions in cam analysis, however, is in the differentiation of

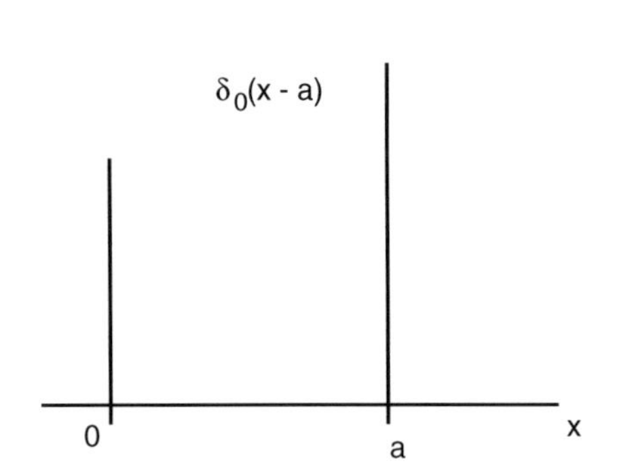

FIGURE 16.10.2
Representation of Dirac's delta function.

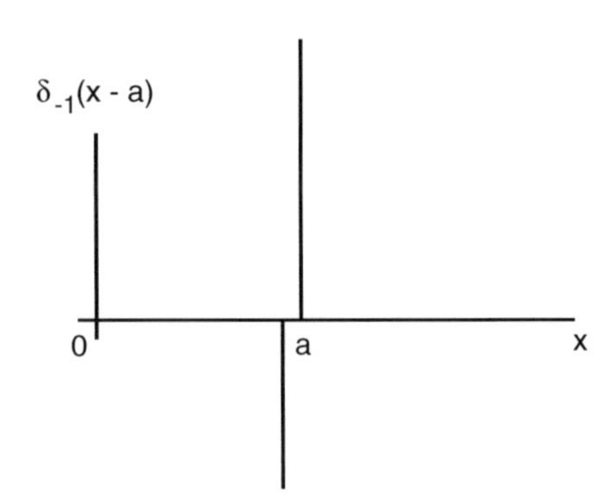

FIGURE 16.10.3
Representation of the derivative of Dirac's delta function.

the follower rise function, providing information about the follower velocity, acceleration, and rate of change of acceleration (*jerk*).

To demonstrate this, consider again the piecewise-linear follower rise function of Figure 16.9.3 and as expressed in Eq. (16.10.4):

$$h(\theta) = h_1 + m(\theta - \theta_1)\left[\delta_1(\theta - \theta_1) - \delta_1(\theta - \theta_2)\right] \tag{16.10.17}$$

where as before m is $(h_2 - h_1)/(\theta_2 - \theta_1)$ (see Figure 16.9.3). The velocity v and the acceleration a of the follower are then:

$$v = \frac{dh}{dt} = \frac{dh}{d\theta}\frac{d\theta}{dt} = \omega\frac{dh}{d\theta} \tag{16.10.18}$$

and

$$a = \frac{d^2h}{dt^2} = \frac{d}{dt}\left(\frac{dh}{dt}\right) = \frac{d}{d\theta}\left(\frac{dh}{dt}\right)\frac{d\theta}{dt} = \left[\frac{d}{d\theta}\left(\omega\frac{dh}{d\theta}\right)\right]\omega = \omega^2\frac{d^2h}{d\theta^2} \tag{16.10.19}$$

where, as before, we have assumed the cam rotation $d\theta/dt$ (= ω) to be constant. Then, by substituting from Eq. (16.10.17) into (16.10.18) and (16.10.19), we find v and a to be:

$$v = \omega m\left[\delta_1(\theta - \theta_1) - \delta_1(\theta - \theta_2)\right] + \omega m(\theta - \theta_1)\left[\delta_0(\theta - \theta_1) - \delta_0(\theta - \theta_2)\right] \tag{16.10.20}$$

and

$$a = 2\omega^2 m\left[\delta_0(\theta - \theta_1) - \delta_0(\theta - \theta_2)\right] + \omega(\theta - \theta_1)\left[\delta_{-1}(\theta - \theta_1) - \delta_{-1}(\theta - \theta_2)\right] \tag{16.10.21}$$

We can readily see that the final terms in both Eqs. (16.10.20) and (16.10.21) are zero. To see this, consider first the expression $x\delta_0(x)$. If x is not zero, then $\delta_0(x)$ and thus $x\delta_0(x)$ are

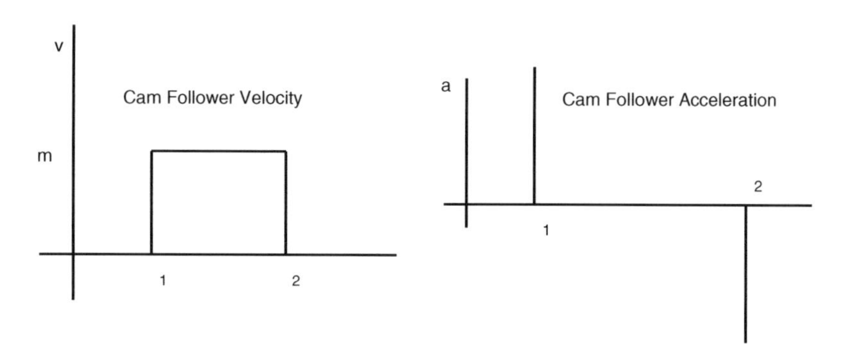

FIGURE 16.10.4
Velocity and acceleration of cam–follower due to linear rise of the follower.

zero. If x is zero, we have an indeterminant form of the type $0 \cdot \infty$. By using l'Hospital's rule of elementary calculus, we then have:

$$\lim_{x\to 0} x\delta_0(x) = \lim_{x\to 0}\frac{\delta_0(x)}{1/x} = \lim_{x\to 0}\frac{\delta_{-1}(x)}{-(1/x)^2} = 0 \tag{16.10.22}$$

Thus, the term $(\theta - \theta_1)\delta_0(\theta - \theta_1)$ in Eq. (16.10.21) is zero. Also, because $\delta_{-1}(x)$ is zero for all x, the last term of Eq. (16.10.21) is zero. Therefore, the velocity and acceleration of the cam–follower become:

$$v = \omega m\left[\delta_1(\theta - \theta_1) - \delta_1(\theta - \theta_2)\right] \tag{16.10.23}$$

and

$$a = 2\omega^2 m\left[\delta_0(\theta - \theta_1) - \delta_0(\theta - \theta_2)\right] \tag{16.10.24}$$

These expressions may be represented graphically as in Figure 16.10.4.

The abrupt change in follower velocity at the beginning and end of the linear follower rise leads to the infinite values of follower acceleration as seen in Figure 16.10.4. These infinite accelerations in turn will theoretically produce infinite forces between the cam and follower. Obviously, this cannot be tolerated in any practical design.

Such problems can be corrected in several ways. One is to adjust the cam profile at the approach and departure of the linear rise so as to remove the abrupt nature of the velocity changes. In addition to this, the followers may be equipped with springs and shock absorbers to keep the follower in contact with the cam and to reduce the forces transmitted. In the following sections we explore ways of "smoothing" the cam profile so as to reduce the accelerations while still accomplishing the desired follower rise and fall.

16.11 Parabolic Rise Function

The principal method for avoiding high follower acceleration is to remove the sharp change or jump in the velocity. To explore this, consider again the linear follower rise

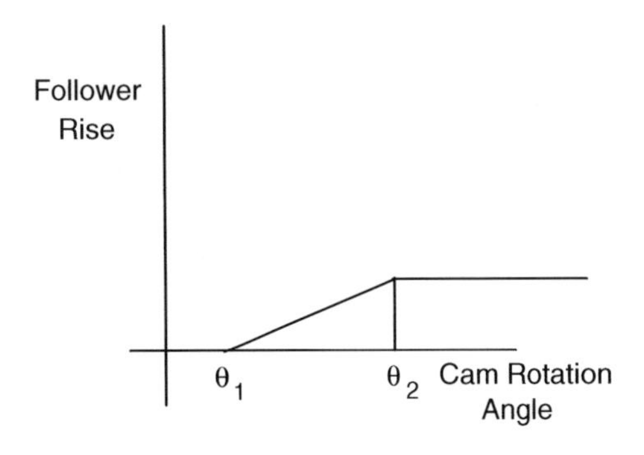

FIGURE 16.11.1
Linear follower rise.

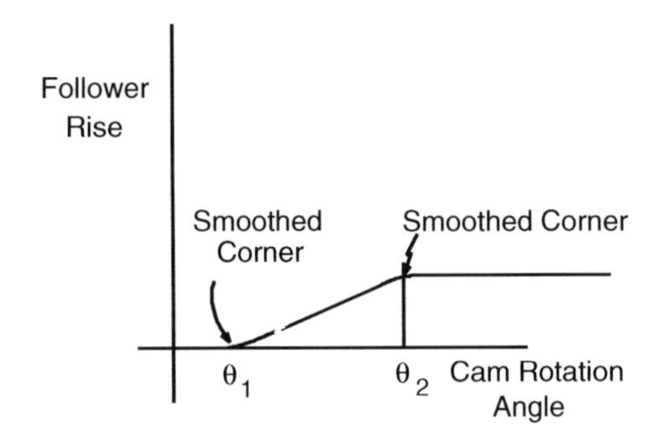

FIGURE 16.11.2
Linear follower rise with smoothes entry (θ_1) and departure (θ_2).

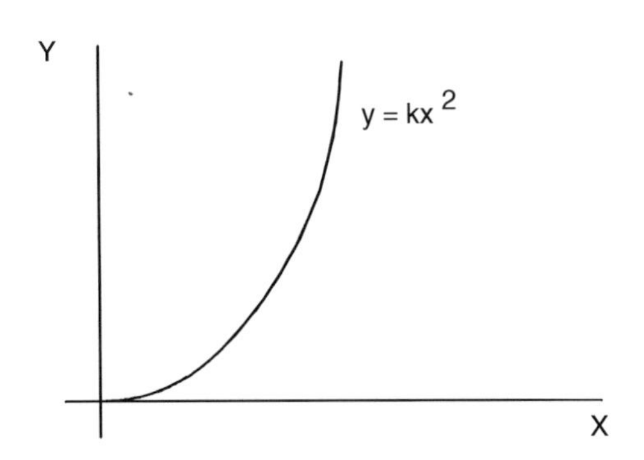

FIGURE 16.11.3
A parabolic curve.

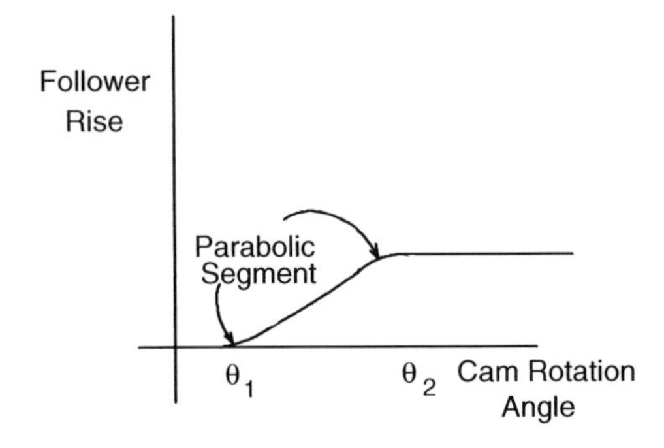

FIGURE 16.11.4
Linear follower rise with parabolic entry and departure.

profile shown in Figure 16.11.1, where the jumps in the velocity occur at the cam angles $\theta = \theta_1$ and $\theta = \theta_2$. To avoid these jumps, we may simply round or smooth the sharp corners at θ_1 and θ_2 as suggested in Figure 16.11.2. To be effective, the smoothing of the corners must be such that the radius of curvature exceeds the radius of the follower wheel.

An analytical approach to this smoothing is to employ either parabolic or sinusoidal arc segments. We will consider parabolic segments in this section and sinusoidal segments in the next section.

Recall that the equation of a parabola may be written in the simple form:

$$y = kx^2 \tag{16.11.1}$$

where k is a constant. Graphically, this equation may be represented as in Figure 16.11.3. The design of the follower rise is then to use arcs of a parabolic curve at the beginning and end of the follower rise function as depicted in Figure 16.11.4.

If the linear segment of the rise function is not critical to the cam–follower applications, a smoother rise can be obtained by eliminating the linear segment with the parabolic entry being made tangent to the parabolic departure. Such a configuration is shown in Figure 16.11.5, where θ is the transition point between the concave and convex parabolic segments.

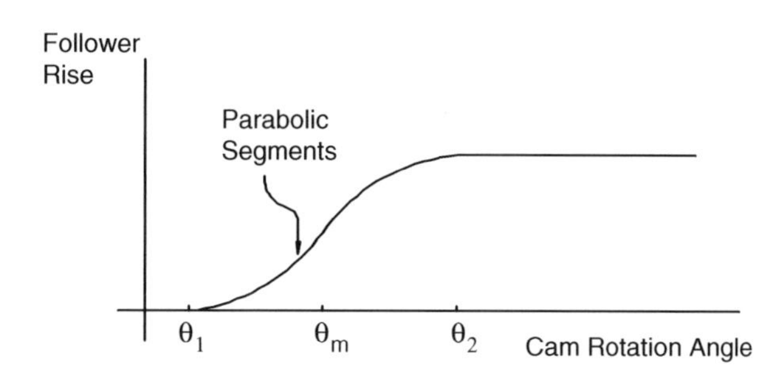

FIGURE 16.11.5
Follower rise with parabolic segments.

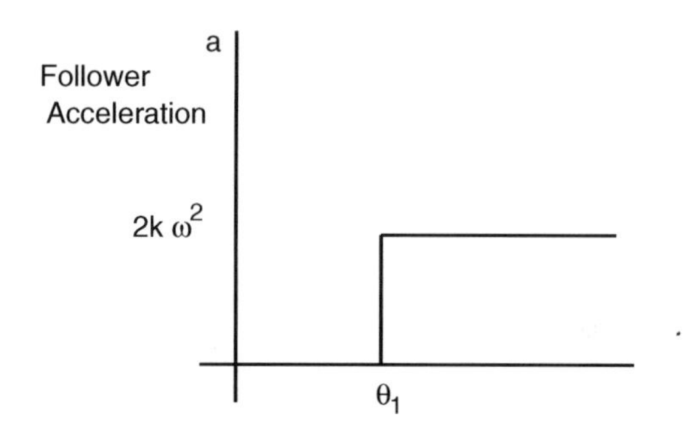

FIGURE 16.11.6
Follower acceleration at transition point to parabolic rise.

A principal advantage of the parabolic profile is that the accelerations at the transition points are finite. To see this, consider that in the vicinity of θ_1 (dwell to rise), the follower rise function may be expressed as:

$$h(\theta) = \begin{cases} 0 & \theta \le \theta_1 \\ k(\theta - \theta_1)^2 & \theta_1 \le \theta \le \theta_m \end{cases} \tag{16.11.2}$$

Alternatively, in view of Eq. (16.10.2), we may express $h(\theta)$ as:

$$h(\theta) = k(\theta - \theta_1)^2 \delta_1(\theta - \theta_1) \quad \theta \le \theta_m \tag{16.11.3}$$

The velocity v and acceleration a of the follower near θ_1 are then

$$v = \frac{dh}{dt} = \frac{dh}{d\theta}\frac{d\theta}{dt} = \omega\frac{dt}{d\theta} = 2k\omega(\theta - \theta_1)\delta_1(\theta - \theta_1) + k\omega(\theta - \theta_1)^2 \delta_0(\theta - \theta_1) \tag{16.11.4}$$

and

$$\begin{aligned} a &= \frac{d^2h}{dt^2} = \frac{d}{d\theta}\left(\omega\frac{dh}{d\theta}\right)\frac{d\theta}{dt} = \omega^2\frac{d^2h}{d\theta^2} \\ &= 2k\omega^2\delta_1(\theta - \theta_1) + 2k\omega^2(\theta - \theta_1)\delta_0(\theta - \theta_1) \\ &\quad + k\omega^2(\theta - \theta_1)^2\delta_{-1}(\theta - \theta_1) \end{aligned} \tag{16.11.5}$$

The acceleration profile in the vicinity of θ_1 then has the form shown in Figure 16.11.6. (Note that $(\theta - \theta_1)\delta_0(\theta - \theta_1)$ and $(\theta - \theta_1)^2\delta_{-1}(\theta - \theta_1)$ are zero at $\theta = \theta_1$.) Similar results are found at the other transition points $\theta = \theta_m$ and $\theta = \theta_2$.

Observe in Figure 16.11.6 that the acceleration is finite in spite of the step jump in its value. This jump, however, produces an infinite value for the acceleration derivative (often called the *jerk*). The jerk (or acceleration derivative) is proportional to the change in inertia forces. An infinite jerk would then (theoretically) produce an infinite change in the inertia force which in turn could produce unwanted vibration and wear in the cam–follower pair. Thus, even though the cam profile is smooth, with no abrupt changes in slope, abrupt accelerations can still occur for the follower.

16.12 Sinusoidal Rise Function

A second approach to smoothing the motion of the follower is through the use of sinusoidal functions. Consider the graph of the cosine function in Figure 16.12.1. To obtain a desired follower rise, we may use the segment of the curve between π and 2π. For example, suppose we want to have a follower rise between elevation h_1 and h_2 for rotation angles between θ_1 and θ_2. We may obtain this by simply inserting the rising segment of the cosine function as shown in Figure 16.12.2.

By comparing Figures 16.12.1 and 16.12.2, we see that to obtain the desired cosine segment fit between (h_1, θ_1) and (h_2, θ_2), the amplitude A must be:

$$A = (h_2 - h_1)/2 \tag{16.12.1}$$

In like manner, the phase and period of the cosine function must be adjusted so that when θ has the values θ_1 and θ_2, the argument ψ of the cosine function has the values π and 2π. Specifically, ψ must have the form:

$$\psi = \pi(\theta + \theta_2 - 2\theta_1)/(\theta_2 - \theta_1) \tag{16.12.2}$$

Then, it is readily seen that the desired cosine function for the rise segment of Figure 16.12.2 may be represented by the function $\phi(\theta)$ given by:

$$\phi = \left(\frac{h_1 + h_2}{2}\right) + \left(\frac{h_2 - h_1}{2}\right)\cos\psi \tag{16.12.3}$$

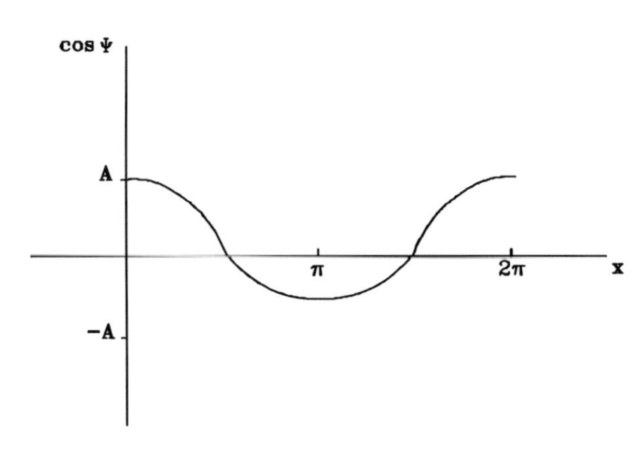

FIGURE 16.12.1
Cosine function.

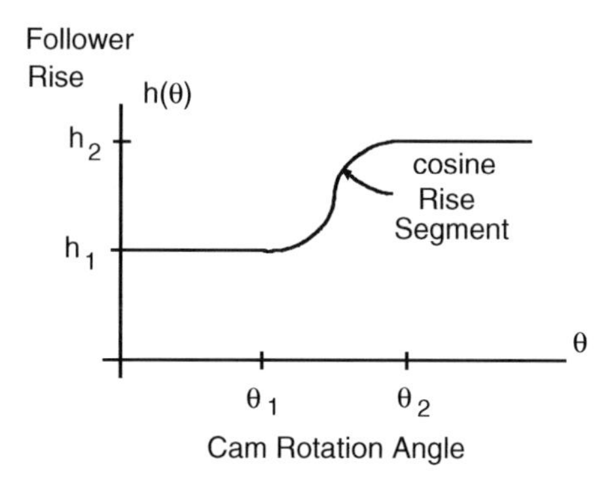

FIGURE 16.12.2
Fitting a cosine segment to obtain a desired follower rise.

Finally, to fit the rise between the dwell segments of Figure 16.12.2, we may employ the step functions of Section 16.10. Specifically, for a dwell at h_1 to θ_1, a rise to h_2 to θ_2, and a dwell at h_2 beyond θ_2, the follower function $h(\theta)$ is:

$$h(\theta) = h_1\left[1 - \delta_1(\theta - \theta_1)\right] + \phi\,\left[\delta_1(\theta - \theta_1) - \delta_1(\theta - \theta_2)\right] + h_2\delta_1(\theta - \theta_2) \tag{16.12.4}$$

A casual inspection of Figure 16.12.2 might suggest that there is a "smooth" transition between the dwell and rise segments. A closer examination, however, shows that we have finite changes in the acceleration of the transition points. This in turn means that we have infinite jerk of these points, producing sudden changes in inertia loading. We can see this by examining the derivatives of $h(\theta)$. Specifically, from Eq. (16.12.4) we have:

$$\begin{aligned}\frac{dh}{dt} = \Big\{ & -h_1\delta_0(\theta - \theta_1) + \frac{d\phi}{d\theta}\left[\delta_1(\theta - \theta_1) - \delta_1(\theta - \theta_2)\right] \\ & + \phi\left[\delta_0(\theta - \theta_1) - \delta_0(\theta - \theta_2)\right] + h_2\delta_0(\theta - \theta_2)\Big\}\omega\end{aligned} \tag{16.12.5}$$

$$\begin{aligned}\frac{d^2h}{dt^2} = \Big\{ & -h_1\delta_{-1}(\theta - \theta_1) + \frac{d^2\phi}{d\theta^2}\left[\delta_1(\theta - \theta_1) - \delta_1(\theta - \theta_2)\right] \\ & + 2\frac{d\phi}{d\theta}\left[\delta_0(\theta - \theta_1) - \delta_0(\theta - \theta_2)\right] + \phi\left[\delta_{-1}(\theta - \theta_1) - \delta_{-1}(\theta - \theta_2)\right] \\ & + h_2\delta_{-1}(\theta - \theta_1)\Big\}\omega^2\end{aligned} \tag{16.12.6}$$

and

$$\begin{aligned}\frac{d^3h}{dt^3} = \Big\{ & -h_1\delta_{-2}(\theta - \theta_1) + \frac{d^3\phi}{d\theta^3}\left[\delta_1(\theta - \theta_1) - \delta_1(\theta - \theta_2)\right] \\ & + 3\frac{d^2\phi}{d\theta^2}\left[\delta_0(\theta - \theta_1) - \delta_0(\theta - \theta_2)\right] + 3\frac{d\phi}{d\theta}\left[\delta_{-1}(\theta - \theta_1) - \delta_{-1}(\theta - \theta_2)\right] \\ & + \phi\left[\delta_{-2}(\theta - \theta_1) - \delta_{-2}(\theta - \theta_2)\right] + h_2\delta_{-2}(\theta - \theta_2)\Big\}\omega^3\end{aligned} \tag{16.12.7}$$

where, as before, ω is $d\theta/dt$ and where, from Eqs. (16.12.2) and (16.12.3), $d\phi/d\theta$, $d^2\phi/d\theta^2$, and $d^3\phi/d\theta^3$ are:

$$\frac{d\phi}{d\theta} = -\left(\frac{h_2 - h_1}{2}\right)\left(\frac{\pi}{\theta_2 - \theta_1}\right)\sin\psi \tag{16.12.8}$$

$$\frac{d^2\phi}{d\theta^2} = -\left(\frac{h_2 - h_1}{2}\right)\left(\frac{\pi}{\theta_2 - \theta_1}\right)^2 \cos\psi \tag{16.12.9}$$

and

$$\frac{d^3\phi}{d\theta^3} = \left(\frac{h_2 - h_1}{2}\right)\left(\frac{\pi}{\theta_2 - \theta_1}\right)^3 \sin\psi \tag{16.12.10}$$

From Eq. (16.12.2) we see that when θ is θ_1, then ψ is π, and when θ is θ_2, ψ is 2π. Hence, from Eq. (16.12.3), we have:

$$\begin{gathered}\text{when } \theta = \theta_1\text{:} \quad \phi = h_1, \quad \frac{d\phi}{d\theta} = 0, \quad \frac{d^2\phi}{d\theta^2} = \left(\frac{h_2 - h_1}{2}\right)\left(\frac{\pi}{\theta_2 - \theta_1}\right)^2, \\ \text{and} \quad \frac{d^3\phi}{d\theta^3} = 0\end{gathered} \tag{16.12.11}$$

and

$$\begin{gathered}\text{when } \theta = \theta_2\text{:} \quad \phi = h_2, \quad \frac{d\phi}{d\theta} = 0, \quad \frac{d^2\phi}{d\theta^2} = -\left(\frac{h_2 - h_1}{2}\right)\left(\frac{\pi}{\theta_2 - \theta_1}\right)^2, \\ \text{and} \quad \frac{d^3\phi}{d\theta^3} = 0\end{gathered} \tag{16.12.12}$$

Then, by inspection of Eqs. (16.12.5), (16.12.6), and (16.12.7), we see that dh/dt is zero at both θ_1 and θ_2, but d^2h/dt^2 has the values:

$$\pm\left(\frac{h_2 - h_1}{2}\right)\left(\frac{\pi}{\theta_2 - \theta_1}\right)^2,$$

at $\theta = \theta_1$ (plus sign) and $\theta = \theta_2$ (minus sign), and d^3h/dt^3 is infinite at both $\theta = \theta_1$ and $\theta = \theta_2$. In the following section, we examine the cycloidal profile which has finite values of d^3h/dt at the transition points.

16.13 Cycloidal Rise Function

In an attempt to avoid sudden changes in a cam–follower pair, many designers have used a cycloidal follower rise profile. The cycloidal profile, while producing slightly higher accelerations than the sinusoidal profile, has the advantage of very gradual changes of the transition points, thus reducing sudden changes in the forces.

A cycloidal curve is commonly developed as the locus of a point on the rim of a rolling vertical disk. Consider, for example, Figure 16.13.1, where we have a circular disk D with radius r rolling in a vertical plane. Let Q be the center of D, let P be a point on the rim, and let θ be the roll angle as shown.

Suppose P is at the origin 0 of an X–Y coordinate system when θ is zero. Then, we can locate P relative to 0 by the position vector **OP** given by:

$$\mathbf{OP} = x\mathbf{n}_x + y\mathbf{n}_y \tag{16.13.1}$$

where (x, y) are the X–Y coordinates of P, and $\mathbf{n}_x$ and $\mathbf{n}_y$ are unit vectors parallel to X and Y as shown. From Figure 16.13.1, we see that **OP** may also be expressed as:

$$\mathbf{OP} = \mathbf{OQ} + \mathbf{QP} \tag{16.13.2}$$

where the position vectors **OQ** and **QP** are:

$$\mathbf{OQ} = r\theta\mathbf{n}_x + r\mathbf{n}_y \tag{16.13.3}$$

and

$$\mathbf{QP} = -r\sin\theta\mathbf{n}_x - r\cos\theta\mathbf{n}_y \tag{16.13.4}$$

Hence, **OP** becomes:

$$\mathbf{OP} = r(\theta - \sin\theta)\mathbf{n}_x + r(1 - \cos\theta)\mathbf{n}_y \tag{16.13.5}$$

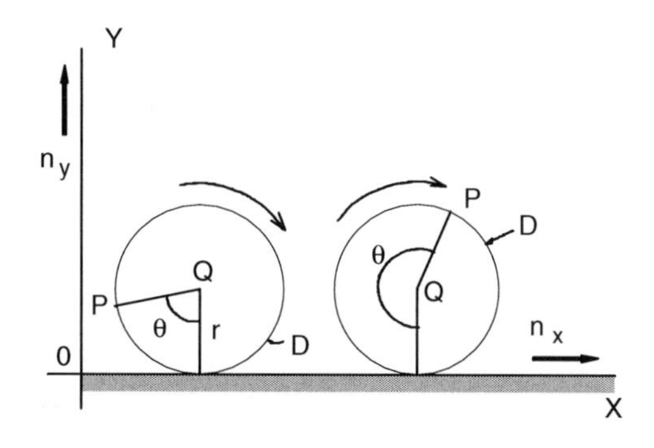

FIGURE 16.13.1
A circular disk with radius r rolling in a vertical X–Y plane.

and then from Eq. (16.13.1) we have:

$$x = r(\theta - \sin\theta) \quad \text{and} \quad y = r(1 - \cos\theta) \tag{16.13.6}$$

Equations (16.13.6) are parametric equations of a cycloid with θ being the parameter. Figure 16.13.2 shows a sketch of a cycloid. Observe that near the origin *O* the horizontal movement of *P* is very gradual. It is this property that makes the cycloid attractive for developing a follower rise function. Specifically, we will use the first of Eqs. (16.13.6) to define our follower rise function $h(\theta)$. That is, let $h(\theta)$ have the general form:

$$h(\theta) = c_1 + c_2\left[c_3(\theta - c_4) - \sin c_3(\theta - c_4)\right] \tag{16.13.7}$$

where c_1, c_2, c_3, and c_4 are constants to be determined so that the curve of the rise function will pass through the points (h_1, θ_1) and (h_2, θ_2) with zero slope at these points as indicated in Figure (16.13.3). It is readily seen that we can obtain the desired curve segment between (h_1, θ_1) and (h_2, θ_2) by assigning c_1, c_2, c_3, and c_4 with the values:

$$\begin{aligned} &c_1 = h_1, \quad c_2 = (h_2 - h_1)/2\pi \\ &c_3 = 2\pi/(\theta_2 - \theta_1), \quad c_4 = \theta_1 \end{aligned} \tag{16.13.8}$$

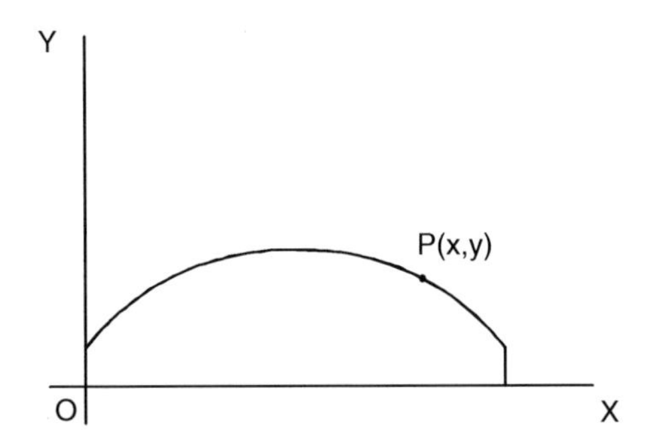

FIGURE 16.13.2
Sketch of a cycloid.

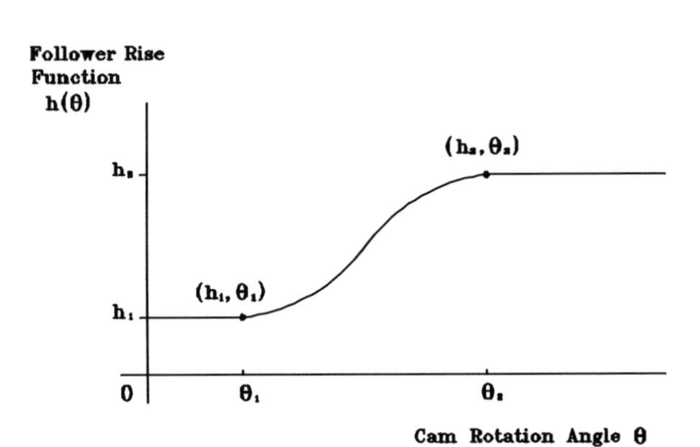

FIGURE 16.13.3
Sketch of cycloidal follower rise function.

Hence, $h(\theta)$becomes:

$$h(\theta) = h_1 + \left(\frac{h_2 - h_1}{2\pi}\right)\left[\left(\frac{2\pi}{\theta_2 - \theta_1}\right)(\theta - \theta_1) - \sin\left(\frac{2\pi}{\theta_2 - \theta_1}\right)(\theta - \theta_1)\right] \tag{16.13.9}$$

or

$$h(\theta) = h_1 + \left(\frac{h_2 - h_1}{\theta_2 - \theta_1}\right)(\theta - \theta_1) - \left(\frac{h_2 - h_1}{2\pi}\right)\sin\left[2\pi\left(\frac{\theta - \theta_1}{\theta_2 - \theta_1}\right)\right] \tag{16.13.10}$$

Then, by differentiation, we obtain the follower velocity as:

$$v = \frac{dh}{dt} = \frac{dh}{d\theta}\frac{d\theta}{dt} = \omega\frac{dh}{d\theta} \tag{16.13.11}$$

where ω is the (constant) angular speed of the cam. Hence,

$$v = \omega\left[\left(\frac{h_2 - h_1}{\theta_2 - \theta_1}\right) - \left(\frac{h_2 - h_1}{\theta_2 - \theta_1}\right)\cos 2\pi\left(\frac{\theta - \theta_1}{\theta_2 - \theta_1}\right)\right] \tag{16.13.12}$$

Similarly, for the follower acceleration we have:

$$a = \frac{d^2h}{dt^2} = \omega^2\frac{d^2h}{d\theta^2} = \omega^2\left[\left(\frac{h_2 - h_1}{\theta_2 - \theta_1}\right)\left(\frac{2\pi}{\theta_2 - \theta_1}\right)\sin 2\pi\left(\frac{\theta - \theta_1}{\theta_2 - \theta_1}\right)\right] \tag{16.13.13}$$

Finally, for the jerk we obtain:

$$jerk = \frac{d^3h}{dt^3} = \omega^3\frac{d^3h}{d\theta^3} = \omega^3\left[\left(\frac{h_2 - h_1}{\theta_2 - \theta_1}\right)^2\left(\frac{2\pi}{\theta_2 - \theta_1}\right)^2\cos 2\pi\left(\frac{\theta - \theta_1}{\theta_2 - \theta_1}\right)\right] \tag{16.13.14}$$

Observe that at the transition points ($\theta = \theta_1$ and $\theta = \theta_2$) the jerk has the finite value:

$$\omega^3\left(\frac{h_2 - h_1}{\theta_2 - \theta_1}\right)\left(\frac{2\pi}{\theta_2 - \theta_1}\right)^2$$

Observe further from Eq. (16.13.13) that the maximum values of the follower acceleration occur at cam rotation angles where the sine argument $2\pi(\theta - \theta_1)/(\theta_2 - \theta_1)$ has values $\pi/4$ and $3\pi/4$. At these angles, the magnitude of the maximum acceleration a_{max} is:

$$a_{\max} = 2\pi\omega^2(h_2 - h_1)/(\theta_2 - \theta_1)^2 \tag{16.13.15}$$

From Eqs. (16.12.11) and (16.12.12) we see that for the sinusoidal rise function the maximum follower acceleration is $(\pi^2/2)(h_2 - h_1)/(\theta_2 - \theta_1)^2$. Because 2π is greater than $\pi^2/2$, we see that the maximum follower acceleration of the cycloidal rise function is larger than that of the sinusoidal rise function, even though the jerk remains finite for the cycloidal function but has infinite values for the sinusoidal function.

Finally, we may obtain an expression for the cycloidal rise function beyond the cam rotation interval $[\theta_1, \theta_2]$ by using the singularity functions of Section 16.10. Specifically, from Eq. (16.13.10), we may express $h(\theta)$ as:

$$h(\theta) = h_1\delta(\theta_1 - \theta) + \left[h_1 + \left(\frac{h_2 - h_1}{\theta_2 - \theta_1}\right)(\theta - \theta_1) - \left(\frac{h_2 - h_1}{2\pi}\right)\sin 2\pi\left(\frac{\theta - \theta_1}{\theta_2 - \theta_1}\right)\right] \tag{16.13.16}$$

$$\left[\delta_1(\theta - \theta_1) - \delta_1(\theta - \theta_2)\right] + h_2\delta_1(\theta - \theta_2)$$

16.14 Summary: Listing of Follower Rise Functions

To summarize the results of the foregoing sections we provide here a tabular listing of the follower rise functions we have developed. Recall that for each case we have developed functions that serve as connections between two dwell regions, as shown in Figure 16.14.1. Specifically, we have the following follower rise functions:

1. Linear rise function (Eq. (16.10.16)):

$$h(\theta) = h_1 + \left(\frac{h_2 - h_1}{\theta_2 - \theta_1}\right)(\theta - \theta_1)\left[\delta_1(\theta - \theta_1) - \delta_1(\theta - \theta_2)\right] \tag{16.14.1}$$

2. Parabolic rise function (Eq. (16.11.3)):

$$h(\theta) = k(\theta - \theta_1)^2\delta_1(\theta - \theta_1) \quad (\text{k} = \text{constant}) \tag{16.14.2}$$

3. Sinusoidal rise function (Eq. (16.12.4)):

$$h(\theta) = h_1\left[1 - \delta_1(\theta - \theta_1)\right] + \phi\left[\delta_1(\theta - \theta_1) - \delta_1(\theta - \theta_2)\right] + h_2\delta_1(\theta - \theta_2) \tag{16.14.3}$$

where (see Eq. (16.12.3)):

$$\phi = \left(\frac{h_1 + h_2}{2}\right) + \left(\frac{h_2 - h_1}{2}\right)\cos\psi \tag{16.14.4}$$

where (see Eq. (16.12.2)):

$$\psi = \pi(\theta + \theta_2 - 2\theta_1)/(\theta_2 - \theta_1) \tag{16.14.5}$$

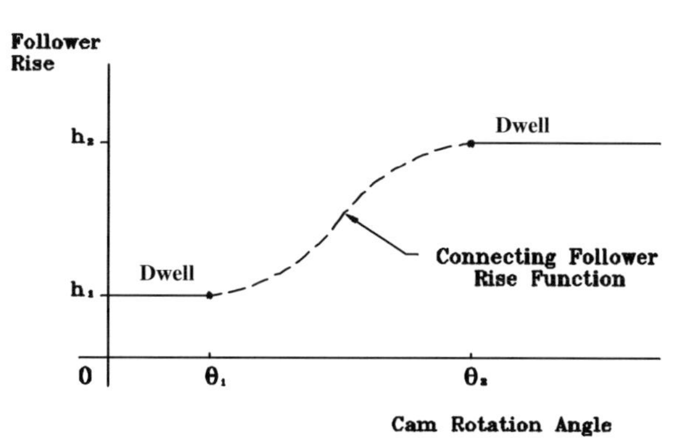

FIGURE 16.14.1
Follower rise connection between dwell positions.

4. Cycloidal rise function (Eq. (16.13.10)):

$$h(\theta) = h_1 + \left(\frac{h_2 - h_1}{\theta_2 - \theta_1}\right)(\theta - \theta_1) - \left(\frac{h_2 - h_1}{2\pi}\right)\sin\left[2\pi\frac{\theta - \theta_1}{\theta_2 - \theta_1}\right] \tag{16.14.6}$$

where the step function $\delta(\theta - \theta_1)$ is defined in Eq. (16.10.2) as:

$$\delta_1(\theta - \theta_1) = \begin{cases} 0 & \theta < \theta_1 \\ 1 & \theta > \theta_1 \end{cases} \tag{16.14.7}$$

Of interest (and concern) with these functions are those values of the acceleration and jerk at the transition points. Table 16.14.1 provides a listing of these quantities. Observe in Table 16.14.1 that the cycloidal rise function is the only function with finite values of the jerk at the transition points. Observe further that, even though the accelerations for the cycloidal rise functions at the transition points are zero, the acceleration is not zero for all angles of the function. Indeed, the maximum value of the acceleration for the cycloidal rise function occurs at cam rotation angles $(\theta_2 - 7\theta_1)/8$ and $(3\theta_2 - 5\theta_1)/8$ with value $2\pi\omega^2(h_2 - h_1)/(\theta_2 - \theta_1)^2$ (see Eq. (16.13.5)). Finally, observe that this maximum acceleration value slightly exceeds the maximum acceleration values of the sinusoidal rise function.

TABLE 16.14.1

Acceleration and Jerk at Transition Points for Various Follower Rise Functions

Follower Rise Function	Acceleration	Position	Jerk	Position
Linear	$\pm\infty$	θ_1, θ_2	$\pm\infty$	θ_1, θ_2
Parabolic	$2k\omega^2$ $(\omega = d\theta/dt)$	θ_1, θ_2	$\pm\infty$	θ_1, θ_2
Sinusoidal	$\pm\left(\frac{h_2 - h_1}{2}\right)\left(\frac{\pi}{\theta_2 - \theta_1}\right)^2$	θ_1, θ_2	$\pm\infty$	θ_1, θ_2
Cycloidal	0	θ_1, θ_2	$\omega^3\left(\frac{h_2 - h_1}{\theta_2 - \theta_1}\right)\left(\frac{2\pi}{\theta_2 - \theta_1}\right)^2$	θ_1, θ_2

16.15 Closure

This concludes our relatively brief introduction to cams and to cam–follower pairs. Although we have focused our attention upon two-dimensional movements, the same principles may readily be applied with cams and followers moving in three dimensions. The difference is simply geometry. The kinematic principles are the same.

A principal concern in cam and follower design is that unwanted accelerations and jerks may be generated. These accelerations and jerks may then produce unwanted and harmful forces, especially with high-speed systems. In the immediate foregoing sections, we have considered various follower rise functions and their effects upon the follower acceleration and jerk. We have seen that the cycloidal function reduces the jerk but at the cost of slightly higher acceleration.

Another way of alleviating the effects of large acceleration and jerk is to use large cams. The disadvantage of large cams, of course, is that they may not be practical for use with small and compact machines. Still another concern in cam–follower design is that the pressure angle between the cam and follower should be kept as small as possible to reduce contact forces.

Finally, for efficient and relatively low-stress cam–follower systems, it is important that precision manufacturing be used to ensure adherence to the desired design geometry. Also, regular maintenance and inspection are essential for long wear and long-lived systems. Readers interested in more advanced aspects of cam–follower design may wish to consult the references.

In the following chapter, we consider gears, another widely used component in machine design. A gear tooth may be viewed as a specialized cam.

References

16.1. Paul, B., *Kinematics and Dynamics of Planar Machinery,* Prentice Hall, Englewood Cliffs, NJ, 1979, chap. 4.

16.2. Graustein, W. C., *Differential Geometry,* Dover, New York, 1962, pp. 64–67.

16.3. Davis, S. A., *Feedback and Control Systems,* rev. ed., Simon & Schuster Technical Outlines, New York, 1974, pp. 44–46.

16.4. Tuma, J. J., *Engineering Mathematics Handbook,* 2nd ed., McGraw-Hill, New York, 1979, p. 230.

16.5. Butkov, E., *Mathematical Physics,* Addison-Wesley, Reading, MA, 1968, chap. 6.

16.6. Roark, R. J., and Young, W. C., *Formulas for Stress and Strain,* 5th ed., McGraw-Hill, New York, 1975, pp. 11, 94.

16.7. Oberg, E., Jones, F. D., and Horton, H. L., in *Machinery's Handbook,* 23rd ed., Ryftel, H. H., Ed., Industrial Press, New York, 988, pp. 2049–2075.

16.8. Erdman, A. G., and Sandov, G. N., *Mechanism Design: Analysis and Synthesis,* Vol. I, 2nd ed., Prentice Hall, Englewood Cliffs, NJ, 1984, chap. 6.

16.9. Wilson, C. F., Sadler, J. P., and Michels, W. J., *Kinematics and Dynamics of Machinery,* Harper & Row, New York, 1983, chap. 5.

Problems

Section 16.4 Graphical Constructions: The Follower Rise Function

P16.4.1: Consider a cam in the shape of an ellipse rotating about a focus point. Let the ellipse have semi-major axis length a and semi-minor axis length b as in Figure P16.4.1. Let c be the distance from the ellipse center to the focus. Thus, from elliptic geometry a, b, and c are related by:

$$a^2 = b^2 + c^2$$

Following the procedure outlined in Section 16.4, obtain a graphical representation of the follower rise function. Let a = 2 in. and b = 1.25 in.

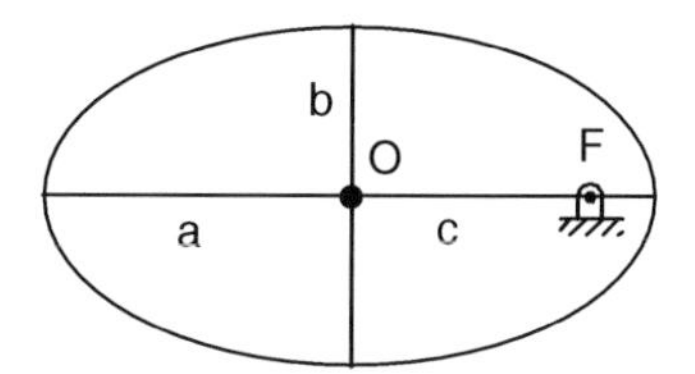

FIGURE P16.4.1
A cam in the shape of an ellipse.

P16.4.2: Repeat Problem P16.4.1 if a = 5 cm and b = 3.25 cm.

Section 16.5 Graphical Construction: Cam Profiles

P16.5.1: Let a follower rise function have the shape of a cosine curve and be defined as:

$$h(\theta) = A[1 + \cos(\theta + \pi)]$$

Use the procedure outlined in Section 16.5 to determine the cam profile if A = 1.5 inches.

P16.5.2: Repeat Problem P16.5.1 if A = 5 cm.

Section 16.6 Graphical Construction: Effect of Cam–Follower Design

P16.6.1: Repeat Problem P16.4.1 if the follower axis is offset from the cam axis by a distance d as represented in Figure P16.6.1. Let a, b, and d have the values a = 2 in., b = 1.25 in., and d = 0.25 in.

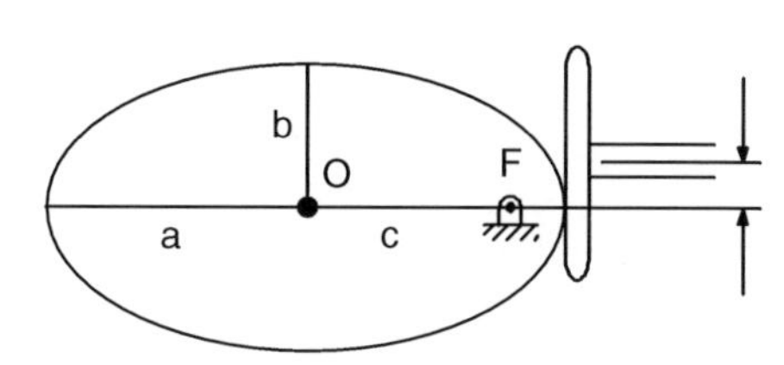

FIGURE P16.6.1
A cam–follower axis offset from the cam rotation axis.

P16.6.2: Repeat Problem P16.6.1 if a, b, and d have the values $a = 5$ cm, $b = 3.25$ cm, and $d = 0.75$ cm.

P16.6.3: Repeat Problem P16.5.1 if the follower axis is offset from the cam rotation axis by 3/8 in.

P16.6.4: Repeat Problem P16.5.2 if the follower axis is offset from the cam rotation axis by 1.0 cm.

Section 16.8 Analytical Construction of Cam Profiles

P16.8.1: Consider the follower rise function represented graphically in Figure P16.8.1. Find an analytical expression for the function.

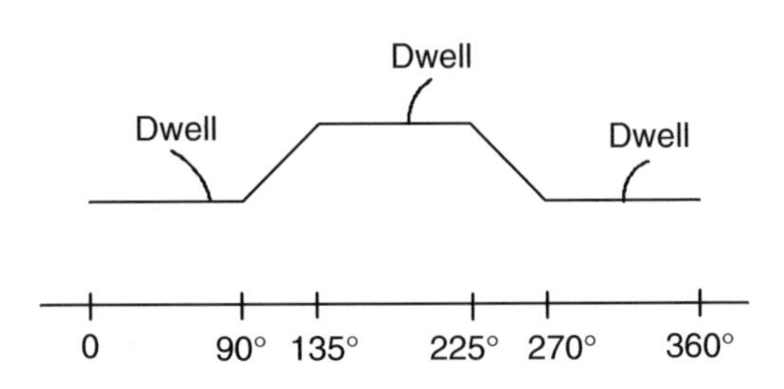

FIGURE P16.8.1
Graphical representation of a follower rise function.

P16.8.2: Consider the follower rise function represented graphically in Figure P16.8.2. Find an analytical expression for the function.

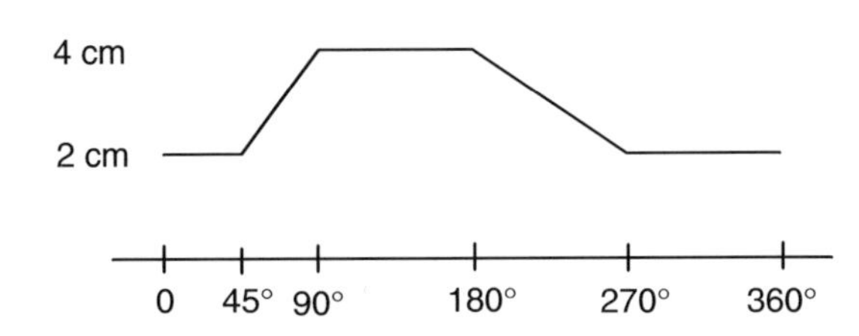

FIGURE P16.8.2
Graphical representation of a follower rise function.

Section 16.10 Use of Singularity Functions

P16.10.1: See Problem P16.8.1. Use singularity functions to describe the follower rise function of Figure P16.8.1.

P16.10.2: See Problem P16.8.2. Use singularity functions to describe the follower rise function of Figure P16.8.2.

Section 16.11 Parabolic Rise Function

P16.11.1: Consider the follower rise function shown graphically in Figure P16.11.1. Let the transitions between the dwell regions and the rise/fall regions be accomplished with parabolic shapes of the same size. That is, let the form of $h(\theta)$ at A, B, C, and D be parabolic. Determine an analytical representation of $h(\theta)$.

P16.11.2: Repeat Problem P16.11.1 if the difference in elevations of the dwell regions of Figure P16.11.1 is 8 cm instead of 3 in.

P16.11.3: Consider the use of a circular arc to smooth the transition between a dwell and a linear rise as suggested in Figure P16.11.3. (A circular arc is easier to fabricate than a

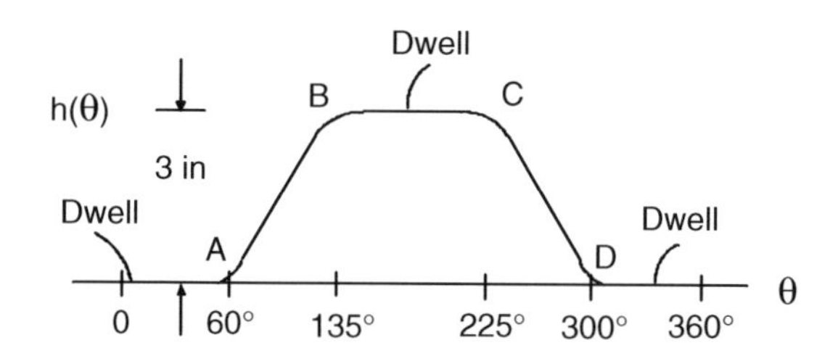

FIGURE P16.11.1
A follower rise function with parabolic transition regions *A*, *B*, *C*, *D*.

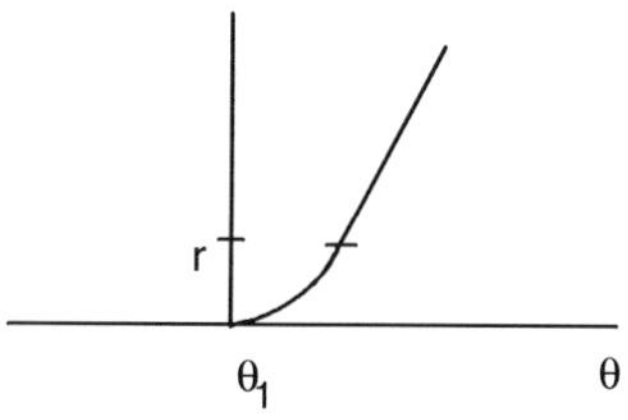

FIGURE P16.11.3
A circular arc transition.

parabolic curve.) Specifically, let the circular arc have a radius r as in Figure P16.11.3 beginning at an angle θ_1. Determine an analytical representation of the rise function $h(\theta)$.

P16.11.4: See Problem P16.11.3. Determine the follower velocity $h'(\theta)$ and acceleration $h''(\theta)$ at $\theta = \theta_1$.

P16.11.5: Verify that $(\theta - \theta_1)\delta_0(\theta - \theta_1)$ and $(\theta - \theta_1)^2\delta_{-1}(\theta - \theta_1)$ are zero at $\theta = \theta_1$ (see Eq. (16.11.5) and the statement following the equation).

P16.11.6: Show that the follower acceleration for the circular arc transition has finite follower rise acceleration.

Section 16.12 Sinusoidal Rise Function

P16.12.1: Consider the follower rise function $h(\theta)$ shown graphically in Figure P16.12.1. Let the transitions between the rises (and falls) and dwells at 45, 135, 225, and 315° be accomplished by sinusoidal functions. Using singularity functions determine an analytical expression for $h(\theta)$.

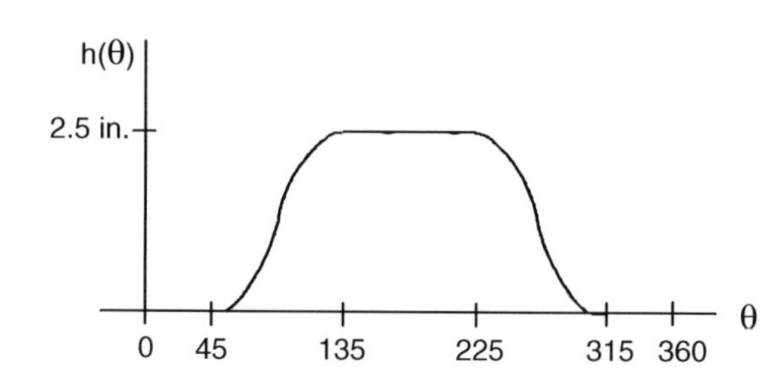

FIGURE P16.12.1
Follower rise function with sinusoidal transitions.

P16.12.2: Repeat Problem P16.12.1 if the difference in the dwell elevations is 6.75 cm instead of 2.5 in.

Section 16.13 Cycloidal Rise Functions

P16.13.1: See Problems P16.11.1, P16.11.3, and P16.12.1. Show that the parabolic, sinusoidal, and circular rise functions all have infinite acceleration derivatives at the transition points.

P16.13.2: Consider the follower rise function $h(\theta)$ shown graphically in Figure P16.13.2. Let the transition between the rises (and falls) and dwells at 45, 135, 225, and 315° be accomplished by cycloidal functions. Using singularity functions determine an analytical expression for $h(\theta)$.

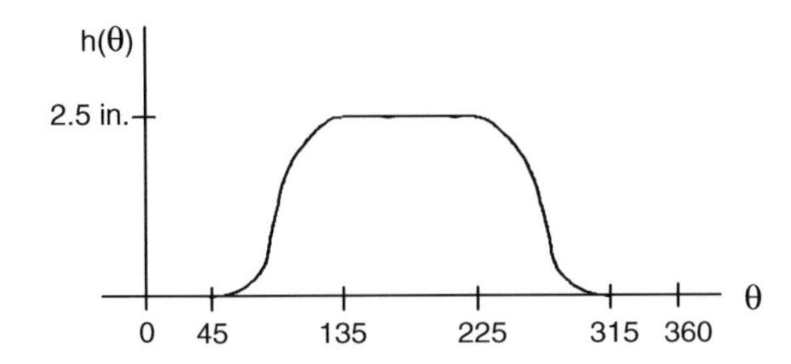

FIGURE P16.13.2
Follower rise function with cycloidal transitions.

P16.13.3: Repeat Problem P16.13.2 if the difference in the dwell elevations is 6.75 cm instead of 2.5 in.

P16.13.4: Show that the cycloidal rise function of Problem P16.13.3 has finite values of the derivatives of the acceleration rise function.

17

Mechanical Components: Gears

17.1 Introduction

As with cams, gears and gearing systems are fundamental mechanical components. Indeed, in the design of machines and mechanical systems, gears are even more common than cams. Gears are perhaps the most common of all mechanical components. From another perspective, however, gears may be viewed as special kinds of cams. Gears and gearing systems have two purposes: (1) to transmit motion, and (2) to transmit forces during that motion. Hence, gearing systems are often called *transmissions*.

Gear design is a very old subject dating back to antiquity with the use of cog-type devices as early as 330 BC or earlier [17.6]. Numerous books, monographs, articles, and research reports have been written about gears. (The references at the end of the chapter represent a small sampling of writings that could be of interest and use for further study.) In recent years, the technical literature has grown dramatically as a result of increased numbers of research efforts stimulated by a quest to obtain higher precision, stronger, more efficient, and longer-lived gearing systems.

The large body of information and knowledge about gears cannot be fully documented or even summarized in a single book chapter, or for that matter even in a single volume. Hence, in this chapter, we will simply attempt to summarize the most fundamental aspects of gearing systems, with an emphasis on gear geometry and kinematics. We will not discuss in any detail such topics as gear vibration, gear strength or deformation, or gear wear and life. Readers interested in those subjects may want to consult the references.

We begin our discussion in the next section with a brief review of the fundamental concepts of gearing and of conjugate action (uniform motion transmission). In subsequent sections, we discuss gear tooth geometry, gear nomenclature, gearing kinematics, and gear trains. Section 17.16 has a glossary of commonly used gearing terms.

17.2 Preliminary and Fundamental Concepts: Rolling Wheels

The transmission of motion and forces is commonly called *power transmission*. With gears, power transmission almost always occurs between rotating shafts. For the most part, these shafts are parallel (as with spur and helical gears), but they may also be intersecting (as with bevel and spiral bevel gears) or even nonparallel and nonintersecting (as with hypoid and worm gears). In the sequel, we will develop our analysis with parallel shaft gears and specifically with spur gears. The analysis of other types of gears is fundamentally the same but somewhat more detailed due to their more complex geometries.

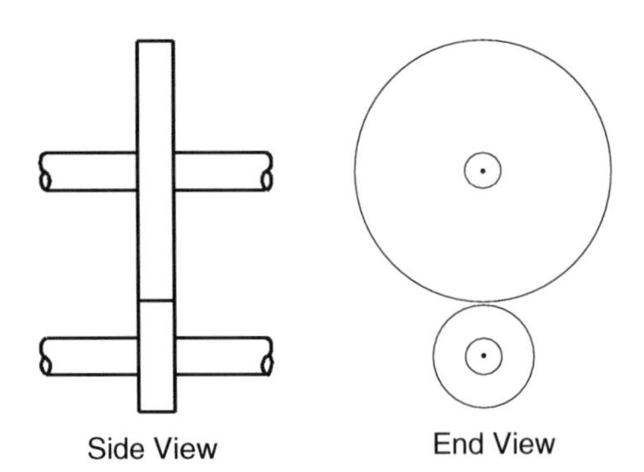

FIGURE 17.2.1
Rolling wheels.

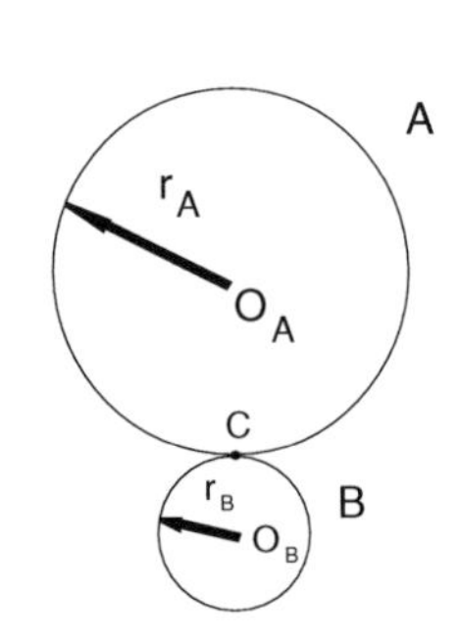

FIGURE 17.2.2
End view of rolling wheels.

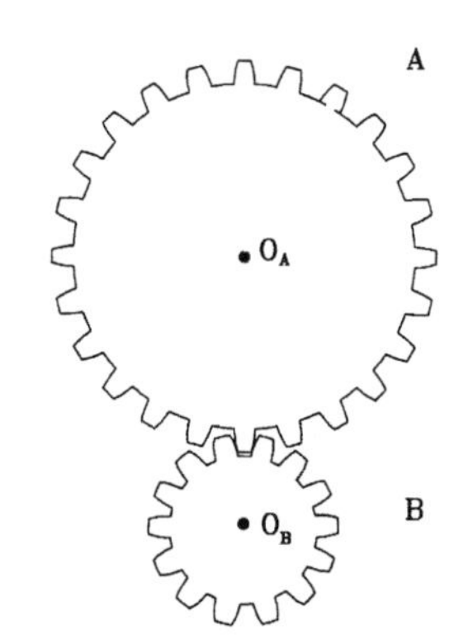

FIGURE 17.2.3
Teeth on the rolling wheels.

For parallel shaft gears, we may consider the motion transmitted between the gears to be the same as that transmitted between rolling disks or rolling wheels as depicted in Figure 17.2.1. Generally, these wheels have different diameters. The smaller wheel is usually called the *pinion* and the larger the *gear.* Also, the disk providing the motion or forces is often called the *driver,* while the wheel receiving the motion or forces is called the *follower.*

Consider again the end view of the rolling wheels as in Figure 17.2.2. Let the wheels be called A and B with centers O_A and O_B and radii r_A and r_B as shown. Let C be the contact point between A and B. Then, from elementary rolling analysis (see Section 4.11), if there is no slipping between A and B then we have:

$$r_A\omega_A = r_B\omega_B \quad \text{or} \quad \omega_A/\omega_B = r_B/r_A \tag{17.2.1}$$

where ω_A and ω_B are the angular speeds of A and B. Equation (17.2.1) shows that the angular speed ratio ω_A/ω_B is inversely equal to the radius ratio r_B/r_A.

A key condition needed to establish Eq. (17.2.1) is that there be no slipping between the wheels A and B. That is, rolling must occur without slipping, implying perfectly rough rolling surfaces. Because, in practice, it is impossible to manufacture perfectly rough rolling surfaces, machine designers have resorted to the use of *finite roughness* in the form of wheel projections or teeth as in Figure 17.2.3. The teeth ensure that there is no slipping between the wheels and that the angular speed relation of Eq. (17.2.1) is maintained.

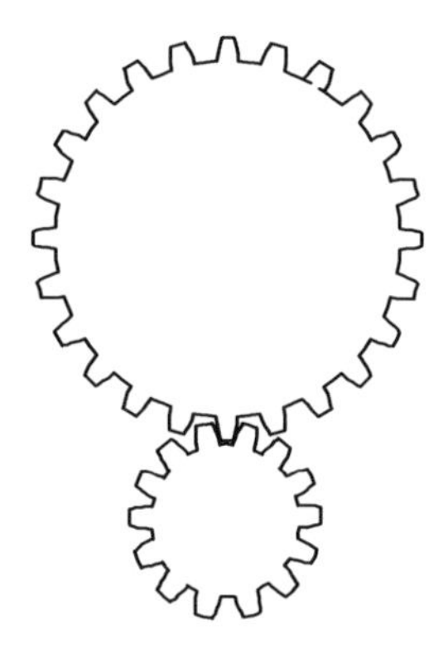

FIGURE 17.3.1
A meshing gear pair.

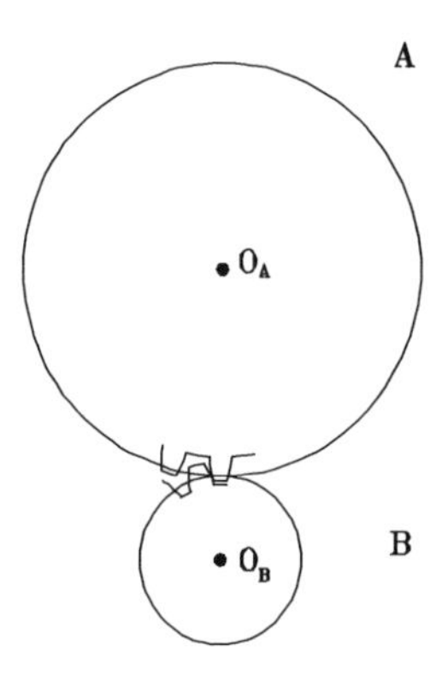

FIGURE 17.3.2
Rolling pitch circles and a typical pair of engaging teeth.

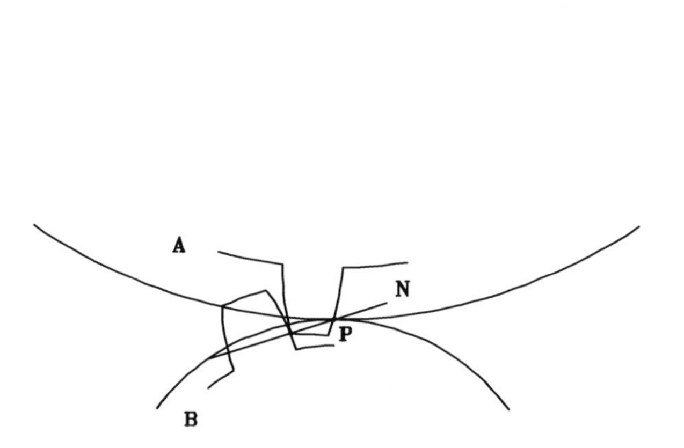

Figure 17.3.3
Enlarged view of mesh region.

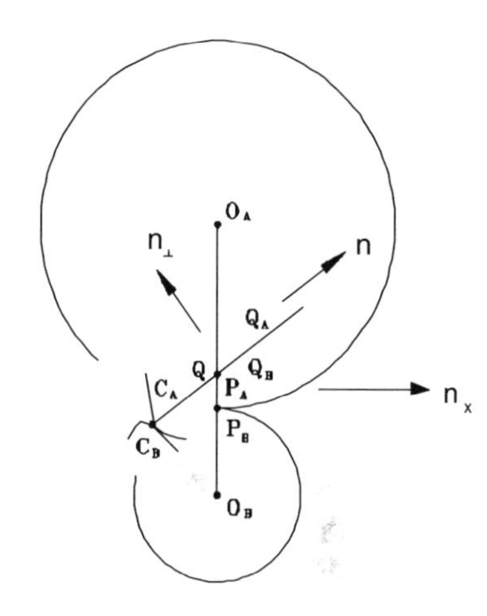

Figure 17.3.4
An enlarged view (not to scale) of engaging gears with a typical pair of contacting teeth surfaces.

17.3 Preliminary and Fundamental Concepts: Conjugate Action

When a pair of gear wheels roll together so that their angular speed ratio remains constant throughout the motion, the associated gears are said to be *conjugate*. When this occurs, the relative motion (or kinematics) of the gears is called *conjugate action*. Conjugate action between meshing gears is desirable to maintain smooth operation of the gears. That is, without conjugate action a constant angular speed rotation of the driver gear will produce nonsteady rotation of the follower gear. In this section, we will discuss the features of gear tooth design required to have conjugate action.

Consider again a pair of gears rolling together (that is, with teeth in mesh) as depicted in Figure 17.3.1. Let the movement of the gears be modeled by a pair of rolling wheels (or circles), as in Figure 17.3.2. Consider a typical pair of contacting teeth as shown. Let O_A and O_B be the centers of the gears (and, hence, of the rolling circles) and let P be the point of contact between the circles. Then, P is called the *pitch point* and the rolling circles are called *pitch circles*.

Consider an enlarged view of the pitch circles and the mating tooth surfaces in the mesh region as in Figure 17.3.3. Let N be a line normal to the mating teeth at their point of contact. In the following paragraphs, we will show that for conjugate action to be maintained N must pass through the pitch point P. This point intersection requirement for N is called the *law of conjugate action*.

To develop this concept, consider an enlarged view of the engaging gears and of the two typical contacting teeth, as in Figure 17.3.4. For conjugate action, there must be no slipping between the pitch circles of the engaging gears. This in turn means that the respective contact points of the gears at the pitch circles will have identical velocities. That is,

$$\mathbf{V}^{P_A} = \mathbf{V}^{P_B} \tag{17.3.1}$$

where P_A and P_B are the respective contact points (this expression is equivalent to Eq. (17.2.1)).

To have contact between the teeth at C, and without penetration, we must have:

$$\mathbf{V}^{C_A} \cdot \mathbf{n} = \mathbf{V}^{C_B} \cdot \mathbf{n} \tag{17.3.2}$$

where C_A and C_B are the respective tooth contact points and where $\mathbf{n}$ is a unit vector normal to the contacting teeth at the point of contact (observe that Eq. (17.3.2) may be satisfied even if $\mathbf{V}^{C_A} \neq \mathbf{V}^{C_B}$).

Because P_A and C_A are both on the same body (gear A), we have from Eq. (4.94) that:

$$\mathbf{V}^{C_A} - \mathbf{V}^{P_A} = \mathbf{V}^{C_A/P_A} = \boldsymbol{\omega}^A \times \mathbf{P}_A\mathbf{C}_A = \alpha\mathbf{n}_T \tag{17.3.3}$$

where $\boldsymbol{\omega}^A$ is the angular velocity of A, $\mathbf{P}_A\mathbf{C}_A$ is a position vector locating C_A relative to P_A, $\mathbf{n}_T$ is a unit vector parallel to the plane of the gears and perpendicular to the line passing through P_A and C_A, and α is a scalar. Similarly, for C_B and P_B we have:

$$\mathbf{V}^{C_B} - \mathbf{V}^{P_B} = \mathbf{V}^{C_B/P_B} = \boldsymbol{\omega}^B \times \mathbf{P}_B\mathbf{C}_B = \beta\mathbf{n}_T \tag{17.3.4}$$

where β is a scalar. Hence, we have:

$$\mathbf{V}^{C_A} \cdot \mathbf{n} = \mathbf{V}^{P_A} \cdot \mathbf{n} \quad \text{and} \quad \mathbf{V}^{C_B} \cdot \mathbf{n} = \mathbf{V}^{P_B} \cdot \mathbf{n} \tag{17.3.5}$$

where here $\mathbf{n}$ is a unit vector parallel to $\mathbf{P}_A\mathbf{C}_A$ and $\mathbf{P}_B\mathbf{C}_B$.

In view of Eqs. (17.3.2) and (17.3.5), we must have:

$$\mathbf{V}^{C_A} \cdot \mathbf{n} = \mathbf{V}^{P_A} \cdot \mathbf{n} = \mathbf{V}^{C_B} \cdot \mathbf{n} = \mathbf{V}^{P_B} \cdot \mathbf{n} \tag{17.3.6}$$

Suppose that the normal line N at the tooth contact point C intersects the line of gear centers at a point Q distinct from the pitch point P. Let Q_A and Q_B be points of A and B (or A or B extended) that coincide with Q. Then, the velocities of Q_A and Q_B are perpendicular to the line of gear centers, or in the $\mathbf{n}_x$ direction (see Figure 17.3.4). Also, these velocities will not be equal to one another. That is,

$$\mathbf{V}^{Q_A} = \mathbf{V}^{Q_A}\mathbf{n}_x \quad \text{and} \quad \mathbf{V}^{Q_B} = \mathbf{V}^{Q_B}\mathbf{n}_x \tag{17.3.7}$$

and

$$\mathbf{V}^{Q_A} \neq \mathbf{V}^{Q_B} \tag{17.3.8}$$

where V^{Q_A} and V^{Q_B} are the magnitudes of $\mathbf{V}^{Q_A}$ and $\mathbf{V}^{Q_B}$. Hence, we also have:

$$V^{Q_A} \neq V^{Q_B} \tag{17.3.9}$$

Then we have:

$$\mathbf{V}^{Q_A} \cdot \mathbf{n} = V^{Q_A}(\mathbf{n}_x \cdot \mathbf{n}) \quad \text{and} \quad \mathbf{V}^{Q_B} \cdot \mathbf{n} = V^{Q_B}(\mathbf{n}_x \cdot \mathbf{n}) \tag{17.3.10}$$

and, therefore, in view of Eq. (17.3.9) we have:

$$\mathbf{V}^{Q_A} \cdot \mathbf{n} \neq \mathbf{V}^{Q_B} \cdot \mathbf{n} \tag{17.3.11}$$

Because C_A and Q_A are both fixed on A and because C_B and Q_B are both fixed on B, we have equations analogous to Eqs. (17.3.3) and (17.3.4) in the forms:

$$\mathbf{V}^{C_A} - \mathbf{V}^{Q_A} = \mathbf{V}^{C_A/Q_A} = \boldsymbol{\omega}^A \times \mathbf{Q}_A\mathbf{C}_A = \hat{\alpha}\mathbf{n}_T \tag{17.3.12}$$

and

$$\mathbf{V}^{C_B} - \mathbf{V}^{Q_B} = \mathbf{V}^{C_B/Q_B} = \boldsymbol{\omega}^B \times \mathbf{Q}_B\mathbf{C}_B = \hat{\beta}\mathbf{n}_T \tag{17.3.13}$$

where $\hat{\alpha}$ and $\hat{\beta}$ are scalars. Then,

$$\mathbf{V}^{C_A} \cdot \mathbf{n} - \mathbf{V}^{Q_A} \cdot \mathbf{n} = 0 \quad \text{and} \quad \mathbf{V}^{C_B} \cdot \mathbf{n} - \mathbf{V}^{Q_B} \cdot \mathbf{n} = 0 \tag{17.3.14}$$

Then, finally, in view of Eq. (17.3.11) we have:

$$\mathbf{V}^{C_A} \cdot \mathbf{n} \neq \mathbf{V}^{C_B} \cdot \mathbf{n} \tag{17.3.15}$$

This relation, however, violates the noninterpenetrability of the contacting surfaces (see Eq. (17.3.2)); therefore, unless Q is at P, the pitch point, the conjugate action will be lost.

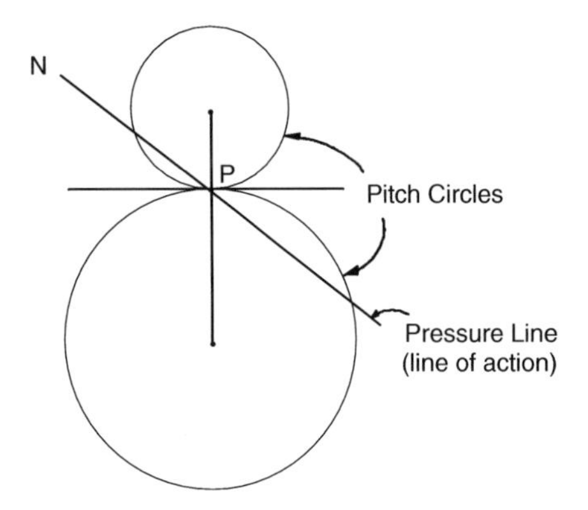

FIGURE 17.4.1
Rolling wheels, pitch point, and pressure line.

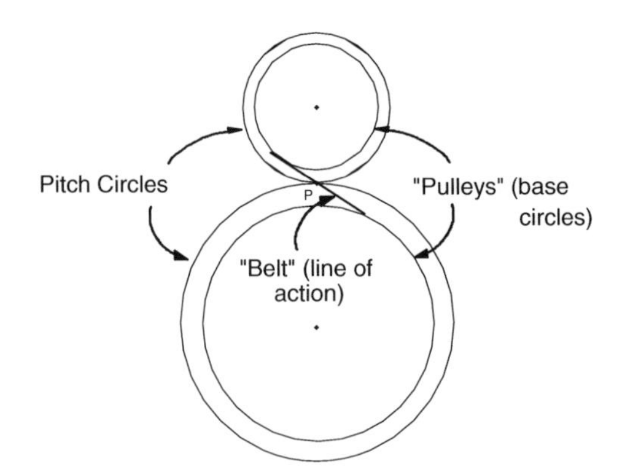

FIGURE 17.4.2
Line of action viewed as a belt connecting pulleys (base circles) of rolling wheels.

17.4 Preliminary and Fundamental Concepts: Involute Curve Geometry

The law of conjugate action places restrictions on the gear tooth profile shape. Of all tooth profiles producing conjugate action, the most widely used has the form of an involute of a circle. In this section, we consider the geometry of such involute curves.

Consider Figure 17.4.1 showing a sketch of rolling wheels (pitch circles). In view of the law of conjugate action, let the inclined line *N* represent the normal of the gear tooth surfaces at their points of contact. Thus, *N* passes through the pitch point *P*.

For smooth gear teeth, the forces transmitted between the teeth will be directed along *N*. For this reason *N* is often called the *pressure line* or *line of action* of the gear tooth.

The line of action *N* may be thought of as describing a crossed belt connecting two pulleys attached to the rolling wheels, as depicted in Figure 17.4.2. In Figure 17.4.3, we see that the radii of these pulleys are related to the radii of the rolling wheels through the inclination angle θ of the line of action (pressure line). The profile circles of the pulleys are called *base circles*, and the pressure line inclination angle θ is called the *pressure angle*. Specifically the base and pitch circle radii r_b and r_p are related by the expressions:

$$r_{Ab} = r_{Ap}\cos\theta \quad \text{and} \quad r_{Bb} = r_{Bp}\cos\theta \tag{17.4.1}$$

or

$$r_b / r_p = \cos\theta \tag{17.4.2}$$

Viewed still another way, the line of action as shown in Figures 17.3.4 and 17.4.3 is a crossed tangent to the base circles. Because the line of action is normal to the gear tooth surfaces, we can construct the gear tooth profile as the locus of points on the gear wheel occupied by a typical point on the line of action. Specifically, suppose we consider again the line of action modeled as a cable or belt wrapped around the base circle pulley. Then, instead of thinking of the belt as moving along a fixed line away from the rotating pulley, let us consider the pulley to be fixed with the belt being unwrapped around the pulley.

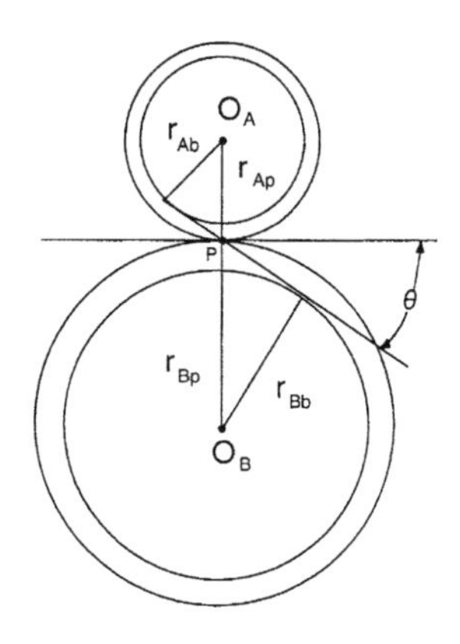

FIGURE 17.4.3
Pitch circle and base circle radii.

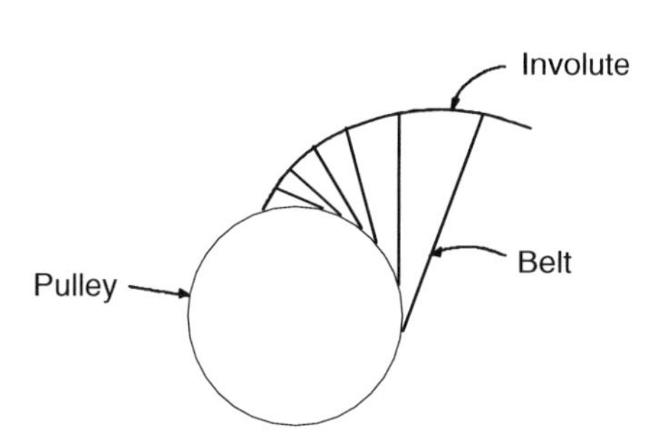

FIGURE 17.4.4
Belt unwrapping around a fixed pulley.

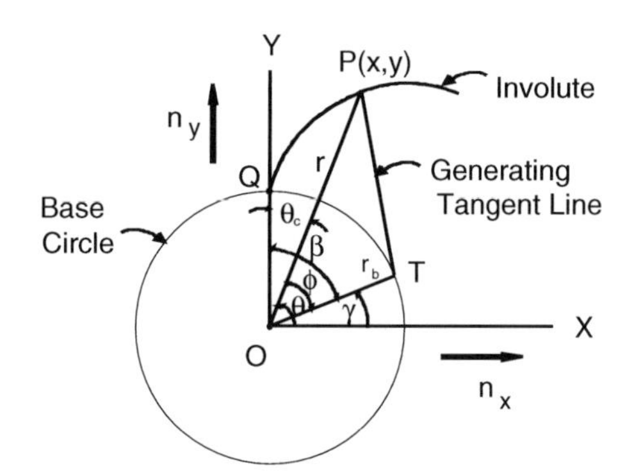

FIGURE 17.4.5
Involute curve geometry.

Such a process might be envisioned as in Figure 17.4.4. The curve traced out by the end point of the unwrapping belt is commonly called the *involute* of the circle perimeter of the pulley. The involute is used to define the gear tooth profile.

To further examine the geometric and analytical properties of an involute curve, consider the sketch of Figure of 17.4.5 where the base circle having radius r_b is centered at the origin of a Cartesian axis system. Let P be a typical point on the involute curve with rectangular (Cartesian) coordinates (x, y) and polar coordinates (r, θ). Let T be the point on the base circle at the base end of the unwrapping line segment as shown in Figure 17.4.5. Let O be the origin of the coordinate system; let γ be the angle between **OT** and the X-axis; let ϕ be the angle between **OT** and **OP**; let θ_C be the angle between **OP** and the Y-axis (θ_C is the complement of the polar angle θ); and let β be the angle between **OT** and the Y-axis. Finally, let $\mathbf{n}_x$ and $\mathbf{n}_y$ be unit vectors parallel to the X- and Y-axes as shown.

By inspection of Figure 17.4.5 we obtain the relations:

$$\theta = \gamma + \phi \tag{17.4.3}$$

$$\theta + \theta_c = \gamma + \phi + \theta_c = \pi/2 \tag{17.4.4}$$

$$\gamma + \beta = \pi/2 \tag{17.4.5}$$

$$\beta = \phi + \theta_c \tag{17.4.6}$$

Let Q be the intersection point of the Y-axis and the base circle. Q is then at the origin of the involute curve. The arc $\widehat{TQ}$ then has the same length as the generating line segment TP. Hence, we have:

$$|\mathbf{TP}| = |\widehat{TQ}| = r_b\beta \tag{17.4.7}$$

The angle β may be considered as the *unwrapping angle*. It is often helpful to express the coordinates of P in terms of β. From Figure 17.4.5 we have:

$$r^2 = |OP|^2 = |OT|^2 + |TP|^2 = r_b^2 + r_b^2\beta^2 = r_b^2(1+\beta^2) \tag{17.4.8}$$

or

$$r = r_b\sqrt{1+\beta^2} \tag{17.4.9}$$

Also, from Figure 17.4.5 we can obtain the coordinates x and y as follows:

$$\mathbf{OP} = \mathbf{OT} + \mathbf{TP} \tag{17.4.10}$$

But, **OT** may be expressed as:

$$\mathbf{OT} = r_b \cos\gamma \mathbf{n}_x + r_b \sin\gamma \mathbf{n}_y \tag{17.4.11}$$

and as:

$$\mathbf{OT} = r_b \sin\beta \mathbf{n}_x + r_b \cos\beta \mathbf{n}_y \tag{17.4.12}$$

Similarly, **TP** may be expressed as:

$$\mathbf{TP} = r_b\beta(-\sin\gamma \mathbf{n}_x + \cos\gamma \mathbf{n}_y) \tag{17.4.13}$$

and as:

$$\mathbf{TP} = r_b\beta(-\cos\beta \mathbf{n}_x + \sin\beta \mathbf{n}_y) \tag{17.4.14}$$

Then, from Eq. (17.4.10) we have:

$$\begin{aligned}\mathbf{OP} &= r_b \sin\beta \mathbf{n}_x + r_b \cos\beta \mathbf{n}_y - r_b\beta\cos\beta \mathbf{n}_x + r_b\beta\sin\beta \mathbf{n}_y \\ &= r_b(\sin\beta - \beta\cos\beta)\mathbf{n}_x + r_b(\cos\beta + \beta\sin\beta)\mathbf{n}_y \\ &= x\mathbf{n}_x + y\mathbf{n}_y\end{aligned} \tag{17.4.15}$$

or

$$x = r_b(\sin\beta - \beta\cos\beta) \quad \text{and} \quad y = r_b(\cos\beta + \beta\sin\beta) \tag{17.4.16}$$

Then, the polar angle θ is:

$$\theta = \tan^{-1} y/x$$
$$= \tan^{-1}\left(\frac{\cos\beta + \beta\sin\beta}{\sin\beta - \beta\cos\beta}\right) \tag{17.4.17}$$

Finally, from Figure 17.4.5, the angle ϕ is related to β by the relation:

$$\tan\phi = r_b\beta/r_b = \beta = \phi + \theta_c \tag{17.4.18}$$

Then, θ_c is:

$$\theta_c = (\tan\phi) - \phi \stackrel{D}{=} Inv\phi \tag{17.4.19}$$

where Invϕ is commonly called the *involute function.*

17.5 Spur Gear Nomenclature

As noted earlier, we will focus our attention upon parallel shaft gears and specifically upon spur gears. In this section, we will introduce and define the notation and nomenclature commonly used with spur gears. In the next section, we will consider the kinematics of these gears.

Figure 17.5.1 shows a profile of an involute spur gear tooth together with some identifying nomenclature. We have already introduced the pitch and base circles. The addendum and dedendum represent the gear tooth height and depth above and below the pitch circle. The tooth profile extends below the base circle with the involute curve being replaced by a fillet. The remainder of the notation should be evident by inspection.

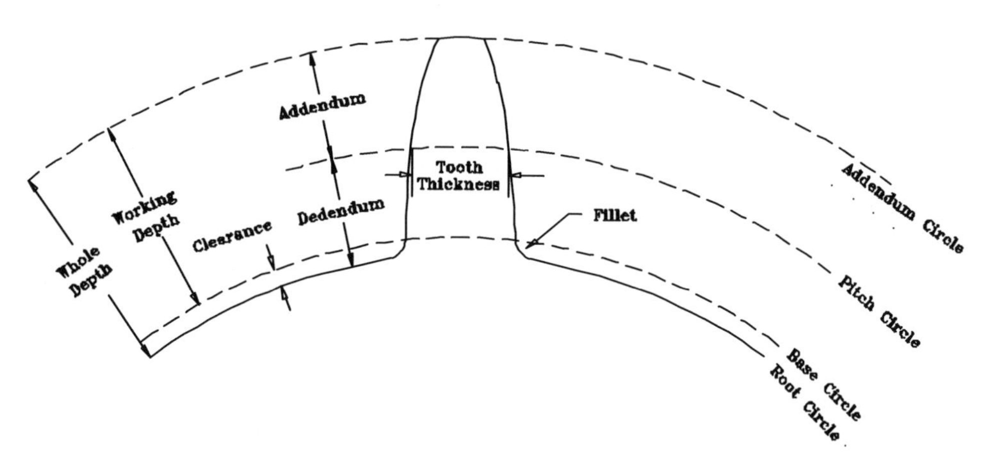

FIGURE 17.5.1
An involute spur gear tooth.

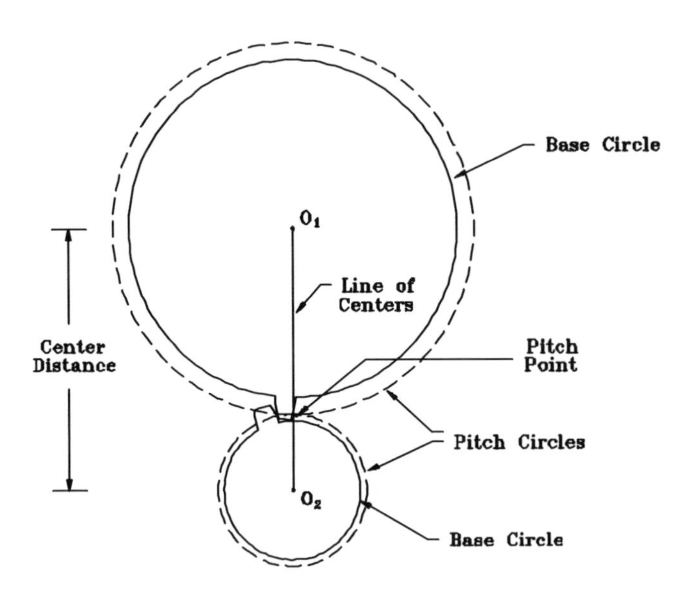

FIGURE 17.5.2
Meshing involute spur gears determining the pitch circles.

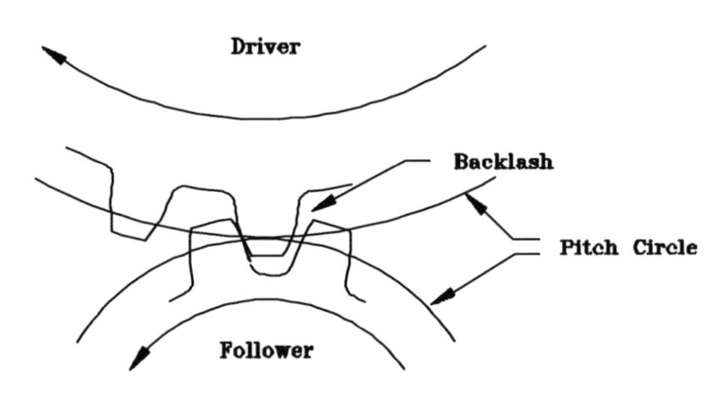

FIGURE 17.5.3
Profile of meshing gear teeth.

Unlike the root circle, the base circle, and the addendum circle, the pitch circle is not fixed relative to the tooth. Instead, the pitch circle is determined by the center distance between the mating gears. Hence, the addendum, dedendum, and tooth thickness may vary by a small amount depending upon the specific location of the pitch circle. The location of the pitch circle on a gear tooth is determined by the location of the point of contact between meshing teeth on the line connecting the gear centers, as illustrated in Figure 17.5.2.

An advantageous property of involute spur gear tooth geometry is that the gears will operate together with conjugate action at varying center distances with only the provision that contact is always maintained between at least one pair of teeth.

Next, consider a closer look at the interaction of the teeth of meshing gears as in Figure 17.5.3. If the gears are not tightly pressed together, there will be a separation between noncontacting teeth. This separation or looseness is called the *backlash*.

The distance measured along the pitch circle between corresponding points on adjacent teeth is called the *circular pitch*. Circular pitch is commonly used as a measure of the size of a gear. From Figure 17.5.3 we see that, even with backlash, unless two gears have the

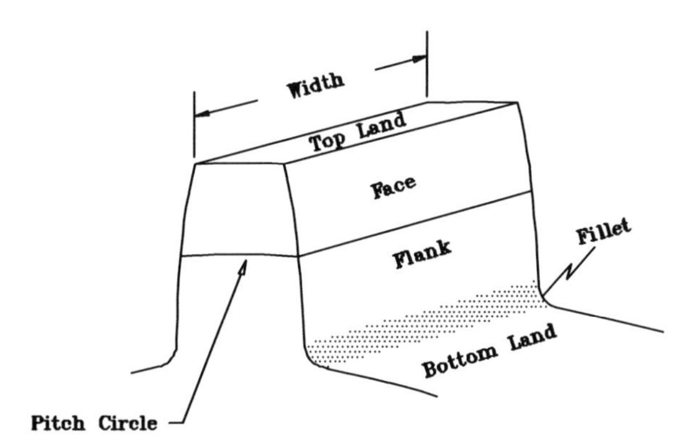

Figure 17.5.4
Perspective view of a spur gear tooth.

same circular pitch they will not mesh or operate together. If a gear has N teeth and a pitch circle with diameter d, then the circular pitch p is:

$$p = \pi d/N \tag{17.5.1}$$

Another commonly used measure of gear size is *diametral pitch P*, defined as the number of teeth divided by the pitch circle diameter. That is,

$$P = N/d \tag{17.5.2}$$

Usually the diametral pitch is computed with the pitch circle diameter d measured in inches.

Still another measure of gear size is the *module*, m, which is the reciprocal of the diametral pitch. With the module, however, the pitch circle diameter is usually measured in millimeters. Then, the module and diametral pitch are related by the expression:

$$m = 25.4/P \tag{17.5.3}$$

From Eqs. (17.5.1) and (17.5.2) we also have the relations:

$$pP = \pi \quad \text{and} \quad p/m = \pi \tag{17.5.4}$$

For terminology regarding spur gear depth, consider Figure 17.5.4, which shows a three-dimensional representation of a gear tooth illustrating the width, top land, face, flank, and bottom land.

Finally, involute spur gear tooth designers have the following tooth proportions depending upon the diametral pitch P [17.3]:

$$\text{Addendum} = 1/P \tag{17.5.5}$$

$$\text{Dedendum} = 1.157/P \tag{17.5.6}$$

$$\text{Clearance} = 0.157/P \tag{17.5.7}$$

$$\text{Fillet radius} = 0.157/P \tag{17.5.8}$$

17.6 Kinematics of Meshing Involute Spur Gear Teeth

In this section, we consider the fundamentals of the kinematics of meshing involute spur gear teeth. We will focus upon ideal gears — that is, gears with exact geometry. In practice, of course, gear geometry is not exact, but the closer the geometry is to the theoretical form, the more descriptive our analysis will be. Lent [3] provides an excellent elementary description of spur gear kinematics. We will follow his outline here, and the reader may want to consult the reference itself for additional details.

In our discussion, we will consider the interaction of a pair of mating teeth from the time they initially come into contact, then as they pass through the mesh (the region of contact), and, finally, as they separate. We will assume that the gears always have at least one pair of teeth in contact and that each pair of mating teeth is the same as each other pair of mating teeth.

With involute gear teeth, the locus of points of contact between the teeth is a straight line (see Figures 17.3.4, 17.4.2, and 17.5.2). This line is variously called the *line of contact*, *path of contact*, *pressure line*, or *line of action*. The angle turned through by a gear as a typical tooth travels along the path of contact is called the *angle of contact*. (Observe that for gears with different diameters, thus having different numbers of teeth, that their respective angles of contact will be different — even though their paths of contact are the same.)

Figure 17.6.1 illustrates the path and angles of contact for a typical pair of meshing spur gears. The angle of contact is sometimes called the *angle of action*. The path of contact *AB* is formed from the point of initial contact *A* to the point of ending contact *B*. The path of contact passes through the pitch point *P*. The angle turned through during the contact, measured along the pitch circle, is the *angle of contact*. The *angle of approach* is the angle turned through by the gear up to contact from the pitch point *P*. The *angle of recess* is the angle turned through by the gear during tooth contact at the pitch point to the end of contact at *B*.

Next, consider Figure 17.6.2, which shows an outline of addendum circles of a meshing spur gear pair. Contact between teeth will occur only within the region of overlap of the circles. The extent of the overlap region is dependent upon the gear radii. The location of the pitch point within the overlap region is dependent upon the size or length of the addendum. Observe further that the location of the pitch point *P* determines the lengths of the paths of approach and recess. Observe moreover that the paths of approach and recess are generally not equal to each other.

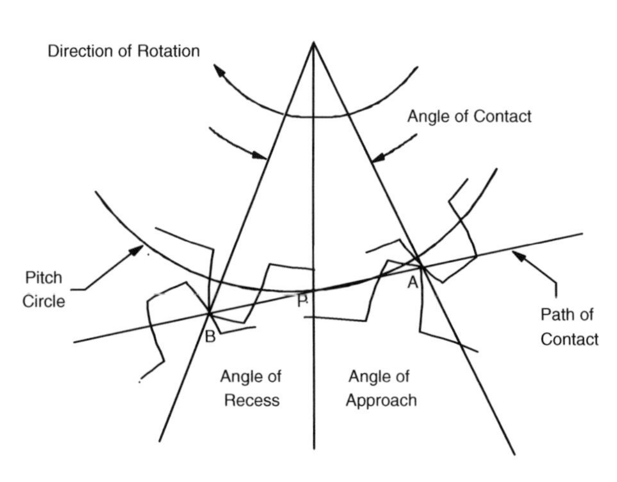

FIGURE 17.6.1
Path and angle of contact for a pair of meshing involute teeth.

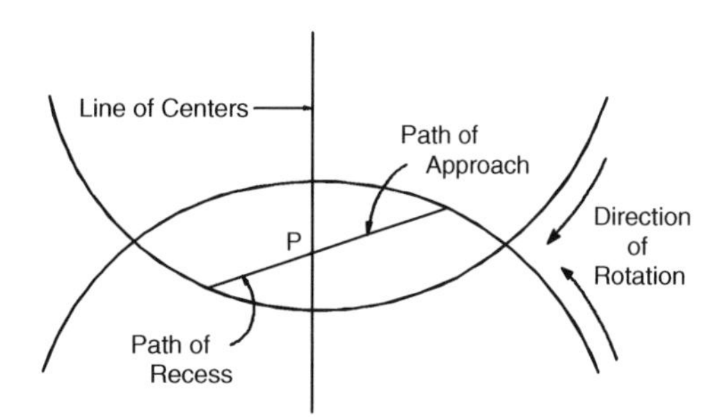

FIGURE 17.6.2
Overlap of addendum circles of meshing gear teeth.

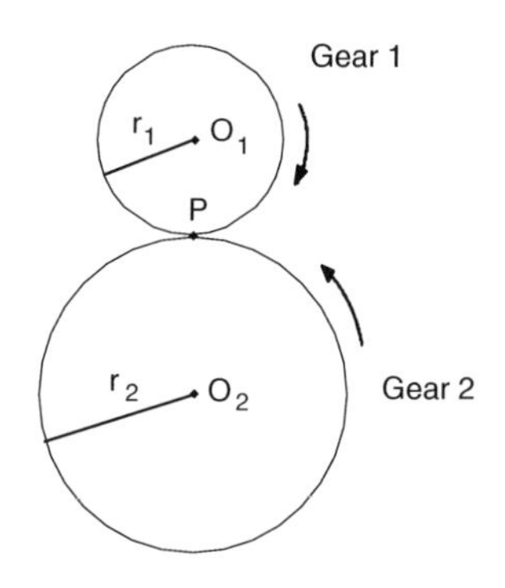

FIGURE 17.6.3
Rolling pitch circles.

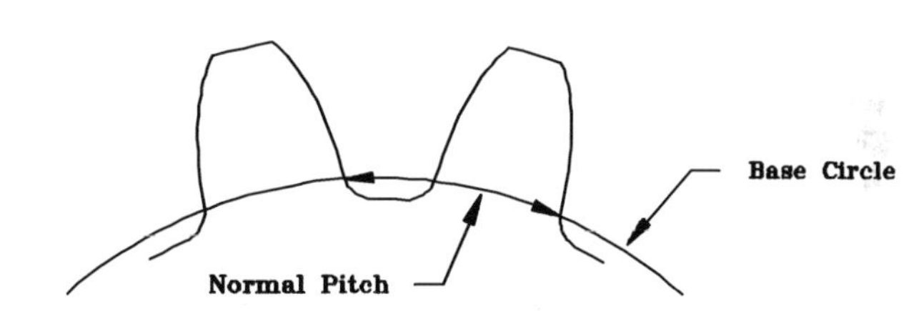

FIGURE 17.6.4
Normal pitch.

As noted in the foregoing sections, the rotation of meshing gears may be modeled by rolling wheels whose profiles are determined by the pitch circles as in Figure 17.6.3. Then, for there to be rolling without slipping, we have from Eq. (17.2.1):

$$r_1\omega_1 = r_2\omega_2 \tag{17.6.1}$$

Then,

$$\omega_1/\omega_2 = r_2/r_1 = d_2/d_1 \tag{17.6.2}$$

where r_1, r_2 and d_1, d_2 are the radii and diameters, respectively, of the pitch circles of the gears. From Eq. (17.5.2) we may express the angular speed ratio as:

$$\omega_1/\omega_2 = N_2/N_1 \tag{17.6.3}$$

where, as before, N_1 and N_2 are the numbers of teeth in the gears. Observe that the angular speed ratio is inversely proportional to the tooth ratio.

Observe that for meshing gears to maintain conjugate action it is necessary that at least one pair of teeth be in contact at all times (otherwise there will be intermittent contact with kinematic discontinuities). The average number of teeth in contact at any time is called the *contact ratio*.

The contact ratio may also be expressed in geometric terms. To this end, it is helpful to introduce the concept of *normal pitch*, defined as the distance between corresponding points on adjacent teeth measured along the base circle, as shown in Figure 17.6.4. From the property of involute curves being generated by the locus of the end point of a belt or

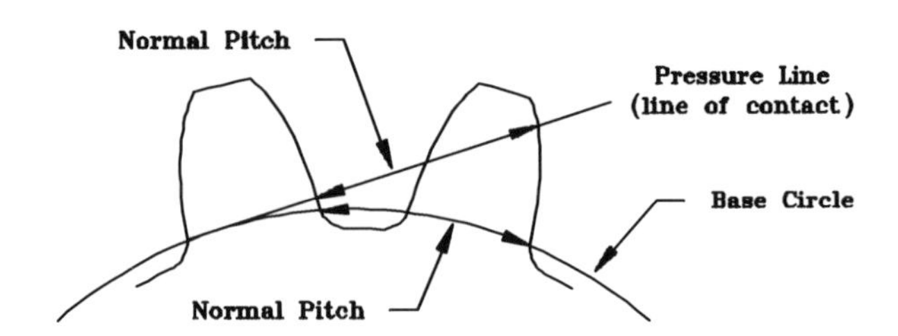

FIGURE 17.6.5
Normal pitch measured along the pressure line.

cord being unwrapped about the base circle (see Figure 17.4.4), we see that the normal pitch may also be expressed as the distance between points of adjacent tooth surfaces measured along the pressure line as in Figure 17.6.5. Observing that the path of contact also lies along the pressure line (see Figure 17.6.1) we see that the length of the path of contact will exceed the normal pitch if more than one pair of teeth are in contact. Indeed, the length of the path of contact is proportional to the average number of teeth in contact. Hence, we have the verbal equation:

$$\text{Contact Ratio} = \frac{\text{Path of Contact}}{\text{Normal Pitch}} \tag{17.6.4}$$

Observe further that the definitions of normal pitch and circular pitch are similar. That is, the normal pitch is defined relative to the base circle, whereas the circular pitch is defined relative to the pitch circle. Hence, from Eq. (17.5.1), we may express the normal pitch p_n as:

$$p_n = \pi d_b / N \tag{17.6.5}$$

where d_b is the diameter of the base circle. Then, from Eqs. (17.5.1) and (17.4.2), the ratio of the normal pitch to the circular pitch is:

$$p_n / p = d_b / d = \cos\theta \tag{17.6.6}$$

where θ is the pressure angle.

In view of Figures 17.6.2 and 17.6.5, we can use Eq. (17.6.4) to obtain an analytical representation of the contact ratio. To see this, consider an enlarged and more detailed representation of the addendum circle and the path of contact as in Figure 17.6.6, where P is the pitch point and where the distance between A_1 and A_2 (pressure line/addendum circle intersections) is the length of the path of contact. Consider the segment A_2P; let Q be at the intersection of the line through A_2 perpendicular to the line through the gear centers as in Figure 17.6.7. Then A_2QO_2 is a right triangle whose sides are related by:

$$(O_2Q)^2 + (QA_2)^2 = (O_2A_2)^2 = r_{2a}^2 \tag{17.6.7}$$

where r_{2a} is the addendum circle radius of gear 2.

Let ℓ_2 be the length of segment A_2P (the path of approach of Figure 17.6.2). Then, from Figure 17.6.7, we have the relations:

$$QA_2 = \ell_2 \cos\theta \tag{17.6.8}$$

$$PQ = \ell_2 \sin\theta \tag{17.6.9}$$

$$O_2Q = r_{2p} + PQ = r_{2p} + \ell_2 \sin\theta \tag{17.6.10}$$

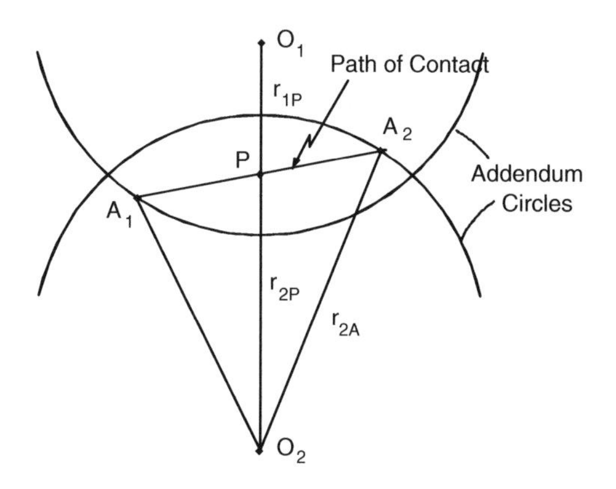

FIGURE 17.6.6
Addendum circles and path of contact for meshing gears.

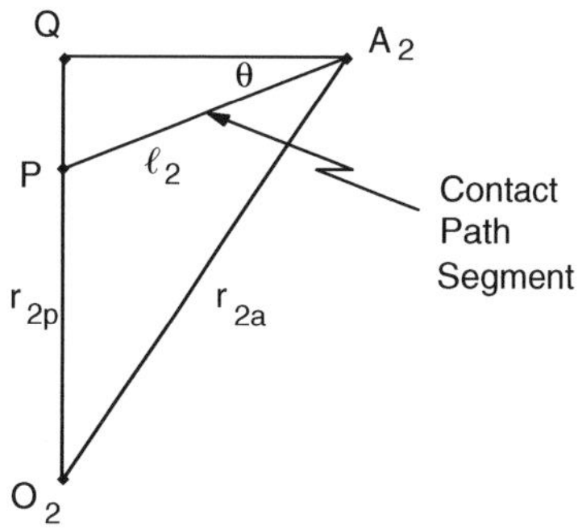

Q
A2
θ
P
ℓ2
Contact
Path
Segment
r2p
r2a
O2

FIGURE 17.6.7
Geometry surrounding the contact path.

where, as before, θ is the pressure angle, and r_{2p} is the pitch circle radius of gear 2. Then, by substituting from Eqs. (17.6.8) and (17.6.10) into (17.6.7), we have:

$$\left(r_{2p} + \ell_2 \sin\theta\right)^2 + \left(\ell_2 \cos\theta\right)^2 = r_{\theta a}^2 \tag{17.6.11}$$

Solving for ℓ_2, we obtain:

$$\ell_2 = -r_{2p} \sin\theta + \left[r_{2a}^2 - r_{2p}^2 \cos^2\theta\right]^{1/2} \tag{17.6.12}$$

Similarly, for gear 1 we have:

$$\ell_1 = -r_{1p} \sin\theta + \left[r_{1a}^2 - r_{1p}^2 \cos^2\theta\right]^{1/2} \tag{17.6.13}$$

where ℓ_1 is the length of the contact path segment from P to A_1 (the path of recess in Figure 17.6.2), and r_{1p} and r_{1a} are the pitch and addendum circle radii for gear 1.

From Eqs. (17.6.4) and (17.6.6), the contact ratio is then:

$$\text{Contact Ratio} \overset{D}{=} CR = \frac{\ell_1 + \ell_2}{p\cos\theta} \tag{17.6.14}$$

where p is the circular pitch, and the lengths of the contact path segments ℓ_1 and ℓ_2 are given by Eqs. (17.6.12) and (17.6.13). Spotts and Shoup [17.10] state that for smooth gear operation the contact ratio should not be less than 1.4.

To illustrate the use of Eq. (17.6.14), suppose it is desired to transmit angular motion with a 2-to-1 speed ratio between axes separated by 9 inches. If 20° pressure angle gears with diametral pitch 7 are chosen, what will be the contact ratio? To answer this question, let 1 and 2 refer to the pinion and gear, respectively. Then, from Eq. (17.6.2), we have:

$$r_{2p}/r_{1p} \quad \text{and then} \quad r_{1p} + 2r_{1p} = 9 \tag{17.6.15}$$

or

$$r_{1p} = 3 \text{ in.} \quad \text{and} \quad r_{2p} = 6 \text{ in.} \tag{17.6.16}$$

From Eqs. (17.5.2) and (17.5.5), we then also have:

$$N_1 = 7(2r_{1p}) = 42 \quad \text{and} \quad N_2 = 7(2r_{2p}) = 84 \tag{17.6.17}$$

and

$$r_{1a} = r_{1p} + 1/7 = 3.143 \text{ in.} \quad \text{and} \quad r_{2a} = r_{2p} + 1/2 = 6.143 \tag{17.6.18}$$

Hence, from Eqs. (17.6.12) and (17.6.13) the contact path segment lengths are found to be:

$$\ell_1 = 0.363 \text{ in.} \quad \text{and} \quad \ell_2 = 0.387 \text{ in.} \tag{17.6.19}$$

Finally, from Eq. (17.5.4), the circular pitch p is:

$$p = \pi/7 = 0.449 \text{ in.} \tag{17.6.20}$$

The contact ratio is then:

$$CR = (\ell_1 + \ell_2)/p\cos\theta = (0.363 + 0.387)/(0.449)(0.940) = 1.777 \tag{17.6.21}$$

17.7 Kinetics of Meshing Involute Spur Gear Teeth

As noted earlier, gears have two purposes: (1) transmission of motion, and (2) transmission of forces. In this section, we consider the second purpose by studying the forces transmitted between contacting involute spur gear teeth. To this end, consider the contact of two teeth as they pass through the pitch point as depicted in Figure 17.7.1. It happens that at the pitch point there is no sliding between the teeth. Hence, the forces transmitted between the teeth are equivalent to a single force **N**, normal to the contacting surfaces and thus

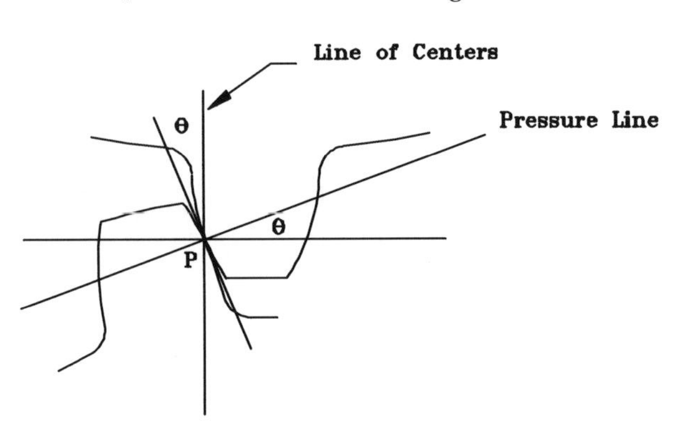

FIGURE 17.7.1
Contacting gear teeth at the pitch point.

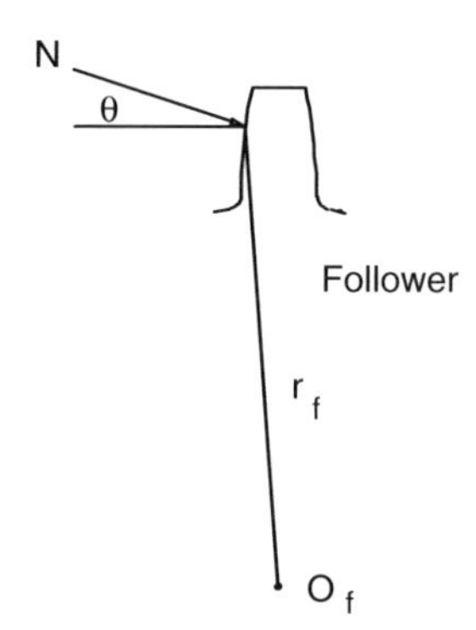

FIGURE 17.7.2
Force acting onto the follower gear.

along the pressure line, which is also the path of contact. (This is the reason the path of contact, or line of action, is also called the *pressure line*.)

As we noted earlier, the angle θ between the pressure line and a line perpendicular to the line of centers is called the *pressure angle* (see Section 17.4). The pressure angle is thus the angle between the normal to the contacting gear surfaces (at their point of contact) and the tangent to the pitch circles (at their point of contact). We also encountered the concept of the pressure angle in our previous chapter on cams (see Section 16.3).

The pressure angle determines the magnitude of component N_T of the normal force **N** tangent to the pitch circle, generating the moment about the gear center. That is, for the follower gear, the moment M_f about the gear center O_f is seen from Figure 17.7.2 to be:

$$M_f = r_f N_T = r_f N \cos\theta \tag{17.7.1}$$

where N is the magnitude of **N**.

Equation (17.7.1) shows the importance of the pressure angle in determining the magnitude of the driving moment. For most gears, the pressure angle is designed to be either 14.5 or 20°, with a recent trend toward 20°. The pressure angle, however, is also dependent upon the gear positioning. That is, because the position of the pitch circles depends upon the gear center separation (see Section 17.5 and Figure 17.5.2), the pressure angle will not, in general, be exactly 14.5 or 20°, as designed.

17.8 Sliding and Rubbing between Contacting Involute Spur Gear Teeth

When involute spur gear teeth are in mesh (in contact), if the contact point is not at the pitch point the tooth surfaces slide relative to each other. This sliding (or "rubbing") can lead to wear and degradation of the tooth surfaces. As we will see, this rubbing is greatest at the tooth tip and tooth root, decreasing monotonically to the pitch point. The effect of the sliding is different for the driver and the follower gear.

To see all this, consider Figure 17.8.1 showing driver and follower gear teeth in mesh. Let P_1 be the point of initial mesh (contact) of the gear teeth and let P_2 be the point of final contact. If P_{1f} is that point of the follower gear coinciding with P_1, then the velocity of P_{1f} may be expressed as:

$$\mathbf{V}^{P_{1f}} = \boldsymbol{\omega}_f \times \mathbf{O}_f\mathbf{P}_{1f} \tag{17.8.1}$$

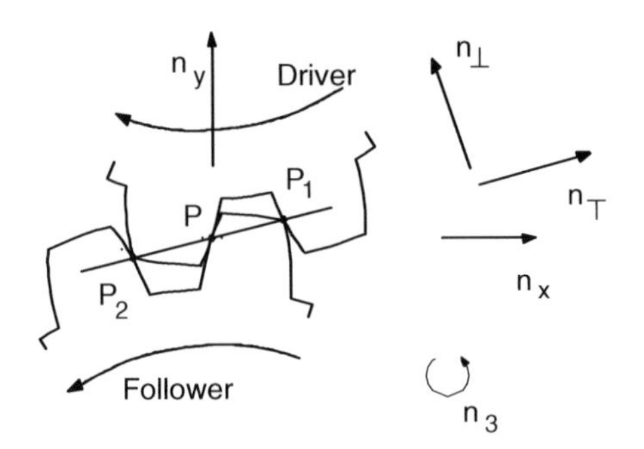

FIGURE 17.8.1
Driver and follower gears in mesh.

where ω_f is the angular velocity of the follower gear and O_fP_{1f} is the position vector locating P_{1f} relative to the follower gear center O_f. Then, in terms of unit vectors shown in Figure 17.8.1, O_fP_{1f} may be expressed as:

$$\mathbf{O}_f\mathbf{P}_{1f} = r_f\mathbf{n}_y + \xi\mathbf{n} \tag{17.8.2}$$

where r_f is the pitch circle radius of the follower gear and ξ is the distance from the pitch point P to P_{1f}. The unit vector $\mathbf{n}$ is parallel to the line of contact (or pressure line) and is thus normal to the contacting gear tooth surfaces. Also, $\boldsymbol{\omega}_f$ may be expressed as:

$$\boldsymbol{\omega}_f = \omega_f\mathbf{n}_3 \tag{17.8.3}$$

where $\mathbf{n}_3$ is normal to the plane of the gears.

By substituting from Eqs. (17.8.2) and (17.8.3) into (17.8.1) we obtain:

$$\begin{aligned}\mathbf{V}^{P_{1f}} &= \omega_f\mathbf{n}_3 \times \left(r_f\mathbf{n}_y + \xi\mathbf{n}\right)\\ &= -r_f\omega_f\mathbf{n}_x + \omega_f\xi\mathbf{n}_\perp\end{aligned} \tag{17.8.4}$$

where $\mathbf{n}_\perp$ is perpendicular to $\mathbf{n}$ and parallel to the plane of the gear as in Figure 17.8.1.

Similarly, if P_{1d} is that point of the driver gear coinciding with P_{1f}, the velocity of P_{1d} is:

$$\mathbf{V}^{P_{1d}} = -r_d\omega_d\mathbf{n}_x - \omega_d\xi\mathbf{n}_\perp \tag{17.8.5}$$

The difference in these velocities is the sliding (or rubbing) velocity. From Eq. (17.2.1) we see that:

$$r_f\omega_f = r_d\omega_d \tag{17.8.6}$$

Hence, the rubbing (or sliding) velocity $\mathbf{V}_s$ is:

$$\mathbf{V}_s = \mathbf{V}^{P_{1f}} - \mathbf{V}^{P_{1d}} = \left(\omega_f + \omega_d\right)\xi\mathbf{n}_\perp \tag{17.8.7}$$

Equation (17.8.7) shows that the rubbing is greatest at the points of initial and final contact (maximum ξ) and that the rubbing vanishes at the pitch point ($\xi = 0$).

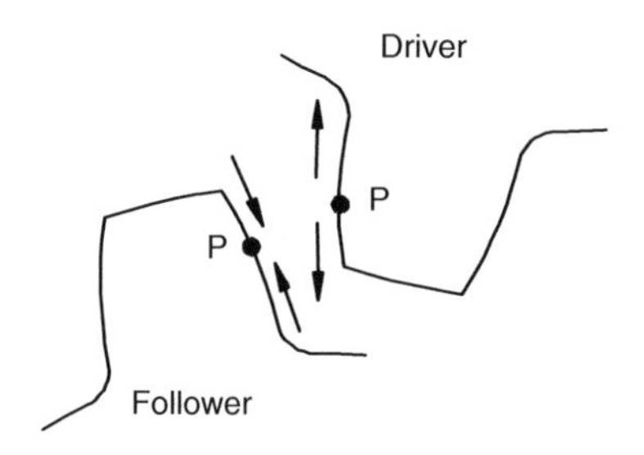

FIGURE 17.8.2
Sliding or rubbing direction for meshing drives and follower gear teeth.

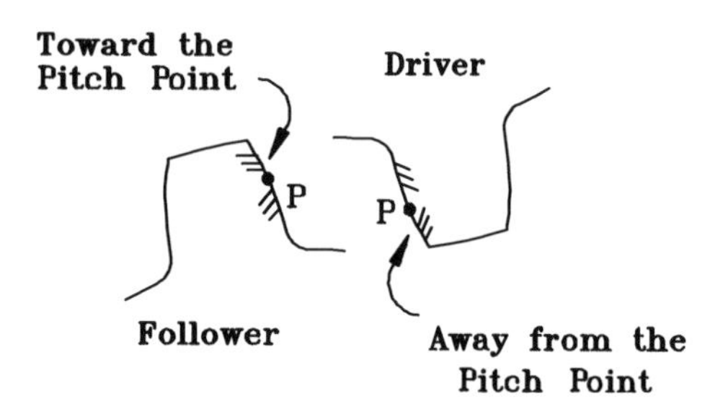

FIGURE 17.8.3
Fracture pattern for meshing driver and follower gear teeth.

Consider now the rubbing itself: First, for the follower gear, as the meshing begins the contact point is at the tip P_{1f}. The contact point then moves down the follower tooth to the pitch point P. During this movement we see from Eq. (17.8.7) that with $\xi > 0$ the sliding velocity $\mathbf{V}_s$ is in the $\mathbf{n}_\perp$ direction. This means that, because $\mathbf{V}_s$ is the sliding velocity of the follower gear relative to the driver gear, the upper portion of the follower gear tooth has the rubbing directed toward the pitch point.

Next, after reaching the pitch point, the contact point continues to move down the follower gear tooth to the root point P_{2f}. During this movement, however, ξ is negative; thus, we see from Eq. (17.8.7) that the sliding velocity $\mathbf{V}_s$ is now in the $-\mathbf{n}_\perp$ direction. This in turn means that the rubbing on the lower portion of the follower gear tooth is also directed toward the pitch point.

Finally, for the driver gear tooth the rubbing is in the opposite directions. When the contact is on the lower portion of the tooth (below the pitch point), the rubbing is directed away from the pitch point and toward the root. When the contact is on the upper portion of the tooth (above the pitch point), the rubbing is also directed away from the pitch point, but now toward the tooth tip.

Figure 17.8.2 shows this rubbing pattern on the driver and follower teeth. For tooth wear (or degradation), this rubbing pattern has the tendency to pull the driver tooth surface away from the pitch point. On the follower gear tooth the rubbing tends to push the tooth surface toward the pitch point. If the gear teeth are worn to the point of fracture, the fracture will initiate as small cracks directed as shown in Figure 17.8.3.

17.9 Involute Rack

Many of the fundamentals of involute tooth geometry can be understood and viewed as being generated by the *basic rack* gear. A basic rack is a gear of infinite radius as in Figure 17.9.1. For an involute spur gear the basic rack has straight-sided teeth. The inclination of the tooth then defines the pressure angle as shown.

A rack can be used to define the gear tooth geometry by visualizing a plastic (perfectly deformable) wheel, or gear blank, rolling on its pitch circle over the rack, as in Figure 17.9.2. With the wheel being perfectly plastic the rack teeth will create impressions, or *footprints* on the wheel, thus forming involute gear teeth, as in Figure 17.9.3.

The involute rack may also be viewed as a reciprocating cutter forming the gear teeth on the gear blank as in Figure 17.9.4. Indeed, a reciprocating rack cutter, as a *hob*, is a common procedure for involute spur gear manufacture [17.2]. The proof that the

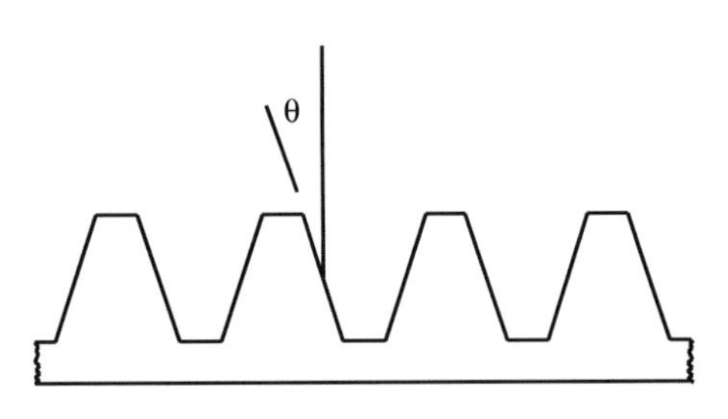

FIGURE 17.9.1
Basic involute rack.

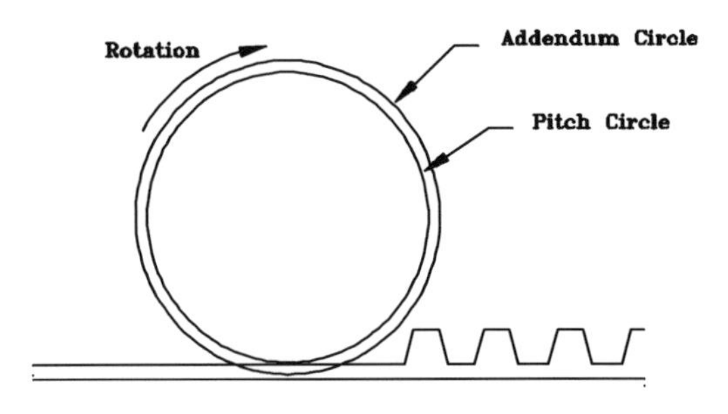

FIGURE 17.9.2
A plastic gear blank rolling on its pitch circle over a rack gear.

FIGURE 17.9.3
Involute teeth formed on a plastic rolling wheel.

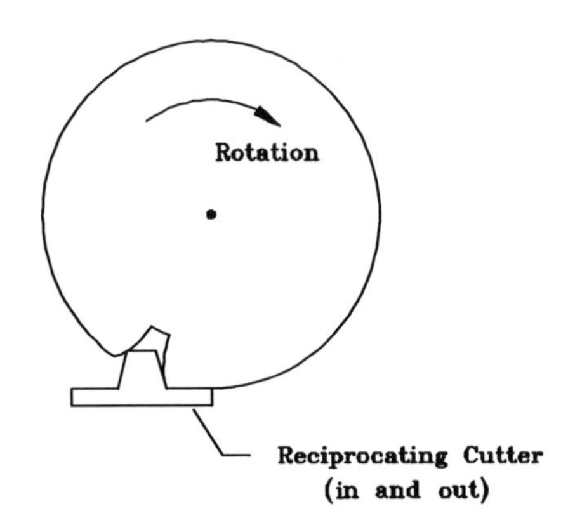

FIGURE 17.9.4
A reciprocating hob cutter cutting a tooth on a gear blank.

straight-sided rack cutter generates an involute spur gear tooth is somewhat beyond our scope; however, a relatively simple proof using elementary procedures of differential geometry may be found in Reference 17.12.

17.10 Gear Drives and Gear Trains

As we have noted several times, gears are used for the transmission of forces and motion. Thus, a pair of meshing gears is called a *transmission*. Generally speaking, however, a transmission usually employs a series of gears, and is sometimes called a *gear train*. Figure 17.10.1 depicts a gear train of parallel shaft gears. If the pitch diameter of the first gear is d_1 and the pitch diameter of the nth gear is d_n, then by repeated use of Eq. (17.2.1) we find the angular speed ratio to be:

$$\text{Speed Ratio} = \omega_1/\omega_n = d_1/d_n \tag{17.10.1}$$

Theoretically, there is no limit to this speed ratio; however, the larger the number of gears, the greater is the friction loss. Also, from a practical viewpoint, the number of gears is often limited by the space available in a given mechanical system.

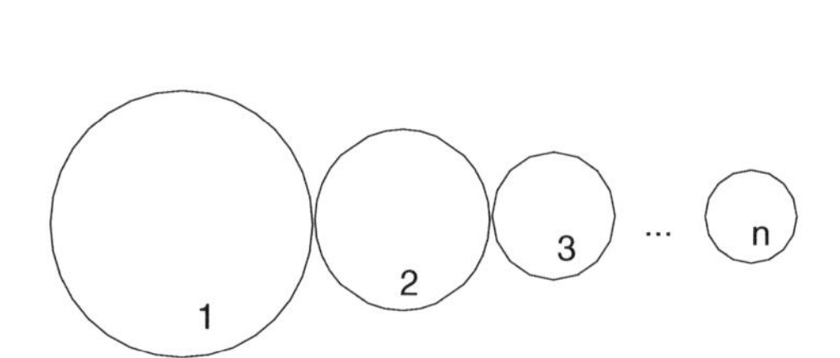

FIGURE 17.10.1
A gear train with n gears.

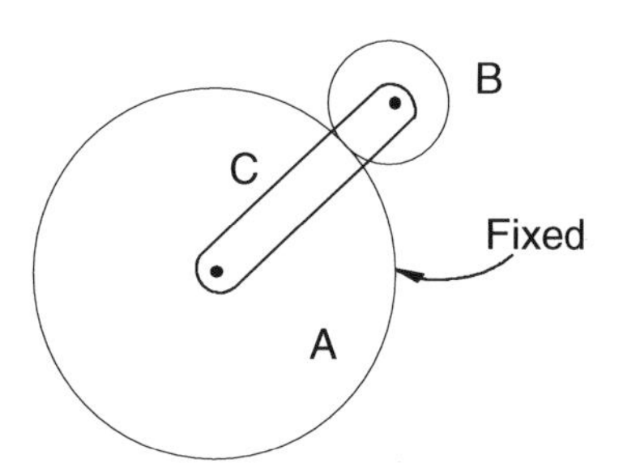

FIGURE 17.10.2
A simple planetary gear system.

The gears of a gear train producing an angular speed ratio, as in Eq. (17.10.1), thus produce either an angular speed increase or an angular speed reduction between the shafts of the first and the last gears. In an ideal system, with no friction losses, there will be a corresponding reduction or increase in the moments applied to the first and last shafts. That is, if the moment applied to the shaft of the first gear is M_1 and if the moment produced at the last shaft is M_n, then we have the ratios:

$$M_n/M_1 = \omega_1/\omega_n = d_1/d_n \tag{17.10.2}$$

An efficient method of speed and moment reduction (or increase) may be obtained by using a *planetary gear system* — so called because one or more of the gears does not have a fixed axis of rotation but instead has an axis that rotates about the other gear axis. That is, although the axes remain parallel, the axis of one or more of the gears is itself allowed to rotate. To illustrate this, consider the system of Figure 17.10.2 consisting of two gears A and B, whose axes are connected by a link C. If gear A is fixed, then as gear B engages gear A in mesh, the axis of B will move in a circle about the axis of A. The connecting link C will rotate accordingly. Because B moves around A, B is often called a *planet gear* and then A is called the *sun gear.*

Using our principles of elementary kinematics, we readily discover that the angular speeds of B and C are related by the expressions:

$$\omega_B = \omega_C[1 + r_A/r_B] \tag{17.10.3}$$

or

$$\omega_B/\omega_C = 1 + (r_A/r_B) \tag{17.10.4}$$

where r_A and r_B are the pitch circle radii of A and B.

To see this, let O_A and O_B be the centers of gears A and B. Then because O_A is fixed and because C rotates about O_A, the speed of O_B is:

$$V^{O_B} = r_C\omega_C = (r_A + r_B)\omega_C \tag{17.10.5}$$

where r_C is the effective length of the connecting link C. However, because the pitch circle of B rolls on the pitch circle of A, we also have (see Eq. (4.11.5)):

$$V^{O_B} = r_B\omega_B \tag{17.10.6}$$

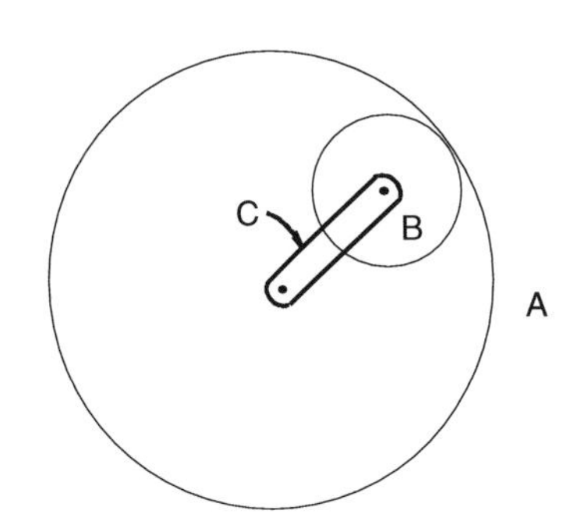

FIGURE 17.10.4
A simple planetary gear system with an external (or ring) gear.

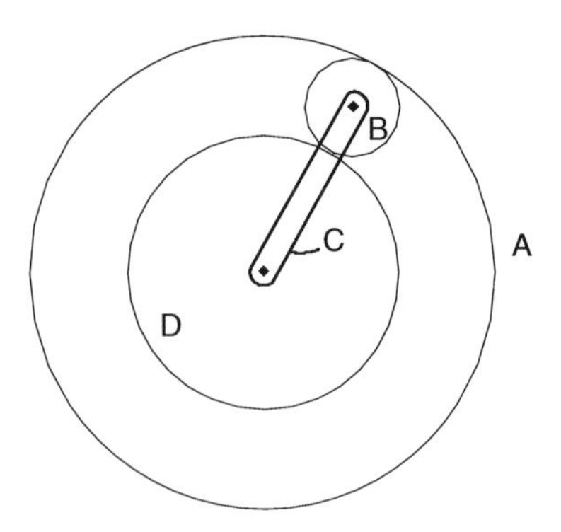

FIGURE 17.10.5
A planetary gear system with both a sun gear and a ring gear.

Hence, we have:

$$r_B\omega_B = (r_A + r_B)\omega_C \tag{17.10.7}$$

or

$$\omega_B/\omega_C = (r_A + r_B)/r_B = 1 + (r_A/r_B) \tag{17.10.8}$$

As a second illustration, consider the system of Figure 17.10.4, where the fixed sun gear *A* is external to the planet gear *B*. The external sun gear is often called a *ring gear*. Again, using the principles of elementary kinematics as above, we readily see that, if the ring gear *A* is fixed, the angular speeds of the planet gear *B* and the connecting link *C* are related by the expression:

$$\omega_B/\omega_C = 1 - (r_A/r_B) \tag{17.10.9}$$

Planetary gear systems generally have both a sun gear and a ring gear as represented in Figure 17.10.5. In this case, the system has four members, *A*, *B*, *C*, and *D*, with *C* being a connecting link between the centers of the sun gear *D* and the planet gear *B*.

Generally, in applications, either the sun gear or the ring gear is fixed. If the sun gear *D* is fixed, we can, by again using the principles of elementary kinematics, find relations between the angular velocities of the ring gear *A*, the planet gear *B*, and the connecting link *C*. For example, the angular velocities of *A* and *C* are related by the expression:

$$\omega_A = [1 + (r_D/r_A)]\omega_C \tag{17.10.10}$$

Similarly, if the ring gear *A* is fixed, the angular velocities of the sun gear *D* and the connecting link *C* are related as:

$$\omega_D = [1 + (r_A/r_D)]\omega_C \tag{17.10.11}$$

17.11 Helical, Bevel, Spiral Bevel, and Worm Gears

Thus far we have focused our attention and analyses on parallel shaft gears and specifically on spur gears. Although these are the most common gears, there are many other kinds of gears and other kinds of gear tooth forms. In the following sections, we will briefly consider some of the more common of these other types of gears. Detailed analyses of these gears, however, is beyond our scope. Indeed, the geometry of these gears makes their analyses quite technical and complex. In fact, comprehensive analyses of many of these gears have not yet been developed, and research on them is continuing. Nevertheless, the fundamental principles (conjugate action, pitch points, rack forms, etc.) are the same or very similar to those for involute spur gears.

17.12 Helical Gears

Helical gears, like spur gears, are gears for parallel shafts. They may be viewed as a modification of spur gears where the teeth are curved in the axial direction. The objective of the curved tooth is to provide a smoother and quieter mesh than is obtained with spur gears. Specifically, the teeth form helical curves along the cylinder of the pitch circles. Figure 17.12.1 shows a sketch of a helical gear.

To explore the geometry of helical gears a bit further, consider the basic rack with inclined teeth as shown in Figure 17.12.2. If the rack is deformed into a cylindrical shape, the gear form of Figure 17.12.1 is obtained. When formed in this way, the teeth have an involute form in planes parallel to the plane of the gear.

If we visualize the rack being deformed and wrapped into a helical gear, the teeth form helix segments along the gear cylinder. To see this, consider a rectangular sheet as in Figure 17.12.3 having a diagonal line AB as shown. If the sheet is wrapped, or spindled, into a cylinder, as in Figure 17.12.4, the line AB becomes a circular helix.

The inclination angle of the rack gear teeth of Figure 17.12.2a is called the *helix angle*. In practice, the helix angle can range from just a few degrees up to 45° [17.6]. The greater the helix angle, the greater the gear tooth length and thus the greater the time of contact.

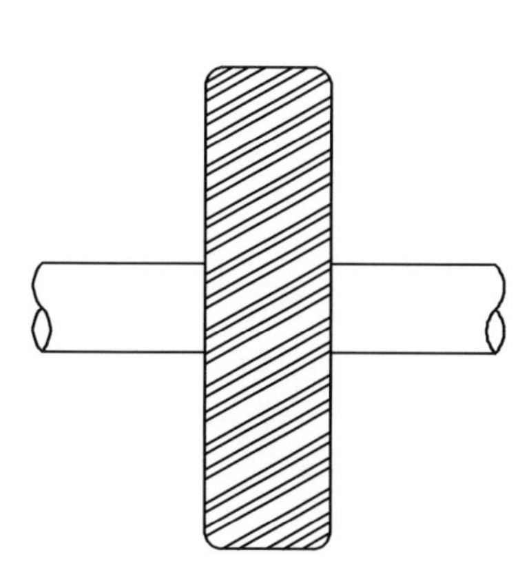

FIGURE 17.12.1
A helical gear.

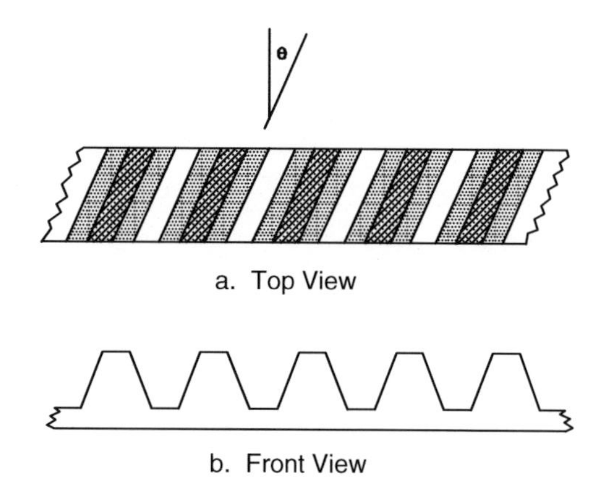

FIGURE 17.12.2
Basic rack of a helical gear.

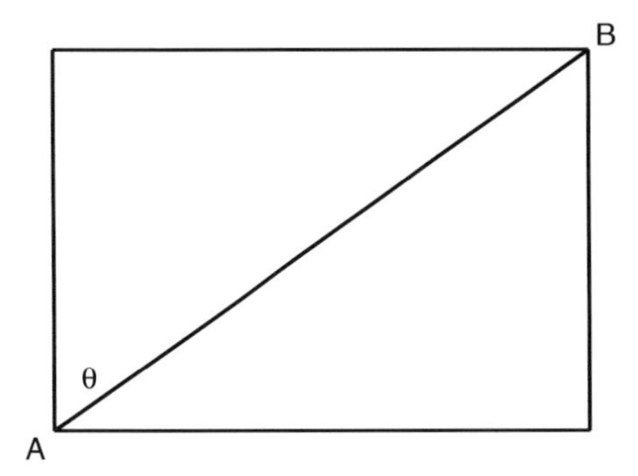

FIGURE 17.12.3
A rectangular sheet with diagonal line *AB*.

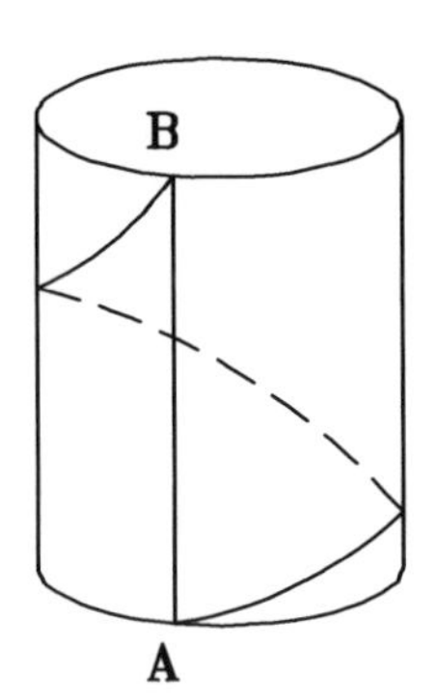

FIGURE 17.12.4
A circular helix.

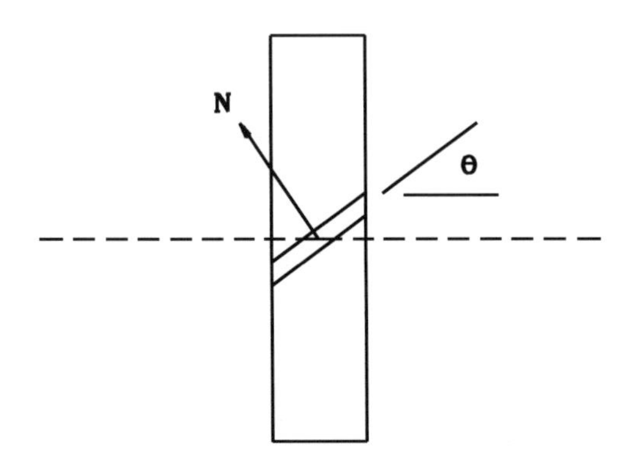

FIGURE 17.12.5
Normal force N on helix gear tooth surface.

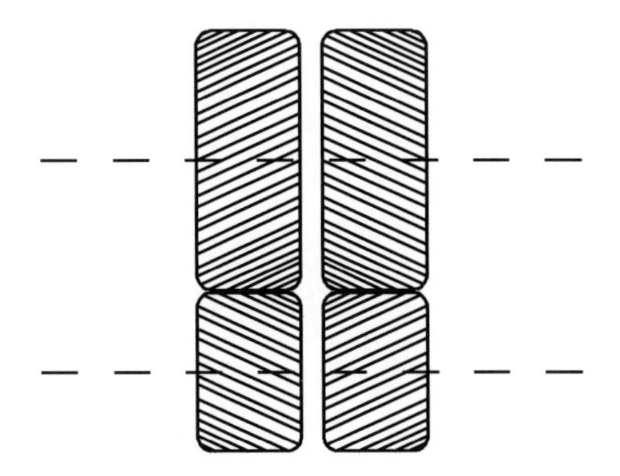

FIGURE 17.12.6
A pair of meshing helical gears — herringbone gears.

Unfortunately, however, the inclined tooth surface creates an axial thrust for gears in mesh, as depicted in Figure 17.12.5. Specifically, the normal force N will have an axial component $N\sin\theta$. This axial force can then in turn reduce the efficiency of meshing helical gears by introducing friction forces to be overcome by the driving gear.

Also, in the absence of thrust-bearing constraint, the axial force component can cause the meshing helical gears to tend to separate axially. To eliminate this separation tendency, helical gears are often used in pairs with opposite helix angles as in Figure 17.12.6. Such gears are generally called *herringbone gears*.

17.13 Bevel Gears

Unlike spur and helical gears, bevel gears transmit forces and motion between nonparallel shafts. With bevel gears, the shaft axes intersect, usually at 90°. Figure 17.13.1 depicts a typical bevel gear pair. As is seen in the figure, bevel gears have a conical shape. Their geometrical characteristics are thus somewhat more complex than those of spur or helical gears. The kinematic principles, however, are essentially the same.

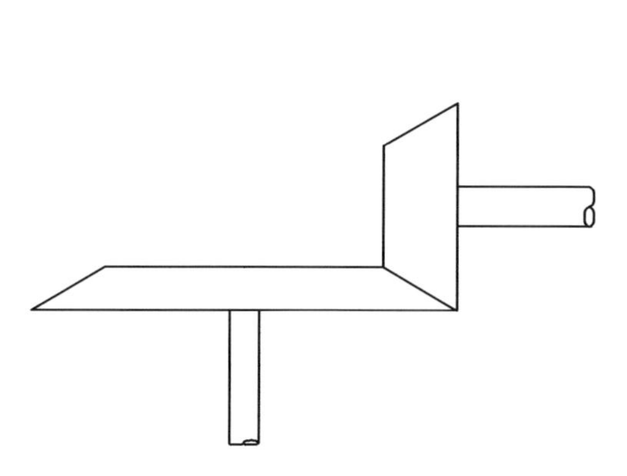

FIGURE 17.13.1
A bevel gear pair.

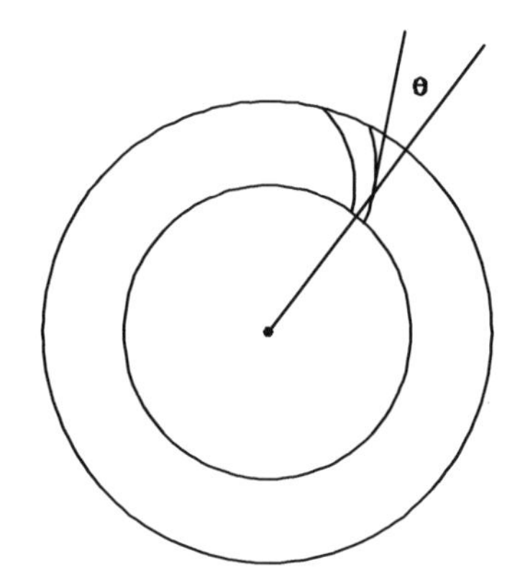

Figure 17.13.2
Spiral angle.

Bevel gears have tapered teeth. The geometric properties of these teeth (for example, pressure angle, pitch, backlash, etc.) are generally measured at the mean cone position — that is, halfway between the large and small ends ("heel and toe") of the cone frustrum.

In their profile, bevel gear teeth are like involute spur gear teeth. The teeth themselves may be either straight or curved. Bevel gears with curved teeth are commonly called *spiral bevel gears*. The curvature of a spiral bevel gear tooth may vary somewhat, but typically at the mid-tooth position the tangent line to the tooth will make an angle θ with a conical element, as depicted in Figure 12.13.2. This angle, which is typically 35°, is called the *spiral angle*.

Straight bevel gears are analogous to involute spur gears, whereas spiral bevel gears are analogous to helical gears. The advantages of spiral bevel gears over straight bevel gears are analogous to the advantages of helical gears over spur gears — that is, stronger and smoother acting teeth with longer tooth contact. The principal disadvantage of spiral bevel and helical gear teeth is their need for precision manufacture. Also, if their precise geometry is distorted under load, the kinematics of the gears can be adversely affected.

The geometry of bevel gears makes their analysis and design more difficult than for parallel shaft gears. Also, bevel gears are generally less efficient than parallel shaft gears. As a consequence, engineers and designers use parallel shaft gears wherever possible (as, for example, with front-wheel-drive cars with engines mounted parallel to the drive axles). Details of the geometry and kinematics of bevel gears are beyond our scope, but the interested reader may want to contact the references for additional information.

17.14 Hypoid and Worm Gears

Hypoid and worm gears are used to transmit forces and motion between nonparallel *and* nonintersecting shafts. Generally, the shafts are perpendicular. In this sense, hypoid and worm gears are similar to bevel gears.

Figure 17.14.1 depicts a hypoid gear set. As with other gear pairs, the larger member is called the *gear* and the smaller is called the *pinion*. Hypoid gears have the same form and shape as spiral bevel gears; however, the nonintersecting shafts produce hyperbolic, as opposed to conical, gear shapes.

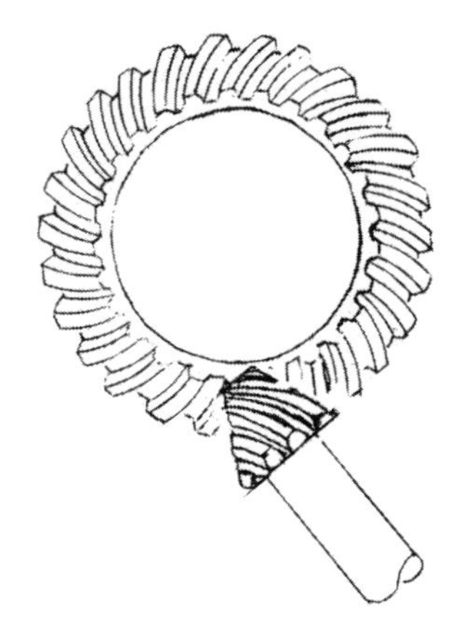

FIGURE 17.14.1
A hypoid gear set.

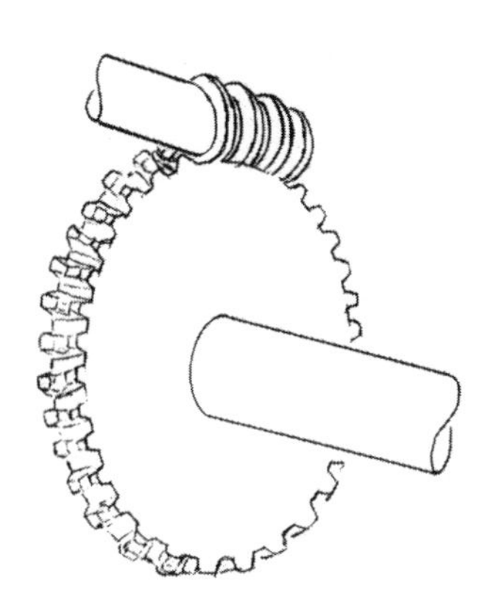

FIGURE 17.14.2
A worm gear set.

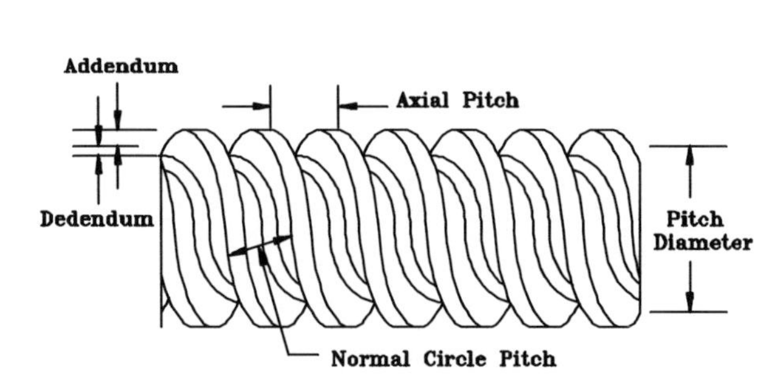

FIGURE 17.14.3
A profile sketch of a worm.

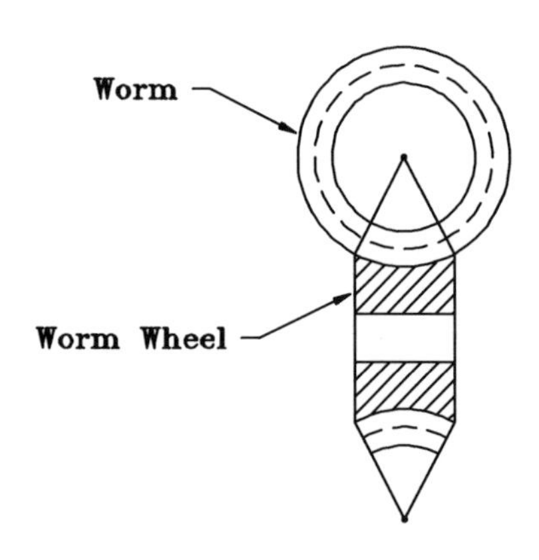

FIGURE 17.14.4
Profile of a worm gear set in mesh.

Figure 17.14.2 depicts a worm gear set. With worm gears, the smaller member is called the *worm* and the larger the *worm wheel*. A worm is similar to a screw, and the worm wheel is similar to a helical gear. The geometry of a worm is analogous to that of a helical gear rack. Figure 17.14.3 provides a profile view of a worm and common terminology [17.10]. Similarly, Figure 17.14.4 provides a profile of a worm gear set in mesh.

Hypoid and worm gears are generally used when a large speed-reduction ratio is needed with smooth action, or with little or no backlash. Hypoid and worm gears are employed when it is impractical or impossible to use intersecting shafts. Indeed, a principal advantage of hypoid and worm gears — in addition to their high speed reduction and smooth operation — is that bearings for both shafts may be used on both sides of the gear elements, thus providing structural rigidity and stability. On the other hand, a disadvantage of hypoid and worm gears (as with bevel gears) is that they are not nearly as efficient as parallel shaft gears. The inclined and curved surfaces, while strong, also induce sliding between the mating surfaces, leading to friction losses. Readers interested in additional details should consult the references.

17.15 Closure

In this chapter, we have briefly considered the fundamentals of gearing, with a focus upon spur gear geometry and kinematics. The dynamic principles we have developed in earlier chapters are directly applicable with gearing systems. In spite of the relative simplicity of spur gear geometry, the complex geometry of other gear forms (that is, helical, bevel, spiral bevel, hypoid, and worm gears) makes elementary analyses impractical, even though the basic principles are essentially the same as those as spur gears. Research on gears and gearing systems is continuing and expanding in response to increasing demands for greater precision and longer-lived transmission systems. Although details of this research are beyond our scope, interested readers may want to consult the references for additional information about this work. We conclude our chapter in the following section with a glossary of gearing terms.

17.16 Glossary of Gearing Terms

The following is a partial listing of terms commonly used in gearing technology together with a brief definition and the chapter section where the term is first discussed:

Addendum — height of a spur gear tooth above the pitch circle (see Figure 17.5.1) [17.5]

Addendum circle — external or perimeter circle of a gear (see Figure 17.5.1) [17.5]

Angle of action — angle turned through by a gear as a typical tooth passes through the path of contact (see also *angle of contact*) [17.6]

Angle of approach — angle turned through by a gear from the position of initial contact of a pair of teeth up to contact at the pitch point (see Figure 17.6.1) [17.6]

Angle of contact — angle turned through by a gear as a typical tooth travels through the path of contact (see also *angle of action*) [17.6]

Angle of recess — angle turned through by a gear from contact of a pair of teeth at the pitch point up to the end of contact (see Figure 17.6.1) [17.6]

Axial pitch — distance between corresponding points of adjacent screw surfaces of a worm, measured axially (see Figure 17.14.3) [17.3]

Backlash — looseness or rearward separation of meshing gear teeth (see Figure 17.5.3) [17.5]

Base circle — circle of rolling wheel pulley (see Figure 17.4.2) or generating involute curve circle (see Figures 17.4.4 and 17.4.5) [17.4]

Bevel gear — gear in the shape of a frustrum of a cone and used with intersecting shaft axes [17.12]

Bottom land — inside or root surface of a spur gear tooth (see Figure 17.5.5) [17.5]

Circular pitch — distance, measured along the pitch circle, between corresponding points of adjacent teeth (see Figure 17.5.3 and Eq. (17.5.1)) [17.5]

Clearance — difference between root circle and base circle radii; the elevation of the support base of a tooth (see Figure 17.5.1) [17.5]

Conjugate action — constant angular speed ratio between meshing gears [17.3]

Conjugate gears — pair of gears that have conjugate action (constant angular speed ratio) when they are in mesh [17.3]

Contact ratio — average number of teeth in contact for a pair of meshing gears [17.6]

Dedendum — depth of a spur gear tooth below the pitch circle (see Figure 17.5.1) [17.5]

Diametral pitch — number of gear teeth divided by the diameter of the pitch circle (see Eq. (17.5.2)) [17.5]

Driver — gear imparting or providing the motion or force [17.2]

Face — surface of a spur gear tooth above the pitch circle (see Figure 17.5.4) [17.5]

Fillet — gear tooth profile below the pitch circle (see Figure 17.5.1) [17.5]

Flank — surface of a spur gear tooth below the pitch circle (see Figure 17.5.5) [17.5]

Follower — gear receiving the motion or force [17.2]

Gear — larger of two gears in mesh [17.2]

Gear train — transmission usually employing several gears in mesh in a series [17.10]

Heel — large end of a bevel gear or of a bevel gear tooth [17.12]

Helical gear — parallel shaft gear with curved teeth in the form of a helix (see Figure 17.12.1) [17.12]

Helix angle — inclination of a helix gear tooth (see Figure 17.12.2a) [17.12]

Herringbone gears — pair of meshing helical gears with opposite helix angles (see Figure 17.12.6) [17.12]

Hob — reciprocating cutter in the form of a rack gear tooth [17.9]

Involute function — function Invϕ defined as $(\tan\phi) - \phi$ (see Eq. (17.4.17)) [17.4]

Involute of a circle — curve formed by the end of an unwrapping cable around a circle (see Figure 17.4.4) [17.4]

Law of conjugate action — requirement that the normal line of contacting gear tooth surfaces passes through the pitch point [17.3]

Line of action — line normal to contacting gear tooth surfaces at their point of contact (see also *line of contact, pressure line,* and *path of contact*) [17.4]

Line of contact — line passing through the locus of contact points of meshing spur gear teeth (see also *line of action, pressure line,* and *path of contact*) [17.6]

Mesh — interaction and engaging of gear teeth

Module — pitch circle diameter divided by the number of teeth of a gear

Normal circular pitch — distance between corresponding points of adjacent screw surfaces of a worm, measured perpendicular or normal to the screw (see Figure 17.13.4) [17.3]

Normal pitch — distance between corresponding points on adjacent teeth measured along the base circle (see Figure 17.6.4) [17.6]

Path of contact — locus of points of contact of a pair of meshing spur gear teeth (see also *line of action, pressure line,* and *line of contact*) [17.6]

Pinion — smaller of two gears in mesh [17.2]

Pitch circles — perimeters of rolling wheels used to model gears in mesh [17.3]

Pitch point — point of contact between rolling pitch circles [17.3]

Planet gear — gear of a planetary gear system whose center moves in a circle about the center of the sun gear [17.10]

Planetary gear system — gear train or transmission where one or more of the gears rotate on moving axes [17.10]

Pressure angle — inclination of line of action (see Figure 17.4.3) [17.4]

Pressure line — line normal to contacting gear tooth surfaces at this point of contact (see also *line of action, line of contact,* and *path of contact*) [17.4]

Rack gear — gear with infinite radius [17.9]

Ring gear — external sun gear of a planetary gear system (see Figure 17.10.4) [17.10]

Root circle — boundary of the root, or open space, between spur gear teeth (see Figure 17.5.1) [17.5]

Spiral angle — angle between the tangent to a spiral bevel gear tooth and a conical element (see Figure 17.12.2) [17.12]

Spiral bevel gear — bevel gear with curved teeth [17.12]

Sun gear — central gear of a planetary gear system with a fixed center [17.10]

Toe — small end of a bevel gear or of a bevel gear tooth [17.12]

Tooth thickness — distance between opposite points at the pitch circle for an involute spur gear tooth (see Figure 17.5.1) [17.5]

Top land — outside surface of a spur gear tooth (see Figure 17.5.4) [17.5]

Transmission — pair of gears in mesh [17.1]

Whole depth — total height of a spur gear tooth (see Figure 17.6.1) [17.3]

Width — axial thickness of a spur gear tooth (see Figure 17.5.4) [17.5]

Working depth — height of the involute portion of a spur gear tooth (see Figure 17.5.1) [17.5]

Worm — smaller member of a worm gear set, in the form of a screw (see Figure 17.13.2) [17.3]

Worm wheel — larger member of a worm gear set, similar to a helical gear (see Figure 17.3.2) [17.3]

References

17.1. Buckingham, E., *Analytical Mechanics of Gears*, Dover, New York, 1963.

17.2. Townsend, D. P., Ed., *Dudley's Gear Handbook*, 2nd ed., McGraw-Hill, New York, 1991.

17.3. Lent, D., *Analysis and Design of Mechanisms*, 2nd ed., Prentice Hall, Englewood Cliffs, NJ, 1970, chap. 6.

17.4. Oberg, E., Jones, F. D., and Horton, H. L., *Machinery's Handbook*, 23rd ed., Industrial Press, New York, pp. 1765–2076.

17.5. Dudley, D. W., *Handbook of Practical Gear Design*, McGraw-Hill, New York, 1984.

17.6. Drago, R. J., *Fundamentals of Gear Design*, Butterworth, Stoneham, MA, 1988.

17.7. Michalec, G. W., *Precision Gearing Theory and Practice*, Wiley, New York, 1966.

17.8. Jones, F. D., and Tyffel, H. H., *Gear Design Simplified*, 3rd ed., Industrial Press, New York, 1961.

17.9. Litvin, F. L., *Gear Geometry and Applied Theory*, Prentice Hall, Englewood Cliffs, NJ, 1994.

17.10. Spotts, M. F., and Shoup, T. E., *Design of Machine Elements*, 7th ed., Prentice Hall, Englewood Cliffs, NJ, 1998, p. 518.

17.11. Coy, J. J., Zaretsky, E. V., and Townsend, D. P., *Gearing*, NASA Reference Publication 1152, AVSCOM Technical Report 84-C-15, 1985.
17.12. Chang, S. H., Huston, R. L., and Coy, J. J., A Computer-Aided Design Procedure for Generating Gear Teeth, ASME Paper 84-DET-184, 1984.

Problems

Section 17.5 Spur Gear Nomenclature

P17.5.1: The gear and pinion of meshing spur gears have 35 and 25 teeth, respectively. Let the pressure angle be 20° and let the teeth have involute profiles. Let the diametral pitch be 5. Determine the following:

a. Pitch circle radii for the gear and pinion
b. Base circle radii for the gear and pinion
c. Center-to-center distance
d. Circular pitch
e. Module
f. Addendum
g. Dedendum
h. Clearance
i. Fillet radius

P17.5.2: Repeat Problem P17.5.1 if the gear and pinion have 40 and 30 teeth, respectively.

P17.5.3: Repeat Problems P17.5.1 and 17.5.2 if the module m, given by Eq. (17.5.3), is 5. Express the answers in millimeters.

P17.5.4: Suppose two meshing spur gears are to have an angular speed ratio of 3 to 2. Suppose further that the distance between the centers of the gears is to be 5 inches. For a diametral pitch of 6, determine the number of teeth in each gear. (*Hint:* The angular speed ratio is the inverse of the ratio of the pitch circle radii [see Eq. (17.6.2)].)

P17.5.5: See Problem P17.5.4. Let the pressure angle of the gears be 20°. Determine the circular pitch and the base circle radii of the gears.

P17.5.6: Repeat Problem P17.5.5 if the pressure angle is 14.5°.

P17.5.7: See Problem P17.5.4. Determine the module of the gears and pitch radii in both inches and centimeters.

Section 17.6 Kinematics of Meshing Involute Spur Gear Teeth

P17.6.1: Consider three parallel shaft spur gears in mesh as represented in Figure P17.6.1. Develop expressions analogous to Eqs. (17.6.2) and (17.6.3) for the angular speed ratio of gear 1 to gear 3.

P17.6.2: Generalize the results of Problem P17.6.1 for n gears in mesh.

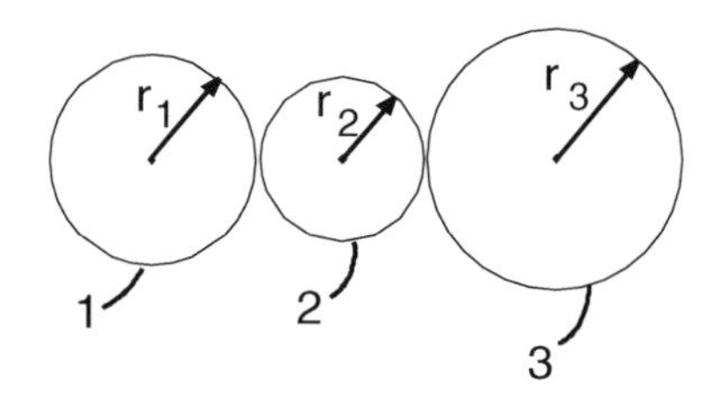

FIGURE P17.6.1
A train of three spur gears.

P17.6.3: Two meshing spur gears have an angular speed ratio of 2.0 to 1. Suppose the pinion (smaller gear) has 32 teeth and that the distance separating the gear centers is 12 inches. Determine: (a) the number of teeth in the gear (larger gear); (b) the diametral pitch; (c) the circular pitch; and (d) the pitch circle radii of the gears.

P17.6.4: Repeat Problem P17.6.3 if the center-to-center distance is 30 cm. Instead of the diametral pitch, find the module.

P17.6.5: Verify Eqs. (17.6.12) and (17.6.13).

P17.6.6: Meshing 20° pressure angle spur gears with 25 and 45 teeth, respectively, have diametral pitch 6. Determine the contact ratio.

P17.6.7: Repeat Problem P17.6.6 if the gears have 30 and 50 teeth, respectively.

P17.6.8: Repeat Problem P17.6.6 if the diametral pitch is (a) 7 and (b) 8.

Section 17.8 Sliding and Rubbing between Contacting Involute Spur Gear Teeth

P17.7.1: See Problem P17.6.6 where 20° pressure angle gears with 25 and 45 teeth and diametral pitch 6 are in mesh. Determine the maximum distance from the pitch point to a point of contact between the teeth.

P17.7.2: Repeat Problem P17.7.1 for the data of Problem P17.6.6.

P17.7.3: See Problem P17.7.1. If the pinion is the driving gear with an angular speed of 350 rpm, determine the magnitude of the maximum sliding velocity between the gears.

P17.7.4: Repeat Problem P17.7.3 for the data and results of Problems P17.5.5 and P17.7.2.

Section 17.10 Gear Drives and Gear Trains

P17.10.1: In the simple planetary gear system of Figure P17.10.1 the connecting arm C has an angular speed of 150 rpm, rotating counterclockwise. If the stationary gear A has 85 teeth and if gear B has 20 teeth, determine the angular speed of gear B.

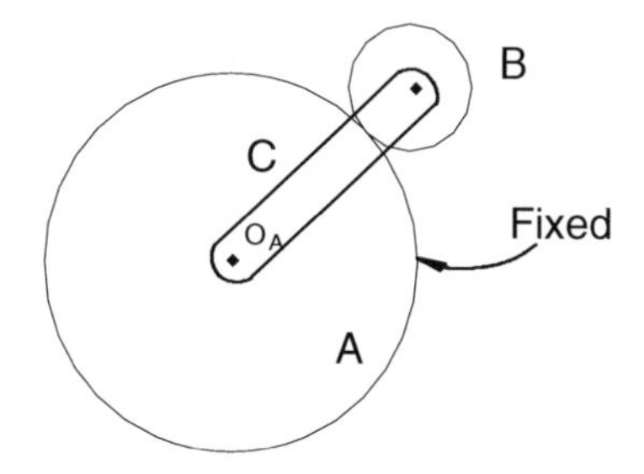

FIGURE P17.10.1
A simple planetary gear system.

P17.10.2: Repeat Problem P17.10.1 if gear A, instead of being stationary, is rotating (a) counterclockwise at 45 rpm, and (b) clockwise at 45 rpm.

P17.10.3: Repeat Problems P17.10.1 and P17.10.2 if the connecting arm C is rotating clockwise at 100 rpm.

P17.10.4: Consider the planetary gear systems of Figure P17.10.4. Let the angular speed of gear B be the angular speed of the connecting arm C if the fixed ring gear A has 100 teeth and gear B has 35 teeth.

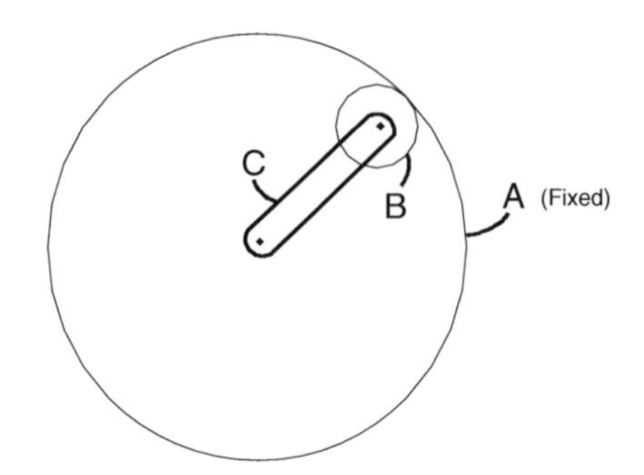

FIGURE P17.10.4
A planetary gear system with a fixed external (ring) gear.

P17.10.5: Repeat Problem P17.10.4 if gear A, instead of being stationary, is rotating (a) clockwise at 25 rpm, and (b) counterclockwise at 25 rpm.

P17.10.6 Repeat Problems P17.10.4 and P17.10.5 if the angular speed of gear B is rotating counterclockwise at 75 rpm.

18

Introduction to Multibody Dynamics

18.1 Introduction

In the foregoing chapters — indeed, in the major portions of this book — we have focused upon relatively simple mechanical systems. Our objective has been to illustrate fundamental principles. Often our systems have been as simple as a single particle or body. Mechanical systems of practical importance, however, are generally far more complex than such simple systems. Nevertheless, the procedures of analysis for the more complex systems are essentially the same as those of the simple systems. In this chapter, we develop procedures for extending our analysis techniques to multibody systems. As the name implies, multibody systems contain many bodies. Multibody systems may be used to model virtually all physical systems. Although the fundamental principles used with simple systems can also be used with multibody systems, it is necessary to develop procedures for organizing the complex geometry of the systems. These organizational procedures are the focus of this chapter.

Technically, a multibody system is simply a collection of bodies. The bodies themselves may be either connected to each other or free to translate (or separate) relative to each other. The bodies may be either rigid or flexible. They may or may not form closed loops.

Multibody systems consisting entirely of connected rigid bodies and without closed loops are called *open-chain* or *open-tree* systems. Figure 18.1.1 shows such a system. Alternatively, a multibody system may have large separation between the bodies, closed loops, and flexible members, as depicted in Figure 18.1.2.

As we noted earlier, multibody systems may be used to model many physical systems of interest and of practical importance. In the next two chapters, we will consider two such systems that have recently received considerable attention from analysts — specifically, robots and biosystems.

To keep our analysis simple, and at least moderate in length, we will focus our attention upon open-chain systems with connected rigid bodies. Extension to other systems having flexible bodies, closed loops, and relative separation between the bodies may be considered by using procedures documented in References 18.1 to 18.3.

18.2 Connection Configuration: Lower Body Arrays

A characteristic of multibody systems, particularly large systems, is that the multitude of bodies creates unwieldy geometric complexity. An effective way to work with this complexity is to use a *lower body array,* which is an array of body numbers (or labels) that

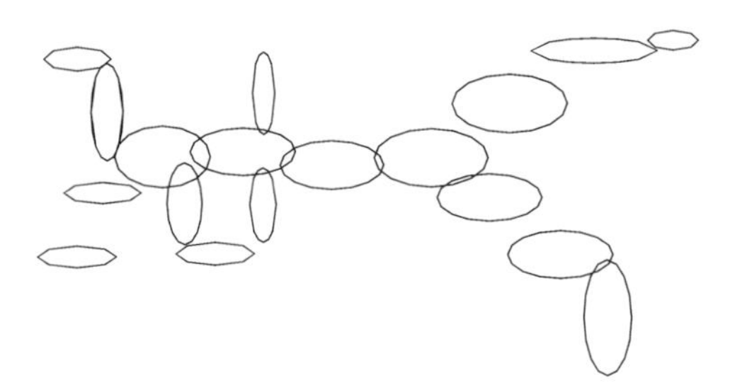

FIGURE 18.1.1
An open-chain, open-tree multibody system.

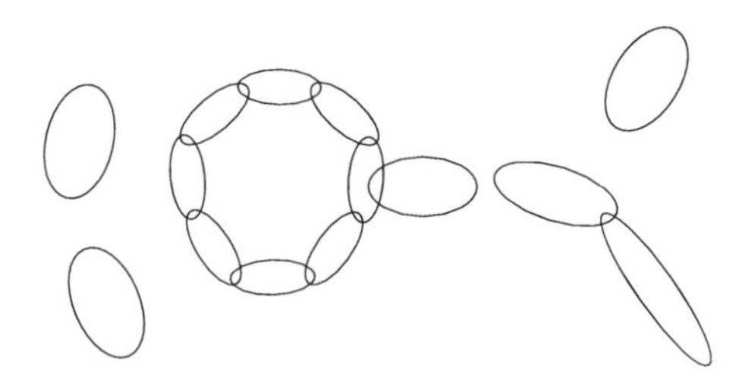

FIGURE 18.1.2
A multibody system with a closed loop, separation between the bodies, and long flexible members.

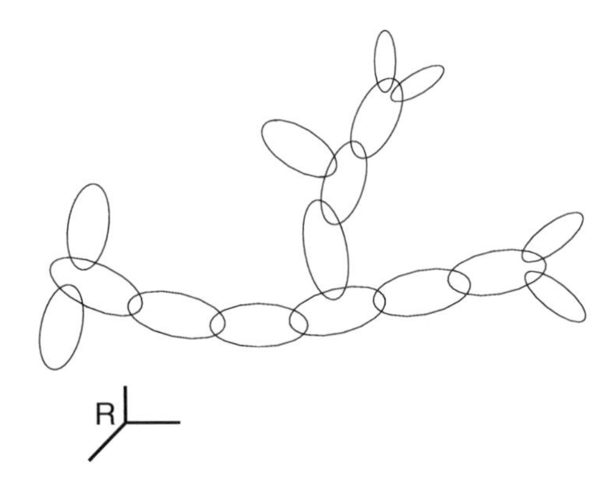

FIGURE 18.2.1
A multibody system.

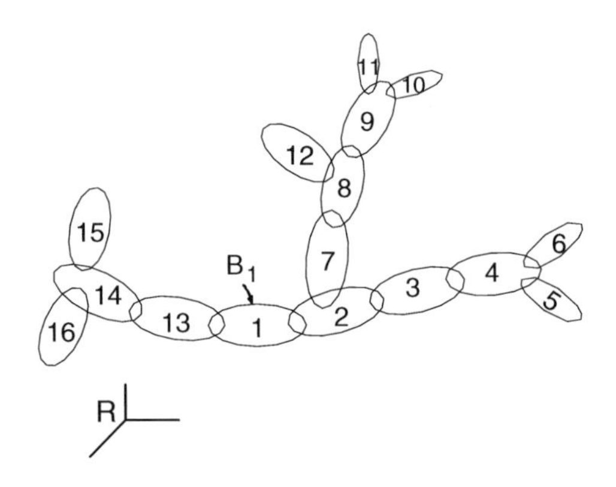

FIGURE 18.2.2
Reference body B_1 for the system of Figure 18.2.1.

define the connection configuration of the multibody system. The lower body arrays may be used to develop a simplified approach to the kinematics of the system.

To define and develop the lower body array, consider the multibody system of Figure 18.2.1. This is an open-chain (or open-tree) multibody system moving in an inertial reference frame R. Let the bodies of the system be numbered and labeled as follows: arbitrarily select a body, perhaps one of the larger bodies, as a reference body and call it Body 1, or B_1, and label it 1 as in Figure 18.2.2. Next, number and label the other bodies of the system in ascending progression away from B_1 through the branches of the tree system. Specifically, consider the representation of the multibody system of Figures 18.2.1 and 18.2.2 as a projection of the images of the bodies onto a plane. Next, select a body adjacent to B_1, call it B_2, and label it 2. Then, continue to number the bodies in a serial manner through the branch of bodies containing B_2 until the extremity of the branch (B_5) is reached, as in Figure 18.2.3. Observe that two extremity bodies branch off of B_4. Let the other body be called B_6 and label it 6. Next, return to B_2, which is also a branching body, and number the bodies in the other branch of B_2 in a similar manner, as in Figure 18.2.4. Then label the remaining bodies in the second branch in ascending progression, moving clockwise in the projected image of the system. Finally, return to B_1 and number and label the bodies in the remaining branch, leading to a complete numbering and labeling of the system, as in Figure 18.2.5.

Once the bodies of the system are numbered and labeled, the connection configuration may be described by a lower body array defined as follows: First, observe that except for B_1 each body of the system has a *unique* adjacent lower numbered body. (A body may have more than one adjacent higher numbered bodies [such as B_2, B_4, B_8, B_9, and B_{14} of

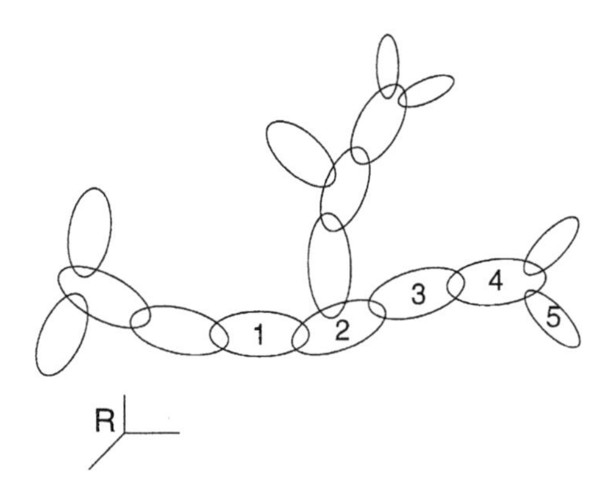

FIGURE 18.2.3
A numbering and labeling of the bodies in a branch of the multibody system.

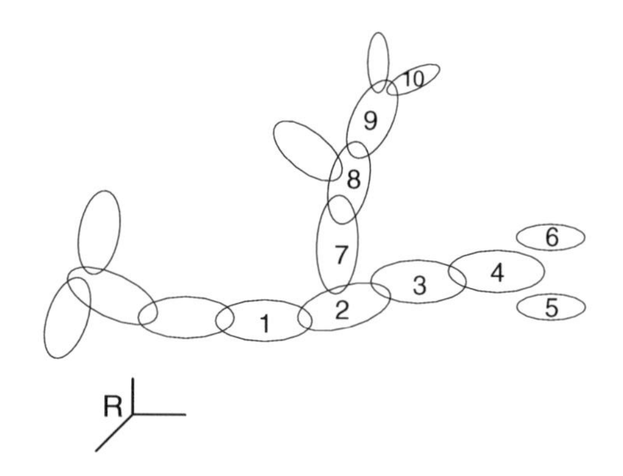

FIGURE 18.2.4
A numbering and labeling of the bodies of two branches of the multibody system.

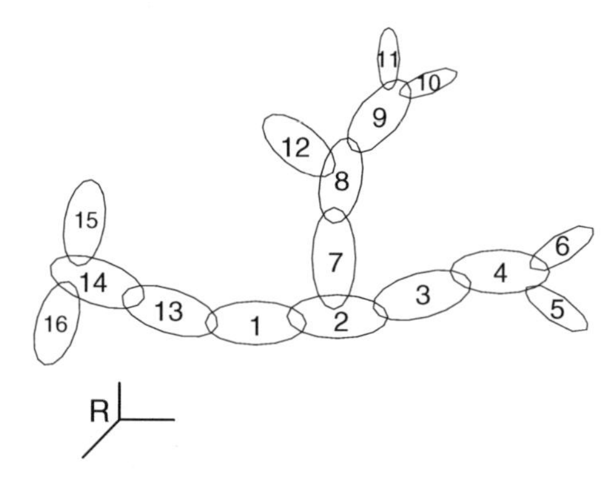

FIGURE 18.2.5
A complete numbering and labeling of the multibody system.

Figure 18.2.5] but only one adjacent lower numbered body.) Next, consider the inertial frame R to be the adjacent lower body for B_1 and let the number or label of R be 0. Finally, form a listing of labels of the adjacent lower numbered bodies. This listing is the lower body array for the system. For the numbering and labeling of the system of Figure 18.2.5, the lower body array $L(K)$ is:

$$
\begin{array}{lcccccccccccccccc}
K: & 1 & 2 & 3 & 4 & 5 & 6 & 7 & 8 & 9 & 10 & 11 & 12 & 13 & 14 & 15 & 16 \\
L(K): & 0 & 1 & 2 & 3 & 4 & 4 & 2 & 7 & 8 & 9 & 9 & 8 & 1 & 13 & 14 & 14
\end{array}
\tag{18.2.1}
$$

where K is an index corresponding to the body number — that is, B_k with $k = K$.

The labeling and numbering of a multibody system is somewhat arbitrary in that the reference body B_1 may be arbitrarily selected. Also, the order of the choice of branching away from B_1 is arbitrary. However, once the system is numbered and labeled, the lower body array is unique. Moreover, the listing of the lower body array, $L(K)$, is a numerical description and definition of the connection configuration of the multibody system. That is, the graphical representation (as in Figure 18.2.5) and the lower body array, as in Eq. (18.2.1), are equivalent.

The lower body array also provides information about the bodies themselves. If a body number does not appear in the $L(K)$ listing, then that body is an extremity of the system.

That is, the body is an ending body of a branch. For the system of Figure 18.2.5, bodies B_5, B_6, B_{10}, B_{11}, B_{12}, B_{15}, and B_{16} are seen to be extremities or ending bodies. Correspondingly, the numbers 5, 6, 10, 11, 12, 15, and 16 do not appear in $L(K)$ of Eq. (18.2.1).

Next, a body number that occurs more than once in the $L(K)$ array designates a branching body. In Figure 18.2.5, we see that bodies B_2, B_4, B_8, B_9, and B_{14} are branching bodies. As a consequence, we see in Eq. (18.2.1), that the numbers 1, 2, 4, 8, 9, and 14 are repeated in the $L(K)$ array.

Finally, if a body number occurs once but only once in the $L(K)$ array, the corresponding body is an intermediate body (that is, neither a branching nor an ending body). In Eq. (18.2.1), body numbers 3, 7, and 13 occur singly in the $L(K)$ array, and from Figure 18.2.5 we see that B_1, B_7, and B_{13} are intermediate bodies. Observe that each of the bodies of the system may be classified as being a branching body, an intermediate body, or an ending body.

In dynamical analyses of multibody systems we will see that it is useful to know not only the lower body array but also the array of lower bodies of the lower body array itself. For example, for the system of Figure 18.2.5, the lower body array of the lower body array, written as $L(L(K))$ is:

$$\begin{array}{cccccccccccccccccc} L(K) & 0 & 1 & 2 & 3 & 4 & 4 & 2 & 7 & 8 & 9 & 9 & 8 & 1 & 13 & 14 & 14 \\ L(L(K)) & 0 & 0 & 1 & 2 & 3 & 3 & 1 & 2 & 7 & 8 & 8 & 7 & 0 & 1 & 13 & 13 \end{array} \tag{18.2.2}$$

where we have assigned $L(0)$ to be 0.

Observe that once $L(K)$ is known, it is possible to obtain $L(L(K))$ without referring to the figure depicting the system (Figure 18.2.5) but instead by simply using $L(K)$ and the (K) array as in Eq. (18.2.1). In this sense $L(K)$ could be considered as an operator on K (as a differential operator). Thus, $L(L(K))$ might be written as $L^2(K)$. In this context, $L(K)$ is $L^1(K)$, and (K) is $L^0(K)$. We can continue to develop lower body arrays until all zeros occur. Again using the system of Figure 18.2.5 as an example, we obtain the listing of Table 18.2.1.

To see how these higher order lower body arrays might be useful in kinematic analyses, consider a typical extremity body of the system of Figure 18.2.5, say B_{10}. The numbers in the 10th column of Table 18.2.1 (that is, 10, 9, 8, 7, 2, 1, and 0) designate a listing of the bodies (B_{10}, B_9, B_8, B_7, B_2, B_1, and R) in the branch of the system containing B_{10}, from R out to B_{10}. Then, from Figure 18.2.5 if, for example, we want to obtain the angular velocity of B_{10} in R we could use the addition theorem for angular velocity (see Eq. (4.7.6)), which leads to the expression:

$$^R\boldsymbol{\omega}^{B_{10}} = {}^{B_9}\boldsymbol{\omega}^{B_{10}} + {}^{B_8}\boldsymbol{\omega}^{B_9} + {}^{B_7}\boldsymbol{\omega}^{B_8} + {}^{B_2}\boldsymbol{\omega}^{B_7} + {}^{B_1}\boldsymbol{\omega}^{B_2} + {}^R\boldsymbol{\omega}^{B_1} \tag{18.2.3}$$

(Observe how the sequence of numbers in the 10th column of Table 18.2.1 occurs in this expression.)

We may simplify the form of Eq. (18.2.3) by adopting the notation:

$$^R\boldsymbol{\omega}^{B_k} \overset{\text{D}}{=} \boldsymbol{\omega}_k \quad \text{and} \quad {}^{B_j}\boldsymbol{\omega}^{B_k} = \hat{\boldsymbol{\omega}}_k \tag{18.2.4}$$

where, as before, B_j is the adjacent lower numbered body of B_k. Thus, Eq. (18.2.3) becomes:

$$\boldsymbol{\omega}_{10} = \hat{\boldsymbol{\omega}}_{10} + \hat{\boldsymbol{\omega}}_9 + \hat{\boldsymbol{\omega}}_8 + \hat{\boldsymbol{\omega}}_7 + \hat{\boldsymbol{\omega}}_2 + \hat{\boldsymbol{\omega}}_1 \tag{18.2.5}$$

(Observe again that the subscripts on the right side of this equation occur in the 10th column of Table 18.2.1.)

TABLE 18.2.1

Higher Order Lower Body Arrays for the Multibody System of Figure 18.2.5

$L^0(K)$	1	2	3	4	5	6	7	8	9	10	11	12	13	14	15	16
$L^1(K)$	0	1	2	3	4	4	2	7	8	9	9	8	1	13	14	14
$L^2(K)$	0	0	1	2	3	3	1	2	7	8	8	7	0	1	13	13
$L^3(K)$	0	0	0	1	2	2	0	1	2	7	7	2	0	0	1	1
$L^4(K)$	0	0	0	0	1	1	0	0	1	2	2	1	0	0	0	0
$L^5(K)$	0	0	0	0	0	0	0	0	0	1	1	0	0	0	0	0
$L^6(K)$	0	0	0	0	0	0	0	0	0	0	0	0	0	0	0	0

18.3 A Pair of Typical Adjoining Bodies: Transformation Matrices

Consider a typical pair of adjoining bodies, say B_j and B_k, of a multibody system, as in Figure 18.3.1. Let B_j be the lower numbered body. Let $\mathbf{n}_{jm}$ and $\mathbf{n}_{kw}$ (m = 1, 2, 3) be unit vectors fixed in B_j and B_k. Then by following the procedures of Sections 2.11 and 4.3 we can define the orientation of B_k relative to B_j by the relative inclinations of the $\mathbf{n}_{km}$ relative to the $\mathbf{n}_{jm}$. That is, we can bring B_k into a general orientation relative to B_j by aligning the unit vectors and then successively rotating B_k about axes parallel to $\mathbf{n}_{k1}$, $\mathbf{n}_{k2}$, and $\mathbf{n}_{k3}$ through the dextral (or Bryan) angles α, β, and γ. When this is done, the $\mathbf{n}_{jm}$ and $\mathbf{n}_{km}$ may be expressed in terms of each other by using configuration graphs as defined in Section 4.3. Specifically, in matrix notation we have:

$$\mathbf{n}_j = ABC\mathbf{n}_k \quad \text{and} \quad \mathbf{n}_k = C^{-1}B^{-1}A^{-1}\mathbf{n}_j \tag{18.3.1}$$

where $\mathbf{n}_j$ and $\mathbf{n}_k$ are the column arrays:

$$\mathbf{n}_j = \begin{bmatrix} \mathbf{n}_{j1} \\ \mathbf{n}_{j2} \\ \mathbf{n}_{j3} \end{bmatrix} \quad \text{and} \quad \mathbf{n}_k = \begin{bmatrix} \mathbf{n}_{k1} \\ \mathbf{n}_{k2} \\ \mathbf{n}_{k3} \end{bmatrix} \tag{18.3.2}$$

and where A, B, and C are the orthogonal matrices:

$$A = \begin{bmatrix} 1 & 0 & 0 \\ 0 & c_\alpha & -s_\alpha \\ 0 & s_\alpha & c_\alpha \end{bmatrix} \quad B = \begin{bmatrix} c_\beta & 0 & s_\beta \\ 0 & 1 & 0 \\ -s_\beta & 0 & c_\beta \end{bmatrix} \quad C = \begin{bmatrix} c_\gamma & -s_\gamma & 0 \\ s_\gamma & c_\gamma & 0 \\ 0 & 0 & 1 \end{bmatrix} \tag{18.3.3}$$

where, as before, s_α, c_α, … are abbreviations for sinα, cosα, … .

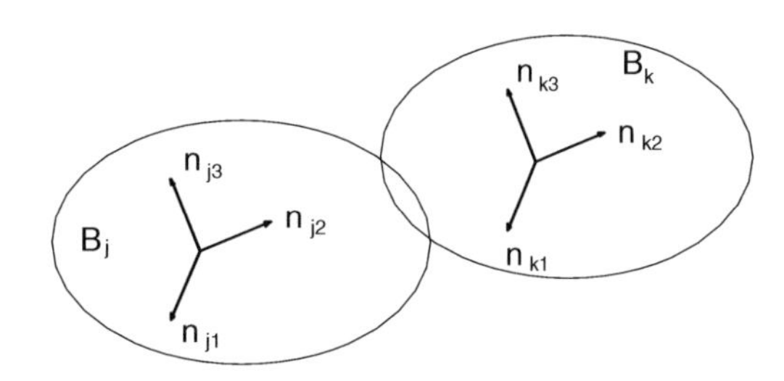

FIGURE 18.3.1
Two typical adjoining bodies.

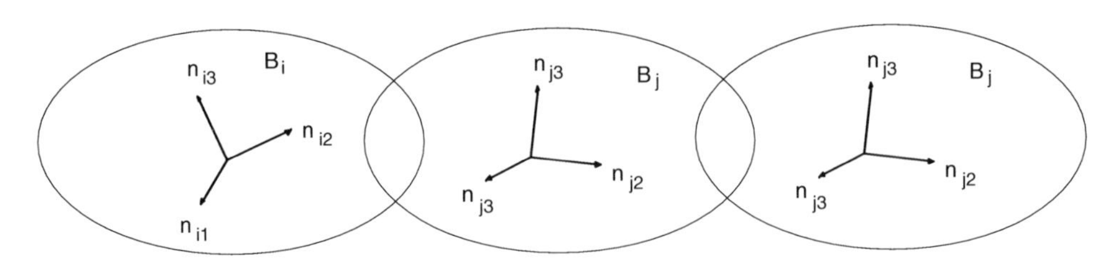

FIGURE 18.3.2
Three typical adjoining bodies.

The matrix product ABC is also an orthogonal matrix given by:

$$ABC = S = \begin{bmatrix} c_\beta c_\gamma & -c_\beta s_\gamma & s_\beta \\ (c_\alpha s_\gamma + s_\alpha s_\beta c_\gamma) & (c_\alpha c_\gamma - s_\alpha s_\beta s_\gamma) & -s_\alpha c_\beta \\ (s_\alpha s_\gamma - c_\alpha s_\beta c_\gamma) & (s_\alpha c_\gamma + c_\alpha s_\beta s_\gamma) & c_\alpha c_\beta \end{bmatrix} \tag{18.3.4}$$

Recall that an "orthogonal" matrix has the property that its inverse is also its transpose; that is,

$$S^{-1} = (ABC)^{-1} = C^{-1}B^{-1}A^{-1} = C^T B^T A^T = (CBA)^T = S^T \tag{18.3.5}$$

Hence, from Eq. (18.3.4) we have:

$$S^{-1} = S^T = \begin{bmatrix} c_\beta c_\gamma & (c_\alpha s_\gamma + s_\alpha s_\beta s_\gamma) & (s_\alpha s_\gamma - c_\alpha s_\beta c_\gamma) \\ -c_\beta s_\gamma & (c_\alpha c_\gamma - s_\alpha s_\beta s_\gamma) & (s_\alpha c_\gamma + c_\alpha s_\beta s_\gamma) \\ s_\beta & -s_\alpha c_\beta & c_\alpha c_\beta \end{bmatrix} \tag{18.3.6}$$

In view of Eqs. (18.3.4), (18.3.5), and (18.3.6), we can rewrite Eq. (18.3.1) in the simple forms:

$$\mathbf{n}_j = s\mathbf{n}_k \quad \text{and} \quad \mathbf{n}_k = S^T \mathbf{n}_j \tag{18.3.7}$$

The elements S_{mn} of the transformation matrix may be expressed in terms of the unit vectors $\mathbf{n}_{jm}$ and $\mathbf{n}_{km}$ as:

$$S_{mn} = \mathbf{n}_{jm} \cdot \mathbf{n}_{kn} \tag{18.3.8}$$

Then, in index notation, we can write Eqs. (18.3.7) in the forms:

$$\mathbf{n}_{jm} = S_{mn}\mathbf{n}_{kn} \quad \text{and} \quad \mathbf{n}_{kn} = S_{mn}\mathbf{n}_{jm} \tag{18.3.9}$$

where as before repeated indices (such as the m and n) designate a sum over the range (in this case from 1 to 3) of the index.

Recall further from Section 2.11 that a principal use of the transformation matrices is to obtain expressions relating the components of a vector referred to the different unit vector sets. Specifically, if a vector V is expressed as:

$$\mathbf{V} = V_m^{(j)}\mathbf{n}_{jm} \quad \text{and as} \quad V_n^{(k)}\mathbf{n}_{kn} \tag{18.3.10}$$

then the components $V_m^{(j)}$ and $V_n^{(k)}$ may be related as:

$$V_m^{(j)} = S_{mn}V_n^{(k)} \quad \text{and} \quad V_n^{(k)} = S_{mn}V_m^{(j)} \tag{18.3.11}$$

Observe the similarity in the pattern of the indices of Eqs. (18.3.9) and (18.3.11).

Finally, regarding notation, observe that S is a transformation matrix between the unit vectors of the typical adjoining bodies B_j and B_k. To keep an account of this, let S be replaced by:

$$S = SJK \tag{18.3.12}$$

Similarly, let the matrices A, B, and C and the angles α, β, and γ be replaced by:

$$A = AJK \quad B = BJK \quad C = CJK \tag{18.3.13}$$

and

$$\alpha = \alpha_k \quad \beta = \beta_k \quad \gamma = \gamma_k \tag{18.3.14}$$

Consider next a series of three adjoining bodies B_i, B_j, and B_k as in Figure 18.3.2. As before, let B_i be the adjacent lower numbered body of B_j. Let $\mathbf{n}_{im}$, $\mathbf{n}_{jm}$, and $\mathbf{n}_{km}$ ($m = 1, 2, 3$) be unit vector sets fixed in B_i, B_j, and B_k as in Figure 18.3.2. Let SJK be the transformation matrix between $\mathbf{n}_{jm}$ and $\mathbf{n}_{km}$ and let SIJ be the analogous transformation matrix between $\mathbf{n}_{im}$ and $\mathbf{n}_{jm}$. Then, from Eq. (18.3.7), we have:

$$\mathbf{n}_i = SIJ\mathbf{n}_j \quad \text{and} \quad \mathbf{n}_j = SJK\mathbf{n}_k \tag{18.3.15}$$

Hence, by substitution for $\mathbf{n}_j$, we also have:

$$\mathbf{n}_i = SIJ\, SJK\mathbf{n}_k = SIK\mathbf{n}_k \tag{18.3.16}$$

That is,

$$SIK = SIJ\, SJK \tag{18.3.17}$$

Observe the transitive nature of Eq. (18.3.17). By repeated use of this expression, we can obtain a transformation matrix between the unit vectors of a typical body, say B_k, and the unit vectors of the inertial frame R. For example, consider again the multibody system of

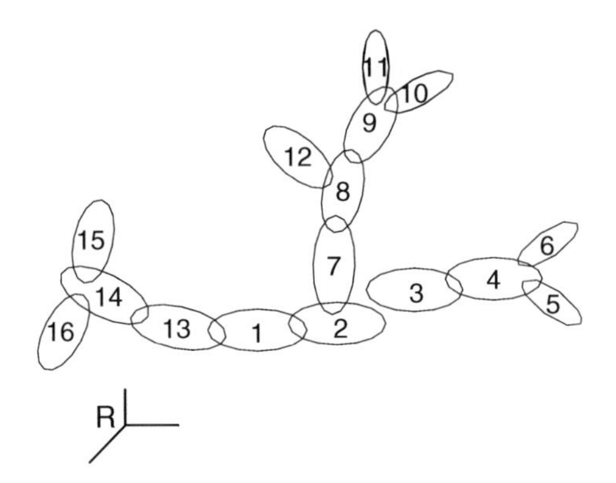

FIGURE 18.3.3
A typical multibody system.

Figure 18.2.5 and as shown again in Figure 18.3.3. A transformation matrix relating the unit vectors of say B_{10} to those of R may be obtained from Equation (18.3.17) as:

$$S010 = S01\ \ S12\ \ S27\ \ S78\ \ S89\ \ S910 \tag{18.3.18}$$

Then the unit vector relationships are:

$$\mathbf{n}_{0i} = S010_{im}\mathbf{n}_{10m} \quad \text{and} \quad \mathbf{n}_{10i} = S010_{mi}\mathbf{n}_{0m} \tag{18.3.19}$$

Finally, observe the pattern of the numbers in Eq. (18.3.18): 1, 2, 7, 8, 9, 10. These are the same indices as occur in the column of B_{10} in Table 18.2.1 and also in Eq. (18.2.5).

18.4 Transformation Matrix Derivatives

In developing the kinematics of multibody systems it is useful to have expressions for derivatives of the transformation matrices. To obtain these derivatives, consider the transformation matrix SOK of a typical body B_k relative to the inertia frame R. From Eq. (18.3.8), we see that the elements of SOK are:

$$SOK_{mn} = \mathbf{n}_{0m} \cdot \mathbf{n}_{kn} \tag{18.4.1}$$

Because the $\mathbf{n}_{0m}$ are fixed in R, they are constants in R. Hence, we have:

$$\frac{d}{dt} SOK_{mn} = \frac{d}{dt}\left(\mathbf{n}_{0m} \cdot \mathbf{n}_{kn}\right) = \mathbf{n}_{0m} \cdot \frac{{}^R d\mathbf{n}_{kn}}{dt} \tag{18.4.2}$$

Because the $\mathbf{n}_{kn}$ are fixed in B_k, their derivatives are simply:

$$\frac{{}^R d\mathbf{n}_{kn}}{dt} = \boldsymbol{\omega}_k \times \mathbf{n}_{kn} \tag{18.4.3}$$

where, as before, $\boldsymbol{\omega}_k$ is the angular velocity of B_k in R. Then, by substituting into Eq. (18.4.2), we have:

$$\frac{d}{dt}SOK_{mn} = \mathbf{n}_{0m} \cdot \boldsymbol{\omega}_k \times \mathbf{n}_{kn} \tag{18.4.4}$$

To develop this expression, let $\boldsymbol{\omega}_k$ and $\mathbf{n}_{kn}$ be expressed in terms of the $\mathbf{n}_{0m}$ by again using the transformation matrices. That is, let $\boldsymbol{\omega}_k$ and $\mathbf{n}_{kn}$ be written as:

$$\boldsymbol{\omega}_k = \omega_{km}\mathbf{n}_{0m} \quad \text{and} \quad \mathbf{n}_{kn} = SOK_{mn}\mathbf{n}_{0m} \tag{18.4.5}$$

Then, Eq. (18.4.4) becomes:

$$\begin{aligned}\frac{d}{dt}SOK_{mn} &= \mathbf{n}_{0m} \cdot \omega_{kr}\mathbf{n}_{0r} \times SOK_{sn}\mathbf{n}_{0s} \\ &= \omega_{kr}SOK_{sn}\mathbf{n}_{0m} \cdot \mathbf{n}_{0r} \times \mathbf{n}_{0s} \\ &= e_{mrs}\omega_{kr}SOK_{sn} \\ &= WK_{ms}SOK_{sn}\end{aligned} \tag{18.4.6}$$

where as before the e_{mrs} are elements of the permutation array and where the WK_{ms} are the elements of the matrix WK defined as:

$$WK_{ms} = -e_{mrs}\omega_{kr} = \begin{bmatrix} 0 & -\omega_{k3} & \omega_{k2} \\ \omega_{k3} & 0 & -\omega_{k1} \\ -\omega_{k2} & \omega_{k1} & 0 \end{bmatrix} \tag{18.4.7}$$

Then, in index-free notation, Eq. (18.4.6) becomes:

$$\frac{d}{dt}SOK = WK\,SOK \tag{18.4.8}$$

Equation (18.4.8) provides the desired differentiation expression. Observe the "algorithmic" nature of the expression: the derivative is produced by a multiplication. Observe further that the multiplier matrix WK is developed from components of the angular velocity matrix. In this context, $\boldsymbol{\omega}_k$ is the *dual vector* [18.4] of WK.

18.5 Euler Parameters

With large multibody systems, the solution of the governing dynamical equations will generally require the use of numerical procedures and computational techniques. The objective of such procedures is to obtain a time history of the dependent variables of the system, most of which are orientation angles of the bodies of the system.

A problem that arises in these numerical solutions is that, for some orientations of the bodies, singularities can occur that will disrupt the solution procedure. To see this, consider

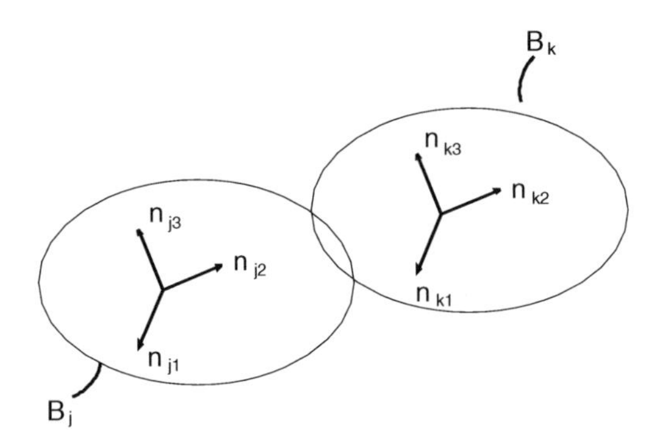

FIGURE 18.5.1
Two typical adjoining bodies.

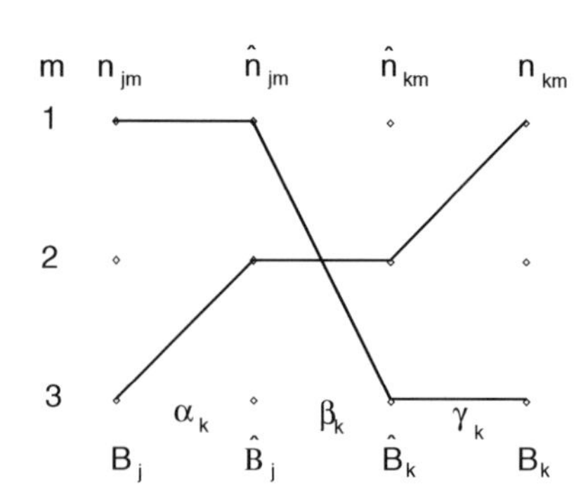

FIGURE 18.5.2
Configuration graph with dextral angles.

again two typical adjoining bodies as in Figure 18.5.1. As before, let the bodies be called B_j and B_k, and let $\mathbf{n}_{jm}$ and $\mathbf{n}_{km}$ be unit vectors fixed in B_j and B_k, respectively. If α_k, β_k, and γ_k are the relative orientation angles as described in the foregoing sections, we can construct a configuration graph (see Section 4.3) relating the unit vectors $\mathbf{n}_{jm}$ and $\mathbf{n}_{km}$, as well as intermediate unit vector sets $\hat{\mathbf{n}}_{jm}$ and $\hat{\mathbf{n}}_{km}$, as in Figure 18.5.2. In this graph, the intermediate unit vector sets $\hat{\mathbf{n}}_{jm}$ and $\hat{\mathbf{n}}_{km}$ are fixed in intermediate reference frames $\hat{B}_j$ and $\hat{B}_k$ formed in the process of defining the α_k, β_k, and γ_k as described in Sections 4.3 and 18.3. Then, following the procedure of Section 4.7, the angular velocity of B_k relative to B_j may be expressed as:

$$^{B_j}\boldsymbol{\omega}^{B_k} = \hat{\boldsymbol{\omega}}_k = \dot{\alpha}_k \mathbf{n}_{j1} + \dot{\beta}_k \hat{\mathbf{n}}_{j2} + \dot{\gamma}_k \hat{\mathbf{n}}_{k3} \tag{18.5.1}$$

where from the configuration graph of Figure 18.5.1 the unit vectors $\mathbf{n}_{j1}$, $\hat{\mathbf{n}}_{j2}$, and $\hat{\mathbf{n}}_{k3}$ are the unit vectors parallel to the rotation axes, defining α_k, β_k, and γ_k. By referring to the configuration graph, we see that $\hat{\mathbf{n}}_{j2}$ and $\hat{\mathbf{n}}_{k3}$ may be expressed in terms of $\mathbf{n}_{j1}$, $\mathbf{n}_{j2}$, and $\mathbf{n}_{j3}$ as:

$$\hat{\mathbf{n}}_{j2} = \cos\alpha_k \mathbf{n}_{j2} + \sin\alpha_k \mathbf{n}_{j3} \tag{18.5.2}$$

and

$$\hat{\mathbf{n}}_{k3} = \sin\beta_k \mathbf{n}_{j1} - \cos\beta_k \sin\alpha_k \mathbf{n}_{j2} + \cos\beta_k \cos\alpha_k \mathbf{n}_{j3} \tag{18.5.3}$$

Hence, $\hat{\boldsymbol{\omega}}_k$ becomes:

$$\begin{aligned}
\hat{\boldsymbol{\omega}}_k &= \dot{\alpha}_k \mathbf{n}_{j1} + \dot{\beta}_k \cos\alpha_k \mathbf{n}_{j2} + \dot{\beta}_k \sin\alpha_k \mathbf{n}_{j3} \\
&\quad + \dot{\gamma}_k \sin\beta_k \mathbf{n}_{j1} - \dot{\gamma}_k \cos\beta_k \sin\alpha_k \mathbf{n}_{j2} + \dot{\gamma}_k \cos\beta_k \cos\alpha_k \mathbf{n}_{j3} \\
&= (\dot{\alpha}_k + \dot{\gamma}_k \sin\beta_k)\mathbf{n}_{j1} + (\dot{\beta}_k \cos\alpha_k - \dot{\gamma}_k \cos\beta_k \sin\alpha_k)\mathbf{n}_{j2} \\
&\quad + (\dot{\beta}_k \sin\alpha_k + \dot{\gamma}_k \cos\beta_k \cos\alpha_k)\mathbf{n}_{j3} \\
&= \hat{\omega}_{k1}\mathbf{n}_{j1} + \hat{\omega}_{k2}\mathbf{n}_{j2} + \hat{\omega}_{k3}\mathbf{n}_{j3}
\end{aligned} \tag{18.5.4}$$

where the $\hat{\omega}_{km}$ (m = 1, 2, 3) are defined by inspection as:

$$\hat{\omega}_{k1} = \dot{\alpha}_k + \dot{\gamma}_k \sin\beta_k \tag{18.5.5}$$

$$\hat{\omega}_{k2} = \dot{\beta}_k \cos\alpha_k - \dot{\gamma}_k \cos\beta_k \sin\alpha_k \tag{18.5.6}$$

$$\hat{\omega}_{k3} = \dot{\beta}_k \sin\alpha_k + \dot{\gamma}_k \cos\beta_k \cos\alpha_k \tag{18.5.7}$$

Equations (18.5.5), (18.5.6), and (18.5.7) are nonlinear in α_k and β_k but are linear in $\dot{\alpha}_k$, $\dot{\beta}_k$, and $\dot{\gamma}_k$. Hence, they may be solved for $\dot{\alpha}_k$, $\dot{\beta}_k$, and $\dot{\gamma}_k$ in terms of the $\hat{\omega}_{km}$, leading to the expressions:

$$\dot{\alpha}_k = \hat{\omega}_{k1} + \sin\beta_k\left(\hat{\omega}_{k2}\sin\alpha_k - \hat{\omega}_{k3}\cos\alpha_k\right)/\cos\beta_k \tag{18.5.8}$$

$$\dot{\beta}_k = \hat{\omega}_{k2}\cos\alpha_k + \hat{\omega}_{k3}\sin\alpha_k \tag{18.5.9}$$

$$\dot{\gamma}_k = \left(-\hat{\omega}_{k2}\sin\alpha_k + \hat{\omega}_{k3}\cos\alpha_k\right)/\cos\beta_k \tag{18.5.10}$$

Equations (18.5.8), (18.5.9), and (18.5.10) form a series of first-order nonlinear ordinary differential equations for α_k, β_k, and γ_k. Observe that if β_k is 90° there are singularities in Eqs. (18.5.8) and (18.5.10), and then $\dot{\alpha}_k$ and $\dot{\gamma}_k$ are undefined. A numerical integration of Eqs. (18.5.8), (18.5.9), and (18.5.10) will thus become disrupted as β_k gets close to or is at 90°. Similar problems occur if β_k gets close to or is at 270°.

We can avoid these difficulties if we are studying a system where we have a general understanding of what the motion will be in a given analysis or application. In those cases, we can simply configure the unit vectors so that the singularities do not occur.

For example, suppose we are studying a system where we know (or reasonably expect) that between two adjoining bodies (say B_j and B_k) the orientation angle β_k will pass through 90°, as depicted in Figure 18.5.3. We can then avoid the singularity by reorienting the unit vectors $\mathbf{n}_{ki}$ as in Figure 18.5.4.

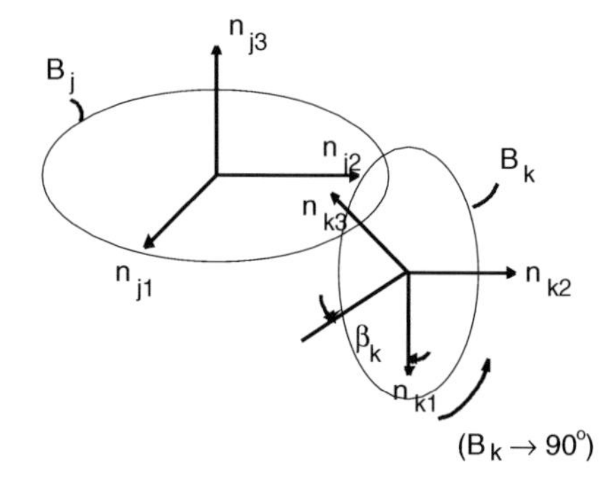

FIGURE 18.5.3
Relative rotation of typical bodies where the orientation angle β_k is passing through 90°.

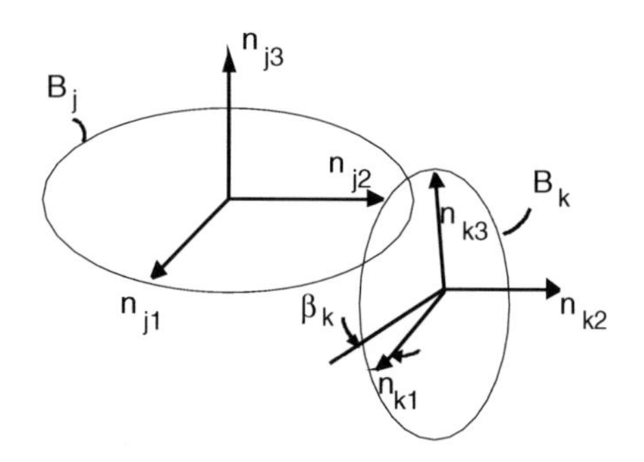

FIGURE 18.5.4
Reorientation of unit vectors $\mathbf{n}_{ki}$ to avoid the singularity of Figure 18.5.2.

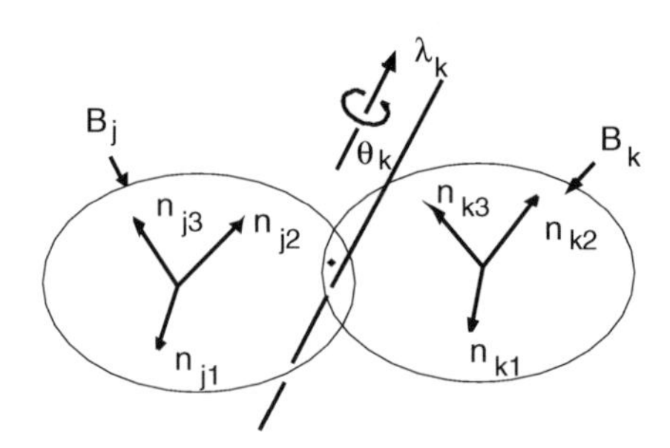

FIGURE 18.5.5
Two typical adjoining bodies.

Such adjustments, however, may not always be convenient or even possible. Indeed, in analyses of large multibody systems, it will not generally be known in advance when singularities might occur. Therefore, it is desirable to have a procedure that will not be disrupted by such singularities.*

In the early 1970s, it was discovered that a little known procedure attributed to Leonard Euler could be used to eliminate the singularity problem (see References 18.5 and 18.6). This procedure uses four parameters (called *Euler parameters*) to define the relative orientation of adjoining bodies. Because only three angles are needed to define orientation, the use of four Euler parameters introduces a redundancy and, as a consequence, when the governing equations are developed, an additional equation must be solved for each body. The benefit in exchange for this redundancy and additional governing equations is that the singularities do not occur.

The method proceeds as follows. Consider again two typical adjoining bodies of a multibody system such as B_j and B_k of Figure 18.5.5. It has been shown (as will be discussed in the following section) that B_k may be brought into a general orientation relative to B_j by a single rotation about an axis fixed in both B_j and B_k. Let $\boldsymbol{\lambda}_k$ be a unit vector parallel to this axis, and let θ_k be the rotation angle. Then, four Euler parameters ε_{ki} ($i = 1,\ldots, 4$) are defined as:

$$
\begin{aligned}
\varepsilon_{k1} &\overset{D}{=} \lambda_{k1}\sin(\theta_k/2) \\
\varepsilon_{k2} &\overset{D}{=} \lambda_{k2}\sin(\theta_k/2) \\
\varepsilon_{k3} &\overset{D}{=} \lambda_{k3}\sin(\theta_k/2) \\
\varepsilon_{k4} &\overset{D}{=} \cos(\theta_k/2)
\end{aligned}
\tag{18.5.11}
$$

where the $\boldsymbol{\lambda}_{km}$ ($m = 1, 2, 3$) are components of $\boldsymbol{\lambda}_k$ relative to the $\mathbf{n}_{jm}$ ($m = 1, 2, 3$), unit vectors fixed in B_j (see Figure 18.5.4).

Observe that the four Euler parameters are not independent. That is, they are related by the expression:

$$
\varepsilon_{k1}^2 + \varepsilon_{k2}^2 + \varepsilon_{k3}^2 + \varepsilon_{k4}^2 = 1 \tag{18.5.12}
$$

* One might wonder whether a different choice of orientation angles might not produce singularities, but it happens that singularities as in Eqs. (18.5.8) and (18.5.10) occur no matter how the orientation angles are defined (see Problems P18.5.2 and P18.5.3).

The Euler parameters take the place of orientation angles in defining the orientation, rotation, and angular velocity. In this regard, we will see (Section 18.7) that the transformation matrix SJK between unit vectors of B_j and B_k may be expressed in terms of the Euler parameters ε_{ki} as:

$$SJK = \begin{bmatrix} \left(\varepsilon_{k1}^2 - \varepsilon_{k2}^2 - \varepsilon_{k3}^2 + \varepsilon_{k4}^2\right) & 2\left(\varepsilon_{k1}\varepsilon_{k2} - \varepsilon_{k3}\varepsilon_{k4}\right) & 2\left(\varepsilon_{k1}\varepsilon_{k3} + \varepsilon_{k2}\varepsilon_{k4}\right) \\ 2\left(\varepsilon_{k1}\varepsilon_{k2} + \varepsilon_{k3}\varepsilon_{k4}\right) & \left(-\varepsilon_{k1}^2 + \varepsilon_{k2}^2 - \varepsilon_{k3}^2 + \varepsilon_{k4}^2\right) & 2\left(\varepsilon_{k2}\varepsilon_{k3} - \varepsilon_{k1}\varepsilon_{k4}\right) \\ 2\left(\varepsilon_{k1}\varepsilon_{k3} - \varepsilon_{k2}\varepsilon_{k4}\right) & 2\left(\varepsilon_{k2}\varepsilon_{k3} + \varepsilon_{k1}\varepsilon_{k4}\right) & \left(-\varepsilon_{k1}^2 - \varepsilon_{k2}^2 + \varepsilon_{k3}^2 + \varepsilon_{k4}^2\right) \end{bmatrix} \tag{18.5.13}$$

Also, we will see that the angular velocity of B_k relative to B_j may be expressed in terms of the ε_{ki} as:

$$^{B_j}\boldsymbol{\omega}^{B_k} = \hat{\boldsymbol{\omega}}_k = \hat{\omega}_{km}\mathbf{n}_{jm} \tag{18.5.14}$$

where the $\hat{\omega}_{km}$ are:

$$\begin{aligned} \hat{\omega}_{k1} &= 2\left(\varepsilon_{k4}\dot{\varepsilon}_{k1} - \varepsilon_{k3}\dot{\varepsilon}_{k2} + \varepsilon_{2}\dot{\varepsilon}_{k3} - \varepsilon_{k1}\dot{\varepsilon}_{k4}\right) \\ \hat{\omega}_{k2} &= 2\left(\varepsilon_{k3}\dot{\varepsilon}_{k1} + \varepsilon_{k4}\dot{\varepsilon}_{k2} - \varepsilon_{k1}\dot{\varepsilon}_{k3} - \varepsilon_{k2}\dot{\varepsilon}_{k4}\right) \\ \hat{\omega}_{k3} &= 2\left(-\varepsilon_{k2}\dot{\varepsilon}_{k1} - \varepsilon_{k1}\dot{\varepsilon}_{k2} + \varepsilon_{k4}\dot{\varepsilon}_{k3} - \varepsilon_{k3}\dot{\varepsilon}_{k4}\right) \end{aligned} \tag{18.5.15}$$

Finally, by using Eq. (18.5.12) we may solve Eq. (18.5.15) for the $\dot{\varepsilon}_{km}$ in terms of the $\hat{\omega}_{km}$ as:

$$\begin{aligned} \dot{\varepsilon}_{k1} &= (1/2)\left(\varepsilon_{k4}\hat{\omega}_{k1} + \varepsilon_{k3}\hat{\omega}_{k2} - \varepsilon_{k2}\hat{\omega}_{k3}\right) \\ \dot{\varepsilon}_{k2} &= (1/2)\left(-\varepsilon_{k3}\hat{\omega}_{k1} + \varepsilon_{k4}\hat{\omega}_{k2} + \varepsilon_{k1}\hat{\omega}_{k3}\right) \\ \dot{\varepsilon}_{k3} &= (1/2)\left(\varepsilon_{k2}\hat{\omega}_{k1} - \varepsilon_{k1}\hat{\omega}_{k2} + \varepsilon_{k4}\hat{\omega}_{k3}\right) \\ \dot{\varepsilon}_{k4} &= (1/2)\left(-\varepsilon_{k1}\hat{\omega}_{k1} - \varepsilon_{k2}\hat{\omega}_{k2} - \varepsilon_{k3}\hat{\omega}_{k3}\right) \end{aligned} \tag{18.5.16}$$

Equation (18.5.16) is analogous to Eqs. (18.5.8), (18.5.9), and (18.5.10), but due to the linear nature of Eq. (18.5.16) no singularities occur.

In the following section we will explore the basis for these equations in greater detail.

18.6 Rotation Dyadics

The fundamental premise providing the basis for the development and use of Euler parameters is that a body may undergo any given orientation change by a single rotation about an axis through an appropriate angle. This axis is fixed in both the body and the

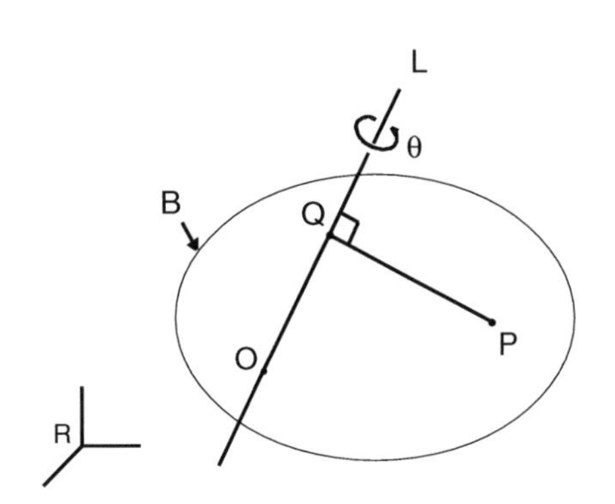

FIGURE 18.6.1
Rotation of a body about a line.

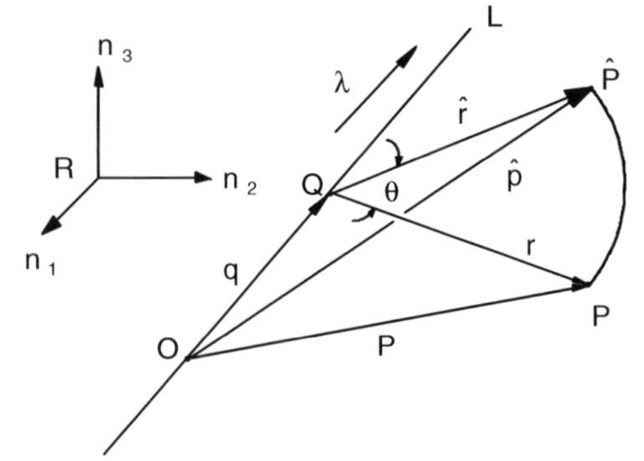

FIGURE 18.6.2
Rotation of point P of B to point $\hat{P}$ of R.

reference frame or the adjacent body, in which the rotation is measured. In this section, we will explore the basis for this fundamental premise.

Consider first the simple rotation of a body about a fixed line as in Figure 18.6.1. Specifically, let the body, called B, be rotated about a line L through an angle where L is fixed in both B and a reference frame R as shown. Let O be a point on L and let P be a typical point of B. Then as B rotates about L, P will rotate about L, moving on a circle whose center Q is at the intersection with L of the line through P that is perpendicular to L.

Let $\hat{P}$ be the point of R to which P is rotated as B rotates through the angle θ, as depicted in Figure 18.6.2. Let $\mathbf{p}$ locate P relative to O, and let $\hat{\mathbf{p}}$ locate $\hat{P}$ relative to O. Similarly, let $\mathbf{r}$ locate P relative to Q, let $\hat{\mathbf{r}}$ locate $\hat{P}$ relative to Q, let $\mathbf{q}$ locate Q relative to O, and let $\mathbf{n}_1$, $\mathbf{n}_2$, and $\mathbf{n}_3$ be unit vectors fixed in R.

From Figure 18.6.2, we see that $\mathbf{p}$ and $\hat{\mathbf{p}}$ may be expressed as:

$$\mathbf{p} = \mathbf{q} + \mathbf{r} \tag{18.6.1}$$

and

$$\hat{\mathbf{p}} = \mathbf{q} + \hat{\mathbf{r}} \tag{18.6.2}$$

Observe that $\mathbf{q}$ is the projection of $\mathbf{p}$ onto L and that therefore $\mathbf{q}$ may be expressed as:

$$\mathbf{q} = (\mathbf{p} \cdot \boldsymbol{\lambda})\boldsymbol{\lambda} \tag{18.6.3}$$

Observe also that $\mathbf{r}$ is the component of $\mathbf{p}$ that is perpendicular to L. That is:

$$\mathbf{r} = \mathbf{p} - \mathbf{q} = \mathbf{p} - (\mathbf{p} \cdot \boldsymbol{\lambda})\boldsymbol{\lambda} \tag{18.6.4}$$

Next, consider a point view of line L showing the plane of Q, P, and $\hat{P}$ as in Figure 18.6.3. Let $\boldsymbol{\mu}$ be a unit vector parallel to QP (and, hence, parallel to $\mathbf{r}$), and let $\mathbf{v}$ be a unit vector generated by the vector product $\boldsymbol{\lambda} \times \boldsymbol{\mu}$. Let A be the point at the base of the perpendicular to QP through $\hat{P}$. Then, from Figure 18.6.3, $\hat{\mathbf{r}}$ may be expressed as:

$$\hat{\mathbf{r}} = \mathbf{QA} + \mathbf{A\hat{P}} \tag{18.6.5}$$

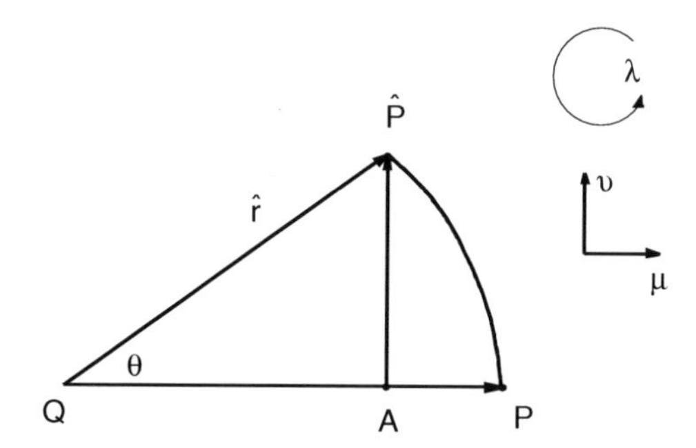

FIGURE 18.6.3
View of the plane of $QP\hat{P}$.

But, also from Figure 18.6.3, we can express **QA** and $\mathbf{A}\hat{\mathbf{P}}$ in the forms:

$$\mathbf{QA} = |\hat{\mathbf{r}}|\cos\theta\boldsymbol{\mu} \quad \text{and} \quad \mathbf{A}\hat{\mathbf{P}} = |\hat{\mathbf{r}}|\sin\theta\mathbf{v} \tag{18.6.6}$$

Because $\hat{\mathbf{r}}$ is the rotation of **r** through θ about *L*, we have:

$$|\hat{\mathbf{r}}| = |\mathbf{r}| \tag{18.6.7}$$

Hence, **QA** and $\mathbf{A}\hat{\mathbf{P}}$ may be written as:

$$\mathbf{QA} = |\mathbf{r}|\cos\theta\boldsymbol{\mu} = \mathbf{r}\cos\theta \tag{18.6.8}$$

and

$$\begin{aligned} \mathbf{A}\hat{\mathbf{P}} &= |\mathbf{r}|\sin\theta\mathbf{v} = |\mathbf{r}|\sin\theta\boldsymbol{\lambda}\times\boldsymbol{\mu} \\ &= \boldsymbol{\lambda}\times(|\mathbf{r}|\sin\theta)\boldsymbol{\mu} = \boldsymbol{\lambda}\times\mathbf{r}\sin\theta \end{aligned} \tag{18.6.9}$$

Then, by substituting from Eqs. (18.6.4), (18.6.8), and (18.6.9) into Eq. (18.6.5), we can express $\hat{\mathbf{r}}$ as:

$$\begin{aligned} \hat{\mathbf{r}} &= \cos\theta\mathbf{r} + \sin\theta\boldsymbol{\lambda}\times\mathbf{r} \\ &= \cos\theta\left[\mathbf{p} - (\mathbf{p}\times\boldsymbol{\lambda})\boldsymbol{\lambda}\right] + \sin\theta\boldsymbol{\lambda}\times\mathbf{p} \end{aligned} \tag{18.6.10}$$

Finally, by substituting from Eq. (18.63) and (18.6.10) into Eq. (18.6.2) we can express $\hat{\mathbf{p}}$ as:

$$\begin{aligned} \hat{\mathbf{p}} &= \mathbf{q} + \hat{\mathbf{r}} = (\mathbf{p}\cdot\boldsymbol{\lambda})\boldsymbol{\lambda} + \cos\theta\left[\mathbf{p} - (\mathbf{p}\cdot\boldsymbol{\lambda})\boldsymbol{\lambda}\right] + \sin\theta\boldsymbol{\lambda}\times\mathbf{p} \\ &= (1-\cos\theta)\boldsymbol{\lambda}\boldsymbol{\lambda}\cdot\mathbf{p} + \cos\theta\mathbf{p} + \sin\theta\boldsymbol{\lambda}\times\mathbf{p} \end{aligned} \tag{18.6.11}$$

or as:

$$\hat{\mathbf{p}} = R\cdot\mathbf{p} \tag{18.6.12}$$

where R is the dyadic defined as:

$$\boldsymbol{R} = (1 - \cos\theta)\boldsymbol{\lambda\lambda} + \cos\theta\mathbf{I} + \sin\theta\boldsymbol{\lambda} \times \mathbf{I} \tag{18.6.13}$$

where **I** is the identity dyadic.

It is helpful to express R in component form as:

$$\boldsymbol{R} = R_{ij}\mathbf{n}_i\mathbf{n}_j \tag{18.6.14}$$

where the $\mathbf{n}_i$ (i = 1, 2, 3) are unit vectors fixed in reference frame R (see Figure 18.6.2). In terms of the $\mathbf{n}_i$, $\boldsymbol{\lambda}$ and **I** may be expressed as:

$$\boldsymbol{\lambda} = \lambda_i\mathbf{n}_i \tag{18.6.15}$$

and

$$\mathbf{I} = \delta_{ij}\mathbf{n}_i\mathbf{n}_j \tag{18.6.16}$$

where, as before, δ_{ij} is Kronecker's delta function (see Eq. (2.6.7), Section 2.6). Then, by substituting from Eqs. (18.6.15) and (18.6.16) into Eq. (18.6.13), the R_{ij} of Eq. (18.6.14) are seen to be:

$$R_{ij} = (1 - \cos\theta)\lambda_i\lambda_j + \cos\theta\delta_{ij} - \sin\theta e_{ijk}\lambda_k \tag{18.6.17}$$

where, as before, the e_{ijk} are components of the permutation symbol (see Eq. (2.7.7), Section 2.7).

$\boldsymbol{R}$ is called a *rotation dyadic*. Viewed as an operator, we see from Eq. (18.6.12) that $\boldsymbol{R}$ "rotates" the vector **p** about L, through the angle θ, into the vector $\hat{\mathbf{p}}$. Because P is an arbitrary point of body B, **p** is an arbitrary vector fixed in B. Hence, if **V** is any vector fixed in B, $\boldsymbol{R}$ will rotate **V** about L through θ into a vector $\hat{\mathbf{V}}$. That is,

$$\hat{\mathbf{V}} = \boldsymbol{R} \cdot \mathbf{V} \tag{18.6.18}$$

Equation (18.6.18) might be visualized as in Figure 18.6.4. In this regard, suppose a mutually perpendicular unit vector set $\mathbf{b}_i$ (i = 1, 2, 3) fixed in B coincides with the $\mathbf{n}_i$ of R before B is rotated. Then, when B is rotated about L through θ the $\mathbf{b}_i$ will be rotated away

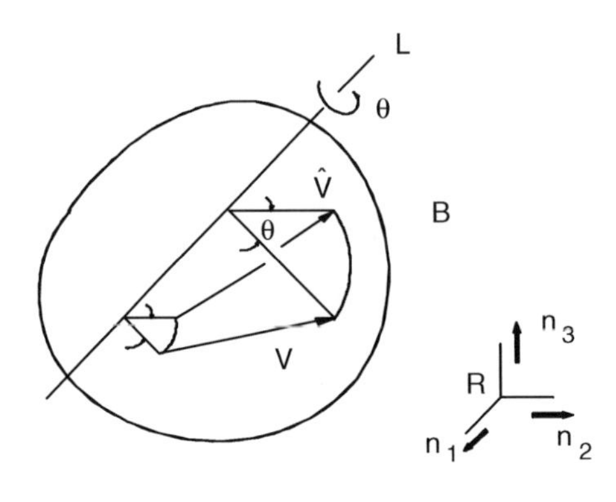

FIGURE 18.6.4
Rotation of body B and imbedded vector V about L.

from the $\mathbf{n}_i$. Let $\hat{\mathbf{n}}_i$ be unit vectors fixed in R that coincide with the $\mathbf{b}_i$ after rotation. Then, from Eq. (8.16.18), we have:

$$\hat{\mathbf{n}}_i = \boldsymbol{R} \cdot \mathbf{n}_i \tag{18.6.19}$$

By substituting from Eq. (18.6.14), the $\hat{\mathbf{n}}_i$ may be expressed as:

$$\hat{\mathbf{n}}_i = R_{kj}\mathbf{n}_k\mathbf{n}_j \cdot \mathbf{n}_i = R_{kj}\mathbf{n}_k\delta_{ji}$$

or

$$\hat{\mathbf{n}}_i = R_{ki}\mathbf{n}_k \tag{18.6.20}$$

Equation (18.6.20) has exactly the same form as Eq. (2.11.7) of Section 2.11, which means that the elements of the transformation matrix S_{ij} may be obtained by direct identification with the elements of the rotation tensor dyadic. That is,

$$S_{ij} = R_{ij} \tag{18.6.21}$$

As we will see, Eq. (18.6.21) is useful for establishing the validity of Eq. (18.5.13). One word of caution, however. Eq. (18.6.21) does *not* mean that the rotation dyadic and the transformation matrix are the same. The R_{ij} of Eq. (18.6.21) are referred to the $\mathbf{n}_i$ whereas the S_{ij} are referred to both $\mathbf{n}_i$ and $\hat{\mathbf{n}}_i$. The S_{ij} are elements of the identity dyadic referred to the two different unit vector sets.

From Eq. (18.6.17) we can readily see that, like the transformation matrix, the matrix of the rotation dyadic is orthogonal; that is, the inverse is the transpose. Specifically,

$$R_{ij}R_{\ell j} = \delta_{i\ell} \tag{18.6.22}$$

Equation (18.6.22) may be verified by direct substitution from Eq. (18.6.17). In the process of this verification it is helpful to recall that the permutation symbol e_{ijk} is antisymmetric in any pair of its indices so that forms of the type $e_{ijk}\lambda_j\lambda_k$ are zero (that is, $\boldsymbol{\lambda} \times \boldsymbol{\lambda} = 0$). Specifically,

$$e_{ijk} = -e_{ikj} \tag{18.6.23}$$

Also, because $\boldsymbol{\lambda}$ is a unit vector, we have:

$$\lambda_k\lambda_k = 1 \tag{18.6.24}$$

Finally, products of permutation symbols may be expressed in terms of Kronecker delta symbols as:

$$e_{ijk}e_{\ell mk} = \delta_{i\ell}\delta_{jm} - \delta_{im}\delta_{j\ell} \tag{18.6.25}$$

Then, by substituting from Eq. (18.6.17) and by using Eqs. (18.6.23), (18.6.24), and (18.6.25), we immediately obtain Eq. (18.6.22).

Equation (18.6.22) shows that we can reverse a rotation (that is, find its inverse) simply by using the transpose. Specifically,

$$\boldsymbol{R}^{-1} = \boldsymbol{R}^T \tag{18.6.26}$$

Hence, from Eq. (18.6.18), we have:

$$\mathbf{V} = \boldsymbol{R}^{-1} \cdot \hat{\mathbf{V}} = \boldsymbol{R}^T \cdot \hat{\mathbf{V}} \tag{18.6.27}$$

We can use Eq. (18.6.17) to obtain an explicit form of the rotation dyadic matrix as:

$$R_{ij} = \begin{bmatrix} [\lambda_1^2(1-\cos\theta)+\cos\theta] & [\lambda_1\lambda_2(1-\cos\theta)-\lambda_3\sin\theta] & [\lambda_1\lambda_3(1-\cos\theta)+\lambda_2\sin\theta] \\ [\lambda_1\lambda_2(1-\cos\theta)+\lambda_3\sin\theta] & [\lambda_2^2(1-\cos\theta)+\cos\theta] & [\lambda_2\lambda_3(1-\cos\theta)-\lambda_1\sin\theta] \\ [\lambda_3\lambda_1(1-\cos\theta)-\lambda_2\sin\theta] & [\lambda_2\lambda_3(1-\cos\theta)+\lambda_1\sin\theta] & [\lambda_3^2(1-\cos\theta)+\cos\theta] \end{bmatrix} \tag{18.6.28}$$

Observe that if we know the rotation angle θ and the components λ_1, λ_2, λ_3 of a unit vector parallel to the rotation axis we can immediately obtain the rotation dyadic matrix from Eq. (18.6.25). Alternatively, if we know the elements of the rotation dyadic matrix we can use Eq. (18.6.25) to obtain the rotation angle θ and the unit vector components λ_1, λ_2, λ_3. That is, from the sum of the diagonal elements (the *trace*), we have:

$$R_{kk} = 1 + 2\cos\theta \tag{18.6.29}$$

or

$$\cos\theta = (R_{kk} - 1)/2 \quad \text{and} \quad \theta = \cos^{-1}[(R_{kk} - 1)/2] \tag{18.6.30}$$

Then, knowing θ, and if $\sin\theta$ is not zero, we can determine λ_1, λ_2, λ_3 from the off-diagonal elements as:

$$\lambda_1 = (R_{32} - R_{23})/2\sin\theta \tag{18.6.31}$$

$$\lambda_2 = (R_{13} - R_{31})/2\sin\theta \tag{18.6.32}$$

and

$$\lambda_3 = (R_{21} - R_{12})/2\sin\theta \tag{18.6.33}$$

If $\sin\theta$ is zero, then θ is zero or π. If θ is zero, then there is no rotation, the rotation dyadic is the identity dyadic, and $\boldsymbol{\lambda}$ is undefined and need not be determined. If θ is π, then the rotation dyadic matrix has the simplified form:

$$R_{ij} = \begin{bmatrix} (2\lambda_1^2 - 1) & 2\lambda_1\lambda_2 & 2\lambda_1\lambda_3 \\ 2\lambda_1\lambda_2 & (2\lambda_2^2 - 1) & 2\lambda_2\lambda_3 \\ 2\lambda_1\lambda_3 & 2\lambda_2\lambda_3 & (2\lambda_3^2 - 1) \end{bmatrix} \tag{18.6.34}$$

Then, the diagonal elements λ_1, λ_2, and λ_3 are:

$$\lambda_1 = \pm\left[(1 + R_{11})/2\right]^{1/2} \tag{18.6.35}$$

$$\lambda_2 = \pm\left[(1 + R_{22})/2\right]^{1/2} \tag{18.6.36}$$

and

$$\lambda_3 = \pm\left[(1 + R_{33})/2\right]^{1/2} \tag{18.6.37}$$

where the signs are chosen to be consistent with the off diagonal terms.*

Equations (18.6.21) and Eqs. (18.6.30) through (18.6.37) establish the fundamental premise; that is, if a body B undergoes any orientation change, that same orientation change may be obtained by a single rotation about a fixed axis through an angle θ where θ is given by Eq. (18.6.30) and the rotation axis is defined by the components of a parallel unit vector $\boldsymbol{\lambda}$ given by Eqs. (18.6.31), (18.6.32), and (18.6.33).

18.7 Transformation Matrices, Angular Velocity Components, and Euler Parameters

If Euler parameters are used to define the orientation of a body then it is necessary to be able to express the angular velocity of the body in terms of the Euler parameters. Also, it is necessary to be able to express transformation matrices relating unit vectors of the body to unit vectors of the observer's reference frame in terms of the Euler parameters. Equations (18.5.14) and (18.5.16) provide these expressions. In this section, we will establish the basis for these equations.

To this end, consider again a body B moving in a reference frame R as in Figure 18.7.1 where as before $\boldsymbol{\lambda}$ is a unit vector parallel to the axis of rotation defining the orientation of B in R. Let the $\mathbf{n}_i$ and the $\mathbf{N}_i$ (i = 1, 2, 3) be unit vectors of B and R as shown. Then, the Euler parameters ε_i (i = 1, 2, 3, 4) defining the orientation of B in R are defined as (see Eq. (18.5.12)):

$$\begin{aligned} \varepsilon_1 &= \lambda_1 \sin\theta/2 \\ \varepsilon_2 &= \lambda_2 \sin\theta/2 \\ \varepsilon_3 &= \lambda_3 \sin\theta/2 \\ \varepsilon_4 &= \cos\theta/2 \end{aligned} \tag{18.7.1}$$

where, as before, θ is the rotation angle bringing B into its desired orientation and the λ_i (i = 1, 2, 3) are the components of $\boldsymbol{\lambda}$ relative to the $\mathbf{N}_i$.

* Observe that when θ is π it is not necessary to uniquely determine the sense of $\boldsymbol{\lambda}$.

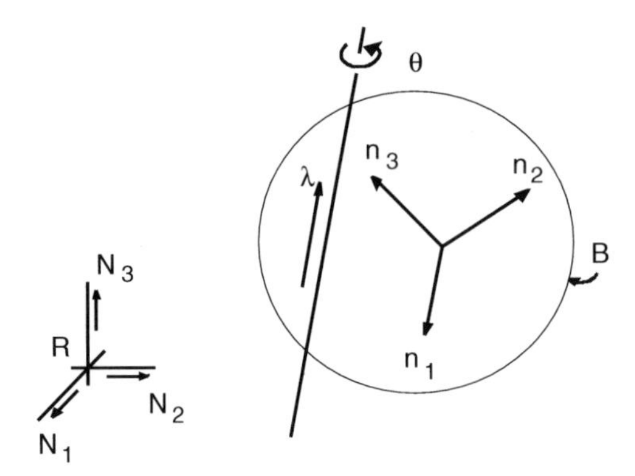

FIGURE 18.7.1
Orientation of a body B in a reference frame R.

Recall from Eqs. (18.6.21) and (18.6.28) of the foregoing section that the elements of the transformation matrix relating the $\mathbf{N}_i$ and the $\mathbf{n}_i$ may be expressed in terms of the λ_i and θ as:

$$S_{ij} = \begin{bmatrix} [\lambda_1^2(1-\cos\theta)+\cos\theta] & [\lambda_1\lambda_2(1-\cos\theta)-\lambda_3\sin\theta] & [\lambda_1\lambda_3(-\cos\theta)+\lambda_2\sin\theta] \\ [\lambda_1\lambda_2(1-\cos\theta)+\lambda_3\cos\theta] & [\lambda_2^2(1-\cos\theta)+\cos\theta] & [\lambda_2\lambda_3(1-\cos\theta)-\lambda_1\sin\theta] \\ [\lambda_3\lambda_1(1-\cos\theta)-\lambda_2\sin\theta] & [\lambda_2\lambda_3(1-\cos\theta)+\lambda_1\sin\theta] & [\lambda_3^2(1-\cos\theta)+\cos\theta] \end{bmatrix} \tag{18.7.2}$$

Then, by solving Eq. (18.7.1) for λ_1, λ_2, and λ_3 and by using trigonometric identities we can express the elements of S_{ij} in terms of the ε_i. Consider S_{11}. From Eq. (18.7.2), we have:

$$S_{11} = \lambda_1^2(1-\cos\theta)+\cos\theta \tag{18.7.3}$$

But, from Eq. (18.7.1), we have:

$$\varepsilon_1^2 = \lambda_1^2\sin^2\theta/2 \equiv \lambda_1^2\left(\frac{1}{2}-\frac{1}{2}\cos\theta\right)$$

or

$$\lambda_1^2(1-\cos\theta) = 2\varepsilon_1^2 \tag{18.7.4}$$

Also, from Eq. (18.2.1), we have:

$$\varepsilon_4^2 = \cos^2\theta/2 \equiv \frac{1}{2}+\frac{1}{2}\cos\theta$$

or

$$\cos\theta = 2\varepsilon_4^2 - 1 \tag{18.7.5}$$

Next, recall the identity of Eq. (18.5.13):

$$\varepsilon_1^2 + \varepsilon_2^2 + \varepsilon_3^2 + \varepsilon_4^2 = 1 \tag{18.7.6}$$

By combining Eqs. (18.7.5) and (18.7.6), we have:

$$\cos\theta = \varepsilon_4^2 - \varepsilon_1^2 - \varepsilon_2^2 - \varepsilon_3^2 \tag{18.7.7}$$

Finally, by substituting from Eqs. (18.7.4) and (18.7.7) into Eq. (18.7.3), we can eliminate both λ_1 and θ and thus express S_{11} as:

$$S_{11} = \varepsilon_1^2 - \varepsilon_2^2 - \varepsilon_3^2 + \varepsilon_4^2 \tag{18.7.8}$$

Consider next S_{12}. From Eq. (18.7.2) we have:

$$S_{12} = \lambda_1\lambda_2(1 - \cos\theta) - \lambda_3 \sin\theta \tag{18.7.9}$$

From Eq. (18.7.1) we have:

$$\varepsilon_1\varepsilon_2 = \lambda_1\lambda_2 \sin^2\theta/2 \equiv \lambda_1\lambda_2(1/2 - 1/2\cos\theta)$$

or

$$\lambda_1\lambda_2(1 - \cos\theta) = 2\varepsilon_1\varepsilon_2 \tag{18.7.10}$$

By recalling the trigonometric identity:

$$\sin\theta \equiv 2\sin(\theta/2)\cos(\theta/2) \tag{18.7.11}$$

and, by using Eq. (18.7.1), we can express λ_3 sinθ as:

$$\lambda_3 \sin\theta = 2\lambda_3 \sin(\theta/2)\cos(\theta/2) = 2\varepsilon_3\varepsilon_4 \tag{18.7.12}$$

Finally, by substituting from Eqs. (18.7.10 and (18.7.12) into Eq. (18.7.9), we can express S_{12} as:

$$S_{12} = 2\varepsilon_1\varepsilon_2 - 2\varepsilon_3\varepsilon_4 \tag{18.7.13}$$

In like manner, we can eliminate λ_1, λ_2, and λ_3 and θ from the remaining elements of the S_{ij} matrix, so that S_{ij} may be expressed as:

$$S_{ij} = \begin{bmatrix} (\varepsilon_1^2 - \varepsilon_2^2 - \varepsilon_3^2 + \varepsilon_4^2) & 2(\varepsilon_1\varepsilon_2 - \varepsilon_3\varepsilon_4) & 2(\varepsilon_1\varepsilon_3 + \varepsilon_2\varepsilon_4) \\ 2(\varepsilon_1\varepsilon_2 + \varepsilon_3\varepsilon_4) & (-\varepsilon_1^2 + \varepsilon_2^2 - \varepsilon_3^2 + \varepsilon_4^2) & 2(\varepsilon_2\varepsilon_3 - \varepsilon_1\varepsilon_4) \\ 2(\varepsilon_1\varepsilon_3 - \varepsilon_2\varepsilon_4) & 2(\varepsilon_2\varepsilon_3 + \varepsilon_1\varepsilon_4) & (-\varepsilon_1^2 - \varepsilon_2^2 + \varepsilon_3^2 + \varepsilon_4^2) \end{bmatrix} \tag{18.7.14}$$

This therefore establishes Eq. (18.5.14).

Next, concerning the angular velocity, recall from Eq. (18.4.8) that the transformation matrix derivative may be expressed as:

$$dS/dt = WS \tag{18.7.15}$$

or as:

$$dS_{mn}/dt = W_{ms}S_{sn} \tag{18.7.16}$$

where the elements W_{ms} of the array W are:

$$W_{ms} = -e_{msr}\omega_r \tag{18.7.17}$$

where the ω_r (r = 1, 2, 3) are the components of the angular velocity of body B in the reference frame R in which B moves. Hence, we can also express the transformation matrix element derivatives as:

$$dS_{mn}/dt = -\omega_r e_{rms}S_{sn} \tag{18.7.18}$$

We may use Eq. (18.7.18) to obtain the individual transformation matrix element derivatives as:

$$\begin{aligned} dS_{11}/dt &= -\omega_3 S_{21} + \omega_2 S_{31} \\ dS_{12}/dt &= -\omega_3 S_{22} + \omega_2 S_{32} \\ dS_{13}/dt &= -\omega_3 S_{23} + \omega_2 S_{33} \end{aligned} \tag{18.7.19}$$

$$\begin{aligned} dS_{21}/dt &= -\omega_1 S_{31} + \omega_3 S_{11} \\ dS_{22}/dt &= -\omega_1 S_{32} + \omega_3 S_{12} \\ dS_{23}/dt &= -\omega_1 S_{33} + \omega_3 S_{13} \end{aligned} \tag{18.7.20}$$

$$\begin{aligned} dS_{31}/dt &= -\omega_2 S_{11} + \omega_1 S_{21} \\ dS_{32}/dt &= -\omega_2 S_{12} + \omega_1 S_{22} \\ dS_{33}/dt &= -\omega_2 S_{13} + \omega_1 S_{23} \end{aligned} \tag{18.7.21}$$

Recall that the transformation matrix is orthogonal; the inverse and the transpose are the same. That is,

$$S_{mn}S_{mr} = \delta_{nr} \quad \text{and} \quad S_{mn}S_{rn} = S_{mr} \tag{18.7.22}$$

By using these expressions we may solve Eqs. (18.7.19), (18.7.20), and (18.7.21) for ω_1, ω_2, and ω_3. If we multiply the expressions in Eq. (18.7.19) by S_{31}, S_{32}, and S_{33}, respectively, and then add the equations, we have:

$$\omega_2 = S_{31}dS_{11}/dt + S_{32}dS_{12}/dt + S_{33}dS_{13}/dt \tag{18.7.23}$$

Similarly, by multiplying the expressions in Eqs. (18.7.20) by S_{11}, S_{12}, and S_{13}, respectively, and adding we obtain:

$$\omega_3 = S_{11}dS_{21}/dt + S_{12}dS_{22}/dt + S_{13}dS_{23}/dt \tag{18.7.24}$$

Finally, by multiplying the expressions in Eq. (18.7.22) by S_{21}, S_{22}, and S_{23}, respectively, and adding we obtain:

$$\omega_1 = S_{21}dS_{31}/dt + S_{22}dS_{32}/dt + S_{23}dS_{33}/dt \tag{18.7.25}$$

By substituting for the S_{ij} elements from Eq. (18.7.14) into Eqs. (18.7.23), (18.7.24), and (18.7.25), we obtain the angular velocity components in terms of the Euler parameters and their derivatives as:

$$\begin{aligned}
\omega_1 &= 2(\varepsilon_4\dot{\varepsilon}_1 - \varepsilon_3\dot{\varepsilon}_2 + \varepsilon_2\dot{\varepsilon}_3 - \varepsilon_1\dot{\varepsilon}_4) \\
\omega_2 &= 2(\varepsilon_3\dot{\varepsilon}_1 + \varepsilon_4\dot{\varepsilon}_2 - \varepsilon_1\dot{\varepsilon}_3 - \varepsilon_2\dot{\varepsilon}_4) \\
\omega_3 &= -2(\varepsilon_2\dot{\varepsilon}_1 - \varepsilon_1\dot{\varepsilon}_2 + \varepsilon_4\dot{\varepsilon}_3 - \varepsilon_3\dot{\varepsilon}_4)
\end{aligned} \tag{18.7.26}$$

This therefore establishes Eq. (18.5.16).

Finally, in Section 18.5 it was stated that with the use of Eq. (18.5.12) we could solve Eq. (18.5.15) (or, equivalently, Eq. (18.7.26)) for the Euler parameter derivatives in terms of the angular velocity components. To see this, recall from Eqs. (18.5.12) and (18.7.6) that the Euler parameters satisfy the identity:

$$\varepsilon_1^2 + \varepsilon_2^2 + \varepsilon_3^2 + \varepsilon_4^2 \equiv 1 \tag{18.7.27}$$

Then, by differentiating, we have:

$$2(\varepsilon_1\dot{\varepsilon}_1 + \varepsilon_2\dot{\varepsilon}_2 + \varepsilon_3\dot{\varepsilon}_3 + \varepsilon_4\dot{\varepsilon}_4) = 0 \tag{18.7.28}$$

Observe the similarity in form of this expression with the expressions of Eq. (18.7.26). Indeed, if we define the left side of Eq. (18.7.28) to be ω_4 and then append this expression to Eq. (18.7.26), we have the set of four equations:

$$\begin{aligned}
\omega_1 &= 2(\varepsilon_4\dot{\varepsilon}_1 - \varepsilon_3\dot{\varepsilon}_2 + \varepsilon_2\dot{\varepsilon}_3 - \varepsilon_1\dot{\varepsilon}_4) \\
\omega_2 &= 2(\varepsilon_3\dot{\varepsilon}_1 + \varepsilon_4\dot{\varepsilon}_2 - \varepsilon_1\dot{\varepsilon}_3 - \varepsilon_2\dot{\varepsilon}_4) \\
\omega_3 &= 2(-\varepsilon_2\dot{\varepsilon}_1 + \varepsilon_1\dot{\varepsilon}_2 + \varepsilon_4\dot{\varepsilon}_3 - \varepsilon_3\dot{\varepsilon}_4) \\
\omega_4 &= 2(\varepsilon_1\dot{\varepsilon}_1 + \varepsilon_2\dot{\varepsilon}_2 + \varepsilon_3\dot{\varepsilon}_3 + \varepsilon_4\dot{\varepsilon}_4)
\end{aligned} \tag{18.7.29}$$

where ω_4 is zero.

The expressions in Eq. (18.7.29) are linear equations in both the Euler parameters and their derivatives. They may therefore be written in the matrix form:

$$\begin{bmatrix} \omega_1 \\ \omega_2 \\ \omega_3 \\ \omega_4 \end{bmatrix} = 2\begin{bmatrix} \varepsilon_4 & -\varepsilon_3 & \varepsilon_2 & -\varepsilon_1 \\ \varepsilon_3 & \varepsilon_4 & -\varepsilon_1 & -\varepsilon_2 \\ -\varepsilon_2 & \varepsilon_1 & \varepsilon_4 & -\varepsilon_3 \\ \varepsilon_1 & \varepsilon_2 & \varepsilon_3 & \varepsilon_4 \end{bmatrix}\begin{bmatrix} \dot{\varepsilon}_1 \\ \dot{\varepsilon}_2 \\ \dot{\varepsilon}_3 \\ \dot{\varepsilon}_4 \end{bmatrix} \tag{18.7.30}$$

or more succinctly as:

$$\omega = 2\mathrm{E}\dot{\varepsilon} \tag{18.7.31}$$

where the arrays ω, E, and $\dot{\varepsilon}$ are defined by inspection and comparison of the two equations. Then, $\dot{\varepsilon}$ may be written as:

$$\dot{\varepsilon} = 1/2\,\mathrm{E}^{-1}\omega \tag{18.7.32}$$

The solution of Eq. (18.7.29) for the Euler parameter derivatives is thus at hand once we know the inverse E^{-1} of the matrix E. By inspection, however, we can see that E is orthogonal so that its inverse is its transpose. That is,

$$\mathrm{E}^{-1} = \begin{bmatrix} \varepsilon_4 & \varepsilon_3 & -\varepsilon_2 & \varepsilon_1 \\ -\varepsilon_3 & \varepsilon_4 & \varepsilon_1 & \varepsilon_2 \\ \varepsilon_2 & -\varepsilon_1 & \varepsilon_4 & \varepsilon_3 \\ -\varepsilon_1 & -\varepsilon_2 & -\varepsilon_3 & -\varepsilon_4 \end{bmatrix} \tag{18.7.33}$$

Therefore, from Eq. (18.7.32), we immediately obtain (with $\omega_4 = 0$) the desired expressions:

$$\begin{aligned} \dot{\varepsilon}_1 &= 1/2(\varepsilon_4\omega_1 + \varepsilon_3\omega_2 - \varepsilon_2\omega_3) \\ \dot{\varepsilon}_2 &= 1/2(-\varepsilon_3\omega_1 + \varepsilon_4\omega_2 + \varepsilon_1\omega_3) \\ \dot{\varepsilon}_3 &= 1/2(\varepsilon_2\omega_1 - \varepsilon_1\omega_2 + \varepsilon_4\omega_3) \\ \dot{\varepsilon}_4 &= 1/2(-\varepsilon_1\omega_1 - \varepsilon_2\omega_2 - \varepsilon_3\omega_3) \end{aligned} \tag{18.7.34}$$

This establishes Eq. (18.5.17).

18.8 Degrees of Freedom, Coordinates, and Generalized Speeds

Returning our attention to multibody systems, consider the open-chain system of Figure 18.8.1. Suppose the bodies are connected to one another by spherical (ball-and-socket) joints. Such joints will allow three rotational degrees of freedom of each body relative to its adjacent lower numbered body (see Sections 11.2 and 18.2). Hence, a system with N bodies will have $3N$ rotational degrees of freedom. If the system is moving in a reference frame R, then R (or B_0) is the lower numbered body for B_1. The translation of the system in R may then be measured by the coordinates of an arbitrarily selected reference point (say, O_1) of B_1 in R. Thus, in addition to the rotational degrees of freedom, there are three translation degrees of freedom leading to a total of $3N + 3$ degrees of freedom.

Following the procedures of Chapter 11, we may describe these degrees of freedom with variables (or "coordinates") representing the translation of O_1 in R and the rotation of each body relative to its adjacent lower numbered body. While the choice of these variables is somewhat arbitrary, it is natural to let Cartesian coordinates (x, y, z) describe the translation of O_1 in R, and to let orientation angles (say, α, β, γ) describe the rotations of

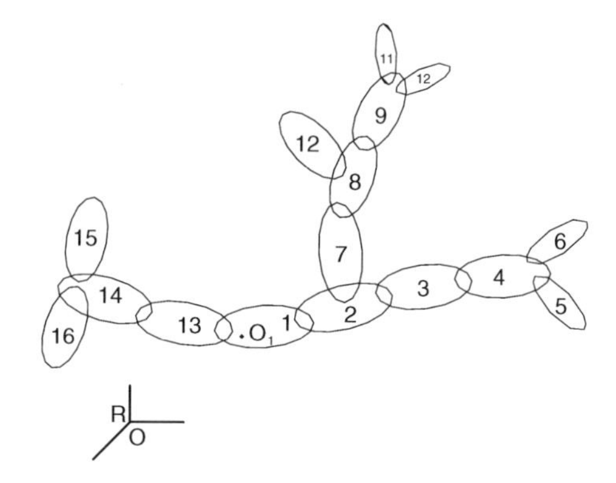

FIGURE 18.8.1
An open-chain multibody system with spherical joints.

the bodies. (To avoid singularities with these orientation angles, we may want to use Euler parameters as discussed in the previous section.)

If the movement (or anticipated movement) of the system is such that singularities are not likely to occur, then the coordinates may be represented by variables x_ℓ ($\ell = 1,\ldots, 3N + 3$), defined as:

$$\begin{array}{lllllll} x_1 = x & x_4 = \alpha_1 & x_7 = \alpha_2 & x_{10} = \alpha_3 & & x_{3N+1} = \alpha_N \\ x_2 = y & x_5 = \beta_1 & x_8 = \beta_2 & x_{11} = \beta_3 & \cdots & x_{3N+2} = \beta_N \\ x_3 = z & x_6 = \gamma_1 & x_9 = \gamma_2 & x_{12} = \gamma_3 & & x_{3N+3} = \gamma_N \end{array} \tag{18.8.1}$$

From our analysis of the triple pendulum (see Section 12.4), it would appear from an analytical perspective that it would be simpler to use absolute orientation angles as opposed to relative orientation angles to define the orientation of the bodies. (Absolute angles would orient the bodies directly in the inertia frame R, whereas relative angles orient the bodies with respect to one another and thereby indirectly in R.) While the absolute angles do indeed simplify the analysis, the relative angles are easier to visualize. Hence, if the analysis is to be conducted on a computer it is to the analyst's benefit to simplify the visualization of the system and let the computer do the work of the computation. It happens that in the development of computer algorithms and software that the preferred approach is to use absolute angles for the numerical procedures and relative angles for the input/output procedures. Thus, expressions enabling the conversion from one to the other are needed. Such expressions may be developed from the transformation matrices as discussed in the following section.

In either approach, Eqs. (18.8.1) or analogous expressions define the coordinates of the system. Recall from Chapter 11, however, that the velocity functions (that is, the mass center velocities and the angular velocities of the bodies) will be linear functions of the coordinate derivatives (see Section 11.4). It is from these velocity functions that the partial velocity and partial angular velocity vectors are obtained. To this end, we immediately see from Eq. (18.8.1) that the coordinate derivatives are:

$$\begin{array}{lllllll} \dot{x}_1 = \dot{x} & \dot{x}_4 = \dot{\alpha}_1 & \dot{x}_7 = \dot{\alpha}_2 & \dot{x}_{10} = \dot{\alpha}_3 & & \dot{x}_{3N+1} = \dot{\alpha}_N \\ \dot{x}_2 = \dot{y} & \dot{x}_5 = \dot{\beta}_1 & \dot{x}_8 = \dot{\beta}_2 & \dot{x}_{11} = \dot{\beta}_3 & \cdots & \dot{x}_{3N+2} = \dot{\beta}_N \\ \dot{x}_3 = \dot{z} & \dot{x}_6 = \dot{\gamma}_1 & \dot{x}_9 = \dot{\gamma}_2 & \dot{x}_{12} = \dot{\gamma}_3 & & \dot{x}_{3N+3} = \dot{\gamma}_N \end{array} \tag{18.8.2}$$

The coordinate derivatives are the fundamental variables in dynamical analyses, and in numerical solutions they become dependent variables to be determined by the solution procedure. From our discussions in the foregoing sections, however, we know that with the use of orientation angles singularities can occur in solution expressions for the coordinate derivatives. We saw that we can avoid these singularities with the use of Euler parameters. To facilitate the use of Euler parameters, it is helpful to group or arrange the coordinate derivatives into sets called *generalized speeds.*

A generalized speed is a linear combination of coordinate derivatives. Specifically, if a system has n degrees of freedom, represented by coordinates x_r $(r = 1,\ldots, n)$ with coordinate derivatives $\dot{x}_r$ $(r = 1,\ldots, n)$, then n generalized speeds y_s $(s = 1,\ldots, n)$ may be introduced as:

$$\begin{aligned} y_1 &= a_{11}\dot{x}_1 + a_{12}\dot{x}_2 + \cdots + a_{1n}\dot{x}_n + b_1 \\ y_2 &= a_{21}\dot{x}_1 + a_{22}\dot{x}_2 + \cdots + a_{2n}\dot{x}_n + b_2 \\ &\vdots \\ y_n &= a_{n1}\dot{x}_1 + a_{n2}\dot{x}_2 + \cdots + a_{nn}\dot{x}_n + b_n \end{aligned} \tag{18.8.3}$$

or as:

$$y_s = \sum_{r=1}^{n} a_{sr}\dot{x}_r + b_s \tag{18.8.4}$$

where the a_{sr} and the b_s $(r, s = 1,\ldots, n)$ may be functions of the coordinates x_r and time t. The a_{sr} and b_s are arbitrary, provided only that Eq. (18.8.4) may be solved for the $\dot{x}_r$ in terms of the y_s. That is, the a_{sr} and b_s may be selected in any way provided only that:

$$\det a_{sr} \neq 0 \tag{18.8.5}$$

The coordinate derivatives $\dot{x}_r$ may then be expressed in terms of these generalized speeds as:

$$\dot{x}_r = \sum_{s=1}^{n} a_{sr}^{-1}(y_s - b_s) \tag{18.8.6}$$

where the a_{sr}^{-1} are the elements of the inverse of the matrix whose elements are a_{sr}.

Generalized speeds may be conveniently used in dynamical analyses in place of the coordinate derivatives. When this is done, the partial velocities, the partial angular velocities, and the generalized forces (see Chapter 11) may each be expressed in terms of the generalized speeds. Specifically, if P is a typical particle of a mechanical system S and if B is a typical body of S, then the partial velocity of P and the partial angular velocity of B relative to the generalized speeds Y_s are defined as:

$$\mathbf{V}_{ys}^{P} \overset{\text{D}}{=} \partial\mathbf{V}^{P}/\partial y_s \quad \text{and} \quad \boldsymbol{\omega}_{y_s}^{B} \overset{\text{D}}{=} \partial\boldsymbol{\omega}^{B}/\partial y_s \tag{18.8.7}$$

By using Eq. (18.8.6) and by recalling our original definitions of the partial velocity and partial angular velocity vectors (see Eqs. (11.4.4) and (11.4.16)), we immediately obtain the relations:

$$\mathbf{V}_{y_s}^{P} = \sum_{r=1}^{n} a_{sr}^{-1}\mathbf{V}_{\dot{x}_r}^{P} \quad \text{and} \quad \boldsymbol{\omega}_{y_s}^{B} = \sum_{r=1}^{n} a_{sr}^{-1}\boldsymbol{\omega}_{\dot{x}_r}^{B} \tag{18.8.8}$$

Suppose P is now a point of B and suppose the force system on B is replaced by an equivalent force system consisting of a single force **F** passing through P together with a couple with torque **M**. Then, the generalized applied force (or *active force*) acting on B for the generalized speed y_s may be expressed as:

$$F_s = \mathbf{V}_{y_s}^P \cdot \mathbf{F} + \boldsymbol{\omega}_{y_s}^B \cdot \mathbf{M} \tag{18.8.9}$$

Similarly, if the inertia forces acting on B are replaced by a force **F*** passing through P together with a couple with torque **M***, the generalized inertia force (or *passive force*) acting on B for the generalized speed y_s may be expressed as:

$$F_s^* = \mathbf{V}_{y_s}^P \cdot \mathbf{F}^* + \boldsymbol{\omega}_{y_s}^B \cdot \mathbf{M}^* \tag{18.8.10}$$

At this point it might be helpful to make several comments about the actual use of general speeds. First, in practice, it is usually convenient to express velocities and angular velocities in terms of the generalized speeds, then the partial velocities and partial angular velocities may be obtained directly from Eq. (18.8.7) as opposed to using the more cumbersome chain rule of Eq. (18.8.8).

Next, we see that even though the expressions of Eq. (18.8.4) are invertible and may be solved for the coordinate derivatives $\dot{x}_r$ in terms of the general speeds y_s, the individual generalized speeds are not generally integrable in terms of elementary functions. That is, even though the $\dot{x}_r$ may be integrated to produce x_r, there are in general no elementary functions (say, z_s) such that $\dot{z}_s = y_s$. Hence, with the use of generalized speeds there are in general no corresponding generalized coordinates for which derivatives produce the generalized speeds. (This nonexistence of corresponding generalized coordinates has sometimes led analysts to refer to formulations using generalized speeds as formulations based upon *quasi-coordinates*.)

Finally, in practice the most widely used generalized speeds are components of the angular velocity vectors. Specifically, for a typical multibody system, as in Figure 18.8.1, it is usually convenient to define the first three generalized speeds as derivatives of the translation coordinates of the reference point O_1 of Body 1. The remaining generalized speeds are then defined as the components of the relative angular velocities of the bodies. That is,

$$\begin{array}{lllll} y_1 = \dot{x} & y_4 = \hat{\omega}_{11} & y_7 = \hat{\omega}_{21} & & y_{3n-2} = \hat{\omega}_{N1} \\ y_2 = \dot{y} & y_5 = \hat{\omega}_{12} & y_8 = \hat{\omega}_{22} & \cdots & y_{3n-1} = \hat{\omega}_{N2} \\ y_3 = \dot{z} & y_6 = \hat{\omega}_{13} & y_9 = \hat{\omega}_{23} & & y_{3n} = \hat{\omega}_{N3} \end{array} \tag{18.8.11}$$

where, as before, N is the number of bodies of the system, n is the number of degrees of freedom, and the angular velocity components are defined relative to unit vectors fixed in the adjacent lower numbered bodies. That is, the angular velocity components of Eq. (18.8.11) are defined in terms of the relative angular velocity vectors expressed in the form:

$$\hat{\boldsymbol{\omega}}_k = \hat{\omega}_{ki}\mathbf{n}_{ji} = \hat{\omega}_{ki}\mathbf{n}_{j1} + \hat{\omega}_{k2}\mathbf{n}_{j2} + \hat{\omega}_{k3}\mathbf{n}_{j3} \tag{18.8.12}$$

where $j = L(k)$.

18.9 Transformations between Absolute and Relative Coordinates

As noted in the previous section, it is desirable to use absolute orientation angles in computer analyses of multibody system dynamics and relative orientation angles for input and output data. Absolute angles provide for efficient numerical analysis, but relative angles provide a simpler geometric description of the system. Thus, it is desirable to have a means for readily converting from one to the other. Such conversions may be obtained using the transformation matrices.

To develop this, consider a typical multibody system S as in Figure 18.9.1. Let B_k be a typical body of the system, and let B_j be its adjacent lower numbered body. As before, let R be an inertial reference frame in which S moves. Let $\mathbf{n}_{si}$, $\mathbf{n}_{ji}$, and $\mathbf{n}_{ki}$ ($i = 1, 2, 3$) be unit vector sets fixed in R, B_j, and B_k as shown in Figure 18.9.2. Let α_k, β_k, and γ_k be dextral orientation angles defining the orientation of B_k in R, and let $\hat{\alpha}_k$, $\hat{\beta}_k$, and $\hat{\gamma}_k$ be dextral orientation angles defining the orientation of B_k in B_j. Recall from Section 18.3 that dextral orientation angles (or *Bryan angles*) may be defined as follows: imagine the unit vectors $\mathbf{n}_{ki}$ of B_k to be mutually aligned with the unit vectors $\mathbf{n}_{0i}$ of R. Then, B_k may be brought into a general orientation (dextral rotation) of B_k about axes parallel to $\mathbf{n}_{k1}$, $\mathbf{n}_{k2}$, and $\mathbf{n}_{k3}$ through the angles α_k, β_k, and γ_k. The relative orientation angles $\hat{\alpha}_k$, $\hat{\beta}_k$, and $\hat{\gamma}_k$ of B_k and B_j are defined similarly with the $\mathbf{n}_{0i}$ replaced by the $\mathbf{n}_{ji}$.

If we know the absolute orientation angles α_k, β_k, and γ_k, we may use Eq. (18.3.4) to obtain the transformation matrix SOK relating the $\mathbf{n}_{0i}$ and the $\mathbf{n}_{ki}$; that is,

$$SOK_{mn} = \begin{bmatrix} C\beta_k C\gamma_k & -C\beta_k S\gamma_k & S\beta_k \\ (C\alpha_k S\gamma_k + S\alpha_k SB_k C\gamma_k) & (C\alpha_k C\gamma_k - S\alpha_k S\beta_k S\gamma_k) & -S\alpha_k C\beta_k \\ (S\alpha_k S\gamma_k - C\alpha_k S\beta_k C\gamma_k) & (S\alpha_k C\gamma_k + C\alpha_k S\beta_k S\gamma_k) & C\alpha_k C\beta_k \end{bmatrix} \tag{18.9.1}$$

where, as before, the SOK_{mn} are defined as:

$$SOK_{mn} = \mathbf{n}_{0m} \cdot \mathbf{n}_{kn} \tag{18.9.2}$$

Also, as before, S and C are abbreviations for sine and cosine.

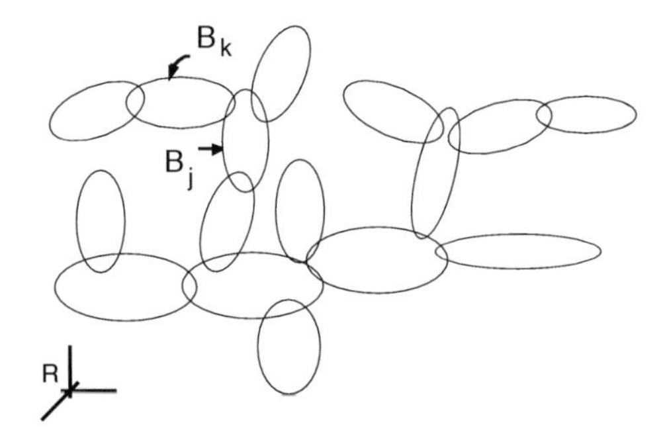

FIGURE 18.9.1
A typical multibody system with typical adjacent bodies B_j and B_k.

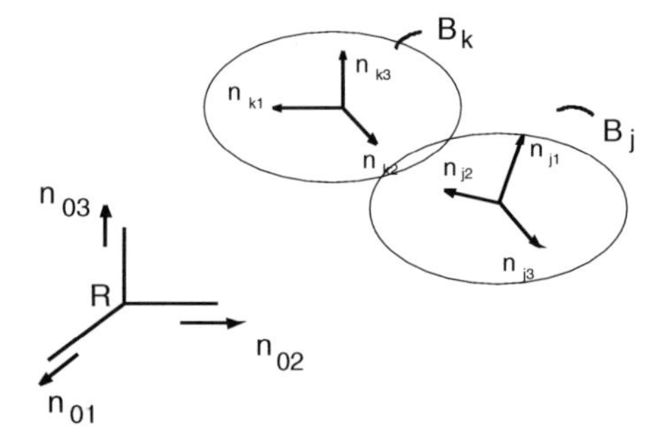

FIGURE 18.9.2
Typical adjoining bodies, inertial frame and unit vector sets.

Similarly, if we know the relative orientation angles $\hat{\alpha}_k$, $\hat{\beta}_k$, and $\hat{\gamma}_k$, we obtain the transformation matrix SJK relating the $\mathbf{n}_{ji}$ and $\mathbf{n}_{ki}$ as:

$$SJK_{mn} = \begin{bmatrix} C\hat{\beta}_k C\hat{\gamma}_k & -C\hat{\beta}_k S\hat{\gamma}_k & S\hat{\beta}_k \\ \left(C\hat{\alpha}_k S\hat{\gamma}_k + S\hat{\alpha}_k S\hat{\beta}_k C\hat{\gamma}_k\right) & \left(C\hat{\alpha}_k C\hat{\gamma}_k - S\hat{\alpha}_k S\hat{\beta}_k S\hat{\gamma}_k\right) & -S\hat{\alpha}_k C\hat{\beta}_k \\ \left(S\hat{\alpha}_k S\hat{\gamma}_k - C\hat{\alpha}_k S\hat{\beta}_k C\hat{\gamma}_k\right) & \left(S\hat{\alpha}_k C\hat{\gamma}_k + C\hat{\alpha}_k S\hat{\beta}_k S\hat{\gamma}_k\right) & C\hat{\alpha}_k C\hat{\beta}_k \end{bmatrix} \tag{18.9.3}$$

Recall from Eqs. (18.3.17) and (18.3.18) that the absolute transformation matrix SOK may be expressed as products of relative transformation matrices as:

$$SOK = SO1\ S12\ \ldots\ SIJ\ SJK \tag{18.9.4}$$

where the products are taken over the relative transformation matrices of the bodies in the branches of the system containing B_k. Therefore, if we know the relative transformation matrices, we can obtain the absolute transformation matrix.

Recall further from Eq. (18.3.7) that SOK may be expressed as:

$$SOK = SOJ\ SJK \tag{18.9.5}$$

Hence, SJK may be expressed as:

$$SJK = SOJ^{-1}\ SOK \tag{18.9.6}$$

Therefore, if we know the absolute transformation matrices we can readily obtain the relative transformation matrices.

In summary, we see from Eqs. (18.9.1) and (18.9.3) that if we know the orientation angles, be they absolute or relative, we can immediately obtain the absolute and relative transformation matrices. What remains to be shown is that if we know the transformation matrices we can obtain the orientation angles. To develop this, consider again the elements of the transformation matrices. Specifically, from Eq. (18.9.1), we have:

$$SOK_{13} = S\beta_k \tag{18.9.7}$$

thus β_k is:

$$\beta_k = \sin^{-1}\left[SOK_{13}\right] \tag{18.9.8}$$

Similarly, once β_k is known, α_k and γ_k may be obtained from SOK_{33} and SOK_{11}, respectively. That is,

$$\alpha_k = \cos^{-1}\left[SOK_{33}/C\beta_k\right] \quad \text{and} \quad \gamma_k = \cos^{-1}\left[SOK_{11}/C\beta_k\right] \tag{18.9.9}$$

Therefore, if we know the elements of the absolute transformation matrix, we can obtain the absolute orientation angles. In exactly the same manner, if we know the elements of the relative transformation matrix we can obtain the relative orientation angles. Thus, because we can relate the absolute and relative transformation matrices to each other we have the desired relations between the absolute and relative orientation angles.

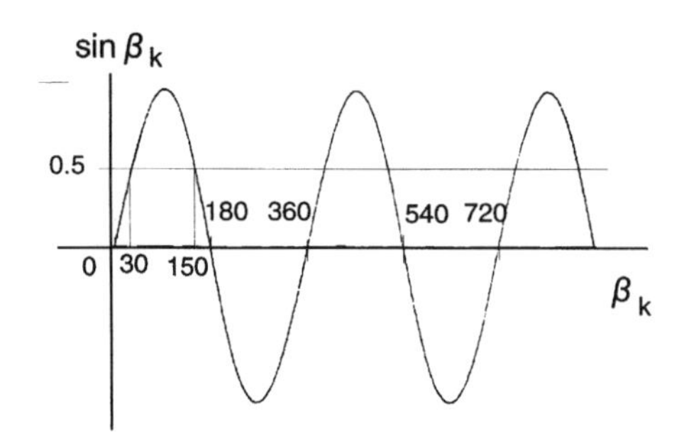

FIGURE 18.9.3
Multiple values of the inverse sine function.

Two questions arise in view of Eqs. (18.9.8) and (18.9.9): Because β_k is not uniquely determined by its sine function, which value should be chosen? What if β_k is 90°, or close to 90°? Will that not create singularities in Eq. (18.9.9)? To answer these questions, consider first that the principal purpose of obtaining relations between the absolute and relative orientation angles is for computational efficiency and for ease of data interpretation in numerical procedures. Because numerical procedures employ small incremental steps and because from physical considerations the angle values must be continuous, the choice of values of the inverse sine function is simply that value closest to the immediate preceding values. For example, suppose in Eq. (18.9.8) that SOK_{13} has the value 0.5. Is β_k then 30° or 150°, or even a multiple of 360° greater or less than 30 or 150°? (See Figure 18.9.3.) The answer is that β_k is to be chosen between 0 and 360° (or, alternatively, between –180 and 180°) so that its value is closest to the previous incrementally determined value. That is, if for the immediate previous time step β_k was 29° and if $\sin\beta_k$ is 0.5, then β_k will be assigned the value 30°.

Next, regarding a singularity in Eq. (18.9.9), if S_{β_k} is zero, then C_β is ±1. Then from Eq. (18.9.1) SOK_{21} and SOK_{22} are:

$$SOK_{21} = \sin(\gamma_k \pm \alpha_k) \quad \text{and} \quad SOK_{22} = \cos(\gamma_k \pm \alpha_k) \tag{18.9.10}$$

where, as before, the plus or minus sign will be apparent from the previous time step. Then, the values of α_k and γ_k are determined by simultaneous solution of Eq. (18.9.10) in view of the previous time step.

Finally, suppose Euler parameters are used in the analysis. The conversions back and forth to orientation angles may also be obtained using transformation matrices. To see this, consider Eq. (18.7.14) showing transformation matrix elements in terms of Euler parameters:

$$S_{ij} = \begin{bmatrix} (\varepsilon_1^2 - \varepsilon_2^2 - \varepsilon_3^2 + \varepsilon_4^2) & 2(\varepsilon_1\varepsilon_2 - \varepsilon_3\varepsilon_4) & 2(\varepsilon_1\varepsilon_3 + \varepsilon_2\varepsilon_4) \\ 2(\varepsilon_1\varepsilon_3 + \varepsilon_2\varepsilon_4) & (-\varepsilon_1^2 + \varepsilon_2^2 - \varepsilon_3^2 + \varepsilon_4^2) & 2(\varepsilon_2\varepsilon_3 - \varepsilon_1\varepsilon_4) \\ 2(\varepsilon_1\varepsilon_3 - \varepsilon_2\varepsilon_4) & 2(\varepsilon_2\varepsilon_3 + \varepsilon_1\varepsilon_4) & (-\varepsilon_1^2 - \varepsilon_2^2 + \varepsilon_3^2 + \varepsilon_4^2) \end{bmatrix} \tag{18.9.11}$$

In this expression the S_{ij} provides a transformation between unit vectors fixed in a reference frame *R* and a body *B*. The Euler parameters ε_i ($i = 1,\ldots, 4$) then define the orientation of *B* in *R*. In this regard, recall from Eq. (18.7.1) that the Euler parameters are defined as:

$$\varepsilon_1 = \lambda_1 \sin\theta/2, \quad \varepsilon_2 = \lambda_2 \sin\theta/2, \quad \varepsilon_3 = \lambda_3 \sin\theta/2, \quad \varepsilon_4 = \cos\theta/2 \tag{18.9.12}$$

where B is considered to be brought into its general orientation by a single rotation through the angle θ about an axis L fixed in both B and R and where the λ_i are the components relative to unit vectors in R of a unit vector λ parallel to L.

If we know the λ_i and θ we may immediately obtain the Euler parameters using Eq. (18.9.12) and then the transformation matrix elements using Eq. (18.9.11). Then, as before, Eqs. (18.9.8) and (18.9.9) provide the orientation angles.

What remains to be shown is how to obtain the Euler parameters if we know the orientation angles. To do this, consider first that in knowing the orientation angles we may use Eq. (18.9.1) to obtain the transformation matrix elements. Next, from Eq. (18.9.11) we see that the sum of the diagonal elements S_{kk} of the transformation matrix is:

$$S_{kk} = S_{11} + S_{22} + S_{33} = -\varepsilon_1^2 - \varepsilon_2^2 - \varepsilon_3^2 + 3\varepsilon_4^2 = -1 + 4\varepsilon_4^2 \tag{18.9.13}$$

where the last equality is obtained using the identity of Eq. (18.7.6). Then, from the last expression of Eq. (18.9.12) we immediately obtain θ as:

$$\theta = 2\cos^{-1}(S_{kk} + 1)/4 \tag{18.9.14}$$

Finally, consider the off-diagonal elements of the transformation matrix. From Eqs. (18.9.11) and (18.9.12) we see that:

$$\begin{aligned} S_{32} - S_{23} &= 4\varepsilon_1\varepsilon_4 = 4\lambda_1 \sin\theta/2\cos\theta/2 = 2\lambda_1 \sin\theta \\ S_{13} - S_{31} &= 4\varepsilon_2\varepsilon_4 = 4\lambda_2 \sin\theta/2\cos\theta/2 = 2\lambda_2 \sin\theta \\ S_{21} - S_{12} &= 4\varepsilon_3\varepsilon_4 = 4\lambda_3 \sin\theta/2\cos\theta/2 = 2\lambda_3 \sin\theta \end{aligned} \tag{18.9.15}$$

Then, the λ_i ($i = 1, 2, 3$) are:

$$\begin{aligned} \lambda_1 &= (S_{32} - S_{23})/2\sin\theta \\ \lambda_2 &= (S_{13} - S_{31})/2\sin\theta \\ \lambda_3 &= (S_{21} - S_{12})/2\sin\theta \end{aligned} \tag{18.9.16}$$

Hence, if we know the transformation matrix elements S_{ij} we can obtain the rotation angle θ from Eq. (18.9.14) and then the components of $\boldsymbol{\lambda}$ from Eq. (18.9.16). The Euler parameters are then immediately obtained from Eq. (18.9.12).

18.10 Angular Velocity

Consider again a typical multibody system as in Figure 18.10.1. Let B_k be a typical body of the system and let G_k be its mass center. The movement of G_k, and of all the particles of B_k, is known (or can be determined) if we know the velocity and acceleration of G_k and the angular velocity and angular acceleration of B_k itself. Of these four kinematical quantities, it is the angular velocity that is the most fundamental. We will see that knowledge

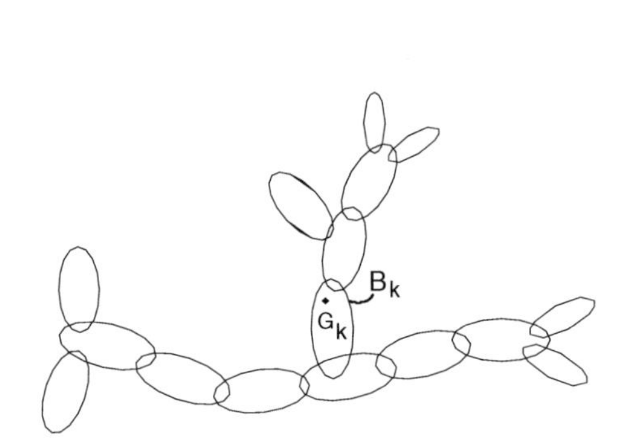

FIGURE 18.10.1
Typical multibody system.

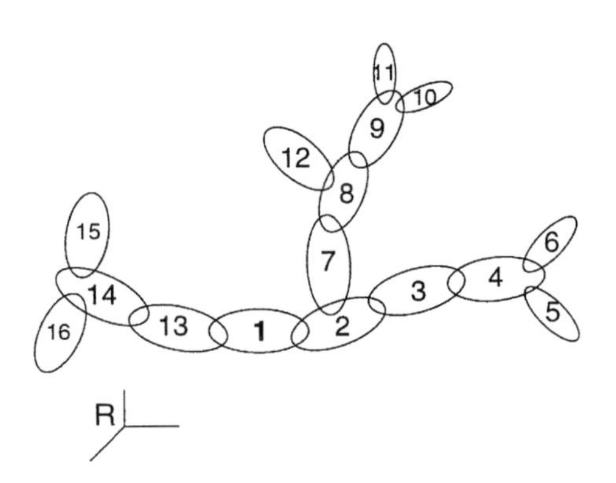

FIGURE 18.10.2
A numbered multibody system.

of the angular velocity of B_k enables us to determine the angular acceleration of B_k and then the velocity and acceleration of the mass center G_k.

Remarkably, the angular velocity of B_k is readily obtained and can be expressed in terms of the transformation matrices and lower body arrays. To see this, recall again from the addition theorem for angular velocity, Eq. (4.7.6), that the angular velocity of B_k in an inertial reference frame R may be expressed as:

$$ {}^{R}\boldsymbol{\omega}^{B_k} = {}^{R}\boldsymbol{\omega}^{B_1} + {}^{B_1}\boldsymbol{\omega}^{B_2} + \ldots + {}^{B_i}\boldsymbol{\omega}^{B_j} + {}^{B_j}\boldsymbol{\omega}^{B_k} \tag{18.10.1} $$

where the implied summation is taken over the bodies in the branch of the multibody system that contains B_k, where B_j is the adjacent lower numbered body of B_k, and where B_i is the adjacent lower numbered body of B_j.

Using the notation of Section 18.2 we can simplify the form of the terms of Eq. (18.10.1) by the definition:

$$ \boldsymbol{\omega}_k \overset{\text{D}}{=} {}^{R}\boldsymbol{\omega}^{B_k} \quad \text{and} \quad \hat{\boldsymbol{\omega}}_k = {}^{B_j}\boldsymbol{\omega}^{B_k} \tag{18.10.2} $$

Then, Eq. (18.10.1) may be written as:

$$ \boldsymbol{\omega}_k = \boldsymbol{\omega}_1 + \boldsymbol{\omega}_2 + \ldots + \hat{\boldsymbol{\omega}}_j + \hat{\boldsymbol{\omega}}_k \tag{18.10.3} $$

For the system of Figure 18.10.1, let the bodies be labeled and numbered as in Figure 18.10.2 (this is the same system we considered in Section 18.2; see Figure 18.2.5). Then, for, say, Body 10, Eq. (18.10.3) takes the form:

$$ \boldsymbol{\omega}_{10} = \hat{\boldsymbol{\omega}}_1 + \hat{\boldsymbol{\omega}}_2 + \hat{\boldsymbol{\omega}}_7 + \hat{\boldsymbol{\omega}}_8 + \hat{\boldsymbol{\omega}}_9 + \hat{\boldsymbol{\omega}}_{10} \tag{18.10.4} $$

or, equivalently,

$$ \boldsymbol{\omega}_{10} = \hat{\boldsymbol{\omega}}_{10} + \hat{\boldsymbol{\omega}}_9 + \hat{\boldsymbol{\omega}}_8 + \hat{\boldsymbol{\omega}}_7 + \hat{\boldsymbol{\omega}}_2 + \hat{\boldsymbol{\omega}}_1 \tag{18.10.5} $$

TABLE 18.10.1

Higher Order Lower Body Arrays for the Multibody System of Figure 18.10.2 (See Table 18.2.1)

$L^0(K)$	1	2	3	4	5	6	7	8	9	10	11	12	13	14	15	16
$L^1(K)$	0	1	2	3	4	4	2	7	8	9	9	8	1	13	14	14
$L^2(K)$	0	0	1	2	3	3	1	2	7	8	8	7	0	1	13	13
$L^3(K)$	0	0	0	1	2	2	0	1	2	7	7	2	0	0	1	1
$L^4(K)$	0	0	0	0	1	1	0	0	1	2	2	1	0	0	0	0
$L^5(K)$	0	0	0	0	0	0	0	0	0	1	1	0	0	0	0	0
$L^6(K)$	0	0	0	0	0	0	0	0	0	0	0	0	0	0	0	0

Recall also from Section 18.2 that the subscripts of Eq. (18.10.5) may be recognized as the entries of the 10th column of the table of higher order lower body arrays as in Table 18.10.1. Hence, by examining the sequencing of the indices we can write Eq. (18.10.5) as:

$$\boldsymbol{\omega}_{10} = \sum_{p=0}^{5} \boldsymbol{\omega}_8, \quad q = L^p(10) \tag{18.10.6}$$

In general, we can write:

$$\boldsymbol{\omega}_k = \sum_{p=0}^{r} \hat{\boldsymbol{\omega}}_q, \quad q = L^p(k) \tag{18.10.7}$$

where r is the index such that

$$L^r(k) = 1 \tag{18.10.8}$$

We may determine r by computing $L^p(k)$ and comparing the result to 1. The index r is the number of bodies from B_1 to B_k in the branch of the system containing B_k. Equation (18.10.7) may be viewed as an algorithm for finding $\boldsymbol{\omega}_k$ once the $\hat{\boldsymbol{\omega}}_k$ and $L(k)$ are known.

Let the bodies of the system of Figure 18.10.2 be connected by spherical joints. Then, because the system has 16 bodies, it has $(16 \times 3) + 3$, or 51, degrees of freedom (three in rotation for each body and three in translation [in R] for B_1). Then, by following the procedure outlined in Section 18.8, let us describe these degrees of freedom with generalized speeds. For the translation degrees of freedom, let the generalized speeds be the derivatives of the Cartesian coordinates of a reference point O_1 of B_1 in R. For the rotation degrees of freedom, let the generalized speeds be components of the relative angular velocity vectors.

Let the generalized speeds be labeled as y_r ($r = 1,\ldots, 51$). Then, from their definitions, the y_r may be divided into 17 sets, each containing three generalized speeds, as shown in Table 18.10.2, where the relative angular velocity components $\hat{\omega}_{ki}$ are defined relative to unit vectors fixed in the adjacent lower numbered body. That is, for typical body B_k we have:

$$\begin{aligned}\hat{\boldsymbol{\omega}}_k &= \hat{\omega}_{k1}\mathbf{n}_{j1} + \hat{\omega}_{k2}\mathbf{n}_{j2} + \hat{\omega}_{k3}\mathbf{n}_{j3} = \hat{\omega}_{ki}\mathbf{n}_{ji} \\ &= y_{3k+1}\mathbf{n}_{j1} + y_{3k+2}\mathbf{n}_{j2} + y_{3k+3}\mathbf{n}_{j3} = y_{3k+i}\mathbf{n}_{ji}\end{aligned} \tag{18.10.9}$$

or

$$y_r = \hat{\omega}_{ki} \quad \text{where} \quad r = 3k + i \quad (r \geq 4) \tag{18.10.10}$$

An inspection and comparison of Eqs. (18.10.3) and (18.10.9) shows that each term of Eq. (18.10.3) contains one and only one generalized speed. In this regard, suppose the unit vectors of Eq. (18.10.9) are expressed in terms of the unit vectors of the inertia frame R as:

$$\mathbf{n}_{jm} = SOJ_{nm}\mathbf{n}_{on} \tag{18.10.11}$$

Then, substitution from Eqs. (18.10.10) and (18.10.11) into (18.10.9) puts the relative angular velocity in the form:

$$\hat{\boldsymbol{\omega}}_k = y_r SOJ_{ni} n_{on} \quad (r = 3k + i) \tag{18.10.12}$$

TABLE 18.10.2

Triplets of General Speeds for the System of Figure 18.10.2

$y_1 = \dot{x}$	$y_4 = \hat{\omega}_{11}$	$y_7 = \hat{\omega}_{21}$	$y_{10} = \hat{\omega}_{31}$
$y_2 = \dot{y}$	$y_5 = \hat{\omega}_{12}$	$y_8 = \hat{\omega}_{22}$	$y_{11} = \hat{\omega}_{32}$
$y_3 = \dot{z}$	$y_6 = \hat{\omega}_{13}$	$y_9 = \hat{\omega}_{23}$	$y_{12} = \hat{\omega}_{33}$
$y_{13} = \hat{\omega}_{41}$	$y_{16} = \hat{\omega}_{51}$	$y_{19} = \hat{\omega}_{61}$	$y_{22} = \hat{\omega}_{71}$
$y_{14} = \hat{\omega}_{42}$	$y_{17} = \hat{\omega}_{52}$	$y_{20} = \hat{\omega}_{62}$	$y_{23} = \hat{\omega}_{72}$
$y_{15} = \hat{\omega}_{43}$	$y_{18} = \hat{\omega}_{53}$	$y_{21} = \hat{\omega}_{63}$	$y_{24} = \hat{\omega}_{73}$
$y_{25} = \hat{\omega}_{81}$	$y_{28} = \hat{\omega}_{91}$	$y_{31} = \hat{\omega}_{10,1}$	$y_{34} = \hat{\omega}_{11,1}$
$y_{26} = \hat{\omega}_{82}$	$y_{29} = \hat{\omega}_{92}$	$y_{32} = \hat{\omega}_{10,2}$	$y_{35} = \hat{\omega}_{11,2}$
$y_{27} = \hat{\omega}_{83}$	$y_{30} = \hat{\omega}_{93}$	$y_{33} = \hat{\omega}_{10,3}$	$y_{36} = \hat{\omega}_{11,3}$
$y_{37} = \hat{\omega}_{12,1}$	$y_{40} = \hat{\omega}_{13,1}$	$y_{43} = \hat{\omega}_{14,1}$	$y_{46} = \hat{\omega}_{15,1}$
$y_{38} = \hat{\omega}_{12,2}$	$y_{41} = \hat{\omega}_{13,2}$	$y_{44} = \hat{\omega}_{14,2}$	$y_{47} = \hat{\omega}_{15,2}$
$y_{39} = \hat{\omega}_{12,3}$	$y_{42} = \hat{\omega}_{13,3}$	$y_{45} = \hat{\omega}_{14,3}$	$y_{48} = \hat{\omega}_{15,3}$
$y_{49} = \hat{\omega}_{16,1}$			
$y_{50} = \hat{\omega}_{16,2}$			
$y_{51} = \hat{\omega}_{16,3}$			

Finally, by substituting into Eq. (18.10.3), we see that the angular velocity of B_k in R may be expressed in the form:

$$\boldsymbol{\omega}_k = \omega_{k\ell m} y_\ell \mathbf{n}_{om} \tag{18.10.13}$$

where, as before, there is a sum on repeated indices over the range of the index (in this case, m from 1 to 3 and ℓ from 1 to 51).

In Eq. (18.10.12), the $\omega_{k\ell m}$ may be recognized as scalar components (relative to the $\mathbf{n}_{om}$) of the partial angular velocity vectors (see Section 11). That is,

$$\partial \boldsymbol{\omega}_k / \partial y_\ell = \omega_{k\ell m} \mathbf{n}_{om} \tag{18.10.14}$$

By comparing Eqs. (18.10.12) and (18.10.14) we see that the $\omega_{k\ell m}$ may be expressed in terms of the transformation matrices.

To illustrate this, consider again B_{10} of Figure 18.10.2. Recall from Eq. (18.10.4) that the angular velocity of B_{10} in the inertia frame R may be written as:

$$\boldsymbol{\omega}_{10} = \hat{\boldsymbol{\omega}}_1 + \hat{\boldsymbol{\omega}}_2 + \hat{\boldsymbol{\omega}}_7 + \hat{\boldsymbol{\omega}}_8 + \hat{\boldsymbol{\omega}}_9 + \hat{\boldsymbol{\omega}}_{10} \tag{18.10.15}$$

Consider a typical term (say, $\hat{\boldsymbol{\omega}}_8$) of this expression. Because B_7 is the adjacent lower numbered body of B_8, we have:

$$\hat{\boldsymbol{\omega}}_8 = \hat{\omega}_{81} \mathbf{n}_{71} + \hat{\omega}_{82} \mathbf{n}_{72} + \hat{\omega}_{83} \mathbf{n}_{73} = \hat{\omega}_{8n} \mathbf{n}_{7n} \tag{18.10.16}$$

Then, from Table 18.10.2, we have:

$$\hat{\boldsymbol{\omega}}_8 = y_{25} \mathbf{n}_{71} + y_{26} \mathbf{n}_{72} + y_{27} \mathbf{n}_{73} = y_{24+n} \mathbf{n}_{7n} \tag{18.10.17}$$

From Eq. (18.10.11), we have:

$$\mathbf{n}_{7n} = SO7_{mn} \mathbf{n}_{om} \tag{18.10.18}$$

Therefore, $\hat{\boldsymbol{\omega}}_8$ may be written as:

$$\hat{\boldsymbol{\omega}}_8 = y_{24+n} SO7_{mn} \mathbf{n}_{om} \tag{18.10.19}$$

By similar analyses we can readily obtain the expressions:

$$\hat{\boldsymbol{\omega}}_1 = y_{3+n} \delta_{mn} \mathbf{n}_{om} \tag{18.10.20}$$

$$\hat{\boldsymbol{\omega}}_2 = y_{6+n} SO1_{mn} \mathbf{n}_{om} \tag{18.10.21}$$

$$\hat{\boldsymbol{\omega}}_7 = y_{21+n} SO2_{mn} \mathbf{n}_{om} \tag{18.10.22}$$

$$\hat{\boldsymbol{\omega}}_9 = y_{27+n} SO8_{mn} \mathbf{n}_{om} \tag{18.10.23}$$

$$\hat{\boldsymbol{\omega}}_{10} = y_{30+n} SO9_{mn} \mathbf{n}_{om} \tag{18.10.24}$$

Hence, if we write $\boldsymbol{\omega}_{10}$ in the form:

$$\boldsymbol{\omega}_{10} = \omega_{10\ell m} y_\ell \mathbf{n}_{om}$$

we can by inspection of Eqs. (18.10.19) to (18.10.24) identify the $\omega_{10\ell m}$ as:

$$\omega_{10\ell m} = \begin{cases} 0 & \ell = 1,2,3 \\ \delta_{m\ell-3} & \ell = 4,5,6 \\ SO1_{m\ell-6} & \ell = 7,8,9 \\ 0 & \ell = 10,\ldots,21 \\ SO2_{m\ell-21} & \ell = 22,23,24 \\ SO7_{m\ell-24} & \ell = 25,26,27 \\ SO8_{m\ell-27} & \ell = 28,29,30 \\ SO9_{m\ell-30} & \ell = 31,32,33 \\ 0 & \ell \geq 34 \end{cases} \quad (m = 1,2,3) \tag{18.10.25}$$

By similar analyses, we may obtain the other $\omega_{k\ell m}$. The results may be tabulated as in Table 18.10.3.

Observe that most of the entries of Table 18.10.3 are zero. Observe also the nonzero entries in the rows and columns. The entries in any given column are all the same. Finally, note that the pattern in the rows follows that of the columns of the higher-order lower body arrays of Table 18.10.1.

18.11 Angular Acceleration

Once we know the angular velocities of the bodies, we can obtain their angular accelerations by differentiation. Specifically, from Eq. (18.10.13), let the angular velocity of a typical body B_k of the system be expressed as:

$$\boldsymbol{\omega}_k = \omega_{k\ell m} y_\ell \mathbf{n}_{om} \tag{18.11.1}$$

Then, recalling that the $\mathbf{n}_{om}$ are unit vectors fixed in the inertial frame R, the angular acceleration of B_k in R is:

$$\boldsymbol{\alpha}_k = d\boldsymbol{\omega}_k/dt = \left(\omega_{k\ell m} \dot{y}_\ell + \dot{\omega}_{k\ell m} y_\ell\right) \mathbf{n}_{om} \tag{18.11.2}$$

where as before the overdot represents time differentiation.

TABLE 18.10.3

Partial Angular Velocity Components $\omega_{k\ell m}$ for the System of Figure 18.10.2

y_ℓ	1	4	7	10	13	16	19	22	25	28	31	34	37	40	43	46	49
	2	5	8	11	14	17	20	23	26	29	32	35	38	41	44	47	50
B_k	3	6	9	12	15	18	21	24	27	30	33	36	39	42	45	48	51
1	0	I	0	0	0	0	0	0	0	0	0	0	0	0	0	0	0
2	0	I	S01	0	0	0	0	0	0	0	0	0	0	0	0	0	0
3	0	I	S01	S02	0	0	0	0	0	0	0	0	0	0	0	0	0
4	0	I	S01	S02	S03	0	0	0	0	0	0	0	0	0	0	0	0
5	0	I	S01	S02	S03	S04	0	0	0	0	0	0	0	0	0	0	0
6	0	I	S01	S02	S03	0	S04	0	0	0	0	0	0	0	0	0	0
7	0	I	S01	0	0	0	0	S02	0	0	0	0	0	0	0	0	0
8	0	I	S01	0	0	0	0	S02	S07	0	0	0	0	0	0	0	0
9	0	I	S01	0	0	0	0	S02	S07	S08	0	0	0	0	0	0	0
10	0	I	S01	0	0	0	0	S02	S07	S08	S09	0	0	0	0	0	0
11	0	I	S01	0	0	0	0	S02	S07	S08	0	S09	0	0	0	0	0
12	0	I	S01	0	0	0	0	S02	S07	0	0	0	S08	0	0	0	0
13	0	I	0	0	0	0	0	0	0	0	0	0	0	S01	0	0	0
14	0	I	0	0	0	0	0	0	0	0	0	0	0	S01	S013	0	0
15	0	I	0	0	0	0	0	0	0	0	0	0	0	S01	S013	S014	0
16	0	I	0	0	0	0	0	0	0	0	0	0	0	S01	S013	0	S014

TABLE 18.11.1

Derivatives of Partial Angular Velocity Components $\dot{\omega}_{k\ell m}$ for the System of Figure 18.10.2

y_ℓ	1	4	7	10	13	16	19	22	25	28	31	34	37	40	43	46	49
	2	5	8	11	14	17	20	23	26	29	32	35	38	41	44	47	50
B_k	3	6	9	12	15	18	21	24	27	30	33	36	39	42	45	48	51
1	0	0	0	0	0	0	0	0	0	0	0	0	0	0	0	0	0
2	0	0	$S\dot{0}1$	0	0	0	0	0	0	0	0	0	0	0	0	0	0
3	0	0	$S\dot{0}1$	$S\dot{0}2$	0	0	0	0	0	0	0	0	0	0	0	0	0
4	0	0	$S\dot{0}1$	$S\dot{0}2$	$S\dot{0}3$	0	0	0	0	0	0	0	0	0	0	0	0
5	0	0	$S\dot{0}1$	$S\dot{0}2$	$S\dot{0}3$	$S\dot{0}4$	0	0	0	0	0	0	0	0	0	0	0
6	0	0	$S\dot{0}1$	$S\dot{0}2$	$S\dot{0}3$	0	$S\dot{0}4$	0	0	0	0	0	0	0	0	0	0
7	0	0	$S\dot{0}1$	0	0	0	0	$S\dot{0}2$	0	0	0	0	0	0	0	0	0
8	0	0	$S\dot{0}1$	0	0	0	0	$S\dot{0}2$	$S\dot{0}7$	0	0	0	0	0	0	0	0
9	0	0	$S\dot{0}1$	0	0	0	0	$S\dot{0}2$	$S\dot{0}7$	$S\dot{0}8$	0	0	0	0	0	0	0
10	0	0	$S\dot{0}1$	0	0	0	0	$S\dot{0}2$	$S\dot{0}7$	$S\dot{0}8$	$S\dot{0}9$	0	0	0	0	0	0
11	0	0	$S\dot{0}1$	0	0	0	0	$S\dot{0}2$	$S\dot{0}7$	$S\dot{0}8$	0	$S\dot{0}9$	0	0	0	0	0
12	0	0	$S\dot{0}1$	0	0	0	0	$S\dot{0}2$	$S\dot{0}7$	0	0	0	$S\dot{0}8$	0	0	0	0
13	0	0	0	0	0	0	0	0	0	0	0	0	0	$S\dot{0}1$	0	0	0
14	0	0	0	0	0	0	0	0	0	0	0	0	0	$S\dot{0}1$	$S\dot{0}13$	0	0
15	0	0	0	0	0	0	0	0	0	0	0	0	0	$S\dot{0}1$	$S\dot{0}13$	$S\dot{0}14$	0
16	0	0	0	0	0	0	0	0	0	0	0	0	0	$S\dot{0}1$	$S\dot{0}13$	0	$S\dot{0}14$

The $\dot{\omega}_{k\ell m}$ are derivatives of the partial angular velocity components. They may be obtained by differentiation in expressions for the $\omega_{k\ell m}$. For example, for Body 10 of the system of Figure 18.10.2, the $\omega_{10\ell m}$ are listed in Table 18.10.3, and the $\dot{\omega}_{k\ell m}$ are as listed in Table 18.11.1. The derivatives of the transformation matrices are given by Eq. (18.4.8): That is,

$$S\dot{O}K = WK\ SOK \tag{18.11.3}$$

where from Eq. (18.4.7) the elements of the WK matrix may be expressed in terms of the angular velocity components as:

$$WK_{ms} = -e_{msr}\omega_{kr} \tag{18.11.4}$$

where the e_{msr} are elements of the permutation symbol. Observe in Table 18.11.1 that most of the $\dot{\omega}_{k\ell m}$ are zero.

18.12 Joint and Mass Center Positions

In the previous section, we saw that knowledge of the angular velocities of the bodies of a multibody system enabled us to readily determine the angular accelerations of the bodies. In the following sections, we will see that angular velocities of the bodies will also enable us to find the mass center velocities and accelerations. In this section, we develop expressions for the joint and mass center positions. Differentiating these expressions then leads to the desired mass center velocities and accelerations.

To begin the analysis, consider a typical series of adjoining bodies such as B_i, B_j, and B_k as in Figure 18.12.1. Let O_i, O_j, and O_k be at the centers of the spherical connecting joints, and let G_i, G_j, and G_k be the mass centers of the bodies. Let O_i, O_j, and O_k also serve as origins of the reference frames fixed in B_i, B_j, and B_k, respectively.

Let $\mathbf{r}_i$, $\mathbf{r}_j$, and $\mathbf{r}_k$ locate G_i, G_j, and G_k relative to O_i, O_j, and O_k as shown in Figure 18.12.2. Then, $\mathbf{r}_i$, $\mathbf{r}_j$, and $\mathbf{r}_k$ are fixed in B_i, B_j, and B_k.

Let $\boldsymbol{\xi}_j$ and $\boldsymbol{\xi}_k$ locate O_j and O_k relative to O_i and O_j as shown in Figure 18.12.3. Then, $\boldsymbol{\xi}_j$ and $\boldsymbol{\xi}_k$ are fixed in B_i and B_j, respectively.

Consider again the typical multibody system of Figure 18.12.4. As before, let the system be numbered and labeled as shown. Let O_1 be the reference point of the reference body B_1, and let $\boldsymbol{\xi}_1$ locate O_1 relative to the origin O of the inertial frame R. Then, $\boldsymbol{\xi}_1$ may be expressed as:

$$\boldsymbol{\xi}_1 = x\mathbf{n}_{01} + y\mathbf{n}_{02} + z\mathbf{n}_{03} \tag{18.12.1}$$

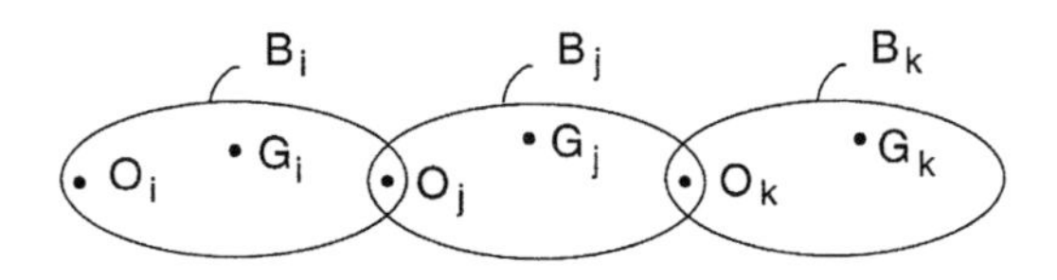

FIGURE 18.12.1
Three typical adjoining bodies of a multibody system.

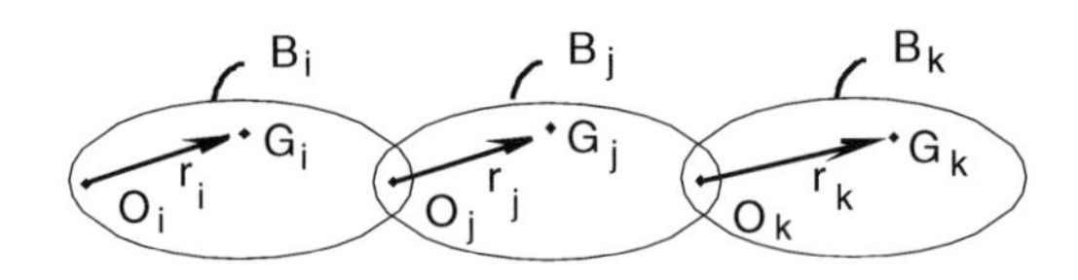

FIGURE 18.12.2
Mass center position vectors.

where, as before, $\mathbf{n}_{01}$, $\mathbf{n}_{02}$, and $\mathbf{n}_{03}$ are unit vectors fixed in R and x, y, and z are coordinates of O_1 in R determining the first three generalized speeds of Table 18.10.2.

Consider a typical body (say, B_{10}) of this system. Using the position vector definitions and notation of Figures 18.10.2 and 18.10.3, the position vector locating mass center G_{10} relative to O may be expressed as (see Figure 18.12.5):

$$\mathbf{p}_{10} = \boldsymbol{\xi}_1 + \boldsymbol{\xi}_2 + \boldsymbol{\xi}_7 + \boldsymbol{\xi}_8 + \boldsymbol{\xi}_9 + \boldsymbol{\xi}_{10} + \mathbf{r}_{10} \tag{18.12.2}$$

Observe the pattern of the indices of Eq. (18.12.2), and observe that the indices are the same as the numbers in column 10 of Table 18.10.1. Observe further that each $\boldsymbol{\xi}_k$ of Eq. (18.12.2), except $\boldsymbol{\xi}_1$, is fixed in B_j where B_j is the adjacent lower numbered body of B_k ($j = L(k)$). Then, in terms of the $\mathbf{n}_{om}$, we may express Eq. (18.12.2) as:

$$\begin{aligned} \mathbf{p}_{10} = \big(&\xi_{1m} + S01_{mn}\xi_{2n} + S02_{mn}\xi_{7n} + S07_{mn}\xi_{8n} \\ &+ S08_{mn}\xi_{9n} + S09_{mn}\xi_{10n} + S010_{mn}r_{10n}\big)\mathbf{n}_{om} \end{aligned} \tag{18.12.3}$$

In view of Eq. (18.10.6), we may express Eq. (18.12.3) in the compact form:

$$\mathbf{p}_{10} = \left(\xi_{1m} + \sum_{p=0}^{4} S0Q_{mn}\xi_{qn} + S010_{mn}r_{10n} \right)\mathbf{n}_{om} \tag{18.12.4}$$

where here we have:

$$q = L^p(10) \quad Q = L(8) = L^{p+1}(10) \tag{18.12.5}$$

Then, in general, we have:

$$\mathbf{p}_k = \left(\xi_{1m} + \sum_{s=0}^{r-1} SOP_{mn}\xi_{qn} + SOK_{mn}r_{kn} \right)\mathbf{n}_{om} \tag{18.12.6}$$

where now q and P are:

$$q = L^s(K) \quad P = L(q) = L^{s+1}(K) \tag{18.12.7}$$

and r is the index such that

$$L^r(K) = 1 \tag{18.12.8}$$

and where $K = k$.

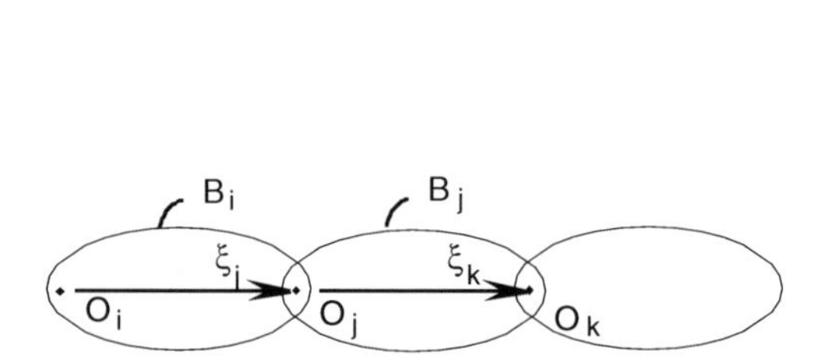

FIGURE 18.12.3
Joint position vectors.

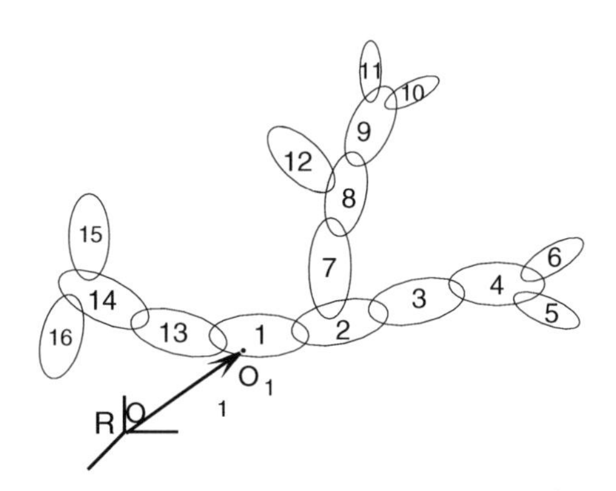

FIGURE 18.12.4
Typical multibody system.

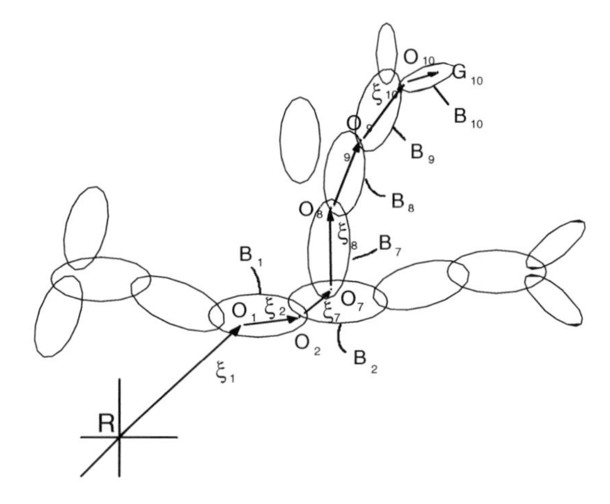

FIGURE 18.12.5
Position vectors locating the mass center of body B_{10}.

18.13 Mass Center Velocities

The mass center velocities may be obtained by differentiation in Eq. (18.12.6). Specifically, we have:

$$
\begin{aligned}
{}^{R}\mathbf{v}^{G_k} &\overset{\text{D}}{=} \mathbf{v}_k = d\mathbf{p}_k/dt \\
&= \left(\dot{\xi}_{1m} + \sum_{s=0}^{r-1} S\dot{O}P_{mn}\xi_{qn} + S\dot{O}K_{mn}r_{kn} \right)\mathbf{n}_{om}
\end{aligned}
\tag{18.13.1}
$$

where, as before, q and P are:

$$
q = L^s(k) \quad P = L(q) = L^{s+1}(k) \tag{18.13.2}
$$

and r is the index such that:

$$L^r(K) = 1 \tag{18.13.3}$$

The $\dot{SOK}$ in Eq. (18.13.1) may be obtained from Eq. (18.4.8) as:

$$\dot{SOK} = WK\ SOK \tag{18.13.4}$$

or in component form from Eq. (18.4.6) as:

$$\dot{SOK}_{mn} = WK_{ms} SOK_{sn} \tag{18.13.5}$$

where the WK_{ms} are:

$$WK_{ms} = -e_{msr}\omega_{kr} \tag{18.13.6}$$

where the ω_{kr} are the $\mathbf{n}_{or}$ components of the $\boldsymbol{\omega}_k$. From Eq. (18.10.13), we have:

$$\omega_{kr} = \omega_{k\ell r} y_\ell \tag{18.13.7}$$

Thus, by substitution, Eq. (18.13.5) may be written in the form:

$$\dot{SOK}_{mn} = -e_{msr}\omega_{k\ell r} y_\ell SOK_{sn} \stackrel{D}{=} UK_{m\ell n} y_\ell \tag{18.13.8}$$

where by inspection the $UK_{m\ell n}$ are defined as:

$$UK_{m\ell n} \stackrel{D}{=} -e_{msr}\omega_{k\ell r} SOK_{sn} \tag{18.13.9}$$

Then, from Eq. (18.13.1), the $\mathbf{v}_k$ in turn become:

$$\mathbf{v}_k = \left(\dot{\xi}_{1m} + \sum_{s=0}^{r-1} UP_{m\ell n} y_\ell \xi_{qn} + UK_{m\ell n} y_\ell r_{kn} \right) \mathbf{n}_{om} \tag{18.13.10}$$

By examining Eq. (18.13.10), we see that, analogous to Eq. (18.10.13) for angular velocity, we can express the mass center velocities in the form:

$$\mathbf{v}_k = v_{k\ell m} y_\ell \mathbf{n}_{om} \tag{18.13.11}$$

where by inspection of Eq. (18.13.10) the $v_{k\ell m}$ are seen to be:

$$v_{k\ell m} = \delta_{km} \quad k = 1,\ldots,N; \quad \ell = 1,2,3; \quad m = 1,2,3 \tag{18.13.12}$$

and

$$v_{k\ell m} = \sum_{s=0}^{r-1} UP_{m\ell n}\xi_{qn} + UK_{m\ell n}r_{kn} \quad k=1,\ldots,N; \quad \ell=4,\ldots,3N+3; \quad m=1,2,3 \tag{18.13.13}$$

Observe that the $v_{k\ell m}$ depend directly on the $\omega_{k\ell m}$ through the $UK_{m\ell n}$.

18.14 Mass Center Accelerations

With the mass center velocities given by Eq. (18.13.11), we may obtain the mass center accelerations by differentiations. Specifically, the accelerations of the G_k in R are:

$$^{R}\mathbf{a}^{G_k} = \mathbf{a}_k = \left(v_{k\ell m}\dot{y}_\ell + \dot{v}_{k\ell m}y_\ell\right)\mathbf{n}_{om} \tag{18.14.1}$$

where the $v_{k\ell n}$ are given by Eqs. (18.13.12) and (18.13.13) and where by differentiation the $\dot{v}_{k\ell n}$ are:

$$\dot{v}_{k\ell m} = 0 \quad k=1,\ldots,N; \quad \ell=1,2,3; \quad m=1,2,3 \tag{18.14.2}$$

and

$$\dot{v}_{k\ell m} = \sum_{p=0}^{r-1} U\dot{Q}_{m\ell n}\xi_{qn} + U\dot{K}_{m\ell n}r_{kn} \quad k=1,\ldots,N; \quad \ell=4,\ldots 3N+3; \quad m=1,2,3 \tag{18.14.3}$$

where from Eqs. (18.13.8) and (18.13.9) the $U\dot{K}_{m\ell n}$ are seen to be:

$$U\dot{K}_{m\ell n} = -e_{msr}\left(\dot{\omega}_{k\ell r}SOK_{sn} + \omega_{k\ell r}UK_{spn}y_p\right) \tag{18.14.4}$$

Observe again the direct dependence upon the $\omega_{k\ell m}$.

18.15 Kinetics: Applied (Active) Forces

Consider again a multibody system S. Let S be subjected to an applied force field as depicted in Figure 18.15.1. The force field could occur external to S as gravity and contact forces. Contact forces could also occur internal to S, between the bodies themselves.

The procedures for studying the effect of the force field on S are the same as those discussed in Chapters 6 and 11. Because multibody systems are generally relatively large systems with many degrees of freedom, it is efficient to use the procedures of Chapter 11 in Section 11.5 concerning generalized forces.

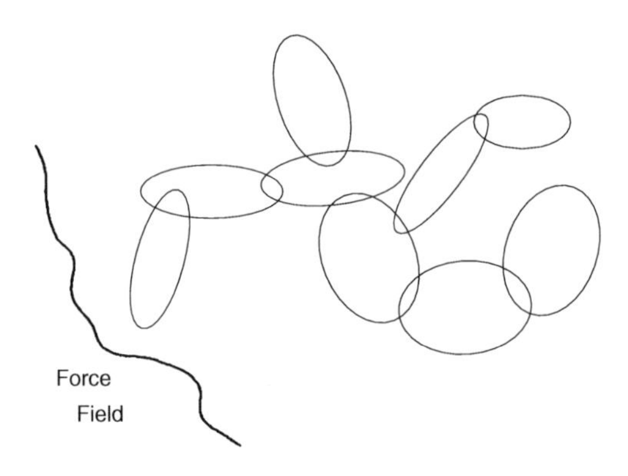

FIGURE 18.15.1
A multibody system in a force field.

Specifically, let the forces applied to a typical body B_k of S be represented by an equivalent force system (see Section 6.5) consisting of a single force $\mathbf{F}_k$ passing through the mass center G_k together with a couple with torque $\mathbf{T}_k$. Then, from Eq. (11.5.7), the generalized active force $F_{k\ell}$ acting on B_k for the generalized speed y_ℓ is:

$$F_\ell^k = \mathbf{F}_k \cdot \partial \mathbf{v}_k / \partial y_\ell + \mathbf{T}_k \cdot \partial \boldsymbol{\omega}_k / \partial y_\ell \quad \text{(no sum on } k\text{)} \tag{18.15.1}$$

where, as before, $\mathbf{v}_k$ and $\boldsymbol{\omega}_k$ are the velocity of the mass center G_k of B_k and the angular velocity of B_k.

Let $\mathbf{n}_{om}$ $(m = 1, 2, 3)$ form a mutually perpendicular unit vector set fixed in an inertia frame R, and let $\mathbf{F}_k$ and $\mathbf{T}_k$ be expressed in terms of the $\mathbf{n}_{om}$ as:

$$\mathbf{F}_k = F_{km}\mathbf{n}_{om} \quad \text{and} \quad \mathbf{T}_k = T_{km}\mathbf{n}_{om} \tag{18.15.2}$$

Also, from Eqs. (18.10.13) and (18.13.11), let $\mathbf{v}_k$ and $\boldsymbol{\omega}_k$ be expressed in the forms:

$$\mathbf{v}_k = v_{k\ell m} y_\ell \mathbf{n}_{om} \quad \text{and} \quad \boldsymbol{\omega}_k = \omega_{k\ell m} y_\ell \mathbf{n}_{om} \tag{18.15.3}$$

Then, by substituting into Eq. (18.15.1), the generalized force has the form:

$$F_\ell^k = F_{km} v_{s\ell m} + T_{km} \omega_{k\ell m} \tag{18.15.4}$$

Observe that in Eq. (18.15.1) there is no sum on k. If we omit that restriction, however, we have the generalized force for y_ℓ for the entire system.

18.16 Kinetics: Inertia (Passive) Forces

In a similar and directly analogous manner we can develop expressions for the generalized inertia forces. Consider again a typical body B_k of a multibody system as in Figure 18.16.1. Following the procedures of Chapters 6 and 8, let the inertia forces on B_k be represented

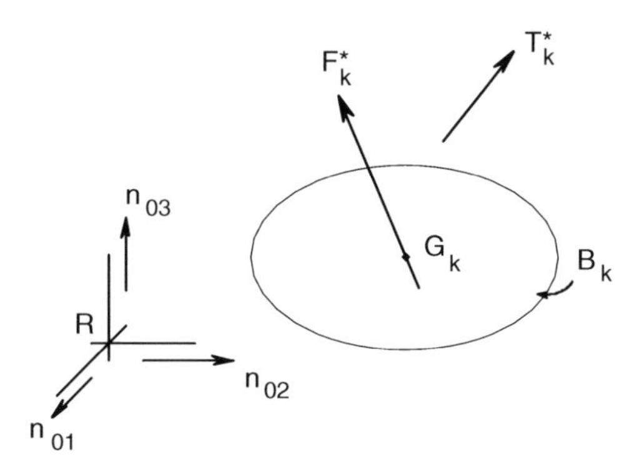

FIGURE 18.16.1
Equivalent inertia force system on a typical body B_k of a multibody system.

by a single force $\mathbf{F}_k^*$ passing through the mass center G_k together with a couple with torque $\mathbf{T}_k^*$, as in Figure 18.16.1. Then, analogous to Eq. (18.15.4), the generalized inertia force on B_k for the generalized speed y_ℓ is:

$$F_\ell^{*k} = F_{km}^* v_{k\ell m} + T_{km}^* \omega_{k\ell m} \tag{18.16.1}$$

where F_{km}^* and T_{km}^* are the $\mathbf{n}_{om}$ components of $\mathbf{F}_k^*$ and $\mathbf{T}_k^*$.

Observe in Eq. (18.16.1), as in Eq. (18.15.4), if we add the forces for each body (that is, sum over k), we have the generalized inertia force for y_ℓ for the entire system.

By using Eqs. (8.6.5) and (8.6.6) of Chapter 8, we can obtain a more detailed and explicit expression for the generalized inertia forces. Specifically, from Eqs. (8.6.5) and (8.6.6) we can express $\mathbf{F}_k^*$ and $\mathbf{T}_k^*$ in the forms:

$$\mathbf{F}_k^* = -m_k \mathbf{a}_k \quad \text{(no sum on } k\text{)} \tag{18.16.2}$$

and

$$\mathbf{T}_k^* = -\mathbf{I}_k \cdot \boldsymbol{\alpha}_k - \boldsymbol{\omega}_k \times (\mathbf{I}_k \cdot \boldsymbol{\omega}_k) \quad \text{(no sum on } k\text{)} \tag{18.16.3}$$

where m_k is the mass of B_k, and $\mathbf{I}_k$ is the central inertia dyadic of B_k. From Eqs. (18.10.13), (18.11.2), and (18.14.1), we can express $\boldsymbol{\omega}_k$, $\boldsymbol{\alpha}_k$, and $\mathbf{a}_k$ as:

$$\boldsymbol{\omega}_k = \omega_{k\ell m} y_\ell \mathbf{n}_{om} \tag{18.6.4}$$

$$\boldsymbol{\alpha}_k = (\omega_{k\ell m} \dot{y}_\ell + \dot{\omega}_{k\ell m} y_\ell)\mathbf{n}_{om} \tag{18.6.5}$$

and

$$\mathbf{a}_k = (\mathrm{v}_{k\ell m} \dot{y}_\ell + \dot{\mathrm{v}}_{k\ell m} y_\ell)\mathbf{n}_{om} \tag{18.6.6}$$

Then, by substituting into Eqs. (18.16.2) and (18.16.3), we can express $\mathbf{F}_k^*$ and $\mathbf{T}_k^*$ as:

$$\mathbf{F}_k^* = -m_k (v_{k\ell m} \dot{y}_\ell + \dot{v}_{k\ell m} y_\ell)\mathbf{n}_{om} \tag{18.6.7}$$

and

$$\mathbf{T}_k^* = -\big[I_{kmn}\big(\omega_{k\ell n}\dot{y}_\ell + \dot{\omega}_{k\ell n}y_\ell\big) \\ + e_{rsm}\omega_{k\ell r}\omega_{kpn}I_{ksn}y_\ell y_p\big]\mathbf{n}_{om} \tag{18.6.8}$$

where I_{kmn} are the $\mathbf{n}_{om}$ and $\mathbf{n}_{on}$ components of $\mathbf{I}_k$, and the e_{rsm} are the components of the permutation symbol. Finally, by substituting into Eq. (18.16.1), we can express the generalized inertia force on B_k in the form:

$$\mathbf{F}_{k\ell}^* = -m_k v_{k\ell m}\big(v_{kpm}\dot{y}_p + \dot{v}_{kpm}y_p\big) \\ - I_{kmn}\omega_{k\ell m}\big(\omega_{kpn}\dot{y}_p + \dot{\omega}_{kpn}y_p\big) \\ - I_{ksn}\omega_{k\ell m}e_{rsm}\omega_{kqr}\omega_{kpn}y_q y_p \tag{18.6.9}$$

Observe again the dependence upon the partial velocity and partial angular velocity arrays $v_{k\ell m}$, $\dot{v}_{k\ell m}$, $\omega_{k\ell m}$, and $\dot{\omega}_{k\ell m}$.

18.17 Multibody Dynamics

Once we know the generalized forces we can immediately obtain the governing dynamical equations by using Kane's equation (see Chapter 12, Section 12.2). Specifically, from Eq. (12.2.1) we have:

$$F_\ell + F_\ell^* = 0 \quad \ell = 1,\ldots n \tag{18.17.1}$$

where as in the foregoing sections F_ℓ and F_ℓ^* represent the generalized applied (active) and inertia (passive) forces on a multibody system for the generalized speed y_ℓ and where n is the number of degrees of freedom. Hence, by substituting from Eqs. (18.15.4) and (18.16.9), we have:

$$F_{km}v_{k\ell m} + T_{km}\omega_{k\ell m} - m_k v_{k\ell m}\big(v_{kpm}\dot{y}_p + \dot{v}_{kpm}y_p\big) \\ - I_{kmn}\omega_{k\ell m}\big(\omega_{kpn}\dot{y}_p + \dot{\omega}_{kpn}y_p\big) \\ - I_{ksn}\omega_{k\ell m}e_{rsm}\omega_{bqr}\omega_{kpn}y_q y_p = 0 \tag{18.17.2}$$

Equation (18.17.2) may be written in the simplified form:

$$a_{\ell p}\dot{y}_p = f_\ell \tag{18.17.3}$$

where the $a_{\ell p}$ and f_ℓ are:

$$a_{\ell p} = m_k v_{k\ell m}v_{kpm} + I_{kmn}\omega_{k\ell m}\omega_{kpn} \tag{18.17.4}$$

and

$$
\begin{aligned}
f_\ell = f_{km}v_{k\ell m} + T_{km}\omega_{k\ell m} - \big(m_k v_{k\ell m}\dot{v}_{kpm}y_p + I_{kmn}\omega_{k\ell m}\omega_{kpn}\dot{y}_p \\
+ I_{kmn}\omega_{k\ell m}\dot{\omega}_{kpn}y_p + e_{rsm}I_{ksn}\omega_{k\ell m}\omega_{kqr}\omega_{kpn}y_p y_q\big)
\end{aligned} \tag{18.17.5}
$$

where the $a_{\ell p}$ are sometimes called *inertia coefficients*.

Observe from Eq. (18.17.4) that $a_{\ell p} = a_{p\ell}$ so the inertia coefficients form the elements of a symmetric $n \times n$ matrix A. In this context, Eq. (18.17.3) may be written in the matrix form:

$$A\dot{y} = f \tag{18.17.6}$$

where y and f are $n \times 1$ column arrays with elements y_ℓ and f_ℓ, respectively.

Finally, observe the direct dependence of A and f upon the four partial velocity arrays $\omega_{k\ell m}$, $\dot{\omega}_{k\ell m}$, $v_{k\ell m}$, and $\dot{v}_{k\ell m}$. By determining these arrays, using the procedures of the foregoing sections, we can readily obtain the multibody dynamical equations.

18.18 Closure

In this relatively long chapter we have attempted to summarize the modern analyses procedures for studying multibody systems. Because our objective was to summarize the procedures, we restricted our discussions to open-chain systems with spherical joints. Systems with closed loops or with other constraints, systems with other kinds of connecting joints, systems allowing translation or separation between adjoining bodies, and systems having flexible members may be studied using procedures similar to those discussed here. Interested readers may want to consult the references for details on analysis of such systems.

To a large extent, multibody dynamics is an evolving field of research with new analyses and numerical solution procedures still being developed (see References 18.2 and 18.3). Nevertheless, the governing equations will have the form of Eq. (18.17.3) or (18.17.6). Moreover, these equations are readily developed once the partial velocity arrays $\omega_{k\ell m}$, $\dot{\omega}_{k\ell m}$, $v_{k\ell m}$, and $\dot{v}_{k\ell m}$ are known.

Finally, observe that, through the use of generalized speeds, the governing equations form a set of first-order ordinary differential equations. As such, they are in a form ideally suited for numerical integration. To conduct the integration with a numerical integrator, we need to append to the dynamical equations a set of first-order equations such as Eq. (18.8.6) defining the generalized coordinate derivatives in terms of the generalized speeds. In the remaining two chapters, we will look at application of these concepts in the analyses of robotic and biosystems.

References

18.1. Huston, R. L., *Multibody Dynamics*, Butterworth-Heinemann, Stoneham, MA, 1990.

18.2. Huston, R. L., Multibody dynamics: modeling and analysis methods, *Appl. Mech. Rev.*, 44(3), 109, 1991.

18.3. Huston, R. L., Multibody dynamics since 1990, *Appl. Mech. Rev.*, 44(3), S35, 1996.
18.4. Brand, L., *Vector and Tensor Analysis*, Wiley, New York, 1947.
18.5. Huston, R. L., and Passerello, C. E., Eliminating singularities in governing equations of mechanical systems, *Mech. Res. Commun.*, 3, 361, 1976.
18.6. Whittaker, E. T., *Analytical Dynamics*, Cambridge University Press, London, 1937, pp. 8, 16.
18.7. Kane, T. R., and Wang, C. F., On the derivation of equations of motion, *J. Soc. Ind. Appl. Math.*, 13, 487–492, 1965.
18.8. Schiehen, W., Ed., *Multibody Systems Handbook*, Springer-Verlag, New York, 1990.
18.9. Schiehen, W., Ed., *Advanced Multibody System Dynamics: Simulation and Software Tools*, Kluwer Academic Publishers, Dordrecht, 1993.
18.10. Amirouche, F. M. L., *Computational Methods in Multibody Dynamics*, Prentice Hall, Englewood Cliffs, NJ, 1992.
18.11. Xie, M., *Flexible Multibody System Dynamics: Theory and Applications*, Taylor & Francis, London, 1994.
18.12. Pereira, M. F. O. S., and Ambrosio, J. A. C., *Computer-Aided Analysis of Rigid and Flexible Mechanical Systems*, Kluwer, Dordrecht, 1994.
18.13. Garcia de Jalon, J., and Bayo, E., *Kinematic and Dynamic Simulation of Multibody Systems: The Real Time Challenges*, Springer-Verlag, New York, 1994.
18.14. Sinha, S. C., Waites, H. B., and Book, W. J., *Dynamics of Flexible Multibody Systems: Theory and Experiment*, ASME Publication AMD-141, DSC-37, New York, 1992.
18.15. Lagnese, J. E., Leugering, G., and Schmidt, E. J. P. G., *Modeling, Analysis and Control of Dynamic Elastic Multi-Link Structures*, Birkhauser, Boston, MA, 1994.
18.16. Pfeiffer, F., and Glocker, C. H., *Multibody Dynamics with Unilateral Contacts*, John Wiley & Sons, New York, 1996.

Problems

Section 18.2 Connection Configuration: Lower Body Arrays

P18.2.1: Consider the multibody system depicted in Figure P18.2.1. Let the bodies of the system be numbered and labeled as shown. Determine the lower body array $L(K)$.

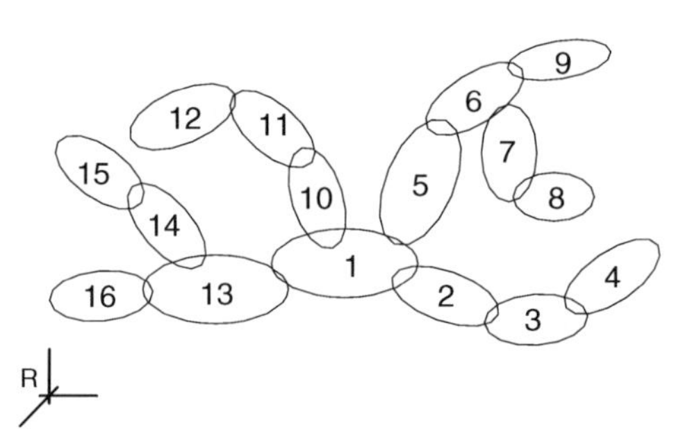

FIGURE P18.2.1
A numbered multibody system.

P18.2.2: See Problem P18.2.1. Determine the higher order lower body arrays as shown in Table 18.2.1 for the system of Figure P18.2.1.

P18.2.3: See Problem P18.2.1. For the system of Figure P18.2.1 determine: (a) the branching bodies, (b) the intermediate bodies, and (c) the extremities. Observe that (a) the body numbers for the branching bodies appear more than once in the lower body array $L(K)$; (b) the body numbers for the intermediate bodies appear once, and only once, in $L(K)$; and (c) the body numbers of the extremities do not appear in $L(K)$.

P18.2.4: Suppose a multibody system has a lower body array $L(K)$ given by:

$$L(K)\text{: } 0\ 1\ 2\ 3\ 4\ 2\ 6\ 7\ 6\ 9\ 6\ 11\ 1\ 13\ 13\ 15$$

Draw a sketch of the multibody system.

P18.2.5: See Problem P18.2.4. Determine the higher-order lower body arrays as shown in Table 18.2.1 for the system defined by the lower body array $L(K)$.

P18.2.6: See Problem P18.2.4. Determine: (a) the branching bodies; (b) the intermediate bodies; and (c) the extremities for the system defined by the lower body array $L(K)$.

P18.2.7: See Problems P18.2.1, P18.2.2, and P18.2.3. Determine expressions for the angular velocities of the extremity bodies of the system of Figure P18.2.1. Write the expressions in the form of Eq. (18.2.5).

Section 18.3 Transformation Matrices

P18.3.1: Verify that the product of the matrices A, B, and C of Eq. (18.3.3) give the matrix in Eq. (18.3.4).

P18.3.2: Verify that the product of the matrices of Eqs. (18.3.5) and (18.3.6) is the identity matrix.

P18.3.3: In the matrix of Eq. (18.3.4) let α, β, and γ have the values 30, 45, and 60°, respectively. Show that the determinant of the matrix is 1.

P18.3.4: See Problems P18.2.1 and P18.2.3. Determine expressions for the transformation matrices in a form similar to Eq. (18.3.18) for the extremity bodies of the system of Figure 18.3.3.

Section 18.4 Transformation Matrix Derivatives

P18.4.1: Suppose a body B is moving in a reference frame R. Let $\mathbf{n}_i$ ($i = 1, 2, 3$) be mutually perpendicular unit vectors fixed in B and let N_i ($i = 1, 2, 3$) be mutually perpendicular unit vectors fixed in R. Let α, β, and γ be dextral orientation angles defining the orientation of B in R, as in Section 18.3. Suppose the angular velocity of B in R may be expressed as:

$$\omega = 4N_1 - 3N_2 + 5N_3 \text{ rad/sec}$$

Suppose further that at an instant of interest α, β, and γ are 30, 45, and 60°, respectively. Determine: (a) the transformation matrix S between B and R, and (b) the derivative dS/dt of S.

P18.4.2: Repeat Problem P18.4.1 if α, β, and γ have the values 15, –30, and –45°, respectively.

P18.4.3: Repeat Problem P18.4.1 if the angular velocity of B in R is expressed as:

$$\boldsymbol{\omega} = -5\mathbf{n}_1 + 7\mathbf{n}_2 + 6\mathbf{n}_3 \text{ rad/sec}$$

P18.4.4: Let a body B be moving in a reference frame R. Let the orientation of B in R be determined by the inclination of mutually perpendicular unit vectors $\mathbf{n}_i$ ($i = 1, 2, 3$) fixed in B relative to analogous unit vectors $\mathbf{N}_i$ ($i = 1, 2, 3$) fixed in R. Let the transformation matrix S relating the $\mathbf{N}_i$ and $\mathbf{n}_i$ have elements S_{ij} given by:

$$S_{ij} = \begin{bmatrix} \cos t & -\sin t & 0 \\ \sin t \cos t & \cos^2 t & -\sin t \\ \sin^2 t & \sin t \cos t & \cos t \end{bmatrix}$$

where t is time.

a. Determine dS/dt by differentiation of the elements S_{ij}.
b. From the definition of S and its elements (that is, $S_{ij} = \mathbf{N}_i \cdot \mathbf{n}_j$; see Eq. (2.11.3)), express the $\mathbf{N}_i$ in terms of the $\mathbf{n}_i$.
c. Similarly, express the $\mathbf{n}_i$ in terms of the $\mathbf{N}_i$.
d. Verify that the results of (b) and (c) satisfy the relations:

$$\mathbf{N}_i = S_{ij}\mathbf{n}_j \quad \text{and} \quad \mathbf{n}_i = S_{ji}\mathbf{N}_j$$

P18.4.5: See Problem P18.4.4. From the results of (c) determine the derivatives ${}^R d\mathbf{n}_1/dt$, ${}^R d\mathbf{n}_2/dt$, and ${}^R d\mathbf{n}_3/dt$.

P18.4.6: See Problems P18.4.4 and P18.4.5. From the results of Problem P18.4.5 determine the angular velocity $\boldsymbol{\omega}$ of B in R. Recall from Eq. (4.5.1) that $\boldsymbol{\omega}$ may be expressed as:

$$\boldsymbol{\omega} = \left[(d\mathbf{n}_2/dt)\cdot\mathbf{n}_3\right]\mathbf{n}_1 + \left[(d\mathbf{n}_3/dt)\cdot\mathbf{n}_1\right]\mathbf{n}_2 + \left[(d\mathbf{n}_1/dt)\cdot\mathbf{n}_2\right]\mathbf{n}_3$$

Express $\boldsymbol{\omega}$ both in terms of the $\mathbf{n}_i$ and the $\mathbf{N}_i$.

P18.4.7: See Problem P18.4.6. Let $\boldsymbol{\omega}$ be expressed in the forms:

$$\boldsymbol{\omega} = \Omega_i \mathbf{N}_i = \omega_j \mathbf{n}_j$$

Show that the $\boldsymbol{\omega}_i$ and Ω_j as determined in Problem P18.4.6 are related by the expressions:

$$\Omega_i = S_{ij}\omega_j \quad \text{and} \quad \omega_j = S_{ij}\Omega_i$$

where, as before, S_{ij} is given in Problem P18.4.4.

P18.4.8: Review Problems P18.4.4 to P18.4.7. Use the results of Problem P18.4.6 to determine the angular velocity array W of Eq. (18.4.7).

P18.4.9: See Problem P18.4.4 and P18.4.8. Use W as determined in Problem P18.4.8 to find dS/dt as in Eq. (18.4.8). Compare the results with those of Problem P18.4.4.

Section 18.5 Euler Parameters

P18.5.1: Consider the configuration graph shown in Figure P18.5.1 relating unit vector sets $\mathbf{n}_{jm}$ and $\mathbf{n}_{km}$ ($m = 1, 2, 3$) of adjoining bodies B_j and B_k of a multibody system. Determine expressions for ${}^{B_j}\boldsymbol{\omega}^{B_k}$, the angular velocity of B_k relative to B_j. Express the results in terms of both the $\mathbf{n}_{jm}$ and the $\mathbf{n}_{km}$.

P18.5.2: For the results of Problem P18.5.1, find expressions for $\dot{\alpha}_k$, $\dot{\beta}_k$, and $\dot{\gamma}_k$ in terms of the angular velocity components relative to both the $\mathbf{n}_{jm}$ and the $\mathbf{n}_{km}$ unit vector sets. Identify the singularities.

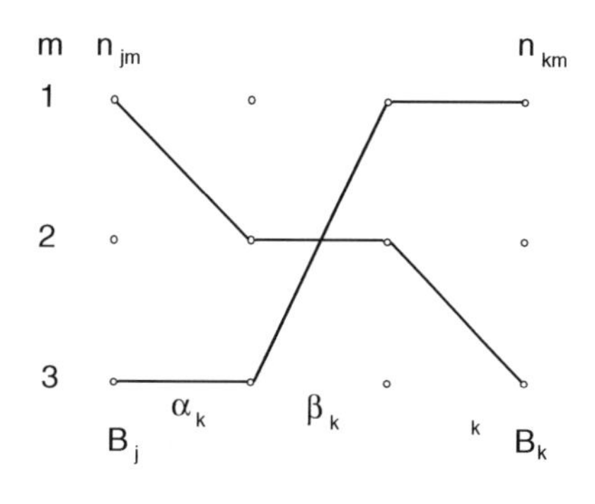

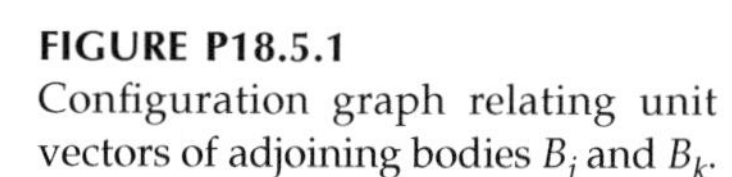

FIGURE P18.5.1
Configuration graph relating unit vectors of adjoining bodies B_j and B_k.

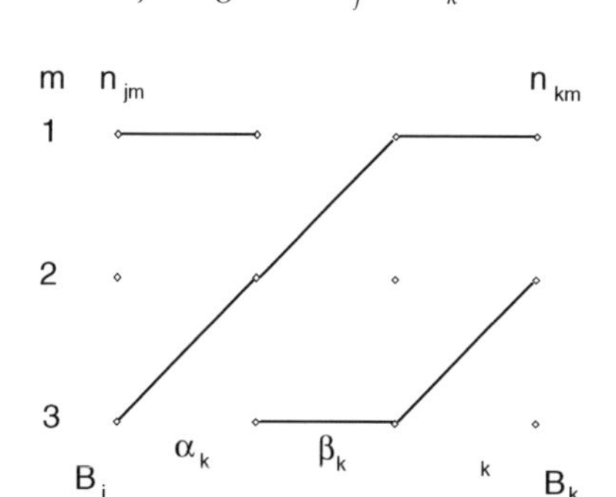

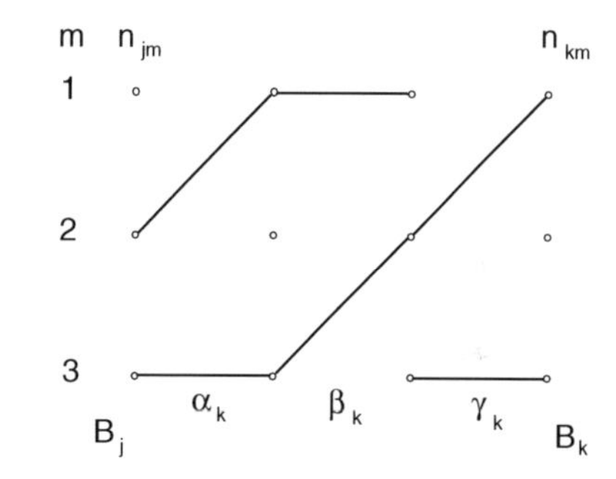

FIGURE P18.5.3
(A) Configuration graph relating unit vectors of adjoining bodies. (B) Configuration graph relating unit vectors of adjoining bodies.

P18.5.3: Repeat Problems P18.5.1 and P18.5.2 for the configuration graphs of Figures P18.5.3A and B.

P18.5.4: Let a body *B* be moving in a reference frame *R*. Suppose the movement of *B* in *R* may be described as a rotation through an angle θ about an axis fixed in both *B* and *R* and parallel to a unit vector $\boldsymbol{\lambda}$. If θ is 60° and if $\boldsymbol{\lambda}$ is given by:

$$\boldsymbol{\lambda} = (3/13)\mathbf{N}_1 + (4/13)\mathbf{N}_2 + (12/13)\mathbf{N}_3$$

where $\mathbf{N}_1$, $\mathbf{N}_2$, and $\mathbf{N}_3$ are mutually perpendicular unit vectors fixed in *R*, determine the Euler parameters expressed by Eq. (18.5.11).

P18.5.5: Using the results of Problem P18.5.4, determine the transformation matrix of Eq. (18.5.13).

P18.5.6: Verify that the matrix obtained in Problem P18.5.5 is orthogonal (that is, the inverse is the same as the transpose).

Section 18.6 Rotation Dyadics

P18.6.1: Show that the matrix of components of a rotation dyadic is orthogonal. Specifically, verify Eq. (18.6.22) using the components defined by Eq. (18.6.17).
P18.6.2: Consider the matrix T given by:

$$T = \begin{bmatrix} 0.5266 & -0.7639 & 0.3729 \\ 0.8349 & 0.8494 & -0.0578 \\ -0.1599 & 0.3418 & 0.9260 \end{bmatrix}$$

Show that to a reasonable degree of accuracy, T is orthogonal.

P18.6.3: See Problem P18.6.2. If T contains the components of a rotation dyadic, determine the rotation angle θ and the components of a unit vector parallel to the axis of rotation.

P18.6.4: Determine the components of the rotation dyadic if θ is 60° and $\boldsymbol{\lambda}$ is given by:

$$\boldsymbol{\lambda} = (3/13)\mathbf{N}_1 + (4/13)\mathbf{N}_2 + (12/13)\mathbf{N}_3$$

where $\mathbf{N}_1$, $\mathbf{N}_2$, and $\mathbf{N}_3$ are mutually perpendicular unit vectors. Compare the results with those of Problem P18.5.4.

Section 18.7 Transformation Matrices, Angular Velocity Components, and Euler Parameters

P18.7.1: Verify the correctness of the elements of Eq. (18.7.14) by following the procedure outlined in Eqs. (18.7.3) to (18.7.13).

P18.7.2: Let a body B moving in a reference frame R rotate about a line L fixed in both B and R. Let $\boldsymbol{\lambda}$ be a unit vector parallel to L given by:

$$\boldsymbol{\lambda} = (1/3)\mathbf{n}_1 + (2/3)\mathbf{n}_2 + (2/3)\mathbf{n}_3$$

where $\mathbf{n}_1$, $\mathbf{n}_2$, and $\mathbf{n}_3$ are mutually perpendicular unit vectors fixed in R. Let the rotation angle θ be 30°. Use Eq. (18.7.2) to determine the transformation matrix S between the $\mathbf{n}_i$ (i = 1, 2, 3) and mutually perpendicular unit vectors $\hat{\mathbf{n}}_i$ (i = 1, 2, 3) fixed in B and which coincide with the $\mathbf{n}_i$ before rotation.

P18.7.3: See Problem P18.7.2. Use Eq. (18.5.12) to determine the Euler parameters for the 30° rotation about L.

P18.7.4: See Problems P18.7.2 and P18.7.3. Use the results of Problems P18.7.3 and Eq. (18.7.4) to determine the elements of the transformation matrix. Compare these results with those of Problem P18.7.2.

P18.7.5: Verify that matrix E of Eq. (18.7.30) is orthogonal (that is, $E^{-1} = E^T$).

Section 18.8 Degrees of Freedom, Coordinates, and Generalized Speeds

P18.8.1: See Problem 11.2.1. Determine the number of degrees of freedom of the following systems.

a. Pair of eyeglasses
b. Pair of pliers or a pair of scissors
c. Child's tricycle rolling on a flat horizontal surface; let the tricycle be modeled by a frame, two rear wheels, and a front steering wheel
d. Human arm model consisting of three rigid bodies representing the upper arm, the lower arm, and the hand; let there be spherical joints at the wrist and shoulder and a hinge joint at the elbow

P18.8.2: Consider the relatively simple six-body system shown in Figure P18.8.2. (This system might represent the end effector of a robot arm.) Develop explicit expressions analogous to Eqs. (18.8.1) to (18.8.11) for this system. For simplicity, assume that the bodies are connected by spherical joints and that the entire system may move in an inertial frame *R*.

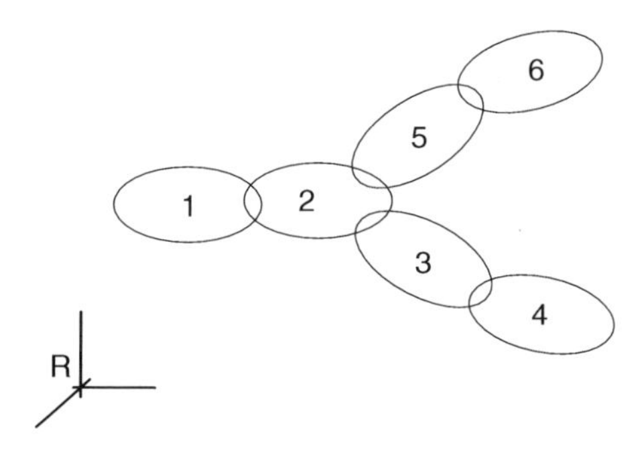

FIGURE P18.8.2
A six-body system.

Section 18.10 Angular Velocity

P18.10.1: See Problems P18.2.1 and P18.2.2. Consider the multibody system depicted and shown in Figure P18.2.1 and shown again in Figure P18.10.1. Let the system be connected by spherical joints and let the system be moving in a reference frame *R*. Determine the triplets of generalized speeds as in Table 18.10.2.

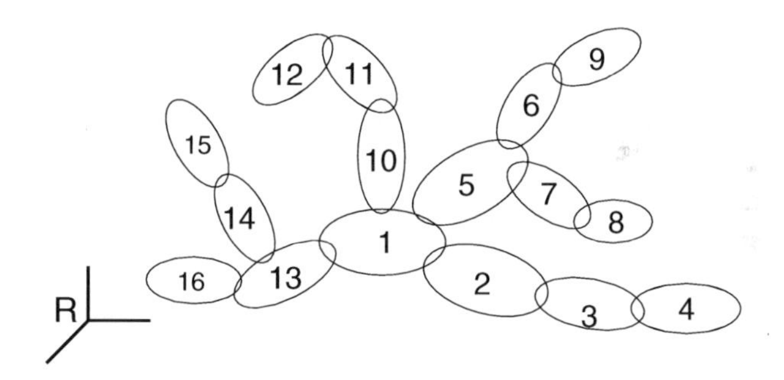

FIGURE P18.10.1
A numbered multibody system.

P18.10.2: See Problems P18.10.1. Using the results of Problems P18.2.1 and P18.2.2, determine a table of partial angular velocity components for the system of Figure P18.10.1 analogous to Table 18.10.3.

P18.10.3: See Problems P18.2.4 and P18.2.5. Repeat Problems P18.10.1 and P18.10.2 for the system defined by the lower body array $L(K)$ given by:

$$L(K)\text{: } 0\ 1\ 2\ 3\ 4\ 2\ 6\ 7\ 6\ 9\ 6\ 11\ 1\ 13\ 13\ 15$$

Section 18.11 Angular Acceleration

P18.11.1: See Problems P18.10.1, P18.10.2, and P18.10.3. Determine the derivation of the partial angular velocity components for the systems of Problems P18.10.1 and P18.10.3. List the results in tables analogous to Table 18.11.1.

Section 18.12 Joint and Mass Center Positions

P18.12.1: Consider again the multibody systems of Problems P18.2.1 and P18.10.1 and as shown again in Figure P18.12.1. Using the notation of Section 18.12, let position vectors locating the origins O_j (connecting joints) of the bodies relative to the origin O_i of the adjacent lower numbered bodies be ξ_j ($j = 1,\ldots, 16$), and let the position vectors locating the mass centers G_j relative to the O_j be $\mathbf{r}_j$ ($j = 1,\ldots, 16$), as shown selectively in Figure P18.12.1. Determine explicit expressions, analogous to Eq. (18.12.2), for the position vectors $\mathbf{p}_j$ ($j = 1,\ldots, 16$) of the mass centers G_j relative to the origin O of reference frame R.

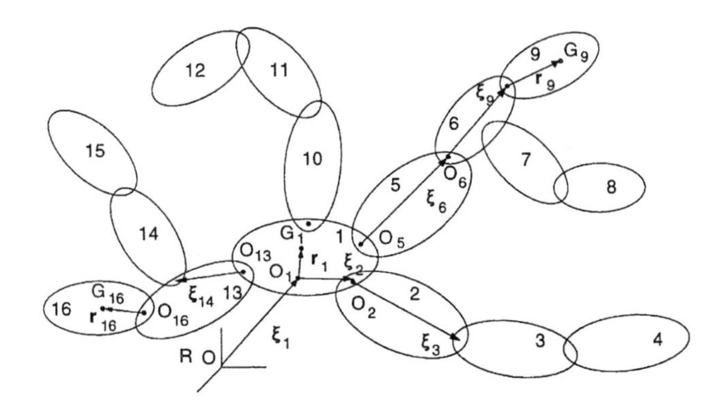

FIGURE P18.12.1
A numbered multibody system.

P18.12.2: See Problem P18.12.1. Using the results of Problem P18.12.1, express the position vectors $\mathbf{p}_j$ ($j = 1,\ldots, 16$) in terms of mutually perpendicular unit vectors $\mathbf{n}_{om}$ ($m = 1, 2, 3$) fixed in R, analogous to Eq. (18.12.3).

P18.12.3: See Problems P18.2.1, P18.2.2, and P18.12.2. Use the results of Problems P18.2.1 and P18.2.2 for the lower body arrays together with Eqs. (18.12.6), (18.12.7), and (18.12.8) to determine the position vectors $\mathbf{p}_j$ ($j = 1,\ldots, 16$) in terms of the $\mathbf{n}_{om}$. Compare the results with those of Problem P18.12.2.

P18.12.4: See Problems P18.2.4, P18.2.5, and P18.12.3. Consider a multibody system whose lower body array $L(K)$ is:

$$L(K)\text{: } 0\ 1\ 2\ 3\ 4\ 2\ 6\ 7\ 6\ 9\ 6\ 11\ 1\ 13\ 13\ 15$$

as in Problem P18.2.4. Using the results of Problem P18.2.5 for the higher-order lower body arrays, together with Eqs. (18.12.6), (18.12.7), and (18.12.8), determine expressions for the position vectors $\mathbf{p}_j$ ($j = 1,\ldots, 16$) locating the mass centers G_j relative to the origin O of a reference frame R in which the system moves.

P18.12.5: See Problems P18.2.4 and P18.12.4. Use the sketch obtained in Problem P18.2.4 to determine expressions for the position vectors $\mathbf{p}_j$ ($j = 1,\ldots, 16$) analogous to Eq. (18.12.2). Then, express these vectors in terms of mutually perpendicular unit vectors $\mathbf{n}_{om}$ ($m = 1, 2,$ 3) fixed in R as in Eq. (18.12.3). Compare the results with those of Problem P18.12.4.

Section 18.13 Mass Center Velocities

P18.13.1: Consider the relatively simple multibody system depicted in Figure P18.13.1, consisting of seven spherical-joint connected bodies moving in a reference frame R. Let the bodies be numbered and labeled as shown. Develop a table of lower body arrays for the system, analogous to Table 18.2.1.

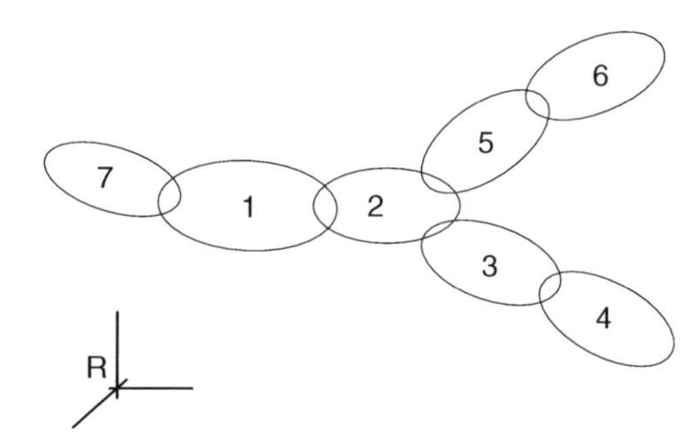

FIGURE P18.13.1
A simple illustrative multibody system with spherical-joint connections.

P18.13.2: See Problem P18.13.1. With spherical joints show that the system of Figure P18.13.1 has 24° of freedom. Let these degrees of freedom be represented by translation and rotation coordinates x_ℓ ($\ell = 1,\dots, 24$) as in Eq. (18.8.1). Make a listing of these x_ℓ ($\ell = 1,\dots, 24$).

P18.13.3: See Problems P18.13.1 and P18.13.2. Let the movement of the system of Figure P18.13.1 be described by generalized speeds y_ℓ ($\ell = 1,\dots, 24$) as in Eq. (18.8.10). Make a listing of these y_ℓ ($\ell = 1,\dots, 24$).

P18.13.4: See Problems P18.13.1, P18.3.2, and P18.13.3. Let position vectors p_k ($k = 1,\dots, 7$) locating the mass centers G_k of the bodies be constructed in terms of joint position vectors $\boldsymbol{\xi}_j$ and local mass center position vectors $\mathbf{r}_k$ ($k = 1,\dots, 7$) as in Eq. (18.12.2). Make a listing of these $\mathbf{p}_k$ ($k = 1,\dots, 7$).

P18.13.5: See Problems P18.13.1 to P18.13.4. Let $\mathbf{n}_{om}$ ($m = 1, 2, 3$) be mutually perpendicular unit vectors fixed in reference frame R. Using the results of Problem P18.13.4, express the mass center position vectors $\mathbf{p}_k$ ($k = 1,\dots 7$) in terms of the $\mathbf{n}_{om}$ and transformation matrices SOJ as in Eq. (18.12.3).

P18.13.6: See Problems P18.13.1 to P18.13.5. Using the lower body array table developed in Problem P18.13.1, use Eqs. (18.12.6), (18.12.7), and (18.12.8) to determine the mass center position vectors $\mathbf{p}_k$ ($k = 1,\dots, 7$). Compare the results with those of Problem P18.13.5.

P18.13.7: See Problem P18.13.1 to P18.13.4. Using the results of Problem P18.13.4 for the mass center position vectors $\mathbf{p}_k$ ($k = 1,\dots, 7$), determine the mass center velocities $\mathbf{v}_k$ ($k = 1,\dots, 7$) R, by term-by-term differentiation and by using vector products of angular velocities to calculate the derivatives (see Eq. (4.5.2)). Then, using transformation matrices, express the results in terms of the $\mathbf{n}_{om}$ unit vectors fixed in R.

P18.3.8: See Problems P18.13.1 to P18.13.7. Using the results of Problem P18.13.5, with the mass center position vectors $\mathbf{p}_k$ ($k = 1,\dots, 7$) expressed in terms of the unit vectors $\mathbf{n}_{om}$, determine the mass center velocities $\mathbf{v}_k$ ($k = 1,\dots, 7$) in R, by term-by-term differentiation, using Eqs. (18.13.5) and (18.13.6) to obtain derivatives of the transformation matrices. Compare the results with those of Problem P18.13.7.

P18.13.9: See Problems P18.13.1 to P18.3.8. Use Eq. (18.13.10) together with the supporting equations to determine the mass center velocities $\mathbf{v}_k$ $(k = 1,\ldots, 7)$ in R. Compare the results with these of Problems P18.13.7 and P18.13.8.

Section 18.14 Mass Center Acceleration

P18.14.1: See Problems P18.13.1 to P18.13.9. Continue the process of differentiation to obtain the mass center accelerations of the bodies of the simple system of Figure P18.13.1. Specifically, obtain the acceleration expressions by direct differentiation of the results of Problems P18.3.8 and P18.3.9. Then, obtain acceleration expressions by using Eqs. (18.14.1) to (18.14.4).

Section 18.16 Kinetics: Inertia (Passive) Forces

P18.16.1: By substituting from Eqs. (18.16.4), (18.16.5), and (18.16.6) into Eqs. (18.16.2) and (18.16.3) verify the results of Eqs. (18.16.7), (18.16.8), and (18.16.9).

19

Introduction to Robot Dynamics

19.1 Introduction

A robot is a system of intelligently controlled connected bodies whose objective is to accomplish a given task. As such, robots come in a wide variety of sizes and shapes. In manufacturing settings, however, robots typically have the forms shown in Figure 19.1.1 — a series of connected bodies attached to a fixed base at one end with a tool (or *end effector*) at the other end. The end effector is sometimes called a *gripper,* and the connected series of bodies is generally called a *robot arm*. The task or job of a robot is usually to produce a movement or to produce a force through its end effector.

Around 1980 there began an outpouring of interest in robots. Robots were viewed as machines of the future that promised to provide dramatic changes in the workplace and even in the home. Since then, as robots have become more commonplace, the excitement has ebbed a bit. For analysts, however, the interest is still strong, as these machines are becoming increasingly sophisticated with an ever-increasing range of applications. Some of the more sophisticated machines may have more than one arm, so-called *multi-arm robots*, and some have a moveable base, so-called *mobile robots*.

In this chapter we will apply the procedures of multibody dynamics analysis in the study of robotic systems. These procedures are ideally suited for studying all kinds of robotic systems including mobile robots and multi-arm robots. For simplicity, however, we will focus our attention upon single-arm robots with a fixed base, as in Figure 19.1.1. For additional simplicity, we will consider the bodies to be connected with hinge (or revolute) joints. We will, however, allow the hinge joints to be inclined relative to one another so that the end effector may move in three dimensions. Also, we will consider robots with an arbitrary number of bodies so that the system may have more degrees of freedom than needed to accomplish a given task (so-called *redundant robots*). Finally, the end effector may vary considerably from one machine to another and may contain a number of parts. For simplicity, however, we will initially think of the end effector as a relatively small body at the end of the robot arm.

19.2 Geometry, Configuration, and Degrees of Freedom

Consider the robot system depicted in Figures 19.1.1 and 19.2.1. Let the system consist of a chain of n pin-connected links L_k ($k = 1,\ldots, n$) connected at one end to a base B and at the other end to a gripper, or end effector tool E. Let the base B be able to pivot or swivel about a vertical axis in a fixed (or inertial) reference frame R. Thus, aside from the end

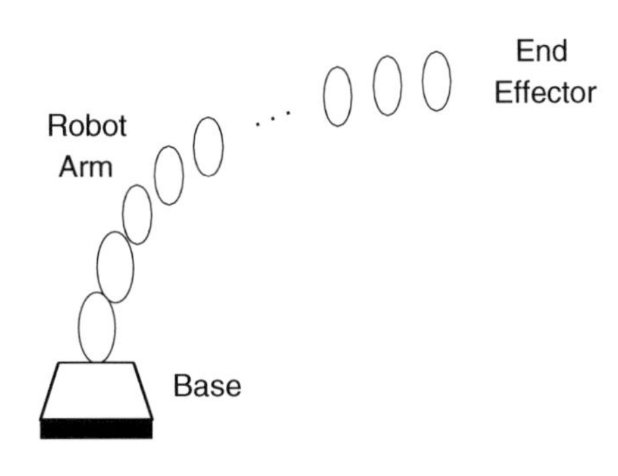

FIGURE 19.1.1
A robotic system.

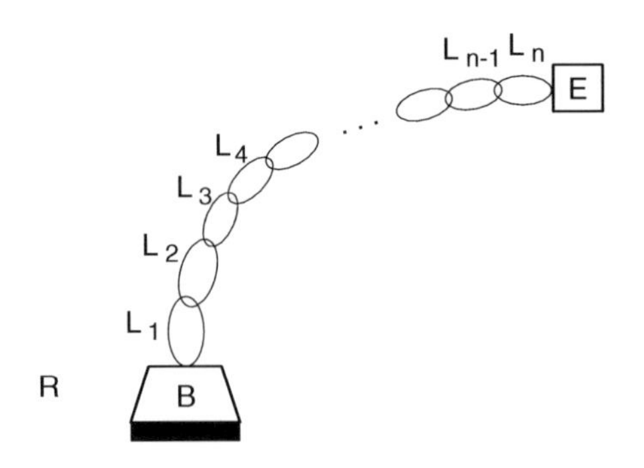

FIGURE 19.2.1
A robot system.

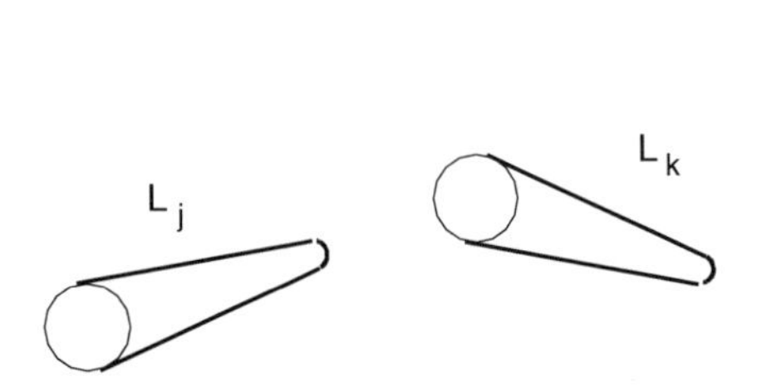

FIGURE 19.2.2
Two typical adjoining links.

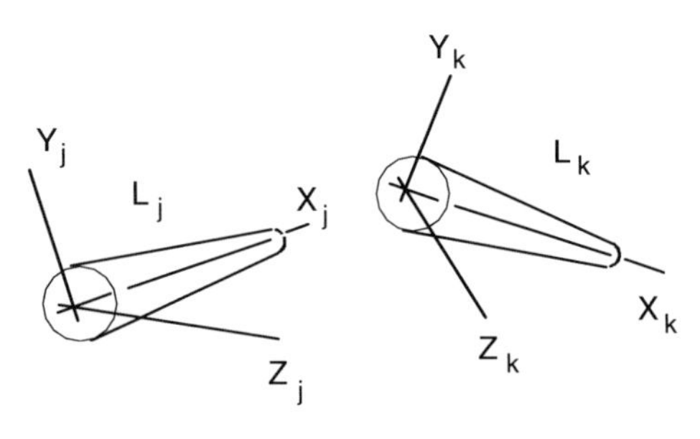

FIGURE 19.2.3
Axis systems of typical adjoining links.

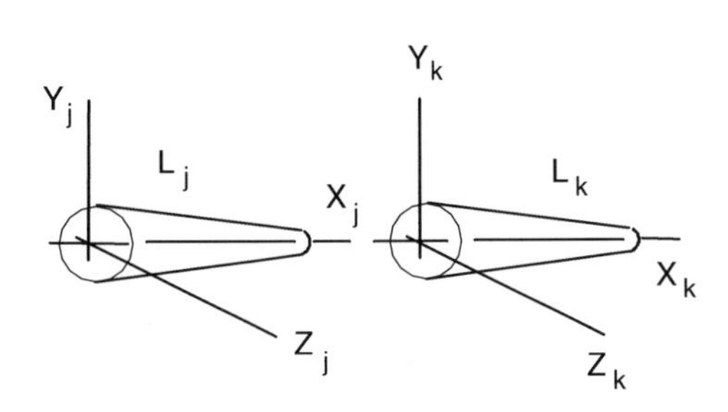

FIGURE 19.2.4
Adjoining links with aligned axes.

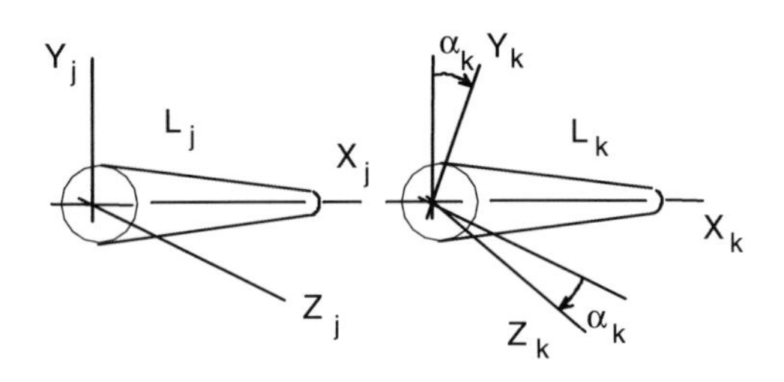

FIGURE 19.2.5
Axial rotation of link L_k through an angle α_k.

effector E, the system has $n + 1$ degrees of freedom (one for each link and one for the base). Let these degrees of freedom be represented by an angle α_B defining the rotation of the base B in R and n angles γ_i ($i = 1,\ldots, n$) defining the relative rotations of the links. These angles are sometimes called *articulation* angles.

Consider two typical adjoining links such as L_j and L_k as in Figure 19.2.2. For the purpose of keeping the geometry relatively simple, but without loss of generality, let the links be represented by frustrums of cylindrical cones as shown. To define the relative orientation of the links, let the links have X-, Y-, Z-axes systems imbedded in them as shown in Figure 19.2.3, where the X-axis is aligned with the cone axis. Imagine the links and their axes to be mutually aligned as in Figure 19.2.4. Next, let link L_k be rotated about its X_k-axis through an angle α_k as in Figure 19.2.5. Finally, let L_k be rotated about its Z_k-axis through the angle γ_k as in Figure 19.2.6.

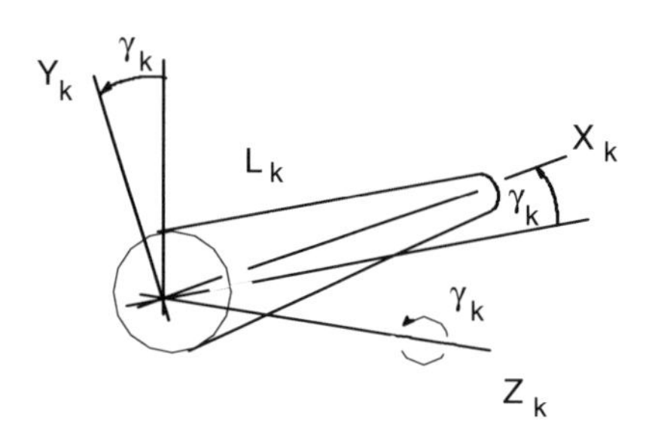

FIGURE 19.2.6
Definition of articulation angle γ_k.

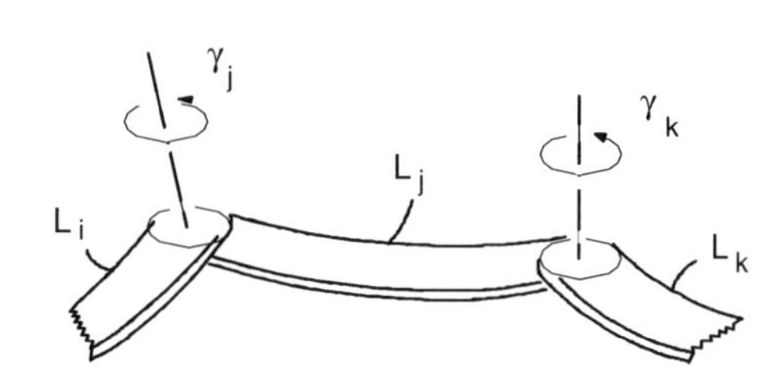

FIGURE 19.2.7
Adjoining links with arbitrary shape.

Although we have used frustrums of circular cones to represent the robot links, we could have used bodies with arbitrary geometry as in Figure 19.2.7. There is, however, no significant loss in generality in using cones; therefore, in the sequel for simplicity we will use conical links.

19.3 Transformation Matrices and Configuration Graphs

Consider again two typical adjoining robot links and let unit vectors n_{ji} and n_{ki} (i = 1, 2, 3) be introduced parallel to the link axis systems as in Figure 19.3.1. The orientation and movement of the robot links may be described in terms of the relative inclination of these unit vector sets. To develop this concept, recall that transformation matrix elements between the unit vector sets may be defined as (see Eq. (2.11.3)):

$$SJK_{mn} = \mathbf{n}_{jm} \cdot \mathbf{n}_{kn} \tag{19.3.1}$$

Then, the respective unit vectors are related by the expressions (see Eqs. (2.11.6) and (2.11.7)):

$$\mathbf{n}_{jm} = SJK_{mn}\mathbf{n}_{kn} \quad \text{and} \quad \mathbf{n}_{kn} = SJK_{mn}\mathbf{n}_{jm} \tag{19.3.2}$$

Recall that the transformation matrices are useful for relating vector components referred to the unit vector sets. Specifically, if a vector **V** is expressed in the forms:

$$\mathbf{V} = v_{jm}\mathbf{n}_{jm} = v_{kn}\mathbf{n}_{kn} \quad (\text{no sum on } j \text{ or } k) \tag{19.3.3}$$

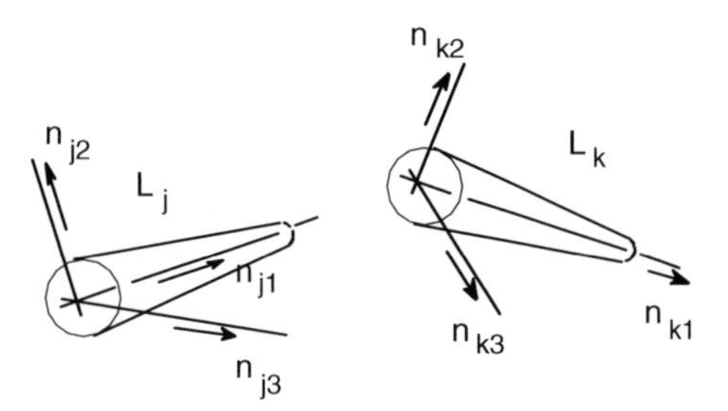

FIGURE 19.3.1
Unit vectors parallel to axes of adjoining links.

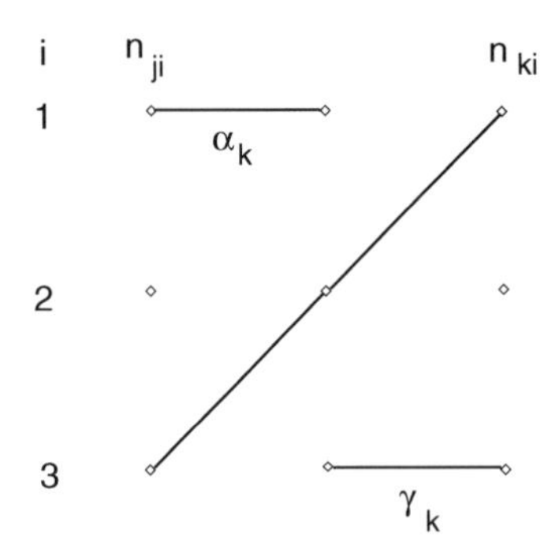

FIGURE 19.3.2
Configuration graph relating the unit vector sets of adjoining links L_j and L_k.

then the components v_{jm} and v_{kn} are related by the expressions:

$$v_{jm} = SJK_{mn} v_{kn} \quad \text{and} \quad v_{kn} = SJK_{mn} v_{jm} \tag{19.3.4}$$

(Observe the identical patterns of Eqs. (19.3.2) and (19.3.4).)

Equations (19.3.1), (19.3.2), and (19.3.4) need not be restricted for application only with adjoining links. Indeed, they may be used with any pair of links or between a typical link and the inertial frame R. Specifically, if $\mathbf{n}_{0i}$ (i = 1, 2, 3) are mutually perpendicular unit vectors fixed in R, then Eqs. (19.3.1), (19.3.2), and (19.3.4) take the forms:

$$SOK_{mn} = \mathbf{n}_{om} \cdot \mathbf{n}_{kn} \tag{19.3.5}$$

$$\mathbf{n}_{om} = SOK_{mn} \mathbf{n}_{kn} \quad \text{and} \quad \mathbf{n}_{kn} = SOK_{mn} \mathbf{n}_{om} \tag{19.3.6}$$

and

$$v_{om} = SOK_{mn} v_{kn} \quad \text{and} \quad v_{kn} = SOK_{mn} v_{om} \tag{19.3.7}$$

Recall also from Eq. (2.11.18) that the transformation matrix SOK may be expressed as a product of relative transformation matrices as:

$$SOK = SOB\ SB1\ S12\ S23 \ldots SIJ\ SJK \tag{19.3.8}$$

The elements of the relative transformation matrices may be efficiently obtained by using the configuration graphs of Section 4.3. Specifically, from Figures 19.2.4, 19.2.5, and 19.2.6, we see that for the typical adjoining links, their unit vectors $\mathbf{n}_{ji}$ and $\mathbf{n}_{ki}$ are related by the graph of Figure 19.3.2. Then, following the procedures of Section 4.3, we see that SJK may be expressed as:

$$SJK = AJK\ CJK \tag{19.3.9}$$

where the AJK and CJK are:

$$AJK = \begin{bmatrix} 1 & 0 & 0 \\ 0 & c\alpha_k & -s\alpha_k \\ 0 & s\alpha_k & c\alpha_k \end{bmatrix} \quad \text{and} \quad CJK = \begin{bmatrix} c\gamma_k & -s\gamma_k & 0 \\ s\gamma_k & c\gamma_k & 0 \\ 0 & 0 & 1 \end{bmatrix} \tag{19.3.10}$$

where as before s and c are abbreviations for sine and cosine. Then, by multiplication, we see that the elements of SJK are:

$$SJK_{mn} = \begin{bmatrix} c\gamma_k & -s\gamma_k & 0 \\ c\alpha_k s\gamma_k & c\alpha_k c\gamma_k & -s\alpha_k \\ s\alpha_k s\gamma_k & s\alpha_k c\gamma_k & c\alpha_k \end{bmatrix} \tag{19.3.11}$$

19.4 Angular Velocity of Robot Links

Consider again the robot system of Figures 19.1.1 and 19.2.1 and as shown again in Figure 19.4.1. In this system, each link is connected to its adjoining links by hinge (pin), or revolute, joints. Hence, relative angular velocities between adjoining links are *simple angular velocities* (see Section 4.4). This means that we can readily obtain simple expressions for the angular velocities of the respective links. To see this, consider first the base B moving in the inertial frame R as in Figure 19.4.2. For convenience, let the axis systems of B and R be as shown so that B has simple angular velocity in R about the X_0, or X_B, axis through the rotation angle α_B. Specifically,

$$^R\boldsymbol{\omega}^B = \boldsymbol{\omega}_B = \dot{\alpha}_B \mathbf{n}_{01} \tag{19.4.1}$$

where, as before, $\mathbf{n}_{01}$ is a unit vector parallel to X_0.

Next, consider the movement of the first link L_1 relative to the base B, as depicted in Figure 19.4.3. Then, the angular velocity of L_1 in B may be expressed as:

$$^B\boldsymbol{\omega}^{L_1} \stackrel{D}{=} \hat{\boldsymbol{\omega}}_1 = \dot{\gamma}_1 \mathbf{n}_{B3} \tag{19.4.2}$$

where γ_1 is the rotation angle and $\mathbf{n}_{B3}$ is a unit vector parallel to Z_B. Using Eq. (19.3.2), we may express $\mathbf{n}_{B3}$ in terms of the $\mathbf{n}_{0i}$ ($i = 1, 2, 3$) of R as:

$$\mathbf{n}_{B3} = SOB_{m3}\mathbf{n}_{0m} \tag{19.4.3}$$

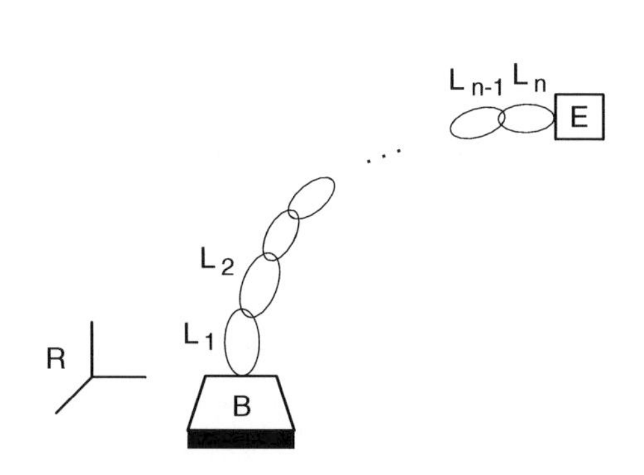

FIGURE 19.4.1
Example robot system.

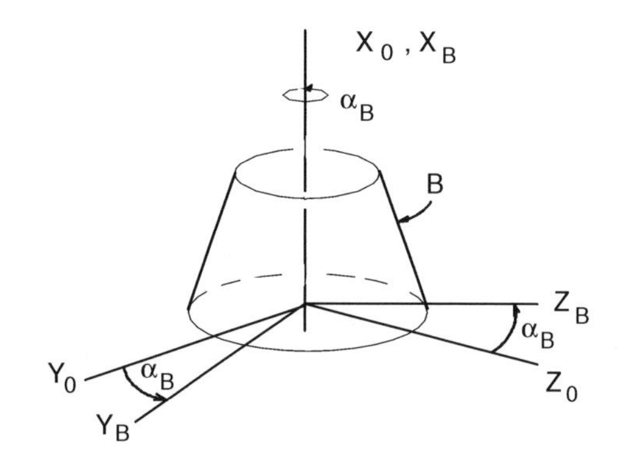

FIGURE 19.4.2
Rotation of base B in reference frame R.

where from Figure 19.4.2 the elements of S0B are seen to be:

$$S0B_{mn} = \begin{bmatrix} 1 & 0 & 0 \\ 0 & c\alpha_B & -s\alpha_B \\ 0 & s\alpha_B & c\alpha_B \end{bmatrix} \tag{19.4.4}$$

Then, by substituting into Eq. (19.4.2), $\hat{\boldsymbol{\omega}}_1$ becomes:

$$\hat{\boldsymbol{\omega}}_1 = \dot{\gamma}_1 S0B_{m3}\mathbf{n}_{0m} = -\dot{\gamma}_1 s\alpha_B \mathbf{n}_{02} + \dot{\gamma}_1 c\alpha_B \mathbf{n}_{03} \tag{19.4.5}$$

Finally, consider two typical adjoining links such as L_j and L_k as in Figure 19.4.4. The angular velocity of L_k relative to L_j is then seen to be:

$$^{L_j}\boldsymbol{\omega}^{L_k} = \hat{\boldsymbol{\omega}}_k = \dot{\gamma}_k \mathbf{n}_{k3} \quad \text{(no sum on } k) \tag{19.4.6}$$

Then, by using Eq. (19.3.6) we have:

$$\hat{\boldsymbol{\omega}}_k = \dot{\gamma}_k S0K_{m3}\mathbf{n}_{0m} \tag{19.4.7}$$

Observe and recall that the axes z_k and z_j in Figure 19.4.4 are not necessarily parallel to each other. Indeed, in general they are inclined at an angle α_k relative to each other as in Figure 19.2.5. Therefore, it also follows that $\mathbf{n}_{k3}$ is not in general parallel to $\mathbf{n}_{j3}$.

From the addition theorem for angular velocity (see Eq. (4.7.6)) we see that the angular velocity of a typical link L_k in the inertia frame R may be expressed as:

$$^{R}\boldsymbol{\omega}^{L_k} \stackrel{\text{D}}{=} \boldsymbol{\omega}_k = \boldsymbol{\omega}_B + \hat{\boldsymbol{\omega}}_1 + \hat{\boldsymbol{\omega}}_2 + \ldots + \hat{\boldsymbol{\omega}}_j + \hat{\boldsymbol{\omega}}_k \tag{19.4.8}$$

Then, by substituting from Eq. (19.4.7), we see that $\boldsymbol{\omega}_k$ may be expressed as:

$$\begin{aligned} \boldsymbol{\omega}_k &= \boldsymbol{\omega}_B + \dot{\gamma}_1 S01_{m3}\mathbf{n}_{0m} + \dot{\gamma}_2 S02_{m3}\mathbf{n}_{0m} \\ &\quad + \ldots + \dot{\gamma}_j S0J_{m3}\mathbf{n}_{0m} + \dot{\gamma}_k S0K_{m3}\mathbf{n}_{0m} \\ &= \dot{\alpha}_B \mathbf{n}_{01} + \sum_{i=1}^{k} \dot{\gamma}_i S0I_{m3}\mathbf{n}_{0m} \end{aligned} \tag{19.4.9}$$

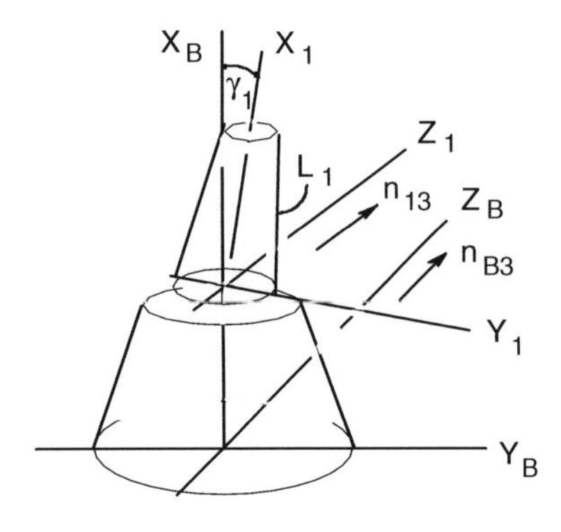

FIGURE 19.4.3
Rotation of link L_1 relative to the base B.

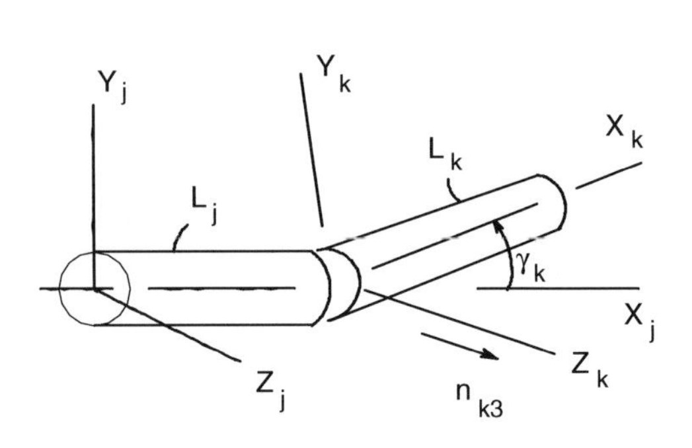

FIGURE 19.4.4
Two typical adjoining robot links.

19.5 Partial Angular Velocities

Upon inspecting the angular velocity expression of Eq. (19.4.9) we see that each term is a linear function of an angle derivative. The angles themselves (the articulation angles) describe the degrees of freedom of the system as discussed in Section 19.2. We can therefore define the generalized coordinates of the system in terms of these angles. Specifically, let the robot system, aside from the end effector, have $n + 1$ generalized coordinates q_r $(r = 1,\ldots, n + 1)$ defined as:

$$\begin{aligned} &q_1 = \alpha_B, \quad q_2 = \gamma_1, \quad q_3 = \gamma_2, \quad q_4 = \gamma_3, \ldots, \\ &q_r = \gamma_{r-1}, \ldots, \quad q_n = \gamma_{n-1}, \quad q_{n+1} = \gamma_n \end{aligned} \tag{19.5.1}$$

Then, the link angular velocities of Eq. (19.4.9) may be expressed as:

$$\boldsymbol{\omega}_B = \dot{q}_1 \mathbf{n}_{01} \quad \text{and} \quad \boldsymbol{\omega}_k = \dot{q}_1 \mathbf{n}_{01} + \sum_{r=1}^{k} \dot{q}_{r+1} SOR_{m3} \mathbf{n}_{03} \quad (k = 1,\ldots,n) \tag{19.5.2}$$

or as:

$$\boldsymbol{\omega}_B = \omega_{Brm} \dot{q}_r \mathbf{n}_{0m} \quad \text{and} \quad \boldsymbol{\omega}_k = \omega_{krm} \dot{q}_r \mathbf{n}_{0m} \quad (k = 1,\ldots,n) \tag{19.5.3}$$

where the ω_{Brm} and ω_{krm} are components of the partial angular velocities (see Section 11.4) defined as $\partial\boldsymbol{\omega}_B/\partial\dot{q}_r$ and $\partial\boldsymbol{\omega}_k/\partial\dot{q}_r$. Specifically, the ω_{Brm} and ω_{krm} are seen to be:

$$\begin{aligned}
\omega_{Brm} &= \begin{cases} 0 & r = 1, \quad m = 1 \text{ or } 2 \\ 1 & r = 1, \quad m = 3 \\ 0 & r > 1 \end{cases} \\
\omega_{1rm} &= \begin{cases} \omega_{Brm} & r = 1 \\ S01_{m3} & r = 2 \\ 0 & r > 0 \end{cases} \\
\omega_{2rm} &= \begin{cases} \omega_{1rm} & r = 1,2 \\ S01_{m3} & r = 2 \\ 0 & r > 0 \end{cases} \\
&\vdots \\
\omega_{krm} &= \begin{cases} \omega_{jrm} & r = 1,\ldots,k;\ j = k-1 \\ SOK_{m3} & r = k+1 \\ 0 & r > k+1 \end{cases} \\
&\vdots \\
\omega_{nrm} &= \begin{cases} \omega_{n-1,rm} & r = 1,\ldots,n \\ SON_{m3} & r = n+1 \end{cases}
\end{aligned} \tag{19.5.4}$$

The partial angular velocity components are useful in the sequel as we develop the governing dynamics equations.

19.6 Transformation Matrix Derivatives

By knowing the angular velocities of the links, we can follow the procedures of Section 18.4 to obtain expressions for derivatives of the transformation matrices. Specifically, from Eq. (19.3.5), we have:

$$SOK_{mn} = \mathbf{n}_{0m} \cdot \mathbf{n}_{kn} \tag{19.6.1}$$

Then, the derivative of SOK_{mn} may be developed as:

$$\begin{aligned} d(SOK_{mm})/dt &= d(\mathbf{n}_{0m} \cdot \mathbf{n}_{kn})/dt \\ &= \mathbf{n}_{0m} \cdot d\mathbf{n}_{kn}/dt \\ &= \mathbf{n}_{0m} \cdot \boldsymbol{\omega}_k \times \mathbf{n}_{kn} \text{ (no sum on } k) \\ &= \mathbf{n}_{0m} \cdot \omega_{kr}\mathbf{n}_{0r} \times SOK_{sn}\mathbf{n}_{0s} \\ &= \omega_{kr} SOK_{sn} \mathbf{n}_{0m} \cdot \mathbf{n}_{0r} \times \mathbf{n}_{0s} \\ &= \omega_{kr} e_{mrs} SOK_{sn} \\ &= WK_{ms} SOK_{sn} \end{aligned} \tag{19.6.2}$$

where the WK_{ms} are the elements of an array W defined as:

$$WK_{ms} \stackrel{D}{=} -e_{msr}\omega_{kr} \tag{19.6.3}$$

Then, Eq. (19.6.2) may be written in the simple matrix form:

$$S\dot{O}K = WK\ SOK \tag{19.6.4}$$

19.7 Angular Acceleration of the Robot Links

We can immediately obtain expressions for the angular acceleration of the links by differentiating the angular velocity expressions. Specifically, from Eq. (19.5.3), we have:

$$\boldsymbol{\alpha}_B = d\boldsymbol{\omega}_B/dt = (\dot{\omega}_{Brm}\dot{q}_r + \omega_{Brm}\ddot{q}_r)\mathbf{n}_{0m} \tag{19.7.1}$$

and

$$\boldsymbol{\alpha}_k = d\boldsymbol{\omega}_k/dt = \left(\dot{\omega}_{krm}\dot{q}_r + \omega_{krm}\ddot{q}_r\right)\mathbf{n}_{0m} \quad (k = 1,\ldots,n) \tag{19.7.2}$$

where, from Eqs. (19.5.4), the $\dot{\omega}_{Brm}$ and the $\dot{\omega}_{krm}$ are:

$$\begin{aligned}
&\dot{\omega}_{Brm} = 0 \\
&\dot{\omega}_{1rm} = \begin{cases} 0 & r = 1 \\ S\dot{0}1_{m3} & r = 2 \\ 0 & r > 0 \end{cases} \\
&\dot{\omega}_{2rm} = \begin{cases} \dot{\omega}_{1rm} & r = 1,2 \\ S\dot{0}2_{m3} & r = 3 \\ 0 & r > 3 \end{cases} \\
&\quad\vdots \\
&\dot{\omega}_{krm} = \begin{cases} \dot{\omega}_{jrm} & r = 1,\ldots,k;\ j = k-1 \\ S\dot{0}K_{m3} & r = k+1 \\ 0 & r > k+1 \end{cases} \\
&\quad\vdots \\
&\dot{\omega}_{nrm} = \begin{cases} \dot{\omega}_{n-1,rm} & r = 1,\ldots,n \\ S\dot{0}1_{m3} & r = n+1 \end{cases}
\end{aligned} \tag{19.7.3}$$

where the $S\dot{0}\mathrm{K}$ are given by Eq. (19.6.4).

19.8 Joint and Mass Center Position

In our example robot system of Figure 19.4.1 and shown again in Figure 19.8.1, let the connecting joints serve as origins of the axis system of the links. Specifically, for typical link L_k let O_k be the origin of an axis system, X_k, Y_k, and Z_k, fixed in L_k. Let O_k be located on the pin axis connecting links L_j and L_k, and as before let the X_k axis be aligned with the link axis, as in Figure 19.8.2. Let G_k be the mass center of L_k and let $\mathbf{r}_k$ locate G_k relative to O_k. Then, $\mathbf{r}_k$ is fixed in L_k.

Next, consider again two typical adjoining links L_j and L_k as in Figure 19.8.3. Let X_j, Y_j, and Z_j be axes fixed in L_j with origin O_j as shown. Let $\boldsymbol{\xi}_k$ locate O_k relative to O_j. Then $\boldsymbol{\xi}_k$ is fixed in L_j.

Finally, let the axis systems and origins for the base B and end effector E be as shown in Figures 19.8.4 and 19.8.5. In these figures, X_B, Y_B, and Z_B and X_E, Y_E, and Z_E are axes fixed in B and E, respectively, with O_B and O_E being the respective origins. The mass centers G_B and G_E are located relative to O_B and O_E by the vectors $\mathbf{r}_B$ and $\mathbf{r}_E$ as shown.

As before, we are representing the end effector E by a generic block. We will consider a more detailed representation of E in Section 19.12.

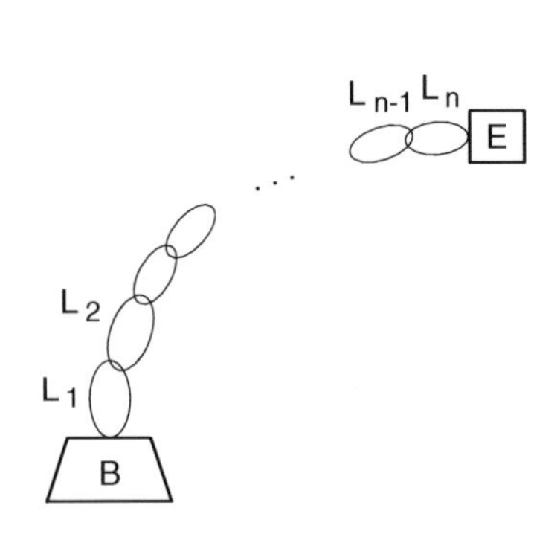

FIGURE 19.8.1
Example robot system.

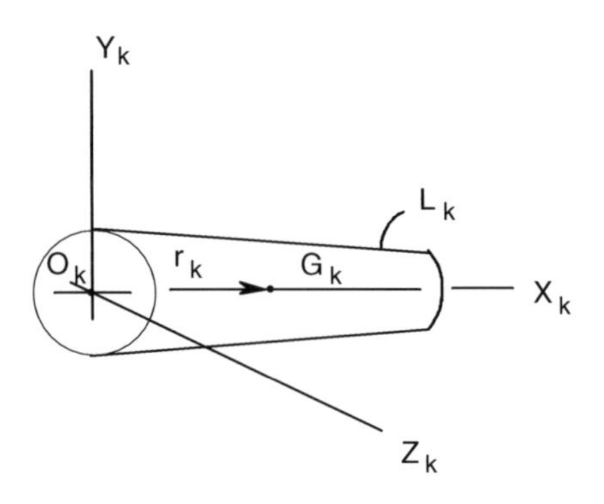

FIGURE 19.8.2
Typical link L_k with imbedded axis system X_k, Y_k, Z_k; origin O_k; and mass center G_k.

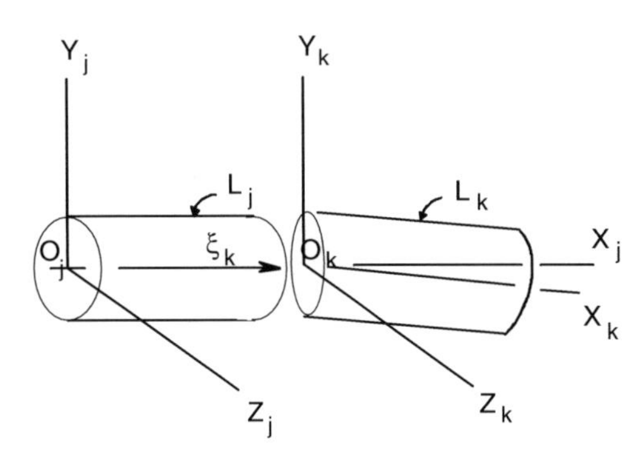

FIGURE 19.8.3
Two typical adjoining links.

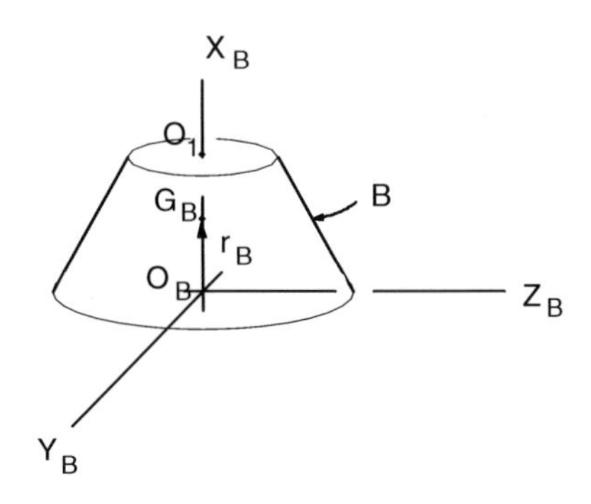

FIGURE 19.8.4
Base axis system and mass center G_B.

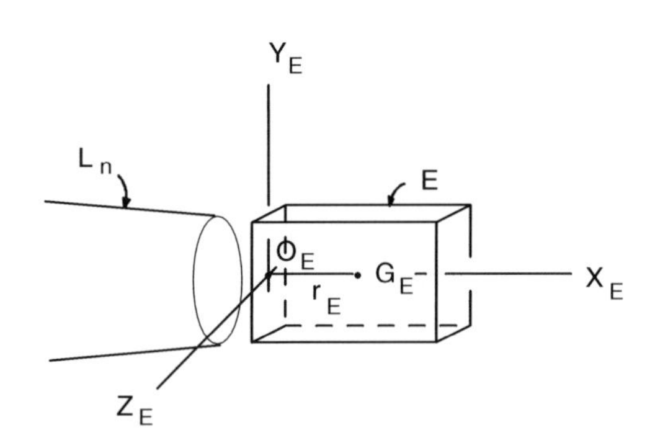

FIGURE 19.8.5
Axis system and mass center G_E of end effector E.

Using the foregoing notation, a position vector $\mathbf{p}_k$ locating mass center G_k of typical link L_k relative to a fixed point O of an inertial frame R may be expressed as:

$$\mathbf{p}_k = \boldsymbol{\xi}_1 + \boldsymbol{\xi}_2 + \ldots + \boldsymbol{\xi}_j + \boldsymbol{\xi}_k + \mathbf{r}_k = \sum_{i=1}^{k} \boldsymbol{\xi}_i + \mathbf{r}_k \tag{19.8.1}$$

Then, for end effector E, the position vector locating G_E relative to O is:

$$\mathbf{p}_E = \sum_{i=1}^{n} \xi_i + \mathbf{r}_E \tag{19.8.2}$$

Because the end effector E (or *gripper*) may itself contain a number of bodies depending upon its purpose, $\mathbf{r}_E$ will generally not be fixed in E. However, if we regard E to be "small" compared to the robot links, we may obtain a reasonable representation of the dynamical effects of E by considering $\mathbf{r}_E$ to be fixed in E.

19.9 Mass Center Velocities

By differentiating in Eqs. (19.8.1) and (19.8.2), we may immediately obtain expressions for the velocities of the mass centers of the robot links. Specifically, the velocities of G_B, G_k, and G_E in inertia frame R are:

$$^R\mathbf{V}^{G_B} \stackrel{D}{=} \mathbf{V}_B = {}^R d\mathbf{r}_B/dt = 0 \tag{19.9.1}$$

$$^R\mathbf{V}^{G_k} \stackrel{D}{=} \mathbf{V}_k = {}^R d\mathbf{p}_k/dt \tag{19.9.2}$$

and

$$^R\mathbf{V}^{G_E} \stackrel{D}{=} \mathbf{V}_E = {}^R d\mathbf{P}_E/dt \tag{19.9.3}$$

where in Eq. (19.9.1) the velocity of G_B is found to be zero because, from Figure 19.8.4, G_B is seen to be fixed in R as B has axial rotation about the vertical axis X_B. Then, from Eqs. (19.8.1) and (19.8.2), we have:

$$\begin{aligned} \mathbf{V}_k &= {}^R d\xi_1/dt + {}^R d\xi_2/dt + \ldots + {}^R d\xi_j/dt + {}^R d\xi_k/dt + {}^R d\mathbf{r}_k/dt \\ &= \sum_{i=1}^{k} {}^R d\xi_i/dt + {}^R d\mathbf{r}_k/dt \end{aligned} \tag{19.9.4}$$

and

$$\mathbf{V}_E = \sum_{i=1}^{n} {}^R d\xi_i/dt + {}^R d\mathbf{r}_E/dt \tag{19.9.5}$$

Observe that $\boldsymbol{\xi}_k$ is fixed in L_j, the adjacent lower numbered link, and that $\mathbf{r}_k$ is fixed in L_k. Then, the derivatives of $\boldsymbol{\xi}_k$ and r_k may be expressed as:

$$^R d\boldsymbol{\xi}_k/dt = \boldsymbol{\omega}_j \times \boldsymbol{\xi}_k \quad \text{and} \quad ^R d\mathbf{r}_k/dt = \boldsymbol{\omega}_k \times \mathbf{r}_k \quad (\text{no sum on } k) \tag{19.9.6}$$

If the end effector E is relatively small compared to the robot links, we may regard $\mathbf{r}_E$ as being fixed in E. Then, $Rd\mathbf{r}_E/dt$ is approximately:

$$^R d\mathbf{r}_E/dt = \boldsymbol{\omega}_E \times \mathbf{r}_E \tag{19.9.7}$$

From Eqs. (19.9.4) and (19.9.6), we see that $\mathbf{V}_k$ may thus be expressed as:

$$\mathbf{V}_k = \boldsymbol{\omega}_B \times \boldsymbol{\xi}_1 + \boldsymbol{\omega}_1 \times \boldsymbol{\xi}_2 + \ldots + \boldsymbol{\omega}_i \times \boldsymbol{\xi}_j + \boldsymbol{\omega}_k \times \mathbf{r}_k \tag{19.9.8}$$

$\mathbf{V}_E$ may be expressed in a similar form.

Recall from Eqs. (19.5.3) that the angular velocities $\boldsymbol{\omega}_B$ and $\boldsymbol{\omega}_k$ may be expressed as:

$$\boldsymbol{\omega}_B = \omega_{Brm}\dot{q}_r\mathbf{n}_{0m} \quad \text{and} \quad \boldsymbol{\omega}_k = \omega_{krm}\dot{q}_r\mathbf{n}_{0m} \tag{19.9.9}$$

where the partial angular velocity components ω_{Brm} and ω_{krm} are given by Eq. (19.5.4) and the $\dot{q}_r$ are defined in Eq. (19.5.1). As before, the $\mathbf{n}_{0m}$ ($m = 1, 2, 3$) are mutually perpendicular unit vectors fixed in R. Also, from Figures 19.8.2 and 19.8.3 we see that $\boldsymbol{\xi}_k$ and $\mathbf{r}_k$ may be expressed as:

$$\boldsymbol{\xi}_k = \ell_j\mathbf{n}_{j1} \quad (\text{no sum on } j) \quad \text{and} \quad \mathbf{r}_k = r_k\mathbf{n}_{k1} \quad (\text{no sum on } k) \tag{19.9.10}$$

where ℓ_j is the length of link L_j and as before $\mathbf{n}_{ji}$ and $\mathbf{n}_{ki}$ ($i = 1, 2, 3$) are unit vectors fixed in L_j and L_k with $\mathbf{n}_{j1}$ and $\mathbf{n}_{k1}$ being parallel to the axes of L_j and L_k. In terms of the $\mathbf{n}_{0m}$, $\boldsymbol{\xi}_k$ and $\mathbf{r}_k$ may be expressed as:

$$\boldsymbol{\xi}_k = \ell_j SOJ_{m1}\mathbf{n}_{0m} \quad \text{and} \quad \mathbf{r}_k = r_k SOK_{m1}\mathbf{n}_{0m} \tag{19.9.11}$$

Finally, recall from Figure 19.4.2 that the base B rotates about its own axis. Hence, $\boldsymbol{\omega}_B$ is parallel to $\boldsymbol{\xi}_1$ so the first term of Eq. (19.9.6) is zero. Therefore, $\mathbf{V}_k$ may be expressed in the form:

$$V_k = \sum_{r=1}^{k-1} \boldsymbol{\omega}_r \times \boldsymbol{\xi}_s + \boldsymbol{\omega}_k \times r_k \quad (s = r+1) \tag{19.9.12}$$

Then, by substituting from Eqs. (19.9.7) and (19.9.9), we obtain:

$$\begin{aligned} V_k &= \sum_{r=1}^{k-1} \omega_{r\ell m}\dot{q}_\ell\mathbf{n}_{0m} \times \ell_r SOR_{n1}\mathbf{n}_{0m} + \omega_{k\ell m}\dot{q}_\ell\mathbf{n}_{0m} \times r_k SOK_{n1}\mathbf{n}_{0n} \\ &= \sum_{r=1}^{k-1} e_{mnp}\omega_{r\ell m}SOR_{n1}\ell_r\dot{q}_\ell\mathbf{n}_{0p} + e_{mnp}\omega_{k\ell m}SOK_{n1}r_k\dot{q}_\ell\mathbf{n}_{0p} \end{aligned} \tag{19.9.13}$$

19.10 Mass Center Partial Velocities

From Eq. (19.9.13) we see that the mass center velocities may be expressed in the compact form:

$$\mathbf{V}_k = V_{k\ell p}\dot{q}_\ell \mathbf{n}_{0p} \quad k = 1,\ldots,n \tag{19.10.1}$$

where, by inspection of Eq. (19.9.13), the $V_{k\ell p}$ are seen to be:

$$V_{k\ell p} = \sum_{r=1}^{k-1} e_{mnp}\omega_{r\ell m}SOR_{n\ell}\ell_r + e_{mnp}\omega_{k\ell m}SOK_{n\ell}r_k \tag{19.10.2}$$

The $V_{k\ell p}$ are the $\mathbf{n}_{0p}$ components of the partial velocity vectors of the link mass centers with respect to the generalized coordinate q_ℓ (see Eq. (19.5.1)).

Similarly, the velocities of the base and end effector mass centers may be expressed as:

$$\mathbf{V}_B = V_{B\ell p}\dot{q}_\ell \mathbf{n}_{0p} \quad \text{and} \quad \mathbf{V}_E = V_{E\ell p}\dot{q}_\ell \mathbf{n}_{0p} \tag{19.10.3}$$

where from Eq. (19.9.1) the $V_{B\ell p}$ are seen to be zero, and from Eq. (19.9.4) and (19.9.5) the $V_{E\ell p}$ are seen to have expressions similar in form to those of Eq. (19.10.2).

19.11 Mass Center Accelerations

By differentiating in Eqs. (19.10.1) and (19.10.2) we immediately obtain expressions for the accelerations $\mathbf{a}_k$ of the link mass centers in an inertial frame R. Specifically, we have:

$$\mathbf{a}_k = \left(V_{k\ell p}\ddot{q}_\ell + \dot{V}_{k\ell p}\dot{q}_\ell\right)\mathbf{n}_{0p} \quad k = 1,\ldots,n \tag{19.11.1}$$

where, from Eq. (19.10.2), the $\dot{V}_{k\ell p}$ are seen to be:

$$\dot{V}_{k\ell p} = \sum_{r=1}^{k-1} e_{mnp}\left(\dot{\omega}_{r\ell m}SOR_{n1} + \omega_{r\ell m}S\dot{O}R_{n1}\ell_r\right) + e_{mnp}\left(\dot{\omega}_{k\ell m}SOK_{n1} + \omega_{k\ell m}S\dot{O}K_{n1}\right)r_k \tag{19.11.2}$$

where the $\dot{\omega}_{r\ell m}$ and $\dot{\omega}_{k\ell m}$ are given by Eq. (19.7.3) and where the $S\dot{O}R_{n1}$ and $S\dot{O}K_{n1}$ may be obtained from Eq. (19.6.4).

From Eqs. (19.9.1) we see that the acceleration $\mathbf{a}_B$ of the base mass center is:

$$\mathbf{a}_B = 0 \tag{19.11.3}$$

Finally, if the end effector is relatively small compared to the robot links, its mass center acceleration $\mathbf{a}_E$ may be expressed as:

$$\mathbf{a}_E = \left(V_{E\ell p}\ddot{q}_\ell + \dot{V}_{E\ell p}\dot{q}_\ell \right)\mathbf{n}_{0p} \tag{19.11.4}$$

where the $V_{E\ell p}$ and $\dot{V}_{E\ell p}$ have expressions similar to those of Eqs. (19.10.2) and (19.11.2), respectively. We will discuss end effector kinematics in more detail in the following section.

19.12 End Effector Kinematics

For an illustration of an elementary kinematical analysis of an end effector, consider the simple device sketched in Figure 19.12.1. It consists of a main body called the *hand* and two *fingers* consisting of slender members pinned to each other and to the hand as shown. The hand is affixed to the nth link, but it has the ability to rotate axially as represented in Figure 19.12.2. The movement of the hand and fingers is accomplished through actuators at the joints. Although the design of actuators is beyond our scope, interested readers may want to consult the references for information about actuators, drivers, and stepping motors.

Despite its simplicity, the end effector of Figure 19.12.1 could be used to grasp a variety of objects, as depicted in Figures 19.12.3 and 19.12.4. In practice, a robot and effector may

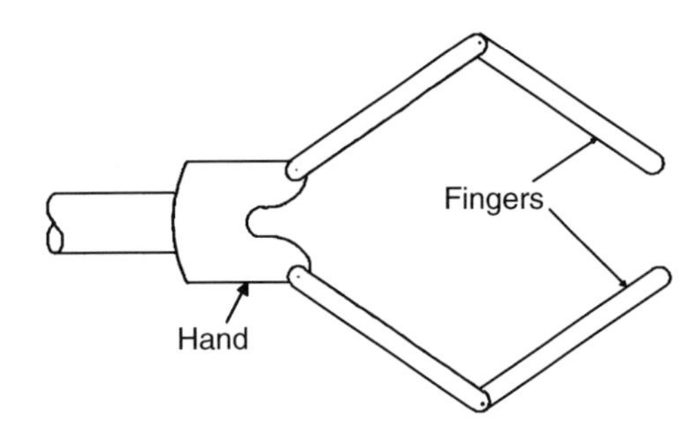

FIGURE 19.12.1
A simple end effector.

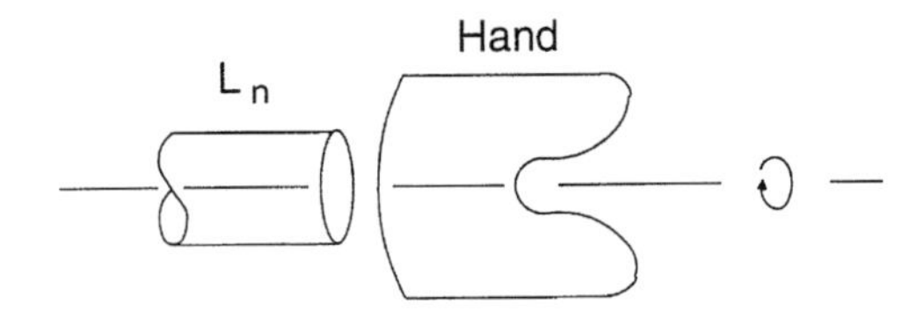

FIGURE 19.12.2
Attachment of the hand to the nth link allowing for axial rotation.

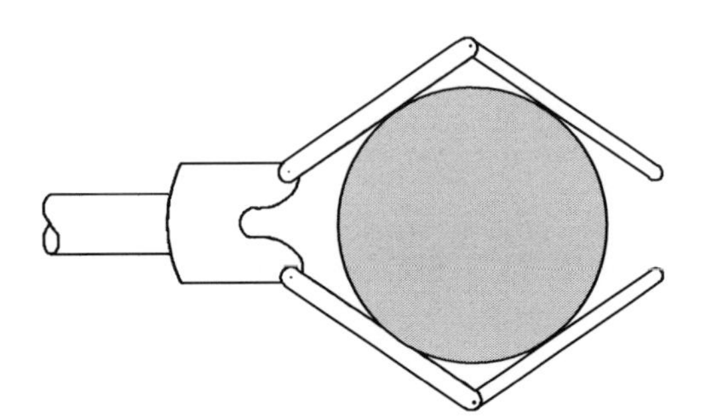

FIGURE 19.12.3
Grasping a circular object.

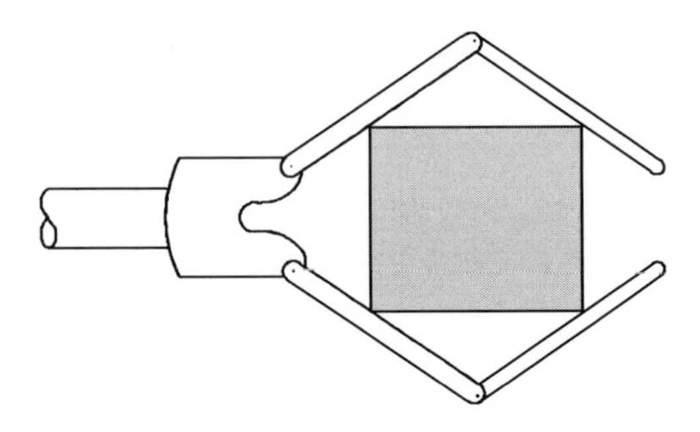

FIGURE 19.12.4
Grasping a rectangular object.

be used to grasp both light and heavy objects, as well as objects that may be durable or fragile; hence, precise control of the hand and fingers is a necessity. Such control in turn requires an analysis of the forces involved in lifting various objects.

A force analysis for the hand and fingers may be developed using the procedures of Chapter 6. To illustrate such an analysis, let the bodies and joints of the simple end effector of Figure 19.12.1 be labeled as in Figure 19.12.5, where P and Q are points at the extremities (or fingertips). To simplify the analysis, let the upper and lower fingers be identical, symmetrically configured about an axis through the hand H as in Figure 19.12.6. Specifically, let both the geometry and the movement of the upper and lower fingers be the same, with ℓ_1 and ℓ_2 being the lengths of D_1,D_3 and D_2,D_4, and θ_1 and θ_2 the rotative angles at the joints O_1,O_3 and O_2,O_4, respectively, as shown. With this symmetry, we may study the motion and forces in the fingers by considering only one of the fingers, say the upper finger. Therefore, consider a sketch of the upper finger as in Figure 19.12.7. Let X and Y be a Cartesian axis system fixed in the hand H with origin O as shown. Let (x_p, y_p) be the coordinates of P relative to X and Y. Then, in terms of ℓ_1, ℓ_2, θ_1, and θ_2, x_p and y_p are:

$$x_p = \ell_1 \cos\theta_1 + \ell_2 \cos\theta_2 \tag{19.12.1}$$

$$y_p = \eta + \ell_1 \sin\theta_1 + \ell_2 \sin\theta_2 \tag{19.12.2}$$

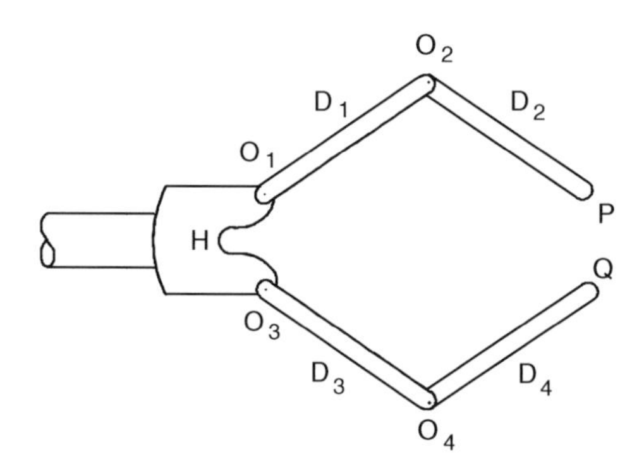

FIGURE 19.12.5
Simple illustrative end effector.

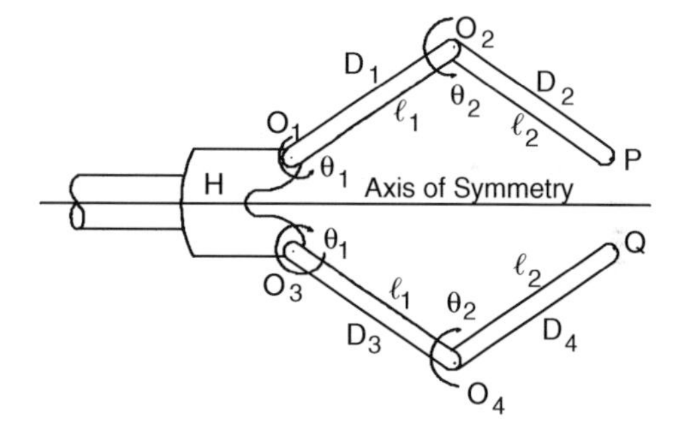

FIGURE 19.12.6
Symmetrical illustrative end effector.

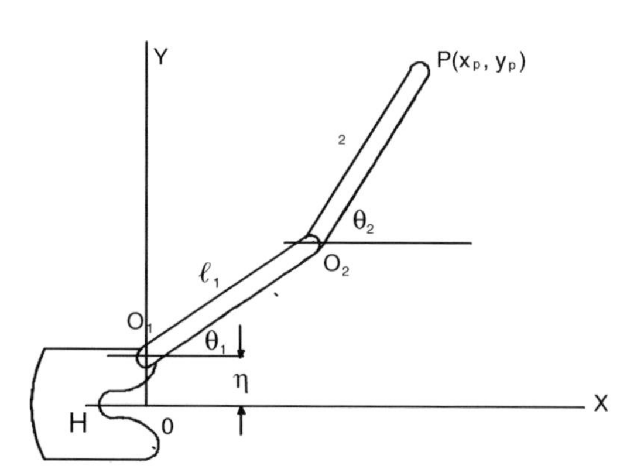

FIGURE 19.12.7
Upper finger of end effector.

where η is the separation of O and O_1 as shown in Figure 19.12.7. The X–Y components of the velocity and acceleration of P in H are then simply:

$$\dot{x}_p = -\ell_1\dot{\theta}_1 \sin\theta_1 - \ell_2\dot{\theta}_2 \sin\theta_2 \tag{19.12.3}$$

$$\dot{y}_p = \ell_1\dot{\theta}_1 \cos\theta_1 + \ell_2\dot{\theta}_2 \cos\theta_2 \tag{19.12.4}$$

and

$$\ddot{x}_p = -\ell_1\left(\ddot{\theta}_1 \sin\theta_1 + \dot{\theta}_1^2 \cos\theta_1\right) - \ell_2\left(\ddot{\theta}_2 \sin\theta_2 + \dot{\theta}_2^2 \cos\theta_2\right) \tag{19.12.5}$$

$$\ddot{y}_p = \ell_1\left(\ddot{\theta}_1 \cos\theta_1 - \dot{\theta}_1^2 \sin\theta_1\right) + \ell_2\left(\ddot{\theta}_2 \sin\theta_2 - \dot{\theta}_2^2 \sin\theta_2\right) \tag{19.12.6}$$

Similarly, by symmetry, the coordinates, the velocity, and the acceleration of tip Q of D_4 are:

$$x_Q = \ell_1 \cos\theta_1 + \ell_2 \cos\theta_2 \tag{19.12.7}$$

$$y_Q = -\eta - \ell_1 \sin\theta_1 - \ell_2 \sin\theta_2 \tag{19.12.8}$$

and

$$\dot{x}_Q = -\ell_1\dot{\theta}_1 \sin\theta_1 - \ell_2\dot{\theta}_2 \sin\theta_2 \tag{19.12.9}$$

$$\dot{y}_Q = -\ell_1\dot{\theta}_1 \cos\theta_1 - \ell_2\dot{\theta}_2 \cos\theta_2 \tag{19.12.10}$$

and

$$\ddot{x}_Q = -\ell_1\left(\ddot{\theta}_1 \sin\theta_1 + \dot{\theta}_1^2 \cos\theta_1\right) - \ell_2\left(\ddot{\theta}_2 \sin\theta_2 + \dot{\theta}_2^2 \cos\theta_2\right) \tag{19.12.11}$$

$$\ddot{y}_Q = -\ell_1\left(\ddot{\theta}_1 \cos\theta_1 - \dot{\theta}_1^2 \sin\theta_1\right) - \ell_2\left(\ddot{\theta}_2 \cos\theta_2 - \dot{\theta}_2^2 \sin\theta_2\right) \tag{19.12.12}$$

From Eqs. (19.12.2) and (19.12.7), we see that the separation between P and Q (the gap at the finger tips) h is:

$$h = y_p - y_Q = 2\eta + 2\ell_1 \sin\theta_1 + 2\ell_2 \sin\theta_2 \tag{19.12.13}$$

Finally, consider the end effector gripping an object between its finger tips as in Figure 19.12.8. The gripping forces will in general be large compared with the inertia forces in the gripper fingers. Hence, in most analyses, the inertia forces may be neglected. The contact forces at the finger tips P and Q are then directly related to the joint moments caused by the joint actuators. Specifically, the relation between the gripping force G at P and the moments M_1 and M_2 at the upper finger joints may be obtained from a free-body

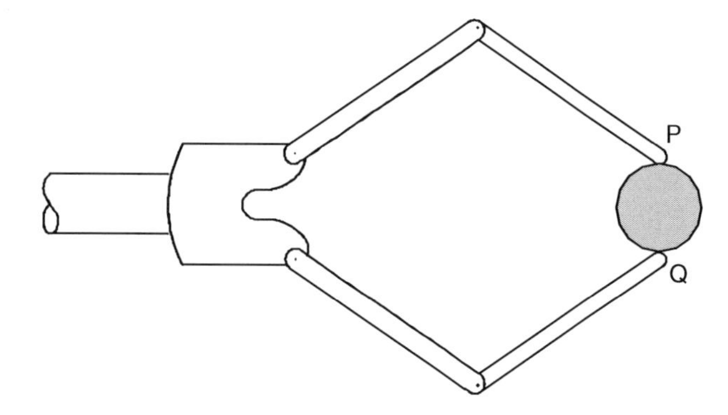

FIGURE 19.12.8
End effector grasping an object at the finger tips.

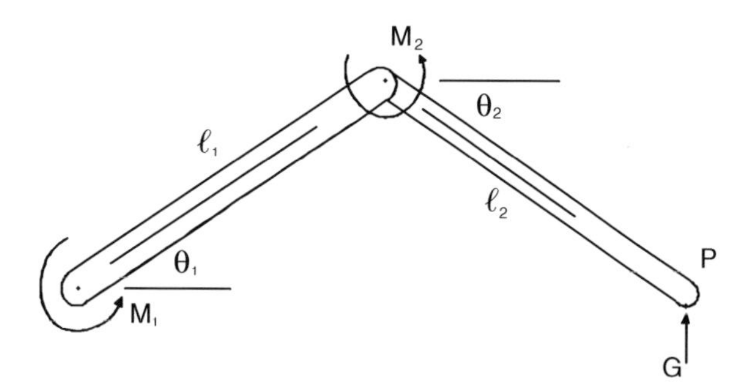

FIGURE 19.12.9
Free-body diagram of upper fingers.

diagram of the upper figure as in Figure 19.12.9. From this sketch, we immediately obtain the expressions:

$$M_2 = -G\ell_2 \cos\theta_2 \tag{19.12.14}$$

and

$$M_1 = -G(\ell_1 \cos\theta_1 + \ell_2 \cos\theta_2) \tag{19.12.15}$$

This analysis is, of course, greatly simplified compared to what might be needed in studying an actual end effector. Nevertheless, the underlying principles of a more detailed analysis will be the same.

19.13 Kinetics: Applied (Active) Forces

With knowledge of the kinematics of a robot arm, we are in a position to consider and discuss the kinetics (the forces) acting on the system. To this end, consider again two typical adjoining robot links as in Figure 19.13.1. The forces exerted on these links occur due to gravity (from the weights of the links), interactive forces and moments between the links (from actuators at the joint), and inertia (from acceleration of the links). In this section, we will consider the gravity and interactive forces, and in the next section we will study the inertia forces. As before, our objective will be to formulate the analysis in terms of generalized forces.

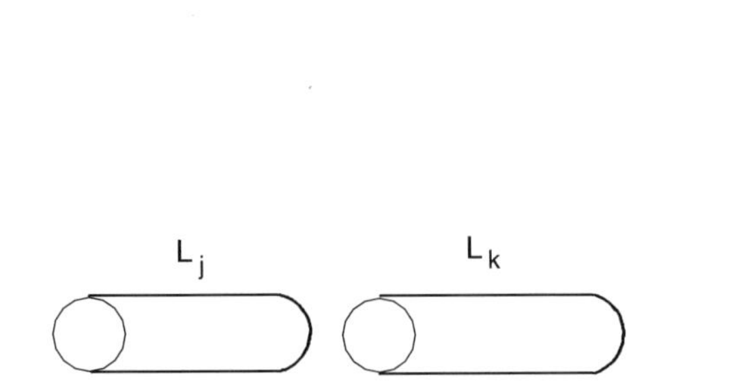

FIGURE 19.13.1
Two typical adjoining robot links.

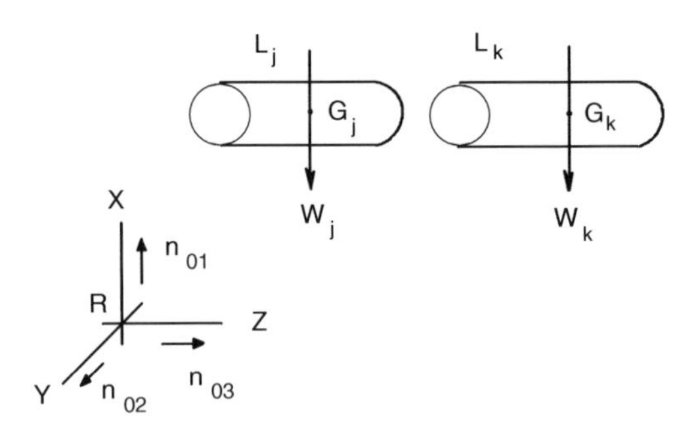

FIGURE 19.13.2
Gravitational forces and robot links.

The gravitational forces on L_j and L_k may be represented by weight forces $\mathbf{W}_j$ and $\mathbf{W}_k$ acting through the mass centers G_j and G_k of the links as in Figure 19.13.2, where, as before, R is the inertial reference frame. Then, the contribution $\overset{G}{F}_\ell$ from these weight forces to the generalized applied (or *active*) forces for each generalized coordinate q_2 may be expressed in the form (see Eq. (11.5.3)):

$$\overset{G}{F}_\ell = \frac{\partial \mathbf{V}_j}{\partial \dot{q}_\ell} \cdot \mathbf{W}_j + \frac{\partial \mathbf{V}_k}{\partial \dot{q}_\ell} \cdot \mathbf{W}_k \quad \ell = 1, \ldots, n+1 \tag{19.13.1}$$

where, as before, $\mathbf{V}_j$ and $\mathbf{V}_k$ are the velocities of G_j and G_k in R. Recall from Eq. (19.10.1) that $\mathbf{V}_j$ and $\mathbf{V}_k$ may be expressed as:

$$\mathbf{V}_j = v_{j\ell p} \dot{q}_\ell \mathbf{n}_{0p} \quad \text{and} \quad \mathbf{V}_k = v_{k\ell p} \dot{q}_\ell \mathbf{n}_{0p} \tag{19.13.2}$$

where the $v_{j\ell p}$ and the $v_{k\ell p}$ are the components of the partial velocity vectors of G_j and G_k, as given by Eq. (19.10.2). If we let the X-axis be vertical, as in Figure 19.13.2, we can express $\mathbf{W}_j$ and $\mathbf{W}_k$ as:

$$\mathbf{W}_j = -m_j g \mathbf{n}_{01} \quad \text{and} \quad \mathbf{W}_k = -m_k g \mathbf{n}_{01} \tag{19.13.3}$$

where m_j and m_k are the masses of L_j and L_k. Hence, from Eqs. (19.13.1, 19.13.2, and 19.13.3), the $\overset{G}{F}_\ell$ become:

$$\overset{G}{F}_\ell = -m_j g v_{j\ell 1} - m_k g v_{k\ell 1} \tag{19.13.4}$$

The interactive forces and moments transmitted between the links at the connecting joint, represented by O_k, may be represented by opposite equivalent force systems as in Figure 19.13.3. Specifically, let the force system exerted on L_k and L_j be represented by a force $\mathbf{F}_j$ passing through O_k together with a couple with torque $\mathbf{M}_j$. Then, by the law of action and reaction (see Reference 19.1), we have:

$$\mathbf{F}_j = -\mathbf{F}_k \quad \text{and} \quad \mathbf{M}_j = -\mathbf{M}_k \tag{19.13.5}$$

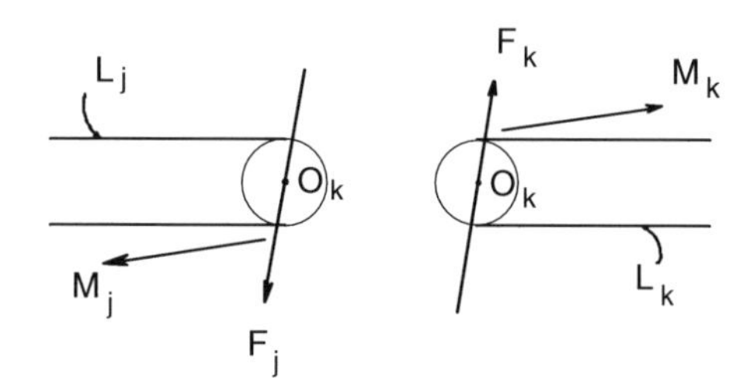

FIGURE 19.13.3
Force systems acting on robot links at a connecting joint.

Then, from Eq. (11.5.7), we see that the contribution $\overset{J}{F}_\ell$ of these joint force systems to the generalized forces is:

$$\overset{J}{F}_\ell = \frac{\partial \mathbf{V}^{O_k}}{\partial \dot{q}_\ell} \cdot \mathbf{F}_j + \frac{\partial \mathbf{V}^{O_k}}{\partial \dot{q}_\ell} \cdot \mathbf{F}_k + \frac{\partial \boldsymbol{\omega}_j}{\partial \dot{q}_\ell} \cdot \mathbf{M}_j + \frac{\partial \boldsymbol{\omega}_k}{\partial \dot{q}_\ell} \cdot \mathbf{M}_k \tag{19.13.6}$$

But, from Eq. (19.13.5), we see that the first two terms cancel one another, and from the addition theorem for angular velocity, we see that the link angular velocities are related as (see Eq. (19.4.6)):

$$\boldsymbol{\omega}_k = \boldsymbol{\omega}_j + \hat{\boldsymbol{\omega}}_k \tag{19.13.7}$$

Hence, the last two terms of Eq. (19.13.6) become:

$$\begin{aligned} \frac{\partial \boldsymbol{\omega}_j}{\partial \dot{q}_\ell} \cdot \mathbf{M}_j + \frac{\partial \boldsymbol{\omega}_k}{\partial \dot{q}_\ell} \cdot \mathbf{M}_k &= \frac{\partial \boldsymbol{\omega}_j}{\partial \dot{q}_\ell} \cdot \mathbf{M}_j + \frac{\partial \boldsymbol{\omega}_j}{\partial \dot{q}_\ell} \cdot \mathbf{M}_k + \frac{\partial \hat{\boldsymbol{\omega}}_k}{\partial \dot{q}_\ell} \cdot \mathbf{M}_k \\ &= \frac{\partial \boldsymbol{\omega}_j}{\partial \dot{q}_\ell} \cdot (\mathbf{M}_j + \mathbf{M}_k) + \frac{\partial \hat{\boldsymbol{\omega}}_k}{\partial \dot{q}_\ell} \cdot \mathbf{M}_k \\ &= \frac{\partial \hat{\boldsymbol{\omega}}_k}{\partial \dot{q}_k} \cdot \mathbf{M}_k \end{aligned} \tag{19.13.8}$$

From a review of Section 19.5 we see that $\partial \dot{\boldsymbol{\omega}}_k / \partial \dot{q}_\ell$ is:

$$\frac{\partial \dot{\boldsymbol{\omega}}_k}{\partial \dot{q}_\ell} = \begin{cases} 0 & k \neq \ell - 1 \\ \mathbf{n}_{k3} & k = \ell - 1 \end{cases} \tag{19.13.9}$$

Therefore, $\overset{J}{F}_\ell$ becomes:

$$\overset{J}{F}_\ell = M_{k3} \quad (k = \ell - 1) \tag{19.13.10}$$

where M_{k3} is the $\mathbf{n}_{k3}$ component of $\mathbf{M}_k$ and thus represents the actuator moment exerted at 0_k by L_j on L_k.

Finally, for the end effector E, let us suppose that the forces exerted on E may be represented by a force $\mathbf{F}_E$ passing through the mass center G_E, together with a couple having torque $\mathbf{M}_E$. Then, the contribution $\overset{E}{F}_\ell$ to the generalized forces from $\mathbf{F}_E$ and $\mathbf{M}_E$ may be expressed as:

$$\begin{aligned}\overset{E}{F}_\ell &= \frac{\partial \mathbf{V}_E}{\partial \dot{q}_\ell} \cdot \mathbf{F}_E + \frac{\partial \boldsymbol{\omega}_E}{\partial \dot{q}_\ell} \cdot \mathbf{M}_E \\ &= \mathbf{v}_{\mathbf{E}\ell m} F_{Em} + \omega_{E\ell m} M_{Em}\end{aligned} \tag{19.13.11}$$

where $\mathbf{V}_E$ and $\boldsymbol{\omega}_E$ are the end effector mass center velocity and angular velocity; $v_{E\ell m}$ and $\omega_{E\ell m}$ are the corresponding partial velocity and partial angular velocity components relative to unit vectors $\mathbf{n}_{0m}$ in the inertia frame R; and F_{Em} and M_{Em} are the $\mathbf{n}_{0m}$ components of $\mathbf{F}_E$ and $\mathbf{M}_E$.

By combining Eqs. (19.13.4), (19.13.10) and (19.13.11), we see that the generalized applied force F_R becomes:

$$F_\ell = \overset{G}{F}_\ell + \overset{J}{F}_\ell + \overset{E}{F}_\ell = -m_j g v_{j\ell 1} + M_{k3} + \overset{E}{F}_\ell \tag{19.13.12}$$

where $\overset{E}{F}_\ell$ is given by Eq. (19.13.11), where time is a sum over j from 1 to n, and where k is $\ell - 1$.

19.14 Kinetics: Passive (Inertia) Forces

The inertia forces on the robot links may be obtained by following the procedures of Section 18.16. Specifically, let the inertia forces exerted on a typical link L_k be represented by an equivalent force system consisting of a force $\mathbf{F}_k^*$ passing through mass center G_k together with a torque $\mathbf{M}_k^*$. Then, $\mathbf{F}_k^*$ and $\mathbf{M}_k^*$ may be expressed as (see Eqs. (18.16.2) and (18.16.3)):

$$\mathbf{F}_k^* = -m_k \mathbf{a}_k \quad (\text{no sum on } k) \tag{19.14.1}$$

and

$$\mathbf{M}_k^* = -\mathbf{I}_k \cdot \boldsymbol{\alpha}_k - \boldsymbol{\omega}_k \times (\mathbf{I}_k \cdot \boldsymbol{\omega}_k) \quad (\text{no sum on } k) \tag{19.14.2}$$

where, as before, m_k and $\mathbf{I}_k$ are the mass and central inertia dyadic of L_k, and $\boldsymbol{\omega}_k$, $\boldsymbol{\alpha}_k$, and $\mathbf{a}_k$ are the angular velocity, angular acceleration, and mass center acceleration, respectively, of L_k given by Eqs. (19.5.3), (19.7.2), and (19.11.1) as:

$$\boldsymbol{\omega}_k = \omega_{k\ell m} \dot{q}_\ell \mathbf{n}_{0m} \tag{19.14.3}$$

$$\boldsymbol{\alpha}_k = (\dot{\omega}_{k\ell m} \dot{q}_\ell + \omega_{k\ell m} \ddot{q}_\ell) \mathbf{n}_{0m} \tag{19.14.4}$$

and

$$\mathbf{a}_k = (\dot{v}_{k\ell m}\dot{q}_\ell + v_{k\ell m}\ddot{q}_\ell)\mathbf{n}_{0m} \tag{19.14.5}$$

where the partial velocity and partial angular velocity components and their derivatives are given by Eqs. (19.10.2), (19.11.2), (19.5.4), and (19.7.3). Then, by substituting from Eqs. (19.14.3), (19.14.4), and (19.14.5) into Eqs. (19.14.1) and (19.14.2), $\mathbf{F}_k^*$ and $\mathbf{M}_k^*$ become:

$$\mathbf{F}_k^* = -m_k(v_{k\ell m}\ddot{q}_\ell + \dot{v}_{k\ell m}\dot{q}_\ell)\mathbf{n}_{0m} \tag{19.14.6}$$

and

$$\mathbf{M}_k^* = -\left[I_{kmn}(\omega_{k\ell n}\ddot{q}_\ell + \dot{\omega}_{k\ell n}\dot{q}_\ell) + e_{rsm}\omega_{k\ell r}\omega_{kpn}I_{ksn}\dot{q}_\ell\dot{q}_p\right]\mathbf{n}_{0m} \tag{19.14.7}$$

where, as before, the I_{kmn} are the $\mathbf{n}_{0m}$ and $\mathbf{n}_{0n}$ components of $\mathbf{I}_k$, and the e_{rsm} are permutation symbol components.

The generalized inertia force on L_k, for the generalized coordinate q_ℓ, is given by Eq. (18.16.1) as:

$$F_\ell^* = F_{km}^* v_{k\ell m} + M_{km}^* \omega_{k\ell m} \quad \text{(no sum on } k) \tag{19.14.8}$$

where the F_{km}^* and M_{km}^* are the $\mathbf{n}_{0m}$ components of $\mathbf{F}_k^*$ and $\mathbf{M}_k^*$. Then, by substituting from Eqs. (19.14.6) and (19.14.7) into (19.14.8), we may express the F_ℓ^* as:

$$\begin{aligned} F_\ell^* = &-m_k v_{k\ell m}(v_{kpm}\ddot{q}_p + \dot{v}_{kpm}\dot{q}_p) \\ &-I_{kmn}\omega_{k\ell m}(\omega_{kpn}\ddot{q}_p + \dot{\omega}_{kpn}\dot{q}_p) \\ &-I_{ksn}\omega_{k\ell m}e_{rsm}\omega_{kqr}\omega_{kpn}\dot{q}_q\dot{q}_p \quad \text{(no sum on } k) \end{aligned} \tag{19.14.9}$$

Observe again the major dependence of the generalized inertia forces upon the partial velocity and partial angular velocity vector components and their derivatives. Observe also that if we relax the no-sum restriction, we obtain the generalized inertia forces for all of the links.

19.15 Dynamics: Equations of Motion

By knowing the generalized applied and inertia forces we can use Kane's equations (see Chapter 12) to obtain the governing equations of motion. Specifically, from Eq. (12.2.1) we have:

$$F_\ell + F_\ell^* = 0 \quad (\ell = 1, \ldots, n+1) \tag{19.15.1}$$

Then, by substituting from Eqs. (19.13.11) and (19.14.8), we obtain:

$$-m_k v_{k\ell m}\left(v_{kpm}\ddot{q}_p + \dot{v}_{kpm}\dot{q}_p\right) - I_{kmn}\omega_{k\ell m}\left(\omega_{kpn}\ddot{q}_p + \dot{\omega}_{kpn}\dot{q}_p\right)$$

$$-I_{ksn}\omega_{k\ell m}e_{rsm}\omega_{kqr}\omega_{kpn}\dot{q}_q\dot{q}_p - m_j g v_{j\ell 1} + M_{k3} + \overset{E}{F_\ell} = 0 \tag{19.15.2}$$

$$(\ell = 1,\ldots,n+1) \quad (k = \ell - 1)$$

Equation (19.15.2) may be written in the compact matrix form as:

$$A\ddot{q} = f + F \tag{19.15.3}$$

where A is a symmetrical $(n + 1) \times (n + 1)$ generalized mass array, q is an $(n + 1) \times 1$ column array of generalized coordinates, f is an $(n + 1) \times 1$ column array of inertia forces, and F is an $(n + 1) \times 1$ column array of generalized applied forces. The elements $a_{\ell p}$ of A are:

$$a_{\ell p} = m_k v_{k\ell m} v_{kpm} + I_{kmn}\omega_{k\ell m}\omega_{k\ell n} \tag{19.5.4}$$

The elements f_ℓ of f are:

$$f_\ell = -\left(m_k v_{k\ell m}\dot{v}_{kpm}\dot{q}_p + I_{kmn}\omega_{k\ell m}\dot{\omega}_{kpn}\dot{q}_p + e_{rsm}I_{ksn}\omega_{k\ell m}\omega_{kqr}\omega_{kpn}\dot{q}_p\dot{q}_q\right) \tag{19.15.5}$$

The elements f_ℓ of F are:

$$F_\ell = -m_j g v_{j\ell 1} + M_{\ell - h3} + \overset{E}{F_\ell} \tag{19.15.6}$$

where the $\overset{E}{F_\ell}$ are given by Eq. (19.13.11).

19.16 Redundant Robots

Redundant robots have more degrees of freedom than are needed to accomplish a given task. These extra degrees of freedom pose special problems for dynamics analysis. Decisions need to be made as to how the superfluous degrees of freedom are to be used. Expressed another way, constraints need to be imposed upon redundant robots for the robots to accomplish a given task. These constraints may take a variety of forms such as minimum joint torques, minimum energy consumption, obstacle avoidance, minimum jerk, or least time of operation. Methods of analysis to accommodate such constraints are still being developed. Difficulties of these analyses have not all been resolved.

To avoid these difficulties, early manufacturers of robots have restricted their machines to at most six degrees of freedom; however, without redundancy, the machines cannot accomplish their given tasks in an optimal manner. In this and the following section we will explore some of the issues in the dynamics and control of redundant robot systems.

Consider the general system of Figure 19.16.1. Suppose we have a desired movement of the end effector E. For simplicity (and without loss of generality), let us consider E to

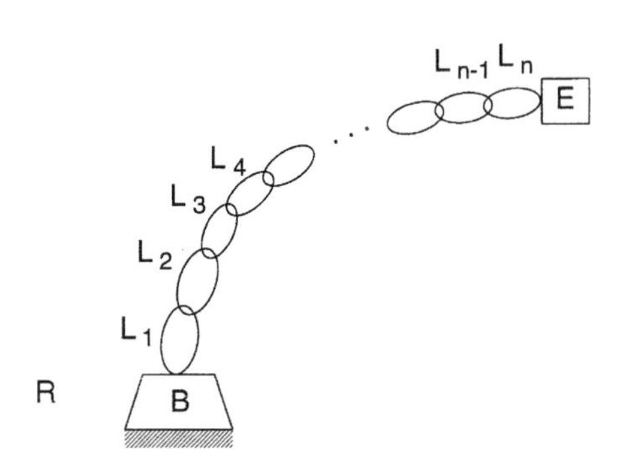

FIGURE 19.16.1
Generic robot system.

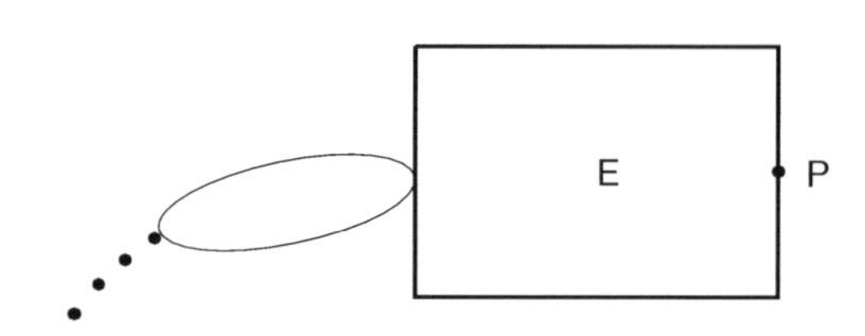

FIGURE 19.16.2
Representation of an end effector.

be a single body (that is, suppose E is a gripper firmly holding an object or workpiece). Let P be a typical point, perhaps at an extremity, of E as in Figure 19.16.2. For a given desired motion of E, we will know both the angular velocity of E and the velocity of P.

Let us represent the movement of the system by a series of articulation angles γ_i ($i = 1,\ldots, n$), as in Section 19.2. For simplicity (and again without loss of generality) let us assume a fixed base angle α_B. Then, we can express the angular velocity of E and the velocity of P in the forms:

$$\boldsymbol{\omega}^E = \omega_{E\ell m}\dot{\gamma}_\ell \mathbf{n}_{0m} \quad \text{and} \quad \mathbf{V}^P = v_{p\ell m}\dot{\gamma}_\ell \mathbf{n}_{0m} \tag{19.16.1}$$

where, as before, the $\mathbf{n}_{0m}$ ($m = 1, 2, 3$) are unit vectors fixed in the inertia frame R, and the $v_{p\ell m}$ and $\omega_{E\ell m}$ are the partial velocity and partial angular velocities, respectively, of P and E. Then, with the movement and P and E being known, we will know the $\mathbf{n}_{0m}$ components of $\boldsymbol{\omega}^E$ and V^P. These components may then be expressed as:

$$\omega_{E\ell m}\dot{\gamma}_\ell = g_m(t) \quad \text{and} \quad V_{p\ell m}\dot{\gamma}_\ell = g_{m+3}(t) \quad m = 1,2,3 \tag{19.16.2}$$

where the $g_i(t)$ ($i = 1,\ldots, 6$) are known functions of time.

Equation (19.16.2) may be written in the compact form:

$$B\dot{\gamma} = g \tag{19.16.3}$$

where γ is a column array of the articulation angles, g is a column array of functions on the right sides of Eq. (19.16.2), and B is a block array of partial angular velocity and partial velocity array components of Eq. (19.16.2). If there are n articulation angles γ_i ($i = 1,\ldots, n$), then γ is an ($n \times 1$) array, g is a (6×1) array, and B is a ($6 \times n$) array. B is sometimes called the *constraint matrix*.

The articulation angles are governed by the dynamics equations (Eqs. (19.15.3)):

$$A\ddot{\gamma} = f + F \tag{19.16.4}$$

If there are n articulation angles γ_i, Eq. (19.16.4) is equivalent to n scalar equations. Then, Eqs. (19.16.3) and (19.16.4) form a total of ($6 + n$) equations for the n γ_i and the n joint moments M_i ($i = 1,\ldots, n$). Thus, we have ($n + 6$) equations for $2n$ unknowns and the system is undetermined. There are ($n - 6$) fewer equations than needed to uniquely specify the motion of the system. Expressed another way, the robot has $(n - 6)$ more degrees of freedom than needed to uniquely determine the movement of the system. These extra degrees of freedom represent the redundancy in the system, thus we have a *redundant robot*.

To resolve the redundancy, we need $(n - 6)$ additional equations or *constraints* on the system. As noted earlier, these equations may be obtained in a number of ways, by specifying some of the articulation angles or by specifying some of the joint torques, or a combination thereof, or by imposing more general requirements such as having minimum joint torques or minimum kinetic energy for the system. We will further discuss the form and solutions of the governing equations in the following sections.

19.17 Constraint Equations and Constraint Forces

Before we look for solutions of the governing equation let us attempt to obtain further insight into the nature of the equations themselves by considering the following problem. Suppose we have an n-link robot arm with a desired motion of the nth link as in Figure 19.17.1. As before, for simplicity, let us assume that we have a fixed base B. Suppose this system is initially at rest in an arbitrary configuration. Suppose further that, for this discussion, the system is in a weightless and force-free environment and that the moments between adjoining links at the joints are all zero. Imagine further that we are somehow able to move the nth link through its desired motion. Under these rather specialized conditions and with the given motion of the nth link, we could ask the question: What are the resulting motions of the first $n - 1$ links? That is, in the absence of gravity, with free movement at the joints, and with a specified movement of the last link, what are the movements of the first $n - 1$ links?

To answer this question, consider again the governing dynamical equations (Eqs. (19.16.4)):

$$A\ddot{\gamma} = f + F \tag{19.17.1}$$

where, as before, A is the $(n \times n)$ generalized mass array with elements as in Eq. (19.15.4); γ is the array of articulation angles; f is an array of inertia force terms with elements as in Eq. (19.15.5); and F is the array of generalized applied forces with elements F_ℓ as in Eq. (19.15.6). In this rather specialized case with no gravity and with zero joint moments, the F_ℓ have the reduced form:

$$F_\ell = F'_\ell \tag{19.17.2}$$

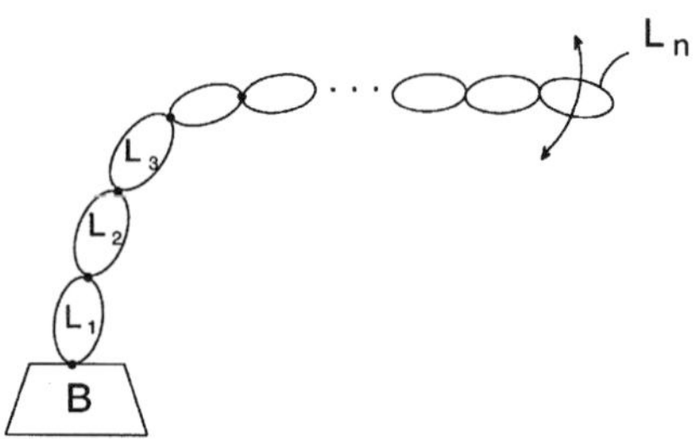

FIGURE 19.17.1
An n-link robot with a desired motion of the last link.

where the F'_ℓ are generalized forces arising due to the specified motion of the nth link. That is, the F'_ℓ are generalized forces due to the applied forces and movements needed to drive the nth link through the specified motion. These forces are constraint forces and the resulting generalized forces of Eqs. (19.17.2) form an array F'_ℓ that might be called a *generalized constraint force array.*

The specified motion of the nth link also generates a set of constraint equations of the form of Eq. (19.16.3). That is,

$$B\dot{\gamma} = g \tag{19.17.3}$$

Remarkably, the generalized constraint force array F' is directly related to the constraint matrix array B by the expression:

$$F' = B^T\lambda \tag{19.17.4}$$

where λ is an array of constraint force and moment components and B^T is the transpose of B. To see this, suppose a point P of the nth link (perhaps a tool point of an end effector) has a specified motion (see Figure 19.17.2). Suppose further that the angular motion of the link L_n is specified. Suppose still further that the specified motions of P and L_n are given in the form of velocities or in a form such that the velocity of P and the angular velocity of L_n may be determined. Then, the velocity of G_n, the mass center of L_n, will be known or can readily be determined. Let these velocities be expressed in the forms:

$$^R\mathbf{V}^{G_n} = \mathbf{V}_n = \mathbf{v}(t) \tag{19.17.5}$$

and

$$^R\boldsymbol{\omega}^{L_n} = \boldsymbol{\omega}_n = \boldsymbol{\Omega}(t) \tag{19.17.6}$$

where $\mathbf{v}(t)$ and $\boldsymbol{\Omega}(t)$ are known (specified) vector functions of time. Referring to the unit vectors $\mathbf{n}_{0m}$ of R it is convenient to express $\mathbf{V}_n$ and $\boldsymbol{\omega}_n$ in the component forms:

$$\mathbf{V}_n = V^{G_n}_{m\ell}\dot{\gamma}_\ell\mathbf{n}_{0m} \quad \text{and} \quad \boldsymbol{\omega}_n = \omega^{L_n}_{m\ell}\dot{\gamma}_\ell\mathbf{n}_{0m} \tag{19.17.7}$$

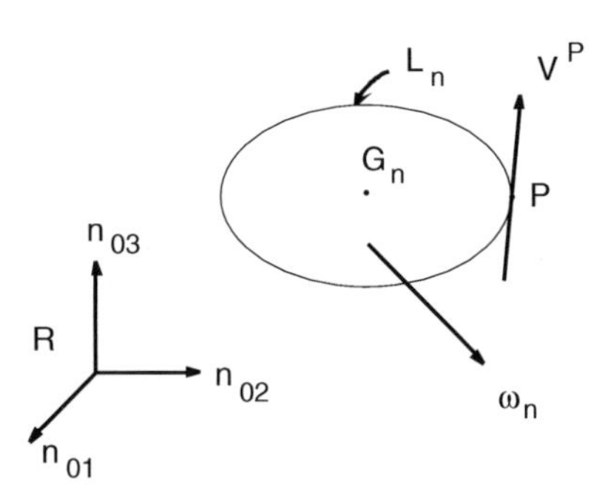

FIGURE 19.17.2
Specified motion of link L_n.

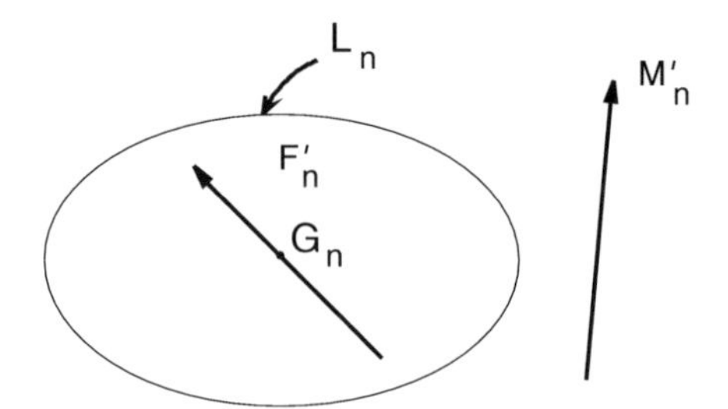

FIGURE 19.17.3
Equivalent constraint force system on link L_n.

(Observe that this is different than our usual notation of $V_{n\ell m}\mathbf{n}_{0m}$ and $\omega_{n\ell m}\mathbf{n}_{0m}$. The change is introduced to simplify the matrix analysis of the following paragraphs.) In scalar form, Eqs. (19.17.5) and (19.17.6) become:

$$V_{m\ell}^{G_n}\dot{\gamma}_\ell = \text{v}_m(t) \quad \text{and} \quad \omega_{m\ell}^{L_n}\dot{\gamma}_\ell = \Omega_m(t) \quad m = 1,2,3 \tag{19.17.8}$$

where $\text{v}_m(t)$ and $\Omega_m(t)$, the $\mathbf{n}_{0m}$ components of $\mathbf{v}$ and $\mathbf{\Omega}$, are specified functions of time.

The expressions in Eq. (19.17.8) are constraint equations in the form of Eqs. (19.16.3) and (19.17.3). Specifically, the constraint matrix B is a $(6 \times n)$ array whose elements are the partial velocity and partial angular velocity components. That is,

$$B = \begin{bmatrix} V_{11}^{G_n} & V_{12}^{G_n} & V_{13}^{G_n} & \cdots & V_{1n}^{G_n} \\ V_{21}^{G_n} & V_{22}^{G_n} & V_{23}^{G_n} & \cdots & V_{2n}^{G_n} \\ V_{31}^{G_n} & V_{32}^{G_n} & V_{33}^{G_n} & \cdots & V_{3n}^{G_n} \\ \omega_{21}^{L_n} & \omega_{12}^{L_n} & \omega_{13}^{L_n} & \cdots & \omega_{1n}^{L_n} \\ \omega_{21}^{L_n} & \omega_{22}^{L_n} & \omega_{13}^{L_n} & \cdots & \omega_{2n}^{L_n} \\ \omega_{32}^{L_n} & \omega_{32}^{L_n} & \omega_{33}^{L_n} & \cdots & \omega_{3n}^{L_n} \end{bmatrix} \tag{19.17.9}$$

Then, by comparison of Eqs. (19.17.3), (19.17.8), and (19.17.4), we see that the array g is:

$$g(t) = \begin{bmatrix} \text{v}_1(t) \\ \text{v}_2(t) \\ \text{v}_3(t) \\ \Omega_1(t) \\ \Omega_2(t) \\ \Omega_3(t) \end{bmatrix} \tag{19.17.10}$$

Next, let the constraining force system on L_n be represented by an equivalent force system consisting of a force $\mathbf{F}_n'$ passing through G_n together with a couple with torque $\mathbf{M}_n'$ as in Figure 19.17.3. Then, the generalized constraint for force F_ℓ' for $\dot{\gamma}_\ell$ is (see Eq. (11.5.7)):

$$F_\ell' = F_{nm}' V_{m\ell}^{G_n} + M_{nm}' \omega_{m\ell}^{L_n} \quad (\ell = 1, \ldots, n) \tag{19.17.11}$$

where the F'_{nm} and M'_{nm} are the $\mathbf{n}_{0m}$ components of $\mathbf{F}'_n$ and $\mathbf{M}'_n$. Let λ be the (6×1) column array of these components. That is,

$$\lambda = \begin{bmatrix} F'_{n1} \\ F'_{n2} \\ F'_{n3} \\ M'_{n1} \\ M'_{n2} \\ M'_{n3} \end{bmatrix} \tag{19.17.12}$$

Observe from Eq. (19.17.9) that the transpose of B is:

$$B^T = \begin{bmatrix} V_{11}^{G_n} & V_{21}^{G_n} & V_{31}^{G_n} & \omega_{11}^{L_n} & \omega_{21}^{L_n} & \omega_{31}^{L_n} \\ V_{12}^{G_n} & V_{22}^{G_n} & V_{32}^{G_n} & \omega_{12}^{L_n} & \omega_{22}^{L_n} & \omega_{32}^{L_n} \\ V_{13}^{G_n} & V_{23}^{G_n} & V_{33}^{G_n} & \omega_{13}^{L_n} & \omega_{23}^{L_n} & \omega_{33}^{L_n} \\ \vdots & \vdots & \vdots & \vdots & \vdots & \vdots \\ V_{1n}^{G_n} & V_{2n}^{G_n} & V_{3n}^{G_n} & \omega_{1n}^{L_n} & \omega_{2n}^{L_n} & \omega_{3n}^{L_n} \end{bmatrix} \tag{19.17.13}$$

Then, by inspection of Eq. (19.17.11) and by comparison with Eqs. (19.17.12) and (19.17.13), we see that:

$$F' = B^T\lambda \tag{19.17.14}$$

thus establishing Eq. (19.17.4).

In view of Eq. (19.17.14) the governing equations, Eqs. (19.17.1) and (19.17.3), may be written as:

$$A\ddot{\gamma} = f + B^T\lambda \tag{19.17.15}$$

and

$$B\dot{\gamma} = g \tag{19.17.16}$$

We will consider solutions to these equations in the following section.

19.18 Governing Equation Reduction and Solution: Use of Orthogonal Complement Arrays

Recall that Eqs. (19.17.15) and (19.17.16) are the governing equations for an n-link robot whose nth link is driven with a prescribed motion. The joints are moment free, and no gravitational or externally applied forces are present, other than the constraint forces needed to drive the nth link through its desired motion. Also, the base is fixed. Under these specialized conditions, Eq. (19.17.15) is equivalent to n scalar equations involving the n articulation angles γ_i ($i = 1,\ldots, n$) and the six constraint (or *driving*) force and moment components. Correspondingly, Eq. (19.17.16) is equivalent to six scalar constraint equations involving the articulation angles. Therefore, taken together, Eqs. (19.17.15) and (19.17.16) are equivalent to $(n + 6)$ scalar equations for the $(n + 6)$ variables consisting of the n articulation angles and the six driving force and moment components.

Equations (19.17.15) and (19.17.16) are nonlinear, coupled differential–algebraic equations. As such, we cannot expect to obtain an analytical solution. Instead, we are left to seek numerical solutions. Generally, we are more interested in knowing the system motion than in knowing the constraint force and moment components. Hence, the question arising is can we reduce and thus simplify the equations by eliminating the λ array? To answer this question, observe that λ may be eliminated if we could find an $(n - 6) \times n$ array (say, C^T), with rank $(n - 6)$ such that $C^T B^T$ is zero. That is, λ may be eliminated from Eq. (19.17.15) by premultiplying by C^T if $C^T B^T$ is zero. Equation (19.17.15) then becomes:

$$C^T A \ddot{\gamma} = C^T f \tag{19.18.1}$$

Equation (19.17.16) may then be differentiated and cast into the same format as:

$$B\ddot{\gamma} = \dot{g} - \dot{B}\dot{\gamma} \tag{19.18.2}$$

Equations (19.18.1) and (19.18.2) may be combined into a single matrix equation of the form:

$$Q\ddot{\gamma} = h \tag{19.18.3}$$

where Q is an $(n \times n)$ array which has the partitioned form:

$$Q = \begin{bmatrix} C^T A \\ \hline B \end{bmatrix} \tag{19.18.4}$$

and h is an $(n \times 1)$ column array with the partitioned form:

$$h = \begin{bmatrix} C^T f \\ \hline \dot{g} - \dot{B}\dot{\gamma} \end{bmatrix} \tag{19.18.5}$$

Then, with Q being nonsingular, Eq. (19.18.3) may be solved for the $\ddot{\gamma}$ as:

$$\ddot{\gamma} = Q^{-1} h \tag{19.18.6}$$

Equation (19.18.6) is equivalent to n scalar differential equations for the n articulation angles. They are in a format that is ideally suited for numerical integration using any of a number of numerical integration algorithms (*solvers*).*

* A discussion of such software is beyond our scope here, but interested readers may want to consider numerical analysis writings, such as References 19.2 to 19.5, for information about such software.

The key to developing Eq. (19.18.6) is in obtaining the $(n - 6) \times n$ array C^T with rank $(n - 6)$ such that:

$$C^T B^T = 0 \tag{19.18.7}$$

Taking the transpose of Eq. (19.18.7), we have:

$$BC = 0 \tag{19.18.8}$$

In this context, C is often called an *orthogonal complement* of B.

One way of obtaining an orthogonal complement of B is to use a rather ingenious zero-eigenvalues theorem as developed by Walton and Steeves [19.6]. In this procedure, the $(6 \times n)$ array B is premultiplied by its transpose B^T, forming an $(n \times n)$ symmetric array B^TB of rank 6. Then, in the eigenvalue problem:

$$B^T Bx = \mu x \tag{19.18.9}$$

there are $(n - 6)$ zero eigenvalues μ. Let x_i $(i = 1,\ldots, n - 6)$ be the eigenvectors x associated with the zero eigenvalues. Next, let C be an $n \times (n - 6)$ array whose columns are the $(n - 6)$ eigenvectors x_i, associated with the zero eigenvalues. Then, with the eigenvalues μ being zero, we have:

$$B^T BC = 0 \tag{19.18.10}$$

Finally, by premultiplying by C^T we have:

$$\begin{gathered} C^T B^T BC = 0 \quad \text{or} \quad (BC)^T BC = 0 \quad \text{or} \quad (BC)^2 = 0 \\ \text{or} \quad BC = 0 \quad \text{and} \quad C^T B^T = 0 \end{gathered} \tag{19.18.11}$$

Thus, C is the desired orthogonal complement array. Once C is known, the arrays Q and h may be immediately calculated from Eqs. (19.18.4) and (19.18.5), and the articulation angles γ_i may then be obtained by integrating Eq. (19.18.6).

Observe that Eq. (19.18.3) represents a reduced form of the governing equations obtained by eliminating the constraint force and moment component array λ from Eq. (19.17.15). If one is interested in knowing these constraint force and moment components, they may be obtained by back substitution into Eq. (19.17.15). Specifically, Eq. (19.17.15) may be solved for λ as:

$$\lambda = (BB^T)^{-1}(A\ddot{\gamma} - f) \tag{19.18.12}$$

19.19 Discussion, Concluding Remarks, and Closure

The foregoing analysis, although somewhat specialized by our simplifying assumptions, is nevertheless typical of the kinds of analyses used when studying the dynamics of large

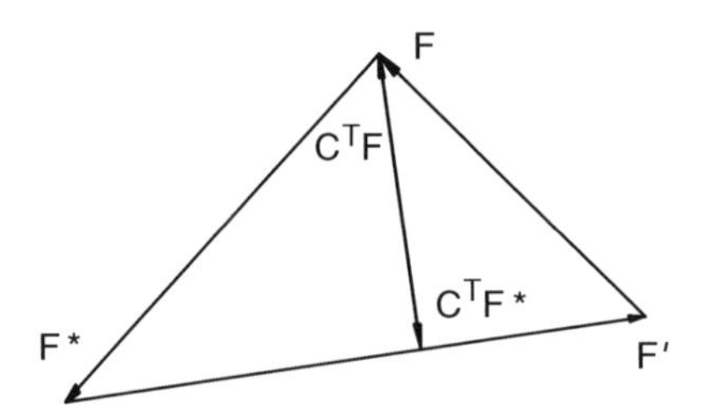

FIGURE 19.19.1
Generalized force triangle.

systems. For an actual robot, the analysis would of course be more detailed, but the basic principles would be the same. In robot hardware design, analysts are concerned with end effector mechanisms, increased mobility, inverse kinematic problems (determining the articulation angles to obtain a desired end effector movement), multiple arm systems, control problems, location sensing, and effects of flexibility and compliance.

The immediate foregoing analysis illustrates a procedure for working with constraint equations. To discuss this further, consider a matrix form of Kane's equations for a specified-motion constrained system (a so-called *acatastatic system* [19.7]):

$$F + F^{*} + F' = 0 \tag{19.19.1}$$

where F is a column array of generalized applied forces, F^* is the corresponding array of generalized inertia forces, and F' is an array of constraint forces. The constraint force array arises as a result of the specified motion of the end effector. The constraint force array is actually a specialized applied force array due to the forces needed to obtain the specified end effector motion. In terms of these force components we have seen that F' has the form:

$$F' = B^{T}\lambda \tag{19.19.2}$$

where B is the array of constraint equation coefficients and λ is the array of constraint force and moment components.

In many cases, the values of the constraint force and moment components are not of interest. In those cases, their array λ may be eliminated from the analysis by premultiplying by the transpose of an orthogonal complement C of B. That is, because BC and C^TB^T are zero, Eq. (19.19.1) may be reduced to the form:

$$C^{T}F + C^{T}F^{*} = 0 \tag{19.19.3}$$

Equations (19.19.1) and (19.19.3) may be depicted graphically as in Figure 19.19.1, where F, F^*, and F' form a generalized force triangle representing Eq. (19.19.1). The projection of F and F^* perpendicular to F', in the form of C^TF and C^TF^* as shown, represents Eq. (19.19.3). In this sense, the system may be viewed as being able to move in directions orthogonal to the constraints.

As we saw in the foregoing analysis, once Eq. (19.19.3) is solved we can use the solution with back substitution into Eq. (19.19.1) to obtain the force and moment components of λ, if they are desired.

The references provide a source for more specific information and analysis of robot systems. As with many other current technologies, robotics is rapidly developing. It is beyond our scope, however, to attempt to be more detailed about specific systems than we have been. Our objective has been simply to present an application of our dynamical procedures. Biomechanics represents another and somewhat related application area that we will discuss in the following chapter.

References

19.1. Kane, T. R., *Analytical Elements of Mechanics*, Vol. 1, Academic Press, New York, 1959, p. 128.
19.2. Champion, E. R., Jr., *Numerical Methods for Engineering Applications*, Marcel Dekker, New York, 1993.
19.3. Hornbeck, R. W., *Numerical Methods*, Quantum Publishers, New York, 1975.
19.4. Griffiths, D. V., and Smith, I. M., *Numerical Methods for Engineers*, CRC Press, Boca Raton, FL, 1991.
19.5. Rice, J. R., *Numerical Methods, Software, and Analysis*, McGraw-Hill, New York, 1993.
19.6. Walton, W. C., Jr., and Steeves, E. C., A New Matrix Theorem and Its Application for Establishing Independent Coordinates for Complex Dynamical Systems with Constraints, NASA Technical Report TR-326, 1986.
19.7. Pars, L. A., *A Treatise on Analytical Dynamics*, Ox Box Press, Woodbridge, CT, 1979, p. 24.
19.8. Angeles, J., *Fundamentals of Robotic Mechanical Systems: Theory, Methods, and Algorithms*, Springer-Verlag, New York, 1997.
19.9. Craig, J. J., *Introduction to Robotics: Mechanics and Control*, Addison-Wesley, Reading, MA, 1989.
19.10. Duffy, J., *Analysis of Mechanisms and Robot Manipulators*, Edward Arnold, London, 1980.
19.11. Engelberger, J. F., *Management and Application of Industrial Robots*, American Management Association, New York, 1980.
19.12. Featherstone, R., *Robot Dynamics Algorithms*, Kluwer Academic, Dordrecht, 1987.
19.13. Koivo, A. J., *Fundamentals for Control of Robotic Manipulators*, Wiley, New York, 1989.
19.14. Mason, M. T., and Salisbury, J. K., *Robot Hands and the Mechanics of Manipulation*, MIT Press, Cambridge, MA, 1985.
19.15. Murray, R. M., Li, Z., and Sastry, S. S., *A Mathematical Introduction to Robotic Manipulation*, CRC Press, Boca Raton, FL, 1993.
19.16. Nakamura, Y., *Advanced Robotics: Redundancy and Optimization*, Addison-Wesley, Reading, MA, 1991.
19.17. Paul, R. P., *Robot Manipulators, Mathematics, Programming, and Control*, MIT Press, Cambridge, MA, 1981.
19.18. Spong, M. W., and Vidyasagar, M. M., *Dynamics and Control of Robot Manipulators*, Wiley, New York, 1989.

Problems

Section 19.2 Geometry, Configuration, and Degrees of Freedom

P19.2.1: Refer to the simple planar robots shown in Figures P19.2.1a and b consisting of two- and three-link arms, respectively, and consider the following questions:

1. How many degrees of freedom does each system have?
2. What are the hinge inclination angles for these systems?
3. From an analysis perspective, what are the relative advantages and disadvantages of these systems?
4. From an applications perspective, what are the relative advantages and disadvantages of these systems?

P19.2.2: Consider the simple end effector, or gripper, shown in Figure P19.2.2. If this end effector were attached to the extremities of the systems of Problem P19.2.1, what would

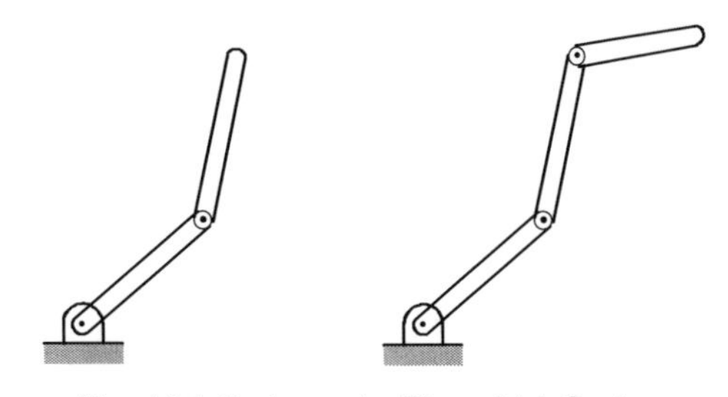

FIGURE P19.2.1
Two- and three-link planar robots.

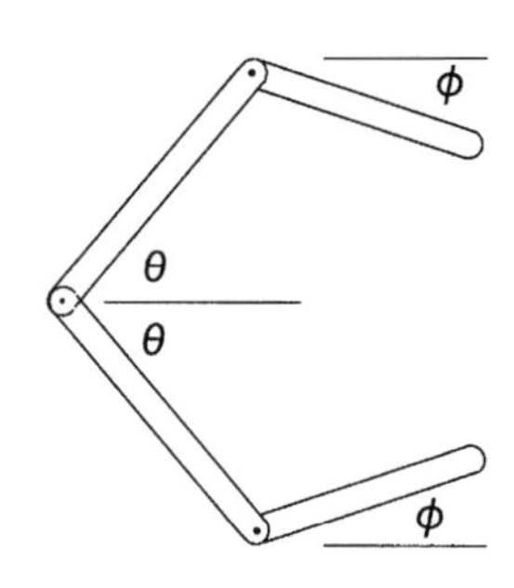

FIGURE P19.2.2
An end effector.

be the resulting degrees of freedom of each system? Would the presence of the end effector affect the relative advantages and disadvantages of each system?

P19.2.3: Consider a robot arm consisting of four identical links each having length 0.5 m (Figure P19.2.3). Let the links be pin-connected to each other with articulation angles γ_i ($i = 1,\ldots, 4$) as defined in Section 19.2. Also, let the pin axes be oriented relative to each other with inclination angles α_i ($i = 1,\ldots, 4$), as defined in Section 19.2. Let the α_i and γ_i have the values:

$$\alpha_1 = \alpha_2 = \alpha_3 = \alpha_4 = 45^\circ$$

$$\gamma_1 = 15^\circ, \quad \gamma_2 = 30^\circ, \quad \gamma_3 = 45^\circ, \quad \gamma_4 = 60^\circ$$

Determine the coordinates of the extremity E of the fourth link relative to an axis system fixed in the base B with origin O_1.

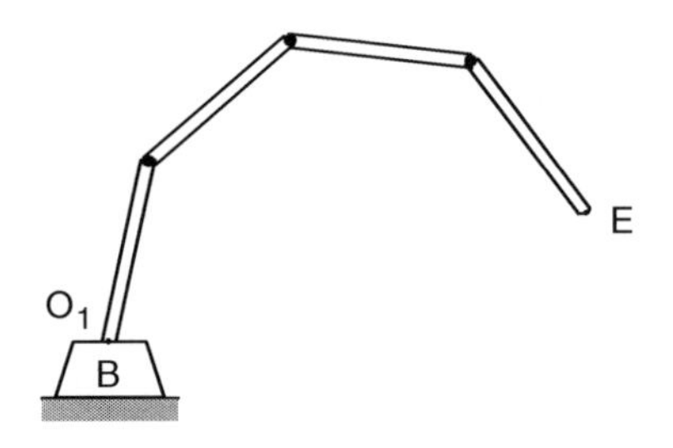

FIGURE P19.2.3
A robot arm consisting of four identical pin-connected links.

P19.2.4: Repeat Problem P19.2.3 if the link lengths are 12 inches long and the inclination and articulation angles are:

$$\alpha_1 = \alpha_2 = \alpha_3 = \alpha_4 = 30^\circ$$

$$\gamma_1 = 60^\circ, \quad \gamma_2 = 45^\circ, \quad \gamma_3 = 30^\circ, \quad \gamma_4 = 15^\circ$$

Section 19.3 Transformation Matrices and Configuration Graphs

P19.3.1: See Problem P19.2.3. Let the robot arm links be numbered outward from the base B. Using the angles listed in the problem statement, determine the transformation matrices, S01, S02, S03, and S04. Assume that the base B is fixed in the inertial reference frame R (that is, S0B = I).

P19.3.2: Repeat Problem P19.3.1 for the orientation angles listed in Problem P19.2.4.

Section 19.4 Angular Velocity of Robot Links

P19.4.1: Consider a robot arm with six links L_k ($k = 1,\ldots, 6$) connected to each other by hinge joints as in Figure 19.4.4 and with a rotating base B as in Figure 19.4.2, as represented in Figure P19.4.1. Let the orientation of the links be defined by the angles α_k ($k = 1,\ldots, 6$) as in Section 19.4. How many degrees of freedom does this system have? Using the notation and unit vector directions of Section 9.4, develop expressions for the transformation matrices S0K (K = 1,…, 6).

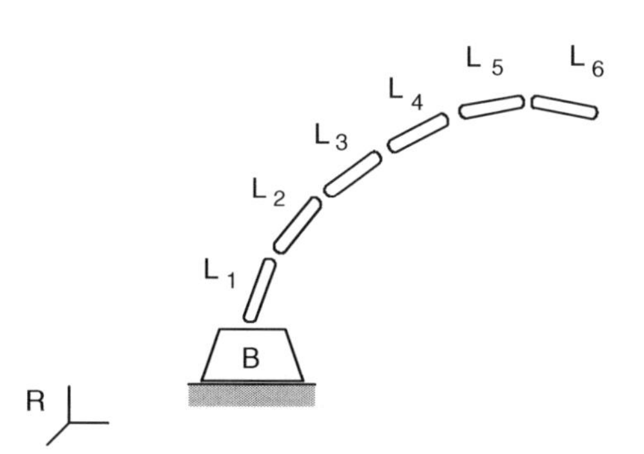

FIGURE P19.4.1
A robot with six links and a rotating base.

P19.4.2: See Problem P19.4.1. Let the orientation angles α_k ($k = 1,\ldots, 6$) of the link axes and the articulation angles α_B and γ_k ($k = 1,\ldots, 6$) and be:

$$\alpha_1 = 0, \quad \alpha_2 = 90°, \quad \alpha_3 = 0, \quad \alpha_4 = -90°, \quad \alpha_5 = 0, \quad \alpha_6 = 90°$$

$$\gamma_1 = 30°, \quad \gamma_2 = 45°, \quad \gamma_3 = 60°, \quad \gamma_4 = 45°, \quad \gamma_5 = 30°, \quad \gamma_6 = 45°$$

$$\alpha_B = 30°$$

Determine the numerical values of the elements of the transformation matrices S0K (K = 1,…, 6).

P19.4.3: See Problems P19.4.1 and P19.4.2. Develop equations analogous to Eq. (19.4.9) for the robot links of the system of Figure P19.4.1 with the angles α_k ($k = 1,\ldots, 6$) of Problem P19.4.2.

P19.4.4: See Problem P19.4.3. Develop more explicit expressions for the link angular velocities if at an instant of interest the articulation angles α_B and γ_k ($k = 1,\ldots, 6$) are:

$$\alpha_B = 30°, \quad \gamma_1 = 30°, \quad \gamma_2 = 45°, \quad \gamma_3 = 60°$$

$$\gamma_4 = 45°, \quad \gamma_5 = 30°, \quad \gamma_6 = 45°$$

P19.4.5: See Problem P19.4.4. Let the articulation angle rates all be equal to 30 rpm. Determine the link angular velocities in terms of unit vectors $\mathbf{n}_{0i}$ fixed in the inertia frame R. Assume the orientation angles α_k ($k = 1,\ldots, 6$) and the articulation angles α_B and γ_k ($k = 1,\ldots, 6$) are the same as those of Problems P19.14.2 and P19.14.4.

P19.4.6: See Problems P19.4.1 and P19.4.5. Suppose all the orientation angles α_k ($k = 1,\ldots, 6$) as well as α_B are zero. We then have a planar robot. Repeat Problems P19.4.1 to P19.4.5 for this planar system.

Section 19.5 Partial Angular Velocities

P19.5.1: See Problems P19.4.1 to P19.4.6. Develop expressions for the partial angular velocity components as in Eq. (19.5.4) for the system of Figure P19.4.1 as described in Problems P19.4.1 to P19.4.6.

Section 19.6 Transformation Matrix Derivative

P19.6.1: See Problems P19.3.1 and P19.3.2. Determine, by direct differentiation, the derivative of the transformation matrices.

P19.6.2: See Problems P19.2.3 and P19.2.4. Determine expressions for the link angular velocities.

P19.6.3: See Problem P19.6.2. Use the results of Problem P19.6.2 to determine the *WK* ($K = 1,\ldots, 4$) as defined by Eq. (19.6.3) for the robot links. Then, use Eq. (19.6.4) to determine the transformation matrix derivatives. Compare the results with those of Problem P19.6.1.

P19.6.4: Repeat the procedure of Problems P19.6.1, P19.6.2, and P19.6.3 for the robot system of Problem P19.4.1.

Section 19.7 Angular Acceleration of Robot Links

P19.7.1: Consider again the robot system of Problem P19.2.3 consisting of four identical pin-connected links and as shown again in Figure P19.7.1. Recall that the hinge inclination angles α_k ($k = 1,\ldots, 4$) are all 45°. At an instant of interest, let the articulation angles γ_k ($k = 1,\ldots, 4$) and their derivatives be:

$$\begin{array}{llll}
\gamma_1 = 15^\circ, & \gamma_2 = 30^\circ, & \gamma_3 = 45^\circ, & \gamma_4 = 60^\circ \\
\dot{\gamma}_1 = \frac{\pi}{2}\ \text{rad/sec}, & \dot{\gamma}_2 = \pi\ \text{rad/sec}, & \dot{\gamma}_3 = \frac{3\pi}{2}\ \text{rad/sec}, & \dot{\gamma}_4 = 2\pi\ \text{rad/sec} \\
\ddot{\gamma}_1 = -1\ \text{rad/sec}^2, & \ddot{\gamma}_2 = 0, & \ddot{\gamma}_3 = 1\ \text{rad/sec}^2, & \ddot{\gamma}_4 = 0
\end{array}$$

Let the base B be, and remain, at rest in an inertial frame R. Using the results of Problems P19.3.1, P19.6.1, and P19.6.3, determine the angular acceleration of the robot links. Express the results in terms of unit vectors fixed in R.

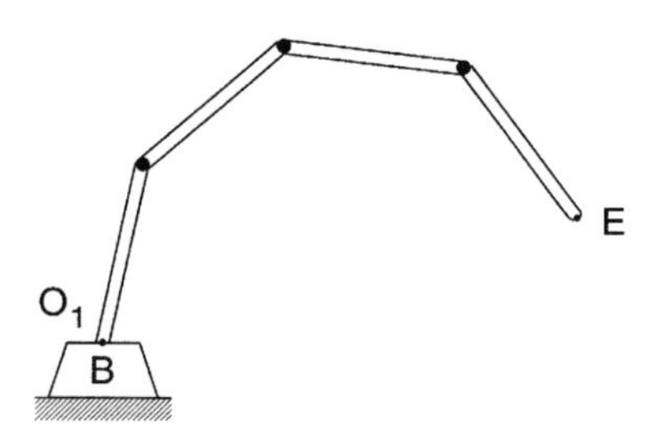

FIGURE P19.7.1
The robot arm of Problem P19.2.3 with four identical pin-connected links.

P19.7.2: Repeat Problem P19.7.1 for the following data:

$$\begin{array}{llll}
\gamma_1 = 60^\circ, & \gamma_2 = 45^\circ, & \gamma_3 = 30^\circ, & \gamma_4 = 15^\circ \\
\dot{\gamma}_1 = 2\pi\ \text{rad/sec}, & \dot{\gamma}_2 = \frac{3\pi}{2}\ \text{rad/sec}, & \dot{\gamma}_3 = \pi\ \text{rad/sec}, & \dot{\gamma}_4 = \frac{\pi}{2}\ \text{rad/sec} \\
\ddot{\gamma}_1 = 0, & \ddot{\gamma}_2 = 1\ \text{rad/sec}^2, & \ddot{\gamma}_3 = 0, & \ddot{\gamma}_4 = -1\ \text{rad/sec}^2
\end{array}$$

P19.7.3: Consider again the robot system of Problem P19.4.1 consisting of six pin-connected links and a rotating base as shown again in Figure P19.7.3. As in Problems P19.4.2 and P19.4.5, let the orientation and articulation angles and their derivatives have the values:

$$\alpha_1 = 0, \quad \alpha_2 = 90°, \quad \alpha_3 = 0, \quad \alpha_4 = -90°, \quad \alpha_5 = 0, \quad \alpha_6 = 90°$$

$$\gamma_1 = 30°, \quad \gamma_2 = 45°, \quad \gamma_3 = 60°, \quad \gamma_4 = 45°, \quad \gamma_5 = 30°, \quad \gamma_6 = 45°$$

$$\alpha_B = 30°$$

$$\dot{\gamma}_1 = \dot{\gamma}_2 = \dot{\gamma}_3 = \dot{\gamma}_4 = \dot{\gamma}_5 = \dot{\gamma}_6 = \dot{\alpha}_B = \pi \text{ rad/sec}^2$$

Let the second derivatives of the articulation angles have the values:

$$\ddot{\gamma}_1 = 1 \text{ rad/sec}^2, \quad \ddot{\gamma}_2 = 0, \quad \ddot{\gamma}_3 = -1 \text{ rad/sec}^2, \quad \ddot{\gamma}_4 = 0$$

$$\ddot{\gamma}_5 = 1 \text{ rad/sec}^2, \quad \ddot{\gamma}_6 = 0, \quad \ddot{\alpha}_B = 0$$

Let the second derivatives of the articulation angles have the values:

$$\ddot{\gamma}_1 = 1 \text{ rad/sec}^2, \quad \ddot{\gamma}_2 = 0, \quad \ddot{\gamma}_3 = -1 \text{ rad/sec}^2, \quad \ddot{\gamma}_4 = 0$$

$$\ddot{\gamma}_5 = 1 \text{ rad/sec}^2, \quad \ddot{\gamma}_6 = 0, \quad \ddot{\alpha}_B = 0$$

Using the results of Problems P19.4.2, P19.4.3, P19.4.5, and P19.6.4, determine the angular accelerations of the links and the base. Express the results in terms of unit vectors fixed in the inertial frame R.

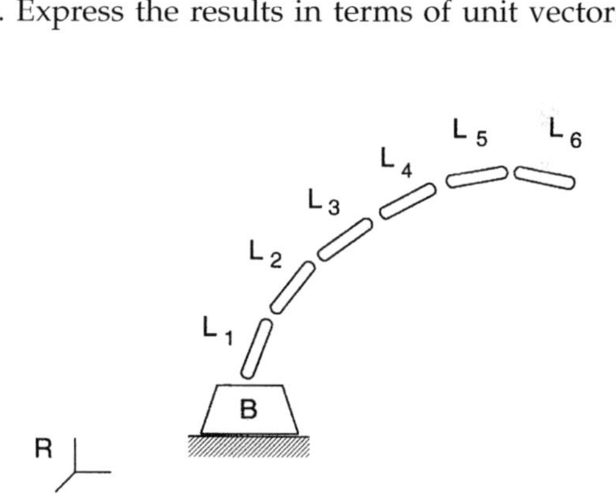

FIGURE P19.7.3
The robot of Problem P19.4.1 with six links and a rotating base.

P19.7.4: See Problem P19.4.6. Suppose that all the orientation angles of the robot arms of Problems P19.7.1 and P19.7.3 are zero so that they are planar robots. Using the data of Problems P19.7.1 and P19.7.3, determine the angular accelerations of the links for each system.

Section 19.8 Joint and Mass Center Position

P19.8.1: See Problems P19.2.3 and P19.3.1. Consider again the robot arm of Problem P19.2.3 as shown in Figure P19.2.3 and as shown again in Figure P19.18.1. Let the length of each link be 12 in., let the origin O_1 of the first link be at the top of a fixed base B as shown, and let O_1 be 6 in. above a fixed point O in an inertial frame R. Let the orientation and articulation angles be as listed in Problem P19.2.3. That is,

$$\alpha_1 = \alpha_2 = \alpha_3 = \alpha_4 = 45°$$

$$\gamma_1 = 15°,\quad \gamma_2 = 30°,\quad \gamma_3 = 45°,\quad \gamma_4 = 60°$$

Determine an expression for the position vector $\mathbf{P}_E$ locating the end effector origin E relative to O.

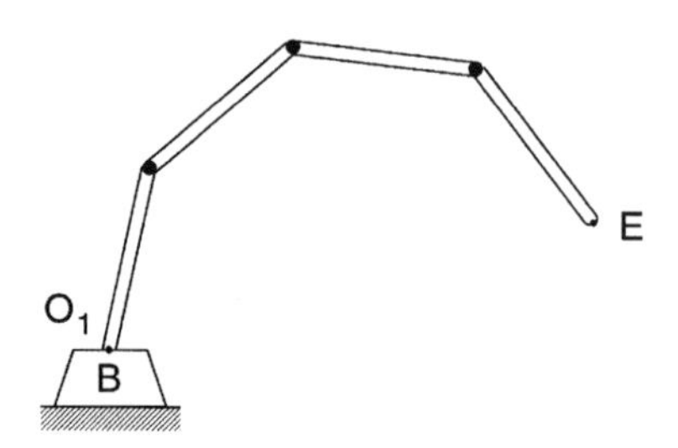

FIGURE P19.8.1
A robot arm with four identical pin-connected links.

P19.8.2: See Problems P19.8.1, P19.2.4, and P19.5.2. Repeat Problem P19.8.1 if the robot link lengths are 0.25 m and if O_1 is 0.125 m above O. Let the orientation and articulation angles be those of Problem P19.2.4. That is,

$$\alpha_1 = \alpha_2 = \alpha_3 = \alpha_4 = 30°$$

$$\gamma_1 = 60°,\quad \gamma_2 = 45°,\quad \gamma_3 = 30°,\quad \gamma_4 = 15°$$

P19.8.3: See Problems P19.4.1, P19.4.2, and 19.4.3. Consider again the robot arms of Problem P19.4.1 as shown in Figure P19.4.1 and as shown again in Figure P19.8.3. Use the following lengths of the robot links L_k: L_1 and L_2, 6 in.; L_3 and L_4, 8 in.; L_5 and L_6, 12 in. Let the origin O_1 of link L_1 be 6 in. above a fixed point O of the inertial frame R. Finally, let the orientation and articulation angles be those listed in Problem P19.4.2. That is,

$$\alpha_1 = 0,\quad \alpha_2 = 90°,\quad \alpha_3 = 0,\quad \alpha_4 = -90°,\quad \alpha_5 = 0,\quad \alpha_6 = 90°$$

$$\gamma_1 = 30°,\quad \gamma_2 = 45°,\quad \gamma_3 = 60°,\quad \gamma_4 = 45°,\quad \gamma_5 = 30°,\quad \gamma_6 = 45°,\quad \alpha_B = 30°$$

Determine an expression for the position vector $\mathbf{P}_E$ locating the origin of an end effector at the extremity of link L_6 relative to O.

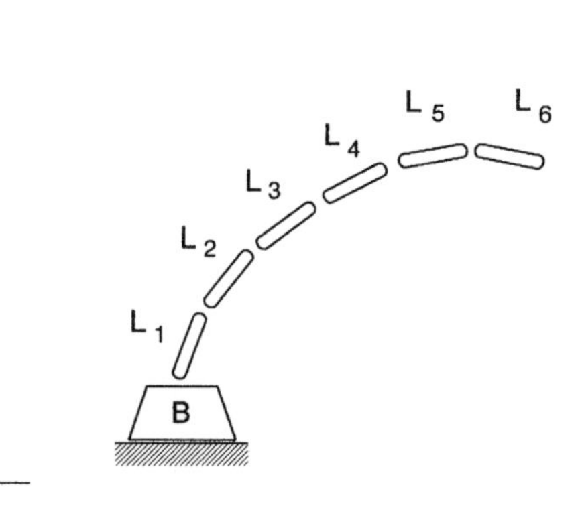

FIGURE P19.8.3
A robot with six pin-connected links and a rotating base B.

P19.8.4: Repeat Problem P19.8.3 if the link lengths are L_1 and L_2, 12 in.; L_3 and L_4, 8 in.; L_5 and L_6, 6 in.

P19.8.5: Repeat Problems P19.8.1 to P19.8.4 if all the articulation angles α_k including α_B are zero. That is, repeat the problems for planar robots.

P19.8.6: See Problems 19.8.1 and 19.8.2. Let the robot links be uniform in cross section so that the mass center is at mid-link. Determine expressions for the mass center position vectors using the data of Problems P19.8.1 and P19.8.2.

P19.8.7: See Problems P19.8.3 and P19.8.4. Let the robot links be uniform so that the mass center is at mid-link. Determine expressions for the mass center position vectors using the data of Problems P19.8.3 and P19.8.4.

P19.8.8: See Problem P19.8.5. Repeat Problems P19.8.6 and P19.8.7 for the robots with a planar configuration.

Section 19.9 Mass Center Velocities

P19.9.1: See Problem P19.8.6. Determine expressions analogous to Eq. (19.9.3) for the robot link mass center velocities for the robot arm of Figure P19.8.1. Let the articulation angle derivatives $\dot{\gamma}_k$ ($k = 1,\ldots, 4$) be arbitrary.

P19.9.2: See Problem P19.8.7. Determine expressions analogous to Eq. (19.9.3) for the robot link mass center velocities for the robot arm of Figure P19.8.3. Let the articulation angle derivatives $\dot{\alpha}_B$ and $\dot{\gamma}_k$ ($k = 1,\ldots, 4$) be arbitrary.

P19.9.3: See Problem P19.8.8. Determine expressions analogous to Eqs. (19.9.3) for the robot link mass center velocities for the robot with planar arm configurations. Let the articulation angle derivatives be arbitrary.

P19.9.4: Repeat Problems P19.9.1, P19.9.2, and P19.9.3 if the articulation angle rates are all 30 rpm.

Section 19.10 Mass Center Partial Velocities

P19.10.1: Use the results of Problems P19.9.1, P19.9.2, and P19.9.3 to identify the partial velocity vectors for the systems studied.

Section 19.11 Mass Center Accelerations

P19.11.1: Using the data and results of Problems P19.9.1, P19.9.2, and P19.9.3, determine expressions for the link mass center acceleration in the inertial frame *R* for the systems studied. Let the articulation angle derivatives and second derivatives be arbitrary.

P19.11.2: Consider again the robot system of Figure P19.11.2 and as studied in Problems P19.2.3, P19.2.4, P19.3.1, P19.3.2, P19.9.1, P19.9.3, and P19.10.1. For the data given in these problems, determine the accelerations of the link mass centers in the inertial frame *R*.

P19.11.3: Consider again the robot system of Figure P19.11.3 and as studied in Problems P19.4.1 to P19.4.6, P19.5.1, P19.6.4, P19.7.3, P19.7.4, P19.8.3, P19.8.4, P19.8.5, P19.8.7, P19.8.8, P19.9.2, P19.9.3, and P19.10.1. For the data given in these problems, determine the acceleration of the link mass centers in the inertial frame *R*.

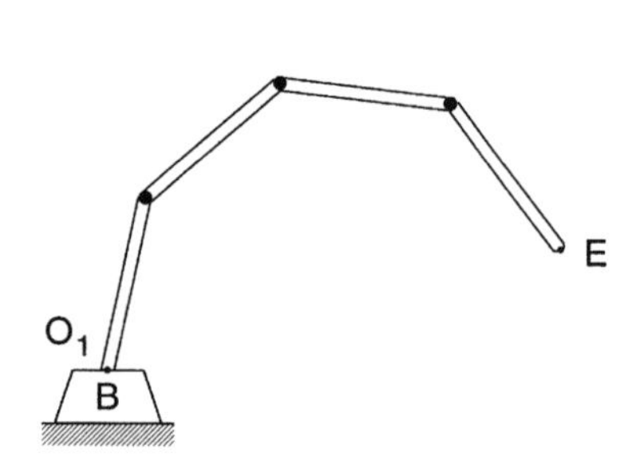

FIGURE P19.11.2
A robot arm with four identical, uniform, pin-connected links.

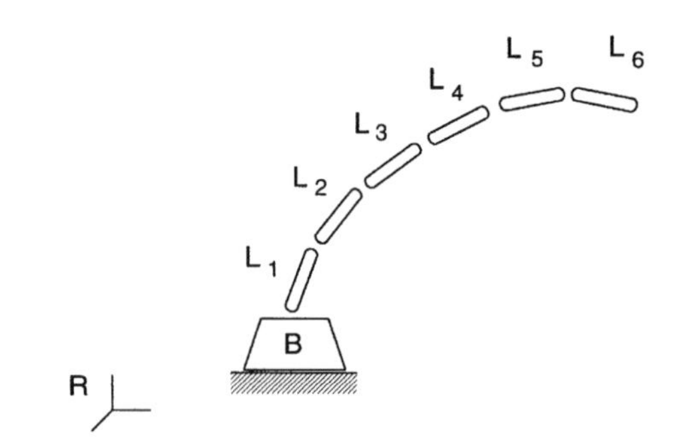

FIGURE P19.11.3
A robot with six uniform pin-connected links and a rotating base.

Section 19.12 End Effector Kinematics

P19.12.1: Design an end effector able to firmly grasp a sphere.

P19.12.2: Let an end effector consist of three two-link fingers spaced 120° from one another as shown in Figure P19.12.2. Let the finger links have equal lengths, and let the connecting and supporting joints be revolute, or hinge, joints with axes oriented at 120° relative to one another from finger to finger. Develop the kinematics of this system.

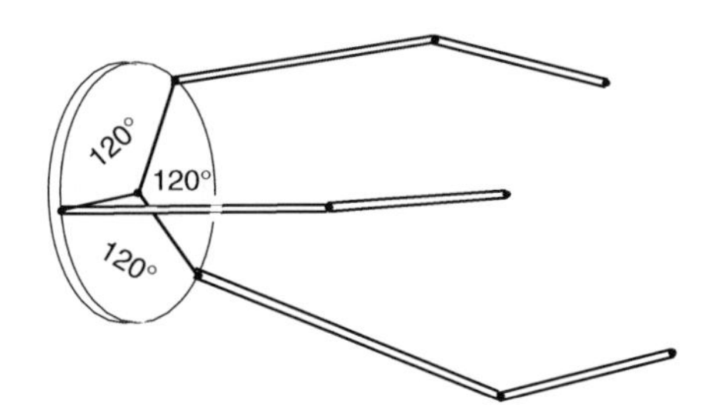

FIGURE P19.12.2
A three-fingered end effector.

Section 19.13 Kinetics: Applied (Active) Forces

P19.13.1: Illustrate the use of Eqs. (19.13.4), (19.13.10), and (19.13.11) with the relatively simple three-link planar robot arm shown in Figure P19.13.1. Specifically, let the arm consist of three pin-connected links L_1, L_2, and L_3 which are rotated relative to their adjacent lower numbered links by joint located actuators exerting torques T_1, T_2, and T_3, as shown. Let the links have masses m_1, m_2, and m_3 and lengths ℓ_1, ℓ_2, and ℓ_3. Let there be an end effector E with mass M at the extremity of L_3. Let **F** be an externally applied force on E with horizontal and vertical components F_x and F_g. Finally, let β_1, β_2, and β_3 be orientation angles measured between the links as shown. Determine the generalized applied forces on the system using β_1, β_2, and β_3 as generalized coordinates.

Section 19.14 Kinetics: Passive (Inertia) Forces

P19.14.1: Using Eqs. (19.4.1) to (19.4.5), verify Eqs. (19.14.6) to (19.14.9).

P19.14.2: Consider again the simple three-link robot arm of Problem P19.13.1. Using the same data as in Problem P19.13.1 and using β_1, β_2, and β_3 as generalized coordinates, determine the generalized inertia forces on the system.

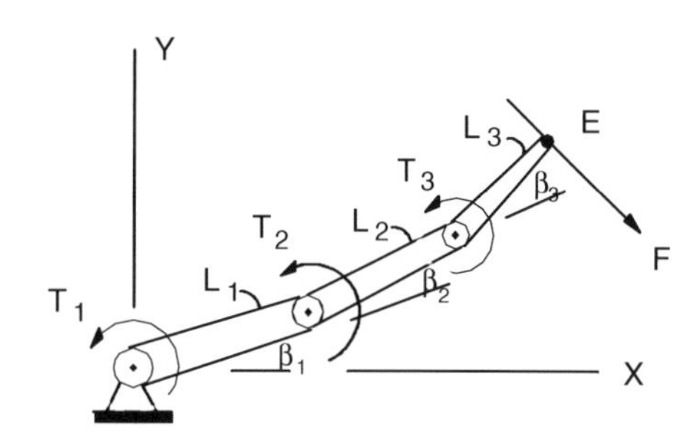

FIGURE P19.13.1
A three-link planar joint actuated robot arm.

Section 19.15 Dynamics: Equations of Motion

P19.15.1: Using Eqs. (19.13.11), (19.14.8), and (19.15.1), verify Eqs. (19.15.2) through (19.15.6).

P19.5.2: See Problems P19.13.1 and P19.14.2. Determine the equations of motion of the simple three-link robot shown in Figure P19.13.1. (Use the same data and parameters as in Problems P19.13.1 and P19.14.2.)

Sections 19.16 Redundant Robots

P19.16.1: Consider the system of three identical pin-connected rods moving in a horizontal plane as represented in Figure P19.16.1. Let the rods each have length ℓ and mass m, and let the connecting pins be frictionless. Let the orientation of the rods in the plane be defined by the relative angles γ_1, γ_2, and γ_3 as shown. Finally, let the movement of the extremity E be specified. Specifically, let the plane of movement of the system be the X–Y plane as shown and let the coordinates (x, y) of E be specified functions of time as:

$$x = \phi(t) \quad \text{and} \quad y = \psi(t)$$

Develop the governing equations for this system. That is, determine the constraint equations as in Eq. (19.16.3) and the dynamic equations as in Eq. (19.16.4).

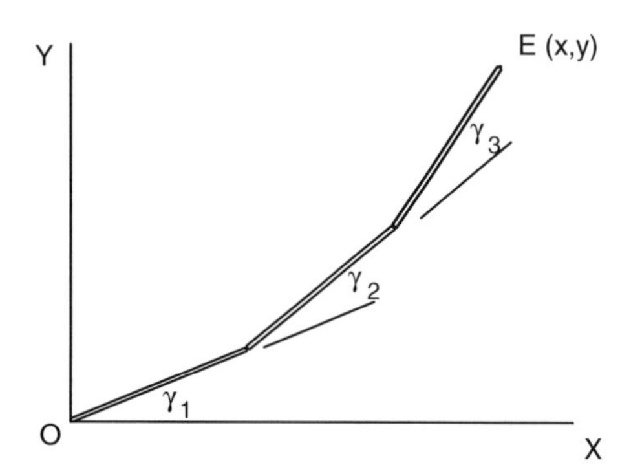

FIGURE P19.16.1
A three-link system with prescribed motion of the extremity.

P19.16.2: See Problem 19.16.1. Repeat Problem P19.16.1 if E is required to move on a circle of radius r with center C (x_O, y_O) where r, x_O, and y_O are compatible with the geometry of the system.

P19.16.3: See Problems P19.16.1 and P19.16.2. Repeat Problems P19.16.1 and P19.16.2 if the orientation of the rods is defined by the absolute angles θ_1, θ_2, and θ_3 as in Figure P19.16.3.

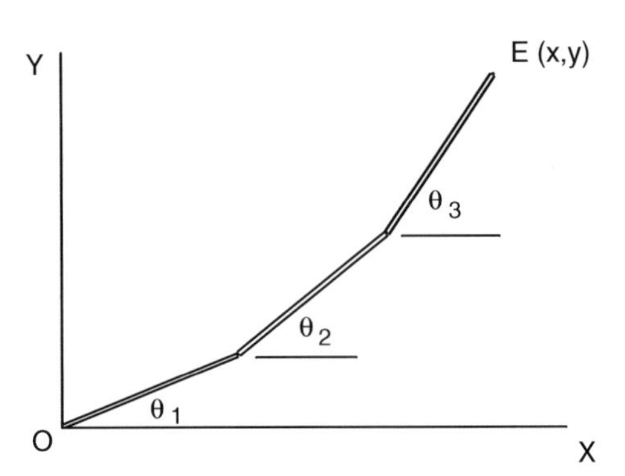

FIGURE P19.16.3
A three-link system with prescribed extremity motion described by absolute orientation angles.

Section 19.17 Constraint Equations and Constraint Forces

P19.17.1 See Problems P19.16.1, P19.16.2, and P19.16.3. Use Eq. (19.17.4) to determine the generalized constraint array F'_ℓ. Identify the components of the λ array.

Section 19.18 Governing Equation Reduction and Solution: Use of Orthogonal Complement Arrays

P19.18.1: See Problems P19.16.1, P19.16.2, P19.16.3, and P19.17.1. Determine an orthogonal complement array C of B.

P19.18.2: See Problem P19.18.1. Use Eq. (19.18.12) to determine an expression for the components of the constraint force array λ.

20

Application with Biosystems, Human Body Dynamics

20.1 Introduction

In this final chapter, we apply the foregoing procedures with biosystems, principally the human body. Interest in this application is growing faster than for any other area of dynamics analysis. This interest is stimulated by unanswered questions about human body dynamics arising in regard to the workplace, accidents, sport activities, space exploration, and routine daily activities. The principal areas of interest are performance enhancement, understanding causes of injury, and rehabilitation from injury and illness.

As with robotics, the application of dynamics procedures with biosystems is more extensive than can be covered, or even summarized, in a single chapter. Hence, our purpose is to simply introduce the application, to discuss specific problems, and to propose methods of analysis. Readers interested in more in-depth analyses may want to consult the references at the end of the chapter.

Unlike a robot, a biosystem — specifically, the human body — is more complex and less well defined. The human form is not composed of simple geometrical shapes, and, even if the body is studied in small parts such as with arms, hands, or fingers, the analysis is not simple. Indeed, even the location of the mass center of elemental parts is imprecise, and comprehensive analyses of joint kinematics are extremely difficult.

These difficulties, while apparently formidable, are nevertheless intriguing and challenging to dynamics analysts. As such they provide motivation for the development of modeling and analysis procedures. As with other mechanical systems, the objective is to find simple models that are sufficiently representative of the physical system to provide useful information.

To this end, Figure 20.1.1 shows a typical whole-body or gross-motion model of the human body. It consists of a series of bodies representing the main parts of a human frame. These bodies have relatively simple geometric shapes: ellipsoids, elliptical cylinders, and frustums of elliptical cones. They are assembled in the form of the human frame and are connected by pin and spherical joints simulating the human joints. Springs and dampers are used to represent the restraining and articulation effects of the ligaments, muscles, and other connective tissues.

This model is well suited for studying overall gross motion of the human body as might occur in the workplace, in sport activities, and due to accidents or injuries. The model is also suitable for health and fitness studies as in analyses of exercise maneuvers and in the design of rehabilitative devices such as wheelchairs, crutches, and braces. We will discuss some of these applications in the sequel, but, as before, our focus will be upon dynamic analysis and, specifically, the methods for obtaining the governing dynamical equations.

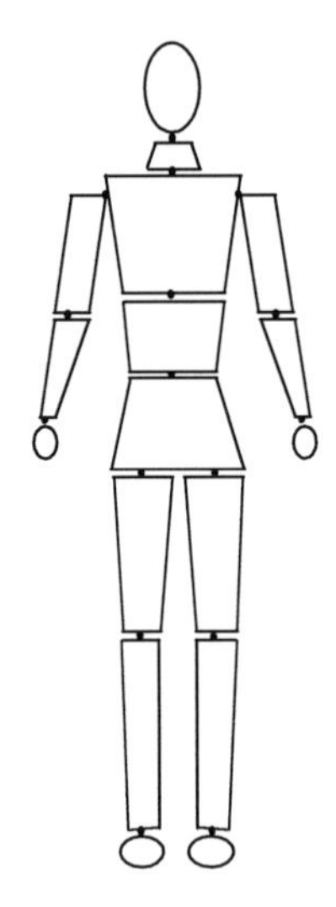

FIGURE 20.1.1
A gross-motion human body model.

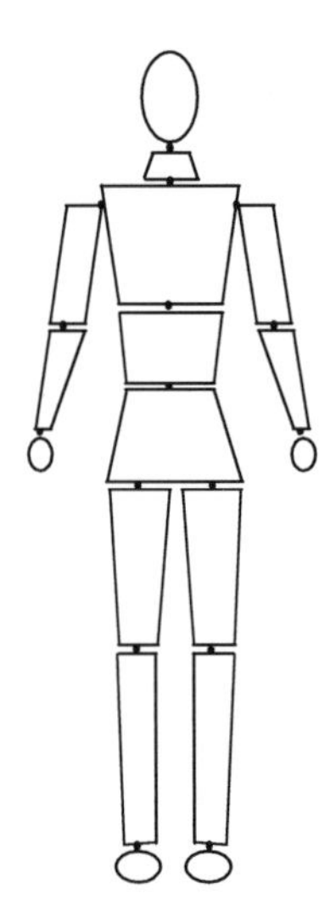

FIGURE 20.2.1
A global representation of the human frame.

20.2 Human Body Modeling

The model of Figure 20.1.1 and shown again in Figure 20.2.1 is a *global* or gross representation of the human frame. The segments represent the major human links. They are connected with pin and spherical joints. The actuating muscles and ligaments are represented by springs and dampers. The segments themselves are in the form of elementary geometrical shapes such as ellipsoids, elliptical cylinders, cones, and frustums of elliptical cones. As such, the modeling system is a rather simplified and idealized representation of the human body. It is best used for gross-motion simulation. It is not particularly useful for studying the movement of individual body parts such as the hands, the feet, or even the head/neck system. Also, a close examination of the human joints shows that they are not pins or spherical joints. Instead they are somewhat more complex, having both translation and rotation between adjoining bodies. Therefore, if we are interested in a more detailed representation of the human than that of Figure 20.2.2, we will need more detailed models.

Figure 20.2.2 shows a more detailed model of the head and neck system. Here, the individual bodies represent the vertebrae, the upper torso, and the head. In head and neck studies, the effects of the discs, muscles, and ligaments connecting and separating the vertebrae are of particular interest. These may be effectively modeled by nonlinear springs and dampers. For example, suppose θ represents a relative rotation of a pair of adjoining vertebrae. Then, the moment exerted by the discs, muscles, and ligaments on the vertebrae may be represented by an equivalent moment M of the form [20.10, 20.14, 20.16]:

$$\begin{aligned}
&M = 0 \quad \text{for} \quad \theta_{min} \le \theta \le \theta_{max} \\
&M = -k_1\theta - c_1(\theta - \theta_{max})\dot{\theta} \quad \text{for} \quad \theta > \theta_{max} \quad \text{and} \quad \dot{\theta} > 0 \\
&M = -k_2\theta - c_2(\theta - \theta_{min})\dot{\theta} \quad \text{for} \quad \theta < \theta_{min} \quad \text{and} \quad \dot{\theta} < 0 \\
&M = 0 \quad \text{for} \quad \theta > \theta_{max} \quad \text{and} \quad \dot{\theta} < 0 \\
&M = 0 \quad \text{for} \quad \theta < \theta_{min} \quad \text{and} \quad \dot{\theta} > 0
\end{aligned} \tag{20.2.1}$$

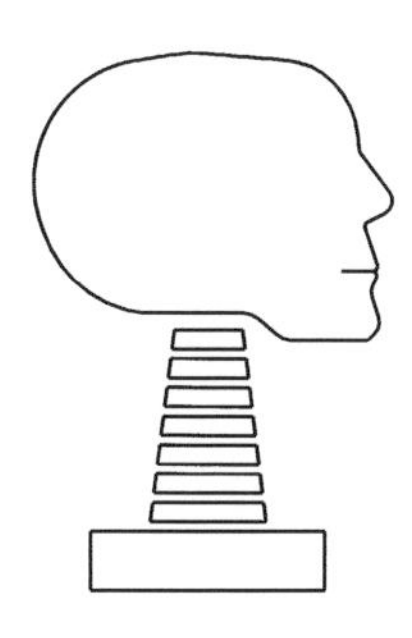
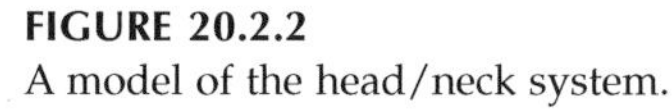

FIGURE 20.2.2
A model of the head/neck system.

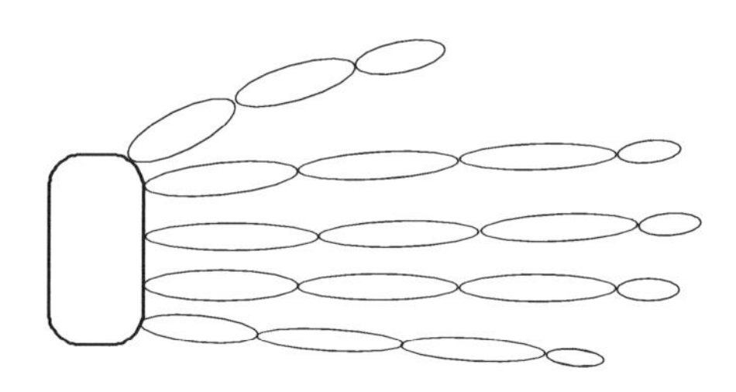

FIGURE 20.2.3
A hand model.

where θ_{min} and θ_{max} are minimum and maximum rotations, respectively, that establish limits of normal movement before pain is experienced and where k_1, k_2, c_1, and c_2 are constants.

Observe in Eq. (20.2.1) that the moment is nonlinearly dependent upon the rate of movement beyond the limiting values θ_{min} and θ_{max}. Observe further that the "restoring" aspect of M is zero when the vertebrae are returning to their normal movement range. This nonlinearity is based upon empirical data [20.10, 20.14, 20.16, 20.18, 20.19]. Finally, if our interest is in hand movement we might use a model as in Figure 20.2.3.

In many cases of interest, the models of Figures 20.2.1, 20.2.2, and 20.2.3 are useful in providing information about both the movement and forces experienced during various activities. With smaller bodies such as the fingers and vertebrae, the applied forces, such as the muscle and ligament forces, greatly exceed the inertia forces. Also, in many instances such as hand modeling, interest in the movement (that is, the kinematics) is greater than interest in the forces.

A question arising in the development of these models is what are the values of the physical and geometrical parameters (that is, the masses, the inertia matrix components, the sizes, and the movement limits) for the various members of the models? The somewhat unsatisfying answer is these values depend upon the particular individual and may vary considerably. Banks of data, however, are available for individuals of varying sizes. References 20.9, 20.20, and 20.21 are good sources for such data.

In the sequel for simplicity, but without loss of generality, we will focus our discussion upon the whole-body model of Figure 20.2.1. As with robotic systems, we will initially consider the kinematics, kinetics, and dynamics. We will then briefly discuss applications.

20.3 A Whole-Body Model: Preliminary Considerations

Consider again the model of Figure 20.2.1 and as redrawn here in Figure 20.3.1. Let the individual bodies of the model be numbered as shown, consistent with the procedures used with lower body arrays (see Section 18.2). As before, let R be an inertial reference frame in which the system moves. Table 20.3.1 then provides a listing and labeling of the various links and segments of the model. By inspection of Figure 20.3.1, we then immediately obtain the lower body array $L(K)$ as follows:

K:	0	1	2	3	4	5	6	7	8	9	10	11	12	13	14	15	16	17
L(K):	0	0	1	2	3	4	5	3	7	3	9	10	1	12	13	1	15	16

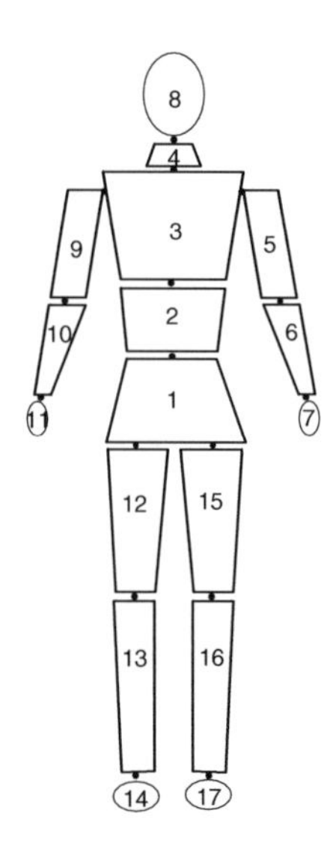

FIGURE 20.3.1
A numbered whole-body model.

TABLE 20.3.1

A Listing and Labeling of the Segments of Figure 20.3.1

Body Number	Label	Name
0	R	Inertial reference frame
1	B_1	Lower torso/pelvis
2	B_2	Mid-torso/abdomen
3	B_3	Upper torso/chest
4	B_4	Upper left arm
5	B_5	Lower left arm
6	B_6	Left hand
7	B_7	Neck
8	B_8	Head
9	B_9	Upper right arm
10	B_{10}	Lower right arm
11	B_{11}	Right hand
12	B_{12}	Upper right leg/right thigh
13	B_{13}	Lower right leg
14	B_{14}	Right foot
15	B_{15}	Upper left leg/left thigh
16	B_{16}	Lower left leg
17	B_{17}	Left foot

By following the procedures of Section 18.2, we then obtain the higher-order lower body arrays as listed in Table 20.3.2.

Observe in the lower body array $L(K)$ that the repeated numbers are 1 and 3, designating the branching bodies. The missing numbers are 6, 8, 11, 14, and 17, designating extremities. The numbers appearing only once are 2, 4, 5, 7, 9, 10, 12, 13, 15, and 16, designating intermediate bodies. Table 20.3.3 provides a summary listing of these.

Consider two typical adjoining bodies of the system such as B_j and B_k as in Figure 20.3.2. Let B_j be the lower numbered body. As in our previous analyses, let the bodies be connected by a spherical joint. Let $\mathbf{n}_{ji}$ and $\mathbf{n}_{ki}$ ($i = 1, 2, 3$) be mutually perpendicular unit vectors fixed in B_j and B_k as shown. Then, as before, the orientation of B_k relative to B_j may be defined in terms of a transformation matrix (or direction cosine matrix) SJK whose elements SJK_{min} are:

$$SJK_{mn} = \mathbf{n}_{jm} \cdot \mathbf{n}_{kn} \tag{20.3.1}$$

Recall also from our previous analyses that we can describe the orientation of B_k relative to B_j in terms of orientation angles such as dextral orientation angles (or Bryan angles) as follows: let the unit vectors $\mathbf{n}_{ji}$ and $\mathbf{n}_{ki}$ be mutually aligned — that is, with the $\mathbf{n}_{ki}$ parallel to the $\mathbf{n}_{ji}$ (i = 1, 2, 3), respectively. Then, B_k may be brought into a general orientation relative to B_j by three successive rotations of B_k about axes parallel to $\mathbf{n}_{k1}$, $\mathbf{n}_{k2}$, and $\mathbf{n}_{k3}$ through angles α_k, β_k, and γ_k, which are positive when the rotation is dextral (or right-handed). When this is done, the transformation matrix SJK has the form (see Eq. (18.3.4)):

$$SJK = \begin{bmatrix} c\beta_k c\gamma_k & -c\beta_k s\gamma_k & s\beta_k \\ (c\alpha_k s\gamma_k + s\alpha_k s\beta_k c\gamma_k) & (c\alpha_k c\gamma_k - s\alpha_k s\beta_k s\gamma_k) & -s\alpha_k c\beta_k \\ (s\alpha_k s\gamma_k - c\alpha_k s\beta_k c\gamma_k) & (s\alpha_k c\gamma_k + c\alpha_k s\beta_k s\gamma_k) & c\alpha_k c\beta_k \end{bmatrix} \tag{20.3.2}$$

where as before s and c are abbreviations for sine and cosine.

Recall from Section 18.5 that the use of orientation angles, although conceptually simple, allows singularities to occur under certain orientations of the bodies. Specifically, for the

TABLE 20.3.2

Higher-Order Lower Body Arrays for the Model of Figure 20.3.1

$L^0(K)$	0	1	2	3	4	5	6	7	8	9	10	11	12	13	14	15	16	17
$L^1(K)$	0	0	1	2	3	4	5	3	7	3	9	10	1	12	13	1	15	16
$L^2(K)$	0	0	0	1	2	3	4	2	3	2	3	9	0	1	12	0	1	15
$L^3(K)$	0	0	0	0	1	2	3	1	2	1	2	3	0	0	1	0	0	1
$L^4(K)$	0	0	0	0	0	1	2	0	1	0	1	2	0	0	0	0	0	0
$L^5(K)$	0	0	0	0	0	0	1	0	0	0	0	1	0	0	0	0	0	0
$L^6(K)$	0	0	0	0	0	0	0	0	0	0	0	0	0	0	0	0	0	0

TABLE 20.3.3

Branching Bodies, Intermediate Bodies and Extremities for the Model of Figure 20.3.1

Branching Bodies	Intermediate Bodies	Extremities
Lower torso (B_1)	Mid-torso (B_2)	Left hand (B_6)
Upper torso (B_3)	Upper left arm (B_4)	Head (B_8)
	Lower left arm (B_5)	Right hand (B_{11})
	Neck (B_2)	Right foot (B_{14})
	Upper right arm (B_9)	Left foot (B_{17})
	Lower right arm (B_{10})	
	Upper right leg (B_{12})	
	Lower right leg (B_{13})	
	Upper left leg (B_{15})	
	Lower left leg (B_{16})	

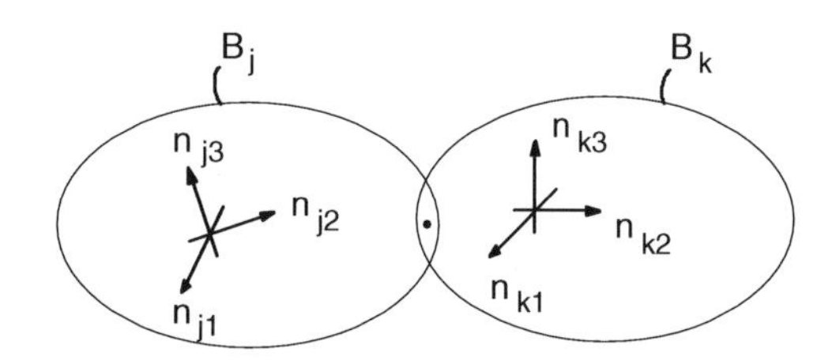

FIGURE 20.3.2
Two typical adjoining bodies.

Bryan angles, there is a singularity where β_k is either 90° or 270°. From a computational (or numerical analysis) viewpoint, it is necessary to avoid these singularities if the range of motion of the bodies allows them to occur. As we saw in Section 18.5, the singularities may be avoided through use of Euler parameters, where B_k is envisioned to be brought into a general orientation relative to B_j by a rotation about an axis fixed in both B_j and B_k. Then, if $\boldsymbol{\lambda}_k$ is a unit vector parallel to this axis and if θ_k is the rotation angle, the Euler parameters may be defined as:

$$\begin{aligned}
\varepsilon_{k1} &= \lambda_{k1}\sin(\theta_k/2) \\
\varepsilon_{k2} &= \lambda_{k2}\sin(\theta_k/2) \\
\varepsilon_{k3} &= \lambda_{k3}\sin(\theta_k/2) \\
\varepsilon_{k4} &= \cos(\theta_k/2)
\end{aligned} \tag{20.3.3}$$

where the λ_{ki} (i = 1, 2, 3) are the $\mathbf{n}_{ji}$ components of $\boldsymbol{\lambda}_k$. The transformation matrix SJK may then be written as (see Eq. (18.5.14)):

$$SJK = \begin{bmatrix} \left(\varepsilon_{k1}^2 - \varepsilon_{k2}^2 - \varepsilon_{k3}^2 + \varepsilon_{k4}^2\right) & 2\left(\varepsilon_{k1}\varepsilon_{k2} - \varepsilon_{k3}\varepsilon_{k4}\right) & 2\left(\varepsilon_{k1}\varepsilon_{k3} + \varepsilon_{k2}\varepsilon_{k4}\right) \\ 2\left(\varepsilon_{k1}\varepsilon_{k2} + \varepsilon_{k3}\varepsilon_{k4}\right) & \left(-\varepsilon_{k1}^2 + \varepsilon_{k2}^2 - \varepsilon_{k3}^2 + \varepsilon_{k4}^2\right) & 2\left(\varepsilon_{k2}\varepsilon_{k3} - \varepsilon_{k1}\varepsilon_{k4}\right) \\ 2\left(\varepsilon_{k1}\varepsilon_{k3} - \varepsilon_{k2}\varepsilon_{k4}\right) & 2\left(\varepsilon_{k2}\varepsilon_{k3} + \varepsilon_{k1}\varepsilon_{k4}\right) & \left(-\varepsilon_{k1}^2 - \varepsilon_{k2}^2 + \varepsilon_{k3}^2 + \varepsilon_{k4}^2\right) \end{bmatrix} \tag{20.3.4}$$

Observe in these two methods for describing the relative orientation of adjoining bodies that the orientation angles are somewhat easier to visualize in terms of actual limb configuration and movement. The Euler parameters, however, have the computation advantage. As we saw in Section 18.9, it is possible in computer analyses to take advantage of the desirable features of each method by making transformations between the two. Then computations may be made using Euler parameters while the relative body orientation and movement are described by the orientation angles.

Observe further that from a strict anatomical perspective it may not be accurate to use spherical joints to represent the body joints. For example, a spherical joint does not allow translation to occur between adjoining bodies (such as neck stretching). Also, knees and elbows might better be modeled as revolute (or hinge) joints, and even shoulders and hips are not exactly spherical joints. For gross motion simulations, however, these are generally unimportant concerns in that joint translation between adjoining bodies is relatively small. Also, revolute or hinge joints may be regarded as a specialization of a spherical joint where two of the angles are assigned zero values.

Finally, in Figure 20.3.1 we will regard the model to be in its reference configuration, where it is standing erect with legs together and arms at the sides with palms facing forward. The axes of each body are then all respectively parallel, with the X-axis being forward, the Z-axis vertically up, and the Y-axis to the left.

20.4 Kinematics: Coordinates

In the foregoing chapters, we have developed methods for obtaining kinematical expressions for multibody systems. Those same methods are applicable here. Indeed, obtaining

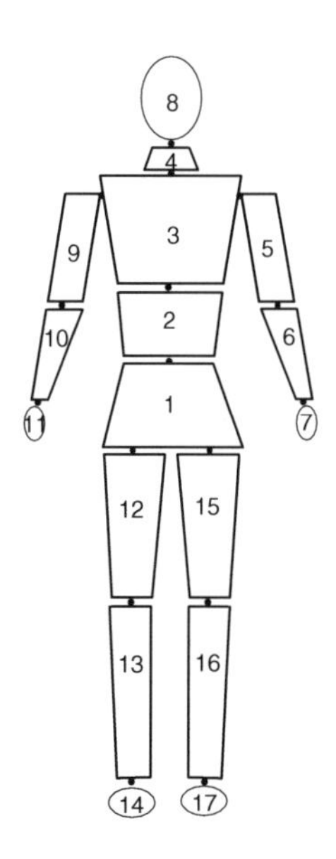

FIGURE 20.4.1
Human body model.

the kinematics of human body models has been a motivating factor in the development of many of these methods.

As before, we seek to find for each body of the model the velocity and acceleration of the mass center and the angular velocity and angular acceleration of the body itself. Because the methods and procedures are the same as we used earlier, we will simply outline them again here. The interested reader may then want to develop the details.

Consider again the global human model of Figure 20.4.1. If the model has 17 bodies as shown and if these bodies are connected by spherical joints, the model will have $(17 \times 3) + 3$, or 54, degrees of freedom. These degrees of freedom may be represented by three translation variables (say, x, y, z) for a reference point (say, O_1) of body B_1 (the pelvis), and by three orientation angles (say, α_k, β_k, λ_k) for each of the bodies B_k ($k = 1,\ldots, 17$) of the model.

If the model is to be used for large gross motions, we will want to use Euler parameters and generalized speeds to describe the angular movement of the bodies, so as to avoid singularities that may occur with orientation angles (see Section 18.5). It is convenient to let the generalized speed be the components of the relative angular velocities of the bodies. As such, they have the dimension of time derivatives of angular variables, but they cannot individually be integrated to obtain angular variables or orientation angles. Hence, when using generalized speed to describe the angular movements of the bodies, the body orientations are said to be described by *quasi-coordinates*.

Table 20.4.1 provides a listing of generalized speeds for the model of Figure 20.4.1. In Table 20.4.1, we are using the same notation for relative angular velocity components that we used earlier. Specifically, for two typical adjoining bodies B_j and B_k, the overhat (^) designates the angular velocity of B_k relative to B_j. That is,

$$^{B_j}\boldsymbol{\omega}^{B_k} = \hat{\boldsymbol{\omega}}_k = \hat{\omega}_{km}\mathbf{n}_{jm} \tag{20.4.1}$$

Questions may arise as a result of choosing relative angular velocity components as the fundamental variables. Will these variables produce excessive computational effort compared to the use of, say, absolute orientation angles? Will the use of Euler parameters obscure the visualization, or interpretation, of the movement of the model? To answer these questions, consider that human movement is easiest to describe in terms of relative

TABLE 20.4.1

Generalized Speeds and Variable Names for the Model of Figure 20.4.1

Body	Body Name	Generalized Speeds	Variable Names
B_1	Lower torso	$\hat{x}, \hat{y}, \hat{z}$ (translation)	y_1, y_2, y_3
		$\hat{\omega}_{11}, \hat{\omega}_{12}, \hat{\omega}_{13}$ (rotation)	y_4, y_5, y_6
B_2	Mid-torso	$\hat{\omega}_{21}, \hat{\omega}_{22}, \hat{\omega}_{23}$	y_7, y_8, y_9
B_3	Upper torso	$\hat{\omega}_{31}, \hat{\omega}_{32}, \hat{\omega}_{33}$	y_{10}, y_{11}, y_{12}
B_4	Upper left arm	$\hat{\omega}_{41}, \hat{\omega}_{42}, \hat{\omega}_{43}$	y_{13}, y_{14}, y_{15}
B_5	Lower left arm	$\hat{\omega}_{51}, \hat{\omega}_{52}, \hat{\omega}_{53}$	y_{16}, y_{17}, y_{18}
B_6	Left hand	$\hat{\omega}_{61}, \hat{\omega}_{62}, \hat{\omega}_{63}$	y_{19}, y_{20}, y_{21}
B_7	Neck	$\hat{\omega}_{71}, \hat{\omega}_{72}, \hat{\omega}_{73}$	y_{22}, y_{23}, y_{24}
B_8	Head	$\hat{\omega}_{81}, \hat{\omega}_{82}, \hat{\omega}_{83}$	y_{25}, y_{26}, y_{27}
B_9	Upper right arm	$\hat{\omega}_{91}, \hat{\omega}_{92}, \hat{\omega}_{93}$	y_{28}, y_{29}, y_{30}
B_{10}	Lower right arm	$\hat{\omega}_{10,1}, \hat{\omega}_{10,2}, \hat{\omega}_{10,3}$	y_{31}, y_{32}, y_{33}
B_{11}	Right hand	$\hat{\omega}_{11,1}, \hat{\omega}_{11,2}, \hat{\omega}_{11,3}$	y_{34}, y_{35}, y_{36}
B_{12}	Upper right leg	$\hat{\omega}_{12,1}, \hat{\omega}_{12,2}, \hat{\omega}_{12,3}$	y_{37}, y_{38}, y_{39}
B_{13}	Lower right leg	$\hat{\omega}_{13,1}, \hat{\omega}_{13,2}, \hat{\omega}_{13,3}$	y_{40}, y_{41}, y_{42}
B_{14}	Right foot	$\hat{\omega}_{14,1}, \hat{\omega}_{14,2}, \hat{\omega}_{14,3}$	y_{43}, y_{44}, y_{45}
B_{15}	Upper left leg	$\hat{\omega}_{15,1}, \hat{\omega}_{15,2}, \hat{\omega}_{15,3}$	y_{46}, y_{47}, y_{48}
B_{16}	Lower left leg	$\hat{\omega}_{16,1}, \hat{\omega}_{16,2}, \hat{\omega}_{16,3}$	y_{49}, y_{50}, y_{51}
B_{17}	Left foot	$\hat{\omega}_{17,1}, \hat{\omega}_{17,2}, \hat{\omega}_{17,3}$	y_{52}, y_{53}, y_{54}

variables. For example, hand orientation and movement are easier to define and describe in terms of wrist and elbow angles than in terms of orientation angles of the hand in space. Although it is true that the use of absolute orientation angles will produce simpler coefficients in the governing equations of motion (see, for example, the triple pendulum of Section 12.4), the computational burden is borne by the computer. It is preferable to have a simplified description of the movement with an increased computational burden than to use an obscure description of the movement to ease the burden on the computer. Expressed another way, it is preferable to shift the burden of complexity away from the analyst and to the computer. Even then the computational burden for the computer need not be excessive, as we can convert from relative to absolute variables (as in Section 18.9) to obtain computational efficiency.

Regarding the use of Euler parameters, observe that human joints are usually best modeled as either revolute (hinge) or spherical (ball-and-socket) joints. For hinges, as at the elbows and knees, the use of Euler parameters involves only the rotation angle itself so that two of the four parameters are zero throughout the analysis. For spherical joints, as at the shoulders and hips, if the rotations are large, then orientation angles (such as Bryan angles or Euler angles) are not necessarily more explicit descriptors than Euler parameters.

20.5 Kinematics: Velocities and Acceleration

Consider once again the gross human body model in Figure 20.5.1. In developing the kinematics of this system, we are (as with other multibody systems) interested in knowing the position, velocity, and acceleration of the mass center of each body of the system and the angular velocity and angular acceleration of each body. In addition, depending upon the application, we may be interested in knowing the kinematics of other parts of the model as well. For example, if we are modeling the throwing of a ball we may be interested in knowing the details of the finger tip movements.

Because the kinematics can be developed following the same procedures that we used with robots and multibody systems in Chapters 18 and 19, we will omit some of the details of the development and primarily outline the procedures leading to the principal results.

Consider again the lower body arrays for the model as listed in Table 20.3.2 and reproduced in Table 20.5.1. Consider a typical body of the system, say B_{11}, the right hand. The position vector $\mathbf{p}_{11}$ locating the mass center G_{11} in the inertia frame R is:

$$\mathbf{p}_{11} = \xi_1 + \xi_2 + \xi_3 + \xi_9 + \xi_{10} + \xi_{11} + \mathbf{r}_{11} \qquad (20.5.1)$$

Observe that the indices of the ξ_i (that is, 1, 2, 3, 9, 10, and 11) are precisely those of the column of B_{11} in Table 20.5.1.

By knowing the mass center positions we can immediately determine the mass center velocities by differentiation. To this end, observe that each of the ξ_i vectors (except for $\boldsymbol{\xi}_1$) is fixed in a body. Specifically, $\boldsymbol{\xi}_i$ is fixed in B_h, where h is $L(i)$. Also, $\mathbf{r}_k$ is fixed in B_k. Hence,

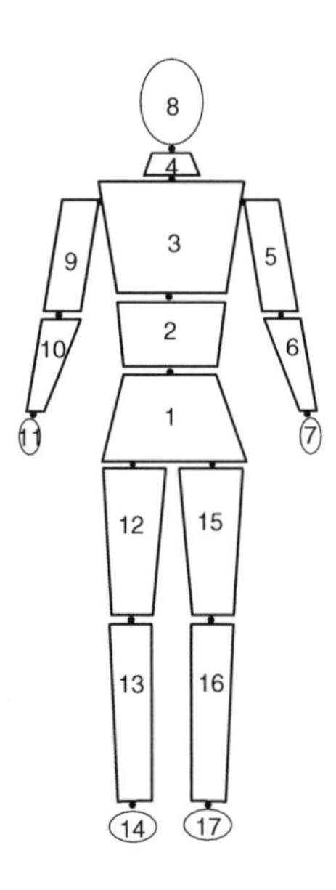

FIGURE 20.5.1
Gross human body model.

TABLE 20.5.1

Higher-Order Lower Body Arrays for the Model of Figure 20.5.1

$L^0(K)$	0	1	2	3	4	5	6	7	8	9	10	11	12	13	14	15	16	17
$L^1(K)$	0	0	1	2	3	4	5	3	7	3	9	10	1	12	13	1	15	16
$L^2(K)$	0	0	0	1	2	3	4	2	3	2	3	9	0	1	12	0	1	15
$L^3(K)$	0	0	0	0	1	2	3	1	2	1	2	3	0	0	1	0	0	1
$L^4(K)$	0	0	0	0	0	1	2	0	1	0	1	2	0	0	0	0	0	0
$L^5(K)$	0	0	0	0	0	0	1	0	0	0	0	1	0	0	0	0	0	0
$L^6(K)$	0	0	0	0	0	0	0	0	0	0	0	0	0	0	0	0	0	0

in differentiating the mass center position vectors we need simply compute vector products by angular velocity vectors. Thus, for the velocity of G_{11} we have:

$$\mathbf{V}_{11} = {}^R d\mathbf{p}_{11}/dt = \dot{\boldsymbol{\xi}}_1 + \boldsymbol{\omega}_1 \times \boldsymbol{\xi}_2 + \boldsymbol{\omega}_2 \times \boldsymbol{\xi}_3 + \boldsymbol{\omega}_3 \times \boldsymbol{\xi}_9 + \boldsymbol{\omega}_9 \times \boldsymbol{\xi}_{10} + \boldsymbol{\omega}_{10} \times \boldsymbol{\xi}_{11} + \boldsymbol{\omega}_{11} \times \mathbf{r}_{11} \tag{20.5.2}$$

where $\dot{\boldsymbol{\xi}}_1$ is simply:

$$\dot{\boldsymbol{\xi}}_1 = \dot{x}\mathbf{n}_{01} + \dot{y}\mathbf{n}_{02} + \dot{z}\mathbf{n}_{03} \tag{20.5.3}$$

where (x, y, z) are the coordinates of the origin O_1 of B_1 relative to a Cartesian axis system fixed in R.

In like manner, the mass center accelerations may be obtained by differentiating the expressions for the mass center velocities. For example, the acceleration of G_{11} is:

$$\begin{aligned}\mathbf{a}_{11} = {}^R d\mathbf{V}_{11}/dt &= \ddot{\boldsymbol{\xi}}_1 + \boldsymbol{\alpha}_1 \times \boldsymbol{\xi}_1 + \boldsymbol{\omega}_1 \times (\boldsymbol{\omega}_1 \times \boldsymbol{\xi}_2) + \boldsymbol{\alpha}_2 \times \boldsymbol{\xi}_3 \\ &+ \boldsymbol{\omega}_2 \times (\boldsymbol{\omega}_2 \times \boldsymbol{\xi}_3) + \boldsymbol{\alpha}_3 \times \boldsymbol{\xi}_4 + \boldsymbol{\omega}_3 \times (\boldsymbol{\omega}_3 \times \boldsymbol{\xi}_4) \\ &+ \boldsymbol{\alpha}_9 \times \boldsymbol{\xi}_{10} + \boldsymbol{\omega}_9 \times (\boldsymbol{\omega}_9 \times \boldsymbol{\xi}_{10}) + \boldsymbol{\alpha}_{10} \times \boldsymbol{\xi}_{11} \\ &+ \boldsymbol{\omega}_{10} \times (\boldsymbol{\omega}_{10} \times \boldsymbol{\xi}_{11}) + \boldsymbol{\alpha}_{11} \times \mathbf{r}_{11} + \boldsymbol{\omega}_{11} \times (\boldsymbol{\omega}_{11} \times \mathbf{r}_{11})\end{aligned} \tag{20.5.4}$$

where $\ddot{\xi}_1$ is:

$$\ddot{\boldsymbol{\xi}}_1 = \ddot{x}\mathbf{n}_{01} + \ddot{y}\mathbf{n}_{02} + \ddot{z}\mathbf{n}_{03} \tag{20.5.5}$$

Observe in both Eqs. (20.5.2) and (20.5.4) that we need to know the angular velocities and the angular acceleration of the bodies to complete the indicated computations. Recall that we can obtain the angular velocities of the bodies by using the addition theorem for angular velocity (see Eq. (4.7.5)) with the terms determined by the indices of the higher-order lower body arrays of Table 20.5.1 (see Section 18.2). For example, for the right hand, the angular velocity is simply:

$$\boldsymbol{\omega}_{11} = \hat{\boldsymbol{\omega}}_{11} + \hat{\boldsymbol{\omega}}_{10} + \hat{\boldsymbol{\omega}}_9 + \hat{\boldsymbol{\omega}}_3 + \hat{\boldsymbol{\omega}}_2 + \hat{\boldsymbol{\omega}}_1 \tag{20.5.6}$$

where the indices (11, 10, 9, 8, 3, 2, 1) are the entries in column 11 of Table 20.5.1. Regarding notation, as before, the overhat (ˆ) designates relative angular velocity with respect to the adjacent lower numbered body. That is,

$$\boldsymbol{\omega}_{11} = {}^{R}\boldsymbol{\omega}^{B_{11}} \quad \text{and} \quad \hat{\boldsymbol{\omega}}_{11} = {}^{B_{10}}\boldsymbol{\omega}^{B_{11}} \tag{20.5.7}$$

Then, in view of the generalized speeds listed in Table 20.4.1, we see that the relative angular velocity terms may be expressed in terms of the generalized speeds. For example, $\hat{\boldsymbol{\omega}}_{11}$ and $\hat{\boldsymbol{\omega}}_{3}$ may be expressed as:

$$\begin{aligned} \hat{\boldsymbol{\omega}}_{11} &= y_{34}\mathbf{n}_{10,1} + y_{35}\mathbf{n}_{10,2} + y_{36}\mathbf{n}_{10,3} \\ &= y_{3k+n}\mathbf{n}_{10,n} \\ &= S010_{mn}y_{3k+n}\mathbf{n}_{0m} \quad (k = 11) \end{aligned} \tag{20.5.8}$$

and

$$\begin{aligned} \hat{\boldsymbol{\omega}}_{3} &= y_{10}\mathbf{n}_{21} + y_{11}\mathbf{n}_{22} + y_{12}\mathbf{n}_{23} \\ &= y_{3k+n}\mathbf{n}_{2n} \\ &= S02_{mm}y_{3k+n}\mathbf{n}_{0m} \quad (k = 3) \end{aligned} \tag{20.5.9}$$

where, as before, the S0K are transformation matrices (see Section 18.3). Then, in view of Eqs. (20.5.6), (20.5.8), and (20.5.9), we see that the angular velocities of the bodies may be expressed as:

$$\boldsymbol{\omega}_{k} = \omega_{k\ell m}y_{\ell}\mathbf{n}_{0m} \tag{20.5.10}$$

where the $\omega_{k\ell m}$ are components of the partial angular velocity vectors (see Sections 11.4 and 18.10). As we see in Eq. (20.5.9) and in Section 18.10, the $\omega_{k\ell m}$ may be expressed in terms of the transformation matrices and that most of them are zero. Specifically, for the model of Figure 20.5.1 the nonzero $\omega_{k\ell m}$ are listed in Table 20.5.2.

Observe the pattern of the indices in Table 20.5.2. Observe that the numbers follow the sequences of the higher-order lower body arrays as listed in Table 20.5.1.

By knowing the angular velocities of the bodies we can immediately obtain the angular accelerations of the bodies by differentiating in Eq. (20.5.10). For typical body B_k the angular acceleration is:

$$\boldsymbol{\alpha}_{k} = \left(\omega_{k\ell m}\dot{y}_{\ell} + \dot{\omega}_{k\ell m}y_{\ell}\right)\mathbf{n}_{0m} \tag{20.5.11}$$

where $\dot{\omega}_{k\ell m}$ the are immediately obtained by differentiating the $\omega_{k\ell m}$ listed in Table 20.5.2. As with the $\omega_{k\ell m}$, most of the $\dot{\omega}_{k\ell m}$ are zero. Table 20.5.3 provides a listing of the nonzero $\dot{\omega}_{k\ell m}$ for the model of Figure 20.5.2.

Observe that the entries in Table 20.5.3 consist of derivatives of the transformation matrices. Interestingly, through Eq. (18.4.8) and the procedures of Section 18.4, we see that

TABLE 20.5.2

Nonzero Partial Angular Velocity Vector Components () for the Model of Figure 20.5.1

y_ℓ	1	4	7	10	13	16	19	22	25	28	31	34	37	40	43	46	49	52
	2	5	8	11	14	17	20	23	26	29	32	35	38	41	44	47	50	53
B_k	3	6	9	12	15	18	21	24	27	30	33	36	39	42	45	48	51	54
1		*I*																
2		*I*	S01															
3		*I*	S01	S02														
4		*I*	S01	S02	S03													
5		*I*	S01	S02	S03	S04												
6		*I*	S01	S02	S03	S04	S05											
7		*I*	S01	S02				S03										
8		*I*	S01	S02				S03	S07									
9		*I*	S01	S02						S03								
10		*I*	S01	S02						S03	S09							
11		*I*	S01	S02						S03	S09	S010						
12		*I*											S01					
13		*I*											S01	S012				
14		*I*											S01	S012	S013			
15		*I*														S01		
16		*I*														S01	S015	
17		*I*														S01	S015	S016

TABLE 20.5.3

Non-Zero Derivatives of the Partial Angular Velocity Vector Components $\dot{\omega}_{k\ell m}$ for the Model of Figure 20.5.1

y_ℓ	1	4	7	10	13	16	19	22	25	28	31	34	37	40	43	46	49	52
	2	5	8	11	14	17	20	23	26	29	32	35	38	41	44	47	50	53
B_k	3	6	9	12	15	18	21	24	27	30	33	36	39	42	45	48	51	54
1																		
2			$S\dot{O}1$															
3			$S\dot{O}$	$S\dot{O}$														
4			$S\dot{O}$	$S\dot{O}$	$S\dot{O}$													
5			$S\dot{O}$	$S\dot{O}$	$S\dot{O}$	$S\dot{O}$												
6			$S\dot{O}$	$S\dot{O}$	$S\dot{O}$	$S\dot{O}$	$S\dot{O}$											
7			$S\dot{O}$	$S\dot{O}$				$S\dot{O}$										
8			$S\dot{O}$	$S\dot{O}$				$S\dot{O}$	$S\dot{O}$									
9			$S\dot{O}$	$S\dot{O}$						$S\dot{O}$								
10			$S\dot{O}$	$S\dot{O}$						$S\dot{O}$	$S\dot{O}$							
11			$S\dot{O}$	$S\dot{O}$						$S\dot{O}$	$S\dot{O}$	$S\dot{O}$						
12													$S\dot{O}$					
13													$S\dot{O}$	$S\dot{O}$				
14													$S\dot{O}${E	$S\dot{O}${E	$S\dot{O}${E			
15																$S\dot{O}$		
16																$S\dot{O}${E	$S\dot{O}${E	
17																$S\dot{O}$	$S\dot{O}$	$S\dot{O}$

these derivatives may be expressed in terms of the angular velocity components. Specifically, the $S\dot{0}K$ are:

$$S\dot{0}K = d\ S0K/dt = WK\ S0K \tag{20.5.12}$$

where from Eq. (18.4.7) the WK elements are:

$$WK_{mn} = -e_{msr}\omega_{kr} = \begin{bmatrix} 0 & -\omega_{k3} & \omega_{k2} \\ \omega_{k3} & 0 & -\omega_{k1} \\ -\omega_{k2} & \omega_{k1} & 0 \end{bmatrix} \tag{20.5.13}$$

Alternatively, by expressing the ω_{kr} in terms of the partial angular velocity components as $\omega_{k\ell r}y_\ell$, we see from Eq. (18.3.6) that the elements of the $S\dot{0}K$ arrays may be expressed as:

$$S\dot{0}K_{mn} = -e_{msr}\omega_{k\ell r}y_\ell S0K_{sn} = UK_{m\ell n}y_\ell \tag{20.5.14}$$

where the $UK_{m\ell n}$ are:

$$UK_{m\ell n} = -e_{msr}\omega_{k\ell r}S0K_{sn} \tag{20.5.15}$$

Returning now to our mass center velocity expressions, as in Eq. (20.5.2), we see that these velocities may also be expressed in terms of angular velocity components and thus in terms of transformation matrices. Specifically, from Eq. (18.3.1), we have:

$$\mathbf{V}_k = d\mathbf{p}_k/dt = \left(\dot{\xi}_{1m} + \sum_{p=0}^{r-1} S\dot{0}Q_{mn}\xi_{qn} + S\dot{0}K_{mn}r_{kn}\right)\mathbf{n}_{0m} \tag{20.5.16}$$

Then, by substituting from Eq. (20.5.14), we see that the $\mathbf{V}_k$ may be expressed as (see Eq. (18.3.8)):

$$\mathbf{V}_k = \left(\dot{\xi}_{1m} + \sum_{p=0}^{r-1} UQ_{m\ell n}y_\ell\xi_{qn} + UK_{m\ell n}y_\ell r_{kn}\right)\mathbf{n}_{0m} \tag{20.5.17}$$

By examining Eq. (20.5.17) in view of Eq. (20.5.3) and Table 20.4.1, we see that the $\mathbf{V}_k$ may be expressed in the compact form as:

$$\mathbf{V}_k = v_{k\ell m}y_\ell\mathbf{n}_{0m} \tag{20.5.18}$$

where the $v_{k\ell m}$ (the partial velocity components) are:

$$v_{k\ell m} = \begin{cases} \delta_{km} & \ell = 1,2,3 \\ \sum_{p=0}^{r-1} UQ_{m\ell n}\xi_{qn} + UK_{m\ell n}r_{kn} & \ell = 4,\ldots,54 \end{cases} \tag{20.5.19}$$

Observe from Eq. (20.5.15) that through the $UK_{m\ell n}$ the $v_{k\ell m}$ depend directly upon the $\omega_{k\ell m}$.

Finally, by differentiating in Eq. (20.5.18), we see that the mass center accelerations may be expressed as:

$$\mathbf{a}_k = \left(v_{k\ell m}\dot{y}_\ell + \dot{v}_{k\ell m}y_\ell\right)\mathbf{n}_{0m} \tag{20.5.20}$$

where, by differentiating in Eq. (20.5.19), the $\dot{v}_{k\ell m}$ are:

$$\dot{v}_{k\ell m} = \begin{cases} 0 & \ell = 1,2,3 \\ \sum_{p=0}^{r-1} U\dot{Q}_{m\ell n}\xi_{qn} + U\dot{K}_{m\ell n}r_{kn} & \ell = 4,\ldots,54 \end{cases} \tag{20.5.21}$$

where from Eqs. (20.5.15) and (20.5.14) the $U\dot{K}_{m\ell n}$ are seen to be [see Eq. (18.14.3)]:

$$U\dot{K}_{m\ell n} = -e_{msr}\left(\dot{\omega}_{k\ell r}SOK_{sn} + \omega_{k\ell r}UK_{s\ell n}y_\ell\right) \tag{20.5.22}$$

Observe again the direct dependence upon the $\omega_{k\ell m}$.

20.6 Kinetics: Active Forces

We can study the kinetics of biosystems by following the same procedures that we developed and applied in Chapters 5, 18, and 19. To illustrate this, consider a typical body of the model of Figure 20.5.1 — say, the right forearm, B_{10}, as in Figure 20.6.1. The forces exerted on the forearm will generally consist of externally applied forces such as gravity and possibly contact forces, forces applied at the joints (the elbow and wrist), muscle forces, and inertia forces due to movement of forearm. In this section, we will consider the applied, or active, forces in general terms. In the next section, we will consider the muscle and joint forces in greater detail. We will consider inertia forces in Section 20.8.

Let the gravity, or weight, forces on the forearm be represented by a single vertical force $\mathbf{W}_{10}$ passing through the mass center G_{10} (see Figure 20.6.2). Let the other applied forces

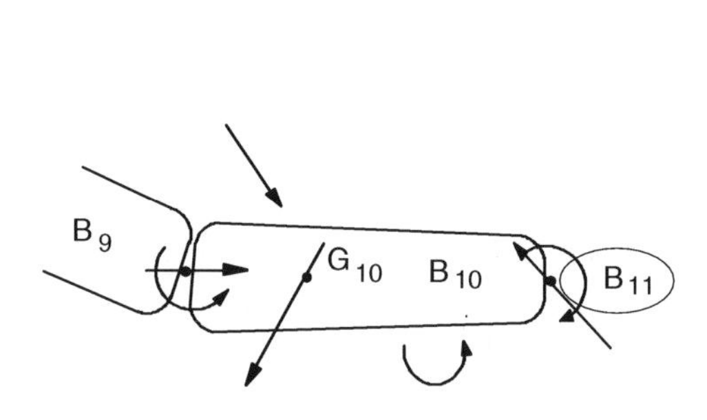

FIGURE 20.6.1
Forces and moments applied to the right forearm.

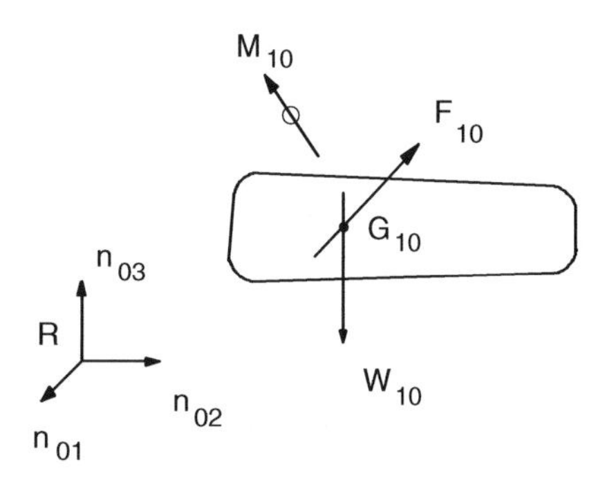

FIGURE 20.6.2
A representation of the applied forces on the forearm.

(the joint, muscle, and contact forces) be represented by a single force $\mathbf{F}_{10}$ passing through G_{10} together with a couple with torque $\mathbf{M}_{10}$. Then, the contribution F_ℓ^{10} of these forces to the generalized active force F_ℓ for the generalized speed y_ℓ is (see Eq.11.5.7):

$$F_\ell^{10} = (\mathbf{F}_{10} + \mathbf{W}_{10}) \cdot \frac{\partial \mathbf{V}_{10}}{\partial y_\ell} + \mathbf{M}_{10} \cdot \frac{\partial \boldsymbol{\omega}_{10}}{\partial y_\ell} \tag{20.6.1}$$

where, as before, $\mathbf{V}_{10}$ and $\boldsymbol{\omega}_{10}$ are the velocity of the mass center G_{10} and the angular velocity, respectively, of B_{10}, the forearm.

As before, it is convenient to express the vectors of Eq. (20.6.1) in terms of unit vectors $\mathbf{n}_{0i}$ (i = 1, 2, 3) fixed in the inertia frame R. Specifically, let $\mathbf{F}_{10}$, $\mathbf{M}_{10}$, $\mathbf{W}_{10}$, $\mathbf{V}_{10}$, and $\boldsymbol{\omega}_{10}$ be expressed as:

$$\begin{gathered} \mathbf{F}_{10} = F_{10m}\mathbf{n}_{0m}, \quad \mathbf{M}_{10} = M_{10m}\mathbf{n}_{0m}, \quad \mathbf{W}_{10} = -m_{10}g\mathbf{n}_{03} \\ \mathbf{V}_{10} = V_{10\ell m}y_\ell\mathbf{n}_{0m}, \quad \boldsymbol{\omega}_{10} = \omega_{10\ell m}y_\ell\mathbf{n}_{0m} \end{gathered} \tag{20.6.2}$$

where m_{10} is the mass of the forearm. Then, the generalized force contribution F_ℓ^{10} of Eq. (20.6.1) becomes:

$$F_\ell^{10} = F_{10m}V_{10\ell m} - m_{10}gV_{10\ell 3} + M_{10m}\omega_{10\ell m} \tag{20.6.3}$$

Finally, if we consider the other bodies of the model, we see that the contribution to the generalized forces from the external forces acting on these bodies has the same form as in Eq. (20.6.3). Then, the total generalized active force F_ℓ for y_ℓ is simply the sum of the contributions from each of the bodies. That is,

$$F_\ell = \sum_{k=1}^{17} \left(F_{km}V_{k\ell m} - m_k g V_{k\ell 3} + M_{km}\omega_{k\ell m} \right) \tag{20.6.4}$$

20.7 Kinetics: Muscle and Joint Forces

Of all the active forces, the muscle and joint forces are usually of greatest interest. For the most part, the muscles are connected to tendons, which in turn are connected to the bones. We can model the tendons as cables and the muscles as one-way actuators which create tension in the tendons. Anatomically, there are often sets of muscles/tendons working in parallel to accomplish a given body movement. For each of these muscle/tendon sets, there are also opposing muscle/tendon sets that provide for negation of the body movement. In the following paragraphs, we will consider a model of a single muscle/tendon actuation. The bones may be regarded as rigid bodies, although in actuality they are living tissue resisting forces primarily in compression.

Consider a simple model of the right upper arm/forearm/biceps as represented in Figure 20.7.1. In this model, we represent the muscle (biceps) as a force-producing actuator creating tension on its attaching tendons. The upper arm and forearm are represented as single rigid bodies. We are thus greatly simplifying the anatomical representation by

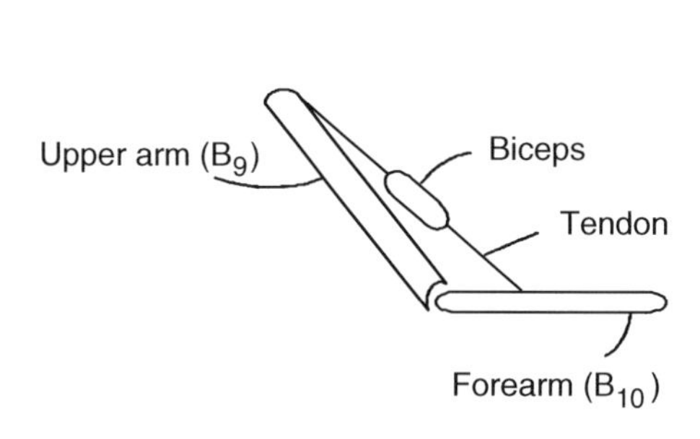

FIGURE 20.7.1
Simple model of the right upper arm/forearm/biceps.

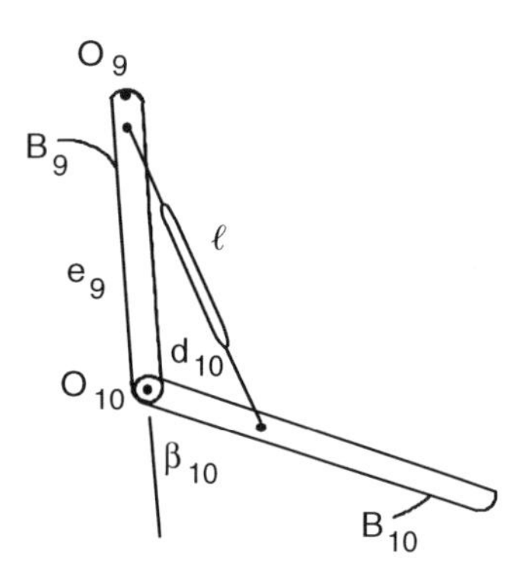

FIGURE 20.7.2
Upper arm/forearm/biceps geometry.

lumping the soft tissue, fluids, and bones into single units. (Recall also that the forearm has two bones.)

To make this model more specific, let the lower tendon be attached to the forearm (B_{10}) a distance d_{10} from the elbow (O_{10}), and let the upper tendon be attached to the upper arm (B_9) at a distance e_9 above the elbow, as in Figure 20.7.2. Let ℓ be the muscle tendon length at any instant, and let $\hat{F}_{10}$ be the magnitude of the muscle force. Then, by an elementary analysis, we see that the moment M_9 of the muscle force about the elbow O_{10} is:

$$M_9 = \left(\hat{F}_9 e_9 d_{10}/\ell\right)\sin\beta_{10} \tag{20.7.1}$$

where $\hat{F}_{10}$ is the magnitude of $\hat{\mathbf{F}}_{10}$ and β_{10} is the angle between the upper arm and forearm as shown. By further consideration of the geometry of Figure 20.7.2, we see that ℓ may be expressed in terms of d_{10}, e_9, and β_{10} as:

$$\ell = \left(d_{10}^2 + 2d_{10}e_9\cos\beta_{10} + e_9^2\right)^{1/2} \tag{20.7.2}$$

The muscle force $\hat{\mathbf{F}}_9$ may also be a function of the length ℓ and thus a function of β_{10} as well. Therefore, the elbow moment M_9 is a highly nonlinear function of β_{10}.

To simplify the analysis, we may represent the effect of the muscle force exerted on B_9 by a force $\hat{\mathbf{F}}_9$ passing through O_{10} together with a couple with torque M_9, as in Figure 20.7.3. Similarly, the muscle force exerted on B_{10} may be represented by a force $-\hat{\mathbf{F}}_9$ passing through O_{10} together with a couple with torque $-M_9$ as in Figure 20.7.4.

The reduction of the muscle forces to a single force passing through the joint together with a couple provides for a simplified determination of the contribution to the generalized forces. Recall from Section 11.5 that the joint force does not contribute to the generalized force and that the couple torque components contribute singly to the generalized forces. To develop this, consider two typical adjoining bodies such as B_j and B_k as in Figure 20.7.5, where O_k is a shared, common point at the joint. Let the forces exerted between the bodies on each other be represented for forces $\mathbf{F}_{k/j}$ and $\mathbf{F}_{j/k}$ passing through O_k together with couples with torques $\mathbf{M}_{k/j}$ and $\mathbf{M}_{j/k}$ as in Figure 20.7.6. From the law of action and reaction, the forces and couple torques have equal magnitudes but opposite directions. That is,

$$\boldsymbol{F}_{k/j} = -\boldsymbol{F}_{j/k} \quad \text{and} \quad \boldsymbol{M}_{k/j} = -\boldsymbol{M}_{j/k} \tag{20.7.3}$$

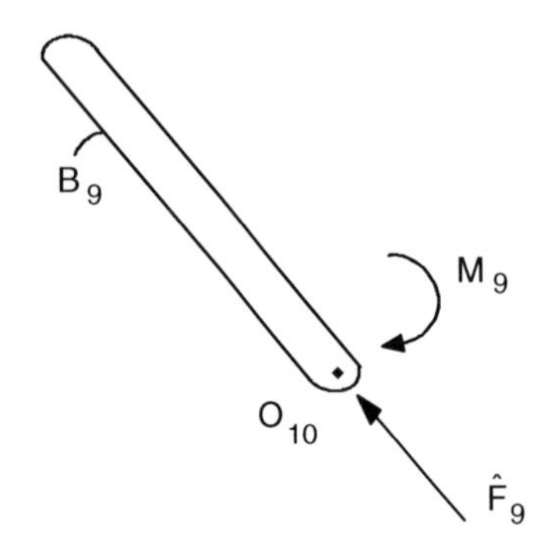

FIGURE 20.7.3
Representation of the biceps muscle force on the upper arm.

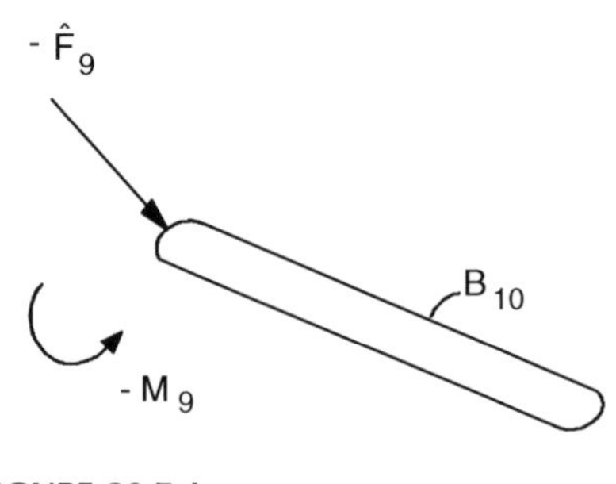

FIGURE 20.7.4
Representation of the biceps muscle force on the forearm.

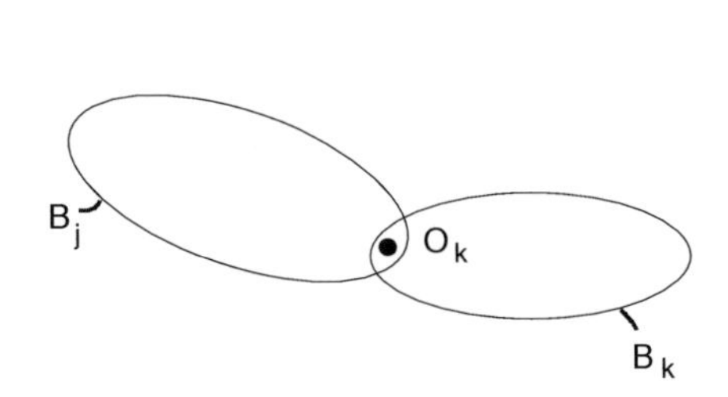

FIGURE 20.7.5
Two typical adjoining bodies.

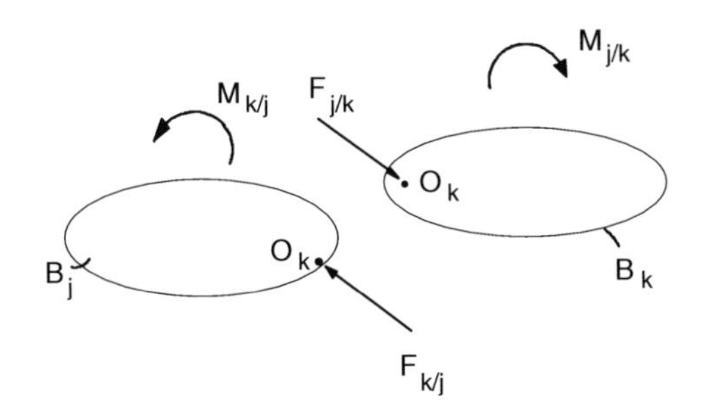

FIGURE 20.7.6
Representation of force systems between two typical adjoining bodies.

Let $\hat{F}_\ell$ be the contributions of these forces and moments to the generalized force F_ℓ, associated with the generalized speed y_ℓ. Then, from Eq. (11.5.7), $\hat{\mathbf{F}}_\ell$ is:

$$\hat{F}_\ell = \frac{\partial \boldsymbol{\omega}_j}{\partial y_\ell} \cdot \mathbf{M}_{k/j} + \frac{\partial \mathbf{V}_{O_k}}{\partial y_\ell} \cdot \mathbf{F}_{k/j} + \frac{\partial \mathbf{U}_{O_k}}{\partial y_\ell} \cdot \mathbf{F}_{j/k} + \frac{\partial \omega_k}{\partial y_\ell} \cdot \mathbf{M}_{j/k} \tag{20.7.4}$$

Recall that $\boldsymbol{\omega}_j$ and $\boldsymbol{\omega}_k$ are related by the expression:

$$\boldsymbol{\omega}_k = \boldsymbol{\omega}_j + \hat{\boldsymbol{\omega}}_k \tag{20.7.5}$$

Then,

$$\frac{\partial \boldsymbol{\omega}_k}{\partial y_\ell} = \frac{\partial \boldsymbol{\omega}_j}{\partial y_\ell} + \frac{\partial \hat{\boldsymbol{\omega}}_k}{\partial y_\ell} \tag{20.7.6}$$

Then, by substituting from Eqs. (20.7.3) and (20.7.6) into (20.7.4), we have:

$$\hat{F}_\ell = \frac{\partial \hat{\boldsymbol{\omega}}_k}{\partial y_\ell} \cdot \mathbf{M}_{j/k} \tag{20.7.7}$$

Recall that $\hat{\boldsymbol{\omega}}_k$ is the angular velocity of B_k relative to B_j. Then, $\partial \hat{\boldsymbol{\omega}}_k / \partial y_\ell$ is:

$$\frac{\partial \hat{\boldsymbol{\omega}}_k}{\partial y_\ell} = \begin{cases} \mathbf{n}_{j1} & \ell = 3k+1 \\ \mathbf{n}_{j2} & \ell = 3k+2 \\ \mathbf{n}_{j3} & \ell = 3k+3 \end{cases} \tag{20.7.8}$$

and

$$\frac{\partial \hat{\boldsymbol{\omega}}_k}{\partial y_\ell} = 0 \quad \ell \le 3k \quad \text{or} \quad \ell > 3k+3 \tag{20.7.9}$$

Let the moment $\mathbf{M}_{j/k}$ be expressed as:

$$\mathbf{M}_{j/k} = \overset{(k)}{M_1}\mathbf{n}_{j1} + \overset{(k)}{M_2}\mathbf{n}_{j2} + \overset{(k)}{M_3}\mathbf{n}_{j3} \tag{20.7.10}$$

Then, from Eqs. (20.7.7) through (20.7.10), $\hat{F}_\ell$ becomes:

$$\hat{F}_\ell = \begin{cases} \overset{(k)}{M^1} & \ell = 3k+1 \\ \overset{(k)}{M^2} & \ell = 3k+2 \\ \overset{(k)}{M^3} & \ell = 3k+3 \end{cases} \tag{20.7.11}$$

and

$$\hat{F}_\ell = 0 \quad \ell \le 3k \quad \text{or} \quad \ell > 3k+3 \tag{20.7.12}$$

Observe the convenience of Eqs. (20.7.11) and (20.7.12). Most of the $\hat{F}_\ell$ are zero, and the contributions of the moments $\mathbf{M}_{j/k}$ occur singly. Observe further that the forces $\mathbf{F}_{j/k}$ and $\mathbf{F}_{k/j}$ do not contribute to the $\hat{F}_\ell$.

20.8 Kinetics: Inertia Forces

We can develop the generalized inertia forces in the same manner as we did with robotic systems and as we did with multibody systems in Chapter 18. Specifically, consider again the gross-motion, whole-body model of Figure 20.8.1. Consider a typical body B_k represented by a force $\mathbf{F}_k^*$ passing through the mass center G_k together with a couple with torque $\mathbf{M}_k^*$ as in Figure 20.8.2. Then, $\mathbf{F}_k^*$ and $\mathbf{M}_k^*$ as given by Eqs. (8.6.5) and (8.6.6) are:

$$\mathbf{F}_k^* = -m_k \mathbf{a}_k \quad \text{(no sum on } k\text{)} \tag{20.8.1}$$

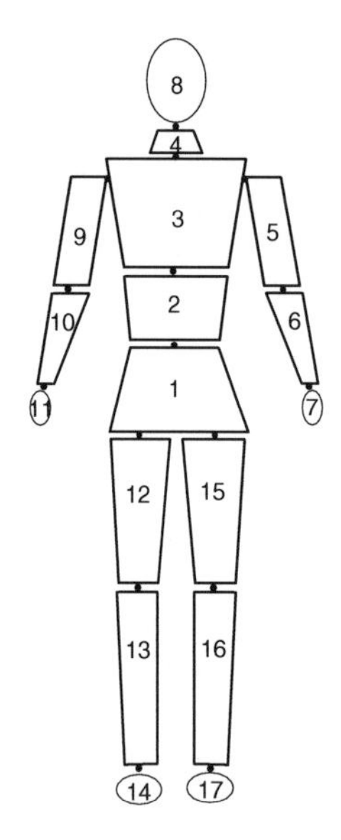

FIGURE 20.8.1
Gross-motion whole-body model.

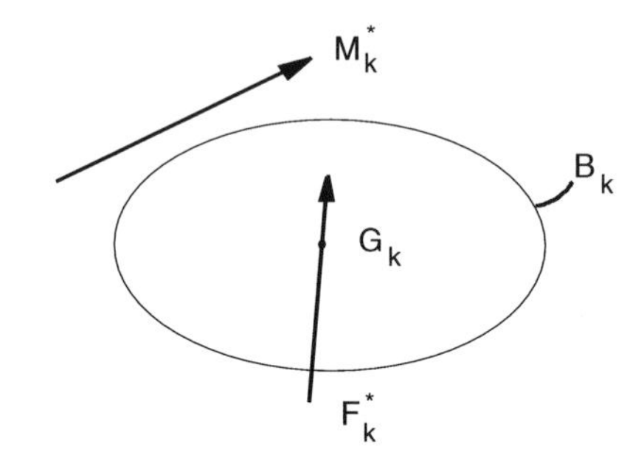

FIGURE 20.8.2
Equivalent inertia force system exerted on typical body B_k.

and

$$\mathbf{M}_k^* = -\mathbf{I}_k^* \cdot \boldsymbol{\alpha}_k - \boldsymbol{\omega}_k \times (\mathbf{I}_k \cdot \boldsymbol{\omega}_k) \quad \text{(no sum on } k) \tag{20.8.2}$$

where m_k is the mass of B_k, and $\boldsymbol{I}_k$ is the central inertia dyadic of B_k.

From Eqs. (20.5.10), (20.5.11), and (20.5.20) $\boldsymbol{\omega}_k$, $\boldsymbol{\alpha}_k$, and $\mathbf{a}_k$ may be expressed as:

$$\boldsymbol{\omega}_k = \omega_{k\ell m} y_\ell \mathbf{n}_{0m} \tag{20.8.3}$$

$$\boldsymbol{\alpha}_k = (\omega_{k\ell m} \dot{y}_\ell + \dot{\omega}_{k\ell m} y_\ell) \mathbf{n}_{0m} \tag{20.8.4}$$

and

$$\mathbf{a}_k = (v_{k\ell m} \dot{y}_\ell + \dot{v}_{k\ell m} y_\ell) \mathbf{n}_{0m} \tag{20.8.5}$$

where as before $v_{k\ell m}\mathbf{n}_{0m}$ and $\omega_{k\ell m}\mathbf{n}_{0m}$ are the partial velocities and partial angular velocities of G_k and B_k for the generalized speeds y_ℓ and the $\mathbf{n}_{0m}$ (m = 1, 2, 3) are mutually perpendicular unit vectors fixed in the inertial reference frame R.

By substituting from Eqs. (20.8.3), (20.8.4), and (20.8.5) into Eqs. (20.8.1) and (20.8.2), we see that $\mathbf{F}_k^*$ and $\mathbf{M}_k^*$ may be expressed in the forms:

$$\mathbf{F}_k^* = F_{km}^* \mathbf{n}_{0m} \quad \text{and} \quad \mathbf{M}_k^* = M_{km}^* \mathbf{n}_{0m} \tag{20.8.6}$$

where $\mathbf{F}_k^*$ and $\mathbf{M}_k^*$ are:

$$F_{km}^* = (-\mathbf{M}_k v_{k\ell m} \dot{y}_\ell + \dot{v}_{k\ell m} y_\ell) \quad \text{(no sum on } k) \tag{20.8.7}$$

and

$$M^*_{km} = -\left[I_{kmn}\left(\omega_{k\ell m}\dot{y}_\ell + \dot{\omega}_{k\ell n}y_\ell\right) + e_{rsm}\omega_{k\ell r}\omega_{kpn}I_{ksn}y_\ell y_p\right] \tag{20.8.8}$$

(See Eqs. (18.16.7) and (18.16.8).)

Finally, from Eq. (18.16.1), the generalized inertia force $F^*_{k\ell}$ on B_k for the generalized speed y_ℓ is:

$$F^*_{k\ell} = \frac{\partial \mathbf{v}_k}{\partial y_\ell}\cdot \mathbf{F}^*_k + \frac{\partial \boldsymbol{\omega}_k}{\partial y_\ell}\cdot \mathbf{M}^*_k = v_{k\ell m}F^*_{km} + \omega_{k\ell m}M^*_{km} \quad \text{(no sum on } k) \tag{20.8.9}$$

Hence, by substituting from Eqs. (20.8.7) and (20.8.8), $F^*_{k\ell}$ becomes:

$$\begin{aligned} F^*_{k\ell} &= -m_k v_{k\ell m}\left(v_{kpm}\dot{y}_p + \dot{v}_{kpm}y_p\right) \\ &\quad -I_{kmn}\omega_{k\ell m}\left(\omega_{kpn}\dot{y}_p + \dot{\omega}_{kpn}y_p\right) \\ &\quad -I_{ksn}\omega_{k\ell m}e_{rsm}\omega_{kqr}\omega_{kpn}y_q y_p \quad \text{(no sum on } k) \end{aligned} \tag{20.8.10}$$

Let F^*_ℓ be the total generalized inertia force on the model for the generalized speed y_ℓ. Then, F^*_ℓ is simply equal to the sum of the generalized inertia forces for the individual bodies. Thus, F^*_ℓ is simply the same as $F^*_{k\ell}$ of Eq. (20.8.10) with the deletion of the no-sum condition. That is,

$$\begin{aligned} F^*_\ell &= -m_k v_{k\ell m}\left(v_{kpm}\dot{y}_p + \dot{v}_{kpm}y_p\right) \\ &\quad -I_{kmn}\omega_{k\ell m}\left(\omega_{kpn}\dot{y}_p + \dot{\omega}_{kpn}y_p\right) \\ &\quad -I_{ksn}\omega_{k\ell m}e_{rsm}\omega_{kqr}\omega_{kpn}y_q y_p \end{aligned} \tag{20.8.11}$$

20.9 Dynamics: Equations of Motion

With knowledge of the generalized forces, the governing dynamical equations are readily obtained using Kane's equations (see Chapter 12, Section 12.2). Recall that Kane's equations simply state that the sums of the generalized active (applied) forces and passive (inertia) forces are zero, for each generalized speed y_ℓ. That is, from Eq. (12.2.1), we have:

$$F_\ell + F^*_\ell = 0 \quad \ell = 1,\ldots,n \tag{20.9.1}$$

where n is the number of degrees of freedom. (The gross-motion model of the foregoing sections [see, for example, Figure 20.8.1], has 54 degrees of freedom.)

By substituting from Eqs. (20.6.4), (20.7.11), (20.7.12), and (20.8.11), the dynamics equations take the form:

$$a_{\ell p}\dot{y}_p = f_\ell \quad (\ell, p = 1,\ldots n) \tag{20.9.2}$$

where the $a_{\ell p}$ and f_ℓ are:

$$a_{\ell p} = m_k v_{k\ell m} v_{kpm} + I_{kmn} \omega_{k\ell m} \omega_{kpn} \quad (\ell, p = 1, \ldots n) \tag{20.9.3}$$

$$\begin{aligned} f_\ell = {} & F_{km} v_{k\ell m} + M_{k\ell} \omega_{km} - m_k g v_{k\ell 3} + \hat{F}_\ell \\ & - \big(m_k v_{k\ell m} \dot{v}_{kpm} y_p + I_{kmn} \omega_{k\ell m} \dot{\omega}_{kpn} y_p \\ & + e_{rsm} I_{ksn} \omega_{k\ell m} \omega_{kqr} \omega_{kpn} y_p y_q \big) \quad (\ell = 1, \ldots, n) \end{aligned} \tag{20.9.4}$$

Equations (20.9.2) may be written in the compact matrix format as:

$$A\dot{y} = f \tag{20.9.5}$$

where A is the $(n \times n)$ square array with elements $a_{\ell p}$, and y and f are the $(n \times 1)$ column arrays with elements y_ℓ and f_ℓ.

20.10 Constrained Motion

Almost all applications of human body modeling will have constraints on the movement of the model. Generally, these constraints are position, or geometric (holonomic), constraints, as opposed to motion, or velocity (nonholonomic), constraints. There are, however, many instances when motion constraints are also required. For example, if we are studying an automobile operator, a typical constraint would be that the operator keep his or her hands on the steering wheel. When the vehicle is going straight at a constant speed, the steering wheel constraint is geometric. When the vehicle is turning, however, the constraint is kinematic. Examples where no constraints are imposed would be a sky diver or a swimmer.

The modeling task is to express the application constraint in terms of constraint equations such as those we developed in Section 19.17 for robots. To illustrate this, suppose the model's right hand has a specified position as depicted in Figure 20.10.1. (Such a condition might model a worker in a manufacturing environment or even a vehicle

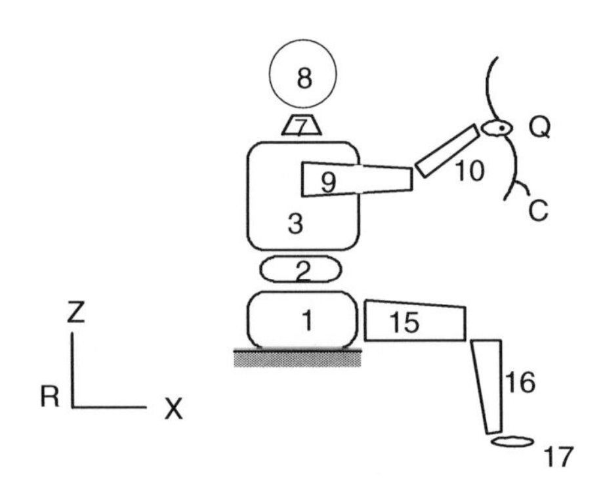

FIGURE 20.10.1
Desired position of right hand of a seated model.

operator.) Then, the constraint equation is immediately obtained by equating position vectors. That is,

$$\mathbf{P}_{11} = \mathbf{P}_Q(t) \tag{20.10.1}$$

where Q is a point of the right hand moving on a curve C, and the movement of Q on C is a given function of time.

Then, using the notation of Section 20.5 and, specifically, from Eq. (20.5.1), we have the constraint:

$$\boldsymbol{\xi}_1 + \boldsymbol{\xi}_2 + \boldsymbol{\xi}_3 + \boldsymbol{\xi}_9 + \boldsymbol{\xi}_{10} + \boldsymbol{\xi}_{11} + \mathbf{r}_{11} = \mathbf{P}_Q \tag{20.10.2}$$

To develop this further, suppose that we express these position vectors in terms of components relative to the unit vectors $\mathbf{n}_{0m}$ (m = 1, 2, 3) fixed in the inertial frame R. Specifically, let the vectors of Eq. (20.10.2) be expressed as:

$$\begin{aligned}
\boldsymbol{\xi}_1 &= x\mathbf{n}_{01} + y\mathbf{n}_{02} + z\mathbf{n}_{03} \\
\boldsymbol{\xi}_2 &= \xi_{2n}\mathbf{n}_{1n} = \xi_{2n}S01_{mn}\mathbf{n}_{0m} \\
\boldsymbol{\xi}_3 &= \xi_{3n}\mathbf{n}_{2n} = \xi_{3n}S02_{mn}\mathbf{n}_{0m} \\
\boldsymbol{\xi}_9 &= \xi_{9n}\mathbf{n}_{3n} = \xi_{9n}S03_{mn}\mathbf{n}_{0m} \\
\boldsymbol{\xi}_{10} &= \xi_{10n}\mathbf{n}_{9n} = \xi_{10n}S09_{mn}\mathbf{n}_{0m} \\
\boldsymbol{\xi}_{11} &= \xi_{11n}\mathbf{n}_{10n} = \xi_{11n}S010_{mn}\mathbf{n}_{0m} \\
\mathbf{r}_{11} &= r_{11n}\mathbf{n}_{11n} = r_{11n}S011_{mn}\mathbf{n}_{0m} \\
\mathbf{P}_Q &= x_Q(t)\mathbf{n}_{01} + y_Q(t)\mathbf{n}_{02} + z_Q(t)\mathbf{n}_{03}
\end{aligned} \tag{20.10.3}$$

where ξ_{kn} and r_{11n} (n = 1, 2, 3) are geometric constants. Then, by substituting into Eq. (20.10.2), we obtain the following scalar constraint equations:

$$x + \xi_{2n}S01_{1n} + \xi_{3n}S02_{1n} + \xi_{9n}S03_{1n} + \xi_{10n}S09_{1n} + \xi_{11n}S010_{1n} + r_{11n}S011_{1n} = x_Q(t) \tag{20.10.4}$$

$$y + \xi_{2n}S01_{2n} + \xi_{3n}S02_{2n} + \xi_{9n}S03_{2n} + \xi_{10n}S09_{2n} + \xi_{11n}S010_{2n} + r_{11n}S011_{2n} = y_Q(t) \tag{20.10.5}$$

$$z + \xi_{2n}S01_{3n} + \xi_{3n}S02_{3n} + \xi_{9n}S03_{3n} + \xi_{10n}S09_{3n} + \xi_{11n}S010_{3n} + r_{11n}S011_{3n} = z_Q(t) \tag{20.10.6}$$

In Eqs. (20.10.4), (20.10.5), and (20.10.6), the transformation matrices depend upon the orientation of the bodies and are thus time dependent. Recall from Eqs. (20.4.12) and (20.5.13) that the derivatives of the transformation matrices may be expressed in terms of angular velocity components and thus in terms of the generalized speeds. Moreover, from Eq. (20.5.13) we see that the derivatives are linear functions of the generalized speeds.

Therefore, by differentiating Eqs. (20.10.4), (20.10.5), and (20.10.6) we obtain equations that are linear in the generalized speeds and can be expressed in the matrix form:

$$By = g \tag{20.10.7}$$

where g is the (3×1) array:

$$g = \begin{bmatrix} \dot{x}_Q \\ \dot{y}_Q \\ \dot{z}_Q \end{bmatrix} \tag{20.10.8}$$

Equation (20.10.7) together with Eq. (20.9.5) form the governing equations of the constrained system. In the following section, we will discuss methods for solving these equations.

20.11 Solutions of the Governing Equations

For the unconstrained human body model as in Figure 2.8.1, the governing dynamical equations are given in matrix form by Eq. (20.9.5) as:

$$A\dot{y} = f \tag{20.11.1}$$

When the system is constrained as in Figure 20.10.1, there needs to be appended to Eq. (20.11.1) a constraint equation as in Eq. (20.10.7):

$$By = g \tag{20.11.2}$$

Observe in comparing Eqs. (20.11.1) and (20.11.2) that the constraint equations are expressed in terms of the generalized speed array y, whereas the dynamics equations are expressed in terms of the array of derivatives of the generalized speeds $\dot{y}$. Therefore, for numerical integration purposes, it is useful to differentiate in Eq. (20.11.2), leading to the expression:

$$B\dot{y} = \dot{g} - \dot{B}y \tag{20.11.3}$$

If the unconstrained human model has n degrees of freedom, then the matrix expression of Eq. (20.11.1) is equivalent to n scalar equations. If the movement of the model is constrained so that m of the degrees of freedom are constrained, then Eq. (20.11.2) represents m scalar equations. Thus, for the constrained system, Eqs. (20.11.1) and (20.11.2) form $(m + n)$ scalar equations. The unknowns are the n y_ℓ and the n muscle torques. There are thus more unknowns $(2n)$ than there are equations $(n + m)$. The excess unknowns $(n - m)$ are the degrees of freedom. These are often determined so as to optimize a movement to provide for the least effort or minimum energy expenditure.

In our example of the seated model with the specified right-hand motion, as in Figure 20.10.1 and redrawn in Figure 20.11.1, there are three constraints for the specified hand

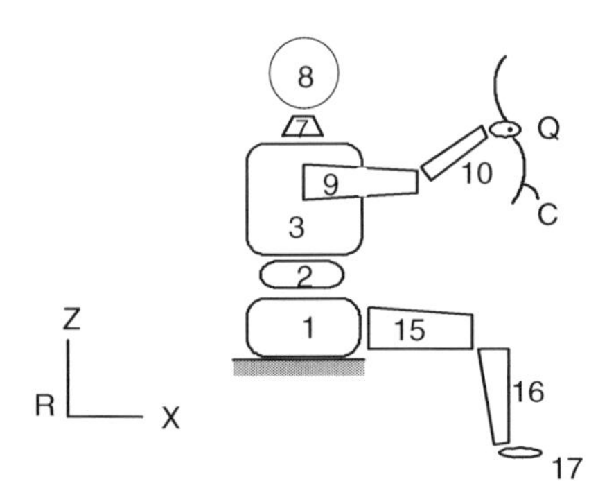

FIGURE 20.11.1
Specified motion for the right hand of a seated model.

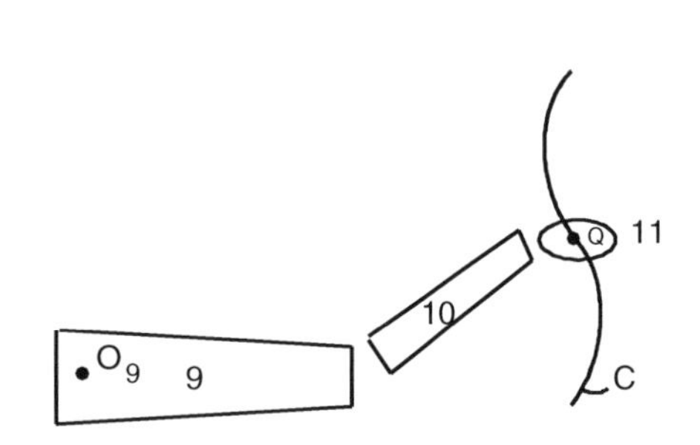

FIGURE 20.11.2
Specified movement of the right arm with fixed right shoulder.

motion. In addition, there are position constraints on Body 1 at the seat and Bodies 14 and 17 at the feet. To simplify this and to illustrate the optimization concept of the previous paragraph, we can consider the right shoulder joint as being fixed in the inertial frame R and focus upon the right-arm movement as represented in Figure 20.11.2. If the right-hand movement was not specified, the right-arm model could be considered as having seven degrees of freedom: three at the shoulder, one at the elbow, and three at the wrist. In this case, Eq. (20.11.1) would be equivalent to seven equations. If now the hand movement is constrained, Eq. (20.11.2), or, equivalently, Eq. (20.11.3), would be equivalent to three constraint equations. Then, in the notation of the foregoing paragraph $n = 7$ and $m = 3$.

Continuing this discussion about the right-arm movement, there are seven generalized speeds and seven muscle moments, for a total of 14 unknowns. Equations (20.11.1) and (20.11.2) produce ten equations; hence, four additional equations are needed. Otherwise, the solution will not be unique.

The non-uniqueness of the solution may be physically interpreted by observing that the model of Figure 20.11.1 can accommodate the desired hand movement with a variety of arm configurations. For example, the elbow could be up or down, in or out, etc. Some of these configurations would doubtless be more comfortable than others. Some would be more efficient. Some would be less fatiguing. The questions arising then, in view of the large number of solution configurations, are which one should be chosen and how is that selection to be made? The answers to these questions are dependent upon the task objective. That is, how the arm is to be configured depends upon whether the task is to be accomplished with the least effort, with the most comfort, in the least time, or under some other optimizing criteria.

To illustrate the use of an optimization criterion in solving Eqs. (20.11.1) and (20.11.2), consider that the comfort or ease of a given motion will depend largely upon the stress created in the muscles driving and controlling the arm movement. Because muscles vary in size, it is reasonable to expect that the larger muscles will bear the greater burden in accomplishing a desired task. The muscle stress is simply the muscle force divided by the muscle cross-section area. For the purposes of illustration, let us assume that the muscle cross-section area is proportional to the arm cross-section area. Then, for uniform muscle stress in the arm, the muscle moment at the shoulder would exceed the moment at the elbow which in turn would exceed the moment at the wrist. Hence, as a person is carrying a burden, as a waiter or waitress carrying a tray, the arm will assume a configuration to obtain comfort by having the muscle moments at the joints proportional to the cross-section areas of the joints. Table 20.11.1 provides a listing of typical segment weights and joint cross-section areas for a human arm, as taken from Reference 20.23.

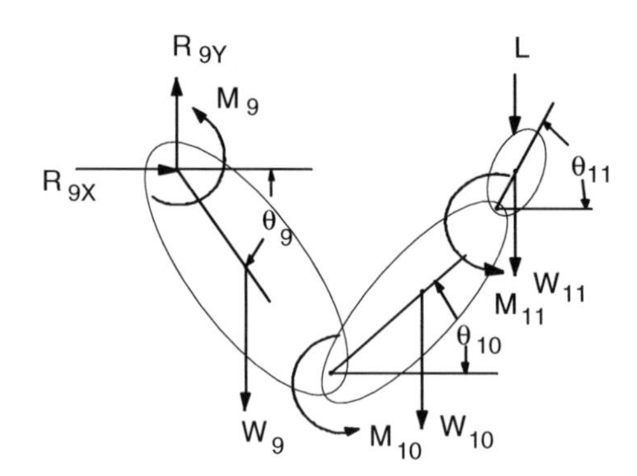

FIGURE 20.11.3
Free-body diagram of the arm.

TABLE 20.11.1

Segment Weights and Joint Cross-Section Areas of a Typical Human Arm

Segment/Joint	Weight (lb.)	Cross-Section Area (in.²)
Upper arm (B_9)/shoulder	4.68	17.9
Forearm (B_{10})/elbow	2.71	8.0
Hand (B_{11})/wrist	1.18	3.4

Source: Adapted from Huston, R. L. and Liu, Y. S., *Ohio J. Sci.*, 96, 93, 1996.

Next, consider a free-body diagram of the arm as in Figure 20.11.3. Let the orientations of the segments (the upper arm, the forearm, and the hand) be defined by the angles θ_9, θ_{10}, and θ_{11} as shown. Let the segment weights be W_9, W_{10}, and W_{11}, and let the joint moments be M_9, M_{10}, and M_{11}. Let L be the load, or weight, supported by the hand, and let R_{9X} and R_{9Y} be forces exerted at the shoulder joint. Then, by successively setting moments about the joints equal to zero in Figure 20.11.3, we readily obtain the equations:

$$M_{11} - W_{11} r_{11} \cos\theta_{11} - L\ell_{11} \cos\theta_{11} = 0 \tag{20.11.4}$$

$$M_{10} - M_{11} - W_{10} r_{10} \cos\theta_{10} - (W_{11} + L)\ell_{10} \cos\theta_{10} = 0 \tag{20.11.5}$$

$$M_9 - M_{10} - W_9 r_9 \cos\theta_9 - (W_{10} + W_{11} + L)\ell_9 \cos\theta_9 = 0 \tag{20.11.6}$$

where ℓ_9, ℓ_{10}, and ℓ_{11} are the lengths of the upper arm, the forearm, and hand, respectively, and r_9, r_{10}, and r_{11} are the distances of the mass centers of the upper arm, forearm, and hand from the shoulder, elbow, and wrist, respectively. Table 20.11.2 provides a listing of these lengths and distances for a typical human arm [20.23].

Equations (20.11.4), (20.11.5), and (20.11.6) are three equations for six unknowns: M_9, M_{10}, M_{11}, θ_9, θ_{10}, and θ_{11}. Hence, additional equations are needed. Our uniform muscle stress criterion stating that the joint moments are proportional to the joint cross-section areas leads to the equations:

$$M_9 = cA_9, \quad M_{10} = cA_{10}, \quad M_{11} = cA_{11} \tag{20.11.7}$$

where c is a constant, and A_9, A_{10}, and A_{11} are the cross-section areas of the shoulder, elbow, and wrist as listed in Table 20.11.1.

TABLE 20.11.2

Segment Lengths and Mass Center Locations for a Typical Human Arm

Segment	Length (ℓ) (in.)	Joint–Mass Center Distance (r) (in.)
Upper arm	11.78	4.47
Forearm	10.35	5.08
Hand	2.62	2.62

Source: Adapted from Huston, R. L. and Liu, Y. S., *Ohio J. Sci.*, 96, 93, 1996.

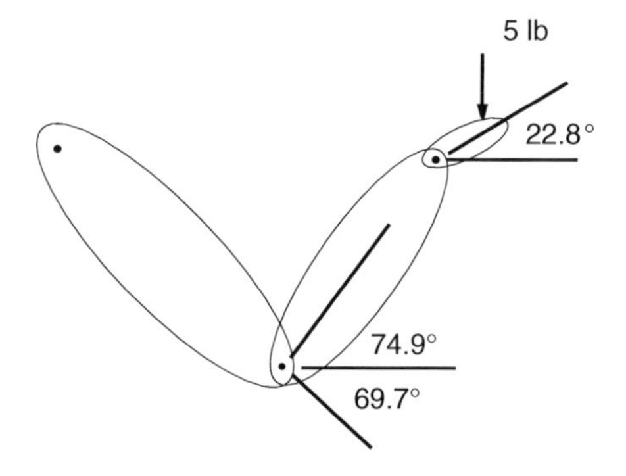

FIGURE 20.11.4
Arm configuration to support a 5-lb load.

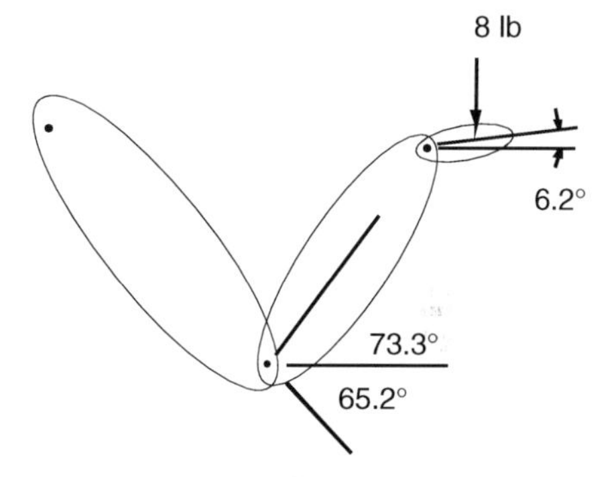

FIGURE 20.11.5
Arm configuration to support an 8-lb load.

Although Eqs. (20.11.7) provide three additional equations, they also introduce an additional unknown, c; hence, another equation is needed. This may be obtained by envisioning the burden carrier (the waiter or waitress) carrying the tray or burden at eye level, or such that the hand is held at the shoulder level. This leads to the expression:

$$\ell_9 \sin\theta_9 + \ell_{10} \sin\theta_{10} + \ell_{11} \sin\theta_{11} = 0 \tag{20.11.8}$$

Equations (20.11.4) through (20.11.8) provide seven equations for the seven unknowns: M_9, M_{10}, M_{11}, θ_9, θ_{10}, θ_{11}, and c. Using the data of Tables 20.11.1 and 20.11.2, these equations have been solved (see Reference 20.23) for two burden weights (5 lb and 8 lb), providing the solutions depicted in Figures 20.11.4 and 20.11.5. Observe the similarity of the computed configurations and the actual posture of waiters and waitresses.

20.12 Discussion: Application and Future Development

The foregoing sections of this chapter are intended to show how the principles of dynamics can be applied with biosystems and specifically the human body. Our focus has been in modeling the human body so that the dynamics principles are applicable. The utility of it all, however, is in the use of the models to study problems of practical importance. As noted earlier, the three main areas where problems must be addressed are (1) performance and enhancement of performance as in athletics, in the workplace, and in the home; (2) injury analysis and prevention; and 3) health care and rehabilitation. In this closing section, we will briefly outline a few problems and opportunities in each of these areas. These problems are typical of many that remain to be fully addressed.

First, consider performance. We often think of performance in terms of optimal achievement, as in sport maneuvers, but performance is also of great interest in the workplace, and in the home. A common activity for which performance has been studied extensively is walking. Walking (or gait) has been documented experimentally with large banks of data. Numerous analytical studies have also been conducted. Still, we have much to learn and many details to develop.

A difficulty with gait studies is the extreme complexity of movement of the links, the large number of bodies or segments involved (virtually the entire body), and the variability of gait from one person to another. It is believed that the procedures developed herein can be of assistance in obtaining a better understanding of this common, yet quite complex, activity. References 20.24 to 20.36 also provide a basis for further studies.

In addition to walking numerous other common performance activities include standing and sitting, as well as lifting, carrying, and holding objects. As with walking, these activities are very commonplace but quite complex from a dynamics perspective. Even simplified models of these movements lead to analyses that are often intractable, particularly if they are developed without using the procedures of multibody dynamics as outlined in the previous sections.

Beyond these routine movements are the more complex movements of running, jumping, tumbling, throwing, and kicking — movements for which there is great interest in optimizing, especially when they occur in sport maneuvers.

As we saw in our earlier waiter/waitress model, the human body generally has more degrees of freedom available than are needed to accomplish a given task. The question is then how best to use the degrees of freedom to optimize a given movement. For example, is it more advantageous to throw a baseball overhand or sidearm? Answers to such questions have seldom been addressed from a comprehensive dynamics perspective.

Regarding throwing, recall from our previous discussions that a simple model of the human arm has seven degrees of freedom (three at the shoulder, one at the elbow, and three at the wrist), whereas only six degrees of freedom are required to determine the motion of a thrown body (for example, a baseball). A closer examination, however, shows that such a simple model may not be satisfactory because in the throw of, say, a baseball, the fingers, the upper body, the pelvis, and the legs all contribute. This occurs with virtually all sport throwing: baseball, football, basketball, bowling, horseshoes, field events (shot put, javelin, discus), and perhaps to a lesser extent darts. Similar comments could be made for kicking in sports. A comprehensive dynamical analysis of throwing and kicking could well lead to performance improvement by identifying optimal techniques and corresponding exercises and practice maneuvers needed to acquire the optimal techniques.

Next, consider injury analyses. The increasing costs of injuries and the increasing rate of injury occurrence have stimulated interest in biomechanics to a greater extent than any other area of inquiry. As a result, more biomechanics research has been devoted to injury analysis than to any other area; consequently, more technical writings are available on injury than in any other area of biomechanics.

Vehicle accidents are the greatest single cause of injuries. As many as 42,000 people die annually in the U.S. alone from vehicle accidents, and hundreds of thousands more suffer serious permanent disablement from vehicle accidents [20.37].

Of all injuries, the most serious are those of the head and neck system, for which a gross model is shown in Figure 20.12.1. The model consists of a single large body representing the head together with a series of smaller bodies representing the neck vertebrae. Springs and dampers may be used to connect these bodies and thus model the soft tissues (discs, muscles, and ligaments).

If the model of Figure 20.12.1 is used in conjunction with a gross-motion model such as that of Figure 20.8.1, reasonable representations of head and neck motions can be obtained in the high-acceleration environments as occur in vehicle accidents. Specifically, the gross-motion model is used to obtain chest movement, and this movement in turn is used to derive the base of the neck.

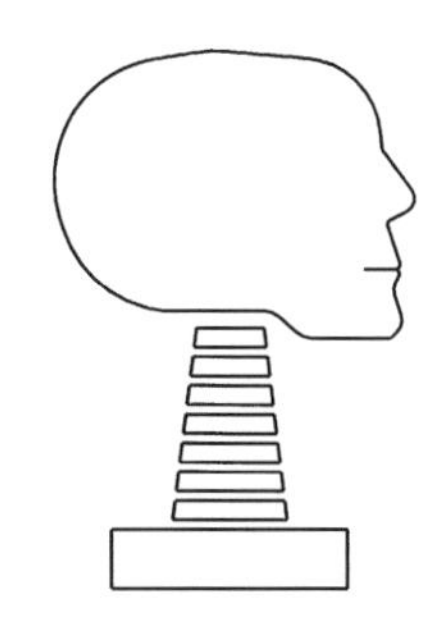

FIGURE 20.12.1
A gross-motion head/neck model.

A number of drawbacks and deficiencies are inherent with this approach, however. First, the model simulates primarily inertia loading on the head and neck members, as opposed to direct impact loading. Thus, the modeling is better suited for studying neck injury than head injury. Next, even for neck injuries, the relative movement of the vertebrae may not be sufficient to accurately model spinal injury, as such injuries can occur with only minor shifts in vertebral and disc positions and geometry. Third, the large disparity in size between the head and vertebrae can cause numerical difficulties and inefficiencies in solving the governing equation. Therefore, to study more detail about head and neck injuries, more detailed models than those of Figures 20.12.1 and 20.8.1 are needed. Just as the gross-motion, whole-body model of Figure 20.8.1 is used to provide input motion data for the head and neck model of Figure 20.12.1, so also can the head and neck model be used for input data for a more refined model of either the head or the neck.

The gross-motion, whole-body model and the head and neck model of Figure 20.12.1 and 20.8.1 may also be used to study and design protective devices for injury prevention. In motor vehicles, such devices include seat belts, air bags, helmets, head rests, seats, doors, flooring, and roofs. Also, the models can be used to study the effects of vehicle steering wheels, dashboards, and windshields upon occupant dynamics during crashes.

Aside from vehicle accidents, major sources of injuries are slips and falls. Unfortunately, the modeling of slip-and-fall accidents lags behind that of vehicle accidents. Nevertheless, the procedures outlined in this and the preceding chapters would seem to be ideally suited for modeling slips and falls. The difficulty is that slipping and falling are complex, three-dimensional motions involving the entire human frame. Thus, considerable care and insight are needed in the modeling to obtain accurate simulation.

Finally, consider rehabilitation. Here we have perhaps the greatest challenges and opportunities in dynamic modeling. Whereas performance and injury analyses require skill in studying a given system, rehabilitation requires the ingenuity to create the designs of the rehabilitative devices, be they crutches, braces, implants, or prosthetics. The problems, and hence the opportunities, for the design and analysis of these devices are manifold. Most of these devices must be designed and fabricated individually to meet the particular requirements of the person in need of rehabilitation. Complicating the design and analysis is the complex geometry of both the devices and the human body itself. For example, at first glance a shoulder or hip joint may appear to be simply a spherical joint. A closer examination, however, reveals a nonspherical geometry that is extremely difficult to duplicate in an implant. Similarly, the knee has much greater flexibility than that offered by a single hinge. Indeed, a knee provides for rolling, sliding, and even to some extent a translation between the tibia and femur. Hence, simple designs of artificial joints are almost certain to be deficient and the opportunity for design optimization is great.

In a related area, there is increasing evidence that biological tissue heals best and remains healthy under alternating periods of stimulation (or loading) and rest. Without loading, healing is greatly retarded. Alternatively, too much loading can aggravate a wound or a fracture. Therefore, the challenge is to determine optimum loads and appropriate exercises and maneuvers to apply those loads.

In summary, unsolved dynamic problems in healing, in rehabilitation, and in biomechanics in general, are undoubtedly more numerous and have greater complexity than those in any other area of dynamics analysis. It is our belief that the methods presented herein are ideally suited for studying these problems.

References

20.1. Smith, P. G., and Kane, T. R., On the dynamics of the human body in free fall, *J. Appl. Mech.*, 35, 167, 1968.

20.2. Kane, T. R., and Scher, M. P., Human self-rotation by means of limb movements, *J. Biomech.*, 3, 39, 1970.

20.3. Passerello, C. E., and Huston, R. L., Human attitude control, *J. Biomech.*, 4, 95, 1971.

20.4. Huston, R. L., and Passerello, C. E., On the dynamics of a human body model, *J. Biomech.*, 4, 369, 1971.

20.5. Gallenstein, J., and Huston, R. L., Analysis of swimming motions, *Hum. Factors*, 15, 91, 1973.

20.6. Huston, R. L., Hessel, R. E., and Winget, J. W., Dynamics of a crash victim — a finite segment model, *AIAA J.*, 14(2), 173, 1976.

20.7. King, A. I., and Chou, C. C., Mathematical modelling, simulation and experimental testing of biomechanical system crash response, *J. Biomech.*, 9, 301, 1976.

20.8. King, A. I., A review of biomechanical models, *J. Biomech. Eng.*, 106, 97, 1984.

20.9. Hanavan, E. P., A Mathematical Model of the Human Body, Report AMRL-TR-64- 102, Aerospace Medical Research Laboratory, Wright Patterson Air Force Base, Dayton, OH, 1964.

20.10. Advani, S., Huston, J., Powell, W., and Cook, W., Human head–neck dynamics response: analytical models and experimental data, *Aircraft Crashworthiness*, University Press of Virginia, Charlottesville, VA, 1975, p. 197.

20.11. Merrill, T., Goldsmith, W., and Deng, Y. C., Three-dimensional response of a lumped parameter head–neck model due to impact and impulsive loading, *J. Biomech.*, 17(2), 81, 1984.

20.12. Kabo, J. M., and Goldsmith, W., Response of a human head–neck model to transient sagittal plane loading, *J. Biomech.*, 16, 313, 1983.

20.13. Reber, J. A., and Goldsmith, W., Analysis of large head–neck motion, *J. Biomech.*, 3, 211, 1979.

20.14. Huston, R. L., Harlow, M. W., and Huston, J. C., Comprehensive three-dimensional head–neck model for impact and high acceleration studies, *Aviation, Space, and Environ. Med.*, 49, 205, 1978.

20.15. Huston, R. L., and Sears, J., Effect of protective helmet mass on head/neck dynamics, *J. Biomech. Eng.*, 109, 163, 1987.

20.16. Tein, C. S., and Huston, R. L., Numerical advances in gross-motion simulation of head/neck dynamics, *J. Biomech. Eng.*, 109, 163, 1987.

20.17. Huston, R. L., A summary of three-dimensional gross-motion, crash-victim simulators, *Structural Mechanics Software Series*, Vol. I, University Press of Virginia, Charlottesville, VA, 1977, p. 611.

20.18. Ewing, C. L., and Thomas, D. J., Human Head and Neck Response to Impact Acceleration, Army–Navy Joint Report, Naval Aerospace Medical Research Laboratory, NAMRL Monograph 21, 1972.

20.19. Young, R. D., A Three-Dimensional Mathematical Model of an Automobile Passenger, Research Report 140-2, NTIS Report No. PB 197 159, Texas Transportation Institute, Texas A&M University, College Station, TX, 1970.

20.20. Web Associates, Anthropometry for designers, in *Anthropometric Source Book*, Vol. I, NASA Reference Publication 1024, 1978.

20.21. Tilley, A. R., *The Measure of Man and Woman — Human Factors in Design*, Whitney Library of Design, New York, 1993.

20.22. Kane, T. R., *Analytical Elements of Mechanics*, Vol. 2, Academic Press, New York, 1961, p. 229.

20.23. Huston, R. L., and Liu, Y. S., Optimal human posture — analysis of a waitperson holding a tray, *Ohio J. Sci.*, 96, 93, 1996.

20.24. Abdelnour, T. A., Passerello, C. E., and Huston, R. L., An analytical analysis of walking, ASME Paper 75-WA/Bio-4, American Society of Mechanical Engineers, New York, 1975.
20.25. Kirtley, C., Shittle, M. W., and Jefferson, R. J., Influence of walking speed on gait parameters, *J. Biomed. Eng.*, 7, 282, 1985.
20.26. McMahon, T. A., Mechanics of locomotion, *Int. J. Robotics Res.*, 3(2), 4, 1984.
20.27. Aleshinsky, S. Y., and Zatsiorsky, V. M., Human locomotion in space analyzed biomechanically through a multi-link chain model, *J. Biomech.*, 11, 101, 1978.
20.28. Hardt, D. E., and Mann, R. W., A fine body — three-dimensional dynamic analysis of walking, *J. Biomech.*, 13, 455, 1980.
20.29. Zernicke, R. F., and Roberts, E. M., Lower extremity forces and torques during systematic variation of non-weight-bearing motion, *Med. Sci. Sports*, 10(1), 21, 1978.
20.30. Cheng, M. Y., and Lin, C. S., Genetic algorithm for control design of biped locomotion, *J. Robotic Syst.*, 14(5), 365, 1997.
20.31. Levens, A. S., Berkeley, C. E., Inman, V. T., and Blosser, J. A., Transverse rotation of the segments of the lower extremity in locomotion, *J. Bone Joint Surg.*, 30-4(4), 859, 1948.
20.32. Winter, D. A., Quanburg, A. O., Hobson, D. A., Sidwall, H. G., Reimer, G., Trenholm, B. G., Steinke, T., and Shlosser, H., Kinematics of normal locomotion — a statistical study based on T.V. data, *J. Biomech.*, 7, 479, 1974.
20.33. Alexander, R. M., Walking and running, *Am. Scientist*, 72, 348, 1984.
20.34. Rohrle, H., Scholten, R., Sigolotto, C., and Sollbach, W., Joint forces in the human pelvis–leg skeleton during walking, *J. Biomech.*, 17(6), 409, 1984.
20.35. Garcia, M., Chatterjee, A., Ruina, A., and Coleman, M., The simplest walking model: stability, complexity, and scaling, *J. Biomech. Eng.*, 120, 281, 1998.
20.36. Alexander, R. M., Simple models of human motion, *Appl. Mech. Rev.*, 48, 461, 1995.
20.37. National Safety Council, *Traffic Safety*, Jan./Feb., 24, 1998.

Problems

Section 20.2 Human Body Modeling

P20.2.1: Develop a model of the human arm (upper arm and forearm) analogous to that of the hand of Figure 20.2.3.

Section 20.3 A Whole-Body Model: Preliminary Consideration

P20.3.1: Consider the model of a hand as in Figure 20.2.3 and shown again in Figure P20.4.1. Make a listing of the bodies of this system and construct a lower body array for the system.

P20.3.2: See Problem P20.3.1. Develop a table of higher-order lower body arrays for the hand model analogous to Table 20.3.2.

Section 20.4 Kinematics: Coordinates

P20.4.1: Consider the model of a hand of Figure 20.2.3 and as shown again in Figure P20.4.1. Let there be spherical joints between the wrist and the fingers and thumb. Let the joints within the fingers and thumb, however, be pin (hinge) joints. Determine the number of degrees of freedom of this system.

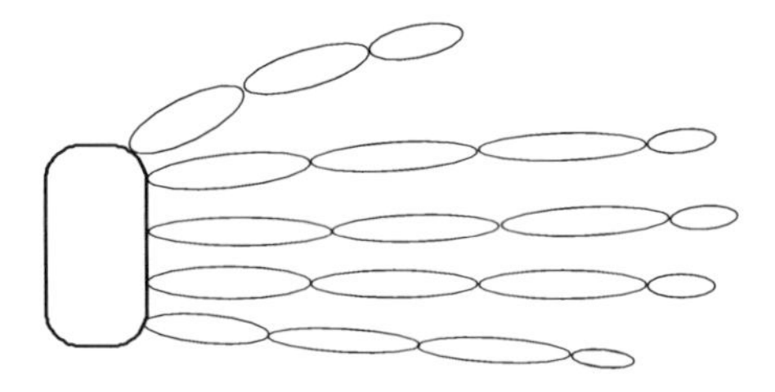

FIGURE P20.4.1
A model of the human hand.

P20.4.2: See Problem P20.4.1. Construct a table analogous to Table 20.4.1 for the hand model, showing the body numbers, the body names, the generalized speeds, and the variable names.

Section 20.5 Kinematics: Velocities and Accelerations

P20.5.1: Review the analysis of Section 20.5 relating to the example for the kinematics of the right hand. Develop an analogous analysis for the kinematics of the right foot.

P20.5.2: See Problems P20.3.1, P20.3.2, P20.4.1, and P20.4.2. Consider again the model of the hand shown in Figure 20.2.3 and again in Figure P20.4.1. Develop the kinematics of the hand model analogous to the gross-motion development in Section 20.5.

Section 20.6 Kinetics: Active Forces

P20.6.1: Consider a simple planar model of the arm as shown in Figure P20.6.1. It consists of three bodies representing the upper arm, the forearm, and the hand, labeled and numbered as 1, 2, and 3, respectively, as shown. Let the shoulder, elbow, and wrist joints be O_1, O_2, and O_3, respectively. Let the lengths of the bodies be ℓ_1, ℓ_2, and ℓ_3; let the weights of the bodies be w_1, w_2, and w_3; and let the mass centers each be one third of the body length distal from the upper joint. Let the hand support a mass with weight W at the finger tips as shown. Finally, let the orientation of the bodies be defined by the relative angles β_1, β_2, and β_3. If O_1 is a fixed point in a reference frame R, develop the kinematics of this system using β_1, β_2, and β_3 as generalized coordinates.

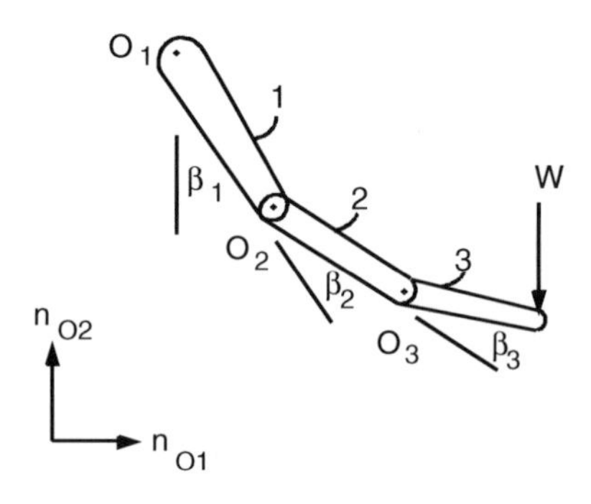

FIGURE P20.6.1
A model of the arm.

P20.6.2: See Problem P20.6.1. Determine the contribution to the generalized active forces of the weights of the bodies.

P20.6.3: See Problem P20.6.1. Determine the contribution to the generalized active forces of the weight of the mass at the finger tips.

P20.6.4: See Problem P20.6.1 and Figure P20.6.1. Let the movement of the system be defined by the absolute angles θ_1, θ_2, and θ_3 as in Figure P20.6.4. Repeat Problems P20.6.1, P20.6.2, and P20.6.4 using θ_1, θ_2, and θ_3 as generalized coordinates.

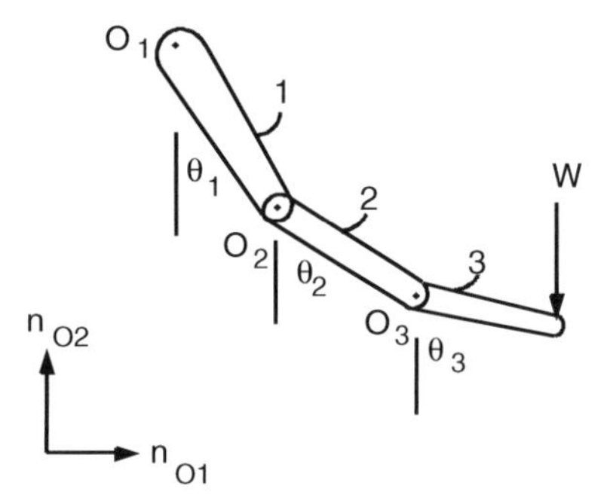

FIGURE P20.6.4
A model of the arm with absolute orientation angles.

Section 20.7 Kinetics: Muscle and Joint Forces

P20.7.1: See Problems P20.6.1 to P20.6.4. Let the forces exerted between the bodies due to the muscles be equivalent to and represented by single forces passing through the joint centers together with couples. For example, at a typical joint — say, the elbow (joint 2) — let the muscle forces exerted by B_1 (the upper arm) on B_2 (the forearm) be represented by a single force $\mathbf{F}_{1/2}$ passing through O_2 together with a couple with torque $\mathbf{M}_{1/2}$. Similarly, let the muscle forces exerted by the forearm on the upper arm be represented by a single force $\mathbf{F}_{2/1}$ passing through O_2 together with a couple with torque $\mathbf{M}_{2/1}$. Let these forces and moments be negative to one another; that is,

$$\mathbf{F}_{1/2} = -\mathbf{F}_{2/1} \quad \text{and} \quad \mathbf{M}_{1/2} = -\mathbf{M}_{2/1}$$

Determine the contribution of the muscle forces to the generalized active forces using both the relative angles β_1, β_2, and β_3 and the absolute angles θ_1, θ_2, and θ_3 as generalized coordinates.

Section 20.8 Kinetics: Inertia Forces

P20.8.1: Verify again the validity of Eq. (20.8.11).

P20.8.2: Consider again the model of the arm of Problem P20.6.1 shown in Figure P20.6.1 and as shown again in Figure P20.8.2. Let the bodies be modeled by frustrums of cones and let mutually perpendicular unit vectors $\mathbf{n}_{ki}$ ($k = 1, 2, 3$; $i = 1, 2, 3$) be fixed in the bodies with the $\mathbf{n}_{k2}$ being along the axes of the bodies (and cones) and the $\mathbf{n}_{k3}$ being normal to the plane of motion. Let the $\mathbf{n}_{ki}$ be parallel to principal inertia axes of the bodies, and let the corresponding moments and products of inertia then be:

$$I_{k11} = I_{k33} = \overset{(k)}{I_3}, \quad I_{k22} = \overset{(k)}{I_2}$$
$$I_{k12} = I_{k13} = I_{k23} = 0$$

Determine the generalized inertia forces on the system.

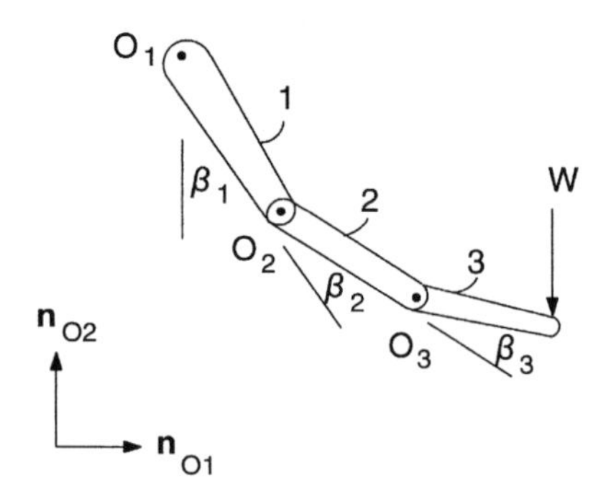

FIGURE P20.8.2
A model of the arm.

P20.8.3: See Problems P20.6.4 and P20.8.2. Let the movement of the arm model be defined by the absolute angles θ_1, θ_2, and θ_3 as shown in Figure P20.6.4 and as shown again in Figure P20.8.3. Find the generalized inertia forces corresponding to θ_1, θ_2, and θ_3 using the inertia data of Problem P20.8.2.

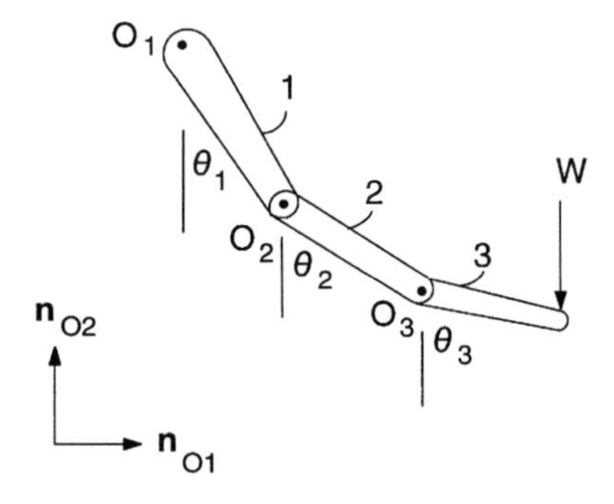

FIGURE P20.8.3
A model of the arm with absolute orientation angles.

Section 20.9 Dynamics, Equations of Motion

P20.9.1: See Problems P20.6.1, P20.6.2, P20.6.3, P20.7.1, and P20.8.2. Using the results obtained in these problems, determine the governing equations of motion for the arm model using the relative orientation angles β_1, β_2, and β_3 as generalized coordinates.

P20.9.2: See Problems P20.6.4, P20.7.1, and P20.8.3. Using the results obtained in these problems, determine the governing equations of motion for the arm model using the absolute orientation angles θ_1, θ_2, and θ_3 as generalized coordinates.

Section 20.10 Constrained Motion

P20.10.1: See Problems P20.9.1 and P20.9.2. Let the finger tips (extremity of body 3) have a desired motion — say, movement on a circle at constant speed v (let the circle be in the vertical plane with center (x_O, y_O) and with radius a). Determine the governing equations of motion for the model with this constraint. Use both relative and absolute orientation angles.

Appendix I*

Centroid and Mass Center Location for Commonly Shaped Bodies with Uniform Mass Distribution

I. Curves, wires, thin rods

1. Straight line, rod:

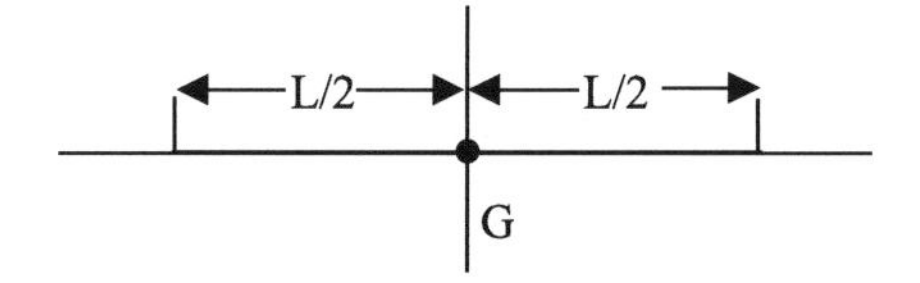

2. Circular arc, circular rod:

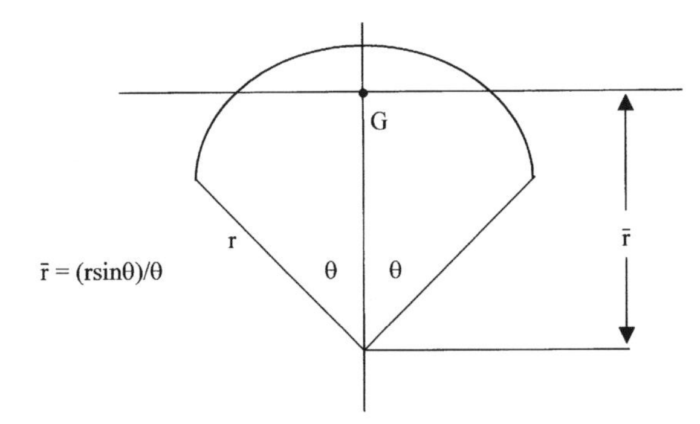

3. Semicircular arc, semicircular rod:

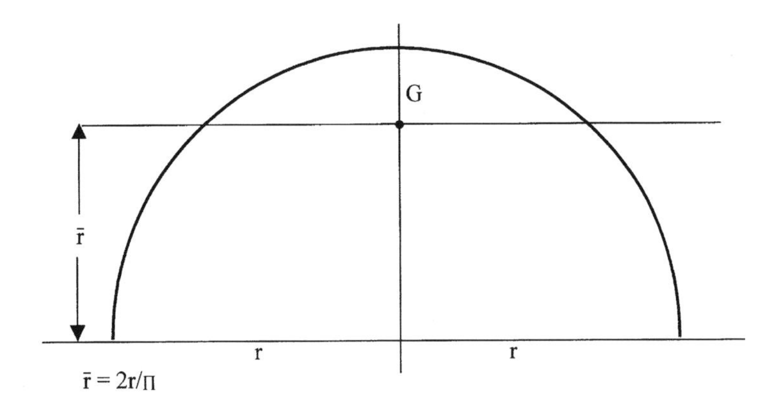

* Reprinted from Huston, R. L., and Liu, C. Q., *Formulas for Dynamic Analysis*, pp. 303–310, by courtesy of Marcel Dekker, New York, 2001.

4. Circle, hoop:

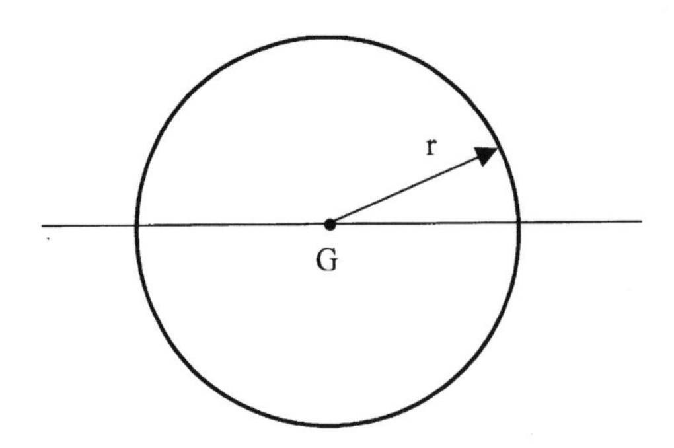

II. Surfaces, thin plates, shells

1. Triangle, triangular plate:

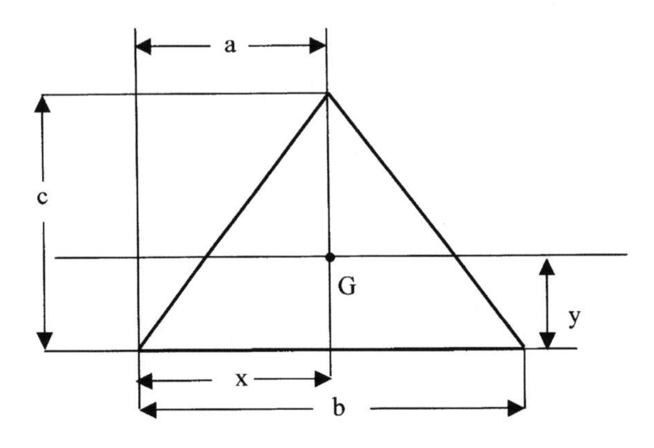

$x = (a + b)/3$

$y = c/3$

2. Rectangle, rectangular plate:

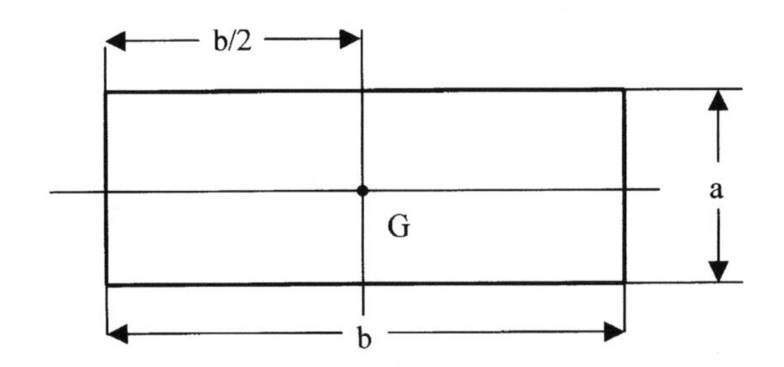

3. Circular sector, circular section plate:

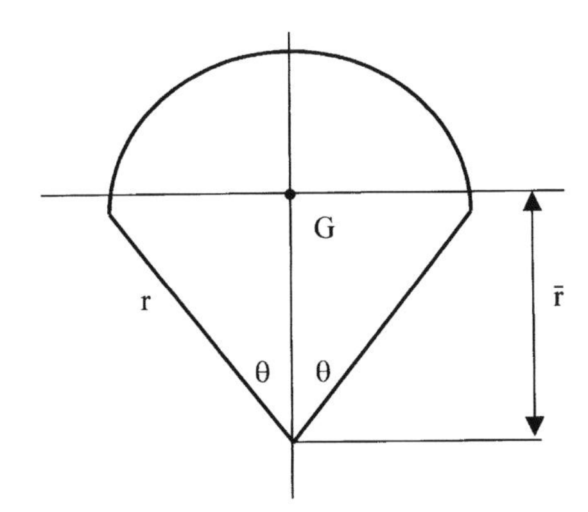

$\bar{r} = (2r/3)\,(\sin\theta)/\theta$

4. Semicircle, semicircular plate:

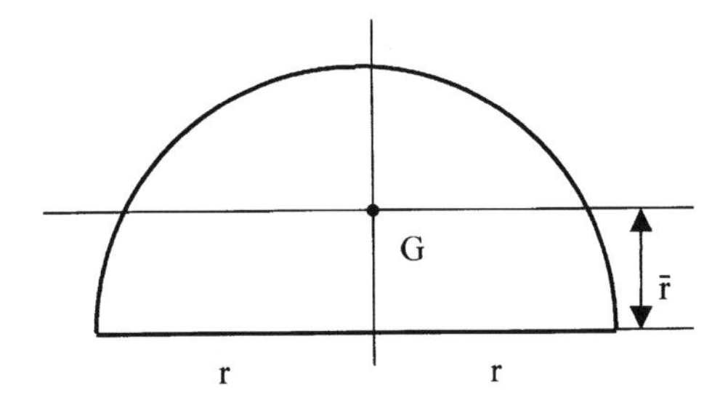

$\bar{r} = 4r/\Pi$

5. Circle, circular plate:

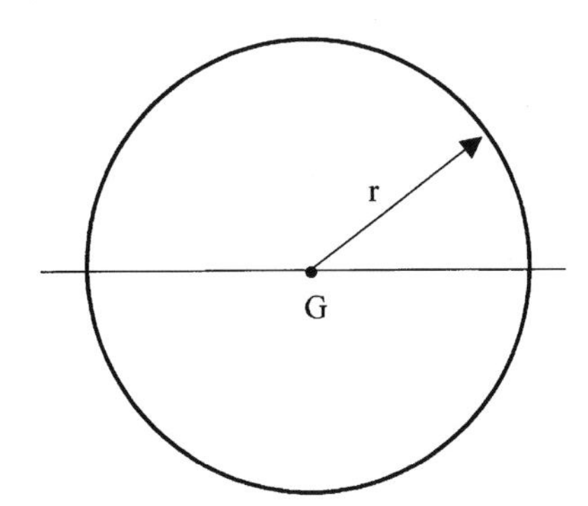

6. Circular segment, circular segment plate:

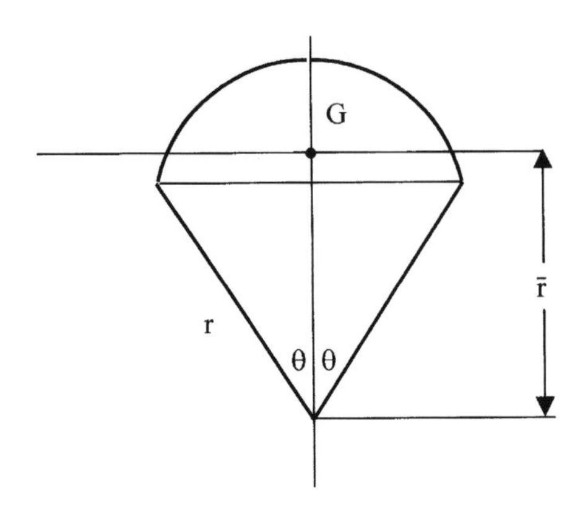

$\bar{r} = (2r/3)(\sin^3\theta)/(\theta - \sin\theta\cos\theta)$

7. Cylinder, cylindrical shell:

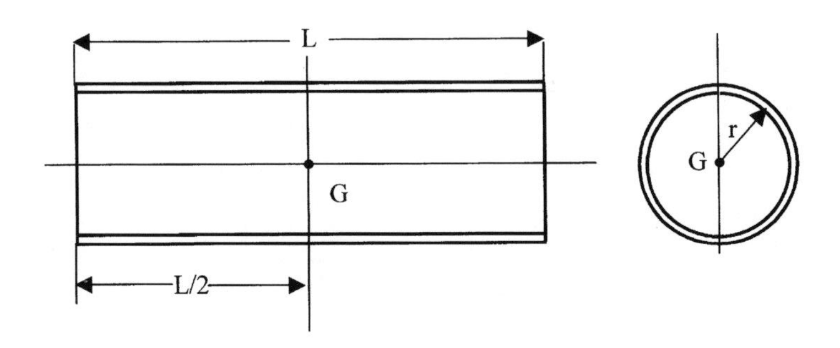

8. Semicylinder, semicylindrical shell:

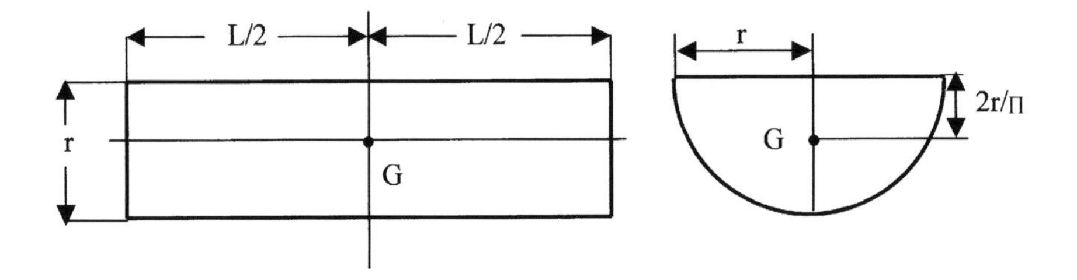

9. Sphere, spherical shell:

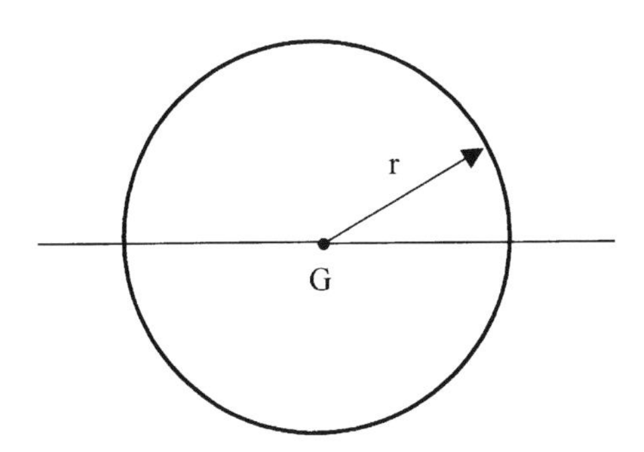

10. Hemisphere, hemispherical shell:

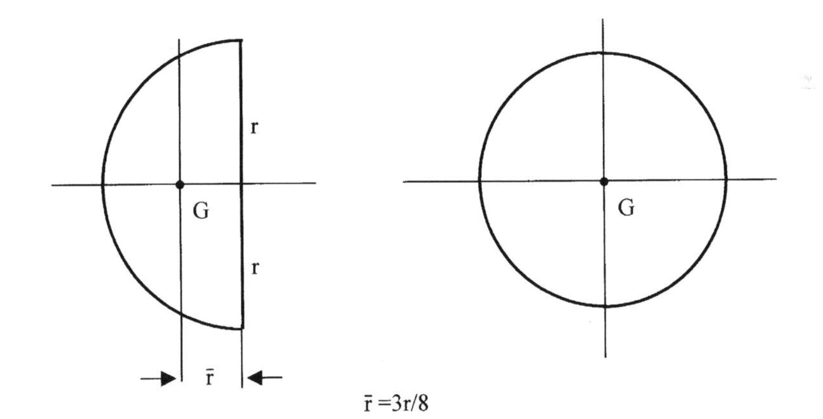

11. Cone, conical shell:

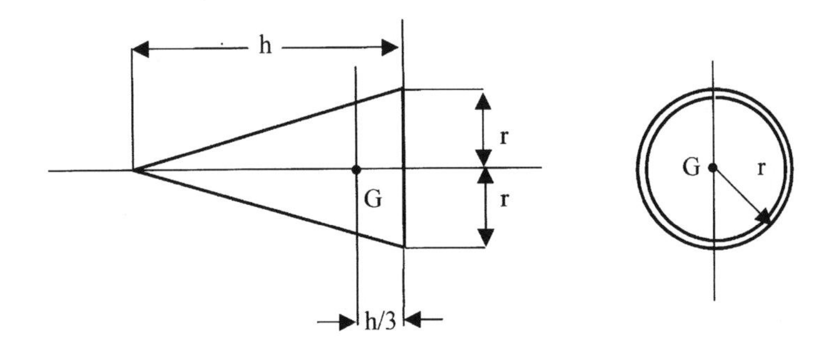

12. Half cone, half-conical shell:

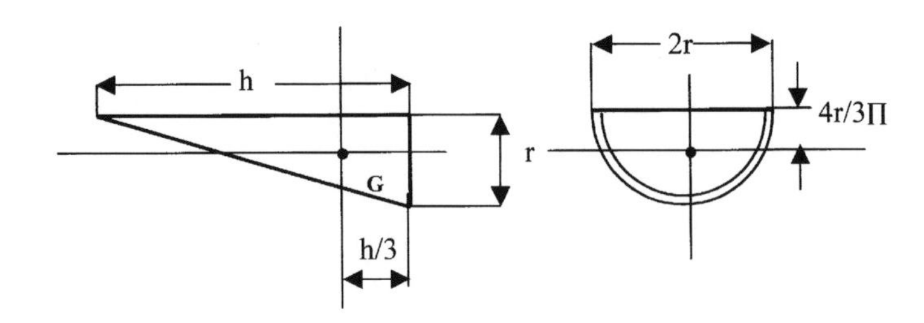

III Solids, bodies

1. Parallelepiped, block:

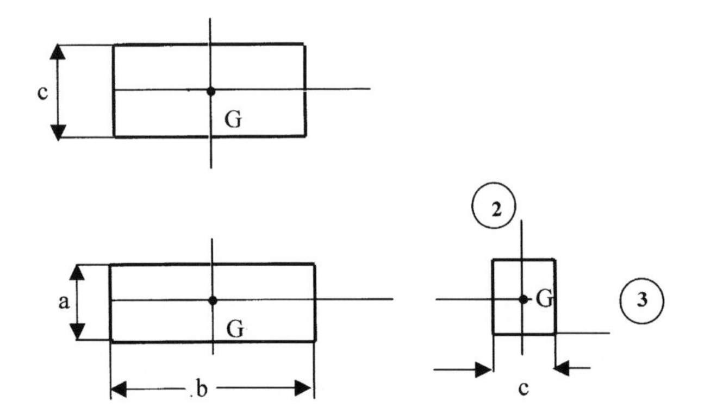

2. Cylinder:

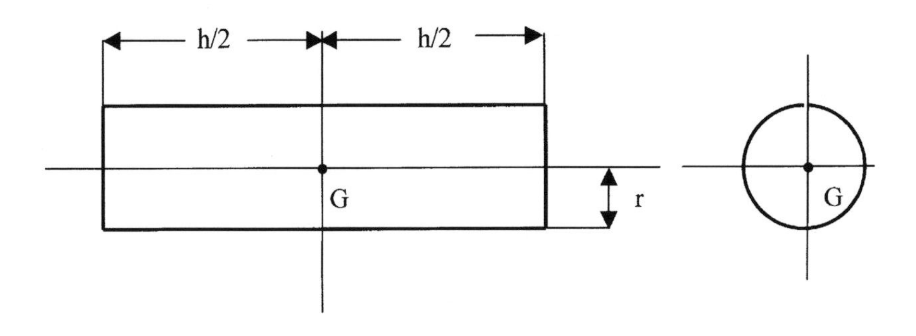

3. Half cylinder:

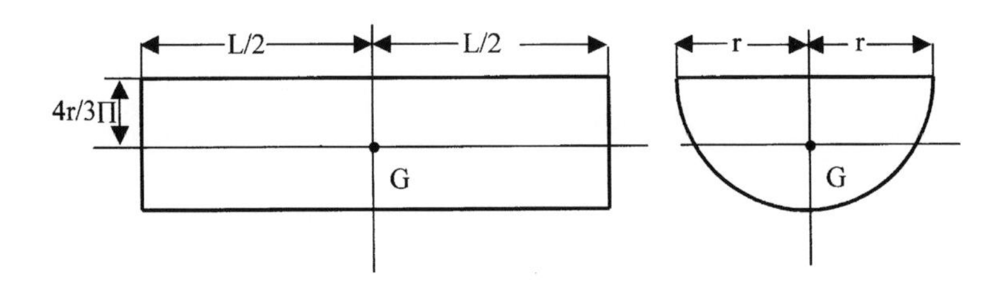

4. Cone:

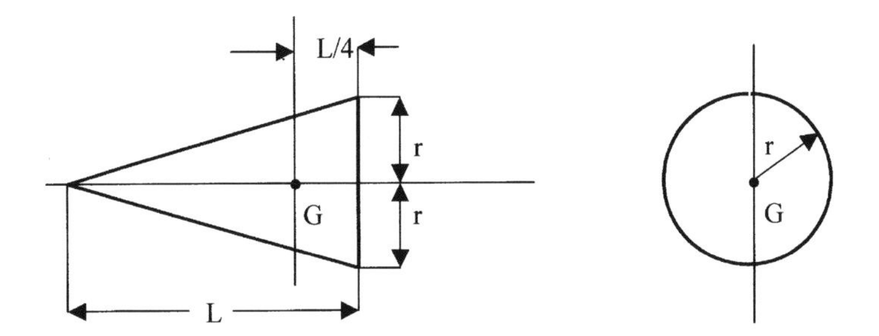

5. Half cone:

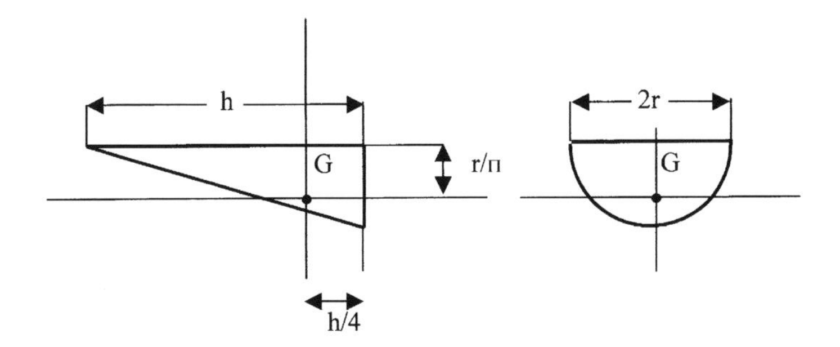

6. Sphere:

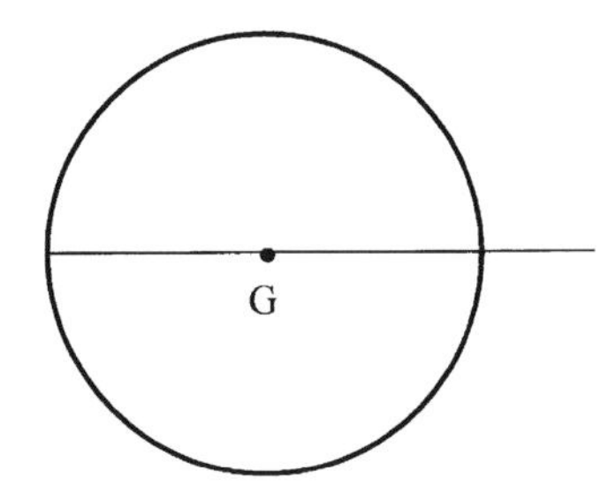

7. Hemisphere:

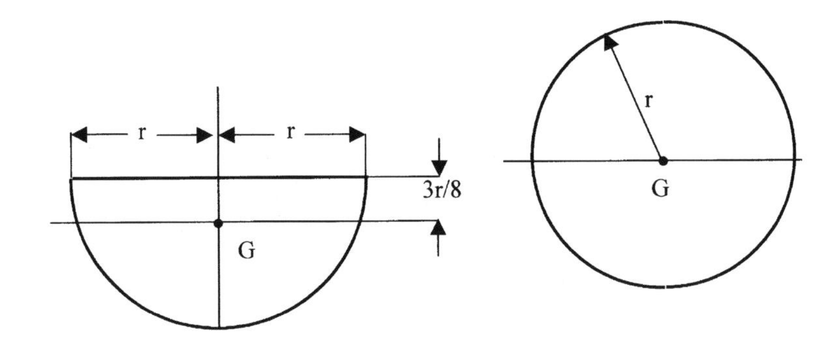

Appendix II*

Inertia Properties (Moments and Products of Inertia) for Commonly Shaped Bodies with Uniform Mass Distribution

I. Curves, wires, thin rods

1. Straight line, rod:

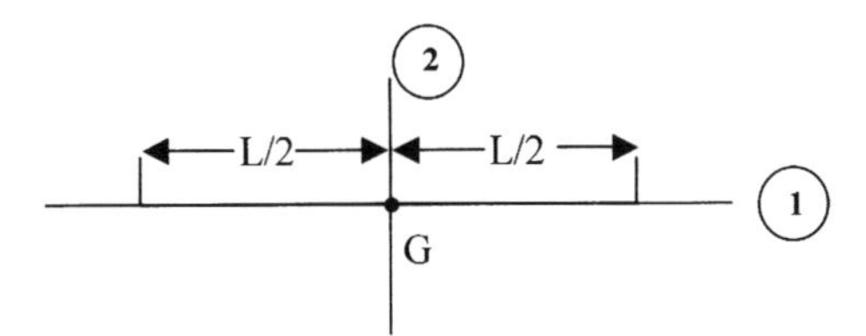

$I_{11} = 0$

$I_{22} = ML^2/12$

2. Circular arc, circular rod:

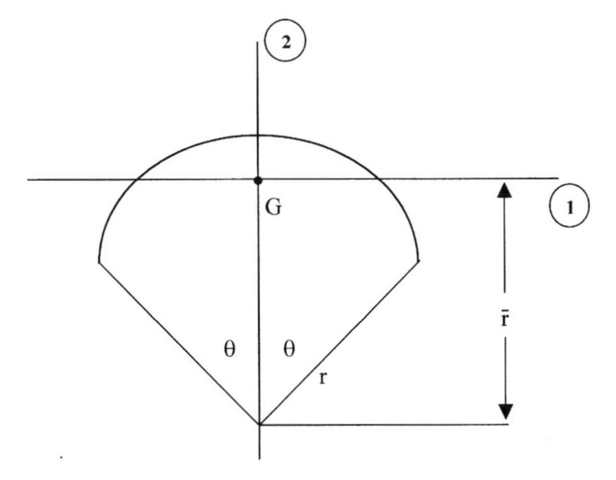

$\bar{r} = (r\sin\theta)/\theta$

$I_{11} = mr^2(1 + \sin2\theta/2\theta - 2\sin^2\theta/\theta^2)/2$

$I_{22} = mr^2(1 - \sin2\theta/2\theta)/2$

* Reprinted from Huston, R. L., and Liu, C. Q., *Formulas for Dynamic Analysis*, pp. 303–310, by courtesy of Marcel Dekker, New York, 2001.

3. Semicircular arc, semicircular rod:

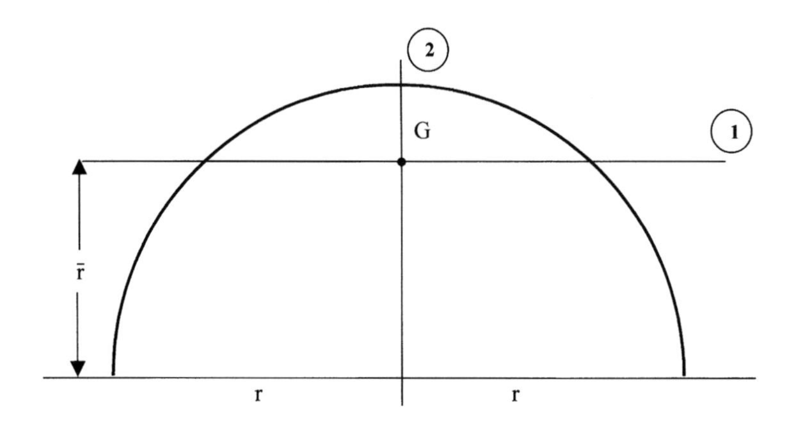

$I_{11} = mr^2(1 - 8/\Pi^2)/2$
$I_{22} = mr^2$
$\bar{r} = 2r/\Pi$

4. Circle, hoop:

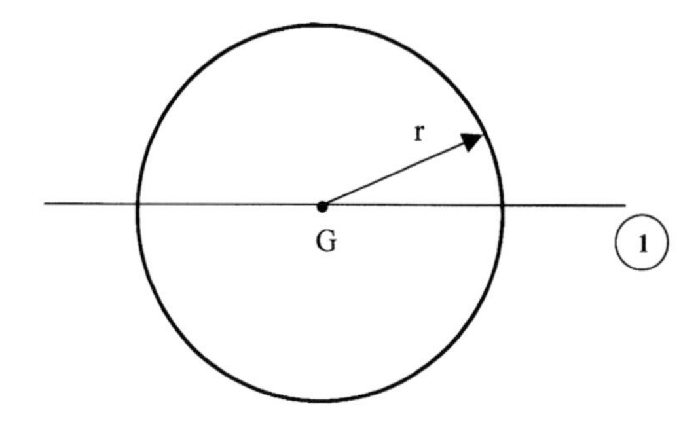

$I_{11} = mr^2/2$

II. Surfaces, thin plates, shells

1. Triangle, triangular plate:

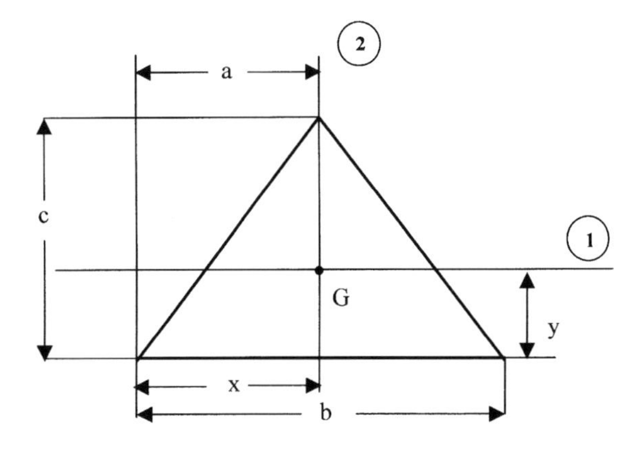

$x = (a + b)/3$

$y = c/3$

$I_{11} = mc^2/18$

$I_{22} = m(b^2 - ab + a^2)/18$

$I_{12} = mc(2a - b)/18$

2. Rectangle, rectangular plate:

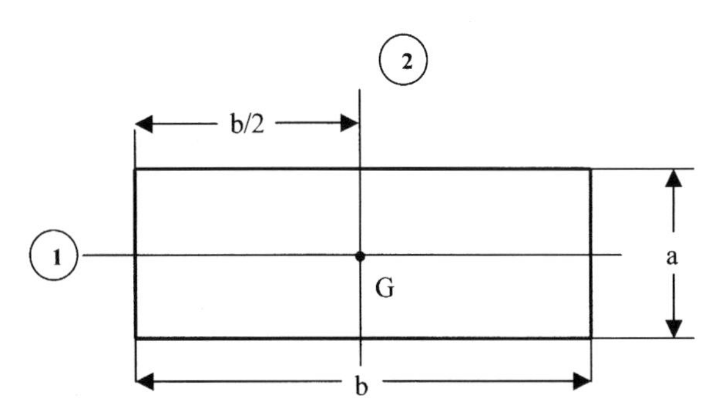

$I_{11} = ma^2/12$

$I_{22} = mb^2/12$

3. Circular sector, circular section plate:

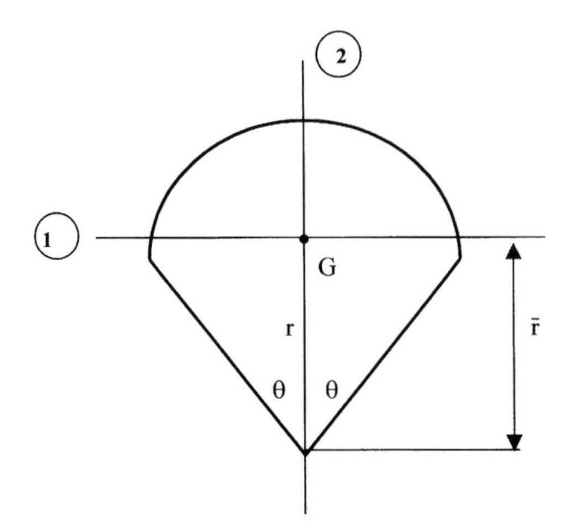

$\bar{r} = (2r/3)(\sin\theta)/\theta$

$I_{11} = (mr^2/4)(1 + \sin\theta\cos\theta/\theta - 16\sin^2\theta/9\theta^2)$

$I_{22} = (mr^2/4)(1 - \sin\theta\cos\theta/\theta)$

4. Semicircle, semicircular plate:

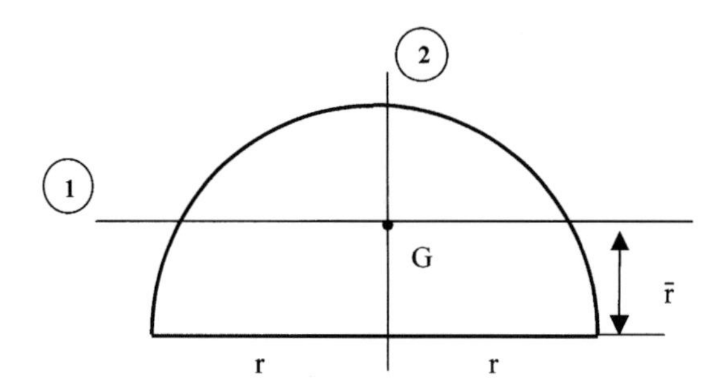

$\bar{r} = 4r/3\Pi$

$I_{11} = mr^2(9\Pi^2 - 64)/36\Pi^2$

$I_{22} = mr^2/4$

5. Circle, circular plate:

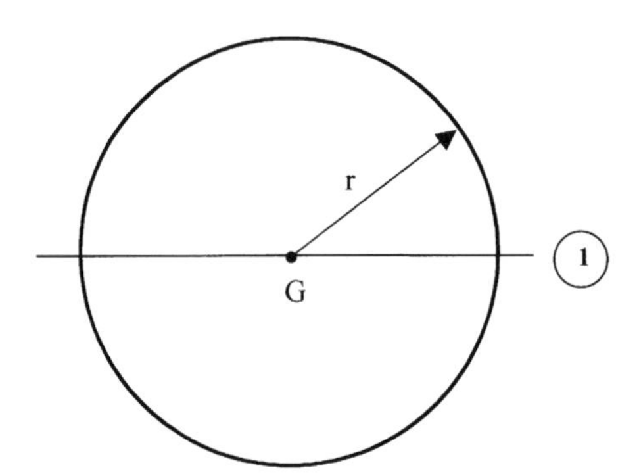

$I_{11} = mr^2/4$

6. Circular segment, circular segment plate:

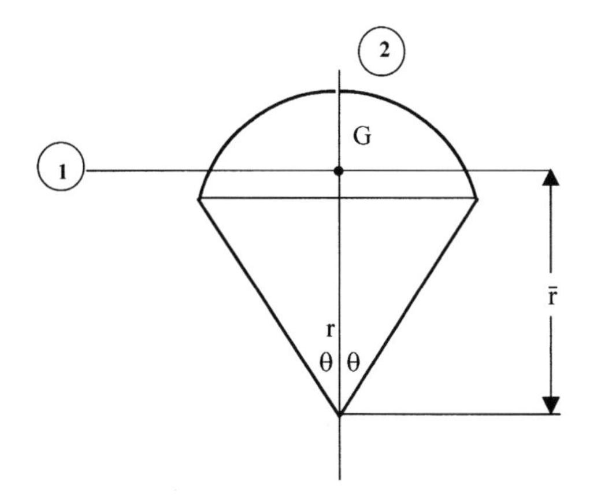

$\bar{r} = (2r/3)(\sin^3\theta)/(\theta - \sin\theta\cos\theta)$

$I_{11} = mr^2(9\theta^2 + 9\sin^2\theta\cos^2\theta - 36\sin^4\theta\cos^4\theta - 18\theta\sin\theta\cos\theta + 36\sin^3\theta\cos\theta - 8\sin^6\theta)/18(\theta - \sin\theta\cos\theta)^2$

$I_{22} = mr^2[1 - 2\sin^3\theta\cos\theta/3(\theta - \sin\theta\cos\theta)]/4$

7. Cylinder, cylindrical shell:

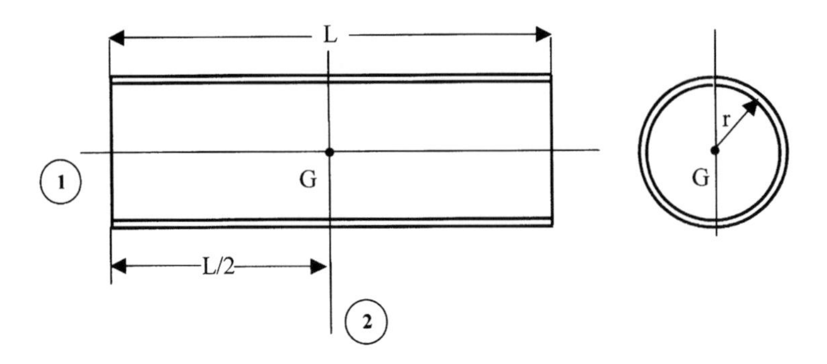

$I_{11} = mr^2$
$I_{22} = m(6r^2 + L^2)/12$

8. Semicylinder, semicylindrical shell:

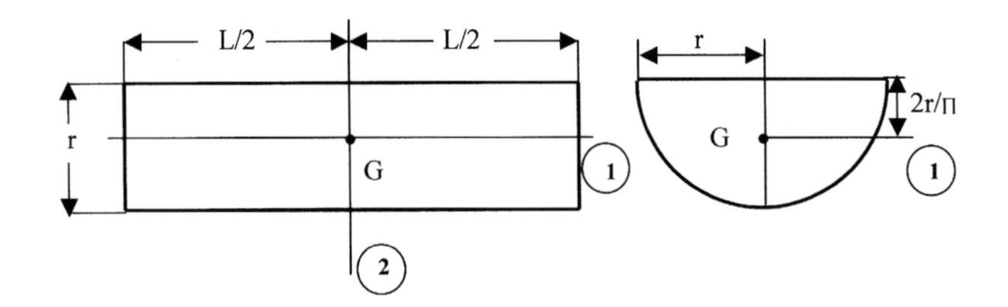

$I_{11} = mr^2(1 - 4/\Pi^2)$
$I_{22} = m(r^2 + L^2/6)/2$
$I_{33} = mr^2(1/2 - 4/\Pi^2) + mL^2/12$

9. Sphere, spherical shell:

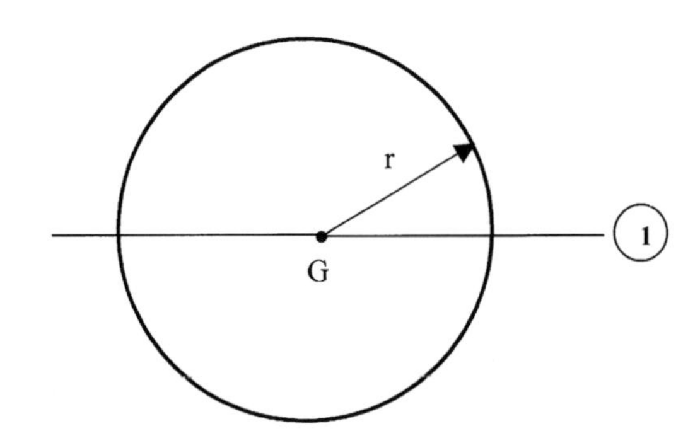

$I_{11} = 2mr^2/3$

10. Hemisphere, hemispherical shell:

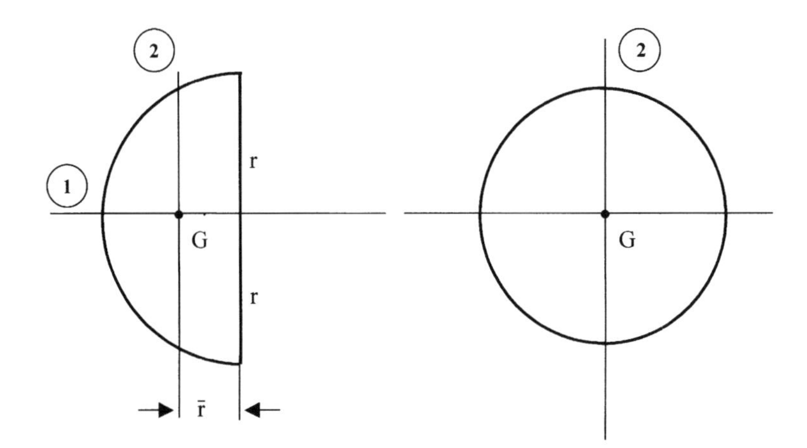

$\bar{r} = 3r/8$

$I_{11} = 2mr^2/3$

$I_{22} = 5mr^2/12$

11. Cone, conical shell:

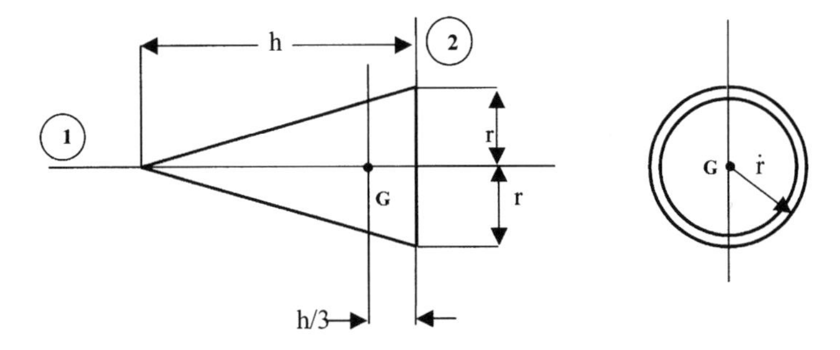

$I_{11} = mr^2/2$

$I_{22} = m(r^2/2 + h^2/9)/2$

12. Half cone, half-conical shell:

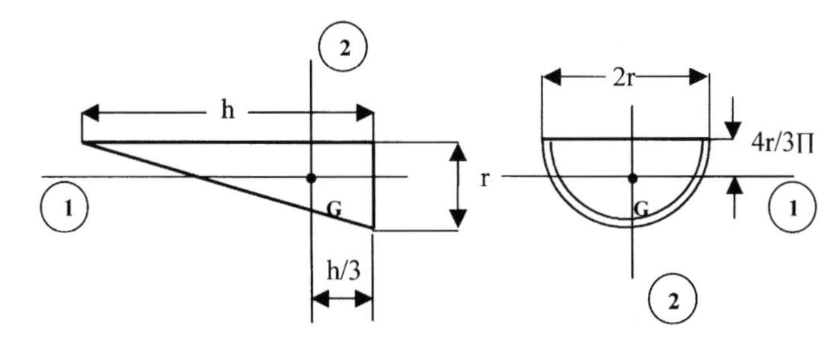

$I_{11} = mr^2(1 - 8/9\Pi^2)/2$

$I_{22} = (m/36)\,(9r^2 + 2h^2)$

$I_{33} = mh^2/18 + mr^2(1 - 16/9\Pi^2)/4$

$I_{12} = -mrh/9\Pi$

III. Solids, bodies

1. Parallelepiped, block:

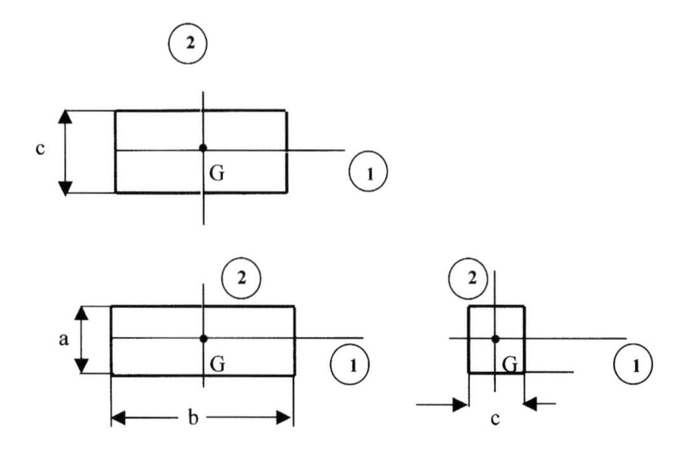

$I_{11} = m(a^2 + c^2)/12$
$I_{22} = m(b^2 + c^2)/12$
$I_{33} = m(a^2 + b^2)/12$

2. Cylinder:

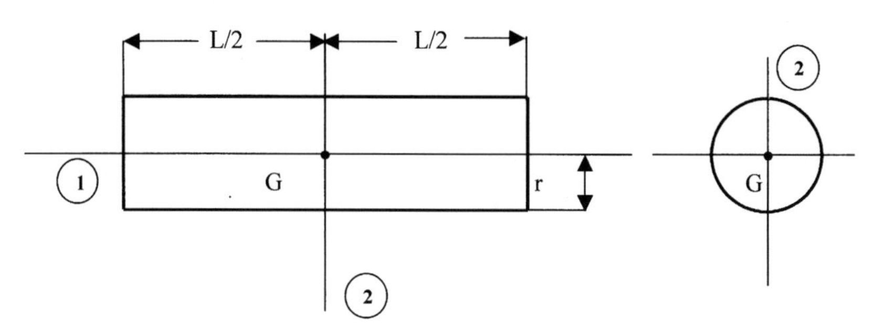

$I_{11} = mr^2/2$
$I_{22} = mr^2/4 + mL^2/12$

3. Half cylinder:

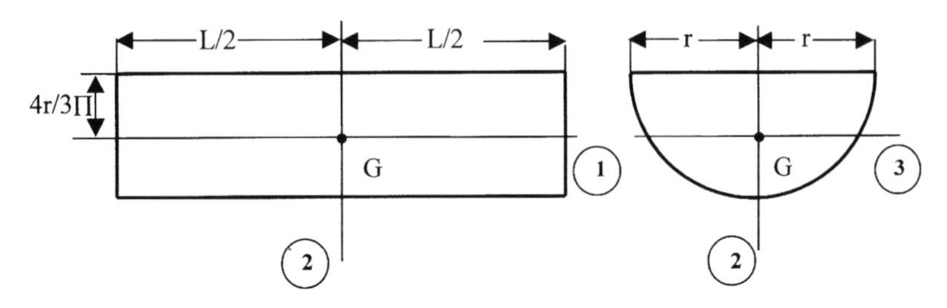

$I_{11} = mr^2(9 - 32/\Pi^2)/18$
$I_{22} = m(3r^2 + L^2)/12$
$I_{33} = mL^2/12 + mr^2(9 - 64/\Pi^2)/36$

4. Cone:

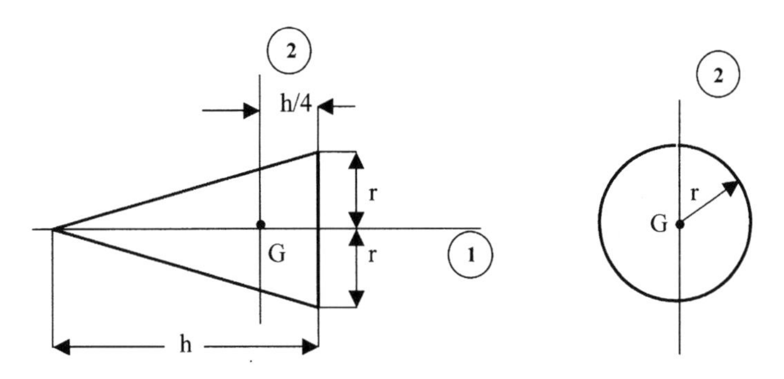

$I_{11} = 3mr^2/10$
$I_{22} = 3m(4r^2 + h^2)/80$

5. Half cone:

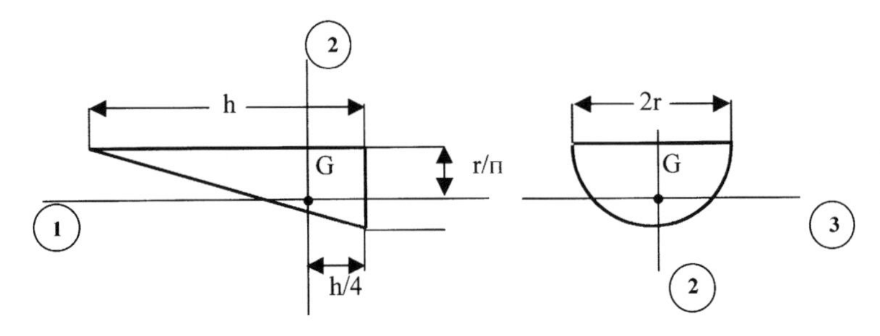

$I_{11} = mr^2(3/10 - 1/\Pi^2)$
$I_{22} = 3m(4r^2 + h^2)/80$
$I_{33} = mr^2(3/20 - 1/\Pi^2) + 3mh^2/80$
$I^{12} = -mrh/20\Pi$

6. Sphere:

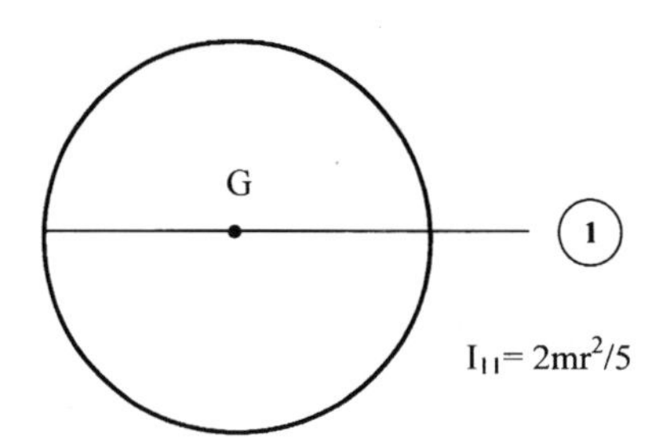

7. Hemisphere:

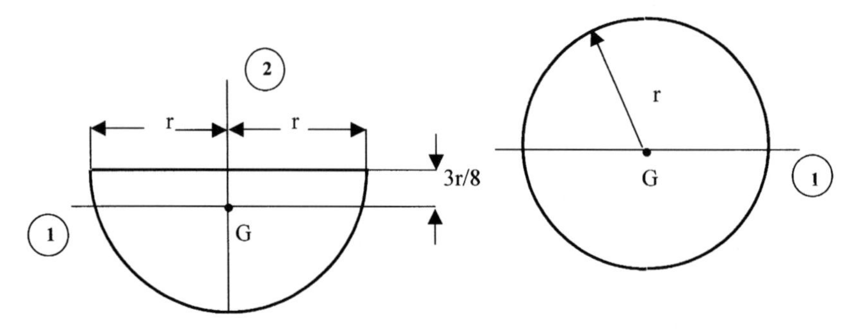

$I_{11} = 83mr^2/320$

$I_{22} = 2mr^2/5$

Index

A

Absolute orientation angles, 142
Acatastatic system, 690
Acceleration, 5, 59, 66
Accident reconstruction, 338
Action–reaction, 181, 241
Active forces, 2, 177, 244, 379, 647, 677, 715
Addendum, 581, 599
Addition theorem for angular velocity, 90
Adjoining bodies, 609
Adjoint, 40
Amplitude, 440
Angle, 9
Angle between two vectors, 23, 28
Angle of action, 584, 599
Angle of approach, 584, 599
Angle of contact, 584, 599
Angular acceleration, 83, 93, 640
Angular impulse, 280
Angular momentum, 282
Angular speed, 83
Angular velocity, 9, 83, 85, 635, 667
Anticyclic indices, 30
Antisymmetric matrix, 39
Applied forces, 177, 677
Articulation angles, 662, 688
Associative law, 18, 40
Axial pitch, 599

B

Backlash, 582, 599
Balancing, 513
Ball-and-socket joint, 4
Base circle, 578, 599
Basic rack, 591
Bevel gears, 596, 599
Biceps, 716, 717
Biosystem, 701
Boltzmann-Hamel equations, 242, 243
Branching body, 608
Buridan, John, 241

C

Cam-pair, 539
Cam profiles, 544
Cams, 5, 15, 539
Cam systems, 3
Cartesian coordinate system, 6, 8
Center of percussion, 298
Centroid, 735
Chord vector, 60
Circular frequency, 440
Circular pitch, 582, 599
Clearance, 583, 599
Closed loops, 606
Coefficient of restitution, 303
Column matrix, 39
Commutative law, 16, 23
Complete elliptic integral, 462
Components of vectors, 16, 19
Compression stroke, 529
Configuration graphs, 79, 614, 664
Conformable matrix, 39
Conjugate action, 575, 600
Connection configuration, 605
Conservation of momentum, 294, 301
Constrained motion, 722
Constraint equations, 125, 126, 684
Constraint forces, 684
Constraint matrix, 683
Constraints, 3, 125, 353
Contact forces, 180
Contact ratio, 585, 600
Coordinates, 6, 125, 353, 628
Couples, 170
Cross product, 29
Cyclic indices, 30
Cycloidal rise function, 563
Cylindrical coordinate system, 7

D

d'Alembert's principle, 185, 242, 243, 262, 279, 290
Damped linear oscillator equation, 442
Dedendum, 581, 600
Degrees of freedom, 3, 125, 353, 628, 661
Derivative of transformation matrices, 612, 668

Determinant, 40
Dextral rotation, 43
Dextral vectors, 30
Diagonal matrix, 39
Diametral pitch, 583
Dirac's delta function, 554
Directed line segment, 5
Direct impact, 306
Direction cosines, 21
Direction of a vector, 5
Distributive law, 17, 25, 32, 40
Dot product, 23
Double-rod pendulum, 258, 381, 396, 418, 426
Driver, 3, 539
Driver gear, 574
Dwell, 542
Dyad, 203
Dyadic, 203
Dynamic balancing, 514
Dynamics, 1
Dynamic unbalance, 516

E

Earth rotation effect, 89
Eigen unit vector, 209
Eigenvalue of inertia, 209
Elastic collision, 304
Elements of a matrix, 39
Elliptic integral 460
End effector, 661, 674, 692, 698
Ending body, 607-608
Energy, 9, 10
Equality of vectors, 15
Equivalent force systems, 170
Euler angles, 82
Euler parameters, 613, 707-709
Euler torque, 230
Exhaust stroke, 529
Extremity body, 607

F

Fillet radius, 583
Finite segment model, 142
Firing order of internal combustion engines, 530
First integral, 459
First moments, 182
Fixed stars, 244
Fixed vector, 15
Follower, 3, 539
Follower gear, 574
Force, 2, 5, 9, 163
Forced vibration, 446, 449
Forcing function, 442
Four-bar linkage, 136
Four-stroke engines, balancing of, 528
Free-body diagram, 245-246
Free index, 38
Free vector, 15
Frequency, 440

G

Gear drive, 592
Gear glossary, 599-601
Gears, 539, 573
Gear systems, 3
Gear train, 592
Generalized active force, 363
Generalized applied force, 363
Generalized coordinates, 242, 353
Generalized forces, 360
Generalized inertia forces, 360, 377
Generalized passive force, 377
Generalized speeds, 628
General plane motion, 129, 130
Gibbs equations, 243
Gibbs function, 243
Gravity forces, 177
Gripper, 661
Gross-motion model, 702

H

Hamilton's principle, 242
Helical gears, 595
Helix joint, 4
Holonomic constraint, 357
Human body dynamics, 702
Human body model, 704
Hypoid gears, 597

I

Identity matrix, 39
Imbalance, 513
Impact, 303
Impulse, 279
Impulse-momentum, 242
Incomplete elliptic integral, 462
Inertia, 1, 2, 199, 241
Inertia coefficients, 651
Inertia ellipsoid, 228
Inertia forces, 177, 184, 243, 244, 248, 648, 680, 719
Inertial reference frame, 2, 185, 244
Inertia properties (common shapes), 743
Inertia torque, 228, 249-250
Infinitesimal stability, 479
Inside unit vector, 80
Instant center of zero acceleration, 150
Instant center of zero velocity, 133, 147

Intake stroke, 529
Integration algorithms, 688
Intermediate body, 608
Inverse matrix, 40
Involute curve, 578
Involute function, 579-581
Involute rack, 591

J

Joint, 3
Joint forces, 716
Jourdain's principle, 243

K

Kane's equations, 242, 243, 263, 415, 422, 435
Kane, T. R., 416, 460
Kinematic chain, 4
Kinematics, 1, 2, 57, 241
Kinetic energy, 327
Kinetics, 163, 241
Kronecker's delta function, 24, 38, 42
Krylov and Bogoliuboff method, 463

L

Lagrange multiplier, 224
Lagrange's equations, 242, 262, 423, 435
Lagrange's form of d'Alembert's principle, 243, 416
Lagrangian, 242, 424
Lanchester balancing mechanism, 525
Law of action and reaction, 717
Law of conjugate action, 576
Linear impulse, 279
Linear momentum, 280
Linear oscillator equation, 246, 439
Linear rise function, 551
Line of action, 578
Line of centers, 585
Line of contact, 584
Link, 3
Linkage, 3
Logarithmic decrement, 471
Loop closure equation, 137
Lower body arrays, 605

M

Machine, 3
Magnitude of a vector, 5, 15, 27
Mass, 2, 10, 241
Mass center, 177
Mass center locations (common shapes), 735
Mass density, 10
Mass-spring system, 336
Matrix, 39
Matrix inverse, 40
Maximum moment of inertia, 223
Mechanism, 3
Mesh, 575
Minimum moment of inertia, 223
Minimum moments, 175
Minor, 40
Mobile robot, 661
Modes of vibration, 455
Module, 583
Moment, 10, 163
Moment of inertia, 200, 743
Moment of momentum, 282
Momentum, 280
Motion on a circle, 66
Motion on a plane, 68
Multi-arm robot, 662
Multibody system, 258, 605
Muscle forces, 716

N

Natural modes of vibration, 456
Newton, I., 241
Newton's laws, 2, 241, 285, 287
Nonholonomic constraint, 357
Nonlinear vibrations, 458
Normal pitch, 585
N-rod pendulum, 260, 433

O

Oblique impact, 306
Open-chain system, 606
Open-tree system, 606
Orientation angles, 79
Orientation of a vector, 5, 15
Orientation of bodies, 77, 84
Orthogonal complement arrays, 687, 689
Orthogonal matrix, 40
Orthogonal transformation, 42, 77
Outside unit vector, 80

P

Parabolic rise function, 557
Parallel axis theorem, 206, 207
Parallelogram law, 16
Partial angular velocity, 359, 667
Partial angular velocity array, 639
Partial velocity, 359

Partial velocity arrays, 651
Particle, 3, 57
Passive forces, 2, 177, 184, 244, 377, 680
Period, 440
Permutation symbol, 30
Perturbation, 479
Phase, 440
Pinion, 539
Pinion gear, 574, 597
Pitch circle, 576
Pitch point, 582
Pivoting, 107
Planar joint, 4
Planar motion, 128
Planetary gear system, 593
Planet gear, 593
Plastic collision, 304
Position, 58
Position vector, 8
Potential energy, 389
Power, 10, 327
Power stroke, 529
Power transmission, 573
Pressure, 10
Pressure angle, 542, 578
Pressure line, 578, 584
Principal axis of inertia, 208, 215
Principal moment of inertia, 208, 209
Principal unit vector, 209
Principle of angular momentum, 289
Principle of linear momentum, 285
Product of inertia, 200, 743
Projectile motion, 251
Projection of a vector, 24, 28, 31
Pure rolling, 107
Pythagorean theorem, 7

R

Radius of gyration, 202
Reciprocating machines, balancing of, 520
Reduction of a force system, 171
Redundant robots, 661, 684
Reference frame, 3, 6, 41
Relative acceleration, 63, 97
Relative orientation angles, 137, 142
Relative velocity, 61, 97
Resultant, 16, 165
Right-hand rule, 29
Rigid body, 3
Ring gear, 594
Rise of cam follower, 543
Robot, 663
Robot arm, 663
Rod pendulum, 255, 380, 396, 418, 425
Rolling, 106
Rolling circular disk, 267
Rolling disk, 107, 357, 385, 399, 421, 488
Root circle, 581
Rotating pinned rod, 263
Rotating unit vectors, 63
Rotation, 129, 130
Rotation dyadics, 617
Routh-Hurwitz stability criteria, 505
Row–column product of matrices, 39
Row matrix, 39

S

Scalar, 5, 9, 15
Scalar product, 23, 27
Scalar triple product, 33
Screw joint, 4
Second-moment vectors, 199
Sense of a vector, 5, 15
Simple angular velocity, 83, 87, 665
Simple chains, 4
Simple pendulum, 245, 324, 333, 365, 379, 395, 417, 424, 445, 459, 479
Singularity functions, 553
Singular matrix, 39
Sinistral vectors, 30
Sinusoidal rise function, 560
Sliding joint, 4
Sliding vector, 15
Solver, 688
Space, 2, 3
Speed, 63, 64
Spherical coordinate system, 7
Spherical joint, 4
Spiral angle, 597
Spiral bevel gears, 597
Spring forces, 178
Spur gear, 4, 581
Stability, 479
Static balancing, 513
Statics, 1
Stress, 10
Stroke, 528
Substitution symbol, 38
Summation convention, 37
Sun gear, 593
Symmetric matrix, 39
System of forces, 165

T

Tensor, 204
Thrust bearing, 110
Time, 2
Torque, 10, 170
Transformation matrices, 78, 609, 663, 704
Transformation matrix derivatives, 612, 668
Translation, 129
Transmission, 540, 573
Transpose of a matrix, 39

Triple-rod pendulum, 260, 429
Tzu, Mo, 242

U

Units, 9
Unit vector, 6, 15
Universal joint, 4
Unstable system, 482

V

Vector, 5, 15
Vector addition, 16
Vector characteristics, 5, 57
Vector components, 16, 19
Vector differentiation, 57
Vector multiplication, 23, 28
Vector product, 28, 33
Vector space, 5
Vector subtraction, 17
Vector triple product, 33
Velocity, 5, 10, 59
Vibration, 439
Virtual work, 242

W

Weight, 177
Wheel rolling over a step, 302, 341
Whole-body model, 701, 703
Work, 321
Work-energy, 242
Work-energy principles, 329
Worm gears, 597
Worm wheel, 598
Wrench, 173

Z

Zero force systems, 170
Zero vector, 6, 15